McGRAW-HILL BOOK COMPANY

New York
St. Louis
San Francisco
Auckland
Düsseldorf
Johannesburg
Kuala Lumpur
London
Mexico
Montreal
New Delhi
Panama
Paris
São Paulo
Singapore
Sydney
Tokyo
Toronto

J. O. HINZE

Turbulence

SECOND EDITION

This book was set in Times New Roman.
The editor was B. J. Clark;
the production supervisors were Alan Chapman and Thomas J. LoPinto.
The drawings were done by Reproduction Drawings Ltd.
Kingsport Press, Inc., was printer and binder.

Library of Congress Cataloging in Publication Data
Hinze, J O
Turbulence.

(McGraw-Hill series in mechanical engineering)
Includes bibliographical references.
1. Turbulence.
QA913.H5 1975 532′.0527 75-5765
ISBN 0-07-029037-7

TURBULENCE

234567890KPKP79876

CONTENTS

Preface to the Second Edition viii
Preface to the First Edition ix

Chapter 1 General Introduction and Concepts 1

1-1 Definition of Turbulence and Introductory Concepts 1
1-2 Equations of Motion for Turbulent Flow; Reynolds Stresses 15
Cylindrical Coordinates 26
1-3 Equation for the Conservation of a Transferable Scalar Quantity in a Turbulent Flow 29
1-4 Double Correlations between Turbulence-Velocity Components 30
1-5 Change in Double Velocity Correlations with Time; Introduction of Triple Velocity Correlations 35
1-6 Features of the Double Longitudinal and Lateral Correlations in a Homogeneous Turbulence 39
1-7 Integral Scale of Turbulence 43
1-8 Other Eulerian Correlations 44
1-9 Turbulent Diffusion of Fluid Particles; Lagrangian Correlations 48
1-10 Recapitulation of Correlations 56
Eulerian Space Correlations 56
Eulerian Time Correlation 57
Lagrangian Time Correlation 57
1-11 Empirical Formulas for Double-Correlation Curves 58
1-12 Taylor's One-dimensional Energy Spectrum 61
1-13 Energy Relations in Turbulent Flows 68

References 79
Nomenclature for Chapter 1 80

Chapter 2 Principles of Methods and Techniques in the Measurement of Turbulent Flows 83

2-1 Introduction 83
2-2 The Hot-Wire Anemometer 85
Effect of Compressibility 91
2-3 Constant-Temperature Method 96
Cooling Effect of the Wire Supports 98
Dynamic Behavior with Heat Loss to Supports 100
Effect of Large Turbulence Fluctuations on the Response of a Hot Wire 105
Effect of Compressibility 111
2-4 Constant-Current Method 113
Effect of Large Turbulence Fluctuations 115
Effect of Nonlinearity Due to Temperature Dependence of Factors 117
2-5 Measurement of Turbulence Characteristics with the Hot-Wire Anemometer 119
Measurement of Turbulence Intensities 123
Measurement of Double and Triple Velocity Correlations 127
Measurement of Spectrum and Scale of Turbulence 131
2-6 Measurement of Temperature and Concentration Fluctuations with the Hot-Wire Anemometer: Constant-Current Method 135
Measurement of Concentration Fluctuations 139
2-7 Limitations of the Hot-Wire Anemometer 139
2-8 Other Turbulence Measuring Probes 144
Thermistor 145
Mechanical Probes 145
Electrical Probes 149
Electrochemical Probe 153
2-9 Methods Based upon Flow Visualization 154
Hydrogen Bubble Technique 155
Light Refraction by Density Variations 159
Methods Based on Scattering and Absorption of Light 160
2-10 Measurement of Mean Values of Static Pressure and Velocity 164
Measurement of Mean Velocity 166
References 168
Nomenclature for Chapter 2 172

Chapter 3 Isotropic Turbulence **175**

3-1 Introduction 175
3-2 Correlation Tensors 178
First-Order Correlation Tensor 178
Second-Order Correlation Tensor 181
Third-Order Correlation Tensor 189
3-3 Differential Equation for the Dynamic Behavior of an Isotropic Turbulence 195
3-4 The Three-dimensional Energy Spectrum 202
3-5 The Dynamic Equation for the Energy Spectrum 211
The "Physical Closure" Approximations in the Light of Experimental Evidence 250
3-6 The Decay of an Isotropic Turbulence 259
3-7 Extension to an Isotropic-Turbulent Scalar Field 278
Correlations 278
The Dynamic Equation for $\mathbf{Q}_{\gamma,\gamma}(r,t)$ 282
The Spectral Distribution of a Scalar Quantity 283
The Dynamic Equation for $\mathbf{E}_{\gamma}(k,t)$ 286
The Decay of an Isotropic Scalar Field 300
3-8 Pressure Fluctuations in Isotropic Turbulence 305
3-9 Some Remarks on the Effect of Compressibility 310
References 315
Nomenclature for Chapter 3 319

Chapter 4 Homogeneous, Shear-Flow Turbulence **321**

4-1 Introduction 321
Dynamics of the One-Point Velocity Correlation $\overline{u_i u_j}$ 323
Dynamics of the Two-Point Velocity Correlation $\overline{(u_i)_A (u_j)_B}$ 328
Homogeneous Turbulence 331
4-4 The Dynamic Equation for the Energy Spectrum 334
Experimental Results 350
References 356
Nomenclature for Chapter 4 356

Chapter 5 Transport Processes in Turbulent Flows **358**

5-1 Introduction 358
5-2 Mixing-Length and Phenomenological Theories 360
5-3 Analogies in Turbulent Transport 376
5-4 Transport by Turbulent Diffusion 384

	Relative Diffusion of Two Fluid Particles	406
	Deformation of a Fluid Element	412
	Relationship Between Eulerian and Lagrangian Correlations	416
5-5	Diffusion from a Fixed Source in a Uniform Flow	427
5-6	Diffusion from a Fixed Source in a Turbulent Shear Flow	451
5-7	Diffusion of Discrete Particles in a Homogeneous Turbulence	460
5-8	Effect of Compressibility	471
	References	477
	Nomenclature for Chapter 5	480

Chapter 6	**Free Turbulent Shear Flows**	**483**
6-1	Introduction	483
6-2	Similarity Considerations	486
6-3	Approximations Applied to the Equations of Motion	489
6-4	Velocity Distribution Behind a Cylinder According to the Classical Theories	496
	Axi-symmetric Wake	502
6-5	The Transport of a Scalar Quantity in the Wake Flow of a Cylinder	503
6-6	Measurements of Mean-Velocity and Mean-Temperature Distribution in the Wake Flow of a Cylinder	505
	Axi-symmetric Wake	509
6-7	Measurements of Turbulence Quantities in the Wake Flow of a Cylinder	510
	Axi-symmetric Wake	519
6-8	Velocity Distribution in a Round Free Jet According to the Classical Theories	520
	Prandtl's Mixing-Length Theory	527
	Reichardt's Inductive Theory	529
6-9	The Transport of a Scalar Quantity in a Round Free Jet	531
6-10	Measurements of Mean-Velocity and Mean-Temperature Distribution in a Round Free Jet	534
6-11	Measurements of Turbulence Quantities in a Round Free Jet	546
6-12	The Structure of Free Turbulent Shear Flow and Transport Processes	558
	References	581
	Nomenclature for Chapter 6	583

Chapter 7 "Wall" Turbulent Shear Flows 586

7-1 Introduction
7-2 Approximations to the Equations of Motion, and Their 586
Integral Relations for a Plane Boundary Layer 588
Boundary-Layer Approximations 589
Integral Relations 593
7-3 The Laminar Boundary Layer and Transition 597
7-4 Further Considerations on Transition to Turbulence of a Laminar Boundary Layer 600
Mathematical Models 613
7-5 Turbulent Boundary Layer Along a Flat Plate. Classical Theories 614
7-6 Measurements on Mean-Velocity Distribution 626
7-7 Measurements on Turbulence Quantities in a Boundary Layer 638
Effect of Fluid Compressibility 653
7-8 More Detailed Information on the Structure of the Turbulent Boundary Layer 656
Mechanism of the Turbulence in the Wall Region 659
Wall-Pressure Fluctuations 668
Convection Velocities 673
Active and Inactive Motions 680
Dependency Between the Turbulence in the Outer Region and that in the Wall Region 681
7-9 The Structure of the Outer Region 684
7-10 Nonclassical Theories on the Distribution of Mean Properties in a Plane Boundary Layer 690
"Model" Theories 703
7-11 Transition to Turbulence of a Laminar Poiseuille Flow 707
7-12 Turbulent Flow through a Straight Circular Pipe. Mean-velocity Distribution 715
7-13 Measurements of Turbulence Quantities in Pipe Flow 724
7-14 Transport of a Scalar Quantity in Wall Turbulence 742
Effect of Wall Roughness on Heat Transfer 757
Effect of Free-Stream Turbulence on Boundary-Layer Transport 760
References 762
Nomenclature for Chapter 7 769

Appendix—Elements of Cartesian Tensors 771

Name Index 781
Subject Index 785

PREFACE TO THE SECOND EDITION

During the fifteen years or so which have elapsed since the publication of the first edition of this book, a lot of new material resulting from experimental and theoretical investigations on turbulence has become available. The new experimental material has broadened and deepened our insight into the turbulent flow phenomena, both with respect to the large-scale and the fine-scale structure. The concurrent theoretical investigations have been most useful in this respect, though we are still some way from a complete theoretical solution of the turbulence problem.

All this new material has made it necessary to revise drastically the contents of the first edition, parts of which have even become entirely obsolete.

When preparing this second edition, the author was confronted with the basic problem: what material to retain from the first edition, what could be or had to be dropped, and what new material had to be included at least. Considering the abundant amount of new material available in the literature, at the same time taking into account an upper limit for the allotted extra space in the second edition, a selection was imperative. Consequently much important published material had to be disregarded. The literature references which have been given with respect to a specific subject, should therefore be considered as not being complete. In general the author was guided by the idea that the book is still intended as an introduction, notwithstanding the increased size of the new edition. Parts containing new material but which possibly may be skipped are printed in small type.

Since the figures presented should be considered more as illustrating material rather than showing the most recent and accurate experimental evidence, many figures of the first edition have been retained, as long as they still do elucidate sufficiently well the subject matter treated in the text.

The same method of presentation from an instructional point of view has been followed as in the first edition. The same coverage has been held, and the same number of chapters. However, depending on the subject covered by a chapter, the material has been more or less changed and extended. Consequently, Chap. 2 which deals with measuring methods and instruments, has been almost entirely rewritten. In contrast to the first edition much attention has been paid to the transition process from laminar to turbulent flow in Chaps. 6 and 7, dealing with free turbulent and wall turbulent shear flows, respectively.

Due to the restriction imposed by the allotted extra space not much more attention has been given to the effect of compressibility than in the first edition, though a lot of new information has become available since 1959. For the same reason, such subjects as magnetohydrodynamics and non-Newtonian fluids have been entirely left out of consideration.

The author feels much indebted to a number of referees for their critical and constructive comments concerning an earlier version of the manuscript. He also wants to acknowledge his gratitude to his collaborator, Mr P. J. H. Builtjes for his help during the preparation of the manuscript and in correcting the proofs. The same acknowledgement applies to Mrs. A. I. H. Sijnja-Teillers and to Mrs H. J. Van der Brugge-Peeters for typing the manuscript, and to Messrs C. J. de Kat and L. B. Kok in preparing the drawings for a number of new figures.

J. O. HINZE

PREFACE TO THE FIRST EDITION

The subject matter of the present book originates from a series of lectures given by the author before a group of chemical engineers of the Royal/Dutch-Shell laboratories at Amsterdam and Delft, during 1950.

The aim of this course was to make the participant familiar with current notions and theories about turbulent fluid flow and, by so doing, to give him a sufficient theoretical basis not only for studying the specialized literature on turbulence but also for theoretical investigations on those chemical engineering problems in which turbulence plays an essential part, such as mixing and heat and mass transfer.

The book follows essentially the same outline, but the subject matter has been extended, modified, and brought up to date. It should be of interest to research engineers not only in the field of chemical engineering, but also in the field of mechanical engineering, since special attention has been paid to discussing the mechanism of turbulence in relation to flow resistance and to heat and mass transfer.

The choice of subjects to be treated has been made accordingly. Not all questions about turbulence have been considered; a complete textbook on turbulence was not intended. For instance, instability theories and transition from laminar to turbulent flow have not been considered, and the effect of compressibility of the fluid on turbulence phenomena has been touched upon only incidentally.

The author has tried to present the subject matter in such a way as to make it accessible to readers who are not already experts in fluid flow. However, although the treatment starts from first principles and the use of mathematics has been reduced to a minimum, a fair knowledge of mathematics and physics is a prerequisite if the reader is to succeed in making himself familiar with the contents of the book.

The material of Chapter 1 serves, in the first place, as a general introduction through which the reader may become familiar with various notions specific to turbulence theories; in the second place, general basic formulas are derived here that will be used as a starting point in other chapters.

Chapter 2 deals with the methods and instruments commonly applied in measuring turbulence quantities. This subject has been treated more extensively than might perhaps be expected. But the underlying idea was that it is certainly useful to make the research engineer familiar with existing methods and their merits, so as to enable him to criticize and interpret empirical data.

In Chapter 3 the generally accepted theories of isotropic turbulence are expounded. It is believed that these theories have been treated here fully enough to serve as a starting point for further theoretical studies on turbulence.

In sharp contrast with isotropic turbulence, theoretical knowledge of non-isotropic turbulence is much less advanced. Some of the most important attempts to arrive at a statistical theory of this kind of turbulence are considered in Chapter 4.

Typical of turbulent flows is the diffusive character of transport processes, owing to the randomness of the turbulent motions. This feature of turbulence is considered to be so basic as to justify treatment of theoretical and experimental investigations on transport processes in a separate chapter, Chapter 5.

The last two chapters, 6 and 7, deal, respectively, with nonisotropic free turbulence and with turbulent flow along fixed walls.

An Appendix contains a short introduction to Cartesian tensors, valuable in view of the extensive use that has been made of tensor notation and tensor calculus.

A complete nomenclature for each chapter is included at the end of the chapter.

The author would like to express his gratitude to Dr. T. Baron of Shell-Development Co., Emeryville, whose encouragement induced him to write the book, and to a number of his former collaborators at the Royal/Dutch-Shell laboratory at Delft for the help they gave him in many respects during the preparation of the manuscript. Dr. H. J. Merk especially deserves mention, for he has carefully worked through the whole manuscript and has made many important contributions by his critical and instructive remarks. Finally he wishes to acknowledge his debt to his collaborator, Mr. G. F. Buisman, for his valuable help in correcting the proofs.

J. O. HINZE

1
GENERAL INTRODUCTION AND CONCEPTS

1-1 DEFINITION OF TURBULENCE AND INTRODUCTORY CONCEPTS

The notion of turbulence is generally accepted nowadays, and, broadly speaking, its meaning is understood, at least by technical people. Yet it is curious to note that the use of the word "turbulent" to characterize a certain type of flow, namely, the counterpart of streamline motion, is comparatively recent. Osborne Reynolds, one of the pioneers in the study of turbulent flows, named this type of motion "sinuous motion."

Turbulence is rather a familiar notion; yet it is not easy to define in such a way as to cover the detailed characteristics comprehended in it and to make the definition agree with the modern view of it held by professionals in this field of applied science.

According to Webster's "New International Dictionary," turbulence means: agitation, commotion, disturbance. . . . This definition is, however, too general, and does not suffice to characterize turbulent fluid motion in the modern sense. In

1937 Taylor and Von Kármán[1]† gave the following definition: "Turbulence is an irregular motion which in general makes its appearance in fluids, gaseous or liquid, when they flow past solid surfaces or even when neighboring streams of the same fluid flow past or over one another." According to this definition, the flow has to satisfy the condition of irregularity.

Indeed, this irregularity is a very important feature. Because of irregularity, it is impossible to describe the motion in all details as a function of time and space coordinates. But, fortunately, turbulent motion is irregular in the sense that it is possible to describe it by laws of probability. It appears possible to indicate distinct average values of various quantities, such as velocity, pressure, temperature, etc., and this is very important.

Therefore, it is not sufficient just to say that turbulence is an irregular motion and to leave it at that. Perhaps a *definition* might be formulated somewhat more precisely as follows: "Turbulent fluid motion is an irregular condition of flow in which the various quantities show a random variation with time and space coordinates, so that statistically distinct average values can be discerned."

The addition "with time and space coordinates" is necessary; it is not sufficient to define turbulent motion as irregular in time alone. Take, for instance, the case in which a given quantity of a fluid is moved bodily in an irregular way; the motion of each part of the fluid is then irregular with respect to time to a stationary observer, but not to an observer moving with the fluid. Nor is turbulent motion a motion that is irregular in space alone, because a steady flow with an irregular flow pattern might then come under the definition of turbulence. Though the two cases of irregular motions may be useful for studying theoretically certain aspects of turbulence. It may be remarked that in the first case the Eulerian velocity at a point with respect to a stationary coordinate system is a random function of time, in the second case it is the Lagrangian velocity of a fluid particle that is a random function of time.

As Taylor and Von Kármán have stated in their definition, turbulence can be generated by friction forces at fixed walls (flow through conduits, flow past bodies) or by the flow of layers of fluids with different velocities past or over one another. As will be shown in what follows, there is a distinct difference between the kinds of turbulence generated in the two ways. Therefore it is convenient to indicate turbulence generated and continuously affected by fixed walls by the designation "wall turbulence" and to indicate turbulence in the absence of walls by "free turbulence," the generally accepted term.

In the case of real viscous fluids, viscosity effects will result in the conversion of kinetic energy of flow into heat; thus turbulent flow, like all flow of such fluids, is dissipative in nature. If there is no continuous external source of energy for the continuous generation of the turbulent motion, the motion will decay. Other effects of viscosity are to make the turbulence more homogeneous and to make it less

† Numbers refer to References at the end of each chapter.

dependent on direction. In the extreme case, the turbulence has quantitatively the same structure in all parts of the flow field; the turbulence is said to be homogeneous. The turbulence is called *isotropic* if its statistical features have no preference for any direction, so that perfect disorder reigns. As we shall see later, no average shear stress can occur and, consequently, no gradient of the mean velocity. This mean velocity, if it occurs, is constant throughout the field. A more complete definition of isotropic turbulence will be given later.

In all other cases where the mean velocity shows a gradient, the turbulence will be nonisotropic, or anisotropic. Since this gradient in mean velocity is associated with the occurrence of an average shear stress, the expression "shear-flow turbulence" is often used to designate this class of flow. Wall turbulence and anisotropic free turbulence fall into this class.

Von Kármán[2] has introduced the concept of homologous turbulence for the case of constant average shear stress throughout the field, for instance, in plane Couette flow.

Frequently the expression "pseudo turbulence" is used; this refers to the hypothetical case of a flow field with a regular pattern that shows a distinct constant periodicity in time and space. The difference between pseudo and real turbulence becomes clear if we compare pictures made of the two types. The first picture shows a regular flow pattern with constant periodicities throughout the field, whereas the second can show the condition only at one instant—the next instant the pattern may have changed in shape and magnitude. Pseudo-turbulent flow fields may be very useful for simulating real turbulent fields, for they can be made more accessible to theoretical treatment; it is relatively easy, for instance, to calculate the dissipation of kinetic energy by viscous effects in such a field. In his book "The Structure of Turbulent Shear Flow," Townsend suggests a few types of pseudo-turbulent flows that are suitable for studying various characteristics typical of real turbulent flows. On the other hand, in using a pseudo turbulence in a theoretical study to show some of the features of real turbulence, one often has to be very careful in interpreting the results. For instance, serious errors might result if one calculated transport and diffusion by turbulence from an assumed pseudo-turbulent flow pattern, since these processes are mainly, if not entirely, determined by the irregularity and randomness of the real turbulent motions.

This disorderliness and randomness of turbulence is clearly shown by the following case. Consider an oscillogram of the velocity fluctuations at a point in a flow field. If we determine from this oscillogram the number of amplitudes that have an assigned value, and so the probability of amplitudes, for isotropic turbulence a Gaussian distribution is obtained. For turbulent shear flow, generally, the distribution will be more or less skew.

As we pointed out in connection with our definition of turbulence, average values of quantities exist with respect to time and space. Mere observation of turbulent flows and of oscillograms of quantities varying turbulently shows that these average values exist, because:

1. At a given point in the turbulent domain a distinct pattern is repeated more or less regularly in time.

2. At a given instant a distinct pattern is repeated more or less regularly in space; so turbulence, broadly speaking, has the same over-all structure throughout the domain considered.

If we compare different turbulent motions in each of which a distinct pattern can be discerned, we shall observe differences, for instance, in the size of the patterns. This means that, to describe a turbulent motion quantitatively, it is necessary to introduce the notion of *scale of turbulence*: a certain scale in time and a certain scale in space. The magnitude of these scales will be determined by the dimensions of and the velocities within the apparatus in which the turbulent flow occurs. For turbulent flow through a pipe, for instance, one may expect a time scale of the order of magnitude of the ratio between pipe diameter and mean-flow velocity and a space scale of the order of magnitude of the diameter of the pipe.

It is apparent that it is insufficient to characterize a turbulent motion by its scale alone, since to do so does not tell us anything about the violence of the motion. One cannot take the average value of the velocity as a measure of this violence, because the violence of the fluctuations with respect to this average velocity is just what one wants to know.

If the momentary value of the velocity is written

$$U = \bar{U} + u$$

where the overscore denotes the average value, so that by definition $\bar{u} = 0$, it might be possible to take the average of the absolute values of the fluctuation, i.e., $\overline{|u|}$, as a measure of this violence. However, it is not usual to do it in this way. For reasons which will become obvious later on, it has been usual, since Dryden and Kuethe[36] introduced this definition in 1930, to define the violence or *intensity* of the turbulence fluctuations by the root-mean-square value

$$u' = \sqrt{\overline{u^2}}$$

The *relative intensity* will then be defined by the ratio

$$\frac{u'}{\bar{U}} \dagger$$

Average values can be determined in various ways. If the turbulence flow field is quasi-steady, or stationary random, averaging with respect to time can be used. In the case of a homogeneous turbulence flow field, averaging with respect to space can be considered. It is not always possible, however, to take time-mean or space-mean values if the flow field is neither steady nor homogeneous. In such cases we may assume that an average is taken over a large number of experiments

† Many investigators, including Dryden and Kuethe,[36] use the designation "intensity" or "degree of turbulence" or "turbulence level" for the relative intensity just defined.

that have the same initial and boundary conditions. We then speak of an ensemble-mean value.

If we use the Eulerian description of the flow field, one of the above three methods of averaging may have to be applied to a varying quantity at any point in the flow field.

If we want to study turbulent transport or diffusion processes, it is often convenient to apply the Lagrangian description of the paths of separate fluid particles. In this case averaging may be carried out with respect to a large number of particles that have either the same starting time but different origins (this requires on the average a homogeneous flow field) or the same origin but different starting times (this requires on the average a quasi-steady flow field). Of course we may also consider an ensemble average.

Expressed in mathematical form, the three methods of averaging applied, for instance, to the Eulerian velocity U, are:

Time average for a stationary turbulence:

$$\overset{t}{\bar{U}}(x_0) = \lim_{T\to\infty} \frac{1}{2T}\int_{-T}^{+T} dt\, U(x_0,t)$$

Space average for a homogeneous turbulence

$$\overset{s}{\bar{U}}(t_0) = \lim_{X\to\infty} \frac{1}{2X}\int_{-X}^{+X} dx\, U(x,t_0)$$

Ensemble average of N identical experiments:

$$\overset{e}{\bar{U}}(x_0,t_0) = \frac{\sum_{1}^{N} nU_n(x_0,t_0)}{N}$$

By introducing a probability density function $\mathfrak{P}(U)$, which in normalized form satisfies

$$\int_{-\infty}^{+\infty} dU\, \mathfrak{P}(U) = 1$$

the ensemble average may be also expressed as follows

$$\overset{e}{\bar{U}}(x_0,t_0) = \int_{-\infty}^{+\infty} dU \cdot U\,\mathfrak{P}(U)$$

For a stationary and homogeneous turbulence we may expect and assume that the three averaging procedures lead to the same result.

$$\overset{t}{\bar{U}} = \overset{s}{\bar{U}} = \overset{e}{\bar{U}}$$

This assumption is known as the *ergodic hypothesis*.

Actual turbulent flows are neither really stationary nor homogeneous. Therefore, for practical reasons, we cannot carry out the averaging procedures with respect to time or with respect to space for infinite values of T or X respectively, but only for finite values. However, certain conditions then have to be satisfied.

Let us for instance, consider averaging with respect to time of the Eulerian velocity of a turbulent flow. The flow may contain very slow variations that we do not wish to regard as belonging to the turbulent motion of the flow. Take, for instance, the case where the turbulent flow through a duct shows a slight pulsation of low frequency—or, in meteorology, where we wish to distinguish between the average wind speed during certain periods of the day and the average speed during much longer periods.

Therefore we take T to be a finite time interval. This interval must be sufficiently large compared with the time scale T_1 of the turbulence or else, since this corresponds to a certain quasi periodicity, with the main period of change in flow pattern. On the other hand it must be small compared with the period T_2 of any slow variations in the field of flow that we do not wish to regard as belonging to the turbulence. It is clear that there is a certain arbitrariness in the choice of the fluctuations that we do wish to consider. Fortunately, in practice, such a choice can be made without too much difficulty. If we take an oscillogram of a turbulent flow, it is usually easy to discern some average main period of the change in flow pattern. Furthermore it may be helpful to keep in mind that, for a given mean velocity, the order of magnitude of such main periods corresponds to the size of the turbulence-generating object or of the apparatus in which the turbulent flow is studied.

Taking for T a finite value, we now define the average value by

$$\bar{U} = \frac{1}{T}\int_0^T d\tau \, U(t + \tau)$$

with the condition $T_1 \ll T \ll T_2$.

The average value should be independent of the origin t of the averaging procedure, provided $t < T_2$. Thus $\partial\bar{U}/\partial t$ should be either zero or, in the case of a slightly varying main flow, negligibly small.

In the foregoing an average value has been designated by an overscore. In this book any averaging procedure will be denoted by such an overscore. In the study of turbulence we often have to carry out an averaging procedure not only on single quantities but also on products of quantities. Here the overscores have the following properties.

Let $A = \bar{A} + a$ and $B = \bar{B} + b$. In any further averaging procedure $\bar{A}$ and $\bar{B}$ may be treated as constants. Thus,

$$\bar{A} = \overline{\bar{A} + a} = \bar{\bar{A}} + \bar{a} = \bar{A} + \bar{a} \qquad \text{whence } \bar{a} = 0$$

$$\overline{\bar{A}\bar{B}} = \bar{\bar{A}}\bar{\bar{B}} = \bar{A}\bar{B}$$

$$\overline{\bar{A}b} = \bar{\bar{A}}\bar{b} = \bar{A}\bar{b} = 0 \qquad \text{since } \bar{b} = 0$$

Similarly,

$$\overline{\bar{B}a} = \bar{\bar{B}}\bar{a} = \bar{B}\bar{a} = 0 \qquad \text{since } \bar{a} = 0$$

$$\overline{AB} = \overline{(\bar{A} + a)(\bar{B} + b)} = \overline{\bar{A}\bar{B}} + \overline{\bar{A}b} + \overline{\bar{B}a} + \overline{ab} = \bar{A}\bar{B} + \overline{ab}$$

We have already mentioned the concepts of space scale and time scale and, corresponding to them, the quasi periodicity in turbulence. A study of photographs of turbulent flows or of oscillograms of velocity fluctuations will reveal that, properly speaking, it is not permissible to speak of the quasi periodicity or the scale of turbulence. It is possible to speak of an average maximum quasi periodicity or scale determined principally by the dimensions of the apparatus. But besides this there are many smaller quasi periodicities and others smaller still. Turbulence consists of many superimposed quasi-periodic motions.

The counterpart of the scale or quasi periodicity is the quasi frequency. Hence many quasi frequencies are present.

The characteristic features of turbulence: irregularity and disorderliness, involve the impermanence of the various frequencies and also of the various periodicities and scales. For this reason we have used the adjective "quasi." Henceforward, if we keep this impermanent character in mind, we can for convenience leave out the "quasi."

It is said that turbulence consists of the superposition of ever-smaller periodic motions—or, since a periodicity in velocity distribution involves the occurrence of velocity gradients which correspond to a certain vortex motion, the extent of which is determined by the periodicity, we may also say that turbulence consists of the superposition of eddies of ever-smaller sizes. But can this go on indefinitely? Intuitively, one would expect not. In real fluids viscosity effects prevent this from happening. The smaller an eddy, the greater also in general the velocity gradient in the eddy and the greater the viscous shear stress that counteracts the eddying motion. Thus, in each turbulent flow, there will be a statistical lower limit to the size of the smallest eddy; there is a minimum scale of turbulence that corresponds to a maximum frequency in the turbulent motion.

All these various-sized eddies of which a turbulent motion is composed have a certain kinetic energy, determined by their vorticity or by the intensity of the velocity fluctuation of the corresponding frequency. An interesting question which soon arises when the more detailed features of turbulence are being studied is how the kinetic energy of turbulence will be distributed according to the various frequencies. Although, as stated, in real turbulence a distinct frequency is not permanently present, yet it is possible on the average to allocate a certain amount of the total energy to a distinct frequency. Such a distribution of the energy between the frequencies is usually called an *energy spectrum*. It can be established by means of suitable instruments. Though a harmonic analysis of the velocity fluctuations can be carried out, this fact is no proof that, conversely, the turbulent fluctuations are composed of these harmonics. Compare the similar problem in the case of sound, where one may distinguish between noise (turbulent) and note (composed of a number of harmonics). Burgers[3] has drawn attention to the similar controversy in the case of light, an old one in the theory of optics, namely, whether the colors of the spectrum can be said to be present originally in white light or whether they are produced by the spectroscope.

In the foregoing we have spoken about turbulent motion, which can be assumed to consist of the superposition of eddies of various sizes and vorticities with distinguishable upper and lower limits. The upper size limit of the eddies is determined mainly by the size of the apparatus, whereas the lower limit is determined by viscosity effects and decreases with increasing velocity of the average flow, other conditions remaining the same. Within these smallest eddies the flow is of a strong viscous nature, where molecular effects are dominant. Now the reader may wonder whether these smallest eddies might not become so small that the flow within them could no longer be treated as a continuum flow. In other words, what is the size of these smallest eddies compared with the mean free path of the molecules? The following figures may help to convey an idea of the problem.

For moderate flow velocities, that is, not much greater than, say, 100 m/s, the smallest space scale or eddy will hardly be less than about 1 mm; this value is still very large compared with the mean free path in gases under atmospheric conditions, which is of the order of 10^{-4} mm. One cubic millimeter of air under atmospheric conditions contains roughly 2.7×10^{16} molecules. Thus gases under atmospheric conditions and certainly liquids also may be treated as continua in the study of turbulent flow of moderate speed.

Relevant values of turbulent fluctuations are roughly 10 per cent of average velocity and are between, say, 0.01 and 10 m/s. These values must be compared with the mean velocity of molecules, which for air is of the order of 500 m/s. Turbulence frequencies vary between, say, 1 and 10,000 s^{-1}, whereas molecular-collision frequencies for air are about 5×10^9 s^{-1}.

The domain of turbulent magnitudes is, therefore, sufficiently far away from the domain of molecular magnitudes.

We will conclude this first introduction by discussing a few photographs of fluid motion which will serve to elucidate the specific character of turbulent flow.

Figure 1-1 shows the flow pattern just downstream of a circular cylinder at low values of the Reynolds number.[4] The general flow pattern is so regular that it hardly falls within the definition of real turbulence, that is, the condition in which randomness prevails. At most this might be considered a pseudo turbulence.

Figure 1-2 shows a similar flow pattern, but one pertaining to a higher value of the Reynolds number.[5] Up to downstream distances 30 to 40 times the cylinder diameter, the general flow pattern is still fairly regular; the more detailed patterns—and, beyond this distance, the general flow pattern also—gradually become more and more turbulent. The detailed patterns become more turbulent as the Reynolds number increases; this is seen clearly in Fig. 1-3, which shows a close-up of the flow pattern close behind the cylinder.[4] Within the region of the large regular eddies

FIGURE 1-1
Flow pattern downstream of a cylinder. Low Reynolds numbers.

the flow pattern is distinctly turbulent, with a space scale much smaller than those of the large regular eddies.

The regularity and irregularity of the flow in the wake of a cylinder are well illustrated by velocity oscillograms taken at different locations in the wake flow. Figure 1-4 shows such oscillograms together with one taken in the turbulent flow through a windtunnel. An oscillogram taken at a point on the line through the centers of the vortices of each row (that is, at a point eccentric with respect to the center of the wake flow) shows a preference for a distinct frequency; an oscillogram taken centrally behind the cylinder (that is, on the center line of the wake flow) also shows a preference for a distinct frequency, which is, however, equal to twice the previous frequency (this is the effect of the vortices, which are separated from either side of the cylinder alternately). Compare these oscillograms with that for the real turbulent flow through a windtunnel, and the difference is clear.

Figure 1-5*a* to *f* shows a series of flow patterns corresponding to increasing distances downstream from a grid (mesh $M = 45$ mm, rod diameter $d = 15$ mm) which is towed at a speed of 66.5 mm/s through a stagnant liquid. Whereas

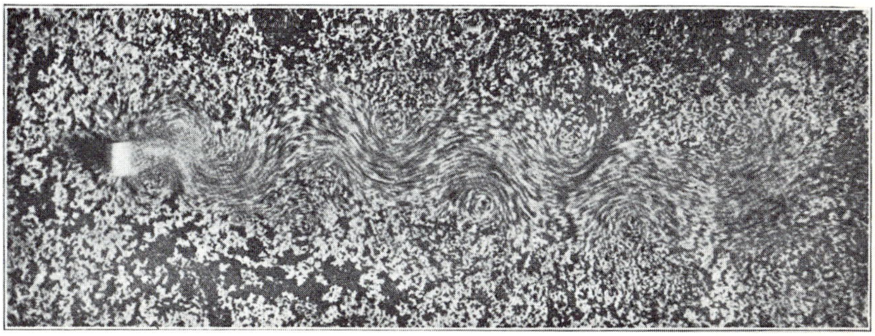

FIGURE 1-2
Flow pattern downstream of a cylinder. Moderate Reynolds number.

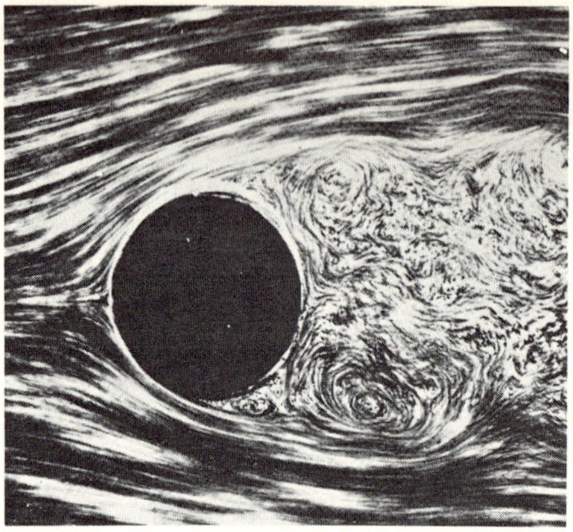

FIGURE 1-3
Flow pattern close behind a cylinder. High Reynolds number.

Fig. 1-5*a* shows a distinct and regular pattern, the patterns shown subsequently become less and less distinct. Figure 1-5*d* to *f* shows, in addition to the gradually increasing turbulent character, the decay of the eddies; the contours become less sharp.

The foregoing photographs have shown the generation of turbulence through the flow of a fluid relative to a solid body. The following photographs show this generation and development of turbulence when two neighboring streams of the same fluid flow past each other.

Figure 1-6 shows the initial vortices produced at the boundary of a free jet (half-jet boundary). During the development of these vortices farther downstream, there is a gradual transition into irregular turbulent flow. Figure 1-7 shows the

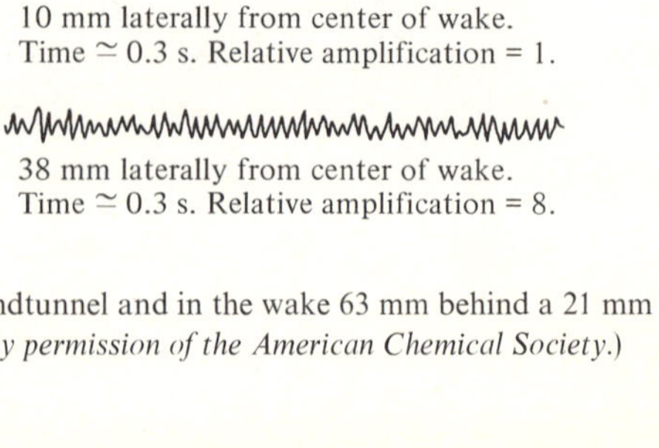
Turbulence in windtunnel, $u'/\overline{U} = 0.0085$.
Time $\simeq$ 0.4 s. Relative amplification = 64.
Center of wake. 63 mm behind 21 mm cylinder.
Time $\simeq$ 0.3 s. Relative amplification = 1.
10 mm laterally from center of wake.
Time $\simeq$ 0.3 s. Relative amplification = 1.
38 mm laterally from center of wake.
Time $\simeq$ 0.3 s. Relative amplification = 8.

FIGURE 1-4
Oscillograms of turbulence in a windtunnel and in the wake 63 mm behind a 21 mm cylinder. (From: *Dryden, H. L.,*[6] *by permission of the American Chemical Society.*)

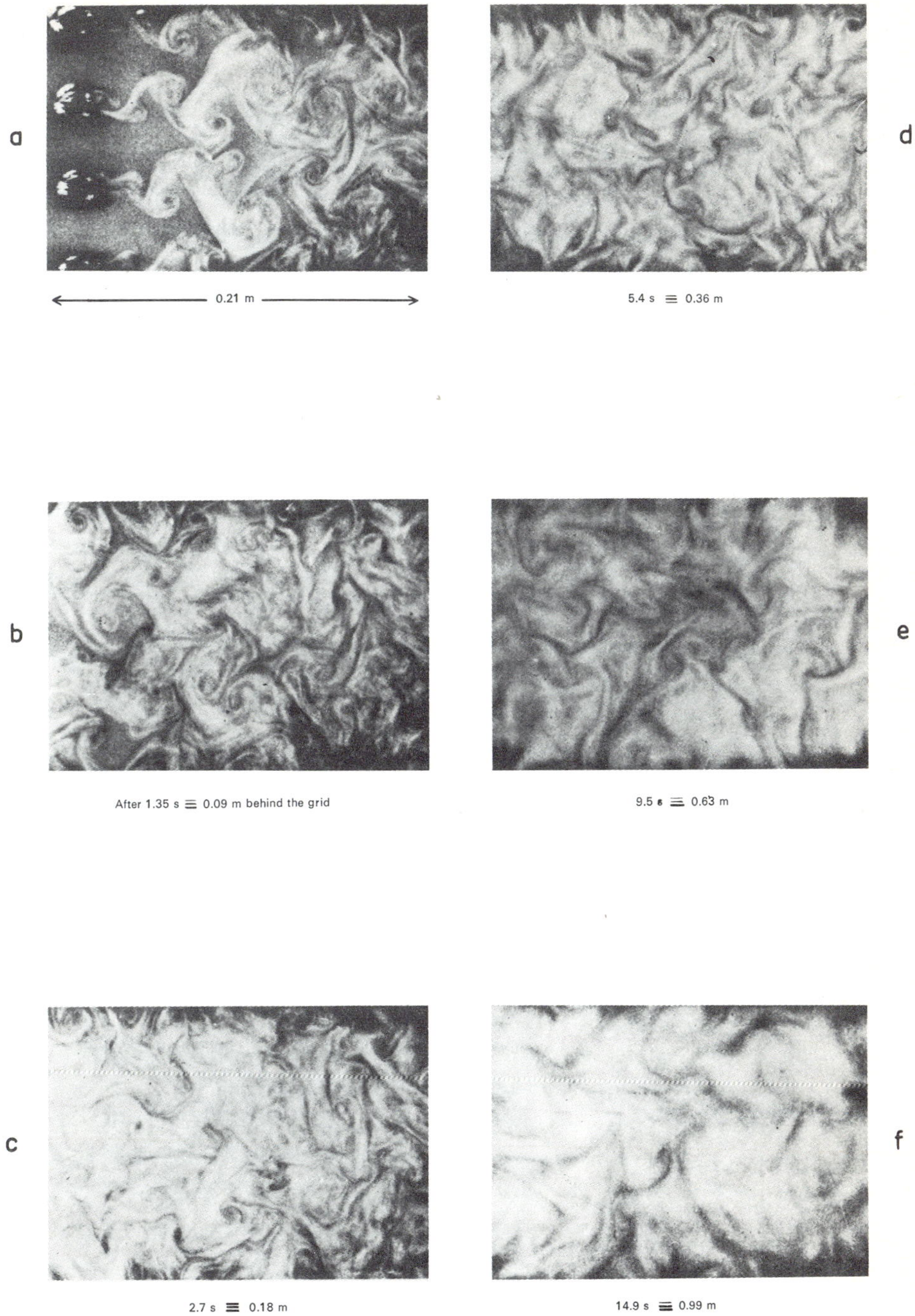

FIGURE 1-5
Flow pattern behind a grid. $U = 66.5$ mm/s, $M = 45$ mm, and $d = 15$ mm. (From: *Prandtl, L.*[7])

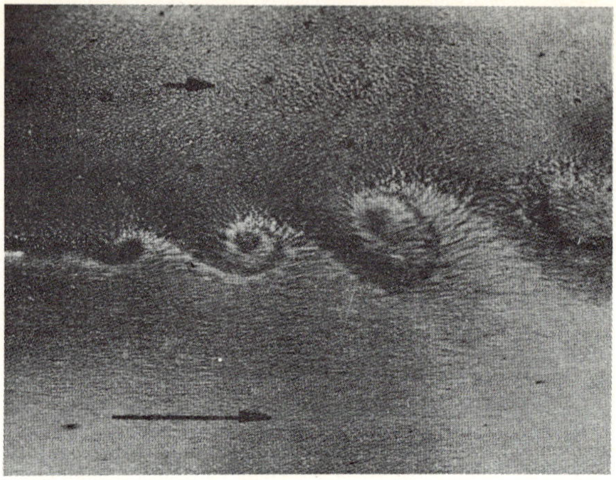

FIGURE 1-6
Vortices at boundary of a half-jet. (From: *Flügel, G.,*[8] *by permission of the Verlag des Vereins Deutscher Ingenieure.*)

turbulent character of the flow in a free jet, with the separate eddying domains at the boundary region still distinguishable.

Figure 1-8 shows a "Schlieren" photograph of a jet at just the moment when it was issuing.[9] In the first stages distinct separate vortices are seen, in the later stages complete turbulent flow.

As Fig. 1-7 shows, the flow in the central region of a jet is different from the flow in the boundary region: the boundary-region flow is not continuous but becomes more and more intermittent toward the outside. This difference in character will of course find expression also in oscillograms taken at different points in the jet. This is shown by Fig. 1-9. This difference in character will be discussed

FIGURE 1-7
Flow pattern in a free jet. (*Photograph by Van der Hegge Zijnen; taken at Royal/Dutch-Shell Laboratory, Delft.*)

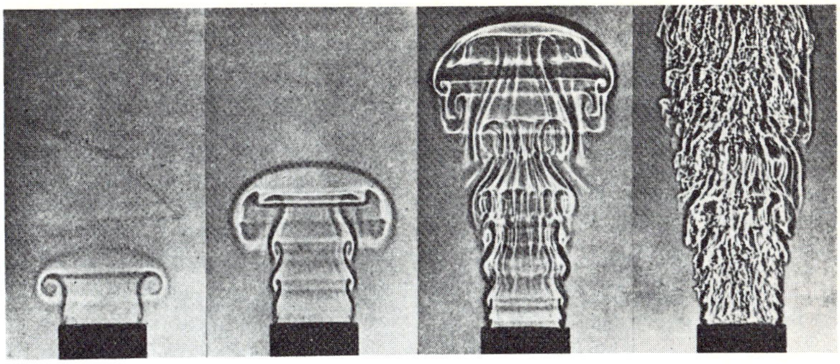

FIGURE 1-8
"Schlieren" photographs of a starting free jet. (From: *Garside, J. E., et al.*[9])

extensively in Chap. 6. At the moment it will be sufficient to mention that this intermittence, which is typical of turbulence in free boundaries, indicates the presence of large-scale eddies.

Finally, Fig. 1-10*a* and *b* shows the flow through a channel as photographed with a moving camera.[4] With a view to a more precise definition of scale of

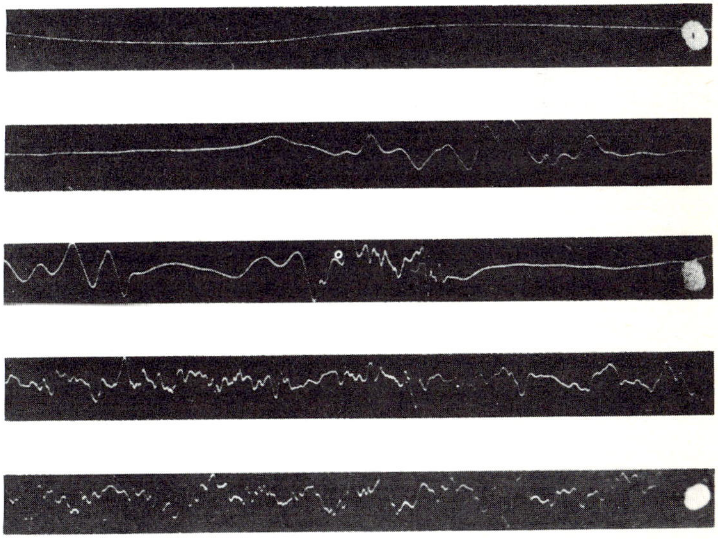

FIGURE 1-9
Oscillograms of velocity fluctuations in a free jet at different distances from the axis. (From: *Corrsin, S.,*[10] *by permission of the National Advisory Committee for Aeronautics.*)

(a)

(b)

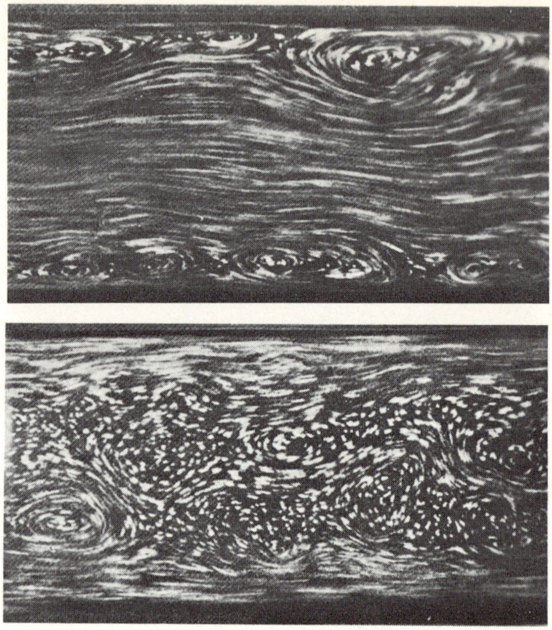

FIGURE 1-10
Flow pattern in a channel. (*Nikuradse, in Prandtl, L. and O. Tietjens.*[4])

turbulence, to be given later, it is important to note that there seems to be a correlation between the velocities within a region which extends from the center of the channel to about midway between the center and the wall (this is particularly strongly suggested by Fig. 1-10*b*).

In the foregoing we have mentioned and discussed briefly certain features of turbulent flows. It may be useful to summarize them.

Thus, turbulence is a random phenomenon which shows a quasi-permanency and quasi-periodicity both in time and in space. In a sense this may be considered as being an assumption, due to the fact that our information concerning the turbulent flow at any instant is incomplete, so that henceforth it is impossible to predict in detail the future behavior of the flow field.

Another assumption is that turbulence is a continuum phenomenon, an assumption applicable to liquids and gases under atmospheric conditions. But caution is dictated in the case of ultra-high supersonic or hypersonic turbulent flows.

The turbulence is characterized by a strong diffusive nature with respect to any transferable property. And as all flows of real fluids it is dissipative due to viscous actions; and therefore rotational. This dissipation takes place mainly in the region of the smallest eddies of the hierarchy of eddies of many different sizes of which the turbulence may be composed. There is a strong interaction between these eddies due to the nonlinear and three-dimensional character of turbulence. For, a nonhomogeneous vortex flow pattern can only be either one-dimensional or three-dimensional.

These nonlinear and spatial interactions of vortices resulting in transfer of kinetic energy between them, the diffusive nature of turbulence, and another important property not mentioned hitherto, namely the memory behavior, will be discussed in more detail in later chapters.

1-2 EQUATIONS OF MOTION FOR TURBULENT FLOW; REYNOLDS STRESSES

In an Eulerian description of the flow field the balance of a transferable property $\mathscr{P}$, defined per unit of mass, in a unit control volume reads

$$\frac{\partial}{\partial t}(\rho\mathscr{P}) + \frac{\partial}{\partial x_j}(U_j\rho\mathscr{P}) = -\frac{\partial}{\partial x_j}(\mathfrak{J}_{\mathscr{P}})_j + F_{\mathscr{P}} \tag{1-1}$$

Here, and in the following, Cartesian tensor notation will be used, and the summation convention with respect to repeated indices.

The left-hand side shows the change with time of the property in a unit volume, and the change due to a divergence in convective transport by the flow through the boundaries of the control volume.

The first term on the right-hand side describes the divergence of the transport $(\mathfrak{J}_{\mathscr{P}})_j$ through these boundaries by molecular effects. It is defined positive if the transport is in the positive x_j-direction. The last term stands for any internal or external process, or source, that contributes to the change of $\mathscr{P}$ in the control volume.

The expression of the left-hand side is general, independent of $\mathscr{P}$ and of the process concerned. The first term on the right-hand side depends on the nature of the property, while the last term depends on both the nature of the property and the process considered.

If we consider the mass as the transferable property, thereby making no distinction between possible components of which the fluid is composed, then $\mathscr{P}$ becomes equal to unity, being the mass per unit of mass, and $\mathfrak{J}_{\mathscr{P}} = 0$. If we further assume no sources or sinks of mass present in the flow, also $F_{\mathscr{P}} = 0$. We then obtain the equation for the conservation of mass

$$\frac{\partial\rho}{\partial t} + \frac{\partial}{\partial x_i}\rho U_i = 0 \tag{1-2}$$

An alternative expression is obtained if we introduce the dilatation

$$\Theta = \frac{\partial U_i}{\partial x_i}$$

viz.

$$\frac{\partial \rho}{\partial t} + U_i \frac{\partial \rho}{\partial x_i} + \rho \frac{\partial U_i}{\partial x_i} = \frac{D\rho}{Dt} + \rho\Theta = 0$$

Or

$$\frac{1}{\rho}\frac{D\rho}{Dt} = -\Theta \qquad \text{(1-2a)}$$

Here

$$\frac{D}{Dt} = \frac{\partial}{\partial t} + U_i \frac{\partial}{\partial x_i}$$

In turbulent flows this equation must hold at any instant, but also on the average. How will the equation read for the average values $\bar{\rho}$ and $\bar{U}_i$? For this write

$$\rho = \bar{\rho} + \tilde{\rho} \qquad \text{and} \qquad U_i = \bar{U}_i + u_i$$

Substitute these expressions in Eq. (1-2) and carry out the averaging procedure

$$\overline{\frac{\partial(\bar{\rho} + \tilde{\rho})}{\partial t}} + \frac{\partial}{\partial x_i} \overline{(\bar{\rho} + \tilde{\rho})(\bar{U}_i + u_i)} = 0$$

Apply the properties of overscores given in the first section. Then we obtain

$$\frac{\partial \bar{\rho}}{\partial t} + \frac{\partial}{\partial x_i}(\bar{\rho}\bar{U}_i + \overline{\tilde{\rho}u_i}) = 0 \qquad \text{(1-3)}$$

Thus we can see that we cannot simply apply an equation like (1-2) to the average values $\bar{\rho}$ and $\bar{U}_i$, but that an additional term occurs when there is a correlation between the density fluctuations and the velocity fluctuations.

As was first shown by Osborne Reynolds, the equations of motion for the average values in turbulent flows can be derived similarly from the Navier-Stokes equations, which apply for the instantaneous values. But before doing this, it might be worth while to give first a short derivation of these equations in order to bring to the fore the meaning of the separate terms.

Since these equations are, in fact, nothing but equations for the conservation of momentum, we may start from the above general balance equation, where in this case we take for $\mathscr{P}$ any of the three components U_i of momentum per unit of mass.

The transfer of momentum by molecular effects results in a stress, so that for the balance equation in the x_i-direction we have

$$(\mathfrak{J}_{U_i})_j = -\sigma_{ji}$$

The minus sign results from the definition of σ_{ji}. Assume a control-surface element perpendicular to the x_j-direction. Then, σ_{ji} is the stress in the x_i-direction exerted by the fluid at the positive side of the surface element on the fluid at the negative side of the surface element. The stress is defined positive if it is directed in the positive direction of x_i.

Let F_i be an external force working on a unit volume of the fluid in the x_i-direction. The balance equation for the momentum U_i then reads

$$\frac{\partial}{\partial t}\rho U_i + \frac{\partial}{\partial x_j}\rho U_j U_i = \frac{\partial}{\partial x_j}\sigma_{ji} + F_i \tag{1-4}$$

With Eq. (1-2) the left-hand side can be written in an alternative way

$$\frac{\partial}{\partial t}\rho U_i + \frac{\partial}{\partial x_j}\rho U_j U_i = \rho\left(\frac{\partial U_i}{\partial t} + U_j\frac{\partial U_i}{\partial x_j}\right) = \rho\frac{DU_i}{Dt}$$

so that Eq. (1-4) becomes

$$\rho\frac{DU_i}{Dt} = \frac{\partial}{\partial x_j}\sigma_{ji} + F_i \tag{1-4a}$$

Of course, these equations can be obtained in a direct way from the dynamical equilibrium of a cube of fluid. Consider for instance the x_1-direction. On a cubical fluid element we have the following divergence of surface forces (see Fig. 1-11).

$$\frac{\partial}{\partial x_1}\sigma_{11}\,dx_1\,dx_2\,dx_3 \qquad \frac{\partial}{\partial x_2}\sigma_{21}\,dx_2\,dx_1\,dx_3 \qquad \frac{\partial}{\partial x_3}\sigma_{31}\,dx_3\,dx_1\,dx_2$$

With an external force F_1 working on a unit volume of the fluid in the x_1-direction, Newton's second law applied to this unit volume results in the balance equation for the x_1-direction:

$$\rho\frac{DU_1}{Dt} = \frac{\partial}{\partial x_1}\sigma_{11} + \frac{\partial}{\partial x_2}\sigma_{21} + \frac{\partial}{\partial x_3}\sigma_{31} + F_1$$

Similar equations are obtained for the equilibrium in the x_2- and x_3-directions, respectively.

Using Cartesian tensor notation, these equations can be written in the condensed form of Eq. (1-4a).

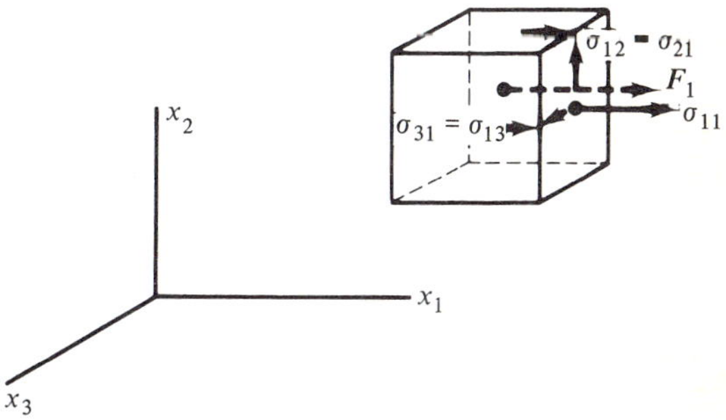

FIGURE 1-11
Stresses on a cubic element of the fluid.

The nine stress components σ_{ji} form a second-order tensor. This stress tensor can be divided into a part that corresponds to the average value of the normal stresses for all directions, that is, a spherically-symmetric part, which is invariant under rotation of the coordinate system, and an anti-symmetric part.

The spherically-symmetric part is equal to $\frac{1}{3}\sigma_{ii}$. It contains the thermodynamic pressure P and, in general, an additional term proportional with Θ. So:

$$\tfrac{1}{3}\sigma_{ii} = -P + \kappa\Theta \tag{1-5}$$

We shall not give the derivation of the result $\frac{1}{3}\sigma_{ii}$ for the symmetric part; the reader will find it in any textbook on advanced stress analysis. Moreover, a similar case, the distribution of turbulence stresses, will be considered in greater detail in Sec. 1-4.

With Eq. (1-5) the expression for the stress tensor can be written

$$\sigma_{ji} = [-P + \kappa\Theta]\,\delta_{ji} + \sigma_{ji}^* \tag{1-6}$$

where δ_{ji} is the Kronecker delta, a unit second-order tensor; σ_{ji}^* is the anti-symmetrical part, so that $\sigma_{ii}^* = 0$.

The stresses σ_{ji} will cause deformations of the fluid elements. Since these deformations are determined by the spatial variations $\partial U_i/\partial x_j$ of the velocities U_i, it is possible to relate the stresses σ_{ji} to these spatial variations.

The spatial variations $\partial U_i/\partial x_j$ also form a second-order tensor, which can be divided into a symmetrical and an antisymmetrical part.

$$\frac{\partial}{\partial x_j}U_i = \frac{1}{2}\left(\frac{\partial U_i}{\partial x_j} + \frac{\partial U_j}{\partial x_i}\right) + \frac{1}{2}\left(\frac{\partial U_i}{\partial x_j} - \frac{\partial U_j}{\partial x_i}\right)$$

$$= \tfrac{1}{2}D_{ij} + \tfrac{1}{2}\Omega_k \varepsilon_{ijk}$$

where ε_{ijk} is the alternating, third-order tensor.

The symmetrical part D_{ij} determines the deformation of the fluid and is called the deformation tensor. The antisymmetrical part $\Omega_k \varepsilon_{ijk}$ determines a rotation without deformation; it defines the vorticity of the motion. Ω_k is called the vector of the antisymmetrical tensor. It is identified with the vorticity, and an equivalent relation is

$$\Omega_k = \frac{\partial U_i}{\partial x_j}\varepsilon_{ijk}$$

As stated, the stresses cause a deformation of the fluid elements. Hence it is logical to assume a relation between the stress tensor and the deformation tensor. Throughout the book we will deal only with Newtonian fluids, for which a linear relationship between σ_{ji}^* and D_{ji} is assumed

$$\sigma_{ji}^* = AD_{ji} + B\,\delta_{ji} \tag{1-7}$$

Since σ_{ji}^* is the anti-symmetrical part of the stress tensor, and $\sigma_{ii}^* = 0$, we must have

$$B = -\tfrac{1}{3}AD_{ii} = -\tfrac{2}{3}A \cdot \Theta$$

By considering a simple one-dimensional shear flow, A is identified as being the dynamical viscosity μ of the fluid.

The expression for σ_{ji} now reads

$$\sigma_{ji} = (-P + \kappa\Theta)\,\delta_{ji} + \mu D_{ji} - \tfrac{2}{3}\mu\Theta\,\delta_{ji} \tag{1-8}$$

The constant κ occurring in Eq. (1-5) is usually referred to as the volume viscosity of the fluid.

For most Newtonian fluids $\kappa \simeq 0$, and its effect may therefore be neglected. In that case Eq. (1-8) can be written

$$\sigma_{ji} = -P\,\delta_{ji} + \mu(D_{ji} - \tfrac{2}{3}\Theta\,\delta_{ji}) \tag{1-9}$$

Substitution of this equation for σ_{ji} in Eq. (1-4a) yields

$$\rho\frac{DU_i}{Dt} = \frac{\partial}{\partial x_j}(-P\,\delta_{ji} + \mu D_{ji} - \tfrac{2}{3}\mu\Theta\,\delta_{ji}) + F_i$$

or

$$\rho\frac{DU_i}{Dt} = -\frac{\partial P}{\partial x_i} + \frac{\partial}{\partial x_j}\mu D_{ji} - \frac{2}{3}\frac{\partial}{\partial x_i}\mu\Theta + F_i \tag{1-10}$$

This is the general expression for the Navier-Stokes equation for a compressible fluid and a viscosity μ that may still be a function of the space coordinates.

In general μ is a function of both temperature and pressure, though the latter dependence is very slight. Through its temperature dependence the relative variations of the dynamic viscosity in compressible flows are of the same order of magnitude as the relative variations in density. Except, of course, when the flow is isothermal.

Equation (1-10) may also be written in a different way, by making use of Eq. (1-2a). For an isentropic process

$$\frac{1}{\rho}\frac{D\rho}{Dt} = \frac{1}{\kappa P}\frac{DP}{Dt} = \frac{1}{\rho c^2}\frac{DP}{Dt}$$

where $\kappa = c_p/c_v$ and c is the velocity of sound.

Thus

$$\Theta = -\frac{1}{\rho c^2}\frac{DP}{Dt} \tag{1-11}$$

and Eq. (1-10) may be written

$$\rho\frac{DU_i}{Dt} = \frac{\partial}{\partial x_j}\mu D_{ji} - \frac{\partial}{\partial x_i}\left(P - \frac{2}{3}\frac{\mu}{\rho c^2}\frac{dP}{dt}\right) + F_i \tag{1-12}$$

Two cases are of special interest, namely:

1. *Constant Viscosity;* $\mu =$ *Constant*. In the case of isothermal flow we have

$$\frac{\partial}{\partial x_j}(\mu D_{ji}) = \mu\frac{\partial}{\partial x_j}D_{ji} = \mu\frac{\partial}{\partial x_j}\left(\frac{\partial U_i}{\partial x_j} + \frac{\partial U_j}{\partial x_i}\right)$$

$$= \mu \frac{\partial^2}{\partial x_j \partial x_j} U_i + \mu \frac{\partial}{\partial x_j} \frac{\partial U_j}{\partial x_i}$$

$$= \mu \frac{\partial^2}{\partial x_j \partial x_j} U_i + \mu \frac{\partial}{\partial x_i} \Theta$$

so Eq. (1-10) becomes

$$\rho \frac{DU_i}{Dt} = -\frac{\partial P}{\partial x_i} + \mu \frac{\partial^2}{\partial x_j \partial x_j} U_i + \frac{1}{3} \mu \frac{\partial \Theta}{\partial x_i} + F_i \tag{1-13}$$

and Eq. (1-12) becomes

$$\rho \frac{DU_i}{Dt} = \mu \frac{\partial^2}{\partial x_j \partial x_j} U_i - \frac{\partial}{\partial x_i}\left(P + \frac{1}{3} \frac{\mu}{\rho c^2} \frac{dP}{dt}\right) + F_i \tag{1-14}$$

where c is now the isothermal speed of propagation.

2. *Incompressible Fluid;* $\rho =$ *Constant;* $\Theta = 0$.

$$\rho \frac{DU_i}{Dt} = -\frac{\partial P}{\partial x_i} + \frac{\partial}{\partial x_j} \mu D_{ji} + F_i \tag{1-15}$$

or, since

$$\frac{\partial}{\partial x_j} \mu D_{ji} = \mu\left(\frac{\partial^2 U_i}{\partial x_j \partial x_j} + \frac{\partial^2 U_j}{\partial x_i \partial x_j}\right) + \left(\frac{\partial U_i}{\partial x_j} + \frac{\partial U_j}{\partial x_i}\right) \frac{\partial \mu}{\partial x_j}$$

$$= \frac{\partial}{\partial x_j}\left(\mu \frac{\partial U_i}{\partial x_j}\right) + \frac{\partial U_j}{\partial x_i} \frac{\partial \mu}{\partial x_j}$$

$$\rho \frac{DU_i}{Dt} = -\frac{\partial P}{\partial x_i} + \frac{\partial}{\partial x_j}\left(\mu \frac{\partial U_i}{\partial x_j}\right) + \frac{\partial U_j}{\partial x_i} \frac{\partial \mu}{\partial x_j} + F_i \tag{1-16}$$

The above equations apply to nonturbulent as well as turbulent flows. In order to show the turbulent motions and their interference with the mean motion more explicitly, we follow a procedure due to Osborne Reynolds. Let us apply this procedure to Eq. (1-13) for the flow of a compressible fluid with constant viscosity, i.e. for isothermal flow.

Put

$$U_i = \bar{U}_i + u_i \qquad \rho = \bar{\rho} + \tilde{\rho} \qquad P = \bar{P} + p$$

where $\bar{A} = 1/T \int_0^T A(t + \tau)\, d\tau$, with T large compared with the time scale of the turbulent motions.

Substitution of these expressions in Eq. (1-13) and averaging with respect to time must give the required equations for turbulent flow. When we apply the averaging procedure, we have to use the properties of overscores given in Sec. 1-1. We have also to keep in mind that

$$\frac{DU_i}{Dt} = \frac{\partial U_i}{\partial t} + U_j \frac{\partial U_i}{\partial x_j}$$

We then obtain for the left-hand side of Eq. (1-13)

$$\overline{\rho\frac{DU_i}{Dt}} = \bar{\rho}\frac{\partial \bar{U}_i}{\partial t} + \bar{\rho}\bar{U}_j\frac{\partial \bar{U}_i}{\partial x_j} + \bar{\rho}\overline{u_j\frac{\partial u_i}{\partial x_j}} + \overline{\tilde{\rho}\frac{\partial u_i}{\partial t}} + \overline{\tilde{\rho}u_j}\frac{\partial \bar{U}_i}{\partial x_j} + \bar{U}_j\overline{\tilde{\rho}\frac{\partial u_i}{\partial x_j}} + \overline{\tilde{\rho}u_j\frac{\partial u_i}{\partial x_j}}$$

For the right-hand side we obtain

$$-\frac{\partial \bar{P}}{\partial x_i} + \mu\frac{\partial^2}{\partial x_j \partial x_j}\bar{U}_i + \frac{1}{3}\mu\frac{\partial \bar{\Theta}}{\partial x_i} + \bar{F}_i$$

So Eq. (1-13) becomes

$$\bar{\rho}\left(\frac{\partial \bar{U}_i}{\partial t} + \bar{U}_j\frac{\partial \bar{U}_i}{\partial x_j}\right) = -\frac{\partial \bar{P}}{\partial x_i} + \mu\left(\frac{\partial^2 \bar{U}_i}{\partial x_j \partial x_j} + \frac{1}{3}\frac{\partial \bar{\Theta}}{\partial x_i}\right) + \bar{F}_i - \bar{\rho}\overline{u_j\frac{\partial u_i}{\partial x_j}} - \overline{\tilde{\rho}\frac{\partial u_i}{\partial t}}$$
$$- \overline{\tilde{\rho}u_j}\frac{\partial \bar{U}_i}{\partial x_j} - \bar{U}_j\overline{\tilde{\rho}\frac{\partial u_i}{\partial x_j}} - \overline{\tilde{\rho}u_j\frac{\partial u_i}{\partial x_j}} \qquad (1\text{-}17)$$

Thus we see that, besides the well-known terms referring to the mean motion, there occur on the right-hand side additional terms due to the turbulent motion; four are determined by the double and even triple correlations between density and velocity fluctuations and one by the double correlation $\overline{u_i u_j}$.

By applying the equation of continuity (1-2) to the turbulence terms on the right-hand side of Eq. (1-17), one obtains

$$-\left(\frac{\partial}{\partial x_j}\bar{\rho}\overline{u_i u_j} + \frac{\partial}{\partial t}\overline{\tilde{\rho}u_i} + \overline{\tilde{\rho}u_j}\frac{\partial}{\partial x_j}\bar{U}_i + \frac{\partial}{\partial x_j}\overline{\tilde{\rho}u_i}\bar{U}_j + \frac{\partial}{\partial x_j}\overline{\tilde{\rho}u_i u_j}\right)$$

Hence, in fact the effect of the turbulence is determined by the three correlations $\overline{u_i u_j}$, $\overline{\tilde{\rho}u_i}$, and $\overline{\tilde{\rho}u_i u_j}$.

A considerable simplification and reduction in length of the equations for turbulent flow can be obtained if *density-weighted mean velocities* are introduced as suggested and worked out by Favre.[12] The density-weighted mean velocity $\bar{U}_i^0$ is defined by

$$\bar{\rho}\bar{U}_i^0 = \overline{\rho U_i}$$

where $\tilde{\rho} = \rho - \bar{\rho}$, and $\bar{\tilde{\rho}} = 0$.

If we put

$$U_i = \bar{U}_i + u_i = \bar{U}_i^0 + u_i^0$$

it follows that in contrast with $\bar{u}_i = 0$, $\bar{u}_i^0 \neq 0$. Furthermore $\overline{\rho u_i^0} = 0$.

The averaged expression for the mass-balance equation reads

$$\frac{\partial \bar{\rho}}{\partial t} + \frac{\partial}{\partial x_i}\bar{\rho}\bar{U}_i^0 = 0$$

which has to be compared with Eq. (1-3).

The momentum-balance equation corresponding to Eq. (1-17) becomes

$$\bar{\rho}\left(\frac{\partial \bar{U}_i^0}{\partial t} + \bar{U}_j^0\frac{\partial \bar{U}_i^0}{\partial x_j}\right) = \bar{\rho}\frac{\mathring{D}\bar{U}_i^0}{Dt} = -\frac{\partial \bar{P}}{\partial x_i} + \frac{\partial}{\partial x_j}\left\{\mu\left(\frac{\partial \bar{U}_i}{\partial x_j} + \frac{\partial \bar{U}_j}{\partial x_i} - \frac{2}{3}\bar{\Theta}\right) - \overline{\rho u_i^0 u_j^0}\right\} + \bar{F}_i$$

Apart from being more simple, the last equation has only one turbulence term, which can be understood more easily than many of the turbulence terms in Eq. (1-17). Also, as will be discussed in Chap. 2, many measuring methods determine density-weighted velocities rather than the non-density-weighted velocities. However, there is the other side to the picture. We have to distinguish the Lagrangian differential operator D/Dt from $\mathring{D}/Dt$ which takes into account changes in volume and mass. Second, the momentum-balance equation contains the non-density-weighted velocities, determining the deformations of the mean motion, in addition to the density-weighted velocities. The same holds true for the energy dissipation term occurring in the balance equation for the mechanical energy.

The compressibility effect in the turbulent motions may be neglected if $\tilde{\rho}/\bar{\rho} \ll 1$. For these small values of $\tilde{\rho}/\bar{\rho}$ we have the approximation $\tilde{\rho}/\bar{\rho} \simeq p/\bar{\rho}c^2$. Since p is of the order of $\bar{\rho}u^2$, $\tilde{\rho}/\bar{\rho}$ becomes of the order of u^2/c^2, that is, of the square of the Mach number of the turbulence.

An appreciable simplification is then obtained in Eq. (1-17), because four out of the five turbulence terms may be neglected.

Thus, for an *incompressible* fluid and *constant viscosity*, Eq. (1-17) reduces to

$$\rho\left(\frac{\partial \bar{U}_i}{\partial t} + \bar{U}_j \frac{\partial \bar{U}_i}{\partial x_j}\right) = -\frac{\partial \bar{P}}{\partial x_i} + \mu \frac{\partial^2}{\partial x_j \partial x_j} \bar{U}_i - \rho \overline{u_j \frac{\partial u_i}{\partial x_j}} + \bar{F}_i$$

This equation may be written in another form which shows more clearly the physical meaning of the turbulence term on the right-hand side.

Because of the assumption of incompressibility, we have

$$0 = \overline{U_i \frac{\partial U_j}{\partial x_j}} = \bar{U}_i \frac{\partial \bar{U}_j}{\partial x_j} + \overline{u_i \frac{\partial u_j}{\partial x_j}} = \overline{u_i \frac{\partial u_j}{\partial x_j}}$$

If we add $-\overline{u_i\, \partial u_j/\partial x_j} = 0$ to the right-hand side of the equation of motion, it becomes

$$\rho\left(\frac{\partial \bar{U}_i}{\partial t} + \bar{U}_j \frac{\partial \bar{U}_i}{\partial x_j}\right) = -\frac{\partial \bar{P}}{\partial x_i} + \frac{\partial}{\partial x_j}\left(\mu \frac{\partial \bar{U}_i}{\partial x_j} - \rho \overline{u_i u_j}\right) + \bar{F}_i \qquad \text{(1-18)}$$

Comparing this equation with Eq. (1-4a), we come to the conclusion that the turbulence terms $\rho\overline{u_i u_j}$ can be interpreted as stresses on an element of the fluid in addition to the stresses determined by the pressure P and the viscous stresses.

With these turbulence stresses the complete stress components read

$$\sigma_{ij} = -\bar{P}\,\delta_{ij} + \mu \bar{D}_{ij} - \rho \overline{u_i u_j} \qquad \text{(1-19)}$$

Because Reynolds was the first to give the equation for turbulent flow in the form of Eq. (1-18), the turbulence stresses $\rho\overline{u_i u_j}$ are often called Reynolds stresses.

Reynolds stresses have normal as well as tangential components. The normal components $\rho\overline{u_n{}^2}$ are obtained by putting $i = j$. That this may be interpreted as a normal stress can be demonstrated in the following way.

Consider a surface element dS perpendicular, for instance, to the x_1-direction. Consider further only the turbulent motion through that element, so that there is no average motion $\bar{U}_1$. Then $\rho\overline{u_1{}^2}\, dS$ denotes the average flow of momentum through that surface element, namely $\overline{(\rho u_1)u_1}\, dS$. Because this flow of momentum causes a reaction (in this case a pressure) on the surface dS, the term $\rho\overline{u_1^2}$ in the expression for the positive stress σ_{11} has a negative sign.

Similarly we can show that the terms $-\rho\overline{u_i u_j}$ for $i \neq j$ may be interpreted as tangential stresses. Consider, for instance, $-\rho\overline{u_2 u_1}$. We now take the surface element dS perpendicular to the x_2-direction. Fluid elements passing this plane have both a u_1 and a u_2 velocity component. If, on the average, most of the particles with a positive value of u_2 have also a positive value of u_1 and vice versa, then on the average $\overline{u_2 u_1}$ must be positive. In this case a force in the positive-x_1-direction is exerted on the fluid at the positive-x_2-side of the plane dS; and, conversely, a force in the negative-x_1-direction will be exerted on the fluid at the negative-x_2-side of the plane dS. Consequently, $\sigma_{21} = -\rho\overline{u_2 u_1}$.

Though, according to the above, we may interpret in a "macroscopic" consideration the terms $-\rho\overline{u_i u_j}$ as stresses, we still have to appreciate that they stem from the convective accelerations in the turbulent motion, a point which has been stressed in particular by Irmay.[51,52]

In comparing the turbulence stresses in the equations of motion with the corresponding stresses caused by viscosity effects, it is tempting to assume that the turbulence stresses act like the viscous stresses, that is, that they are directly proportional to the velocity gradient. This assumption was made by Boussinesq,[11] who introduced the concept of an "apparent," or "turbulence," or "eddy" viscosity ϵ_m, such that the stresses σ_{ij} read

$$\sigma_{ij} = -\bar{P}\,\delta_{ij} + (\mu + \rho\,\epsilon_m)\bar{D}_{ij} \tag{1-20}$$

If this expression is compared with the corresponding relation (1-19), we obtain the relation

$$-\overline{u_i u_j} = \epsilon_m \bar{D}_{ij} = \epsilon_m\left(\frac{\partial \bar{U}_i}{\partial x_j} + \frac{\partial \bar{U}_j}{\partial x_i}\right) \tag{1-21}$$

According to Boussinesq's concept, the eddy viscosity ϵ_m has a scalar value. Originally Boussinesq assumed only directional constancy but, in the application of the theory to turbulent flows through channels, he assumed that ϵ_m was spatially constant also. Such a constant value can be expected to occur only if the turbulent flow field is at least homogeneous. Although, as we shall discuss later (see Chap. 6), in a few cases of free turbulent flows which are not homogeneous and which show a pronounced velocity gradient and, hence, a shear stress in one direction only, it is possible to describe the over-all flow field in a satisfactory way on the assumption of a constant eddy viscosity, a constant value cannot be expected as a general rule. In turbulent shear flow near walls, for instance, eddy viscosity is not constant in the boundary layer.

Moreover, if ϵ_m were a constant scalar, a general relation like Eq. (1-21) could not be correct. For, if we apply a contraction to this relation, we obtain

$$-\overline{u_i u_i} = 2 \times \text{kinetic energy per unit of mass}$$

$$= 2\epsilon_m \frac{\partial \bar{U}_i}{\partial x_i}$$

For an incompressible fluid, the right-hand side is zero for finite values of ϵ_m; but the left-hand side can be zero only if there is no turbulence at all.

If we still want to stick to the concept of a scalar eddy viscosity, a more correct procedure would be to extract the average turbulence pressure from the turbulence stresses as a separate term (see Sec. 1-4).

$$\bar{P}_t = \underset{\substack{\text{Average} \\ \text{over all } n}}{\rho\langle\overline{u_n^2}\rangle} = \tfrac{1}{3}\rho\overline{u_i u_i}$$

Instead of the relations (1-20) and (1-21) we must then consider

$$\bar{\sigma}_{ij} = -(\bar{P} + \bar{P}_t)\,\delta_{ij} + (\mu + \rho\,\epsilon_m^*)\bar{D}_{ij} \tag{1-22}$$

and

$$-\overline{u_i u_j} = -\tfrac{1}{3}\overline{u_k u_k}\cdot\delta_{ij} + \epsilon_m^*\bar{D}_{ij} \tag{1-23}$$

A contraction applied to the latter relation now makes the two sides of the equation identical for an incompressible fluid.

Of course the turbulence pressure $\bar{P}_t$ has to be clearly distinguished from the static pressure $\bar{P}$. There is no similar relation between $\bar{P}_t$ and the velocity head of the mean flow as occurs for the static pressure $\bar{P}$. A safe attitude would be to consider $\bar{P}_t$ as a second parameter in addition to the "eddy viscosity" ϵ_m^* for describing the stress-strain relationship for the mean motion.

Instead of using the relation (1-23) it is possible to express the complete turbulence stress tensor $-\overline{u_i u_j}$ in terms of the deformation tensor $\bar{D}_{ij}$ for the mean motion, like Eq. (1-21), however assuming a second-order tensor for the eddy viscosity

$$-\overline{u_i u_j} = (\epsilon_m)_{ik}\left(\frac{\partial \bar{U}_j}{\partial x_k} + \frac{\partial \bar{U}_k}{\partial x_j}\right) \tag{1-24}$$

Or, even assuming a fourth-order tensor for the eddy viscosity

$$-\overline{u_i u_j} = (\epsilon_m)_{ijkl}\left(\frac{\partial \bar{U}_k}{\partial x_l} + \frac{\partial \bar{U}_l}{\partial x_k}\right) \tag{1-25}$$

The components of the second-order tensor $(\epsilon_m)_{ik}$, and of the fourth-order term $(\epsilon_m)_{ijkl}$ have to satisfy the conditions imposed by the incompressibility of a fluid and by the symmetry of $\overline{u_i u_j}$ with respect to i and j.

The relations (1-24) and (1-25) are invariant with respect to a coordinate transformation, but in contrast to the relations (1-21) and (1-23) they are not objective. For, though they are invariant with respect to a Galilean transformation, they are not invariant with respect to a rotation of the coordinate system with a constant angular velocity. This condition of objectivity, introduced in theories on rheology, has to be satisfied by any constitutive equation of a fluid. If the condition is made that a relation between the stress tensor $\overline{u_i u_j}$ and the deformation tensor D_{ij} must be invariant with respect to the position of the coordinates, to a Galilean transformation and to a constant rotation of the coordinate system, so that the relation must be objective, a necessary and sufficient condition for objectivity is, that the relation must be of the form

$$-\overline{u_i u_j} = \epsilon_m \delta_{ij} + \epsilon_m^* \overline{D_{ij}} + \epsilon_m^{**} \overline{D_{ik}}\,\overline{D_{ij}}$$

where the scalars ϵ_m, ϵ_m^* and ϵ_m^{**} are functions of the principal invariants of $\overline{D_{ij}}$. This nonlinear relation is similar to the constitutive equation for a viscous fluid of Stokes.

Still there is an unsatisfactory situation with this latter relation when it is applied to actual turbulent flows. Consider a simple one-dimensional mean-shear flow in the x_1-direction, so that $\bar{U}_1$ is a function of x_2 alone. Experimental evidence yields for the three turbulence intensities $\overline{u_1^2} \neq \overline{u_2^2} \neq \overline{u_3^2}$, the axial component being the largest of the three. Now, neither the latter (Stokes's type) relation, nor the relations (1-23) and (1-24) produce results which can be made to agree with this experimental evidence. Only the relation (1-25) can yield this agreement, but as stated, this relation is not objective.

However, in the case of turbulence, when its properties are considered as those of a non-Newtonian fluid, the condition of invariancy with respect to a rotation of the coordinate system may certainly be questioned. For, as may be assumed to be well known, in the case of a rotating coordinate system, Coriolis and centrifugal effects on the fluid motion are present, the first proportional with the angular velocity, the second proportional with the square of this velocity. Lumley,[55] when considering the possibility of a constitutive equation for turbulence in a fluid, has shown that in contrast to actual fluids considered in rheology, the spatial and time scales occurring in turbulent flows are such that already the Coriolis effects may not be neglected. Which means that one has to drop the extra invariancy condition. In other words, a constitutive equation for turbulence need not be objective in the sense used in rheology.

It may be concluded that the introduction of an eddy-viscosity concept does not yield a complete solution, that is satisfactory in all respects. The correct way should be to solve the dynamic equations for the stress-tensor $-\rho\overline{u_i u_j}$ as given in Sec. 4-2. But owing to the occurrence of higher-order correlations in these dynamic equations no solution has been obtained hitherto, unless some more or less realistic assumption concerning these higher-order correlations is introduced. Agreement between theoretical solutions and experimental results may then be obtained. As an example we may refer to solutions obtained by Hanjalić and Launder[53] for the case of asymmetric, turbulent boundary layers, notwithstanding the fact that not all of the assumptions made are basically correct (see also Sec. 7-10).

Some objection may be raised to the concept of eddy viscosity if it is used in the same way as the "molecular" viscosity of the fluid. For molecular viscosity is a real property of the fluid, present also when the fluid is at rest, whereas the eddy viscosity can become effective only if there is some flow of the fluid and is clearly not a property of the fluid as such. However, there is nothing against the formal introduction of the term "eddy viscosity" if we are fully aware of its specific office, namely, to express the behavior of the turbulence stresses in terms of the mean-velocity gradients in a flowing fluid.

This latter, however, implies the assumption that the turbulence transport is of the gradient-type. As will be discussed later in more detail in Sec. 5-2, this may lead to conflicting and unacceptable results in parts of the flow, where large-scale turbulent motions give a not-inconsiderable contribution to the transport. For instance, a negative value of the eddy-viscosity may be obtained.

Notwithstanding the above objections against the introduction of an eddy viscosity, for many turbulent flows which are mainly thin shear flows, it does appear possible to obtain a satisfactory description of mean properties, by assuming the turbulent flows to behave as a hypothetical Newtonian fluid with a scalar eddy-viscosity. Though, at the same time, it should be realized that such an assumption may yield a result that can be only approximately true.

Earlier it was said that we would deal only with Newtonian fluids. However, considered as a hypothetical fluid turbulent flow is not Newtonian in its behavior. A nonlinear stress-strain relationship may be assumed as, e.g., the one mentioned earlier which satisfies the condition of objectivity.

Another possibility is to consider turbulent behavior as that of a visco-elastic fluid,[13] including a memory behavior.

$$-\rho\overline{u_i u_j} = -\bar{P}_t\delta_{ij} + A\bar{P}_t\int_0^\infty d\tau\, M(\tau)\left[\frac{\partial U_i^*}{\partial x_j}(t-\tau) + \frac{\partial U_j^*}{\partial x_i}(t-\tau)\right]$$

Here $M(\tau)$ is a memory function characteristic for the turbulent flow considered. The relation may refer to the smaller-scale motions, where U_i^* stands for the large-scale motions,[50] or to the total turbulent field where then U_i^* stands for the mean velocity. As will be discussed later, certain features of turbulent boundary layer can be reproduced theoretically by ascribing visco-elastic properties to the turbulence in the boundary layer.

A variation of the above relation has been proposed by Hinze, et. al.[54] By assuming an exponential function for $M(\tau)$ with a characteristic or relaxation time equal to the streamwise Lagrangian integral time scale (see Sec. 1-9) an eddy-viscosity including a memory behavior is obtained. Such an eddy-viscosity appeared useful for describing a turbulent shear flow with a relatively rapid streamwise change of the transverse mean-velocity gradient $\partial\bar{U}_1/\partial x_2$.

Cylindrical Coordinates. Since in this book we shall deal with axisymmetrical flow problems, which can be described most conveniently in terms of cylindrical coordinates, for the reader's convenience we give here the equation for the conservation of mass and the equations of motion in cylindrical coordinates. These equations can be derived in a way similar to that shown in the previous sections for Cartesian coordinates.

Let U_r, U_φ, and U_z be the velocity components in the directions of the three cylindrical coordinates r, φ, and z.

The equation for the conservation of mass reads

$$\frac{\partial \rho}{\partial t} + \frac{\partial}{\partial r}(\rho U_r) + \rho\frac{U_r}{r} + \frac{\partial}{r\,\partial\varphi}(\rho U_\varphi) + \frac{\partial}{\partial_z}(\rho U_z) = 0$$

or

$$\frac{D\rho}{Dt} + \rho\Theta = 0 \tag{1-26}$$

where

$$\Theta = \frac{\partial U_r}{\partial r} + \frac{U_r}{r} + \frac{\partial U_\varphi}{r\,\partial\varphi} + \frac{\partial U_z}{\partial z} \tag{1-27}$$

and

$$\frac{D}{Dt} = \frac{\partial}{\partial t} + U_r\frac{\partial}{\partial r} + \frac{U_\varphi}{r}\frac{\partial}{\partial\varphi} + U_z\frac{\partial}{\partial z}$$

The equations of motion in the three coordinate directions, assuming constant viscosity, are

$$\rho\left(\frac{DU_r}{Dt} - \frac{U_\varphi{}^2}{r}\right) = -\frac{\partial P}{\partial r} + \frac{1}{3}\mu\frac{\partial\Theta}{\partial r} + \mu\left(\nabla^2 U_r - \frac{U_r}{r^2} - \frac{2\partial U_\varphi}{r^2\,\partial\varphi}\right) + F_r \tag{1-28a}$$

$$\rho\left(\frac{DU_\varphi}{Dt} + \frac{U_r U_\varphi}{r}\right) = -\frac{\partial P}{r\,\partial\varphi} + \frac{1}{3}\mu\frac{\partial\Theta}{r\,d\varphi} + \mu\left(\nabla^2 U_\varphi - \frac{U_\varphi}{r^2} + \frac{2}{r^2}\frac{\partial U_r}{\partial\varphi}\right) + F_\varphi \tag{1-28b}$$

$$\rho\frac{DU_z}{Dt} = -\frac{\partial P}{\partial z} + \frac{1}{3}\mu\frac{\partial\Theta}{\partial z} + \mu\nabla^2 U_z + F_z \tag{1-28c}$$

where

$$\nabla^2 = \frac{\partial^2}{\partial r^2} + \frac{\partial}{r\,\partial r} + \frac{\partial^2}{r^2\,\partial\varphi^2} + \frac{\partial^2}{\partial z^2}$$

The stress components read:

$$\sigma_{rr} = -P + 2\mu\frac{\partial U_r}{\partial r} - \frac{2}{3}\mu\Theta \qquad \sigma_{r\varphi} = \mu\left(\frac{\partial U_\varphi}{\partial r} + \frac{\partial U_r}{r\,\partial\varphi} - \frac{U_\varphi}{r}\right)$$

$$\sigma_{\varphi\varphi} = -P + 2\mu\left(\frac{\partial U_\varphi}{r\,\partial\varphi} + \frac{U_r}{r}\right) - \frac{2}{3}\mu\Theta \qquad \sigma_{\varphi z} = \mu\left(\frac{\partial U_z}{r\,\partial\varphi} + \frac{\partial U_\varphi}{\partial z}\right)$$

$$\sigma_{zz} = -P + 2\mu\frac{\partial U_z}{\partial z} - \frac{2}{3}\mu\Theta \qquad \sigma_{zr} = \mu\left(\frac{\partial U_r}{\partial z} + \frac{\partial U_z}{\partial r}\right) \tag{1-29}$$

The Reynolds equations of motion for turbulent flow may be obtained again by substituting for ρ, P, U_r, U_φ, and U_z:

$$\rho = \bar{\rho} + \tilde{\rho}$$

$$P = \bar{P} + p$$

$$U_r = \bar{U}_r + u_r$$

$$U_\varphi = \bar{U}_\varphi + u_\varphi$$

$$U_z = \bar{U}_z + u_z$$

and by carrying out an averaging procedure, as shown in the foregoing. It would be a good exercise for the reader to do this. As a result he will obtain for the mean motion essentially the same expressions as those of (1-28) with, in addition, many new expressions resulting from double and triple correlations among density fluctuations and turbulence components of the velocities. For instance, in the first equation of (1-28), the terms with the turbulence fluctuations, when put on the right-hand side of the equation, are

$$-\bar{\rho}\left(\overline{u_r\frac{\partial u_r}{\partial r}+u_\varphi\frac{\partial u_r}{r\,\partial\varphi}+u_z\frac{\partial u_z}{\partial z}-\frac{u_\varphi{}^2}{r}}\right)-\overline{\tilde{\rho}\frac{\partial u_r}{\partial t}}$$

$$-\overline{\tilde{\rho}u_r}\frac{\partial\bar{U}_r}{\partial r}-\overline{\tilde{\rho}u_\varphi}\left(\frac{\partial\bar{U}_r}{r\,\partial\varphi}-\frac{2\bar{U}_\varphi}{r}\right)-\overline{\tilde{\rho}u_z}\frac{\partial\bar{U}_z}{\partial z}$$

$$-\overline{\tilde{\rho}\frac{\partial u_r}{\partial r}}\bar{U}_r-\overline{\tilde{\rho}\frac{\partial u_r}{r\,\partial\varphi}}\bar{U}_\varphi-\overline{\tilde{\rho}\frac{\partial u_z}{\partial z}}\bar{U}_z$$

$$+\frac{\overline{\tilde{\rho}u^2}}{r}-\overline{\tilde{\rho}u_r\frac{\partial u_r}{\partial r}}-\overline{\tilde{\rho}u_\varphi\frac{\partial u_r}{r\,\partial\varphi}}-\overline{\tilde{\rho}u_z\frac{\partial u_z}{\partial z}}$$

Similar results are obtained for the second and third equations of (1-28).

The reader may notice that most of these turbulence terms are due to the correlations that include density fluctuations, so the number of these terms is appreciably reduced when the fluid is incompressible. For the case of an incompressible fluid we give the equations in full.

$$\rho\left(\frac{D\bar{U}_r}{Dt}-\frac{\bar{U}_\varphi{}^2}{r}\right)=-\frac{\partial\bar{P}}{\partial r}+\mu\left(\nabla^2\bar{U}_r-\frac{\bar{U}_r}{r^2}-2\frac{\partial\bar{U}_\varphi}{r^2\,d\varphi}\right)-\frac{\rho}{r}\frac{\partial}{\partial r}r\overline{u_r{}^2}$$

$$-\rho\frac{\partial}{r\,\partial\varphi}\overline{u_ru_\varphi}-\rho\frac{\partial}{\partial z}\overline{u_ru_z}+\rho\frac{\overline{u_\varphi{}^2}}{r}+\bar{F}_r$$

$$\rho\left(\frac{D}{Dt}\bar{U}_\varphi+\frac{\overline{U_rU_\varphi}}{r}\right)=-\frac{\partial\bar{P}}{r\,\partial\varphi}+\mu\left(\nabla^2\bar{U}_\varphi-\frac{\bar{U}_\varphi}{r^2}+\frac{2}{r^2}\frac{\partial\bar{U}_r}{\partial\varphi}\right)$$

$$-\rho\frac{\partial}{r\,\partial\varphi}\overline{u_\varphi{}^2}-\rho\frac{\partial}{\partial r}\overline{u_\varphi u_r}-\rho\frac{\partial}{\partial z}\overline{u_\varphi u_z}$$

$$-2\rho\frac{\overline{u_\varphi u_r}}{r}+\bar{F}_\varphi$$

$$\rho\frac{D\bar{U}_z}{Dt}=-\frac{\partial\bar{P}}{\partial z}+\mu\nabla^2\bar{U}_z-\rho\frac{\partial}{\partial z}\overline{u_z{}^2}-\frac{\rho}{r}\frac{\partial}{\partial r}r\overline{u_ru_z}$$

$$-\rho\frac{\partial}{r\,\partial\varphi}\overline{u_\varphi u_z}+\bar{F}_z \qquad \text{(1-30)}$$

The stress components read:

$$\sigma_{rr} = -\bar{P} + 2\mu \frac{\partial \bar{U}_r}{\partial r} - \rho \overline{u_r^2}$$

$$\sigma_{\varphi\varphi} = -\bar{P} + 2\mu\left(\frac{\partial \bar{U}_\varphi}{r\,\partial \varphi} + \frac{\bar{U}_r}{r}\right) - \rho \overline{u_\varphi^2}$$

$$\sigma_{zz} = -\bar{P} + 2\mu \frac{\partial \bar{U}_z}{\partial z} - \rho \overline{u_z^2}$$

$$\sigma_{r\varphi} = \mu\left(\frac{\partial \bar{U}_\varphi}{\partial r} + \frac{\partial \bar{U}_r}{r\,\partial \varphi} - \frac{\bar{U}_\varphi}{r}\right) - \rho \overline{u_r u_\varphi}$$

$$\sigma_{zr} = \mu\left(\frac{\partial \bar{U}_r}{\partial z} + \frac{\partial \bar{U}_z}{\partial r}\right) - \rho \overline{u_r u_z}$$

$$\sigma_{\varphi z} = \mu\left(\frac{\partial \bar{U}_z}{r\,\partial \varphi} + \frac{\partial \bar{U}_\varphi}{\partial z}\right) - \rho \overline{u_\varphi u_z} \qquad (1\text{-}31)$$

On comparing the expressions of (1-31) with the corresponding ones of (1-19), the reader may notice that the expressions for the turbulence stresses are the same.

1-3 EQUATION FOR THE CONSERVATION OF A TRANSFERABLE SCALAR QUANTITY IN A TURBULENT FLOW

We shall consider now the transport of a scalar quantity, such as heat, energy, or matter, in a turbulent flow of an incompressible fluid. More particularly, we shall consider the transport by the turbulent motions apart from transport by molecular diffusion or by forced convection by the mean flow velocities.

Application of the balance Eq. (1-1) to the scalar quantity Γ per unit of mass gives

$$\frac{\partial \Gamma}{\partial t} + U_i \frac{\partial \Gamma}{\partial x_i} = \frac{\partial}{\partial x_i}\left(\mathfrak{k}\frac{\partial \Gamma}{\partial x_i}\right) + F_\gamma \qquad (1\text{-}32)$$

where $\mathfrak{k}$ denotes the molecular-transport coefficient. F_γ is a "driving force" or source; for instance, if heat is the transported quantity, F_γ is the heat generation by dissipation of the kinetic energy of the flow.

If we assume that $\mathfrak{k}$ is constant and if we put

$$\Gamma = \bar{\Gamma} + \gamma \qquad U_i = \bar{U}_i + u_i$$

after we carry out the usual averaging procedure, for the left-hand side of the equation we obtain

$$\frac{\partial \bar{\Gamma}}{\partial t} + \bar{U}_i \frac{\partial \bar{\Gamma}}{\partial x_i} + \overline{u_i \frac{\partial \gamma}{\partial x_i}} = \frac{\partial \bar{\Gamma}}{\partial t} + \bar{U}_i \frac{\partial \bar{\Gamma}}{\partial x_i} + \frac{\partial}{\partial x_i} \overline{u_i \gamma}$$

since, for an incompressible fluid, $\overline{\gamma(\partial u_i/\partial x_i)} = 0$.

Substitution in Eq. (1-32), after some rearranging, gives

$$\frac{\partial \bar{\Gamma}}{\partial t} + \bar{U}_i \frac{\partial \bar{\Gamma}}{\partial x_i} = \frac{\partial}{\partial x_i}\left(\mathfrak{k} \frac{\partial \bar{\Gamma}}{\partial x_i} - \overline{u_i \gamma}\right) + \bar{F}_\gamma \qquad (1\text{-}33)$$

Thus the distribution of $\bar{\Gamma}$ by the mean motion is affected by the molecular diffusion and by the turbulence convective motions, the latter effect being determined by the correlation $\overline{u_i \gamma}$ between the turbulent fluctuations of the velocity and the scalar quantity. Also we may here formally introduce a turbulence- or eddy-transport coefficient ϵ_γ, just as Boussinesq did for the transport of momentum.

$$-\overline{u_i \gamma} = \epsilon_\gamma \frac{\partial \bar{\Gamma}}{\partial x_i} \qquad (1\text{-}34)$$

Because $\overline{u_i \gamma}$ and $\partial \bar{\Gamma}/\partial x_i$ are vector quantities, the turbulence-transport coefficient must be either a tensor of the second order or a scalar; to the scalar case the relations (1-34) apply. If the transport coefficient is a tensor of the second order, the relation between $\overline{u_i \gamma}$ and $\partial \bar{\Gamma}/\partial x_i$ should read

$$-\overline{u_i \gamma} = (\epsilon_\gamma)_{ij} \frac{\partial \bar{\Gamma}}{\partial x_j} \qquad (1\text{-}35)$$

1-4 DOUBLE CORRELATIONS BETWEEN TURBULENCE-VELOCITY COMPONENTS

Here and in the following sections, we assume an incompressible fluid and a constant viscosity. In Sec. 1-2 we have shown that the turbulence components at a point show up in the equations of motion in the form of the double correlation $\overline{u_i u_j}$ and that $\rho \overline{u_i u_j}$ can be interpreted as normal (when $i = j$) and tangential (when $i \neq j$) turbulence stresses.

We shall now apply to these turbulence stresses the analysis that is customary for stresses due to viscosity and we shall see what comes of it. Consider, for instance, a small element dS of a surface in the fluid, whose normal N makes angles ni with the coordinate axes x_i.

Let $\sigma_{Nn} = -\rho \overline{u_n^2}$ be the normal component of the total turbulence stress σ_N on this surface element; then the stress analysis just mentioned will show how this normal component of the turbulence stress is related to the stresses $\sigma_{ij} = -\rho \overline{u_i u_j}$. The total turbulence stress σ_N may be resolved into three components σ_{Ni} along the directions i of the coordinate axes. Between σ_{Nn} and σ_{Ni} exists the relation (see Fig. 1-12)

$$\sigma_{Nn} = \sigma_{Ni} \cos ni = \sigma_{Ni} e_{ni}$$

where $\cos ni$, the direction cosine of the normal N with respect to the axis i, is written e_{ni} for short.

The equilibrium of the fluid element $OABC$ requires that

$$\sigma_{Ni} = \sigma_{ij} e_{nj}$$

Hence

$$\sigma_{Nn} = \sigma_{ij} e_{ni} e_{nj}$$

or

$$\overline{u_n{}^2} = \overline{u_i u_j} e_{ni} e_{nj} \qquad (1\text{-}36)$$

When we introduce $u'_n = \sqrt{\overline{u_n{}^2}}$, the intensity of the turbulence velocity normal to the surface element dS, we may rewrite Eq. (1-36)

$$\overline{u_i u_j} \frac{e_{ni} e_{nj}}{u'^2_n} = 1 \qquad (1\text{-}37)$$

Since $\overline{u_i u_j}$ is a fixed quantity for a given coordinate system, it cannot be affected by a change in the orientation of the surface element dS; only e_{ni}, e_{nj}, and u'_n change with this orientation. So (1-37) is the equation for a quadratic surface, namely, an ellipsoid, with coordinates e_{ni}/u'_n. The intersections of this surface with the coordinate axes are $1/u'_i$, respectively.

Assume that the coordinate axes coincide with the three principal axes of the ellipsoid, and let this new coordinate system be designated i^*. The equation of the ellipsoid with respect to this new coordinate system then reads

$$\overline{u_n{}^2} = \overline{u_{i^*}{}^2} e_{ni^*}{}^2 \qquad (1\text{-}38)$$

This equation does not contain the mixed or product terms $\overline{u_{i^*} u_{j^*}}$ for $i^* \neq j^*$;

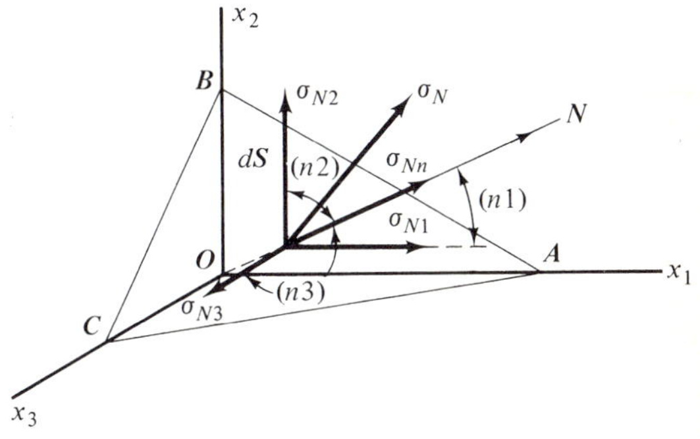

FIGURE 1-12
Stresses on a surface element dS.

this means that the planes perpendicular to the principal axes of the ellipsoid have no shear stresses. This is a well-known theorem in the theory of stresses.

In fluid flow, the static pressure at a point is defined as the average value of the normal stress for all values of the orientation of a plane through this point.

Similarly, the turbulence pressure can be defined by taking the average value of Eq. (1-36) for all values of the direction N. Since

$$\underset{\substack{\text{average for}\\ \text{all } n}}{\langle e_{ni} e_{nj} \rangle} = \tfrac{1}{3}\delta_{ij}$$

from Eq. (1-36) we have

$$\underset{\substack{\text{average for}\\ \text{all } n}}{\langle \overline{u_n{}^2} \rangle} = \overline{u_i u_j} \tfrac{1}{3} \delta_{ij} = \tfrac{1}{3}\overline{u_i u_i} = \tfrac{2}{3} \times \text{(kinetic energy)}$$

In the general case considered hitherto, the stresses $\rho\overline{u_i{}^2}$ are different according to whether $i = 1$, 2, or 3. What will be the physical meaning when $\overline{u_i{}^2}$ is the same for $i = 1$, 2, and 3? The ellipsoid then reduces to a sphere, and Eq. (1-38) holds with respect to any coordinate system; in other words, this equation is invariant under rotation of the coordinate system and under reflection with respect to the coordinate planes. A flow field that can be described by such an equation is called *isotropic*.† Since terms $\overline{u_i u_j}$ with $i \neq j$ do not occur with respect to either coordinate system, in isotropic turbulent flow no average turbulent shear stress can exist.

In addition to the correlations between velocity components in a fixed point of the flow field, which have a distinct physical meaning (they may be interpreted as stresses), there are correlations between velocities at two different points in the flow field that are important, for these can be used in studying the spatial configurations of the field of flow. For instance, it is possible to determine those regions of the field in which the fluid motions have a certain connection or correlation with each other.

Let A and B be two points in the field of flow, and let a and b be two given directions at the points A and B, respectively. The velocity components along these directions are U_a and U_b, respectively. What is the correlation between these velocities?

The velocities U_a and U_b consist of the mean velocity and the turbulence velocity at the points A and B, respectively. We assume that, in the region of the turbulence flow field considered, the mean velocity has a constant value. Let us assume further, for simplicity, that this mean velocity is along the direction of the x_1-axis. Thus, at any instant,

$$U_a = \bar{U}_1 e_{a1} + u_a \qquad \text{at } A$$

$$U_b = \bar{U}_1 e_{b1} + u_b \qquad \text{at } B$$

where e_{a1} and e_{b1} are the direction cosines of the directions a and b, respectively, with respect to the direction of the x_1-axis.

† A more complete definition of isotropy will be given in Chap. 3.

The correlation product of $(U_a)_A$ and $(U_b)_B$

$$\overline{(U_a)_A(U_b)_B} = \bar{U}_1{}^2 e_{a1} e_{b1} + \overline{(u_a)_A(u_b)_B}$$

consists of two parts; only the second, namely, $\overline{(u_a)_A(u_b)_B}$, which refers to the turbulence velocities, is of interest in describing the turbulence flow field. Therefore, in what follows, let us consider this part only, i.e.,

$$\mathbf{Q}_{A,B} = \overline{(u_a)_A(u_b)_B} \tag{1-39}$$

For a quasi-steady flow field, that is, a field where the average flow pattern does not change with time, the correlation product is a function only of the locations of the points A and B and of the directions a and b. Hence, for given locations and directions, $\mathbf{Q}_{A,B}$ has a fixed value.

Now $(u_a)_A$ and $(u_b)_B$ can be expressed in terms of the velocity components $(u_i)_A$ and $(u_i)_B$, respectively, along the coordinate axes.

$$u_a = u_i e_{ai}$$

$$u_b = u_i e_{bi} \tag{1-40}$$

Substitution of Eq. (1-40) in Eq. (1-39) then gives

$$\mathbf{Q}_{A,B} = \overline{(u_a)_A(u_b)_B} = \overline{(u_i)_A(u_j)_B e_{ai} e_{bj}}$$

Since the averaging procedure is taken with respect to time and since e_{ai} and e_{bj} are the constant direction cosines of a and b, respectively, the overscore in fact applies only to $(u_i)_A(u_j)_B$.

$$\mathbf{Q}_{A,B} = \overline{(u_a)_A(u_b)_B} = \overline{(u_i)_A(u_j)_B} e_{ai} e_{bj}$$

$$= (\mathbf{Q}_{i,j})_{A,B} e_{ai} e_{bj} \tag{1-41}$$

Thus $\mathbf{Q}_{A,B}$ can be considered the product of $(\mathbf{Q}_{i,j})_{A,B}$ and $e_{ai}e_{bj}$. The correlation

$$(\mathbf{Q}_{i,j})_{A,B} = \overline{(u_i)_A(u_j)_B} \tag{1-42}$$

is a tensor of the second order, since under rotation of the coordinate system it transforms like a tensor. According to the rule of transformation for a tensor [see the Appendix] we must have

$$(\mathbf{Q}^*_{p,q} = e_{ip} e_{jq} \mathbf{Q}_{i,j})_{A,B}$$

where $(\mathbf{Q}^*_{p,q})_{A,B}$ is the tensor according to a new coordinate system, whose p-axis makes an angle ip with the i-axis of the original coordinate system, and where $e_{ip} = \cos ip$ and $e_{jq} = \cos jq$. This relation is easily proved because

$$(e_{ip} e_{jq} \mathbf{Q}_{i,j})_{A,B} = e_{ip} e_{jq} \overline{(u_i)_A(u_j)_B} = \overline{(u_p)_A(u_q)_B} = (\mathbf{Q}^*_{p,q})_{A,B}$$

As already mentioned, $\mathbf{Q}_{A,B}$ has a fixed value and, according to Eq. (1-41), may be considered as a scalar formed by the scalar product of the tensor $(\mathbf{Q}_{h,k})_{A,B}$ and the tensor or dyadic $e_{ai}e_{bj}$, as follows:

$$\mathbf{Q}_{A,B} = \delta_{hi} \delta_{kj} (Q_{h,k})_{A,B} e_{ai} e_{bj} = (\mathbf{Q}_{i,j})_{A,B} e_{ai} e_{bj}$$

Equation (1-41) shows that the correlation product $\mathbf{Q}_{A,B} = \overline{(u_a)_A(u_b)_B}$ can be described in terms of the nine correlations $\overline{(u_i)_A(u_j)_B}$.

Considerable simplification results when the turbulent flow is isotropic and homogeneous. As we shall show later, the correlation $\mathbf{Q}_{A,B}$ can then be described in terms of two special types of correlation.

By way of introducing to the reader the characteristic features of an isotropic turbulence, we shall give here a proof simplified by the assumption of a special position of the coordinate system, namely, where the x_1-axis passes through the two points A and B (see Fig. 1-13). According to the definition of isotropy, any relation between turbulence quantities must be invariant under rotation of the coordinate system and under reflection with respect to the coordinate planes.

We have to show that $\mathbf{Q}_{A,B} = \overline{(u_a)_A(u_b)_B}$ can be expressed in terms of two correlations. This can easily be done if one makes use of the definition of isotropy and shows that

$$(\mathbf{Q}_{i,j})_{A,B} = \overline{(u_i)_A(u_j)_B} = 0$$

for $i \neq j$.

Consider, for instance, $(u_1)_A$ and $(u_2)_B$. Rotation of the coordinate system through 180° about the x_1-axis must, because of isotropy, give

$$\overline{(u_1)_A(u_2)_B} = \overline{(u_1)_A[-(u_2)_B]} = -\overline{(u_1)_A(u_2)_B}$$

which can be true only when $\overline{(u_1)_A(u_2)_B} = 0$.

Hence we have

$$\mathbf{Q}_{A,B} = \overline{(u_a)_A(u_b)_B} = \overline{(u_i)_A(u_j)_B}e_{ai}e_{bj} \tag{1-43}$$

Of the three correlations

$$\overline{(u_1)_A(u_1)_B} \qquad \overline{(u_2)_A(u_2)_B} \qquad \overline{(u_3)_A(u_3)_B}$$

the last two must be the same, because the coordinate system is invariant under rotation about the x_1-axis. Thus, $\mathbf{Q}_{A,B}$ is indeed determined by two correlations only, namely, the longitudinal velocity correlation $\overline{(u_1)_A(u_1)_B}$ and either of the lateral velocity correlations $\overline{(u_2)_A(u_2)_B}$ and $\overline{(u_3)_A(u_3)_B}$.

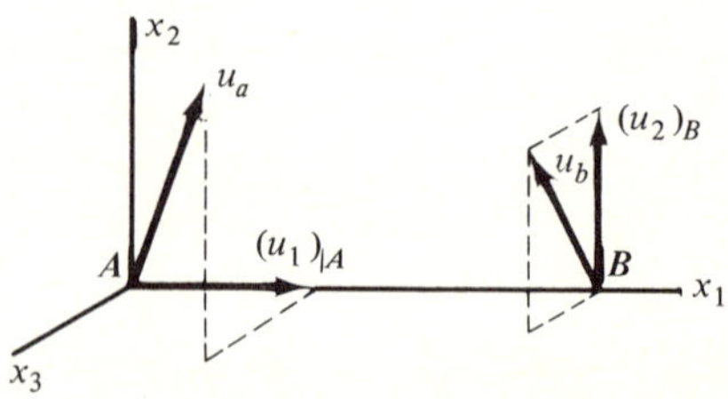

FIGURE 1-13
Correlation between the velocities u_a and u_b.

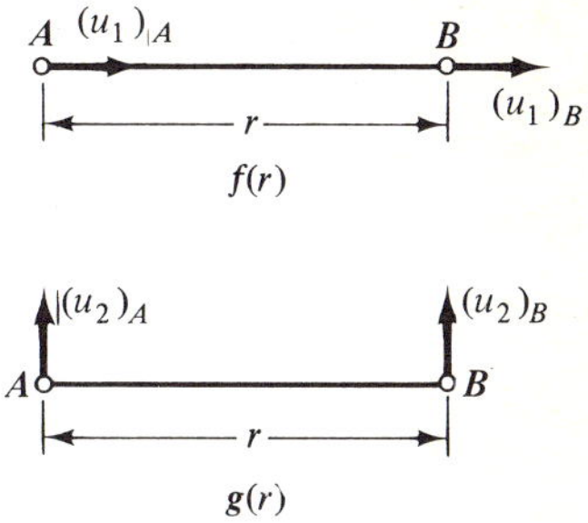

FIGURE 1-14
Longitudinal velocity correlation coefficient **f** and lateral velocity correlation coefficient **g**.

Instead of using the correlation products $\overline{(u_a)_A(u_b)_B}$ and $\overline{(u_i)_A(u_j)_B}$, it is very common to use correlation coefficients defined by

$$\mathbf{R}_{A,B} = \frac{\overline{(u_a)_A(u_b)_B}}{(u'_a)_A(u'_b)_B} \qquad (\mathbf{R}_{i,j})_{A,B} = \frac{\overline{(u_i)_A(u_j)_B}}{(u'_i)_A(u'_j)_B} \quad \text{(no summation)} \tag{1-44}$$

In the case of homogeneous isotropic turbulence we have

$$(u'_a)_A = (u'_i)_A = (u'_b)_B = (u'_j)_B;$$

we may write u' for short.

The coefficients of the two special types of correlation in which the correlation $\mathbf{Q}_{A,B}$ or its coefficient $\mathbf{R}_{A,B}$ can be expressed are usually denoted by the symbols **f** and **g** for the coefficients of the longitudinal and lateral velocity correlations, respectively (see Fig. 1-14). The relation corresponding to Eq. (1-43) reads

$$\mathbf{R}_{A,B} = \mathbf{f} e_{a1} e_{b1} + \mathbf{g}(e_{a2} e_{b2} + e_{a3} e_{b3})$$

Later, in Chap. 3, we will discuss how to describe the correlation $\mathbf{R}_{A,B}$ with respect to an arbitrary coordinate system in terms of the two types **f** and **g** and in terms of the distance r between the points A and B.

1-5 CHANGE IN DOUBLE VELOCITY CORRELATIONS WITH TIME; INTRODUCTION OF TRIPLE VELOCITY CORRELATIONS

In real fluids the viscous stresses in turbulent motions will cause the kinetic energy of the motions to dissipate in heat. If there are no external effects present to supply energy continuously for maintaining the turbulent motions, these will decay in the course of time. An interesting problem to investigate is how the flow pattern and the relations between velocities change during decay. Since these relations can be described by the tensor $(\mathbf{Q}_{i,j})_{A,B}$ of the double velocity correlations, we have to consider the change in this tensor with time.

To this end we start with the equations of motion, because they describe the turbulent motions. These equations have to be transformed in such a way as to obtain other equations containing the double velocity correlation $\overline{(u_a)_A(u_b)_B}$. $\partial/\partial t\, \overline{(u_a)_A(u_b)_B}$ is, then, the term in which we are interested.

The procedure for obtaining this term from the equations of motion is as follows.

Since $\overline{(u_a)_A(u_b)_B} = \overline{(u_i)_A(u_j)_B} e_{ai} e_{bj}$, it is sufficient to consider only the changes in the terms $\overline{(u_i)_A(u_j)_B}$ separately, namely,

$$\frac{\partial}{\partial t}\overline{(u_i)_A(u_j)_B} = \overline{(u_i)_A\frac{\partial}{\partial t}(u_j)_B} + \overline{(u_j)_B\frac{\partial}{\partial t}(u_i)_A}$$

The value of each term on the right-hand side of this equation can be obtained by using the equations of motion for U_j at the point B and for U_i at the point A.

Now we assume that the mean velocity $\bar{U}_i$ is constant throughout the region considered and independent of time and we put

$$(U_i = \bar{U}_i + u_i)_A \qquad (U_j = \bar{U}_j + u_j)_B$$

The equation of motion for u_i at the point A reads

$$\left[\frac{\partial}{\partial t}u_i + (\bar{U}_k + u_k)\frac{\partial u_i}{\partial x_k} = -\frac{1}{\rho}\frac{\partial p}{\partial x_i} + \nu\frac{\partial^2}{\partial x_l\,\partial x_l}u_i\right]_A$$

Multiply this equation by $(u_j)_B$.

$$(u_j)_B\frac{\partial}{\partial t}(u_i)_A + [\bar{U}_k + (u_k)_A]\left(\frac{\partial}{\partial x_k}\right)_A(u_i)_A(u_j)_B$$

$$= -\frac{1}{\rho}\left(\frac{\partial}{\partial x_i}\right)_A p_A(u_j)_B + \nu\left(\frac{\partial^2}{\partial x_l\,\partial x_l}\right)_A(u_i)_A(u_j)_B$$

In this procedure we have to keep in mind that $(u_j)_B$ can be treated as a constant in a differentiation process at point A.

In the same way we can multiply the equation of motion for u_j at point B by $(u_i)_A$ and obtain

$$(u_i)_A\frac{\partial}{\partial t}(u_j)_B + [\bar{U}_k + (u_k)_B]\left(\frac{\partial}{\partial x_k}\right)_B(u_j)_B(u_i)_A$$

$$= -\frac{1}{\rho}\left(\frac{\partial}{\partial x_j}\right)_B p_B(u_i)_A + \nu\left(\frac{\partial^2}{\partial x_l\,\partial x_l}\right)(u_j)_B(u_i)_A$$

For an incompressible fluid we have

$$u_i\frac{\partial u_k}{\partial x_k} = 0 \qquad \text{and} \qquad u_j\frac{\partial u_k}{\partial x_k} = 0$$

and thus also

$$(u_j)_B\left(u_i\frac{\partial u_k}{\partial x_k}\right)_A = 0 \qquad \text{and} \qquad (u_i)_A\left(u_j\frac{\partial u_k}{\partial x_k}\right)_B = 0$$

If we add the first term to the transformed equation of motion at point A and the second term to the transformed equation of motion at point B and, finally, if we add the two equations, we obtain

$$\frac{\partial}{\partial t}(u_i)_A(u_j)_B + \left[\left(\frac{\partial}{\partial x_k}\right)_A(u_i)_A(u_k)_A(u_j)_B + \left(\frac{\partial}{\partial x_k}\right)_B(u_i)_A(u_k)_B(u_j)_B\right]$$
$$+ \bar{U}_k\left[\left(\frac{\partial}{\partial x_k}\right)_A(u_i)_A(u_j)_B + \left(\frac{\partial}{\partial x_k}\right)_B(u_i)_A(u_j)_B\right]$$
$$= -\frac{1}{\rho}\left[\left(\frac{\partial}{\partial x_i}\right)_A p_A(u_j)_B + \left(\frac{\partial}{\partial x_j}\right)_B p_B(u_i)_A\right]$$
$$+ \nu\left[\left(\frac{\partial^2}{\partial x_k\,\partial x_k}\right)_A + \left(\frac{\partial^2}{\partial x_k\,\partial x_k}\right)_B\right](u_i)_A(u_j)_B$$

We are interested in the relation of the turbulent motions at point B to those at point A. Thus it will make no difference if we take one of the points B or A as the origin of the coordinate system. Let us take point A as the origin and write

$$\xi_k = (x_k)_B - (x_k)_A$$

We then have

$$\left(\frac{\partial}{\partial x_k}\right)_A = -\frac{\partial}{\partial \xi_k} \qquad \left(\frac{\partial}{\partial x_k}\right)_B = \frac{\partial}{\partial \xi_k}$$
$$\left(\frac{\partial^2}{\partial x_k\,\partial x_k}\right)_A = \left(\frac{\partial^2}{\partial x_k\,\partial x_k}\right)_B = \frac{\partial^2}{\partial \xi_k\,\partial \xi_k}$$

If we make use of these relations, after an averaging procedure has been carried out with respect to time the above equation becomes

$$\frac{\partial}{\partial t}\overline{(u_i)_A(u_j)_B} - \frac{\partial}{\partial \xi_k}\overline{(u_i)_A(u_k)_A(u_j)_B} + \frac{\partial}{\partial \xi_k}\overline{(u_i)_A(u_k)_B(u_j)_B}$$
$$= -\frac{1}{\rho}\left[-\frac{\partial}{\partial \xi_i}\overline{p_A(u_j)_B} + \frac{\partial}{\partial \xi_j}\overline{p_B(u_i)_A}\right] + 2\nu\frac{\partial^2}{\partial \xi_k\,\partial \xi_k}\overline{(u_i)_A(u_j)_B} \qquad (1\text{-}45)$$

Notice that the term with $\bar{U}_k$ has vanished. The resulting Eq. (1-45) describes the relations between the turbulence motions exactly as if we had considered the turbulence motions with respect to a coordinate system moving with the mean velocity $\bar{U}_k$.

We notice further that Eq. (1-45) contains, apart from the terms with the double velocity correlation $\overline{(u_i)_A(u_j)_B}$, terms with double correlations such as $\overline{p_A(u_j)_B}$ and terms with triple correlations such as $\overline{(u_i)_A(u_k)_A(u_j)_B}$.

Since the pressure is a scalar quantity, the correlations $\overline{p_A(u_j)_B}$ and $\overline{p_B(u_i)_A}$

form tensors of the first order; the triple correlations $\overline{(u_i)_A(u_k)_A(u_j)_B}$ and $\overline{(u_i)_A(u_k)_B(u_j)_B}$ form tensors of the third order.

In what follows we shall designate these first-order and third-order correlations $(\mathbf{K}_{p,i})_{A,B}$ and $(\mathbf{S}_{ij,k})_{A,B}$ respectively.

Thus

$$(\mathbf{K}_{i,p})_{A,B} = \overline{(u_i)_A p_B} \qquad (\mathbf{K}_{p,j})_{A,B} = \overline{p_A(u_j)_B} \tag{1-46}$$

and

$$(\mathbf{S}_{ik,j})_{A,B} = \overline{(u_i)_A(u_k)_A(u_j)_B} \qquad (\mathbf{S}_{i,kj})_{A,B} = \overline{(u_i)_A(u_k)_B(u_j)_B} \tag{1-47}$$

Notice that in $\mathbf{K}_{i,p}$ the index p refers to the pressure p and is not an index like i or j, so that the summation convention does not apply to p.

With these notations Eq. (1-45) reads

$$\frac{\partial}{\partial t}\mathbf{Q}_{i,j} - \frac{\partial}{\partial \xi_k}\mathbf{S}_{ik,j} + \frac{\partial}{\partial \xi_k}\mathbf{S}_{i,kj} = -\frac{1}{\rho}\left(-\frac{\partial}{\partial \xi_i}\mathbf{K}_{p,j} + \frac{\partial}{\partial \xi_j}\mathbf{K}_{i,p}\right) + 2\nu\frac{\partial^2}{\partial \xi_k\,\partial \xi_k}\mathbf{Q}_{i,j} \tag{1-48}$$

where all correlations refer to the two points A and B.

It is not our intention to give here a solution of this equation. Later, in Chap. 3, a solution for the case of isotropic turbulence will be given of this equation which then can be appreciably simplified.

It will suffice at the moment to mention that, in order to solve the equation for the double-velocity-correlation tensor $\mathbf{Q}_{i,j}$, it is necessary to know the other tensors $\mathbf{K}_{p,i}$ and $\mathbf{S}_{ik,j}$ and $\mathbf{S}_{i,kj}$. As we shall see later, this constitutes a fundamental obstacle to solving the turbulence problem in this way.

As in the case of the double correlations, it is usual to introduce for the double pressure-velocity correlations and for the triple velocity correlations their coefficients, defined by

$$(\mathbf{L}_{i,p})_{A,B} = \frac{1}{(u_i')_A p_B'}(\mathbf{K}_{i,p})_{a,B} \qquad (\mathbf{L}_{p,j})_{A,B} = \frac{1}{p_A'(u_j')_B}(\mathbf{K}_{p,j})_{A,B} \tag{1-49}$$

$$(\mathbf{T}_{ik,j})_{A,B} = \frac{(\mathbf{S}_{ik,j})_{A,B}}{(u_i')_A(u_k')_A(u_j')_B} \qquad (\mathbf{T}_{i,kj})_{A,B} = \frac{(\mathbf{S}_{i,kj})_{A,B}}{(u_i')_A(u_k')_B(u_j')_B} \tag{1-50}$$

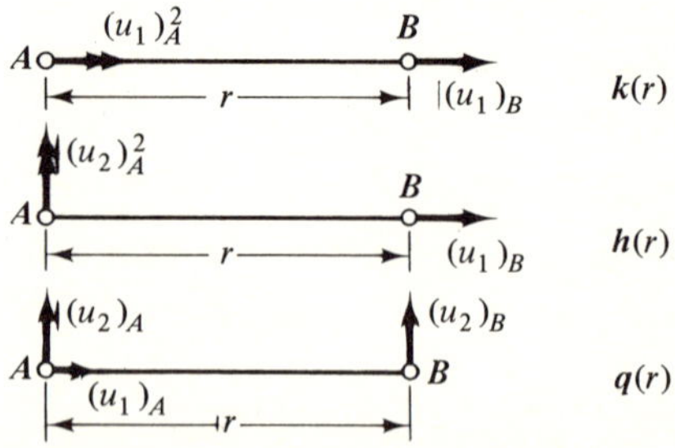

FIGURE 1-15
Triple-velocity-correlation coefficients **k**, **h**, and **q**.

In the previous section we have made it appear reasonable that, for homogeneous isotropic turbulence, the correlation coefficient $\mathbf{R}_{A,B}$ can be expressed in terms of only two specific correlations $\mathbf{f}(r,t)$ and $\mathbf{g}(r,t)$. It will be shown later, in Chap. 3, that for this type of turbulence the triple-velocity-correlation coefficient can be expressed in terms of only three specific correlation coefficients designated $\mathbf{k}(r,t)$, $\mathbf{h}(r,t)$, and $\mathbf{q}(r,t)$ (see Fig. 1-15):

$$\mathbf{k}(r,t) = \frac{\overline{(u_1{}^2)_A(u_1)_B}}{u'^3} = \mathbf{T}_{11,1}$$

$$\mathbf{h}(r,t) = \frac{\overline{(u_2{}^2)_A(u_1)_B}}{u'^3} = \mathbf{T}_{22,1} \qquad (1\text{-}51)$$

$$\mathbf{q}(r,t) = \frac{\overline{(u_2)_A(u_1)_A(u_2)_B}}{u'^3} = \mathbf{T}_{21,2}$$

1-6 FEATURES OF THE DOUBLE LONGITUDINAL AND LATERAL CORRELATIONS IN A HOMOGENEOUS TURBULENCE

In this section we shall discuss a few important features of the longitudinal and lateral correlation coefficients $\mathbf{f}(r,t)$ and $\mathbf{g}(r,t)$, respectively, which were introduced in Sec. 1-4. Since these features are the same for both correlation coefficients, it will suffice to consider here only one of them; we take the lateral correlation coefficient $\mathbf{g}$ between the lateral velocity components at the two points A and B. For simplicity we take the two points A and B on the x_2-axis, A with the coordinate ξ_2 and B with the coordinate $\xi_2 + x_2$.

By definition we have

$$\mathbf{g}(x_2) = \frac{\overline{u_1(\xi_2)u_1(\xi_2 + x_2)}}{u_1'(\xi_2)u_1'(\xi_2 + x_2)} \qquad (1\text{-}52)$$

Here the averaging procedure can be carried out with respect to time. But, because the turbulence is assumed to be homogeneous in its average statistical properties throughout the whole flow field, it is also possible to apply the averaging procedure with respect to ξ_2. Then we consider a large number of points A and B on the x_2-axis, at constant intervals x_2, and determine the average value of the correlation for all those points. Thus

$$\overline{u_1(\xi_2)u_1(\xi_2 + x_2)} = \frac{1}{\xi_2'' - \xi_2'}\int_{\xi_2'}^{\xi_2''} d\xi_2 u_1(\xi_2)u_1(\xi_2 + x_2)$$

where $\xi_2'' - \xi_2'$ is the range of values of ξ_2 considered in the averaging procedure.

Because of the assumed homogeneity of the flow field, we have

$$u_1'(\xi_2) = u_1'(\xi_2 + x_2) = u_1'$$

From its definition it follows that

$$\mathbf{g}(0) = 1$$

We show first that this value is also the maximum value of $\mathbf{g}(x_2)$, thus that

$$\mathbf{g}(x_2) \leq 1 \qquad (1\text{-}53)$$

To this end consider the square of the velocity difference

$$\overline{[u_1(\xi_2 + x_2) - u_1(\xi_2)]^2} \geq 0$$

from which follows

$$2u_1'^2 - \overline{2u_1(\xi_2 + x_2)u_1(\xi_2)} \geq 0$$

and, consequently, the relation (1-53).

Furthermore, it follows from the homogeneity of the flow field that the correlation must be a symmetrical function of x_2. For, because of this homogeneity, the value of the correlation should not change when we decrease the coordinates of A and B by the value x_2. But this is the same as if we had changed the positive sign of x_2 in Eq. (1-53) into a negative sign. So

$$\mathbf{g}(x_2) = \mathbf{g}(-x_2) \qquad (1\text{-}54)$$

Thus $\mathbf{g}$ is a symmetrical function of x_2, with a maximum value equal to unity for $x_2 = 0$ and decreasing with increasing x_2. Intuitively one would expect $\mathbf{g}$ to decrease and ultimately to become zero, as x_2 increased to infinity. How it decreased with increasing x_2 would depend on the character of the turbulence. It may decrease to zero either monotonously or oscillating. Therefore, in general, there is very little to say about the shape of the correlation function. Only about the shape in the immediate neighborhood of $x_2 = 0$ is it possible to say anything. As a matter of fact it is possible to derive certain interesting relations between the velocity and its various derivatives as a consequence of the homogeneity of the flow field, and these relations can be used to describe the shape of the correlation function at $x_2 = 0$.

Thus from homogeneity follows

$$\overline{u_1(\xi_2)u_1(\xi_2 + x_2)} = \overline{u_1(\xi_2 - x_2)u_1(\xi_2)}$$

whence

$$\overline{u_1(\xi_2)\frac{\partial u_1(\xi_2 + x_2)}{\partial(\xi_2 + x_2)}} = -\overline{\frac{\partial u_1(\xi_2 - x_2)}{\partial(\xi_2 - x_2)}u_1(\xi_2)}$$

Hence

$$u_1'^2\left[\frac{\partial \mathbf{g}}{\partial x_2}\right]_{x_2=0} = \overline{\left[u_1\frac{\partial u_1}{\partial x_2}\right]}_{x_2=0} = -\overline{\left[\frac{\partial u_1}{\partial x_2}u_1\right]}_{x_2=0} = 0$$

Again because of homogeneity we have

$$\overline{u_1(\xi_2 - x_2)\frac{\partial u_1(\xi_2)}{\partial x_2}} = \overline{u_1(\xi_2)\frac{\partial u_1(\xi_2 + x_2)}{\partial x_2}}$$

whence

$$-\overline{\frac{\partial u_1(\xi_2 - x_2)}{\partial(\xi_2 - x_2)}\frac{\partial u_1(\xi_2)}{\partial x_2}} = \overline{u_1(\xi_2)\frac{\partial^2 u_1(\xi_2 + x_2)}{\partial x_2 \partial(\xi_2 + x_2)}}$$

Hence

$$u_1'^2\left[\frac{\partial^2 \mathbf{g}}{\partial x_2^2}\right]_{x_2=0} = \overline{\left[u_1\frac{\partial^2 u_1}{\partial x_2^2}\right]_{x_2=0}} = -\overline{\left[\frac{\partial u_1}{\partial x_2}\right]^2_{x_2=0}}$$

By a similar procedure these relations can be extended to the higher derivatives. The result will be

$$\left[\frac{\partial^{n+m}\mathbf{g}}{\partial x_2^{n+m}}\right]_{x_2=0} = \frac{1}{u_1'^2}\overline{\left[\frac{\partial^n u_1}{\partial x_2^n}\frac{\partial^m u_1}{\partial x_2^m}\right]_{x_2=0}} = 0 \quad \text{for} \quad n+m = \text{uneven} \tag{1-55}$$

and

$$\left[\frac{\partial^{2n}\mathbf{g}}{\partial x_2^{2n}}\right]_{x_2=0} = \frac{1}{u_1'^2}\overline{\left[u_1\frac{\partial^{2n} u_1}{\partial x_2^{2n}}\right]_{x_2=0}} = (-1)^n\frac{1}{u_1'^2}\overline{\left[\frac{\partial^n u_1}{\partial x_2^n}\right]^2_{x_2=0}} \tag{1-56}$$

As mentioned above, it is possible to express the shape of the correlation function at $x_2 = 0$ in terms of the velocity derivatives at that point. To this end expand $\mathbf{g}(x_2)$ in a Taylor series, at the same time taking into account that $\mathbf{g}(x_2)$ is symmetrical with respect to x_2.

$$\mathbf{g}(x_2) = 1 + \frac{1}{2!}x_2^2\left[\frac{\partial^2\mathbf{g}}{\partial x_2^2}\right]_{x_2=0} + \frac{1}{4!}x_2^4\left[\frac{\partial^4\mathbf{g}}{\partial x_2^4}\right]_{x_2=0} + \cdots$$

If we introduce into this series the relations (1-56) we obtain

$$\mathbf{g}(x_2) = 1 - \frac{x_2^2}{2!}\frac{1}{u_1'^2}\overline{\left[\frac{\partial u_1}{\partial x_2}\right]^2_{x_2=0}} + \frac{x_2^4}{4!}\frac{1}{u_1'^2}\overline{\left[\frac{\partial^2 u_1}{\partial x_2^2}\right]^2_{x_2=0}} - \cdots \tag{1-57}$$

For very small values of x_2, $\mathbf{g}(x_2)$ approaches a parabolic function of x_2.

Introduce a length λ_g so that, for very small values of x_2 we have

$$\mathbf{g}(x_2) \simeq 1 - \frac{x_2^2}{\lambda_g^2} \tag{1-58}$$

From this definition of λ_g follows

$$\frac{1}{\lambda_g^2} = -\frac{1}{2}\left[\frac{\partial^2\mathbf{g}}{\partial x_2^2}\right]_{x_2=0} \tag{1-59}$$

and

$$\frac{1}{\lambda_g^2} = \frac{1}{2u_1'^2}\overline{\left[\frac{\partial u_1}{\partial x_2}\right]^2_{x_2=0}} \tag{1-60}$$

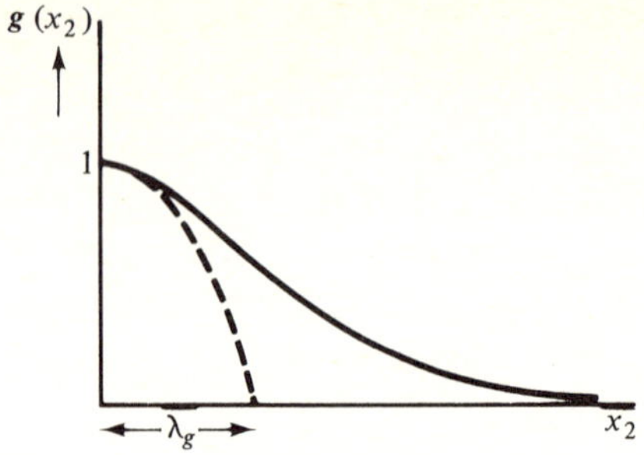

FIGURE 1-16
Osculation parabola of the lateral correlation coefficient **g**, and dissipation length-scale λ_g.

Since $\overline{[\partial u_1/\partial x_2]^2_{x_2=0}}$ is the mean square of the rate of local change of u_1, λ_g may be regarded as a measure of this change. We may imagine that this local change in u_1 is caused by the smallest eddies present in the turbulent flow field. Then λ_g may also be considered a measure of the average dimension of the smallest eddies. As we shall show later, dissipation of kinetic energy of the flow field into heat by the effect of molecular viscosity is determined by the value of $\overline{[\partial u_1/\partial x_2]^2_{x_2=0}}$. Hence, one may also say that λ_g is a measure of the average dimension of the eddies that are mainly responsible for dissipation.† Depending on whether it is being regarded as a measure of the smallest eddies or as a measure of the size of eddy mainly responsible for dissipation, λ_g is called either the micro scale or the dissipation scale. In the following we will use the expression *dissipation scale.*

Since Eq. (1-58) defines a parabola with vertex at the point $x_2 = 0$, $\mathbf{g} = 1$ where it osculates the correlation curve, λ_g is found from the intersection of the parabola with the x_2-axis (see Fig. 1-16).

We mentioned in the beginning of this section that for the longitudinal correlation coefficient

$$\mathbf{f}(x_1) = \frac{\overline{u_1(\xi_1)u_1(\xi_1 + x_1)}}{u_1'^2}$$

the same considerations will apply and will lead to the same results. Thus we may introduce a dissipation scale λ_f also. But since in general

$$\overline{\left[\frac{\partial u_1}{\partial x_1}\right]^2_{x_1=0}} \neq \overline{\left[\frac{\partial u_1}{\partial x_2}\right]^2_{x_2=0}}$$

$\lambda_f \neq \lambda_g$ also. See, for instance, in Chap. 3 the relation between λ_f and λ_g for isotropic turbulence.

† More correctly stated, λ_g is a measure of the dimension of eddies which at the same intensity produce the same dissipation as the turbulence considered.

1-7 INTEGRAL SCALE OF TURBULENCE

There is another length, in addition to the length λ_g or λ_f, that is an important characteristic of the structure of turbulence. This new length is to a certain extent a measure of the longest connection, or correlation distance, between the velocities at two points of the flow field. It is reasonable to expect that the degree of correlation will decrease with increasing distance and that, beyond some finite distance, this correlation will be practically zero. Thus $\mathbf{g} \simeq 0$ for $x_2 > x_2^*$, where x_2^* is the correlation distance.

The new length introduced, however, is not taken equal to this correlation distance x_2^*, but is defined by

$$\Lambda_g = \int_0^\infty \mathbf{g}(x_2)\, dx_2 \qquad (1\text{-}61)$$

That is why this length is also called the integral scale, which name will be further used in the present textbook.

This value of Λ_g will be comparable to that of x_2^* if the $\mathbf{g}$ curve does not deviate too much from the rectangular. The difference becomes substantial if $\mathbf{g}$ obtains a negative value for great values of x_2, which will certainly occur if, for instance, there is a certain constant periodicity in the flow pattern, so that the $\mathbf{g}$ curve oscillates about the x_2-axis at high values of x_2.

To illustrate the effect of periodicity, consider an exact periodic function of x_2

$$u_1 = u_1'\sqrt{2}\sin 2\pi\frac{x_2}{X}$$

Then, by definition,

$$\begin{aligned}\mathbf{g}(x_2) &= \frac{\overline{u_1(\xi_2)u_1(\xi_2 + x_2)}}{u'^2} \\ &= \frac{2}{\Xi}\int_0^\Xi d\xi_2 \sin 2\pi\frac{\xi_2}{X}\sin 2\pi\frac{\xi_2 + x_2}{X} \\ &= \frac{2}{\Xi}\left(\cos\frac{2\pi x_2}{X}\int_0^\Xi d\xi_2\sin^2 2\pi\frac{\xi_2}{X} + \sin\frac{2\pi x_2}{X}\int_0^\Xi d\xi_2 \sin 2\pi\frac{\xi_2}{X}\cos 2\pi\frac{\xi_2}{X}\right)\end{aligned}$$

If we make $\Xi \to \infty$, we obtain for $\mathbf{g}(x_2)$ a pure periodic function of x_2,

$$\mathbf{g}(x_2) = \cos\frac{2\pi x_2}{X}$$

With this periodic correlation function, the integral scale Λ_g can be determined only if we consider a finite domain for the integral. Assume this domain to be equal to $\frac{1}{2}nX$, where n is a large number. We then obtain $\Lambda_g = (X/2\pi)\sin \pi n$; so Λ_g is zero if n is an integer, and varies between zero and $\pm X/2\pi$ if n is not an integer.

Hence, strictly speaking, the definition according to Eq. (1-61) makes sense only if the correlation function shows no pure periodicity, which will be the case if the turbulent motions are irregular.

Periodicity in the flow pattern is not essential in order that $\mathbf{g}$ shall have negative values. As will be shown later (in Chap. 3), in isotropic turbulence, where the turbulent motions are certainly completely random, $\mathbf{g}$ must become negative for high values of x_2. Also, in the following case we shall see that $\mathbf{g}$ becomes negative for large x_2.

Consider the turbulent flow through a two-dimensional channel with breadth $2b$. The condition of continuity requires that

$$\int_{-b}^{+b} dx_2(\bar{U} + u) = \int_{-b}^{+b} dx_2\bar{U} = \text{const. with respect to time}$$

Hence the turbulence fluctuation $u(x_2)$ must be such that, at any instant,

$$\int_{-b}^{+b} dx_2 u(x_2) = 0$$

At any instant, then, we have also

$$u(0)\int_{-b}^{+b} u(x_2)\,dx_2 = \int_{-b}^{+b} u(0)u(x_2)\,dx_2 = 0$$

If we take the time-average value and introduce

$$\overline{u(0)u(x_2)} = u'(0)u'(x_2)\mathbf{g}(x_2)$$

then

$$u'(0)\int_{-b}^{+b} dx_2\, u'(x_2)\mathbf{g}(x_2) = 0$$

or

$$\int_{-b}^{+b} dx_2\, u'(x_2)\mathbf{g}(x_2) = 0$$

Because $u'(x_2)$ is positive, $\mathbf{g}(x_2)$ must become negative for large x_2.

Similarly, in the case of a circular tube we shall find that

$$\int_{0}^{D/2} dr\, ru'(r)\mathbf{g}(r) = 0$$

Taylor[14] has found that experimental results obtained by Simmons confirm this relation.

1-8 OTHER EULERIAN CORRELATIONS

In the preceding sections we have considered correlations between velocity components at a fixed point in the flow field and correlations between velocity components at two points in the flow field all at the same instant; the latter correlations then appear to be functions of the distance between the two points. In this section we shall discuss the correlation between the values of a fluctuating quantity—such as the velocity component in a given direction—at a fixed point in the flow field but at two different instants t' and $t' - t$ or $t' + t$. For obvious reasons, this correlation will be called an Eulerian time correlation.

Consider the velocity component $u_1(x_j,t)$ for constant x_j. Assume that the average statistical character of u_1 is homogeneous with respect to time, that is, that the flow is quasi-steady. We want to know the features of the Eulerian correlation

$$\overline{u_1(t')u_1(t'-t)}$$

or those of its coefficient

$$\mathbf{R}_E(t) = \frac{\overline{u_1(t')u_1(t'-t)}}{u_1'^2} \tag{1-62}$$

where the average is taken with respect to time t'.

Because of the assumed homogeneity, we may deduce, exactly as shown in Sec. 1-6, that $\mathbf{R}_E$(t) is a symmetrical function of t and that the following relations hold:

$$\left[\frac{\partial^{n+m}\mathbf{R}_E}{\partial t^{n+m}}\right]_{t=0} = \frac{1}{u_1'^2}\overline{\left[\frac{\partial^n u_1}{\partial t^n}\frac{\partial^m u_1}{\partial t^m}\right]_{t=0}} = 0 \qquad \text{for } n+m = \text{uneven} \tag{1-63}$$

$$\left[\frac{\partial^{2n}\mathbf{R}_E}{\partial t^{2n}}\right]_{t=0} = \frac{1}{u_1'^2}\overline{\left[u_1\frac{\partial^{2n}u_1}{\partial t^{2n}}\right]_{t=0}} = \frac{1}{u_1'^2}(-1)^n\overline{\left[\frac{\partial^n u_1}{\partial t^n}\right]^2_{t=0}} \tag{1-64}$$

Thus the shape of the $\mathbf{R}_E$ curve in the neighborhood of $t = 0$ can be expressed in terms of the velocity derivatives with respect to time,

$$\mathbf{R}_E(t) = 1 - \frac{t^2}{2u_1'^2}\overline{\left[\frac{\partial u_1}{\partial t}\right]^2_{t=0}} + \frac{t^4}{4!u_1'^2}\overline{\left[\frac{\partial^2 u_1}{\partial t^2}\right]^2_{t=0}} - \cdots \tag{1-65}$$

The equation for the osculating parabola in the vertex of the $\mathbf{R}_E$ curve reads

$$\mathbf{R}_E(t) \simeq 1 - \frac{t^2}{\tau_E^2}$$

where

$$\frac{1}{\tau_E^2} = \frac{1}{2u_1'^2}\overline{\left[\frac{\partial u_1}{\partial t}\right]^2_{t=0}} = -\frac{1}{2}\left[\frac{\partial^2\mathbf{R}_E}{\partial t^2}\right]_{t=0} \tag{1-66}$$

Here the Eulerian time scale τ_E is a measure of the most rapid changes that occur in the fluctuations of $u_1(t)$.

Because of the close connection between this time scale τ_E and the dissipation scales λ_g or λ_f (see, e.g., Eq. (1-71a)), we shall refer to τ_E as the *Eulerian dissipation time scale.*

Like the integral scale Λ defined by Eq. (1-61), the scale

$$\mathfrak{J}_E = \int_0^\infty dt\, \mathbf{R}_E(t) \qquad (1\text{-}67)$$

may be considered to be a rough measure of the longest connection in the turbulent behavior of $u_1(t)$.

The question arises whether there is a relation between $\mathbf{R}_E$ and $\mathbf{f}$ or $\mathbf{g}$. If the turbulent field is homogeneous in its average statistical structure, it is reasonable to expect a certain relationship. Yet, what this relationship will be for the general case it is very difficult to say. Very little is known about this at the present juncture.

Only if the homogeneous field has a constant mean velocity, say $\bar{U}_1 = \bar{U}$ in the x_1-direction, is a relation known and, even then, only an approximate one. In this case it is usually assumed (Taylor's hypothesis) that

$$\frac{\partial}{\partial t} = -\bar{U}\frac{\partial}{\partial x_1} \qquad (1\text{-}68)$$

The negative sign is necessary, because a positive value of $\partial/\partial x_1$ at a point in the space corresponds to a negative value of $\partial/\partial t$.

Closer consideration soon reveals that the relation (1-68) holds approximately but only if u_1 is very small with respect to $\bar{U}$. This can be shown by using the equation of motion,

$$\frac{DU_1}{Dt} = \frac{Du_1}{Dt} = \frac{\partial u_1}{\partial t} + (\bar{U} + u_i)\frac{\partial u_1}{\partial x_i} = -\frac{1}{\rho}\frac{\partial p}{\partial x_1} + \nu\frac{\partial^2}{\partial x_k\,\partial x_k}u_1$$

or

$$\frac{\partial u_1}{\partial t} + \bar{U}\frac{\partial u_1}{\partial x_1} = -u_i\frac{\partial u_1}{\partial x_i} - \frac{1}{\rho}\frac{\partial p}{\partial x_1} + \nu\frac{\partial^2}{\partial x_k\,\partial x_k}u_1$$

Taylor's hypothesis is, then, valid only if all terms on the right-hand side of the equation are very small compared with the terms on the left-hand side. This certainly occurs if $u_1/\bar{U} \ll 1$, since the pressure fluctuation p is of the order of $\rho u_1{}^2$.

A more rigorous treatment of this problem has been presented by Corrsin[15] and by Lin.[16] Lin showed that, for isotropic turbulence and large Reynolds numbers, an estimate of the accuracy of Taylor's hypothesis can be made from

$$\frac{\overline{(du_1/dt)^2}}{\bar{U}^2\overline{(\partial u_1/\partial x_1)^2}} \simeq 5\frac{u_1'^2}{\bar{U}^2}$$

but that, for shear flow, the validity of Taylor's hypothesis is less clear.

The physical interpretation of Eq. (1-68) is that, if $\bar{U} \gg u_1$, the fluctuations at a fixed point of the field may be imagined to be caused by the whole turbulent flow field passing that point as a "frozen" field. The oscillogram of the velocity fluctuations at that point will then be nearly identical with the instantaneous distribution of the velocity u_1 along the x_1-axis through that point. The correlation $\overline{u_1(t')u_1(t'-t)}$

averaging with respect to t' must then be identical with the correlation $\overline{u_1(\xi_1)u_1(\xi_1 - x_1)}$ averaging with respect to ξ_1; so

$$\mathbf{f}(x_1) \equiv \mathbf{R}_E(t) \qquad (1\text{-}69)$$

where the relation between x_1 and t is given by

$$|x_1 - x_1{}^0| = \bar{U}|(t - t^0)| \qquad (1\text{-}70)$$

Finally, between the Eulerian integral scale $\mathfrak{I}_E$ as defined by Eq. (1-67) and the space integral scale Λ_f defined by

$$\Lambda_f = \int_0^\infty dx_1\, \mathbf{f}(x_1)$$

[equivalent to Eq. (1-61)], the simple relation

$$\Lambda_f = \bar{U}\mathfrak{I}_E \qquad (1\text{-}71)$$

holds.

Similarly the Eulerian dissipation scales λ_f and τ_E are connected by the relation

$$\lambda_f = \bar{U}\tau_E \qquad (1\text{-}71a)$$

As we shall see in Chap. 2, the relation (1-70) or (1-68) may be a very welcome means of simplifying measuring procedures; but it must be kept in mind that these relations are only approximate.

For turbulent flows occurring at sufficiently large distances downstream of turbulence-producing grids, the mean velocity $\bar{U}$ is usually great compared with the turbulent fluctuations; so the approximation in the relation (1-70) will be sufficient. The decay there is a relatively slow process; so the field may be considered sufficiently homogeneous for this relation to be applied. But if the same relation is used for measurements in turbulent shear flows with relatively large fluctuations, such as occur in free jets, in which, moreover, the turbulent flow pattern is certainly not homogeneous, the results of such measurements must be viewed with reserve. In fact, this reservation applies to all relations given in the foregoing that are based on the assumption of homogeneity of the flow field.

The reader may get an idea of the validity of Taylor's hypothesis (1-68) from Fig. 1-17, which shows the correlations $\mathbf{f}(x_1)$ and $\mathbf{R}_E(t)$ measured separately by Favre and coworkers[39] in the turbulent flow at $x_1 = 40M$ downstream of a grid ($\mathbf{Re}_M = \bar{U}M/\nu = 21{,}500$; $u'_1/\bar{U} = 0.01$ to 0.02).

An extension to the above purely spatial and purely time correlations is obtained if separations in space and in time are considered, i.e., Eulerian space-time correlations. Thus the correlation is considered, for instance, between a velocity component at A at a given instant, and a velocity component at B some time later.

For a stationary turbulence

$$(\mathbf{Q}_{i,j})_{A,B}(x_k,t) = \overline{u_{i_A}(\xi_k,t')u_{j_B}(\xi_k + x_k, t' + t)}$$

where ξ_k is the position vector of A, and $\xi_k + x_k$ that of B.

It goes without saying that still further extensions can be given, by considering the correlations between any number of velocity components at different points of the turbulence field and at different time instants.

1-9 TURBULENT DIFFUSION OF FLUID PARTICLES; LAGRANGIAN CORRELATIONS

We know that in real turbulence the motions of fluid particles are randomly distributed. In turbulent motions, as in motions of molecules in a gas, which also have a random character, two arbitrary fluid particles or two molecules will move in such a way that statistically the distance between them increases with time. If we consider a number of neighboring particles or molecules at one instant and if we observe the position of the various particles or molecules at subsequent instants, we observe a gradual spread of the particles throughout the space. This is the basic idea of diffusion.

In the kinetic theory of gases, relations for the diffusion of molecules are derived on the basis of the concept of random motion of the molecules. For a short comprehensive treatment see, for instance, the "Introduction to the Kinetic Theory of Gases" by Jeans.[17]

In order to study diffusion phenomena it is often useful to introduce a diffusion constant. Consider, for instance, a plane in the (x_1,x_3) direction, perpendicular to x_2, and let Γ be a transferable quantity. Then the transport $\mathfrak{J}_\gamma$ of this quantity across a unit area of this plane is proportional to the gradient of Γ in the x_2-direction.

$$\mathfrak{J}_\gamma = -D_\gamma \frac{\partial \Gamma}{\partial x_2}$$

The diffusion constant for the quantity Γ is defined as the proportionality constant D_γ.

Consider now a second plane parallel to the first and at a short distance dx_2 from it. The law of conservation applied to the quantity Γ in the space between the two planes gives the equation

$$\frac{\partial \Gamma}{\partial t} = -\frac{\partial \mathfrak{J}_\gamma}{\partial x_2} = D_\gamma \frac{\partial^2 \Gamma}{\partial {x_2}^2} \tag{1-72}$$

The simplest solution of this equation,

$$\Gamma(x_2,t) = \frac{\text{const.}}{\sqrt{t}} \exp\left(-{x_2}^2/4D_\gamma t\right) \tag{1-73}$$

refers to the case in which molecules diffuse gradually from the plane $x_2 = 0$. Equation (1-73) gives the value of Γ at x_2 after the lapse of a time t since the molecules started from the plane $x_2 = 0$.

If we take for the transferable quantity the mass of the molecules themselves, we obtain

$$n(x_2,t) = \frac{\text{const.}}{\sqrt{t}} \exp(-x_2{}^2/4Dt) \qquad (1\text{-}74)$$

where D is the constant of self-diffusion for the molecules. Here $n(x_2,t)$ is the number of molecules that have reached the plane x_2 after a time t.

Let us now consider the random motion of molecules and let us assume that all the free paths of the molecules are the same, constant, and parallel to the direction x_2. It can be shown[17] that the probability $\mathfrak{P}$ that a molecule has advanced a distance x_2 after N free paths is equal to

$$\mathfrak{P}(x_2,N) = \sqrt{\frac{2}{\pi N}} \exp(-x_2{}^2/2l^2N) \qquad (1\text{-}75)$$

where l is the free path of the molecule. We can express this probability also in terms of t by introducing the molecular velocity c,

$$N = \frac{ct}{l}$$

with which Eq. (1-75) becomes

$$\mathfrak{P}(x_2,t) = \sqrt{\frac{2l}{\pi ct}} \exp(-x_2{}^2/2clt)$$

This is exactly the same expression as (1-74). Indeed, $\mathfrak{P}(x_2,t)$ is identical with $n(x_2,t)$. Hence, apparently, we have

$$D = \tfrac{1}{2}cl = \frac{Nl^2}{2t} \qquad (1\text{-}76)$$

Consider next the case of successive displacements Δy_i of the molecules in time intervals Δt in the x_2-direction. These displacements are assumed to be large compared with l and not equal, but still independent and random. After N time intervals, the resultant displacement of a molecule is

$$y_2 = \sum_1^N \Delta y_i$$

If we assume a homogeneous field, the average value over a large number of molecules is zero: $\overline{y_2} = 0$. However, the mean-square value, or variance, of y_2 reads

$$\overline{y_2{}^2} = \overline{\left(\sum_1^N \Delta y_i\right)^2} = \overline{\sum_1^N \sum_1^N \Delta y_i \, \Delta y_j} = N\,\overline{\Delta y^2}$$

since the average value of the product terms vanishes because of the independence of the displacements.

For random motion of molecules, we have derived Eq. (1-76) for the diffusion constant where Nl^2 is the sum of the squares of the displacements of the molecules between successive collisions.

If for the last case considered we take instead of Nl^2 the sum of the squares $N\,\overline{\Delta y^2}$, the expression for the diffusion constant reads

$$D = \frac{N\,\overline{\Delta y^2}}{2N\,\Delta t} = \frac{\overline{y_2{}^2}}{2t} \tag{1-77}$$

This relation shows, just as Eq. (1-76) did, that the root mean square $\sqrt{\overline{y_2{}^2}}$ is proportional to $\sqrt{t}$ as diffusion proceeds.

It must be remarked that the above results hold only if the time of diffusion is long compared with the time required to traverse the mean free path or the displacement Δy_i. Furthermore, the motions of the molecules are discontinuous, and in the rough theory given above they preserve their properties completely during collisions.

It is important to note that at the beginning of each next step after a collision there is no after-effect or pre-history effect. There is no memory behavior of the molecules. A random process where at any instant the further course of the process is entirely determined by its state at that instant and independent of its prehistory or, in other words, where the future is independent of the past for a known present, is usually referred to as a *Markov process*.[49]

Let us now apply similar considerations to the diffusion of fluid particles by the turbulent motion of the fluid. This process is distinct from the diffusion process according to the kinetic theory of gases at least for the following reasons. First, the motions of the fluid particles are continuous. Second, because of the intensive interactions between the fluid particles, there may occur a continuous exchange of a transferable property. Third, as will be shown in the following, there is a correlation in time between properties of a fluid particle at subsequent instants. Because of this memory behavior the turbulence diffusion process may not be considered as a Markov process. We will return to this point later in Chap. 5.

Taylor[18] extended the above considerations to the diffusion in turbulent flow, taking into account the continuous movements of the fluid particles, by considering the path of a marked fluid particle during its motion through the flow field.

Let $v_2(t)$ be the (turbulent) velocity of a particle in the x_2-direction. The distance $y_2(t) - y_2(0)$ traveled by the marked particle in this direction after a time t reads

$$y_2(t) = y_2(0) + \int_0^t dt'\, v_2(t') \tag{1-78}$$

The reader may have noticed that we distinguish between the Lagrangian coordinate y_2 of the particle and the Eulerian coordinate x_2 of the space.

If $v_2(t)$ were a periodic function, the average value of $y_2(t)$ would be equal to

$y_2(0)$. In this periodic motion the size and number of positive "steps" would be equal to the size and number of negative "steps." This is not precisely so in turbulent motion, any more than in the other cases of random motions considered. The intervals with positive v_2 may be greater or smaller than the intervals with negative values. The result is that, after a certain time, the particle will have acquired a positive or negative displacement with respect to $y_2(0)$.

Consider now a large number of particles that are assumed to start in succession from a fixed point $0(x_1,x_2,x_3)$. After a certain time has elapsed since the start of each particle, some of them will have acquired a resultant displacement in one direction, others in another direction. If the structure of the field is isotropic and homogeneous with respect to both time and space, the spread of the particles will be (spherically) symmetrical with respect to 0.

Assume next that the whole field has a constant velocity in the x_1-direction. If at point 0 coloring matter is supplied continuously to the flowing field, downstream of this point it will be observed that a colored strip of the fluid broadens with increasing distance from 0 and is symmetrical with respect to the x_1-axis.

This symmetry obtains also if the flow field is not homogeneous in the flow direction but is homogeneous in the other directions. Homogeneity with respect to time remains a necessary condition, because otherwise it would not be permissible to average over a great number of particles that have started at consecutive moments.

If the field is not homogeneous with respect to time (unsteady flow), it is still possible to carry out an averaging procedure with respect to a great number of particles by considering the simultaneous start of these particles from as many points throughout the field, provided that the field is homogeneous in space.

In order to study in more detail the diffusion by continuous turbulent movement first considered by Taylor, we assume a turbulent flow field that is homogeneous in space and time. We will treat it slightly differently than Taylor did in his original paper.[18] We take the simplest case of diffusion in one direction only, say the x_2-direction. Equation (1-78) gives the displacement from the point $y_2(0)$ of a particle whose velocity is v_2 after time t. For simplicity, we take this point as the origin, so that $y_2(0) = 0$. Since $v_2(t)$ is a random quantity, the average value $\overline{y_2(t)} = 0$ for a large number of particles.

Let $y_2(t_0 + t)$ be the distance traveled by a marked fluid particle with starting time t_0 during the time interval t.

$$y_2(t_0 + t) = \int_0^t dt' \, v_2(t_0 + t')$$

Making use of the assumed homogeneity of the flow field, for the mean-square value or variance of y_2 we obtain

$$\overline{y_2{}^2(t)} = \frac{1}{T}\int_0^T dt_0 \, y_2{}^2(t_0 + t)$$

$$= \frac{1}{T}\int_0^T dt_0 \int_0^t dt' \int_0^t dt''\, v_2(t_0+t')v_2(t_0+t'')$$

$$= \int_0^t dt' \int_0^t dt'' \frac{1}{T}\int_0^T dt_0\, v_2(t_0+t')v_2(t_0+t'')$$

$$= \int_0^t dt' \int_0^t dt''\, \overline{v_2(t_0+t')v_2(t_0+t'')}$$

Averaging has been carried out with respect to a large number of particles with different starting times t_0.

The integration in the (t',t'') plane is over a square with the limits 0 and t. Instead of this square we may take twice the triangle formed by the half square, since the integrand is symmetrical with respect to t' and t''. Then we have

$$\int_0^t \int_0^t dt'\, dt'' = 2\int_0^t dt' \int_0^{t'} dt''$$

Thus

$$\overline{y_2{}^2}(t) = 2\int_0^t dt' \int_0^{t'} dt''\, \overline{v_2(t_0+t')v_2(t_0+t'')} \tag{1-79}$$

In this case—in contrast with the corresponding case for random motion of molecules—the motions of the fluid particles in two instances t' and t'' are not independent but are more or less correlated. Therefore we introduce the Lagrangian auto-correlation coefficient

$$\mathbf{R}_L(\tau) = \frac{\overline{v_2(t)v_2(t+\tau)}}{\overline{v_2'^2}} \tag{1-80}$$

In order to introduce this Lagrangian correlation into Eq. (1-79) we put

$$t'' - t' = \tau$$

so that Eq. (1-79) becomes

$$\overline{y_2{}^2}(t) = 2\int_0^t dt' \int_{-t'}^0 d\tau\, \overline{v_2(t')v_2(t'+\tau)}$$

$$= 2\int_0^t dt' \int_0^{t'} d\tau\, \overline{v_2(t')v_2(t'-\tau)}$$

$$= 2\overline{v_2'^2}\int_0^t dt' \int_0^{t'} d\tau\, \mathbf{R}_L(\tau) \tag{1-81}$$

This is the relation obtained by Taylor.[18]

This expression may be written somewhat differently by carrying out an integration by parts

$$\int_0^t dt' \int_0^{t'} d\tau\, \mathbf{R}_L(\tau) = \left| t' \int_0^{t'} d\tau\, \mathbf{R}_L(\tau) \right|_0^t - \int_0^t dt'\, t'\mathbf{R}_L(t')$$

$$= t \int_0^t d\tau\, \mathbf{R}_L(\tau) - \int_0^t d\tau\, \tau\mathbf{R}_L(\tau)$$

Then Eq. (1-81) reads

$$\overline{y_2{}^2}(t) = 2\overline{v_2'{}^2} \int_0^t d\tau\, (t-\tau)\, \mathbf{R}_L(\tau) \qquad \text{(1-82)}$$

This relation was first given by Kampé de Fériet.[19]

The Lagrangian correlation $\mathbf{R}_L(\tau)$ has properties similar to those of the previously treated Eulerian correlations, namely: $\mathbf{R}_L(0) = 1$; $\mathbf{R}_L(\tau)$ is symmetrical with respect to τ, because of the homogeneity of the field, and will decrease with increasing τ; so, for large τ, $\mathbf{R}_L(\tau)$ will eventually become zero.

Consider more specifically very small values of the time t and very large values.

For very small values of τ the correlation $\mathbf{R}_L(\tau)$ is nearly equal to unity and may be treated as a constant in either of the relations (1-81) and (1-82). After carrying out the integration we obtain, for $t =$ small, $\mathbf{R}_L(\tau) \simeq 1$,

$$\overline{y_2{}^2}(t) \simeq \overline{v_2'{}^2}t^2$$

or

$$y_2'(t) \simeq v_2't \qquad \text{(1-83)}$$

Hence diffusion proceeds proportionally with time.

If, on the other hand, we consider very long periods of time, such that $t \gg t^*$, where t^* is the time for which $\mathbf{R}_L(t^*) \simeq 0$, the relation (1-82) gives

$$\int_0^t d\tau\, (t-\tau)\mathbf{R}_L(\tau) = t\int_0^{t^*} d\tau\, \mathbf{R}_L(\tau) - \int_0^{t^*} d\tau\, \tau\mathbf{R}_L(\tau)$$

For $t \gg t^*$, the second term on the right-hand side will become very small with respect to the first term; so it may be neglected. For the constant value of the integral we write

$$\mathfrak{I}_L = \int_0^{t^*} d\tau\, \mathbf{R}_L(\tau) \qquad \text{(1-84)}$$

$\mathfrak{I}_L$ is usually considered a measure of the longest time during which, on the average, a particle persists in a motion in a given direction. We will call it the *Lagrangian integral time scale.*

We may also introduce a time scale τ_L defined in a similar way as the Eulerian dissipation time scale τ_E, namely:

$$\frac{1}{\tau_L{}^2} = -\frac{1}{2}\left(\frac{\partial^2 \mathbf{R}_L}{\partial t^2}\right)_{t=0}$$

Though the link with the dissipation scale λ_f is not so close as in the case of τ_E, except that the magnitude of τ_L too is determined by the smaller eddies, we will call it the *Lagrangian dissipation time scale.*

With the introduction of $\mathfrak{I}_L$, for $t =$ large, $t \gg t^*$, so that $\mathbf{R}_L(t^*) \simeq 0$, or $t \gg \mathfrak{I}_L$, we obtain from Eq. (1-82)

$$\overline{y_2{}^2(t)} \simeq 2v_2'{}^2\mathfrak{I}_L t$$

or

$$y_2' \simeq v_2'\sqrt{2\mathfrak{I}_L t} \qquad (1\text{-}85)$$

In this case diffusion is proportional to the square root of time.

For these long periods of time we may introduce a coefficient of eddy diffusion, defined by Eq. (1-86), which is similar to Eq. (1-77),

$$\epsilon = \frac{\overline{y_2{}^2(t)}}{2t} = v_2'^2\mathfrak{I}_L = v_2'^2\int_0^{t^*} d\tau\, \mathbf{R}_L(\tau) = v_2'^2\int_0^{\infty} d\tau\, \mathbf{R}_L(\tau) \qquad (1\text{-}86)$$

since the correlation $\mathbf{R}_L(t) \simeq 0$ for $t > t^*$.

This relation may also be written

$$\epsilon = v_2'\Lambda_L \qquad (1\text{-}87)$$

where

$$\Lambda_L = v_2'\mathfrak{I}_L = v_2'\int_0^{\infty} d\tau\, \mathbf{R}_L(\tau) \qquad (1\text{-}88)$$

Λ_L may be interpreted as a space scale in which the particle moves substantially in only one direction.

Since Λ_L appears to be a measure of eddy diffusion, it is often called the scale of eddy diffusion. In the above relation we may put the intensity of the turbulent fluctuations of the particles v_2' equal to u_2', because of the assumed homogeneity of the flow field.

Again the question arises whether there is a relation between the Lagrangian and Eulerian correlations. In the case of an isotropic, or even a homogeneous, turbulent flow field one might expect a certain relationship. Yet it has not been possible to find it on purely theoretical grounds. As we shall show later, empirical relationships have been found which show that Λ_L is roughly of the same magnitude as Λ_f.

The value of ϵ given by Eq. (1-86) would be valid also for the effective turbulent-transport coefficient of a quantity Γ if there were no exchange of this quantity with the surrounding fluid during the path of the particle. However, because the individual fluid particle to be considered in turbulent flow consists of a very large complex of molecules, exchange of at least a molecular nature will take place. The rate of this exchange will be different according to whether the transported quantity is momentum, heat, or mass. This means that the values of ϵ will be different for momentum, heat, and mass. It is reasonable to expect that the higher the degree of exchange the sooner the particle will adapt its transferable property to its surroundings, and the smaller will be the value of ϵ.

If α is a kind of exchange coefficient, we may expect for $(\Lambda_L)_\gamma$ an expression of the form

$$(\Lambda_L)_\gamma = v'_2 \int_0^\infty d\tau\, f(\alpha,\tau)\mathbf{R}_L(\tau)$$

where $f(\alpha,\tau)$ accounts for the effect of the exchange during the path of the particle. Consequently

$$\epsilon(\alpha) = v_2'^2 \int_0^\infty d\tau\, f(\alpha,\tau)\mathbf{R}_L(\tau)$$

A description of turbulence in terms of Eulerian spatial and time correlations, or of Lagrangian correlations cannot be complete due to the memory effects in the dynamics of the turbulence. What occurs in a volume element is not determined by local conditions alone, since the fluid particles contained in this volume element may have arrived from different parts of the flow field. Therefore Kraichnan[37] introduced a kind of a *mixed Eulerian-Lagrangian correlation.* To this end he defined a *generalized velocity* $\mathfrak{u}_i$, namely

$$\mathfrak{u}_i(x_l,t|t_m)$$

which is the velocity of a fluid particle at time t_m which was, or will be, in the point x_l at an earlier, or later, time t.

When $t = t_m$ the velocity reduces to the Eulerian velocity $\mathfrak{u}_i(x_l,t|t) = u_i(x_l,t)$, while for a fixed value of (x_l,t) the velocity becomes the Lagrangian velocity $v_i(t_m)$. The velocity is further defined to satisfy

$$\left(\frac{\partial}{\partial t} + u_k\frac{\partial}{\partial x_k}\right)\mathfrak{u}_i(x_l,t|t_m) = 0$$

for $t \neq t_m$, while for $t = t_m$ the Eulerian velocity has to satisfy the momentum balance equation.

Also, for an incompressible fluid, where $\partial u_i/\partial x_i = 0$, this zero divergence does not apply to the generalized velocity when $t \neq t_m$.

With this generalized velocity the following generalized double velocity correlation can be introduced.

$$(\mathbf{Q}_{i,j})_{\text{gen}}(x_l,t|t_m;\, x_l + \xi_l,t + \tau|t_m + \tau_m) = \overline{\mathfrak{u}_i(x_l,t|t_m)\mathfrak{u}_j(x_l + \xi_l,t + \tau|t_m + \tau_m)}^{\,e}$$

where an ensemble average is taken. Thus it is the correlation between the velocity of a fluid particle at time t_m that was in x_l at time t, and the velocity of another fluid particle at time $t_m + \tau_m$ that was in $x_l + \xi_l$ at time $t + \tau$.

1-10 RECAPITULATION OF CORRELATIONS

In the preceding sections we have introduced various kinds of correlations. It may be convenient to the reader to summarize them.

EULERIAN SPACE CORRELATIONS

Double Correlations between a Velocity Component and Pressure or Other Scalar Quantity. First-order tensor:

$$(\mathbf{K}_{i,p})_{A,B} = \overline{(u_i)_A p_B} \qquad (\mathbf{K}_{i,\gamma})_{A,B} = \overline{(u_i)_A \gamma_B}$$

Correlation coefficient:

$$(\mathbf{L}_{i,p})_{A,B} = \frac{(\mathbf{K}_{i,p})_{A,B}}{(u_i')_A p_B'}$$

(No summation)

Special case when $A \equiv B$:

$$\mathbf{K}_{i,\gamma}(0) = \overline{\gamma u_i} = -\epsilon_\gamma \frac{\partial \bar{\Gamma}}{\partial x_i}$$

Measure of "eddy" transport

Double Correlations between Two Velocity Components (see Fig. 1-13). Second-order tensor:

$$(\mathbf{Q}_{i,j})_{A,B} = \overline{(u_i)_A (u_j)_B}$$

Correlation coefficient:

$$(\mathbf{R}_{i,j})_{A,B} = \frac{\overline{(u_i)_A (u_j)_B}}{(u_i')_A (u_j')_B}$$

Special case when $A \equiv B$:

$$\mathbf{Q}_{i,j}(0) = \overline{u_i u_j} = -(\epsilon_m)_{ik}\left(\frac{\partial \bar{U}_k}{\partial x_j} + \frac{\partial \bar{U}_j}{\partial x_k}\right)$$

Measure of turbulence stresses, eddy "viscosity," or eddy-transport coefficient of momentum

Isotropic homogeneous turbulence (see Fig. 1-14):

Longitudinal correlation coefficient:

$$\mathbf{f}(r) = \frac{\overline{(u_r)_A (u_r)_B}}{u'^2}$$

Transverse correlation coefficient:

$$\mathbf{g}(r) = \frac{\overline{(u_n)_A (u_n)_B}}{u'^2}$$

Integral scale:

$$\Lambda_f = \int_0^\infty dr\, \mathbf{f}(r) \qquad \Lambda_g = \int_0^\infty dr\, \mathbf{g}(r)$$

Dissipation scale:

$$\lambda_f \text{ from } \frac{1}{\lambda_f^2} = -\frac{1}{2}\left[\frac{\partial^2 \mathbf{f}}{\partial r^2}\right]_{r=0}$$

$$\lambda_g \text{ from } \frac{1}{\lambda_g^2} = -\frac{1}{2}\left[\frac{\partial^2 \mathbf{g}}{\partial r^2}\right]_{r=0}$$

Triple Correlations between Velocity Components. Third-order tensor:

$$(\mathbf{S}_{ik,j})_{A,B} = \overline{(u_i)_A(u_k)_A(u_j)_B}$$

Correlation coefficient:

$$(\mathbf{T}_{ik,j})_{A,B} = \frac{\overline{(u_i)_A(u_k)_A(u_j)_B}}{(u_i')_A(u_k')_A(u_j')_B}$$

Isotropic homogeneous turbulence (see Fig. 1-15):

$$\mathbf{k}(r) = \mathbf{T}_{rr,r} = \frac{\overline{(u_r)_A{}^2(u_r)_B}}{u'^3}$$

$$\mathbf{h}(r) = \mathbf{T}_{nn,r} = \frac{\overline{(u_n)_A{}^2(u_r)_B}}{u'^3}$$

$$\mathbf{q}(r) = \mathbf{T}_{nr,n} = \frac{\overline{(u_n)_A(u_r)_A(u_n)_B}}{u'^3}$$

EULERIAN TIME CORRELATION

$$\mathbf{R}_E(t) = \frac{\overline{u(\tau)u(\tau - t)}}{u'^2} \quad \text{for velocity } u \text{ at a fixed point}$$

Steady flow field; averaging with respect to time:

Eulerian integral time scale:

$$\mathfrak{I}_E = \int_0^\infty dt\, \mathbf{R}_E$$

Eulerian dissipation time scale:

$$\tau_E \text{ from } \frac{1}{\tau_E{}^2} = -\frac{1}{2}\left[\frac{\partial^2 \mathbf{R}_E}{\partial t^2}\right]_{t=0}$$

If the field has a uniform mean velocity $\bar{U}$, so that $\bar{U} \gg u$, there is a simple relation between $\mathfrak{I}_E$ and Λ_f: because of $x = \bar{U}t$,

$$\Lambda_f = \bar{U}\mathfrak{I}_E$$

and

$$\mathbf{f}(x) \equiv \mathbf{R}_E(t)$$

LAGRANGIAN TIME CORRELATION

$$\mathbf{R}_L(t) = \frac{\overline{v(\tau)v(\tau - t)}}{v'^2}$$

for the velocity components of a fluid particle. Homogeneous flow field. Averaging over a large number of particles.

Lagrangian or diffusion-integral time scale:

$$\mathfrak{I}_L = \int_0^\infty dt\, \mathbf{R}_L$$

Lagrangian dissipation time scale:

$$\tau_L \text{ from } \frac{1}{\tau_L{}^2} = -\frac{1}{2}\left(\frac{\partial^2 \mathbf{R}_L}{\partial t^2}\right)_{t=0}$$

Lagrangian or diffusion-integral length scale: $\Lambda_L = v'\mathfrak{I}_L$

Coefficient of eddy diffusion for fluid particles:

$$\epsilon = v'\Lambda_L = v'^2 \int_0^\infty dt\, \mathbf{R}_L$$

Similarly, for transport of Γ by fluid particles:

$$\epsilon_\gamma(\alpha) = v'^2 \int_0^\infty dt\, f(\alpha,t)\mathbf{R}_L(t)$$

where $f(\alpha,t)$ accounts for the effect of exchange during the path of the particles.

1-11 EMPIRICAL FORMULAS FOR DOUBLE-CORRELATION CURVES

In view of the importance of the correlation coefficient in studying a turbulent flow field, it would be very convenient if it were possible to represent the correlation curve by a suitable formula. It is known that in homogeneous turbulence the double-correlation function is symmetrical, with a parabolic shape in its vertex. But for the rest rather little is known, theoretically, about the exact shape of the other parts of the curve for the general case of turbulent flow. Von Kármán[20] succeeded in obtaining theoretically a formula for the two correlation coefficients $\mathbf{f}(r)$ and $\mathbf{g}(r)$ for isotropic turbulent flow at large Reynolds numbers. Batchelor[21] has obtained a solution, but not in closed form, by means of which the shape of the curves for $\mathbf{f}(r)$ and $\mathbf{g}(r)$ can be calculated, again for isotropic turbulent flow, but at lower Reynolds numbers. For still lower Reynolds numbers, such that the viscous effect in the turbulent motion becomes predominant, a simple formula can be derived for these correlation coefficients. We shall come back to these theoretical treatments of correlations later, in Chap. 3. These theories, as well as experimental evidence, have shown that there is no universal formula for these correlation coefficients but that the different formulas to be applied depend on the type, condition, and stage of the turbulent flow.

Since at the moment formulas are known only for a few restricted cases and since, moreover, most of those formulas are very complicated, it may be useful to work with empirical formulas that approximate measured correlation curves pertinent to different types, conditions, and stages of turbulence.

In many cases, particularly at large Reynolds numbers, it appears possible to approach the shape of such curves roughly by means of an exponential function. Attempts in this direction have been made by Dryden[22] and coworkers with reference to measured correlation curves for the isotropic turbulent flow downstream of grids.

They found that

$$\mathbf{g}(x_2) = \exp(-x_2/\Lambda_g) \qquad (1\text{-}89)$$

The same expression was obtained by Kalinske and Pien[23] for the coefficient of the Lagrangian correlation $\mathbf{R}_L(t)$. Taylor, too, in his publication on diffusion by continuous movements,[18] has used such an exponential function for the Lagrangian time correlation. In this case of Lagrangian correlation, calculations by Doob[24] concerning Brownian movements appear to lend support to the assumption of an exponential curve. Doob comes to the conclusion that correlations pertinent to these movements are exponential functions.

Yet it is readily seen that the exponential function of Eq. (1-89), although it may describe satisfactorily the over-all shape of a correlation, cannot be correct (1) because this function is not parabolic in its vertex, (2) because this function is essentially positive for all values of its argument whereas, at least in the case of the g-correlation, the correct curve must have negative values for large distances between the points concerned.

Now let us assume, notwithstanding the above, that such an exponential function is correct; it is then very convenient for use in further theoretical analyses. Consider, for instance, the expression (1-81) for the diffusion of fluid particles, into which we introduce

$$\mathbf{R}_L(t) = \exp(-t/\mathfrak{I}_L)$$

This function satisfies Eq. (1-84). We then obtain

$$\overline{y_2^2}(t) = 2\overline{v_2'^2}\int_0^t dt' \int_0^{t'} d\tau \exp(-\tau/\mathfrak{I}_L)$$

$$= 2\overline{v_2'^2}\,\mathfrak{I}_L^2\left\{\frac{t}{\mathfrak{I}_L} - [1 - \exp(-t/\mathfrak{I}_L)]\right\}$$

This, again, shows clearly that diffusion is proportional, for short times, to the time but, for long periods, to the square root of the time.

We have said that the exponential function Eq. (1-89) does not satisfy the actual correlation function, especially in the neighborhood of its vertex. In the search for other functions we may ask whether there are certain minimum requirements imposed by the actual correlation functions that have to be met by the functions sought for. There are indeed a number of such requirements. Thus, for homogeneous turbulence, for instance, the correlation coefficients g or f have to satisfy at least the following conditions:

$$-1 \leq \mathbf{g}(x_2) \leq 1$$

$$\lim_{x_2 \to \infty} \mathbf{g}(x_2) = 0$$

$$\lim_{x_2 \to 0} \mathbf{g}(x_2) = 1$$

$$\lim_{x_2 \to 0} \frac{d\mathbf{g}}{dx_2} = 0$$

$$\lim_{x_2 \to 0} \frac{d^2\mathbf{g}}{dx_2{}^2} = \text{finite} = -\frac{2}{\lambda_g{}^2}$$

We may try a more general function of the form

$$\mathbf{g}(x_2) = \exp\left(-|cx_2|^m\right) \qquad (1\text{-}90)$$

It is easy to verify that all the requirements listed above are satisfied if $m \geq 2$. If $m = 2$ we have

$$\left[\frac{d^2\mathbf{g}}{dx_2{}^2}\right]_{x_2=0} = -2c^2$$

so $c = 1/\lambda_g$. On the other hand, according to the definition of Λ_g, (1-61), the value of c is related to the integral scale Λ_g by

$$c = \tfrac{1}{2}\Gamma\left(\tfrac{1}{2}\right)\frac{1}{\Lambda_g} = \frac{\sqrt{\pi}}{2}\frac{1}{\Lambda_g} \simeq \frac{0.88}{\Lambda_g}$$

Hence, for the case $m = 2$, where we have the Gaussian error curve, the ratio λ_g/Λ_g has a constant value, namely, $2/\sqrt{\pi} \simeq 1.14$. As we will show later, in Chap. 3, such a Gaussian curve does indeed apply to actual correlation curves in an exact way, but only during the final stages of the decay period of an isotropic turbulence. It does not apply to the earlier stages of decay, where, moreover, it has been found that the ratio λ_g/Λ_g does not always remain constant during decay.

Frenkiel[25] has applied the function (1-90) to various measured correlation curves of turbulence generated by grids in a windtunnel. Though this function can give a satisfactory solution, including the part near the vertex of the correlation curve, only when $m \geq 2$, yet Frenkiel found that an adequate approximation to the whole correlation curve is often achieved with $1 < m < 2$ (in most cases closer to 1). The conclusion is that the function (1-90) is still too simple an expression for most of the actual correlation functions.

For this reason, and also because calculations including exponential functions of the type (1-90) are not very convenient when m is not an integer, Frenkiel[26] tried various other functions. His conclusion was that, in general, a fairly close agreement is obtained with functions of the type

$$\varphi(x_2)\exp\left(-|cx_2|\right)$$

where

$$\varphi(x_2) = a_0 + \Sigma a_n \cos m_n c x_2$$

or

$$\varphi(x_2) = 1 + \Sigma a_n c^n |x_2|^m$$

1-12 TAYLOR'S ONE-DIMENSIONAL ENERGY SPECTRUM

In Sec. 1-1 we referred to the possibility of analyzing the kinetic energy of the turbulence fluctuations according to its distribution over the various frequencies occurring in these fluctuations. Consider, for instance, the component u_1 of the turbulence fluctuation of velocity in a fixed point of the flow field. We assume that this field is quasi-steady, so that it is statistically homogeneous with respect to time. There then exists a constant average value of $u_1{}^2$, which can be considered to consist of the sum of the contributions of all the frequencies n.

Let

$$\mathbf{E}_1(n)\,dn$$

be the contribution to $\overline{u_1{}^2}$ of the frequencies between n and $n + dn$; the distribution function $\mathbf{E}_1(n)$ then has to satisfy the condition

$$\int_0^\infty dn\,\mathbf{E}_1(n) = \overline{u_1{}^2} \qquad (1\text{-}91)$$

When discussing the Eulerian time correlation of the velocity fluctuation u_1 at a fixed point, we found a relation between the dissipation time scale τ_E and the shape of the $\mathbf{R}_E$-correlation curve at its vertex [see Eq. (1-66)]. This time τ_E was found to be a measure of the most rapid changes of the fluctuation $u_1(t)$; hence it is reasonable to expect a direct relation between the value of τ_E and the highest frequencies occurring in the fluctuations.

Such a relation between the shape of the correlation curve and the frequencies need not be restricted to the vertex of the curve and to the highest frequencies. The integral scale was also found to be a measure of the average size of the largest eddies. Imagine that the whole flow field has a translation with uniform velocity $\bar{U}$ in the x_1-direction. Then, at a stationary point, the largest eddies will cause fluctuations of low frequencies, whereas the smallest eddies will cause fluctuations

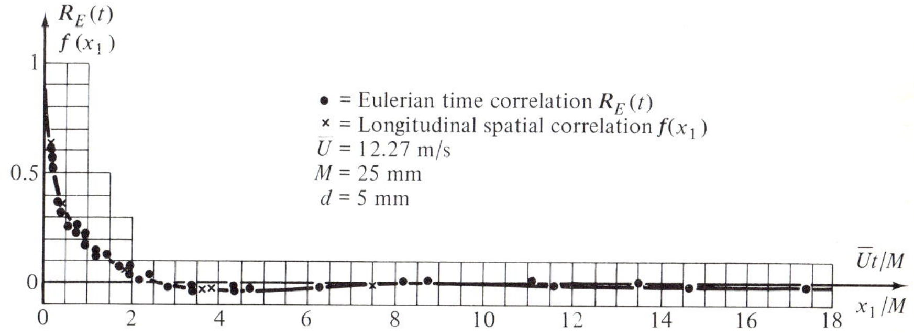

FIGURE 1-17
Comparison between time and spatial correlations according to Taylor's hypothesis. (From: *Favre, A. et. al.*[39])

of high frequencies. If the turbulence contains only large eddies, the distribution function $\mathbf{E}_1(n)$ will exist mainly in the region of low frequencies; if there are only small eddies, $\mathbf{E}_1(n)$ will exist mainly in the region of high frequencies. Hence one may expect that, when the $\mathbf{R}_E$-curve decreases rapidly to zero, the spectrum will have high values in the region of high frequencies.

The relation that must exist between the correlation function and the spectrum function was first put into mathematical form by Taylor.[27] In what follows we shall give the derivation of this relation, but in a somewhat different way from Taylor's.

We assume that the velocity fluctuation $u_1(t)$ can be represented by a Fourier integral

$$u_1(t) = 2\pi \int_{-\infty}^{+\infty} dn\, a(n)\, e^{-i2\pi nt} \qquad (1\text{-}92)$$

where

$$a(n) = \frac{1}{2\pi} \int_{-\infty}^{+\infty} dt\, u_1(t)\, e^{+i2\pi nt}$$

In the theory of Fourier integrals it is shown[28] that the integral for the complex coefficient $a(n)$ exists if $u_1(t)$ satisfies the condition

$$\int_{-\infty}^{+\infty} dt\, |u_1(t)| = \text{finite}$$

Because of the permanent statistical character of the turbulence, so that $u_1(t)$ is a stationary random function of time, the average $\overline{u_1{}^2}$ has a constant value, and the above condition cannot be satisfied. To overcome this difficulty, it is usually assumed that the duration of $u_1(t)$ is finite, so that $u_1(t)$ is different from zero only in the region bounded by $-T$ and $+T$. Since T can be given any large value, though finite, this assumption is no real restriction from a physical point of view.

For the stationary random function $u_1(t)$, we define the correlation function:

$$\mathbf{Q}_E = \overline{u_1(\tau)u_1(\tau - t)} = \lim_{T\to\infty} \frac{1}{2T} \int_{-T}^{+T} d\tau\, u_1(\tau)u_1(\tau - t)$$

With the above assumption concerning the finite duration of $u_1(t)$, we put

$$(2T - t)\overline{u_1(\tau)u_1(\tau - t)} = \int_{-T_1}^{+T_1} d\tau\, u_1(\tau)u_1(\tau - t)$$

where $T_1 > T$.

Substitution of the expression (1-92) for $u_1(\tau - t)$ yields

$$(2T - t)\mathbf{Q}_E(t) = 2\pi \int_{-T_1}^{+T_1} d\tau\, u_1(\tau) \int_{-\infty}^{+\infty} dn\, a(n)\, e^{-i2\pi n(\tau - t)}$$

$$= 2\pi \int_{-\infty}^{+\infty} dn\, a(n)\, e^{+\imath 2\pi nt} \int_{-T_1}^{+T_1} d\tau\, u_1(\tau)\, e^{-\imath 2\pi n\tau}$$

$$= 4\pi^2 \int_{-\infty}^{+\infty} dn\, a(n) a^*(n)\, e^{+\imath 2\pi nt}$$

since T and hence T_1 can be taken sufficiently large. $a^*(n)$ is the complex conjugate of $a(n)$.

Hence

$$\mathbf{Q}_E(t) = \frac{4\pi^2}{2T - t} \int_{-\infty}^{+\infty} dn\, a(n) a^*(n)\, e^{+\imath 2\pi nt}$$

Since T can be taken much larger than any value of t for which $\mathbf{Q}_E(t)$ is still finite, we may write

$$\mathbf{Q}_E(t) = \int_{-\infty}^{+\infty} dn\, 2\pi^2 \frac{|a(n)|^2}{T}\, e^{+\imath 2\pi nt}$$

When $t = 0$, $\mathbf{Q}_E(0) = \overline{u_1^2}$.

We now define an energy or power density spectrum $\mathbf{E}_1(n)$, such that

$$\overline{u_1^2} = \frac{1}{2} \int_{-\infty}^{+\infty} dn\, \mathbf{E}_1(n) \qquad (1\text{-}93)$$

Thus $\mathbf{E}_1(n) = 4\pi^2 |a(n)|^2/T$, and the integral for $\mathbf{Q}_E(t)$ becomes

$$\mathbf{Q}_E(t) = \frac{1}{2} \int_{-\infty}^{+\infty} dn\, \mathbf{E}_1(n)\, e^{\imath 2\pi nt} \qquad (1\text{-}94)$$

For the corresponding correlation coefficient $\mathbf{R}_E(t)$, we obtain

$$\mathbf{R}_E(t) = \frac{1}{2\overline{u_1^2}} \int_{-\infty}^{+\infty} dn\, \mathbf{E}_1(n)\, e^{\imath 2\pi nt} \qquad (1\text{-}95)$$

In order to express $\mathbf{E}_1(n)$ in terms of $\mathbf{Q}_E(t)$, we multiply both parts of Eq. (1-94) by $\exp(-\imath 2\pi mt)$, and integrate with respect to t between the limits $-T_1$ and $+T_1$.

$$\int_{-T_1}^{+T_1} dt\, \mathbf{Q}_E(t)\, e^{-\imath 2\pi mt} = \frac{1}{2} \int_{-T_1}^{+T_1} dt\, e^{-\imath 2\pi mt} \int_{-\infty}^{+\infty} dn\, \mathbf{E}_1(n)\, e^{\imath 2\pi nt}$$

$$= \frac{1}{2} \int_{-\infty}^{+\infty} dn\, \mathbf{E}_1(n) \int_{-T_1}^{+T_1} dt\, e^{\imath 2\pi (n-m)t}$$

$$= \int_{-\infty}^{+\infty} dn\, \mathbf{E}_1(n) \frac{\sin 2\pi(n-m)T_1}{2\pi(n-m)}$$

$$= \frac{1}{2\pi} \int_{-\infty}^{+\infty} dz \frac{\sin z}{z}\, \mathbf{E}_1\!\left(m + \frac{z}{2\pi T_1}\right)$$

where $z = 2\pi(n-m)T_1$.

Since (sin z)/z only gives a contribution to the integral where z is not large, we may increase T_1 indefinitely.

$$\int_{-\infty}^{+\infty} dt\, \mathbf{Q}_E(t)\, e^{-i2\pi mt} = \frac{1}{2\pi}\int_{-\infty}^{+\infty} dz \frac{\sin z}{z}\, \mathbf{E}_1(m) = \tfrac{1}{2}\mathbf{E}_1(m)$$

Hence

$$\mathbf{E}_1(m) = 2\int_{-\infty}^{+\infty} dt\, \mathbf{Q}_E(t)\, e^{-i2\pi mt} \tag{1-96}$$

and

$$\mathbf{E}_1(m) = 2\overline{u_1{}^2}\int_{-\infty}^{+\infty} dt\, \mathbf{R}_E(t)\, e^{-i2\pi mt} \tag{1-97}$$

Now the correlation $\mathbf{Q}_E(t)$ of a stationary random function is an even function of t. Hence we may also write the expressions (1-94) and (1-95)

$$\mathbf{Q}_E(t) = \int_0^{\infty} dn\, \mathbf{E}_1(n) \cos 2\pi nt \tag{1-94a}$$

and

$$\mathbf{R}_E(t) = \frac{1}{\overline{u_1{}^2}}\int_0^{\infty} dn\, \mathbf{E}_1(n) \cos 2\pi nt \tag{1-95a}$$

Similarly $\mathbf{E}_1(n)$ must be an even function of n,

$$\mathbf{E}_1(n) = 4\int_0^{\infty} dt\, \mathbf{Q}_E(t) \cos 2\pi nt \tag{1-96a}$$

and

$$\mathbf{E}_1(n) = 4\overline{u_1{}^2}\int_0^{\infty} dt\, \mathbf{R}_E(t) \cos 2\pi nt \tag{1-97a}$$

From these relations we conclude that the correlation function and the energy density spectrum are Fourier cosine transforms, a fact which has already been pointed out by Taylor.

At this point it may be worth while to remark that the correlation and the energy density spectrum are insensitive to phase differences between Fourier components, or to the value of initial phases.

This is because any phase shift or initial phase that may still be different for the different Fourier components may be included in the complex coefficients $a(n)$, and will therefore drop out in the product $a(n)a^*(n)$ occurring in the expression for the correlation function $\mathbf{Q}_E(t)$, and defining the energy density spectrum $\mathbf{E}_1(n)$. So formally we may have an infinite number of stationary turbulence functions which only differ in different initial phases of their Fourier components, which all give the same correlation functions and energy density spectra.

Thus if a turbulence is described in terms of an infinite number of Fourier components, it has also an infinite number of degrees of freedom corresponding to the infinite number of initial phases of the Fourier components.[45]

If we assume that the turbulent flow field has a uniform mean velocity $\bar{U}$ in the x_1-direction and that $\bar{U}$ is large compared with the turbulence velocities, so that we may apply the approximate relation between the time t and the distance x_1 as given by Eq. (1-70), and if we take account of the identity between $\mathbf{f}(x_1)$ and $\mathbf{R}_E(t)$, we obtain from Eqs. (1-95a) and (1-97a):

$$\mathbf{f}(x_1) = \frac{1}{\overline{u_1^2}} \int_0^\infty dn\, \mathbf{E}_1(n) \cos \frac{2\pi n x_1}{\bar{U}} \tag{1-98}$$

$$\mathbf{E}_1(n) = \frac{4\overline{u_1^2}}{\bar{U}} \int_0^\infty dx_1\, \mathbf{f}(x_1) \cos \frac{2\pi n x_1}{\bar{U}} \tag{1-99}$$

From the relations between $\mathbf{R}_E(t)$ or $\mathbf{f}(x_1)$ and $\mathbf{E}_1(n)$ just given, we can obtain the following interesting results.

1. Determine in the expressions (1-97a) and (1-99) the limit for $n = 0$.

$$\lim_{n\to 0} \frac{1}{\overline{u_1^2}} \mathbf{E}_1(n) = 4 \int_0^\infty dt\, \mathbf{R}_E(t) = 4\mathfrak{I}_E \tag{1-100}$$

$$\lim_{n\to 0} \frac{1}{\overline{u_1^2}} \mathbf{E}_1(n) = \frac{4}{\bar{U}} \int_0^\infty dx_1\, \mathbf{f}(x_1) = \frac{4}{\bar{U}} \Lambda_f \tag{1-101}$$

These results show that the integral scales $\mathfrak{I}_E$ and Λ_f can be obtained from the intersection of the $[\mathbf{E}_1(n), n]$ curve with the $\mathbf{E}_1(n)$ axis. This provides a means for determining these scales.

2. According to Eq. (1-66) we have, by Eq. (1-95a),

$$\frac{1}{\tau_E^2} = -\frac{1}{2}\left[\frac{\partial^2 \mathbf{R}_E}{\partial t^2}\right]_{t=0} = \frac{2\pi^2}{\overline{u_1^2}} \int_0^\infty dn\, n^2 \mathbf{E}_1(n)$$

and, similarly,

$$\frac{1}{\lambda_f^2} = -\frac{1}{2}\left[\frac{\partial^2 \mathbf{f}}{\partial x_1^2}\right]_{x_1=0} = \frac{2\pi^2}{\bar{U}^2 \overline{u_1^2}} \int_0^\infty dn\, n^2 \mathbf{E}_1(n) \tag{1-102}$$

These results show again that the size of the smaller eddies, of which the dissipation scales τ_E and λ_f are a measure, corresponds mainly to the value of $\mathbf{E}_1(n)$ for high values of the frequency n.

Measured correlation curves of $\mathbf{f}$ show that their rough shape appears to be independent of the velocity $\bar{U}$, apart from the curvature at the vertex of the $\mathbf{f}$-curve. Thus one might infer from Eq. (1-99) that $\bar{U}\mathbf{E}_1(n)$ is a function of $n/\bar{U}$ alone. Yet this can be true only for relatively low values of n. For, if $\bar{U}\mathbf{E}_1(n)$ were a function of $n/\bar{U}$ alone, so also would be the size of the dissipation scale λ_f, according to Eq. (1-102). But this conclusion is not confirmed by experiment, which, on the contrary, shows that λ_f decreases with increasing $\bar{U}$. Hence, for high values of n, where the value of $\mathbf{E}_1(n)$ is decisive for the value of λ_f, $\bar{U}\mathbf{E}_1(n)$ must certainly depend on both $n/\bar{U}$ and $\bar{U}$. A closer analysis of measured $\mathbf{E}_1(n)$-curves for high values of n has shown that this is so.[29]

Since $\mathbf{E}_1(n)$ and $\mathbf{f}(x_1)$ appear to be Fourier cosine transforms, the one can be found from the other by an integration process, either analytically or graphically. It can, of course, be found analytically only if an analytical equation of the function to be integrated is known. Just as, if no theoretical equation is known, we may try to approximate $\mathbf{f}(x_1)$-curves by an empirical equation, we make the same attempt with $\mathbf{E}_1(n)$-curves.

Theoretically this is all correct; so it would be very attractive to be able to make do with measuring only one of the two functions. In reality, however, with our present techniques the experimental determination of spectrum functions as well as of correlation functions is still not free from inaccuracies. In general, a correlation function is measured more accurately when the distance between the two points is not too small, and spectrum functions are measured more accurately in the higher frequency range. Thus a further analytical investigation using, for instance, a spectrum function that has been obtained entirely by integration of a measured correlation curve is a rather risky matter. In our opinion it is better to consider the relation between spectrum and correlation function merely a welcome means of checking the experimental determination of both.

Yet it may be of some interest to show what expression is obtained for the spectrum function if the correlation curve is approximated by the exponential function $\exp(-x_1/\Lambda_f)$.

Substitution of this expression for $\mathbf{f}(x_1)$ in the integral (1-99) yields

$$\frac{1}{\overline{u_1^2}}\mathbf{E}_1(n) = \frac{4}{\bar{U}}\int_0^\infty dx_1 \cos\frac{2\pi n x_1}{\bar{U}}\exp(-x_1/\Lambda_f)$$

$$= \frac{4}{\bar{U}}\frac{\Lambda_f}{1+(4\pi^2 n^2/\bar{U}^2)\Lambda_f^2}$$

or

$$\frac{\bar{U}\mathbf{E}_1(n)}{4\overline{u_1^2}\Lambda_f} = \frac{1}{1+(4\pi^2 n^2/\bar{U}^2)\Lambda_f^2} \tag{1-103}$$

Just as $\exp(-x_1/\Lambda_f)$ appears to be a satisfactory approximation of the rough shape of many $\mathbf{f}(x_1)$-curves, so also Eq. (1-103) appears to approximate measured spectrum curves satisfactorily, except of course for that part of the $\mathbf{E}_1(n)$-curve that pertains to high values of n. When n approaches zero, the value $\bar{U}_1\mathbf{E}_1(n)/\overline{u_1^2}\Lambda_f = 4$ agrees very satisfactorily even with experimental data if those are obtained from extrapolation of the measured parts of the $\mathbf{E}_1(n)$-curve.[29]

The reader can form an idea of the degree of approximation from Fig. 1-18, which is due to Favre.[39] The conditions used were the same as for the measurements

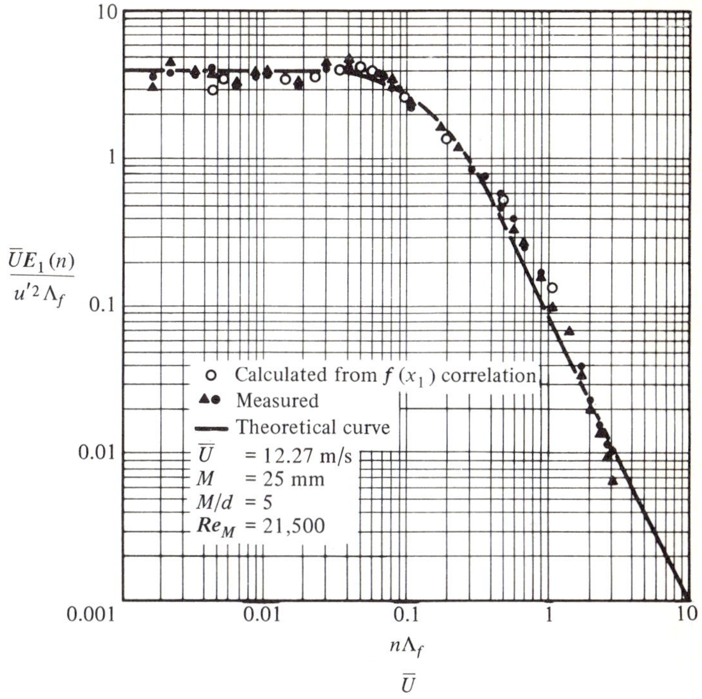

FIGURE 1-18
Spectral distribution $\mathbf{E}_1(n)$ of the longitudinal turbulence velocity component. (From: *Favre, A., et. al.*[39])

shown in Fig. 1-17. The spectral distribution of the longitudinal turbulence component was obtained in two ways, by direct measurement and by calculation from the corresponding correlation curve.

It is possible to treat the components $u_2(t)$ and $u_3(t)$ of the turbulence velocity at a fixed point in the same way. Also, from the corresponding Eulerian time correlations, the Fourier transforms $\mathbf{E}_2(n)$ and $\mathbf{E}_3(n)$ can be determined. If, again, we assume that the flow field has a uniform mean velocity $\bar{U}$ in the x_1-direction, so that $x_1 = \bar{U}t$, the Eulerian space correlations of $u_2(t)$ and $u_3(t)$ in the x_1-direction are the $\mathbf{g}(x_1)$ correlations, which in the case of isotropic turbulence are the same for u_2 and u_3. Consequently their Fourier transforms $\mathbf{E}_2(n)$ and $\mathbf{E}_3(n)$ will also be the same.

The next step is to consider, instead of two points in the x_1-direction, two arbitrary points in the flow field. We have already shown that the correlation between the velocity fluctuations at the two points can be described by means of the correlation tensor $(\mathbf{R}_{i,j})_{A,B}$. To the nine components of this tensor correspond nine spectrum functions, which form the components of a spectral tensor. Each component of this tensor is the Fourier transform of the corresponding component of the correlation tensor.

We shall come back to this matter in Chap. 3. There it will be shown that Taylor's spectrum function is a one-dimensional section of what is in fact the three-dimensional spectrum function. This three-dimensional spectrum function is not the same as the above spectral tensor but, as a matter of fact, is closely related to it.

In this connection we may mention the work by Kampé de Fériet,[30] which is based on the features of the spectral tensor. Kampé de Fériet has shown that working with the spectral tensor has mathematical advantages over working with the correlation tensor. One of the advantages from a physical point of view is that various relations can be deduced without the restrictive assumption of isotropy; the field need only be homogeneous.

1-13 ENERGY RELATIONS IN TURBULENT FLOWS

We propose now to consider the energy relations that occur in turbulent flows. We know that turbulent flow of real fluids is of a dissipative nature. Because of this dissipation of turbulence kinetic energy and conversion of it into heat, a continuous supply of energy is necessary to maintain the turbulence, certainly so for a steady flow. At the same time, owing to this turbulent motion, diffusion of fluid particles together with their kinetic energy takes place. Thus an average steady state can only exist, if there is equilibrium between the energy supplied to the turbulent motion, the transport by the mean motion, and the diffusion plus dissipation of turbulence energy.

For the general case of fluids with variable properties the mechanical energy considered above is interrelated to the internal energy $\mathscr{E}_{\text{int}}$ of the fluid, due to compressibility effects, in addition to the above dissipation. Therefore we will start by giving the balance equation for the total energy $\mathscr{E}$ per unit of mass of the fluid. If, as we will do, no external effects as those due, e.g., to gravity or electrical fields, are assumed, the mechanical energy is essentially the kinetic energy of the flowing fluid. Thus we may write

$$\mathscr{E} = \mathscr{E}_{\text{kin}} + \mathscr{E}_{\text{int}} = \frac{U_i U_i}{2} + \mathscr{E}_{\text{int}}$$

It is possible to obtain the balance equations separately for the total energy, the internal energy, and the mechanical energy. The three equations then have to satisfy the condition that the sum of the latter two must be equal to the first, of course. Or, stated in a different way, it is sufficient to derive only two of the three equations, which two equations then determine the third equation.

Let us first consider the total energy $\mathscr{E}$, and start from Eq. (1-1), assuming the term for the external effects zero. The transport through the boundaries of the volume element due to molecular effects consists of two parts. The first part is due to molecular thermal conductivity, the second part is due to the work done per unit of time by the stresses acting on the boundaries. Thus writing $\mathscr{E}$ for $\mathscr{P}$ in Eq. (1-1) we obtain:

$$\frac{\partial}{\partial t}\rho\mathscr{E} + \frac{\partial}{\partial x_j}(U_j\rho\mathscr{E}) = \rho\frac{D\mathscr{E}}{Dt} = \frac{\partial}{\partial x_j}\left(\kappa\frac{\partial\theta}{\partial x_j} + \sigma_{ji}U_i\right) \tag{1-104}$$

where θ is the temperature of the fluid, and κ its coefficient of thermal conductivity.

Introducing the expression (1-9) for σ_{ji}, we obtain after some elaboration

$$\underset{\text{I}}{\rho\frac{D\mathscr{E}}{Dt}} = \underset{\text{II}}{\frac{\partial}{\partial x_j}\left(\kappa\frac{\partial\theta}{\partial x_j}\right)} - \underset{\text{III}}{P\Theta} - \underset{\text{IV}}{U_j\frac{\partial P}{\partial x_j}} + \underset{\text{V}}{\frac{\partial}{\partial x_j}\left[\mu\left(\frac{\partial U_i}{\partial x_j} + \frac{\partial U_j}{\partial x_i}\right)U_i - \tfrac{2}{3}\mu\Theta U_j\right]} \tag{1-105}$$

The third term (III) can be written in an alternative way, using the mass-balance equation (1-2a)

$$-P\Theta = \frac{P}{\rho}\frac{D\rho}{Dt} = -P\rho\frac{D}{Dt}\frac{1}{\rho}$$

and shows the work done by compression or expansion depending on whether

$$\Theta = \rho\frac{D}{Dt}\frac{1}{\rho}$$

is negative or positive.

The meaning of the different terms in Eq. (1-105) may be described as follows,

I = local change with time plus change due to convective transport per unit of volume and time of the total energy.

II = change per unit of volume and of time due to thermal conductivity.

III = change per unit of volume and of time due to work by compression or expansion.

IV = change per unit of volume and of time due to work by the pressure gradient.

V = change per unit of volume and of time due to work by the viscous stresses.

The terms (III) and (IV) have been obtained from the term $\partial/\partial x_j(U_jP)$, giving the contribution to the total energy of the work by the pressure per unit of volume and of time. The part III is converted into heat, whereas the part IV gives the contribution to the kinetic energy. For, if we multiply the momentum balance equation (1-10) by U_i, we may easily conclude that the term $U_i\,\partial P/\partial x_i$ gives the abovementioned contribution to the kinetic energy.

In a similar way we may write the term (V) of Eq. (1-105) as follows:

$$\frac{\partial}{\partial x_j}\left[\mu\left(\frac{\partial U_i}{\partial x_j} + \frac{\partial U_j}{\partial x_i}\right)U_i - \tfrac{2}{3}\mu\Theta U_j\right]$$
$$= \left\{U_i\left[\frac{\partial}{\partial x_j}\mu\left(\frac{\partial U_i}{\partial x_j} + \frac{\partial U_j}{\partial x_i}\right)\right] - U_j\frac{\partial}{\partial x_j}(\tfrac{2}{3}\mu\Theta)\right\}$$

$$+\left\{\mu\left(\frac{\partial U_i}{\partial x_j}+\frac{\partial U_j}{\partial x_i}\right)\frac{\partial U_i}{\partial x_j}-\tfrac{2}{3}\mu\Theta^2\right\}$$

The first term on the right-hand side gives the contribution to the kinetic energy, as may again be concluded from Eq. (1-10) after its multiplication by U_i. The second term on the right-hand side is the dissipation term, for it is the work of deformation per unit of volume and of time by the viscous stresses, which for a Newtonian fluid is completely converted into heat.

So considering the kinetic energy and the internal energy separately, the balance equations read:

For the kinetic energy,

$$\rho\frac{D}{Dt}\frac{U_iU_i}{2}=U_i\frac{\partial}{\partial x_j}\left[-P\,\delta_{ji}+\mu\left(\frac{\partial U_i}{\partial x_j}+\frac{\partial U_j}{\partial x_i}\right)-\tfrac{2}{3}\mu\Theta\right]$$

or

$$\rho\frac{D}{Dt}\frac{U_iU_i}{2}=-U_j\frac{\partial P}{\partial x_j}+\frac{\partial}{\partial x_j}\left[\mu\left(\frac{\partial U_i}{\partial x_j}+\frac{\partial U_j}{\partial x_i}\right)U_i-\tfrac{2}{3}\mu\Theta U_j\right]$$

$$-\mu\left[\left(\frac{\partial U_i}{\partial x_j}+\frac{\partial U_j}{\partial x_i}\right)\frac{\partial U_i}{\partial x_j}-\tfrac{2}{3}\Theta^2\right] \qquad (1\text{-}106)$$

For the internal energy,

$$\underset{\text{I}}{\rho\frac{D}{Dt}\mathscr{E}_{\text{int}}}=\underset{\text{II}}{\frac{\partial}{\partial x_j}\left(\kappa\frac{\partial\theta}{\partial x_j}\right)}-\underset{\text{III}}{P\Theta}+\underset{\text{IV}}{\mu\left[\left(\frac{\partial U_i}{\partial x_j}+\frac{\partial U_j}{\partial x_i}\right)\frac{\partial U_i}{\partial x_j}-\tfrac{2}{3}\Theta^2\right]} \qquad (1\text{-}107)$$

Which equation, of course, is nothing else but the first law of thermodynamics, where the terms (II) and (IV) are the heat supplied to the unit of volume per unit of time, and the term (III) may either be interpreted as the external work done, or as the heat of compression or expansion contributing to the change in internal energy given by the term (I).

If we want to apply these balance equations to turbulent motions, making use of the Reynolds procedure for each turbulently varying quantity, we run into the same difficulties as mentioned earlier, when we considered the momentum-balance equations for a compressible fluid. For, mainly due to the temperature variations associated with the density variations, the dynamic viscosity μ and the thermal conductivity κ undergo relative variations of the same order of magnitude as the density variations. Even if we follow Favre,[12] and introduce density-weighted velocities, the equations still remain very complicated, while at the same time they contain the non-weighted velocities in the dissipation term.

Since in this textbook we shall confine ourselves mainly to incompressible fluids, we shall not trouble to give the formal expressions that will be obtained after applying Reynolds's procedure to the Eqs. (1-105), (1-106) and (1-107). Such expressions have been given by Favre.[12] He used density-weighted velocities, in which case the equations become less complicated.

For a fluid with constant properties, only the balance equation for the kinetic energy remains of interest. So, with $\Theta = 0$, Eq. (1-106) becomes

$$\frac{D}{Dt}\frac{U_iU_i}{2} = -U_j\frac{\partial}{\partial x_j}\frac{P}{\rho} + \nu\frac{\partial}{\partial x_j}U_i\left(\frac{\partial U_i}{\partial x_j} + \frac{\partial U_j}{\partial x_i}\right) - \nu\left(\frac{\partial U_i}{\partial x_j} + \frac{\partial U_j}{\partial x_i}\right)\frac{\partial U_i}{\partial x_j}$$

Or,

$$\underset{\text{I}}{\frac{\partial}{\partial t}\frac{U_iU_i}{2}} = \underset{\text{II}}{-\frac{\partial}{\partial x_j}U_j\left(\frac{P}{\rho} + \frac{U_iU_i}{2}\right)} + \underset{\text{III}}{\nu\frac{\partial}{\partial x_j}U_i\left(\frac{\partial U_i}{\partial x_j} + \frac{\partial U_j}{\partial x_i}\right)} - \underset{\text{IV}}{\nu\left(\frac{\partial U_i}{\partial x_j} + \frac{\partial U_j}{\partial x_i}\right)\frac{\partial U_i}{\partial x_j}} \qquad (1\text{-}108)$$

The first term on the right-hand side can be interpreted as the work done per unit mass and time by the total dynamic pressure $P + (\rho/2)U_jU_j$. But this term may also be interpreted as the change in transport of the energy $P/\rho + \frac{1}{2}U_jU_j$ per unit mass through convection by the velocity U_i.

Thus for the meaning of the various terms in the energy equation (1-108) we have:

I = local change of kinetic energy per unit mass and time
II = change in convective transport of the pressure and kinetic energy per unit mass and time, or work done per unit mass and time by the total dynamic pressure
III = work done per unit mass and time by the viscous stresses
IV = dissipation per unit mass

Now apply Reynolds's procedure to Eq. (1-108) by dividing each quantity into its mean and its fluctuating part.

$$U_i = \bar{U}_i + u_i$$

$$P = \bar{P} + p$$

$$U_iU_i = \bar{U}_i\bar{U}_i + 2\bar{U}_iu_i + u_iu_i = \bar{U}_i\bar{U}_i + 2\bar{U}_iu_i + q^2$$

Substitution in Eq. (1-108) and averaging with respect to time gives

$$\frac{1}{2}\frac{\partial}{\partial t}\bar{U}_j\bar{U}_j + \frac{1}{2}\frac{\partial\overline{q^2}}{\partial t} = -\frac{\partial}{\partial x_i}\bar{U}_i\left(\frac{\bar{P}}{\rho} + \tfrac{1}{2}\bar{U}_j\bar{U}_j\right) + \nu\frac{\partial}{\partial x_i}\bar{U}_j\left(\frac{\partial\bar{U}_i}{\partial x_j} + \frac{\partial\bar{U}_j}{\partial x_i}\right)$$
$$-\nu\left(\frac{\partial\bar{U}_i}{\partial x_j} + \frac{\partial\bar{U}_j}{\partial x_i}\right)\frac{\partial\bar{U}_j}{\partial x_i} - \frac{\partial}{\partial x_i}\overline{u_i\left(\frac{p}{\rho} + \tfrac{1}{2}q^2\right)}$$
$$-\frac{\partial}{\partial x_i}\bar{U}_j\overline{u_iu_j} - \frac{1}{2}\frac{\partial}{\partial x_i}\bar{U}_i\overline{q^2}$$
$$+\nu\frac{\partial}{\partial x_i}\overline{u_j\left(\frac{\partial u_i}{\partial x_j} + \frac{\partial u_j}{\partial x_i}\right)} - \nu\overline{\left(\frac{\partial u_i}{\partial x_j} + \frac{\partial u_j}{\partial x_i}\right)\frac{\partial u_j}{\partial x_i}}$$

Multiplication by $\bar{U}_i$ of the equation of motion (1-18) (F_i are assumed zero) gives

$$\frac{1}{2}\frac{\partial}{\partial t}\bar{U}_j\bar{U}_j = -\frac{\partial}{\partial x_i}\bar{U}_i\left(\frac{\bar{P}}{\rho}+\tfrac{1}{2}\bar{U}_j\bar{U}_j\right)+\nu\bar{U}_i\frac{\partial^2}{\partial x_j\partial x_j}\bar{U}_i-\bar{U}_i\frac{\partial}{\partial x_j}\overline{u_iu_j}$$

Or,

$$\underset{\text{I}}{\frac{1}{2}\frac{\partial}{\partial t}\bar{U}_j\bar{U}_j}+\underset{\text{II}}{\frac{\partial}{\partial x_i}\bar{U}_i\left(\frac{\bar{P}}{\rho}+\tfrac{1}{2}\bar{U}_j\bar{U}_j\right)} = \underset{\text{III}}{-(-\overline{u_iu_j})\frac{\partial\bar{U}_i}{\partial x_j}}+\underset{\text{IV}}{\frac{\partial}{\partial x_j}(-\overline{u_iu_j}\bar{U}_i)}$$

$$+\underset{\text{V}}{\nu\frac{\partial}{\partial x_i}\bar{U}_j\left(\frac{\partial\bar{U}_i}{\partial x_j}+\frac{\partial\bar{U}_j}{\partial x_i}\right)}-\underset{\text{VI}}{\nu\left(\frac{\partial\bar{U}_i}{\partial x_j}+\frac{\partial\bar{U}_j}{\partial x_i}\right)\frac{\partial\bar{U}_i}{\partial x_j}} \qquad (1\text{-}109)$$

This is the balance equation for the kinetic energy of the mean motion. The terms I, II, V, and VI have the same meaning as the corresponding terms in Eq. (1-108), though now pertaining to the mean motion. The term III is the work of deformation by the turbulence stresses per unit of mass and of time When $i \neq j$, so that $-\overline{u_iu_j}$ is a shear stress, usually this stress has the same sign as $\partial\bar{U}_i/\partial x_j$. Hence for $i \neq j$, the term III gives a negative contribution to the kinetic energy of the mean motion; or it extracts energy from the mean motion. When $i = j$, the contribution may be either positive or negative, depending on the sign of $\partial\bar{U}_i/\partial x_j$ for $i = j$. The term IV is the work per unit of mass and of time by the turbulence stresses.

Subtraction of Eq. (1-109) from the previous one yields

$$\underset{\text{I}}{\frac{D}{Dt}\frac{\overline{q^2}}{2}} = \underset{\text{II}}{-\frac{\partial}{\partial x_i}\overline{u_i\left(\frac{p}{\rho}+\frac{q^2}{2}\right)}}\;\underset{\text{III}}{-\overline{u_iu_j}\frac{\partial\bar{U}_j}{\partial x_i}}+\underset{\text{IV}}{\nu\frac{\partial}{\partial x_i}\overline{u_j\left(\frac{\partial u_i}{\partial x_j}+\frac{\partial u_j}{\partial x_i}\right)}}$$

$$\underset{\text{V}}{-\nu\overline{\left(\frac{\partial u_i}{\partial x_j}+\frac{\partial u_j}{\partial x_i}\right)\frac{\partial u_j}{\partial x_i}}} \qquad (1\text{-}110)$$

This is the turbulence-energy equation, which states:

The change (I) in kinetic energy of turbulence per unit of mass and of time including the convective transport by the mean motion, is equal to: (II) the convective diffusion by turbulence of the total turbulence mechanical energy, or the work by the total dynamic pressure of turbulence, plus (III) the work of deformation of the mean motion by the turbulence stresses, plus (IV) the work by the viscous shear stresses of the turbulent motion, plus (V) the viscous dissipation by the turbulent motion. all per unit of mass and of time.

We may note that the term (III) in Eq. (1-110) has an opposite sign to the same term (III) in Eq. (1-109). Thus, again when $i \neq j$, in Eq. (1-110) the work of deformation of the mean motion by the turbulence shear stresses usually gives a positive contribution to the turbulence energy. For this reason this term is also referred to as a turbulence production term. Through extraction of energy from the mean motion, it transfers this energy to the turbulent motion. Physically one may visualize this transfer of energy as taking place through a stretching process of turbulence vortices due to the mean motion.†

When $i \neq j$ in the term (III) in Eq. (1-110), the following interesting inferences can be made.

Consider a flow with a velocity that increases in the x_1-direction so that $\partial \bar{U}_1/\partial x_1 > 0$. The term in (1-110) or (1-111) corresponding to this main motion, namely, $-\overline{u_1^2}\, \partial \bar{U}_1/\partial x_1$, is then negative; thus it promotes a decrease in the turbulence energy. Conversely, the turbulence energy will have a tendency to increase if $\partial \bar{U}_1/\partial x_1 < 0$. Hence we have the following rule. In a flow with a positive velocity gradient in the flow direction (accelerated flow in space), there is a tendency for a relative turbulence intensity to decrease; in a flow with a negative velocity gradient (retarded flow in space), the tendency is toward an increase in turbulence.

Because, in general, an increase in velocity is associated with a decrease in static pressure and vice versa, the above may also be formulated as follows. A decrease in static pressure in the flow direction decreases turbulence and an increase in static pressure promotes turbulence, in so far as these pressure gradients are not caused by frictional effects.

The form of the term (III) of Eq. (1-109) when compared with the term VI, and that of IV when compared with V suggest a similar behavior of viscous and turbulence stresses. The terms (III) and (VI) give the work of deformation of the mean motion by the turbulence and viscous stresses respectively. In the latter case, namely the term (VI), which is the viscous dissipation, this work is converted into heat, i.e., in chaotic motions of the molecules. Similarly we may interpret the term (III) as a "dissipation" term for the mean motion, where the work of deformation by the turbulence stresses is converted into chaotic turbulent motions. Provided, of course, the usual case is considered that turbulence shear stress and the corresponding mean-velocity gradient have the same sign. The expressions for the dissipation and for the work done by the stresses may attain exactly the same form, if we introduce the modified Boussinesq concept of a turbulence viscosity

$$-\overline{u_i u_j} = -\bar{P}_t\, \delta_{ij} + \epsilon_m^* \left(\frac{\partial \bar{U}_i}{\partial x_j} + \frac{\partial \bar{U}_j}{\partial x_i} \right)$$

With this expression for $-\overline{u_i u_j}$ the terms (III) and (IV) of Eq. (1-109) read

$$-\epsilon_m^* \left(\frac{\partial \bar{U}_i}{\partial x_j} + \frac{\partial \bar{U}_j}{\partial x_i} \right) \frac{\partial \bar{U}_i}{\partial x_j} + \frac{\partial}{\partial x_j} \epsilon_m^* \bar{U}_i \left(\frac{\partial \bar{U}_i}{\partial x_j} + \frac{\partial \bar{U}_j}{\partial x_i} \right) - \bar{U}_j \frac{\partial \bar{P}_t}{\partial x_j}$$

† This interaction between mean and turbulent motion only holds good as long as the turbulence shear stresses and the corresponding gradients of the mean velocities have the same sign. But an opposite sign may not be impossible. In fact there is experimental evidence that certain turbulent flows show local regions where this does occur. However, before concluding that this would imply a transfer of energy back from the turbulent to the mean motion, the contribution of the term (IV) in Eq. (1-102) should also be considered.

Which, apart from the turbulence pressure term, have to be compared with terms (VI) and (V) respectively.

With the introduction of the eddy viscosity ϵ_m^*, the ratio of the local "dissipation" of mean flow energy by the turbulence and the local viscous dissipation of the mean flow becomes equal to the ratio between the eddy and the molecular viscosity. The same ratio thus also exists between the production of turbulence and the viscous dissipation of the mean motion. Based upon this relation, which is in fact an identity, inferences concerning the possibility of a constant eddy viscosity in turbulent flow regions have been made by Kline.[46]

Looking now at the turbulence kinetic-energy equation (1-110) the turbulence production term (III) is the "dissipation" term (III) in Eq. (1-108) for the mean motion, and has its counterpart only in the viscous dissipation term (V) of the turbulent motion. Hence one may visualize the following flow of energy. Due to the interaction between mean motion and turbulent motion, energy is extracted from the mean motion through work of deformation by the turbulence stresses, converted into turbulence energy which ultimately is converted through work of deformation by the viscous stresses in the turbulent motion into heat. Of course, this does not imply that the local production of turbulence will always be equal to the local viscous dissipation, and thus that the turbulence is gaining energy from the mean motion at the same rate as it is losing energy through viscous dissipation. If that were the case, the turbulence could then be considered as being locally in *mechanical-energy equilibrium.* That this cannot be true in general, can be concluded from the occurrence of the convection due to the mean motion and the turbulence transport or diffusion in the complete balance equation (1-110). Thus the turbulence at a point is in general dependent on upstream conditions and on turbulent conditions outside of it.

Frequently the two viscous terms in Eq. (1-110) are written in a different way, namely,

$$\frac{\partial}{\partial x_i}\overline{u_j\left(\frac{\partial u_i}{\partial x_j}+\frac{\partial u_j}{\partial x_i}\right)}-\overline{\left(\frac{\partial u_i}{\partial x_j}+\frac{\partial u_j}{\partial x_i}\right)\frac{\partial u_j}{\partial x_i}}$$

$$=\frac{\partial}{\partial x_i}\overline{u_j\frac{\partial u_i}{\partial x_j}}+\frac{\partial}{\partial x_i}\frac{1}{2}\frac{\partial}{\partial x_i}\overline{q^2}-\overline{\frac{\partial u_i}{\partial x_j}\frac{\partial u_j}{\partial x_i}}-\overline{\frac{\partial u_j}{\partial x_i}\frac{\partial u_j}{\partial x_i}}$$

$$=\frac{1}{2}\frac{\partial^2}{\partial x_i\,\partial x_i}\overline{q^2}-\overline{\frac{\partial u_j}{\partial x_i}\frac{\partial u_j}{\partial x_i}}$$

since, for an incompressible fluid, $\overline{u_j(\partial^2 u_i/\partial x_i\,\partial x_j)}=0$. Equation (1-110) then reads

$$\frac{D}{Dt}\frac{\overline{q^2}}{2}=-\frac{\partial}{\partial x_i}\overline{u_i\left(\frac{p}{\rho}+\frac{q^2}{2}\right)}-\overline{u_iu_j}\frac{\partial \bar{U}_j}{\partial x_i}+\tfrac{1}{2}\nu\frac{\partial^2}{\partial x_i\,\partial x_i}\overline{q^2}-\nu\overline{\frac{\partial u_j}{\partial x_i}\frac{\partial u_j}{\partial x_i}} \qquad \text{(1-111)}$$

It must be remarked that the two viscous terms in Eq. (1-111) do not have the same meaning as the two (IV and V) in (1-110); thus the last term in (1-111) is not the dissipation term, as is term V in (1-110).[38] Only in homogeneous turbulence is this last term equal to the dissipation.

Denoting the viscous dissipation of the turbulent motions per unit of mass by ε, we thus have for a homogeneous turbulence,

$$\varepsilon = \nu \overline{\frac{\partial u_i}{\partial x_j}\left(\frac{\partial u_i}{\partial x_j} + \frac{\partial u_j}{\partial x_i}\right)} = \nu \overline{\frac{\partial u_i}{\partial x_j}\frac{\partial u_i}{\partial x_j}} \tag{1-112}$$

On the same grounds we also have

$$\overline{\frac{\partial u_i}{\partial x_j}\frac{\partial u_i}{\partial x_j}} = \frac{1}{2}\overline{\left(\frac{\partial u_i}{\partial x_j} - \frac{\partial u_j}{\partial x_i}\right)\left(\frac{\partial u_i}{\partial x_j} - \frac{\partial u_j}{\partial x_i}\right)} = \overline{\omega_k \omega_k} \tag{1-113}$$

where

$$\omega_k = -\varepsilon_{ijk}\frac{\partial u_i}{\partial x_j}$$

is the vorticity of the turbulence. Hence, for a homogeneous turbulence, and only then, we have that the viscous dissipation per unit of mass is equal to the viscosity times the mean-square of the rate of strain, and to the viscosity times the mean-square vorticity.

Now in a homogeneous turbulence it is not only the viscous dissipation that takes a simplified form. Since all spatial derivatives of mean turbulence quantities become zero, we obtain either from Eq. (1-110) or from Eq. (1-111), for a *homogeneous turbulence*

$$\frac{\partial}{\partial t}\frac{\overline{q^2}}{2} = -\overline{u_i u_j}\frac{\partial \bar{U}_j}{\partial x_i} - \nu\overline{\left(\frac{\partial u_i}{\partial x_j} + \frac{\partial u_j}{\partial x_i}\right)\frac{\partial u_j}{\partial x_i}} \tag{1-114}$$

When the turbulence is stationary, the local production of turbulence energy is just equal to the local viscous dissipation of this turbulence energy. The turbulence is then locally in mechanical-energy equilibrium.

Hitherto we have considered only an unlimited flow field. Let us now take the case in which the flow field is bounded by fixed walls. The velocities must then be zero at these boundaries. Hence, if Eq. (1-110) is integrated over the entire domain, the terms II and IV will become zero at the boundaries (mathematically this follows immediately from Gauss' theorem). This means that the effects of convective diffusion and of work by viscous shear stresses of the turbulent motion disappear in the integrated equation.

Let

$$\mathscr{E}_t = \tfrac{1}{2}\rho \textstyle\int\int\int dx_1\, dx_2\, dx_3\, \overline{q^2}$$

be the total kinetic energy of turbulence in the entire bounded domain.

Since

$$\rho \iiint dx_1\, dx_2\, dx_3 \frac{D}{Dt}\overline{q^2} = \rho \iiint dx_1\, dx_2\, dx_3 \left(\frac{\partial}{\partial t}\overline{q^2} + \bar{U}_i \frac{\partial \overline{q^2}}{\partial x_i}\right)$$

$$= \rho \iiint dx_1\, dx_2\, dx_3 \frac{\partial}{\partial t} \overline{q^2}$$

$$= \rho \frac{d}{dt} \iiint dx_1\, dx_2\, dx_3\, \overline{q^2} = 2 \frac{d\mathscr{E}_t}{dt}$$

Eq. (1-110), integrated, reads

$$\frac{d\mathscr{E}_t}{dt} = -\rho \iiint dx_1\, dx_2\, dx_3 \left(\overline{u_i u_j} \frac{\partial \bar{U}_j}{\partial x_i} \right) - \mu \iiint dx_1\, dx_2\, dx_3 \overline{\left(\frac{\partial u_i}{\partial x_j} + \frac{\partial u_j}{\partial x_i} \right) \frac{\partial u_i}{\partial x_j}} \tag{1-115}$$

Here the second term on the right-hand side (the dissipation term) is always positive. The first term can be either positive or negative. If the first term is negative and if its absolute value is greater than the second term, then $d\mathscr{E}_t/dt$ may become positive.

It is worth noting that the pressure term does not occur in this integrated energy equation. This means that there is no net effect of the turbulence pressure fluctuations on the total kinetic energy of the turbulence in the bounded domain. Of course, there may be an effect on the transfer of energy from one region to another within this bounded domain, as well as an effect on the transfer of energy between components of the turbulence velocity. We shall come back to this point later in Chap. 4.

Equation (1-115) has been given by Reynolds and by Lorentz.[31] It has been used as the starting point for calculating the critical value of the Reynolds number below which any disturbance will be damped by the effect of dissipation. Consider, for example, a laminar viscous flow, and assume a small disturbance superimposed on the main viscous motion. If we take a suitable analytical function for this disturbance, the two terms on the right-hand side of Eq. (1-115) can be calculated. If such a calculation shows that the second term is greater than the first, the disturbance is damped out and the flow remains laminar. If, on the other hand, the energy transferred to the disturbance by the main motion is greater than the dissipation, a growth of the disturbance will result, the laminar flow will be unstable, and transition into turbulent flow will occur. In the limiting case $d\mathscr{E}_t/dt = 0$; so the ratio between the two terms on the right-hand side of Eq. (1-115) is just equal to 1. From the assumed flow pattern for the disturbance the turbulence terms in Eq. (1-115) can be calculated. By increasing the main velocity $\bar{U}_i$, the velocity gradient $\partial \bar{U}_i/\partial x_j$ is increased. At a certain value of this gradient, that is, at a certain value of $\bar{U}_i$ and thus also at a certain value of the Reynolds number, $d\mathscr{E}_t/dt$ becomes zero and the critical Reynolds number has been reached. This shows why transition from laminar flow to turbulent flow occurs when the velocity of the main motion increases beyond a certain critical value. Lorentz has calculated in this way the critical Reynolds number for the flow through a cylindrical tube. The same has been done by Orr,[32] who calculated this critical number for Couette flow also. Reynolds himself made the calculation for the pressure flow between two parallel walls. The results of all these calculations show a critical value of the Reynolds number appreciably lower than the experimental value. Thus, Reynolds obtained a critical Reynolds number $\bar{U}_{av}h/\nu = 516$ for the pressure flow between two parallel walls (distance h), whereas experiments yielded a value of roughly 1,100 to 1,200. Lorentz calculated $U_0h/\nu = 288$ for Couette flow, whereas measurements by Couette yielded a critical value of 1,900.

In this connection it may be of interest to refer to results obtained by Lindgren,[44] who studied experimentally the transition-process in the flow through a straight tube with highly disturbed entry conditions. He found that, at Reynolds numbers based on average flow velocity and tube diameter as low as 200, the initial entry disturbances were first amplified before they eventually were damped a certain distance from the inlet. He observed the occurrence of turbulence slugs under these conditions, whose length and frequency of occurrence increased with increasing Reynolds number. Similar observations had been reported earlier by Reynolds and by others.[47,48] It was only after the Reynolds number increased beyond the value of 2,200 to 2,400 that these turbulence slugs did not decay eventually, and a continuous turbulence flow condition was maintained. We will consider these experiments in more detail in Sec. 7-11.

Another criticsm that can be made with respect to the actual application of the above Reynolds and Lorentz method is that the assumed analytical function for the disturbance has only been made to satisfy the equation of continuity, but it has not in all cases been checked whether the disturbance is consistent with the dynamics of the motion, i.e., whether it is a solution of the Navier-Stokes equations.

It may be questioned whether it is correct to start from the integrated form of the energy equation. It is known, for instance, that in actual flows turbulence originates more or less locally. The starting point for the calculation of local instabilities of the main flow in response to certain disturbances should, therefore, be the much more complicated equation (1-110) or (1-111).

In this connection it may be remarked that later investigations of the stability of laminar boundary layers and so on start, not from the energy equations (1-110) or (1-111), but from the equations of motion. The former method, which is due to Reynolds, is often called the energy-balance method; the second method is usually designated the perturbation method. Rayleigh,[33] too, has considered the transition to turbulence as a local phenomenon; Rayleigh has shown that a flow is unstable when the velocity profile shows an inflection point.

Theories concerning perturbation methods, which have been confirmed by many experimental investigations, have shown that, in the case of instability of originally laminar flow in response to a certain type of sinusoidal disturbance, the sinusoidal waves increase in amplitude downstream. How such a single wave develops ultimately into a fully turbulent flow is not yet fully understood, but there is certainly a connection with the nonlinear character of the equations of motion. The nonlinear terms in these equations cause interaction between the motions of various frequencies, so that kinetic energy is transferred from the motions with lower frequencies to the motions with higher frequencies. In other words, the effect of the nonlinear terms in the equations of motion is to increase the amplitude of the motions with higher frequencies. An indication of this may be found in the theoretical investigation by Burgers[34] on a simplified model of turbulent flow.

Later investigations[40,41] have indicated that development into turbulence of these sinusoidal waves occurs via the generation of spots of turbulence through a secondary instability of the flow. These turbulent spots grow and spread in the downstream direction, as acoustic disturbances do in supersonic flow, until the whole field of flow is turbulent. There are further indications that the secondary waves form a more or less regular system of vortices with axes along concave curved streamlines, like the three-dimensional vortices shown by Görtler[42] to occur in the flow over a concave surface.

A characteristic feature of the transition into turbulence is its sudden behavior: relatively weak and slow disturbances change in a very short time into a spot of turbulence of high intensity and with a wide spectrum. The spot grows quickly not only in the axial direction but also in the lateral direction parallel to the wall, which indicates that the turbulence of the spot is three-dimensional. This occurs even if the original disturbance is only two-dimensional.

Mention may also be made of interesting results obtained by Lin[43] and Lindgren.[44] Lin showed that viscosity effects at the wall lead to a phase shift between the axial velocity component u_1 and the normal velocity component u_2 of a two-dimensional disturbance, and consequently to a positive value of the turbulence shear stress $-\rho\overline{u_1 u_2}$. Since at the wall $\partial\bar{U}_1/\partial x_2 > 0$, we have a positive value of the turbulence production term III of Eq. (1-110). This result finds support in a conclusion obtained by Lindgren from his experiments on transition phenomena in tube flow, namely, "turbulence should originate from wall effects acting on disturbed flow in the immediate vicinity of—or in contact with—the boundary walls."

Another problem, hitherto disregarded in investigations on turbulence and also closely related to the nonlinear character of the equations of motion, is how turbulence is maintained. Munk[35] has made an attempt to visualize a mechanism of turbulence that might serve as a basis for theoretical work leading to an explanation. He started from the concept that turbulence is an essentially three-dimensional motion in which randomly distributed vortex rings turn, stretch or shrink, accumulate, and diffuse. Bigger and stronger vortices are formed by merging, but at the same time vortex rings break up into smaller ones, whose intensity is then reduced by viscosity or diffusion. In this respect we may refer also to the above-mentioned results by Lin and Lindgren concerning the effect of a wall not only on the generation but also on the self-maintenance of turbulence near the wall.

A much more detailed consideration concerning the mechanisms of transition processes in connection with those of fully developed continuous turbulence will be given in Chap. 6, and especially in Chap. 7, for more recent investigations on the transition processes and on the resulting fully developed wall turbulence have revealed that similar mechanisms occur during these transition processes and in the continuous generation of turbulence in the fully developed turbulent flow.

REFERENCES

1. *J. Roy. Aeronaut. Soc.,* **41,** 1109 (1937). See also Ref. 6.
2. Kármán, Th. von: *J. Aeronaut. Sci.,* **4,** 131 (1937).
3. Burgers, J. M.: *Proc. Koninkl. Ned. Akad. Wetenschap.,* **51,** 1073 (1948).
4. Prandtl, L., and O. Tietjens: "Fundamentals of Hydro- and Aeromechanics," vol. 2, McGraw-Hill Book Company, Inc., New York, 1934.
5. "Handbuch der Experimental-Physik," vol. 4, p. 1, Akademische Verlag Gesellschaft, Leipzig, 1931.
6. Dryden, H. L.: *Ind. Eng. Chem.,* **31,** 416 (1939).
7. Prandtl, L.: *Proc. 5th Intern. Congr. Appl. Mech.,* Cambridge, Mass., 1938, p. 340.
8. Flügel, G.: *VDI-Forschungsheft No.* 395, 1939.
9. Garside, J. E., A. R. Hall, and D. T. A. Townend: *Nature,* **152,** 748 (1943).
10. Corrsin, S.: *Natl. Advisory Comm. Aeronaut. Wartime Repts. No.* ACR 3L23, 1943.
11. Boussinesq, J.: *Mém. prés. par div. savant à l'acad. sci. Paris,* **23,** 46 (1877).
12. Favre, A.: *J. de Mécanique,* **4,** 361 (1965).
13. Rivlin, R. S.: *Quart. Appl. Math.,* **15,** 212 (1957).
14. Taylor, G. I.: *Proc. Roy. Soc. London,* **157A,** 537 (1936).
15. Corrsin, S., and M. S. Uberoi: *Natl. Advisory Comm. Aeronaut. Tech. Repts. No.* 1142, 1953.
16. Lin, C. C.: *Quart. Appl. Math.,* **10,** 295 (1953).
17. Jeans, J. H.: "An Introduction to the Kinetic Theory of Gases," Cambridge University Press, New York, 1946.
18. Taylor, G. I.: *Proc. London Math. Soc.,* **20,** 196 (1921).
19 Kampé de Fériet, J.: *Ann. soc. sci. Bruxelles, Ser. I,* **59,** 145 (1939).
20. Kármán, Th. von: *Proc. Natl. Acad. Sci. U.S.,* **34,** 530 (1948).
21. Batchelor, G. K.: *Quart. Appl. Math.,* **6,** 97 (1948).
22. Dryden, H. L., G. B. Schubauer, W. C. Mick, and H. K. Skramstad: *Natl. Advisory Comm. Aeronaut. Tech. Repts. No.* 581, 1937.
23. Kalinske, A. A., and C. L. Pien: *Ind. Eng. Chem.,* **36,** 220 (1944).
24. Doob, J. L.: *Ann. Math.,* **43,** 352 (1946).
25. Frenkiel, F. N.: *J. Aeronaut. Sci.,* **15,** 57 (1948).
26. Frenkiel, F. N.: *O.N.E.R.A. Rapp. tech. No.* 34, 1948.
27. Taylor, G. I.: *Proc. Roy. Soc. London,* **164A,** 476 (1938).
28. See, for instance, Jeffreys, H., and B. S. Jeffreys: "Methods of Mathematical Physics," p. 452, Cambridge University Press, New York, 1950.
29. Dryden, H. L.: *Quart. Appl. Math.,* **1,** 7 (1943).
30. Kampé de Fériet, J.: *Compt. rend.,* **227,** 760 (1948); see also *Proc. 7th Intern. Congr. Appl. Mech., Public Addresses No. 1,* London, 1948.
31. Lamb, H.: "Hydrodynamics," 6th ed., p. 677, Cambridge University Press, New York, 1932.
32. Orr, W. M. F.: *Proc. Roy. Irish Acad.,* **27A,** 69 (1906–1907).
33. Rayleigh, Lord: *Proc. Math. Soc. London,* **11,** 57 (1880); **19,** 67 (1887).
34. Burgers, J. M.: *Proc. Koninkl. Ned. Akad. Wetenschap.,* **17,** no. 2 (1939). See also Mises, R. von, and Th. von Kármán: "Advances in Applied Mechanics," vol. 1, p. 171, New York, 1948.
35. Munk, M.: *Aero Digest,* June, 1951, p. 100; July, 1952, p. 32.
36. Dryden, H. L., and A. M. Kuethe: *Natl. Advisory Comm. Aeronaut. Tech. Repts. No.* 342, 1930.
37. Kraichnan, R. H.: *Phys. Fluids,* **8,** 575 (1965).

38. Corrsin, S.: *J. Aeronaut. Sci.*, **20,** 853 (1953).
39. Favre, A., J. Gaviglio, and R. Dumas: *Recherche aéronaut. No.* 32, p. 21, 1953.
40. Schubauer, G. B., and P. S. Klebanoff: *Natl. Advisory Comm. Aeronaut. Tech. Notes No.* 3489, 1955.
41. Dryden, H. L.: *Z. Flugwissenschaften,* **4,** 89–95 (1956).
42. Görtler, H.: *Z. angew. Math. u. Mech.*, **21,** 250–252 (1949).
43. Lin, C. C.: "The Theory of Hydrodynamic Stability," p. 61, Cambridge University Press, New York, 1955.
44. Lindgren, E. R.: *Arkiv Fysik,* **12,** 1 (1957).
45. See, for instance, Landau, L. D. and E. M. Lifshitz: "Fluid Mechanics," p. 106, Pergamon Press, New York, 1959.
46. Kline, S. J.: *Proc. Inst. Mech. Engrs.*, **180,** Part 3J, 222 (1965–1966).
47. Couette, M.: *Ann. Chim. Phys.*, **21,** 433 (1890).
48. Rotta, J.: *Ing. Archiv,* **24,** 258 (1956).
49. See, for instance, Dynkin, E. B.: "Markov Processes," Vol. 1, p. 2, Springer-Verlag, Berlin-Göttingen, 1965.
50. Crow, S. C.: *J. Fluid Mech.*, **33,** 1 (1968).
51. Irmay, S.: *J. Basic Engng., Trans. A.S.M.E.*, **82D,** 961 (1960).
52. Irmay, S.: "Int. Symposium on Two-Phase Systems," Haifa, 1971.
53. Hanjalić, K., and B. E. Launder: *Report* TM/TN/A/8, Imp. College of Science and Technology, London, 1971.
54. Hinze, J. O., R. E. Sonnenberg, and P. J. H. Builtjes: *Appl. Sci. Res.*, **29,** 1 (1974).
55. Lumley, J. L.: *J. Fluid Mech.*, **41,** 413 (1970).

NOMENCLATURE FOR CHAPTER 1

c velocity of sound; molecular velocity.
D coefficient of molecular diffusion.
D_{ij} deformation tensor.
d diameter of cylinder or rod.
$\mathbf{E}$ spectrum function of turbulence kinetic energy.
e_{ij} direction cosines.
F force; driving force.
$\mathbf{f}$ coefficient of spatial, longitudinal velocity correlation.
$\mathbf{g}$ coefficient of spatial, lateral velocity correlation.
$\mathbf{h}$ coefficient of spatial, triple velocity correlation.
$\mathbf{K}_i$ first order correlation tensor.
$\mathbf{k}$ coefficient of spatial, triple velocity correlation.
κ coefficient of thermal conductivity
$\mathbf{L}_i$ coefficient of first order correlation tensor.
l mean free path of molecules.
M mesh length; $M(\tau)$, memory function.
N normal to a plane.
n frequency; number of molecules.

P static pressure; $\bar{P}$, time-mean value; p, turbulent fluctuation of static pressure.
$\bar{P}_t$ time-mean value of turbulence "pressure."
$\mathscr{P}$ transferable property
$\mathbf{Q}_{ij}$ second order velocity-correlation tensor.
$\mathbf{q}$ coefficient of spatial, triple velocity correlation.
q^2 $u_i u_i$ = twice energy of turbulence; $\overline{q^2}$, time-mean value.
$\mathbf{R}_{ij}$ coefficient of second-order velocity-correlation tensor.
$\mathbf{R}_E$ coefficient of Eulerian time correlation.
$\mathbf{R}_L$ coefficient of Lagrangian time correlation.
r cylindrical polar coordinate in radial direction.
$\mathbf{S}_{i,kj}$ or $\mathbf{S}_{ik,j}$ third-order correlation tensor.
T time period.
$\mathbf{T}_{i,kj}$ or $\mathbf{T}_{ik,j}$ coefficient of third-order correlation tensor.
t time.
U Eulerian velocity; $\bar{U}$, time-mean value; u, turbulence component; $u' = \sqrt{\overline{u^2}}$, root-mean-square turbulence-velocity component; subscripts 1, 2, 3, i, j, k refer to Cartesian coordinates, n to normal component, r, φ, z to cylindrical polar coordinates.
v Langrangian turbulence velocity; $v' = \sqrt{\overline{v^2}}$.
x_i, x_1, x_2, x_3 Eulerian Cartesian coordinates.
y_i, y_1, y_2, y_3 Lagrangian Cartesian coordinates.
z cylindrical polar coordinate in axial direction.
α exchange coefficient
∂ partial differential operator.
δ_{ij} Kronecker delta.
ε_{ijk} alternating tensor.
ϵ coefficient of eddy diffusion; ϵ_m, ditto for momentum; ϵ_γ, ditto for a scalar quantity.
$\mathscr{E}$ energy per unit of mass
$\mathscr{E}_t$ total turbulence kinetic energy in a finite domain.
φ cylindrical polar coordinate in angular direction.
Γ scalar quantity; $\bar{\Gamma}$ time-mean value.
γ turbulent fluctuation of a scalar quantity.
$\mathfrak{J}_E$ Eulerian integral time scale.
$\mathfrak{J}_L$ Lagrangian integral time scale.
$\mathfrak{J}$ flux per unit of area.
$\varkappa$ c_p/c_v = ratio of specific heats.
$\mathfrak{k}$ molecular transport coefficient.
Λ spatial integral scale; Λ_f, longitudinal integral scale; Λ_g, lateral integral scale.
Λ_L Lagrangian integral length scale.
λ spatial dissipation scale.
μ dynamic viscosity.
ν kinematic viscosity.

∇^2 Laplacian operator $\partial^2/\partial x_i\,\partial x_i$.
Ω_k vorticity.
ω_k vorticity of the turbulence.
π 3.14159...
$\mathfrak{P}$ probability.
ρ density; $\bar{\rho}$, time-mean value; $\tilde{\rho}$, turbulent fluctuation.
σ_{ij} stress tensor.
τ time.
τ_E Eulerian dissipation time scale.
τ_L Lagrangian dissipation time scale.
Θ dilatation.
$\mathfrak{u}$ generalized velocity.
$\Xi_i, \Xi_1, \Xi_2, \Xi_3$ Eulerian Cartesian coordinates.
$\xi_i, \xi_1, \xi_2, \xi_3$ Eulerian Cartesian coordinates.

2
PRINCIPLES OF METHODS AND TECHNIQUES IN THE MEASUREMENT OF TURBULENT FLOWS

2-1 INTRODUCTION

During the experimental investigations of fluid flow a great number of methods, techniques, and instruments have been developed and used; so today a choice among them is available, one being more suitable for a particular kind of measurement than the other. Most of these methods and instruments have been developed and are used for measuring velocities in flows that are either nonturbulent or assumed to be nonturbulent; in fact only a few are suitable for making reliable measurements in turbulent flows or, more specifically, for measuring the turbulence itself.

The main difficulties in measuring turbulence are caused by the fact that turbulence is a random, fluctuating flow and three-dimensional. Moreover, the high frequencies of the fluctuations occurring in the turbulent flows in which we are interested make it very difficult for a measuring instrument to satisfy in every respect the basic requirement that recordings of the quantity to be measured must be as free as possible from distortion.

In measuring turbulent flows we have to distinguish between measurement of the mean flow and measurement of the turbulence proper. The problems connected with these two types of measurement are to a certain extent related, yet the requirements which the methods and instruments used must fulfill are different. For instance, the result of a measurement of the mean velocity at a given point is more or less affected by the turbulence present in the flow, and it is necessary to know what correction must be made in the readings of the measuring instrument, but in measuring the turbulence itself we cannot tolerate any influence of the mean velocity that produces an error in the turbulence velocities recorded.

In this chapter we shall consider first the principles of various methods of measuring the turbulence proper, and later on of measuring mean values. It might be wondered why the reverse order has not been chosen, for this might seem to be more logical. The answer is that the problems associated with the measurement of the mean flow will be more fully appreciated if we are well acquainted with the problems presented by the measurement of the turbulence. Moreover, most of the current methods for measuring mean values are affected by turbulence; so corrections have to be applied to the experimental data. These corrections can be made only if the turbulence values are known; so it is necessary to measure these turbulence quantities in addition to the mean values.

Broadly speaking, we may divide the various methods, techniques, and instruments into two groups.

In the first group, use is made of a tracer or other indicator which is introduced into the fluid to make the flow pattern visible (photographic recording) or observable by a suitable detecting apparatus outside the field of flow.

In the second group, a detecting element is introduced into the flowing fluid, and the turbulence quantities are measured by the changes of a mechanical, physical, or chemical nature that occur in this element.

When we apply methods of the first group for measuring turbulence quantities, we are immediately confronted with the difficulties associated with very rapid changes according to time and place, so that practically instantaneous recordings are necessary, frequently at very short intervals. Moreover, the three-dimensionality of the turbulent motions does not make the interpretation of such recordings any simpler—on the contrary.

As regards the second group, there are a number of requirements that must be satisfied by the detecting element and the rest of the measuring apparatus before turbulence can be measured reliably:

1 The detecting element introduced into the flowing field must be so small that it causes only the minimum admissible disturbance of the flow pattern.
2 The instantaneous velocity distribution must be uniform in the region occupied by the element. This means that the detecting element must be smaller than the dimensions of the smallest eddies of the turbulence. If we confine our measurements to flows of low or moderate velocities, the size of the detecting element should not exceed, say, 1 mm.

3 The inertia of the instrument must be low, so that response to even the most rapid fluctuations is practically instantaneous. For flow velocities that are not too high, frequencies up to 5,000 s^{-1} may be expected.

4 The instrument must be sufficiently sensitive to record small differences in the fluctuations; these differences are often only a few per cent of the mean value.

5 The instrument must be stable, so that no noticeable change occurs in the calibration parameters during at least one test run.

6 The instrument must be sufficiently strong and sufficiently rigid to exclude vibrations or motions caused by the turbulent flow.

Accordingly, the normal-sized Pitot, or total-head, tube, which is employed so successfully in the measurement of nonturbulent flows, cannot be used for measuring turbulence (except perhaps turbulence on a very large scale and of very low frequencies), mainly because of its excessive inertia. More modern designs of such tubes which can be successfully applied to measuring turbulent variations of the total head use tiny hypodermic-needle tubes in conjunction with a sensitive transducer.

There is one instrument whose development and application for measuring turbulent flow have far outstripped those of other instruments up to now, namely, the hot-wire anemometer. Its popularity for making turbulence measurements will be easily understood if it is realized that this is an instrument that reasonably satisfies all the above-mentioned requirements—although, of course, it has its limitations.

Since the hot-wire anemometer is to date still the most important instrument for measuring turbulence, the reader will understand if an apparently disproportionate part of this chapter is devoted to a description (still rather restricted) of this instrument and of the relevant methods of measuring various important quantities in turbulence, such as intensities, correlations, integral and dissipation scales, etc.

Since the main idea of this chapter is to familiarize the reader to some extent with the various measuring methods and their relative merits and so to enable him to criticize and interpret empirical data, we will restrict ourselves to the principles only, without considering the technical details of the instrumentation. The reader should not expect to be given a complete account of measuring techniques such as hc would find in an inclusivc tcxtbook on thc subjcct. For instancc, wc shall discuss neither the construction details of the sensing elements proper, nor the electronic equipment, while also data processing will be left out of consideration. Such a discussion would go far beyond the scope of the present book.

2-2 THE HOT-WIRE ANEMOMETER

The detecting element of a hot-wire anemometer consists of a very fine short metal wire, which is heated by an electric current. The wire is cooled by the flowing fluid, causing the temperature to drop and, consequently, the electric resistance of the wire to diminish. For turbulence measurements in gases, wires of 1 to 5 μm diameter are used. The usual materials are platinum, platinum-iridium, and tungsten.

The total amount of heat transferred depends on:

1 The flow velocity
2 The difference in temperature between the wire and the fluid
3 The physical properties of the fluid
4 The dimensions and physical properties of the wire

Generally (2) and (4) are known. If (3) is known or kept constant, the value of (1) can be measured; or if (1) is known or constant, (3) can be measured.

The wire is cooled by heat conduction, free and forced convection, and radiation. In general, the effect of radiation is neglected, and that of free convection may also be neglected.

Since the hot-wire anemometer is used mostly to measure air flows, we will confine ourselves here to its use in this connection. In Sec. 2-8 we will consider its use in turbulence measurements in liquids.

The heat transferred per unit time to the ambient gas from a wire with length l and diameter d is

$$h\pi\, dl(\Theta_w - \Theta_g)$$

Here, h is the heat-transfer coefficient, Θ_w the wire temperature and Θ_g the gas temperature. If the temperature of the wire is not uniform along its length, read for h and Θ_w average values.

For thermal-equilibrium conditions, this heat loss per unit time must be equal to the heat generated per unit time by the electric current through the wire; i.e., it must be equal to I^2R_w, where I is the electric heating current and R_w the total electric resistance of the wire.

With the introduction of the dimensionless group of Nusselt, $\mathbf{Nu} = hd/\kappa_g$, where κ_g is the heat conductivity of the gas at the temperature Θ_g, the thermal equilibrium of the wire gives the following equations

$$I^2R_w = h\pi\, dl(\Theta_w - \Theta_g) = \pi l\kappa_g(\Theta_w - \Theta_g)\cdot\mathbf{Nu} \tag{2-1}$$

The heat-transfer coefficient, and so the Nusselt number, depend on a large number of geometrical and physical parameters, which may conveniently be included in dimensionless groups. Since under the usual operating conditions wire temperatures do not exceed 100 to 150°C, say, radiation is relatively small and its effect will be neglected. We will also neglect, right from the beginning, the *thermo-electrical Thomson effect* and the effect due to a departure from the continuum conditions. The Thomson effect occurs whenever there is an electric current in the direction of a temperature gradient. It may make an originally symmetric temperature distribution along the hot wire slightly skew. However, even in the case of an oblique position of the hot wire with respect to the flow, when this Thomson effect will be more pronounced, the integrated effect over the whole hot wire turns out to be negligibly small, as shown by Davis and Davies.[20,164] The departure from the continuum condition is usually expressed by the value of the *Knudsen number*, which is the ratio between the mean free path of the molecules and a suitable length dimension of the object. It has been assumed that for the wire the effect on the heat transfer can be neglected when the ratio mean-free-path to wire diameter is 0.01 or smaller. However, calculations by Sauer and Drake,[21] as well as experimental results obtained by Davis and Davies[20,164] indicate that the limit for a continuum condition, where the Nusselt number is independent of the Knudsen number, is only reached at still lower values of the Knudsen number for larger values of the Reynolds number. Nevertheless, we will neglect in the following a possible effect of the Knudsen number.

The dependence of the Nusselt number on the other dimensionless groups of importance may be formally expressed as follows

$$\mathbf{Nu} = f\left[\mathbf{Re}, \mathbf{Pr}, \mathbf{Gr}, \frac{\Delta\Theta_w}{\Theta_g}, \frac{U^2}{c_p\Delta\Theta_w}, \frac{l}{d}, \varphi\right] \tag{2-2}$$

where

$$\mathbf{Re} = \frac{\rho_g U d}{\mu_g} \qquad \mathbf{Pr} = \frac{c_p \mu_g}{\kappa_g} \qquad \mathbf{Gr} = \frac{g\rho_g{}^2 d^3 \beta \Delta\Theta_w}{\mu_g{}^2}$$

$\Delta\Theta_w = \Theta_w - \Theta_g$.

μ_g = dynamic viscosity of the gas at the temperature Θ_g.
ρ_g = density of the gas at the temperature Θ_g.
c_p = specific heat of the gas at constant pressure.
β = coefficient of thermal expansion of the gas.
g = gravitational acceleration.
U = velocity of the gas.
φ = angle between wire axis and flow direction.

The *overheat ratio* $\Delta\Theta_w/\Theta_g$ expresses the effect of variable gas properties due to the non-uniform temperature distribution of the gas in the vicinity of the hot wire. The dimensionless group $U^2/c_p\Delta\Theta_w$, often referred to as the *Eckert number,* is of importance when temperature differences caused by dynamic flow effects become comparable with the temperature difference $\Delta\Theta_w$.

By rendering the balance equation for the internal energy, Eq. (1-99), dimensionless it can be shown that the Eckert number occurs in both the compressibility term (III), and the viscous dissipation term (IV), where it shows up combined with the Reynolds number, namely $\mathbf{Re} \cdot U^2/c_p\Delta\Theta_w$. As far as the compressibility effect is concerned, this can be expressed in terms of the adiabatic or stagnation-temperature difference $\Delta\Theta_s$, since $U^2 = 2c_p\Delta\Theta_s$. Because for perfect gases $\Delta\Theta_s$ is directly related to the Mach number, we may express the Eckert number $U^2/c_p\Delta\Theta_w$ in terms of the Mach number as well, making use of the relation $c^2 = c_p(\varkappa - 1)\Theta_g$ for the velocity of sound of a perfect gas. Here $\varkappa$ is the ratio of the specific heats.

Hence

$$\frac{U^2}{c_p\Delta\Theta_w} = 2\frac{\Delta\Theta_s}{\Delta\Theta_w} = \mathbf{Ma}^2(\varkappa - 1)\frac{\Theta_g}{\Delta\Theta_w} \tag{2-3}$$

So, instead of the Eckert number we may use in the general relation (2-2) the Mach number.

The complete relation (2-2) will be too complicated for practical use in hot-wire anemometry. Therefore simplifying approximations are made, depending on the actual conditions. One of the most important conditions stems from the flow pattern around the wire, as determined by the values of the pertinent dimensionless groups, of which the Reynolds number is the most important. Considering current fluid velocities in turbulence measurements, and wire diameters ranging from 1 μm up to 50 μm, the Reynolds number may vary from a value as low as 0.01 up to several hundreds, in the case of supersonic flows. Since the flow pattern around a cylinder at very low values of the Reynolds number differs appreciably from the one at much higher Reynolds numbers, no single and simple relation between the Nusselt number and the Reynolds number can be expected.

Let us consider the current relations for the heat transfer from circular cylinders to a fluid by forced convection alone, and used in hot-wire anemometry. Thereby we will first consider an *infinitely long cylinder,* and neglect the effect of **Gr**, **Ma** and the over-heat ratio $\Delta\Theta_w/\Theta_g$.

One of the first relations used in hot-wire anemometry was suggested by King,[3] namely

$$\mathbf{Nu} = 1 + \sqrt{2\pi\,\mathbf{P\acute{e}}} \qquad (2\text{-}4)$$

valid for the Péclet number

$$\frac{\rho_g c_p U d}{\kappa_g} > 0.08$$

It was derived theoretically on the assumption of two-dimensional potential flow around a circular cylinder, and of unacceptable heat-flow distribution along the circumference of the cylinder. Since $\mathbf{P\acute{e}} = \mathbf{Pr} \cdot \mathbf{Re}$, the essential result is a square-root dependence of the Nusselt number on the Reynolds number. This relation has been used, and is often still being used, in a slightly modified form by maintaining the square-root dependence on the Reynolds number but introducing two constants instead of the number 1 and $\sqrt{2\pi}$, which constants then are determined experimentally. In the absence of free-convection and compressibility effects the constants are mainly functions of the Prandtl number. So

$$\mathbf{Nu} = C_1(\mathbf{Pr}) + C_2(\mathbf{Pr})\,\mathbf{Re}^{1/2} \qquad (2\text{-}5)$$

An empirical relation of this kind, and said to give satisfactory results for many gases and liquids, was evolved by Kramers[2]:

$$\mathbf{Nu} = 0.42\,\mathbf{Pr}^{0.20} + 0.57\,\mathbf{Pr}^{0.33}\,\mathbf{Re}^{0.50} \qquad (2\text{-}6)$$

For air and diatomic gases it was proved valid in the range (see, e.g., McAdams' book[1]):

$$0.01 < \mathbf{Re} < 10{,}000$$

The values of the gas properties have to be taken at the so-called "*film temperature*"

$$\Theta_f = \frac{\Theta_w + \Theta_g}{2} = \Theta_g + \frac{\Delta\Theta_w}{2} \tag{2-7}$$

However, in actual application of hot-wire anemometry, it turns out that Kramers's relation does not satisfactorily describe the observed heat transfer rates, in particular at low values of **Re**. Indeed a plot of

$$\frac{\mathbf{Nu} - 0.42\,\mathbf{Pr}^{0.20}}{0.57\,\mathbf{Pr}^{0.33}} \quad \text{versus} \quad \mathbf{Re}^{0.50}$$

shows for values of $\mathbf{Re} < 0.5$ a slight but definite S-shape of the curve instead of a straight line. Delleur, Toebes, and Liu[22] made a careful check of Kramers's relation and obtained a satisfactory agreement in the range $0.5 < \mathbf{Re} < 70$, though the experimental values were slightly lower than predicted, while the above plot showed a slightly convex behavior.

Since hot-wire anemometers are frequently used at much smaller values of **Re** than 0.5, a better knowledge of the (**Nu**, **Re**) relationship at these low values of **Re** is of utmost importance. Now, theoretically it can be shown that a square-root relationship can certainly not be correct when $\mathbf{Re} \ll 1$.

By applying an Oseen approximation to the convective heat transfer from a cylinder, Cole and Roshko[23] derived the following relation

$$\mathbf{Nu} \simeq \frac{2}{\ln(8/\mathbf{P\acute{e}}) - \Gamma}$$

where $\Gamma = 0.577\ldots$, the Euler number. Further improvements to this relation, by either taking higher order approximations or using asymptotic expansion methods, have been made by others.[24,25,26,30] Very careful measurements in the low **Re**-range have been made by Collis and Williams.[31] For air and $\mathbf{Re} < 0.5$ they arrived at the empirical relation

$$\mathbf{Nu}\left(\frac{\Theta_f}{\Theta_g}\right)^{-0.17} = (1.18 - 1.10\log_{10}\mathbf{Re})^{-1}$$

However, for practical applications, where also much higher values of **Re** have to be considered, they suggested a power-law relationship, thereby making a distinction between the range $\mathbf{Re} < 44$ and the range $\mathbf{Re} > 44$. At $\mathbf{Re} \simeq 44$ vortex shedding begins to take place. These relations are

$$\mathbf{Nu}\left(\frac{\Theta_f}{\Theta_g}\right)^{-0.17} = 0.24 + 0.56\,\mathbf{Re}^{0.45} \quad \text{for} \quad 0.02 < \mathbf{Re} < 44 \tag{2-8a}$$

$$= 0.48\,\mathbf{Re}^{0.51} \quad \text{for} \quad 44 < \mathbf{Re} < 140 \tag{2-8b}$$

Again, the air properties have to be taken at the film temperature. Note, that a correction for the effect of the temperature ratio Θ_f/Θ_g is still included.

The first relation (2-8a) refers to Reynolds numbers most frequently encountered in subsonic flow measurements.

Later experiments by Ahmed[32] have shown that the relation (2-8a) does not apply to the case where the heat transfer is in the opposite direction, namely from a hot fluid to a relatively cool cylinder. For helium, nitrogen, and mixtures of these gases with carbon dioxide, all having a Prandtl number close to that of air, Ahmed obtained a relation similar to Eq. (2-8a), however containing the viscosity ratio instead of the temperature ratio

$$\mathbf{Nu}\left(\frac{\mu_f}{\mu_g}\right)^{-0.15} = 0.21 + 0.50\,\mathbf{Re}^{0.45}$$

At low values of **Re**, below the value of vortex shedding, it may be expected that a close relationship exists between the heat transfer and the frictional resistance of the wire. Therefore Davies and Fisher[63] suggested the following relationship, based on similarity between transport of heat and that of momentum

$$\mathbf{Nu} = \frac{c_f}{\pi\varkappa}\,\mathbf{Pr}\cdot\mathbf{Re}$$

Here c_f is the skin friction coefficient. In the range $0.1 < \mathbf{Re} < 50$, they suggest

$$c_f = 2.6\,\mathbf{Re}^{-2/3}$$

In contrast to Eq. (2-8a) by Collis and Williams, **Nu** is now proportional with $\mathbf{Re}^{1/3}$. Furthermore, in the range $0.1 < \mathbf{Re} < 10$ and taking air, the values of **Nu** are approximately half the values according to Collis and Williams. So a better agreement is obtained if in Davies and Fisher's relation c_f is replaced by $2c_f$. Yet, according to Bradbury and Castro,[166] the semi-empirical relation (2-8a) gives a better representation still.

At low Reynolds numbers free convection may become of importance. Its effect depends on the value of $\mathbf{Gr}\cdot\mathbf{Pr}$, which is the Rayleigh number. According to measurements by Van der Hegge Zijnen,[35] this effect may be neglected for values of $\mathbf{Re} > 0.5$, when

$$\mathbf{Ra} = \mathbf{Gr}\cdot\mathbf{Pr} = \frac{gc_p\rho_f^{\,2}\beta d^3\Delta\Theta}{\mu_f\kappa_f} < 10^{-4} \qquad (2\text{-}9)$$

Collis and Williams concluded from their low Reynolds number experiments in air, that buoyancy effects may be neglected when

$$\mathbf{Gr} < \mathbf{Re}^3 \qquad (2\text{-}10)$$

where they took the fluid properties at ambient temperature. So for $\mathbf{Re} \simeq 10^{-2}$, **Gr** should be smaller than 10^{-6}. For air this corresponds with a wire of 5 μm diameter.

When $\mathbf{Re} \to 0$ the above theoretical relations for the two-dimensional flow around a cylinder, i.e., for an infinitely long cylinder, show that also $\mathbf{Nu} \to 0$. For a cylinder of finite length, where the effect of three-dimensionality can no longer be

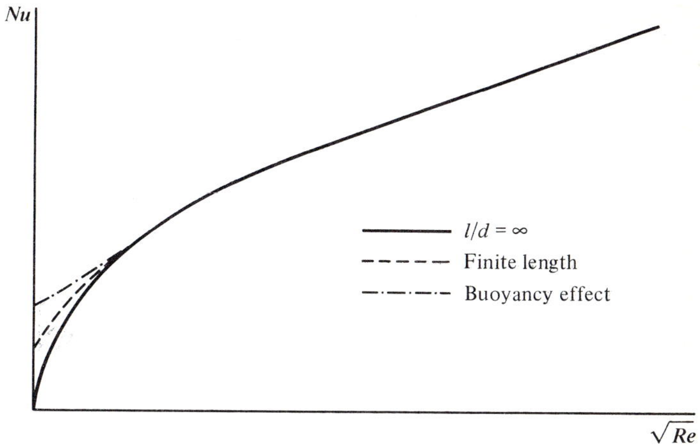

FIGURE 2-1
Relationship between **Nu** and **Re**, applicable to hot-wire anemometers.

neglected, this will no longer hold true. For the extreme case of a cylinder whose length is equal to its diameter a finite value for **Nu** when **Re** $\rightarrow 0$ will be obtained, roughly of the magnitude of that for a sphere, namely the value 2.

Accepting a square-root dependence for high values of the Reynolds number, the shape of the heat-loss characteristic in a (**Nu**, $\sqrt{\mathbf{Re}}$) plot is shown in Fig. 2-1.

Effect of Compressibility

At high flow velocities, say, higher than 100 m/s, compressibility effects on the flow around the wire may become appreciable. Particularly at supersonic speeds, the flow pattern differs from the pattern that occurs at low, subsonic speeds. Because of this difference in flow pattern and because of direct compressibility effects, the temperature distribution in the flow field around the hot wire is different, too. For instance, in the conversion of kinetic energy into pressure, work of compression is done on a fluid element and this work, through conversion into heat, leads to a temperature increase. Accordingly, at a stagnation point the temperature will be higher than the "static" temperature of the flowing fluid.

To take the compressibility effects into account, it is necessary to retain in the relation (2-2) at least the Eckert number, or better the Mach number. So we have to consider

$$\mathbf{Nu} = f(\mathbf{Re}, \mathbf{Pr}, \mathbf{Ma})$$

Kovasznay,[15] Lowell,[7] and Laufer and McClellan[82] have studied the effect of flowing-fluid compressibility on the response of a hot wire. Kovasznay has experimented with wires in supersonic flows in the range of Mach numbers from 1.15 to 2.05. He found that, for small temperature differences between wire and air, a relation similar to (2-6) applies to supersonic flows provided that velocity and density are referred to free-stream conditions but viscosity and conductivity are referred to stagnation conditions. A unique relation is then obtained, which is independent of the Mach number. For air this relation reads

$$\mathbf{Nu} = 0.58\,\mathbf{Re}^{0.5} - 0.795$$

For calculating the heat loss, the temperature difference $\Theta_w - \Theta_e$ has to be taken, where Θ_e is the equilibrium temperature that the unheated wire would attain in the supersonic flow.

With large temperature loadings $(\Theta_w - \Theta_e)/\Theta_e$ of the wire, a nonlinearity appears; this effect, however, is different from that which would occur at subsonic speeds. In contrast to the incompressible case, the Nusselt number decreases with an increase in $(\Theta_w - \Theta_e)/\Theta_{\text{stagn}}$. The empirical correlation suggested by Kovasznay[15] for air reads

$$\mathbf{Nu} = (0.58\sqrt{\mathbf{Re}} - 0.795)\left(1 - 0.18\frac{\Theta_w - \Theta_e}{\Theta_{\text{stagn}}}\right)$$

For estimating the value of Θ_e, use can be made of the empirical relation $\Theta_e/\Theta_{\text{stagn}} \simeq 0.93 - 0.98$. Eckert (see, e.g., ref. 14) gives the relation

$$\Theta_e/\Theta_{\text{stagn}} \simeq \mathbf{Pr}^{0.5}$$

if **Pr** does not differ too much from unity.

However, according to Laufer and McClellan,[82] who also carried out a systematic investigation with fine platinum-rhodium wires (platinum 90–rhodium 10 per cent) in a supersonic airstream, Kovasznay's values are about 20 to 25 per cent too high. Within the Mach-number range from 1.3 to 4.5, Laufer and McClellan also found that the Nusselt number varies as the square root of the Reynolds number, provided that both are referred to the flow conditions behind the detached shockwave in front of the wire. For values of this Reynolds number greater than 20, the equilibrium temperature Θ_e had a constant value of $0.95\Theta_{\text{stagn}}$, independent of Reynolds number and free-stream Mach number. From their graphs one might conclude a relationship

$$\mathbf{Nu} = a\,\mathbf{Re}^{0.5} - b$$

where a varied from 0.465 to 0.49 for a range of temperature loadings $(\Theta_w - \Theta_e)/\Theta_e$ between 0 and 1, and b varied from 0.55 to 0.8. For $\mathbf{Re} < 20$, the ratio $\Theta_e/\Theta_{\text{stagn}}$ rises sharply with decreasing Reynolds number, and the above square-root relationship no longer holds.

Later investigations by others[73,81] have confirmed a linear relation between **Nu** and $\sqrt{\mathbf{Re}}$ for a fixed value of **Ma** and the overheat ratio $\Delta\Theta_w/\Theta_{\text{stagn}}$, for transonic and supersonic flows, up to high Mach number values. Most of these measurements were made at Reynolds numbers greater than 20, say.

A more recent study of published data on the heat transfer to thin wires in compressible flow has been made by Dewey.[83] He presented the correlations obtained in the following form

$$\mathbf{Nu}(\mathbf{Re},\mathbf{Ma}) = \mathbf{Nu}(\mathbf{Re},\infty)\cdot\Phi(\mathbf{Re},\mathbf{Ma})$$

where all fluid properties are again taken at the free-stream stagnation temperature. For larger values of **Ma**, the function Φ reduces to 1, whereas at large **Re**-values, the square-root dependence is obtained.

So for large **Ma**, its effect becomes negligible, and the resulting $\mathbf{Nu} - \sqrt{\mathbf{Re}}$ relation shows that the hot-wire anemometer effectively measures the mass velocity, just as in the incompressible case.

Let us now return to the heat-balance equation (2-1) for the *incompressible case.* The fluid properties occurring in this relation are still functions of the temperature, for which usually the film temperature is taken. These temperature dependences give rise to effects which are considered of second order, though use can be made of these effects in certain turbulence investigations. We will return to this later on.

The *temperature dependence of the electric resistance* of the wire gives the effect on which the use of a hot wire as an anemometer is based. This temperature dependence may be expressed as follows:

$$R_w = R_0[1 + b(\Theta_w - \Theta_0) + b_1(\Theta_w - \Theta_0)^2 + \cdots] \qquad (2\text{-}11)$$

where R_0 = resistance at a reference temperature Θ_0, say 270 K, b, b_1 = temperature coefficients of the electric resistivity of the wire.

Usually, under the normal operating conditions of hot-wire anemometers, the nonlinear terms may be disregarded, because their coefficients are so much smaller than b. For example,

For platinum:

$$b = 3.5 \times 10^{-3}(^\circ\text{C})^{-1} \qquad b_1 = -5.5 \times 10^{-7}(^\circ\text{C})^{-2}$$

For tungsten:

$$b = 5.2 \times 10^{-3}(^\circ\text{C})^{-1} \qquad b_1 = 7.0 \times 10^{-7}(^\circ\text{C})^{-2}$$

Disregarding the quadratic term in Eq. (2-11), the temperature difference $(\Theta_w - \Theta_g)$ in Eq. (2-1) can easily be expressed in terms of $R_w - R_g$, where R_g denotes the electric resistance of the wire at the fluid temperature Θ_g. We obtain

$$\Theta_w - \Theta_g = \frac{R_w - R_g}{bR_0}$$

The relation (2-1) can then be written

$$I^2 R_w = \frac{\pi l \kappa_f}{b}\frac{R_w - R_g}{R_0}\cdot\mathbf{Nu}$$

For sufficiently high Reynolds number, say greater than 40, we may either use the Kramers's relation (2-6), or Collis and Williams' relation (2-8b). Taking the Kramers's relation we obtain

$$I^2R_w = \frac{\pi l \kappa_f}{b}\frac{R_w - R_g}{R_0}[0.42(\mathbf{Pr})_f{}^{0.20} + 0.57(\mathbf{Pr})_f{}^{0.33}(\mathbf{Re})_f{}^{0.50}] \qquad (2\text{-}12)$$

valid for $\mathbf{Re}_f > 40$, say.

For purposes of hot-wire anemometry it is convenient and usual to write this relation in the form

$$\frac{I^2R_w}{R_w - R_g} = A + B\sqrt{U} \qquad (2\text{-}13)$$

where

$$A = 0.42\frac{\pi l \kappa_f}{bR_0}(\mathbf{Pr})_f{}^{0.20} \qquad (2\text{-}14)$$

$$B = 0.57\frac{\pi l \kappa_f}{bR_0}\left(\frac{\rho_f d}{\mu_f}\right)^{0.50}(\mathbf{Pr})_f{}^{0.33} \qquad (2\text{-}15)$$

For a hot wire of 5 μm diameter and normal atmospheric working conditions, this square-root relationship should only be applicable for relatively high velocities, namely in excess of 150 m/s, say. Since most turbulence measurements where compressibility effects may be neglected are carried out at much lower speeds, a more suitable relationship should be used. For instance, the empirical relation (2-8a) suggested by Collis and Williams. For practical applications this relation may be written

$$\frac{I^2R_w}{R_w - R_g} = A + BU^{0.45} \qquad (2\text{-}16)$$

The temperature factor occurring in the original relation (2-8a) can be included in the coefficients A and B, which have to be determined experimentally anyway for the pertinent operating conditions. A better and safer procedure still is to use a relation

$$\frac{I^2R_w}{R_w - R_g} = A + BU^n \qquad (2\text{-}17)$$

and to determine the values of A, B, and n by a suitable calibration method under the operating conditions just before and, as a check, also just after a series of measurements. The reason for this is that these constants, and in particular the exponent n, are rather sensitive to slight changes in the condition of the hot wire, as, e.g., by contamination due to the deposition of extremely fine dust particles.

For carrying out measurements of turbulence velocities two basically different methods can be applied: the *constant-current* method, and the *constant-temperature* method.

Historically speaking, the first method used is that where the electric current I is kept constant. In the second method the wire temperature and so its electric resistance are kept constant. In the first case the temperature of the wire and hence its electric resistance changes due to the fluctuating cooling effects of the fluctuating fluid velocity, in the second case it is the electric current that fluctuates. The hot wire is built into a Wheatstone bridge, as shown schematically, together with the main electronic equipment, in Fig. 2-2. Obviously this equipment differs basically for the two methods.

As mentioned, the earlier applications of hot-wire anemometers for turbulence measurements used the constant-current method. In the course of years it has been developed into an almost foolproof tool. This, notwithstanding the fact that the constant-temperature method has basic advantages over the other with respect to accuracy of measurements. However, the inherent instability of the feedback system required for keeping the wire resistance and temperature constant has for a long time been considered an almost insuperable difficulty. Owing to important developments in electronics it is nowadays possible to build stably operating feedback systems with the same degree of reliability as the electronic circuits used in the constant-current method. Because of its advantages the constant-temperature method is now widely adopted. Complete equipments using this method, and operating to a satisfactory degree of reliability, are commercially available. It is only in special and exceptional cases, that the constant-current method will be used.

Therefore, in what follows we will consider first the constant-temperature method, and in much greater detail than the constant-current method.

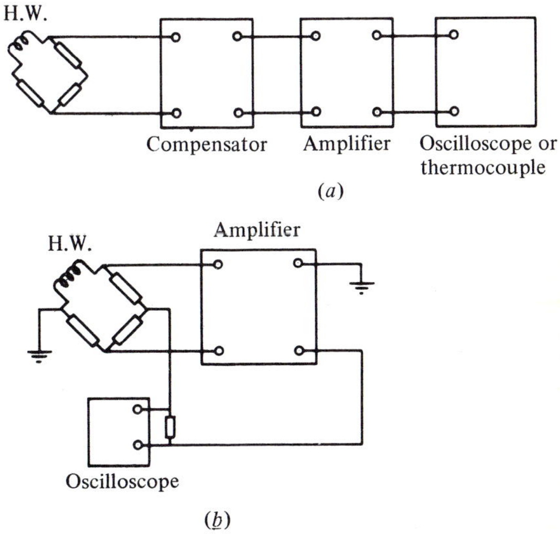

FIGURE 2-2
Block diagrams of (a) constant-current method and (b) constant-temperature method.

2-3 CONSTANT-TEMPERATURE METHOD

The electric resistance of the wire and, accordingly, its temperature are kept constant, as far as possible. Any slight variation in temperature and thus in electric resistance due to the cooling effect of the turbulent-velocity fluctuations is immediately compensated for by means of an electronic feedback system. This feedback system changes the electric current through the wire as soon as a variation in temperature, and hence, in electric resistance occurs. Such a feedback system operates almost instantaneously.

Let r_w be the—slight—variation in electric resistance, and g_{tr} the transconductance of the electronic circuit. The relation between the compensating electronic current i and the resistance variation r_w is then given by

$$i = -g_{tr} I r_w \tag{2-18}$$

The variation in electric resistance r_w is that of the total wire. Since, mainly due to the heat loss by conduction along the wire to its supports, the temperature of the wire is not uniform along its length, also the local electric resistance per unit of length is not uniform.

In order to study the dynamic behavior of the hot wire we have to consider this non-uniform temperature and electric resistance distribution.

We assume that at each section of the wire the heat loss to the ambient flowing fluid is by forced convection alone according to Eq. (2-17).

Let $\mathscr{R}_w$ and $\mathscr{R}_g$ denote the electric resistance per unit length at the local wire temperature and at the fluid temperature, respectively. The heat-balance equation for a differential element of the wire, considering heat exchange only by forced convection and assuming an incompressible fluid, reads

$$I^2\mathscr{R}_w = (\mathscr{R}_w - \mathscr{R}_g)(A + BU^n) + \rho_w c_w \frac{\pi}{4} d^2 \frac{\partial \Theta_w}{\partial t} - \kappa_w \frac{\pi}{4} d^2 \frac{\partial^2 \Theta_w}{\partial x^2} \tag{2-19}$$

where ρ_w = density of wire material,
c_w = specific heat of wire material per unit mass.

The coordinate x is taken along the wire, with the origin midway between its ends.

With the assumed linear dependence of the electric resistance on the temperature, the two temperature terms in the above equation can be expressed in the local electric resistance

$$I^2\mathscr{R}_w = (\mathscr{R}_w - \mathscr{R}_g)(A + BU^n) + \frac{\rho_w c_w \pi d^2}{4b\mathscr{R}_0} \frac{\partial \mathscr{R}_w}{\partial t} - \frac{\kappa_w \pi d^2}{4b\mathscr{R}_0} \frac{\partial^2 \mathscr{R}_w}{\partial x^2} \tag{2-20}$$

Owing to the electronic feedback system, variation in electric resistance $\mathfrak{r}_w$ and, provided $g_{tr} r_w \ll 1$, also variations in electric current i will be small, so that

$$\frac{\mathfrak{r}_w}{\overline{\mathscr{R}}_w} \ll 1 \quad \text{and} \quad \frac{i}{\overline{I}} \ll 1$$

Overscores again denote time-mean values. These small values make possible a linearization procedure in order to keep relations simple. However, we have also to reckon with variations of the fluid velocity U, which variations are not necessarily small. In order to extend the linearization procedure to the term containing the fluid velocity, we will assume that also

$$\frac{u}{\bar{U}} \ll 1$$

So putting

$$U = \bar{U} + u$$

$$\mathscr{R}_w = \bar{\mathscr{R}}_w + \mathfrak{r}_w$$

$$I = \bar{I} + i$$

substituting these quantities in the heat-balance Eq. (2-20), and carrying out the linearization procedure, yields the following relations for the time-mean condition and for the fluctuating part respectively

$$\bar{I}^2\bar{\mathscr{R}}_w = (\bar{\mathscr{R}}_w - \mathscr{R}_g)(A + B\bar{U}^n) - \frac{\kappa_w \pi d^2}{4b\mathscr{R}_0}\frac{d^2\bar{\mathscr{R}}_w}{dx^2} \tag{2-21}$$

and

$$2\bar{I}\bar{\mathscr{R}}_w i + \bar{I}^2\mathfrak{r}_w = (A + B\bar{U}^n)\mathfrak{r}_w + (\bar{\mathscr{R}}_w - \mathscr{R}_g)nB\bar{U}^{n-1}u$$
$$+ \frac{\rho_w c_w \pi d^2}{4b\mathscr{R}_0}\frac{\partial \mathfrak{r}_w}{\partial t} - \frac{\kappa_w \pi d^2}{4b\mathscr{R}_0}\frac{\partial^2 \mathfrak{r}_w}{\partial x^2} \tag{2-22}$$

The fluctuation i of the electric current may be expressed in terms of the fluctuation of the electric resistance, by making use of the relation (2-18). Namely

$$i = -g_{tr}\bar{I}r_w = -g_{tr}\bar{I}\int_{-l/2}^{+l/2} dx\, \mathfrak{r}_w$$

Equation (2-22) then may be written

$$\frac{\partial \mathfrak{r}_w}{\partial t} - \frac{\kappa_w}{\rho_w c_w}\frac{\partial^2 \mathfrak{r}_w}{\partial x^2} + \frac{4b\mathscr{R}_0(A + B\bar{U}^n - \bar{I}^2)}{\rho_w c_w \pi d^2}\mathfrak{r}_w + \frac{8b\mathscr{R}_0\bar{\mathscr{R}}_w\bar{I}^2 g_{tr}}{\rho_w c_w \pi d^2} r_w$$
$$= -\frac{4b\mathscr{R}_0(\bar{\mathscr{R}}_w - \mathscr{R}_g)nB\bar{U}^{n-1}}{\rho_w c_w \pi d^2}u \tag{2-23}$$

From this equation it may be concluded that not only is there a phase shift between the velocity fluctuation and the fluctuation of the electric resistance, but also that this phase shift will be a function of x, depending on the degree of nonuniformity of the distribution of temperature along the wire. So the variation in phase shift will be more pronounced close to the supports, where there is a strong temperature drop along the wire. On the other hand we may expect that for sufficiently long wires the phase shift will be almost constant along a large part of the wire length.

A similar equation, but for a hot wire operating according to the constant current method, has been studied extensively by Betchov.[6] He arrived at a similar conclusion.

In the above Eq. (2-23) the electric resistance per unit length as well as the total electric resistance occur. Since we are interested in the dynamic behavior of the total wire, we shall integrate this equation with respect to x from $-l/2$ to $+l/2$. With

$$\bar{R}_w = \int_{-l/2}^{+l/2} dx\, \bar{\mathscr{R}}_w, \quad R_g = \mathscr{R}_g l, \quad R_0 = \mathscr{R}_0 l \quad \text{and} \quad C_w = \frac{\pi}{4} d^2 l \rho_w c_w$$

we obtain:

$$\frac{dr_w}{dt} + bR_0 \frac{A + B\bar{U}^n - \bar{I}^2 + 2g_{tr}\bar{I}^2\bar{R}_w}{C_w} r_w - \frac{2\kappa_w}{\rho_w c_w}\left[\frac{\partial \mathfrak{r}_w}{\partial x}\right]_0^{l/2}$$

$$= -\frac{bR_0(\bar{R}_w - R_g)nB\bar{U}^{n-1}}{C_w} u \qquad \text{(2-24)}$$

The last term on the left-hand side of this equation accounts for the heat loss due to conduction from the wire to the relatively cold supports. Thus the dynamic behavior of the wire will be affected by this fluctuating heat loss. In order to take account of this effect, we will first consider the temperature distribution and so the heat loss to the supports under steady conditions.

Cooling Effects of the Wire Supports

Usually, for reasons explained later on, the wire supports are thick compared with the wire. This means that the heat flow from the wire to the supports will only increase in a negligible way the temperature of the supports with respect to that of the ambient air. So the thermal inertia of the supports may be assumed to be infinitely large. Hence, also under non-steady temperature conditions of the wire, as caused by the fluctuating velocity, we may assume the supports to have the (constant) ambient temperature.

Instead of the temperature distribution we may consider the distribution of the electric resistance along the wire. For steady thermal and flow conditions, this distribution is described by Eq. (2-21). Since R_g remains constant, we write this equation as follows.

$$\frac{d^2(\bar{\mathscr{R}}_w - \mathscr{R}_g)}{dx^2} - \frac{4b\mathscr{R}_0(A + B\bar{U}^n - \bar{I}^2)}{\kappa_w \pi d^2}(\bar{\mathscr{R}}_w - \mathscr{R}_g) + \frac{4b\mathscr{R}_0\mathscr{R}_g\bar{I}^2}{\kappa_w \pi d^2} = 0 \qquad \text{(2-25)}$$

The boundary conditions are:

$$\bar{\mathscr{R}}_w - \mathscr{R}_g = 0 \quad \text{for} \quad x = \pm l/2$$

$$\frac{d}{dx}(\bar{\mathscr{R}}_w - \mathscr{R}_g) = 0 \quad \text{for} \quad x = 0$$

The solution of the differential equation that satisfies the boundary condition reads

$$\bar{\mathscr{R}}_w - \mathscr{R}_g = \frac{\bar{I}^2 \mathscr{R}_g}{A + B\bar{U}^n - \bar{I}^2}\left[1 - \frac{\cosh(x/l_c)}{\cosh(l/2l_c)}\right] \qquad (2\text{-}26)$$

where

$$l_c = \frac{d}{2}\sqrt{\frac{\pi \kappa_w}{b\mathscr{R}_0(A + B\bar{U}^n - \bar{I}^2)}} \qquad (2\text{-}27)$$

This quantity l_c has the dimension of a length. It accounts for the cooling effect of the supports. It may be interpreted as the length of the part of the wire, adjoining the support, that is effectively cooled by the supports. For this reason, Betchov[6] called it the "cold length."

The following relations can be obtained from the solution for the distribution of the electric resistance

$$[\bar{\mathscr{R}}_w]_{\text{aver}} - \mathscr{R}_g = \frac{\bar{I}^2 \mathscr{R}_g}{A + B\bar{U}^n - \bar{I}^2}\left[1 - \frac{\tanh l^*}{l^*}\right] \qquad (2\text{-}28)$$

and

$$\frac{[\bar{\mathscr{R}}_w]_{\max} - \mathscr{R}_g}{[\bar{\mathscr{R}}_w]_{\text{aver}} - \mathscr{R}_g} = \frac{l^* \cosh l^* - l^*}{l^* \cosh l^* - \sinh l^*} \qquad (2\text{-}29)$$

where $l^* = l/2l_c$.

The heat loss to the supports is determined by $d(\bar{\Theta}_w - \Theta_g)/dx$ at $x = \pm l/2$. For the corresponding gradient of the electric resistance we obtain from Eq. (2-26)

$$\left[\frac{d(\bar{\mathscr{R}}_w - \mathscr{R}_g)}{dx}\right]_{-l/2}^{+l/2} = -\frac{2\bar{I}^2 \mathscr{R}_g}{A + B\bar{U}^n - \bar{I}^2}\frac{\tanh l^*}{l_c}$$

Or, expressed in terms of the average electric resistance, given by Eq. (2-28)

$$\left[\frac{d(\bar{\mathscr{R}}_w - \mathscr{R}_g)}{dx}\right]_{-l/2}^{+l/2} = -2\frac{[\bar{\mathscr{R}}_w]_{\text{aver}} - \mathscr{R}_g}{l_c}\frac{\tanh l^*}{1 - (\tanh l^*)/l^*} = -4\mathfrak{C}\frac{\bar{R}_w - R_g}{l^2} \qquad (2\text{-}30)$$

where

$$\mathfrak{C} = \frac{l^{*2} \tanh l^*}{l^* - \tanh l^*} \qquad (2\text{-}31)$$

The relations (2-26) and (2-29) show that the distribution of the electric resistance becomes more uniform with increase of l^*, or of the ratio l/d, an obvious result. If the total resistance $\bar{R}_w$ is kept constant, l^* increases with increasing $\bar{U}$. The gradient in electric resistance at the supports increases, almost proportional with l^*. We may calculate the minimum value of l^* required to obtain a reasonably uniform resistance distribution. Such a calculation shows that l^* should not be much smaller than 5, say. According to Eq. (2-26) for $l^* = 5$, the resistance is almost uniform along the central 60 per cent of the wire length. So, for platinum-iridium wires, l/d should preferably be greater than 200. For tungsten wire this should be higher still, since κ_w for tungsten is roughly 2.5 times the value for platinum.

The cooling effect of the supports was studied by Lowell[7] also, though in a somewhat different way. He determined the fractional end loss, defined as the ratio between the heat loss by conduction to the supports and the heat loss to the airstream. More recently Davies and Fisher[63] made similar calculations. They determined the ratio of the Nusselt number for a wire with and without end losses.

Dynamic Behavior with Heat Loss to Supports

The differential equation for the fluctuation of the electrical resistance including this heat loss is given by Eq. (2-24). When discussing the partial differential equation (2-23), we mentioned the non-uniform distribution of the phase shift between velocity fluctuations and resistance fluctuations. This implies that the distribution of the electric resistance and temperature will not remain similar when it fluctuates about the time-mean or "steady-state"-distribution. However, we may expect that the deviation from similarity, i.e., from the "steady-state" distribution at any instant will be smaller when this "steady-state" distribution is more uniform. This requires a large value of l^*. Calculations show that for the dynamic sensitivity of a hot wire the error made by assuming similarity is less than 5 per cent for $l^* = 5$. We may also consider this question of similarity in a different way. When the wire is not long enough, the wire will have no time to adapt itself to the "steady-state" temperature distribution. Since the deviation becomes of importance in the parts of the wire where the distribution is not uniform, that is in the parts adjoining the supports, an estimate of this adaption time can be made from $\rho_w c_w l_c^2/\kappa_w$. This should be small compared with the characteristic time of the smallest eddies. If we assume a wire length l equal to the size of the smallest eddies, we obtain the condition

$$\frac{\rho_w c_w l \bar{U}}{4\kappa_w} \ll l^{*2}$$

The reader may convince himself that for usual laboratory conditions and the current hot-wire anemometers, this condition can hardly be satisfied for a value of $l^* \simeq 5$.

However, since calculations concerning the dynamic behavior of the hot wire are greatly simplified when the assumption of similarty of the temperature distribution is made, and since it is our intention with the following theoretical consideration to show the main aspects of the dynamic behavior without aiming at accurate quantitative results, we will make this assumption of similarity. So, in Eq. (2-24) we assume for the last term on the left-hand side a similar expression to the steady-state relation (2-30)

$$-\frac{2\kappa_w}{\rho_w c_w}\left[\frac{\partial \mathfrak{r}_w}{\partial x}\right]_0^{l/2} = \frac{4\kappa_w}{\rho_w c_w}\mathfrak{C}\frac{r_w}{l^2} = \frac{\kappa_w \pi d^2 \mathfrak{C}}{C_w}\cdot\frac{r_w}{l}$$

Equation (2-24) then reads

$$\frac{dr_w}{dt} + bR_0 \frac{A + B\bar{U}^n - \bar{I}^2 + 2g_{tr}\bar{I}^2\bar{R}_w}{C_w} r_w + \frac{\kappa_w \pi d^2 \mathfrak{C}}{C_w l} r_w = - \frac{bR_0(\bar{R}_w - R_g)nB\bar{U}^{n-1}}{C_w} u \tag{2-32}$$

Integration with respect to x of Eq. (2-21) gives

$$(\bar{R}_w - R_g)(A + B\bar{U}^n - \bar{I}^2) = \bar{I}^2 R_g + \frac{\kappa_w \pi d^2}{4b\mathscr{R}_0}\left[\frac{d\bar{\mathscr{R}}_w}{dx}\right]_{-l/2}^{+l/2} = \bar{I}^2 R_g - \frac{\kappa_w \pi d^2 \mathfrak{C}}{bR_0 l}(\bar{R}_w - R_g) \tag{2-33}$$

Elimination of $A + B\bar{U}^n - \bar{I}^2$ from Eqs. (2-32) and (2-33) yields the dynamic equation for r_w in a suitable form

$$\frac{dr_w}{dt} + \frac{bR_0\bar{I}^2}{C_w}\left(\frac{R_g}{\bar{R}_w - R_g} + 2g_{tr}\bar{R}_w\right) r_w = - \frac{bR_0(\bar{R}_w - R_g)nB\bar{U}^{n-1}}{C_w} u \tag{2-34}$$

Or

$$\frac{dr_w}{dt} + \frac{1}{M_{\text{c.t.}}} r_w = \varphi(t) \tag{2-35}$$

when, for short, we write

$$\varphi(t) = - \frac{bR_0(\bar{R}_w - R_g)nB\bar{U}^{n-1}}{C_w} u(t) \tag{2-36}$$

and

$$M_{\text{c.t.}} = \frac{C_w}{bR_0\bar{I}^2[R_g/(\bar{R}_w - R_g) + 2g_{tr}\bar{R}_w]} \tag{2-37}$$

This expression for the "time constant" $M_{\text{c.t.}}$ (the subscript c.t. refers to constant temperature) does not contain explicitly a term that accounts for the heat loss to the supports. It is implicitly determined by a higher value of $\bar{I}$ for the same value of $\bar{R}_w$. Thus for the same total electric resistance of the hot wire this heat loss results in a favorable reduction of the time constant. This can also be shown directly by elimination of $\bar{I}$ from Eq. (2-37) and Eq. (2-28), after integration of the latter equation over the total length of the hot wire

$$M_{\text{c.t.}} = \frac{C_w(\bar{R}_w/R_g) - (\tanh l^*)/l^*}{bR_0(A + B\bar{U}^n)[1 + 2g_{tr}(\bar{R}_w - R_g)(\bar{R}_w/R_g)]} \tag{2-38}$$

When $l_c \to 0$, so that $l^* \to \infty$, $(\tanh l^*)/l^* \to 0$, and

$$\frac{M_{\text{c.t.}}}{M_{\text{c.t.}}(l_c = 0)} = 1 - \frac{R_g}{C_w\bar{R}_w}\frac{\tanh l^*}{l^*} \tag{2-39}$$

This result clearly shows that with increasing heat loss to the supports, i.e., with decreasing l^*, the time constant will decrease. This would call for a short wire.

It should be remarked that the expressions (2-37) and (2-38) are for the time constant of the hot-wire including the feedback of the electronic circuit, through the transconductance g_{tr}. The expression for the time constant of the hot wire proper is obtained by omitting the term $2g_{tr}\bar{R}_w$ in Eqs. (2-37) and (2-38).

Let us now consider the response of the system to a fluctuation of the velocity, for which we assume a simple harmonic function with angular frequency ω,

$$u(t) = u^* \exp(\imath\omega t), \quad \text{where} \quad \imath = \sqrt{-1}$$

The corresponding expression for $\varphi(t)$ occurring in the dynamic equation (2-35) becomes

$$\varphi(t) = -\varphi^* \exp(\imath\omega t)$$

with

$$\varphi^* = \frac{bR_0(\bar{R}_w - R_g)nB\bar{U}^n}{C_w}\frac{u^*}{\bar{U}}$$

Or, after introducing the time constant as given by Eq. (2-37)

$$\varphi^* = \frac{(\bar{R}_w - R_g)nB\bar{U}^n}{M_{\text{c.t.}}\bar{I}^2[R_g/(\bar{R}_w - R_g) + 2g_{tr}\bar{R}_w]}\cdot\frac{u^*}{\bar{U}}$$

Now we assume a solution of Eq. (2-35) of the form

$$r_w = r_w^* \exp[\imath(\omega t - \psi)]$$

Substitution in Eq. (2-35) then yields

$$r_w^* = -\frac{\varphi^* M_{\text{c.t.}}}{\sqrt{1 + \omega^2 M_{\text{c.t.}}^2}} \qquad \psi = \arctan \omega M_{\text{c.t.}}$$

The relation between the voltage fluctuation e and the resistance fluctuation is obtained from

$$\bar{E} + e = (\bar{I} + i)(\bar{R}_w + r_w)$$

$$\text{i.e. } e \simeq \bar{R}_w i + \bar{I} r_w = \bar{I} r_w(1 - g_{tr}\bar{R}_w) \simeq -g_{tr}\bar{I}\bar{R}_w r_w$$

since $g_{tr}\bar{R}_w \gg 1$.

With the above solution for r_w we obtain

$$e = \frac{\varphi^* M_{\text{c.t.}}}{\sqrt{1 + \omega^2 M_{\text{c.t.}}^2}} g_{tr}\bar{I}\bar{R}_w \exp[\imath(\omega t - \psi)]$$

$$= \frac{g_{tr}\bar{R}_w(\bar{R}_w - R_g)nB\bar{U}^{n-1}}{\sqrt{1 + \omega^2 M_{\text{c.t.}}^2}\cdot\bar{I}[R_g/(\bar{R}_w - R_g) + 2g_{tr}\bar{R}_w]}\cdot u^* \exp[\imath(\omega t - \psi)]$$

If we introduce the sensitivity $s_{c.t.}$ of the hot wire, defined by

$$\sqrt{\overline{(e^2)}} = s_{c.t.}\sqrt{\overline{(u^2)}} \tag{2-40}$$

then for a purely harmonic fluctuation of the velocity u, this sensitivity reads

$$s_{c.t.}(\omega) = \frac{g_{tr}\bar{R}_w(\bar{R}_w - R_g)nB\bar{U}^{n-1}}{\bar{I}[R_g/(\bar{R}_w - R_g) + 2g_{tr}\bar{R}_w]\sqrt{1+\omega^2 M_{c.t.}^2}} \tag{2-41}$$

Since usually $R_g/(\bar{R}_w - R_g) = \mathcal{O}(1)$, and $g_{tr}\bar{R}_w \gg 1$, the expression (2-41) may be simplified to

$$s_{c.t.}(\omega) = \frac{(\bar{R}_w - R_g)nB\bar{U}^{n-1}}{2\bar{I}\cdot\sqrt{1+\omega^2 M_{c.t.}^2}} \tag{2-42}$$

Consequently, owing to the finite thermal inertia of the hot wire, its sensitivity decreases for large values of the frequency. The value of the frequency above which its effect becomes noticeable depends on the value of the "time constant" $M_{c.t.}$. At a given frequency the sensitivity increases, though slightly so, with decreasing time constant. Due to the electronic feedback system the time-constant can be made very small indeed. This means that even for moderately large frequencies (up to 10^4 s^{-1}, say), $\omega^2 M_{c.t.}^2 \ll 1$ and the sensitivity is frequency independent and becomes equal to the value under "static" conditions

$$s_{c.t.} = \frac{(\bar{R}_w - R_g)nB\bar{U}^{n-1}}{2\bar{I}} \tag{2-43}$$

In this case it will not be necessary, though it is possible, to account for the reduction of the sensitivity at high frequencies due to thermal lag, by including in the electronic circuit a compensator that accounts for the reduction factor $[1 + \omega^2 M_{c.t.}^2]^{-1/2}$.

For an actual turbulence which is composed of many harmonics we have to take the contribution of all harmonics to the sensitivity.

The solution of Eq. (2-35) reads

$$r_w = \int_{-\infty}^{t} d\tau\, \varphi(\tau) \exp\left[-\frac{1}{M_{c.t.}}(t-\tau)\right] \tag{2-44}$$

Or, with the expression (2-36) for $\varphi(\tau)$

$$r_w = A\int_{-\infty}^{t} d\tau\, u(\tau) \exp\left[-\frac{t-\tau}{M_{c.t.}}\right]$$

$$= A\int_{0}^{\infty} dt'\, u(t-t') \exp\left[-t'/M_{c.t.}\right]$$

where

$$A = -\frac{bR_0(\bar{R}_w - R_g)nB\bar{U}^{n-1}}{C_w}$$

Since $e \simeq -g_{tr}\bar{I}\bar{R}_w r_w$, the sensitivity $s_{c.t.}$, defined by Eq. (2-40) has to be obtained from

$$s_{c.t.} = g_{tr}\bar{I}\bar{R}_w \frac{\sqrt{(\overline{r_w^2})}}{\sqrt{(\overline{u^2})}}$$

$$\overline{r_w^2} = A^2 \int_0^\infty dt' \int_0^\infty dt'' \, \overline{u(t-t')u(t-t'')} \exp[-(t'+t'')/M_{c.t.}]$$

$$= A^2\overline{u^2} \int_0^\infty dt' \int_0^\infty dt'' \, \mathbf{R}_E(t''-t') \exp[-(t'+t'')/M_{c.t.}]$$

Since the integrand does not change sign when t' and t'' are changed, the integration in the entire first quadrant of the (t',t'')-plane is equal to twice the integration over half of the first quadrant between the limits $(0 \le t'' \le t'; 0 \le t' \le \infty)$.

$$\overline{r_w^2} = 2A^2\overline{u^2} \int_0^\infty dt' \int_0^{t'} dt'' \, \mathbf{R}_E(t''-t') \exp[-(t'+t'')/M_{c.t.}]$$

In evaluating the double integral we follow a procedure applied by Burgers.[80] We introduce the new variables

$$t_1 = t' + t'' \qquad t_2 = t'' - t'$$

According to the transformation rule we have

$$\int_0^\infty dt' \int_0^{t'} dt'' \, f(t',t'') = \int_0^\infty dt_2 \int_{t_2}^\infty dt_1 \, f(t_1,t_2) \begin{vmatrix} \dfrac{\partial t'}{\partial t_1} & \dfrac{\partial t'}{\partial t_2} \\ \dfrac{\partial t''}{\partial t_1} & \dfrac{\partial t''}{\partial t_2} \end{vmatrix}$$

$$= 1/2 \int_0^\infty dt_2 \int_{t_2}^\infty dt_1 \, f(t_1,t_2)$$

The integration over half of the first quadrant of the (t',t'')-plane corresponds with an integration over half of the first quadrant of the (t_1,t_2)-plane between the limits $(t_2 \le t_1 \le \infty; 0 \le t_2 \le \infty)$. Application to the double integral occurring in $\overline{r_w^2}$ yields

$$\overline{r_w^2} = A^2\overline{u^2} \int_0^\infty dt_2 \int_{t_2}^\infty dt_1 \, \mathbf{R}_E(t_2) \exp(-t_1/M_{c.t.})$$

$$= A^2\overline{u^2} M_{c.t.} \int_0^\infty dt_2 \, \mathbf{R}_E(t_2) \exp(-t_2/M_{c.t.})$$

The expression for the sensitivity $s_{c.t.}$ then becomes

$$s_{c.t.} = \frac{b g_{tr}\bar{I}R_0\bar{R}_w(\bar{R}_w - R_g)nB\bar{U}^{n-1}}{C_w} \left[M_{c.t.} \int_0^\infty dt \, \mathbf{R}_E(t) \exp(-t/M_{c.t.}) \right]^{1/2}$$

With the expression (2-37) for the time constant $M_{c.t.}$, we obtain

$$s_{c.t.} = \frac{g_{tr}\bar{I}\bar{R}_w(\bar{R}_w - R_g)nB\bar{U}^{n-1}}{A + B\bar{U}^n - \bar{I}^2 + 2g_{tr}\bar{I}^2\bar{R}_w} \left[\int_0^\infty d\tau \, \mathbf{R}_E(\tau M_{c.t.}) \exp(-\tau) \right]^{1/2} \qquad (2\text{-}45)$$

where $\tau = t/M_{c.t.}$

For the usual large values of g_{tr} and the corresponding very small values of $M_{c.t.}$, the contribution to the integral is given for such values of τ where $\mathbf{R}_E \simeq 1$, and the integral becomes equal to 1.

Since $A + B\bar{U}^n - \bar{I}^2$ is roughly of the same magnitude as $\bar{I}^2R_g/(\bar{R}_w - R_g)$, for large values of g_{tr}

this term becomes small with respect to $2g_{tr}\bar{I}^2\bar{R}_w$, and the expression (2-45) for $s_{\text{c.t.}}$ reduces to the value for "static" conditions, given by Eq. (2-43).

In considering the cooling effect of the supports on the temperature distribution along the wire, we have assumed that the supports have an infinitely great heat capacity and that their temperature is equal to the ambient-gas temperature. It is obvious that these assumptions hold only if the dimensions of the supports are large compared with the wire diameter and if, moreover, the heat conductivity of the support material is high. A rough calculation will soon show that, if the conductivity of the support material is not lower than that of the wire material, the effect of a finite support diameter is negligible provided that the ratio of cross section of support to cross section of wire exceeds, roughly, 100. Thus, for a wire of 5 μm diameter, this would occur if the support diameter were greater than 0.05 mm, a condition that will always be satisfied. Therefore the requirement of relatively large heat capacity of the supports will not determine the lower limit of the tolerable support diameter.

This lower limit will be determined rather by requirements concerning the rigidity of the supports and the possible vibrations. These vibrations may be of a "seismic" nature or may be caused by the flow past the wire or supports (aeolian effect). These vibrations are undesirable not only from a mechanical point of view, but even more so from an aerodynamical point of view, since the corresponding vibrations of the wire might simulate a nonexistent turbulence in the flow past the wire and might be measured and interpreted as such. The effect may become serious when measurements are made of narrow band-width spectra. Both Van der Hegge Zijnen[8] and Lowell[7] have considered this problem of rigidity.

Effect of Large Turbulence Fluctuations on the Response of a Hot Wire

So far, we have tacitly assumed that the relative turbulence intensity is small, in order to make possible a linearization of the hot-wire response to turbulence fluctuations.

It is of practical interest to know at what turbulence intensities such a linearization is still permitted without incurring too great errors in the measurement of turbulence intensity.

In the flow at some distance from turbulence-producing grids, the relative intensity is seldom higher than 10 per cent. From pure intuition this intensity may be considered sufficiently low to permit a linearization of the hot-wire response. The turbulence occurring in pipe flow and in free jets usually appreciably exceeds the above value of 10 per cent; again, from pure intuition it may be concluded that we may then expect a distortion of the linearized hot-wire response. Many investigators, among them Corrsin,[13] who experimented with turbulent free jets, have appreciated this deviation from a linear response at high turbulence intensities and have tried to estimate the error made if a linear response is still assumed.

A first estimate of the effect of large turbulence fluctuations on the hot-wire response can be made by extending the series expansion for the n-th power of the velocity [see Eq. (2-20)] to the second- and higher-order terms. At the same time it is no longer permissible to consider only the turbulence component in the same direction as the main velocity $\bar{U}$; we must take into account the other turbulence components also. Now the hot wire is sensitive essentially to the velocity components perpendicular to it; only when the velocity component parallel to the wire becomes very large does its effect on the cooling become noticeable vis-à-vis the effect of the components perpendicular to the wire.

This nonlinearity effect, due to large velocity fluctuations, has to be distinguished well from the nonlinearity effect due to temperature dependence of fluid properties, to be discussed later. This latter effect will be increased too with increase of the velocity fluctuations. However, for the constant-temperature operations this effect may still be considered as not serious. It may become of importance when the constant-current operation is applied.

Thus, if u_1 is the turbulence-velocity component in the direction of the main stream $\bar{U}$ (which is assumed to be perpendicular to the wire) and if u_2 is the lateral turbulence-velocity component perpendicular to the wire, then we have to consider as the velocity that influences the cooling of the wire

$$U_{\text{eff}} = \sqrt{(\bar{U} + u_1)^2 + u_2^2} \tag{2-46}$$

A series expansion gives

$$U_{\text{eff}} = \bar{U}\left(1 + \frac{u_1}{\bar{U}} + \frac{u_2^2}{2\bar{U}^2} - \frac{u_1 u_2^2}{2\bar{U}^3} + \cdots\right)$$

from which, for the average value of U_{eff}, we obtain,

$$\bar{U}_{\text{eff}} = \bar{U}\left(1 + \frac{\overline{u_2^2}}{2\bar{U}^2} - \frac{\overline{u_1 u_2^2}}{2\bar{U}^3} + \cdots\right) \tag{2-46a}$$

This, however, is not the value measured by the hot wire. For that value we have to start from U_{eff}^n, according to Eq. (2-17).

$$U_{\text{eff}}^n = \bar{U}^n\left[1 + n\frac{u_1}{\bar{U}} + n(n-1)\frac{u_1^2}{2\bar{U}^2} + n\frac{u_2^2}{2\bar{U}^2} + n(n-1)(n-2)\frac{u_1^3}{6\bar{U}^3}\right.$$
$$\left. + n(n-2)\frac{u_1 u_2^2}{2\bar{U}^3} + \mathcal{O}\left(\frac{u_1^4}{\bar{U}^4}\right)\right] \tag{2-47}$$

If we take the time average, we obtain

$$\overline{U_{\text{eff}}^n} = \bar{U}^n\left[1 + n(n-1)\frac{\overline{u_1^2}}{2\bar{U}^2} + n\frac{\overline{u_2^2}}{2\bar{U}^2} + n(n-1)(n-2)\frac{\overline{u_1^3}}{6\bar{U}^3}\right.$$

$$+n(n-2)\frac{\overline{u_1u_2^2}}{2\bar{U}^3}+\mathcal{O}\left(\frac{\overline{u_1^4}}{\bar{U}^4}\right)\Bigg]$$

Thus we see that what affects the cooling of the wire is a "mixture" of the two turbulence components u_1 and u_2. We may formally equate this expression with $\bar{U}^n_{\text{meas}}$, which then yields

$$\bar{U}_{\text{meas}}=\bar{U}\left[1+(n-1)\frac{\overline{u_1^2}}{2\bar{U}^2}+\frac{\overline{u_2^2}}{2\bar{U}^2}+(n-1)(n-2)\frac{\overline{u_1^3}}{6\bar{U}^3}\right.$$

$$\left.+(n-2)\frac{\overline{u_1u_2^2}}{2\bar{U}^3}+\mathcal{O}\left(\frac{\overline{u_1^4}}{\bar{U}^4}\right)\right] \tag{2-48}$$

With $n=\frac{1}{2}$ we obtain the relation given by Corrsin[13] with which he corrected the measured value to obtain the actual mean velocity in a free jet. The correction may be either negative or positive, depending mainly on the ratio $\overline{u_1^2}/\overline{u_2^2}$. If the departure from isotropy is small, so that $\overline{u_1^2}\simeq\overline{u_2^2}$, the correction is negative; that is, the actual velocity is smaller than the measured value.

A more practical procedure, however, is the following, based on the actual calibration of the hot-wire anemometer.

Let U_{cal} be the calibration velocity. From Eq. (2-20), assuming a sufficiently long hot-wire, so that $l^*>5$, say, we obtain for the steady state after integration with respect to x:

$$I^2R_w=(R_w-R_g)(A+BU^n)-\frac{\kappa_w\pi d^2}{4b\mathcal{R}_0}\frac{\partial\mathcal{R}_w}{\partial x}\Bigg|_{-l/2}^{+l/2}$$

$$=(R_w-R_g)(A_1+BU^n) \tag{2-49}$$

where

$$A_1=A+\frac{\kappa_w\pi d^2}{bR_0l^2}\mathfrak{C}$$

Hence

$$E_{\text{cal}}=I_{\text{cal}}R_w=[R_w(R_w-R_g)(A_1+BU^n_{\text{cal}})]^{1/2} \tag{2-50}$$

Putting the hot-wire anemometer in a turbulent flow results in a measured voltage

$$E_{\text{meas}}=[R_w(R_w-R_g)(A_1+BU^n_{\text{eff}})]^{1/2} \tag{2-51}$$

Its time-average value is

$$\overline{E_{\text{meas}}}=\overline{[R_w(R_w-R_g)(A_1+BU^n_{\text{eff}})]^{1/2}} \tag{2-52}$$

This is put equal to

$$\overline{E_{\text{meas}}}=[\bar{R}_w(\bar{R}_w-R_g)(A_1+B\bar{U}^n_{\text{meas}})]^{1/2} \tag{2-53}$$

Writing $U_{\text{eff}}^n = \bar{U}^n(1 + u^*)$, where

$$u^* = n\frac{u_1}{\bar{U}} + n(n-1)\frac{\overline{u_1^2}}{2\bar{U}^2} + n\frac{\overline{u_2^2}}{2\bar{U}^2} + n(n-1)(n-2)\frac{\overline{u_1^3}}{6\bar{U}^3}$$

$$+ n(n-2)\frac{\overline{u_1u_2^2}}{2\bar{U}^3} + \mathcal{O}\left(\frac{\overline{u_1^4}}{\bar{U}^4}\right) \tag{2-54}$$

the expression (2-51) for E_{meas} becomes

$$E_{\text{meas}} = \left[\bar{R}_w(\bar{R}_w - R_g)(A_1 + B\bar{U}^n)\left(1 + \frac{r_w}{\bar{R}_w}\right)\left(1 + \frac{r_w}{\bar{R}_w - R_g}\right)(1 + \alpha u^*)\right]^{1/2} \tag{2-55}$$

where

$$\alpha = \frac{B\bar{U}^n}{A_1 + B\bar{U}^n}$$

In the *constant-temperature* operation usually the transconductance is taken so large that

$$g_{tr}\bar{R}_w \gg \frac{R_g}{\bar{R}_w - R_g}$$

Hence

$$\frac{r_w}{\bar{R}_w - R_g} <<< 1 \quad \text{and} \quad \frac{r_w}{\bar{R}_w} <<< 1$$

and these terms may be neglected in Eq. (2-55). The time-averaged value of this reduced Eq. (2-55) put equal to Eq. (2-53) gives

$$[\bar{R}_w(\bar{R}_w - R_g)(A_1 + B\bar{U}^n)]^{1/2}\overline{(1 + \alpha u^*)^{1/2}} = [\bar{R}_w(\bar{R}_w - R_g)(A_1 + B\bar{U}^n_{\text{meas}})]^{1/2}$$

whence follows after some calculation

$$\bar{U}_{\text{meas}} = \bar{U}\left[1 + \frac{2(n-1) - \alpha n}{4}\frac{\overline{u_1^2}}{\bar{U}^2} + \frac{\overline{u_2^2}}{2\bar{U}^2}\right.$$

$$+ \frac{2(n-1)(2n - 3\alpha n - 4) + 3\alpha^2 n^2}{24}\frac{\overline{u_1^3}}{\bar{U}^3}$$

$$\left. + \frac{2n - \alpha n - 4}{4}\frac{\overline{u_1u_2^2}}{\bar{U}^3} + \mathcal{O}\left(\frac{\overline{u_1^4}}{\bar{U}^4}\right)\right] \tag{2-56}$$

This result is slightly different from Eq. (2-48) because of the occurrence of terms with α. For current operating conditions a reasonable value is $\alpha \simeq 0.5$ to 0.7.

The parameter α here is still expressed in terms of the actual velocity $\bar{U}$, whereas actually it should better be expressed in terms of the measured velocity $\bar{U}_{\text{meas}}$:

$$\alpha_m = \frac{b\bar{U}^n_{\text{meas}}}{A_1 + B\bar{U}^n_{\text{meas}}}$$

The relation between α and α_m is given by

$$\alpha = \alpha_m \left[1 + \mathcal{O}\left(\frac{\overline{u_1^2}}{\bar{U}^2} \right) \right]$$

So in the expression (2-56) the difference between α and α_m is felt in the term $\mathcal{O}(\overline{u_1^4}/\bar{U}^4)$.

Let us now consider the hot-wire response to the fluctuations of the velocity, when the latter are not small. The fluctuating voltage is obtained from Eq. (2-55) in which again the terms with r_w are neglected

$$e = E_{\text{meas}} - \bar{E}_{\text{meas}} = [\bar{R}_w(\bar{R}_w - R_g)(A_1 + B\bar{U}^n)]^{1/2}[(1 + \alpha u^*)^{1/2} - \overline{(1 + \alpha u^*)^{1/2}}]$$

The mean-square value of this equation yields the relation between $\overline{e^2}$ and $\overline{u_1^2}$. After a lengthy algebraic calculation we obtain

$$\overline{e^2} = \bar{R}_w(\bar{R}_w - R_g)\frac{B^2\bar{U}^{2n}}{A_1 + B\bar{U}^n} \cdot \frac{n^2}{4}[1 - \varphi(u_1,u_2)] \cdot \frac{\overline{u_1^2}}{\bar{U}^2} \tag{2-57}$$

where

$$\varphi(u_1,u_2) = \frac{2 - (2 - \alpha)n}{2} \frac{\overline{u_1^3}}{\bar{U}\overline{u_1^2}} - \frac{\overline{u_1 u_2^2}}{\bar{U}\overline{u_1^2}}$$

$$- \frac{4(n-1)(7n - 9\alpha n - 11) + 15\alpha^2 n^2}{48} \frac{\overline{u_1^4}}{\bar{U}^2\overline{u_1^2}}$$

$$- \frac{1}{4}\frac{\overline{u_2^4}}{\bar{U}^2\overline{u_1^2}} + \frac{\alpha^2 n^2 - 4(n-1)(\alpha n - n + 1)}{16}\frac{\overline{u_1^2}}{\bar{U}^2} - \frac{2 - (2 - \alpha)n}{4}$$

$$\times \frac{\overline{u_2^2}}{\bar{U}^2} + \frac{1}{4}\frac{(\overline{u_2^2})^2}{\bar{U}^2\overline{u_1^2}} + \frac{10 - 3n(2 - \alpha)}{4}\frac{\overline{u_1^2 u_2^2}}{\bar{U}^2\overline{u_1^2}} + \mathcal{O}\left(\frac{\overline{u_1^3}}{\bar{U}^3}\right) \tag{2-58}$$

Now by definition we may put

$$\overline{e^2} = \bar{R}_w(\bar{R}_w - R_g)\frac{B^2\bar{U}^{2n}_{\text{meas}}}{A_1 + B\bar{U}^n_{\text{meas}}} \cdot \frac{n^2}{4}\left(\frac{\overline{u_1^2}}{\bar{U}^2}\right)_{\text{meas}} \tag{2-59}$$

So that

$$\frac{\overline{u_1^2}}{\bar{U}^2} = \left(\frac{\overline{u_1^2}}{\bar{U}^2}\right)_{\text{meas}} \left(\frac{\bar{U}_{\text{meas}}}{\bar{U}}\right)^{2n} \frac{A_1 + B\bar{U}^n}{A_1 + B\bar{U}^n_{\text{meas}}}[1 - \varphi(u_1,u_2)]^{-1} \tag{2-60}$$

Equations (2-57) to (2-60) show that the relation between the measured and actual values is quite involved. For a correct measurement of the turbulence component

u_1 we must not only know the value of u_2, but also the value of the correlations $\overline{u_1u_2^2}, \overline{u_1^2u_2^2}, \ldots$, which are usually unknown for nonisotropic turbulent flows.

In order to estimate the order of magnitude of the error made by applying the linearized theory, even if the turbulence intensities are not small, we shall calculate the value of $\varphi(u_1,u_2)$ for a one-dimensional turbulence ($u_2 = 0$) and for an isotropic turbulence.

For normal operating conditions of a hot wire, the quantity $\alpha \simeq 0.5$ to 0.7. Taking $\alpha = \frac{1}{2}$ and $n = \frac{1}{2}$, the expression (2-58) for $\varphi(u_1,u_2)$ reads for a one-dimensional turbulence

$$\varphi(u_1) = \frac{5}{8} \cdot \frac{\overline{u_1^3}}{\bar{U}\overline{u_1^2}} - \frac{109}{256} \frac{\overline{u_1^4}}{\bar{U}^2\overline{u_1^2}} + \frac{25}{256} \frac{\overline{u_1^2}}{\bar{U}^2} + \mathcal{O}\left(\frac{\overline{u_1^3}}{\bar{U}^3}\right)$$

Since the turbulence velocity u_1 is a random quantity, we will assume a normal Gaussian distribution, in which case we have

$$\overline{u_1^3} = 0 \qquad \overline{u_1^4} = 3(\overline{u_1^2})^2$$

Hence

$$\varphi(u_1) = -\frac{151}{128} \frac{\overline{u_1^2}}{\bar{U}^2} + \cdots$$

For an isotropic turbulence, $\varphi(u_1,u_2)$ becomes

$$\varphi(u_1,u_2) = -\frac{109}{256} \frac{\overline{u_1^4}}{\bar{U}^2\overline{u_1^2}} - \frac{1}{4} \frac{\overline{u_2^4}}{\bar{U}^2\overline{u_1^2}} + \frac{25}{256} \frac{\overline{u_1^2}}{\bar{U}^2} - \frac{5}{16} \frac{\overline{u_2^2}}{\bar{U}^2}$$
$$+ \frac{1}{4} \frac{(\overline{u_2^2})^2}{\bar{U}^2\overline{u_1^2}} + \frac{31}{16} \frac{\overline{u_1^2u_2^2}}{\bar{U}^2\overline{u_1^2}} + \mathcal{O}\left(\frac{\overline{u_1^3}}{\bar{U}^3}\right)$$

since for an isotropic turbulence $\overline{u_1^3} = \overline{u_1u_2^2} = 0$, as will be shown in Chap. 3.

If we assume not only a Gaussian distribution for u_1 and u_2, but also that these components are normally correlated, we obtain

$$\overline{u_2^2} = \overline{u_1^2} \quad \overline{u_2^4} = \overline{u_1^4} = 3(\overline{u_1^2})^2 \quad \overline{u_1^2u_2^2} = (\overline{u_1^2})^2$$

and the value of $\varphi(u_1,u_2)$ becomes

$$\varphi(u_1,u_2) = -\frac{7}{128} \frac{\overline{u_1^2}}{\bar{U}^2} + \cdots$$

Thus we see that, in the two special cases considered, $\varphi(u_1,u_2)$ has a negative value. This means that the linearized theory would give too high a value for $u_1'/\bar{U}$. The error is of the order of $u_1'^2/\bar{U}^2$ and less for an isotropic turbulence; for $u_1'/\bar{U} = 0.2$, this error would be roughly 5 per cent for a one-dimensional and 0.2 per cent for an isotropic turbulence.

Sometimes the effect of high turbulence intensities on the hot-wire response is studied by vibrating a wire in an air flow of low turbulence intensity, thus simulating a one-dimensional turbulence of high intensity. If we assume that the fluctuating motion relative to the wire is a pure harmonic $u_1 = u_1^* \sin \omega t$, then we obtain, since $\overline{u_1{}^4} = \frac{3}{2} u_1'^4$, $\varphi(u_1, u_2) = -\frac{277}{512} u_1'^2 / \bar{U}^2$, that is, a much smaller value than would be obtained in a one-dimensional randomly fluctuating motion. Hence, such experiments lead to overoptimistic conclusions. At any rate this assumption makes rather dubious the application to actual turbulent flows of high intensity of correction factors obtained in this way.

Summarizing, it may be concluded that the correct measurement of turbulence of high intensity is nearly impossible.

In principle it is possible to apply a theory in which the curvature in the basic calibration curve has been accounted for. However, to apply such a theory, a knowledge of the turbulence structure of the flow is still required.

But if errors of 10 per cent, say, in the measurement of turbulence are accepted, it is permissible to use the linear relations to measure flows with a turbulence intensity of up to 20 or 25 per cent.

Part of the error is due to the nonlinear characteristic, given by Eq. (2-47), of the hot wire. It is possible, in principle, to include a linearizer in the electronic circuit, which accounts for the curvature in the basic calibration curve. However, as Parthasarathy and Tritton[85] have pointed out, such a linearizer does not always improve the measurement.

For instance, with a perfect linearizer, what is measured is the instantaneous effective velocity U_{eff}, given by Eq. (2-46) or Eq. (2-46a). On the other hand without a linearizer Eq. (2-56) would give, assuming $n = \frac{1}{2}$ and $\alpha = \frac{1}{2}$

$$\bar{U}_{\text{meas}} = \bar{U}\left[1 - \frac{5}{16}\frac{\overline{u_1{}^2}}{\bar{U}^2} + \frac{\overline{u_2{}^2}}{\bar{U}^2} + \mathcal{O}\left(\frac{\overline{u_1{}^3}}{\bar{U}^3}\right)\right]$$

which may compare favorably with the measured value according to Eq. (2-46a) with a perfect linearizer.

Also no better result may be obtained with a linearizer in measuring turbulent quantities.

Effect of Compressibility

Some remarks will be made here on the effect of compressibility of the fluid on the measurements of turbulence velocity fluctuations with the hot-wire anemometer. We have to be brief because of the very complexity of the processes affecting the response of the hot-wire anemometer, while on the other hand, due to this complexity, the problem has not yet been solved experimentally in a satisfactory way.

The basic problem arises from the fact that the hot-wire anemometer is sensitive essentially to variations in heat transfer. In high-speed turbulent flow, the velocity fluctuations may be accompanied by density variations and temperature variations of sufficient intensity to affect the response of the hot-wire anemometer. So the instantaneous voltage fluctuation, e, across the wire may be composed of three contributions given respectively by the variations of velocity, density, and temperature separately

$$e = e_u + e_\rho + e_\theta$$

The velocity, density, and temperature fluctuations are not independent, being related, also through the pressure fluctuations, by the mass balance equation, the energy balance equation or the first law of thermodynamics, and the constitutive equation of the fluid. Now any velocity field may be decomposed locally into an irrotational mode, a solenoidal vorticity mode, and a harmonic mode. The harmonic mode is irrotational and solenoidal, but for the case at hand, Morkovin[81,88] suggested including it in the effect of the vorticity mode, since it will be indistinguishable from rotational flow by means of a single hot-wire anemometer.

The irrotational mode appears to be identical with the isentropic sound field. So, as early as 1953 Kovasznay[87] proposed for a correct interpretation of the response of a hot-wire anemometer to make a distinction between a sound mode, a vorticity mode, and an entropy mode. If we denote by p the pressure fluctuations associated with the sound mode, by ω the vorticity fluctuation, and by ς the entropy fluctuation, we then may write for the instantaneous value of the voltage e:

$$e = s_p p + s_\omega \omega + s_\varsigma \varsigma$$

where s_p, s_ω, and s_ς are the corresponding sensitivities of the hot-wire anemometer.

s_p includes the effect of velocity, temperature, and density, s_ω that of velocity and temperature, and s_ς that of temperature and density. They depend in general on the parameters occurring in the general relation (2-2), and on the variable, mainly temperature sensitive, fluid properties. Since with the electronic equipment mean-square values of random variables are measured, we conclude that the mean-square value of the voltage fluctuations $\overline{e^2}$ contains six unknowns, namely the three mean-square values $\overline{p^2}$, $\overline{\omega^2}$, and $\overline{\varsigma^2}$, and the three mean crossproducts $\overline{p\omega}$, $\overline{p\varsigma}$, and $\overline{\omega\varsigma}$, or their corresponding correlation coefficients. It is possible, in principle, to obtain these six unknowns experimentally, e.g., by making measurements at six different overheat temperatures. But this is not a practical procedure, and in fact hitherto no such complete set of calibrations have been made. Fortunately, under certain conditions and depending on the type of flow it will be possible to neglect some of the terms, as has been pointed out by Morkovin.[81,88] This is because fluctuation amplitudes of one mode are much smaller than those of the other modes. Or because correlations between two modes are sufficiently small to warrant their neglect. This, for instance, is often the case between the sound mode and the other two modes.

We will not go further into this matter. For a very complete discussion of the application of hot-wire anemometry to measurements in high-speed turbulent flow, based upon the above concepts, the reader is referred to the AGARDograph No. 24, by Morkovin.[81]

2-4 CONSTANT-CURRENT METHOD

Here the electric current I is kept constant, and it is the change in temperature or electric resistance which is used for measuring a change of the fluid velocity. First we assume again that the fluctuations of the fluid velocity are relatively small, and consequently also the fluctuations of the electric resistance of the hot wire. So we assume

$$\frac{u}{\bar{U}} \ll 1 \quad \text{and} \quad \frac{\mathfrak{r}_w}{\mathscr{R}_w - \mathscr{R}_g} \ll 1$$

The equation for the heat balance of a differential element of the wire is obtained from Eq. (2-22) by putting $i = 0$. The dynamic behavior of the hot wire is obtained by integrating this equation with respect to x. This integrated equation reads

$$\frac{dr_w}{dt} + bR_0 \frac{A + B\bar{U}^n - I^2}{C_w} r_w - \frac{2\kappa_w}{\rho_w c_w}\left[\frac{\partial \mathfrak{r}_w}{\partial x}\right]_0^{l/2} = -\frac{bR_0(\bar{R}_w - R_g)nB\bar{U}^{n-1}}{C_w} \cdot u \qquad (2\text{-}61)$$

Here again we make the approximation

$$-\frac{2\kappa_w}{\rho_w c_w}\left[\frac{\partial \mathfrak{r}_w}{\partial x}\right]_0^{l/2} = \frac{\kappa_w \pi d^2 \mathfrak{C}\, r_w}{C_w\, l}$$

The error made in this way is greater than for the constant-temperature method. But for $l^* = 5$ the error made in the sensitivity of the hot-wire is still smaller than 10 per cent. The advantage of making the above approximation is, as mentioned earlier, that the expressions obtained for the time constant and the sensitivity remain simple, and still applicable for relatively long hot wires ($l^* > 5$, say).

With this approximation Eq. (2-61) becomes

$$\frac{dr_w}{dt} + \frac{bR_0 I^2}{C_w} \frac{R_g}{\bar{R}_w - R_g} r_w = -\frac{bR_0(\bar{R}_w - R_g)nB\bar{U}^{n-1}}{C_w} u \qquad (2\text{-}62)$$

So the expression for the time constant reads

$$M_{\text{c.c.}} = \frac{C_w}{bR_0 \bar{I}^2} \frac{\bar{R}_w - R_g}{R_g} \qquad (2\text{-}63)$$

This expression can also be obtained from the corresponding Eq. (2-37) for $M_{\text{c.t.}}$ by putting $g_{tr} = 0$. Thus the two time constants are related by the equation

$$M_{\text{c.c.}} = \left[1 + 2\frac{\bar{R}_w - R_g}{R_g} \bar{R}_w g_{tr}\right] \cdot M_{\text{c.t.}} \qquad (2\text{-}64)$$

The sensitivity of the hot wire, defined again by Eq. (2-40) in the case of a purely harmonic fluctuation of the velocity becomes

$$S_{\text{c.c.}}(\omega) = \frac{(\bar{R}_w - R_g)^2 nB\bar{U}^{n-1}}{IR_g\sqrt{1 + \omega^2 M_{\text{c.c.}}{}^2}} \qquad (2\text{-}65)$$

Since $M_{\text{c.c.}} \gg M_{\text{c.t.}}$ the effect of the frequency on the sensitivity $s_{\text{c.c.}}$ is certainly not negligibly small. Comparing it with the corresponding Eq. (2-42) for $s_{\text{c.t.}}$, we obtain, if the same value of the electric current I is assumed,

$$s_{\text{c.c.}} = 2\frac{\bar{R}_w - R_g}{R_g}\sqrt{\frac{1 + \omega^2 M_{\text{c.t.}}{}^2}{1 + \omega^2 M_{\text{c.c.}}{}^2}} \cdot s_{\text{c.t.}} \tag{2-66}$$

So for $\bar{R}_w/R_g > 3/2$ at low frequencies the sensitivity at the constant-current operation is higher than the sensitivity at constant-temperature operation. But at higher frequencies the constant-temperature operation soon gives a much higher sensitivity, as may be expected. It must be noted that the simple expression (2-66) has been obtained by making the simplifying assumption concerning similarity, or "steady-state" distributions of temperature along the wire. When this assumption cannot be made, i.e., for relatively short hot wires then the expressions become much more involved.

As stated above, since for the constant current operation the time constant $M_{\text{c.c.}}$ is not very small, the effect of the frequency on the sensitivity cannot be neglected, even at moderate frequencies. So for a tungsten wire of 5 μm diameter, under normal operating conditions the amplitude of the velocity fluctuations with a frequency as low as 250 s^{-1} will be measured 10 per cent too low and the phase shift would amount to $\sim 25°$.

For exact measurement of these fluctuations and fluctuations with higher frequencies, an electronic compensator must be used that compensates at least for the reduction factor $(1 + \omega^2 M^2)^{-1/2}$ due to thermal inertia.

Dryden and Kuethe[5] were among the first to realize the importance of compensation for thermal-inertia lag and to use an electronic compensation circuit in their equipment.

Usually compensation for the amplitude only is provided for, and this is sufficient not only for measuring the turbulence intensity and the turbulence-energy spectrum, but also for measuring the correlation functions that are Fourier transforms of the energy-spectrum functions.

However, if an undistorted oscillogram of the velocity fluctuations is required, the phase shift also must be reduced to zero by proper compensation.

As already noted, without compensation the intensity will be measured too low. Moreover, it appears that measurement of spatial correlations by means of two hot wires without compensation yields too flat a correlation curve, particularly in the neighborhood of the vertex, so that too large a value of the dissipation scale λ is obtained.[4] Since the intensity divided by the dissipation scale is a measure of the dissipation, too low a value of the dissipation will be obtained.

The integral scale is also affected and is also measured too large, but this effect is much less pronounced than the exaggeration of the dissipation scale.

Effect of Large Turbulence Fluctuations

We will first consider the effect on the measurement of the mean velocity $\bar{U}$. To this end we start from Eq. (2-49), where now the electric current I is taken to maintain a constant value, and U_{eff} has to be read for U. Putting

$$R_w = \bar{R}_w + r_w \quad \text{and} \quad U_{\text{eff}}^n = \bar{U}^n(1 + u^*)$$

where u^* is given by Eq. (2-54), Eq. (2-49) becomes

$$I^2(\bar{R}_w + r_w) = (\bar{R}_w - R_g + r_w)[A_1 + B\bar{U}^n(1 + u^*)]$$

Take the average value of this equation, and subtract this averaged equation from the original non-averaged equation. The following result for r_w then is obtained

$$r_w = [\overline{u^* r_w} - (\bar{R}_w - R_g)(u^* - \overline{u^*})] \frac{\beta}{1 - \beta u^*} \tag{2-67}$$

where

$$\beta = \frac{B\bar{U}^n}{A_1 + B\bar{U}^n - I^2}$$

From this equation the value of $\overline{u^* r_w}$ can be obtained

$$\overline{u^* r_w} = (\bar{R}_w - R_g)[A_1 + B\bar{U}^n(1 + \overline{u^*} - \beta\overline{u^*}^2 + \beta\overline{u^{*2}} + 2\beta^2\overline{u^{*3}} - 2\beta^2\overline{u^*}\,\overline{u^{*2}} + \cdots)]$$

Now

$$\bar{E}_{\text{meas}} = I\bar{R}_w = [\bar{R}_w(\bar{R}_w - R_g)\{A_1 + B\bar{U}^n(1 + \overline{u^*})\} + B\bar{U}^n\,\overline{u^* r_w}]^{1/2}$$

This is put equal to

$$[\bar{R}_w(\bar{R}_w - R_g)(A_1 + B\bar{U}^n_{\text{meas}})]^{1/2}$$

Whence follows after some calculation, and making use of the relation (2-54) for u^*:

$$\bar{U}_{\text{meas}} = \bar{U}\left[1 + \left(\frac{n-1}{2} - \beta n\right)\frac{\overline{u_1^2}}{\bar{U}^2} + \frac{\overline{u_2^2}}{2\bar{U}^2} + \left\{\frac{(n-1)(n-2)}{6} - \beta n(n-1) + 2\beta^2 n^2\right\}\frac{\overline{u_1^3}}{\bar{U}^3} + \left(\frac{n-2}{2} - \beta n\right)\frac{\overline{u_1 u_2^2}}{\bar{U}^3} + \mathcal{O}\left(\frac{\overline{u_1^4}}{\bar{U}^4}\right)\right] \tag{2-68}$$

In the expression between brackets, $\bar{U}$ and the parameter β are not yet known. However, since the correction to $\bar{U}_{\text{meas}}/\bar{U}$ is of the $\mathcal{O}(\overline{u_1^2}/\bar{U}^2)$, and also

$$\beta = \beta_m\left[1 + \mathcal{O}\left(\frac{\overline{u_1^2}}{\bar{U}^2}\right)\right]$$

where

$$\beta_m = \frac{B\bar{U}^n_{\text{meas}}}{A_1 + B\bar{U}^n_{\text{meas}} - I^2}$$

the difference obtained when in the expression between brackets $\bar{U}_{\text{meas}}$ and β_m are taken instead of $\bar{U}$ and β respectively, is of the

$$\mathcal{O}\left(\frac{\overline{u_1^4}}{\bar{U}^4}\right)$$

The mean square of the voltage fluctuations

$$\overline{e^2} = I^2\overline{r_w^2}$$

is obtained by taking the mean square of Eq. (2-67), thereby expressing again u^* in u_1 and u_2, according to Eq. (2-54).

The result is

$$\overline{e^2} = \beta^2 I^2(\bar{R}_w - R_g)^2 n^2 \frac{\overline{u_1^2}}{\bar{U}^2}[1 - \varphi_1(u_1,u_2)] \qquad \text{(2-69)}$$

where

$$\begin{aligned}\varphi_1(u_1,u_2) = {} & \{2\beta n - (n-1)\}\frac{\overline{u_1^3}}{\bar{U}\overline{u_1^2}} - \frac{\overline{u_1u_2^2}}{\bar{U}\overline{u_1^2}} - \left\{\frac{(n-1)(7n-11)}{12}\right. \\ & \left. - 3\beta(n-1) + 3\beta^2 n^2\right\}\frac{\overline{u_1^4}}{\bar{U}^2\overline{u_1^2}} + \left\{\frac{(n-1)^2}{4} - 2\beta(n-1)\right. \\ & \left. + 3\beta^2 n^2\right\}\frac{\overline{u_1^2}}{\bar{U}^2} - \frac{1}{4}\frac{\overline{u_2^4}}{\bar{U}^2\overline{u_1^2}} + \frac{1}{4}\frac{(\overline{u_2^2})^2}{\bar{U}^2\overline{u_1^2}} + \left\{\frac{n-1}{2} - 2\beta n + 3\beta\right\}\frac{\overline{u_2^2}}{\bar{U}^2} \\ & - \frac{3n-5}{2}\frac{\overline{u_1^2u_2^2}}{\bar{U}^2\overline{u_1^2}} + \mathcal{O}\left(\frac{\overline{u_1^3}}{\bar{U}^3}\right)\end{aligned} \qquad \text{(2-70)}$$

By definition we put Eq. (2-69) equal to

$$\overline{e^2} = \beta_m^2 I^2(\bar{R}_w - R_g)^2 n^2\left(\frac{\overline{u_1^2}}{\bar{U}^2}\right)_{\text{meas}}$$

Whence follows:

$$\left(\frac{\overline{u_1^2}}{\bar{U}^2}\right)_{\text{meas}} = \frac{\beta^2}{\beta_m^2}\frac{\overline{u_1^2}}{\bar{U}^2}[1 - \varphi_1(u_1,u_2)]$$

For β/β_m the following relation can be obtained

$$\frac{\beta}{\beta_m} = 1 - (1-\beta_m)\overline{u^*} + \beta_m(1-\beta_m)\overline{u^{*2}}\ldots$$

$$= 1 - (1 - \beta_m)\left\{\frac{n(n-1)}{2} - \beta_m n^2\right\}\frac{\overline{u_1^2}}{\bar{U}^2} - (1 - \beta_m)n\frac{\overline{u_2^2}}{2\bar{U}^2} + \mathcal{O}\left(\frac{\overline{u_1^3}}{\bar{U}^3}\right)$$

Hence

$$\left(\frac{\overline{u_1^2}}{\bar{U}^2}\right)_{\text{meas}} = \frac{\overline{u_1^2}}{\bar{U}^2}\left[1 - \varphi_1(u_1,u_2) - (1 - \beta_m)\{n(n-1) - 2\beta_m n^2\}\frac{\overline{u_1^2}}{\bar{U}^2} - (1 - \beta_m)n\frac{\overline{u_2^2}}{\bar{U}^2} + \mathcal{O}\left(\frac{\overline{u_1^3}}{\bar{U}^3}\right)\right] \tag{2-71}$$

where, again, in the expression between brackets we may take $\bar{U}_{\text{meas}}$ for $\bar{U}$.

In order to estimate the order of magnitude of the error involved when using the linearized theory, we will again consider a one-dimensional turbulence and an isotropic turbulence assuming Gaussian distributions for u_1 and u_2. Taking $n = \frac{1}{2}$ and $\beta_m = \frac{3}{4}$, the values for $\varphi_1(u_1,u_2)$ obtained are

for the one-dimensional turbulence:

$$\varphi_1(u_1) = -\frac{139}{32}\frac{\overline{u_1^2}}{\bar{U}^2}$$

for the isotropic turbulence:

$$\varphi_1(u_1,u_2) = -\frac{59}{32}\frac{\overline{u_1^2}}{\bar{U}^2}$$

A comparison with the corresponding values of $\varphi(u_1,u_2)$ for the constant-temperature method, shows that in this respect the constant-temperature method yields smaller errors.

Though more complicated, it is also possible to build a linearizer into the electronic circuit used in the constant-current method. But, here too, just as in the case of the constant-temperature method this does not always yield an improvement. For instance, measuring the mean-velocity a smaller error is made according to Eq. (2-68) when the turbulence is nearly isotropic than according to Eq. (2-46a), with a perfect linearizer.

Another aspect not considered hitherto, and on which Comte-Bellot and Schon[163] have focused attention, is that thermal lag and large-amplitude velocity fluctuations generate higher harmonics in the hot-wire response at constant-current operations. So large errors may be obtained in the measured energy spectra, in particular in the high-wavenumber, viscous, range.

Effect of Nonlinearity Due to Temperature Dependence of Factors

In the Eq. (2-12) there are several factors whose temperature dependence may give rise to nonlinearity effects.

The electric resistance R_w of the wire is only approximately linear in its temperature dependence. If we take into account the third, quadratic, term on the right-hand side of the relation (2-11) for R_w, the Eq. (2-12) becomes nonlinear in $\Theta_w - \Theta_g$.

Fortunately, the second coefficient b_1 is very small compared with b for the materials used in hot-wire anemometers (see, for instance, the values given for tungsten and platinum); so, under the normal operating temperatures of the hot wire, the quadratic term in Eq. (2-11) is still very small compared with the linear term and may certainly be neglected.

The coefficients A and B in Eq. (2-17) have the form

$$A = \frac{\pi l \kappa_f}{bR_0} f_1(\mathbf{Pr}_f)$$

$$B = \frac{\pi l \kappa_f}{bR_0}\left(\frac{\rho_f d}{\mu_f}\right)^n f_2(\mathbf{Pr}_f)$$

while in addition they may still be a function of the temperature ratio Θ_f/Θ_g (see, e.g., Eq. (2-8a)).

The specific heat c_p is practically independent of the temperature, but the heat conductivity, the viscosity, and the density are very much dependent upon it. Since the heat conductivity and the viscosity vary in essentially the same way with temperature, namely, approximately linearly within a rather large range of temperatures, the Prandtl group is practically independent of temperature. For an exponent n not very much different from the value 0.5, the group $\kappa_f(\rho_f/\mu_f)^n$ is only slightly dependent on the temperature; for diatomic gases, and air, very slightly so.

Hence, of the two factors A and B of the relation (2-17), it is practically the factor A, which is proportional to κ_f, that is sensitive to temperature changes.

If we approximate the temperature dependence of κ_f by introducing a linear relation, we may write

$$\kappa_f = a + a_1\Theta_f$$

For air

$$a = 5 \times 10^{-3} \text{ watt/(m)(°C)}$$

$$a_1 = 7 \times 10^{-5} \text{ watt/(m)(°C)}^2$$

In order to be more specific, consider Kramers's Eq. (2-6).

The temperature dependence of A_f then reads

$$A_f = 0.42\frac{\pi l}{bR_0}\mathbf{Pr}^{0.2}(a + a_1\Theta_f)$$

$$= A_g\left[1 + \frac{a_1}{2(a + a_1\Theta_g)}(\Theta_w - \Theta_g)\right]. \qquad (2\text{-}72)$$

where

$$A_g = 0.42 \frac{\pi l}{bR_0} \mathbf{Pr}^{0.2} (a + a_1 \Theta_g)$$

For air and $\Theta_g = 300$ K, we obtain

$$A_f = A_g[1 + 0.0013(\Theta_w - \Theta_g)]$$

With the relation (2-72) for A_f, instead of (2-17), we obtain

$$\frac{I^2 R_w}{R_w - R_g} = A_g \left[1 + \frac{a_1}{2(a + a_1 \Theta_g)} (\Theta_w - \Theta_g) \right] + B\sqrt{U}$$

$$= A_g \left[1 + \frac{a_1}{2(a + a_1 \Theta_g) b R_0} (R_w - R_g) \right] + B\sqrt{U} \qquad (2\text{-}73)$$

Betchov and Welling[11,12] investigated theoretically as well as experimentally the effect of this nonlinearity for the constant-current operation, where it is of more importance than for the constant-temperature operations. The result of Betchov's calculations is to obtain a more uniform temperature distribution along the wire than is obtained according to the "linear" theory. This sounds reasonable, since the quadratic term causes greater heat-transfer rates from the wire to the flowing fluid at the parts of higher temperature than at those of lower temperature. Because of the more uniform temperature distribution the time-constant will be smaller. The few experiments performed have confirmed the nonlinear effect, at least qualitatively.

2-5 MEASUREMENT OF TURBULENCE CHARACTERISTICS WITH THE HOT-WIRE ANEMOMETER

In this section we will consider methods for measuring the intensities of the three components of the turbulence velocity, their one-point and two-point correlations, dissipation scales, and integral scales. But first we will consider the effect of finite wire length with respect to the size of the eddies, or in other words the effect of a *non-uniform velocity distribution along the wire*. For, all the foregoing considerations have presupposed a uniform velocity distribution along the wire. Now we know that in turbulent flow the velocity distributions are not uniform in regions down to the smallest scale of turbulence. Hence, for making true "point" measurements, the wire length of the hot-wire anemometer should not exceed, say, 0.5 mm, even for turbulent flows of moderate average velocity. Accordingly, for aerodynamic efficiency of the hot-wire anemometer, the wires should be as short as possible. On the other hand, we have shown that, owing to the cooling effect of the supports, the wire cannot be too short or the temperature distribution along the wire will vary too widely. A length-diameter ratio of not less than 200 was recommended for platinum-iridium wires and an even higher ratio for tungsten wires.

In order to meet this latter requirement, wires shorter than 0.5 mm must have a diameter smaller than 2.5 μm. This diameter, however, already approaches the minimum value for the minimum strength required, though wires of 1 μm diameter have already been used with success. Another point is, that such short wires require a special design, in order that the supports do not cause an adverse aerodynamic interference with the flow around the wire. For instance, the supports may be placed much wider apart than 0.5 mm, and a coated, and therefore stronger, wire used, of which only a central part of 0.5 mm length is left uncoated.

Now if the velocity distribution in regions smaller than the wire length deviates appreciably from uniformity—which it does quite soon with increasing velocities of turbulent flow—errors will be made in the measurement of turbulence characteristics. Skramstad[9] was the first not only to appreciate this effect but also to give correction formulas for the intensity of turbulence and for the correlation coefficients when two hot wires are used.

Consider the case in which the intensity of the axial component u_1 of the turbulence velocity has to be measured; then the wire is placed perpendicular to the average velocity $\bar{U}_1$. If the velocity distribution were uniform along the wire, the value of the mean square of the voltage across the wire would read

$$\overline{e^2} = K^2 l^2 \overline{u_1{}^2} \tag{2-74}$$

where K is a constant depending on the wire characteristics and on the operating conditions.

Conversely, the value of $\overline{u_1{}^2}$ calculated with this expression for $\overline{e^2}$ would be correct if a true "point" measurement were made by the wire.

Assuming that the turbulent flow was homogeneous and isotropic, Skramstad showed that, because of the actually nonuniform velocity distribution for which, for instance, the double Eulerian correlation coefficient $\mathbf{g}(x_2)$ (see Sec. 1-6) is a measure, the measured value of the square of the voltage must read

$$(\overline{e^2})_{\text{meas}} = 2K^2 \overline{u_1'^2} \int_0^l dx_2\,(l - x_2)\mathbf{g}(x_2) \tag{2-75}$$

where x_2 is the coordinate along the wire.

We may introduce a correction factor $\mathscr{C}$, defined by

$$\overline{e^2} = \mathscr{C}(\overline{e^2})_{\text{meas}} \tag{2-76}$$

From (2-74) and (2-75) we then obtain

$$\frac{1}{\mathscr{C}} = \frac{2}{l^2}\int_0^l dx_2\,(l - x_2)\mathbf{g}(x_2) \tag{2-77}$$

For evaluating $\mathscr{C}$ with this formula we may consider three cases: (1) l very large compared with the integral scale Λ_g, (2) l comparable to Λ_g, and (3) l very small compared with Λ_g.

If $l \gg \Lambda_g$, it follows from (2-77) that $\mathscr{C}$ increases indefinitely as l/Λ_g increases, since $1/\mathscr{C}$ then becomes proportional to Λ_g/l. An infinitely long wire would measure no turbulence at all!

Before considering the case $l \simeq \Lambda_g$, let us first consider the case $l \ll \Lambda_g$, which may correspond to the case in which l is smaller than the dissipation scale λ_g. The shape of the correlation curve for such small values of x_2 is parabolic (see Sec. 1-6):

$$\mathbf{g}(x_2) \simeq 1 - \frac{x_2^2}{\lambda_g^2} \tag{1-58}$$

The integral (2-77) then yields

$$\mathscr{C} = \left(1 - \frac{l^2}{6\lambda_g^2}\right)^{-1}$$

From this result one might conclude that no correction is needed when $l < 0.5\lambda_g$, say. However, this does not mean that an effective point measurement can be made. For, though the dissipation scale may be considered as a measure for the smaller eddies responsible for the dissipation, it is by no means a measure for the smallest eddies, i.e., for the finest structure of the turbulence. For that the so-called Kolmogoroff length scale, defined by $\eta = [\nu^3/\varepsilon]^{1/4}$, is a more appropriate measure. (See Chap. 3.) So the condition of a point measurement would better be approached by putting $l < \eta$. By making use of the relation between η and λ_g, given in Chap. 3, the above expression for the correction factor $\mathscr{C}$ can be modified accordingly.

For the remaining case to be considered, $l \simeq \Lambda_g$, the correction factor will depend on the shape of the correlation curve. Skramstad had already considered an exponential function for $\mathbf{g}(x_2)$. In addition we will also consider a Gaussian error function for it. When

$$\mathbf{g}(x_2) = \exp\left(-\frac{|x_2|}{\Lambda_g}\right)$$

we obtain

$$\mathscr{C} = 0.5\frac{l^2}{\Lambda_g l - \Lambda_g^2[1 - \exp(-l/\Lambda_g)]}$$

By developing the exponential function in a Taylor series, this expression can be approximated by the simpler expression

$$\mathscr{C} \simeq 1 + \frac{1}{3}\frac{l}{\Lambda_g} - \cdots$$

when $l/\Lambda_g < 1$.

When $\mathbf{g}(x_2) = \exp(-\pi x_2^2/4\Lambda_g^2)$, the correction factor reads

$$\frac{1}{\mathscr{C}} = 2\frac{\Lambda_g}{l}\operatorname{erf}\left(\frac{\sqrt{\pi}}{2}\frac{l}{\Lambda_g}\right) - \frac{4}{\pi}\left(\frac{\Lambda_g^2}{l^2}\right)\left[1 - \exp\left(-\frac{\pi l^2}{4\Lambda_g^2}\right)\right]$$

where

$$\operatorname{erf}\left(\frac{\sqrt{\pi}}{2}\frac{l}{\Lambda_g}\right) = \frac{2}{\sqrt{\pi}}\int_0^{(\sqrt{\pi}/2)(l/\Lambda_g)} dy \exp(-y^2)$$

For the measured value of the lateral-correlation coefficient between the velocity fluctuation u_1 at two points at a distance of x_3 from each other, Skramstad obtains

$$[\mathbf{g}(x_3)]_{\text{meas}} = \frac{2\mathscr{C}}{l^2}\int_0^l dx_2 (l - x_2)\mathbf{g}(\sqrt{x_2{}^2 + x_3{}^2})$$

This integral can be evaluated when $\mathbf{g}$ follows either an exponential function or the Gaussian error function.

If $\mathbf{g}(x_3)$ follows a Gaussian error function, the curious result is obtained that

$$[\mathbf{g}(x_3)]_{\text{meas}} = \mathbf{g}(x_3)$$

and so the correlation function is correctly measured.

In the case in which $\mathbf{g}(x_3) = \exp(-x_3/\Lambda_g)$, a rather complicated expression is obtained for $[\mathbf{g}(x_3)]_{\text{meas}}$. A very simple approximate expression, however, is obtained if it is assumed that $l/x_3 \ll 1$, so that $l/\Lambda_g \ll 1$ also; i.e.,

$$\mathbf{g}(x_3) \simeq \frac{1}{\mathscr{C}}[g(x_3)]_{\text{meas}} \simeq \left(1 - \frac{1}{3}\frac{l}{\Lambda_g}\right)[\mathbf{g}(x_3)]_{\text{meas}}$$

From this relation we obtain

$$\frac{\Lambda_g}{(\Lambda_g)_{\text{meas}}} \simeq 1 - \frac{1}{3}\frac{l}{(\Lambda_g)_{\text{meas}}} - \cdots$$

If, instead of the lateral-correlation coefficient $\mathbf{g}$, the longitudinal-correlation coefficient $\mathbf{f}$ is measured, for a homogeneous, isotropic turbulent flow we may make use of the relation between $\mathbf{f}$ and $\mathbf{g}$, which will be given in Chap. 3, namely, Eq. (3-9).

Instead of Eq. (2-77) we then obtain

$$\frac{1}{\mathscr{C}} = \frac{1}{l}\int_0^l dx_2\, \mathbf{f}(x_2)$$

and

$$\mathbf{f}(x_1) = \frac{\mathscr{C}}{l}\int_0^l dx_2\, \mathbf{f}(\sqrt{x_1{}^2 + x_2{}^2})$$

The same expressions as those obtained by Skramstad were deduced by Frenkiel, though in a different way. Moreover, Frenkiel extended the calculations considerably by considering the more general case of nonisotropic but still homogeneous turbulence. The reader may find a full account of Frenkiel's calculations in Ref. 10.

Turning now to the methods for turbulence measurements mentioned above, we assume a uniform mean velocity $\bar{U}$ with the turbulence component u_1 in the same direction, u_2 and u_3 in the lateral directions.

We further assume that the turbulence fluctuations are small compared with $\bar{U}$ and that there are no compressibility effects, so that the simplified linear relations for the heat loss hold good.

Measurement of Turbulence Intensities

The intensity of the axial turbulence-velocity component $u_1' = \sqrt{\overline{u_1^2}}$ can be measured quite simply by leading the voltage signal $e = \pm s u_1$ via an amplifier, with or without compensation for any thermal lag, to a thermocouple; this thermocouple measures directly the mean-square value of the current that corresponds to e.

The plus sign refers to the constant-temperature operation. The minus sign refers to the constant-current operation. No distinction in notations by means of subscripts will be made here between the constant-temperature and the constant-current operation. Because, once introducing the sensitivity, in what follows no distinction is necessary between the two methods of operation. So when the constant-temperature operation is used the expression (2-42), or (2-43), for s has to be taken, when the constant-current operation is used the expression (2-65).

The intensity of the lateral turbulence-velocity components u_2' and u_3' can be measured by making use of the direction-sensitivity of the hot wire.

The rate of cooling of a hot wire is a maximum for a given fluid velocity when the wire is placed perpendicular to the direction of this velocity. Thus, if the wire makes an angle φ with the velocity vector (see Fig. 2-3a) we may write

$$U_{\text{eff}} = U f(\varphi) \qquad (2\text{-}78)$$

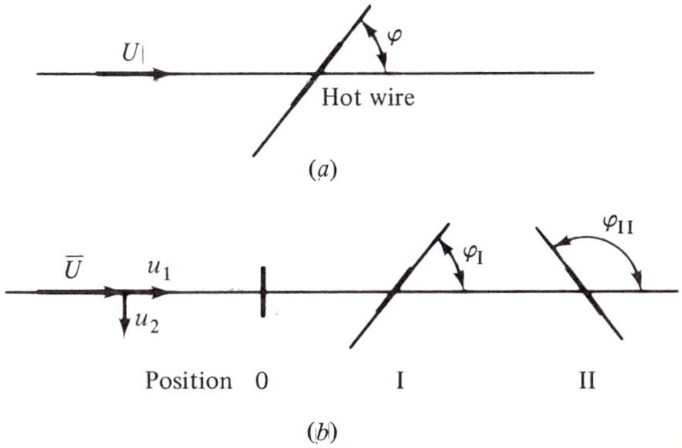

FIGURE 2-3
Measurement of turbulence velocities u_1 and u_2 with a hot-wire anemometer.

where $f(\pi/2) = 1$. The direction characteristic of the hot wire $f(\varphi)$ decreases with decreasing value of φ. Since we consider relatively small turbulence fluctuations only, so that we may restrict ourselves to first-order effects, the hot-wire anemometer placed in the (x_1, x_2) plane in oblique position (see Fig. 2-3b) will only respond to the u_1 and u_2 components of the turbulence.

The voltage fluctuation, e, then is given by

$$e = \frac{\overline{\partial E}}{\partial U} u_1 + \frac{\overline{\partial E}}{\partial \varphi} d\varphi = \frac{\overline{\partial E}}{\partial U} u_1 + \frac{\overline{\partial E}}{\partial \varphi} \frac{u_2}{U}$$

$$= \pm s_1(\varphi) u_1 \pm s_2(\varphi) u_2 \qquad (2\text{-}79)$$

where again the minus sign refers to the constant-current operation.

Introduction of the direction-sensitivity characteristic $f(\varphi)$, yields for the sensitivities $s_1(\varphi)$ and $s_2(\varphi)$ respectively:

$$s_1(\varphi) = s\left(\frac{\pi}{2}\right) \cdot f^n \qquad (2\text{-}80a)$$

$$s_2(\varphi) = s\left(\frac{\pi}{2}\right) \cdot f^{n-1} \frac{df}{d\varphi} \qquad (2\text{-}80b)$$

So if by means of a calibration $f(\varphi)$ is known, the sensitivities $s_1(\varphi)$ and $s_2(\varphi)$ for the u_1 and u_2 component respectively can be determined.

Now for a hot wire the cooling is mainly caused by the component $U \sin \varphi$. Only for small values of φ does the component $U \cos \varphi$ also assume importance. The effective value of U can be obtained approximately from the relation

$$U_{\text{eff}}^2 = U^2 (\sin^2 \varphi + A^2 \cos^2 \varphi) \qquad (2\text{-}81)$$

where the factor A has a small value.

Early experimental results, mentioned in the first edition of this book, indicated that A varies between 0.1 and 0.3, increasing with decreasing velocity. Since then several investigators have conducted more or less systematic and extensive experiments on the direction sensitivity of the hot-wire anemometer. Though the values of A obtained are of the same, small, order of magnitude as the values mentioned above, no general agreement was reached concerning the effect of the different operating parameters concerned. Here we will mention only a few results obtained. Webster[89] experimented with 2.5 μm platinum wires with l/d between 86 and 1456, and air velocities varying between 4 and 6.5 m/s. He obtained a value of $A \simeq 0.2$, with no systematic effect of either the velocity or the l/d ratio.

In contrast to this Champagne[90,159] concluded from his experiments that A depends on those parameters that affect the temperature distribution along the wire, the l/d ratio being the most important parameter. For a 5 μm platinum wire and an air velocity of 30 m/s, A decreased from 0.2 at $l/d = 200$ to almost zero at $l/d = 600$. But also much higher values of A have been observed. Hill[160] reports values up to 0.9 for short cylindrical quartz-coated hot-film probes in mercury. Recent systematic investigations by Leijdens[91] confirm the rather strong dependence of A on l/d, though this dependence did not yet reach a zero value at $l/d = 1000$, A still having a value of roughly 0.04 for a 5 μm tungsten wire. Moreover, a marked effect of the air velocity, i.e., of the Reynolds number was found, A decreasing with increasing velocity, while also A increased with increasing overheat ratio $\Delta\Theta_w/\Theta_g$. Depending on l/d ratio, Reynolds number, and overheat ratio, A varied between roughly 0.04 and 0.3. For these, small, values of A, we may write for $f(\varphi)$ defined by Eq. (2-78), and according to the relation (2-81)

$$f(\varphi) \simeq \sin\varphi \cdot (1 + \tfrac{1}{2}A^2 \operatorname{cotan}^2 \varphi)$$

Fujita and Kovasznay[92] proposed a different function, namely

$$f(\varphi) = \sin\varphi \cdot \left[1 + A_1\left(1 + \frac{\cos 2\varphi}{\sin\varphi}\right)\right]$$

for the range $30° < |\varphi| \le 90°$. The coefficient A_1 is slightly dependent on the velocity, but in contrast to the former coefficient A, increases with increasing velocity. For the low speed range between 2 m/s and 12 m/s it varied between 0.01 and 0.05, roughly. Since $\operatorname{cotan}^2 \varphi$ varies more with φ than $(1 + \cos 2\varphi/\sin\varphi)$, the correction to be made when using the simple $\sin\varphi$-relation is slightly greater according to the first function of φ.

There are at least three causes which affect the non-zero heat transfer by forced convection in a position of the hot wire parallel to the flow. First, for a wire of finite length the axial flow component causes a boundary layer along the wire with the result that the downstream part of the wire tends to have a lower rate of heat transfer, which will be compensated by a higher local wire temperature. Thus an asymmetric temperature distribution. Second, due to the skewed temperature distribution there is, though small still, an integrated Thomson effect. Third, the skewness of the temperature distribution may be increased by the shadow and wake effect of the upstream support of the hot wire. The two supports should be really very wide apart in order to avoid this effect. In general the supports and the holder of a hot wire have a much greater influence on the total heat-transfer characteristic than assumed by many investigators. Even in the much more favorable situation of a hot wire placed perpendicular to the flow direction there can be an effect of the supports and holder on the total heat-transfer rate of the wire depending on the angle between the supports and the flow direction.[93]

When the angle φ is taken not smaller than $|45°|$, the effect of the tangential velocity component at an oblique position of the wire on U_{eff} may be small. Though as Champagne[90,161] has pointed out, with $\varphi = 45°$ and $A = 0.2$ errors up to 20 per cent may be made in the measurement of $\overline{u_2{}^2}/\bar{U}^2$ and up to 10 per cent in the measurement of $\overline{u_1u_2}/\bar{U}^2$ by assuming $A = 0$ instead of the value 0.2. However, by employing hot wires with $l/d > 200$, and suitable, low, overheat ratios, A can be kept sufficiently small in the usual operating range of $\mathbf{Re}_d$ between 2 and 15 say, so as to make it permissible to consider only the component $U \sin \varphi$ in the practical range $45° < \varphi \leq 90°$. Hence, with $f(\varphi) = \sin \varphi$ in Eq. (2-78) the expressions for the sensitivities $s_1(\varphi)$ and $s_2(\varphi)$ in Eq. (2-79), and according to the Eqs. (2-80a) and (2-80b), read

$$s_1(\varphi) = s\left(\frac{\pi}{2}\right) |\sin^n \varphi| \tag{2-82a}$$

$$s_2(\varphi) = s\left(\frac{\pi}{2}\right) |\sin^n \varphi| \operatorname{cotan} \varphi \tag{2-82b}$$

According to Eq. (2-79) we measure a mixture of the effects of the two components u_1 and u_2 when the wire is in an oblique position φ. In order to separate these effects we determine the hot-wire responses in positions 0, I, and II, as shown in Fig. 2-3b. We have, for the *constant-current* operation, but with similar results for the constant-temperature operation:

In position 0:

$$(e)_0 = -(s_1)_0 u_1$$

In position I:

$$(e)_{\text{I}} = -(s_1)_{\text{I}} u_1 - (s_2)_{\text{I}} u_2$$

In position II:

$$(e)_{\text{II}} = -(s_1)_{\text{II}} u_1 + (s_2)_{\text{II}} u_2$$

If we lead the signals obtained in the various positions to a thermocouple, we obtain

$$\begin{aligned}
(\overline{e^2})_0 &= (s_1{}^2)_0 \overline{u_1{}^2} \\
(\overline{e^2})_{\text{I}} &= (s_1)_{\text{I}}{}^2 \overline{u_1{}^2} + (s_2)_{\text{I}}{}^2 \overline{u_2{}^2} + 2(s_1)_{\text{I}}(s_2)_{\text{I}} \overline{u_1 u_2} \\
(\overline{e^2})_{\text{II}} &= (s_1)_{\text{II}}{}^2 \overline{u_1{}^2} + (s_2)_{\text{II}}{}^2 \overline{u_2{}^2} - 2(s_1)_{\text{II}}(s_2)_{\text{II}} \overline{u_1 u_2}
\end{aligned} \tag{2-83}$$

These equations can be solved for the components u_1' and u_2' and, moreover, for the quantity $\overline{u_1 u_2}$.

Fujita and Kovasznay[92] have successfully applied a method where a single hot wire is uniformly rotated by a motor with an angular velocity of, say 2°/s. Continuous recordings are made with two $X - Y$ recorders of $\bar{E}$ and $\overline{e^2}$ as a function of φ.

Since

$$\overline{e^2}(\varphi) = s^2\left(\frac{\pi}{2}\right)\left[f^{2n}(\varphi) \cdot \overline{u_1{}^2} + 2f^{2n-1} \frac{df}{d\varphi} \overline{u_1 u_2} + f^{2n-2}\left(\frac{df}{d\varphi}\right)^2 \overline{u_2{}^2}\right]$$

a large number of equations can be obtained with coefficients given by the pertinent values of φ, for solving $\overline{u_1{}^2}$, $\overline{u_2{}^2}$, and $\overline{u_1 u_2}$, and applying the least-squares method.

Instead of using one wire and setting it in various positions, it is possible to use two wires either in X-array or in V-array. If the wires have the same features, are identical, we have $(s_1)_{\rm I} = (s_1)_{\rm II}$ and $(s_2)_{\rm I} = (s_2)_{\rm II}$. The intensities u'_1 and u'_2 can then be obtained in a simple, direct way from

$$\overline{(e_{\rm I} + e_{\rm II})^2} = 4s_1{}^2\overline{u_1{}^2}$$

and (2-84)

$$\overline{(e_{\rm I} - e_{\rm II})^2} = 4s_2{}^2\overline{u_2{}^2}$$

The intensity of the lateral turbulence component u'_3 can be similarly determined.

Producing two identical wires, however, is far from an easy job; it is hardly feasible and requires very skilled labor. Yet, to obtain reliable values of the turbulence velocities the two wires should differ only very slightly. If $s_{\rm I} = (1 + n)s_{\rm II}$, we obtain an error of about $100n$ per cent in the measured values of $\overline{u_1{}^2}$, $\overline{u_2{}^2}$, and $\overline{u_3{}^2}$.

Measurement of Double and Triple Velocity Correlations

We have already shown that the double correlation $\overline{u_1u_2}$ can be obtained by solving Eqs. (2-83). This is very simple if we assume that $(s_1)_{\rm I} = (s_1)_{\rm II}$ and $(s_2)_{\rm I} = (s_2)_{\rm II}$, since we then have

$$(\overline{e^2})_{\rm I} - (\overline{e^2})_{\rm II} = 4s_1s_2\overline{u_1u_2}$$

Similarly, we may determine the correlation $\overline{u_1u_3}$.

To determine the correlation $\overline{u_2u_3}$, we place the X-wires, say, in such a position that one wire coincides with the $(\bar{U},u_2)$ plane and the other with the $(\bar{U},u_3)$ plane, and both wires make equal angles φ with the $\bar{U}$-direction. If the two wires are again assumed identical,

$$e_{\rm I} = -s_1u_1 - s_2u_2 \qquad e_{\rm II} = -s_1u_1 - s_2u_3$$

so $\overline{u_2u_3}$ can be calculated from $\overline{(e_{\rm I} - e_{\rm II})^2} = s_2{}^2(\overline{u_2{}^2} + \overline{u_3{}^2} - 2\overline{u_2u_3})$.

The spatial Eulerian correlations **g** and **f** can be obtained with two wires.

For determining the lateral correlations coefficient $\mathbf{g}(x_2)$ we locate the wires at the two points A and B considered at a distance x_2 from each other in the same plane perpendicular to U. With the wires we have to measure the axial turbulence components at the points A and B; this is done by placing the wires perpendicular to U. The voltage signals of the wires "A" and "B" are

$$e_A = (s_1)_A(u_1)_A \qquad e_B = (s_1)_B(u_1)_B$$

We then have

$$\overline{(e_A + e_B)^2} = (s_1{}^2\overline{u_1{}^2})_A + (s_1{}^2\overline{u_1{}^2})_B + 2(s_1)_A(s_1)_B\overline{(u_1)_A(u_1)_B}$$

$$\overline{(e_A - e_B)^2} = (s_1{}^2\overline{u_1{}^2})_A + (s_1{}^2\overline{u_1{}^2})_B - 2(s_1)_A(s_1)_B\overline{(u_1)_A(u_1)_B}$$

The correlation $\overline{(u_1)_A(u_1)_B}$ can then be calculated either from

$$\frac{\overline{(e_A + e_B)^2} - \overline{(e_A - e_B)^2}}{\overline{(e_A + e_B)^2} + \overline{(e_A - e_B)^2}}$$

provided that the wires are sufficiently alike to assume $(s_1)_A = (s_1)_B$, or from

$$\frac{\overline{(e_A + e_B)^2} - \overline{(e_A - e_B)^2}}{e'_A e'_B}$$

in which case it is not necessary to have $(s_1)_A = (s_1)_B$, since these quantities cancel out in this expression.

If the wires are not identical and we nevertheless want to apply the first method, the equation can still be solved when the turbulence flow field is homogeneous, so that $(u'_1)_A = (u'_1)_B = u'_1$. Assuming that $(s_1)_B = (1 + n)(s_1)_A$, we have

$$\frac{\overline{(e_A + e_B)^2} - \overline{(e_A - e_B)^2}}{\overline{(e_A + e_B)^2} + \overline{(e_A - e_B)^2}} = \frac{2(s_1)_A(s_1)_B}{(s_1{}^2)_A + (s_1{}^2)_B} \frac{\overline{(u_1)_A(u_1)_B}}{u_1'^2} \simeq \left(1 - \frac{n^2}{2}\right)\mathbf{g}(x_2)$$

Even if $(s_1)_A$ and $(s_1)_B$ differ by as much as 20 per cent, the error made in the measured value of $\mathbf{g}(x_2)$ is about 2 per cent. Hence we conclude that the wires need not be exactly identical but may have somewhat different features.

A similar procedure can be applied to determine the longitudinal correlation coefficient $\mathbf{f}$.

Difficulties may arise if one wire is very close behind the other, because then the wire located downstream will be affected by the wake of the other wire. In order to get around this difficulty, it is possible to place the wires not exactly in a plane parallel to the mean flow $\bar{U}$ but with the second wire slightly to one side of the first wire. It is obvious, however, that in this case some effect of the lateral correlation $\mathbf{g}(x_2)$ must be felt. Thus the first difficulty is more or less replaced by another. Consequently, measurements made in either way will be somewhat unreliable for small values of the axial distance x_1. Though in the latter case some corrections can be made for the effect of $\mathbf{g}(x_2)$.

There is, however, another method that makes it possible to determine the longitudinal correlation coefficient $\mathbf{f}(x_1)$. This method can be applied when the whole field has a constant and uniform average velocity and when the relative turbulence intensity $u'_1/\bar{U}$ is very small. We have shown in Chap. 1 that in this case the Eulerian space correlation $\mathbf{f}(x_1)$ is approximately identical with the Eulerian time correlation $\mathbf{R}_E(t)$, because we have approximately $x_1 = \bar{U}t$ [see Eqs. (1-69) and (1-70)]. Thus $\mathbf{f}(x_1)$ can be determined from measurements of the Eulerian time correlation $\mathbf{R}_E(t)$.

Favre[16,17] has developed a method and technique for measuring time correlations at a point in the flow field. In principle, this method consists in recording the velocity fluctuations as a function of time on a magnetic tape. This tape is then run off at a known velocity V along two pickups with a distance a. The variations in electric voltage produced by the two pickups correspond to the turbulence fluctuations with a time interval $\tau = a/V$. For determining the correlation coefficient $\mathbf{R}_E(t)$, the outputs of the two pickups can be handled like the outputs of two hot wires, described earlier.

With this method it is also possible to measure correlations according to both time and space by using two hot wires placed at two points in the flow field and recording the fluctuations detected by the two hot wires on two magnetic tapes. Favre and collaborators have made such measurements in the turbulent flow field downstream of turbulence-producing grids and in the turbulent boundary layer along a flat plate.

Another method for obtaining correlations from oscillograms is described by Tucker.[79] The records of the oscillograms are prepared photographically in the form of black-and-white silhouettes on a transparent strip. Cross correlations are obtained by scanning the two records with their corresponding light beams and directing the two light beams, after they pass the transparent strip, onto the same photocell. The sum of the two voltages produced by the photocell is passed via an amplifier to a squaring circuit. If $e_1(t)$ is the voltage function of one record and $e_2(t + \tau)$ that of the other, we have

$$\overline{[e_1(t) + e_2(t + \tau)]^2} = \overline{[e_1(t)]^2} + \overline{[e_2(t + \tau)]^2} + \overline{2e_1(t)e_2(t + \tau)}$$

The third term on the right-hand side is proportional to the correlation required. Of course, one record can be scanned by two light beams at a distance V_τ, where V is the scanning speed; in this way time correlations can be obtained.

Instead of using a thermocouple to determine correlations, it is also possible to supply the outputs of two hot wires to the two pairs of deflection plates of a cathode-ray tube, so that the deflection of the light spot on the screen of the tube that corresponds to one of the hot wires is perpendicular to the deflection of the light spot corresponding to the other hot wire. The result is that the light spot makes so-called "Lissajous" figures on the screen. A photograph taken with a long exposure then shows an ellipse, the correlation ellipse. Let $2a$ and $2b$ denote, respectively, the long and the short axis of the ellipse. It can then be shown that the correlation coefficient $-(a^2 - b^2)/(a^2 + b^2)$.

The deviation from the circular will be very small when the correlation coefficient is nearly zero. In that case it will be very difficult to measure accurately the values of a and b and, consequently, small values of the correlation coefficient. However, since the exact shape of the correlation curve for small values is of only minor importance, the drawback inherent in this method is not very serious. What is more serious is that the same drawback also occurs where the correlation coefficient becomes nearly equal to unity. The ellipse is then reduced almost to a straight line of varying thickness. And, since the curves recorded on the tube screen already have a finite thickness, it will be very difficult to measure slight deviations from the "normal" thickness of these lines. Even an error of 1 per cent in the determination of the correlation coefficient becomes large if this error is referred to the difference from unity. Since the shape of the Eulerian space correlation in the neighborhood of its vertex is of great importance for determining the size of the dissipation scale, it is desirable to modify this method so as to give a more accurate value of the correlation coefficient in the neighborhood of perfect correlation.

It may perhaps be recalled that the other methods described earlier also have this same drawback of providing a relatively inaccurate determination of the correlation coefficient at values close to unity.

Taylor[18] gives a method with which it is possible to determine the correlation coefficient $\mathbf{R}$ more accurately in the neighborhood of $\mathbf{R} = 1$. In this method the value of $1 - \mathbf{R}^2$ is measured directly.

As has been shown in Chap. 1, the triple correlations $\mathbf{S}_{ik,j}$ must be known in order to describe the change in the double correlations with time. It has also been mentioned that, in the case of a homogeneous isotropic turbulence, only 3 out of the 27 components of the triple-correlation tensor are of importance; the coefficients of these three correlations are $\mathbf{k}(r)$, $\mathbf{h}(r)$, and $\mathbf{q}(r)$ [see (1-51)]. As will be shown later in Chap. 3, there is a relationship between these three correlations, so that it is sufficient to measure only one of them. The simplest one to measure is the longitudinal triple-correlation coefficient $\mathbf{k}(r)$. To do this two hot wires are placed parallel to each other, perpendicular to the main velocity U, and at a distance $r = x_1$ in the main-flow direction. Let A and B denote the two points, and $(e)_A$ and $(e)_B$ the voltage fluctuations caused by the wires at A and B, respectively. Then

$$\mathbf{k}(r) = \frac{\overline{e_A{}^2 e_B}}{e'^3}$$

The triple product $\overline{e^3}$ can easily be obtained by leading the signal of one of the hot wires to a squaring circuit, then the squared signal plus the original signal to a multiplier, and the output of this multiplier to an integrator in order to obtain the mean value. The same procedure can be applied to obtaining $\overline{e_A{}^2 e_B}$. The squared signal of hot wire A is led together with the signal e_B of hot wire B to the multiplier.

The procedure can be generalized in order to obtain other triple-correlations. Consider two hot wires, one in A and the other in B. The two wires are placed in oblique positions, for instance $\varphi = \pm 45°$, with respect to the main stream.

With the hot wires in the (x_1,x_2) plane we obtain

$$e_A = s_{1A} u_{1A} \pm s_{2A} u_{2A} \quad \text{and} \quad e_B = s_{1B} u_{1B} \pm s_{2B} u_{2B}$$

where the plus sign refers to the $+45°$ position, the minus sign to the $-45°$ position.

For the value $\overline{e_A{}^2 e_B}$ there is obtained

$$\overline{e_A{}^2 e_B} = s_{1A}{}^2 s_{1B} \overline{u_{1A}{}^2 u_{1B}} + s_{2A}{}^2 s_{1B} \overline{u_{2A}{}^2 u_{1B}} + 2 s_{1A} s_{2A} s_{2B} \overline{u_{1A} u_{2A} u_{2B}}$$
$$\pm 2 s_{1A} s_{2A} s_{1B} \overline{u_{1A} u_{2A} u_{1B}} \pm s_{1A}{}^2 s_{2B} \overline{u_{1A}{}^2 u_{2B}} \pm s_{2A}{}^2 s_{2B} \overline{u_{2A}{}^2 u_{2B}}$$

When we put $\varphi = 90°$ for the two hot wires so that $s_{2A} = 0$ and $s_{2B} = 0$, then $\overline{e_A{}^2 e_B} =$

$s_{1A}{}^2 s_{1B} \overline{u_{1A}{}^2 u_{1B}}$, from which also the value for $\mathbf{k}(r)$ may be obtained when A and B are both on the x_1-axis.

When we put $\varphi = \pm 45°$ and consider the one-point triple correlation, using a single hot wire, then

$$\overline{e^3} = s_1{}^3 \overline{u_1{}^3} + 3s_2{}^2 s_1 \overline{u_2{}^2 u_1} \pm 3s_1{}^2 s_2 \overline{u_1{}^2 u_2} \pm s_2{}^3 \overline{u_2{}^3}$$

whence

$$\overline{e^3}(+) - \overline{e^3}(-) = 6s_1{}^2 s_2 \overline{u_1{}^2 u_2} + 2s_2{}^3 \overline{u_2{}^3}$$

The value of $\overline{u_2{}^3}$ can be obtained separately by using two identical hot wires in X-array and applying the above procedure to the difference of the signals of the two wires.

It appears to be of great advantage if a linearizer is used to account for the nonlinear characteristic of the hot wire.

First measurements on the triple correlation $\mathbf{k}(r)$ in grid-produced turbulence have been made by Townsend.[19] His method was based on the quadratic plate characteristic of a triode with a low plate current if the triode operates at the point of maximum curvature of this characteristic. A more recent procedure using digital computational methods has been applied by Frenkiel and Klebanoff,[94,95] to measure the triple and higher order correlations between the longitudinal velocity component u_1 at two points, so $\overline{u_{1A}{}^m(t) \cdot u_{1B}{}^n(t+\tau)}$. To this end the voltage fluctuations e_A and e_B corresponding to u_{1A} and u_{1B} are recorded simultaneously on magnetic tape, at a constant speed (1.524 m/s). At the same time a timing signal of 13,800 Hz is recorded on a third channel. The two channels for the turbulence signals should be well matched, with frequency response characteristics equal within 1 per cent, say. With an analog-to-digital converter the two analog tapes of the turbulence channels are digitized simultaneously, however at a much lower speed (47.6 mm/s), using the timing signal as a trigger for digitizing at each 1/13,800 s interval of real time. The digitized values are put in a computer for further data processing to yield the required correlation. The reading time of the digitizer for each digitized value was about 80 μs, corresponding with 2.5 μs real time.

Measurement of Spectrum and Scale of Turbulence

The spectral analysis of, say, the squared intensity $\overline{u_1{}^2}$ can be accomplished with a suitable narrow-band filtering circuit. The general requirement is that the input signal remain free from distortion and not change appreciably in amplitude. Of course, it would be possible to raise this amplitude back to its original value by means of an amplifier, but then the amplifier must be free from distortion.

The integral scale, for instance the Eulerian integral length-scale

$$\Lambda_f = \int_0^\infty dx_1 \, \mathbf{f}$$

can be obtained by integration of the $\mathbf{f}(x_1)$ curve. Difficulties may arise when the correlation curve oscillates persistently about the 0-axis at great values of the distance. Of course it is possible to interrupt the curve somewhere, but this would be an arbitrary process and it is not correct.

For the specific case of the Eulerian integral time scale $\mathfrak{J}_E = \int_0^\infty dt\, \mathbf{R}_E$, another, very elegant method has been proposed by Townsend,[19] namely, to make use of the energy-spectrum function $\mathbf{E}_1(n)$. From the relation

$$\mathbf{R}_E(t) = \frac{1}{\overline{u_1^2}} \int_0^\infty dn\, \mathbf{E}_1(n) \cos 2\pi nt \qquad \text{(1-95a)}$$

it follows that

$$\int_0^\infty dt\, \mathbf{R}_E(t)\Phi(t) = \frac{1}{\overline{u_1^2}} \int_0^\infty dn\, \mathbf{E}_1(n) \int_0^\infty dt\, \Phi(t) \cos 2\pi nt \qquad \text{(2-85)}$$

where $\Phi(t)$ is an arbitrary function, which, however, must satisfy the conditions:

1 $\int_0^\infty dt\, \mathbf{R}_E(t)\Phi(t) = \int_0^\infty dt\, \mathbf{R}_E(t) = \mathfrak{J}_E$.

2 $\int_0^\infty dt\, \Phi(t) \cos 2\pi nt$ converges.

3 The integration procedure with the function Φ is feasible by means of electronic circuits.

These conditions can be satisfied if the function $\Phi(t)$ is such that the deviation from unity is small for the values of the time where $\mathbf{R}_E(t)$ still has a noticeable value, but decreases to zero when t is large so that the second integral will converge.

Such a function, which also satisfies the third condition, is, for instance,

$$\Phi(t) = \left(1 + \frac{t}{t_0}\right) \exp(-t/t_0)$$

Substitution in Eq. (2-85) yields

$$\mathfrak{J}_E = \int_0^\infty dt\, \mathbf{R}_E(t)\Phi(t) = \frac{1}{\overline{u_1^2}} \int_0^\infty dn\, \mathbf{E}_1(n) \frac{2t_0}{(1 + 4\pi^2 n^2 t_0^2)^2}$$

Now the value of t_0 must be chosen such that

$$\int_0^\infty dt\, \mathbf{R}_E(t)\left(1 + \frac{t}{t_0}\right) \exp(-t/t_0) = \int_0^\infty dt\, \mathbf{R}_E(t)$$

This condition is satisfied if t_0 is large with respect to $\mathfrak{J}_E$. Assume, for instance, that $\mathbf{R}_E(t)$ is given by the exponential function $\exp(-t/\mathfrak{J}_E)$. The right-hand side of the equation then yields the value $\mathfrak{J}_E$, and the left-hand side yields

$$t_0 \frac{2 + t_0/\mathfrak{J}_E}{(1 + t_0/\mathfrak{J}_E)^2}$$

Hence, if $\mathfrak{I}_E/t_0 \ll 1$, the ratio between the two sides of the equation becomes

$$\frac{t_0}{\mathfrak{I}_E}\frac{2 + t_0/\mathfrak{I}_E}{(1 + t_0/\mathfrak{I}_E)^2} = 1 + \text{a term of the order of } \frac{\mathfrak{I}_E{}^2}{t_0{}^2}$$

It is obvious that, when the turbulent field has a constant mean velocity $\bar{U}$ that is large with respect to the turbulence fluctuations, the method yields the Eulerian integral length-scale $\Lambda_f = \bar{U}\mathfrak{I}_E$.

The dissipation scale λ can be determined with the aid of the osculation parabola in the vertex of the correlation curve. A basic requirement is that the shape of the correlation curve at its vertex must be accurately known. In the foregoing section we have shown that an accurate measurement of the correlation is not very simple when the correlation coefficient is close to unity. It would be very pleasant, therefore, if there were another method of determining λ.

Again, it was Townsend[19] who succeeded in finding a fairly simple method for measuring the dissipation scale λ_f, which is quite direct when the turbulent field has a constant main velocity $\bar{U}$ high compared with the turbulence fluctuations. For this he made use of the relations

$$\frac{2}{\lambda_f{}^2} = -\left[\frac{\partial^2 \mathbf{f}}{\partial x_1{}^2}\right]_{x_1=0} = \frac{1}{u'^2}\overline{\left(\frac{\partial u_1}{\partial x_1}\right)^2} = \frac{1}{\bar{U}^2 u'^2}\overline{\left(\frac{\partial u_1}{\partial t}\right)^2}$$

Thus, by introducing a differentiation circuit in the electronic system, λ_f can be measured.

A repeated differentiation would allow us to measure also

$$\left[\frac{\partial^4 \mathbf{f}}{\partial x_1{}^4}\right]_{x_1=0} = \frac{1}{\bar{U}^4 u'^2}\overline{\left(\frac{\partial^2 u_1}{\partial t^2}\right)^2}$$

and so on.

Laufer[27] and Liepmann[28] propose a method for determining λ_f from the average number N_0 of zeros of the u_1-fluctuations per unit time,

$$\frac{1}{\lambda_f} = \frac{\pi N_0}{\bar{U}\sqrt{2}} \tag{2-86}$$

This relation can be proved in a simple way for the case in which u_1 is a pure harmonic

$$u_1 = u_1^* \sin\frac{2\pi t}{T}$$

where T is the period. The number N_0 of zeros of u_1 per unit time is $N_0 = 2/T$. Now it is possible to express T in terms of u_1 and du_1/dt. It is more convenient, however, to express T in terms of $(\overline{u_1{}^2})^{1/2}$ and $[\overline{(du_1/dt)^2}]^{1/2}$, because then a simpler expression is obtained. Moreover, in real turbulence u_1 and du_1/dt are statistical quantities, and we know that the root-mean-square values may be taken for the average values.

Because

$$\sqrt{\overline{u_1^2}} = \frac{u^*}{\sqrt{2}}$$

and

$$\sqrt{\overline{\left(\frac{du_1}{dt}\right)^2}} = \frac{2\pi u^*}{T\sqrt{2}}$$

it follows that

$$N_0 = \frac{2}{T} = \frac{\sqrt{\overline{(du_1/dt)^2}}}{\pi\sqrt{\overline{u_1^2}}} = \bar{U}\frac{\sqrt{\overline{(du_1/dx_1)^2}}}{\pi\sqrt{\overline{u_1^2}}} \tag{2-87}$$

Laufer and Liepmann have shown that exactly the same expression can be obtained for a real turbulence provided that both u_1 and du_1/dx_1 have a Gaussian probability-density distribution and that $\overline{u_1(du_1/dx_1)} = 0$. The latter condition is fulfilled in a homogeneous turbulence (see Chap. 1). Townsend[19] has shown that in isotropic turbulence the probability-density distribution of u_1 is practically Gaussian but that this is not true of du_1/dx_1; the distribution here shows in his experiments a definite though slight skew.

Now we have shown in Chap. 1 that in a homogeneous turbulence we have the relation

$$\frac{2}{\lambda_f^2} = \frac{\overline{(du_1/dx_1)^2}}{\overline{u_1^2}}$$

Substitution in the relation (2-87) then yields

$$\frac{1}{\lambda_f} = \frac{\pi N_0}{\bar{U}\sqrt{2}}$$

In Chap. 3 we shall show that in an isotropic turbulence

$$\lambda_f = \lambda_g\sqrt{2}$$

so that the corresponding relation for the dissipation scale λ_g will read

$$\frac{1}{\lambda_g} = \frac{\pi N_0}{\bar{U}} \tag{2-86a}$$

The experimental technique for determining the number of zeros of u_1 is very simple in principle. The oscillogram of u_1 is shown on the screen of a cathode-ray tube. In front of this tube is a disk with a narrow slit which corresponds to the zero line of the oscillogram. Each time that u_1 passes its zero value the light beam of the cathode-ray tube passes through the slit and falls onto a photoelectric cell. The voltage impulse thus generated is led to an electronic counting circuit.

Liepmann and Robinson[59] have worked out a method and built an apparatus based on this method, in which the probability density of a statistical quantity can be determined by pulse counting. Zero counting is a special case of this; with the probability density it is also possible to determine mean values of various powers of this function from the corresponding moments of the probability-density function. Application to the measurement of the probability density of u_1 and du_1/dt in isotropic turbulence has shown that some statistical dependence of the two probability densities might exist, but not enough to explain any difference between the values of λ_f measured by the zero-counting method and measured otherwise.

2-6 MEASUREMENT OF TEMPERATURE AND CONCENTRATION FLUCTUATIONS WITH THE HOT-WIRE ANEMOMETER; CONSTANT-CURRENT METHOD

The measurement of temperature fluctuations of gas by means of a hot-wire anemometer is possible (1) because the temperature difference between wire and gas is affected by the temperature fluctuations of the gas, but also (2) because the gas properties in the film around the wire vary with the gas temperature. We have demonstrated in Sec. 2-4 that, of the factors A and B in the formula (2-17), it is A in particular that is sensitive to temperature changes. We shall show, however, that, under the normal operating conditions of the hot wire, it is mainly the direct effect of the temperature difference between wire and gas that determines the hot-wire sensitivity to gas-temperature fluctuations rather than the variation of the factor A.

For an assumed linear relation between the gas heat conductivity and the temperature, the following expression was obtained for the factor A (see Sec. 2-4):

$$A_f = A_g\left[1 + \frac{a_1}{2(a + a_1\Theta_g)}(\Theta_w - \Theta_g)\right] \tag{2-72}$$

Since A_g here is also considered to be a fluctuating quantity, we had better introduce a reference temperature Θ_0 and the value A_0 referred to this temperature. We then write A_f as follows:

$$A_f = \frac{A_0}{a + a_1\Theta_0}\left[a + \frac{a_1}{2}(\Theta_w - \Theta_g)\right] + \frac{a_1 A_0}{a + a_1\Theta_0}\Theta_g \tag{2-88}$$

where

$$A_0 = 0.42\frac{\pi l}{bR_0}\mathbf{Pr}^{0.20}(a + a_1\Theta_0)$$

Substitution in Eq. (2-17) yields

$$\frac{I^2R_w}{R_w - R_g} = \frac{A_0}{a + a_1\Theta_0}\left[a + \frac{a_1}{2}(\Theta_w - \Theta_g)\right] + \frac{a_1 A_0}{a + a_1\Theta_0}\Theta_g + B\sqrt{U} \tag{2-89}$$

Again assume

$$R_w = \bar{R}_w + r_w \qquad R_g = \bar{R}_g + r_g \qquad \Theta_g = \bar{\Theta}_g + \theta_g \qquad U = \bar{U} + u$$

From Eq. (2-89), neglecting the nonlinear terms in r_w, r_g, θ_g, and u, we then obtain

$$\frac{I^2 r_w}{\bar{R}_w - \bar{R}_g} = \left[\frac{\bar{A}_f + B\sqrt{\bar{U}}}{\bar{R}_w - \bar{R}_g} + \frac{A_0 a_1}{2b(a + a_1\Theta_0)R_0}\right] r_w$$

$$+ \left[\frac{A_0 a_1}{2(a + a_1\Theta_0)} - \frac{\bar{A}_f + B\sqrt{\bar{U}}}{\bar{R}_w - \bar{R}_g} bR_0\right]\theta_g + B\sqrt{\bar{U}}\,\frac{u}{2\bar{U}}$$

where $\bar{A}_f$ is the average value of (2-88).

Now, under the normal operating conditions of the hot wire, with the dimensions usually applied in turbulence measurements, it turns out that

$$\frac{A_0 a_1}{2b(a + a_1\Theta_0)R_0} \ll \frac{\bar{A}_f + B\sqrt{\bar{U}}}{\bar{R}_w - \bar{R}_g}$$

that is, of the order of 10 per cent or less. Neglecting the smaller term, we simplify the former equation to

$$\frac{I^2 r_w}{\bar{R}_w - \bar{R}_g} = \frac{\bar{A}_f + B\sqrt{\bar{U}}}{\bar{R}_w - \bar{R}_g} r_w - \frac{\bar{A}_f + B\sqrt{\bar{U}}}{\bar{R}_w - \bar{R}_g} bR_0\theta_g + B\sqrt{\bar{U}}\,\frac{u}{2\bar{U}}$$

or

$$\frac{I^2 \bar{R}_g r_w}{(\bar{R}_w - \bar{R}_g)^2} = \frac{bI^2 \bar{R}_w R_0}{(\bar{R}_w - \bar{R}_g)^2}\theta_g - B\sqrt{\bar{U}}\,\frac{u}{2\bar{U}}$$

Hence,

$$e = Ir_w = I\frac{b\bar{R}_w R_0}{\bar{R}_g}\theta_g - \frac{(\bar{R}_w - \bar{R}_g)^2}{2I\bar{R}_g} B\sqrt{\bar{U}}\,\frac{u}{\bar{U}} \tag{2-90}$$

We introduce the sensitivity to velocity fluctuations s defined by Eq. (2-40) and a sensitivity to temperature fluctuations s_θ, defined by

$$\sqrt{\overline{e^2}} = s_\theta\sqrt{\overline{\theta_g{}^2}} \tag{2-91}$$

The relation (2-92) then reads

$$e = -su + s_\theta\theta_g \tag{2-92}$$

where

$$s_\theta = I\frac{b\bar{R}_w R_0}{\bar{R}_g} \tag{2-93}$$

Though the values of s and s_θ can be calculated if we take for B, e.g., the expression given in (2-15), the usual and safe procedure is to determine these values experimentally.

The relation (2-92) forms the basis for the measurement of turbulence temperature fluctuations and of their correlations with velocity fluctuations.

Let us take first the mean-square value of (2-92),

$$\overline{e^2} = s^2\overline{u^2} + s_\theta{}^2\overline{\theta_g{}^2} - 2ss_\theta\overline{u\theta_g} \qquad (2\text{-}94)$$

If it were possible to make $s \ll s_\theta$, then Eq. (2-94) could be reduced to a relation that gives $\overline{\theta_g{}^2}$ directly. If, on the other hand, it were possible to make $s \gg s_\theta$, then we could obtain from (2-94) the value of $\overline{u^2}$. For the first u must be at most of the same order as θ_g; for the second $\theta_g \gtrsim u$. Thus, to make both procedures possible, we may require that u and θ_g be of the same order of magnitude.

By making $s \ll s_\theta$ we obtain $e \simeq s_\theta\theta_g$; so it is possible to obtain an oscillogram of the temperature fluctuations.

Now it appears possible to make $s \ll s_\theta$ when $\bar{R}_w - \bar{R}_g$ is sufficiently small. This cannot be inferred directly from the expressions for s and s_θ, but it can easily be verified when the numerical values of the quantities occurring in these expressions are determined for relevant operating conditions.

When a sufficiently small value for $\bar{R}_w - \bar{R}_g$ is taken, the hot wire operates like a resistance thermometer.

Conversely, s can be made large with respect to s_θ by taking a large value for $\bar{R}_w - \bar{R}_g$. This, however, is not always possible, particularly when $\bar{R}_g$ and $\bar{U}$ are large, because the value of $\bar{R}_w$ is limited by the maximum allowable temperature of the wire.

To get around this difficulty, we may obtain all three unknowns on the right-hand side of (2-94) by giving s and s_θ three values each. This can be done because s and s_θ depend in different ways on $\bar{\Theta}_w/\bar{\Theta}_g$, so that the hot wire may be operated at three temperature levels. Another, but less practicable way, is to use three wires of different diameter, because s and s_θ depend on the wire diameter. It is obvious that the dependence of s and s_θ on $\bar{\Theta}_w/\bar{\Theta}_g$ or on the wire diameter must be known. From the three values of $\overline{e^2}$ obtained by working either with one wire at three different temperature levels or with three wires of different diameter, the values of $\overline{u^2}$, $\overline{\theta_g{}^2}$, and $\overline{u\theta_g}$ can be determined.

By placing the wire obliquely with respect to the main flow $\bar{U}$, the values of $\overline{u_2{}^2}$ and $\overline{u_2\theta_g}$ or of $\overline{u_3{}^2}$ and $\overline{u_3\theta_g}$ can be measured. Actually this can be done, for instance, with two identical wires in X-array. The output of the one wire reads

$$e_{\text{I}} = -(s_1)_{\text{I}}u_1 - (s_2)_{\text{I}}u_2 + s_\theta\theta_g$$

The output of the second wire reads

$$e_{\text{II}} = -(s_1)_{\text{II}}u_1 + (s_2)_{\text{II}}u_2 + s_\theta\theta_g$$

From these relations we obtain $\overline{u_1{}^2}$ and $\overline{u_2{}^2}$ directly [see Eq. (2-84)], since for the identical wires $(s_1)_{\text{I}} = (s_1)_{\text{II}}$ and $(s_2)_{\text{I}} = (s_2)_{\text{II}}$ and since the average of the differences of the squares yields

$$\overline{e_I^2 - e_{II}^2} = -4s_2 s_\theta \overline{u_2\theta_g} + 4s_1 s_2 \overline{u_1 u_2}$$

Again, by making $s_\theta \gg s$, ($\bar{R}_w - \bar{R}_g$ small; resistance thermometer), it is possible to obtain $\overline{u_2\theta_g}$; $\overline{u_1 u_2}$ can be obtained by making $s_\theta \ll s$.

It is also possible to evaluate $\overline{u_2\theta_g}$ and $\overline{u_1 u_2}$ by operating the wires at two different temperature levels, so that two independent equations for $\overline{u_2\theta_g}$ and $\overline{u_1 u_2}$ can be obtained.

If the value of $\overline{\theta_g^2}$ as a direct output of a hot wire can be obtained, a spectral distribution by means of a suitable filter circuit can be determined.

To measure spatial correlations between velocity and temperature fluctuations at two points in the flow field, we may operate either with two single hot wires (namely for the axial u_1-component and θ_g) or with two sets of X-wires (for the lateral u_2- or u_3-component and θ_g). It must then be assumed that the velocity and temperature fields are homogeneous in the average values.

For instance, for measuring the correlations between the u_1-components at the two points A and B and the correlations between the θ_g-fluctuations, we must start from

$$e_A = -s_A(u_1)_A + (s_\theta)_A(\theta_g)_A \qquad e_B = -s_B(u_1)_B + (s_\theta)_B(\theta_g)_B$$

Because of the assumed homogeneity of the flow field and the identity of the two wires, $s_A = s_B$ and $(s_\theta)_A = (s_\theta)_B$.

From these relations we obtain

$$\overline{(e_A + e_B)^2} - \overline{(e_A - e_B)^2} = 4s_\theta^2\overline{(\theta_g)_A(\theta_g)_B} + 4s^2\overline{(u_1)_A(u_1)_B}$$
$$- 4ss_\theta\left[\overline{(\theta_g)_A(u_1)_B} + \overline{(\theta_g)_B(u_1)_A}\right]$$

By operating at three different wire temperatures it is possible to determine $\overline{(\theta_g)_A(\theta_g)_B}$ and $\overline{(u_1)_A(u_1)_B}$. The cross correlations between temperature and velocity at the two points, though obtained automatically by this procedure, have no physical meaning and are, therefore, of no practical importance.

Of course, we may again make either $s_\theta \gg s$ or $s_\theta \ll s$ for obtaining $\overline{(\theta_g)_A(\theta_g)_B}$ or $\overline{(u_1)_A(u_1)_B}$, respectively.

Once the correlation $\overline{(\theta_g)_A(\theta_g)_B}$ has been obtained for different distances between the two points, it is possible to determine the spatial integral scale of the temperature fluctuations.

The measurement of temperature fluctuations in a turbulent flow field was first introduced by Corrsin.[29] Theoretically Corrsin started from the equation given by King for the heat transfer from the hot wire to the flowing gas; consequently, he took for the gas temperature that of the ambient gas outside the boundary layer and not the film temperature. Accordingly, he obtained an expression for s_θ different from that given in Eq. (2-93). Since in the experimental procedure s_θ is determined experimentally, this difference has no consequences of practical importance.

Measurement of Concentration Fluctuations

Mainly because of the dependence of κ_g and ρ_g on the type of gas, it is possible in principle to measure concentration fluctuations with a hot-wire anemometer. It is again possible to derive from Eq. (2-17) a relation such as

$$e = -su - s_\gamma \gamma$$

where s_γ is the sensitivity of the wire to fluctuations γ in the concentration of one gas in another.

At first sight it might be thought that exactly the procedure given for temperature fluctuations would apply in this case. But there is an essential difference. The sensitivities s and s_θ do not depend in exactly the same way on the temperature; so operation at different wire temperatures produces different and independent relations. In concentration measurements, however, both s and s_γ vary proportionally with wire temperature; so no independent equations can be obtained by operating at different temperature levels of the wire. It is possible to obtain independent equations only by using wires of different diameter. Fortunately, the dependence of the two factors A and B on the wire diameter is very strong; so the two wires need differ only slightly in diameter to become suitable for this purpose. It is obvious, however, that measurements with two different wires one after the other are very cumbersome.

In Ref. 29 Corrsin describes methods for measuring statistical quantities in a turbulent concentration field.

2-7 LIMITATIONS OF THE HOT-WIRE ANEMOMETER

Although the hot-wire anemometer is at present still the most useful instrument for measuring turbulence, it has its limitations. Some of these limitations have already become apparent from the preceding sections. These involve the nonlinear character of the heat transfer with respect to velocity and temperature. Both the nonlinear temperature effect and the effect of the curvature in the velocity-response curve can be reduced considerably by applying the constant-temperature method, especially when a linearization circuit is added.

In the case of compressible flows, practical limitations may be set by the complexity of the nature of the heat transfer between wire and fluid.

But there are still more limitations. We have already mentioned the limitation set by the "resolution power in space," i.e., in the direction of the wire, owing to its finite length. Related to this is the resolution power in time, i.e., in the flow direction, due to the finite time constant of the hot wire.

According to Kovasznay,[36] we may introduce a resolution length $l_{res} = U/2n_{max}$, where n_{max} is the maximum frequency of the amplifier at which there is still no appreciable loss of response. This resolution length must be small compared with the size of the micro eddies. Since n_{max} is inversely proportional to the time constant of the hot wire, l_{res} is directly proportional to it.

At very high flow velocities (say, $U > 100$ m/s) extremely small eddies may occur. So that the resolution length may remain small compared with these eddies, small values of the time constant must be used; it is desirable, therefore, to use very fine wires. On the other hand, because of the strength required of the wire in such high-velocity flow, the wire should not be too thin. Moreover, the supports and their aerodynamic interference with the wire may cause disturbances in the flow pattern comparable in size to the smallest eddies.

Vernotte[41] advances another criticism, which is of a fundamental nature. The use of the hot wire is based on a law of heat transfer, operating under steady conditions of fluid flow, whereas in measuring turbulence the wire is exposed to a fluid flow of unsteady character. Now the transfer of heat from a wire to a flowing fluid depends on the flow pattern, including the turbulent motions; these, however, are not yet known. Vernotte recommends a more thorough investigation of this point. In this connection there are two factors of importance. These are (1) the effect of fluctuations of the general flow pattern around the wire, and (2) the effect of fluctuations of the flow in the boundary layer of the wire. The first effect will be unimportant when the turbulence fluctuations are small compared with the mean velocity, since the flow pattern will be determined mainly by this mean flow and so any effect of unsteadiness of the flow around the wire will be of secondary importance as far as the flow pattern is concerned. The second effect, however, may be more serious. Lighthill[84] has studied the effect on the heat transfer of a fluctuating flow in a laminar boundary layer. Application to hot-wire anemometry shows that, particularly in the constant-current method, the time lag due to the thermal inertia of the wire is increased by the thermal-inertia effects in the boundary layer. However, for the usual conditions under which the hot-wire anemometer is used, this increase is negligibly small.

The limitations set by the resolution powers in space and in time, and those following from Vernotte's criticism, can be expressed in terms of the ratios between the corresponding time scales and a characteristic time for the quickest fluctuations in the turbulence considered. For this characteristic time we may take $\tau_c = \eta/\bar{U}$, where η is the Kolmogoroff length scale, a measure for the smallest eddies of the turbulence. The above resolution power in space then is limited by the ratio l/η, or by the ratio τ_1/τ_c, where $\tau_1 = l/\bar{U}$. The resolution power in time is determined by the time scale $\tau_2 = 1/n_{max}$; n_{max} being again the maximum frequency with no appreciable loss of response of the hot wire.

For Vernotte's criticism we have to consider first the time scale for the flow pattern around the wire, for which we may take the time corresponding to the Strouhal number, namely $\tau_3 = d/\bar{U}$. However, since shedding of vortices only takes place when the Reynolds number $\bar{U}d/\nu$ is greater than 40, say, and most hot-wire anemometers are operated at lower Reynolds numbers, instead of τ_3, the time scale $\tau_4 = d^2/\nu$ for the viscous flow closely around the wire will be a more suitable one.

Similarly the time scale for the temperature field around the wire $\tau_5 = \rho_g c_p d^2/\kappa_g$ is of importance in connection with the second part of Vernotte's criticism. For gases with $\mathbf{Pr} = \mathcal{O}(1)$, we have $\tau_4 \simeq \tau_5$.

Corrsin[96] considers two more time scales, one determined by the non-steady heat flow within the hot wire in radial direction, $\tau_6 = \rho_w c_w d^2/\kappa_w$, the other in axial direction, $\tau_7 = \rho_w c_w l^2/\kappa_w$. The latter is similar to the adaptation time $\rho_w c_w l_c^2/\kappa_w$ considered in Sec. 2-3.

All these time scales then should be much smaller than τ_c. It can be easily verified, that for the usual dimensions of the hot wire and the properties of wire materials and of the fluid, the condition that the time scale should be small compared with the characteristic time τ_c, is easily satisfied for τ_6, in a reasonable way for τ_2, τ_3, τ_4, and τ_5, but cannot be satisfied for τ_1 and τ_7, and as discussed earlier hardly for the adaptation time. Since the length l of a hot wire cannot be reduced below a practical minimum value, little can be done to reduce the time scales τ_1 and τ_7. Also, for practical reasons, there is a lower limit for the diameter d of a hot wire. This limit has already been reached in the thinnest hot wires at present in use. So, again very little can be done in this respect with the time scales τ_3, τ_4, and τ_5. Since τ_2 is determined by the time constant of the hot wire, which time constant contains the heat capacity of the hot wire, improvement can be expected from reducing this heat capacity by reducing the mass or volume of the sensing material. Fingerson[97] describes a hot wire composed of a thin glass rod of 50 μm diameter coated with a very thin platinum film. The relatively large value of $d = 50$ μm, however, will adversely affect the time scales which are proportional to d^2. Another promising development is mentioned by Wehrmann,[98] who describes a "whisker" hot wire, where a 0.1 μm thin nickel film is vapor-deposited on a single sapphire Al_2O_3 "whisker" of 5 μm diameter. This way of reducing the time constant of the hot wire has the additional advantage of reducing also its noise level.

Another difficulty, which becomes more pronounced in the case of thin wires, is that the hot wire is sensitive to deposition and impact effects of small particles (dust); consequently, changes may occur in the values of calibration parameters. In particular the hot-wire characteristic is very sensitive in the high-frequency range to even slight deposition of extremely small dust particles. Though the measuring of, for instance, the turbulence velocity intensity may be hardly affected, the measuring of the energy spectrum in the high-frequency range or of the dissipation spectrum may be seriously affected. Hence cleaning and recalibration of a hot wire up into the high-frequency range are unavoidable.

This cleaning and recalibration may become quite a nuisance when the intervals become very short. When turbulence measurements are carried out in a closed-circuit windtunnel, provision may be made to clean the air in the windtunnel. Collis[37] describes a simple means of cleaning the air, namely, by installing very fine woven-wire gauze in a suitable section of the tunnel. Since the air is recirculated, the collecting efficiency need not be high, but nevertheless, particles much smaller than 1 μm must eventually be removed from the air. Even when the air is cleaned in this way it will still be necessary to clean the wire itself after some hours of operation, for there will always be some leakage of impure air and this method of cleaning the air is never 100 per cent efficient.

A common method of removing dust from the wire is to brush it very carefully. This requires some experience, to avoid breaking the wire. The method fails, however, when the deposit is of a tarry nature. Wyatt[38] has successfully applied a technique, soaking the hot wire for half an hour in a liquid with some solvent properties (e.g., methylated spirits) and slightly agitating the liquid acoustically with frequencies of 15 to 60 Hz by means of a loudspeaker. Eighty to ninety per cent of the dust is removed in this way. The sensitivity of a hot wire periodically so treated is said to remain constant to within 2 per cent.

A related difficulty is presented by the effect of changes in wire material due to its over-temperature. In particular its oxidation stability is of importance. In this respect tungsten is a less favorable material. Its oxidation instability calls for a lower over-temperature than for instance with platinum, at the cost of its sensitivity. However, platinum-coated tungsten wire is available and this has the advantage of the strength of tungsten and the oxidation stability of platinum.

Normally the hot-wire anemometer is operated with a transfer of heat from the wire to the surrounding fluid of room temperature. This way of operation is no longer possible in the case of high-temperature gases. It is obvious then to think of the possibility of a sensor as a heat sink instead of a heat source. This idea has led to the design of liquid-cooled tubular sensors, as described by Fingerson[97] and by Ellington and Trottier.[99]

Still another disturbing effect, though not specific for the hot-wire anemometer, is that caused by the vicinity of a wall when measurements very close to a wall have to be made. For the usual hot-wire anemometers operating in air, the effect of a wall becomes non-negligible when the distance of the hot wire to the wall drops below 25 to 50 wire diameters. This value in general decreases with increasing Reynolds number. From the beginning of the application of hot-wire anemometers experimenters have been aware of this effect, which in the case of heat-conducting walls causes an optimistic value of the measured velocity because of the extra heat loss direct to the wall. Van der Hegge Zijnen[100] assumed this loss to be caused only by heat conduction through the gas to the wall, which assumption soon turned out to be incorrect. Piercy, Richardson, and Winney[101] determined the simultaneous effect of forced convection on the correction factor to be applied to the hot-wire anemometer reading by making the calibration measurements with the stationary hot wire close to a rotating plane disk. Much earlier, Reichardt[102] had already calibrated a hot-wire anemometer for the wall effect in a two-dimensional Poiseuille flow. Much more recent and more complete investigations have been made by Wills.[103] He used the same method as Reichardt. In general the heat transfer from a hot wire to the flowing fluid in the vicinity of a wall depends on the dimensionless groups mentioned in Eq. (2-2) and, in addition, on the relative distance to the wall, the velocity distribution close to the wall and the heat transporting properties of the wall. Wills used a 4 μm tungsten wire with $l/d = 625$. He found that he could correlate his experimental results with the relation (2-8a) by Collis and Williams, with a correction factor that appeared to be only a function of the relative distance s/d, when he wrote for the Reynolds number term $\mathbf{Re}_f{}^{0.45} - K(s/d)$; K decreases uniformly from 0.41 at $s/d = 5$, via the values 0.22 at $s/d = 10$ and 0.1 at $s/d = 20$ to 0.02 at $s/d = 50$. Since Wills experimented with only one hot wire with $l/d = 625$ he could not determine a possible additional effect of l/d. For one may expect a decrease of the wall effect with decreasing l/d. Thus, the problem has not yet been completely solved in a satisfactory way. So far, all these calibration experiments have been done in steady, laminar flow conditions. Still unknown is whether the corrections so obtained may also be applied to turbulent flow conditions. Wills reported for turbulent flow in the same two-dimensional duct a value of 0.5 ($K_{\mathrm{lam.flow}}$).

Hitherto, we have considered the application of the hot-wire anemometer to turbulence measurements only in air or other gases. For many years its application has been practically limited to gases, since the hot wire had proven to be much less suitable for *measurements in liquids*. A low operating temperature is necessary because of the possibility of evaporation and the formation of scale.

Overheat ratios $\Delta\Theta_w/\Theta_f$ not much greater than 0.1 should be applied. Closely related to the boiling problem is the occurrence of cavitation. In order to avoid cavitation de-gassed liquids have to be used and, for water, limitations of velocities to below 15 m/s, say. Other troubles are caused by contamination and coating of the wire due to electrochemical and physical processes, algae, lint, and minerals, conductivity of the liquid and, when operating at a $\mathbf{Re} > 40$, the effect of vortex shedding, which effect according to Fabula[104] seems to be rather sensitive to contamination of the wire. Middlebrook and Piret[39] state that electrolysis is responsible for many troubles.

Another disadvantage is that a much sturdier construction with a thicker and longer wire is required to give the required strength, with adverse effects on the sensitivity and time constant of the hot wire. Yet in the early days the hot-wire anemometer in its original design was used for measurements in liquids, but then only to measure mean velocities (for instance, in boundary layers), transition phenomena, or turbulence of rather low frequencies.[40]

Of course the difficulties originating from the liquid can be reduced or even prevented by cleaning the liquid. In the case of water, degassing and de-ionization and the use of filters and chemicals will be of great help.

Ling and Hubbard[86] introduced a new anemometer, which showed a great similarity to the hot-wire anemometer. A heated, very thin, platinum film was used as the sensing element. This film, which had a length of 1 mm and a width of 0.2 mm, was fused onto the wedge-shaped end of a glass or ceramic support. Its operation is similar to that of the hot-wire anemometer. Because of its superior mechanical characteristics, it was thought to be much better suited for measurements in liquids and in either supersonic or high-temperature flow of gases. Ling and Hubbard claimed less sensitivity to surface contamination (e.g., dust) and a superior signal-to-noise ratio under similar operating conditions.

The first designs of Ling and Hubbard's wedge-shaped hot-film sensors were not very successful for a general application, because in practice they did not come up to expectations, giving a lot of troubles of various nature. However, since then there have been important developments. Shapes other than the wedge have been introduced, namely the cone and the cylinder with a body of glass (quartz) or glass fiber. Moreover, in order to protect the platinum film against the abrasive action of minute solid particles and also to eliminate the detrimental effects of conductivity of the water, the film is coated with a thin layer of quartz. The conical-shape sensors have proven to eliminate almost completely deposition of small fiber-shaped particles in the liquid. They have even been used successfully for measurements in high-polymer solutions. The wedge-shape sensor was much less successful, while the use of a cylindrical sensor was out of the question. On the other hand the conical-shape probe is limited in its application because it is insensitive to the direction of the velocity component in a plane normal to the cone axis.

For more detailed descriptions of these hot-film sensors the reader may be referred to the more recent publications by Fingerson,[97] Fabula,[104] and Delleur, Toebes, and Liu.[22,105] Bellhouse and Schultz[106] report the necessity of making a dynamic calibration of a wedge-shape hot-film sensor for measurements in airflows, but that a static calibration may suffice for measurements in water at low speeds because of the much lower turbulence frequencies involved.

A special application of a hot-film sensor is described by Armistead and Keyes,[107] consisting of an 8 μm thin strip of gold on an epoxy substrate and flush with the wall for making turbulence measurements very close to the wall, within the so-called viscous sublayer (see Chap. 7). The idea of a built-in wall sensor is not new. Ludwieg[108] describes a sensor flush with the wall for measuring the wall shear-stress via the rate of heat-transfer to the flowing fluid. It has been applied with success by Ludwieg and Tillmann[109] to turbulent boundary layers. Bellhouse and Schultz[110] extended this method to the measurement of the fluctuating wall shear-stress of a flat plate oscillating in its own plane in the direction of an air stream.

2-8 OTHER TURBULENCE MEASURING PROBES

In this and the following section we will describe briefly a number of methods which have been applied more or less successfully to the measurement of turbulence quantities. They are considered only briefly because they have not yet reached the stage of more general applicability as obtained with the hot-wire anemometer. The reason is that they either have only a limited applicability (as, e.g., the mechanical probes) or are still in the development stage. Only the very few workers who have devoted a lot of time to development work on such new methods have obtained some success, while others not familiar with the many subtle tricks in the correct design, manufacture, and operation of the probe will only obtain no or disappointing results, if they do not put in sufficient effort to acquire the required knowledge and experience.

Thermistor

A probe whose operation principle shows a great similarity with the hot-wire or hot-film probe, is that where use is made of the *temperature dependency of the electric resistance of a thermistor* on its temperature. When a small electrically heated thermistor bead is placed in a flowing fluid, either gas or liquid, its temperature, for a constant electric current, will depend on the velocity of the fluid past the bead, and on the thermal properties of the fluid. Relations for the dynamic behavior of the thermistor similar to those for hot-wire anemometry are obtained. Rasmussen[111] describes the application of a thermistor operating with a constant current and with a constant temperature. Later Pigott and Strum[112] showed that a third method is possible which is neither strictly constant current nor strictly constant temperature. The resistance of the thermistor bead is one leg of a Wheatstone bridge and the unbalance of the bridge is a measure of the flow velocity along the bead.

For such a thermistor bead there is no or a poor direction sensitivity. In this respect it is comparable with the conical hot-film anemometer. Also the same problems of hot-wire anemometry concerning the time-constant, nonlinearity, stability, noise-level, sensitivity to contamination, etc., apply to the thermistor probe. Velocity gradients, and thus shear-stresses at the wall can be measured with a thermistor built-in flush with the wall and thermally insulated from it.[113]

The other probes to be considered here may be classified in two groups, one where the probes operate according to a mechanical principle, and the other where the operation is based on an electrical principle.

Mechanical Probes

We may begin with the special pressure probes for measuring either the dynamic velocity pressure or the static pressure. They have been specially designed for reducing inertia effects. A *total-head tube* of the hypodermic needle type can be made sufficiently small (outer diameter less than 1 mm) to satisfy sufficiently well this condition and that for a "point" measurement. Broer and De Haan,[114] and De Haan[115] describe such a tiny total-head tube. Mechanically it consists of the hypodermic-needle tube and a pressure transducer, for which a thin membrane plus a piezo-electric crystal of Rochelle salt has been chosen. The whole system operates as a coupled linear system of acoustical, mechanical, and electrical elements. One of the main problems is to get rid of or to compensate for the resonance, or "organ-pipe," effect. De Haan has succeeded in obtaining for the whole system an almost constant response in a frequency range up to 3,000 s^{-1}. It has been successfully applied to measuring the intensity of the axial turbulence component in a flow of low relative turbulence intensity, so that the nonlinear turbulent terms in the total-head expression may be neglected. No correction for effects of the lateral turbulence components are made, which is only permitted at low values of the relative turbulence intensity. By taking two equally long hypodermic needles, placed parallel and as close as possible to each other, it is possible to measure also a lateral component of turbulence.[116] To this end the tips of the two tubes are chamfered. By giving this twin-tube a yawed position with respect to the main-flow direction the pressure readings of the two tubes separately make it possible to determine the intensity of the axial component of the turbulence, that of the lateral component in the plane of the two tubes, and their correlations, following a procedure similar to that used in hot-wire anemometry.

Several investigators[117,118,119] have studied the possibility of measuring the fluctuations in pressure in a turbulent flow, by means of *static-pressure tubes*. Because of the many problems presented as, for instance, how to eliminate or to correct for the dynamic effects of the lateral turbulence velocity components, the correct measurement of the pressure fluctuations has not yet been possible. These problems will be appreciated if we consider the many factors that influence the pressure at the static-pressure holes. There is the geometric factor, which accounts for the design of the probe. Of importance are the nose curvature, the distance of the holes (or annular slit) from the tip and from the downstream stem or holder. The second factor is the Reynolds number (viscosity) effect. The third factor is connected with the nature of the turbulence; its intensity, scales and correlations, etc. Usually in the studies made it is assumed that the scale, which is taken as that of the eddies with the largest kinetic energy, is large compared with the diameter of the cylindrical probe. Furthermore it is assumed that the instantaneous flow pattern around the probe may be considered as quasi-steady, so that the nonsteady effects may be neglected. At each instant the flow pattern is well established, which assumption at least requires that the time constant occurring in the Strouhal number is small compared with the periodicity of the turbulence fluctuations. When the relative turbulence intensity is rather small, as a first approximation use can be made of the direction sensitivity of the static-pressure tube for steady flows. The difference in measured fluctuating pressure p_{meas} and the actual value p then is given by

$$p_{\text{meas}} - p \simeq -C\rho[u_n^2 - \overline{u_n^2}]$$

where $u_n^2 = u_2^2 + u_3^2$.

The pressure tube is assumed to be aligned in the main flow direction, coinciding with the x_1-coordinate. Calibration under static conditions yields a value of ~ 0.5 for the numerical constant C. This is the value which would occur for a potential flow around a cylinder. Siddon[119] obtained under dynamic conditions values varying from 0.3 to 0.5, depending on the nature of the fluctuating flow.

From the above relation we obtain

$$\left[\overline{(p_{\text{meas}} - p)^2}\right]^{1/2} \simeq C\rho\overline{u_n^2}\left[\frac{\overline{u_n^4}}{(\overline{u_n^2})^2} - 1\right]^{1/2}$$

The flatness factor $\overline{u_n^4}/(\overline{u_n^2})^2$ depends on the nature of the turbulence. But for current values, and considering the above values of C, it is concluded that the R.M.S. of the error in measured pressure is of the same magnitude as that of the turbulence pressure.

On the other hand we may obtain for the difference in mean-square values

$$\overline{p^2} - \overline{p_{\text{meas}}^2} \simeq 2C\rho\overline{p_{\text{meas}}u_n^2} + (C\rho\overline{u_n^2})^2\left[\frac{\overline{u_n^4}}{(\overline{u_n^2})^2} - 1\right]$$

For a flatness factor greater than 1, which is usually the case, the second term on the right-handside is positive. Since $\overline{p_{\text{meas}}u_n^2}$ may have negative values, it is possible that the measured value of $\overline{p^2}$ becomes equal to the actual value. Indeed, Siddon[119] found for the turbulent flow considered in his experiments that the differences between the two values were less than 20 per cent.

At a wall the fluctuations of the pressure can be measured with a tiny hole and by applying De Haan's method, discussed above. The hole in the wall must be carefully designed and made. In order to reduce any effect of the hole dimensions, its length-diameter ratio should be smaller than 1, while its diameter should be so small, when possible, that u^*d/ν does not exceed the value 100. Here u^* is the wall shear-stress velocity (see Chap. 7). If this condition cannot be satisfied, too high a pressure will be measured. The error in pressure measured becomes greater than the wall shear-stress when $u^*d/\nu > 500$, roughly.[120,162]

Instead of a hole in the wall, Willmarth and collaborators[121,122,123] use a piezo-electric crystal (barium-titanate, later lead-zirconate) built in flush with the wall surface. By making the diameter of the sensitive area as small as 1.5 mm, a sufficient spatial resolution power in a boundary layer with a displacement thickness (see Chap. 7) about eight times this diameter is obtained.[123]

Preston[124] used with success a *hypodermic needle tube* to determine the *local wall shear-stress* from the dynamic pressure indicated by the tube. To this end a tube with an outer diameter of 1 mm and internal diameter to outer diameter ratio of 0.6, was placed parallel to the main flow and against the wall. The diameter of the tube is still large enough to ensure that the main part of the tube diameter is in the fully turbulent part of the flow along the wall. The dynamic pressure indicated by the tube depends on the local velocity distribution close to the wall. As will be discussed extensively in Chap. 7, there is close to the wall a universal velocity distribution, which for a smooth wall is only determined by the kinematic viscosity ν of the fluid, the wall shear-stress σ_w and the distance to the wall. So for a tube with outer diameter d_o the dynamic pressure ΔP indicated by the tube depends on ν, σ_w, ρ, d_o, and the inner diameter d_i. When d_i/d_o is kept close to 0.6, dimensional analysis shows that for the universal velocity distribution (so-called law of the wall) $\sigma_w d_o^2/\rho\nu^2$ is a function of $\Delta P d_o^2/\rho\nu^2$. This function can be obtained by calibration.

Patel[125] investigated the effect of the diameter ratio d_i/d_o. When this ratio is kept close to the recommended value of 0.6, the effect may be neglected. In the case of flows with favorable and in particular with adverse gradients, where the law of the wall for zero pressure gradient no longer applies, of course the calibration curve for the Preston tube fails.

Quarmby and Das[126] experimented with flattened Preston tubes. The mouth of the tube had a flat elliptical shape. The same calibration curve as for a circular tube could be used, if the short principal axis of the outer "ellipse," perpendicular to the wall is taken instead of d_o for a circular tube. The corresponding principal axis of the inner "ellipse" should then be taken 0.6 times that of the outer "ellipse."

Transverse components of turbulence have been measured through the deflection of an *elastically suspended small thin plate*. Siddon and Ribner[127] describe such an aerofoil type of probe. They used either a thin round disk of 1.8 mm diameter or a rectangular plate of the same area and an aspect ratio 2. The fluctuating forces in the suspension rod of the plate corresponding with the deflections of the plate due to the transverse component of the turbulence, are recorded by means of a piezo-electric quartz transducer. For not too high relative intensities of the turbulence (say, smaller than 30 per cent), a linear relationship is obtained between the voltage signal and the transverse velocity component. Comparison with measurements in a free turbulent air jet by means of a hot-wire anemometer showed very satisfactory agreement in the frequency range between 200 and 2,500 s^{-1}.

For measuring the transverse turbulence component in water Hartung and Csallner[128] applied the same principle of a deflecting plate in two instruments. In one of the instruments the deflecting plate causes a change in the electric field around two electrodes, feeded with a high-frequency (10^5 Hz) alternating current, and positioned close to the plate. The bending elasticity of the plate, fixed at one end, produces the spring action. In the other instrument the plate is connected at one end to a long rod which is mounted rotatably along its axis. The alternating rotary motion, due to the forces on the plate, is counteracted by a spring. The deflection of the plate then is recorded through an induction coil on the rod in conjunction with a concentric annular magnet. The latter instrument had been made quite robust, with the adverse effect on the time constant. This instrument was recommended only for measuring the large-scale properties of the turbulence.

Siddon[165] describes a tiny probe consisting essentially of a thin, hollow rod with a flexibly mounted frontal part, which utilizes aerodynamic lift effects due to the transverse turbulence velocity-component. Comparison of the spectral distribution of this velocity component measured in this way in a turbulent free jet, with that obtained with a hot-wire anemometer showed satisfactory agreement in the frequency range up to 2,000 Hz.

Simple *vane and propeller-type anemometers* are well known, and have long been used for measuring mean-flow velocities.

For measuring the vorticity of a steady component of streamwise vortices in a turbulent flow, Trutt[129] and Elder[130] used a simple, straight thin vane (10 × 10 mm) rotatable along an axis parallel to the mean flow. The rotational speed of the vane was measured with a stroboscope. Due to its simplicity of construction no high accuracy can be expected.

A much more refined construction of a propeller flowmeter, in this case for measuring turbulence velocity fluctuations, has been used satisfactorily in waterflows.[131] It consists of a four-bladed propeller, made of Luran, with an outer diameter of 15 mm and a pitch of 50 mm. Around the tips of the four blades is mounted a perforated ring of 0.5 × 1.2 mm, with 60 holes of 0.5 mm. Close to the outer surface of the ring, with a gap of 0.1 mm, are two electrodes at a separation distance equal to 0.75 times the distance between two holes. When the propeller rotates in water a small variation in electric resistance between each of the electrodes and the support occurs each time a hole passes the electrode. Fluctuations of the rotational speed due to turbulence (mainly the axial velocity component) can be recorded through a modification of the number of pulses per unit of time. A detailed analysis of the response of this propeller flowmeter to turbulence is given by Schuyf.[132] Also Plate[133] and Iwasa[134] describe the behavior of a helical type flowmeter in turbulent flow of water.

Since the propeller flowmeter appears to be direction-sensitive also, though within a much more restricted range of angles of yaw than the hot-wire anemometer, in principle it should be possible to measure also transverse velocity components. Limitations to its use are determined by its spatial resolution power, its sensitivity to mechanical damage and to fine particulate material (lint) in the water.

Electrical Probes

A method which potentially offers many advantages as, e.g., its almost inertia-free operation, for measuring turbulence in gases is based on the potential-electric-current characteristics of an *electric discharge* between two electrodes. For electrodes of a given shape and gap, the characteristic depends on the nature of the gas, the pressure, temperature, humidity, and velocity. It is this last dependence—which, moreover, is rather strong—that makes the electric discharge applicable to velocity measurements.

It is probable that Thomas[42] was the first to apply the electric discharge for a technical purpose, namely, for designing a diaphragmless microphone for radio broadcasting.

The first application of such an electric-discharge anemometer to the measurement of turbulence has been made in 1934 by Lindvall,[43] who used the sensitivity of the glow-discharge potential to gas velocity.

According to Lindvall, the best materials to use are platinum and platinum-iridium, because of their relatively stable discharge and good reproducibility, although tungsten also gives satisfactory results.

The glow discharge is characterized by a cathode-glow zone of a few tenths of a micron with a potential difference of 300 volts and by a positive zone with a gradient of 150 volts/mm. The discharge current amounts to about 10 to 30 mA.

To measure turbulence, Lindvall used platinum electrodes of 1.5-mm diameter separated by a distance of about 0.1 to 0.2 mm. The total potential difference across these electrodes amounts to about 300 to 400 volts. As already mentioned, the potential-electric-current characteristic has a negative gradient.

It has been stated that there is a sensitive response to gas velocity. It is possible to obtain a potential difference of 1 volt for a difference in velocity of 1 m/s. This is considerably more than is obtainable with the hot-wire anemometer, where the potential difference is of the order of 0.01 volt. Lindvall considered this a very important advantage, because a less strong amplification is now needed. Another important advantage is that effects corresponding to the thermal-inertia effects of the hot-wire anemometer are very insignificant. A new discharge is built up in less than 10^{-5} s; so velocity fluctuations up to frequencies of 10^5 s^{-1} may be followed by the glow-discharge anemometer.

Lindvall has applied this method to measurements of turbulence in the wake of a circular cylinder and has found satisfactory agreement with measurements with a compensated hot-wire anemometer. During and since World War II, Fucks[44–46] in Germany studied extensively the application of the corona-discharge anemometer for turbulence measurements; the same was done in France by Agostini.[47]

Agostini worked with two small spherical electrodes and also with a spherical anode placed centrally with respect to an annular cathode. Fucks advocates the use of spherical platinum electrodes. Tiny spherical electrodes can be produced quite simply by melting the ends of platinum wires in an electric discharge.

Fucks[44,45] and Werner[48] have also studied experimentally the sensitivity to variations in pressure, temperature, and humidity, and Fucks[44] has investigated the effect (beneficial) of radioactive preionization.

Fucks, Agostini, and Werner have used greater distances between the electrodes. The potential-electric-current characteristic here has a positive gradient.

Franzen, Fucks, and Schmitz[135] showed that satisfactory results, compared with a hot-wire anemometer, can be obtained when measuring the spectral energy distribution of the turbulence produced by a grid.

In Ref. 46 Fucks shows that it is possible to measure not only total turbulence intensities but also the turbulence intensities of the components, separately. This is possible because the corona-discharge anemometer is sensitive to the direction of the gas flow.

When the vector connecting the positive with the negative electrode makes an angle with the velocity vector, the cathode current of the corona discharge has a maximum when the electrode vector coincides with the velocity vector. It becomes zero when the angle is 45° and attains a minimum when the two vectors again coincide but have opposite sense. This direction sensitivity can be used to measure the transverse components of turbulence, following the same hot-wire anemometer procedure.

A number of electrical methods have been developed and used for making turbulence measurements in liquids. Since the hot-wire or hot-film anemometer has not yet given the same satisfactory results as in the case of gases, these other methods would be very welcome if they could be developed to the same degree of accuracy and reliability as the hot-wire anemometer in air. However, although these other methods have very attractive properties, there are apparently many difficulties still to overcome before a general application will be possible.

The method of *electromagnetic induction* has proved its usefulness for turbulence measurements in liquids.

If the liquid is slightly ionized (the liquid must be an electrolytic conductor, but even a poor one such as tap water will do) and has a velocity with respect to a magnetic field, an electric field strength is induced. This induced field strength is determined solely by the velocity component that is perpendicular to the magnetic field and is directly proportional to this component and to the magnetic-field strength.

Let U be the velocity component perpendicular to a magnetic field with strength H; then the strength V of the induced electric field is, according to the law of Maxwell-Faraday,

$$V = \frac{\mu}{c} HU$$

where μ is the relative magnetic permeability of the liquid and c is the velocity of light.

The direction of the induced field is perpendicular to both the velocity component U and the magnetic field. Its sense can be determined according to the well-known left-hand rule.

If two electrodes with gap s are introduced into the flowing liquid on a line parallel to the induced electric field, the potential difference E between the two electrodes reads

$$E = Vs = \frac{\mu}{c} HUs$$

A very important fact is that the voltage E is a linear function of the velocity U and that it is direction-sensitive, since it is affected only by the velocity component perpendicular to the connecting line through the electrodes and to the direction of the electromagnetic field.

Furthermore, the relation between E and U is affected neither by the density, viscosity, temperature, and electric conductivity of the liquid nor by its composition.

It is recommended that the electrodes be made of materials that are unaffected by acids and cannot be polarized. With polarizable materials, an alternating electromagnetic field must be used; this requires a compensation circuit to neutralize the induced alternating potential.

It is possible either to expose the entire flow field to a magnetic field[49,53,54] or to introduce into the flow field a relatively small electromagnet connected to a pair of electrodes in a line perpendicular to the line through the poles.[52]

In the first case the construction is much simpler, but account must be taken of the effect of locally induced currents if the flow distribution of the fluid is not uniform throughout the flow field. Thürlemann[50] has calculated the distribution of such local electric field strengths for the flow through a circular tube in the case of laminar as well as turbulent flow. Kolin and Reiche[51] have shown theoretically that it is possible to make corrections for these induced currents.

Since the potential differences between the electrodes are extremely small, namely of the order of 10^{-4} to 10^{-5} volt, corrections for background noise have to be made.[53,54]

The electromagnetic induction method has been developed to a reliable tool for measuring the average velocity or the total volume flow rate in a cross section of a tube. Complete apparatuses are commercially available.

However, mainly due to the above-mentioned spurious, local, induced currents and the background noise in relation to the smallness of the voltage signal, many experimenters still have difficulties in making reliable turbulence measurements. Though Grossman's[53] measurements of the intensity of turbulence and of the turbulence shear-stress in the flow of water through a straight circular pipe of 43 mm internal diameter gave reasonable agreement with those obtained by other experimenters with hot-wire anemometers in air flow through tubes.

The *conductivity* of the liquid (even though very small) can be used to measure flow velocities. The liquid in the narrow gap between two electrodes is then the sensing element. The path of the electric current from one electrode to the other is distorted by the flow velocity between the electrodes. Eskinazi[136] has shown that this method can in principle be used to measure turbulence velocities. He used a gap between 0.25 and 0.75 mm, and a sufficiently weak current (10 to 30 μA) to prevent hydrogen-bubble formation at the sharp electrode. The dynamic response expressed in the relative voltage changes is almost instantaneous, and linear for small values of the relative intensity of turbulence. More development work is needed to get more information on the direction sensitivity, and to account for the effect of deposition of impurities on the electrodes and for small changes in the electric conductivity of the liquid due to, e.g., temperature and concentration changes, which changes impair the static calibration relation.

If one is interested in measuring fluctuations in temperature and concentrations, a single electrode can be used, the second being formed by the wall of the liquid-containing system. Gibson and Schwarz[137] used to this end a thin platinum wire of 10 to 50 μm thickness, placed in the center of a conical epoxy probe. The slightly protruding end of the wire at the top of the cone was covered with platinum black to minimize polarization and "noise" at the surface of the probe. The "effective" diameter of the cell around the tip was about 10 times the diameter of the electrode tip.

Chuang and Cermak[138,139] introduced an *electro-kinetic* probe for turbulence measurements in water. It is based on the preferential adsorption of ions at a solid-liquid interface, thus forming an electric double layer. This double layer is perturbed and distorted by the local liquid flow, so that at one place of the interface there is an excess of positive charges and at another place an excess of negative charges.

If at these places electrodes are built in flush with the wall surface, the electrodes being electrically insulated from the main body, a potential difference is generated between the two electrodes. This potential difference depends on the degree of the distortion of the double layer and hence is a measure for the local liquid velocity. It is assumed that what is mainly measured, is the velocity component in the direction of the two electrodes, though a proof of this fact has not yet been given.[140]

The original design of the probe contained two electrodes, but later Binder[141] showed that one single wire can be used and a design of the probe similar to that of a hot-wire anemometer. The wire reacts mainly on the component of the liquid velocity perpendicular to it. An excess of negative charges is generated at the forward stagnation point of the flow around the cylindrical wire. The wire probe has a direction sensitivity almost similar to that of a hot-wire probe. A main problem appears to be the establishment of an independent calibration procedure.

Electrochemical Probe

An interesting probe for measuring turbulence velocities in the immediate vicinity of a wall, similar to the hot-film probe of Armistead and Keyes,[107] or the thermistor built in flush with the wall as used by Lambert, Snyder, and Karlson,[113] is the electrochemical probe used by Reiss and Hanratty,[142,143] and later modified by Mitchell and Hanratty.[144,145] The basic idea is the use of a diffusion-controlled electrode (cathode) built in flush with the wall. The concentration of the diffusing species is constant at the electrode surface, and the electric current becomes proportional to the rate of mass transfer to the electrode. The cathode is used as the measuring or test electrode, the anode, which is given a much larger surface area, as the reference electrode. These nickel electrodes form with the aqueous solution of potassium ferri- and ferrocyanide an electrochemical cell. The diffusing species is the ferricyanide ion. Sodium hydroxide is added to the aqueous solution, so that a low-resistance current is ensured in the cell, except near the electrodes. When the concentration of the sodium hydroxide is large relative to that of ferricyanide, the transfer number occurring in the relation between the current and the reaction rate can be neglected. When, moreover, the area of the anode is made much larger than that of the cathode and a sufficiently high voltage is used, the concentration of the ferricyanide ion at the surface of the electrode is practically zero. In that case the electric current will be directly proportional to the value of the mass-transfer coefficient. Very close to the wall this mass-transfer coefficient depends on the velocity gradient at the wall. This dependence is not linear, but for small values of the relative turbulence intensity, a series expansion yields a linear dependence between the turbulence velocity fluctuation or its gradient at the wall and the corresponding fluctuation of the mass-transfer coefficient.

The flow field that can be investigated in this way is restricted to a very short distance from the wall indeed, much shorter than with the hot-film sensor of Armistead and Keyes,[107] described earlier.

In the modified system test electrodes of rectangular shape with an aspect ratio roughly 10 to 1, are used. For measurements in the turbulent flow through a "Lucite" pipe of 25.4 mm internal diameter, test electrodes were used with a length in the direction perpendicular to that of the main flow varying between 0.5 and 1.5 mm, and with a width (in the main-flow direction) varying between 0.075 and 0.5 mm. Since the rectangular test electrode with large aspect ratio is sensitive to velocity fluctuations perpendicular to the length dimension similar to the hot-wire probe, it is direction sensitive. So by applying the same procedure as for the hot-wire probe the lateral turbulence component of the velocity or of the velocity gradient at the wall can be measured. An arrangement of multiple parallel test electrodes provides the means for measuring Eulerian spatial velocity-correlations.

2-9 METHODS BASED UPON FLOW VISUALIZATION

In this section we shall review briefly a number of methods that belong to the first group mentioned in the introductory section 2-1, i.e., methods based on making flow patterns visible or perceptible to a suitable detecting element outside the flow domain.

If very tiny particles that are insoluble in the continuous phase are supplied to the flowing fluid, their paths can be observed cinematographically or with short-flash photographs. These paths will be identical with the flow paths of the fluid particles if the density of the particles is the same as that of the fluid. By taking short-flash photographs with known exposures or by making these photographs in a series of two or three known intervals, it is possible to evaluate the displacement of the particles according to magnitude and direction and, hence, the components of the flow velocity parallel to the photographic plane. To obtain the total velocity vector, two simultaneous photographs must be taken corresponding to two directions perpendicular to each other. Special illumination arrangements can be made so that the suspended particles appear as bright points.[55]

A method for taking a stereoscopic picture with one camera only instead of two is described by Nieuwenhuizen.[146] Here a front view and a side view of the flow pattern are photographed simultaneously by a single still camera or cinecamera via a set of mirrors. From such a set of photographs the spatial position of a tracer particle can be obtained.

An entirely different method for obtaining the velocity component of a tracer particle perpendicular to the photographic plane has been used by Van Meel and Vermij.[147] The plane light beam for illumination is composed of a number of colored "layers." Through the use of a multi-color filter a color photograph of the streak of a tracer particle obtained during the exposure of fixed duration enables one to determine the above-normal velocity component from the length of the colored parts of the streak. The accuracy of this method, in particular when this normal component is relatively large, is not high, while it requires a high-intensity light source (150 W xenon discharge lamp).

Fage and Townend[61] used an "ultramicroscope" to study the turbulent flow in a straight pipe from the streaks produced by fine particles entrained by the liquid.

The particles to be used should be small with respect to the fine-scale of turbulence in order to obtain reliable results. Furthermore, the introducing apparatus must also be very small, so that flow disturbances produced by this apparatus remain negligible.

For measurements in liquids, emulsions may be used. Suitable emulsions that have been used by various experimenters for making measurements in water[55–57] are mixtures of benzene and carbon tetrachloride and of olive oil and ethylene dibromide. These mixtures can be made such as to have the same density as water. It is obvious that the quality of the photographs will be improved by uniform droplets of the emulsions. For making measurements in turbulence of high intensity, therefore, the emulsion introduced should be sufficiently fine so that no drops are split up by the turbulence stresses; for, when drops are split up, secondary droplets of varying size will usually be produced.

Of course no trouble with deformation or splitting up of tracer particles will be encountered when a suspension of solids is used. Micro polystyrene spheres have been used with success. Since the material density is only 1040 kg/m^3, when used in water an addition of a neutral agent to increase the density of the water to the same value, will prevent any inertia effect. On the other hand if the particle is small enough with respect to the turbulence flow conditions, a difference in density may still have a negligible effect. So even gas-bubbles may be utilized.

Hydrogen Bubble Technique

Clutter, Smith, and Brazier[148] suggested a technique, the so-called "Hydrogen Bubble Technique," which method has been further worked out and successfully applied to the study of the flow pattern near the wall during the transition from laminar to turbulent flow in a boundary layer by Kline and his co-workers.[149] A fine wire of 10 to 50 μm diameter is stretched in the water perpendicular to the main flow-direction. This wire, preferably of platinum, forms the negative electrode of a D.C. circuit, the wall or part of it then acting as the positive electrode. Hard water is sufficient as well as a supply voltage between 10 and 250 volts. Hydrogen bubbles with a diameter between one-half and one wire diameter are produced at the negative wire electrode. After some degree of "aging" of the wire under the operating conditions bubbles are produced uniformly. For illumination a 750-watt projector is used to produce a flat light beam downstream of the wire and parallel to it. For sufficient contrast in the pictures of the streaks shown by the moving gas bubbles the direction of the camera lens should make an angle of roughly 65° with the direction of the light beam. Hydrogen dissolves in water, but even at the low water speeds used of 0.03 to 0.3 m/s the half-life time of the bubbles was sufficiently long to cover a distance long enough for observation. Because of turbulence the bubbles may diffuse rapidly so that more than one wire may be needed to show the turbulent flow over an acceptable area. In order to obtain a marking in space and time, the wire has short sections at regular intervals coated for insulation, while the voltage is pulsed at regular intervals to give the marking in time. Limitations are the restriction to low water speeds and to low-frequency fluctuations of the turbulence. For the quantitative evaluation of the streaks there are the usual sources of uncertainty including the accuracy of length and direction measurements of the streaks, the possible displacement of a bubble out of the plane of observation, the response of the bubbles to velocity and pressure fluctuations, velocity defect downstream of the wire and the finite averaging times.

For measurements in air, tiny soap bubbles can be used,[58] or local regions of higher temperature can be produced periodically by means of an electric spark, as applied by Townend.[60] When the soap-bubble technique is used, in turbulence of high intensity a difficulty exists similar to that mentioned in relation to emulsions in liquids; with Townend's method, the heated micro regions are confined to rather short periods, owing to the effects of heat loss, deformation, and diffusion by micro turbulence. It is also possible that, at least initially, the density difference arising from the colder ambient fluid may lead to errors in the quantitative analysis of the flow paths.

For flow visualization in a windtunnel, Bourot[74] has studied extensively the use of micro aluminum flakes (with a diameter of a few microns) as a tracer.

We have already mentioned that the particles must have the same density as that of the flowing fluid; this, however, is not always possible. The question arises, therefore, as to what difference in density is admissible if we are to achieve a certain degree of measuring accuracy. This admissible difference depends on the nature and intensity of the turbulent-flow pattern and on the size of the suspended particles.

If the density difference is great, the error made may soon become appreciable. To illustrate this, we may assume a particle of sand subjected to the turbulent motion of a water flow. Let us consider a particle of 100 μm in diameter and a discrete frequency of 100 s^{-1}, say. Due to the inertia effect the fluctuation of the sand particle at this frequency will be 10 to 20 per cent smaller in magnitude than that of the surrounding water. In general it is recommended to have particles smaller in diameter than 10 μm for studying quantitatively the turbulence of the water and smaller than 1 μm in the case of a gas.[150] In the case of a hydrogen bubble of 50 μm diameter, taken as an upper limit for the bubble size in the hydrogen bubble technique, the fluctuation of the gas bubble would be a few per cent too high at the frequency of 100 s^{-1}, as compared with that of the water.

We shall not go further into this matter here, since the same problem will be discussed later in relation to the diffusion of discrete particles in turbulent flow. But it is necessary to focus the reader's attention on this problem because, when a measuring technique involving suspended particles is applied, it is absolutely necessary to study first the effect of density differences in order to know how to interpret the results of measurement. The reader will derive useful information from the study by Bourot referred to above.

Instead of using discrete particles, it is also possible to introduce continuously a suitable material miscible with the flowing fluid. In the case of a gas, for instance, wood smoke, hexite, vapors of ammonia, hydrochloric acid, iodine, etc., may be used, or even relatively coarse materials, such as fine balsa-wood dust.[62] In the case of liquids, coloring matters such as ink may be used or, particularly, patent blue (a sulphonic acid of phenylated rosaniline), which is very persistent and obtainable in various colors. Because of their persistency, more than one color may even be used to study, for instance, mixing processes.

The interpretation of the visual observations when the tracer is continuously injected at one or more points in the flow presents difficulties. For what is observed are the streaklines of fluid particles that pass the point of injection. In an unsteady flow the streaklines differ from the pathlines of individual particles and from the instantaneous streamlines. A great help to the quantitative interpretation is the application of "time-lines" as suggested, amongst others, by Clutter *et. al*[148] and by Hama and Nutant.[151] Here, a line of fluid particles, taken normal to the flow direction, is marked at a given instant. At any later instant the location of the marked particles is called a "time-line." Yet, as Schraub *et. al*[149] have pointed out, the knowledge of streaklines and time-lines is not sufficient. An improvement has been the employment of the "combined-time-streak" markers in their hydrogen-bubble technique.

By the introduction of either discrete particles or smoke or dyes, etc., it is possible to measure various turbulence quantities.

When the main flow is parallel to the photographic plane, the main velocity can be determined from the streaks made by numerous particles, and the lateral and axial turbulence velocities as well, although these will be much more difficult to obtain accurately. It goes without saying that this is a very lengthy and tedious way of determining flow velocities.

Consider a homogeneous turbulence superimposed on a constant velocity $\bar{U}$ in the x_1-direction. Consider further a large number of pairs of points marked ξ at a distance x_2 from each other and distributed over the entire flow region. If we introduce particles at these points simultaneously we can measure the lateral turbulence velocity $u_2(\xi)$ and $u_2(\xi + x_2)$ at all pairs of points from the streaks of the particles shown on the photographic image (it is assumed that the particles have the same velocity as the fluid). Hence it is possible to calculate the product $u_2(\xi)u_2(\xi + x_2)$ for all the pairs of points and take the average value with respect to all ξ, namely $\overline{u_2(\xi)u_2(\xi + x_2)}$. If the turbulence is assumed to be quasi-steady, we may also consider only one pair of points, take many pictures of the particles introduced successively at instants t, and determine the average with respect to all $t: \overline{u_2(t)u_2(t; x_2)}$. By giving x_2 various values we obtain the longitudinal Eulerian correlation $\mathbf{f}(x_2)$.

From the spatial dispersion of the particles in an x_2-direction downstream of the injection points, the Lagrangian correlation $\mathbf{R}_L(t)$ [see Eq. (1-80)] can be obtained from the relation (1-81) or (1-82), namely,

$$\mathbf{R}_L(t) = \frac{1}{2v'^2} \frac{d^2}{dt^2} \overline{[y_2(t)]^2}$$

and we may make use of the relation $x_1 = \bar{U}t$.

The Lagrangian integral time-scale $\mathfrak{I}_L$ is obtained from

$$\mathfrak{I}_L = \int_0^\infty dt\, \mathbf{R}_L(t) = \frac{1}{2v'^2} \lim_{t\to\infty} \frac{d}{dt} \overline{[y_2(t)]^2}$$

whence the space scale $\Lambda_L = v'\mathfrak{I}_L$ [see Eq. (1-88)].

From the diffusion of heat, smoke, or a dye downstream of a source, the turbulence intensities and the eddy diffusion coefficient ϵ_y can be obtained.

If the turbulence is homogeneous and isotropic, the distribution of Γ in a plane perpendicular to the flow direction of $\bar{U}$ and at a distance x_1 from the source is given by a Gaussian error curve (see Sec. 5-5),

$$\bar{\Gamma} = \bar{\Gamma}_{max} \exp\left(-\bar{U}x_2^2/4x_1\epsilon_y\right) \qquad (2\text{-}96)$$

where $\bar{\Gamma}_{max}$ is the maximum value of $\bar{\Gamma}$ in the plane at a distance x_1 from the source.

Taking $(x_2)_{1/2}$ as the coordinate value where $\bar{\Gamma} = \frac{1}{2}\bar{\Gamma}_{max}$, it follows from Eq. (2-96) that

$$\epsilon_y = \frac{\bar{U}}{4x_1 \ln 2} (x_2)_{1/2}^2 \qquad (2\text{-}97)$$

The intensity u' can be obtained from measurements at very short distances from the source; so a nearly perfect correlation between velocities exists. As will be shown later (Chap. 5) a Gaussian $\bar{\Gamma}$-distribution is again obtained in a plane at a distance x_1 from the source,

$$\bar{\Gamma} = \bar{\Gamma}_{max} \exp\left[(-x_2^2/2x_1)(\bar{U}^2/u'^2)\right]$$

whence

$$\frac{u'}{\bar{U}} = \frac{(x_2)_{1/2}}{x_1\sqrt{2 \ln 2}}$$

Frenkiel[64] describes a similar method. He used ammonia as a tracer in conjunction with a grating composed of cotton threads which were impregnated with mercurous nitrate or with dibromo cresol sulphophthalein. From the degree of discoloration of the grating threads by the tracer gas emitted from the source upstream of the grating, he calculated the eddy diffusion in the turbulent flow.

Hinze and Van der Hegge Zijnen[65] have shown that the diffusion method can also be suitably applied in determining turbulence intensities and the eddy shear stress in shear flow (see Sec. 5-6). Depending on the degree of anisotropy, i.e., on the magnitude of the eddy shear stress, the $\bar{\Gamma}$-distribution just downstream of the source is more or less skew.

There are drawbacks to these diffusion methods; for some of the effects inherent in the character of turbulent flow are not taken—or are very difficult to take—into account. It would be going too far to discuss these effects here; they will be discussed in Secs. 5-5, 5-6, and 5-7, to which the reader may refer.

If one is particularly interested in the distribution of shear stresses in a flow field, use can be made of the orientation of particles suspended in a liquid and made visible by means of a double refraction of polarized light.[66-69] Suitable materials are colloidal clay suspensions (magnesium, bentonite, or hectorite), tobacco mosaic virus, soap solutions, and aged sols of vanadium pentoxide. The degree of orientation depends on the magnitude of the lateral velocity gradient and on the length-diameter ratio of the particle. The particles must not be too small; at any rate they must be large enough so that their motion is unaffected by Brownian movements.

This method has been applied successfully for measuring two-dimensional laminar flows and also for rather slow two-dimensional turbulent flows with a distinct region of high average shear stresses. The method will probably fail in strong turbulent flows, particularly in three-dimensional flows, because of the obscuring effects of the turbulence fluctuations.

Generally, a visual or motion-picture observation gives better results than short-flash photographs.

Light Refraction by Density Variations

By making use of the phenomenon of light refraction caused by variation in density, useful information can be obtained from density fluctuations that may accompany turbulent flows. These density differences may be caused by differences in temperature, by differences in concentration if two gases with widely different densities are mixed during the turbulent motion, or by pressure differences; this last effect is generally appreciable only in flows where the Mach number is no longer small.

The usual optical methods are: the shadowgraph method, the Toepler "Schlieren" method, and Mach's interferometer method. We shall not discuss these methods and their merits here, but note that in recent years experimenters have shown a growing preference for the interferometer over the "Schlieren" method, because of the relatively simple quantitative analysis involved.

Density differences are a direct measure of differences in temperature or concentration. The gas law gives a direct relation among temperature, pressure, and density.

If the compressibility effects become important as in high-speed turbulent gas flows, the density fluctuations accompanying pressure and temperature fluctuations can be used to gain information on the structure of the turbulence.

Since turbulence is a three-dimensional phenomenon, the density pattern will also be three-dimensional. The question now arises: how to interpret the resultant refraction of a light beam that has passed a turbulent region of a certain depth. Kovasznay[70] has discovered a method of obtaining from a refraction image information concerning the turbulence structure. He used the shadowgraph method. He has shown that, if the turbulence is homogeneous throughout the whole turbulent region and if, in addition, the scale of turbulence is very small with respect to the cross section of the turbulent region, the shadow picture must be identical with the turbulent-flow pattern in a longitudinal section of the region. Furthermore, he has suggested a method for determining velocity correlations and, consequently, the turbulence dissipation and integral scales also, with the aid of these shadowgraphs. This can be accomplished as follows. Two transparent pictures are made from the shadowgraph. If one transparent picture is placed on the other so that there is complete coincidence of the pictures, there is maximum transparence to a parallel light beam. If the pictures are now shifted a little with respect to each other, a decrease in transparency is obtained. Now the resultant transparency of the two pictures is equal to the product of the transparencies of each individual picture. This fact offers the possibility of determining the correlation between densities at two neighboring points of the flow field.

Uberoi and Kovasznay[71] have applied this method for making measurements in the turbulent wake of a projectile at supersonic speed and for measuring the turbulence in a free jet of heated air. In the latter case, the flow was incompressible, and the density variations resulted from the temperature variations; for comparison purposes, measurements were also made using the hot-wire technique. The measurements showed that the shadowgraph method can be used with more reliability in the high-wavenumber region, that is, the region of the smaller eddies. Thus, the dissipation scale is obtained more accurately than the integral scale. Comparison of the values obtained with the two methods showed reasonable agreement for the dissipation scale but very poor agreement for the integral scale.

Hubbard and coworkers[72] used the shadowgraph method to obtain information about the turbulence structure of a free jet of carbon dioxide. They made a Debye-Scherrer spectrum from a short-flash shadowgraph. This method gives a kind of Fourier analysis; so, apparently, some information concerning the energy spectrum of the turbulent flow may be gained. It is probable that, if success is to be achieved by this method, the turbulence must be homogeneous. For the turbulence in a free jet this would be a questionable assumption.

Methods Based on Scattering and Absorption of Light

If a transparent fluid contains a fine or colloidal suspension, light passing through the fluid will be scattered and partly absorbed. Turbulence in the flowing fluid may cause turbulence variations in the concentration of the discrete particles, and so variations in the local scattering and absorption of light. When two fluids have to be mixed and one of the two is marked by the addition of a fine suspension, the turbulent mixing of the two fluids can be studied through the concentration fluctuations. Becker, Hottel, and Williams[152] have developed a method based on scattering of a light-beam by the suspension, for the study of mixing of two fluids.

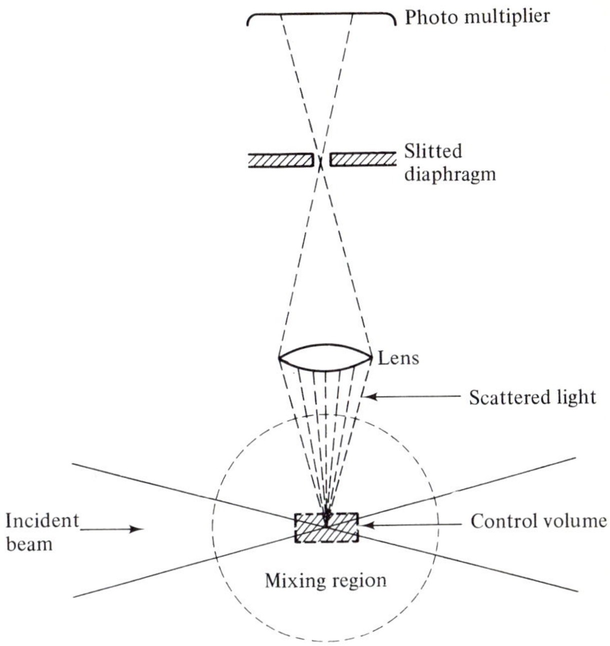

FIGURE 2-4
Principle of light-scatter technique.
(Adapted from: *Becker, H. A., H. C. Hottel and G. C. Williams.*[152])

A size range of the particles between 0.1 and 5 μm appears suitable, with a preference for 0.5 μm particles. A concentration of 10^3 particles/mm^3 in the marked fluid is suggested. Figure 2-4 shows the principle of the light-scatter technique. Through the slit of the diaphragm passes the light scattered by the particles in the control volume, which should be made sufficiently small to obtain "point" values of the concentrations. The wavelength of the light should be at least one order of magnitude larger than the particle diameter. The electric signal from the photomultiplier is proportional to the concentration of particles in the control volume. When molecular diffusion of the fluids and particle diffusion due to Brownian motions are neglected, then the measured concentration of particles gives the concentration of the fluid marked with the particles. Consequently this method gives no information about the mixing of the two fluids on a molecular scale.

The attenuation of a light-beam itself passing a turbulent mixing region containing a fine suspension, due to scattering and absorption, is utilized by Fisher and Krause[153,154] in their *crossed-beam* correlation technique. Two collimated light beams from two light sources, directed perpendicular to each other and in a plane perpendicular to the main-flow direction of the turbulent field, are sent through this field. Before being received by the two photomultiplier detectors the light beams after crossing the flow field pass a monochromator of a suitable wavelength. The signals of the two detectors are sent to a multiplier and a time averaging circuit.

The total attenuation of each light-beam can be expressed in terms of an integral over the total path length, containing the attenuation or extinction coefficient which is a function of time and space. Due to the turbulence fluctuations in the concentration of the suspension, corresponding fluctuations occur in the extinction coefficient. Since the correlation distances in the turbulent flow are finite, the contribution in the integral expressing the total output of the two detectors after multiplication and averaging, is mainly from the volume in the neighborhood of the point of interaction of the two light beams. So what is measured then is the intensity of the turbulent fluctuation of the concentration or of the property causing the attenuation, in a volume containing the point of intersection of the two beams. Two-point space-time correlation can be obtained when the two light-beams are in different parallel planes perpendicular to the main-flow direction at a short, but variable, distance from each other, and if a time-delay is given to the output of one of the detectors before entering the multiplier. There are quite a number of uncontrollable errors, observed as so many sources for "noise" in the final output of the system, which certainly have to be accounted for. Becker, Hottel, and Williams[152] have investigated systematically a number of sources for noise. There is the optical background noise due to possible extraneous light sources, stray light from the lamplight source, re-scatter of scattered light in the mixing field. Noise due to fluctuations in the lamplight source. The effects of optical background noise and of source fluctuations noise have to be determined by calibration. Then there are still a number of potential sources of noise, which, however, are either negligible or can be made negligible. These are: dark noise from the photomultiplier itself, marker shot noise when the number of particles in the control volume is too small, and the white electronic shot noise due to the particulate nature of the electric current. In the light-scatter technique of Becker, Hottel, and Williams there is also the optical attenuation noise because of the attenuation of light along the whole light beam, when the concentration is not small. The figure of 10^3 particles per mm^3 given above is a compromise between the requirements set by keeping both the optical attenuation noise and the marker shot noise negligible.

The last method based on light scatter which we will consider is that for measuring fluid velocities utilizing the *Doppler shift of a scattered laser beam.* First suggested and used by Yeh and Cummins[155] it has been developed further, also for making turbulence measurements, by Goldstein[156,157] and collaborators. Light scattered by a moving particle causes a Doppler shift f_D in frequency, given by the following approximate relation, when the velocity is very small compared with the velocity of light

$$f_D = \frac{n}{\lambda_0} u_j[(\mathbf{e}_j)_{sc} - (\mathbf{e}_j)_{in}]$$

Here n is the index of refraction of the fluid, λ_0 the vacuum wavelength, u_j the fluid velocity vector, $(\mathbf{e}_j)_{sc}$ and $(\mathbf{e}_j)_{in}$ the unit vector in the direction of the scattered beam and of the incident beam respectively. The Doppler shift has a maximum

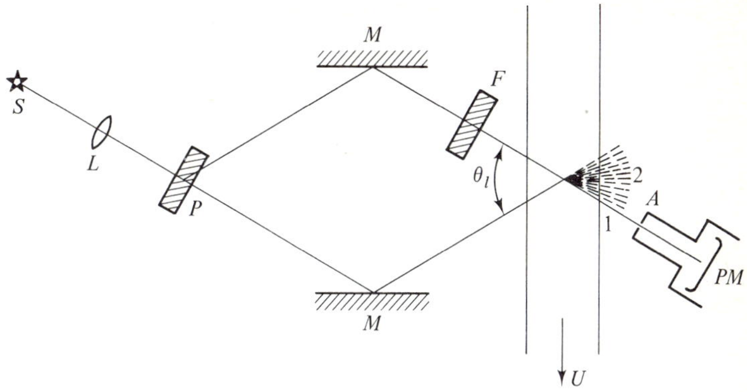

FIGURE 2-5
Doppler-shift of scattered laser beam for measuring velocity in pipe flow.[156]
S = Laser source, L = lens, P = splitter plate, M = mirror, A = aperture,
PM = photomultiplier, F = neutral density filter.
1 = reference beam, 2 = scattered beam.

value when the velocity vector u_j is parallel with the difference vector $(\mathbf{e}_j)_{sc} - (\mathbf{e}_j)_{in}$. The Doppler shift should be small compared with the source frequency, but still large compared with the bandwidth of the beam. Since a laser is a source of an essentially coherent radiation with a very small bandwidth ($\sim 10\ s^{-1}$ for the He–Ne laser), the above conditions are certainly satisfied. The principle of the technique applied and based on the above relation for f_D, is shown in Fig. 2-5. Polystyrene particles of 0.5 μm diameter are added at a concentration of 3×10^{-5} by volume to the liquid flowing through the tube. The light from the incident beam scattered by the particles in the direction of the photomultiplier combines with the non-scattered reference beam. The photomultiplier acts as an amplifier and as a heterodyne receiver. Such a receiver mixes two signals of different frequency and produces an output signal with a frequency equal to the difference in the frequencies of the input signals. The receiver thus generates a current whose A.C. component is equal to the difference, Doppler, frequency. When U is the velocity of the liquid in the tube in the point of interaction of the two beams, the difference frequency reads

$$f_D = \frac{2n}{\lambda_0} U \sin \frac{\Theta_l}{2}$$

Here the angle Θ_l is measured in the liquid. Due to the turbulent motion of the particles the A.C. component of the photomultiplier signal is frequency modulated. Moreover, it is amplitude modulated because of particle concentration fluctuations. Information about the turbulence can be obtained by utilizing a spectrum analyser whose output is slowly scanned and integrated with a R.C.-time constant of 3.5 s. For a given Doppler frequency, i.e., a given value of the fluid velocity, in turbulent flow the intensity of the beat signal is proportional to the fractions of time that the turbulence velocity has a magnitude equal to the given velocity, provided the period of the turbulence fluctuation is small compared with 3.5 s. Hence the probability density distribution of the turbulence velocity is obtained in this way, from which follows the intensity of the turbulence. Also here, measures have to be taken to reduce background noise. The aperture put in front of the photomultiplier, for instance, is intended to minimize noncoherent scattered light.

Goldstein mentions that with a He–Ne laser velocities can be measured in the very wide range between 10^{-6} and 10^{2} m/s.

2-10 MEASUREMENT OF MEAN VALUES OF STATIC PRESSURE AND VELOCITY

We have already discussed the difficulties in measuring turbulent fluctuations of the pressure in a flow. But also the measurement of the mean pressure in a turbulent flow is not a simple matter. If the turbulence velocities are very small relative to a straight mean velocity, usually the assumption is made that the mean static pressure is constant in a plane perpendicular to the mean velocity. It will then suffice to measure the mean static pressure at one point of this plane only. It will be simpler to determine this static pressure when regions of low or no turbulence can be found in this plane. For instance, in the case of free jets, according to the above assumption the static pressure in the turbulent regions of the jet is equal to the pressure in the ambient still fluid outside the jet. And, in the case of turbulent flow through a straight duct, the mean static pressure is assumed constant in a cross section of the duct and equal to the static pressure at the wall. In the first case there is no turbulence in the ambient fluid, and the static pressure can be measured as accurately as pressure measurements can be made. Any possible error in the determination of the static pressure within the jet, therefore, lies in the faultiness of the assumption. Experimental evidence has shown that this assumption may be used to determine mean-velocities in a free-jet with satisfactory results.

The measurement of the static pressure at the wall of a duct is usually accomplished by making several pressure taps in a diametrical plane of the duct, equally spaced along the circumference of the duct; the static pressure is taken as the average of the pressure readings of the pressure taps. The mean pressure in such a tap is affected by the turbulence velocities, normal as well as tangential to the wall. To what extent it is affected is not exactly known. Neither can any general rule be given for possible corrections to be made, since the effect depends on the way in which the turbulence velocities become zero at the wall, on the size of the holes of the pressure taps, and on the roughness of the wall. But it may be expected that the errors made in the measurement of the pressure at the wall are very small, when the wall is smooth and the same measures are taken as discussed earlier, for measuring pressure fluctuations at a wall.

For measuring the mean pressure in the turbulent flow the "static-pressure" holes of a Pitot tube have been utilized. This, however, is an incorrect method, because here the pressure in these holes is too much affected by the turbulence velocities. The static pressure will be measured too low, mainly because of the lateral turbulence velocities perpendicular to the tube. This will become apparent if it is realized that a lateral velocity will cause a pressure distribution along the circumference of the cylindrical tube corresponding to the flow around a cylinder, giving a resultant negative pressure.

As discussed earlier there are many factors, which are difficult to evaluate in an exact way, that affect the pressure in the "static-pressure" holes. Yet, in order to make possible the determination of this effect, in first approximation use is made of the direction sensitivity of the Pitot tube. For relatively small deviations from the main flow direction (corresponding to a relative intensity of the turbulence smaller than 0.15 say) the empirical relation may be approximated by

$$P_{\mathrm{meas}} - P = -A \sin^2 \varphi \cdot \tfrac{1}{2}\rho U^2 \qquad (2\text{-}98)$$

where φ is the angle between the Pitot tube and the velocity.

A static calibration shows that for a standard Pitot tube $A \simeq 1$. However, as discussed earlier, dynamic calibrations show differences with respect to the static calibration, with values depending on the nature of the turbulence and on the design of the tube.[119]

Consider now a turbulent flow with mean velocity parallel to the Pitot tube and with turbulence-velocity components u_1, u_2, and u_3, so that

$$U_{\mathrm{eff}} = \sqrt{(\bar{U}_1 + u_1)^2 + u_2^2 + u_3^2} = \sqrt{(\bar{U}_1 + u_1)^2 + u_n^2} \qquad (2\text{-}99)$$

In Eq. (2-98) we must take for U this U_{eff}; for $\sin \varphi$ we have (see Fig. 2-6)

$$\sin \varphi = \frac{u_n}{U_{\mathrm{eff}}} \qquad (2\text{-}100)$$

Substitution of Eqs. (2-99) and (2-100) in Eq. (2-98), results in

$$P_{\mathrm{meas}} - P = -A \cdot \tfrac{1}{2}\rho u_n^2$$

Hence

$$\bar{P}_{\mathrm{meas}} - \bar{P} = -A \cdot \tfrac{1}{2}\rho \overline{u_n^2} \qquad (2\text{-}101)$$

In an isotropic or axisymmetric turbulence where $\overline{u_2^2} = \overline{u_3^2}$, the error in $(\bar{P}_{\mathrm{meas}} - \bar{P})/\frac{1}{2}\rho\bar{U}_1^2$ will be equal to $-2A\overline{u_2^2}/\bar{U}_1^2$.

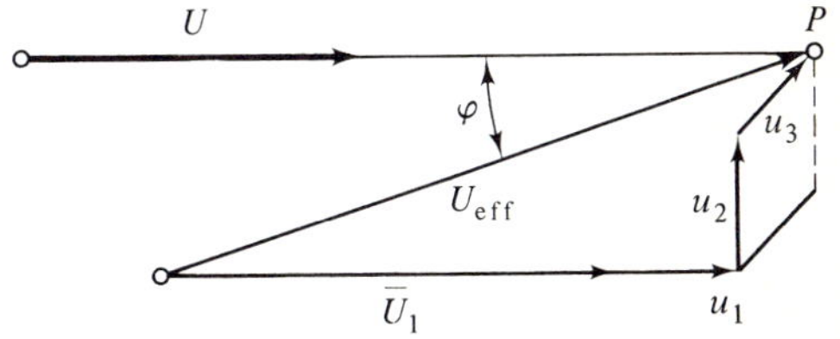

FIGURE 2-6
Instantaneous direction of U_{eff} with respect to $\bar{U}_1$.

Measurement of Mean Velocity

When flow-visualization methods are used, for instance, by introducing discrete particles into the flowing fluid, the value of the mean velocity can be obtained from the average value according to the direction and magnitude of the streaks produced by the particles during the time of illumination or exposure to a recording instrument.[74] The diffusion methods with miscible tracers are not so suitable, since the spread of the tracer downstream of the source depends both on the mean velocity and on the turbulence diffusivity, so that at least one of these factors must be known to measure the other.

Usually the mean velocity is determined directly by means of a detecting element introduced into the flow. For measuring turbulence velocities we have ruled out the standard Pitot tube because of its slowness. But, since we are interested here only in the mean velocity, such an instrument which, by its inertia, gives an average value, might be very useful. In fact, the Pitot tube—or, rather, the total-head tube—is frequently used for measuring mean velocities in turbulent flows. Here, however, we are faced with the same problem as in the measurement of the mean static pressure: how is the measurement of the total head affected by the turbulence velocities?

In a theoretical study Goldstein[75] found that the total-head tube measured the total head of the total velocity vector, since he assumed that the frontal part of the tube might be considered a point of stagnation. In that case we would have

$$\bar{P}_{\text{tot}} = \bar{P} + \tfrac{1}{2}\rho[\overline{(\bar{U} + u_1)^2} + \overline{u_2{}^2} + \overline{u_3{}^2}] \qquad (2\text{-}102)$$

This assumption might be correct if the tube were infinitely small in diameter, so that the frontal area of the tube might be considered a true point.

Owing, however, to the finite diameter of the total-head-tube hole, deviations from the above relation (2-102) may be expected. In the first place, there is always a lateral velocity gradient in turbulent flow, which may give a difference in velocity over a distance equal to the hole diameter. The total pressure measured by the tube is then not equal to the total pressure existing in the center of the tube hole.[76] In the second place, the lateral-velocity fluctuations will not produce an impact pressure as shown in Eq. (2-102) but rather an appreciably smaller one.

However, Goldstein's relation (2-102) may still be used if the tip of the total-head tube is specially designed so as to make the tube much more insensitive to the flow direction than the standard tubes. For instance, by taking a much greater outer to inner diameter ratio and letting the inner hole end at the tip in a conically shaped mouth. According to Gracey[158] deviations from the actual total head within 1 per cent can be obtained in this way for angles $|\varphi| < 27°$. In the absence of basic data to correct for the effect of turbulence on a standard total-head tube, Hinze and Van der Hegge Zijnen[77] neglected the effect of the lateral turbulence velocities and, in Eq. (2-102), considered only the effect of the axial turbulence velocity component. In a few experiments Alexander, Baron, and Comings[78] observed that the total-head-tube reading decreased markedly with increasing relative turbulence intensity, contrary to expectations. Therefore, in using the total-head tube, they further neglected the effect of turbulence and assumed $\bar{P}_{tot} = \bar{P} + \frac{1}{2}\rho\bar{U}^2$. Neither of these two methods is entirely correct, but, as already indicated, at present no better method can be suggested.

We may make use of the direction-sensitivity of a total-head tube to estimate the effect of turbulence on its readings. Instead of Eq. (2-102) we assume the relation

$$(P_{tot})_{meas} = \bar{P} + \tfrac{1}{2}\rho U_{eff}^{\,2} - B\sin^2\varphi\tfrac{1}{2}\rho U_{eff}^{\,2}$$

where $\sin\varphi = u_n/U_{eff}$.

The third term is an approximation, for not too large values of φ, to the empirical direction-sensitivity characteristic.

From static flow measurements a value of roughly 1.2 is obtained for the constant B. But dynamic measurements, where the tube itself is given a transverse vibration with respect to a steady flow, seem to indicate that B may become greater than this value.

Following the procedure already outlined for measuring the mean static pressure, after some calculation we obtain

$$(\bar{P}_{tot})_{meas} = \bar{P} + \tfrac{1}{2}\rho\bar{U}^2\left[1 + \frac{\overline{u_1^{\,2}} + (1 - B)\overline{u_n^{\,2}}}{\bar{U}^2}\right] \tag{2-103}$$

Depending on the value of B, the total head may be measured too high or too low. For a turbulence that is isotropic we obtain

$$(\bar{P}_{tot})_{meas} = \bar{P} + \tfrac{1}{2}\rho\bar{U}^2\left[1 + \frac{u'^2}{\bar{U}^2}(3 - 2B)\right]$$

For $B = 1.2$ a result is obtained from the relation (2-103) that comes quite close to the assumption made by Hinze and Van der Hegge Zijnen.

By combining the expressions (2-101) and (2-103) we may obtain the expression for a Pitot tube,

$$(\bar{P}_{tot})_{meas} - \bar{P}_{meas} = \tfrac{1}{2}\rho\bar{U}^2\left[\frac{\overline{u_1^{\,2}} + (1 + A - B)\overline{u_n^{\,2}}}{\bar{U}^2}\right] \tag{2-104}$$

The last expression is correct only if the velocity fluctuations at the total-pressure hole and at the static-pressure holes are uncorrelated. This requires that the distance between the two points be much greater than the correlation distance or than the integral scale of the turbulence.

More recently an extensive study has been made by Becker and Brown[168] of the response of a Pitot tube in turbulent flows. They obtained the following relation for the directional sensitivity

$$(P_{tot})_{meas} = P + \tfrac{1}{2}\rho U^2[1 - K(\sin^2\varphi)^m]$$

where U is the total velocity vector. K and m depend on the Pitot-tube geometry.

By applying in essence the same procedure as given above, Becker and Brown obtained semi-empirical relations for the response in turbulent shear flows. For a round-nosed probe with a ratio of the inside and outside diameters equal to 0.45 they found the simple relation

$$(\bar{P}_{\text{tot}})_{\text{meas}} \simeq \bar{P} + \tfrac{1}{2}\rho \bar{U}_1^{\,2}$$

to be applicable with satisfactory accuracy.

REFERENCES

1. McAdams, W. H.: "Heat Transmission," 3rd ed., p. 260, McGraw-Hill Book Company, Inc., New York, 1954.
2. Kramers, H.: *Physica,* **12,** 61 (1946).
3. King, L. V.: *Phil Trans. Roy. Soc. London,* **214A,** 373 (1914).
4. Frenkiel, F. N.: *O.N.E.R.A. Rapp. tech. No.* 37, 1948.
5. Dryden, H. L., and A. M. Kuethe: *Natl. Advisory Comm. Aeronaut. Tech. Repts. No.* 320, 1929.
6. Betchov, R.: *Proc. Koninkl. Ned. Akad. Wetenschap.,* **51,** 721 (1948).
7. Lowell, H. H.: *Natl. Advisory Comm. Aeronaut. Tech. Notes No.* 2117, 1950.
8. Van der Hegge Zijnen, B. G.: *Appl. Sci. Research,* **2A,** 351 (1951).
9. Dryden, H. L., G. B. Schubauer, W. C. Mock, Jr., and H. K. Skramstad: *Natl. Advisory Comm. Aeronaut. Tech. Repts. No.* 581, 1937.
10. Frenkiel, F. N.: *Aeronaut. Quart.,* **5,** 1 (1954).
11. Betchov, R.: *Proc. Koninkl. Ned. Akad. Wetenschap.,* **52,** 195 (1949).
12. Betchov, R., and W. Welling: *Proc. Koninkl. Ned. Akad. Wetenschap.,* **53,** 432 (1950).
13. Corrsin, S.: *Natl. Advisory Comm. Aeronaut. Wartime Repts. No.* ACR 3L23, 1943.
14. Eckert, E. R. G.: *Introduction to Heat and Mass Transfer,* p. 173, McGraw-Hill Book Company, Inc., New York, 1963.
15. Kovasznay, L. S. G.: *J. Aeronaut. Sci.,* **17,** 565 (1950).
16. Favre, A.: *Proc. 7th Intern. Congr. Appl. Mech.,* London, 1948, vol. 2, p. 44.
17. Favre, A., J. Gaviglio, and R. Dumas: *Recherche aéronaut. No.* 31, p. 37, 1953.
18. Taylor, G. I.: *Proc. Roy. Soc. London,* **157A,** 537 (1936).
19. Townsend, A. A.: *Proc. Cambridge Phil. Soc.,* **43,** 560 (1947).
20. Davis, M. R., and P. O. A. L. Davies: *Inst. of Sound and Vibration Res., Univ. of Southampton, Techn. Rep.* No. 2, 1968.
21. Sauer, F. M., and R. M. Drake: *J. Aero. Sci.,* **20,** 175 (1953).
22. Delleur, J. W., G. H. Toebes, and C. L. Liu: *Proc. 12th Congr. I.A.H.R.,* Fort Collins, Col., 1967, Vol. 2, p. 227.
23. Cole, J. D., and A. Roshko: *Proc. Heat Transfer and Fluid Mech. Inst.,* Stanford, Calif., 1954, p. 13.
24. Kassoy, D. R.: *Phys. Fluids,* **10,** 938 (1967).
25. Dennis, S. C. R., J. D. Hudson, and N. Smith: *Phys. Fluids,* **11,** 933 (1968).
26. Wood, W. W.: *J. Fluid Mech.,* **32,** 9 (1968).
27. Laufer, J.: Ph.D. thesis, California Institute of Technology, Pasadena, 1948.
28. Liepmann, H. W.: *Helv. Phys. Acta,* **2,** 119 (1949).

29. Corrsin, S.: *Natl. Advisory Comm. Aeronaut. Tech. Notes No.* 1864, 1949.
30. Hieber, C. A., and B. Gebhart: *J. Fluid Mech.*, **32,** 21 (1968).
31. Collis, D. C., and M. J. Williams: *J. Fluid Mech.*, **6,** 357 (1959).
32. Ahmed, A. M.: "Advances in Hot-wire Anemometry," p. 65, AFOSR No. 68-1492, Univ. of Maryland, 1968.
33. Tournier, M., P. Laurenceau, J. J. Huebès, and A. Seigneurin: *Recherche aéronaut. No.* 15, p. 19, 1950; *No.* 19, p. 11, 1951.
34. Laurence, J. P., and L. G. Landes: *Instruments,* **26,** 1890 (1953); or *ISA Journal,* **9,** 128 (1953).
35. Van der Hegge Zijnen, B. G.: *Appl. Sci. Research,* **6A,** 129 (1956).
36. Kovasznay, L. S. G.: *Natl. Advisory Comm. Aeronaut. Tech. Repts. No.* 1209, 1954.
37. Collis, D. C.: *Aeronaut. Quart.*, **4,** 93 (1952).
38. Wyatt, L. A.: *J. Sci. Instr.*, **30,** 13 (1953).
39. Middlebrook, G. B., and E. L. Piret: *Ind. Eng. Chem.*, **42,** 1551 (1950).
40. Richardson, E. G.: *Trans. North East Coast Inst. Engrs. & Shipbuilders,* **64,** 273 (1948).
41. Vernotte, P.: *Compt. rend.*, **230,** 58 (1950).
42. Thomas, P.: *J. Am. Inst. Elec. Engrs.*, **42,** 219 (1923).
43. Lindvall, F. C.: *Trans. Am. Inst. Elec. Engrs.*, **53,** 1068 (1934).
44. Fucks, W.: *Deut. Luftfahrt-Forsch. U. u. M. Nos.* 1202, 1203, 1205, 1431, 1943–1944.
45. Fucks, W.: *Z. Naturforsch.*, **5A,** 89 (1950).
46. Fucks, W.: *Z. Physik,* **137,** 49 (1954).
47. Agostini, L.: *Proc. 7th Intern. Congr. Appl. Mech.*, London, 1948, vol. 2, p. 56.
48. Werner, F. D.: *Rev. Sci. Instr.*, **21,** 61 (1950).
49. Kolin, A.: *Proc. Soc. Exp. Biol. Med.*, **35,** 53 (1936).
50. Thürlemann, B.: *Helv. Phys. Acta,* **14,** 383 (1941).
51. Kolin, A., and F. Reiche: *J. Appl. Phys.*, **25,** 409 (1954).
52. Kolin, A.: *Rev. Sci. Instr.*, **16,** 109 (1945).
53. Grossman, L. M., and E. A. Shay: *Mech. Eng.*, **71,** 744 (1949).
54. Grossman, L. M., and A. F. Charwatt: *Rev. Sci. Instr.*, **23,** 741 (1952).
55. Gaffyn, J. E., and R. M. Underwood: *Nature,* **169,** 239 (1952).
56. Van Driest, E. R.: *J. Appl. Mechanics,* **12,** A91 (1945).
57. Kalinske, A. A., and C. L. Pien: *Ind. Eng. Chem.*, **36,** 220 (1944).
58. Kampé de Fériet, J.: *Proc. 5th Intern. Congr. Appl. Mech.*, Cambridge, Mass., 1938, p. 352.
59. Liepmann, H. W., and M. S. Robinson: *Natl. Advisory Comm. Aeronaut. Tech. Notes No.* 3037 (1953).
60. Townend, H. C. H.: *Proc. Roy. Soc. London,* **145A,** 180 (1934).
61. Fage, A., and H. C. H. Townend: *Proc. Roy. Soc. London,* **135A,** 656 (1932).
62. Taylor, M. K.: *Mech. Eng.*, **72,** 658 (1950).
63. Davies, P. O. A. L., and M. J. Fisher: *Proc. Roy. Soc. London,* **280A,** 486 (1964).
64. Frenkiel, F. N.: *Compt. rend.*, **22,** 1331 (1946).
65. Hinze, J. O., and B. G. van der Hegge Zijnen: General Discussion on Heat Transfer, *Proc. Inst. Mech. Engrs. London,* 188 (1951).
66. Leaf, W.: *Mech. Eng.*, **67,** 586 (1945).
67. Leaf, W., and L. C. Atchinson: *J. Phys. & Colloid Chem.*, **53,** 957 (1949).
68. Weller, R.: *J. Appl. Mechanics,* **14,** 103A (1947).
69. Ullyot, T.: *Trans. ASME,* **69,** 245 (1947).

70. Kovasznay, L. S. G.: *Proc. Heat Transfer and Fluid Mechanics Inst.*, Berkeley, Calif., 1949, p. 211.
71. Uberoi, M. S., and L. S. G. Kovasznay: *J. Appl. Phys.*, **26,** 19 (1955).
72. Hubbard, J. C., J. A. Fitzpatrick, W. J. Thaler, L. Cheng, and R. J. Beeber: *Phys. Rev.*, **74,** 708 (1948).
73. Spangenberg, W. G.: *Natl. Advisory Comm. Aeronaut. Techn. Notes* No. 3381, 1955.
74. Bourot, J. M.: *Publ. sci. et tech. ministère air France No.* 226, 1949.
75. Goldstein, S.: *Proc. Roy. Soc. London,* **155A,** 570 (1936).
76. Young, A. D., and J. N. Maas: *Aeronaut. Research Comm. Repts. & Mem. No.* 1770, 1937.
77. Hinze, J. O., and B. G. van der Hegge Zijnen: *Appl. Sci. Research,* **1A,** 435 (1949).
78. Alexander, L. G., T. Baron, and E. W. Comings: *Univ. Illinois Eng. Exp. Sta. Tech. Rept. No.* 8, 1950.
79. Tucker, M. J.: *J. Sci. Instr.*, **29,** 327 (1952).
80. Burgers, J. M.: *Proc. Koninkl. Ned. Akad. Wetenschap.*, **51,** 1222 (1948).
81. Morkovin, M. V.: *AGARDograph.* No. 24, 1956.
82. Laufer, J., and R. McClellan: *J. Fluid Mech.*, **1,** 276 (1956).
83. Dewey Jr., C. F.: *Int. J. Heat Mass Trans.*, **8,** 245 (1965).
84. Lighthill, M. J.: *Proc. Roy. Soc. London,* **224A,** 1 (1954).
85. Parthasarathy, S. P., and D. J. Tritton: *A.I.A.A. Jl.*, **1,** 1210 (1963).
86. Ling, S. C., and P. G. Hubbard: *J. Aeronaut. Sci.*, **23,** 890 (1956).
87. Kovasznay, L. S. G.: *J. Aeronaut Sci.*, **20,** 657 (1953).
88. Morkovin, M. V.: "Advances in Hot-Wire Anemometry," p. 38, AFOSR No. 68-1492, Univ. Maryland, 1968.
89. Webster, C. A. G.: *J. Fluid Mech.*, **13,** 307 (1962).
90. Champagne, F. H.: *Boeing Sc. Res. Lab. Doc.* D1-82-0491 (1965).
91. Leijdens, H.: Fluid Mech. Lab., Delft Univ. Technology. Private communication.
92. Fujita, H., and L. S. G. Kovasznay: *Res. Inst. Sci.*, **39,** 1351 (1968).
93. Hoole, B. J., and J. R. Calvert: *J. Roy. Aeronaut. Soc.*, **71,** 511 (1967).
94. Frenkiel, F. N., and P. S. Klebanoff: *Phys. Fluids,* **10,** 507 (1967).
95. Frenkiel, F. N., and P. S. Klebanoff: *Phys. Fluids,* **10,** 1737 (1967).
96. Corrsin, S.: "Turbulence, Experimental Methods," in "Encyclopedia of Physics," vol VIII/2, p. 568, Springer-Verlag, Berlin, 1963.
97. Fingerson, L. M.: "Advances in Hot-wire Anemometry," p. 258, AFOSR No. 68-1492, Univ. Maryland, 1968.
98. Wehrmann, O. H.: "Advances in Hot-wire Anemometry," p. 194, AFOSR. No. 68-1492, Univ. Maryland, 1968.
99. Ellington, E., and G. Trottier: "Advances in Hot-wire Anemometry," p. 52, AFOSR. No. 68-1492, Univ. Maryland, 1968.
100. Van der Hegge Zijnen, B. G.: *Ph.D. thesis,* Delft Univ. Technology, Delft, 1924.
101. Piercy, N. A. V., E. G. Richardson, and H. F. Winney: *Proc. Phys. Soc.*, **69B,** 731 (1956).
102. Reichardt, H.: *Z. angew. Math. u. Mech.*, **20,** 297 (1940).
103. Wills, J. A. B.: *J. Fluid Mech.*, **12,** 388 (1962).
104. Fabula, A. G.: "Advances in Hot-wire Anemometry," p. 167, AFOSR No. 68-1492, Univ. Maryland, 1968.
105. Delleur, J. W., G. H. Toebes, and C. L. Liu: *Proc. 12th Congr. I.A.H.R.*, Fort Collins, Col., 1967, vol. 2, p. 153.
106. Bellhouse, B. J., and D. L. Schultz: *J. Fluid Mech.*, **29,** 289 (1967).

107. Armistead Jr., R. A., and J. J. Keyes Jr.: *Rev. Sci. Instr.,* **39,** 61 (1968).
108. Ludwieg, H.: *Ing. Archiv,* **17,** 207 (1949).
109. Ludwieg, H., and W. Tillmann: *Ing. Archiv,* **17,** 288 (1949).
110. Bellhouse, B. J., and D. L. Schultz: *J. Fluid Mech.,* **32,** 675 (1968).
111. Rasmussen, R. A.: *Rev. Sci. Instr.,* **33,** 38 (1962).
112. Pigott, M. T., and R. C. Strum: *Rev. Sci. Instr.,* **38,** 743 (1967).
113. Lambert, R. B., H. A. Snyder, and S. K. F. Karlson: *Rev. Sci. Instr.,* **36,** 924 (1965).
114. Broer, L. J. F., and R. E. de Haan: *Z. angew. Math. u. Phys.,* **9b,** 162 (1958).
115. De Haan, R. E.: *Appl. Sci. Res.,* **22,** 306 (1970).
116. Jezdinsky, V.: *A.I.A.A. Jl.* **4,** 2072 (1966).
117. Strasburg, M.: *David Taylor Model Basin, Res. & Dev. Rep.* 1779 (1963).
118. Kobashi, Y.: *J. Phys. Soc. Japan,* **12,** 533 (1957).
119. Siddon, Th. E.: *Inst. Aerosp. Studies, Univ. Toronto, UTIAS-Rep.* No. 136 (1969).
120. Shaw, R.: *J. Fluid Mech.,* **7,** 550 (1960).
121. Willmarth, W. W.: *Rev. Sci. Instr.,* **29,** 218 (1958).
122. Willmarth, W. W., and C. E. Wooldridge: *J. Fluid Mech.,* **14,** 187 (1962).
123. Willmarth, W. W., and F. W. Roos: *J. Fluid Mech.,* **22,** 81 (1965).
124. Preston, J. H.: *J. Roy. Aeronaut. Soc.,* **58,** 109 (1954).
125. Patel, V. C.: *J. Fluid Mech.,* **23,** 185 (1965).
126. Quarmby, A., and H. K. Das: *Aeronaut J.,* **73,** 228 (1969).
127. Siddon, Th. E., and H. S. Ribner: *A.I.A.A. Jl.,* **3,** 747 (1965).
128. Hartung, F., and K. Csallner: *Proc. 12th Congr. I.A.H.R.,* Fort Collins, Col., 1967, vol. 2, p. 264.
129. Trutt, R. W.: *J. Aeronaut. Sci.,* **23,** 889 (1956).
130. Elder, J. W.: *J. Fluid Mech.,* **9,** 133 (1960).
131. Koppe, H.: *Instr. Practice,* **18,** 143 (1964).
132. Schuyf, J. P.: *J. Hydr. Res.,* **4,** 37 (1966).
133. Plate, E. J.: *Proc. 12th Congr. I.A.H.R.,* Fort Collins, Col., 1967, vol. 2, p. 281.
134. Iwasa, Y.: *Proc. 12th Congr. I.A.H.R.,* Fort Collins, Col., 1967, vol. 2, p. 273.
135. Franzen, B., W. Fucks, and G. Schmitz: *Forschungsber. des Landes Nordrhein-Westfalen,* No. 760, Westdeutscher Verlag, Cologne, 1959.
136. Eskinazi, S.: *Phys. Fluids,* **1,** 161 (1958).
137. Gibson, C. H., and W. H. Schwarz: *J. Fluid Mech.,* **16,** 357 (1963).
138. Chuang, H., and J. E. Cermak: *Proc. A.I.Civ.E., J. Hydr. Div.,* **91** HY6, Paper 4526 (1965).
139. Chuang, H., and J. E. Cermak: *A.I.Ch.E. Jl.,* **13,** 266 (1967).
140, Liu, H.: *A.I.Ch.E. Jl.,* **14,** 983 (1968).
141. Binder, G.: *Proc. 12th Congr., I.A.H.R.,* Fort Collins, Col., 1967, vol. 2, p. 249.
142. Reiss, L. P., and T. J. Hanratty: *A.I.Ch.E. Jl.,* **8,** 245 (1962).
143. Reiss, L. P., and T. J. Hanratty: *A.I.Ch.E. Jl.,* **9,** 154 (1963).
144. Mitchell, J. E., and T. J. Hanratty: *J. Fluid Mech.,* **26,** 199 (1966).
145. Hanratty, T. J.: *Phys. Fluids,* **10,** No. 9 II, S 126 (1967).
146. Nieuwenhuizen, J. K.: *Chem. Eng. Sci.,* **19,** 367 (1964).
147. Van Meel, D. A., and H. Vermij: *Appl. Sci. Res.,* **A10,** 109 (1961).
148. Clutter, D. W., O. M. O. Smith, and J. G. Brazier: *Aerosp. Engrg.,* **20,** Jan. 1961.
149. Schraub, F. A., S. J. Kline, J. Henry, P. W. Rundstadler Jr., and A. Littell: *Trans. A.S.M.E., J. Basic Eng.,* **87D,** 429 (1965).
150. Fauré, J.: *La Houille Blanche,* **18,** 298 (1963).

151. Hama, F. R., and J. Nutant: *Proc. Heat Transfer and Fluid Mech. Inst.*, Pasadena Calif., 1963, p. 77.
152. Becker, H. A., H. C. Hottel, and G. C. Williams: *J. Fluid Mech.*, **30,** 259 (1967).
153. Fisher, M. J., and F. R. Krause: *Proc. 5th Int. Congr. on Acoustics,* Paper K 64, Liege, 1965.
154. Fisher, M. J., and F. R. Krause: *J. Fluid Mech.*, **28,** 705 (1967).
155. Yeh, Y., and H. Cummins: *Appl. Phys. Letters,* **4,** 176 (1964).
156. Goldstein, R. J., and W. F. Hagen: *Phys. Fluids,* **10,** 1349 (1967).
157 Goldstein, R. J., and D. K. Kreid: "Fluid Velocity Measurement from the Doppler Shift of Scattered Laser Radiation," Heat Transfer Laboratory Rep. HTL. TR 85, 1968. University of Minesota.
158. Gracey, W.: *Natl. Advisory Comm. Aeronaut. Tech. Rept.* No. 1303, 1957.
159. Champagne, F. H., C. A. Sleicher, and O. H. Wehrmann: *J. Fluid Mech.*, **28,** 153 (1967).
160. Hill, J. C., and C. A. Sleicher: *Phys. Fluids,* **12,** 1126 (1969).
161. Champagne, F. H., and C. A. Sleicher: *J. Fluid Mech.*, **28,** 177 (1967).
162. Franklin, R. E., and J. M. Wallace: *J. Fluid Mech.*, **42,** 33 (1970).
163. Comte-Bellot, G., and J. P. Schon: *Int. J. Heat Mass Transf.*, **12,** 1661 (1969).
164. Davis, M. R., and P. O. A. L. Davies: *Int. J. Heat Mass Transf.*, **15,** 1659 (1972).
165. Siddon, Th. E.: *Rev. Sci. Instr.*, **42,** 653 (1971).
166. Bradbury, L. J. S., and J. P. Castro: *J. Fluid Mech.*, **51,** 487 (1972).
167. Perry, A. E., and G. L. Morrison: *J. Fluid Mech.*, **50,** 815 (1971).
168. Becker, H. A., and A. P. G. Brown: *J. Fluid Mech.*, **62,** 85 (1974).

NOMENCLATURE FOR CHAPTER 2

A coefficient of hot-wire characteristic; A_g, at gas temperature; A_f, at film temperature.
B coefficient of hot-wire characteristic.
b, b_1 temperature coefficients of electric resistance.
C_w heat capacity of hot wire.
c_w specific heat of hot-wire material.
c_p specific heat at constant pressure.
c velocity of light.
d diameter of hot wire.
E difference in electric potential between electrodes.
$\mathbf{E}$ spectrum function of turbulence kinetic energy.
e voltage across hot wire; $\overline{e^2}$, mean-square value.
f_D Doppler shift in frequency.
$\mathbf{f}$ coefficient of spatial, longitudinal velocity correlation.
Gr Grashof number.
$\mathbf{g}$ coefficient of spatial, lateral velocity correlation.
g gravitational acceleration.
g_{tr} transconductance.
H strength of magnetic field.
h heat-transfer coefficient.

h coefficient of spatial, triple velocity correlation.
I electric heating current; $\bar{I}$, time-mean value; i, turbulent fluctuation of current.
k coefficient of spatial, triple velocity correlation.
l length of hot wire; l_c, cold length.
l_{res} resolution length of hot wire.
M time constant of hot wire.
Ma Mach number.
Nu Nusselt number
P_{stat} static pressure; $\bar{P}_{stat}$, time-mean value.
P_{tot} total pressure; $\bar{P}_{tot}$, time-mean value.
Pr Prandtl number.
q coefficient of spatial, triple velocity correlation.
Re Reynolds number.
R_w electric resistance of hot wire; $\bar{R}_w$, time-mean value; R_0, at reference temperature; R_g, at gas temperature; r_w, turbulent fluctuation.
$\mathbf{R}_E$ coefficient of Eulerian time correlation.
$\mathbf{R}_L$ coefficient of Lagrangian time correlation.
$\mathbf{S}_{ik,j}$ third-order correlation tensor.
s sensitivity of hot wire; distance between electrodes.
t time.
U Eulerian velocity; $\bar{U}$, time-mean value; u, turbulence component; $u' = \sqrt{\overline{u^2}}$, root-mean-square value.
V strength of induced electric field.
x_i Eulerian Cartesian coordinates.
y_i Lagrangian Cartesian coordinates.
α coefficient of heat transfer.
β coefficient of expansion of gas; β_f, at film temperature.
$\mathscr{C}$ correction factor.
$\mathfrak{C}$ "cold-length" parameter of hot-wire anemometer.
Δ difference.
ε turbulence viscous dissipation per unit mass.
ϵ coefficient of eddy diffusion, ϵ_γ, for a scalar quantity.
Γ scalar quantity; $\bar{\Gamma}$, time-mean value; γ, turbulent fluctuation.
$\mathfrak{I}_E$ Eulerian integral time scale.
$\mathfrak{I}_L$ Lagrangian integral time scale.
$\imath$ $\sqrt{-1}$.
$\varkappa$ c_p/c_v.
κ heat conductivity; κ_g, of gas; κ_f, ditto at film temperature
Λ_g spatial, lateral integral length-scale.
λ_g spatial, lateral dissipation length-scale.
μ_g dynamic viscosity; μ_f, at film temperature.
μ relative magnetic permeability.
$\mathcal{O}$ order of magnitude.

Ω unit electric resistance.

π 3.14159.....

$\mathscr{R}_w$ electric resistance of hot wire per unit length; $\mathscr{R}_0$, at reference temperature; $\mathscr{R}_g$, at gas temperature; $\mathfrak{r}_w$, turbulent fluctuation.

ρ density; $\bar{\rho}$, time-mean value; $\tilde{\rho}$, turbulent fluctuation; ρ_g, gas density; ρ_f, ditto at film temperature; ρ_{stagn} ditto at stagnation point; ρ_w, density wire material.

ς entropy fluctuation.

τ time.

Θ temperature; $\bar{\Theta}$, time-mean value; θ, turbulent fluctuation; Θ_w, wire temperature; Θ_g, gas temperature; Θ_f, film temperature $= \frac{1}{2}(\Theta_g + \Theta_w)$; Θ_e, equilibrium temperature of hot wire; Θ_{stagn}, stagnation temperature.

ξ_i Eulerian Cartesian coordinates.

3 ISOTROPIC TURBULENCE

3-1 INTRODUCTION

Isotropic turbulence is the simplest type of turbulence, since no preference for any specific direction occurs and a minimum number of quantities and relations are required to describe its structure and behavior. However, it is also a hypothetical type of turbulence, because no actual turbulent flow shows true isotropy, though conditions may be made such that isotropy is more or less closely approached.

In this chapter we shall consider isotropic turbulence much more extensively than we did in the first chapter. At the same time the assumptions of spatial homogeneity in the mean properties of the turbulence will be made. However, this implies that the turbulence cannot be stationary. For, spatial homogeneity and stationarity are conflicting conditions for any turbulent flow. This can be easily concluded from Eq. (1-110) for the balance of turbulence mechanical energy. In the case of homogeneity all terms on the right-hand side, except the dissipation term, are zero. Also zero are the terms on the left-hand side for the convection by the mean motion. So, only the nonstationary term is left to balance the dissipation term. In other words, a homogeneous turbulent flow field is at the same time a decaying turbulent field. On the other hand, if we assume stationarity of the turbulence, at least one inhomogeneous term in Eq. (1-110) must be nonzero in order to balance the dissipation term.

Usually, for reasons of simplicity, a homogeneous isotropic turbulence is assumed with respect to a coordinate system moving with the constant mean velocity of the flow. Fortunately, the rate of decay of the mean properties is rather slow with respect to the time scale of the smaller eddies. Therefore, the actual state of nonstationarity is considered not to be a serious drawback in the experimental study of the smaller scale turbulence.

For the theoretical study, this makes it possible to apply the concepts and theories of stationary random processes.

Many properties of the turbulence, to be discussed in this chapter, are based on the assumption of homogeneity, without the additional conditions of isotropy. In order to avoid any misunderstanding, homogeneity of the turbulence will be assumed throughout this chapter, and when the additional condition of isotropy is required it will be so mentioned explicitly.

The reader who is concerned primarily with the practical and applied aspects of fluid flow might be inclined to ask why this hypothetical type of turbulence is treated so extensively. The answer is that, notwithstanding its hypothetical character, a knowledge of its characteristics may still form a fundamental basis for the study of actual, nonisotropic turbulent flows. Because of its relative simplicity it is more amenable to fundamental theoretical treatment than any other type of turbulence, and theoretical relations may be checked more easily by suitable experiments carried out in turbulence that is isotropic to a sufficient degree of approximation.

Moreover, the theoretical and experimental results of such a study are of more practical value than one might believe. Certain theoretical considerations concerning the energy transfer through the eddy-size spectrum from the bigger to the smaller eddies lead to the conclusion that the fine structure of nonisotropic turbulent flows is almost isotropic (local isotropy). There is also experimental evidence which supports this conclusion, though it must be admitted that not all experimental results conform in this respect. But if the conclusion is, at least approximately, true many features of isotropic turbulence apply to phenomena in actual turbulence that are determined mainly by the fine-scale structure.

But even if we consider the nonisotropic large-scale structure of an actual turbulence or if this turbulence is nonisotropic through an essential part of its spectrum, it is often possible to treat such a turbulence for purposes of a first approximation as if it were isotropic. Differences between results based upon the assumed isotropy and actual results are often sufficiently small to be negligible in a first approximation and may be even smaller than the spread in the experimental data.

Because of its relative simplicity, isotropic turbulence has been studied most, theoretically as well as experimentally. The geometrical and kinematic relations involved in turbulence have been studied through correlations and spectrum functions, and the dynamic behavior of these correlations and spectrum functions (decay of turbulence).

It has been possible to obtain a satisfactory and nearly complete picture of the formal geometrical relations only when these are based upon the conditions of isotropy and of incompressibility of the fluid. Usually the condition of isotropy is defined, rather briefly, by the invariance under rotation of the coordinate system and under reflection with respect to the coordinate planes of the statistically averaged properties of the turbulence. However, this definition does not express explicitly that the distance vectors between points have also to be taken into account. Therefore, Batchelor[11] gives the following definition. In an isotropic turbulence the joint-probability distribution of the velocities at any arbitrarily chosen n points in space is invariant under arbitrary rotations of the configuration as formed by the n points and by the various direction vectors, and under reflection of the configuration with respect to any plane.

It has not yet been possible to obtain general, complete solutions of the differential equations involved. No complete solutions are known for the various correlations and spectrum functions, nor for their changes with time, except for one single, limiting case. However, by making some additional assumption or postulate that applies more or less approximately, it has been possible to obtain solutions of some restricted validity.

One of the usual assumptions is that the structure of turbulence and the kinematic relations remain similar or, better, self-preserving during decay. Similarity and self-preservation are not identical notions. In order to avoid any misunderstanding we shall use here the distinction between the two suggested by Stewart and Townsend.[17]

Complete or perfect kinematic similarity exists between two flows with geometrically similar boundaries, when one velocity scale and one length scale are sufficient to make the distributions of reduced velocities in terms of reduced coordinates identical. *Incomplete, or imperfect kinematic similarity exists* if two or more velocity scales and two or more length scales are required. For instance, different velocity scales may be required for the $\overline{U}_1$, $\overline{U}_2$, and $\overline{U}_3$ components and a different velocity scale and length scale for the turbulent motions.

Self-preservation occurs in a nonstationary, decaying or developing flow when at any one Reynolds number the structure remains similar, either complete or incomplete, at all times. In a stationary flow this is so when similarity exists at all sections perpendicular to the main flow.

In the following we shall follow the example of most investigators and assume an incompressible fluid. It goes without saying that this assumption results in considerable simplification of the problem. However, at the end of this chapter we shall make some remarks on the effect of compressibility.

3-2 CORRELATION TENSORS

In the first chapter we have shown that in current theories the statistical behavior of turbulence may conveniently be described with the aid of correlations between the pressure fluctuations and the velocity fluctuations of the flow field. For the purely spatial correlations one has to consider at least two points in this flow field. Of these two-point correlations we have introduced the correlation tensors of the first, second, and third order, namely,

$$(\mathbf{K}_{i,p})_{A,B} = \overline{(u_i)_A p_B}$$

$$(\mathbf{Q}_{i,j})_{A,B} = \overline{(u_i)_A (u_j)_B}$$

$$(\mathbf{S}_{ij,k})_{A,B} = \overline{(u_i)_A (u_j)_A (u_k)_B}$$

The index p is fixed, unlike i, j, k, etc., and merely indiates reference to the scalar quantity p.

We shall now consider these correlation tensors in greater detail, especially with regard to features that follow from the condition of isotropy. General expressions will be derived for the isotropic tensors. In this derivation we shall apply the theory of invariants, as expounded by Robertson.[1] This theory is quite formal and does not give direct insight into the geometrical relations that obtain among specific correlations, the distance r between the two points A and B, and their position.

Von Kármán and Howarth[2] found the same relations by applying a more lucid geometrical derivation. For instruction purposes, we shall give two examples in which their derivation is applied, namely, for the double correlation tensors $(\mathbf{K}_{p,i})_{A,B}$ and $(\mathbf{Q}_{i,j})_{A,B}$.

In general the correlation tensors are functions of the space coordinates, the distance between A and B, and the time. As long as we consider only the geometrical relations, we shall not show the time dependence explicitly in the expressions for the tensors, for the sake of simplicity.

First-order Correlation Tensor $(\mathbf{K}_{p,i})_{A,B}$

Before giving Robertson's theory, we will first apply Von Kármán and Howarth's method to the following derivation. This tensor has three components, corresponding

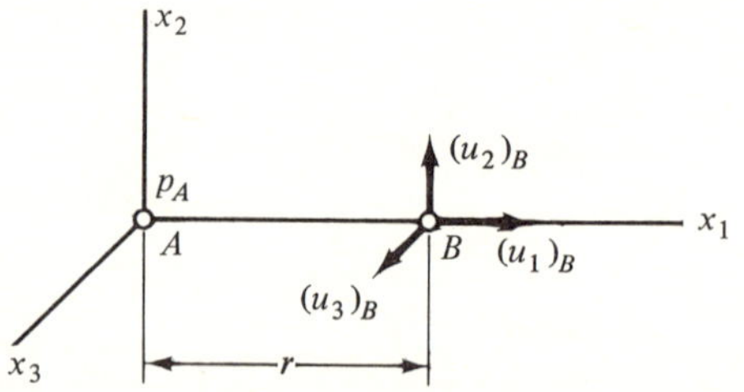

FIGURE 3-1
Correlation between the pressure at A and the velocity components at B.

to the three components into which the velocity at the point B can be resolved. Without departing from generality we may choose the coordinate system in such a way that the x_1-axis coincides with the line AB.

According to this coordinate system, the three components of the tensor are (see Fig. 3-1)

$$\overline{p_A(u_1)_B} \qquad \overline{p_A(u_2)_B} \qquad \overline{p_A(u_3)_B}$$

The condition of isotropy requires that these relations be invariant under reflection with respect to the coordinate planes. Of the three components of the tensor, the following two components can be invariant only when they are zero:

$$\overline{p_A(u_2)_B} = 0 \qquad \overline{p_A(u_3)_B} = 0$$

We shall designate by $\mathbf{K}_1(r)$ the remaining component $\overline{p_A(u_1)_B}$ or $\overline{p_A(u_r)_B}$ and show that this correlation must be an uneven function of r. To this end we expand $\mathbf{K}_1(r)$ in a power series of r:

$$\mathbf{K}_1(r) = [\mathbf{K}_1]_{r=0} + \frac{r}{1!}\left[\frac{\partial \mathbf{K}_1}{\partial r}\right]_{r=0} + \frac{r^2}{2!}\left[\frac{\partial^2 \mathbf{K}_1}{\partial r^2}\right]_{r=0} + \cdots$$

where

$$[\mathbf{K}_1]_{r=0} = [\overline{p_A(u_r)_B}]_{r=0}$$

$$\left[\frac{\partial \mathbf{K}_1}{\partial r}\right]_{r=0} = \left[\overline{p_A\left(\frac{\partial u_r}{\partial r}\right)_B}\right]_{r=0}$$

$$\left[\frac{\partial^2 \mathbf{K}_1}{\partial r^2}\right]_{r=0} = \left[\overline{p_A\left(\frac{\partial^2 u_r}{\partial r^2}\right)_B}\right]_{r=0}$$

. . . .

The coefficients

$$[\mathbf{K}_1]_{r=0}, \left[\frac{\partial \mathbf{K}_1}{\partial r}\right]_{r=0}, \ldots$$

in the series expansion containing the derivatives of the velocity vector in its own direction, may be considered as scalar quantities and must, therefore, be invariant.

Now u_r, $\partial^2 u_r/\partial r^2$, $\partial^4 u_r/\partial r^4, \ldots$ change sign under reflection because u_r and r both change sign. Thus their averaged products with the pressure fluctuation p at $r = 0$, given above, must be zero.

$$[\mathbf{K}_1]_{r=0} = \left[\frac{\partial^2 \mathbf{K}_1}{\partial r^2}\right]_{r=0} = \cdots = 0$$

On the other hand $\partial u_r/\partial r$, $\partial^3 u_r/\partial r^3, \ldots$ remain unchanged under reflection.

Hence the series expansion of $\mathbf{K}_1(r)$ must read

$$\mathbf{K}_1(r) = r\left[\frac{\partial \mathbf{K}_1}{\partial r}\right]_{r=0} + \frac{r^3}{3!}\left[\frac{\partial^3 \mathbf{K}_1}{\partial r^3}\right]_{r=0} + \cdots \tag{3-1}$$

When we have an arbitrary position of the points A and B with respect to a coordinate system, what will be the expression for the tensor $(\mathbf{K}_{p,i})_{A,B}$ in terms of the correlation $\mathbf{K}_1(r)$ and the coordinates ξ_j of B with respect to A?

Let us now follow Robertson's method. Consider the scalar correlation $\mathbf{K}_{A,B}$ between the pressure at A and the velocity component at B according to a given direction b.

$$\mathbf{K}_{A,B} = (\mathbf{K}_{p,i})_{A,B} e_{bi} \tag{3-2}$$

where e_{bi} is the direction cosine of b with respect to the direction of the x_i-axis (see Sec. 1-4). Thus $\mathbf{K}_{A,B}$ is a homogeneous linear function of e_{bi}, and $(\mathbf{K}_{p,i})_{A,B}$ is a function of ξ_j. Since $\mathbf{K}_{A,B}$ is a scalar, $(\mathbf{K}_{p,i})_{A,B}$ must be a function of ξ_j such that the terms in $\mathbf{K}_{A,B}$ in which ξ_j and e_{bi} occur are scalars.

The only scalar combinations that can be made of ξ_j and e_{bi} are $\xi_i\xi_i$, $e_{bi}e_{bi}$, and $\xi_i e_{bi}$.

Since $e_{bi}e_{bi} = 1$, only the first and third combinations are of importance. Consequently $\mathbf{K}_{A,B}$ must be a function of $\xi_i\xi_i = r^2$ and of $\xi_i e_{bi}$. The only combination that is linear and homogeneous with respect to e_{bi} reads

$$\mathbf{K}_{A,B} = \mathbf{K}_1^*(r^2)\xi_i e_{bi} \tag{3-3}$$

From Eqs. (3-2) and (3-3) it follows that

$$(\mathbf{K}_{p,i})_{A,B} = \mathbf{K}_1^*(r^2)\xi_i \tag{3-4}$$

This is the *general expression for an isotropic tensor of the first order.*

If we take the origin of the coordinate system at A and put the x_1-axis through A and B, then $\xi_1 = r$, and it follows from Eq. (3-4) that

$$\mathbf{K}_1(r) = \mathbf{K}_1^*(r^2)r$$

Consequently,

$$(\mathbf{K}_{p,i})_{A,B} = \frac{\mathbf{K}_1(r)}{r}\xi_i \tag{3-5}$$

The former expression, too, shows that $\mathbf{K}_1(r)$ must be an uneven function of r.

Let us now see what the consequences are when the fluid is incompressible. We then have

$$\left(\frac{\partial \mathbf{K}_{p,i}}{\partial x_i}\right)_B = \overline{p_A\left(\frac{\partial u_i}{\partial x_i}\right)_B} = 0$$

so that $(K_{p,i})_{A,B}$ is of vanishing divergence. Since

$$\left(\frac{\partial}{\partial x_i}\right)_B = \frac{\partial}{\partial \xi_i}$$

we obtain from Eq. (3-5)

$$\frac{\partial}{\partial \xi_i}\left(\frac{\mathbf{K}_1(r)}{r}\xi_i\right) = 3\frac{\mathbf{K}_1}{r} - \frac{\mathbf{K}_1}{r^2}\frac{\xi_i\xi_i}{r} + \frac{\xi_i\xi_i}{r^2}\frac{\partial \mathbf{K}_1}{\partial r} = \frac{2\mathbf{K}_1}{r} + \frac{\partial \mathbf{K}_1}{\partial r} = 0$$

whence $\mathbf{K}_1(r) = c/r^2$. Because $\mathbf{K}_1(r)$ cannot become infinite for $r = 0$, and while moreover $\mathbf{K}_1(r)$ can only be an odd function of r, the constant must be zero; consequently, $\mathbf{K}_1(r) = 0$. Hence, in an isotropic turbulence of an *incompressible fluid,*

$$(\mathbf{K}_{p,i})_{A,B} = \overline{p_A(u_i)_B} = 0 \tag{3-6}$$

Note that from isotropy alone we obtain

$$\overline{p_A(u_n)_B} = 0 \qquad \text{(reflection-invariancy condition)}$$

$$\overline{[p_A(u_r)_B]_{r=0}} = 0 \qquad \text{(reflection-invariancy condition)}$$

but

$$\overline{p_A(u_r)_B} \neq 0 \qquad \text{for } r \neq 0$$

and that the requirement of incompressibility also makes the last correlation and, hence, the tensor $(\mathbf{K}_{p,i})_{A,B}$ equal to zero.

Second-order Correlation Tensor $(\mathbf{Q}_{i,j})_{A,B}$

In Chap. 1 we have discussed the correlations $(\mathbf{Q}_{i,j})_{A,B}$ and a few features of two specific correlations, namely, the longitudinal correlation $u'^2\mathbf{f}(r)$ and the lateral correlation $u'^2\mathbf{g}(r)$. We have also shown, but only for a special position of the points A and B with respect to the coordinate system considered, that $(\mathbf{Q}_{i,j})_{A,B}$ can be expressed in terms of the longitudinal and lateral correlations.

We shall now derive a general expression for a second-order isotropic tensor and, from that, the expression for $(\mathbf{Q}_{i,j})_{A,B}$ in terms of $\mathbf{f}(r)$ and $\mathbf{g}(r)$.

Consider the scalar correlation $\mathbf{Q}_{A,B}$ between the velocity component at A in the fixed, arbitrary direction a and the velocity component at B in the fixed, arbitrary direction b.

$$\mathbf{Q}_{A,B} = (\mathbf{Q}_{i,j})_{A,B}e_{ai}e_{bj}$$

$\mathbf{Q}_{A,B}$ is a function of ξ_k, e_{ai}, and e_{bj}, its terms being scalar combinations of ξ_k, e_{ai}, and e_{bj}; the function must, moreover, be homogeneous and bilinear in e_{ai} and e_{bj}.

The scalar combinations of ξ_k, e_{ai}, and e_{bj} up to the terms bilinear in e_{ai} and e_{bj} are:

$$\xi_k\xi_k = r^2 \qquad e_{ai}e_{ai} = 1 \qquad e_{bj}e_{bj} = 1 \qquad \xi_i e_{ai} \qquad \xi_j e_{bj} \qquad \delta_{ij}e_{ai}e_{bj} \qquad \varepsilon_{ijk}\xi_k e_{ai}e_{bj}$$

The homogeneous bilinear function of e_{ai} and e_{bj} must read

$$\mathbf{Q}_{A,B} = \mathbf{Q}_1\xi_i\xi_j e_{ai}e_{bj} + \mathbf{Q}_2\delta_{ij}e_{ai}e_{bj} + \mathbf{Q}_3\varepsilon_{ijk}\xi_k e_{ai}e_{bj}$$

whence

$$(\mathbf{Q}_{i,j})_{A,B} = \mathbf{Q}_1\xi_i\xi_j + \mathbf{Q}_2\delta_{ij} + \mathbf{Q}_3\varepsilon_{ijk}\xi_k$$

where $\mathbf{Q}_1$, $\mathbf{Q}_2$, $\mathbf{Q}_3$ are functions of $\xi_i\xi_i = r^2$.

Because of homogeneity and isotropy, there must be invariance under translation and reflection with respect to an arbitrary point; hence

$$(\mathbf{Q}_{i,j})_{A,B} = (\mathbf{Q}_{j,i})_{A,B}$$

Since $\varepsilon_{ijk} = -\varepsilon_{jik}$, it follows that $\mathbf{Q}_3 = 0$; so

$$(\mathbf{Q}_{i,j})_{A,B} = \mathbf{Q}_1(r^2)\xi_i\xi_j + \mathbf{Q}_2(r^2)\,\delta_{ij} \tag{3-7}$$

This is the *general expression for an isotropic tensor of the second order.*

What is the meaning of $\mathbf{Q}_1(r^2)$ and $\mathbf{Q}_2(r^2)$? Let us consider a special case, where the points A and B are on the x_1-axis, with the origin at A. Then $\xi_1 = r$, $\xi_2 = \xi_3 = 0$, $\mathbf{Q}_{1,1} = u'^2\mathbf{f}(r)$, and $\mathbf{Q}_{2,2} = u'^2\mathbf{g}(r)$; so from Eq. (3-7),

$$\mathbf{Q}_{1,1} = u'^2\mathbf{f}(r) = \mathbf{Q}_1(r^2)r^2 + \mathbf{Q}_2(r^2)$$

and

$$\mathbf{Q}_{2,2} = u'^2\mathbf{g}(r) = \mathbf{Q}_2(r^2)$$

Hence

$$\mathbf{Q}_1(r^2) = u'^2\,\frac{\mathbf{f}(r) - \mathbf{g}(r)}{r^2} \qquad \mathbf{Q}_2(r^2) = u'^2\mathbf{g}(r)$$

Then Eq. (3-7) becomes

$$(\mathbf{Q}_{i,j})_{A,B} = u'^2\left[\frac{\mathbf{f}(r) - \mathbf{g}(r)}{r^2}\,\xi_i\xi_j + \mathbf{g}(r)\,\delta_{ij}\right] \tag{3-8}$$

Geometrical Derivation of Eq. (3-8). We shall consider each term of $\overline{(u_i)_A(u_j)_B}$ separately. It is sufficient to do this only for the two types $i = j$ and $i \neq j$.

Take the origin of the coordinate system at A; r is the distance AB with components ξ_i. Consider $\overline{(u_1)_A(u_1)_B}$. Take a plane through AB and the x_1-axis, as shown in Fig. 3-2. In this plane the velocities at A and B can be resolved into the components $(u_{1*})_A$ and $(u_{1*})_B$, respectively, directed along AB, and the components $(u_{2*})_A$ and $(u_{2*})_B$, respectively, perpendicular to AB; the third components, $(u_{3*})_A$ and $(u_{3*})_B$, respectively, are perpendicular to the plane. These last components provide no contributions to $(u_1)_A$ and $(u_1)_B$, respectively.

From Fig. 3-2 we read

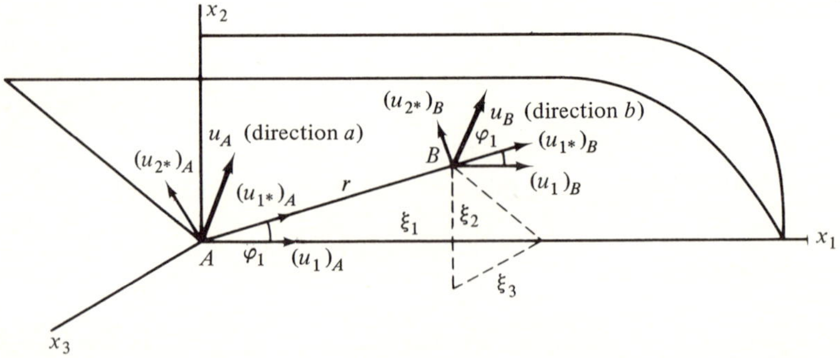

FIGURE 3-2
Relation between $\overline{(u_1)_A(u_1)_B}$ and $\mathbf{f}(r)$ and $\mathbf{g}(r)$.

$$(u_1)_A = (u_{1*})_A \cos\varphi_1 - (u_{2*})_A \sin\varphi_1 = (u_{1*})_A \frac{\xi_1}{r} - (u_{2*})_A \sqrt{1 - \frac{\xi_1^2}{r^2}}$$

$$(u_1)_B = (u_{1*})_B \frac{\xi_1}{r} - (u_{2*})_B \sqrt{1 - \frac{\xi_1^2}{r^2}}$$

If we multiply $(u_1)_A$ by $(u_1)_B$ we obtain the products

$$\overline{(u_{1*})_A(u_{1*})_B} = u'^2\mathbf{f}(r) \qquad \overline{(u_{2*})_A(u_{2*})_B} = u'^2\mathbf{g}(r)$$

and, because of invariance conditions

$$\overline{(u_{1*})_A(u_{2*})_B} = \overline{(u_{2*})_A(u_{1*})_B} = 0$$

Hence

$$\mathbf{Q}_{1,1} = \overline{(u_1)_A(u_1)_B} = u'^2\mathbf{f}(r)\frac{\xi_1^2}{r^2} + u'^2\mathbf{g}(r)\left(1 - \frac{\xi_1^2}{r^2}\right)$$

$$= u'^2\left[\frac{\mathbf{f}(r) - \mathbf{g}(r)}{r^2}\xi_1\xi_1 + \mathbf{g}(r)\right]$$

Similarly, we shall obtain

$$\mathbf{Q}_{2,2} = u'^2\left[\frac{\mathbf{f}(r) - \mathbf{g}(r)}{r^2}\xi_2\xi_2 + \mathbf{g}(r)\right]$$

or, more generally,

$$\mathbf{Q}_{i,j} = u'^2\left[\frac{\mathbf{f}(r) - \mathbf{g}(r)}{r^2}\xi_i\xi_j + \mathbf{g}(r)\right] \qquad \text{for } i = j \tag{a}$$

Let us now consider $\overline{(u_1)_A(u_2)_B}$. Again, take a plane through AB and the x_1-axis (see Fig. 3-3). Just as before it can be seen from this figure that

$$(u_1)_A = (u_{1*})_A \frac{\xi_1}{r} - (u_{2*})_A \sqrt{1 - \frac{\xi_1^2}{r^2}}$$

$$(u_2)_B = (u_{1*})_B \cos\varphi_2 + (u_{2*})_B \frac{\cos\varphi_2}{\tan\varphi_1} + \text{component of } (u_{3*})_B$$

$$= (u_{1*})_B \frac{\xi_2}{r} + (u_{2*})_B \frac{\xi_1\xi_2}{r^2\sqrt{1 - \xi_1^2/r^2}} + \text{component of } (u_{3*})_B$$

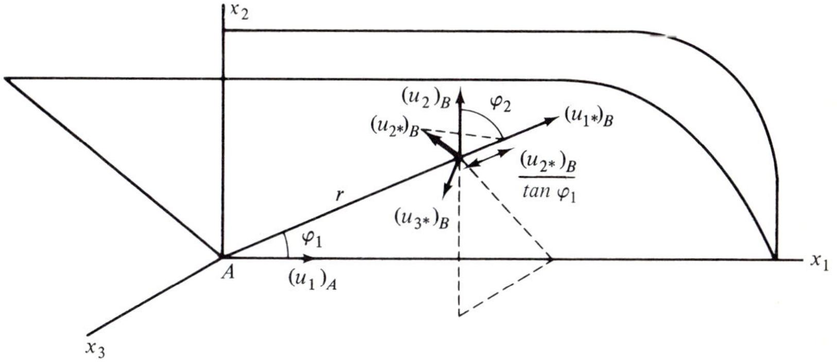

FIGURE 3-3
Relation between $\overline{(u_1)_A(u_2)_B}$ and $\mathbf{f}(r)$ and $\mathbf{g}(r)$.

Because of the condition of isotropy,

$$\overline{(u_{1*})_A(u_{2*})_B} = 0 \quad \overline{(u_{1*})_A(u_{3*})_B} = 0 \quad \overline{(u_{2*})_A(u_{1*})_B} = 0 \quad \overline{(u_{2*})_A(u_{3*})_B} = 0$$

so

$$\overline{(u_1)_A(u_2)_B} = \overline{(u_{1*})_A(u_{1*})_B}\frac{\xi_1\xi_2}{r^2} - \overline{(u_{2*})_A(u_{2*})_B}\frac{\xi_1\xi_2}{r^2}$$

or

$$\mathbf{Q}_{1,2} = u'^2\frac{\mathbf{f}(r) - \mathbf{g}(r)}{r^2}\xi_1\xi_2$$

And, in general,

$$\mathbf{Q}_{i,j} = u'^2\frac{\mathbf{f}(r) - \mathbf{g}(r)}{r^2}\xi_i\xi_j \qquad \text{for } i \neq j \tag{b}$$

Thus (a) and (b) can be combined into

$$\mathbf{Q}_{i,j} = u'^2\left[\frac{\mathbf{f}(r) - \mathbf{g}(r)}{r^2}\xi_i\xi_j + \mathbf{g}(r)\,\delta_{ij}\right]$$

Figure 3-4 shows an example of the longitudinal correlation coefficient $\mathbf{f}(r)$ as measured by Stewart[4] in a grid-generated turbulent flow at three mesh-Reynolds numbers. More recent measurements of the one-point auto-correlation coefficient $\mathbf{R}_E(t)$, also in grid-generated turbulence, and using linearized hot-wire anemometry, digital sampling, and high-speed digital computation, have been made by Frenkiel and Klebanoff,[24] and by Van Atta and Chen.[60] Figure 3-5 shows a result of the measurements by Van Atta and Chen. In order to compare Fig. 3-4 with Fig. 3-5, one has to read for $\bar{U}_1 t = r$, according to Taylor's hypothesis (see Chap. 1, e.g., Eq. 1-68) and to take account of the fact that Stewart did not find an effect on the correlation coefficient of the distance x_1/M from the grid.

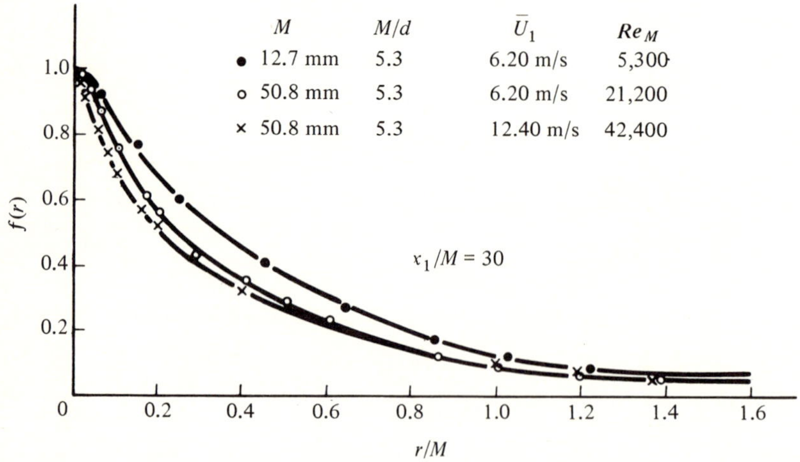

FIGURE 3-4
Double velocity-correlation coefficient $\mathbf{f}(r)$. (From: *Stewart, R. W.*,[4] *by permission of the Cambridge Philosophical Society.*)

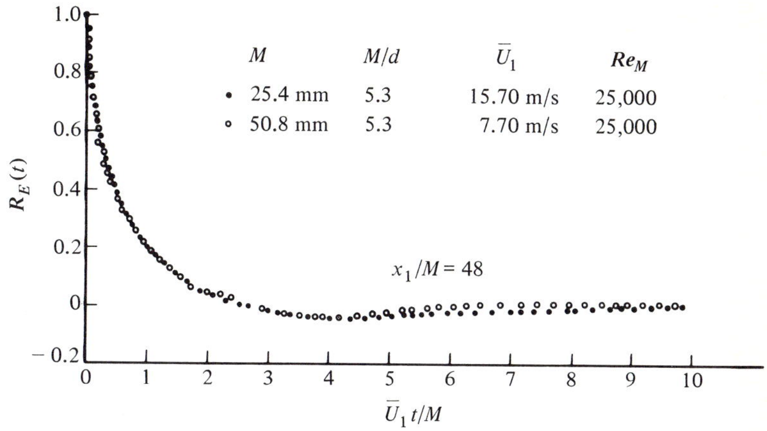

FIGURE 3-5
Auto-correlation coefficient $\mathbf{R}_E(t)$. (From: *Van Atta, C. W., and W. Y. Chen,*[60] *by permission of the Cambridge University Press.*)

We shall now use the *condition of fluid incompressibility* to derive a relation $\mathbf{Q}_1$ and $\mathbf{Q}_2$ or between **f** and **g**.

For an incompressible fluid we obtain

$$\left(\frac{\partial}{\partial x_j}\right)_B (\mathbf{Q}_{i,j})_{A,B} = \overline{(u_i)_A\left(\frac{\partial u_j}{\partial x_j}\right)_B} = 0$$

With $(x_j)_B - (x_j)_A = \xi_j$

$$\frac{\partial}{\partial \xi_j}(\mathbf{Q}_{i,j})_{A,B} = 0$$

Substitution of Eq. (3-8) yields

$$\frac{1}{u'^2}\frac{\partial}{\partial \xi_j}(\mathbf{Q}_{i,j})_{A,B} = 3\,\frac{\mathbf{f}-\mathbf{g}}{r^2}\,\xi_i + \frac{\mathbf{f}-\mathbf{g}}{r^2}\,\xi_j\delta_{ij} - 2\,\frac{\mathbf{f}-\mathbf{g}}{r^3}\,\frac{\xi_j\xi_j}{r}\,\xi_i$$
$$+\frac{1}{r^3}\left(\frac{\partial \mathbf{f}}{\partial r} - \frac{\partial \mathbf{g}}{\partial r}\right)\xi_j\xi_j\xi_i + \delta_{ij}\,\frac{\xi_j}{r}\,\frac{\partial \mathbf{g}}{\partial r} = 0$$

whence

$$\mathbf{f}(r) + \frac{r}{2}\frac{\partial \mathbf{f}(r)}{\partial r} = \mathbf{g}(r) \qquad (3\text{-}9)$$

The equivalent reciprocal integral relation reads

$$\mathbf{f}(r) = \frac{2}{r^2}\int_0^r dr\, r\mathbf{g}(r) \qquad (3\text{-}10)$$

Figure 3-6 shows an experimental verification of the relation (3-9), given by Von Kármán and Howarth.[2]

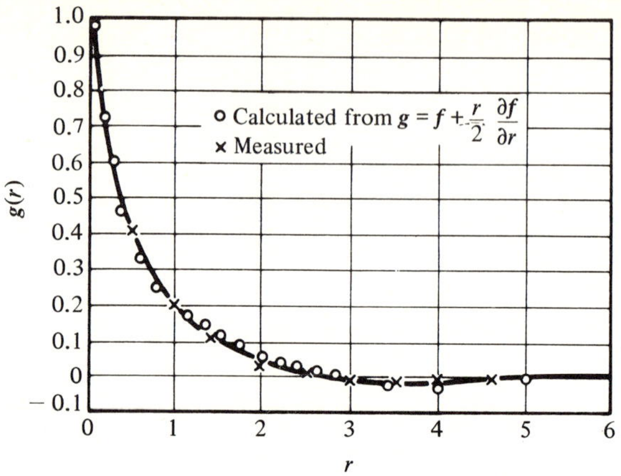

FIGURE 3-6
Confirmation of the relation between $\mathbf{f}(r)$ and $\mathbf{g}(r)$.

From the relation (3-9) between $\mathbf{f}(r)$ and $\mathbf{g}(r)$, which results from the continuity condition for an incompressible fluid, it thus appears that the double-velocity-correlation tensor $(\mathbf{Q}_{i,j})_{A,B}$ may be expressed in terms of one scalar only, either $\mathbf{f}(r)$ or $\mathbf{g}(r)$, for instance,

$$(\mathbf{Q}_{i,j})_{A,B} = u'^2\left\{-\frac{1}{2r}\frac{\partial\mathbf{f}(r)}{\partial r}\xi_i\xi_j + \left[\mathbf{f}(r) + \frac{r}{2}\frac{\partial\mathbf{f}(r)}{\partial r}\right]\delta_{ij}\right\} \tag{3-11}$$

Figure 3-7 shows curves for $\mathbf{Q}_{1,1}$ as a function of ξ_1/M for various values of ξ_2/M according to measurements by Favre, Gaviglio, and Dumas.[3] They made the measurements at a distance $40M$ downstream of a grid with mesh $M = 25$ mm and with an air velocity of 12.27 m/s ($\mathbf{Re}_M = 21{,}500$).

In the curves corresponding to $\xi_2/M = 0.157$ and $\xi_2/M = 0.315$ points are also given that are calculated according to Eq. (3-11) by

$$\frac{1}{u'^2}\mathbf{Q}_{1,1} = -\frac{1}{2r}\frac{\partial\mathbf{f}(r)}{\partial r}\xi_1{}^2 + \mathbf{f}(r) + \frac{r}{2}\frac{\partial\mathbf{f}(r)}{\partial r} = \mathbf{f}(r) + \frac{r}{2}\frac{\partial\mathbf{f}(r)}{\partial r}\left(1 - \frac{\xi_1{}^2}{r^2}\right)$$

where $r^2 = \xi_1{}^2 + \xi_2{}^2$.

The reader may notice that, especially for $\xi_2/M = 0.157$, the agreement between measured and calculated values is very satisfactory.

From Eq. (3-8) or Eq. (3-11) we easily derive the expression for

$$(\mathbf{Q}_{i,i})_{A,B} = \overline{(u_i)_A(u_i)_B}$$

Later on we shall make use of this special value of $(\mathbf{Q}_{i,j})_{A,B}$. From Eq. (3-8)

$$(\mathbf{Q}_{i,i})_{A,B} = u'^2[\mathbf{f}(r) + 2\mathbf{g}(r)]$$

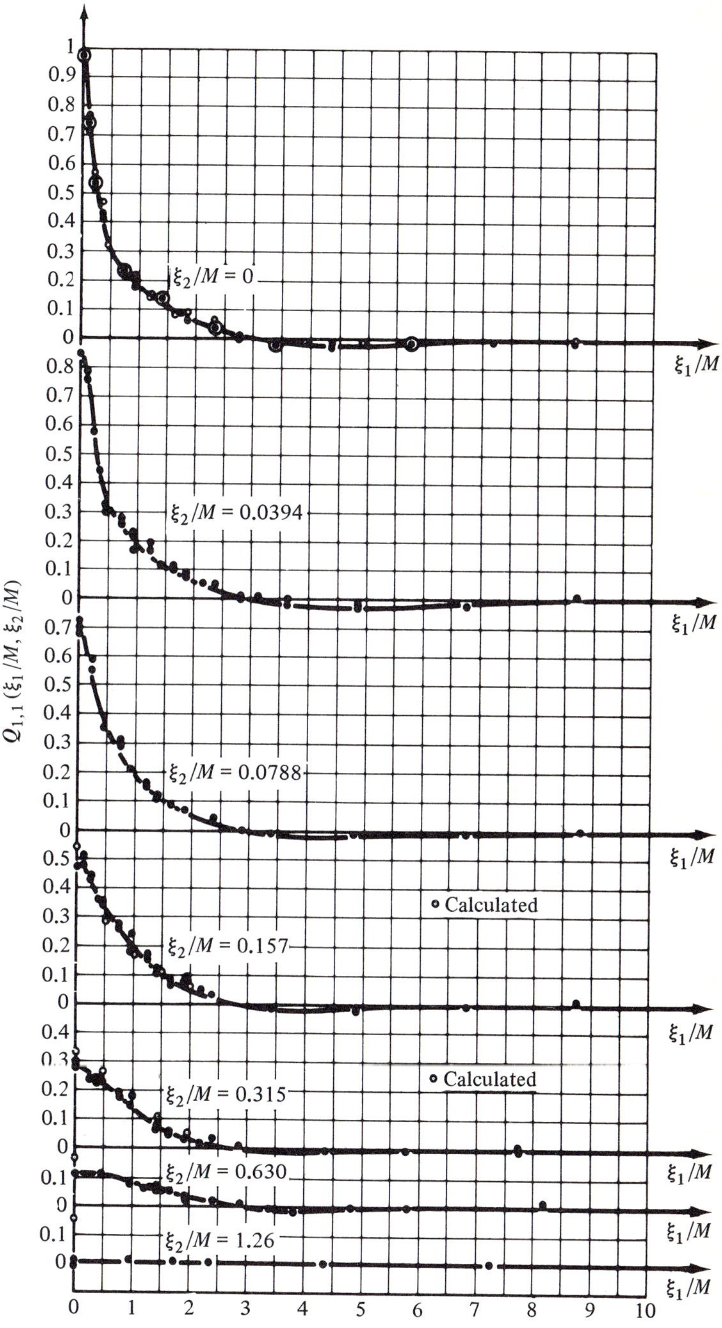

FIGURE 3-7
Spatial correlation $\mathbf{Q}_{1,1} = \overline{(u_1)_A(u_1)_B}$ as a function of ξ_1 and ξ_2. (From: *Favre, A. et. al.*[3])

and from Eq. (3-11)

$$(\mathbf{Q}_{i,i})_{A,B} = u'^2\left[3\mathbf{f}(r) + r\frac{\partial \mathbf{f}(r)}{\partial r}\right] = \frac{u'^2}{r^2}\frac{\partial}{\partial r}\left[r^3\mathbf{f}(r)\right] \tag{3-12}$$

When $A \equiv B$, so $r = 0$, these relations yield

$$\mathbf{Q}_{i,i}(0) = 3u'^2 = 2 \times \text{kinetic energy per unit mass}$$

We have shown in Sec. 1-6 that $\mathbf{f}(r)$ and $\mathbf{g}(r)$ are even functions of r (this also follows directly from the relations with $\mathbf{Q}_1$ and $\mathbf{Q}_2$ given earlier). Their series expansions read

$$\mathbf{g}(r) = 1 + \frac{r^2}{2!}\left[\frac{\partial^2 \mathbf{g}}{\partial r^2}\right]_{r=0} + \frac{r^4}{4!}\left[\frac{\partial^4 \mathbf{g}}{\partial r^4}\right]_{r=0} + \cdots \tag{3-13}$$

and

$$\mathbf{f}(r) = 1 + \frac{r^2}{2!}\left[\frac{\partial^2 \mathbf{f}}{\partial r^2}\right]_{r=0} + \frac{r^4}{4!}\left[\frac{\partial^4 \mathbf{f}}{\partial r^4}\right]_{r=0} + \cdots \tag{3-14}$$

Substitution of Eqs. (3-13) and (3-14) in Eq. (3-9) gives the relations:

$$\left[\frac{\partial^2 \mathbf{g}}{\partial r^2}\right]_{r=0} = 2\left[\frac{\partial^2 \mathbf{f}}{\partial r^2}\right]_{r=0}$$

$$\left[\frac{\partial^4 \mathbf{g}}{\partial r^2}\right]_{r=0} = 3\left[\frac{\partial^4 \mathbf{f}}{\partial r^4}\right]_{r=0}$$

$$\cdots\cdots\cdots\cdots\cdots$$

$$\left[\frac{\partial^{2n} \mathbf{g}}{\partial r^{2n}}\right]_{r=0} = (n+1)\left[\frac{\partial^{2n} \mathbf{f}}{\partial r^{2n}}\right]_{r=0} \tag{3-15}$$

From the first relation we obtain the following relation between the dissipation scales λ_g and λ_f:

$$\lambda_f = \lambda_g\sqrt{2} \tag{3-16}$$

Presently we shall need correlations between derivatives of velocities, namely,

$$\left(\frac{\partial u_i}{\partial x_k}\right)_A \quad \text{and} \quad \left(\frac{\partial u_j}{\partial x_l}\right)_B$$

Since $(u_i)_A$ is not affected by differentiations at the point B, nor $(u_j)_B$ by differentiations at the point A, we may write

$$\overline{\left(\frac{\partial u_i}{\partial x_k}\right)_A\left(\frac{\partial u_j}{\partial x_l}\right)_B} = \left(\frac{\partial}{\partial x_k}\right)_A\left(\frac{\partial}{\partial x_l}\right)_B\overline{(u_i)_A(u_j)_B} = \left(\frac{\partial}{\partial x_k}\right)_A\left(\frac{\partial}{\partial x_l}\right)_B(\mathbf{Q}_{i,j})_{A,B}$$

$$= -\frac{\partial^2}{\partial \xi_k\,\partial \xi_l}(\mathbf{Q}_{i,j})_{A,B} \tag{3-17}$$

where $\xi_k = (x_k)_B - (x_k)_A$, so that $\left(\frac{\partial}{\partial x_k}\right)_A = -\frac{\partial}{\partial \xi_k}$ and $\left(\frac{\partial}{\partial x_l}\right)_B = \frac{\partial}{\partial \xi_l}$. When $A \equiv B$ we obtain

$$\overline{\left(\frac{\partial u_i}{\partial x_k}\right)\left(\frac{\partial u_j}{\partial x_l}\right)} = -\left[\frac{\partial^2}{\partial \xi_k \, \partial \xi_l}(\mathbf{Q}_{i,j})\right]_{r=0} \tag{3-18}$$

By twice differentiating $(\mathbf{Q}_{i,j})_{A,B}$ according to Eq. (3-17) and making use of the series expansion (3-14) for $\mathbf{f}(r)$, after some calculation, we obtain

$$\begin{aligned}\frac{\partial^2}{\partial \xi_k \, \partial \xi_l}(\mathbf{Q}_{i,j})_{A,B} &= u'^2(2\delta_{kl}\,\delta_{ij} - \tfrac{1}{2}\delta_{il}\,\delta_{jk} - \tfrac{1}{2}\delta_{jl}\,\delta_{ik})\left[\frac{\partial^2 \mathbf{f}}{\partial r^2}\right]_{r=0} \\ &\quad - u'^2\left(\frac{\delta_{il}\,\delta_{jk} + \delta_{jl}\,\delta_{ik} - 6\delta_{kl}\,\delta_{ij}}{12}\right. \\ &\quad \left. + \frac{\xi_i\xi_j\delta_{kl} + \xi_j\xi_k\delta_{li} - 6\xi_k\xi_l\delta_{ij} + \xi_l\xi_i\delta_{jk} + \xi_l\xi_j\delta_{ik} + \xi_i\xi_k\delta_{jl}}{6r^2}\right) \\ &\quad \times \left[\frac{\partial^4 \mathbf{f}}{\partial r^4}\right]_{r=0} r^2 + \cdots\end{aligned} \tag{3-19}$$

Hence

$$\overline{\frac{\partial u_i}{\partial x_k}\frac{\partial u_j}{\partial x_l}} = -\left(\frac{\partial^2 \mathbf{Q}_{i,j}}{\partial \xi_k \, \partial \xi_l}\right)_{r=0} = -u'^2(2\delta_{kl}\,\delta_{ij} - \tfrac{1}{2}\delta_{il}\,\delta_{jk} - \tfrac{1}{2}\delta_{jl}\,\delta_{ik})\left[\frac{\partial^2 \mathbf{f}}{\partial r^2}\right]_{r=0} \tag{3-20}$$

From this general relation we obtain for specific values of i, j, k, and l:

$$\frac{1}{u'^2}\left[\frac{\partial^2 \mathbf{Q}_{i,j}}{\partial \xi_k \, \partial \xi_l}\right]_{r=0} = \left[\frac{\partial^2 \mathbf{f}}{\partial r^2}\right]_{r=0} \qquad \text{if } i = j = k = l \tag{3-21a}$$

$$= -\frac{1}{2}\left[\frac{\partial^2 \mathbf{f}}{\partial r^2}\right]_{r=0} \qquad \text{if } j = l \text{ and } i = k \tag{3-21b}$$

$$\text{or } i = l \text{ and } j = k \text{ and } i \neq j$$

$$= 2\left[\frac{\partial^2 \mathbf{f}}{\partial r^2}\right]_{r=0} \qquad \text{if } k = l \text{ and } i = j \text{ and } i \neq k \tag{3-21c}$$

$$= 0 \qquad \text{for all other combinations} \tag{3-21d}$$

Third-order Correlation Tensor $(\mathbf{S}_{ik,j})_{A,B}$

The triple correlation tensor $\overline{(u_i)_A(u_k)_A(u_j)_B}$ has 27 components. From the product $(u_i)_A(u_k)_A$ six different combinations can be made, corresponding to various values of i and k. Consequently, 18 different combinations can be made of the product $(u_i)_A(u_k)_A(u_j)_B$, so that only 18 of the 27 components of the triple correlation tensor are different.

Consider a coordinate system with the x_1-axis through the points A and B. Though we thus have a special position of the coordinate system with respect to these points, the following will nevertheless have general validity.

By reflection with respect to one of the coordinate planes it may be shown that 13 components of the tensor must be zero because of the condition of invariance.

$$\overline{(u_1)_A{}^2(u_2)_B} = \overline{(u_1)_A{}^2(u_3)_B} = 0$$

$$\overline{(u_2)_A{}^2(u_2)_B} = \overline{(u_3)_A{}^2(u_2)_B} = \overline{(u_3)_A{}^2(u_3)_B} = \overline{(u_2)_A{}^2(u_3)_B} = 0$$

$$\overline{(u_1)_A(u_2)_A(u_1)_B} = \overline{(u_1)_A(u_3)_A(u_1)_B} = 0$$

$$\overline{(u_1)_A(u_2)_A(u_3)_B} = \overline{(u_1)_A(u_3)_A(u_2)_B} = 0$$

$$\overline{(u_2)_A(u_3)_A(u_1)_B} = 0$$

$$\overline{(u_2)_A(u_3)_A(u_2)_B} = \overline{(u_2)_A(u_3)_A(u_3)_B} = 0$$

Thus only the following five components are different from zero: $\overline{(u_1)_A{}^2(u_1)_B}$, $\overline{(u_1)_A(u_2)_A(u_2)_B}$, $\overline{(u_1)_A(u_3)_A(u_3)_B}$, $\overline{(u_2)_A{}^2(u_1)_B}$, and $\overline{(u_3)_A{}^2(u_1)_B}$. Of these five the second and the third form the same configuration, just as do the fourth and the fifth. In Sec. 1-5, we have already designated the remaining three different and invariant components of the tensor: $u'^3\mathbf{k}(r) = \mathbf{S}_{11,1}$, $u'^3\mathbf{h}(r) = \mathbf{S}_{22,1}$, and $u'^3\mathbf{q}(r) = \mathbf{S}_{21,2}$.

These correlations are odd functions of r. This can easily be shown by considering the series expansion of these functions. Since the coefficients of the various powers of r in the series expansion may again be considered as scalar quantities, they must be invariant.

Let us consider, for instance, the correlation $\mathbf{k}(r)$:

$$\mathbf{k}(r) = \mathbf{k}(0) + r\left[\frac{\partial \mathbf{k}}{\partial r}\right]_{r=0} + \frac{r^2}{2!}\left[\frac{\partial^2 \mathbf{k}}{\partial r^2}\right]_{r=0} + \frac{r^3}{3!}\left[\frac{\partial^3 \mathbf{k}}{\partial r^3}\right]_{r=0} + \cdots$$

where

$$\mathbf{k}(0) = \frac{1}{u'^3}\overline{(u_1)_A{}^3}$$

$$\left[\frac{\partial \mathbf{k}}{\partial r}\right]_{r=0} = \frac{1}{u'^3}\left[\overline{(u_1)_A{}^2\left(\frac{\partial u_1}{\partial r}\right)_B}\right]_{r=0} = \frac{1}{u'^3}\overline{(u_1)_A{}^2\left(\frac{\partial u_1}{\partial r}\right)_A}$$

$$\left[\frac{\partial^2 \mathbf{k}}{\partial r^2}\right]_{r=0} = \frac{1}{u'^3}\overline{(u_1)_A{}^2\left(\frac{\partial^2 u_1}{\partial r^2}\right)_A}$$

. . . .

Since $\overline{(u_1)_A{}^3}$, $\overline{(u_1)_A{}^2(\partial^2 u_1/\partial r^2)_A}$, $\overline{(u_1)_A{}^2(\partial^4 u_1/\partial r^4)_A}, \ldots$ do change sign under reflection, they must be zero.

Invariant under reflection are: $\overline{(u_1)_A{}^2(\partial u_1/\partial r)_A}$, $\overline{(u_1)_A{}^2(\partial^3 u_1/\partial r^3)_A}, \ldots$ But, because $\overline{(u_1)_A{}^3} = 0$, $\overline{(u_1)_A{}^2\left(\frac{\partial u_1}{\partial r}\right)_A} = 0$ also, and, consequently, $\left[\frac{\partial \mathbf{k}}{\partial r}\right]_{r=0} = 0$.

Hence

$$\mathbf{k}(0) = \left[\frac{\partial \mathbf{k}}{\partial r}\right]_{r=0} = \left[\frac{\partial^2 \mathbf{k}}{\partial r^2}\right]_{r=0} = \left[\frac{\partial^4 \mathbf{k}}{\partial r^2}\right]_{r=0} = \cdots = \left[\frac{\partial^{2n} \mathbf{k}}{\partial r^{2n}}\right]_{r=0} = 0$$

but

$$\left[\frac{\partial^3 \mathbf{k}}{\partial r^3}\right]_{r=0} \neq 0 \quad \cdots \quad \left[\frac{\partial^{2n+1} \mathbf{k}}{\partial r^{2n+1}}\right]_{r=0} \neq 0$$

Therefore, the correlation $\mathbf{k}(r)$ is an uneven function of r and its series expansion starts with the term r^3.

$$\mathbf{k}(r) = \frac{r^3}{3!}\left[\frac{\partial^3 \mathbf{k}}{\partial r^3}\right]_{r=0} + \frac{r^5}{5!}\left[\frac{\partial^5 \mathbf{k}}{\partial r^5}\right]_{r=0} + \cdots \tag{3-22}$$

Similar results will be obtained for $\mathbf{h}(r)$ and $\mathbf{q}(r)$.

Figure 3-8 shows the results of measurements of $\mathbf{k}(r)$ made by Stewart[4] in a turbulent flow at distance $30M$ downstream of a grid. He used two grids (mesh $M = 12.7$ and 50.8 mm) and made the measurements at two velocities (6.2 and 12.4 m/s).

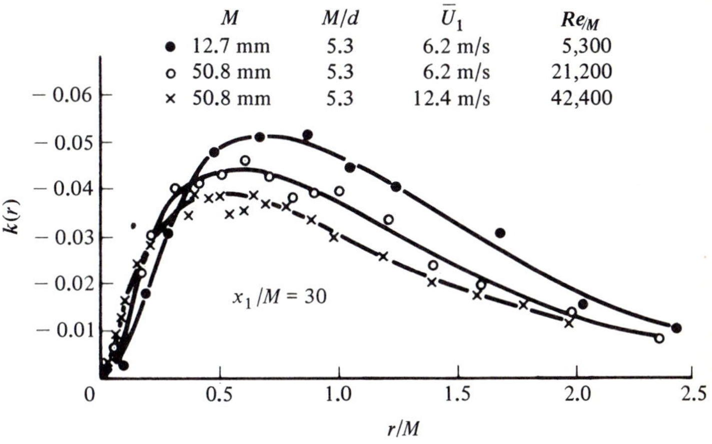

FIG. 3-8
Triple velocity-correlation coefficient $\mathbf{k}(r)$. (From: *Stewart, R. W.,*[4] *by permission of the Cambridge Philosophical Society.*)

For comparison the result of measurements of the one-point triple auto-correlation coefficient $\overset{3}{\mathbf{R}}_E(t)$ by Van Atta and Chen,[60] using the above-mentioned digital techniques is shown in Fig. 3-9. This triple auto-correlation coefficient is defined as

$$\overset{3}{\mathbf{R}}_E(t) = \frac{1}{2}\frac{\overline{u_1{}^2(\tau)u_1(\tau + t) - u_1(\tau)u_1{}^2(\tau + t)}}{u_1'^3}$$

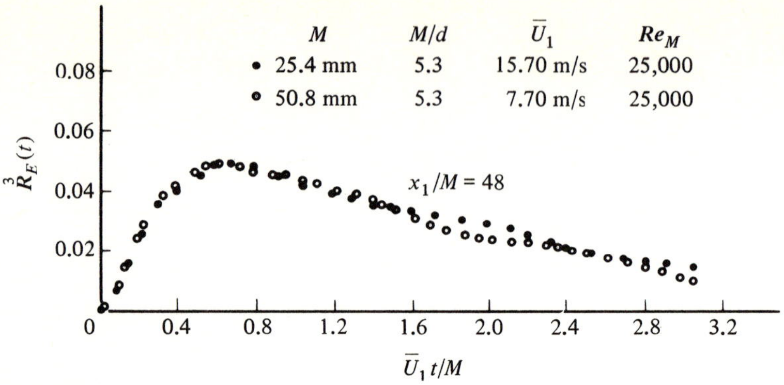

FIGURE 3-9
Composite third-order auto-correlation coefficient $\overset{3}{\mathbf{R}}_E(t)$. (From: *Van Atta, C. W., and W. Y. Chen,*[60] *by permission of the Cambridge University Press.*)

In a homogeneous and isotropic turbulence the two triple-correlation coefficients on the right-hand side should be equal and anti-symmetrical. Actually, the measurements showed slight deviations from this anti-symmetry.

By comparing the longitudinal double-correlation coefficient $\mathbf{f}(r)$ with the longitudinal triple-correlation coefficient $\mathbf{k}(r)$ given in Fig. 3-4 and Fig. 3-8 respectively, measured under the same conditions by Stewart, we may note that though the absolute values of the triple-correlation coefficient are much smaller than those of the double-correlation coefficient, the relative decrease of the triple correlation with increasing separation distance is smaller. So the triple correlation seems to persist longer. At the separation distance where $\mathbf{k}(r)$ shows its maximum, $\mathbf{f}(r)$ has decreased to roughly 0.3.

Of course, the above result that the triple correlation tensor has only three different nonzero components, holds only for the special position of the points A and B with respect to the coordinate system. But, for an arbitrary position of these points, any component of the triple correlation tensor $(\mathbf{S}_{ik,j})_{A,B}$ can be expressed in terms of $\mathbf{k}(r)$, $\mathbf{h}(r)$, and $\mathbf{q}(r)$. This will be shown in the following, by applying Robertson's method.

Consider the scalar quantity (see Fig. 3-10)

$$\mathbf{S}_{A,B} \doteq \overline{u_A u_{\mathring{A}} u_B} = (\mathbf{S}_{ik,j})_{A,B} e_{ai} e_{\mathring{a}k} e_{bj} \tag{3-23}$$

$\mathbf{S}_{A,B}$ is a function of ξ_l, e_{ai}, $e_{\mathring{a}k}$, and e_{bj}, homogeneous and trilinear with respect to $e_{ai} e_{\mathring{a}k} e_{bj}$. The terms of $\mathbf{S}_{A,B}$ consist of scalar combinations of ξ_l, e_{ai}, $e_{\mathring{a}k}$, and e_{bj}. The possible scalar combinations up to the terms trilinear in e_{ai}, $e_{\mathring{a}k}$, and e_{bj} are:

$$\xi_l \xi_l \qquad \xi_i e_{ai} \qquad \xi_k e_{\mathring{a}k} \qquad \xi_j e_{bj}$$

$$\delta_{ik} e_{ai} e_{\mathring{a}k} \qquad \delta_{ij} e_{ai} e_{bj} \qquad \delta_{kj} e_{\mathring{a}k} e_{bj}$$

$$\varepsilon_{ikj} e_{ai} e_{\mathring{a}k} e_{bj} \qquad \varepsilon_{lik} \xi_l e_{ai} e_{\mathring{a}k} \qquad \varepsilon_{lij} \xi_l e_{ai} e_{bj} \qquad \varepsilon_{lkj} \xi_l e_{\mathring{a}k} e_{bj}$$

$$\varepsilon_{lim} \varepsilon_{kjm} \xi_l e_{ai} e_{\mathring{a}k} e_{bj}$$

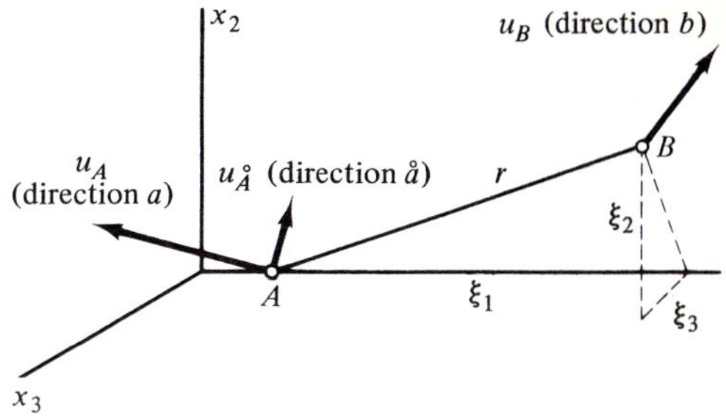

FIGURE 3-10
Triple velocity-correlation $\mathbf{S}_{A,B} = \overline{u_A u_A^0 u_B}$.

Because

$$\varepsilon_{lim}\varepsilon_{kjm}\xi_l = (\delta_{lk}\delta_{ij} - \delta_{lj}\delta_{ik})\xi_l = \delta_{ij}\xi_k - \delta_{ik}\xi_j$$

we obtain

$$\varepsilon_{lim}\varepsilon_{kjm}\xi_l \boldsymbol{e}_{ai}\boldsymbol{e}_{\mathring{a}k}\boldsymbol{e}_{bj} = (\delta_{ij}\xi_k - \delta_{ik}\xi_j)\boldsymbol{e}_{ai}\boldsymbol{e}_{\mathring{a}k}\boldsymbol{e}_{bj}$$

Thus a homogeneous trilinear polynomial in $\boldsymbol{e}_{ai}$, $\boldsymbol{e}_{\mathring{a}k}$, and $\boldsymbol{e}_{bj}$ reads

$$\mathbf{S}_{A,B} = (\mathbf{S}_1\xi_i\xi_k\xi_j + \mathbf{S}_2\xi_j\delta_{ik} + \mathbf{S}_3\xi_k\delta_{ij} + \mathbf{S}_4\xi_i\delta_{kj} + \mathbf{S}_5\varepsilon_{ikj} + \mathbf{S}_6\varepsilon_{lik}\xi_l\xi_j$$
$$+\mathbf{S}_7\varepsilon_{lij}\xi_l\xi_k + \mathbf{S}_8\varepsilon_{lkj}\xi_l\xi_i)\boldsymbol{e}_{ai}\boldsymbol{e}_{\mathring{a}k}\boldsymbol{e}_{bj}$$

Here $\mathbf{S}_1$ to $\mathbf{S}_8$ are functions of r^2.

For the special position of A and B with respect to the coordinate system as considered above, we have shown that the tensor $(\mathbf{S}_{ik,j})_{A,B}$ consists only of the three different correlations $\mathbf{k}(r)$, $\mathbf{h}(r)$, and $\mathbf{q}(r)$. which are odd functions of r. Since the features of a tensor are independent of the coordinate system used, $(\mathbf{S}_{ik,j})_{A,B}$ must be an odd function of r for any coordinate system.

Thus in the above polynomial the terms that are even with respect to r must be zero; that is,

$$\mathbf{S}_5 = \mathbf{S}_6 = \mathbf{S}_7 = \mathbf{S}_8 = 0$$

Since, moreover, $(\mathbf{S}_{ik,j})_{A,B} = (\mathbf{S}_{ki,j})_{A,B}$, we must obtain $\mathbf{S}_3 = \mathbf{S}_4$. Hence

$$(\mathbf{S}_{ik,j})_{A,B} = \mathbf{S}_1\xi_i\xi_k\xi_j + \mathbf{S}_2\xi_j\delta_{ik} + \mathbf{S}_3(\xi_k\delta_{ij} + \xi_i\delta_{kj}) \tag{3-24}$$

This is the *general expression for an isotropic tensor of the third order.*

If, again, the coordinate system is taken with the x_1-axis through A and B, so that $\xi_1 = r$, $\xi_2 = \xi_3 = 0$, it can be verified easily that from Eq. (3-24) it follows that

$$u'^3\mathbf{k}(r) = \mathbf{S}_{11,1} = \mathbf{S}_1 r^3 + r(\mathbf{S}_2 + 2\mathbf{S}_3)$$
$$u'^3\mathbf{h}(r) = \mathbf{S}_{22,1} = \mathbf{S}_2 r$$
$$u'^3\mathbf{q}(r) = \mathbf{S}_{21,2} = \mathbf{S}_3 r$$

or

$$\mathbf{S}_1 = u'^3\left[\frac{\mathbf{k}(r)}{r^3} - \frac{\mathbf{h}(r)}{r^3} - \frac{2\mathbf{q}(r)}{r^3}\right]$$

$$\mathbf{S}_2 = u'^3\frac{\mathbf{h}(r)}{r}$$

$$\mathbf{S}_3 = u'^3\frac{\mathbf{q}(r)}{r}$$

Consequently,

$$(\mathbf{S}_{ik,j})_{A,B} = u'^3\left[(\mathbf{k} - \mathbf{h} - 2\mathbf{q})\frac{\xi_i\xi_k\xi_j}{r^3} + \delta_{ik}\mathbf{h}\frac{\xi_j}{r} + \mathbf{q}\left(\delta_{ij}\frac{\xi_k}{r} + \delta_{kj}\frac{\xi_i}{r}\right)\right] \tag{3-25}$$

For an incompressible fluid $(\mathbf{S}_{ik,j})_{A,B}$ must also have a vanishing divergence

$$\frac{\partial}{\partial\xi_j}(\mathbf{S}_{ik,j})_{A,B} = 0 \tag{3-26}$$

Substitution of Eq. (3-25) in Eq. (3-26), after differentiation, yields,

$$\frac{1}{u'^3}\frac{\partial}{\partial\xi_j}(\mathbf{S}_{ik,j})_{A,B} = \frac{\xi_i\xi_k}{r^2}\left(\frac{\partial\mathbf{k}}{\partial r} - \frac{\partial\mathbf{h}}{\partial r} + 2\frac{\mathbf{k} - \mathbf{h} - 3\mathbf{q}}{r}\right) + \delta_{ik}\left(\frac{\partial\mathbf{h}}{\partial r} + \frac{2\mathbf{h}}{r} + \frac{2\mathbf{q}}{r}\right) = 0$$

Whence

$$\frac{\partial\mathbf{h}}{\partial r} + \frac{2\mathbf{h}}{r} + \frac{2\mathbf{q}}{r} = 0$$

and

$$\frac{\partial\mathbf{k}}{\partial r} - \frac{\partial\mathbf{h}}{\partial r} + 2\frac{\mathbf{k} - \mathbf{h} - 3\mathbf{q}}{r} = 0$$

Or, after some rearranging,

$$\mathbf{q} = -\frac{1}{2r}\frac{\partial}{\partial r}(\mathbf{h}r^2) \tag{3-27a}$$

$$\mathbf{q} = \frac{1}{4r}\frac{\partial}{\partial r}(\mathbf{k}r^2) \tag{3-27b}$$

$$\mathbf{h} = -\tfrac{1}{2}\mathbf{k} \tag{3-27c}$$

Hence, in the case of an *incompressible fluid,* the isotropic triple correlation tensor can be expressed in terms of one single scalar, say, $\mathbf{k}(r)$:

$$(\mathbf{S}_{ik,j})_{A,B} = u'^3\left[\left(\mathbf{k} - r\frac{\partial\mathbf{k}}{\partial r}\right)\frac{\xi_i\xi_k\xi_j}{2r^3} - \frac{\mathbf{k}}{2}\delta_{ik}\frac{\xi_j}{r} + \frac{1}{4r}\frac{\partial}{\partial r}(r^2\mathbf{k})\left(\delta_{ij}\frac{\xi_k}{r} + \delta_{kj}\frac{\xi_i}{r}\right)\right] \tag{3-28}$$

3-3 DIFFERENTIAL EQUATION FOR THE DYNAMIC BEHAVIOR OF AN ISOTROPIC TURBULENCE

In Sec. 1-5 we derived the differential equation (1-48) for $(\mathbf{Q}_{i,j})_{A,B}$, an equation which also contains terms with space derivatives of the correlation tensors $(\mathbf{K}_{i,p})_{A,B}$, $(\mathbf{K}_{p,j})_{A,B}$, $(\mathbf{S}_{ik,j})_{A,B}$, and $(\mathbf{S}_{i,kj})_{A,B}$.

In the preceding section we have shown, for an isotropic turbulence of an incompressible fluid, that the double pressure-velocity correlations are zero and that the correlation tensors $(\mathbf{Q}_{i,j})_{A,B}$ and $(\mathbf{S}_{ik,j})_{A,B}$ can each be expressed in terms of one defining scalar, for instance, $\mathbf{f}(r,t)$ and $\mathbf{k}(r,t)$, respectively.

Now in an isotropic turbulence it follows from the condition of invariance under reflection with respect to point A that

$$\overline{(u_i)_A(u_k)_B(u_j)_B} = -\overline{(u_k)_A(u_j)_A(u_i)_B}$$

or

$$(\mathbf{S}_{i,kj})_{A,B} = -(\mathbf{S}_{kj,i})_{A,B} \tag{3-29}$$

For an isotropic turbulence of an incompressible fluid the differential equation (1-48) simplifies into

$$\frac{\partial}{\partial t}(\mathbf{Q}_{i,j})_{A,B} - \frac{\partial}{\partial \xi_k}[(\mathbf{S}_{ik,j})_{A,B} + (\mathbf{S}_{kj,i})_{A,B}] = 2\nu \frac{\partial^2}{\partial \xi_l\,\partial \xi_l}(\mathbf{Q}_{i,j})_{A,B} \tag{3-30}$$

The second term on the left-hand side is a tensor of the second order, which we shall designate $\mathbf{S}_{i,j}$

$$(\mathbf{S}_{i,j})_{A,B} = \frac{\partial}{\partial \xi_k}[(\mathbf{S}_{ik,j})_{A,B} + (\mathbf{S}_{kj,i})_{A,B}] \tag{3-31}$$

so that

$$\frac{\partial}{\partial t}\mathbf{Q}_{i,j} - \mathbf{S}_{i,j} = 2\nu \frac{\partial^2}{\partial \xi_l\,\partial \xi_l}\mathbf{Q}_{i,j} \tag{3-32}$$

Since $\mathbf{S}_{ik,j}$ is defined by the single scalar $\mathbf{k}(r,t)$, according to Eq. (3-28), $\mathbf{S}_{i,j}$ is also defined by this scalar.

From Eq. (3-28) we obtain

$$\frac{\partial}{\partial \xi_k}\mathbf{S}_{ik,j} = u'^3\left[\left(-\frac{1}{4r}\frac{\partial^2 \mathbf{k}}{\partial r^2} - \frac{1}{r^2}\frac{\partial \mathbf{k}}{\partial r} + \frac{\mathbf{k}}{r^3}\right)\xi_i\xi_j + \left(\frac{r}{4}\frac{\partial^2 \mathbf{k}}{\partial r^2} + \frac{3}{2}\frac{\partial \mathbf{k}}{\partial r} + \frac{\mathbf{k}}{r}\right)\delta_{ij}\right]$$

and, for $(\partial/\partial \xi_k)\mathbf{S}_{kj,i}$, exactly the same result from the equation for $\mathbf{S}_{kj,i}$ that corresponds to Eq. (3-28).

Hence

$$\mathbf{S}_{i,j} = u'^3\left[\left(-\frac{1}{2r}\frac{\partial^2 \mathbf{k}}{\partial r^2} - \frac{2}{r^2}\frac{\partial \mathbf{k}}{\partial r} + \frac{2\mathbf{k}}{r^3}\right)\xi_i\xi_j + \left(\frac{r}{2}\frac{\partial^2 \mathbf{k}}{\partial r^2} + 3\frac{\partial \mathbf{k}}{\partial r} + \frac{2\mathbf{k}}{r}\right)\delta_{ij}\right] \tag{3-33}$$

If we apply a contraction, the expression can be reduced to

$$\mathbf{S}_{i,i} = u'^3\left(r\frac{\partial^2 \mathbf{k}}{\partial r^2} + 7\frac{\partial \mathbf{k}}{\partial r} + 8\frac{\mathbf{k}}{r}\right) = u'^3\frac{1}{r^2}\frac{\partial}{\partial r}\left(r^3\frac{\partial \mathbf{k}}{\partial r} + 4r^2\mathbf{k}\right) \tag{3-34}$$

For this case the differential equation for $\mathbf{Q}_{i,i}$ and $\mathbf{S}_{i,i}$ that corresponds to Eq. (3-32) reads

$$\frac{\partial}{\partial t}\mathbf{Q}_{i,i}(r,t) - \mathbf{S}_{i,i}(r,t) = 2\nu\frac{1}{r^2}\frac{\partial}{\partial r}\left[r^2\frac{\partial}{\partial r}\mathbf{Q}_{i,i}(r,t)\right] \tag{3-35}$$

It is interesting to note that this equation, considered as an equation for the spread of heat, describes such spread in a three-dimensional space. The term $\mathbf{S}_{i,i}$ should then represent the convective term (or may be interpreted as an exchange function); it may be considered to originate from the interaction of eddies.

From Eq. (3-35) we can easily derive the differential equation for $\mathbf{f}(r,t)$ and $\mathbf{k}(r,t)$ by introducing the expressions (3-12) for $\mathbf{Q}_{i,i}$ and (3-34) for $\mathbf{S}_{i,i}$ into Eq. (3-35), at the same time keeping in mind that u' is a function of time only:

$$\frac{\partial}{\partial t}\left[\frac{u'^2}{r^2}\frac{\partial}{\partial r}(r^3\mathbf{f})\right] - \frac{u'^3}{r^2}\frac{\partial}{\partial r}\left(r^3\frac{\partial \mathbf{k}}{\partial r} + 4r^2\mathbf{k}\right) = 2\frac{\nu}{r^2}\frac{\partial}{\partial r}\left\{r^2\frac{\partial}{\partial r}\left[\frac{u'^2}{r^2}\frac{\partial}{\partial r}(r^3\mathbf{f})\right]\right\}$$

Integration with respect to r gives

$$\frac{\partial}{\partial t}(u'^2\mathbf{f}) - u'^3\left(\frac{\partial \mathbf{k}}{\partial r} + \frac{4}{r}\mathbf{k}\right) = 2\nu u'^2\frac{1}{r}\frac{\partial}{\partial r}\left[\frac{1}{r^2}\frac{\partial}{\partial r}(r^3\mathbf{f})\right]$$

or

$$\frac{\partial}{\partial t}(u'^2\mathbf{f}) - u'^3\frac{1}{r^4}\frac{\partial}{\partial r}(r^4\mathbf{k}) = 2\nu u'^2\frac{1}{r^4}\frac{\partial}{\partial r}\left(r^4\frac{\partial \mathbf{f}}{\partial r}\right) \tag{3-36}$$

The constant of integration must be zero, because $\mathbf{f}(r,t)$ and $\mathbf{k}(r,t)$ and their derivatives must remain finite for $r = 0$.

The differential equation (3-36) was first deduced by Von Kármán and Howarth;[2] accordingly, it is usually called the Kármán-Howarth equation.

If Eq. (3-36) also is regarded as describing the spread of heat or matter, it defines a process in five-dimensional space, in contrast with Eq. (3-35).

Frequently the correlation $\mathbf{h}(r,t)$ is used instead of $\mathbf{k}(r,t)$. Since, according to Eq. (3-27c), $\mathbf{k}(r,t) = -2\mathbf{h}(r,t)$, the equation for $\mathbf{h}(r,t)$ reads

$$\frac{\partial}{\partial t}(u'^2\mathbf{f}) + u'^3\frac{2}{r^4}\frac{\partial}{\partial r}(r^4\mathbf{h}) = 2\nu u'^2\frac{1}{r^4}\frac{\partial}{\partial r}\left(r^4\frac{\partial \mathbf{f}}{\partial r}\right) \tag{3-37}$$

Either Eq. (3-35) or (3-36) is used as a starting point for studying the dynamic behavior of an isotropic turbulence. If we try to solve these equations we are immediately confronted with the difficulty that each equation contains two variables. Thus, Eq. (3-36), considered as an equation for $u'^2\mathbf{f}(r,t)$, contains in addition a term with the triple correlation $u'^3\mathbf{k}(r,t)$, which is still an unknown function. As mentioned in Chap. 1, this specific point is one of the fundamental difficulties in solving the turbulence problem when the present statistical approach is used: there are more unknowns than equations available.

Of course, it is possible, starting again from the equation of motion, to derive an equation that describes the behavior of $\mathbf{k}(r,t)$, but such an equation would contain fourth-order velocity correlations, which would then constitute an unknown function in the equation for $\mathbf{k}(r,t)$. This procedure can be continued by considering the dynamic equation of the fourth-order velocity correlation, which then contains as an unknown new dependent variable the fifth-order two-point velocity correlation; and so on. Eventually we would obtain an infinite number of equations, but still one less than the number of dependent variables.

An additional difficulty is that an ever-increasing number of defining scalars will be required for the ever-higher-order correlations, so that the number of terms in the dynamic equations concerned increases indefinitely. Formally, therefore, the present statistical approach to the dynamics of turbulence defines a divergent process. On the other hand, it may be expected that the higher the order of the velocity correlations, the less they will contribute to the complete description of the statistical behavior of the turbulence. So, as Batchelor[11] has pointed out, it is possible to make this description complete to a sufficient degree of approximation by considering only a finite number of velocity correlations (though this number may be large) and an equal number of dynamic equations, since some appropriate assumption can surely be made concerning the next-higher-order velocity correlation in order to determine its value.

Various attempts have been made to solve this so-called *closure problem*. As said above, it seems reasonable to expect that the higher the order of a velocity correlation, the smaller its effect will be on the dynamical behavior of the turbulence. Logically, this idea then leads to a *truncation approximation*. In a study of the turbulent pressure flow in a two-dimensional channel, Chou[55] showed that the divergence of the quadruple-velocity correlation was small compared with the triple correlation between the pressure-gradient fluctuation and the velocity fluctuation. He therefore made the assumption that the quadruple-velocity correlation term could be neglected.

A truncation approximation has also been made by Deissler,[56] and by Loeffler and Deissler,[61] by neglecting fourth-order correlations in the dynamical equations for the lower-order velocity correlations. Lee,[62] considering the development of a turbulent scalar field, went one step further and neglected the fifth-order correlations. However, apart from the fact that the closure procedure by neglecting the next higher-order correlations, in the dynamical equations for a certain-order correlation may converge insufficiently rapidly at large Reynolds numbers, these truncation approximations appear to produce solutions of the initial-value problem, showing after a certain stage of development of the power spectrum and at sufficiently high values of the Reynolds number, negative parts in the spectral distribution, a physically unacceptable situation.

Another method of solving the closure problem is the *quasi-normal approximation*. Here, instead of neglecting higher-order correlations, higher order cumulants† are put equal to zero, for instance, the fourth-order cumulant. This implies a Gaussian behavior of the fourth-order correlation, i.e. a joint normal-probability distribution. However, a jointly normal probability distribution makes the third-order correlation $(S_{ik,j})_{A,B}$ zero, whereas this correlation has to be retained. Therefore the approximation is called quasi-normal. It is possible to show that even if at some initial instant pure Gaussianity were to occur, nonzero values of the triple correlations are developed at later stages, so that the flow cannot remain Gaussian.[63] Now, Kraichnan[64] proved for a turbulent scalar field, that the assumption of a zero fourth-order cumulant also led to a negative power spectrum at larger times. The same has been shown by Ogura[65,66] to occur for the velocity field. The energy-spectrum obtained a negative part after a certain time of development, while back transfer of energy from smaller to bigger eddies occurred. Lee[62] showed that this anomalous behavior of the energy spectrum is obtained due to a too well organized turbulence with too long a memory behavior. Indeed, the quasi-normal approximation could be improved in this respect by introducing an appropriate response function in the expression describing in the dynamical equations the transfer of energy through the spectrum.

The most satisfactory closure procedure suggested, has been the *direct-interaction approximation* introduced by Kraichnan.[67] In this theory the dynamics of the turbulent motion through the interaction of eddies of different sizes is taken into account. And, further, amongst others, the assumption of maximum randomness of the motions is made. This approximation yields solutions that do not show a negative part in the energy spectrum.

Before these purely theoretical approximations were suggested, the closure problem was attacked by making assumptions based more upon considerations of a physical nature. The usual approach to solving Eqs. (3-35) and (3-36) has been by making some assumption concerning the triple correlation, or its Fourier transform. We shall come back to these "physical" closure approximations in a later section. First we will consider a rigorous solution for the limiting case of very low Reynolds numbers, so that the effect of the "convective" terms, represented by the triple-correlation terms in Eqs. (3-35) and (3-36) may be neglected. This situation describes the *asymptotic behavior of turbulence with dominant viscosity effects,* as occurs for instance during the last stage of a decaying turbulence.

† The n-order cumulant tensor of n velocity components is obtained by subtracting from the n-order mean velocity product the various mean products of lower order that can be formed from the n velocity components. For a turbulence with zero mean velocity the cumulant tensor of lowest order is four:

$$\overline{u_i u_j u_k u_l} - \overline{u_i u_j}\,\overline{u_k u_l} - \overline{u_i u_k}\,\overline{u_j u_l} - \overline{u_i u_l}\,\overline{u_j u_k}$$

It is zero when the joint probability distribution is normal.

The solution for $u'^2\mathbf{f}(r,t)$ can be obtained easily for this limiting case. Before giving this solution, however, we should like to discuss a general feature directly obtainable from the Eqs. (3-35) and (3-36).

Let us take the second moment of each term of the Eq. (3-35)

$$\frac{\partial}{\partial t}\int_0^\infty dr\, r^2\mathbf{Q}_{i,i} - \int_0^\infty dr\, r^2\mathbf{S}_{i,i} = 2\nu\left(r^2\frac{\partial\mathbf{Q}_{i,i}}{\partial r}\right)\Bigg|_0^\infty$$

We shall show later on that, with certain assumptions, a consequence of the continuity condition for an incompressible fluid is

$$\int_0^\infty dr\, r^2\mathbf{Q}_{i,i} = 0 \qquad \text{and} \qquad \int_0^\infty dr\, r^2\mathbf{S}_{i,i} = 0$$

Consequently, it follows from the above equation that

$$\lim_{r\to\infty} r^2\frac{\partial\mathbf{Q}_{i,i}}{\partial r} = 0 \tag{3-38}$$

This shows how quickly at least $\mathbf{Q}_{i,i}$ has to decrease as r increases.

We may now apply a similar procedure to Eq. (3-36), this time, however, taking the fourth moment instead of the second:

$$\frac{\partial}{\partial t}\left(u'^2\int_0^\infty dr\, r^4\mathbf{f}\right) - (u'^3 r^4\mathbf{k})\Big|_0^\infty = 2\nu u'^2\left(r^4\frac{\partial\mathbf{f}}{\partial r}\right)\Bigg|_0^\infty$$

Later it will be shown that with certain assumptions concerning the large-scale structure of the turbulence, it is reasonable to expect that at least

$$\lim_{r\to\infty}\left(r^4\frac{\partial\mathbf{f}}{\partial r}\right) = 0$$

This is a much stronger requirement than the one imposed by Eq. (3-38). For according to Eq. (3-12) the asymptotic behavior of $\mathbf{f}(r)$ and $\mathbf{Q}_{i,i}(r)$ are the same. So if $\mathbf{f}(r)$ were to behave as r^{-n} for large r, the condition (3-38) would require $n > 1$, whereas the latter, stronger, condition would require $n > 3$. As to the term $r^4\mathbf{k}$, it has also ordinarily been assumed that this term approaches zero when r increases indefinitely. In general, it has been customary to assume that all velocity correlations in homogeneous turbulence are exponentially small for large distances between the two points. This assumption has also been made by Loitsianskii;[5] from it he concluded that

$$u'^2\int_0^\infty dr\, r^4\mathbf{f}(r,t) \tag{3-39}$$

must be invariant. This integral is, therefore, usually referred to as *"Loitsianskii's invariant."*

If Eq. (3-36) is considered as an equation for the spread of heat in a five-dimensional space, the result (3-39) may be interpreted as the condition for the conservation of the total amount of heat in the five-dimensional space. The interpretation given by Loitsianskii is that the integral (3-39) gives the total amount of disturbance introduced by the turbulence-generating system. And such an amount of disturbance should remain constant with time.

Now Batchelor and Proudman[50] showed that the assumption concerning the rate of decrease of velocity correlations with increasing distance between the two points does not always hold and that, in particular, $\lim_{r\to\infty}(r^4\mathbf{k})$ is not zero. Hence, if we use the complete equation (3-36) including the inertia term $r^4\mathbf{k}$, the Loitsianskii integral (3-39) is not an invariant. We shall come back to this point later on.

Let us now consider the solution of either Eq. (3-35) or (3-36) for the limiting case in which the viscosity effects become predominant, so that the second term on the left-hand side of the equation may be neglected. It is sufficient to consider only one of the two equations, since the solutions obtained will be alike.

Here we shall consider Eq. (3-35), which for the limiting case in question reads

$$\frac{\partial}{\partial t}\mathbf{Q}_{i,i}(r,t) = 2\nu\frac{1}{r^2}\frac{\partial}{\partial r}\left[r^2\frac{\partial \mathbf{Q}_{i,i}(r,t)}{\partial r}\right] \tag{3-40}$$

This is a well-known equation for the conduction of heat or the diffusion of matter in space, in the case of spherical symmetry. A usual method for solving this differential equation is to assume that $\mathbf{Q}_{i,i}(r,t)$ is a function of $\chi = r/\sqrt{\nu t}$ alone. Indeed, simple solutions expressed in terms of $\exp(-r^2/8\nu t)$ may be obtained. However, as we can see if we evaluate $u'^2\mathbf{f}(r,t)$ from $\mathbf{Q}_{i,i}(r,t)$ according to Eq. (3-12), these solutions give values for $u'^2\mathbf{f}(r,t)$ that are not finite for $r = 0$.

Since, in the relation (3-12), u'^2 is a function of t alone, the assumption that $\mathbf{Q}_{i,i}(r,t)$ is a function of χ alone is not justified. The same assumption can better be made with respect to the correlation coefficient $\mathbf{f}(r,t)$.

Hence it is better to follow Agostini and Bass,[6] who assumed that

$$\mathbf{Q}_{i,i}(r,t) = \varphi(t)\psi(\chi) \tag{3-41}$$

where $\chi = r/\sqrt{8\nu t}$. Substitution of Eq. (3-41) in Eq. (3-40) yields, after separation of variables, the two differential equations:

$$\frac{1}{\varphi}\frac{d\varphi}{dt} = -\frac{\alpha}{t}$$

with the solution $\varphi = \text{const.} \times t^{-\alpha}$, and

$$\chi\frac{d^2\psi}{d\chi^2} + 2(\chi^2 + 1)\frac{d\psi}{d\chi} + 4\alpha\chi\psi = 0$$

Assuming $\alpha = (2p + 1)/2$, where p is an integer, Agostini and Bass obtained the solution

$$\psi_p = \frac{1}{\chi}\exp(-\chi^2)H_{2p-1}(\chi)$$

where $H_n(\chi)$ is the Hermite polynomial

$$H_n(\chi) = (-1)^n \exp(\chi^2) \frac{d^n}{d\chi^n} \exp(-\chi^2)$$

[thus $H_0(\chi) = 1$; $H_1(\chi) = 2\chi$; $H_2(\chi) = 4\chi^2 - 2$; $H_3(\chi) = 8\chi^3 - 12\chi; \ldots$]

The general solution of Eq. (3-40), subject to the conditions given above, then reads

$$\mathbf{Q}_{i,i}(r,t) = \frac{\sqrt{8\nu}}{r} \exp(-r^2/8\nu t) \sum_1^\infty \frac{A_p}{t^p} H_{2p-1}\left(\frac{r}{\sqrt{8\nu t}}\right) \tag{3-42}$$

The constants A_p have to be chosen such that the series converges and that $\mathbf{Q}_{i,i}(0,t) = 3u'^2$.

If we apply the condition

$$\int_0^\infty dr\, r^2 \mathbf{Q}_{i,i}(r,t) = 0$$

we find that all the terms of Eq. (3-42) satisfy this condition except the term where $p = 1$; so A_1 must be zero.

Furthermore, for the corresponding solutions for $\mathbf{f}(r,t)$, we find that:

1. When $p = 1$, the solution does not converge to a finite value for $r = 0$; convergence requires that $A_1 = 0$ also.
2. In Loitsianskii's integral (3-39) only the term where $p = 2$ makes a contribution; all other terms make a zero contribution to the integral.

Hence, if we consider further only the particular solution of Eq. (3-40) where $p = 2$, Eq. (3-42) may be reduced to

$$\mathbf{Q}_{i,i}(r,t) = -\frac{4A_2}{t^{5/2}}\left(3 - \frac{r^2}{4\nu t}\right) \exp(-r^2/8\nu t)$$

The condition $\mathbf{Q}_{i,i}(0,t) = 3u'^2$ yields

$$u'^2 = -4A_2 t^{-5/2} = \text{const.} \times t^{-5/2} \tag{3-43}$$

and, consequently,

$$\begin{aligned}\mathbf{Q}_{i,i}(r,t) &= \text{const.} \times t^{-5/2}\left(3 - \frac{r^2}{4\nu t}\right) \exp\left(-\frac{r^2}{8\nu t}\right) \\ &= u'^2\left(3 - \frac{r^2}{4\nu t}\right) \exp\left(-\frac{r^2}{8\nu t}\right)\end{aligned} \tag{3-44}$$

From Eqs. (3-12) and (3-44) for $\mathbf{f}(r,t)$ we obtain

$$\mathbf{f}(r,t) = \exp(-r^2/8\nu t) \tag{3-45}$$

Thus for the case of dominating viscosity effects, the relation (3-43) shows the decay law for the turbulence; and the double correlation coefficient $\mathbf{f}(r,t)$ has the

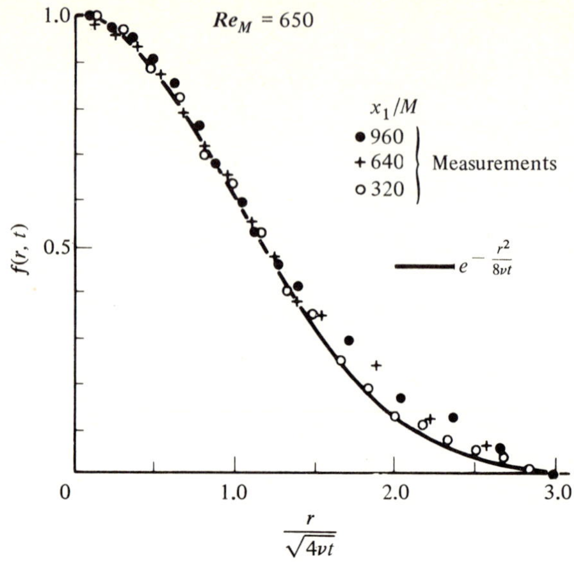

FIGURE 3-11
Longitudinal-velocity-correlation coefficient $\mathbf{f}(r,t)$ in the final period of decay. (From: *Batchelor, G. K., and A. A. Townsend,*[7] *by permission of the Royal Society.*)

shape of a Gaussian error curve and remains self-preserving during the decay. Measurements by Batchelor and Townsend[7] have shown that, in the final period of decay where the turbulent motions are of an entirely viscous nature, the agreement with the Gaussian error curve is, indeed, very close (see Fig. 3-11).

If we use Eqs. (3-43) and (3-45) to calculate the value of Loitsianskii's integral (3-39), we conclude that, in the case of predominant viscosity effects, the integral is an exact invariant with respect to time.

3-4 THE THREE-DIMENSIONAL ENERGY SPECTRUM

In Sec. 1-12 we discussed the one-dimensional energy spectrum $\mathbf{E}_1(n)$ of the velocity fluctuation u_1, as registered by a suitable measuring apparatus, for instance, a hot-wire anemometer. This spectrum function $\mathbf{E}_1(n)$ was defined by Eq. (1-91):

$$\int_0^\infty dn\, \mathbf{E}_1(n) = \overline{u_1^2} \qquad (1\text{-}91)$$

Taylor has shown that $\mathbf{E}_1(n)$ and $u'^2\mathbf{f}(x_1)$ are Fourier cosine transforms, which, if the turbulent flow field has a constant velocity $\bar{U}_1$ that is large compared with u', read [see Eqs. (1-98) and (1-99)]

$$\mathbf{f}(x_1) = \frac{1}{u'^2}\int_0^\infty dn\, \mathbf{E}_1(n) \cos \frac{2\pi n x_1}{\bar{U}_1} \qquad (3\text{-}46)$$

and

$$\mathbf{E}_1(n) = \frac{4u'^2}{\bar{U}_1}\int_0^\infty dx_1 \mathbf{f}(x_1)\cos\frac{2\pi n x_1}{\bar{U}_1} \tag{3-46a}$$

In view of the three-dimensional character of the turbulence, it is obvious that the energy spectrum must also have a three-dimensional character. What is actually measured by means of a hot-wire anemometer is, strictly speaking, a one-dimensional cut of the spatial spectrum. For a more accurate analysis of the spatial structure it will, therefore, be necessary to consider this spatial spectrum more closely.

For this purpose we may consider the spatial structure at a given instant, just as the field presents itself to an observer who is at rest relative to the field. It is then more convenient to consider the wave-number $k_1 = 2\pi n/\bar{U}_1$ instead of the frequency n and to introduce the energy-spectrum function $\mathbf{E}_1(k_1)$ instead of $\mathbf{E}_1(n)$. It appears suitable to define $\mathbf{E}_1(k_1)$ by

$$\mathbf{E}_1(k_1) = \frac{\bar{U}_1}{2\pi}\mathbf{E}_1(n) \tag{3-47}$$

so that

$$\int_0^\infty dk_1\, \mathbf{E}_1(k_1) = \overline{u_1{}^2} \tag{3-48}$$

which is similar to Eq. (1-91). The relation between the energy-spectrum function $\mathbf{E}_1(k_1)$ of the longitudinal velocity fluctuations and the longitudinal velocity correlation $\mathbf{f}(x_1)$ is given by

$$u'^2\mathbf{f}(x_1) = \int_0^\infty dk_1\, \mathbf{E}_1(k_1)\cos k_1 x_1 \tag{3-49}$$

with its Fourier cosine transform

$$\mathbf{E}_1(k_1) = \frac{2}{\pi}u'^2\int_0^\infty dx_1\, \mathbf{f}(x_1)\cos k_1 x_1 \tag{3-50}$$

Since the functions $\mathbf{f}(x_1)$ and $\mathbf{E}_1(k_1)$ are symmetrical, we may also write (see Sec. 1-12)

$$u'^2\mathbf{f}(x_1) = \frac{1}{2}\int_{-\infty}^{+\infty} dk_1\, \mathbf{E}_1(k_1)\exp(\imath k_1 x_1) \tag{3-51}$$

and

$$\mathbf{E}_1(k_1) = \frac{1}{\pi}u'^2\int_{-\infty}^{+\infty} dx_1\, \mathbf{f}(x_1)\exp(-\imath k_1 x_1) \tag{3-51a}$$

where $\imath = \sqrt{-1}$. Similarly, we may introduce the energy-spectrum function $\mathbf{E}_2(k_1)$ of the transverse velocity fluctuations, which is related to the transverse velocity correlation $\mathbf{g}(x_1)$ according to

$$u'^2\mathbf{g}(x_1) = \frac{1}{2}\int_{-\infty}^{+\infty} dk_1\, \mathbf{E}_2(k_1)\exp(\imath k_1 x_1) \tag{3-52}$$

with the Fourier cosine transform

$$\mathbf{E}_2(k_1) = \frac{1}{\pi}u'^2\int_{-\infty}^{+\infty} dx_1\, \mathbf{g}(x_1)\exp(-\imath k_1 x_1) \tag{3-52a}$$

Since the correlations $\mathbf{f}(x_1)$ and $\mathbf{g}(x_1)$ are different, the corresponding one-dimensional spectrum functions $\mathbf{E}_1(k_1)$ and $\mathbf{E}_2(k_1)$ are also different, although

$$\int_0^{\infty} dk_1\, \mathbf{E}_2(k_1) = \int_0^{\infty} dk_1\, \mathbf{E}_1(k_1) = \overline{u^2}$$

The relation between $\mathbf{E}_2(k_1)$ and $\mathbf{E}_1(k_1)$ can easily be obtained from the relation (3-9) between $\mathbf{f}(x_1)$ and $\mathbf{g}(x_1)$, by taking the Fourier transform of this relation

$$\frac{1}{\pi}u'^2\int_{-\infty}^{+\infty} dx_1\, \mathbf{f}(x_1)\exp(-\imath k_1 x_1) + \frac{1}{\pi}u'^2\int_{-\infty}^{+\infty} dx_1\, \frac{x_1}{2}\frac{\partial \mathbf{f}}{\partial x_1}\exp(-\imath k_1 x_1)$$

$$= \frac{1}{\pi}u'^2\int_{-\infty}^{+\infty} dx_1\, \mathbf{g}(x_1)\exp(-\imath k_1 x_1)$$

By integrating by parts the second term on the left-hand side of the equation, and making use of the relations (3-51a) and (3-52a), the following result is obtained

$$\mathbf{E}_2(k_1) = \frac{1}{2}\left[\mathbf{E}_1(k_1) - k_1\frac{\partial \mathbf{E}_1(k_1)}{\partial k_1}\right] \tag{3-53}$$

In the partial integration procedure we have made use of the property of $\mathbf{f}(x_1)$ that

$$\int_{-\infty}^{+\infty} dx_1\, \mathbf{f}(x_1)$$

exists, so that as a consequence

$$\lim_{x_1\to\infty} x_1\,\mathbf{f}(x_1) = 0$$

We will now give the extension to the three-dimensional spectrum function. Instead of the scalar correlations $\mathbf{f}(x_1)$ and $\mathbf{g}(x_1)$ we shall consider the correlation tensor $\mathbf{Q}_{i,j}$ and define its Fourier transform, the energy-spectrum tensor $\mathbf{E}_{i,j}$, so that

$$\mathbf{Q}_{i,j}(x_1,x_2,x_3,t) = \iiint_{-\infty}^{+\infty} dk_1\, dk_2\, dk_3\, \mathbf{E}_{i,j}(k_1,k_2,k_3,t)\exp[\imath(k_1x_1 + k_2x_2 + k_3x_3)] \tag{3-54}$$

Conversely, we have

$$\mathbf{E}_{i,j}(k_1,k_2,k_3,t) = \frac{1}{8\pi^3}\iiint_{-\infty}^{+\infty} dx_1\, dx_2\, dx_3\, \mathbf{Q}_{i,j}(x_1,x_2,x_3,t)\exp[-\imath(k_1x_1 + k_2x_2 + k_3x_3)] \tag{3-55}$$

Hitherto we have only made use of the assumed homogeneity of the turbulence, so that a Fourier transformation in space may be applied.

If, moreover, the turbulence is *isotropic*, then we have

$$\mathbf{Q}_{i,j}(x_1,x_2,x_3,t) = \mathbf{Q}_{i,j}(-x_1,-x_2,-x_3,t)$$

and

$$\mathbf{E}_{i,j}(k_1,k_2,k_3,t) = \mathbf{E}_{i,j}(-k_1,-k_2,-k_3,t)$$

In that case it is convenient to introduce the wavenumber k and the distance r, and to consider instead of $\mathbf{Q}_{i,j}(x_1,x_2,x_3,t)$ and $\mathbf{E}_{i,j}(k_1,k_2,k_3,t)$ the corresponding quantities $\mathbf{Q}_{i,i}(r,t)$ and $\mathbf{E}_{i,i}(k,t)$.

Apply spherical polar coordinates k, φ, and θ in the expression (3-54) with the polar axis along the direction of r. Then k is the radius vector in the wavenumber space, θ the colatitude, and φ the longitude. Thus $dk_1\,dk_2\,dk_3 = k\sin\theta\,d\varphi k\,d\theta\,dk$ and $k_1x_1 + k_2x_2 + k_3x_3 = kr\cos\theta$. The expression (3-54) then becomes

$$\mathbf{Q}_{i,i}(r,t) = \int_0^\infty dk \int_0^{2\pi} d\varphi \int_0^\pi d\theta\,\mathbf{E}_{i,i}(k,t)k^2\sin\theta\exp(\imath kr\cos\theta)$$

whence

$$\mathbf{Q}_{i,i}(r,t) = 4\pi\int_0^\infty dk\,k^2\frac{\sin kr}{kr}\,\mathbf{E}_{i,i}(k,t) \qquad \text{(3-56)}$$

Similarly, from Eq. (3-55) we obtain

$$\mathbf{E}_{i,i}(k,t) = \frac{1}{8\pi^3}\int_0^\infty dr \int_0^{2\pi} d\varphi \int_0^\pi d\theta\,\mathbf{Q}_{i,i}(r,t)r^2\sin\theta\exp(\imath kr\cos\theta)$$

whence

$$\mathbf{E}_{i,i}(k,t) = \frac{1}{2\pi^2}\int_0^\infty dr\,r^2\frac{\sin kr}{kr}\,\mathbf{Q}_{i,i}(r,t) \qquad \text{(3-56a)}$$

Now we define $\mathbf{E}(k,t)$ as the *three-dimensional energy-spectrum function*, so that

$$\mathbf{E}(k,t) = 2\pi k^2\mathbf{E}_{i,i}(k,t) \qquad \text{(3-57)}$$

Since $\mathbf{Q}_{i,i}(0,t) = 3\overline{u^2} = 4\pi\int_0^\infty dk\,k_2\mathbf{E}_{i,i}(k,t)$, we have

$$\int_0^\infty dk\,\mathbf{E}(k,t) = \tfrac{3}{2}\overline{u^2} \qquad \text{(3-58)}$$

Thus, from Eq. (3-56a),

$$\mathbf{E}(k,t) = \frac{1}{\pi}\int_0^\infty dr\,kr\sin kr\,\mathbf{Q}_{i,i}(r,t) \qquad \text{(3-59)}$$

and, conversely,

$$\mathbf{Q}_{i,i}(r,t) = 2\int_0^\infty dk\frac{\sin kr}{kr}\,\mathbf{E}(k,t) \qquad \text{(3-60)}$$

From Eq. (3-59) we may obtain an idea of the shape of the $\mathbf{E}(k,t)$ function for small values of k. Series expansion of sin kr yields

$$\mathbf{E}(k,t) = \frac{k^2}{\pi}\int_0^\infty dr\, r^2\mathbf{Q}_{i,i}(r,t) - \frac{k^4}{3!\pi}\int_0^\infty dr\, r^4\mathbf{Q}_{i,i}(r,t) + \cdots \qquad (3\text{-}61)$$

We shall now show that, provided $\mathbf{E}_{i,j}(k_1,k_2,k_3,t)$ remains finite and analytic at $k = 0$, the integral in the first term on the right-hand side is zero for an incompressible fluid.

Differentiation of Eq. (3-54) with respect to x_j yields

$$\frac{\partial}{\partial x_j}\mathbf{Q}_{i,j}(x_1,x_2,x_3,t) = \iiint_{-\infty}^{+\infty} dk_1\, dk_2\, dk_3\, \imath k_j \mathbf{E}_{i,j}(k_1,k_2,k_3,t) \exp(\imath k_l x_l) = 0$$

whence

$$k_j\mathbf{E}_{i,j}(k_1,k_2,k_3,t) = 0 \qquad (3\text{-}62)$$

Since here $\mathbf{E}_{i,j}(k_1,k_2,k_3,t)$ is an isotropic second-order tensor, its general expression [see Eq. (3-7)] reads

$$\mathbf{E}_{i,j}(k_1,k_2,k_3,t) = C_1(k^2,t)k_ik_j + C_2(k^2,t)\,\delta_{ij}k^2$$

Substitution in Eq. (3-62) yields

$$k_j\mathbf{E}_{i,j}(k_1,k_2,k_3,t) = C_1(k^2,t)k_ik^2 + C_2(k^2,t)k^2k_i = 0$$

or

$$C_2(k^2,t) = -C_1(k^2,t)$$

so

$$\mathbf{E}_{i,j}(k_1,k_2,k_3,t) = C_1(k^2,t)(k_ik_j - k^2\delta_{ij})$$

Consequently, we obtain for $\mathbf{E}(k,t)$, according to Eq. (3-57),

$$\mathbf{E}(k,t) = 2\pi k^2 C_1(k^2,t)(k^2 - 3k^2) = -4\pi k^4 C_1(k^2,t) \qquad (3\text{-}63)$$

Thus

$$\mathbf{E}_{i,j}(k_1,k_2,k_3,t) = -\frac{\mathbf{E}(k,t)}{4\pi k^2}\left(\frac{k_ik_j}{k^2} - \delta_{ij}\right) \qquad (3\text{-}64)$$

For instance, for $\mathbf{E}_{1,1}(k_1,k_2,k_3,t)$ we obtain

$$\mathbf{E}_{1,1}(k_1,k_2,k_3,t) = \frac{\mathbf{E}(k,t)}{4\pi k^2}\left(1 - \frac{{k_1}^2}{k^2}\right)$$

and, for $\mathbf{E}_{1,2}(k_1,k_2,k_3,t)$

$$\mathbf{E}_{1,2}(k_1,k_2,k_3,t) = -\frac{\mathbf{E}(k,t)}{4\pi k^2}\frac{k_1k_2}{k^2}$$

The result (3-63) shows that, if $C_1(k^2,t)$ approaches a constant value at $k = 0$, the series expansion of $\mathbf{E}(k,t)$ starts with a term k^4. This appears to be the case if it is assumed that $\mathbf{E}_{i,j}(k_1,k_2,k_3,t)$ is finite and analytic with respect to k_1, k_2, and k_3 at $k=0$.[11] Let the series expansion of $C_1(k^2,t)$ read

$$C_1(k^2,t) = C_1^{(-1)}(t)k^{-2} + C_1^{(0)}(t) + C_1^{(1)}(t)k^2 + \cdots$$

we then have to show that $C_1^{(-1)} = 0$.

Substitution of the series expansion of $C_1(k^2,t)$ in the above expression for $\mathbf{E}_{i,j}(k_1,k_2,k_3,t)$ yields

$$\mathbf{E}_{i,j}(k_1,k_2,k_3,t) = C_1^{(-1)}(t)\left(\frac{k_ik_j}{k^2} - \delta_{ij}\right) + C_1^{(0)}(t)(k_ik_j - k^2\delta_{ij}) + \cdots$$

Since k_ik_j/k^2 is not analytic at $k=0$, its value depending on the way in which $k=0$ is approached, the assumption that $\mathbf{E}_{i,j}(k_1,k_2,k_3,t)$ is analytic with respect to k_1, k_2, and k_3 at $k=0$ requires that $C_1^{(-1)}(t)=0$. Consequently $\mathbf{E}(k,t) \propto k^4$ at $k=0$.

Comparison of this result for $\mathbf{E}(k,t)$ with Eq. (3-61) shows that, for an incompressible fluid, it is indeed true that

$$\int_0^\infty dr\, r^2\mathbf{Q}_{i,i}(r,t) = 0 \qquad (3\text{-}65)$$

From this result it also follows that at least

$$\lim_{r\to\infty} r^3\mathbf{f}(r,t) = 0 \qquad (3\text{-}66)$$

because, according to Eqs. (3-12) and (3-65),

$$\lim_{r\to\infty} u'^2[r^3\mathbf{f}(r,t)] = \int_0^\infty dr\, r^2\mathbf{Q}_{i,i}(r,t) = 0$$

If the correlation $\mathbf{f}(r,t)$ approaches zero monotonously when $r\to\infty$, it follows from Eq. (3-66) that also

$$\lim_{r\to\infty} r^4\frac{\partial\mathbf{f}(r,t)}{\partial r} = 0$$

This was one of the necessary conditions used to obtain Loitsianskii's invariant (3-39). With Eq. (3-65) the series expansion for $\mathbf{E}(k,t)$ reads

$$\mathbf{E}(k,t) = -\frac{k^4}{3!\pi}\int_0^\infty dr\, r^4\mathbf{Q}_{i,i}(r,t) + \frac{k^6}{5!\pi}\int_0^\infty dr\, r^6\mathbf{Q}_{i,i}(r,t) - \cdots \qquad (3\text{-}67)$$

It should be noted that the above results, given by the Eqs. (3-65), (3-66) and (3-67), are obtained on the assumption of an analytic behavior of $\mathbf{E}_{i,j}(k_1,k_2,k_3,t)$ at $k=0$. In the next section it will be shown that in general this need not necessarily be so.

Relations Among $\mathbf{f}(r,t)$, $\mathbf{g}(r,t)$, $\mathbf{E}_1(k_1,t)$, $\mathbf{E}_2(k_1,t)$, and $\mathbf{E}(k,t)$

In the foregoing sections we have derived relations among $\mathbf{f}(r,t)$, $\mathbf{g}(r,t)$, and $\mathbf{Q}_{i,i}(r,t)$. Since $(1/k)\mathbf{E}(k,t)$ and $r\mathbf{Q}_{i,i}(r,t)$ are Fourier sine transforms, a relation must also exist between $\mathbf{f}(r,t)$ or its transform $\mathbf{E}_1(k_1,t)$ and the function $\mathbf{E}(k,t)$; similarly, between $\mathbf{g}(r,t)$ or its transform $\mathbf{E}_2(k_1,t)$ and $\mathbf{E}(k,t)$. All these relations can be derived from previous formulas. In what follows we shall give these relations, because they may be useful for calculating one function when the other is known.

From Eqs. (3-12) and (3-60) we obtain

$$u'^2\mathbf{f}(r,t) = \frac{1}{r^3}\int_0^r dr\, r^2\mathbf{Q}_{i,i}(r,t) = \frac{2}{r^3}\int_0^r dr\, r^2 \int_0^\infty dk\, \mathbf{E}(k,t)\frac{\sin kr}{kr}$$

or

$$u'^2\mathbf{f}(r,t) = 2\int_0^\infty dk\, \mathbf{E}(k,t)\left(\frac{\sin kr}{k^3r^3} - \frac{\cos kr}{k^2r^2}\right) \tag{3-68}$$

Likewise, with the aid of Eq. (3-9),

$$u'^2\mathbf{g}(r,t) = \int_0^\infty dk\, \mathbf{E}(k,t)\left(\frac{\sin kr}{kr} - \frac{\sin kr}{k^3r^3} + \frac{\cos kr}{k^2r^2}\right) \tag{3-69}$$

Conversely, from Eqs. (3-12) and (3-59),

$$\mathbf{E}(k,t) = u'^2\frac{1}{\pi}\int_0^\infty dr\, kr(\sin kr - kr\cos kr)\mathbf{f}(r,t) \tag{3-70}$$

In order to obtain the relations between $\mathbf{E}_1(k_1,t)$ or $\mathbf{E}_2(k_1,t)$ and $\mathbf{E}(k,t)$, we start from Eq. (3-12) and make use of the relations (3-51) and (3-52).

Thus,

$$\mathbf{Q}_{i,i}(x_1,t) = u'^2[\mathbf{f}(x_1,t) + 2\mathbf{g}(x_1,t)]$$
$$= \frac{1}{2}\int_{-\infty}^{+\infty} dk_1 \exp(\imath k_1x_1)[\mathbf{E}_1(k_1,t) + 2\mathbf{E}_2(k_1,t)]$$

whence $\mathbf{E}_1(k_1,t) + 2\mathbf{E}_2(k_1,t) = \dfrac{1}{\pi}\displaystyle\int_{-\infty}^{+\infty} dx_1 \exp(-\imath k_1x_1)\mathbf{Q}_{i,i}(x_1,t)$

Differentiate with respect to k_1:

$$\frac{\partial\mathbf{E}_1(k_1,t)}{\partial k_1} + 2\frac{\partial\mathbf{E}_2(k_1,t)}{\partial k_1} = \frac{-\imath}{\pi}\int_{-\infty}^{+\infty} dx_1\, x_1 \exp(-\imath k_1x_1)\mathbf{Q}_{i,i}(x_1,t) = -\frac{2}{k_1}\mathbf{E}(k_1,t)$$

when we use the relation (3-59).

With Eq. (3-53),

$$\mathbf{E}(k_1,t) = \tfrac{1}{2}k_1{}^2\frac{\partial^2\mathbf{E}_1(k_1,t)}{\partial k_1{}^2} - \tfrac{1}{2}k_1\frac{\partial\mathbf{E}_1(k_1,t)}{\partial k_1} \tag{3-71}$$

which gives the relation between $\mathbf{E}(k_1,t)$ and $\mathbf{E}_1(k_1,t)$.

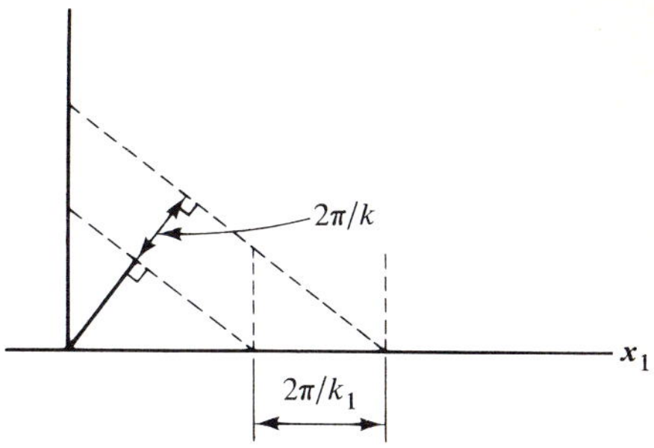

FIGURE 3-12
Contribution of wavelength $2\pi/k$ to the wavelength $2\pi/k_1$ of the one-dimensional spectrum.

The inverse relation can be obtained by solving the differential equations (3-71). Application of the method of variation of constants yields

$$\mathbf{E}_1(k_1,t) = k_1^{\,2}\left[\int_0^{k_1} dk \frac{\mathbf{E}(k,t)}{k^3} + C\right] - \int_0^{k_1} dk \frac{\mathbf{E}(k,t)}{k} + D$$

Since $\lim_{k_1 \to \infty} \mathbf{E}_1(k_1,t) = 0$, it follows that

$$\mathbf{E}_1(k_1,t) = \int_{k_1}^{\infty} dk \frac{\mathbf{E}(k,t)}{k}\left(1 - \frac{k_1^{\,2}}{k^2}\right) \tag{3-72}$$

And with the aid of Eq. (3-53)

$$\mathbf{E}_2(k_1,t) = \frac{1}{2}\int_{k_1}^{\infty} dk \frac{\mathbf{E}(k,t)}{k}\left(1 + \frac{k_1^{\,2}}{k^2}\right) \tag{3-73}$$

The physical meaning of the lower limit occurring in the integration range is that only those wavelengths $2\pi/k$ of the three-dimensional spectrum for which $2\pi/k \le 2\pi/k_1$, i.e., for which $k \ge k_1$, contribute to the value of the wavelength $2\pi/k_1$ of the one-dimensional spectrum[9] (see Fig. 3-12).

Remarks

a. From Eqs. (3-71), (3-72), and (3-73) it may be inferred that if, in the range $k_1 \le k < \infty$, $\mathbf{E}_1(k_1,t)$ can be represented by a power law, $\mathbf{E}(k_1,t)$ will then obey the same power law, and vice versa. This power law should have a negative exponent in order that the spectrum functions may become zero for $k \to \infty$.

b. From Eqs. (3-72) and (3-73) the following relations between the integral scales $\Lambda_f(t)$ and $\Lambda_g(t)$, and $\mathbf{E}(k,t)$ can readily be obtained:

$$\Lambda_f(t) = \frac{\pi}{2u'^2}\mathbf{E}_1(0,t) = \frac{\pi}{2u'^2}\int_0^{\infty} dk \frac{\mathbf{E}(k,t)}{k} \tag{3-74}$$

$$\Lambda_g(t) = \frac{\pi}{2u'^2}\mathbf{E}_2(0,t) = \frac{\pi}{4u'^2}\int_0^{\infty} dk \frac{\mathbf{E}(k,t)}{k} = \tfrac{1}{2}\Lambda_f(t) \tag{3-75}$$

c. Integration of Eq. (3-71), with $\lim_{k_1\to\infty} k_1\mathbf{E}(k_1,t) = 0$, yields

$$\int_0^\infty dk\,\mathbf{E}(k,t) = \frac{3}{2}\int_0^\infty dk\,\mathbf{E}_1(k,t)$$

which is in accord with the definitions (3-48) and (3-58).

It may be of interest to consider what form the spectrum functions take if for the correlation functions we apply expressions considered in previous sections. For example,

1 Assume for the coefficient $\mathbf{f}(x_1,t)$ of the longitudinal velocity correlation the Gaussian error curve given by Eq. (3-45),

$$\mathbf{f}(x_1,t) = \exp(-x_1^2/8\nu t)$$

With Eq. (3-9) we obtain for the coefficient of the transverse velocity correlation

$$\mathbf{g}(x_1,t) = \left(1 - \frac{x_1^2}{8\nu t}\right)\exp(-x_1^2/8\nu t) \tag{3-76}$$

This expression shows that $\mathbf{g}(x_1,t)$ becomes negative for large values of x_1, in contrast with $\mathbf{f}(x_1,t)$. From Eq. (3-50), $\mathbf{E}_1(k_1,t)$ can be calculated:

$$\mathbf{E}_1(k_1,t) = \frac{2}{\pi}u'^2\int_0^\infty dx_1\cos k_1x_1\exp(-x_1^2/8\nu t)$$

$$= u'^2\sqrt{\frac{8\nu t}{\pi}}\exp(-2k_1^2\nu t)$$

whence, $\Lambda_f(t) = (\pi/2u'^2)\mathbf{E}_1(0,t) = \sqrt{2\pi\nu t}$. Thus $\Lambda_f(t)$ increases proportionally with $\sqrt{t}$.

For the dissipation scale of turbulence $\lambda_f(t)$ we obtain

$$\lambda_f(t) = \sqrt{8\nu t}$$

Thus the ratio $\lambda_f(t)/\Lambda_f(t)$ remains constant.

For $\mathbf{E}_2(k_1,t)$ we obtain from Eq. (3-53)

$$\mathbf{E}_2(k_1,t) = \tfrac{1}{2}u'^2\sqrt{\frac{8\nu t}{\pi}}(1 + 4k_1^2\nu t)\exp(-2k_1^2\nu t)$$

With the aid of Eq. (3-71) we calculate the three-dimensional spectrum $\mathbf{E}(k,t)$

$$\mathbf{E}(k,t) = 8u'^2\nu^2t^2\sqrt{\frac{8\nu t}{\pi}}k^4\exp(-2k^2\nu t) = \text{const.}\times k^4\exp(-2k^2\nu t) \tag{3-77}$$

because $u'^2 = \text{const.}\times t^{-5/2}$.

This result shows, just like Eq. (3-63), that $\mathbf{E}(k,t)$ behaves like k^4 when $k\to 0$, but in addition it shows that $\mathbf{E}(k,t)$ becomes independent of t. The reader may easily verify that, in the series expansion (3-61), the term

$$\int_0^\infty dr\, r^2 \mathbf{Q}_{i,i}(r,t) = 0$$

but that the next term $\int_0^\infty dr\, r^4 \mathbf{Q}_{i,i}(r,t) \neq 0$.

What forms do the correlation tensor $\mathbf{Q}_{i,j}(x_1,x_2,x_3,t)$ and the corresponding spectrum tensor $\mathbf{E}_{i,j}(k_1,k_2,k_3,t)$ take? From Eq. (3-8) we obtain for $\mathbf{Q}_{i,j}(x_1,x_2,x_3,t)$

$$\mathbf{Q}_{i,j}(x_1,x_2,x_3,t) = u'^2\left[\frac{r^2}{8vt}\frac{x_i x_j}{r^2} + \left(1 - \frac{r^2}{8vt}\right)\delta_{ij}\right]\exp\left(-r^2/8vt\right)$$

and from Eq. (3-64) we calculate for $\mathbf{E}_{i,j}(k_1,k_2,k_3,t)$

$$\mathbf{E}_{i,j}(k_1,k_2,k_3,t) = -\text{const.}\frac{k^2}{4\pi}\left(\frac{k_i k_j}{k^2} - \delta_{ij}\right)\exp\left(-2k^2 vt\right)$$

2 In Sec. 1-11 we mentioned that for large Reynolds numbers the shape of a velocity-correlation curve can frequently be roughly approximated by an exponential function except close to its vertex. We shall assume here that the coefficient of the longitudinal correlation can be approximated by such a function.

$$\mathbf{f}(x_1,t) = \exp\left[-x_1/\Lambda_f(t)\right]$$

By applying the same method as in the previous example we obtain

$$\mathbf{g}(x_1,t) = \left(1 - \frac{x_1}{2\Lambda_f}\right)\exp\left(-x_1/\Lambda_f\right)$$

$$\mathbf{Q}_{i,i}(x_1,t) = u'^2\left(3 - \frac{x_1}{\Lambda_f}\right)\exp\left(-x_1/\Lambda_f\right)$$

$$\mathbf{E}_1(k_1,t) = \frac{2}{\pi}u'^2\int_0^\infty dx_1 \cos k_1 x_1 \exp\left(-x_1/\Lambda_f\right) = \frac{2}{\pi}u'^2\frac{\Lambda_f}{1 + k_1^2\Lambda_f^2}$$

[compare Eq. (1-103)].

$$\mathbf{E}_2(k_1,t) = \frac{1}{\pi}u'^2\Lambda_f\frac{1 + 3k_1^2\Lambda_f^2}{(1 + k_1^2\Lambda_f^2)^2}$$

$$\mathbf{E}(k,t) = \frac{8}{\pi}u'^2\Lambda_f\frac{k^4\Lambda_f^4}{(1 + k^2\Lambda_f^2)^3}$$

3-5 THE DYNAMIC EQUATION FOR THE ENERGY SPECTRUM

This equation, which describes the behavior of the energy spectrum as a function of time and space coordinates, can be derived by starting from the dynamic equation (3-32) *for an isotropic turbulence*

$$\frac{\partial}{\partial t}\mathbf{Q}_{i,j}(\xi_1,\xi_2,\xi_3,t) - \mathbf{S}_{i,j}(\xi_1,\xi_2,\xi_3,t) = 2\nu\frac{\partial^2}{\partial\xi_l\,\partial\xi_l}\mathbf{Q}_{i,j}(\xi_1,\xi_2,\xi_3,t) \tag{3-78}$$

In the previous section we introduced the Fourier transform $\mathbf{E}_{i,j}(k_1,k_2,k_3,t)$ of the correlation tensor $\mathbf{Q}_{i,j}(\xi_1,\xi_2,\xi_3,t)$:

$$\mathbf{Q}_{i,j}(\xi_1,\xi_2,\xi_3,t) = \iiint_{-\infty}^{+\infty} dk_1\,dk_2\,dk_3\,\mathbf{E}_{i,j}(k_1,k_2,k_3,t)\exp(\imath k_l\xi_l) \tag{3-79}$$

Similarly, we may also introduce the Fourier transform $\mathbf{F}_{i,j}(k_1,k_2,k_3,t)$ of $\mathbf{S}_{i,j}(\xi_1,\xi_2,\xi_3,t)$:

$$\mathbf{S}_{i,j}(\xi_1,\xi_2,\xi_3,t) = \iiint_{-\infty}^{+\infty} dk_1\,dk_2\,dk_3\,\mathbf{F}_{i,j}(k_1,k_2,k_3,t)\exp(\imath k_l\xi_l) \tag{3-80}$$

Because

$$\frac{\partial^2}{\partial\xi_l\,\partial\xi_l}\mathbf{Q}_{i,j}(\xi_1,\xi_2,\xi_3,t) = -\iiint_{-\infty}^{+\infty} dk_1\,dk_2\,dk_3\,k^2\mathbf{E}_{i,j}(k_1,k_2,k_3,t)\exp(\imath k_l\xi_l)$$

after substitution of Eqs. (3-79) and (3-80) in Eq. (3-78) we obtain

$$\iiint_{-\infty}^{+\infty} dk_1\,dk_2\,dk_3\exp(\imath k_l\xi_l)\left(\frac{\partial}{\partial t}\mathbf{E}_{i,j} - \mathbf{F}_{i,j} + 2\nu k^2\mathbf{E}_{i,j}\right) = 0$$

an equation that is satisfied if

$$\frac{\partial}{\partial t}\mathbf{E}_{i,j}(k_1,k_2,k_3,t) = \mathbf{F}_{i,j}(k_1,k_2,k_3,t) - 2\nu k^2\mathbf{E}_{i,j}(k_1,k_2,k_3,t) \tag{3-81}$$

The spectrum function $\mathbf{F}_{i,j}(k_1,k_2,k_3,t)$, just like $\mathbf{E}_{i,j}(k_1,k_2,k_3,t)$, is a symmetrical function of k_1, k_2, k_3. Since for an incompressible fluid

$$\frac{\partial}{\partial\xi_j}\mathbf{S}_{i,j}(\xi_1,\xi_2,\xi_3,t) = 0$$

which can easily be shown by differentiating the expression (3-33) for $\mathbf{S}_{i,j}(\xi_1,\xi_2,\xi_3,t)$ with respect to ξ_j, $\mathbf{F}_{i,j}(k_1,k_2,k_3,t)$ must also satisfy the condition

$$k_j\mathbf{F}_{i,j}(k_1,k_2,k_3,t) = 0$$

Then, just as we have shown for $\mathbf{E}_{i,j}(k_1,k_2,k_3,t)$, it follows that

$$\mathbf{F}_{i,j}(k_1,k_2,k_3,t) = D(k^2,t)(k_ik_j - k^2\delta_{ij}) \tag{3-82}$$

More information concerning $D(k^2,t)$ can be gained if we consider the behavior at small k of the tensor $\mathbf{F}_{ik,j}(k_1,k_2,k_3,t)$, the Fourier transform of $\mathbf{S}_{ik,j}(\xi_1,\xi_2,\xi_3,t)$,

$$\mathbf{S}_{ik,j}(\xi_1,\xi_2,\xi_3,t) = \iiint_{-\infty}^{+\infty} dk_1\, dk_2\, dk_3\, \mathbf{F}_{ik,j}(k_1,k_2,k_3,t) \exp(\imath k_l \xi_l)$$

For an incompressible fluid the equation of continuity yields

$$\frac{\partial}{\partial \xi_j} \mathbf{S}_{ik,j}(\xi_1,\xi_2,\xi_3,t) = 0$$

or

$$k_j \mathbf{F}_{ik,j}(k_1,k_2,k_3,t) = 0 \qquad (3\text{-}83)$$

The general expression for an isotropic tensor of the third order [see Eq. (3-24)] reads

$$\mathbf{F}_{ik,j}(k_1,k_2,k_3,t) = \varphi_1(k^2,t) k_i k_k k_j + \varphi_2(k^2,t) k^2 k_j \delta_{ik} + \varphi_3(k^2,t)(k_k \delta_{ij} + k_i \delta_{kj}) k^2$$

Substitution in Eq. (3-83) yields the relation

$$\varphi_2(k^2,t)\, \delta_{ik} = -\frac{k_i k_k}{k^2} [\varphi_1(k^2,t) + 2\varphi_3(k^2,t)]$$

so

$$\mathbf{F}_{ik,j}(k_1,k_2,k_3,t) = -\varphi_3(k^2,t)(2k_i k_k k_j - k^2 k_k \delta_{ij} - k^2 k_i \delta_{kj}) \qquad (3\text{-}84)$$

$\mathbf{F}_{i,j}(k_1,k_2,k_3,t)$ is the Fourier transform of $\mathbf{S}_{i,j}(\xi_1,\xi_2,\xi_3,t)$. Since

$$\mathbf{S}_{i,j}(\xi_1,\xi_2,\xi_3,t) = \frac{\partial}{\partial \xi_k} (\mathbf{S}_{ik,j} + \mathbf{S}_{kj,i})$$

it follows that

$$\mathbf{F}_{i,j} = \imath k_k (\mathbf{F}_{ik,j} + \mathbf{F}_{jk,i})$$

With the relation (3-84) for $\mathbf{F}_{ik,j}$ and $\mathbf{F}_{jk,i}$ we then obtain

$$\begin{aligned} \mathbf{F}_{i,j}(k_1,k_2,k_3,t) &= -2\imath\varphi_3(k^2,t)(k_i k_j - k^2 \delta_{ij}) k^2 \\ &= \varphi(k^2,t)(k_i k_j - k^2 \delta_{ij}) k^2 \end{aligned}$$

that is, the relation (3-82). Now the expression (3-84) for $\mathbf{F}_{ik,j}(k_1,k_2,k_3,t)$ can be written

$$\mathbf{F}_{ik,j}(k_1,k_2,k_3,t) = \varphi_3(k^2,t) k^2 \left(\delta_{jl} - \frac{k_j k_l}{k^2} \right) (k_i \delta_{lk} + k_k \delta_{li})$$

The term $\delta_{jl} - k_j k_l / k^2$ is nonanalytic at $k = 0$. Hence, if it is assumed that $\mathbf{F}_{ik,j}$ is analytic at $k = 0$, when $k \to 0$, $\varphi_3(k^2,t)$ and $\varphi(k^2,t)$ must be at least of the order k^0, and $D(k^2,t)$, consequently, of the order k^2.

In earlier theories it has been assumed that in homogeneous turbulence the tensors $\mathbf{Q}_{i,j}$ and $\mathbf{S}_{ik,j}$ are exponentially small at large spatial separation r, so that all integral moments converge. Also that $\mathbf{F}_{ik,j}$ is analytic at $k = 0$; accordingly $\mathbf{F}_{i,j}(k_1,k_2,k_3,t)$ should behave like $k^2(k_i k_j - k^2 \delta_{ij})$ at small k.

However, Proudman and Reid[29] showed for a *homogeneous turbulence,* that the assumption of an analytic behavior of $\mathbf{F}_{ik,j}(k_1,k_2,k_3,t)$ at $k = 0$ cannot be correct. The proof was rigorous on the assumption that the joint-probability distribution of velocity components taken simultaneously at any number of points in the turbulence field is normal. More specifically this was assumed to apply to the fourth-order velocity tensor at three different points. Some support to the assumption of a normal joint-probability distribution was given by results of fourth-order and second-order correlation measurements by Uberoi[39,40] in a grid-produced turbulence.

For a homogeneous turbulence a dynamic equation similar to Eq. (3-78) holds, provided the tensor $\mathbf{S}_{i,j}(\xi_1,\xi_2,\xi_3,t)$ contains the pressure-velocity correlations, namely:

$$(\mathbf{S}_{i,j})_{A,B} = \frac{\partial}{\partial \xi_k}\left[(\mathbf{S}_{ik,j})_{A,B} + (\mathbf{S}_{kj,i})_{A,B}\right] + \frac{1}{\rho}\left[\frac{\partial}{\partial \xi_i}\overline{p_A u_{j,B}} + \frac{\partial}{\partial \xi_j}\overline{p_A u_{i,B}}\right] \tag{3-85}$$

The corresponding dynamic equation for $\mathbf{E}_{i,j}$ is also similar to Eq. (3-81), where now $\mathbf{F}_{i,j}$ includes the pressure effects. Proudman and Reid showed that even if $\mathbf{E}_{i,j}$ at some initial instant possessed an analytic expansion, the long-range dynamical action of pressure forces would give rise to a non-analytic form at later instants. To this end they expanded $\overline{p_A u_{j,B}}$ in a series of inverse powers of r. Pressure falls off as an integral power of r^{-1} at large r. So the pressure gradients produce a distribution of acceleration that falls off in the same way. It then turns out that on the ground of a normal joint-probability distribution, the pressure-velocity correlation becomes $\mathcal{O}(r^{-4})$ at large r. Consequently $\mathbf{Q}_{i,j}$ will become $\mathcal{O}(r^{-5})$ and its integral over all space becomes convergent. Thus its Fourier transform $\mathbf{E}_{i,j}$ will be $\mathcal{O}(k^2)$ at small k, and becomes non-analytic at $k = 0$.

These results have been confirmed in a later paper by Batchelor and Proudman,[50] who redeveloped the whole problem. They made the following hypothesis. Homogeneous turbulence, as, e.g., generated approximately by a grid in a uniform flow, has a large-scale structure like that which would develop by dynamical action in a field of turbulence which at some initial instant is homogeneous and has convergent integral moments of cumulant velocity tensors. They showed that the effect of pressure forces, which depends on the velocity distribution in the entire space, is to cause the velocity correlation $\mathbf{Q}_{i,j}$ to decrease at large r not more rapidly as r^{-5}. Just like the pressure-velocity correlation the triple-velocity correlation varies as r^{-4} at large r. This means that $\lim_{r\to\infty} r^4 k$ is finite, so that Loitsianskii's integral (3-39) is not a dynamical invariant. $\mathbf{F}_{i,j}$ will be $\mathcal{O}(k^2)$ at small k and $\mathbf{F}_{ik,j}$ will be $\mathcal{O}(\mathrm{k})$.

When the turbulence is *isotropic,* the above long-range pressure effects vanish, since the pressure-velocity correlations are zero. The leading term in the series expansion of $\mathbf{Q}_{i,j}$ in inverse powers of r, which is $\mathcal{O}(r^{-5})$ then vanishes identically. Thus $\mathbf{Q}_{i,j}$ could not be larger than $\mathcal{O}(r^{-6})$ for large r, and Loitsianskii's integral becomes absolutely convergent. Also $\mathbf{E}_{i,j}$ is no longer singular, but analytic at $k = 0$

and behaves like k^2 at small k, whereas $\mathbf{F}_{ik,j}$ behaves like k and is nonanalytic at $k = 0$. It was impossible to find any term in the expressions for the time derivatives of $\mathbf{Q}_{i,j}$ at the initial instant which remains nonzero. On the other hand it was also not possible to prove that such a nonzero term cannot arise. So the order of magnitude of $\mathbf{f}(r)$ and $\mathbf{k}(r)$ at large r is still an open question and an exponential behavior of $\mathbf{f}(r)$ cannot be excluded. An inference made by Proudman and Reid, namely that $\lim_{r\to\infty} r^4\mathbf{k}(r) \neq 0$, so that Loitsianskii's integral is not a dynamical invariant, was confirmed.

When we apply a contraction, the dynamic equation (3-81) becomes

$$\frac{\partial}{\partial t}\mathbf{E}_{i,i}(k,t) = \mathbf{F}_{i,i}(k,t) - 2\nu k^2\mathbf{E}_{i,i}(k,t)$$

We now introduce the spectrum function $\mathbf{F}(k,t)$, which is defined like $\mathbf{E}(k,t)$ [Eq. (3-57)], that is,

$$\mathbf{F}(k,t) = 2\pi k^2\mathbf{F}_{i,i}(k,t) \tag{3-86}$$

so that, instead of the foregoing equation, we shall consider

$$\frac{\partial}{\partial t}\mathbf{E}(k,t) = \mathbf{F}(k,t) - 2\nu k^2\mathbf{E}(k,t) \tag{3-87}$$

Since $\mathbf{F}_{i,i}(k,t) = -2k^2D(k^2,t)$, the function $\mathbf{F}(k,t)$ should behave like k^4 when $k \to 0$.

$\mathbf{F}(k,t)/k$ is the Fourier transform of $r\mathbf{S}_{i,i}(r,t)$:

$$\mathbf{S}_{i,i}(r,t) = 2\int_0^\infty dk\,\frac{\sin kr}{kr}\,\mathbf{F}(k,t) \tag{3-88}$$

and

$$\mathbf{F}(k,t) = \frac{1}{\pi}\int_0^\infty dr\,kr\sin kr\,\mathbf{S}_{i,i}(r,t) \tag{3-89}$$

By substituting for $\mathbf{S}_{i,i}(r,t)$ the expression (3-34), the transfer function $\mathbf{F}(k,t)$ can be directly expressed in terms of the triple-velocity correlation $\mathbf{k}(r,t)$, viz.

$$\mathbf{F}(k,t) = \frac{u'^3}{\pi}\int_0^\infty dr\,rk^2\left[\frac{3\sin kr}{kr} - 3\cos kr - kr\sin kr\right]\mathbf{k}(r,t) \tag{3-90}$$

Series expansion of sin kr in Eq. (3-89) yields

$$\mathbf{F}(k,t) = \frac{k^2}{\pi}\int_0^\infty dr\,r^2\mathbf{S}_{i,i} - \frac{k^4}{3!\pi}\int_0^\infty dr\,r^4\mathbf{S}_{i,i} + \frac{k^6}{5!\pi}\int_0^\infty dr\,r^6\mathbf{S}_{i,i} - \cdots$$

But since, according to the foregoing, a series expansion of $\mathbf{F}(k,t)$ should start with k^4, the expansion reads

$$\mathbf{F}(k,t) = -\frac{k^4}{3!\pi}\int_0^\infty dr\,r^4\mathbf{S}_{i,i}(r,t) - \cdots \tag{3-91}$$

so

$$\int_0^\infty dr\, r^2 \mathbf{S}_{i,i}(r,t) = 0 \qquad (3\text{-}92)$$

but

$$\int_0^\infty dr\, r^4 \mathbf{S}_{i,i}(r,t) \neq 0 \qquad (3\text{-}93)$$

Now multiply Eq. (3-34) by r^m and integrate with respect to r from 0 to ∞. We then obtain

$$\frac{1}{u'^3}\int_0^\infty dr\, r^m \mathbf{S}_{i,i} = \left[r^{m+1}\frac{\partial \mathbf{k}}{\partial r} - (m-6)r^m \mathbf{k} \right]_0^\infty + (m-2)(m-4)\int_0^\infty dr\, r^{m-1}\mathbf{k}$$

From this equation and from Eqs. (3-92) and (3-93) it may be inferred that

$$\lim_{r\to\infty}\left(r^3 \frac{\partial \mathbf{k}}{\partial r} + 4r^2\mathbf{k} \right) = 0$$

but

$$\lim_{r\to\infty}\left(r^5 \frac{\partial \mathbf{k}}{\partial r} + 2r^4\mathbf{k} \right) \neq 0$$

As discussed earlier, Batchelor and Proudman showed that at large r, the triple correlation $\mathbf{k}(r)$ behaves like r^{-4}. So that

$$\lim_{r\to\infty}\left(r^3 \frac{\partial \mathbf{k}}{\partial r} \right) = \lim_{r\to\infty} (r^2\mathbf{k}) = 0$$

but

$$\lim_{r\to\infty} (r^4\mathbf{k}) \neq 0$$

The reader may remember that a necessary condition for the existence of a Loitsianskii's invariant (3-39) is

$$\lim_{r\to\infty} (r^4\mathbf{k}) = 0$$

According to the Batchelor and Proudman theory, the Loitsianskii's "invariant" (3-39) is not an invariant in the general case. However, in the final stage of a decaying turbulence where the viscous term in the dynamic equation predominates, the Loitsianskii's integral is invariant. This seems to contradict the above result. But one must keep in mind that the nonexistence of a Loitsianskii's "invariant" in the general case results from the fact that the second condition $\lim_{r\to\infty}(r^4\mathbf{k}) = 0$ is not fulfilled, and thus from the effect of the inertia term in the dynamic equation. Since this inertia effect is negligibly small in the final stage of decay, there is no second condition to be fulfilled for the existence of Loitsianskii's invariant. Indeed Proudman and Reid[29] have shown that

$$\frac{d}{dt}\left[u'^3 \lim_{r\to\infty} (r^4\mathbf{k})\right] > 0$$

and that

$$\frac{d^2}{dt^2}\left[u'^2\int_0^\infty dr\, r^4\mathbf{f}(r,t)\right] > 0$$

so that Loitsianskii's integral (3-39) decreases at a diminishing rate during the decay process and becomes constant at the stage where inertia effects are negligible and viscous effects predominate.

Though, as will be shown later, measurements in grid-produced turbulence seem to confirm the above results, the analytic behavior of $\mathbf{E}_{i,j}$ at $k = 0$ in a homogeneous turbulence has still been questioned, and hence the behavior of $\mathbf{E}(k^2)$ as k^4 at small k. Birkhoff,[68] for instance considered the assumption that $\mathbf{E}(k^2)$ behaves like k^2 at small k. He showed that this implies a divergence of Loitsianskii's integral. Saffman[69] reconsidered this problem by taking into account the way in which at some initial instant the turbulence is generated. Now, an initially irrotational flow can only become rotational either by viscous diffusion of vorticity, or by impulsive forces as a force field with rotation. In the case of turbulence it can be generated by a random, rotational external force field or by instability of an initial velocity distribution showing, for instance, at some initial instant concentrated shear layers or layers with velocity discontinuities in transverse direction.

Thus Saffman considered the turbulence vorticity field rather than the turbulence velocity field. Instead of the hypothesis made by Batchelor and Proudman, he examined the consequence of the following hypothesis. Homogeneous turbulence has a large-scale structure like that which would develop by dynamical action in a field of turbulence which at some instant is homogeneous and has convergent integral moments of cumulants of the vorticity distribution, or is generated by random impulsive forces with convergent integral moments of cumulants. Now, in order that the spectral tensor $\mathbf{E}_{i,j}$ of the velocity field be analytic at $k = 0$, the impulsive force field has to satisfy certain conditions. First, it must be solenoidal, so that the divergence of the impulsive forces is zero. Second, the mechanism that generates the turbulence can produce only a finite total linear momentum even if the volume of the flow field increases indefinitely. This might be the case when the turbulence is generated out of a dynamical instability of the fluid motion. Saffman now shows that, if on the other hand the momentum applied to a volume V increases like $V^{1/2}$ when V increases, so that it does not remain bounded, $\mathbf{E}(k^2)$ varies as k^2 at small k. A further consequence is that $\mathbf{Q}_{i,j}$ is $\mathcal{O}(r^{-3})$ at large r. For an isotropic turbulence this implies that Loitsianskii's integral becomes divergent, but that instead

$$\int_0^\infty dr\, r^2\mathbf{Q}_{i,i}(r,t) = \overline{u^2}(t)\lim_{r\to\infty} r^3\mathbf{f}(r,t) \qquad (3\text{-}94)$$

is a dynamic invariant, and not zero. Since, as already mentioned, there are strong experimental indications that grid-produced turbulence shows features in agreement with the Batchelor-Proudman theory, apparently for this turbulence the total linear momentum is bounded and varies to a smaller extent than $V^{1/2}$. This is not unreasonable considering the periodic structure of the individual wakes of the grid rods even at large Reynolds number.

On the other hand Lumley[136] showed that for the flow of an incompressible fluid in an unbounded region, which has a zero net linear momentum and which is globally isotropic in the sense that the total volume integral of $u_i u_j$ is zero when $i \neq j$, the Loitsianskii's integral (3-39) holds true if the average value of $\mathbf{f}(r)$ over the whole space is taken. It also applies to a bounded region when at the boundaries no torque is acting on the fluid.

Let us now return to Eq. (3-87) and consider its integrated form.

$$\frac{\partial}{\partial t}\int_0^k dk\,\mathbf{E}(\mathbf{k},t) = \int_0^k dk\,\mathbf{F}(k,t) - 2\nu\int_0^k dk\,k^2\mathbf{E}(k,t) \tag{3-95}$$

We shall now show that, if the upper limit of the integral is increased to $k = \infty$, the first term on the right-hand side becomes zero.

$$\int_0^\infty dk\,\mathbf{F}(k,t) = 0 \tag{3-96}$$

It follows from Eq. (3-88) that

$$\int_0^\infty dk\,\mathbf{F}(k,t) = \tfrac{1}{2}\mathbf{S}_{i,i}(0,t)$$

We have shown in Sec. 3-2 that, for small values of r, $\mathbf{k}(r,t)$ behaves like r^3 [see Eq. (3-22)], so that, according to Eq. (3-34), $\mathbf{S}_{i,i}(r,t)$ behaves like r^2 for small values of r. Consequently $\mathbf{S}_{i,i}(0,t) = 0$ and Eq. (3-96) must hold.

Hence for $k = \infty$ Eq. (3-95) reads

$$\frac{d}{dt}\int_0^\infty dk\,\mathbf{E}(k,t) = -2\nu\int_0^\infty dk\,k^2\mathbf{E}(k,t) \tag{3-97}$$

According to Eq. (3-58) for $\mathbf{E}(k,t)$, the left-hand side of this equation represents the change of the total kinetic energy of turbulence. Since there are no external energy sources, this change in kinetic energy must equal the dissipation caused by viscous effects. We shall show that the right-hand side of Eq. (3-97) does indeed represent the dissipation of mechanical energy into heat.

To this end we shall consider the expression for the dissipation (term V) given in the energy equation (1-110) of Chap. 1.

$$\text{Dissipation per unit of mass } \varepsilon = \nu\overline{\left(\frac{\partial u_i}{\partial x_j} + \frac{\partial u_j}{\partial x_i}\right)\frac{\partial u_j}{\partial x_i}}$$

Since in isotropic turbulence

$$\overline{\left(\frac{\partial u_1}{\partial x_1}\right)^2} = \overline{\left(\frac{\partial u_2}{\partial x_2}\right)^2} = \overline{\left(\frac{\partial u_3}{\partial x_3}\right)^2}$$

$$\overline{\left(\frac{\partial u_1}{\partial x_2}\right)^2} = \overline{\left(\frac{\partial u_1}{\partial x_3}\right)^2} = \overline{\left(\frac{\partial u_2}{\partial x_1}\right)^2} = \cdots$$

$$\overline{\frac{\partial u_1}{\partial x_2}\frac{\partial u_2}{\partial x_1}} = \overline{\frac{\partial u_1}{\partial x_3}\frac{\partial u_3}{\partial x_1}} = \overline{\frac{\partial u_2}{\partial x_3}\frac{\partial u_3}{\partial x_2}} = \cdots$$

we obtain

$$\varepsilon = 6\nu\left[\overline{\left(\frac{\partial u_1}{\partial x_1}\right)^2} + \overline{\left(\frac{\partial u_1}{\partial x_2}\right)^2} + \overline{\frac{\partial u_1}{\partial x_2}\frac{\partial u_2}{\partial x_1}}\right] \tag{3-98}$$

With the aid of Eqs. (3-20) and (3-21) it is possible to express the dissipation in terms of $[\partial^2\mathbf{f}/\partial r^2]_{r=0}$, viz.,

$$\varepsilon = -15\nu u'^2\left[\frac{\partial^2\mathbf{f}}{\partial r^2}\right]_{r=0} = 30\nu\frac{u'^2}{\lambda_f{}^2} = 15\nu\frac{u'^2}{\lambda_g{}^2} \tag{3-99}$$

From the expression (3-68) for $\mathbf{f}(r,t)$ we obtain

$$u'^2\left[\frac{\partial^2\mathbf{f}}{\partial r^2}\right]_{r=0} = -\frac{2}{15}\int_0^\infty dk\, k^2\mathbf{E}(k,t)$$

so

$$\varepsilon = -15\nu u'^2\left[\frac{\partial^2\mathbf{f}}{\partial r^2}\right]_{r=0} = 2\nu\int_0^\infty dk\, k^2\mathbf{E}(k,t) \tag{3-100}$$

Equation (3-95) may be extended to the case in which energy is supplied continuously to the flow. Let $H_k(k,t)$ be the energy supplied to turbulence in the wavenumber region from 0 to k of the spectrum; then the extension of Eq. (3-95) reads

$$\frac{\partial}{\partial t}\int_0^k dk\,\mathbf{E}(k,t) = \int_0^k dk\,\mathbf{F}(k,t) - 2\nu\int_0^k dk\, k^2\mathbf{E}(k,t) + H_k(k,t) \tag{3-101}$$

The left-hand side of this equation denotes the change of the kinetic energy of and the dissipation in the eddies with wavenumbers between 0 and k. The term $\int_0^k dk\,\mathbf{F}(k,t)$ may then be interpreted as transferring energy to or from the turbulence in this wavenumber region. Now $\mathbf{F}(k,t)$ is the spectrum function of $\mathbf{S}_{i,i}(r,t)$, which is related to the triple correlation $\mathbf{k}(r,t)$ [see Eq. (3-34)]. In considering the dynamic Kármán-Howarth equation (3-36), we interpreted the term with the triple correlation as representing a "convective" action in the spread of the double correlation $\mathbf{f}(r,t)$. This convective action may be caused by the interaction of eddies of different sizes. Similarly, we may interpret the function $\int_0^k dk\,\mathbf{F}(k,t)$ occurring in Eq. (3-101) as representing the interaction of eddies of different wavenumbers, thereby transferring energy by inertial effects to or from the eddies in the wavenumber region 0 to k.

Therefore, $\mathbf{F}(k,t)$ is frequently referred to as the *energy-transfer-spectrum function*, giving the contribution from inertial transfer of energy in the dynamic equation (3-87). [Compare the name energy-spectrum function for $\mathbf{E}(k,t)$].

For obtaining solutions of $\mathbf{E}(k,t)$ we may start either from Eq. (3-87) or from its integral form (3-95). Of course, we are here faced with the same fundamental difficulty we faced in solving the dynamic Kármán-Howarth equation. Here it is the transfer-spectrum function $\mathbf{F}(k,t)$ that is also unknown in Eqs. (3-87) and (3-95). However, in contrast with that case, it appears to be less difficult here to postulate or assume some appropriate form for the function $\mathbf{F}(k,t)$. Indeed, various forms have been suggested to be used in solving the dynamic equation for the spectrum function $\mathbf{E}(k,t)$.

Before considering such solutions, we will consider the simplest case, in which it is assumed that the interaction among eddies of various wavenumbers is negligibly small, so that no transfer of energy between eddies is caused by this interaction. In other words, it is assumed that the term with the transfer-spectrum function $\mathbf{F}(k,t)$ is negligible with respect to the other terms. Equation (3-87) then reduces to

$$\frac{\partial}{\partial t}\mathbf{E}(k,t) = -2\nu k^2 \mathbf{E}(k,t)$$

The solution is readily obtained by integration, and reads

$$\mathbf{E}(k,t) = \mathbf{E}(k,t_0)\exp\left[-2\nu k^2(t - t_0)\right] \qquad (3\text{-}102)$$

Compare the solution (3-77) obtained by means of the solution for the correlation $\mathbf{f}(r,t)$ with reference to the same case of negligible interaction between eddies.

From the solution (3-102) the following interesting result is obtained. The decrease of kinetic energy with time occurs at a higher rate for the eddies with the larger wavenumbers; i.e., the smaller eddies decay at a higher rate than the bigger ones. In other words, the effect of time increases as the wavenumber increases. Conversely, for very small wavenumbers the energy-spectrum function becomes less and less dependent on time. The dependence becomes zero when $k = 0$, i.e.

$$\lim_{k \to 0} \frac{\partial}{\partial t}\mathbf{E}(k,t) = 0$$

Thus, according to Eq. (3-77), the energy-spectrum function increases very rapidly, initially according to k^4, reaches a maximum value, and decreases monotonously to zero as k increases. At the same time, the effect of time on the rate of decay increases, too.

For the more general case in which the viscosity effects are no longer predominant, that is, where the inertial term in the dynamic equation for $\mathbf{E}(k,t)$ has also to be considered, a roughly similar shape is obtained for the energy-spectrum function.

As in all fluid-dynamic processes, the character of the flow is determined by the ratio between the inertial and viscous forces, thus by the Reynolds number.

Depending on the kind of turbulence and upon what region of the turbulence spectrum is being considered, various characteristic lengths and velocities may be considered for defining the Reynolds number of turbulence. We shall return to this point presently. Since the solution (3-102) applies to the case in which the viscous effects are predominant, it obviously refers to low values of the Reynolds number.

The energy spectrum for the more general case of higher Reynolds numbers has been the subject of many theoretical and experimental studies. Stimulated by Kolmogoroff's work,[10] Von Kármán, Lin, Batchelor, Townsend, and others have made important contributions to a better understanding of the mechanism of turbulence.†

Let us consider, then, the energy spectrum of a grid-produced turbulence when the Reynolds number is fairly high, so that the inertial term cannot be neglected in the dynamic equation.

Since turbulence can be generated only by velocity gradients in the flow and since in the domain of this generation the flow is anisotropic, the turbulence at short distances downstream of the grid is neither homogeneous, nor isotropic. It requires some time, that is, some distance from the grid, to become homogeneous in the lateral direction and almost isotropic.

As the turbulence grows to its full development, the larger eddies will produce smaller and smaller eddies through inertial interaction, thereby transferring energy to the smaller eddies. At the same time viscosity effects and, with them, dissipation become more and more important for the smaller eddies.

In the fully developed state it is not the largest eddies that will have the maximum kinetic energy but the eddies in a higher-wavenumber range. The range of the energy spectrum where the eddies make the main contribution to the total kinetic energy of turbulence will be called the range of the *energy-containing eddies*. In this range the energy spectrum shows its maximum. We may associate a wavenumber k_e with this maximum to indicate the range of the energy-containing eddies.

Though the more permanent largest eddies contain much less energy than the energy-containing eddies, their energy is by no means negligibly small and may still amount to as much as 20 per cent of the total kinetic energy.

As mentioned, dissipation by viscous effects increases as the size of the eddy decreases, up to a maximum for a certain size of the smallest eddies. Here too we may associate a wavenumber k_d with the size of the eddies that provide the main contribution to the total dissipation. This value k_d will correspond roughly to the maximum in the $[k^2\mathbf{E}(k,t),k]$ curve. The dissipation will result in a continuous decrease in the total kinetic energy of the turbulence, that is, a decaying turbulence, if no sources of energy are present. This decay takes place at a certain rate. Let us compare this rate of the decay process with the turbulence velocities corresponding to the wavenumbers of the energy spectrum.

† These theories are dealt with by Batchelor in his book on homogeneous turbulence.[11] Much use has been made of this book and of the publications of the aforementioned investigators in the preparation of the following sections.

At the lower extreme of the wavenumber range, corresponding to the largest eddies, the "frequency" $u'_k k$ of these eddies is very small compared with the rate of decay of the total kinetic energy of turbulence expressed by its relative rate of change $(du'^2/dt)/u'^2$. Here u'_k is the root-mean-square of the velocity of eddies with wavenumber equal to k. For a continuous spectral distribution this is an ill-defined quantity since u'_k depends on the bandwidth around k. Therefore we will define it as follows

$$u'_k = \sqrt{k\mathbf{E}(k)} \qquad (3\text{-}103)$$

At the other extreme of the wavenumber range the smallest eddies will have a very high "frequency" $u'_k k_d$ compared with the relative rate of change of the total turbulence.

In between there will be a range of wavenumbers where the "frequencies" $u'_k k$ of the eddies are of the same order as the relative rate of change of the total turbulence. It is logical to expect that this will occur in the range of the energy-containing eddies, since these eddies make the main contribution to the total energy of turbulence and since the relative rate of change of this total energy is then closely related to the "frequency" of these eddies:

$$\frac{1}{u'^2}\frac{du'^2}{dt} \propto -u'_k k_e \propto -u'k_e \qquad (3\text{-}104)$$

Since u'_k for $k = k_e$ is roughly equal to $\frac{1}{2}u'$.

Thus, beyond this range and toward the higher wavenumbers, the decay process may be considered a relatively slow process compared with the motions of the eddies. In other words, in the higher-wavenumber range the turbulence may be considered statistically nearly steady and the rates of change of mean values may be regarded as negligible.

The eddies corresponding to these higher wavenumbers are excited by the transfer of energy by inertia forces from the larger eddies. It may be assumed that, in contrast with the largest eddies, these much smaller eddies are independent of the external conditions producing the forces that generate the initial largest eddies. This assumption is supported by experimental evidence that, even in anisotropic turbulent flow, the high-wavenumber range of the turbulence is close to isotropy.

Thus the turbulence in the higher-wavenumber range may not only be considered statistically steady in its mean values but also independent of external conditions. The character of the turbulence in this range can, therefore, be determined only by parameters resulting from internal conditions. It is found that there are just two parameters possible.

Let us consider so high a value of the Reynolds number that the range of the energy-containing eddies and the range of maximum dissipation are sufficiently wide apart, and hence

$$k_e <<< k_d$$

In the range $k \gg k_e$, the eddies obtain their energy by inertial transfer from the larger eddies, and the larger eddies for their part transfer energy to the smaller eddies. At the same time a part of the energy is dissipated in the eddies. Thus there is a continuous flux of energy through the wavenumber range and a continuous dissipation, the latter increasing with the higher wavenumbers.

In this range the amount of energy transferred through the eddies is large compared with the rate of change of their energy; so these eddies may be considered to be in statistical equilibrium with one another.

Obviously, the character of the turbulence in this range is wholly determined by the energy flux through this range and the rate of dissipation. The energy flux plus the dissipation must be equal to the total energy supply to this range. Neglecting the energy dissipation in the lower-wavenumber range, the total energy supply in the *equilibrium range* is practically equal to the total dissipation

$$\varepsilon(t) = -\frac{3}{2}\frac{du'^2}{dt} = 2\nu \int_0^\infty dk\, k^2 \mathbf{E}(k,t) \qquad (3\text{-}105)$$

Hence the character of the turbulence in this range is determined by $\varepsilon(t)$, which is one of the two parameters, and by the viscosity ν, the other parameter, since this, too, determines the rate of dissipation.

These considerations have led Kolmogoroff to make the following hypothesis:

"At sufficiently high Reynolds numbers there is a range of high wavenumbers where the turbulence is statistically in equilibrium and uniquely determined by the parameters ε and ν. This state of equilibrium is *universal*."

This equilibrium range is termed "universal" because the turbulence in this range is independent of external conditions and any change in the effective length scale and time scale of this turbulence can only be a result of the effect of the parameters ε and ν. Following the example of Kolmogoroff, we shall take a velocity scale instead of the time scale, since this appears to be more convenient. From dimensional reasoning we obtain

For the length scale: $$\eta = \left(\frac{\nu^3}{\varepsilon}\right)^{1/4} \qquad (3\text{-}106a)$$

For the velocity scale: $$v = (\nu\varepsilon)^{1/4} \qquad (3\text{-}106b)$$

Thus the dimensions of the *Kolmogoroff length scale* η and *velocity scale* v are such that the Reynolds number with reference to this length and velocity is

$$\frac{v\eta}{\nu} = 1$$

The wavenumber k_d where the viscous effects become very strong will be of the same order as $1/\eta$. It is usual to define k_d in such a way as to have exactly

$$k_d = \frac{1}{\eta} \qquad (3\text{-}107)$$

However, this then implies that k_d does no longer correspond to the maximum in the dissipation curve. Townsend[52] has shown that most of the viscous dissipation occurs in the range $k\eta < 0.5$.

Measurements by Stewart and Townsend[17] in grid-produced turbulence showed a maximum in the dissipation spectrum at $k\eta \simeq 0.15$. But measurements at a much higher Reynolds number of turbulence by Grant, Stewart, and Moilliet[70] in a tidal channel showed this maximum to be at $k\eta \simeq 0.1$. This latter figure gives a strong support to Kraichnan's Lagrangian-History-Direct-Interaction theory, to be considered later. Kraichnan obtained theoretically a value of $k\eta = 0.09$ in the limiting case of very large Reynolds numbers.

On the other side of the equilibrium range we have the wavenumber k_e marking the range of the energy-containing eddies. We may define a length l_e by

$$k_e = \frac{1}{l_e}$$

so that l_e may be interpreted as the average size of the energy-containing eddies.

We have already introduced as important turbulence-characterizing parameters: the intensity u', the dissipation scale λ_f, and the integral scale Λ_f.

Since it is common practice to follow G. I. Taylor in taking λ_g instead of λ_f, we shall do so in what follows.

The question arises: how are these parameters related to the parameters v, η, and l_e, just introduced?

It is reasonable to assume that Λ_f, which is determined mainly by the size of the larger energy-containing eddies, is of the same order as the average size l_e of these eddies. Furthermore, we may expect that the relations between u' and v and between l_e and η will depend upon the Reynolds number of turbulence.

We have already mentioned the Reynolds number $v\eta/\nu$, which, being exactly equal to unity, marks only the region of strong viscous effects, where the inertial forces are of the same magnitude as the viscous shear forces. This Reynolds number alone is, therefore, not sufficient to characterize the turbulence in the other parts of the wavenumber range. Two other Reynolds numbers that can be obtained from the above parameters present themselves as being suitable, namely,

$$\mathbf{Re}_\lambda = \frac{u'\lambda_g}{\nu} \qquad \mathbf{Re}_l = \frac{u'l_e}{\nu} \tag{3-108}$$

In order to obtain the relations among $\mathbf{Re}_\lambda$, $\mathbf{Re}_l$, u', v, l_e, λ_g, and η, we must keep in mind that these parameters refer partly to the wavenumber range of maximum dissipation, partly to the wavenumber range of the energy-containing eddies. For these two regions the parameter ε is a determining one, since ε is on the one hand practically equal to the dissipation in the highest-wavenumber range and on the other hand practically equal to the work done by the energy-containing eddies, which is the energy supplied to the smaller eddies.

We have already shown that the total dissipation can be expressed in terms of u' and λ_g:

$$\varepsilon = 15\nu \frac{u'^2}{\lambda_g^2} \tag{3-99}$$

and, on the other hand, the work per unit time and mass may be expressed as follows:

$$\varepsilon = A \frac{u'^3}{l_e} \tag{3-109}$$

where A is a numerical constant of the order of unity [compare Eqs. (3-104) and (3-105) from which Eq. (3-109) follows].

Now the reader can easily verify that, according to Eqs. (3-99) and (3-109), the relations among $\mathbf{Re}_\lambda$, $\mathbf{Re}_l$, u', v, l_e, λ_g, and η will read:

$$\mathbf{Re}_l = \frac{A}{15} \mathbf{Re}_\lambda^{\,2} \tag{3-110}$$

$$\frac{u'}{v} = 15^{-1/4}\, \mathbf{Re}_\lambda^{\,1/2} \tag{3-111}$$

$$\frac{\lambda_g}{\eta} = 15^{1/4}\, \mathbf{Re}_\lambda^{\,1/2} \tag{3-112}$$

$$\frac{l_e}{\lambda_g} = \frac{A}{15} \mathbf{Re}_\lambda \tag{3-113}$$

$$\frac{l_e}{\eta} = 15^{-3/4} A\, \mathbf{Re}_\lambda^{\,3/2} \tag{3-114}$$

With the relation (3-114) we are now able to specify more precisely the condition for the existence of the equilibrium range. In this range we have the wavenumbers for which $k \gg k_e$, and, moreover,

$$k_e <<< k_d$$

In terms of l_e and η this condition reads

$$l_e >>> \eta$$

and, by Eq. (3-114),

$$\mathbf{Re}_\lambda^{\,3/2} >>> 1 \tag{3-115}$$

Expressed in terms of the Reynolds number $\mathbf{Re}_l$, the condition reads [see Eq. (3-110)]

$$\mathbf{Re}_l^{\,3/4} >>> 1 \tag{3-116}$$

Later on we shall discuss to what extent this necessary condition is satisfied in the experiments on turbulence.

In the description of the equilibrium range we have said that in this range dissipation not only takes place at all wavenumbers but increases strongly as the wavenumber increases. It may now be expected that, if the Reynolds number of turbulence is very large, the dissipation in the region of wavenumbers very far below the region of maximum dissipation will be negligibly small compared with the flux of energy transferred by inertial effects. In such a subrange the effect of the parameter ν would then vanish, and the turbulence could be determined by the other parameter ε alone. However, this can only be done if there is no direct effect of the energy-containing eddies, that is, of any parameter other than ε that might affect the turbulence in the region of the energy-containing eddies. Hence the condition that the wavenumber k should fall in such a subrange is

$$k_e \ll k \ll k_d \qquad (3\text{-}117)$$

and the turbulence in this range is statistically independent of the range of energy-containing eddies and of the range of strong dissipation.

Kolmogoroff has considered this subrange in his second hypothesis:

"If the Reynolds number is infinitely large, the energy spectrum in the subrange satisfying the condition (3-117) is independent of ν, and is solely determined by one parameter ε."

Since in this subrange the inertial transfer of energy is the dominating factor, this will be called the *inertial subrange.*

The condition (3-117) for the existence of such an inertial subrange shows, in comparison with the condition $k_e <<< k_d$ for the existence of an equilibrium range, that the Reynolds number must be at least one order of magnitude larger. The conditions (3-115) and (3-116), respectively, are just insufficient and should be replaced by

$$\mathbf{Re}_\lambda{}^{3/4} >>> 1 \qquad (3\text{-}118)$$

and

$$\mathbf{Re}_l{}^{3/8} >>> 1 \qquad (3\text{-}119)$$

Let us now consider the range outside the equilibrium range, that is, the range of lower wavenumbers. Adjacent to the equilibrium range, but sufficiently far from it so that there is statistical independence, we have the range of the energy-containing eddies. In this range the "frequency" $u'k_e$ is of the same order as the relative rate of change of the kinetic energy of the total turbulence. Hence it is reasonable to expect that in this range the time t will have to be taken into account as another parameter, in addition to the parameters ε and ν governing the equilibrium range.

We will not yet assume such extremely high values of the Reynolds number that the dissipation and with it the effect of viscosity may be neglected in this range. However, since on dimensional grounds any two of the three parameters ε, ν, and t are already sufficient to render the quantities E and k dimensionless, these three parameters cannot be independent of one another. They form a dimensionless group that must be a constant:

$$\frac{\varepsilon t^2}{\nu} = \text{const.} \qquad (3\text{-}120)$$

As we shall show in the next section, this dimensionless group may be interpreted as a Reynolds number of turbulence. Thus, according to Eq. (3-120), the Reynolds number would remain constant during the decay of turbulence. We shall consider this range in greater detail in the next section on the decay of turbulence.

If, on the other hand, we assume that the Reynolds number is infinitely large, so that in the equilibrium range there is an inertial subrange where the energy spectrum is determined only by one parameter ε and is independent of ν, it is logical to expect that, in the lower-wavenumber range of the energy-containing eddies as well, the effect of viscosity will be negligibly small; so in this lower-wavenumber range the energy spectrum will depend only on the two parameters ε and t.

Finally, in the still lower wavenumber range, we have the largest eddies, which show a strong degree of permanence, and the dependence on time decreases as the wavenumber decreases. This may be inferred from Eq. (3-87) if we assume for $\mathbf{E}(k,t)$ and $\mathbf{F}(k,t)$ their asymptotic forms for $k \to 0$. We already know that in the lowest-wavenumber range the energy spectrum according to the Batchelor-Proudman theory, and probably applicable to grid-produced turbulence, reads

$$\lim_{k \to 0} \mathbf{E}(k,t) = I(t)k^4 \qquad (3\text{-}121)$$

From Eqs. (3-67) and (3-12) we find that the quantity $I(t)$ is equal to

$$I = -\frac{1}{6\pi}\int_0^\infty dr\, r^4 \mathbf{Q}_{i,i}(r,t) = \frac{u'^2}{3\pi}\int_0^\infty dr\, r^4 \mathbf{f}(r,t) \qquad (3\text{-}122)$$

The integral on the right-hand side is Loitsianskii's integral (3-39). [The same result can readily be obtained also by a series expansion of $\sin kr$ and $\cos kr$ in Eq. (3-70).]

Hence we may consider Loitsianskii's integral or I as the parameter determing the lowest part of the wavenumber range.

On the other hand if we follow Saffman[69] and consider as a possibility that

$$\lim_{k \to 0} \mathbf{E}(k,t) = I_1 k^2$$

where $I_1 = \frac{1}{\pi}\int_0^\infty dr\, r^2 \mathbf{Q}_{i,i}(r,t)$ is a dynamical invariant, the quantity I_1 can be considered as the parameter determining the lowest part of the wavenumber range. In what follows, however, we will only consider the result of the Batchelor-Proudman theory.

Now there is another feature that is characteristic of the large eddies in the lower-wavenumber range, namely, the diffusive action of turbulence, which is, as we have indicated in Chap. 1, determined mainly by the larger eddies. Von Kármán and Lin,[12] therefore, assumed that the eddy diffusivity ϵ might be regarded as a parameter determining the character of the turbulence in the lower-wavenumber range.

We have now covered, we believe, the entire wavenumber range. Recapitulating, we have, starting from the lower range:[12]

1 Low range, determined by I and ϵ, where turbulence is strongly permanent
   a. Lowest subrange, determined solely by I
2 Medium range, determined by ϵ and ε
   a. Lower subrange, determined solely by ϵ, where the turbulence may still show some permanence
3 Higher range, determined by ε, ν, and t
   a. This comprises mainly the range of energy-containing eddies
   b. Determined by ε and t only, if $\mathbf{Re}_\lambda^{3/4} >>> 1$
4 Highest range, determined by ε and ν. This is the universal equilibrium range
   a. Lower inertial subrange, determined solely by ε if $\mathbf{Re}_\lambda^{3/4} >>> 1$

It is possible merely on dimensional grounds to indicate the form of the energy spectrum in the various ranges, even quite precisely in the subranges where the turbulence is determined by one parameter only.

$\mathbf{E}(k,t)$ and the parameters considered above have the following dimensions

$$[\mathbf{E}] = [L^3T^{-2}]$$
$$[k] = [L^{-1}]$$
$$[\nu] = [L^2T^{-1}]$$
$$[\epsilon] = [L^2T^{-1}]$$
$$[\varepsilon] = [L^2T^{-3}]$$
$$[I] = [L^7T^{-2}]$$

The results from the form of $\mathbf{E}(k,t)$ in the various wavenumber ranges are given in Table 3-1 and illustrated in Fig. 3-13.

Special attention may be drawn to the form of $\mathbf{E}(k,t)$ in the inertial subrange:

$$\mathbf{E}(k,t) = A\varepsilon^{2/3}k^{-5/3} \qquad (3\text{-}123)$$

This form of $\mathbf{E}(k,t)$, which applies at large Reynolds numbers, is frequently called the *Kolmogoroff spectrum law,* because Kolmogoroff was the first to arrive at this result (see his second hypothesis). Yet both Onsager[47] and Von Weizsäcker[48] have obtained the same result independently of Kolmogoroff and of each other.

It is possible in the inertial subrange to determine the distribution of the intensity of turbulence over the range of wavenumbers or over the range of eddy sizes. Let $u'_{k-\infty}$ be the root mean square of the velocity of all eddies with wavenumbers greater than k; then

$$\tfrac{3}{2}u'^2_{k-\infty} = \int_k^\infty dk\, \mathbf{E}(k,t) \qquad (3\text{-}124)$$

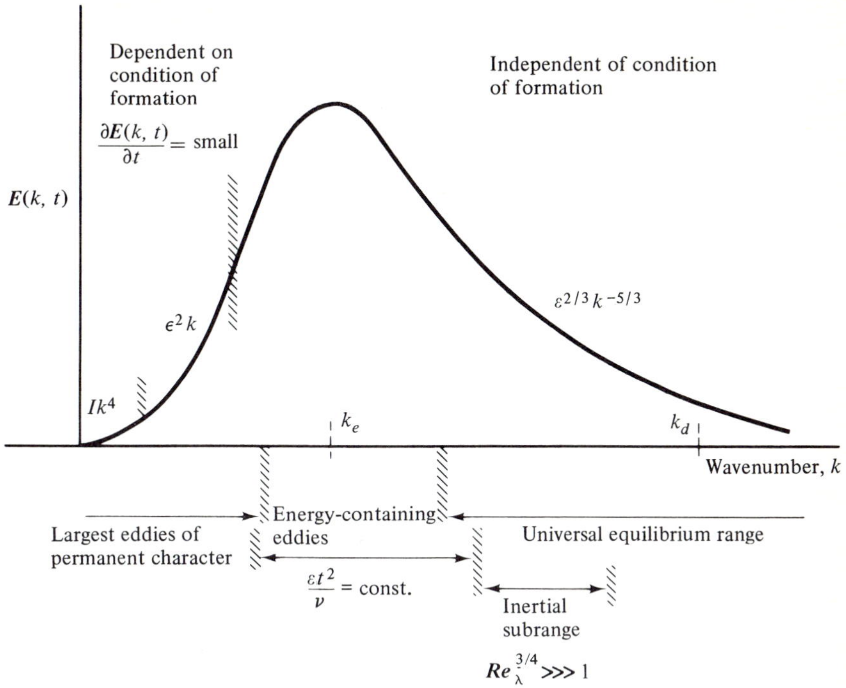

FIGURE 3-13
Form of the three-dimensional spectrum $\mathbf{E}(k,t)$ in the various wavenumber ranges.

If we disregard the contribution to the kinetic energy of the eddies in the wavenumber range beyond the inertial subrange and if we apply for $\mathbf{E}(k,t)$ the above Kolmogoroff law, we obtain

$$u'_{k-\infty} \propto k^{-1/3} \propto l_k^{1/3}$$

A similar result is obtained when u'_k according to the definition (3-103) is used instead of $u'_{k-\infty}$. Hence, the intensity of the turbulent motion is proportional to the third root of the dimension of the region considered (in this case the size of the eddy).

By straightforward reasoning Von Weizsäcker[48] arrived at the same result and then, taking Eq. (3-124) into account, concluded that at large Reynolds numbers the above law (3-123) must follow.

Except for the three subranges shown in Table 3-1, the dimensional analysis does not result in a precise description of the energy spectrum $\mathbf{E}(k,t)$. For that we must return to the dynamic equation of the spectrum function, either in its differential form (3-87) or in its integral form (3-95), and try to solve these equations. At present the only way that has been found to solve these equations is to assume some relation among the transfer function $\mathbf{F}(k,t)$, the energy spectrum $\mathbf{E}(k,t)$, and the wavenumber k.

Various proposals have been made. They are all based more or less on some intuitive physical picture of the mechanism of energy transfer across the wavenumbers. The first one to be considered was made by Obukhoff,[13] who assumed that the energy transfer across the wavenumber k is analogous to the energy transfer from the main motion to the turbulent motion through the turbulence shear stresses, or to the production of turbulence energy [see Chap. 1, Eq. (1-110)]. Here we have to consider the part of the Reynolds stress associated with the wavenumbers larger than k and the part of the mean rate of shear associated with the wavenumbers less than k. Obukhoff put the Reynolds stress proportional to the kinetic energy $\int_k^\infty dk\,\mathbf{E}(k,t)$ and the mean rate of strain for wavenumbers less than k. The mean-square rate of strain of the turbulence [see Eq. (3-20)] is

$$\overline{\frac{\partial u_i}{\partial x_j}\frac{\partial u_i}{\partial x_j}} = -\left[\frac{\partial^2 \mathbf{Q}_{i,i}}{\partial \xi_j \partial \xi_j}\right]_{r=0}$$

By Eq. (3-54) we obtain

$$-\frac{\partial^2 \mathbf{Q}_{i,i}}{\partial \xi_j \partial \xi_j} = \iiint_{-\infty}^{+\infty} dk_1\,dk_2\,dk_3\,k^2\mathbf{E}_{i,i}\exp(\imath k_l \xi_l) = 4\pi\int_0^\infty dk\,k^4\,\frac{\sin kr}{kr}\,\mathbf{E}_{i,i}$$

Hence

$$\overline{\frac{\partial u_i}{\partial x_j}\frac{\partial u_i}{\partial x_j}} = 4\pi\int_0^\infty dk\,k^4\mathbf{E}_{i,i} = 2\int_0^\infty dk\,k^2\mathbf{E}(k,t)$$

and the root-mean-square rate of strain of the turbulence with wavenumbers less than k is

$$\left[2\int_0^k dk\,k^2\mathbf{E}(k,t)\right]^{1/2}$$

Obukhoff's assumption thus reads

$$\int_0^k dk\,\mathbf{F}(k,t) = -\alpha\left[2\int_0^k dk\,k^2\mathbf{E}(k,t)\right]^{1/2}\int_k^\infty dk\,\mathbf{E}(k,t) \qquad (3\text{-}125)$$

where α is an absolute constant.

Kovasznay[14] assumed that $\int_0^k dk\,\mathbf{F}(k,t)$ is a function only of $\mathbf{E}(k,t)$ and k:

$$\int_0^k dk\,\mathbf{F}(k,t) = -\alpha[\mathbf{E}(k)]^{3/2}k^{5/2} \qquad (3\text{-}126)$$

The most useful assumption, perhaps, was made by Heisenberg.[15] This assumption is that the effect of the eddies with wavenumbers larger than k, by which energy is withdrawn from the eddies with wavenumbers less than k, is as if a certain turbulence viscosity were present. It is then possible to write formally

$$\int_0^k dk\,\mathbf{F}(k,t) = -2\epsilon(k,t)\int_0^k dk\,k^2\mathbf{E}(k,t)$$

Table 3-1 FORMS OF E(k,t) IN THE VARIOUS WAVENUMBER RANGES

	Ranges of increasing $k \rightarrow$						
	(1)		(2)		(3)	(4)	
Quantities	$\mathbf{E}, k, I, \epsilon$		$\mathbf{E}, k, \epsilon, \varepsilon$		$\mathbf{E}, k, \varepsilon, \nu, t$	$\mathbf{E}, k, \varepsilon, \nu$	
Basic parameters	I, ϵ		ϵ, ε		ε, ν or ε, t or ν, t	ε, ν	
Charact. length	$(I/\epsilon^2)^{1/3}$		$(\epsilon^3/\varepsilon)^{1/4}$		$(\nu^3/\varepsilon)^{1/4}$	$(\nu^3/\varepsilon)^{1/4}$	
Charact. time	$(I^2/\epsilon^7)^{1/3}$		$(\epsilon/\varepsilon)^{1/2}$		$(\nu/\varepsilon)^{1/2}$	$(\nu/\varepsilon)^{1/2}$	
Charact. velocity	$(\epsilon^5/I)^{1/3}$		$(\epsilon\varepsilon)^{1/4}$		$(\nu\varepsilon)^{1/4}$	$(\nu\varepsilon)^{1/4}$	
$\mathbf{E}(k,t)$	$\epsilon^{8/3}I^{-1/3}\mathbf{E}^*[k(I/\epsilon^2)^{1/3}]$		$\epsilon^{5/4}\varepsilon^{1/4}\mathbf{E}^*[k(\epsilon^3/\varepsilon)^{1/4}]$		$\nu^{5/4}\varepsilon^{1/4}\mathbf{E}^*[k(\nu^3/\varepsilon)^{1/4}, \varepsilon t^2/\nu]$ $\varepsilon^{3/2}t^{5/2}\mathbf{E}^*(k\varepsilon^{1/2}t^{3/2})$ if $\mathbf{Re}_\lambda^{3/4} >>> 1$	$\nu^{5/4}\varepsilon^{1/4}\mathbf{E}^*[k(\nu^3/\varepsilon)^{1/4}]$	
	Subranges						
						$\mathbf{Re}_\lambda^{3/4} >>> 1$	
Basic parameter	I		ϵ			ε	
$\mathbf{E}(k,t)$	Ik^4		const. $\times\ \epsilon^2 k$			const. $\times\ \varepsilon^{2/3}k^{-5/3}$	

where $\epsilon(k,t)$ denotes the kinematic turbulence viscosity caused by the eddies with wavenumbers ranging from k to infinity.

So we may put

$$\epsilon(k,t) = \int_k^\infty dk\, f[k,\mathbf{E}(k,t)]$$

Or

$$\frac{d\epsilon}{dk} = -f[k,\mathbf{E}(k,t)]$$

Dimensional analysis then yields

$$f[k,\mathbf{E}(k,t)] = \text{const.}\sqrt{\frac{\mathbf{E}(k,t)}{k^3}}$$

And hence

$$\epsilon(k,t) = \text{const.}\int_k^\infty dk \sqrt{\frac{\mathbf{E}(k,t)}{k^3}}$$

which is the form given by Heisenberg. Consequently,

$$\int_0^k dk\,\mathbf{F}(k,t) = -2\alpha\int_k^\infty dk\sqrt{\frac{\mathbf{E}(k,t)}{k^3}}\int_0^k dk\,k^2\mathbf{E}(k,t) \qquad (3\text{-}127)$$

The physical picture underlying Heisenberg's assumption seems quite acceptable. Yet, as Batchelor[11] has pointed out, a fundamental objection may be made. From a physical point of view the introduction of a turbulence viscosity to account for the transfer of energy from the larger to the smaller eddies can be correct only if the small eddies responsible for the existence of such a turbulence viscosity are statistically independent of the large eddies. But the form of the Heisenberg expression (3-127) is such that the main contribution to the two integrals on the right-hand side are from the eddies with wavenumber k close to the lower limit of the first integral and close to the upper limit of the second integral.

Intuitively it has been assumed that eddies of widely differing size are statistically independent. Results of measurements have lent support to this assumption because they showed that the small eddies corresponding with the most fine structure of turbulence are not only mutually statistically independent but also independent of the biggest eddies. The idea behind this assumption is that the effect of a big eddy on a small eddy consists of two parts. One is a deformation of the small eddy by the shear flow of the bigger eddy. The other is the entrainment of the small eddy by the flow velocity at its center caused by the big eddy. The assumption then is that it is mainly the second part of the effect that counts when the two eddies differ widely in size. We may recall that the equilibrium theory discussed earlier is based on the statistical independence of the small and the biggest eddies. It has led to the hypothesis[4] that the coupling through nonlinear inertial actions between Fourier components is stronger the closer their frequencies are, and that a Fourier component is statistically independent of Fourier components of much lower and of much higher frequencies. Thus the transfer of energy by inertia forces as represented by the transfer function $\mathbf{F}(k,t)$ is mainly caused by the interaction of eddies of little differing wavenumber. It would be interesting to establish experimentally whether in an actual turbulence the interaction between eddies is a local process in wavenumber space. We shall come back to this point later. On the other hand mathematical theories, as, e.g., Burgers's[71] mathematical model, do not show this local character. A Fourier component interacts through nonlinear actions with other Fourier components in the whole wavenumber space. Also Kraichnan in his direct-interaction theory abandoned the above hypothesis.

The Obukhoff and Heisenberg theories give the integral transfer-function

$$\mathbf{G}(k,t) = \int_0^k dk\,\mathbf{F}(k,t) \qquad (3\text{-}128)$$

as a product of two terms. One includes an integration between 0 and k, the other an integration between k and ∞. The second term then represents the effect of the smaller eddies with wavenumbers $k'' > k$ on the eddy with wavenumber k, the first term the transfer of energy from the bigger eddies with wavenumber $k' < k$ to the eddy with wavenumber k. This suggests an interaction between two eddies with $k'' > k$, and $k' < k$. From a theoretical point of view this is not satisfactory. For, the Fourier cosine transform of the triple correlation between velocities, which describes the inertial transfer of energy, shows that interaction takes place between three Fourier components forming a triangle in wavenumber space, namely the components k, k' and $k - k'$, so that their vectorial sum is zero. This triangle process is, for instance, included in Kraichnan's direct-interaction theory, which will be considered, briefly, later.

We will first discuss a more general expression for $\mathbf{G}(k,t)$ suggested by Von Kármán.[9] The transfer function $\mathbf{F}(k,t)$ describes the interaction of wavenumber k with all other wavenumbers, so that it receives in general energy through dynamical nonlinear interaction from wavenumbers $k' < k$, and loses in general energy to wavenumbers $k'' > k$. We may introduce an *interaction function* $\mathscr{I}(k',k,t)$, describing the positive interaction between k' and k. The function $\mathscr{I}(k,k',t)$ then describes the positive interaction between k and k'. Consequently we have $\mathscr{I}(k',k,t) = -\mathscr{I}(k,k',t)$.

We may then define the interaction function $\mathscr{I}(k',k,t)$ in such a way that

$$\mathbf{F}(k,t) = \int_0^\infty dk\,\mathscr{I}(k',k,t) \qquad (3\text{-}129)$$

We may split the integration in two parts at $k' = k$, so that the part with $k' < k$ describes the contribution from all wavenumbers $k' < k$ to $\mathbf{F}(k,t)$.

$$\mathbf{F}(k,t) = \int_0^k dk'\,\mathscr{I}(k',k,t) + \int_k^\infty dk''\,\mathscr{I}(k'',k,t)$$

$$= \int_0^k dk' \mathscr{I}(k',k,t) - \int_k^\infty dk'' \mathscr{I}(k,k'',t)$$

$$= -\frac{d}{dk}\int_0^k dk' \int_k^\infty dk'' \mathscr{I}(k',k'',t) \qquad (3\text{-}130)$$

Hence

$$\mathbf{G}(k,t) = \int_0^k dk\,\mathbf{F}(k,t) = -\int_0^k dk' \int_k^\infty dk'' \mathscr{I}(k',k'',t) \qquad (3\text{-}131)$$

Now $\mathscr{I}(k',k,t)$ depends on both k' and k, and on the corresponding energy densities $\mathbf{E}(k',t)$ and $\mathbf{E}(k,t)$.

Von Kármán now assumes†

$$\mathscr{I}(k',k,t) = 2\alpha \mathbf{E}^{\varphi'}(k',t)k'^{\psi'}\mathbf{E}^{\varphi}(k,t)k^{\psi} \qquad (3\text{-}132)$$

Whence follows, since $\mathscr{I}$ has the dimension $[L^4T^{-3}]$, from dimensional reasoning

$$\mathscr{I}(k',k,t) = 2\alpha \mathbf{E}^{3/2-\varphi}(k',t)k'^{1/2-\psi}\mathbf{E}^{\varphi}(k,t)k^{\psi} \quad \text{for} \quad k' < k \qquad (3\text{-}133a)$$

And similarly

$$\mathscr{I}(k,k'',t) = 2\alpha \mathbf{E}^{3/2-\varphi}(k,t)k^{1/2-\psi}\mathbf{E}^{\varphi}(k'',t)k''^{\psi} \quad \text{for} \quad k'' > k \qquad (3\text{-}133b)$$

Substitution in the expression $\mathbf{G}(k,t)$ yields

$$\mathbf{G}(k,t) = \int_0^k dk\,\mathbf{F}(k,t) = -2\alpha\left[\int_0^k dk'\,\mathbf{E}^{3/2-\varphi}(k',t)\cdot k'^{1/2-\psi}\right]\cdot\left[\int_k^\infty dk''\,\mathbf{E}^{\varphi}(k'',t)k''^{\psi}\right]$$

$$(3\text{-}134)$$

Since, when passing from $k' < k$ to $k'' > k$, we have $\mathscr{I}(k',k,t) = -\mathscr{I}(k,k'',t)$, there is a change in sign at k, describing a positive contribution by interaction with eddies $k' < k$, and a negative contribution for the eddies $k'' > k$. The gradient $\partial\mathscr{I}/\partial k$ shows a discontinuity at k except when $\varphi = \frac{3}{4}$ and $\psi = \frac{1}{4}$. Furthermore, as has already been pointed out by Batchelor,[11] the expression for $\mathbf{F}(k,t)$ does not incorporate the "triangle" interaction. For in that case, besides $\mathbf{E}(k',t)$ and $\mathbf{E}(k,t)$, it should also include a contribution given by $\mathbf{E}(k' - k,t)$. Depending on the values of φ and ψ, and the energy density $\mathbf{E}(k,t)$, the dynamic interaction between wavenumbers may have a more or less local character. Let us consider the wavenumber range $k > k_e$, where $\mathbf{E}(k,t)$ is a decreasing function of k. When in a restricted wavenumber range $\mathbf{E}(k,t)$ is put proportional with k^{-n}, where $n > 0$, it can easily be concluded that the contribution given by $\mathscr{I}(k',k,t)$ increases with increasing k' when

$$n(\varphi - \tfrac{3}{2}) + \tfrac{1}{2} - \psi > 0$$

The contribution by $\mathscr{I}(k,k'',t)$ increases with decreasing k'' when $\psi - n\varphi < 0$. The

† The same symbol α has been used for the constant of proportionality in all the theories considered. Numerically the value may differ depending on the theory.

interaction becomes more localized around k, when $|n(\varphi - \frac{3}{2}) + \frac{1}{2} - \psi|$ and $|\psi - n\varphi|$ increase.

When we assume $\varphi = 1$, $\psi = 0$, a modified Obukhoff expression is obtained

$$\mathbf{G}(k,t) = -2\alpha\left\{\int_0^k dk' \, [\mathbf{E}(k',t)k']^{1/2}\right\} \cdot \left\{\int_k^\infty dk'' \, E(k'',t)\right\} \tag{3-125a}$$

For $\varphi = \frac{3}{2}$, $\psi = 0$ a modified Kovasznay expression is obtained.

$$\mathbf{G}(k,t) = -2\alpha[\tfrac{2}{3}k^{3/2}] \cdot \left\{\int_k^\infty dk'' \, [\mathbf{E}(k'',t)]^{3/2}\right\} \tag{3-126a}$$

The Heisenberg expression (3-127) is obtained when we put $\varphi = \frac{1}{2}$ and $\psi = -\frac{3}{2}$.

The condition for a localized interaction with the wavenumber k is satisfied for the modified Obukhoff expression only when $n < 1$; so it is not satisfied for the inertial subrange, where according to Eq. (3-123) $n = \frac{5}{3}$. For the modified Kovasznay expression it is satisfied for any value of n, whereas in the case of the Heisenberg expression n must be smaller than 2, which means that for the inertial subrange the value n is already close to this limit.

With the various assumptions made for $\mathbf{F}(k,t)$, solutions of the dynamic equation (3-95) have been obtained for the equilibrium range. In this range where $k \gg k_e$, the term $\partial\mathbf{E}(k,t)/\partial t$ is very small and, consequently, $\dfrac{\partial}{\partial t}\displaystyle\int_k^\infty dk \, \mathbf{E}(k,t)$ is negligible with respect to $\dfrac{\partial}{\partial t}\displaystyle\int_0^\infty dk \, \mathbf{E}(k,t)$. Hence we may put in Eq. (3-95)

$$\frac{\partial}{\partial t}\int_0^k dk \, \mathbf{E}(k,t) \simeq \frac{\partial}{\partial t}\int_0^\infty dk \, \mathbf{E}(k,t) = -\varepsilon = -2\nu\int_0^\infty dk \, k^2\mathbf{E}(k,t)$$

In this case we may write Eq. (3-95) either in the following form:

$$-\varepsilon = \int_0^k dk \, \mathbf{F}(k,t) - 2\nu\int_0^k dk \, k^2\mathbf{E}(k,t) \tag{3-135}$$

or as follows:

$$\int_0^k dk \, \mathbf{F}(k,t) + 2\nu\int_k^\infty dk \, k^2\mathbf{E}(k,t) = 0 \tag{3-136}$$

With Eq. (3-125) Obukhoff obtained a solution of (3-136), not, however, in closed form. For the inertial subrange the solution for $\mathbf{E}(k,t)$ reduced to const. $\times \, k^{-5/3}$, as it should. But for larger values of k the solution decreases at a much slower rate than the continuation of the $k^{-5/3}$-curve until, at a certain cutoff value of k, where $\mathbf{E}(k,t)$ is still finite, the total dissipation is already equal to ε, although this should occur at $\mathbf{E}(k,t) = 0$. Hence, as Batchelor[11] has pointed out, there must be something wrong with Obukhoff's assumption (3-125).

It is remarkable that a slight change in Obukhoff's assumption, as suggested by Ellison[72] removes this objection. Instead of Eq. (3-125a), Ellison assumes

$$\mathbf{G}(k,t) = -\alpha E(k,t)\cdot k\left[2\int_0^k dk'\,k'^2\mathbf{E}(k',t)\right]^{1/2} \tag{3-137}$$

Following Obukhoff's reasoning this would mean that the Reynolds stress interacting with the mean rate of shear or vorticity of the wavenumbers smaller than k, is entirely restricted to the wavenumber k.

For not-small wavenumbers the equation for $\mathbf{E}(k,t)$ is obtained by substituting Eq. (3-137) in Eq. (3-136).

$$k\mathbf{E}(k,t)\left[2\int_0^k dk'\,k'^2\mathbf{E}(k',t)\right]^{1/2} = \frac{2\nu}{\alpha}\int_k^\infty dk''\,k''^2\mathbf{E}(k'',t) \tag{3-138}$$

It can easily be shown that the Kolmogoroff spectrum for the inertial subrange is obtained by considering such relatively small values of k, that the right-hand side of the equation includes the total dissipation, and becomes equal to ε/α. Writing

$$\Theta = \left[2\int_0^k dk'\,k'^2\mathbf{E}(k',t)\right]^{1/2}$$

a simple differential equation for Θ is obtained, that can readily be integrated. The final solution for $\mathbf{E}(k,t)$ reads

$$\mathbf{E}(k,t) = \left(\frac{2}{3\alpha^2}\right)^{1/3}\varepsilon^{2/3}k^{-5/3}$$

Panchev and Kesich[73] have considered Eq. (3-138) without the restriction to relatively small wavenumbers, so including the large wavenumber, dissipation, range. We shall come back to their solution later.

Another way to avoid the above difficulty concerning Obukoff's assumption has been obtained by Eschenroeder,[74] who excluded the high wavenumber, dissipation, range. The dissipation term in the right-hand side of Eq. (3-135) then can be neglected. He used the modified Obukhoff expression (3-125a). The equation for $\mathbf{E}(k,t)$ reads

$$2\alpha\left\{\int_0^k dk'\,[\mathbf{E}(k',t)k']^{1/2}\right\}\cdot\left\{\int_k^\infty dk''\,\mathbf{E}(k'',t)\right\} = \varepsilon$$

Writing for $\int_k^\infty dk''\,\mathbf{E}(k'',t) = \Theta(k,t)$, again a simple first-order differential equation is obtained which can be solved readily. With $\Theta(0,t) = \frac{3}{2}u'^2$, the solution for $\mathbf{E}(k,t)$ is

$$\mathbf{E}(k,t) = \frac{Cu'^2k}{[1+Ck^2]^{4/3}} \tag{3-139}$$

where

$$C = \left(\frac{9\alpha u'^3}{2\varepsilon}\right)^2$$

The solution reduces to the Kolmogoroff spectrum for large k, viz.,

$$\mathbf{E}(k,t) = \left(\frac{2\varepsilon}{9\alpha}\right)^{2/3}k^{-5/3}$$

However, it yields an incorrect, linear behavior when $k \to 0$, which is not surprising since the approximate equation (3-135) no longer holds for small k.

With this assumption for $\int_0^k dk\, F(k,t)$ Kovasznay obtained the solution

$$\mathbf{E}(k,t) = \left(\frac{\varepsilon}{\alpha}\right)^{2/3} k^{-5/3}\left[1 - \frac{\alpha^{-2/3}}{2}\left(\frac{k}{k_d}\right)^{4/3}\right]^2$$

where $k_d(t) = [\varepsilon(t)/v^3]^{1/4}$. This solution is also correct when $k \ll k_d$; like Obukhoff's solution it too has a cutoff value of k, but at this value $\mathbf{E}(k,t) = 0$.

However, the solution no longer applies when k/k_d is not small, for as pointed out by Lumley[137] this solution appears to be the first term in a series expansion in powers of $(k/k_d)^{4/3}$.

Panchev[75] has studied the following generalization of Kovasznay's assumption for the transfer function.

$$\mathbf{G}(k,t) = -\varphi(z) \cdot \alpha[\mathbf{E}(k,t)]^{3/2} \cdot k^{5/2} \tag{3-140}$$

where

$$z = \left(\frac{\alpha}{\varepsilon}\right)^{2/3} \cdot \mathbf{E}(k,t) \cdot k^{5/3}$$

So we have $z = 1$ for the Kolmogoroff spectrum, while Kovasznay's assumption is obtained when $\varphi(z) = 1$. Panchev worked out solutions when $\varphi(z) = z^{-a}$. Only values of $a < \frac{3}{2}$ yield a realistic form of the energy spectrum, whereas for $a < \frac{1}{2}$, the energy spectrum has a cut-off value of k. Kovasznay's solution corresponds with $a = 0$.

Let us now see to what result Heisenberg's assumption (3-127) leads. Substitution of Eq. (3-127) in Eq. (3-135) gives

$$\varepsilon = 2\alpha \int_k^\infty dk \sqrt{\frac{\mathbf{E}(k,t)}{k^3}} \int_0^k dk\, k^2\mathbf{E}(k,t) + 2v \int_0^k dk\, k^2\mathbf{E}(k,t) \tag{3-141}$$

Bass[6] and, independently of him, Chandrasekhar[16] have obtained the solution of Eq. (3-141) as follows. Assume

$$\theta(k) = \int_0^k dk\, k^2\mathbf{E}(k)$$

so

$$\mathbf{E} = \frac{1}{k^2}\frac{d\theta}{dk}$$

Equation (3-141) then reads

$$v + \alpha \int_k^\infty dk \sqrt{\frac{1}{k^5}\frac{d\theta}{dk}} = \frac{\varepsilon}{2\theta}$$

Differentiation with respect to k yields

$$-\alpha \sqrt{\frac{1}{k^5}\frac{d\theta}{dk}} = -\frac{\varepsilon}{2\theta^2}\frac{d\theta}{dk}$$

or

$$\frac{1}{k^5}\frac{d\theta}{dk} = \left(\frac{\varepsilon}{2\alpha}\right)^2 \frac{1}{\theta^4}\left(\frac{d\theta}{dk}\right)^2$$

the solution of which is

$$\theta^3 = \frac{1}{\text{const.} + \frac{3}{4}(2\alpha/\varepsilon)^2 k^{-4}}$$

Since $\theta = \varepsilon/2\nu$ when $k \to \infty$, the constant must be equal to $8\nu^3/\varepsilon^3$. If we further introduce $k_d = (\varepsilon/\nu^3)^{1/4}$ we obtain

$$\theta^3 = \frac{\varepsilon^3}{3\alpha^2\nu^3}\frac{k^4/k_d{}^4}{1 + (8/3\alpha^2)(k/k_d)^4}$$

Hence

$$\mathbf{E}(k,t) = \left(\frac{8}{9\alpha}\right)^{2/3}(\varepsilon\nu^5)^{1/4}\frac{(k/k_d)^{-5/3}}{[1 + (8/3\alpha^2)(k/k_d)^4]^{4/3}} \tag{3-142}$$

When $k/k_d \ll 1$, the expression (3-142) reduces to Eq. (3-123) with

$$A = \left(\frac{8}{9\alpha}\right)^{2/3} \tag{3-143}$$

But when $k/k_d \gg 1$, that is, in the viscous dissipation range, we would obtain

$$\mathbf{E}(k,t) = \left(\frac{\alpha}{2}\right)^2 \cdot (\varepsilon\nu^5)^{1/4} \cdot (k/k_d)^{-7} \tag{3-144}$$

The transition to the region of strong viscous effects is apparently located at $k \simeq k_d$, in accord with the interpretation of $k_d = 1/\eta$ given earlier.

If the various assumptions made concerning the transfer function $\mathbf{G}(k,t)$, written in the general form (3-134), are considered, it may be noticed that they are written as the product of two terms, one involving an integral over the wavenumber range 0 to k, the other involving an integral over the wavenumber range k to ∞. Stewart and Townsend[17] have pointed out that such an assumption must necessarily lead to a power-law spectrum at very large wavenumbers, irrespective of the exact form of the factors.

Earlier in this section we made the objection against Heisenberg's hypothesis of a turbulence viscosity due to the smaller eddies, that is, that it is not consistent with the requirement of statistical independence between these eddies and the large eddies. This objection becomes very serious if we consider the viscous range $k > 0.5k_d$, the range below which most of the viscous dissipation occurs. In this range Heisenberg's hypothesis fails and the energy spectrum $\mathbf{E}(k) \propto k^{-7}$ is not applicable to this range. This has been clearly demonstrated by Townsend.[52] The smallest eddies are produced by a distortion and stretching process caused by the larger eddies, probably at wavenumbers of the order k_d. Now in turbulent flow each fluid element is deformed into a long thin ribbon, a process that will be discussed in more detail in Sec. 5-4. So it is likely that the smallest eddies will form concentrated vortex sheets. Some support to this idea is also given by Burgers.[57] In that case it is certain that the smallest eddies are not statistically independent of the eddies with wavenumber of the order k_d. But there is still another point. Experimental evidence has shown that the vorticity of these smallest eddies is almost negligible with respect to that of the eddies from which they receive their energy, whereas Heisenberg's hypothesis implies the contrary. As a matter of fact, in the main range $k < 0.5k_d$ the energy-supplying eddies in general have a greater energy but a smaller vorticity than the energy-receiving eddies.

We will mention yet another suggestion concerning $\mathbf{G}(k,t)$ made by Dugstad,[76] where an attempt has been made to localize the transfer near the wavenumber k. He assumed in Eq. (3-131), for $k' \leq k''$

$$\mathscr{I}(k',k'',t) = 2\alpha[\mathbf{E}(k',t)]^{3/2}k'^{5/2}\frac{\exp[(k'-k'')/k_d^*]}{k_d^{*2}}$$

whence

$$\mathbf{G}(k,t) = -2\alpha\int_0^k dk'\,[\mathbf{E}(k',t)]^{3/2}\frac{k'^{5/2}}{k_d^*}\exp[(k'-k)/k_d^*]$$

Here k_d^* is a wavenumber characteristic of the dissipation range; and $k_d^* \simeq (0.15 - 0.20)k_d$. Note that the expression for $\mathbf{G}(k,t)$ is not written as a product of two integral terms, like the expression (3-134). A nonlinear first-order differential equation is obtained for $\mathbf{E}(k,t)$, for which no complete analytical solution has been found. However, for relatively small wavenumbers the Kolmogoroff spectrum for the inertial subrange was reproduced, as required dimensionally. For very large wavenumbers an exponential drop-off of $\mathbf{E}(k,t)$ was obtained, viz.,

$$\mathbf{E}(k,t) = A(\varepsilon\nu^5)^{1/4}\left(\frac{k}{k_d^*}\right)^{-2}\exp[-(k/k_d^*)] \qquad (3\text{-}145)$$

The calculations of the average transfer through the spectrum showed that this transfer was more local at low wavenumbers than at large wavenumbers.

An exponential decrease with increasing wavenumber in the dissipation and higher wavenumber range has also been suggested by Corrsin[130] and by Pao.[77] The energy-transfer-spectrum $\mathbf{F}(k,t)$ describes the difference of energy flux by inertial interaction from the eddies in the wavenumber range 0–k into the wavenumber k, and the flux to the smaller eddies in the wavenumber range k–∞. So $\mathbf{G}(k,t)$ is the total flux from the wavenumber range 0–k into the wavenumber range k–∞, that is the energy flux across the wavenumber k.

This energy flux is put proportional to the energy density $\mathbf{E}(k,t)$ at wavenumber k:

$$\mathbf{G} = f \cdot \mathbf{E}(k,t)$$

Here f is the rate at which the energy is transported through a cascade process across the wavenumber range. Pao now postulates that f is determined only by ε and k. So

$$\mathbf{G}(k,t) = f[\varepsilon(t),k] \cdot \mathbf{E}(k,t)$$

Putting $f[\varepsilon,k] = -B\varepsilon^a k^b$, dimensional analysis yields

$$\mathbf{G}(k,t) = -B\varepsilon^{1/3}k^{5/3}E(k,t)$$

whence follows

$$\mathbf{F}(k,t) = -B\frac{\partial}{\partial k}[\varepsilon^{1/3}k^{5/3}\mathbf{E}] = 2\nu k^2\mathbf{E}$$

The solution is

$$\mathbf{E}(k,t) = A \cdot \varepsilon^{2/3}k^{-5/3} \exp\left[-\frac{3}{2B} \cdot \frac{\nu}{\varepsilon^{1/3}} k^{4/3}\right]$$

It too reduces, as it must, to the Kolmogoroff spectrum at smaller k values. The constant B is obtained from the condition that $2\nu \int_0^\infty dk\, k^2\mathbf{E} = \varepsilon$ with the above relation for $\mathbf{E}$, since the dissipation in the range 0–k is negligible. This condition yields $A = B^{-1}$.

Consequently

$$\mathbf{E}(k,t) = A\varepsilon^{2/3}k^{-5/3} \exp\left[-\tfrac{3}{2}A \frac{\nu}{\varepsilon^{1/3}} k^{4/3}\right] \qquad (3\text{-}146a)$$

Or in non-dimensional form

$$\mathbf{E}(k,t) \cdot (\varepsilon\nu^5)^{-1/4} = A\left(\frac{k}{k_d}\right)^{-5/3} \exp\left[-\tfrac{3}{2}A\left(\frac{k}{k_d}\right)^{4/3}\right] \qquad (3\text{-}146b)$$

It may be surmised from Pao's postulate concerning the energy flux described by $\mathbf{G}(k,t)$ and from the evaluation of the constant A, that the relation for $\mathbf{E}(k,t)$ may no longer apply to the extreme high-wavenumber range. In a study of the fine-scale structure of homogeneous, dynamically passive vector fields convected by a turbulent fluid, Saffman[78] considered the transport of a dynamically passive property due to the combined effect of turbulent deformations of fluid particles and of molecular diffusion. The regions considered had dimensions smaller than the Kolmogoroff length scale η. In these small regions the deformation rates may be assumed to be linearly dependent on the rates of strain. Though vorticity is not a dynamically passive property, Saffman assumed the theory still applicable. For the viscous wavenumber range $k > k_d$, he obtained the following expression for $\mathbf{E}(k,t)$:

$$\mathbf{E}(k,t) = \frac{\varepsilon}{\nu}\left(\frac{\nu\sigma}{\alpha}\right)^\sigma \frac{k^{2\sigma-3}}{\Gamma(\sigma)} \cdot \exp\left(-\frac{\nu\sigma k^2}{\alpha}\right) \qquad (3\text{-}147)$$

Here α is an effective rate of strain, being the exponent in the exponential increase in length of line elements. σ is the ratio $|\alpha/\gamma|$, where γ is an average value for the least principal rate of strain. α is proportional to $(\varepsilon/\nu)^{1/2}$, with the proportionality constant depending linearly on the skewness factor of the turbulence velocity field, namely

$$\alpha = -\frac{7}{6\sqrt{15}} S \cdot \left(\frac{\varepsilon}{\nu}\right)^{1/2}$$

where S is the skewness factor of the rate of strain $(\partial u_1/\partial x_1)$. Writing $\alpha = s(\varepsilon/\nu)^{1/2}$, Eq. (3-147) can be rewritten in dimensionless form

$$\mathbf{E}(k,t)(\varepsilon\nu^5)^{-1/4} = \frac{(\sigma/s)^\sigma}{\Gamma(\sigma)} \cdot \left(\frac{k}{k_d}\right)^{2\sigma-3} \exp\left(-\frac{\sigma}{s} \cdot \frac{k^2}{k_d{}^2}\right) \qquad \text{(3-147a)}$$

From experimental evidence Saffman estimated s to lie between 0.09 and 0.15, while for σ he estimated a value of 0.4. Thus Saffman's result indeed shows a much faster fall-off of the energy spectrum at very large wavenumbers than the expression (3-146) according to Pao.

As Panchev and Kesich[73] have shown, Saffman's Gaussian falloff of $\mathbf{E}(k,t)$ at very large wavenumbers, is also obtained as a limiting solution for $k \to \infty$ of Eq. (3-138). For in the case $k \gg k_d$ the equation becomes

$$k \cdot \mathbf{E}(k,t)\left(\frac{\varepsilon}{\nu}\right)^{1/2} = \frac{2\nu}{\alpha} \int_k^\infty dk'' \, k''^2 \mathbf{E}(k'',t)$$

With $\Theta(k,t) = \int_k^\infty dk'' \, k''^2 \mathbf{E}(k'',t)$ the following differential equation is obtained:

$$\frac{1}{k} \frac{d\Theta}{dk} + \frac{2}{\alpha}\left(\frac{\nu^3}{\varepsilon}\right)^{1/2} \Theta = 0$$

With the boundary condition $\Theta(0,t) = \varepsilon/2\nu$, the solution for $\mathbf{E}(k,t)$ reads

$$\mathbf{E}(k,t)(\varepsilon\nu^5)^{-1/4} = \frac{1}{\alpha}\left(\frac{k}{k_d}\right)^{-1} \exp\left[-\frac{1}{\alpha}\left(\frac{k}{k_d}\right)^2\right] \qquad \text{(3-148)}$$

Panchev and Kesich also showed that Pao's relation (3-146) can be obtained from Eq. (3-138) as an intermediate solution between the Kolmogoroff spectrum and the relation (3-148). They proposed the following interpolation formula

$$\mathbf{E}(k,t)(\varepsilon\nu^5)^{-1/4} = \left[\left(\frac{2}{3\alpha^2}\right)^{1/3}\left(\frac{k}{k_d}\right)^{-5/3} + \frac{1}{\alpha}\left(\frac{k}{k_d}\right)^{-1}\right]$$
$$\times \exp\left[-\frac{3}{2}\left(\frac{2}{3\alpha^2}\right)^{1/3}\left(\frac{k}{k_d}\right)^{4/3} - \frac{1}{\alpha}\left(\frac{k}{k_d}\right)^2\right] \qquad \text{(3-149)}$$

According to this interpolation formula Pao's relation (3-146) can be obtained only for $(k/k_d)^{2/3} < 1$.

The strong exponential decrease of the energy spectrum in the dissipation and higher wavenumber range gives a definite indication of the fine-scale structure of the turbulence. Kraichnan[79] has shown that, if all the mean-squares of the velocity derivatives exist, a stronger-than-algebraic decrease of the energy spectrum with increasing wavenumber implies an increasing intermittency of the fluctuations. This shows up in a higher value of the flatness factor of the probability density of velocity derivatives than for a normal distribution. Early measurements by Batchelor and Townsend[30] in grid-produced turbulence and in the wake of a circular cylinder already revealed that the flatness factor T_n of the nth velocity derivative, defined as

$$T_n = \overline{\left(\frac{\partial^n u}{\partial x^n}\right)^4} \Big/ \left[\overline{\left(\frac{\partial^n u}{\partial x^n}\right)^2}\right]^2$$

increased markedly and almost linearly with n. It also increased with increasing Reynolds number. For a normal distribution the flatness factor is equal to 3. Batchelor and Townsend measured in the wake of a cylinder, at a Reynolds number of 4,100 referred to free-stream velocity and cylinder diameter, values of 3, 4.5, 5.7, and 7.3 for $n = 0$, 1, 2, and 3 respectively. If we define an intermittency factor Ω as the fraction of time that turbulence occurs, the value of $T_3 = 7.3$ would correspond with $\Omega \simeq 0.4$. The measurements in grid-produced turbulence showed smaller values of T_n. At a comparable Reynolds number, namely $U_0M/\nu \simeq 5{,}600$, T_n increased from the value 3 for $n = 0$ to $\simeq 5.5$ for $n = 3$. A higher value than 3 of the flatness factor for a distribution with zero mean, means a more peaked distribution around this zero mean than for a normal distribution. So the probability for zero values is higher, while that for nonzero values is smaller, indicating an intermittent nature of the fluctuations. This intermittency indeed shows up in the oscillograms (more, or longer, periods of relative silence). However, as has been pointed out by Kennedy and Corrsin,[124] a high value of the flatness factor does not necessarily imply intermittency. But, if in a separate way intermittency is shown to occur, the value of the flatness factor may be considered a measure for the degree of intermittency.

Batchelor and Townsend[30] suggested the relation $T_n\Omega = 3$, on the assumption that the variable concerned shows a normal distribution during the fraction of time of turbulent state and is zero during the nonturbulent state. Carefully made measurements by Kuo and Corrsin[125] on the fine structure of turbulence in grid-generated turbulence (with and without a slight contraction downstream of the grid to obtain almost isotropic turbulence), and on the axis of round jets where the turbulence is close to isotropy, yielded much larger values of $T\Omega$ than 3, namely as large as 7.5 (in the grid-generated turbulence at $\mathbf{Re}_\lambda = 110$).

Another structural feature connected with this intermittent character of the small-scale motions is that the vorticity, and hence the dissipation, is not homogeneously distributed, but must show a spottiness. Or, better, the dissipation must be present predominantly in small regions, probably formed by concentrated vortex sheets, mentioned earlier. At this small spatial scale the dissipation then must show random fluctuations. Kolmogoroff[80] and Obukhoff[81] therefore modified their theories by taking into account these dissipation fluctuations. They assumed a log-normal probability distribution. Now Gurvich and Yaglom[126] showed that, assuming a cascade process of the breakdown of bigger to smaller eddies, any non-negative quantity governed by small-scale motions of the fine structure of turbulence at large Reynolds numbers must have a log-normal distribution. Saffman[127] obtained the same result in a slightly different way, though. On the other hand Orszag[123] pointed out that a log-normal probability distribution of $\tilde{\varepsilon}$ leads to an inconsistency in the moments of various orders of $\tilde{\varepsilon}$. For, with a log-normal distribution some properties of the very small eddies would not be uniquely defined by the initial values of moments of all orders. Indeed, experimental results obtained by Stewart *et al.*[122] showed substantial deviations from the log-normal distribution at large values of $\tilde{\varepsilon}$, namely smaller values of the probability density than according to the log-normal distribution. Also Kuo and Corrsin[125] found in the aforementioned experiments that the log-normal distribution applied to $u_1{}^2$, $(\partial u_1/\partial t)^2$ and $(\partial^2 u_1/\partial t^2)^2$ only in a limited amplitude range. The agreement improved, however, with increasing order of the derivative and with increasing Reynolds number. At $\mathbf{Re}_\lambda \gtrsim 100$ the log-normal distribution applied to $(\partial^2 u/\partial t^2)^2$ only in the probability range of 0.3 to 0.95.

Corrsin[82] made an estimate of the relative intensity of the dissipation fluctuation $\tilde{\varepsilon}$, by assuming the dissipation to be concentrated in randomly distributed thin sheets or slabs of a characteristic thickness of the order of the Kolmogoroff micro-scale η, surrounding regions of much less activity of a dimension of the order the integral scale Λ. This means that the concentrated-dissipation regions have a volume fraction of the order η/Λ. Corrsin's estimate yields $\tilde{\varepsilon}'/\bar{\varepsilon} \simeq C\,\mathbf{Re}_\lambda{}^{3/4}$, with C of $\mathcal{O}(1)$. Since $\tilde{\varepsilon}$ is determined by $(\partial u_1/\partial x_1)^2$, the above relation corresponds with a flatness factor $T_1 \propto \mathbf{Re}_\lambda{}^{3/2}$. Tennekes[128] considered another model, in which the dissipation is concentrated in randomly distributed vortex tubes with a diameter of $\mathcal{O}(\eta)$ and spaced on the order of the dissipation length-scale λ. This model yields the relation $T_1 \propto \mathbf{Re}_\lambda$, or $\tilde{\varepsilon}'/\bar{\varepsilon} \propto \mathbf{Re}_\lambda{}^{1/2}$. Experiments by Kuo and Corrsin and by others (see ref. 125), however, showed $T_1 \propto \mathbf{Re}_\lambda{}^n$, where n increased gradually from 0.15 for $\mathbf{Re}_\lambda \lesssim 100$ to 0.6 for $\mathbf{Re}_\lambda \gtrsim 500$, i.e., not in agreement with either model. The value of 0.6 is in agreement with that according to Gurvich and Yaglom's theory,[126] which predicts

$$T_1 \propto (\mathbf{Re}_\lambda{}^{3/4})^{2\mu}$$

With a value of 0.4 for the universal constant μ, as estimated by Gurvich and Yaglom[126] from measured spectra of $(\partial u_1/\partial t)^2$, $n = 0.6$ is obtained.

The same value $\mu = 0.4$ has been obtained by Pond and Stewart[131] in a wind blowing over the ground. But in a similar flow Sheih *et al.*[132] obtained $\mu = 0.7$, while Von Atta and Chen[133] found $\mu = 0.5$ in a wind blowing over an open ocean. In a curved mixing-layer flow Wyngaard and Tennekes[134] obtained $\mu = 0.85$ at $\mathbf{Re}_\lambda = 200$, and Ueda and Hinze[135] $\mu = 0.725$ in a windtunnel boundary-layer at $\mathbf{Re}_\lambda \simeq 150$. With the highest value of $\mu = 0.85$ one would obtain according to Gurvich and Yaglom's theory $n \simeq 1.25$, which is between the values 1 and 1.5 according to the models of Tennekes and Corrsin, respectively.

In a second paper Kuo and Corrsin[129] re-examined the models based on concentrated dissipation regions in the form of randomly distributed "sheets" (or "slabs") and "tubes" (or "rods"), and in addition also spherical "blobs" of fluid. The pertinent experiments which they made in a grid-generated turbulence with a slight contraction (acceleration) of the flow, were not conclusive in favor of either model. At the most the dissipation regions were concluded to be more "rod" like, rather than "slab" like or "blob" like. However, as will be discussed in Sec. 5-4, fluid elements are more likely to be deformed by the turbulent motions to long, thin ribbons, and not to slabs or rods. A model based on ribbon-like concentrated-dissipation regions might therefore be better considered.

In an anisotropic shear turbulence (see Chaps. 6 and 7), the bigger eddies have a preferred orientation. With the picture of "ribbon"-like concentrated-dissipation regions, a departure from local isotropy in the most fine-scale structure may occur. In general, in a shear flow, the turbulence in the lower wavenumber range is anisotropic. The smaller eddies, however, will become isotropic with increasing wavenumber, but beyond the inertial subrange, in the dissipation and higher wavenumber range, the structure may become anisotropic again. For, it may be expected that the concentrated-dissipation regions will follow the bigger eddies in their statistically preferred orientation.

We have not yet obtained a solution for $\mathbf{E}(k,t)$ valid in the wavenumber range of the energy-containing eddies and for still lower wavenumbers, except for the lowest range where $\mathbf{E}(k,t)$ varies with k^4. To obtain such a solution we have to consider the complete equations (3-87) or (3-95), because the rate of decay of the total turbulence is of the same order as the "frequency" $u'_k k_e$ of the eddies in these ranges.

Failing a solution valid in the range of the energy-containing eddies we will confine ourselves here to mentioning a formula given by Von Kármán[9] as an interpolation formula covering the range between the k^4-range of the "permanent" largest eddies and the inertial subrange with the $-\frac{5}{3}$-power law:

$$\mathbf{E}(k,t) = \mathbf{E}(k_e,t)2^{17/6} \frac{(k/k_e)^4}{[1 + (k/k_e)^2]^{17/6}} \tag{3-150}$$

where k_e is still a function of time.

The function $\mathbf{E}(k_e,t)$ can be related to the Loitsianskii integral I, since, when $k/k_e \to 0$, we must have

$$Ik^4 = \mathbf{E}(k_e,t)2^{17/6}(k/k_e)^4$$

or

$$\mathbf{E}(k_e,t) = 2^{-17/6} I k_e^{\,4} \tag{3-150a}$$

The formula (3-150) reduces to the $k^{-5/3}$ law for very large values of k/k_e. It has the advantage that it is quite suitable for further use in calculations, since it generally leads to easily integrable functions.

The Von Kármán formula (3-150) applies only to such very large Reynolds numbers that an inertial subrange ($k^{-5/3}$ law) exists and viscosity effects are negligible. Now we have mentioned in Sec. 1-11 that for large Reynolds numbers experimental correlation coefficients may be satisfactorily approximated by an exponential law, except very close to its vertex. In Sec. 3-4 we have shown that, if this exponential law applies to the longitudinal correlation function $\mathbf{f}(r,t)$, there corresponds to it a three-dimensional energy-spectrum function

$$\mathbf{E}(k,t)=\frac{8}{\pi}u'^2\Lambda_f\frac{\Lambda_f{}^4k^4}{(1+\Lambda_f{}^2k^2)^3} \tag{3-151}$$

For large values of k this expression reduces to

$$\mathbf{E}(k,t)\simeq\frac{8}{\pi}\frac{u'^2}{\Lambda_f}k^{-2}$$

which does not differ very much from the $k^{-5/3}$ law. For practical purposes it may sometimes be useful.

It may be of interest to calculate the one-dimensional energy spectra and the correlation functions corresponding to the solutions and formulas just given for $\mathbf{E}(k,t)$. Let us consider first the Heisenberg solution (3-142). With the aid of Eq. (3-72) it would be possible to calculate the one-dimensional energy spectrum $\mathbf{E}_1(k_1,t)$ and, with the aid of Eq. (3-49), the longitudinal correlation coefficient $\mathbf{f}(x_1,t)$. However, the solution of the integrals involved leads to very complex expressions; moreover, the solution obtained for $\mathbf{f}(x_1,t)$ would be valid only in the restricted range $x_1 \ll l_e$, for Heisenberg's solution (3-142) is valid only in the wave-number range $k \gg k_e$. Therefore, we shall not give these solutions here.

In direct contrast with these difficulties, the one-dimensional spectrum and correlation coefficient for the inertial subrange can be obtained very easily. We already know that in this subrange we have the simple power law:

$$\mathbf{E}(k,t)=\left(\frac{8}{9\alpha}\right)^{2/3}(\varepsilon\nu^5)^{1/4}\left(\frac{k}{k_d}\right)^{-5/3}\qquad\text{when } k/k_d\ll 1 \tag{3-152}$$

Substitution in (3-72) will show that $\mathbf{E}_1(k_1,t)$ obeys the same power law.

$$\mathbf{E}_1(k_1,t)=\frac{18}{55}\left(\frac{8}{9\alpha}\right)^{2/3}(\varepsilon\nu^5)^{1/4}\left(\frac{k_1}{k_d}\right)^{-5/3}\qquad\text{when } k/k_d\ll 1 \tag{3-152a}$$

In the viscous subrange $k/k_d \gg 1$, Heisenberg's power law $\mathbf{E}(k) \propto k^{-7}$ would also give the same result for the one-dimensional spectrum $\mathbf{E}_1(k_1)$. But again, owing to the invalidity of Heisenberg's hypothesis in this subrange, we should better consider a spectrum function like Eq. (3-147), suggested by Saffman.

If we want to calculate the longitudinal correlation coefficient $\mathbf{f}(x_1,t)$ we have to bear in mind that the Kolmogoroff spectrum is valid in the range $k_e \ll k \ll k_d$, so that the corresponding solution for $\mathbf{f}(x_1,t)$ is valid only in the range $\eta \ll x_1 \ll l_e$.

Thus from Eq. (3-49)

$$u'^2\mathbf{f}(x_1,t) = C\int_{k_e}^{k_d} dk_1\, k_1^{-5/3}\cos k_1x_1 + \int_0^{k_e} dk_1\, \mathbf{E}_1(k_1,t)\cos k_1x_1 + \int_{k_d}^{\infty} dk_1\, \mathbf{E}_1(k_1,t)\cos k_1x_1$$

where

$$C = \frac{18}{55}\left(\frac{8}{9\alpha}\right)^{2/3}(\varepsilon\nu^5)^{1/4}k_d^{5/3} = \frac{18}{55}\left(\frac{8\varepsilon}{9\alpha}\right)^{2/3}$$

Now we should obtain known definite integrals if the limits of the first integral were 0 and ∞. However, for a zero lower limit this integral is not convergent. Therefore instead of $u'^2\mathbf{f}(x_1,t)$ we had better consider

$$u'^2[1-\mathbf{f}(x_1,t)] = C\int_{k_e}^{k_d} dk_1\, k_1^{-5/3}(1-\cos k_1x_1) + \int_0^{k_e} dk_1\, \mathbf{E}_1(k_1,t)(1-\cos k_1x_1) + \int_{k_d}^{\infty} dk_1\, \mathbf{E}_1(k_1,t)(1-\cos k_1x_1)$$

For small k_1 the integrand becomes zero because $1-\cos k_1x_1$ decreases faster than $k_1^{-5/3}$ increases, but for large k_1 the integrand again becomes zero because of $k_1^{-5/3}$. Hence we may safely replace the limits k_e and k_d by 0 and ∞; we obtain

$$u'^2[1-\mathbf{f}(x_1,t)] = C\int_0^{\infty} dk_1\, k_1^{-5/3}(1-\cos k_1x_1) = Cx_1^{2/3}\int_0^{\infty} dy\, y^{-5/3}(1-\cos y)$$

The integral can be transformed by integration in parts into

$$\frac{3}{2}\int_0^{\infty} dy\, y^{-2/3}\sin y = \tfrac{3}{4}\Gamma(\tfrac{1}{3})$$

so that

$$\mathbf{f}(x_1,t) = 1 - \frac{3}{4}\frac{C}{u'^2}\Gamma(\tfrac{1}{3})x_1^{2/3}$$

or

$$\mathbf{f}(x_1,t) = 1 - \tfrac{18}{55}\Gamma(\tfrac{4}{3})\frac{A}{u'^2}(\varepsilon x_1)^{2/3} \qquad \text{(3-153)}$$

where again $A = (8/9\alpha)^{2/3}$.

The relation (3-153), but for the value of the constant A, may also be obtained in a direct way, by assuming that in the inertial subrange the mean square of the velocity difference between two points, a distance x_1 apart can only be a function of this distance and the dissipation

$$\overline{[u_1(x_1)-u_1(0)]^2} = f(\varepsilon, x_1)$$

Dimensional analysis then results in the relation (3-153).

For the energy spectrum $\mathbf{E}(k,t)$ in the wavenumber range from $k = 0$ up to and including the range of the energy-containing eddies, we have the Von Kármán interpolation formula (3-150) and the empirical formula (3-151). As regards the latter, we have already calculated the corresponding one-dimensional energy spectrum and the correlation coefficients.

Using the Von Kármán interpolation formula, we obtain with Eq. (3-72)

$$\mathbf{E}_1(k_1,t) = \mathbf{E}(k_e,t)2^{17/6}\frac{18}{55}\left[1+\left(\frac{k_1}{k_e}\right)^2\right]^{-5/6}$$

Substitution in Eq. (3-49), after evaluation of the integral, yields

$$u'^2\mathbf{f}(x_1,t) = 2^{7/2}\frac{9}{55}\frac{\sqrt{\pi}}{\Gamma(\frac{5}{6})}k_e\mathbf{E}(k_e,t)(x_1k_e)^{1/3}K_{1/3}(x_1k_e)$$

where $K_{1/3}(y)$ denotes the modified Bessel function of the order $\frac{1}{3}$.

In these expressions $\mathbf{E}(k_e,t)$ is still unknown. Its value, however, can be determined from the condition that the corresponding correlation functions must become unity when $x_1 = 0$. Hence, from the last expression for $\mathbf{f}(x_1,t)$, when $x_1 = 0$,

$$\mathbf{E}(k_e,t) = \frac{u'^2}{k_e}\frac{55}{9}\frac{\Gamma(\frac{5}{6})}{2^{17/6}\sqrt{\pi}\,\Gamma(\frac{1}{3})} \simeq 0.2\frac{u'^2}{k_e}$$

We then obtain for $\mathbf{E}(k,t)$, $\mathbf{E}_1(k_1,t)$, and $\mathbf{f}(x_1,t)$:

$$\mathbf{E}(k,t) = \frac{55}{9}\frac{\Gamma(\frac{5}{6})}{\sqrt{\pi}\,\Gamma(\frac{1}{3})}\frac{u'^2}{k_e}\frac{(k/k_e)^4}{[1+(k/k_e)^2]^{17/6}} \tag{3-154}$$

$$\mathbf{E}_1(k_1,t) = \frac{2}{\sqrt{\pi}}\frac{\Gamma(\frac{5}{6})}{\Gamma(\frac{1}{3})}\frac{u'^2}{k_e}\left[1+\left(\frac{k_1}{k_e}\right)^2\right]^{-5/6} \tag{3-155}$$

and

$$\mathbf{f}(x_1,t) = \frac{2^{2/3}}{\Gamma(\frac{1}{3})}(k_ex_1)^{1/3}K_{1/3}(k_ex_1) \tag{3-156}$$

For the transverse-correlation coefficient we obtain

$$\mathbf{g}(x_1,t) = \frac{2^{2/3}}{\Gamma(\frac{1}{3})}(k_ex_1)^{1/3}\left[K_{1/3}(k_ex_1) - \frac{k_ex_1}{2}K_{-2/3}(k_ex_1)\right] \tag{3-157}$$

For small values of k_ex_1, Eq. (3-156) reduces to

$$\mathbf{f}(x_1,t) \simeq 1 - \frac{\Gamma(\frac{2}{3})}{\Gamma(\frac{4}{3})}\left(\frac{k_ex_1}{2}\right)^{2/3}$$

and Eq. (3-157) reduces to

$$\mathbf{g}(x_1,t) \simeq 1 - \frac{4\,\Gamma(\frac{2}{3})}{3\,\Gamma(\frac{4}{3})}\left(\frac{k_e x_1}{2}\right)^{2/3}$$

Comparison of the result of $\mathbf{f}(x_1,t)$ with Eq. (3-153) shows that

$$k_e = \frac{1}{l_e} = \text{numerical const.} \times \frac{\varepsilon}{u'^3}$$

A closer evaluation of the numerical constant may be obtained by comparing the expression (3-154) for large k/k_e with Eq. (3-123). From Eq. (3-154) for large k/k_e we obtain

$$\mathbf{E}(k,t) = \frac{55}{9}\frac{\Gamma(\frac{5}{6})}{\sqrt{\pi}\,\Gamma(\frac{1}{3})}\frac{u'^2}{k_e}\left(\frac{k}{k_e}\right)^{-5/3}$$

so, by Eq. (3-123),

$$k_e = \frac{8}{9\alpha}\left[\frac{9}{55}\frac{\sqrt{\pi}\,\Gamma(\frac{1}{3})}{\Gamma(\frac{5}{6})}\right]^{3/2}\frac{\varepsilon}{u'^3} = \frac{0.51}{\alpha}\frac{\varepsilon}{u'^3}$$

We have mentioned earlier that l_e will be of the same order of magnitude as the integral scale Λ_f. Now from $\mathbf{E}_1(k_1,t)$ this scale can readily be obtained [see Eq. (3-74)], so that, by Eq. (3-135),

$$\Lambda_f(t) = \frac{\pi}{2u'^2}\mathbf{E}_1(0,t) = \sqrt{\pi}\frac{\Gamma(\frac{5}{6})}{\Gamma(\frac{1}{3})}\frac{1}{k_e} \simeq \frac{0.75}{k_e} = 0.75 l_e \qquad (3\text{-}158)$$

Von Kármán obtained his interpolation formula (3-150) by assuming $\mathbf{E}(k,t)$ to vary with k^4 when $k \to 0$. As discussed earlier, according to Saffman we may consider as well a k^2-behavior of $\mathbf{E}(k,t)$ when $k \to 0$, since there is no experimental evidence in favor of either of the two. The series expansion (3-61) then starts with the first term on the right-hand side, and

$$\lim_{k\to 0} \mathbf{E}(k,t) = \mathbf{I}_1 k^2 \qquad (3\text{-}159)$$

where

$$\mathbf{I}_1 = \frac{1}{\pi}\int_0^\infty dr\, r^2 \mathbf{Q}_{i,i}(r,t)$$

is a dynamical invariant.

The corresponding "Von Kármán" interpolation formula then will read

$$\mathbf{E}(k,t) = \mathbf{E}(k_e,t)2^{11/6}\frac{(k/k_e)^2}{[1+(k/k_e)^2]^{11/6}} \qquad (3\text{-}160)$$

where

$$\mathbf{E}(k_e,t) = 2^{-11/6}\mathbf{I}_1 k_e^{\,2}$$

Since, as will be shown in Sec. 3-7, this is exactly the same relation as for an isotropic-turbulent scalar field, the reader may be referred for further details to that section.

In the "physical" closure approximations considered the assumption is made that the energy transfer between eddies is a very local process in wavenumber space, so that this energy transfer takes place through a cascade process. There is no direct dynamic interaction between big and small eddies, the latter being entrained by the big ones without appreciable distortion when measured in the time scale of the small eddies. In a number of publications Kraichnan[67,83,84,85,86,87] developed a closure theory based on a certain interaction model. We shall give here only a very brief outline of the main ideas underlying his theory, without going into any mathematics. Since the mathematics is very involved and elaborate, its treatment would go far beyond the scope of the present book.

In his *"direct interaction"* theory Kraichnan drops the idea that the direct interaction between big and small eddies may be neglected. Even when there would be no interaction in space at the same instant, so that the purely spatial correlations between the corresponding Fourier components are zero, this does not imply that the space-time correlations too are zero. Therefore Kraichnan considers the dynamic behavior of two-point, two-time velocity correlations, or of their Fourier transforms. We have seen that the transfer of energy in wavenumber space occurs through the interaction of three Fourier components k, k' and k–k'. This interaction may be either direct, or indirect through intermediate interactions with other components. The direct-interaction hypothesis now neglects the contribution of the latter, indirect interactions. But the direct interaction will be affected because each Fourier component will interact with all other Fourier components. This latter interaction may be described as a relaxation process due to an eddy viscosity as the relaxation parameter. For, in the dynamic equation for a Fourier component Kraichnan considers the nonlinear terms as small perturbations (*weak interaction hypothesis*). The direct-interaction approximation then may be considered as the first term in a perturbation expansion procedure. Now, instead of the above relaxation parameter a response function (Green's function) is considered, which function is defined as the statistically average response of the component k at time t to an impulse perturbation applied to it at time t'. This response function occurs in the dynamic equation for the above Fourier transform of the two-time velocity correlation, and represents the effect of the nonlinear terms. Since the response function describes the average response to a small disturbance, it satisfies a similar dynamic equation, which equation then is the supplementary equation needed to make the problem determinate, and so obtain a closure procedure.

When applying this procedure to an isotropic turbulence, and specifically to the inertial subrange of the energy spectrum, a result is obtained which is not in agreement with the Kolmogoroff ($-5/3$) law. The reason is that the time (or memory) integrals occurring in the above dynamic equations, which include the response function, are cut off too effectively by the nonlinear interactions to be consistent with the turbulence in the inertial subrange.[63] The origin of this defect of the theory must be sought in a violation of the invariancy condition for the dynamical relations under random Galilean transformations. The problem of this defect is rather subtle. When we consider the average of a large number of realizations of turbulence, each realization subjected to a uniform convection velocity chosen at random, for this average the dynamical effect of these convection velocities on the *one*-time, or simultaneous, velocity correlations must be zero. Now, the dynamic equation for this simultaneous correlation, or its Fourier transform, according to the direct-interaction theory still contains in the time integrals the two-time velocity correlations and the response function, which are affected by the random convection. So also the simultaneous velocity correlations become affected.

In his *"Lagrangian-History Direct-Interaction"* theory Kraichnan[86,87] modified the direct-interaction theory by considering the integration in space and in Lagrangian time, i.e., along the paths of the fluid particle. In the time integrals occurring in the dynamic equation for the simultaneous velocity correlation he takes the mixed Eulerian-Lagrangian velocity correlation (see Sec. 1-9) and a Lagrangian response function. The modified theory yields equations for these Lagrangian correlations and response function which make the simultaneous velocity correlation no longer affected by random convections. The effective relaxation time for the triads of Fourier components is determined by memory and decay associated with viscous and pressure forces along the particle path. At sufficiently large Reynolds number, so that there exists an inertial subrange, the energy transfer is mainly amongst traids for which

$k \simeq k' \simeq k - k'$. Thus it is local in wavenumber space. But, each component is affected by all other components integrated over its Lagrangian history.

This modified theory yields an inertial-subrange spectrum in agreement with the Kolmogoroff theory. It yields a value for the constant A occurring in Eq. (3-123) equal to 1.77 in the limiting case $\mathbf{Re}_\lambda \to \infty$. This value is close to the values obtained empirically in the high Reynolds number measurements, to be discussed later. And, as mentioned earlier, the theory predicts the optimum wavenumber for maximum dissipation which is, within the experimental accuracy, equal to the value found experimentally.

The "Physical Closure" Approximations in the Light of Experimental Evidence

We have considered various assumptions on the basis of which the dynamic energy-spectrum function can be either partially or completely solved. In a few cases we have shown, too, how the corresponding correlation functions can be calculated. But the value of these results depends entirely on the correctness of the assumptions made. Hence at least some experimental evidence to justify the assumptions made is greatly needed. A direct verification of the assumptions on a relation between the transfer-spectrum function $\mathbf{F}(k,t)$ and the energy-spectrum function $\mathbf{E}(k,t)$ is, however, very difficult to make. The method hitherto adopted is an indirect one, namely, comparing either the calculated energy spectrum or the calculated correlation functions with the measured ones. We have already said in Chap. 1 that correlation functions are more accurately measured in the lower-wavenumber range, whereas spectrum functions are more accurately measured in the higher wavenumber range. This has to be borne in mind in interpreting measurements. But, apart from the accuracy of measurements, there remains the fact that correlation functions calculated by an integration process from the corresponding energy spectra are highly insensitive to the underlying assumptions, particularly since there is usually some adjustable constant available—for instance, the constant α used in the various assumptions in respect of $\int_0^k dk\, \mathbf{F}(k,t)$. Hence a satisfactory agreement between calculated and measured correlation functions is not adequate to prove that an underlying assumption is correct.

Proudman[18] has tried to verify Heisenberg's assumption by such a comparison of computed and measured functions. He found that in general Heisenberg's numerical constant α is not absolute, but that it seems to depend on the Reynolds number, although this dependence is very slight for sufficiently high Reynolds numbers ($\mathbf{Re}_\lambda > 30$), at least in the lower-wavenumber range. Since this lower-wavenumber range determines the general shape of the correlation function, Proudman found that this general shape was in fairly good agreement with results reached by experiment when $\alpha = 0.45$. On the other hand Heisenberg himself suggested a value 0.8. However, if only the large-wavenumber range is considered, α appears to be a function of both $\mathbf{Re}_\lambda$ and the wavenumber k. Proudman mentions unpublished results by Lee, according to which $\alpha = 0.13$ for large wavenumbers. For the final period of decay where viscous effects prevail, it is shown that α may become as low as 0.0005. Obviously, this is too great an extrapolation from Heisenberg's theory, which

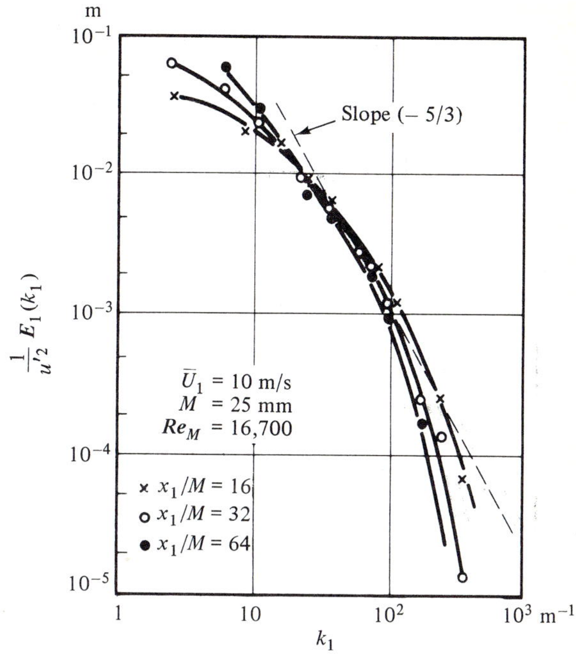

FIGURE 3-14
Comparison of the one-dimensional spectrum $\mathbf{E}_1(k_1)$ with the Kolmogoroff (–5/3) spectrum law. (From: *Sato, H.,*[19] *by permission of the Physical Society of Japan.*)

assumed a sufficiently large Reynolds number for an equilibrium range, including an inertial subrange, to exist. Later, however, Reid[49] showed in a critical discussion of Proudman's paper that the ratio S/α between the skewness factor S of $\partial u_1/\partial x_1$ and α varies only between the limits 0.78 and 1.52 as the Reynolds number $\mathbf{Re}_\lambda$ varies from zero to infinity. Thus, from the observed values of S for various Reynolds numbers, it is concluded that α should vary between 0.62 and 0.20. The value of $\alpha = 0.45$ suggested by Proudman is a good average between these extreme values. A reason why Proudman obtained unreasonably low values of α in the high-wavenumber range where viscous effects prevail may be the failure of Heisenberg's hypothesis, as discussed earlier.

Sato[19] has measured the one-dimensional spectrum $\mathbf{E}_1(k_1,t)$ at distances 0.4, 0.8, and 1.6 m downstream of a square-mesh grid (mesh $M = 25$ mm; bar diameter $d = 5$ mm) at a speed of $\bar{U}_1 = 10$ m/s. $\mathbf{Re}_M = \bar{U}_1 M/\nu \simeq 17{,}000$ and $\mathbf{Re}_\lambda = u'\lambda_g/\nu$ was nearly constant during decay and roughly equal to 60 at the three locations. Sato's results are given in Fig. 3-14. They show that the Kolmogoroff spectrum $\mathbf{E}_1(k_1) \propto k_1^{-5/3}$ applies, if at all, only in a very restricted wavenumber range.

The same conclusion is obtained from similar measurements made by Liepmann[20] for mesh Reynolds number $\mathbf{Re}_M$ of the order of 10^4. Here, too, it is possible only in a very narrow wavenumber range to approximate the measured one-dimensional spectrum by the Kolmogoroff spectrum law. Liepmann has further shown that, in contrast with this unsatisfactory approximation by the Kolmogoroff spectrum law, a closer approximation is obtained over nearly the entire measured wavenumber range by the empirical relation [see Eq. (1-103) or Sec. 3-4]

$$\mathbf{E}_1(k_1,t) = \frac{2}{\pi}u'^2\frac{\Lambda_f}{1+\Lambda_f{}^2k_1{}^2} = \frac{\mathbf{E}_1(0,t)}{1+\Lambda_f{}^2k_1{}^2} \tag{3-161}$$

However, for larger Reynolds numbers $\mathbf{Re}_M = 3 \times 10^5$, a much better agreement with the Kolmogoroff spectrum is obtained in the higher part of the measured wavenumber range. This is shown in Fig. 3-15, which also shows the curve according to Von Kármán's interpolation formula (3-155). Data obtained when $\mathbf{Re}_M = 10^5$ are also given in the same figure. For this slightly lower Reynolds number the agreement is less satisfactory; the empirical relation (3-161) there shows slightly better agreement. The experimental data on the corresponding correlation function are used by Von Kármán[9] for comparison with the theoretical transverse-correlation coefficient given by Eq. (3-157). A satisfactory agreement is obtained, except perhaps for very large values of $k_e x_1$ (see Fig. 3-16). But the same may be said of the calculated correlation curve corresponding to the energy spectrum (3-161).

The unsatisfactory agreement between the measured energy spectrum and the theoretical Kolmogoroff spectrum when $\mathbf{Re}_M < 10^5$ may raise the question: what minimum value of $\mathbf{Re}_M$, or rather of $\mathbf{Re}_\lambda$, is required to give rise to a sufficiently

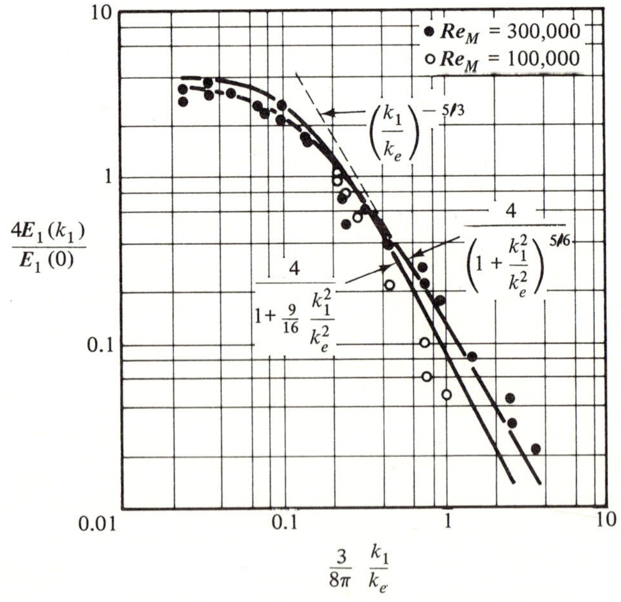

FIGURE 3-15
Comparison of measured and calculated values of the one-dimensional spectrum $\mathbf{E}_1(k_1)$. (From: *Von Kármán, Th.*,[9] *by permission of the University of Chicago Press.*)

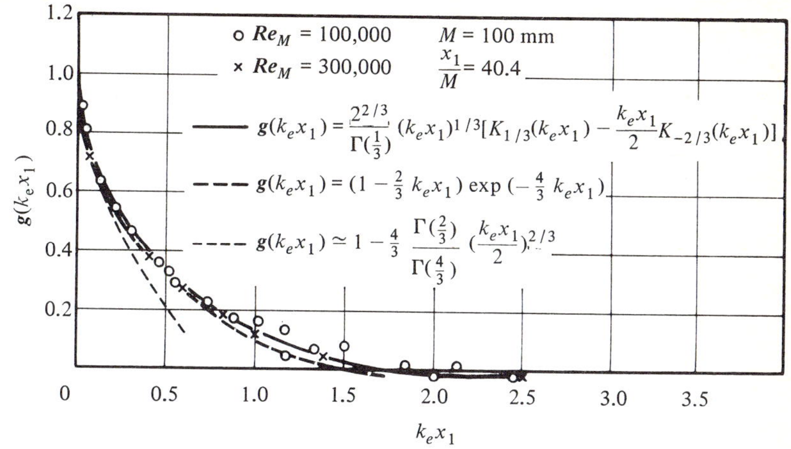

FIGURE 3-16
Comparison of measured and calculated values of the transverse velocity-correlation coefficient $\mathbf{g}(k_e x_1)$. (From: *Von Kármán, Th.*,[9] *by permission of the University of Chicago Press.*)

large inertial subrange? We have already mentioned that the condition for the existence of an equilibrium range is that $\mathbf{Re}_\lambda^{3/2} >>> 1$ [see Eq. (3-115)] and the condition for the existence of an inertial subrange, $\mathbf{Re}_\lambda^{3/4} >>> 1$ [see Eq. (3-118)]. But what we want now is a closer quantitative definition of the order of magnitude. This can be done by first seeing what the values of $\mathbf{Re}_\lambda$ are in laboratory experiments. In general they are not much greater than 100. Stewart and Townsend[17] have found that in such experiments the Kolmogoroff spectrum may exist at most in the range $k/k_d > 0.6$, which is in strong contradiction to the requirement that k/k_d should be much smaller than unity. They further show that if for an inertial subrange we must have at least $k/k_d < 0.1$, then we must have at least $\mathbf{Re}_\lambda > 1{,}500$ and $\mathbf{Re}_M$ should be at least of the order of 10^6.

Also most of the more recent investigations in windtunnel grid-produced turbulent flows, as, e.g., by Uberoi,[88] Comte-Bellot and Corrsin,[89] Uberoi and Wallis,[90] Van Atta and Chen[60,91] have been carried out at too low a Reynolds number to show an inertial subrange in the energy spectra. One of the features of turbulence not showing up in these experiments because of the lack of an inertial subrange, is the zero value of the energy-transfer function $\mathbf{F}(k,t)$ in the inertial subrange. Uberoi and Van Atta and Chen determined this energy-transfer function from the measured decay of energy spectrum $\mathbf{E}(k,t)$ and the local dissipation in wave-number space. For, from Eq. (3-87) we obtain

$$\mathbf{F}(k,t) = \frac{\partial \mathbf{E}(k,t)}{\partial t} + 2\nu k^2 \mathbf{E}(k,t)$$

Now the three-dimensional spectrum $\mathbf{E}(k,t)$ is not measured directly. What can be measured are the one-dimensional spectra $\mathbf{E}_1(k_1,t)$ and $\mathbf{E}_2(k_1,t)$ or $\mathbf{E}_3(k_1,t)$.

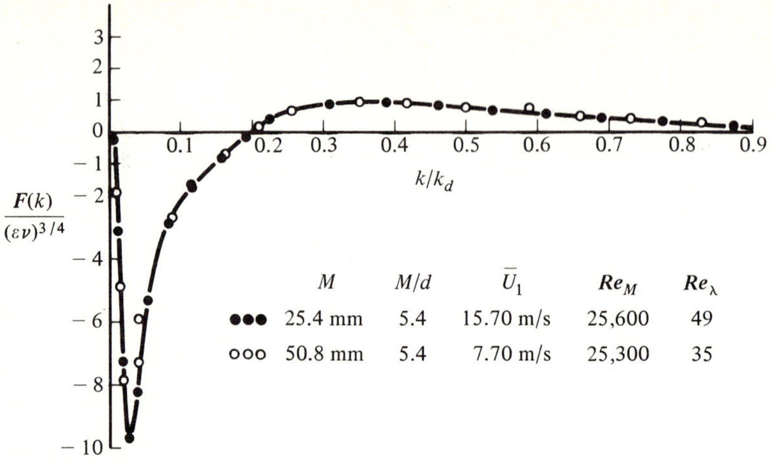

FIGURE 3-17
Normalized energy-transfer spectra $\mathbf{F}(k)$. (From: *Van Atta, C. W., and W. Y. Chen,*[91] *by permission of the Cambridge University Press.*)

If the turbulence shows true isotropy $\mathbf{E}(k,t)$ can be determined by measuring only $\mathbf{E}_1(k,t)$ and using the relation (3-71). However, as will be discussed later, grid-produced turbulence is not isotropic, but axisymmetric. Hence Uberoi and Van Atta and Chen measured $\mathbf{E}_1(k,t)$ and $\mathbf{E}_2(k,t) = \mathbf{E}_3(k,t)$. From the isotropic relation

$$\mathbf{E}(k,t) = -\tfrac{1}{2}k\frac{\partial}{\partial k}[\mathbf{E}_1 + 2\mathbf{E}_2]$$

the three-dimensional spectrum has been obtained. In Uberoi's experiments $\mathbf{Re}_\lambda \simeq 70$, whereas Van Atta and Chen obtained in their experiments even lower values, smaller than 50. Indeed, the measured spectral distributions of the energy did not show an inertial subrange, and consequently no corresponding wavenumber region where $\mathbf{F}(k,t)$ was very small. Figure 3-17 shows this function as determined by Van Atta and Chen. This figure clearly demonstrates the non-existence of an inertial subrange. The function crosses the zero line at $k/k_d \simeq 0.2$. At this wavenumber the dissipation spectrum showed its maximum value. In the range of the energy-containing eddies the dissipation was still roughly 25 per cent of the maximum value. Much higher values of the Reynolds number in grid-produced turbulence have been obtained by Kistler and Vrebalovich.[92] In a windtunnel with a working section 2.6×3.5 m^2 in cross-section, which tunnel could be pressurized, they obtained at a free-stream velocity of 61 m/s grid Reynolds numbers up to $\mathbf{Re}_M = 2.4 \times 10^6$. Figure 3-18 shows the one-dimensional spectra of the axial and of the cross turbulence velocity component, taken with the windtunnel operating at 4 bar absolute pressure. So the kinematic viscosity of the air was reduced to 4.3×10^{-6} m^2/s and a turbulence Reynolds number of $\mathbf{Re}_\lambda \simeq 500$ was obtained. The energy spectrum of the axial velocity component $\mathbf{E}_1(k_1)$ definitely shows an inertial sub-

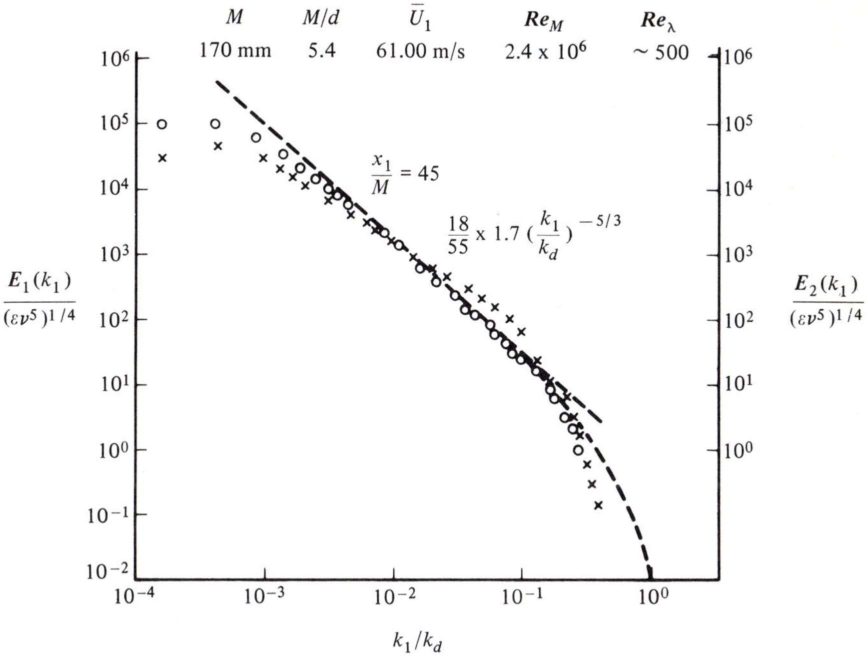

FIGURE 3-18
One-dimensional energy spectra $\mathbf{E}_1(k_1)$ [○○○], and $\mathbf{E}_2(k_1)$ [× × ×].
------- Computed according to Pao.[77]
(After: *Kistler, A. L., and T. Vrebalovich,*[92] *by permission of the Cambridge University Press.*)

range which agrees satisfactorily well with the Kolmogoroff law, Eq. (3-123) with $A \simeq 1.7$. A still higher value of $\mathbf{Re}_\lambda$ has been obtained by Gibson,[93] though not in a windtunnel but in the axis of a free air jet, namely $\mathbf{Re}_\lambda \simeq 780$. Very high values of $\mathbf{Re}_\lambda$ have been obtained by Grant, Stewart and Moilliet[70] in a tidal current, namely up to $\sim$2,000. The results of their measurements of the one-dimensional spectrum $\mathbf{E}_1(k_1)$ have been reproduced in Fig. 3-19, where again the inertial subrange can be represented by Eq. (3-123) with $A = 1.7$.

The variation of Heisenberg's constant α suggested by Reid and mentioned earlier, namely between 0.62 and 0.20 when $\mathbf{Re}_\lambda$ increases from zero to infinity, would correspond according to Eq. (3-143) with a variation of the constant A between 1.27 and 2.7. The measurements by Gibson at $\mathbf{Re}_\lambda = 780$ yield a value of $\alpha = 0.45$, corresponding with $A = 1.58$. The large-scale experiments in a tidal channel by Grant *et al.* resulted in values of A ranging from 1.22 to 1.81 and corresponding values $\alpha = 0.66$ to 0.365, with an average value of $A = 1.43$ and of $\alpha = 0.52$. However, in applying his large-wavenumber theory to Grant's *et al.* experimental results Pao[77] obtained $A = 1.70$, and consequently $\alpha = 0.405$.

Though experimental evidence indicates an effect of $\mathbf{Re}_\lambda$ on α, this effect is not found to be so marked as suggested by Reid. In applying a closure procedure

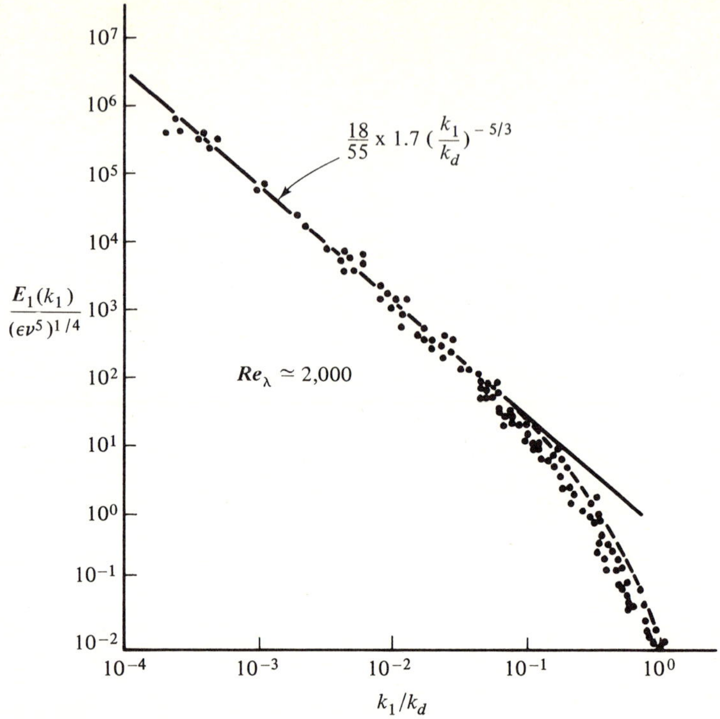

FIGURE 3-19
One-dimensional energy spectrum obtained in a tidal current.
------- Computed according to Pao.[77]
(After: *Grant, H. L., R. W. Stewart and A. Moilliet,*[70] *by permission of the Cambridge University Press.*)

leading in a more direct way to similar results to those obtained by Kraichnan, Orszag[94] found from an approximate calculation that at large Reynolds numbers the constant A should be greater than 1.50, and so $\alpha \leq 0.50$. As mentioned earlier, Kraichnan's Lagrangian-History Direct-Interaction theory predicts an upper limit of $A = 1.77$, and hence a lower limit of $\alpha = 0.38$ when $\mathbf{Re}_\lambda \to \infty$. We have seen from Fig. 3-18, and Fig. 3-19 that the value suggested by Pao, viz., $A = 1.70$ ($\alpha \simeq 0.4$) describes the inertial subrange spectrum satisfactorily at Reynolds numbers sufficiently high to show the existence of this subrange.

Figure 3-18 shows that the one-dimensional energy spectrum of the cross-velocity component, $\mathbf{E}_2(k_1)$ does not follow from the spectrum $\mathbf{E}_1(k_1)$ according to the isotropic relation Eq. (3-53). For, this relation should yield $\mathbf{E}_2(k_1) = \frac{4}{3}\mathbf{E}_1(k_1)$ in the inertial subrange. Actually the values of $\mathbf{E}_2(k_1)$ are in general too low, and they also do not follow a $(-5/3)$ law, as they should. Therefore one may doubt the reliability of these experiments.

Now, it has long since been appreciated that grids cannot produce an isotropic turbulence, because during the generation of the turbulence the axial turbulence velocity component is favored with respect to the lateral components. At the section $x_1/M = 45$, Kistler and Vrebalovich measured $u'_1/u'_2 \simeq 1.25$. This ratio decreased slightly with increasing x_1/M. This trend has been observed by many other experimenters while the values obtained for the ratio u'_1/u'_2 varied between 1.05 and 1.35 depending on the geometry of the grids used, the Reynolds number and the distance x_1/M.

In order to produce an isotropic turbulence out of a grid-generated turbulence a contraction of the windtunnel cross-section downstream of the grids can be applied, since such a contraction has a more reducing effect on the relative intensity of the axial component than on that of the lateral components. Uberoi,[95] Mills and Corrsin,[96] and Comte-Bellot and Corrsin[89] used with success a contraction with an area ratio of 1.27. Comte-Bellot and Corrsin obtained in this way a value $u'_1/u'_2 = 1.00 \pm 0.05$, that is, almost isotropic conditions.

We mentioned earlier that a zero region of the energy-transfer function $\mathbf{F}(k,t)$ corresponding to the inertial subrange did not show up in the low Reynolds number experiments by Van Atta and Chen (see Fig. 3-17). When, on the other hand, the Reynolds number is sufficiently large in the wavenumber range smaller than the inertial subrange, $\mathbf{F}(k,t)$ is negative and just equal to $\partial\mathbf{E}(k,t)/\partial t$, whereas in the wavenumber range larger than the inertial subrange this function is positive and just equal to the dissipation spectrum. Hitherto no determinations have been made of $\mathbf{F}(k,t)$ at large Reynolds numbers to check the theoretical behavior of $\mathbf{F}(k,t)$. Of course, it is possible to calculate this function for the equilibrium range through the dissipation spectrum. It then turns out that, for the results available from high Reynolds number measurements, the dissipation, and hence $\mathbf{F}(k,t)$ is not negligibly small in the inertial subrange as it should be. Grant's *et al.* measurements showed an inertial subrange $5 \times 10^{-4} \gtrsim k_1/k_d \gtrsim 5 \times 10^{-2}$ (see Fig. 3-19). Since Pao's expression, Eq. (3-146b) for $\mathbf{E}(k,t)$ represents this spectrum function satisfactorily well in the inertial subrange up to the dissipation range, we may use it to calculate the dissipation and $\mathbf{F}(k)$. We then obtain

$$\frac{\mathbf{F}(k)}{(\varepsilon\nu)^{3/4}} = \frac{2\nu k^2\mathbf{E}(k)}{(\varepsilon\nu)^{3/4}} = 2A\left(\frac{k}{k_d}\right)^{1/3} \exp\left[-\tfrac{3}{2}A\left(\frac{k}{k_d}\right)^{4/3}\right]$$

Figure 3-20 shows $\mathbf{F}(k)$ calculated in this way, with $A = 1.7$. The function attains its maximum of 1.4 at $k/k_d \simeq 0.1$, in agreement with Grant's *et al.* result. We notice that the transfer functions and the dissipation are still appreciable in the inertial subrange given above.

The function $\mathbf{G}(k) = \int_0^k dk\,\mathbf{F}(k)$ should be constant in the inertial subrange. From Pao's theory follows

$$\frac{\mathbf{G}(k)}{\varepsilon} = -\exp\left[-\tfrac{3}{2}A\left(\frac{k}{k_d}\right)^{4/3}\right]$$

This function too is given in Fig. 3-20. It shows an almost constant behavior in the inertial subrange.

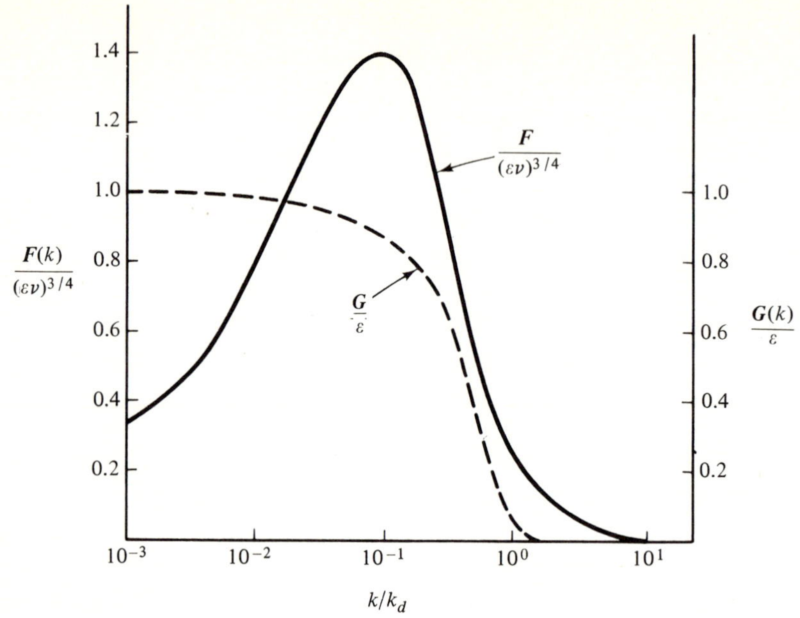

FIGURE 3-20
Calculated $\mathbf{F}(k)$ and $\mathbf{G}(k)$ functions according to Pao's theory.[77]

An interesting point concerning the mechanism of the transfer of energy through the spectrum is the degree of localness in wavenumber space of the interaction between eddies or Fourier components. When discussing the "physical" closure hypotheses we introduced an interaction function $\mathscr{I}(k',k,t)$. In these theories it is assumed that the energy transfer has a local character, which then is shown by the interaction functions having a finite value only in a restricted wavenumber range and increasing when $k' \to k$. Some support from experimental evidence to justify the idea of localness of the interaction resulting in energy transfer would be of importance. Now, the direct determination of $\mathscr{I}(k',k,t)$ from the energy-transfer function $\mathbf{F}(k,t)$ is not possible, the relation between the two being given in the integral form of Eq. (3-130). However, from Von Kármán's general expression Eq. (3-134) for the function $\mathbf{G}(k,t)$ it is possible to determine the values of φ and ψ, and hence from the Eqs. (3-133a) and (3-133b) the form of the interaction function $\mathscr{I}(k',k,t)$. Such a procedure has been followed by Tannenbaum and Mintzer[97] in analyzing the early experimental results of grid-generated turbulence obtained by Stewart and Townsend.[17] From their measurements of the longitudinal correlation $\mathbf{f}(r,t)$ and of the triple correlation $\mathbf{k}(r,t)$ Tannenbaum and Mintzer calculated the energy spectrum function $\mathbf{E}(k,t)$ [see, e.g., Eq. (3-70)] and the transfer function $\mathbf{F}(k,t)$ [see Eq. (3-90)], and hence the function $\mathbf{G}(k,t)$. The value of $\mathbf{G}(k,t)$ calculated in this way should be the same as that calculated from Eq. (3-134), when the assumption is made that Von Kármán's theory holds true. Reasonable agreement could only be obtained with different values of φ and ψ. The closest approximation was found with $\varphi = \frac{1}{4}$, $\psi = -\frac{5}{2}$ and $\alpha = 0.60$. Equations (3-133a) and (3-133b) then yield

$$\mathscr{I}(k',k,t) = 2\alpha \mathbf{E}^{5/4}(k',t)k'^3 \cdot \mathbf{E}^{1/4}(k,t)k^{-5/2} \quad \text{for} \quad k' < k$$

$$\mathscr{I}(k,k'',t) = 2\alpha \mathbf{E}^{5/4}(k,t)k^3 \cdot \mathbf{E}^{1/4}(k'',t)k''^{-5/2} \quad \text{for} \quad k'' > k$$

At the very low Reynolds number measurements [$\mathbf{Re}_M = 5{,}300$, $\mathbf{Re}_\lambda \simeq 20$] made by Stewart and Townsend, with no inertial subrange at all in the energy spectrum, a local behavior of the interaction was only observed in the range of the energy-containing eddies ($k/k_d \gtrsim 0.15$).

3-6 THE DECAY OF AN ISOTROPIC TURBULENCE

The first rather extensive measurements of the decay of turbulence generated by grids were made by Batchelor and Townsend.[7,21,22] From the results of these measurements it follows that different stages or periods of the turbulent flow may be distinguished downstream of the grid. First there is the building-up period or period of establishment of the fully developed, almost homogeneous turbulence. It covers a distance to roughly 20 M from the grid. Depending on the position of the hot-wire anemometer with respect to the rods of the grid the intensity of the turbulence increases more or less strongly to a maximum after which the decay of the turbulence begins. Batchelor and Townsend[21] distinguished in this decay process between an *initial period*, a *transition period*, and a *final period*.

How long the initial period will last is difficult to say, since the transition to other periods is very gradual, and since, moreover, it would depend strongly on the initial value of the Reynolds number. The above-mentioned experimental data by Townsend seem to show that the turbulence may be considered to be in the initial period up to $x_1/M = 100$ to 150, but data obtained by other experimenters have shown that, even at shorter distances, substantial departures from the initial-period decay law may occur. We shall come back to this point later on.

Townsend's experiments have shown that the final period seems to apply to distances greater than 500 M. Of course, this value too should depend on the initial Reynolds number of turbulence. In Townsend's experiments the Reynolds number $\mathbf{Re}_M = \bar{U}_1 M/\nu$ was about 650.

In the initial period the decay, being determined predominantly by the decay of the energy-containing eddies, is caused by the drain of energy from these eddies through inertial interactions towards the small eddies in the dissipation wavenumber range. In the final period the viscous effects dominate even the inertial effects throughout the whole wavenumber range. So in this final period the inertial terms in the dynamic equations may be neglected. We have shown that an exact solution of these equations can then be obtained. They led to a Gaussian function for the double correlation coefficient [Eq. (3-45)]

$$\mathbf{f}(r,t) = \exp(-r^2/8\nu t)$$

and to a similar equation for the spectrum function [Eq. (3-77)]

$$\mathbf{E}(k,t) = \text{const.} \times k^4 \exp(-2\nu k^2 t)$$

The decay law for the turbulence intensity was given by [Eq. (3-43)]

$$u'^2 = \text{const.} \times t^{-5/2}$$

We have shown in Sec. 3-4 that the length scales λ_g and Λ_f vary proportionally with $\sqrt{\nu t}$ and that their ratio remains constant. The Reynolds number of turbulence $\mathbf{Re}_\lambda = u'\lambda_g/\nu$ does not remain constant during decay but varies proportionally with $t^{-3/4}$.

The solutions (3-45) and (3-77) show that the shape of the $\mathbf{f}(r,t)$ and the $\mathbf{E}(k,t)/\mathbf{E}(k,0)$ curves remains similar during decay when r and k are expressed in terms of the length unit $\sqrt{\nu t}$, because $\mathbf{f}(r,t)$ and $\mathbf{E}(k,t)/\mathbf{E}(k,0)$ are functions solely of $r/\sqrt{\nu t}$ and $k\sqrt{\nu t}$, respectively. That λ_g and Λ_f vary proportionally with $\sqrt{\nu t}$ and thus have a constant ratio is a consequence of this similarity of behavior. The maximum value of $\mathbf{E}(k,t)$ is obtained at $k^2\nu t = 1$, or $k = 1/\sqrt{\nu t}$. Furthermore, Loitsianskii's invariant condition (3-39) holds exactly, since with $\sqrt{\nu t}$ as a length unit, this condition yields

$$u'^2(\sqrt{\nu t})^5 = \text{const.}$$

which is precisely the decay law (3-43).

If they are not expressed in terms of the length unit $\sqrt{\nu t}$, the correlation- and energy-spectrum curves do not, of course, remain similar. Thus in absolute terms $\mathbf{E}(k,t)$

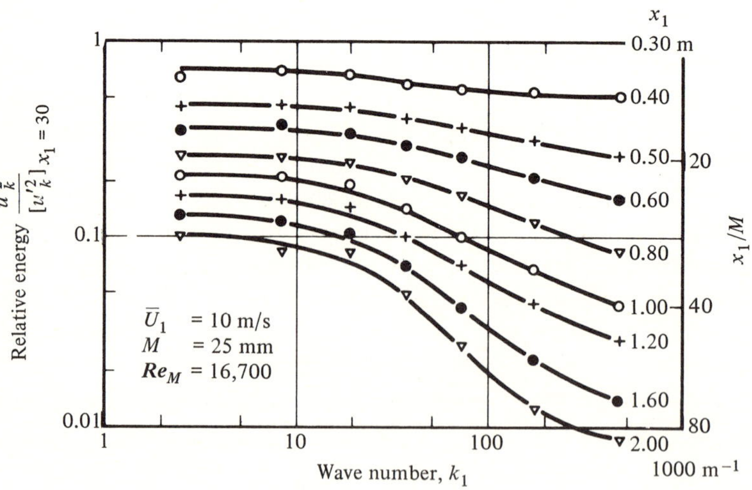

FIGURE 3-21
Decay of energy of the spectral components. (From: *Sato, H.*,[19] *by permission of the Physical Society of Japan.*)

changes more rapidly as the value of the wavenumber k increases; that is, as mentioned earlier, the smaller eddies decay at a higher rate.

This phenomenon of increase in the rate of decay with increase in the wavenumber is not restricted to the case just considered of small Reynolds number of turbulence, but is a general feature of turbulence, as may be inferred from the general dynamic equation (3-87). This is clearly shown by Sato,[19] who measured, at various distances from a grid, the energy $\overline{u_k^2}$ associated with the wavenumber k relative to the value at some arbitrary initial condition [here at a distance 300 mm downstream of the grid (see Fig. 3-21)]. If the decay rate is expressed in terms of a power law against the distance from the grid, the high-wavenumber range reaches a power $-\frac{5}{2}$ in contrast to a power -1 for the total kinetic energy of turbulence. Thus, in the higher-wavenumber range, the "viscous" decay rate (3-43) is reached much earlier. The energy spectrum changes with time in such a way that in the region $k \to 0$ the spectrum becomes more permanent as time increases, but in the higher-wavenumber range the decay rate increases as k increases. Figure 3-22 shows a picture of how the energy spectrum may change with time.

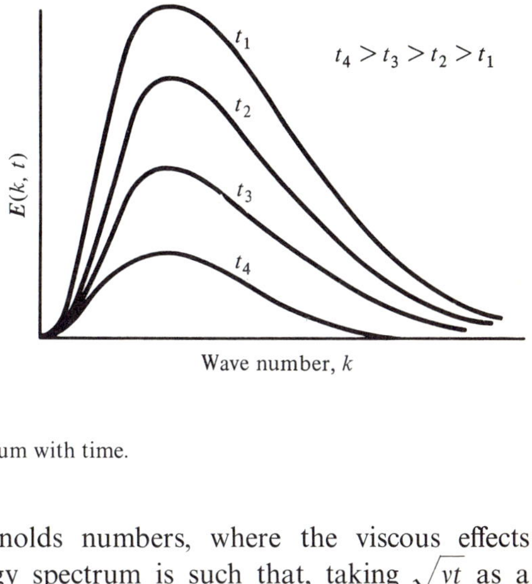

FIGURE 3-22
Change of the energy spectrum with time.

In the case of very small Reynolds numbers, where the viscous effects predominate, the change of the energy spectrum is such that, taking $\sqrt{\nu t}$ as a characteristic length, the energy spectrum as a function of $k\sqrt{\nu t}$ remains similar. It is reasonable to see whether, for larger Reynolds numbers as well, the energy spectrum will remain similar or self-preserving when a suitable characteristic length is chosen. However, we have shown in the preceding section that for different wavenumber ranges different characteristic lengths occur (see Table 3-1). For the equilibrium range we have $\eta = (\nu^3/\varepsilon)^{1/4}$, and the same characteristic length may be taken for the range of the energy-containing eddies, though, for this latter range, the length $\sqrt{\nu t}$ may also be taken, since the relation $\varepsilon t^2/\nu = \text{const.}$ holds, provided $\mathbf{Re}_\lambda$ is not very large. Then we have for the lower-wavenumber ranges the characteristic lengths $(\epsilon^3/\varepsilon)^{1/4}$ and $(I/\epsilon^2)^{1/3}$. Hence it appears to be impossible to take one single characteristic length for the entire wavenumber range to make the energy spectrum self-preserving during decay. It is possible to assume only incomplete self-preservation, that is, self-preservation of only a part of the energy spectrum or of the correlation function. Fortunately, it is possible to assume self-preservation for a substantial part of these functions. According to Table 3-1, we may do so for the wavenumber range of the energy-containing eddies plus the absolute equilibrium range, since they have a common characteristic length $\eta = (v^3/\varepsilon)^{1/4}$.

However, we know that the phenomena in the wavenumber range of the energy-containing eddies are not the same as in the higher-wavenumber range. This is expressed, among other things, by the fact that for the first range the condition $\varepsilon t^2/v = \text{const.}$ should apply, whereas this need not apply with respect to the equilibrium range. Therefore, in order to apply the assumption of self-preservation for the wavenumber range including a major part of that of the energy-containing eddies, Heisenberg[15] made the further assumption that these eddies would be in quasi-equilibrium, so that we may consider them as if they were in equilibrium as far as permissible in view of their finite rate of decay.

The assumption of self-preservation in the region of equilibrium range plus a major part of the range of the energy-containing eddies is supported by the results of measurements made by Stewart and Townsend[17] on the energy spectrum at different stages of decay. Figure 3-23 shows these results. Here, the dissipation scale λ_g is taken instead of η as the characteristic length. The relation between η and λ_g is given by Eq. (3-112). Furthermore, according to Table 3-1 we should have

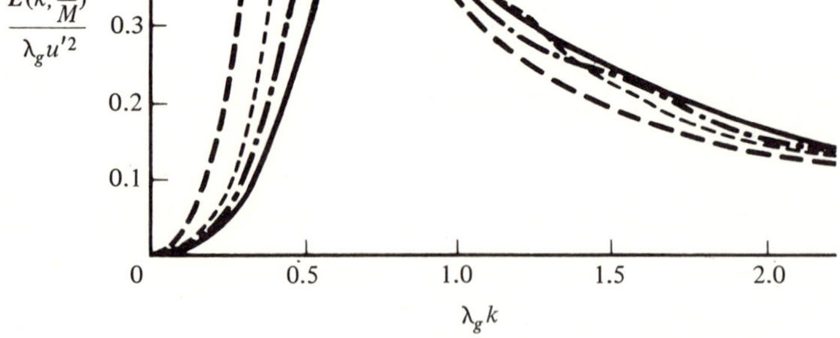

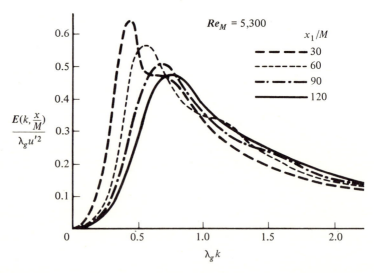

FIGURE 3-23
Energy spectrum measured at various stages of decay. (From: *Stewart, R. W., and A. A. Townsend,*[17] *by permission of the Royal Society.*)

$$\mathbf{E}(k,t) = \nu^{5/4}\varepsilon^{1/4}\mathbf{E}^*\left[k\left(\frac{\nu^3}{\varepsilon}\right)^{1/4}\right] = \eta\upsilon^2\mathbf{E}^*(k\eta)$$

In Fig. 3-23 $\mathbf{E}(k,t)$ has been normalized not by $\eta\upsilon^2$ but by $\lambda_g u'^2$, the relation between u' and υ being given by Eq. (3-111).

It is obvious and, moreover, clearly shown by Fig. 3-23 that no self-preservation obtains in the low-wavenumber range. On the other hand it is possible to assume self-preservation in the low-wavenumber range, by taking, for instance, $(I/\epsilon^2)^{1/3}$ as a characteristic length. But then no self-preservation obtains in the high-wavenumber range. The early measurements on decay were concerned with the turbulence intensity u', which was measured at different distances from a grid.

From Eq. (3-35) or Eq. (3-37), by putting $r = 0$, or from Eq. (3-97) it follows that

$$\frac{d}{dt}u'^2 = -10\nu\frac{u'^2}{\lambda_g^2} \qquad (3\text{-}162)$$

Since λ_g is also a function of time, this relation alone is not sufficient. An additional relation may be obtained by some suitable assumption.

The first assumption that we shall consider is that the decay of the total energy is determined predominantly by that of the energy-containing eddies. In that case we have the additional relation $\varepsilon t^2/\nu = \text{const.}$, that is,

$$\varepsilon = \mathscr{R}\nu t^{-2} \qquad (3.163)$$

Since $\varepsilon = 15\nu u'^2/\lambda_g^2$, it follows from Eqs. (3-162) and (3-163) that

$$\frac{du'^2}{dt} = -\tfrac{2}{3}\mathscr{R}\nu t^{-2}$$

whence, after integration,

$$u'^2 = \tfrac{2}{3}\mathscr{R}\nu\frac{1}{t} + B \qquad (3\text{-}164)$$

For the dissipation scale λ_g we then obtain from Eqs. (3-162) and (3-164)

$$\lambda_g^2 = 10\nu t\left(1 + \frac{3}{2}\frac{Bt}{\nu\mathscr{R}}\right) \qquad (3\text{-}165)$$

and for the Reynolds number $\mathbf{Re}_\lambda$

$$\mathbf{Re}_\lambda = \left(\tfrac{20}{3}\mathscr{R} + \frac{20B}{\nu}t + 15\frac{B^2}{\nu^2\mathscr{R}}t^2\right)^{1/2} \qquad (3\text{-}166)$$

or

$$\mathbf{Re}_\lambda \simeq \sqrt{\tfrac{20}{3}\mathscr{R}}\left(1 + \frac{3}{2}\frac{B}{\nu\mathscr{R}}t\right) \qquad (3\text{-}166a)$$

when $\frac{3}{2}(B/\nu\mathscr{R})t \ll 1$.

It is obvious that, if the constant $B \neq 0$, the solution (3-164) can be true at the most only for a finite time interval, since eventually all turbulence should have decayed to zero when $t = \infty$ and since, moreover, the relation (3-163) does not hold for the final period of decay.

If $B > 0$, then it might be interpreted as the energy of the permanent large eddies. But it would then follow from Eq. (3-165) that λ_g^2 would increase more than linearly, and, according to Eq. (3-166), $\mathbf{Re}_\lambda$ would increase with time. So far as experimental evidence is available, it shows the contrary: if there is a departure from linearity for λ_g it occurs in the opposite direction (eventually λ_g^2 must become equal to $4\nu t$ in the final decay period), and if there is a change in $\mathbf{Re}_\lambda$ with time $\mathbf{Re}_\lambda$ decreases (eventually $\mathbf{Re}_\lambda$ varies proportionally with $t^{-3/4}$ in the final decay period). Thus, if $B \neq 0$, it should be negative; we then arrive at the decay relation proposed by Lin.[23,12] According to Von Kármán and Lin,[12] the negative constant may be interpreted as the negative departure in the low-wavenumber range of the energy spectrum from the curve representing complete self-preservation.

If we assume $B = 0$ then it follows that

$$u'^2 = \tfrac{2}{3}\nu\mathcal{R}t^{-1} \qquad \lambda_g^2 = 10\nu t$$

and

$$\mathbf{Re}_\lambda = \sqrt{\tfrac{20}{3}\mathcal{R}} \quad \text{or} \quad \frac{\varepsilon t^2}{\nu} = 0.15\,\mathbf{Re}_\lambda^2$$

The assumption that the decay of the total turbulence is determined mainly by the decay of the energy-containing eddies has obtained some support from experimental evidence given by Batchelor.[11] According to Eq. (3-104) [which is consistent with Eq. (3-109)], $(l_e/u'^3)(du'^2/dt)$ should be constant during decay; this appears to be roughly the case in the experiments considered by Batchelor.

Another assumption is that u'^2 decays according to some simple power law

$$u'^2 = Ct^{-n}, \quad \text{or} \quad \frac{u'^2}{\bar{U}_1^2} = (At + B)^{-n} \qquad (3\text{-}167)$$

It then follows from Eq. (3-162) that

$$\lambda_g^2 = \frac{10}{n}\nu t, \quad \text{or} \quad \lambda_g^2 = \frac{10}{n}\nu\left(t + \frac{B}{A}\right) \qquad (3\text{-}168)$$

and

$$\mathbf{Re}_\lambda = \sqrt{\frac{10C}{n\nu}}\,t^{(1-n)/2}, \quad \text{or} \quad \mathbf{Re}_\lambda \propto (At + B)^{(1-n)/2} \qquad (3\text{-}169)$$

Since u'^2 and λ_g^2 are measured independently, we have two independent methods of determining the value of n.

We shall presently show that the decay law with $n = 1$ in Eq. (3-167) may correspond with complete preservation of the turbulence during decay. For, let us introduce the dimensionless quantity $\psi = r/\mathscr{L}$ in the dynamic Kármán-Howarth equation (3-36). The length scale $\mathscr{L}$ is a function of time only. In the case of complete similarity $\mathbf{f}(r,t)$ and $\mathbf{k}(r,t)$ are functions of ψ alone. We then obtain from Eq. (3-36):

$$\mathbf{f}\frac{du'^2}{dt} - u'^2\frac{1}{\mathscr{L}}\frac{d\mathscr{L}}{dt}\psi\frac{d\mathbf{f}}{d\psi} - \frac{u'^3}{\mathscr{L}}\frac{1}{\psi^4}\frac{d}{d\psi}(\psi^4\mathbf{k}) - 2\nu\frac{u'^2}{\mathscr{L}^2}\frac{1}{\psi^4}\frac{d}{d\psi}\left(\psi^4\frac{d\mathbf{f}}{d\psi}\right) = 0$$

Or, with Eq. (3-162), and after multiplying all terms by $\lambda_g^2/\nu u'^2$:

$$10\mathbf{f} + \frac{\lambda_g^2}{\nu}\frac{1}{\mathscr{L}}\frac{d\mathscr{L}}{dt}\psi\frac{d\mathbf{f}}{d\psi} + \frac{u'}{\nu}\frac{\lambda_g^2}{\mathscr{L}}\frac{1}{\psi^4}\frac{d}{d\psi}(\psi^4\mathbf{k}) + 2\frac{\lambda_g^2}{\mathscr{L}^2}\frac{1}{\psi^4}\frac{d}{d\psi}\left(\psi^4\frac{d\mathbf{f}}{d\psi}\right) = 0$$

This ordinary differential equation for $\mathbf{f}$ as a function of ψ has coefficients which are functions of t alone. Since the solution for $\mathbf{f}$ must be independent of t, a solution that satisfies this condition can only be obtained when all the coefficients are proportional to one another. Consequently,

$$\frac{\lambda_g}{\mathscr{L}} = \text{const.} \tag{3-170}$$

$$\frac{\lambda_g^2}{\mathscr{L}}\cdot\frac{u'}{\nu} = \text{const.,} \quad \text{or} \quad \frac{u'\mathscr{L}}{\nu} = \text{const.} \tag{3-171}$$

and

$$\frac{\lambda_g^2}{\nu}\frac{1}{\mathscr{L}}\frac{d\mathscr{L}}{dt} = \text{const.,} \quad \text{or} \quad \mathscr{L}\frac{d\mathscr{L}}{dt} = \text{const.} \tag{3-172}$$

Whence follows

$$\lambda_g \propto \mathscr{L} \propto t^{1/2} \quad \text{and} \quad u'^2 \propto t^{-1}$$

We may extend the assumption of complete self-preservation to Loitsianskii's integral Eq. (3-39) and to Saffman's dynamic invariant Eq. (3-94). They yield additional conditions, namely

$$u'^2\mathscr{L}^5\frac{1}{3\pi}\int_0^\infty d\psi\,\psi^4\mathbf{f}(\psi) = I \tag{3-173}$$

and, respectively

$$u'^2\mathscr{L}^3\frac{1}{\pi}\lim_{\psi\to\infty}\psi^3\mathbf{f}(\psi) = I_1 \tag{3-174}$$

When Saffman's dynamic invariant is accepted to be true, so that I_1 is constant, then this additional relation between u' and $\mathscr{L}$ conflicts with a constant value of the Reynolds number during decay as a result of complete self-preservation. Also, when Loitsianskii's integral is assumed to be invariant, a similar conflict arises.

The only possibility to get rid of these conflicting conditions, and if we want to stick to the additional invariant conditions, is to drop one of the other conditions.

For instance we may assume such large values of the Reynolds number that the direct effect of viscosity on the energy-containing eddies is negligibly small, so that the viscous term in Eq. (3-36) may be neglected. Thus we drop the condition $\lambda_g/\mathscr{L} = \text{const.}$, or the condition of complete self-preservation of the correlation and spectrum functions.

We have mentioned that it seems possible to assume self-preservation of the spectrum function in the wavenumber range $k \geq k_e$, which corresponds to the assumption of self-preservation of the correlation function when $r < r_1$, or to assume self-preservation of the spectrum function in the wavenumber range $k < k_e$, which corresponds to the assumption of self-preservation of the correlation function when $r > r_1$.

Experimental studies by Tsuji[51] on the spectrum of turbulence behind two grids of different mesh placed in series have shown that self-preservation of the energy spectrum holds only for the high-wavenumber range and that the low-wavenumber characteristics of the turbulence depend on the initial conditions and show no similarity. Any superimposed disturbances in the high-wavenumber range that would impair the self-preservation of the spectrum decay rapidly.

If self-preservation of the spectrum function in the wavenumber range $k > k_e$ is to be possible, the Reynolds number should be large, so that there is a substantial range of wavenumbers where the eddies are in statistical equilibrium. We might then choose $\eta = (\nu^3/\varepsilon)^{1/4}$ as the characteristic length. Or, we may take the dissipation scale λ_g instead. Doing this, so taking $\mathscr{L} = \lambda_g$, we obtain from Eqs. (3-172) and (3-171)

$$\frac{\lambda_g}{\nu}\frac{d\lambda_g}{dt} = \text{const.} \qquad \text{(3-175)}$$

and

$$\frac{u'\lambda_g}{\nu} = \mathbf{Re}_\lambda = \text{const.} \qquad \text{(3-176)}$$

Equations (3-162) and (3-175) can be solved for $\lambda_g{}^2$ and u'^2. The solutions, subject to the condition (3-176), read

$$\lambda_g{}^2 = 10\nu t + C = 10\nu(t - t_0) \qquad \text{(3-177)}$$

$$u'^2 = \frac{\text{const.}}{C + 10\nu t} = \frac{\text{const.}}{10\nu(t - t_0)} \qquad \text{(3-178)}$$

if the constant C is taken equal to $-10\nu t_0$.

These solutions are consistent with Eqs. (3-167) to (3-169) provided that $n = 1$, and are consistent with Eqs. (3-164) to (3-166) provided that $B = 0$.

Of course the above decay relations for $\lambda_g{}^2$ and u'^2 also hold in the case of complete self-preservation when $\mathscr{L} = \lambda_g$. But the invariance relations cannot be applied. On the other hand, since in the invariance relations higher moments of $\mathbf{f}(\psi)$ are taken, it may be reasonable to extend the self-preservation assumption to including the wavenumber range of the energy-containing eddies. Taking, for instance, in these invariance relations $\mathscr{L} = \Lambda_g$, the integral length scale, one obtains

$$\text{from } I = \text{const.}: \quad \Lambda_g \propto (t - t_0)^{1/5} \qquad (3\text{-}179)$$

$$\text{from } I_1 = \text{const.}: \quad \Lambda_g \propto (t - t_0)^{1/3} \qquad (3\text{-}180)$$

These results are consistent, however, only with a condition of partial self-preservation.

What would be the result if we also take in the relations (3-171) and (3-172) $\mathscr{L} = \Lambda_g$ together with the invariance relations.

With Saffman's invariance relation we obtain:

$$\frac{\lambda_g^2}{\Lambda_g}\frac{u'}{\nu} = \text{const.} \qquad (3\text{-}181)$$

$$\frac{\lambda_g^2}{\nu\Lambda_g}\frac{d\Lambda_g}{dt} = \text{const.} \qquad (3\text{-}182)$$

$$u'^2\Lambda_g^3 = \text{const.} \qquad (3\text{-}183)$$

Whence follows

$$u'^2/\bar{U}_1^2 = (At + B)^{-6/5} \qquad (3\text{-}184)$$

$$\lambda_g^2 = \frac{25}{3}\nu\left(t + \frac{B}{A}\right) \qquad (3\text{-}185)$$

$$\Lambda_g \propto (At + B)^{2/5} \qquad (3\text{-}186)$$

With Loitsianskii's invariance relation, which gives

$$u'^2\Lambda_g^5 = \text{const.} \qquad (3\text{-}187)$$

we obtain

$$u'^2/\bar{U}_1^2 = (At + B)^{-10/7} \qquad (3\text{-}188)$$

$$\lambda_g^2 = 7\nu\left(t + \frac{B}{A}\right) \qquad (3\text{-}189)$$

$$\Lambda_g \propto (At + B)^{2/7} \qquad (3\text{-}190)$$

In both cases we obtain a rate of decay for u'^2 which is greater than according to the (-1) law. So, either there is an initial period with a (-1) decay law and the above faster decay occurs after the initial period, or there is no (-1) decay law at all and in the initial period we have the $(-\frac{6}{5})$ law, or the $(-\frac{10}{7})$ law.

This latter, $(-\frac{10}{7})$ law, has already been obtained by Kolmogoroff.[98] Now this solution for the rate of decay may be ruled out, and it has been so ruled out several times, because of the fundamental objection of Loitsianskii's integral not being a dynamic invariant. However, as has been mentioned earlier, this integral is only a weak function of the time, even at high Reynolds numbers. So it is possible for the time scale of the change of Loitsianskii's integral to be much larger than that of the decay process itself. Hence, measured in the time scale of the decay process, the Loitsianskii integral may be considered, approximately, as a dynamic invariant.

Since, as will be shown later, the decay rate of grid-generated turbulence very seldom shows the value $n = -1$, but a higher absolute value of this exponent, Comte-Bellot and Corrsin[89] reconsidered the applicability of Loitsianskii's "invariant." Instead of this invariant they made the equivalent assumption that the expression (3-121) for $\mathbf{E}(k,t)$ in the lowest wavenumber range is independent of time. Assuming further that this relation for $\mathbf{E}(k,t)$ holds true in the whole range $k \leq k_e$, and that for $k \geq k_e$ Eq. (3-123) may be used, the two relations matching at $k = k_e$, Comte-Bellot and Corrsin obtained with Eq. (3-58) and Eq. (3-105) three relations, from which k_e and $\varepsilon(t)$ could be eliminated. So they obtained the decay law Eq. (3-188), without making any assumption about similarity. On the other hand the same procedure, without making similarity assumptions, can be applied by taking $\mathbf{E}(k,t) = I_1 k^2$ independent of time for $k \leq k_e$, and then arriving at the decay law Eq. (3-184).

What experimental results are available on the decay of an isotropic turbulence? In the course of forty years various investigators have obtained a large amount of data on turbulence produced by grids. Unfortunately, the experiments were not all carried out under the same conditions, in particular with respect to the shape of the grids used. Grids with single and double rows of round and square bars, flat strips, etc., and of different mesh, bar diameter, or bar width, have been used.

From dimensional arguments it may be expected that the turbulence will depend on the relative distance x_1/M, on the mesh-bar width ratio M/d, or its free-area ratio φ (projected free area per unit of total area) on the shape of the rods or slots, and on whether single- or square-mesh grids are taken. It is probable that the effect of rod shape will be less pronounced than that of the other parameters.

It is logical to relate the characteristics of the turbulence to those of the turbulence-producing grid. The grid characteristics comprise the geometry of the grid and its specific drag, since it is the work done by the fluid against this drag that creates the turbulence energy. Hence Batchelor and Townsend[21] proposed the following relation based upon the "linear" decay law:

$$\frac{\bar{U}_1{}^2}{u_1'^2} = \frac{c}{C_D}\left[\frac{x_1}{M} - \left(\frac{x_1}{M}\right)_0\right] \tag{3-191}$$

where C_D is the drag per unit area of the grid and c is a constant depending on the grid geometry. From many measurements on C_D for square-mesh grids of round bars, the following relation has been obtained:

$$C_D = \frac{(d/M)(2 - d/M)}{(1 - d/M)^4}$$

and $c \simeq 106$.

For the square-mesh grids usually applied, that is, grids with $M/d \simeq 5$ to 6, we obtain $c/C_D \simeq 135$.

It may be noted that, at high values of M/d, the length parameter becomes d instead of M.

As mentioned earlier the "linear" decay law has practically never been observed in windtunnel experiments. Only the large Reynolds number experiments by Kistler and Vrebalovich did show a linear decay law. However, the value of c/C_D deduced from their measurements of $u_1'/\bar{U}_1$, turned out to be much smaller than 135, namely ~ 77. Also the drag coefficient C_D of the grid was much lower than the value calculated from the above formula.

We also mentioned that grid-generated turbulence usually becomes practically homogeneous only when $x_1/M > 10$ to 15. The turbulence then is axi-symmetric rather than isotropic, with a ratio u_1'/u_2' varying between 1.05 and 1.35 depending on conditions. One might think that this anisotropy would mainly apply to the bigger, energy-containing eddies, and that isotropy would be approached in the more fine-scale structure of the turbulence. However, some measurements by Uberoi[88] of the terms $\overline{(\partial u_1/\partial x_1)^2}$ and $\overline{(\partial u_2/\partial x_1)^2}$, occurring in the expression for the viscous dissipation, did show noticeable departures from the isotropic relations, up to 40 per cent. So, one has to be very cautious when results of measurements in grid-generated turbulence are to be compared with theories derived for a homogeneous, isotropic turbulence. With this in mind, let us yet consider experiments with grid-generated turbulence.

In the first place, because of the required homogeneity it is safe to consider, for the grids usually applied, at least only measurements with $x_1/M > 20$. The degree of homogeneity, specifically over surfaces normal to the mean flow, should be very high. So the grids must be very carefully manufactured, and of sufficient rigidity in order to prevent any vibration of the bars or strips.[99] Grant and Nisbet[53] reported considerable departures from homogeneity up to $x_1/M = 80$.

But even if the grid itself is sufficiently uniform, still inhomogeneities in the lateral direction of the flow may occur due to "instability" of the "jets-wakes" flow pattern. Due to mutual influence a number of jets from a group of neighboring openings of the grid may form a concentrated bundle, so causing locally a higher velocity than the mean, uniform, value. Investigations by Morgan[100] and Bradshaw[101] with a number of screens have shown that this "instability" may occur when the free-area ratio is within the limits $0.2 \gtrsim \varphi \gtrsim 0.57$ to 0.63. For a square-mesh grid and round bars the upper limit would correspond with $M/d < 4$ to 5. Most of the grids used for turbulence investigations have a $M/d \simeq 5$, so they may be just on the safe side.

Many *experimental results of the decay* of grid-produced turbulence are available. To mention a number: Dryden and coworkers,[26] Hall,[27] Batchelor and Townsend,[7,11,21,22] Baines and Peterson,[28] Sato,[19] Tsuji,[51] Grant and Nisbet,[53] Van der Hegge Zijnen,[54] Uberoi,[88] Comte-Bellot and Corrsin,[89] Kistler and

Vrebalovich,[92] and Tan and Ling.[118] Most of the turbulence intensity measurements concern the axial component $u_1'/\bar{U}_1$, and only a few also the lateral component, e.g. $u_2'/\bar{U}_1$ [Uberoi,[88] Comte-Bellot, and Corrsin,[89] Kistler and Vrebalovich[92]]. Some investigators have also determined the dissipation length scale λ_f or λ_g, and only a very few the integral length scale Λ_f or Λ_g. With the exception of those by Kistler and Vrebalovich, all measurements have been made at low values of the Reynolds number ($\mathbf{Re}_\lambda \gtrsim 50$).

It is now of interest to see whether one of the decay laws discussed can be observed in the actual decay curves. These actual decay curves show a very gradual course and, if an attempt is made to describe the decay by means of the simple function (3-167), also a gradual change in the exponent n. Moreover, the amount of experimental data available for each curve is in most cases too small for reliable and correct fitting of the curve by statistical means. Yet we may try to approximate parts of the decay curve more or less closely by one of the theoretical decay laws.

Since according to all the decay laws with a constant value of n, λ_g^2 varies linearly with x_1/M, of more importance are the results of the decay measurements of the turbulence intensity and of the integral scale. Though the slope of the $\lambda_g^2 - x_1$ line is directly related to the value of n [see Eq. (3-168)]. It will suffice here to show in Fig. 3-24 the results of Batchelor and Townsend's measurements. From the slope of the straight lines one would conclude a value of $n = 1$, though also a value of $n = 1.2$ could be deduced from the data taken at $\mathbf{Re}_M = 11{,}000$. Figure 3-25a and b show the decay of $u_1'/\bar{U}_1$ as measured by Batchelor and Townsend, and by Baines and Peterson[28] respectively. There too, one might conclude to a value of $n \simeq 1$ for

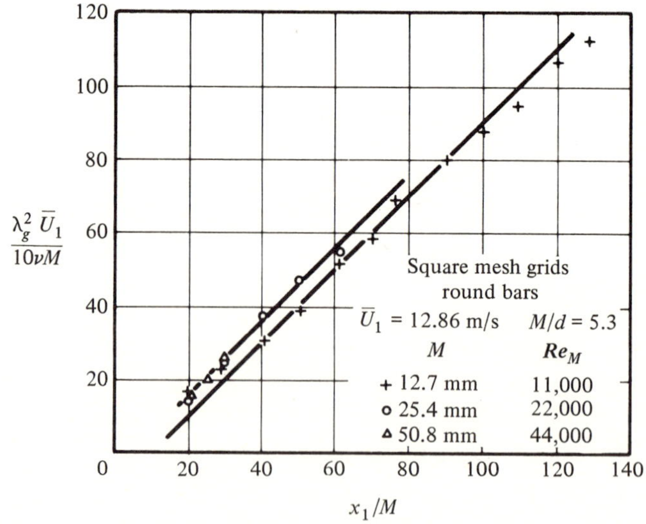

FIGURE 3-24
Increase of λ_g with downstream distance from the grid. (From: *Batchelor, G. K., and A. A. Townsend,*[21] *by permission of the Royal Society.*)

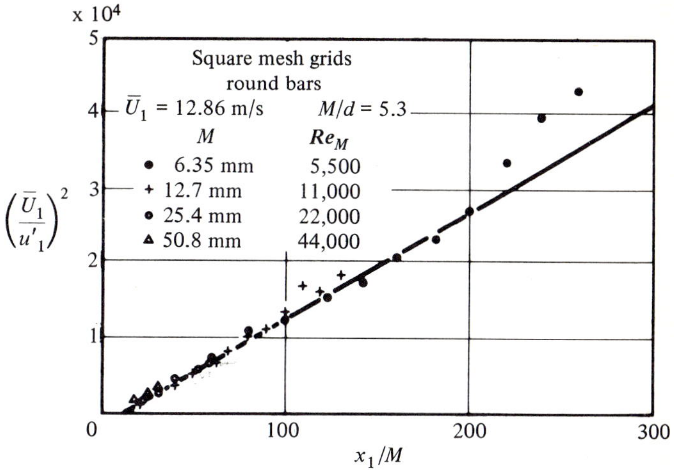

FIGURE 3-25a
Decay of $u'_1/\bar{U}_1$ downstream of grids. (From: *Batchelor, G. K., and A. A. Townsend,*[21] *by permission of the Royal Society.*)

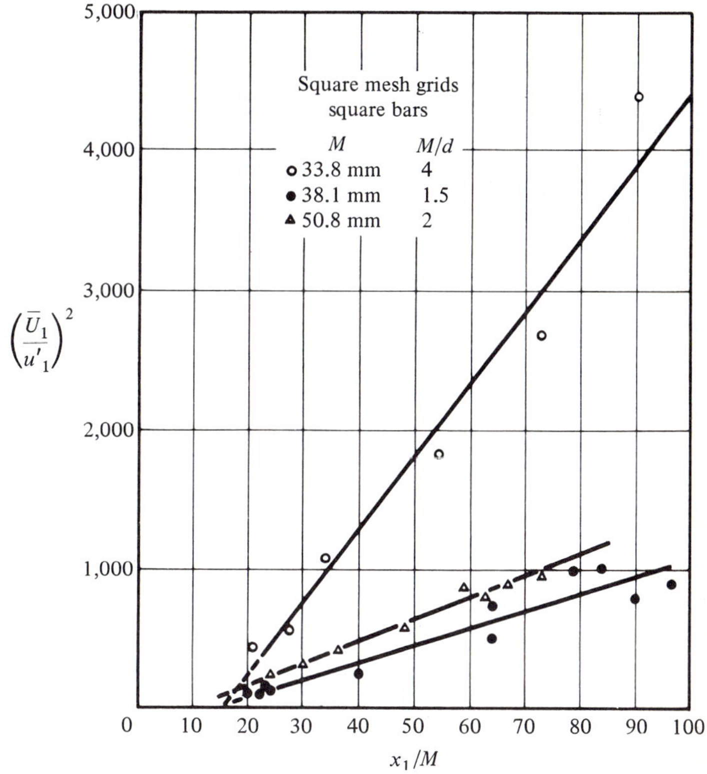

FIGURE 3-25b
Decay of $u'_1/\bar{U}_1$ downstream of grids. (From: *Baines, W. D., and E. G. Peterson.*[28])

the initial period, if one restricts this period to the region $x_1/M < 100$ to 150. However, a reconsideration of Batchelor and Townsend's data including a larger region may lead to a value of 1.25 as well, which value agrees reasonably well with the above value of 1.2 deduced from the $(\lambda_g{}^2 - x_1)$ relation. The data obtained by Baines and Peterson, in particular those at the low value of $M/d = 1.5$, show a lot of scatter. At this low value of M/d the grid has potentially an unfavorable geometry in connection with possible flow instabilities, mentioned above.

When we write the decay relation for the turbulence intensity in the form of Eq. (3-167), then, as mentioned earlier, values of n varying between 1.2 and 1.35 are obtained by most of the investigators, with an overall average value of $\simeq 1.25$. This applies also to the decay of the lateral turbulence velocity component. And also in those cases where, through a slight contraction of the windtunnel, the anisotropy had been reduced to almost isotropy (Uberoi,[88] Comte-Bellot and Corrsin[89]). It is only from the high Reynolds number experiments by Kistler and Vrebalovich[92] that a distinct value of $n = 1$ could be obtained as shown in Fig. 3-26, though it must be noted that their measurements were made in a rather limited region $x_1/M < 60$.

When we write the decay law for the integral scale in a similar way, namely:

$$\frac{\Lambda_g}{M} = (At + B)^m$$

a large variation in the value of A is found, and in the decay rate given by m, though to a smaller degree. The measurement of Λ_g appears not to be a definite

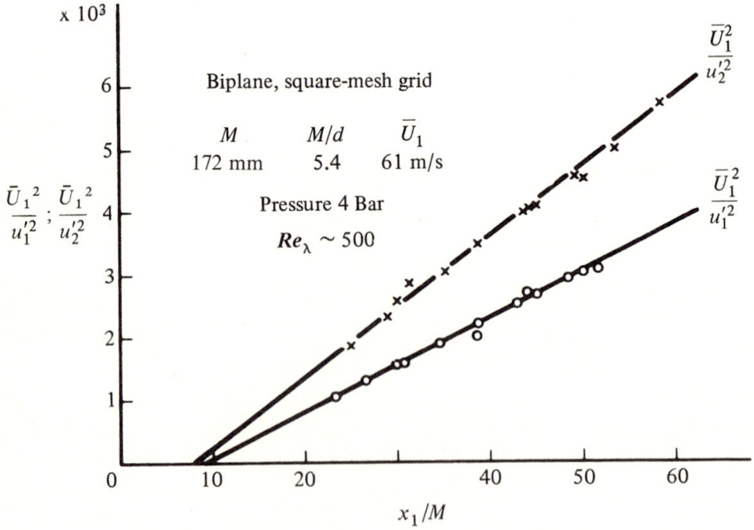

FIGURE 3-26
Decay of $u_1'/\bar{U}_1$ and $u_2'/\bar{U}_1$ downstream of a grid. (From: *Kistler, A. L., and T. Vrebalovich,*[92] *by permission of the Cambridge University Press.*)

procedure, producing consistent results. For, as discussed in Chap. 2, there are several methods for determining the integral scale using hot-wire anemometry. Λ_f and Λ_g can be obtained by an integration of the measured correlation $\mathbf{f}(r)$ or $\mathbf{g}(r)$. Or, the Λ_f can be obtained from the integral time scale $\mathfrak{I}_E$. Or, by measuring the one-dimensional spectrum and making the extrapolation to $k_1 = 0$. The experience is that even for the same experimental conditions it is very difficult to obtain the same value of Λ with the three methods.† This may explain why different experimenters, using different methods and not always under identical test conditions obtain different values of the integral scale. But fortunately the differences obtained in the value of m are less marked. The measurements by Batchelor and Townsend, Uberoi and Corrsin,[102] Mills, Kistler, O'Brien, and Corrsin[103] and by Comte-Bellot and Corrsin reveal a variation $m = 0.30$ to 0.53, with an overall average value of 0.4.

A definite conclusion can still not be drawn concerning which decay law and corresponding theory is supported by the experimental evidence. But the average values of $n \simeq 1.25$ and $m \simeq 0.4$ are very close to the theoretical values based on Saffman's dynamic invariant and the assumption of partial self-preservation, though the applicability of this dynamic invariant to grid-produced turbulence has been doubted.

One of the difficulties in assessing a decay law might be due to the fact that one is inclined to describe the decay by means of a power law with a constant value of n during too long a period.

A different way of describing the decay has been suggested by Naudascher and Farell.[119] From Eqs. (3-167) and (3-168) we obtain, putting $B/A = -t_0$

$$\frac{u'}{\bar{U}_1} = C\left(\frac{t}{t_0} - 1\right)^{-n/2} \qquad \text{(3-167a)}$$

and

$$\frac{\lambda_g^2}{\nu t_0} = \frac{10}{n}\left(\frac{t}{t_0} - 1\right) \qquad \text{(3-168a)}$$

Whence follows the relation:

$$\frac{d\lambda_g}{dt} = \sqrt{\frac{5\nu}{2nt_0}} \cdot \left(\frac{1}{C}\frac{u'}{\bar{U}_1}\right)^{1/n} \qquad \text{(3-192)}$$

We also have the decay relation (3-162):

$$\frac{du'}{dt} = -5\nu\frac{u'}{\lambda_g^2} \qquad \text{(3-162)}$$

Now, instead of the two relations (3-167a) and (3-168a) [or (3-162)] we consider the relations (3-192)

† The neglect of the lowest frequency or wavenumber range caused by a relatively too high cut-off value has an important effect on the measurement of the correlations and consequently on the integral scale. Under laboratory conditions a cut-off frequency lower than 0.05 Hz, say, should be striven after. The procedure using an extrapolation of $\mathbf{E}_1(k_1)$ to $k_1 \to 0$, in general appears to be more reliable, provided the cut-off frequency is sufficiently low to warrant a safe extrapolation.

and (3-162) from which u' and λ_g can be solved. Elimination of u' yields the following differential equation for λ_g:

$$\frac{d^2\lambda_g}{dt^2} + \frac{5\nu}{n\lambda_g^2}\frac{d\lambda_g}{dt} = 0$$

It has tacitly been assumed that n is constant during at least a certain finite interval of the time.

Integrating the differential equation once, yields

$$\frac{1}{\bar{U}_1}\frac{d\lambda_g}{dt} = \frac{5\nu}{n\bar{U}_1}\frac{1}{\lambda_g} + A \qquad (3\text{-}193)$$

When $A = 0$ we obtain again the relation (3-168a) for λ_g^2. However, for $A \neq 0$ another solution for λ_g is possible. An interesting idea is, what decay law would be obtained when we assume the relation (3-192) to hold true during the entire decay process, with a constant value of n? Naudascher and Farell have pursued this idea for the value $n = 1$. Such a restriction is not necessary. So for any, though constant, value of n the solution for λ_g becomes:

$$1 + \frac{A\bar{U}_1 n}{5\nu}\lambda_g - \ln\left(1 + \frac{A\bar{U}_1 n}{5\nu}\lambda_g\right) = A^2\frac{n\bar{U}_1^2}{5\nu}t + B$$

provided $A \neq 0$.

We need two boundary conditions for the constants of integration A and B. For the one boundary condition we may assume $\lambda_g = 0$ when $t = t_0$. This yields

$$B = 1 - A^2\frac{n\bar{U}_1^2}{5\nu}t_0$$

Hence

$$\frac{nA\bar{U}_1}{5\nu}\lambda_g - \ln\left(1 + \frac{nA\bar{U}_1}{5\nu}\lambda_g\right) = A^2\frac{n\bar{U}_1^2}{5\nu}(t - t_0)$$

For small values of λ_g, so that $(nA\bar{U}_1/5\nu)\lambda_g \ll 1$, and hence for small values of $t - t_0$, we obtain again the relation (3-168). According to this relation λ_g increases indefinitely with increasing time. This, however, need not necessarily be so. For a homogeneous, isotropic turbulence, the decay law for the final period does give this indefinite increase of λ_g. But we do know that grid generated turbulence is neither isotropic, nor strictly homogeneous. Therefore Naudascher and Farell assume a limiting, finite value λ_∞ even for the final period in grid-generated turbulence. With this as the secondary boundary condition we obtain $A = -5\nu/n\bar{U}_1\lambda_\infty$, and hence

$$\frac{\lambda_g}{\lambda_\infty} + \ln\left(1 - \frac{\lambda_g}{\lambda_\infty}\right) = -\frac{5\nu}{n\lambda_\infty^2}(t - t_0) \qquad (3\text{-}194)$$

The corresponding relation for $u'/\bar{U}_1$ can conveniently be given in the following form

$$\frac{u'}{\bar{U}_1} = C\left(\frac{10\nu t_0}{n\lambda_\infty^2}\right)^{n/2}\left(\frac{\lambda_\infty}{\lambda_g} - 1\right)^n \qquad (3\text{-}195)$$

As mentioned earlier, for relatively small times the expression (3-194) reduces to the relation (3-168a). Correspondingly the expression for $u'/\bar{U}_1$ reduces to the relation (3-167a). As pointed out by Naudascher and Farell, the advantage of the relations (3-194) and (3-195) relative to the simple power-law relations (3-167a) and (3-168a) is that the new relations may be valid during the entire decay process. They showed, with $n = 1$,† that most of the available experimental data can be represented satisfactorily by the above relations. Though it must be kept in mind, that in the relation (3-194) for λ_g two parameters (λ_∞ and t_0), and in the relation (3-195) for u' three parameters (λ_∞, t_0, and C) are available to adapt the theoretical curves to the experimental data.

† The value $n = 1$ has been assumed by Naudascher and Farell since they concluded from experimental evidence that $\lambda_g \propto t^{1/2}$ and $u' \propto t^{-1/2}$, whence follows $d\lambda_g/dt \propto u'$.

Of course, the whole theory stands and falls by the correctness of the relation (3-192) between u' and λ_g. From the favorable comparison with experimental data which Naudascher and Farell obtained with $n = 1$, one might conclude that there is an indirect confirmation of this relation. Yet a more direct experimental evaluation of the relation would certainly be desirable. Anyway, as we will show presently, the relation seems to fail for the final period of the decay. For large values of $(t - t_0)$, so that $\lambda_g/\lambda_\infty \to 1$, we obtain the following asymptotic relations:

$$\frac{\lambda_g}{\lambda_\infty} = 1 - \exp\left[- \left\{ 1 + \frac{5\nu}{n\lambda_\infty^2}(t - t_0) \right\} \right] \tag{3-196}$$

and

$$\frac{u'}{\bar{U}_1} = C\left(\frac{10\nu t_0}{ne^2\lambda_\infty^2} \right)^{n/2} \exp\left[- \frac{5\nu}{\lambda_\infty^2}(t - t_0) \right] \tag{3-197}$$

That is, an exponential decay law, instead of a power law. Now we do know, that for a homogeneous, isotropic turbulence the decay during the final stage is according to the $(-5/2)$ law, given by Eq. (3-43).

Batchelor and Townsend[7] made measurements at a low mesh Reynolds-number in order to investigate the final period of the decay. Figure 3-27 gives the results, where $[\bar{U}_1/u']^{4/5}$ and λ_g^2 are plotted against x_1/M.

The few data points for $x_1/M > 600$ seem to confirm the theoretical decay law for a homogeneous, isotropic turbulence. But it does not look very convincing. Indeed, Tan and Ling[118] showed, by replotting Batchelor and Townsend's data in a log-log graph, that a decay law $u' \propto (x_1 - x_0)^{-1}$ could also be concluded. Tan and Ling's

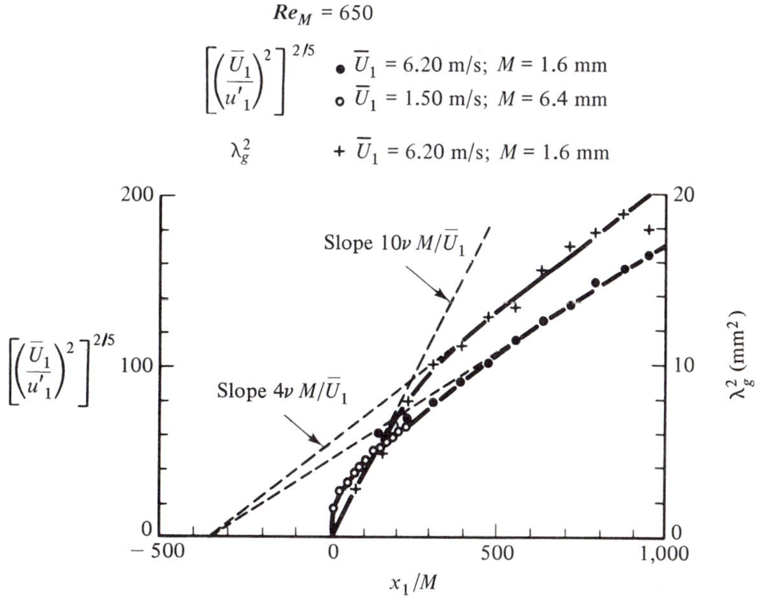

FIGURE 3-27
Decay of $u'_1/\bar{U}_1$ in the final period. (From: *Batchelor, G. K., and A. A. Townsend,*[21] *by permission of the Royal Society.*)

own experiments with grid-generated turbulence in a water channel ($\mathbf{Re}_M = 620$ and 1500), strongly suggested such a decay law. Now the $(-5/2)$ law according to Eq. (3-43) is obtained from the solution of Eq. (3-40), while a (-2) law according to Tan and Ling's experiments could for instance be obtained if we consider during the final period the turbulence to consist of randomly oriented vortices behaving independently of each other and decaying by mere viscous diffusion. For each vortex line ω the diffusion equation

$$\frac{\partial \omega}{\partial t} = \frac{\nu}{r}\frac{\partial}{\partial r}\left(r\frac{\partial \omega}{\partial r}\right)$$

yields a solution

$$\omega = \frac{\text{const.}}{\nu t}\exp\left(-\frac{r^2}{4\nu t}\right)$$

which shows that for very large values of t, ω decays inversely proportional with time, and so does the velocity field. Indeed, experiments by Tan and Ling with dye, clearly showed in the final stage of decay the streaky picture of randomly oriented vortices, which did not interact.

As pointed out earlier more than once, the difference between grid-generated turbulence in a uniform main flow and homogeneous, isotropic turbulence is that the first is axisymmetric, and inhomogeneous in the axial, main flow direction, though the degree of axial inhomogeneity has been assumed sufficiently small to justify the neglect of its effect. However, it might well be that this assumption is not correct, so that the decay law observed by Tan and Ling might be due to the actual axial inhomogeneity of the turbulence.

Now the inverse five-halves decay law for $\overline{u^2}$ *is obtained on the assumption of analytic behavior at* $k = 0$ of the energy-spectrum tensor $\mathbf{E}_{i,j}$. During the final period $\mathbf{E}_{i,j}$ satisfies the equation [see also Eq. (3-81)]

$$\frac{\partial}{\partial t}\mathbf{E}_{i,j}(k_1,k_2,k_3,t) = -2\nu k^2 \mathbf{E}_{i,j}(k_1,k_2,k_3,t)$$

with the solution

$$\mathbf{E}_{i,j}(k_1,k_2,k_3,t) = \mathbf{E}_{i,j}(k_1,k_2,k_3,t_0)\exp\left[-2\nu k^2(t-t_0)\right]$$

For an analytic behavior at $k = 0$, $\mathbf{E}_{i,j}$ must have the form $\mathbf{E}_{i,j} = C[k^2\delta_{ij} - k_ik_j]$ for small k. In the final period of decay we may consider large values of $t - t_0$ so that the contribution for small k only is of importance. Hence, we may consider for the correlation tensor $\mathbf{Q}_{i,j}(x_1,x_2,x_3,t)$ the expression [see Eq. (3-54)]:

$$\mathbf{Q}_{i,j}(x_1,x_2,x_3,t) = C\iiint_{-\infty}^{+\infty} dk_1\,dk_2\,dk_3\,[k^2\delta_{ij} - k_ik_j]\exp\left(-2\nu k^2(t-t_0) + \imath k_jx_j\right]$$

Consequently

$$\overline{u^2}(t) = \tfrac{1}{3}\mathbf{Q}_{i,i}(0,t) = \tfrac{2}{3}C \iiint_{-\infty}^{+\infty} dk_1\, dk_2\, dk_3\, k^2 \exp\left[-2\nu k^2(t-t_0)\right]$$

$$= \tfrac{4}{3}\pi C \int_0^{\infty} dk\, k^4 \exp\left[-2\nu k^2(t-t_0)\right] = \frac{\pi\sqrt{\pi C}}{2[2\nu(t-t_0)]^{5/2}}$$

However, if we drop the condition of analytic behavior at $k = 0$, another decay law may result. Lee[120] considered, for instance, the following expression for $\mathbf{E}_{i,j}$ at small k

$$\mathbf{E}_{i,j}(k_1,k_2,k_3,t_0) = C\left[k\delta_{ij} - \frac{k_i k_j}{k}\right]$$

that is, an odd function of k. With this expression for the initial energy spectrum we obtain

$$\overline{u^2}(t) = \tfrac{1}{3}\mathbf{Q}_{i,i}(0,t) = \tfrac{2}{3}C \iiint_{-\infty}^{+\infty} dk_1\, dk_2\, dk_3 \cdot k \exp\left[-2\nu k^2(t-t_0)\right]$$

$$= \frac{2\pi C}{3[2\nu t - t_0)]^2}$$

Thus the inverse square law of decay, observed by Tan and Ling, is obtained. Notice that the turbulence is still assumed homogeneous. On the other hand if one assumes a non-analytic behavior at $k = 0$, but still an even function in k, so that

$$\mathbf{E}_{i,j}(k_1,k_2,k_3,t_0) = C\left[\delta_{ij} - \frac{k_i k_j}{k^2}\right]$$

one would obtain

$$\overline{u^2}(t) = \frac{\pi^{3/2}C}{3[2\nu(t-t_0)]^{3/2}}$$

which is Saffman's decay law for the final period.

In a later publication Lee and Tan[121] considered the case of an *inhomogeneous turbulence*, more specifically an axisymmetric turbulence, inhomogeneous in the x_1-direction only. By choosing different admissible initial spectra $\mathbf{E}_{i,j}(k_1,k_2,k_3,t_0)$, a considerable variation in decay laws during the final period can be obtained. The spectra were admissible in the sense that they should satisfy the conditions of mass conservation and of symmetry. Again, for a spectrum function that behaved analytically at $k = 0$, the inverse five-halves decay-law was obtained, whereas for another function with non-analytic behavior at $k = 0$, the inverse square law was found.

Another interesting result, worth mentioning, was that the effect of axial inhomogeneity was only perceptible in the final stage of decay, when the grid Reynolds number was very small indeed.

3-7 EXTENSION TO AN ISOTROPIC-TURBULENT SCALAR FIELD

We shall now extend the previous considerations to the case in which the varying quantity is not a vector (like velocity and momentum) but a scalar, such as, for instance, temperature or enthalpy or, in a mixture of miscible fluids, concentration. The study of the behavior of such a scalar field in addition to that of a velocity field is certainly worth while, because it may yield results that may form a basis for studying mixing processes.

In contrast with the case of an isotropic turbulent-velocity field, the study of the corresponding scalar field has only been taken up much later. What are perhaps the first publications are of Soviet origin.[31,32] They have been followed by a few by Corrsin.[33,34] All these papers contain essentially theoretical work. The first experimental evidence of importance has been published by Kistler, O'Brien, and Corrsin,[35] followed soon by a second publication[103] giving additional information. Of the more recent publications we will only consider the theoretical work by Batchelor et al.,[58,59] and by Pao,[77,104] and for experimental evidence the publications by Gibson and Schwarz,[105] by Nye and Brodkey,[106] and by Grant, Hughes, Vogel, and Moilliet.[107]

As proposed in Chap. 1, Γ will denote the total instantaneous value of a scalar quantity and γ, the fluctuation of Γ about an average value $\bar{\Gamma}$. We assume that both the velocity field and the scalar field are isotropic and homogeneous, so $\bar{\Gamma}$ is constant throughout the field.

As Kistler, O'Brien, and Corrsin have shown, such an isotropic scalar field superimposed on an isotropic velocity field may be realized by introducing into the flow field a grid with heated bars. At a certain distance from the heated grid, the temperature field as well as the velocity field will have approached isotropic conditions.

A theoretical study of a turbulent scalar field can be made and the geometrical and dynamical relations obtained in the same way as for a turbulent-velocity field. For this reason it is possible to keep the derivation quite short.

As regards notation, we shall use the same symbols we used for the corresponding quantities in the velocity flow field, marked, however, with a lower index γ to refer to a scalar quantity. Unlike the indices i, j, k, etc., γ then is fixed, and a repetition of the index γ does not mean summation.

Correlations

As in the case of a turbulent-velocity field, the correlations between varying quantities at two points of the field are very useful for describing geometrical relations in the structure of the field.

Let us first consider the double correlation

$$[\mathbf{Q}_{\gamma,\gamma}(r,t)]_{A,B} = \overline{\gamma_A \gamma_B} \qquad (3\text{-}198)$$

with its coefficient

$$[\mathbf{R}_{\gamma,\gamma}(r,t)]_{A,B} = \frac{\overline{\gamma_A \gamma_B}}{\gamma'^2} \tag{3-199}$$

where γ_A and γ_B are the values of γ at the points A and B, respectively, located at a distance r from each other, and where $\gamma' = \sqrt{\overline{\gamma_A{}^2}} = \sqrt{\overline{\gamma_B{}^2}}$ is the root-mean-square value or the intensity of the γ-fluctuations.

This double correlation is a tensor of zero order, thus a scalar, and a function of the distance r and the time t. In order to simplify the notation we shall omit in the following the indication for the time-dependence in all cases where the time as such is of no importance in the relations considered, for instance, in the purely geometrical relations.

Since $\mathbf{Q}_{\gamma,\gamma}(r)$ does not change sign if r is replaced by $-r$, it must be an even function of r, and so must its coefficient $\mathbf{R}_{\gamma,\gamma}(r)$. Hence, a series expansion of $\mathbf{R}_{\gamma,\gamma}(r)$ with respect to r must contain only terms with even powers of r:

$$\mathbf{R}_{\gamma,\gamma}(r) = 1 + \frac{r^2}{2!}\left[\frac{\partial^2 \mathbf{R}_{\gamma,\gamma}}{\partial r^2}\right]_{r=0} + \frac{r^4}{4!}\left[\frac{\partial^4 \mathbf{R}_{\gamma,\gamma}}{\partial r^4}\right]_{r=0} + \cdots \tag{3-200}$$

where

$$\left[\frac{\partial^2 \mathbf{R}_{\gamma,\gamma}}{\partial r^2}\right]_{r=0} = \frac{1}{\gamma'^2} \lim_{r\to 0}\left[\overline{\gamma_A \left(\frac{\partial^2 \gamma}{\partial r^2}\right)_B}\right]$$

$$\left[\frac{\partial^4 \mathbf{R}_{\gamma,\gamma}}{\partial r^4}\right]_{r=0} = \frac{1}{\gamma'^2} \lim_{r\to 0}\left[\overline{\gamma_A \left(\frac{\partial^4 \gamma}{\partial r^4}\right)_B}\right]$$

. . . .

The second derivative $[\partial^2 \mathbf{R}_{\gamma,\gamma}/\partial r^2]_{r=0}$ is a measure of the curvature of the correlation curve at the point $r = 0$. Let λ_γ denote the part of the abscissa between the origin and the point of intersection with the osculation parabola at the vertex of the correlation curve. Then

$$\left[\frac{\partial^2 \mathbf{R}_{\gamma,\gamma}}{\partial r^2}\right]_{r=0} = -\frac{2}{\lambda_\gamma{}^2} \tag{3-201}$$

We may interpret λ_γ, like λ_g or λ_f, as a "dissipation"-scale for the γ-fluctuations. Furthermore, we shall define an integral scale for the γ-fluctuations:

$$\Lambda_\gamma = \int_0^\infty dr\, \mathbf{R}_{\gamma,\gamma}(r) \tag{3-202}$$

Apart from the double correlation between γ at two points, we also have double correlations between γ at one point and a velocity component at another point. If the two points coincide, so that we have the correlation between the γ-fluctuation and velocity fluctuation at the same point, this correlation must then be zero,

$$\overline{(u_i)_A \gamma_A} = 0$$

because this correlation must be invariant under reflection. The physical reason is that $\overline{(u_i)_A \gamma_A}$ is a measure of the Γ-transport by turbulence [see Eq. (1-34)], which must be zero because $\bar{\Gamma}$ is constant.

But the correlation must be zero in the case of two points also.

$$\overline{(u_i)_A \gamma_B} = 0 \qquad (3\text{-}203)$$

which follows from the condition that this tensor of the first order must have a vanishing divergence for an incompressible fluid. This has already been shown for the analogous correlation $\overline{(u_i)_A p_B}$ [see Eq. (3-6)].

Since the relation (3-203) holds for any value of r, it follows that

$$\overline{(u_i)_A \left(\frac{\partial^n \gamma}{\partial r^n}\right)_B} = 0$$

also, as well as

$$\overline{\left(\frac{\partial^n u_i}{\partial r^n}\right)_A \gamma_B} = 0$$

Consequently,

$$\overline{\left(u_i \frac{\partial^n \gamma}{\partial r^n}\right)_A} = \lim_{r \to 0} \overline{(u_i)_A \left(\frac{\partial^n \gamma}{\partial r^n}\right)_B} = 0$$

and

$$\overline{\left(\frac{\partial^n u_i}{\partial r^n} \gamma\right)_B} = \lim_{r \to 0} \overline{\left(\frac{\partial^n u_i}{\partial r^n}\right)_A \gamma_B} = 0 \qquad (3\text{-}204)$$

Triple correlations between γ and the velocity components at two points are

$$\overline{(u_i)_A \gamma_A \gamma_B} \quad \text{and} \quad \overline{(u_i)_A (u_j)_B \gamma_B}$$

They form tensors of the first and second order, respectively. Only the first one is of further interest to us; it occurs in the dynamic equation of $\mathbf{Q}_{\gamma,\gamma}(r,t)$ as will be shown later.

If the two points coincide, the triple correlation

$$\overline{(u_i \gamma^2)_A} = 0 \qquad (3\text{-}205)$$

for the same reason that $\overline{(u_i \gamma)_A} = 0$. And, similar to Eq. (3-203), we have, furthermore,

$$\overline{(u_i)_A \gamma_B^{\,2}} = 0 \qquad (3\text{-}206)$$

We shall now consider the following general triple correlation

$$[\mathbf{S}_\gamma]_{A,B} = \overline{\gamma_B \gamma_A (u_a)_A} = \overline{\gamma_B \gamma_A (u_i)_A} e_{ai} = [\mathbf{S}_{\gamma i,\gamma}]_{A,B} e_{ai} \qquad (3\text{-}207)$$

By applying the theory of invariants, as expounded in Sec. 3-2, to the tensor of the first order $[\mathbf{S}_{\gamma i,\gamma}]_{A,B}$, we obtain [see Eq. (3-5)]

$$[\mathbf{S}_{\gamma i,\gamma}]_{A,B} = \gamma'^2 u' \frac{\mathbf{k}_\gamma(r)}{r} \xi_i \tag{3-208}$$

where $\mathbf{k}_\gamma(r)$ is the coefficient of the spatial correlation between γ_B, γ_A and the velocity component $(u_r)_A$ at A in the direction of r:

$$\mathbf{k}_\gamma(r) = \frac{\overline{\gamma_B \gamma_A (u_r)_A}}{\gamma'^2 u'} \tag{3-209}$$

It is an odd function of r; so its series expansion reads

$$\mathbf{k}_\gamma(r) = r\left[\frac{\partial \mathbf{k}_\gamma}{\partial r}\right]_{r=0} + \frac{r^3}{3!}\left[\frac{\partial^3 \mathbf{k}_\gamma}{\partial r^3}\right]_{r=0} + \cdots \tag{3-210}$$

where

$$\gamma'^2 u' \left[\frac{\partial \mathbf{k}_\gamma}{\partial r}\right]_{r=0} = \lim_{r\to 0}\left[\overline{\gamma_A (u_r)_A \left(\frac{\partial \gamma}{\partial r}\right)_B}\right]$$

$$\gamma'^2 u' \left[\frac{\partial^3 \mathbf{k}_\gamma}{\partial r^3}\right]_{r=0} = \lim_{r\to 0}\left[\overline{\gamma_A (u_r)_A \left(\frac{\partial^3 \gamma}{\partial r^3}\right)_B}\right]$$

. . . .

We shall now show that $[\partial \mathbf{k}_\gamma/\partial r]_{r=0} = 0$. Since $\overline{(u_r)_A \gamma_B{}^2} = 0$ [see (3-206)] we have

$$\overline{(u_r)_A \frac{\partial}{\partial r} \gamma_B{}^2} = \overline{2(u_r)_A \gamma_B \frac{\partial \gamma_B}{\partial r}} = 0$$

Consequently,

$$\overline{\left(u_r \gamma \frac{\partial \gamma}{\partial r}\right)_A} = \lim_{r\to 0} \overline{(u_r)_A \gamma_B \frac{\partial \gamma_B}{\partial r}} = 0$$

so

$$\left[\frac{\partial \mathbf{k}_\gamma}{\partial r}\right]_{r=0} = \frac{1}{\gamma'^2 u'}\left[\overline{\gamma_A (u_r)_A \frac{\partial \gamma_B}{\partial r}}\right]_{r=0} = \frac{1}{\gamma'^2 u'} \overline{\left(\gamma u_r \frac{\partial \gamma}{\partial r}\right)_A} = 0$$

The result is that the series expansion of $\mathbf{k}_\gamma(r)$ starts with r^3, just like the series expansion for the triple velocity correlation $\mathbf{k}(r)$ [see Eq. (3-22)]:

$$\mathbf{k}_\gamma(r) = \frac{r^3}{3!}\left[\frac{\partial^3 \mathbf{k}_\gamma}{\partial r^3}\right]_{r=0} + \frac{r^5}{5!}\left[\frac{\partial^5 \mathbf{k}_\gamma}{\partial r^5}\right]_{r=0} + \cdots \tag{3-211}$$

The Dynamic Equation for $\mathbf{Q}_{\gamma,\gamma}(r,t)$

We shall start from the equation for the diffusion and convection of Γ [see Eq. (1-32)], assuming that the molecular-transport coefficient $\mathfrak{k}$ is constant. If Γ denotes the amount of heat per unit of mass $c_p\Theta$, then $\mathfrak{k} = \kappa/\rho c_p$, where κ is the heat conductivity of the fluid. If Γ is a concentration, $\mathfrak{k}$ denotes the coefficient of molecular diffusion.

$$\frac{\partial \Gamma}{\partial t} + U_i \frac{\partial \Gamma}{\partial x_i} = \mathfrak{k} \frac{\partial^2 \Gamma}{\partial x_i \, \partial x_i}$$

Assume $\Gamma = \bar{\Gamma} + \gamma$ and $U_i = \bar{U}_1 + u_i$, keeping in mind that $\bar{\Gamma}$ and $\bar{U}_1$ are constant throughout the field.

$$\frac{\partial \gamma}{\partial t} + \bar{U}_1 \frac{\partial \gamma}{\partial x_1} + u_i \frac{\partial \gamma}{\partial x_i} = \mathfrak{k} \frac{\partial^2}{\partial x_l \, \partial x_l} \gamma$$

Consider this equation for the two points A and B By multiplying the equation at point A by γ_B and the equation at point B by γ_A, adding the two equations, and averaging with respect to time, after carrying out exactly the same procedure used for deriving Eq. (1-45), we obtain

$$\frac{\partial}{\partial t} \overline{\gamma_A \gamma_B} + 2 \frac{\partial}{\partial \xi_i} \overline{\gamma_A \gamma_B (u_i)_B} = 2\mathfrak{k} \frac{\partial^2}{\partial \xi_l \, \partial \xi_l} \overline{\gamma_A \gamma_B} \tag{3-212}$$

With Eqs. (3-198) and (3-207)

$$\frac{\partial}{\partial t} \mathbf{Q}_{\gamma,\gamma}(\xi_1,\xi_2,\xi_3,t) - 2 \frac{\partial}{\partial \xi_i} \mathbf{S}_{\gamma i,\gamma}(\xi_1,\xi_2,\xi_3,t) = 2\mathfrak{k} \frac{\partial^2}{\partial \xi_l \, \partial \xi_l} \mathbf{Q}_{\gamma,\gamma}(\xi_1,\xi_2,\xi_3,t) \tag{3-213}$$

Introduce the scalar $\mathbf{S}^*_{\gamma,\gamma}$ defined by

$$\mathbf{S}^*_{\gamma,\gamma}(\xi_1,\xi_2,\xi_3,t) = \frac{\partial}{\partial \xi_i} \mathbf{S}_{\gamma i,\gamma}(\xi_1,\xi_2,\xi_3,t) \tag{3-214}$$

and apply Eq. (3-208); then

$$\mathbf{S}^*_{\gamma,\gamma} = \gamma'^2 u' \left[2 \frac{\mathbf{k}_\gamma}{r} + \frac{\partial \mathbf{k}_\gamma}{\partial r} \right] = \gamma'^2 u' \frac{1}{r^2} \frac{\partial}{\partial r} (r^2 \mathbf{k}_\gamma) \tag{3-215}$$

Since $\mathbf{S}^*_{\gamma,\gamma}$ and $\mathbf{Q}_{\gamma,\gamma}$ are functions only of r and t, the Eq. (3-207) may be rewritten

$$\frac{\partial}{\partial t} \mathbf{Q}_{\gamma,\gamma}(r,t) - 2\mathbf{S}^*_{\gamma,\gamma}(r,t) = 2\mathfrak{k} \frac{1}{r^2} \frac{\partial}{\partial r} \left[r^2 \frac{\partial}{\partial r} \mathbf{Q}_{\gamma,\gamma}(r,t) \right] \tag{3-216}$$

and, expressed in terms of the correlation coefficients,

$$\frac{\partial}{\partial t} \gamma'^2 \mathbf{R}_{\gamma,\gamma}(r,t) - 2\gamma'^2 u' \frac{1}{r^2} \frac{\partial}{\partial r} [r^2 \mathbf{k}_\gamma(r,t)] = 2\mathfrak{k} \gamma'^2 \frac{1}{r^2} \frac{\partial}{\partial r} \left[r^2 \frac{\partial \mathbf{R}_{\gamma,\gamma}(r,t)}{\partial r} \right] \tag{3-217}$$

Equations (3-216) and (3-217) describe the behavior of the double correlation in an isotropic turbulence, and Eq. (3-217) is similar to the Kármán-Howarth equation (3-36) for the double velocity correlation.

If, on the other hand, Eq. (3-217) is considered as an equation that describes the spread of a quantity in a space, it is seen to differ essentially from the corresponding equation for the velocity correlation. According to both Eqs. (3-216) and (3-217) we have to deal with a spread in a three-dimensional space, whereas Eq. (3-36) describes the spread in a five-dimensional space. In this respect there is a much higher degree of similarity with Eq. (3-35) for the spread of $\mathbf{Q}_{i,i}(r,t)$, which is, in fact, also a scalar. Like $\mathbf{S}_{i,i}(r,t)$ the terms $\mathbf{S}^*_{\gamma,\gamma}(r,t)$ may be interpreted as an exchange function.

It is of interest to see to what equation Eq. (3-217) reduces when $r \to 0$

Since $\mathbf{k}_\gamma(r,t)$ behaves like r^3 for small r, it follows from the relation (3-215) that

$$\lim_{r \to 0} \mathbf{S}^*_{\gamma,\gamma}(r,t) = 0 \qquad (3\text{-}218)$$

Furthermore,

$$\lim_{r \to 0} \frac{1}{r^2} \frac{\partial}{\partial r} \left[r^2 \frac{\partial \mathbf{R}_{\gamma,\gamma}(r,t)}{\partial r} \right] = -\frac{6}{\lambda_\gamma^2}$$

Hence, for $r \to 0$, the Eq. (3-217) reduces to

$$\frac{d}{dt}\overline{\gamma'^2} = -12 \frac{\mathfrak{f}}{\lambda_\gamma^2} \overline{\gamma'^2} \qquad (3\text{-}219)$$

This result is completely analogous to Eq. (3-162). The decay of the intensity of the γ-fluctuations is, after all, caused only by molecular diffusion. Because of the difference between the numerical constants here and in Eq. (3-162) and because $\mathfrak{f}$ and λ_γ need not have the same values as ν and λ_g, respectively, there is a quantitative difference between the decay rate of $\overline{\gamma'^2}$ and that of $\overline{u'^2}$.

The Spectral Distribution of a Scalar Quantity

In a turbulent scalar field the fluctuating quantity may be assumed to be composed of fluctuations corresponding to different wavenumbers in a continuous wavenumber range. A spectral analysis of the γ-fluctuations will then show how the contributions are distributed over these wavenumbers. As in the case of the velocity fluctuations, we may define spectrum functions that appear to be the Fourier transforms of correlation functions. In space, these spectrum functions will have a three-dimensional character. However, what can be measured with a suitable detecting element is a one-dimensional cut of this spectrum.

Let us first consider this one-dimensional spectrum $\mathbf{E}_{\gamma 1}(k_1,t)$. This will be so defined that

$$\int_0^\infty dk_1 \, \mathbf{E}_{\gamma 1}(k_1,t) = \overline{\gamma'^2} \qquad (3\text{-}220)$$

$\mathbf{E}_{\gamma 1}(k_1,t)$ is the Fourier transform of $\mathbf{Q}_{\gamma,\gamma}(x_1,t)$. Since $\mathbf{Q}_{\gamma,\gamma}(0,t) = \gamma'^2$, the Fourier transform relations must read:

$$\mathbf{Q}_{\gamma,\gamma}(x_1,t) = \frac{1}{2}\int_{-\infty}^{+\infty} dk_1\, \mathbf{E}_{\gamma 1}(k_1,t) \exp{(\imath k_1 x_1)} \qquad \text{(3-221)}$$

and

$$\mathbf{E}_{\gamma 1}(k_1,t) = \frac{1}{\pi}\int_{-\infty}^{+\infty} dx_1\, \mathbf{Q}_{\gamma,\gamma}(x_1,t) \exp{(-\imath k_1 x_1)} \qquad \text{(3-222)}$$

The reader may easily verify that Eq. (3-220) follows from Eq. (3-221) when $x_1 = 0$.

The three-dimensional extension of Eq. (3-222) reads

$$\mathbf{E}_{\gamma,\gamma}(k_1,k_2,k_3,t) = \frac{1}{8\pi^3}\iiint_{-\infty}^{+\infty} dx_1\, dx_2\, dx_3\, \mathbf{Q}_{\gamma,\gamma}(x_1,x_2,x_3,t) \exp{(-\imath k_l x_l)} \qquad \text{(3-223)}$$

However, since $\mathbf{Q}_{\gamma,\gamma}$ is a scalar that depends only on r and t, we may again usefully introduce spherical polar coordinates. We then obtain

$$\mathbf{E}_{\gamma,\gamma}(k,t) = \frac{1}{2\pi^2}\int_0^\infty dr\, r^2 \frac{\sin kr}{kr} \mathbf{Q}_{\gamma,\gamma}(r,t) \qquad \text{(3-224)}$$

with its Fourier transform

$$\mathbf{Q}_{\gamma,\gamma}(r,t) = 4\pi \int_0^\infty dk\, k^2 \frac{\sin kr}{kr} \mathbf{E}_{\gamma,\gamma}(k,t) \qquad \text{(3-225)}$$

From this relation it follows that

$$\gamma'^2 = 4\pi \int_0^\infty dk\, k^2 \mathbf{E}_{\gamma,\gamma}(k,t) \qquad \text{(3-226)}$$

The three-dimensional spectrum function $\mathbf{E}_\gamma(k,t)$ is now defined such that

$$\mathbf{E}_\gamma(k,t) = 4\pi k^2 \mathbf{E}_{\gamma,\gamma}(k,t) \qquad \text{(3-227)}$$

We then have, similar to Eq. (3-220),

$$\int_0^\infty dk\, \mathbf{E}_\gamma(k,t) = \gamma'^2 \qquad \text{(3-228)}$$

Consequently, from Eq. (3-224),

$$\mathbf{E}_\gamma(k,t) = \frac{2}{\pi}\int_0^\infty dr\, kr \sin kr\, \mathbf{Q}_{\gamma,\gamma}(r,t) \qquad \text{(3-229)}$$

and

$$\mathbf{Q}_{\gamma,\gamma}(r,t) = \int_0^\infty dk \frac{\sin kr}{kr} \mathbf{E}_\gamma(k,t) \qquad \text{(3-230)}$$

The three-dimensional spectrum $\mathbf{E}_\gamma(k,t)$ is an even function of k and for small k it behaves like k^2. This immediately follows from a series expansion of Eq. (3-229):

$$\mathbf{E}_\gamma(k,t) = \frac{2}{\pi}\left[k^2 \int_0^\infty dr\, r^2 \mathbf{Q}_{\gamma,\gamma}(r,t) - \frac{k^4}{3!}\int_0^\infty dr\, r^4 \mathbf{Q}_{\gamma,\gamma}(r,t) + \cdots\right]$$

We may compare this expression with the similar relation (3-61) for the three-dimensional energy spectrum $\mathbf{E}(k,t)$. But, unlike the case for $\mathbf{E}(k,t)$, where the first term on the right-hand side is zero, because for an incompressible fluid the condition (3-65) holds, here the first term is not zero.

Let us now consider what relation exists between the one-dimensional and the three-dimensional γ-spectrum. With this relation it is then possible to calculate the three-dimensional spectrum from the measured values of the one-dimensional spectrum.

This relation is easily obtained if we note that Eq. (3-229) is practically equal to the derivative of Eq. (3-222) with respect to k. Thus,

$$\mathbf{E}_{\gamma 1}(k_1,t) = \frac{2}{\pi}\int_0^\infty dx_1\, \mathbf{Q}_{\gamma,\gamma}(x_1,t)\cos k_1 x_1$$

$$\frac{\partial \mathbf{E}_{\gamma 1}(k_1,t)}{\partial k_1} = -\frac{1}{k_1}\frac{2}{\pi}\int_0^\infty dx_1\, k_1 x_1 \sin k_1 x_1 \mathbf{Q}_{\gamma,\gamma}(x_1,t)$$

Hence

$$\mathbf{E}_\gamma(k_1,t) = -k_1 \frac{\partial \mathbf{E}_{\gamma 1}(k_1,t)}{\partial k_1} \tag{3-231}$$

and, conversely,

$$\mathbf{E}_{\gamma 1}(k_1,t) = \int_{k_1}^\infty dk\, \frac{\mathbf{E}_\gamma(k,t)}{k} \tag{3-232}$$

since the integration constant is determined by the condition that $\lim_{k_1\to\infty} \mathbf{E}_{\gamma 1}(k_1,t) - 0$.

The formulas (3-231) and (3-232) show that, if the three-dimensional spectrum $\mathbf{E}_\gamma(k,t)$ follows a negative power law in the range $k_1 \le k < \infty$, the same law holds for the one-dimensional spectrum $\mathbf{E}_{\gamma 1}(k_1,t)$, and vice versa.

The integral scale Λ_γ, defined by Eq. (3-202), and the dissipation scale λ_γ, defined by Eq. (3-201), may be readily obtained from the γ-spectrum functions.

$$\Lambda_\gamma = \frac{1}{\gamma'^2}\int_0^\infty dr\, \mathbf{Q}_{\gamma,\gamma}(r,t) = \frac{\pi}{2}\frac{1}{\gamma'^2}\mathbf{E}_{\gamma 1}(0,t)$$

$$= \frac{\pi}{2}\frac{1}{\gamma'^2}\int_0^\infty dk\, \frac{\mathbf{E}_\gamma(k,t)}{k} \tag{3-233}$$

$$\frac{2}{\lambda_\gamma{}^2} = -\frac{1}{\gamma'^2}\left[\frac{\partial^2 \mathbf{Q}_{\gamma,\gamma}}{\partial r^2}\right]_{r=0} = \frac{1}{\gamma'^2}\int_0^\infty dk\, k^2 \mathbf{E}_{\gamma 1}(k,t)$$

$$= \frac{1}{3}\frac{1}{\gamma'^2}\int_0^\infty dk\, k^2 \mathbf{E}_\gamma(k,t) \tag{3-234}$$

The Dynamic Equation for $\mathbf{E}_\gamma(k,t)$

This equation is obtained from the dynamic equation (3-216) for the correlation $\mathbf{Q}_{\gamma,\gamma}(r,t)$ by taking from each term its Fourier transform. To this end we introduce the spectrum function $\mathbf{F}_{\gamma,\gamma}(k,t)$ as the Fourier transform of $\mathbf{S}^*_{\gamma,\gamma}(r,t)$.

$$\mathbf{S}^*_{\gamma,\gamma}(x_1,x_2,x_3,t) = \iiint_{-\infty}^{+\infty} dk_1\, dk_2\, dk_3\, \mathbf{F}_{\gamma,\gamma}(k_1,k_2,k_3,t) \exp(\imath k_l x_l) \tag{3-235}$$

and

$$\mathbf{F}_{\gamma,\gamma}(k_1,k_2,k_3,t) = \frac{1}{8\pi^3}\iiint_{-\infty}^{+\infty} dx_1\, dx_2\, dx_3\, \mathbf{S}^*_{\gamma,\gamma}(x_1,x_2,x_3,t) \exp(-\imath k_l x_l) \tag{3-236}$$

Since $\mathbf{S}^*_{\gamma,\gamma}$ depends only on r and t, we may write Eq. (3-236) also in spherical polar coordinates. $\mathbf{F}_{\gamma,\gamma}$ then depends only on k and t. Thus.

$$\mathbf{F}_{\gamma,\gamma}(k,t) = \frac{1}{2\pi^2}\int_0^\infty dr\, r^2 \frac{\sin kr}{kr} \mathbf{S}^*_{\gamma,\gamma}(r,t) \tag{3-237}$$

and

$$\mathbf{S}^*_{\gamma,\gamma}(r,t) = 4\pi \int_0^\infty dk\, k^2 \frac{\sin kr}{kr} \mathbf{F}_{\gamma,\gamma}(k,t) \tag{3-238}$$

Substitution of Eqs. (3-238) and (3-225) in Eq. (3-216) yields

$$\frac{\partial}{\partial t}\mathbf{E}_{\gamma,\gamma}(k,t) - 2\mathbf{F}_{\gamma,\gamma}(k,t) = -2\mathfrak{k} k^2 \mathbf{E}_{\gamma,\gamma}(k,t) \tag{3-239}$$

Now put

$$\mathbf{F}_\gamma(k,t) = 8\pi k^2 \mathbf{F}_{\gamma,\gamma}(k,t) \tag{3-240}$$

Then by Eq. (3-227) Eq. (3-239) becomes

$$\frac{\partial}{\partial t}\mathbf{E}_\gamma(k,t) - \mathbf{F}_\gamma(k,t) = -2\mathfrak{k} k^2 \mathbf{E}_\gamma(k,t) \tag{3-241}$$

We have shown that the γ-spectrum functions $\mathbf{E}_{\gamma,\gamma}(k,t)$ and $\mathbf{E}_\gamma(k,t)$ start with k^0 and k^2, respectively, for small k. What will be the behavior of the "transfer" spectrum functions $\mathbf{F}_{\gamma,\gamma}(k,t)$ and $\mathbf{F}_\gamma(k,t)$ at small k?

To this end consider the Fourier transform $\mathbf{F}_{\gamma i,\gamma}(k_1,k_2,k_3,t)$ of $\mathbf{S}_{\gamma i,\gamma}(x_1,x_2,x_3,t)$, viz.

$$\mathbf{S}_{\gamma i,\gamma}(x_1,x_2,x_3,t) = \iiint_{-\infty}^{+\infty} dk_1\, dk_2\, dk_3\, \mathbf{F}_{\gamma i,\gamma}(k_1,k_2,k_3,t) \exp(\imath k_l x_l)$$

For $\mathbf{S}^*_{\gamma,\gamma}(x_1,x_2,x_3,t)$ we then obtain

$$\mathbf{S}^*_{\gamma,\gamma}(x_1,x_2,x_3,t) = \frac{\partial}{\partial x_i}\mathbf{S}_{\gamma i,\gamma} = \imath \iiint_{-\infty}^{+\infty} dk_1\, dk_2\, dk_3\, k_i \mathbf{F}_{\gamma i,\gamma}(k_1,k_2,k_3,t) \exp(\imath k_l x_l)$$

Comparison with Eq. (3-235) shows that $\mathbf{F}_{\gamma,\gamma} = \imath k_i \mathbf{F}_{\gamma i,\gamma}$. Now $\mathbf{F}_{\gamma i,\gamma}(k_1,k_2,k_3,t)$ is a tensor of the first order, whose general expression reads [see (3-4)]

$$\mathbf{F}_{\gamma i,\gamma}(k_1,k_2,k_3,t) = \varphi(k^2,t)k_i$$

Hence, we obtain

$$\mathbf{F}_{\gamma,\gamma}(k,t) = \imath k^2 \varphi(k^2,t) \qquad (3\text{-}242)$$

If now $\mathbf{F}_{\gamma i,\gamma}(k_1,k_2,k_3,t)$ is analytic at $k = 0$, then a series expansion of $\varphi(k^2,t)$ should start with a term k^{2n}, where $n \geq 0$. In that case a series expansion of $\mathbf{F}_{\gamma,\gamma}(k,t)$ starts at least with a term k^2 and that of $\mathbf{F}_\gamma(k,t)$, with a term k^4.

A series expansion of Eq. (3-237), taking account of Eq. (3-242), then yields

$$\mathbf{F}_{\gamma,\gamma}(k,t) = -\frac{1}{2\pi^2}\left[\frac{k^2}{3!}\int_0^\infty dr\, r^4 \mathbf{S}^*_{\gamma,\gamma}(r,t) - \frac{k^4}{5!}\int_0^\infty dr\, r^6 \mathbf{S}^*_{\gamma,\gamma}(r,t) + \cdots\right] \qquad (3\text{-}243)$$

and

$$\int_0^\infty dr\, r^2 \mathbf{S}^*_{\gamma,\gamma}(r,t) = 0 \qquad (3\text{-}244)$$

An equivalent result is

$$\lim_{r\to 0} r^2 \mathbf{k}_\gamma(r,t) = 0$$

From Eq. (3-243) or Eq. (3-242) it follows that

$$\mathbf{F}_{\gamma,\gamma}(0,t) = 0$$

The consequence of this result is that, according to Eq. (3-239),

$$\lim_{k\to 0}\frac{\partial}{\partial t}\mathbf{E}_{\gamma,\gamma}(k,t) = 0$$

and hence, from Eq. (3-224),

$$\frac{\partial}{\partial t}\int_0^\infty dr\, r^2 \mathbf{Q}_{\gamma,\gamma}(r,t) = 0 \qquad (3\text{-}245)$$

or

$$I_\gamma = \frac{2}{\pi}\int_0^\infty dr\, r^2\mathbf{Q}_{\gamma,\gamma}(r,t) = \text{invariant} \qquad (3\text{-}246)$$

This is an expression analogous to the Loitsianskii invariant (3-39), but here it is an exact invariant if $\lim_{r\to\infty} r^2\mathbf{k}_\gamma(r,t) = 0$. Note also the identical expression for Saffman's dynamical invariant for the turbulent-velocity field, Eq. (3-94).

Returning to the dynamic equation (3-241) for $\mathbf{E}_\gamma(k,t)$, the equivalent integrated equation reads

$$\frac{\partial}{\partial t}\int_0^k dk\,\mathbf{E}_\gamma(k,t) - \int_0^k dk\,\mathbf{F}_\gamma(k,t) = -2\mathfrak{k}\int_0^k dk\,k^2\mathbf{E}_\gamma(k,t) \qquad (3\text{-}247)$$

From Eqs. (3-240) and (3-238) it follows that

$$\int_0^\infty dk\,\mathbf{F}_\gamma(k,t) = 8\pi\int_0^\infty dk\,k^2\mathbf{F}_{\gamma,\gamma}(k,t) = 2\mathbf{S}^*_{\gamma,\gamma}(0,t)$$

which is zero, according to Eq. (3-218). Hence when $k = \infty$, Eq. (3-247) reduces to

$$\frac{d}{dt}\int_0^\infty dk\,\mathbf{E}_\gamma(k,t) = -2\mathfrak{k}\int_0^\infty dk\,k^2\mathbf{E}_\gamma(k,t)$$

or, with Eqs. (3-227) and (3-234),

$$\frac{d}{dt}\gamma'^2 = -12\mathfrak{k}\,\frac{\gamma'^2}{\lambda_\gamma^2}$$

This is exactly Eq. (3-219). If we may regard the transformation of turbulent fluctuations into molecular distribution of γ, where further distribution occurs only by molecular-diffusion processes, as "dissipation" of the γ-fluctuations, then $d\gamma'^2/dt$ is the "dissipation" of γ. We shall denote the rate of this "dissipation" of γ by ε_γ; thus,

$$\frac{d}{dt}\gamma'^2 = -\varepsilon_\gamma = -12\mathfrak{k}\,\frac{\gamma'^2}{\lambda_\gamma^2} \qquad (3\text{-}248)$$

A solution of the dynamic equation (3-247) or of the differential form (3-241) for $\mathbf{E}_\gamma(k,t)$ can be obtained only if some postulate concerning the form of the transfer function $\mathbf{F}_\gamma(k,t)$ is made. Since $\mathbf{F}(k,t)$ as well as $\mathbf{F}_\gamma(k,t)$ represents the transfer by interaction of eddies of a transportable quantity, it is logical to assume a direct relationship between the two. Thus we may consider postulates for $\mathbf{F}_\gamma(k,t)$ similar to those that have been made for $\mathbf{F}(k,t)$. For a very small value of the Reynolds number of turbulence, the interaction of eddies is very small and the transfer function $\mathbf{F}(k,t)$ may be assumed to be negligible with respect to the viscous dissipation. Similarly, we may assume for this case that $\mathbf{F}_\gamma(k,t)$ is negligibly small, because $\mathbf{F}_\gamma(k,t)$ also is determined by the interaction of eddies. To the Reynolds number $u'\lambda_g/\nu$ corresponds the dimensionless group $u'\lambda_\gamma/\mathfrak{k}$, which plays the same role in the dynamic equations for $\mathbf{E}_\gamma(k,t)$ and $\mathbf{Q}_{\gamma,\gamma}(r,t)$ as does the Reynolds number $u'\lambda_g/\nu$ in the dynamic equations for $\mathbf{E}(k,t)$ and $\mathbf{Q}_{i,i}(r,t)$.

Physically, the dimensionless group $u'\lambda_\gamma/\mathfrak{k}$ may be interpreted as the ratio between a diffusivity by the smaller eddies $u'\lambda_\gamma$ and the molecular diffusivity.

As we shall show later [see Eq. (3-276)], λ_g/λ_γ is determined by $\nu/\mathfrak{k}$, and, if ν and $\mathfrak{k}$ are of the same order of magnitude, so also will be $u'\lambda_g/\nu$ and $u'\lambda_\gamma/\mathfrak{k}$.

If we neglect the term $\mathbf{F}_\gamma(k,t)$ in Eq. (3-241), we obtain

$$\frac{\partial}{\partial t}\mathbf{E}_\gamma(k,t) = -2\mathfrak{k}k^2\mathbf{E}_\gamma(k,t) \qquad (3\text{-}249)$$

with its solution

$$\mathbf{E}_\gamma(k,t) = \mathbf{E}_\gamma(k,t_0)\exp\left[-2\mathfrak{k}k^2(t-t_0)\right] \qquad (3\text{-}250)$$

Because $\mathbf{E}_\gamma(k,t) \propto k^2$ when $k \to 0$, $\mathbf{E}_\gamma(k,t_0)$ must be equal to a constant times k^2, when $k \to 0$.

This can also be shown by considering the solution for the double correlation $\mathbf{Q}_{\gamma,\gamma}(r,t)$ if the term $\mathbf{S}^*_{\gamma,\gamma}(r,t)$ in Eq. (3-216) is neglected. This solution is identical with Eq. (3-42), and reads

$$\mathbf{Q}_{\gamma,\gamma}(r,t) = \frac{\sqrt{8\mathfrak{k}}}{r}\exp(-r^2/8\mathfrak{k}t)\sum_1^\infty \frac{A_p}{t^p}H_{2p-1}\left(\frac{r}{\sqrt{8\mathfrak{k}t}}\right) \qquad (3\text{-}251)$$

If we now apply to each term of Eq. (3-251) the condition of invariance (3-246)

$$\int_0^\infty dr\, r^2\mathbf{Q}_{\gamma,\gamma}(r,t) = \text{const.}$$

it then appears that, for a finite value of the constant, this condition is satisfied only by the term where $p = 1$; all other terms give zero contribution. If we consider further only the term where $p = 1$, the solution (3-251) reduces to

$$\mathbf{Q}_{\gamma,\gamma}(r,t) = 2A_1\frac{1}{t^{3/2}}\exp(-r^2/8\mathfrak{k}t) \qquad (3\text{-}252)$$

Now $\mathbf{Q}_{\gamma,\gamma}(r,t) = \gamma'^2\mathbf{R}_{\gamma,\gamma}(r,t)$, where γ'^2 is a function of time alone. Thus it may be inferred from Eq. (3-252) that

$$\mathbf{R}_{\gamma,\gamma}(r,t) = \exp(-r^2/8\mathfrak{k}t) \qquad (3\text{-}253)$$

and

$$\frac{\gamma'^2}{\gamma_0'^2} = \left(\frac{t}{t_0}\right)^{-3/2} \qquad (3\text{-}254)$$

Note the difference between this decay rate and the rate for u'^2 given by Eq. (3-43).

From Eqs. (3-201) and (3-253) it follows that

$$\lambda_\gamma^2 = 8\mathfrak{k}t$$

and from Eqs. (3-202) and (3-253) we obtain

$$\Lambda_\gamma^2 = 2\pi\mathfrak{k}t$$

Comparison of these results with the corresponding results for λ_f and Λ_f given in Sec. 3-4 shows that

$$\frac{\lambda_\gamma}{\lambda_f} = \frac{\Lambda_\gamma}{\Lambda_f} = \sqrt{\frac{\mathfrak{k}}{\nu}}$$

With the expression (3-252) for $\mathbf{Q}_{\gamma,\gamma}(r,t)$ and with Eq. (3-229) we can calculate the expression for $\mathbf{E}_\gamma(k,t)$

$$\mathbf{E}_\gamma(k,t) = \frac{4A_1}{\pi}\frac{1}{t^{3/2}}\int_0^\infty dr\, kr \sin kr \exp\left(-r^2/8\mathfrak{k}t\right)$$

$$= \frac{A_1}{\sqrt{\pi}}(8\mathfrak{k})^{3/2}k^2 \exp\left(-2\mathfrak{k}k^2t\right) = \text{const.} \times k^2 \exp\left(-2\mathfrak{k}k^2t\right) \qquad (3\text{-}255)$$

This was for small values of $\mathbf{Re}_\lambda$ and $u'\lambda_\gamma/\mathfrak{k}$. For high values of $\mathbf{Re}_\lambda$ and $u'\lambda_\gamma/\mathfrak{k}$, where the transfer functions $\mathbf{F}(k,t)$ and $\mathbf{F}_\gamma(k,t)$ can no longer be neglected in the corresponding dynamic equations, some postulate must be made about these transfer functions. We have already said that it is reasonable to make similar postulates for the two transfer functions since they are determined by the same turbulence mechanism.

In what follows we shall consider only the case where the Heisenberg assumption (3-127) for $\mathbf{F}(k,t)$ applies. For the transfer function $\mathbf{F}_\gamma(k,t)$ we put

$$\int_0^k dk\, \mathbf{F}_\gamma(k,t) = -2\epsilon_\gamma(k,t)\int_0^k dk\, k^2\mathbf{E}_\gamma(k,t) \qquad (3\text{-}256)$$

where $\epsilon_\gamma(k,t)$ is the turbulence diffusivity caused by the eddies with wavenumbers ranging from k to infinity. According to Heisenberg's assumption for $\mathbf{F}(k,t)$ we write similarly

$$\epsilon_\gamma(k,t) = \beta\int_k^\infty dk\sqrt{\frac{\mathbf{E}(k,t)}{k^3}} \qquad (3\text{-}257)$$

Of course, here, too, the objection to this assumption may be raised as has been made in Sec. 3-5 to the turbulence viscosity.

With Eq. (3-256) and Eq. (3-257), Eq. (3-247) reads

$$\frac{\partial}{\partial t}\int_0^k dk\, \mathbf{E}_\gamma(k,t) = -2\left[\mathfrak{k} + \beta\int_k^\infty dk\sqrt{\frac{\mathbf{E}(k,t)}{k^3}}\right]\int_0^k dk\, k^2\mathbf{E}_\gamma(k,t) \qquad (3\text{-}258)$$

Assume now that $\mathbf{Re}_\lambda$ is so large that an equilibrium range in the turbulence-energy spectrum exists. If $\mathfrak{k}$ is of the same order as ν, $u'\lambda_\gamma/\mathfrak{k}$ will also be very large. For the wavenumbers in this equilibrium range it may also be assumed that

$$\frac{\partial}{\partial t}\int_k^\infty dk\,\mathbf{E}_\gamma(k,t) \ll \frac{\partial}{\partial t}\int_0^\infty dk\,\mathbf{E}_\gamma(k,t)$$

so that

$$\frac{\partial}{\partial t}\int_0^k dk\,\mathbf{E}_\gamma(k,t) \simeq \frac{d}{dt}\int_0^\infty dk\,\mathbf{E}_\gamma(k,t) = -\varepsilon_\gamma$$

For Eq. (3-258) we may then write

$$2\left[\mathfrak{k} + \beta\int_k^\infty dk\sqrt{\frac{\mathbf{E}(k,t)}{k^3}}\right]\int_0^k dk\,k^2\mathbf{E}_\gamma(k,t) = \varepsilon_\gamma \tag{3-259}$$

The solution of this equation can be readily obtained, and reads

$$\mathbf{E}_\gamma(k,t) = \frac{\beta\varepsilon_\gamma\sqrt{\dfrac{\mathbf{E}(k,t)}{k^3}}}{2k^2\left[\mathfrak{k} + \beta\displaystyle\int_k^\infty dk\sqrt{\frac{\mathbf{E}(k,t)}{k^3}}\right]^2} \tag{3-260}$$

Hence $\mathbf{E}_\gamma(k,t)$ is known when $\mathbf{E}(k,t)$ is known.

For the turbulent-velocity field we had accepted Heisenberg's theory only for the not-large wave number range. If there is an equilibrium range with an inertial subrange, in that range $\epsilon \gg \nu$, and the direct viscous effects can be neglected. When we assume the molecular diffusivities of the same order of magnitude, so $\nu/\mathfrak{k} = \mathcal{O}(1)$, also $\epsilon_\gamma \gg \mathfrak{k}$ and we may neglect $\mathfrak{k}$ relative to ϵ_γ in Eq. (3-260).

Hence Eq. (3-260) reduces to

$$\mathbf{E}_\gamma(k,t) = \frac{\varepsilon_\gamma}{2\beta}\frac{\sqrt{\dfrac{\mathbf{E}(k,t)}{k^3}}}{\left[k\displaystyle\int_k^\infty dk\,\frac{\mathbf{E}(k,t)}{k^3}\right]^2}$$

Substitution of Eq. (3-123) for $\mathbf{E}(k,t)$ in the inertial subrange, together with the expression (3-143) for the constant A then yields

$$\mathbf{E}_\gamma(k,t) = \left(\frac{8}{9}\right)^{2/3}\frac{\alpha^{1/3}}{\beta}\varepsilon_\gamma(t)[\varepsilon(t)]^{-1/3}k^{-5/3} \tag{3-261}$$

Or, written in dimensionless form

$$\mathbf{E}_\gamma(k,t)\cdot\varepsilon_\gamma^{-1}\varepsilon^{3/4}\mathfrak{k}^{-5/4} = \left(\frac{8}{9}\right)^{2/3}\frac{\alpha^{1/3}}{\beta}(k/k_{\gamma,d})^{-5/3} \tag{3-261a}$$

Here $k_{\gamma,d} = (\varepsilon/\mathfrak{k}^3)^{1/4}$ is the wavenumber marking the separations between the range where molecular diffusion is negligible from the range in which it is important.

Equation (3-261) is exactly the same law found for the energy-spectrum function.

Corrsin[34] has shown that the same result may be obtained from dimensional reasoning, postulating that, for the higher-wavenumber range considered, the spectrum $\mathbf{E}_\gamma(k,t)$ depends only on the energy flow rate $\varepsilon(t)$ and the flow rate $\varepsilon_\gamma(t)$ through this range.

Equation (3-261) describes the spectral distribution of a scalar in the *inertial-convective subrange* $k_0 \ll k \ll k_{\gamma,d}$. Here k_0 is the wavenumber marking the maximum value of $\mathbf{E}_\gamma(k,t)$. When k is no longer small compared with $k_{\gamma,d}$, so that direct molecular diffusion effects are no longer negligibly small, a faster decrease with increasing k may be expected.

Pao[104] has extended his cascade theory also to the spectral distribution of a scalar in the equilibrium range. Here too he postulates that the flux of $\overline{\gamma^2}$ through a cascading process, i.e., $\mathbf{G}_\gamma(k,t)$, is mainly caused by turbulent convection, and so by the energy flux ε supplied by the large eddies. This scalar flux $\mathbf{G}_\gamma$ is put proportional to the spectral density $\mathbf{E}_\gamma(k,t)$. Dimensional reasoning yields

$$\mathbf{G}_\gamma(k,t) = \int_0^k dk\, \mathbf{F}_\gamma(k,t) = -B_\gamma \varepsilon^{1/3} k^{5/3} \mathbf{E}_\gamma(k,t)$$

Since $\mathbf{F}_\gamma(k,t) = 2\mathfrak{k} k^2 \mathbf{E}_\gamma(k,t)$, the solution for $\mathbf{E}_\gamma(k,t)$ becomes similar to Eq. (3-146a) for the energy spectrum, i.e.:

$$\mathbf{E}_\gamma(k,t) = A_\gamma \varepsilon_\gamma \varepsilon^{-1/3} k^{-5/3} \exp\left[-\tfrac{3}{2} A_\gamma \frac{\mathfrak{k}}{\varepsilon^{1/3}} k^{4/3}\right] \qquad (3\text{-}262)$$

Or, written in dimensionless form

$$\mathbf{E}_\gamma(k,t)\varepsilon_\gamma^{-1}\varepsilon^{3/4}\mathfrak{k}^{-5/4} = A_\gamma \left(\frac{k}{k_{\gamma,d}}\right)^{-5/3} \exp\left[-\tfrac{3}{2} A_\gamma \left(\frac{k}{k_{\gamma,d}}\right)^{4/3}\right] \qquad (3\text{-}262a)$$

One-dimensional spectra $\mathbf{E}_{\gamma,1}(k_1,t)$ have been measured of temperature fluctuations in air flow ($\mathbf{Pr} \simeq 0.74$) by Corrsin and Uberoi[108] and by Kistler, O'Brien, and Corrsin.[35] The first-mentioned measurements were made in the axis of a heated turbulent air jet, where the turbulence is close to isotropy, while Kistler *et al.* made the measurements downstream of a heated grid in a windtunnel.

The one-dimensional spectrum $\mathbf{E}_{\gamma,1}(k_1)$ can be obtained by integrating Eq. (3-262) according to the relation (3-232). Pao carried out the integration numerically, with $A_\gamma = 0.59$, a value which he deduced from Gibson and Schwarz's[105] measurements of temperature fluctuations behind a grid in a water tunnel. Figure 3-28 shows the result of Pao's theoretical $\mathbf{E}_{\gamma,1}(k_1,t)$ compared with the data for the temperature fluctuations in air. It must be remarked that the Reynolds numbers at which the experimental data have been obtained are quite low, so that an inertial subrange can hardly be expected.

Hitherto we have assumed that $\mathfrak{k}$ and ν are of the same order of magnitude. It may be expected that the behavior of $\mathbf{E}_\gamma(k,t)$ at large wavenumbers will largely

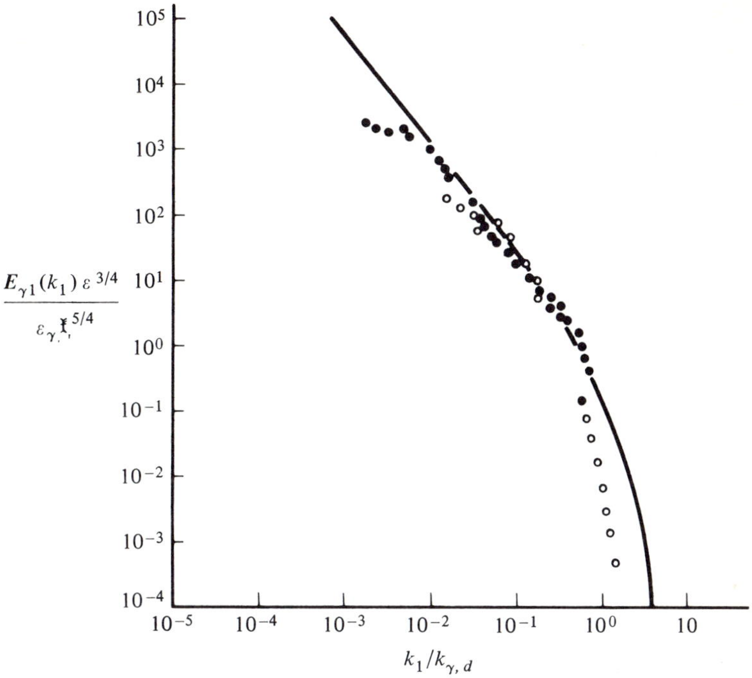

FIGURE 3-28
One-dimensional temperature spectrum.
——— Theoretical according to Pao.
●●●● Heated jet. ○○○○ Heated grid.
(After: *Pao, Y. H.,*[77] *by permission of the American Institute of Physics.*)

depend on the ratio $\nu/\mathfrak{k}$, i.e., the molecular Prandtl or Schmidt number. When $\nu/\mathfrak{k} \ll 1$ the effect of molecular diffusion becomes of importance already at $k \ll k_d$, since $k_{\gamma,d} \ll k_d$. On the other hand when $\nu/\mathfrak{k} \gg 1$, and thus $k_{\gamma,d} \gg k_d$, the effect becomes of importance only at $k \gg k_d$.

Let us first consider *very small values of the Schmidt number*, $\nu/\mathfrak{k} \ll 1$. Then it is possible that the direct effect of molecular diffusion penetrates into the inertial subrange of the energy spectrum, causing the spectrum $\mathbf{E}_\gamma(k,t)$ to fall off more rapidly than according to $k^{-5/3}$.

We may expect the following dependency

$$\mathbf{E}_\gamma(k,t) = f[\mathbf{E}(k,t), \varepsilon_\gamma(t), \mathfrak{k}, k]$$

Since only $\mathbf{E}_\gamma(k,t)$ and $\varepsilon_\gamma(t)$ include the dimension of $\overline{\gamma^2}$, the relation must be of the form

$$\mathbf{E}_\gamma(k,t) = \varepsilon_\gamma(t) \cdot f_1[\mathbf{E}, \mathfrak{k}, k]$$

Batchelor, Howells, and Townsend[59] made the assumption that the contribution to a Fourier component of $\mathbf{E}_\gamma(k,t)$ is determined by the Fourier component of the energy spectrum $\mathbf{E}(k,t)$. This assumption leads to

$$\mathbf{E}_\gamma(k,t) = \varepsilon_\gamma \cdot \mathbf{E} \cdot f_2[\mathfrak{k},k]$$

Dimensional analysis then yields

$$\mathbf{E}_\gamma(k,t) = \text{const.}\ \varepsilon_\gamma \cdot \mathbf{E} \cdot \mathfrak{k}^{-3} k^{-4}$$

Batchelor et al. obtained the value of $\frac{1}{3}$ for the numerical constant. With the Kolmogoroff expression for $\mathbf{E}$ in the inertial subrange we finally obtain the relation, valid for the *inertial-diffusive subrange*: $k_{\gamma,d} < k < k_d$:

$$\mathbf{E}_\gamma(k,t) = \text{const.}\ \varepsilon_\gamma(t) \cdot [\varepsilon(t)]^{2/3} \mathfrak{k}^{-3} k^{-17/3} \qquad (3\text{-}263)$$

Or, written in dimensionless form

$$\mathbf{E}_\gamma(k,t) \cdot \varepsilon_\gamma^{-1} \varepsilon^{3/4} \mathfrak{k}^{-5/4} = \text{const.}\ (k/k_{\gamma,d})^{-17/3} \qquad (3\text{-}263a)$$

Beyond $k = k_d$ the spectrum $\mathbf{E}_\gamma(k,t)$ will decrease even more rapidly with increasing k.

At *large values of the Schmidt number*: $\nu/\mathfrak{k} \gg 1$, there is a range of wavenumbers $k_d < k < k_{\gamma,d}$ where the turbulent motion is in the viscous range and molecular diffusion effects are still negligible. In this range of viscous motions, the main contribution to the convective transport of γ is probably due to the distortion of fluid elements. Consider a fluid element that is sufficiently small, so that its size is smaller than $1/k_d$. As will be shown in Sec. 5-4, owing to the distortion of such a fluid element into a long and thin ribbon, variations in γ remain substantially in the direction of the principal axis of least rate of strain, gradients in γ become steeper, and the effect of molecular diffusion increases. Batchelor[58] showed that convection due to the straining action and molecular diffusion become of the same order of magnitude at $k_{\gamma,c} = (\varepsilon/\nu\mathfrak{k}^2)^{1/4}$. This follows immediately from a dimensional analysis. For, the straining action is determined by $[\overline{(\partial u_1/\partial x_1)^2}]^{1/2}$. Since $\overline{(\partial u_1/\partial x_1)^2} \propto \varepsilon/\nu$, we may put

$$k_{\gamma,c} = f(\varepsilon/\nu,\mathfrak{k})$$

whence follows the above expression for $k_{\gamma,c}$.

In the range $k_d < k < k_{\gamma,c}$, convection is predominant and the distortion process converts Fourier components of lower wavenumber into components of higher wavenumber. Batchelor obtained the following expression for the spectrum of γ-fluctuations:

$$\mathbf{E}_\gamma(k,t) = C\left[\frac{\nu}{\varepsilon(t)}\right]^{1/2} \frac{\varepsilon_\gamma(t)}{k} \exp\left(-C\frac{k^2}{k_{\gamma,c}{}^2}\right) \qquad (3\text{-}264)$$

where C is a numerical constant of order 1, being the constant of proportionality between the effective average least-principal-rate of strain γ and $(\varepsilon/\nu)^{1/2}$, namely $\gamma = -C^{-1}(\varepsilon/\nu)^{1/2}$.

In studying the energy transfer in the viscous wavenumber range $k > k_d$,

Saffman also assumed the same mechanism of transfer through viscous distortion of the extremely small regions smaller than $\eta = (\nu^3/\varepsilon)^{1/4}$. Thus the above result, Eq. (3-264), should also be included in Saffman's more general expression (3-147), which then has to be modified to adapt it to the spectral distribution of a scalar. Indeed Eq. (3-264) can be obtained from Eq. (3-147) when $\mathfrak{k}$ is read for ν in the exponential function, and when we put $\sigma = 1$. Since in Eq. (3-147), $\alpha = s(\varepsilon/\nu)^{1/2}$, we conclude that the constant C in Eq. (3-264) is equal to $1/s$.

Now exactly the same result is obtained when we apply the procedure used by Panchev and Kesich for the energy spectrum in the viscous wavenumber range to the spectral distribution of a scalar. To this end we assume for the total flux from the wavenumber range 0–k, i.e., $\mathbf{G}_\gamma(k,t)$ the same expression as Eq. (3-137).

$$\mathbf{G}_\gamma(k,t) = \int_0^k dk\, \mathbf{F}_\gamma(k,t) = -\beta \mathbf{E}_\gamma(k,t) \cdot k \left[2 \int_0^k dk'\, k'^2 \mathbf{E}(k',t) \right]^{1/2}$$

For large values of k this is approximately equal to

$$\mathbf{G}_\gamma(k,t) \simeq -\beta \mathbf{E}_\gamma(k,t) \cdot k \left(\frac{\varepsilon}{\nu} \right)^{1/2} = -2\mathfrak{k} \int_k^\infty dk'\, k'^2 \mathbf{E}_\gamma(k',t) \qquad \text{(3-265)}$$

since in the equilibrium range

$$-\varepsilon_\gamma = -2\mathfrak{k} \int_0^\infty dk'\, k'^2 \mathbf{E}_\gamma(k',t) = \int_0^k dk'\, \mathbf{F}_\gamma(k',t) - 2\mathfrak{k} \int_0^k dk'\, k'^2 \mathbf{E}_\gamma(k',t)$$

The equation (3-265) is solved in the same way as given for the energy spectrum, taking as the boundary condition

$$\int_0^\infty dk'\, k'^2 \mathbf{E}_\gamma(k',t) = \varepsilon_\gamma / 2\mathfrak{k}$$

The result is an equation identical with Eq. (3-148):

$$\mathbf{E}_\gamma(k,t) = \frac{1}{\beta} \cdot \varepsilon_\gamma \sqrt{\frac{\nu}{\varepsilon}} \cdot \frac{1}{k} \exp\left(-\frac{1}{\beta} k^2 / k_{\gamma,c}{}^2 \right) \qquad \text{(3-266)}$$

We note, when comparing it with Eq. (3-264), that $\beta = 1/C$.

We also note that Eq. (3-266) is obtained without making any assumption concerning the transfer mechanism.

In order to make it possible to compare Eq. (3-264) with Eq. (3-261a) for the inertial-convective subrange we rewrite Eq. (3-264) in the following nondimensional form

$$\mathbf{E}_\gamma(k,t)\varepsilon_\gamma{}^{-1}\varepsilon^{3/4}\mathfrak{k}^{-5/4} = C\mathbf{Sc}^{1/2} \frac{k_{\gamma,d}}{k} \exp\left[-C\mathbf{Sc}^{1/2} \cdot k^2 / k_{\gamma,d}{}^2 \right] \qquad \text{(3-264a)}$$

where we have written the Schmidt number $\mathbf{Sc}$ for $\nu/\mathfrak{k}$. For $k_d < k \ll k_{\gamma,c}$, Eq. (3-264a) reduces to

$$\mathbf{E}_\gamma(k,t)\varepsilon_\gamma{}^{-1}\varepsilon^{3/4}\mathfrak{k}^{-5/4} = C\mathbf{Sc}^{1/2} \cdot \frac{k_{\gamma,d}}{k} \qquad \text{(3-267)}$$

An alternative way is to express the right-hand sides of Eqs. (3-264a) and 3-267) in terms of k/k_d:

$$\mathbf{E}_\gamma(k,t)\varepsilon_\gamma^{-1}\varepsilon^{3/4}\mathfrak{f}^{-5/4} = C\mathbf{Sc}^{5/4}\cdot\frac{k_d}{k}\exp-[C\mathbf{Sc}^{-1}k^2/k_d^{\,2}] \qquad (3\text{-}264b)$$

$$\mathbf{E}_\gamma(k,t)\varepsilon_\gamma^{-1}\varepsilon^{3/4}\mathfrak{f}^{-5/4} \simeq C\mathbf{Sc}^{5/4}\cdot k_d/k \qquad (3\text{-}267a)$$

From the definition of k_d, $k_{\gamma,c}$, and $k_{\gamma,d}$ it follows that

$$\frac{k_{\gamma,c}}{k_d} = \left(\frac{\nu}{\mathfrak{f}}\right)^{1/2} = \mathbf{Sc}^{1/2}; \quad \frac{k_{\gamma,d}}{k_d} = \mathbf{Sc}^{3/4} \quad \text{and} \quad \frac{k_{\gamma,d}}{k_{\gamma,c}} = \mathbf{Sc}^{1/4}$$

The k^{-1} curve intersects the Pao-curve [Eq. (3-262a)] in the vicinity of k_d. We may estimate the position of this intersection, marked by the wavenumber k^*, by taking the $k^{-5/3}$ part instead of the complete relation (3-262a). We then obtain

$$k^*/k_d = (A_\gamma/C)^{3/2} \quad \text{or} \quad k^*/k_{\gamma,d} = (A/C)^{3/2}\cdot\mathbf{Sc}^{-3/4}$$

As we will show later, from experimental evidence it may be concluded that $A/C = \mathcal{O}(0.1)$ so that the viscous convective range practically adjoins the inertial subrange.

Figure 3-29 shows the expected shape of the $\mathbf{E}_\gamma(k,t)$ distribution when $\mathbf{Sc} \gg 1$, so that there is a viscous-convective subrange. According to Eqs. (3-262a)

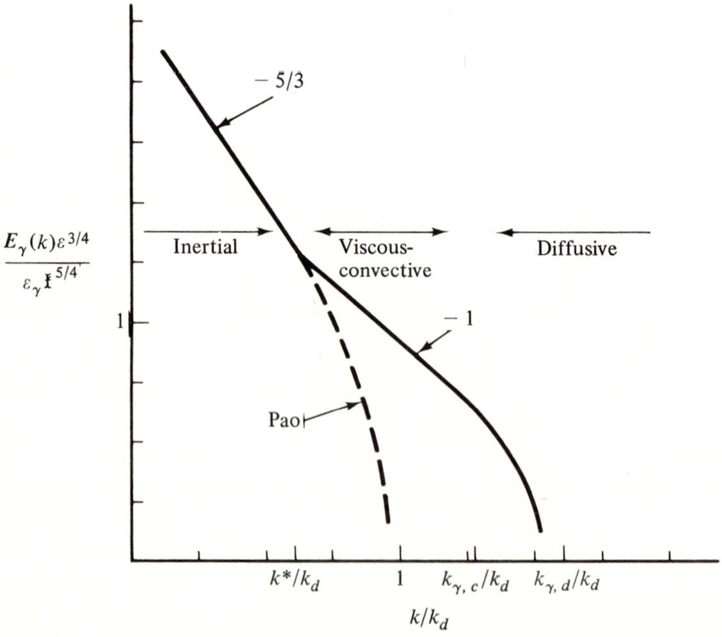

FIGURE 3-29
Spectral distribution $\mathbf{E}_\gamma(k,t)$ for $\nu/\mathfrak{f} \gg 1$ according to theory.

and (3-264b) the value of the normalized spectrum, given by the left-hand side of these equations, depends on $(\nu/\mathfrak{k})$, increasing with it, when k_d is taken to reduce k. So do the values of $k_{\gamma,c}/k_d$ and $k_{\gamma,d}/k_d$. However, k^*/k_d is independent of $(\nu/\mathfrak{k})$.

For comparison with experimental evidence there are available the results of measurements by Gibson and Schwarz,[105] by Nye and Brodkey,[106] and by Grant, Hughes, Vogel, and Moilliet.[107] They determined the one-dimensional spectrum $\mathbf{E}_{\gamma,1}(k,t)$ which is related to the three-dimensional spectrum $\mathbf{E}_\gamma(k,t)$ according to Eq. (3-232). The integral can easily be evaluated when Eq. (3-264) is taken for $\mathbf{E}_\gamma(k,t)$ and an integration by parts is applied, since

$$\int_{x_1}^{\infty} dx \frac{a}{x^2} \exp(-bx^2) = a\left[\frac{1}{x_1}\exp(-bx_1{}^2) - (2\pi b)^{1/2} \cdot \operatorname{erfc}(x_1\sqrt{b})\right]$$

Here

$$\operatorname{erfc} \cdot z = 1 - \frac{2}{\sqrt{\pi}}\int_0^z d\alpha \exp(-\alpha^2)$$

Gibson and Schwarz[105] made the measurements in grid-generated turbulence in a closed-loop water tunnel through which dilute salt water was circulated. Through tiny holes in the hollow round tubes, of which the square-mesh grid was composed, concentrated salt solution was injected for the concentration fluctuation measurements, or heated dilute salt water for temperature fluctuations measurements. Turbulence velocity measurements were carried out with a Lintronic constant-temperature hot-film anemometer, whereas for the measurement of the concentration fluctuations as well as for the temperature fluctuations a conductivity probe was used. See Sec. 2-8 and ref. 137. The Reynolds number $\bar{U}_1 M/\nu$ was varied between 10,000 and 67,000. Only at the highest Reynolds number did the energy spectrum show an inertial subrange (roughly over a decade of wavenumbers). Yet a definite departure from the $(-5/3)$ law could be observed for all experiments at a $k_1/k_d \simeq 0.1$, towards a (-1)-law. A reasonable agreement was obtained with Batchelor's theory with the value $C = 2$ recommended by him. Though for the salt water $\mathbf{Sc} \simeq 700$, and for the heated water $Pr \simeq 7$, no clear distinction was found between the spectra for the concentration fluctuations and the temperature fluctuations respectively.

Nye and Brodkey[106] experimented with dye injection in pipe flow ($D = 78$ mm; $\mathbf{Re}_D \simeq 50{,}000$), using an optical measuring technique for the concentration fluctuations. The dye was injected through a centrally located small coaxial tube. For the dye $\mathbf{Sc} \simeq 4{,}000$. They too found that the spectra of the concentration fluctuations taken in the axis of the pipe at 24 and 36 pipe diameters downstream of the injection tube, showed a (-1)-slope.

Grant *et al.*[107] made measurements at large Reynolds number in a tidal channel. The Reynolds number referred to depth of the channel and the mean speed, was 3×10^8. Measurements of temperature fluctuations caused by the turbulent mixing of a warm top layer with the colder under layers during ebb tide, were made with

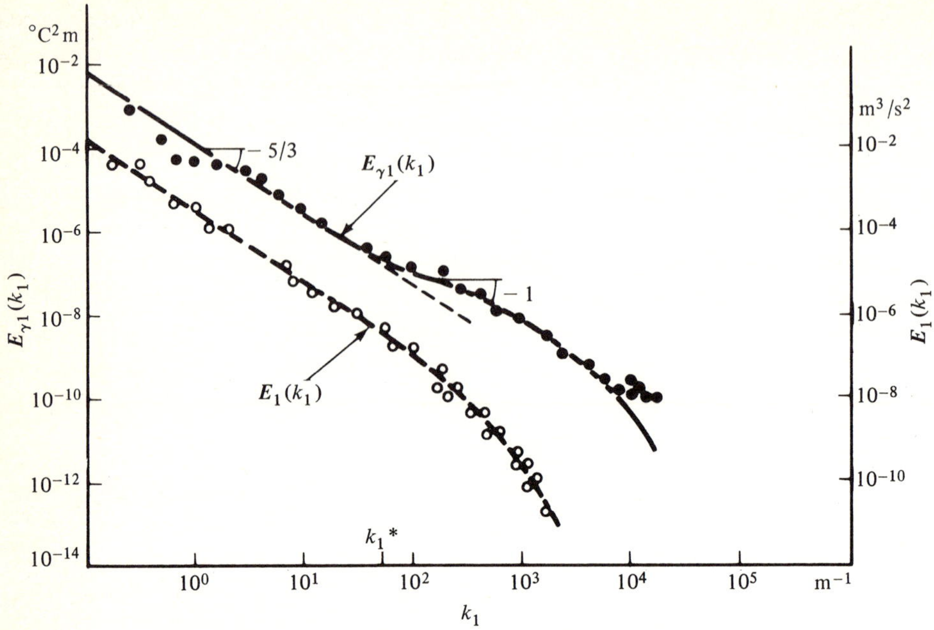

FIGURE 3-30
One-dimensional spectra of temperature and velocity fluctuations measured in a tidal channel. (After: *Grant, H. L., B. A. Hughes, W. M. Vogel, and A. Moilliet,*[107] *by permission of the Cambridge University Press.*)

a conical thin-film resistance thermometer, towed with constant speed through the channel. A typical result of the one-dimensional spectrum $\mathbf{E}_{\gamma,1}(k_1)$ given in dimensional form is shown in Fig. (3-30). The measurements were made during one run at a depth of 23 m and with a towing speed of 1.27 m/s. In the same figure also the one-dimensional energy spectrum is shown, obtained from measurements during the same run.

This figure clearly shows the rather sudden change of the slope of the curve from $(-5/3)$ to (-1) at $k_1 = k_1^*$, which sudden change in slope does not show up in the energy spectrum $\mathbf{E}_1(k_1)$. The (-1)-part of the slope corresponds with a value of $C = 3.6$, a larger value than recommended by Batchelor. However, for the various runs made, quite different values of C had to be taken to adapt the theoretical curve to the experimental data, ranging from 0.72 to 13.2. The average value was $C = 3.9 \pm 1.5$ standard deviation. The value of k_1^*/k_d for the test run represented in Fig. 3-30 was 0.02. This value too differed appreciably for the different test runs, namely from 0.007 to 0.043, with an average value of 0.021. From the relation $k_1^*/k_d = (A_\gamma/C)^{3/2}$ we obtain for the one-dimensional spectrum

$$k_1^*/k_d = \left[\frac{3}{5}\frac{A_\gamma}{C}\right]^{3/2}$$

With $k_1^*/k_d = 0.02$ and $C = 3.9$ we obtain $A_\gamma = 0.48$, which is smaller than the value of 0.59 used by Pao on the basis of Gibson and Schwarz's experimental results. Actually A_γ varied between 0.22 and 0.78 with an average value of 0.48.

By Eqs. (3-262), (3-263), and (3-264), we have obtained expressions for the spectrum function $\mathbf{E}_\gamma(k,t)$ valid in the high inertial subrange of wavenumbers and in the highest-wavenumber range. At the lower extreme of the wavenumber range, $\mathbf{E}_\gamma(k,t)$ is proportional to k^2. No solution has been obtained for the range of the energy-containing eddies where the maximum of the energy spectrum occurs and where it is reasonable to expect the maximum of the $\mathbf{E}_\gamma(k,t)$ spectrum function. But, just as Von Kármán did for the energy spectrum $\mathbf{E}(k,t)$, we suggest an interpolation formula for $\mathbf{E}_\gamma(k,t)$ covering the range from $k = 0$ to the range that includes the inertial subrange, viz.:

$$\mathbf{E}_\gamma(k,t) = \mathbf{E}_\gamma[k_0(t)] \times 2^{11/6} \frac{(k/k_0)^2}{[1 + (k/k_0)^2]^{11/6}} \tag{3-268}$$

Compare the identical expression (3-160) for the energy spectrum when Saffman's dynamical invariant is used. For small k/k_0 the formula reduces to

$$\mathbf{E}_\gamma(k,t) = \mathbf{E}_\gamma(k_0)2^{11/6}\left(\frac{k}{k_0}\right)^2 = I_\gamma k^2$$

so that $\mathbf{E}_\gamma(k_0)2^{11/6} = I_\gamma {k_0}^2$.

On the other hand, for very large k/k_0, the formula yields the $-\frac{5}{3}$ law. With Eq. (3-232) we can calculate the one-dimensional spectrum $\mathbf{E}_{\gamma 1}(k_1,t)$:

$$\mathbf{E}_{\gamma 1}(k_1,t) = \tfrac{3}{5} \times 2^{11/6}\mathbf{E}_\gamma(k_0)\left[1 + \left(\frac{k_1}{k_0}\right)^2\right]^{-5/6} \tag{3-269}$$

The function $\mathbf{E}_\gamma(k_0)$, which also corresponds to $I_\gamma {k_0}^2$, may be determined from the condition that the correlation function $\mathbf{Q}_{\gamma,\gamma}(r,t)$ must become equal to γ'^2 when $r = 0$.

According to Eq. (3-221) with Eq. (3-269), we obtain

$$\mathbf{Q}_{\gamma,\gamma}(r,t) = \tfrac{3}{5} \times 2^{11/6} \frac{\sqrt{\pi}\,k_0}{2^{1/3}\Gamma(\frac{5}{6})} \mathbf{E}_\gamma(k_0)(k_0 r)^{1/3} K_{1/3}(k_0 r)$$

and so the condition $\mathbf{Q}_{\gamma,\gamma}(0,t) = \gamma'^2$ yields

$$\mathbf{E}_\gamma(k_0) = \frac{\Gamma(\frac{5}{6})}{\frac{3}{5}2^{5/6}\sqrt{\pi}\,\Gamma(\frac{1}{3})k_0}\gamma'^2$$

With this expression for $\mathbf{E}_\gamma(k_0)$ we obtain for $\mathbf{Q}_{\gamma,\gamma}(r,t)$, $\mathbf{E}_\gamma(k,t)$, and $\mathbf{E}_{\gamma 1}(k_1,t)$, respectively:

$$\mathbf{Q}_{\gamma,\gamma}(r,t) = \gamma'^2 \frac{2^{2/3}}{\Gamma(\frac{1}{3})}(k_0 r)^{1/3} K_{1/3}(k_0 r) \tag{3-270}$$

$$\mathbf{E}_\gamma(k,t) = \gamma'^2 \frac{10}{3} \frac{\Gamma(\frac{5}{6})}{\sqrt{\pi}\,\Gamma(\frac{1}{3})k_0} \frac{(k/k_0)^2}{[1 + (k/k_0)^2]^{11/6}} \tag{3-271}$$

and

$$\mathbf{E}_{\gamma 1}(k_1,t) = \gamma'^2 \frac{2\Gamma(\frac{5}{6})}{\sqrt{\pi}\,\Gamma(\frac{1}{3})k_0}\left[1 + \left(\frac{k_1}{k_0}\right)^2\right]^{-5/6} \qquad (3\text{-}272)$$

The integral scale Λ_γ, determined from $\mathbf{E}_{\gamma 1}(0,t)$, using Eqs. (3-232) and (3-233), reads

$$\Lambda_\gamma(t) = \frac{\pi}{2}\frac{1}{\gamma'^2}\mathbf{E}_{\gamma 1}(0,t) = \frac{\sqrt{\pi}\,\Gamma(\frac{5}{6})}{\Gamma(\frac{1}{3})k_0} \simeq \frac{0.75}{k_0} \qquad (3\text{-}273)$$

Comparison with the value obtained for Λ_f [see Eq. (3-158)] shows that the values expressed in terms of k_0 and k_e, respectively, are the same. Just as it was possible to express k_e in terms of u' and ε, so it is possible to express k_0 in terms of γ' and ε_γ. From Eqs. (3-261) and (3-271) for high values of k/k_0, we obtain

$$k_0(t) = \frac{8}{9}\left[\frac{3}{10}\frac{\sqrt{\pi}\,\Gamma(\frac{1}{3})}{\Gamma(\frac{5}{6})}\right]^{3/2}\sqrt{\frac{\alpha}{\beta^3}\frac{\varepsilon_\gamma^{\;3/2}}{\varepsilon^{1/2}\gamma'^3}}$$

Incidentally, it may further be remarked that the expressions for $\mathbf{Q}_{\gamma,\gamma}(r,t)$ and $\mathbf{E}_{\gamma 1}(k_1,t)$ are identical with those for $\mathbf{f}(r,t)$ and $\mathbf{E}_1(k_1,t)$.

The Decay of an Isotropic Scalar Field

An isotropic scalar field present in or superimposed upon an isotropic-turbulent velocity field will decay together with this velocity field. Of course, it is possible to imagine an isotropic scalar field superimposed on a nondecaying isotropic-turbulent velocity field, although the latter is hardly possible unless the necessary energy to maintain the turbulence can be so introduced that isotropy is not disturbed. It might be useful to assume such a hypothetical case in order to study some basic properties of mixing.[109,110,111] Here, however, we shall not do this, but rather consider the two isotropic fields while they are decaying simultaneously.

We have already dealt with the decay behavior of a scalar field where both $u'\lambda_\gamma/\mathfrak{k}$ and $\mathbf{Re}_\lambda$ are so small that the convective terms vanish in the dynamic equations (3-254) and (3-255). During decay, complete self-preservation occurs for the spectrum as well as for the correlation functions when k and r, respectively, are expressed in terms of the length $\sqrt{\mathfrak{k}t}$

We do know that for a decaying turbulent velocity field no complete self-preservation of the energy-spectrum function or of the velocity-correlation function is possible when $\mathbf{Re}_\lambda$ is no longer very small.

Let us first assume that during a certain stage of the decay process the decay rate of γ'^2 may be described by a single power law:

$$\frac{\gamma'^2}{\Gamma^2} = (A_1 t + B_1)^{-m} \qquad (3\text{-}274)$$

Equation (3-219) then yields

$$\lambda_\gamma^2 = \frac{12\mathfrak{k}}{m}\left(t + \frac{B_1}{A_1}\right) \tag{3-275}$$

Comparison of this expression for λ_γ^2 with the corresponding expression (3-168) for λ_g^2 shows that

$$\frac{\lambda_\gamma^2}{\lambda_g^2} = \frac{6}{5} \cdot \frac{n}{m} \frac{\mathfrak{k}}{\nu} \cdot \frac{t + B_1/A_1}{t + B/A} \tag{3-276}$$

This is the relation between λ_γ and λ_g to which reference was made earlier. The square of their ratio is directly proportional with $\mathfrak{k}/\nu$, and may change with time, except when the initial instants at which λ_γ and λ_g are zero, are the same for the two.

We may apply to the dynamic equation for $R_{\gamma,\gamma}(r,t)$, i.e., Eq. (3-217), the same procedure for studying a possible behavior of self-preservation of the scalar field as we did for the velocity field. So let again $\psi = r/\mathscr{L}(t)$, and $\mathbf{R}_{\gamma,\gamma}(r,t)$ and $\mathbf{k}_\gamma(r,t)$ be functions of ψ only. With Eq. (3-219) we obtain from Eq. (3-217):

$$12\mathbf{R}_{\gamma,\gamma} + \frac{\lambda_\gamma^2}{\mathfrak{k}} \frac{1}{\mathscr{L}} \frac{d\mathscr{L}}{dt} \psi \frac{dR_{\gamma,\gamma}}{d\psi} + 2\frac{u'\lambda_\gamma^2}{\mathfrak{k}\mathscr{L}} \frac{1}{\psi^2} \frac{d}{d\psi}(\psi^2 \mathbf{k}_\gamma) + 2\frac{\lambda_\gamma^2}{\mathscr{L}^2} \frac{1}{\psi^2} \frac{d}{d\psi}\left(\psi^2 \frac{d\mathbf{R}_{\gamma,\gamma}}{d\psi}\right) = 0$$

Thus, to satisfy the condition of complete self-preservation the following relations have to exist.

$$\frac{\lambda_\gamma}{\mathscr{L}} = \text{const.} \tag{3-277}$$

$$\frac{\lambda_\gamma^2 u'}{\mathfrak{k}\mathscr{L}} = \text{const.,} \quad \text{or} \quad \frac{u'\mathscr{L}}{\mathfrak{k}} = \text{const.} \tag{3-278}$$

$$\frac{\lambda_\gamma^2}{\mathfrak{k}} \frac{1}{\mathscr{L}} \frac{d\mathscr{L}}{dt} = \text{const.,} \quad \text{or} \quad \frac{1}{\mathfrak{k}} \mathscr{L} \frac{d\mathscr{L}}{dt} = \text{const.} \tag{3-279}$$

We have three relations for u', $\mathscr{L}$, and λ_γ. They do not contain γ'. But we still have the invariancy equation (3-246), which yields

$$\gamma'^2 \mathscr{L}^3 = \text{const.} \tag{3-280}$$

From Eq. (3-279) we conclude that

$$\mathscr{L}^2 \propto A_1 t + B_1 \tag{3-281}$$

and, hence, from Eq. (3-278), $u'^2 \propto (A_1 t + B_1)^{-1}$, which is the decay law for a completely self-preserving velocity field. Then, Eq. (3-280) yields

$$\gamma'^2 \propto (A_1 t + B_1)^{-3/2} \tag{3-282}$$

The conclusion is that complete self-preservation of the scalar field is possible, when the turbulence-velocity field decays according to a "linear" decay law, i.e., in the initial period of decay. The decay law for γ'^2 is given by Eq. (3-282), which is the same as for the "viscous" stage according to Eq. (3-254). It is possible that in the transition period of decay, the decay of γ'^2 differs from the $(-\frac{3}{2})$-law, but this does not seem very reasonable. Since in the transition period of decay the velocity field does not follow the "linear" decay law, Eqs. (3-278) and (3-279) become conflicting. No complete self-preservation of the scalar field then is possible.

Now, we have shown that for the velocity field it is not possible to assume complete self-preservation during the decay process, but that we have to admit partial self-preservation. Similarly we may consider partial self-preservation for the scalar field, and consider Λ_γ for the length scale. At the same time we drop the molecular diffusion term in Eq. (3-217), so that the condition $\lambda_\gamma/\Lambda_\gamma = \text{const.}$, does not apply. From the Eqs. (3-278) and (3-279) we obtain:

$$\frac{d\Lambda_\gamma}{dt} = \text{const. } u' \tag{3-283}$$

and

$$\lambda_\gamma{}^2 = \text{const. } \frac{\Lambda_\gamma}{u'} \tag{3-284}$$

With Eq. (3-184) for u', following from the Saffman invariance condition, we obtain

$$\Lambda_\gamma \propto (At + B)^{2/5} \tag{3-285}$$

and with Eq. (3-280)

$$\gamma'^2 \propto (At + B)^{-6/5} \tag{3-286}$$

With Eq. (3-188) for u', following from Loitsianskii's invariance condition, we obtain

$$\Lambda_\gamma \propto (At + B)^{2/7} \tag{3-287}$$

and with Eq. (3-280)

$$\gamma'^2 \propto (At + B)^{-6/7} \tag{3-288}$$

It is possible to give a description of the decay process of a scalar similar to the one of the velocity field suggested by Naudascher and Farell, given earlier. Instead of Eq. (3-274) we consider

$$\frac{d\lambda_\gamma}{dt} = \sqrt{\frac{3k}{mt_0}}\left(\frac{1}{C_\gamma}\frac{\gamma'}{T}\right)^{1/m}.$$

with constant m. Here C_γ is the constant corresponding to C in Eq. (3-167a), and $t_0 = -B_1/A_1$. This relation is obtained from Eqs. (3-274) and (3-275), so that with Eq. (3-275) there are two equations from which λ_γ and γ' can be solved. When, again, a finite $(\lambda_\gamma)_\infty$ is assumed, the solution yields a relation for $\lambda_\gamma/(\lambda_\gamma)_\infty$ similar to Eq. (3-194), when we read $6\mathfrak{k}/m(\lambda_\gamma)_\infty{}^2$ for $5\nu/n\lambda_\infty{}^2$. The decay law of γ' then is obtained from a relation similar to Eq. (3-195). Here, too, the decay law yields an exponential expression for the final period of decay. However, we do know that for this period the decay is according to the $(-\frac{3}{2})$-law, expressed by Eq. (3-254) even for an almost homogeneous turbulent scalar field. For, neglecting triple-correlation effects in the final period, the solution for $\mathbf{E}_{\gamma,\gamma}(k_1,k_2,k_3,t)$ is

$$\mathbf{E}_{\gamma,\gamma}(k_1,k_2,k_3,t) = \mathbf{E}_{\gamma,\gamma}(k_1,k_2,k_3,t_0)\exp\left[-2\mathfrak{k}k^2(t-t_0)\right]$$

Since $\mathbf{E}_{\gamma,\gamma}$ is analytic at $k = 0$, and approaches a constant value when $k \to 0$, the $(-\frac{3}{2})$ decay law for $\overline{\gamma^2} = \mathbf{Q}_{\gamma,\gamma}(0,t)$ is obtained.

First *measurements on the decay of a scalar* were made by Kistler, O'Brien, and Corrsin,[35] followed by those by Mills Jr. and these same authors.[103] In the first-mentioned experiments, carried out in a windtunnel, the turbulence was generated by a square-mesh grid of round wooden dowels ($M = 25.4$ mm; $M/d = 4$). A turbulence temperature field was generated by a grid of heated Nichrome wires placed 37.5 mm upstream of the dowel grid with the wires geometrically in line with the dowels. Later a biplane square-mesh grid with hot rods with the same values of M and d was used.

The turbulence fluctuations were measured with a hot-wire anemometer (platinum, 1.25 μm diameter), acting as a resistance thermometer in the case of the temperature measurements. The spectrum of the temperature fluctuations was determined by setting the wire temperature at the same order of response to velocity and temperature ("mixed" sensitivity). The temperature spectrum was obtained by subtracting the velocity spectrum determined for a cold grid from the mixed spectrum. This procedure is permitted provided all harmonics of the two spectra are uncorrelated $[\overline{u\theta}(k) = 0]$, so that superposition of spectra holds.

All measurements were made at $\bar{U}_1 = 4.2$ m/s, and a $\mathbf{Re}_M \simeq 7{,}000$. Distances up to 100 M downstream of the grid were explored.

No decay laws of either u' and γ' were determined because of insufficient data. However, it was found that the longitudinal velocity correlation $\mathbf{f}(r,t)$ and temperature correlation $\mathbf{R}_{\gamma,\gamma}$ were in close agreement. Furthermore $\mathbf{R}_{\gamma,\gamma}(x_1) \simeq \mathbf{R}_{\gamma,\gamma}(x_2)$ as it should be in case of isotropy. Λ_γ and Λ_f showed the same relationship with x_1/M, with an almost constant ratio $\Lambda_\gamma/\Lambda_f = 0.95$ for $20 < x_1/M < 90$.

The ratio of the relative decay

$$d \ln \gamma'/d \ln u'$$

decreased from roughly 1 at $x_1/M = 20$, to 0.6 at $x_1/M > 80$. If we assume that in the experiments made $\Lambda_g \propto \Lambda_f$, then a constant ratio Λ_γ/Λ_g agrees with any of the decay laws considered above.

If we assume the power law Eq. (3-167) for the energy decay, and Eq. (3-174) for the scalar decay, and assume the same initial instants, then we obtain

$$\frac{d \ln \gamma'}{d \ln u'} = \frac{m}{n}$$

This results in the value 1.5 for the "linear" velocity decay, 1.0 when Saffman's invariant is used, 0.6 when Loitsianskii's invariant is used, and again 0.6 for the final stage of decay. The observed behavior of the above ratio does not agree with the "linear" velocity decay, but it might agree with the predictions based on Saffman's invariant. However, not in agreement with it is the observed behavior of Λ_f, namely proportional with $(At + B)^{0.3}$.

More recent experiments on the decay of a scalar have been made by Gibson and Schwarz.[105] Their experimental setup has been discussed earlier. For the velocity field they assumed a "linear" decay law for the initial period. The result of their

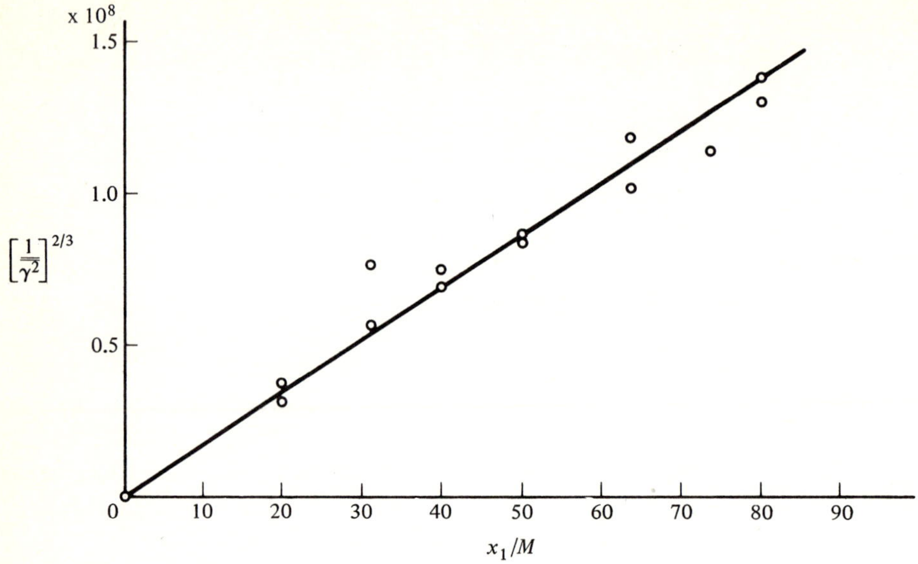

FIGURE 3-31
Decay of concentration fluctuations behind a grid. (After: *Gibson, C. H., and W. H. Schwarz,*[105] *by permission of the Cambridge University Press.*)

measurements of the decay of concentration fluctuations is shown in Fig. 3-31, where $(\gamma'^2)^{2/3}$ is plotted versus x_1/M. Though there is a large scatter and the number of points is rather small, the data are in reasonable agreement with the decay law Eq. (3-282), consistent with complete self-preservation.

A point worth some consideration is the ratio between the eddy viscosity ϵ_m and the eddy diffusivity ϵ_γ. We may recall that for the transfer through the spectrum Heisenberg introduced $\epsilon_\gamma(k,t)$ and $\epsilon_m(k,t)$, defined respectively by

$$\epsilon_m(k,t) = \alpha \int_k^\infty dk \sqrt{\frac{\mathbf{E}}{k^3}} \quad \text{and} \quad \epsilon_\gamma(k,t) = \beta \int_k^\infty dk \sqrt{\frac{\mathbf{E}}{k^3}}$$

According to this concept we have

$$\frac{\epsilon_m(k,t)}{\epsilon_\gamma(k,t)} = \frac{\alpha}{\beta} \quad \text{independent of } k \text{ and } t$$

For sufficiently large Reynolds numbers and Péclet numbers there follows from Eqs. (3-141) and (3-260), when we exclude the viscous and molecular diffusive ranges

$$\varepsilon = 2\nu \int_0^\infty dk\, k^2 \mathbf{E}(k,t) \simeq 2\epsilon_m(k,t) \int_0^k dk\, k^2 \mathbf{E}(k,t) \quad \text{when} \quad k \ll k_d$$

$$\varepsilon_\gamma = 2\mathfrak{k} \int_0^\infty dk\, k^2 \mathbf{E}_\gamma(k,t) \simeq 2\epsilon_\gamma(k,t) \int_0^k dk\, k^2 \mathbf{E}_\gamma(k,t) \quad \text{when} \quad k \ll k_{\gamma,d}$$

Hence

$$\frac{\epsilon_m(k,t)}{\epsilon_\gamma(k,t)} = \frac{\nu}{\mathfrak{k}} \frac{\int_0^\infty dk\, k^2 \mathbf{E}(k,t) \int_0^k dk\, k^2 \mathbf{E}_\gamma(k,t)}{\int_0^\infty dk\, k^2 \mathbf{E}_\gamma(k,t) \int_0^k dk\, k^2 \mathbf{E}(k,t)}$$

Now we assume self-preservation during the simultaneous decay of the velocity field and the scalar field, at least in the range (0–k) considered. Introduce wavenumber scales $\mathscr{K}(t)$ and $\mathscr{K}_\gamma(t)$, and the reduced wavenumbers

$$k^* = k/\mathscr{K}(t) \quad \text{and} \quad k_\gamma{}^* = k/\mathscr{K}_\gamma(t)$$

We then obtain

$$\frac{\epsilon_m(k,t)}{\epsilon_\gamma(k,t)} = \frac{\nu}{\mathfrak{k}} \frac{\mathscr{K}^3 \int_0^\infty dk^*\, k^{*2} \mathbf{E}(k^*) \cdot \mathscr{K}_\gamma{}^3 \int_0^k dk_\gamma^*\, k_\gamma^{*2} \mathbf{E}_\gamma(k_\gamma^*)}{\mathscr{K}_\gamma{}^3 \int_0^\infty dk_\gamma^*\, k_\gamma^{*2} \mathbf{E}_\gamma(k_\gamma^*) \cdot \mathscr{K}^3 \int_0^k dk^*\, k^{*2} \mathbf{E}(k^*)}$$

When $\nu/\mathfrak{k}$ is neither small nor large we may expect that $\mathbf{E}(k^*) \simeq \mathbf{E}_\gamma(k_\gamma^*)$. Hence in this case

$$\frac{\epsilon_m(k,t)}{\epsilon_\gamma(k,t)} \simeq \frac{\nu}{\mathfrak{k}} \tag{3-289}$$

the turbulent Prandtl or Schmidt number is equal to the molecular Prandtl or Schmidt number. Since the ratio $\epsilon_m/\epsilon_\gamma$ is independent of k, the result Eq. (3-289) may also apply to the total eddy diffusivities $\epsilon_m(0,t)$ and $\epsilon_\gamma(0,t)$ for the decay of velocity field and scalar field as a whole, a result which has already been suggested by Ogura.[112]

In a decaying, homogeneous field the diffusivities only effect a leveling process on the turbulent fluctuations. Then Eq. (3-289) holds true. This result is just opposite to what is obtained for the turbulent Prandtl or Schmidt number in a nonhomogeneous scalar field, where there is a gradient of the mean value and consequently a nonzero average transport of the scalar. This will be considered in Chap. 5.

3-8 PRESSURE FLUCTUATIONS IN ISOTROPIC TURBULENCE

For many physical phenomena, not only velocity fluctuations and their correlations are important but also fluctuations in static pressure. For instance, the production of noise by turbulence is directly related to the fluctuating-pressure distribution. Then there is the effect of fluctuating-pressure gradients near a rigid wall on the behavior of the boundary layer. In the mixing of nonmiscible fluids, the breaking up of the dispersed phase caused by turbulence in the continuous phase is another example of the importance of the distribution of pressure fluctuations.

The first contributions to the study of correlations between fluctuating pressures at two neighboring points were made by Heisenberg,[15] Obukhov,[36] and Batchelor.[37] Later, Limber[38] published some numerical results for pressure-velocity correlations, and Uberoi[39,40] gave more detailed theoretical as well as experimental information on quadruple velocity correlations associated with correlations in the fluctuating pressures. All these authors assumed homogeneity of the turbulence.

In what follows we shall confine ourselves to a rather brief study of the double correlation between the pressure fluctuations at two neighboring points.

This double correlation $(\mathbf{Q}_{p,p})_{A,B} = (1/\rho^2)\overline{p_A p_B}$ can be expressed in terms of the velocity correlation, which follows directly from the equations of motion. For an incompressible fluid they may be written

$$\frac{\partial u_i}{\partial t} + \frac{\partial u_i u_j}{\partial x_j} = -\frac{1}{\rho}\frac{\partial p}{\partial x_i} + \nu \frac{\partial^2}{\partial x_l\, \partial x_l} u_i$$

Take the divergence of this equation; we then obtain

$$\frac{1}{\rho}\frac{\partial^2 p}{\partial x_i\, \partial x_i} = -\frac{\partial^2 u_i u_j}{\partial x_i\, \partial x_j}$$

By taking this relation for the pressure at A and multiplying it by the same relation at B, we obtain

$$\frac{1}{\rho^2}\left(\frac{\partial^2 p}{\partial x_i\, \partial x_i}\right)_A \left(\frac{\partial^2 p}{\partial x_l\, \partial x_l}\right)_B = \left(\frac{\partial^2 u_i u_j}{\partial x_i\, \partial x_j}\right)_A \left(\frac{\partial^2 u_k u_l}{\partial x_k\, \partial x_l}\right)_B$$

Introduce $\xi_i = (x_i)_B - (x_i)_A$; then, because of the assumed homogeneity of the turbulence,

$$\frac{1}{\rho^2}\frac{\partial^4 \overline{p_A p_B}}{\partial \xi_i\, \partial \xi_i\, \partial \xi_l\, \partial \xi_l} = \frac{\partial^4 \overline{(u_i u_j)_A (u_k u_l)_B}}{\partial \xi_i\, \partial \xi_j\, \partial \xi_k\, \partial \xi_l}$$

In homogeneous isotropic turbulence the double correlation $\overline{p_A p_B}$ is a function of $r^2 = \xi_i \xi_i$ alone. Write further

$$\frac{\partial^4 \overline{(u_i u_j)_A (u_k u_l)_B}}{\partial \xi_i\, \partial \xi_j\, \partial \xi_k\, \partial \xi_l} = M \qquad \text{(3-290)}$$

Then, with

$$\frac{\partial^4 \mathbf{Q}_{p,p}}{\partial \xi_i\, \partial \xi_i\, \partial \xi_l\, \partial \xi_l} = \left(\frac{\partial^2}{\partial r^2} + \frac{2}{r}\frac{\partial}{\partial r}\right)^2 \mathbf{Q}_{p,p} = \frac{1}{r}\frac{\partial^4 r \mathbf{Q}_{p,p}}{\partial r^4}$$

we obtain

$$\frac{1}{r}\frac{\partial^4 r \mathbf{Q}_{p,p}}{\partial r^4} = M(r,t) \qquad \text{(3-291)}$$

Assume that $\mathbf{Q}_{p,p}(r,t)$ decreases rapidly to zero when r increases indefinitely, so that, for any value of n,

$$\lim_{r\to\infty} \frac{d^n}{dr^n} [r\mathbf{Q}_{p,p}(r,t)] = 0$$

It then follows from Eq. (3-291) that

$$\frac{\partial^3}{\partial r^3} [r\mathbf{Q}_{p,p}(r,t)] = -\int_r^\infty dr'\, r'M(r',t)$$

Consecutively,

$$\begin{aligned}
\frac{\partial^2}{\partial r^2} [r\mathbf{Q}_{p,p}(r,t)] &= \int_r^\infty dr' \int_{r'}^\infty dr''\, r''M(r'',t) \\
&= \left[r' \int_{r'}^\infty dr''\, r''M(r'',t) \right]_r^\infty + \int_r^\infty dr'\, r'^2 M(r',t) \\
&= \int_r^\infty dr'\, r'(r'-r)M(r',t) \\
\frac{\partial}{\partial r} [r\mathbf{Q}_{p,p}(r,t)] &= -\frac{1}{2}\int_r^\infty dr'\, r'(r'-r)^2 M(r',t)
\end{aligned}$$

and, finally,

$$\mathbf{Q}_{p,p}(r,t) = \frac{1}{3!r}\int_r^\infty dr'\, r'(r'-r)^3 M(r',t) \qquad (3\text{-}292)$$

Hence, the double correlation $\mathbf{Q}_{p,p}(r,t)$ can be calculated from $M(r,t)$ by a single integration process.

Now make the following simplifying assumption about the quadruple correlation $\overline{(u_i u_j)_A (u_k u_l)_B}$: that the joint probability distribution of the turbulent velocities is normal. In that case we have the relation

$$\begin{aligned}
\overline{(u_i u_j)_A (u_k u_l)_B} &= \overline{(u_i u_j)_A}\,\overline{(u_k u_l)_B} + \overline{(u_i)_A (u_k)_B}\,\overline{(u_j)_A (u_l)_B} + \overline{(u_i)_A (u_l)_B}\,\overline{(u_j)_A (u_k)_B} \\
&= \mathbf{Q}_{i,j}(0)\mathbf{Q}_{k,l}(0) + \mathbf{Q}_{i,k}(r)\mathbf{Q}_{j,l}(r) + \mathbf{Q}_{i,l}(r)\mathbf{Q}_{j,k}(r) \qquad (3\text{-}293)
\end{aligned}$$

Now we know that, though the probability density of the turbulent velocity may be Gaussian, the joint probability density cannot be Gaussian since a zero value of the fourth-order cumulant would lead to nonpositive parts of the energy spectrum. So the above assumption of a normal joint-probability density can only be used as an approximation, when departures from the normal behavior are small. Experimental evidence for this behavior seems to have been obtained by Uberoi[39,40] who measured the quadruple velocity correlation and the corresponding double velocity correlations simultaneously and found the relation (3-293) at least approximately true within experimental error.

Since for an incompressible fluid

$$\frac{\partial}{\partial \xi_i} \mathbf{Q}_{i,l} = 0$$

etc., it follows that

$$\frac{\partial^4}{\partial \xi_i \, \partial \xi_j \, \partial \xi_k \, \partial \xi_l} \mathbf{Q}_{i,k} \mathbf{Q}_{j,l} = \frac{\partial^2}{\partial \xi_i \, \partial \xi_k} \mathbf{Q}_{j,l} \frac{\partial^2}{\partial \xi_j \, \partial \xi_l} \mathbf{Q}_{i,k}$$

$$\frac{\partial^4}{\partial \xi_i \, \partial \xi_j \, \partial \xi_k \, \partial \xi_l} \mathbf{Q}_{i,l} \mathbf{Q}_{j,k} = \frac{\partial^2}{\partial \xi_j \, \partial \xi_k} \mathbf{Q}_{i,l} \frac{\partial^2}{\partial \xi_i \, \partial \xi_l} \mathbf{Q}_{j,k}$$

hence

$$M(r,t) = 2 \frac{\partial^2}{\partial \xi_i \, \partial \xi_k} \mathbf{Q}_{j,l}(r,t) \frac{\partial^2}{\partial \xi_j \, \partial \xi_l} \mathbf{Q}_{i,k}(r,t) \qquad \text{(3-294)}$$

Now we have shown that the double velocity correlation $\mathbf{Q}_{j,l}(r,t)$ may be expressed in terms of one defining scalar $\mathbf{f}(r,t)$. If that expression (3-11) is introduced in Eq. (3-294), after a lengthy calculation[37] Eq. (3-294) becomes,

$$M(r,t) = 4u'^4 \left[2\left(\frac{\partial^2 \mathbf{f}}{\partial r^2}\right)^2 + 2\left(\frac{\partial \mathbf{f}}{\partial r}\right)\left(\frac{\partial^3 \mathbf{f}}{\partial r^3}\right) + \frac{10}{r}\left(\frac{\partial \mathbf{f}}{\partial r}\right)\left(\frac{\partial^2 \mathbf{f}}{\partial r^2}\right) + \frac{3}{r^2}\left(\frac{\partial \mathbf{f}}{\partial r}\right)^2 \right] \qquad \text{(3-295)}$$

Substituting Eq. (3-295) in the integral (3-292) yields

$$\mathbf{Q}_{p,p}(r,t) = 2u'^4 \int_r^\infty dr' \left(r' - \frac{r^2}{r'}\right)\left(\frac{\partial \mathbf{f}}{\partial r'}\right)^2 \qquad \text{(3-296)}$$

In the specific case $r = 0$, this becomes

$$\overline{p^2} = \rho^2 \mathbf{Q}_{p,p}(0,t) = 2\rho^2 u'^4 \int_0^\infty dr' \, r' \left(\frac{\partial \mathbf{f}}{\partial r'}\right)^2$$

Furthermore, from Eq. (3-296) we may obtain

$$\overline{\left(\frac{\partial p}{\partial x_1}\right)^2} = \overline{\left(\frac{\partial p}{\partial x_2}\right)^2} = \overline{\left(\frac{\partial p}{\partial x_3}\right)^2} = \lim_{B \to A} \overline{\left(\frac{\partial p}{\partial x_1}\right)_A \left(\frac{\partial p}{\partial x_1}\right)_B} = -\lim_{B \to A} \frac{\partial^2 \overline{p_A p_B}}{\partial \xi_1 \, \partial \xi_1}$$

$$= -\lim_{r \to 0} \rho^2 \left[\frac{\partial^2}{\partial r^2} \mathbf{Q}_{p,p}(r,t) \right] = 4\rho^2 u'^4 \int_0^\infty dr' \frac{1}{r'} \left(\frac{\partial \mathbf{f}}{\partial r'}\right)^2 \qquad \text{(3-297)}$$

Instead of $\mathbf{Q}_{p,p}(r,t) = (1/\rho^2)\overline{p_A p_B}$ it may sometimes be of interest to consider $(1/\rho^2)\overline{(p_A - p_B)^2}$, which is related to $\mathbf{Q}_{p,p}(r,t)$ by

$$\frac{1}{\rho^2} \overline{(p_A - p_B)^2} = 2[\mathbf{Q}_{p,p}(0,t) - \mathbf{Q}_{p,p}(r,t)] \qquad \text{(3-298)}$$

The integrals in Eqs. (3-296) and (3-297) can be evaluated if the longitudinal correlation $\mathbf{f}(r,t)$ is known. Since this correlation changes with the Reynolds number of turbulence, different values for the pressure correlations are obtained.

Let us first consider small Reynolds numbers. Then the expression for $\mathbf{f}(r,t)$ reads [see Eq. (3-45)]

$$\mathbf{f}(r,t) = \exp(-r^2/8\nu t) = \exp(-r^2/\lambda_f^2)$$

Substitution in Eq. (3-296) and integration yield

$$\mathbf{Q}_{p,p}(r,t) = u'^4 \exp\left(-\frac{2r^2}{\lambda_f^2}\right)$$

or

$$\frac{\mathbf{Q}_{p,p}(r,t)}{\mathbf{Q}_{p,p}(0,t)} = \exp\left(-\frac{2r^2}{\lambda_f^2}\right)$$

This shows that the double pressure correlation decreases more rapidly with an increase in r than the longitudinal velocity correlation.

For the intensity of the pressure fluctuation we obtain

$$p' = \rho\sqrt{\mathbf{Q}_{p,p}(0,t)} = \rho u'^2$$

Now let us consider large Reynolds numbers and, first, the empirical correlation

$$\mathbf{f}(r,t) = \exp(-r/\Lambda_f)$$

This gives for $\mathbf{Q}_{p,p}(r,t)$

$$\mathbf{Q}_{p,p}(r,t) = 2u'^4\left[\left(\frac{r}{2\Lambda_f} + \frac{1}{4}\right)\exp(-2r/\Lambda_f) - \frac{r^2}{\Lambda_f^2} Ei\left(-\frac{2r}{\Lambda_f}\right)\right]$$

where

$$Ei(-x) = -\int_x^\infty dy \frac{\exp(-y)}{y}$$

Hence

$$p' = \rho\sqrt{\mathbf{Q}_{p,p}(0,t)} = \frac{\rho u'^2}{\sqrt{2}} \simeq 0.7\rho u'^2$$

Batchelor[37] has evaluated the integral (3-296) numerically for a shape of $\mathbf{f}(r,t)$ measured by Proudman[18] in a turbulence of large Reynolds number. Here, too, it is found that $\mathbf{Q}_{p,p}(r,t)/\mathbf{Q}_{p,p}(0,t)$ decreases much more rapidly with an increase in r than $\mathbf{f}(r,t)$. For the intensity of the pressure fluctuations Batchelor obtained

$$p' = \sqrt{0.34}\,\rho u'^2 = 0.58\rho u'^2$$

For large Reynolds numbers, such that an inertial subrange exists in the energy spectrum, we have obtained the expression (3-153) for $\mathbf{f}(r,t)$, valid for small r:

$$\mathbf{f}(r,t) = 1 - \frac{18}{55}\Gamma(4/3)\frac{A}{u'^2}(\varepsilon r)^{2/3}$$

For these values of r we may determine $\overline{(p_A - p_B)^2}$ from Eq. (3-298). With Eq. (3-296) we obtain

$$\overline{(p_A - p_B)^2} = 2\rho^2[\mathbf{Q}_{p,p}(0,t) - \mathbf{Q}_{p,p}(r,t)]$$
$$= 4\rho^2 u'^4 \left[\int_0^r dr'\, r' \left(\frac{\partial \mathbf{f}}{\partial r'}\right)^2 + r^2 \int_r^\infty dr' \frac{1}{r'} \left(\frac{\partial \mathbf{f}}{\partial r'}\right)^2\right]$$

Since the main contribution to the integrand of the second integral is from the value of $\mathbf{f}(r,t)$ for small r, the integral may be evaluated approximately by taking for $\mathbf{f}(r,t)$ the above expression, which is only exact for small r. The result of the calculation then reads

$$\overline{(p_A - p_B)^2} = 4\rho^2 u'^4 \frac{A^2}{u'^4} (\varepsilon r)^{4/3} = 4\rho^2 u'^4 [1 - \mathbf{f}(r,t)]^2 = \rho^2 [\overline{(u_A - u_B)^2}]^2$$

Unfortunately it is not yet possible to make reliable measurements of the pressure fluctuations in a turbulent flow, and so to check in a direct way the above relations. Uberoi,[39,40] however, calculated p' from the measured double velocity-correlation according to Eq. (3-296a). The results are shown in Fig. (3-32). They seem to indicate a value for $p'/\rho u'^2$ closer to 0.7 than to 0.58.

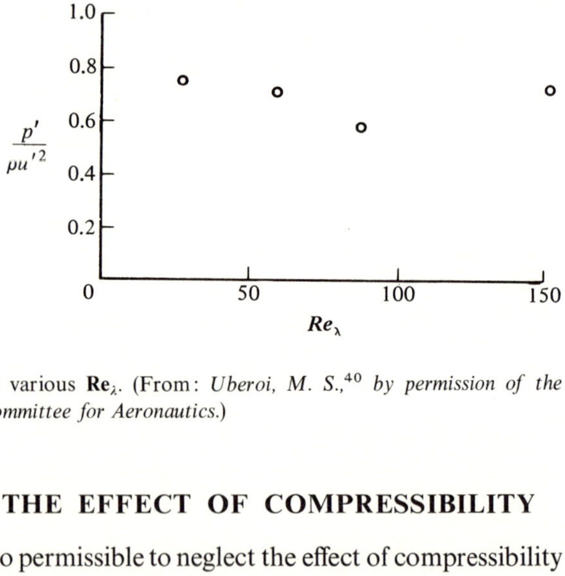

FIGURE 3-32
Values of $p'/\rho u'^2$ for various $\mathbf{Re}_\lambda$. (From: *Uberoi, M. S.,*[40] *by permission of the National Advisory Committee for Aeronautics.*)

3-9 SOME REMARKS ON THE EFFECT OF COMPRESSIBILITY

In fluid dynamics it is usual and also permissible to neglect the effect of compressibility if the Mach number $\mathbf{Ma} \ll 1$. This has been shown in Sec. 1-2. In turbulence we may take the root-mean-square Mach number $\mathbf{Ma}' = u'/c$, where c is the velocity of sound.

Now compressibility effects may only become appreciable when local relative velocities between parts of the fluid cause appreciable pressure fluctuations, and hence appreciable density fluctuations. This requires high values of the turbulence intensities. For main-stream Mach numbers up to 5, say, higher values of the turbulence Mach number **Ma**$'$ than 0.2 have never been observed, even in strong shear-layers. The effect on energy losses due to acoustic radiation is negligible. Only with much higher Mach numbers may compressibility effects change the turbulence, in the sense of decreasing its intensity.

In Chap. 1 we have shown that, if the effect of compressibility is not neglected, the mean motion is affected not only by the turbulence shear stresses but also by stresses originating from double and triple correlations involving the density fluctuation $\tilde{\rho}$ [see Eq. (1-17)].

In the case of an incompressible fluid we have studied extensively the structure of an isotropic turbulence through the features of the double velocity correlation $\overline{(u_i)_A(u_j)_B}$. Krzywoblocki[41,42] has shown that it is possible to obtain relations similar to those obtained for the incompressible case, for the various double, triple, and higher-order correlations: $\overline{(\rho u_i)_A(\rho u_j)_B}$, $\overline{p_A(u_i)_A p_B}$, $\overline{(\rho u_i)_A(u_k)_A(\rho u_j)_B}$, ... These correlations are obtained from the equations of motion for the mass flux ρu_i relative to the two points A and B of the flow field, resulting in a dynamic equation for the correlation $\overline{(\rho u_i)_A(\rho u_j)_B}$.

Now for the incompressible case it was already impossible to obtain complete solutions of the dynamic equations for the double correlation $\overline{(u_i)_A(u_j)_B}$. Needless to say, it will be hopeless for the compressible case.

The reader may appreciate the difficult nature of the problem involved, by considering the many simplifications obtained in turbulence relations through the incompressibility condition, when the fluid is incompressible. For instance, the relation (3-6) for $\overline{p_A(u_i)_B}$. This simplification cannot be obtained, even when we consider the mass velocity ρu_i, or the density-weighted velocity u_i^0 introduced by Favre (see ref. 12, Chap. 1).

In an attempt to gain some insight into the effect of compressibility on turbulence, Chandrasekhar[43] has considered the correlation between the density fluctuations at two points $\overline{\tilde{\rho}_A\tilde{\rho}_B}$. He showed that, as a consequence of the condition of mass conservation (equation of continuity), $\overline{\tilde{\rho}_A\tilde{\rho}_B}$ must satisfy the invariance relation

$$\int_0^\infty dr\, r^2\overline{\tilde{\rho}_A\tilde{\rho}_B}(r,t) = \text{const.}$$

provided that the isotropic first-order tensors $\overline{\tilde{\rho}_A(u_i)_B}$ and $\overline{\tilde{\rho}_A\tilde{\rho}_B(u_i)_B}$ decrease sufficiently rapidly as r increases.

From the equations of motion, Chandrasekhar derived a dynamic equation for $\overline{\tilde{\rho}_A\tilde{\rho}_B}$. He showed that for a small root-mean-square Mach number $u'/c \ll 1$, the solution for $\overline{\tilde{\rho}_A\tilde{\rho}_B}$ can be expressed as a superposition of spherical waves which are propagated with a velocity $\sqrt{2}\,c$. Thus any change in the pressure distribution, i.e., in the density distribution, is also propagated with a finite velocity. The energy radiated in this way is ultimately converted into heat by viscous effects. Thus compressibility acts as a source of energy dissipation in addition to the viscous shear stresses. This effect also follows immediately from the equations of motion, since they contain a viscous term involving the compressibility of the fluid (or the dilatation Θ).

In order to demonstrate in a more general way the effect of compressibility, let us briefly consider the basic equations describing compressible flows. These equations are the balance equations for mass, momentum and energy, and the constitutive equation

$$\frac{\partial \rho}{\partial t} + \frac{\partial}{\partial x_i} \rho U_i = 0; \quad \text{or} \quad \frac{D\rho}{Dt} + \rho\Theta = 0 \tag{1-2}$$

$$\rho \frac{DU_i}{Dt} = -\frac{\partial P}{\partial x_i} + \mu\left[\frac{\partial^2 U_i}{\partial x_l \partial x_l} + \frac{1}{3}\frac{\partial}{\partial x_i}\Theta\right] \tag{1-10}$$

$$\rho \frac{D\mathscr{E}_{\text{int}}}{Dt} = \mathfrak{k} \frac{\partial^2 \Theta}{\partial x_l \partial x_l} - P\Theta + \mu\left[\left(\frac{\partial U_i}{\partial x_j} + \frac{\partial U_j}{\partial x_i}\right)\frac{\partial U_i}{\partial x_j} - \frac{2}{3}\Theta^2\right] \tag{1-107}$$

$$P = \Re \rho \Theta \tag{3-299}$$

We assumed a constant value of the molecular diffusivity and of the dynamic viscosity, though, as explained in Chap. 1, a constant value of the dynamic viscosity may conflict with a variable density and temperature of the fluid. Also we assumed the constitutive equation for a perfect gas.

We transform the equation for the momentum balance to the one for the vorticity balance by taking the rotation of the first-mentioned equation. This yields for the vorticity $\Omega_k = -\varepsilon_{ijk}\,\partial U_i/\partial x_j$ the equation

$$\rho \frac{D\Omega_k}{Dt} = \mu \frac{\partial^2 \Omega_k}{\partial x_l \partial x_l} + \frac{DU_i}{Dt}\varepsilon_{ijk}\frac{\partial \rho}{\partial x_j} + \rho\Omega_l \frac{\partial U_k}{\partial x_l} - \rho\Omega_k\Theta \tag{3-300}$$

Also, instead of Eq. (1-107) we consider the equation for the entropy S, making use of the first law of thermodynamics

$$\frac{D\mathscr{E}_{\text{int}}}{Dt} = \Theta \frac{DS}{Dt} - \Re\Theta\Theta$$

Substitution in Eq. (1-107) gives

$$\rho\Theta \frac{DS}{Dt} = \mathfrak{k} \frac{\partial^2 \Theta}{\partial x_l \partial x_l} + \mu\left[\left(\frac{\partial U_i}{\partial x_j} + \frac{\partial U_j}{\partial x_i}\right)\frac{\partial U_i}{\partial x_j} - \frac{2}{3}\Theta^2\right] \tag{3-301}$$

Interesting features of the flow can already be obtained for the simple case of a turbulence superimposed on a uniform mean velocity. With respect to a coordinate system moving with the mean velocity we have

$$U_i = u_i, \quad \text{so} \quad \Theta = \tilde{\Theta} = \partial u_i/\partial x_i \quad \text{and} \quad \Omega_k = \omega_k$$

Putting

$$\rho = \rho_0 + \tilde{\rho}, \quad P = P_0 + p, \quad \Theta = \Theta_0 + \tilde{\Theta}, \quad S = S_0 + \varsigma,$$

the Eqs. (1-2), (3-300), and (3-301) become

$$\frac{\partial \tilde{\rho}}{\partial t} + \frac{\partial}{\partial x_i} \rho u_i = 0 \tag{3-302}$$

$$\rho \frac{\partial \omega_k}{\partial t} = \mu \frac{\partial^2 \omega_k}{\partial x_l \, \partial x_l} - \rho u_l \frac{\partial \omega_k}{\partial x_l} + \rho \omega_l \frac{\partial u_k}{\partial x_l} + \left(\frac{\partial u_i}{\partial t} + u_l \frac{\partial u_i}{\partial x_l} \right) \varepsilon_{ijk} \frac{\partial \tilde{\rho}}{\partial x_j} - \rho \omega_k \tilde{\Theta} \tag{3-303}$$

and

$$\rho \Theta \frac{\partial \varsigma}{\partial t} = \mathfrak{k} \frac{\partial \tilde{\Theta}}{\partial x_l \, \partial x_l} - \rho \Theta u_l \frac{\partial \varsigma}{\partial x_l} + \mu \left[\left(\frac{\partial u_i}{\partial x_j} + \frac{\partial u_j}{\partial x_i} \right) \frac{\partial u_i}{\partial x_j} - \frac{2}{3} \tilde{\Theta}^2 \right]$$

Or, with

$$c_p \, d\tilde{\Theta} = \Theta \, d\varsigma + \frac{1}{\rho} dp = \Theta \, d\varsigma + \frac{c^2}{\rho} d\tilde{\rho}$$

where $c^2 = \partial p / \partial \tilde{\rho}$, the velocity of sound.

$$\frac{\partial \varsigma}{\partial t} = \frac{\mathfrak{k}}{\rho c_p} \frac{1}{\Theta} \frac{\partial}{\partial x_l} \left[\Theta \frac{\partial s}{\partial x_l} + \frac{c^2}{\rho} \frac{\partial \tilde{\rho}}{\partial x_l} \right] - u_l \frac{\partial \varsigma}{\partial x_l} + \frac{\nu}{\Theta} \left[\left(\frac{\partial u_i}{\partial x_j} + \frac{\partial u_j}{\partial x_i} \right) \frac{\partial u_i}{\partial x_j} - \frac{2}{3} \tilde{\Theta}^2 \right] \tag{3-304}$$

We can eliminate the momentum ρu_i from Eqs. (3-302) and (1-10), which latter equation is written

$$\frac{\partial \rho u_i}{\partial t} + \frac{\partial}{\partial x_l} \rho u_l u_i = - \frac{\partial p}{\partial x_i} + \mu \left[\frac{\partial^2 u_i}{\partial x_l \, \partial x_l} + \frac{1}{3} \frac{\partial}{\partial x_i} \tilde{\Theta} \right]$$

For this elimination process differentiate Eq. (3-302) with respect to t, and the latter equation with respect to x_i, and subtract it from the differentiated Eq. (3-302). Again with $c^2 = \partial p / \partial \tilde{\rho}$, there is obtained

$$\frac{\partial^2 \tilde{\rho}}{\partial t^2} - c^2 \frac{\partial^2 \tilde{\rho}}{\partial x_l \, \partial x_l} = - \frac{4}{3} \mu \frac{\partial^2 \tilde{\Theta}}{\partial x_l \, \partial x_l} + \frac{\partial^2}{\partial x_i \, \partial x_l} \rho u_l u_i$$

or

$$\frac{\partial^2 \tilde{\rho}}{\partial t^2} - c^2 \frac{\partial^2 \tilde{\rho}}{\partial x_l \, \partial x_l} = \frac{4}{3} \mu \frac{\partial^2}{\partial x_l \, \partial x_l} \left(\frac{1}{\rho} \frac{\partial \tilde{\rho}}{\partial t} \right) + \frac{\partial^2}{\partial x_i \, \partial x_l} \rho u_l u_i \tag{3-305}$$

For a weak turbulence, so that the fluctuations are relatively small, and nonlinear effects may be neglected, in the absence of molecular effects Eqs. (3-303), (3-304), and (3-305) reduce respectively to

$$\left. \begin{aligned} \frac{\partial \omega_k}{\partial t} &= 0 \\ \frac{\partial \varsigma}{\partial t} &= 0 \\ \frac{\partial^2 \tilde{\rho}}{\partial t^2} - c_0{}^2 \frac{\partial^2 \tilde{\rho}}{\partial x_l \, \partial x_l} &= 0 \end{aligned} \right\} \tag{3-306}$$

The first two equations show that vorticity and entropy are conserved. This is equivalent to the Taylor hypothesis of a "frozen" turbulence. The last equation is the well-known wave equation in the theory of sound. So we see that the first effect of compressibility is that of sound caused by the turbulence fluctuations.

If we do not disregard molecular effects, but still assume such a weak turbulence that nonlinear effects may be neglected, the equations for ω_k, ς and $\tilde{\rho}$ become

$$\frac{\partial \omega_k}{\partial t} = \nu_0 \frac{\partial^2 \omega_k}{\partial x_l \partial x_l}$$

$$\frac{\partial \varsigma}{\partial t} = \frac{\mathfrak{k}}{\rho_0 c_p} \frac{\partial^2 \varsigma}{\partial x_l \partial x_l} \tag{3-307}$$

$$\frac{\partial^2 \tilde{\rho}}{\partial t^2} - c_0{}^2 \frac{\partial^2 \tilde{\rho}}{\partial x_l \partial x_l} = \frac{4}{3} \nu_0 \frac{\partial^2}{\partial x_l \partial x_l} \frac{\partial \tilde{\rho}}{\partial t}$$

The equations for ω_k and ς being parabolic, describe a dispersion of these quantities due to molecular effects. Also in the last equation viscosity causes a dispersion of the quantity $\partial\tilde{\rho}/\partial t$.

The nonlinear terms cause higher-order effects. Chu and Kovasznay[113] have given a complete evaluation of the second-order effects on the vorticity mode, the entropy mode and the harmonic or sound mode of compressible turbulence, and of the bilinear interactions between these modes.

In Eq. (3-305) the last, nonlinear, term is the sound generating term. It stems from the fluctuations of the turbulence stress tensor $\rho u_l u_i$. They form two pairs of opposing forces or dipoles. When the axes of the dipoles coincide as in the case of the normal stresses (e.g., $\rho u_1{}^2$), they form a longitudinal quadrupole. When the axes are perpendicular, as in the case of the shear stresses (e.g., $\rho u_1 u_2$), then they form a lateral quadrupole. So for this simple type of turbulence with no average shear, and in the absence of internal sources of mass and heat, and of body forces, sound generation is by a continuous field of quadrupoles. By including in the basic equations also source terms for mass, heat, and external foces f_i, it can be shown that they too give rise to sound generation caused, however, by monopoles and dipoles respectively.[114,115]

For an isotropic turbulence, in the absence of these extra source terms, Eq. (3-305) applies. Neglecting the viscous term the formal solution of this equation reads

$$\frac{\tilde{\rho}}{\rho_0} = \frac{1}{4\pi c_0{}^2} \frac{\partial^2}{\partial x_i \partial x_j} \iiint d\xi_1\, d\xi_2\, d\xi_3\, u_j u_i \left[\xi_k, t - \frac{r}{c_0} \right] \cdot \frac{1}{r} \tag{3-308}$$

where $r = |x_k - \xi_k|$.

Lighthill,[45,114] Ffowcs Williams,[116] and others,[115,117] have studied this solution also for the more general case of a turbulent flow with mean shear. In this case the right-hand side of Eq. (3-305) contains additional terms resulting from the mean flow, and cross-coupling of turbulence with the mean flow. Ribner[117] and others, refer to the noise so generated as *shear noise*. Its radiation shows a marked direction effect in contrast to the self-noise caused by the turbulence itself, whose radiation is isotropic.

It can be shown from Eq. (3-308), and also from dimensional analysis,[115] that the acoustic efficiency defined as the ratio between the acoustic power radiated and the energy supplied is proportional to $\mathbf{Ma}^5$. Lighthill[44] already showed for **Ma** not greater than ~ 0.5, that the rate of energy dissipation is increased by a factor $1 + 40\,\mathbf{Ma}^5$.

Lighthill[45] and Proudman[46] studied in greater detail the generation of sound in isotropic turbulence. The associated increase of the density and pressure fluctuations with increasing Mach number is small. It may be estimated from the relation

$$\overline{p^2} = \rho_0{}^2 u'^4 \left(A + 40 \frac{\mathbf{Ma}^4}{e\Lambda} \right)$$

where A is a numerical constant of the order 1, e is the rate of dissipation of acoustic energy per unit distance [the sound energy being attenuated by a factor $\exp(-er)$], and Λ is a length scale of turbulence.

From the expressions for the effect of compressibility on the turbulence-energy dissipation and on p', it may be inferred that for small root-mean-square Mach numbers the effect on the turbulence is very small indeed. This, however, may no longer be so at higher Mach numbers. It is known that at higher Mach numbers the expansion and compression waves propagate with increased steepening of the fronts, resulting in shock waves. The interaction between shock waves creates additional vorticity. Thus Lighthill[44] concluded that, since such a system of shock waves can generate new turbulence, the properties of turbulence when $\mathbf{Ma} \simeq 1$ or greater must differ substantially from those relative to the incompressible case.

REFERENCES

1. Robertson, H. P.: *Proc. Cambridge Phil. Soc.*, **36,** 209 (1940).
2. Kármán, Th. von, and L. Howarth: *Proc. Roy. Soc. London,* **164A,** 192 (1938).
3. Favre, A., J. Gaviglio, and R. Dumas: *Recherche aéronaut. No.* 32, p. 21, 1953.
4. Stewart, R. W.: *Proc. Cambridge Phil. Soc.*, **47,** 146 (1951).
5. Loitsianskii, L. G.: *Natl. Advisory Comm. Aeronaut. Tech. Mem. No.* 1079, 1945.
6. Agostini, L., and J. Bass: *Publ. sci. et tech. ministère air No.* 237, 1950.
7. Batchelor, G. K., and A. A. Townsend: *Proc. Roy. Soc. London,* **194A,** 527 (1948).
8. Batchelor, G. K.: *Proc. Roy. Soc. London,* **195A,** 513 (1949).
9. Kármán, Th. von: *Proc. Natl. Acad. Sci. U.S.*, **34,** 530 (1948).
10. Kolmogoroff, A. N.: *Compt. rend. acad. sci. U.R.S.S.*, **30,** 301 (1941); **32,** 16 (1941).

11. Batchelor, G. K.: "The Theory of Homogeneous Turbulence," Cambridge University Press, New York, 1953.
12. Kármán, Th. von, and C. C. Lin: *Advances in Appl. Mechanics,* **2,** 1 (1951).
13. Obukhoff, A. M.: *Compt. rend. acad. sci. U.R.S.S.,* **32,** 19 (1941).
14. Kovasznay, L. S. G.: *J. Aeronaut. Sci.,* **15,** 745 (1948).
15. Heisenberg, W.: *Z. Physik,* **124,** 628 (1948).
16. Chandrasekhar, S.: *Proc. Roy. Soc. London,* **200A,** 20 (1949). Also *Phys. Rev.,* **75,** 896 (1949).
17. Stewart, R. W., and A. A. Townsend: *Phil. Trans. Roy. Soc. London,* **243A,** 359 (1951).
18. Proudman, I.: *Proc. Cambridge Phil. Soc.,* **47,** 158 (1951).
19. Sato, H.: *J. Phys. Soc. Japan,* **6,** 387 (1951); **7,** 392 (1952).
20. Liepmann, H. W., J. Laufer, and K. Liepmann: *Natl. Advisory Comm. Aeronaut. Tech. Notes No.* 2473, 1951.
21. Batchelor, G. K., and A. A. Townsend: *Proc. Roy. Soc. London,* **193A,** 539 (1948).
22. Batchelor, G. K., and A. A. Townsend: *Proc. Roy. Soc. London,* **190A,** 534 (1947).
23. Lin, C. C.: *Proc. Natl. Acad. Sci. U.S.,* **34,** 760 (1948).
24. Frenkiel, F. N., and P. S. Klebanoff: *Phys. Fluids,* **10,** 507 (1967).
25. Frenkiel, F. N.: *J. Appl. Mechanics,* **15,** 311 (1948).
26. Dryden, H. L., G. B. Schubauer, W. C. Mock, and H. K. Skramstad: *Natl. Advisory Comm. Aeronaut. Tech. Repts. No.* 581, 1937.
27. Hall, A. A.: *Aeronaut. Research Comm. Repts. & Mem. No.* 1842, 1938.
28. Baines, W. D., and E. G. Peterson: *Trans. ASME,* **73,** 467 (1951).
29. Proudman, I., and W. H. Reid: *Phil. Trans. Roy. Soc. London,* **247A,** 926 (1954).
30. Batchelor, G. K., and A. A. Townsend: *Proc. Roy. Soc. London,* **199A,** 238 (1949).
31. Obukhoff, A. M.: *Bull. Acad. Sci. U.R.S.S. Geograph. and Geophys. Ser.,* **13,** 58 (1949).
32. Yaglom, A. M.: *Compt. rend. acad. sci. U.R.S.S.,* **69,** 743 (1949).
33. Corrsin, S.: *J. Aeronaut. Sci.,* **18,** 417 (1951).
34. Corrsin, S.: *J. Appl. Phys.,* **22,** 469 (1951).
35. Kistler, A. L., V. O'Brien, and S. Corrsin: *Natl. Advisory Comm. Aeronaut. Research Mem. No.* RM 54D19, 1954.
36. Obukhov, A. M.: *Compt. rend. acad. sci. U.R.S.S.,* **66,** 17 (1949).
37. Batchelor, G. K.: *Proc. Cambridge Phil. Soc.,* **47,** 359 (1951).
38. Limber, D. N.: *Proc. Natl. Acad. Sci. U.S.,* **37,** 230 (1951).
39. Uberoi, M. S.: *J. Aeronaut. Sci.,* **20,** 197 (1953).
40. Uberoi, M. S.: *Natl. Advisory Comm. Aeronaut. Tech. Notes No.* 3116, 1954.
41. Krzywoblocki, M. Z. E.: *Proc. 1st U.S. Natl. Conf. Appl. Mech.,* p. 827, 1951.
42. Krzywoblocki, M. Z. E.: *Proc. 2nd Midwest. Conf. Fluid Dynamics,* p. 35, 1952.
43. Chandrasekhar, S.: *Proc. Roy. Soc. London,* **210A,** 18 (1951).
44. Lighthill, M. J.: "Gas Dynamics of Cosmic Clouds," p. 121, Symposium, Cambridge, England, 1953.
45. Lighthill, M. J.: *Proc. Roy. Soc. London,* **211A,** 564 (1952); **222A,** 1 (1954).
46. Proudman, I.: *Proc. Roy. Soc. London,* **214A,** 119 (1952).
47. Onsager, L.: *Phys. Rev.,* **68,** 286 (1945).
48. Weizsäcker, G. F. von: *Z. Physik,* **124,** 628 (1948); *Proc. Roy. Soc. London,* **195A,** 402 (1948).
49. Reid, W. H.: *Quart. Appl. Math.,* **14,** 201 (1956).
50. Batchelor, G. K., and I. Proudman: *Phil. Trans. Roy. Soc. London,* **248A,** 369 (1956).

51. Tsuji, H.: *J. Phys. Soc. Japan,* **11,** 1096 (1956).
52. Townsend, A. A.: *Proc. Roy. Soc. London,* **208A,** 534 (1951).
53. Grant, H. L., and I. C. T. Nisbet: *J. Fluid Mech.,* **2,** 263 (1957).
54. Van der Hegge Zijnen, B. G.: *Appl. Sci. Research,* **7A,** 149 (1958).
55. Chou, P. Y.: *Quart. Appl. Math.,* **3,** 38 (1945).
56. Deissler, R. G.: *Physics of Fluids,* **1,** 111 (1958).
57. Burgers, J. M.: *Proc. Koninkl. Ned. Akad. Wetenschap.,* **53,** 122 (1950).
58. Batchelor, G. K.: *J. Fluid Mech.,* **5,** 113 (1959).
59. Batchelor, G. K., I. D. Howells, and A. A. Townsend: *J. Fluid Mech.,* **5,** 134 (1959).
60. Van Atta, C. W., and W. Y. Chen: *J. Fluid Mech.,* **34,** 497 (1968).
61. Loeffler, A. L., and R. G. Deissler: *Int. J. Heat Mass Transf.,* **1,** 312 (1961).
62. Lee, J.: *Phys. Fluids,* **9,** 363 (1966).
63. Orszag, S. A.: *J. Fluid Mech.,* **41,** 363 (1970).
64. Kraichnan, R. H.: *Proc. Symp. Appl. Math.,* **13,** 199 (1962).
65. Ogura, Y.: *J. Geophys. Res.,* **67,** 3143 (1962).
66. Ogura, Y.: *J. Fluid Mech.,* **16,** 33 (1963).
67. Kraichnan, R. H.: *J. Math. Phys.,* **2,** 124 (1961).
68. Birkhoff, G.: *Commun. Pure Appl. Math.,* **7,** 19 (1954).
69. Saffman, P. G.: *J. Fluid Mech.,* **27,** 581 (1967).
70. Grant, H. L., R. W. Stewart, and A. Moilliet: *J. Fluid Mech.,* **12,** 241 (1962).
71. Burgers, J. M.: *Proc. Koninkl. Ned. Akad. Wetenschap.,* **17,** No. 2 (1939); also *Advances in Appl. Mechanics,* **1,** 171 (1948).
72. Ellison, T. H.: "Mécanique de la Turbulence," p. 113, Symposium, Marseille, France, 1961.
73. Panchev, S., and D. Kesich: *Compt. rend. acad. Bulgare sci.,* **22,** 627 (1969).
74. Eschenroeder, A. Q.: *Phys. Fluids,* **8,** 589 (1965).
75. Panchev, S.: "Random Functions and Turbulence," p. 214, Pergamon Press, New York, 1971.
76. Dugstad, I.: *Meteorologiske Annaler, Norske Meteor. Inst.,* **4,** No. 17, 441 (1962).
77. Pao, Y. H.: *Phys. Fluids,* **8,** 1063 (1965).
78. Saffman, P. G.: *J. Fluid Mech.,* **16,** 545 (1963).
79. Kraichnan, R. H.: *Phys. Fluids,* **10,** 2080 (1967).
80. Kolmogoroff, A. N.: *J. Fluid Mech.,* **13,** 82 (1962).
81. Obukhoff, A. M.: *J. Fluid Mech.,* **13,** 77 (1962).
82. Corrsin, S.: *Phys. Fluids,* **5,** 1301 (1962).
83. Kraichnan, R. H.: *Phys. Rev.,* **109,** 1407 (1958).
84. Kraichnan, R. H.: *J. Fluid Mech.,* **5,** 497 (1959).
85. Kraichnan, R. H.: *Phys. Fluids,* **7,** 1030 (1964).
86. Kraichnan, R. H.: *Phys. Fluids,* **8,** 575 (1965).
87. Kraichnan, R. H.: *Phys. Fluids,* **9,** 1728 (1966).
88. Uberoi, M. S.: *Phys. Fluids,* **6,** 1048 (1963).
89. Comte-Bellot, G., and S. Corrsin: *J. Fluid Mech.,* **25,** 657 (1966).
90. Uberoi, M. S. and S. Wallis: *Phys. Fluids,* **12,** 1355 (1969).
91. Van Atta, C. W., and W. Y. Chen: *J. Fluid Mech.,* **38,** 743 (1969).
92. Kistler, A. L., and T. Vrebalovich: *J. Fluid Mech.,* **26,** 37 (1966).
93. Gibson, M. M.: *Nature, Lond.,* **195,** 1281 (1962).
94. Orszag, S. A.: *Phys. Fluids,* **10,** 454 (1967).

95. Uberoi, M. S.: *J. Aeronaut. Sci.*, **23,** 754 (1956).
96. Mills, R. R., and S. Corrsin: *N.A.S.A. Memo.* 5-5-59W.
97. Tannenbaum, B. S., and D. Mintzer: *Phys. Fluids,* **3,** 529 (1960).
98. Kolmogoroff, A. N.: *Compt. rend. acad. sci. U.R.S.S.*, **31,** 538 (1941).
99. Levin, L.: *La Houille Blanche,* **22,** 271 (1967).
100. Morgan, P. G.: *J. Roy. Aeronaut. Soc.*, **64,** 359 (1960).
101. Bradshaw, P.: *J. Roy. Aeronaut. Soc.*, **68,** 198 (1964).
102. Uberoi, M. S., and S. Corrsin: *N.A.C.A. Techn. Repts. No.* 1142 (1953).
103. Mills, R. R., A. L. Kistler, V. O'Brien, and S. Corrsin: *N.A.C.A. Techn. Notes.* No. 4288 (1958).
104. Pao, Y. H.: *Phys. Fluids,* **11,** 1371 (1968).
105. Gibson, C. H., and W. H. Schwarz: *J. Fluid Mech.*, **16,** 365 (1963).
106. Nye, J. O., and R. S. Brodkey: *J. Fluid Mech.*, **29,** 151 (1967).
107. Grant, H. L., B. A. Hughes, W. M. Vogel, and A. Moilliet: *J. Fluid Mech.*, **34,** 423 (1968).
108. Corrsin, S., and M. S. Uberoi: *N.A.C.A. Techn. Reports.* No. 998 (1950).
109. Corrsin, S.: *A.I.Ch.E.Jl.*, **3,** 329 (1957).
110. Corrsin, S.: *A.I.Ch.E.Jl.*, **10,** 870 (1964).
111. Rosenzweig, R. E.: *A.I.Ch.E.Jl.*, **10,** 91 (1964).
112. Ogura, Y.: "Advances in Geophysics," Vol. **6,** p. 175, Academic Press, New York, 1959.
113. Chu, B. T., and L. S. G. Kovasznay: *J. Fluid Mech.*, **3,** 494 (1958).
114. Lighthill, M. J.: *Proc. Roy. Soc. London,* **267A,** 147 (1962).
115. Richards, E. J., and D. J. Mead: "Noise and Acoustic Fatigue in Aeronautics," Ch. 5, John Wiley, New York, 1968.
116. Ffowcs Williams, J. E.: *Phil. Trans. Roy. Soc. London,* **255A,** 469 (1963).
117. Ribner, H. S.: *Advances in Appl. Mechanics,* **8,** 103 (1964).
118. Tan, H. S., and S. C. Ling: *Phys. Fluids,* **6,** 1693 (1963).
119. Naudascher, E., and C. Farell: *J. Eng. Mechanics Div., Am. Soc. Civ. Engrs.,* **96,** 121 (1970).
120. Lee, D. A.: *Phys. Fluids,* **8,** 1911 (1965).
121. Lee, D. A., and H. S. Tan: *Phys. Fluids,* **10,** 1224 (1967).
122. Stewart, R. W., J. R. Wilson, and R. W. Burling: *J. Fluid Mech.*, **41,** 141 (1970).
123. Orszag, S. A.: *Phys. Fluids,* **13,** 2211 (1970).
124. Kennedy, D. A., and S. Corrsin: *J. Fluid Mech.*, **10,** 366 (1961).
125. Kuo, A. Y. S., and S. Corrsin: *J. Fluid Mech.*, **50,** 285 (1971).
126. Gurvich, A. S., and A. M. Yaglom: *Phys. Fluids,* Suppl., Part II, **10,** S. 59 (1967).
127. Saffman, P. G.: *Phys. Fluids,* **13,** 2193 (1970).
128. Tennekes, H.: *Phys. Fluids,* **11,** 669 (1968).
129. Kuo, A. Y. S., and S. Corrsin: *J. Fluid Mech.*, **56,** 447 (1972).
130. Corrsin, S.: *Phys. Fluids,* **7,** 1156 (1964).
131. Pond, S., and R. W. Stewart: *Irv. Acad. Sci. USSR Atmo and Oceanic,* Ser. 1, 1914 (1965).
132. Sheih, C. M., H. Tennekes, and J. L. Lumley: *Phys. Fluids,* **14,** 201 (1971).
133. Van Atta, C. W., and W. Y. Chen: *J. Fluid Mech.*, **44,** 145 (1970).
134. Wyngaard, J. C., and H. Tennekes: *Phys. Fluids,* **13,** 1962 (1970).
135. Ueda, H., and J. O. Hinze: *J. Fluid Mech.*, **67,** 125 (1975).
136. Lumley, J. L.: *Phys. Fluids,* **9,** 2111 (1966).
137. Lumley, J. L.: *Phys. Fluids,* **10,** 855 (1967).

NOMENCLATURE FOR CHAPTER 3

c_p specific heat at constant pressure.
c_v specific heat at constant volume.
$E_{i,j}$ spectrum tensor of turbulence kinetic energy.
$\mathbf{E}(k,t)$ $2\pi k^2\mathbf{E}_{i,i}(k,t)$, three-dimensional energy-spectrum function.
e_{ij} direction cosines.
$\mathbf{F}_{ik,j}$ Fourier transform of $\mathbf{S}_{ik,j}$.
$\mathbf{F}_{i,j}$ Fourier transform of $\mathbf{E}_{i,j}$.
$\mathbf{F}(k,t)$ $2\pi k^2\mathbf{F}_{i,i}(k,t)$, three-dimensional transfer spectrum function.
$\mathbf{f}$ coefficient of spatial longitudinal velocity correlation.
$\mathbf{G}$ total energy-transfer function.
$\mathbf{g}$ coefficient of spatial lateral velocity correlation.
H energy supplied to turbulence.
$\mathbf{h}$ coefficient of spatial, triple velocity correlation.
I $\dfrac{u'^2}{3\pi}\displaystyle\int_0^\infty dr\, r^4\mathbf{f}(r,t)$, Loitsianskii's integral.
$\mathbf{K}_i$ first-order correlation tensor.
$\mathbf{k}$ coefficient of spatial, triple velocity correlation.
k $2\pi n/\bar{U}$, wavenumber; k_i, component of wavenumber vector; k_e, wavenumber range of energy-containing eddies; k_d, wavenumber range of main dissipation.
l_e average size of energy-containing eddies.
M mesh length.
Ma Mach number.
n frequency.
P static pressure.
p turbulent fluctuation of static pressure.
$\mathbf{Q}_{i,j}$ second-order velocity-correlation tensor.
$\mathbf{q}$ coefficient of spatial, triple velocity correlation.
Re Reynolds number: $\mathbf{Re}_\lambda = u'\lambda_g/\nu$; $\mathbf{Re}_M = \bar{U}_1 M/\nu$.
r distance between two points.
S cntropy.
$\mathbf{S}_{i,kj}$ or $\mathbf{S}_{ik,j}$ third-order correlation tensor.
$\mathbf{S}_{i,j}$ $\dfrac{\partial}{\partial \xi_k}[\mathbf{S}_{ik,j} + \mathbf{S}_{kj,i}]$.
T flatness factor.
t time.
U_i Eulerian velocity; $\bar{U}_i$, time-mean value; u_i, turbulence component; $u_i' = \sqrt{\overline{u_i^2}}$, root-mean-square turbulence-velocity component; u_r, radial component; u_n, normal component.
x_i Eulerian Cartesian coordinates.
α Heisenberg constant in relation to the eddy viscosity ϵ.

β	Heisenberg constant in relation to the eddy diffusivity ϵ_γ.
δ_{ij}	Kronecker delta.
$\mathscr{E}$	energy per unit of mass.
ε_{ijk}	alternating unit tensor.
ε	dissipation by turbulence per unit of mass.
ε_γ	"dissipation" of γ-fluctuations.
ϵ	eddy viscosity; ϵ_γ, eddy diffusivity.
η	$(\nu^3/\varepsilon)^{1/4}$. Kolmogoroff micro length-scale.
Γ	scalar quantity; $\bar{\Gamma}$, time-mean value; γ, turbulent fluctuation.
Γ	gamma function.
$\mathscr{I}$	interaction function.
$\imath$	$\sqrt{-1}$.
$\varkappa$	c_p/c_v.
κ	heat conductivity.
$\mathfrak{k}$	$\kappa/c_p\rho$.
$\mathscr{L}$	similarity length scale.
Λ	spatial integral scale; Λ_f, longitudinal scale; Λ_g, lateral scale.
λ	spatial dissipation scale; λ_f, longitudinal; λ_g, lateral.
ν	kinematic viscosity.
π	3.14159....
$\mathscr{O}$	order of magnitude.
$\mathscr{R}$	$\varepsilon t^2/\nu$.
$\mathfrak{R}$	gas constant.
ρ	density; $\bar{\rho}$, time-mean value; $\tilde{\rho}$, turbulent fluctuation.
ς	entropy fluctuation.
σ_{ij}	stress tensor.
Θ	dilatation.
Θ	temperature; $\bar{\Theta}$, time-mean value; θ, turbulent fluctuation.
υ	$(\nu/\varepsilon)^{1/4}$.
ξ_i	Eulerian Cartesian coordinates.

4
HOMOGENEOUS, SHEAR-FLOW TURBULENCE

4-1 INTRODUCTION

The number of contributions in which there is developed a theory on nonisotropic turbulence similar to the theories on isotropic turbulence as expounded in the previous chapter is small. The main reason for this is the extreme complexity of the problems once nonisotropy is brought into the picture.

Axisymmetrical turbulence, a type second in simplicity to isotropic turbulence, was first treated in a few papers by Batchelor,[1] and later by Chandrasekhar.[2,3] In contrast to isotropic turbulence, where the symmetry is spherical, axisymmetrical turbulence has symmetry about a given direction. The main flow is assumed to be in this direction and uniform, and the turbulence is assumed to be homogeneous. The average value of any function of the velocities and of their derivatives is invariant under rotations about an axis in the given direction and under reflections with respect to planes through this axis and perpendicular to it. Batchelor and Chandrasekhar have treated the kinematics of this type of turbulence almost as completely as has been possible with isotropic turbulence and have derived explicit expressions for the various correlation tensors in terms of a minimum number of defining scalars.

In this chapter we will not consider this case of axisymmetric turbulence, notwithstanding that grid-produced turbulence in a uniform main flow is closer to it than to isotropic turbulence. But we will restrict ourselves to considering mainly some contributions to the theory of homogeneous, turbulent shear flow. Though the results are still very meager, they do assist in explaining a few features of shear-flow turbulence and may contribute to some better understanding of it.

In what follows it is assumed that the fluid is *incompressible* and the *main motion steady*.

Because of the existence of a mean shear, there is a continuous production of turbulence. When it just compensates the local dissipation, a steady-state turbulence is possible when at the same time the main motion is steady. Yet it may be of interest to assume a turbulence state that at some initial instant is not in equilibrium, so that we have to do with a time-dependent turbulence in a steady mean shear-flow. It will be assumed that the time-scale of the changes of the statistical relations is small compared with the time-scale of the turbulent fluctuations, so that an averaging procedure with respect to time may still be performed and no recourse to an ensemble average has to be taken.

We shall consider first the dynamics of correlations between velocity components at one point of the turbulent flow field, and then a generalization to correlations between velocities at two points of the field. In the latter case an appreciable simplification is obtained by the assumption of homogeneity. Already in Chap. 1 we brought to light many characteristics of velocity correlations merely by assuming homogeneity. A homogeneous turbulence with shear can be imagined only if the main flow has a constant lateral velocity gradient throughout the entire flow field, and the restriction to this flow type contributes appreciably to the simplification of the problem.

As Burgers and Mitchner[4] have pointed out, in this type of turbulent shear flow the absence of confining walls means that there is no length scale to determine the large-scale structure of the turbulence, in sharp contrast to actual flows. According to Von Kármán's similarity hypothesis (see Chap. 5), in any actual turbulence flow field the large-scale structure is determined by a length scale which follows from $(\partial\bar{U}_i/\partial x_j)/(\partial^2\bar{U}_i/\partial x_j^2)$ for $i \neq j$ (no summation convention); this ratio becomes infinite for a constant value of the velocity gradient.

A steady, homogeneous turbulent shear-flow is, strictly speaking, also a hypothetical type of flow. For, when attempting to produce experimentally such a type of turbulent flow, one is faced with the problem of introducing by external means sufficient energy into the flow that it just compensates the continuous dissipation of mechanical energy in order to keep the whole flow steady. Also, this external energy supply has to be performed without adversely affecting the homogeneity of the flow. Yet Rose,[10] and later Champagne, Harris, and Corrsin,[12] seem to have succeeded in generating in a windtunnel a shear flow that was homogeneous to a satisfactory degree of approximation, and practically steady, at least towards the end of the test section. Finite length-scales of the turbulence were present due to the upstream device generating the shear flow. They must increase downstream so that actually no true steady state of turbulence can occur.[20] Moreover, due to the continuous generation of turbulence by the mean shear flow, the latter also cannot maintain a constant shear in the main flow direction but must continuously decrease in value. This then calls for a compensating $\bar{U}_2$ velocity. However, under the experimental conditions the change of the mean flow was sufficiently small to justify a quasi-homogeneous condition, while a $\bar{U}_2$ velocity appeared too weak to be of any influence.

4-2 DYNAMICS OF THE ONE-POINT VELOCITY CORRELATION $\overline{u_i u_j}$

The dynamic equation for $\overline{u_i u_j}$ is obtained as shown in previous chapters.

The equation of motion of an incompressible fluid for the case of a steady main motion reads

$$\frac{\partial u_i}{\partial t} + (\bar{U}_k + u_k)\frac{\partial}{\partial x_k}(\bar{U}_i + u_i) = -\frac{1}{\rho}\frac{\partial \bar{P}}{\partial x_i} - \frac{1}{\rho}\frac{\partial p}{\partial x_i} + \nu \frac{\partial^2}{\partial x_l\, \partial x_l}(\bar{U}_i + u_i)$$

or, by adding $u_i\, \partial u_k/\partial x_k = 0$,

$$\frac{\partial u_i}{\partial t} + \bar{U}_k \frac{\partial \bar{U}_i}{\partial x_k} + u_k \frac{\partial \bar{U}_i}{\partial x_k} + \bar{U}_k \frac{\partial u_i}{\partial x_k} + \frac{\partial}{\partial x_k} u_i u_k$$

$$= -\frac{1}{\rho}\frac{\partial \bar{P}}{\partial x_i} - \frac{1}{\rho}\frac{\partial p}{\partial x_i} + \nu \frac{\partial^2}{\partial x_l\, \partial x_l}(\bar{U}_i + u_i)$$

Subtract from this its average value, and we obtain

$$\frac{\partial u_i}{\partial t} + u_k \frac{\partial \bar{U}_i}{\partial x_k} + \bar{U}_k \frac{\partial u_i}{\partial x_k} + \frac{\partial}{\partial x_k}(u_i u_k - \overline{u_i u_k}) = -\frac{1}{\rho}\frac{\partial p}{\partial x_i} + \nu \frac{\partial^2 u_i}{\partial x_l\, \partial x_l} \tag{4-1}$$

Write the same equation for the velocity component u_j. Multiply Eq. (4-1) by u_j and the equation for u_j by u_i, and add the two equations. After averaging, the result is

$$\frac{\partial}{\partial t}\overline{u_i u_j} + \overline{u_j u_k}\frac{\partial \bar{U}_i}{\partial x_k} + \overline{u_i u_k}\frac{\partial \bar{U}_j}{\partial x_k} + \bar{U}_k \frac{\partial}{\partial x_k}\overline{u_i u_j}$$

$$= -\left(\overline{u_j \frac{\partial}{\partial x_k} u_i u_k} + \overline{u_i \frac{\partial}{\partial x_k} u_j u_k}\right) - \frac{1}{\rho}\left(\overline{u_j \frac{\partial p}{\partial x_i}} + \overline{u_i \frac{\partial p}{\partial x_j}}\right)$$

$$+ \nu\left(\overline{u_j \frac{\partial^2 u_i}{\partial x_l\, \partial x_l}} + \overline{u_i \frac{\partial^2 u_j}{\partial x_l\, \partial x_l}}\right) \tag{4-2}$$

Because

$$\overline{u_j \frac{\partial p}{\partial x_i}} = \frac{\partial}{\partial x_i}\overline{pu_j} - \overline{p\frac{\partial u_j}{\partial x_i}} \qquad \text{and} \qquad \overline{u_i \frac{\partial p}{\partial x_j}} = \frac{\partial}{\partial x_j}\overline{pu_i} - \overline{p\frac{\partial u_i}{\partial x_j}}$$

$$\overline{u_j\frac{\partial}{\partial x_k}u_iu_k} + \overline{u_i\frac{\partial}{\partial x_k}u_ju_k} = \overline{u_ju_k\frac{\partial u_i}{\partial x_k}} + \overline{u_i\frac{\partial}{\partial x_k}u_ju_k} = \overline{\frac{\partial}{\partial x_k}u_iu_ju_k}$$

$$\frac{\partial^2}{\partial x_l\,\partial x_l}\overline{u_iu_j} = \overline{u_i\frac{\partial^2 u_j}{\partial x_l\,\partial x_l}} + \overline{u_j\frac{\partial^2 u_i}{\partial x_l\,\partial x_l}} + 2\overline{\frac{\partial u_i}{\partial x_l}\frac{\partial u_j}{\partial x_l}}$$

Eq. (4-2) may be rewritten as follows:

$$\frac{\partial}{\partial t}\overline{u_iu_j} + \overline{u_ju_k}\frac{\partial \bar{U}_i}{\partial x_k} + \overline{u_iu_k}\frac{\partial \bar{U}_j}{\partial x_k} + \bar{U}_k\frac{\partial}{\partial x_k}\overline{u_iu_j}$$
$$= -\frac{\partial}{\partial x_k}\overline{u_iu_ju_k} - \frac{1}{\rho}\left(\frac{\partial}{\partial x_i}\overline{pu_j} + \frac{\partial}{\partial x_j}\overline{pu_i}\right) + \frac{1}{\rho}\overline{p\left(\frac{\partial u_j}{\partial x_i} + \frac{\partial u_i}{\partial x_j}\right)}$$
$$+ \nu\frac{\partial^2 \overline{u_iu_j}}{\partial x_l\,\partial x_l} - 2\nu\overline{\frac{\partial u_i}{\partial x_l}\frac{\partial u_j}{\partial x_l}} \tag{4-3}$$

A contraction of these equations yields

$$\frac{\partial}{\partial t}\overline{u_iu_i} + 2\overline{u_iu_k}\frac{\partial \bar{U}_i}{\partial x_k} + \bar{U}_k\frac{\partial}{\partial x_k}\overline{u_iu_i} = -2\frac{\partial}{\partial x_k}\overline{\left(\frac{p}{\rho} + \frac{u_iu_i}{2}\right)u_k} + \nu\frac{\partial^2\overline{u_iu_i}}{\partial x_l\,\partial x_l} - 2\nu\overline{\frac{\partial u_i}{\partial x_l}\frac{\partial u_i}{\partial x_l}}$$

or

$$\frac{\partial}{\partial t}\frac{\overline{q^2}}{2} + \bar{U}_k\frac{\partial}{\partial x_k}\frac{\overline{q^2}}{2} + \overline{u_iu_k}\frac{\partial \bar{U}_i}{\partial x_k} = -\frac{\partial}{\partial x_k}\overline{\left(\frac{p}{\rho} + \frac{q^2}{2}\right)u_k} + \nu\frac{\partial^2}{\partial x_l\,\partial x_l}\frac{\overline{q^2}}{2} - \nu\overline{\frac{\partial u_i}{\partial x_l}\frac{\partial u_i}{\partial x_l}} \tag{4-4}$$

which is the turbulence-energy equation (1-111) already given in Chap. 1, but derived in a different way. For the interpretation of the various terms the reader is referred to that chapter. But we may note that in the contracted equation the pressure-velocity-gradient correlations drop out, since $\overline{p\,\partial u_i/\partial x_i} = 0$ for an incompressible fluid.

From Eq. (4-3) emerge some interesting properties connected with the effects of diffusion, of interferences between the velocity components through their mutual correlations and the pressure-velocity correlations, and of viscosity, on the changes in the individual terms of the correlation tensor $\overline{u_iu_j}$. In order to simplify the argument we take the case of a two-dimensional main flow $\bar{U}_1 = f(x_2)$, $\bar{U}_2 = \bar{U}_3 = 0$, homogeneous in the x_1- and x_3-directions but not necessarily in the x_2-direction.

Because of the symmetry of the flow with respect to planes perpendicular to the x_3-direction, all correlations involving u_3 and uneven derivatives with respect to x_3 are zero. Hence all terms of Eq. (4-3) for the correlations $\overline{u_1u_3}$ and $\overline{u_2u_3}$, respectively, are zero.

The remaining equations of (4-3), that is, those for $\overline{u_1^2}$, $\overline{u_2^2}$, $\overline{u_3^2}$, and $\overline{u_1u_2}$, read, respectively,

$$\frac{\partial}{\partial t}\overline{u_1^2} = -2\overline{u_1u_2}\frac{\partial \bar{U}_1}{\partial x_2} - \frac{\partial}{\partial x_2}\overline{u_1^2u_2} + \frac{2}{\rho}\overline{p\frac{\partial u_1}{\partial x_1}} + \nu\frac{\partial^2\overline{u_1^2}}{\partial x_2^2} - 2\nu\overline{\frac{\partial u_1}{\partial x_l}\frac{\partial u_1}{\partial x_l}} \tag{4-5}$$

$$\frac{\partial}{\partial t}\overline{u_2^2} = -\frac{\partial}{\partial x_2}\overline{u_2^3} - \frac{2}{\rho}\frac{\partial}{\partial x_2}\overline{pu_2} + \frac{2}{\rho}\overline{p\frac{\partial u_2}{\partial x_2}} + \nu\frac{\partial^2}{\partial x_2^2}\overline{u_2^2} - 2\nu\overline{\frac{\partial u_2}{\partial x_l}\frac{\partial u_2}{\partial x_l}} \tag{4-6}$$

$$\frac{\partial}{\partial t}\overline{u_3^2} = -\frac{\partial}{\partial x_2}\overline{u_3^2u_2} + \frac{2}{\rho}\overline{p\frac{\partial u_3}{\partial x_3}} + \nu\frac{\partial^2}{\partial x_2^2}\overline{u_3^2} - 2\nu\overline{\frac{\partial u_3}{\partial x_l}\frac{\partial u_3}{\partial x_l}} \tag{4-7}$$

$$\frac{\partial}{\partial t}\overline{u_1u_2} = -\overline{u_2^2}\frac{\partial \bar{U}_1}{\partial x_2} - \frac{\partial}{\partial x_2}\overline{u_1\left(u_2^2 + \frac{p}{\rho}\right)} + \overline{\frac{p}{\rho}\left(\frac{\partial u_2}{\partial x_1} + \frac{\partial u_1}{\partial x_2}\right)}$$

$$+ \nu\frac{\partial^2}{\partial x_2^2}\overline{u_1u_2} - 2\nu\overline{\frac{\partial u_1}{\partial x_l}\frac{\partial u_2}{\partial x_l}} \tag{4-8}$$

Thus, to recapitulate, the temporal changes in the quantities $\overline{u_1^2}$, $\overline{u_2^2}$, $\overline{u_3^2}$, and $\overline{u_1u_2}$ are determined by convective diffusion and viscous effects due to inhomogeneities of the flow field in the x_2-direction, by production or energy transfer from the main motion through the turbulence shear stresses, by effects due to the correlations between the pressure fluctuations and derivatives of the velocity fluctuations, and by viscous effects mainly of a dissipative nature.

The contributions due to inhomogeneities of the flow field may be either negative or positive, depending on the distribution.

The viscous terms $2\nu\overline{(\partial u_1/\partial x_l)(\partial u_1/\partial x_l)}$, etc., make a negative contribution to the respective components of the turbulence intensity.

As to the production term $\overline{u_1u_2}\,\partial \bar{U}_1/\partial x_2$, the reader may notice that it occurs only in the equation for the $\overline{u_1^2}$ component.

If we consider further the case $\partial \bar{U}_1/\partial x_2 > 0$, $\overline{u_1u_2}$ will be negative. The production term then makes a positive contribution to the temporal change in $\overline{u_1^2}$ but not to the other turbulence components. Hence the production of turbulence results in nonisotropy, in favor of the axial component of the turbulence velocity.

The mean shear $\partial \bar{U}_1/\partial x_2$ also occurs in Eq. (4-8) for $\overline{u_1u_2}$, giving either a positive or negative contribution depending on whether $\partial \bar{U}_1/\partial x_2$ is negative or positive.

What will be the effect of the pressure-velocity-gradient correlation $\overline{p\,\partial u_1/\partial x_1}$? To study this effect we shall only make it appear plausible that, on the average, this correlation will make a negative contribution to $\partial\overline{u_1^2}/\partial t$, if $\overline{u_1^2}$ is much greater than $\overline{u_2^2}$ and $\overline{u_3^2}$. This is, of course, not an exact proof.

Take a point **O** in the flow field moving with the local mean velocity $\bar{U}_1$, and consider the turbulence motion relative to it. At a given instant, let the turbulence velocities be directed as shown in Fig. 4-1a. Assume that $2\overline{u_1^2} > \overline{u_2^2} + \overline{u_3^2}$, owing, for instance, to the effect of the production term that occurs only in the equation for $\overline{u_1^2}$. The spatially decelerated inward motion with velocity u_1 is then associated with a positive value of the local pressure p. Thus $\partial u_1/\partial x_1 < 0$ and $p > 0$; so $p\,\partial u_1/\partial x_1 < 0$. But at the same time $\partial u_2/\partial x_2 > 0$ and $\partial u_3/\partial x_3 > 0$; so $p\,\partial u_2/\partial x_2 > 0$

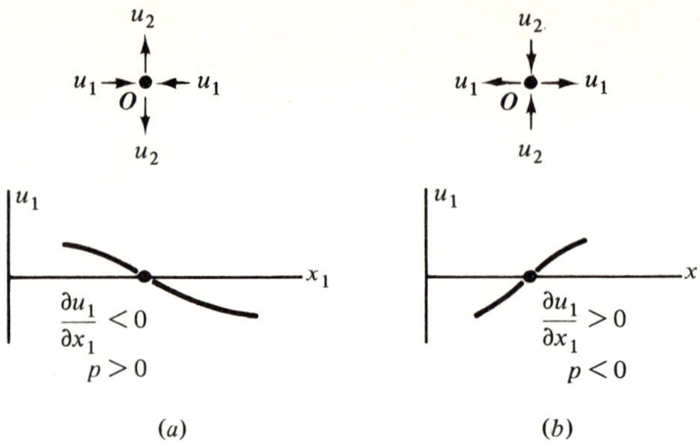

FIGURE 4-1
Concerning the velocity-pressure correlation $\overline{p\,\partial u_1/\partial x_1}$.

and $p\,\partial u_3/\partial x_3 > 0$. This is a consequence of the continuity equation: since $p(\partial u_1/\partial x_1 + \partial u_2/\partial x_2 + \partial u_3/\partial x_3) = 0$, it follows from $p\,\partial u_1/\partial x_1 < 0$ that $p(\partial u_2/\partial x_2 + \partial u_3/\partial x_3) > 0$.

The converse case (Fig. 4-1b), that of a spatially accelerated outward motion with velocity u_1, has a negative value of p associated with it; so $\partial u_1/\partial x_1 > 0$ and $p < 0$, and again $p\,\partial u_1/\partial x_1 < 0$, $p\,\partial u_2/\partial x_2 > 0$, and $p\,\partial u_3/\partial x_3 > 0$.

Hence, if on the average $2\overline{u_1^2} > \overline{u_2^2} + \overline{u_3^2}$, we see that on the average $\overline{p\,\partial u_1/\partial x_1} < 0$, $\overline{p\,\partial u_2/\partial x_2} > 0$, and $\overline{p\,\partial u_3/\partial x_3} > 0$. The pressure-velocity-gradient correlation $\overline{p\,\partial u_1/\partial x_1}$, then, makes a negative contribution to $\partial\overline{u_1^2}/\partial t$, but $\overline{p\,\partial u_2/\partial x_2}$ and $\overline{p\,\partial u_3/\partial x_3}$ make positive contributions to $\partial\overline{u_2^2}/\partial t$ and $\partial\overline{u_3^2}/\partial t$, respectively. The final result is that the lateral components u_2 and u_3 are increased at the expense of the axial component u_1, owing to the transfer of energy via the pressure-velocity-gradient correlation. The effect of this correlation is to make the turbulence less non-isotropic. However, at the same time one may conclude that, according to this mechanism of transfer of energy between velocity components through pressure effects, no further equalization seems possible when $2\overline{u_1^2} \simeq \overline{u_2^2} + \overline{u_3^2}$. Unless other effects are present, as for instance a slight, favorable main-pressure gradient in axial direction which reduces the ratio $\overline{u_1^2}/\overline{u_2^2} \simeq \overline{u_1^2}/\overline{u_3^2}$.

It is reasonable to assume[5] that the transfer of energy from the higher-intensity to the lower-intensity component through the pressure-velocity-gradient correlation is proportional to the difference in intensity, thus that

$$\overline{p\frac{\partial u_1}{\partial x_1}} \propto -[\overline{u_1^2} - \tfrac{1}{2}(\overline{u_2^2} + \overline{u_3^2})] = -\tfrac{1}{2}(3\overline{u_1^2} - \overline{q^2}) \qquad (4\text{-}9)$$

Let us now consider Eq. (4-8) for the turbulence shear stress $\overline{u_1u_2}$, restricting ourselves again to the production term and the term for the pressure-velocity-gradient correlation.

If again we take the case $\partial\bar{U}_1/\partial x_2 > 0$, then the production term $\overline{u_2^2}\,\partial\bar{U}_1/\partial x_2$ makes a negative contribution to $\partial/\partial t\,\overline{u_1u_2}$. But since for $\partial\bar{U}_1/\partial x_2 > 0$ we have

$\overline{u_1u_2} < 0$, the negative contribution results in an increase in the above value of $\overline{u_1u_2}$. The same result will be obtained if $\partial\bar{U}_1/\partial x_2 < 0$, in which case $\overline{u_1u_2} > 0$.

We shall now make it appear reasonable that the term

$$\frac{1}{\rho}\overline{p\left(\frac{\partial u_2}{\partial x_1} + \frac{\partial u_1}{\partial x_2}\right)}$$

makes a positive contribution to $\partial\overline{u_1u_2}/\partial t$ if $\overline{u_1u_2} < 0$ and a negative contribution if $\overline{u_1u_2} > 0$. To this end we express the correlation $\overline{u_1u_2}$ in terms of the turbulence velocity components u_{1*} and u_{2*} along the axes x_{1*} and x_{2*}, respectively, which make an angle φ with the axes x_1 and x_2 (see Fig. 4-2). Again **O** is a point in the fluid moving with the main flow at velocity $\bar{U}_1$.

$$u_1 = u_{1*}\cos\varphi - u_{2*}\sin\varphi$$

$$u_2 = u_{1*}\sin\varphi + u_{2*}\cos\varphi$$

whence $\overline{u_1u_2} = (\overline{u_{1*}^2} - \overline{u_{2*}^2})(\sin 2\varphi)/2 + \overline{u_{1*}u_{2*}}\cos 2\varphi$.

Take $\varphi = 45°$, so that the axes x_{1*} and x_{2*} become the principal axes in the turbulence-stress distribution. Then

$$2\overline{u_1u_2} = \overline{u_{1*}^2} - \overline{u_{2*}^2} \tag{4-9a}$$

From a transformation of the coordinate axes we obtain the relation

$$\frac{\partial u_2}{\partial x_1} + \frac{\partial u_1}{\partial x_2} = \left(\frac{\partial u_1^*}{\partial x_1^*} - \frac{\partial u_2^*}{\partial x_2^*}\right)\sin 2\varphi + \left(\frac{\partial u_1^*}{\partial x_2^*} + \frac{\partial u_2^*}{\partial x_1^*}\right)\cos 2\varphi$$

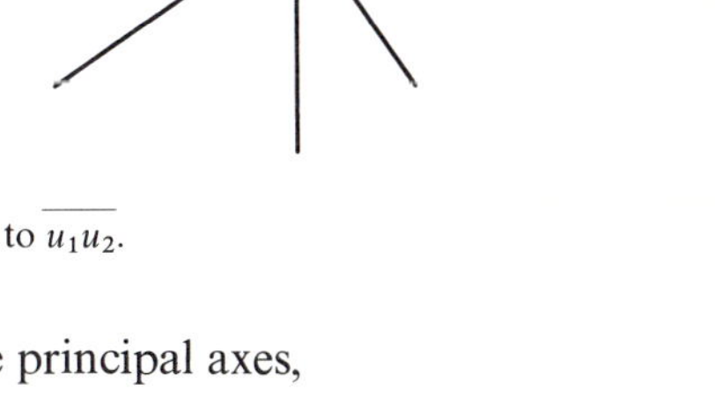

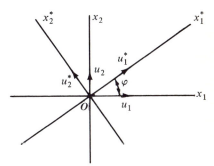

FIGURE 4-2
Coordinate transformation applied to $\overline{u_1u_2}$.

For $\varphi = 45°$, that is, for a transformation on the principal axes,

$$\frac{\partial u_2}{\partial x_1} + \frac{\partial u_1}{\partial x_2} = \frac{\partial u_1^*}{\partial x_1^*} - \frac{\partial u_2^*}{\partial x_2^*}$$

Consequently,

$$\overline{p\left(\frac{\partial u_2}{\partial x_1} + \frac{\partial u_1}{\partial x_2}\right)} = \overline{p\left(\frac{\partial u_{1*}}{\partial x_{1*}} - \frac{\partial u_{2*}}{\partial x_{2*}}\right)}$$

If we apply the relation (4-9) and assume the same proportionality constants, by (4-9a) we obtain

$$\overline{p\left(\frac{\partial u_{1*}}{\partial x_{1*}} - \frac{\partial u_{2*}}{\partial x_{2*}}\right)} \propto -\left(\overline{u_{1*}{}^2} - \overline{u_{2*}{}^2}\right) \propto -\overline{u_1 u_2}$$

and

$$\overline{p\left(\frac{\partial u_2}{\partial u_1} + \frac{\partial u_1}{\partial x_2}\right)} \propto -\overline{u_1 u_2}$$

Hence $\overline{p(\partial u_2/\partial x_1 + \partial u_1/\partial x_2)}$ and $\overline{u_1 u_2}$ do have opposite signs, and the effect of this pressure-velocity-gradient correlation is to decrease the absolute value of $\overline{u_1 u_2}$.

Rotta[5] has made a quantitative estimate of the pressure-velocity-gradient correlation, and arrives at the expression

$$\overline{p\left(\frac{\partial u_i}{\partial x_j} + \frac{\partial u_j}{\partial x_i}\right)} = -A\rho \frac{\sqrt{\overline{q^2}}}{\Lambda}\left(\overline{u_i u_j} - \tfrac{1}{3}\delta_{ij}\overline{q^2}\right) \qquad (4\text{-}10)$$

where A is a numerical constant of the order unity and Λ is an integral scale of turbulence.

Burgers and Mitchner[4] have shown that, for small wavenumber k, the Fourier transform of the pressure-velocity-gradient correlation increases rapidly with k (for very small k, with k^4). Hence the effect of this correlation increases with increasing wavenumber.

We may summarize the above results as follows:

The turbulence shear stress $\overline{u_1 u_2}$ can be produced only if the main flow is not uniform.

The pressure-velocity-gradient correlations tend to decrease the nonisotropy by equalizing the three turbulence velocity components and decreasing the turbulence shear stress.

This tendency to isotropy is greater in the smaller scale range of turbulence.

Since the viscosity effect, through dissipation, increases with increasing intensity of turbulence, its result, as far as nonisotropic turbulence is concerned, is to damp out the greater-intensity components at a higher rate than the smaller; that is, it also tends to equalize. Here, too, the effect is greater in the higher-wavenumber range. However, such a process due to viscosity is a relatively slow one.

4-3 DYNAMICS OF THE TWO-POINT VELOCITY CORRELATION $\overline{(u_i)_A(u_j)_B}$

We shall first derive the dynamic equation for this correlation for the general case of inhomogeneous, nonisotropic turbulence. Though this equation is entirely intractable, it is useful to give it here, together with its derivation, before considering the simplified case of homogeneous turbulence.

We start from Eq. (4-1), referring to point A, and multiply it by $(u_j)_B$, referring to point B:

$$(u_j)_B\left(\frac{\partial u_i}{\partial t}\right)_A + (u_j)_B(u_k)_A\left(\frac{\partial \bar{U}_i}{\partial x_k}\right)_A + (u_j)_B(\bar{U}_k)_A\left(\frac{\partial u_i}{\partial x_k}\right)_A$$

$$= -(u_j)_B\left[\frac{\partial}{\partial x_k}(u_iu_k - \overline{u_iu_k})\right]_A - \frac{1}{\rho}(u_j)_B\left(\frac{\partial p}{\partial x_i}\right)_A + \nu(u_j)_B\left(\frac{\partial^2 u_i}{\partial x_l\,\partial x_l}\right)_A$$

Similarly, we multiply Eq. (4-1) referring to point B by a velocity component referring to point A. Addition of the two equations yields

$$\frac{\partial}{\partial t}(u_i)_A(u_j)_B + (u_j)_B(u_k)_A\left(\frac{\partial \bar{U}_i}{\partial x_k}\right)_A + (u_i)_A(u_k)_B\left(\frac{\partial \bar{U}_j}{\partial x_k}\right)_B$$

$$+ (\bar{U}_k)_A\left(\frac{\partial}{\partial x_k}\right)_A(u_i)_A(u_j)_B + (\bar{U}_k)_B\left(\frac{\partial}{\partial x_k}\right)_B(u_i)_A(u_j)_B$$

$$= -\left(\frac{\partial}{\partial x_k}\right)_A(u_j)_B(u_i)_A(u_k)_A - \left(\frac{\partial}{\partial x_k}\right)_B(u_i)_A(u_j)_B(u_k)_B$$

$$+ \left(\frac{\partial}{\partial x_k}\right)_A(u_j)_B(\overline{u_iu_k})_A + \left(\frac{\partial}{\partial x_k}\right)_B(u_i)_A(\overline{u_ju_k})_B$$

$$- \frac{1}{\rho}\left[\left(\frac{\partial}{\partial x_i}\right)_A p_A(u_j)_B + \left(\frac{\partial}{\partial x_j}\right)_B p_B(u_i)_A\right]$$

$$+ \nu\left[\left(\frac{\partial^2}{\partial x_l\,\partial x_l}\right)_A + \left(\frac{\partial^2}{\partial x_l\,\partial x_l}\right)_B\right](u_i)_A(u_j)_B$$

Because of the inhomogeneity of the turbulence, the mean values of various products of the turbulence velocities occurring in this equation are functions not only of the distance between the points A and B but also of the location of these points in the flow field. In order to differentiate between the effects of distance and of location, we introduce as new independent variables[4]

$$\xi_k = (x_k)_B - (x_k)_A \qquad \text{(4-11a)}$$

$$(x_k)_{AB} = \tfrac{1}{2}[(x_k)_A + (x_k)_B] \qquad \text{(4-11b)}$$

Thus, for any quantity that is a function of $(x_k)_{AB}$ and ξ_k, we have

$$\left(\frac{\partial}{\partial x_k}\right)_A = \left(\frac{\partial}{\partial x_k}\right)_{AB}\frac{\partial(x_k)_{AB}}{(\partial x_k)_A} + \frac{\partial}{\partial \xi_k}\frac{\partial \xi_k}{(\partial x_k)_A} = \frac{1}{2}\left(\frac{\partial}{\partial x_k}\right)_{AB} - \frac{\partial}{\partial \xi_k} \qquad \text{(4-12a)}$$

$$\left(\frac{\partial}{\partial x_k}\right)_B = \left(\frac{\partial}{\partial x_k}\right)_{AB}\frac{\partial(x_k)_{AB}}{(\partial x_k)_B} + \frac{\partial}{\partial \xi_k}\frac{\partial \xi_k}{(\partial x_k)_B} = \frac{1}{2}\left(\frac{\partial}{\partial x_k}\right)_{AB} + \frac{\partial}{\partial \xi_k} \qquad \text{(4-12b)}$$

$$\left(\frac{\partial^2}{\partial x_l\,\partial x_l}\right)_A = \frac{1}{4}\left(\frac{\partial^2}{\partial x_l\,\partial x_l}\right)_{AB} + \frac{\partial^2}{\partial \xi_l\,\partial \xi_l} - \left(\frac{\partial}{\partial x_k}\right)_{AB}\frac{\partial}{\partial \xi_k}$$

$$\left(\frac{\partial^2}{\partial x_l\,\partial x_l}\right)_B = \frac{1}{4}\left(\frac{\partial^2}{\partial x_l\,\partial x_l}\right)_{AB} + \frac{\partial^2}{\partial \xi_l\,\partial \xi_l} + \left(\frac{\partial}{\partial x_k}\right)_{AB}\frac{\partial}{\partial \xi_k}$$

and

$$\left(\frac{\partial^2}{\partial x_l\,\partial x_l}\right)_A + \left(\frac{\partial^2}{\partial x_l\,\partial x_l}\right)_B = \frac{1}{2}\left(\frac{\partial^2}{\partial x_l\,\partial x_l}\right)_{AB} + 2\frac{\partial^2}{\partial \xi_l\,\partial \xi_l} \tag{4-13}$$

If we apply this procedure to the quantities in the dynamic equation that are functions of $(x_k)_{AB}$ and ξ_k, after carrying out an averaging procedure with respect to time, with

$$\overline{(u_i)_A(u_j)_B} = \mathbf{Q}_{i,j}(x_1,x_2,x_3,\xi_1,\xi_2,\xi_3,t)$$
$$\overline{(u_i)_A(u_j)_B(u_k)_B} = \mathbf{S}_{i,kj}(x_1,x_2,x_3,\xi_1,\xi_2,\xi_3,t)$$
$$\overline{(u_j)_B(u_i)_A(u_k)_A} = \mathbf{S}_{ik,j}(x_1,x_2,x_3,\xi_1,\xi_2,\xi_3,t)$$
$$\overline{p_A(u_j)_B} = \mathbf{K}_{p,j}(x_1,x_2,x_3,\xi_1,\xi_2,\xi_3,t)$$
$$\overline{p_B(u_i)_A} = \mathbf{K}_{i,p}(x_1,x_2,x_3,\xi_1,\xi_2,\xi_3,t)$$

we obtain

$$\frac{\partial}{\partial t}\mathbf{Q}_{i,j} + \mathbf{Q}_{k,j}\left(\frac{\partial \bar{U}_i}{\partial x_k}\right)_A + \mathbf{Q}_{i,k}\left(\frac{\partial \bar{U}_j}{\partial x_k}\right)_B + \tfrac{1}{2}[(\bar{U}_k)_A + (\bar{U}_k)_B]\left(\frac{\partial}{\partial x_k}\right)_{AB}\mathbf{Q}_{i,j}$$
$$+ [(\bar{U}_k)_B - (\bar{U}_k)_A]\frac{\partial}{\partial \xi_k}\mathbf{Q}_{i,j}$$
$$= -\frac{1}{2}\left(\frac{\partial}{\partial x_k}\right)_{AB}(\mathbf{S}_{i,kj} + \mathbf{S}_{ik,j}) - \frac{\partial}{\partial \xi_k}(\mathbf{S}_{i,kj} - \mathbf{S}_{ik,j})$$
$$- \frac{1}{2\rho}\left[\left(\frac{\partial}{\partial x_i}\right)_{AB}\mathbf{K}_{p,j} + \left(\frac{\partial}{\partial x_j}\right)_{AB}\mathbf{K}_{i,p}\right] + \frac{1}{\rho}\left[\frac{\partial}{\partial \xi_i}\mathbf{K}_{p,j} - \frac{\partial}{\partial \xi_j}\mathbf{K}_{i,p}\right]$$
$$+ \tfrac{1}{2}\nu\left(\frac{\partial^2}{\partial x_l\,\partial x_l}\right)_{AB}\mathbf{Q}_{i,j} + 2\nu\frac{\partial^2}{\partial \xi_l\,\partial \xi_l}\mathbf{Q}_{i,j} \tag{4-14}$$

This is the complete dynamic equation for the double velocity correlation $(\mathbf{Q}_{i,j})_{A,B}$ for the general case of nonisotropic inhomogeneous turbulence.

It is possible to express the pressure-velocity correlations in terms of the two-point velocity correlations $\mathbf{Q}_{i,j}$ and $\mathbf{S}_{ik,j}$ or $\mathbf{S}_{i,kj}$. To this end we start from Eq. (4-1) and take its divergence $\partial/\partial x_i$. Since $\partial u_i/\partial x_i = 0$ for an incompressible fluid, the acceleration and viscous terms drop out. We obtain

$$\frac{1}{\rho}\frac{\partial^2 p}{\partial x_i\,\partial x_i} = -2\frac{\partial u_k}{\partial x_i}\cdot\frac{\partial \bar{U}_i}{\partial x_k} - \frac{\partial^2}{\partial x_k\,\partial x_i}(u_k u_i - \overline{u_k u_i})$$

This equation is used to obtain

$$\frac{1}{\rho}\left(\frac{\partial^2}{\partial x_i\,\partial x_i}\right)_A p_A(u_j)_B = -2\left(\frac{\partial u_k}{\partial x_i}\right)_A(u_j)_B\left(\frac{\partial \bar{U}_i}{\partial x_k}\right)_A - \left(\frac{\partial^2}{\partial x_k\,\partial x_i}\right)_A[(u_k u_i)_A(u_j)_B - (\overline{u_k u_i})_A(u_j)_B]$$

Whence follows, after averaging

$$\frac{1}{\rho}\left(\frac{\partial^2}{\partial x_i\,\partial x_i}\right)_A \mathbf{K}_{p,j} = -2\left(\frac{\partial}{\partial x_i}\right)_A \mathbf{Q}_{k,j}\left(\frac{\partial \bar{U}_i}{\partial x_k}\right)_A - \left(\frac{\partial^2}{\partial x_k\,\partial x_i}\right)_A \mathbf{S}_{ki,j}$$

After introducing again the new variables ξ_k and $(x_k)_{A,B}$ according to Eqs. (4-11a) and (4-11b), the relation for $\mathbf{K}_{p,j}$ becomes

$$\frac{1}{\rho}\left[\frac{1}{4}\left(\frac{\partial^2}{\partial x_i\,\partial x_i}\right)_{AB} - \left(\frac{\partial}{\partial x_i}\right)_{AB}\frac{\partial}{\partial \xi_i} + \frac{\partial^2}{\partial \xi_i\,\partial \xi_i}\right]\mathbf{K}_{p,j}$$
$$= -\left(\frac{\partial \bar{U}_i}{\partial x_k}\right)_A\left[\left(\frac{\partial}{\partial x_i}\right)_{AB} - 2\frac{\partial}{\partial \xi_i}\right]\mathbf{Q}_{k,j}$$
$$-\left[\frac{1}{4}\left(\frac{\partial^2}{\partial x_k\,\partial x_i}\right)_{AB} + \frac{\partial^2}{\partial \xi_k\,\partial \xi_i} - \frac{1}{2}\left\{\left(\frac{\partial}{\partial x_k}\right)_{AB}\frac{\partial}{\partial \xi_i} + \left(\frac{\partial}{\partial x_i}\right)_{AB}\frac{\partial}{\partial \xi_k}\right\}\right]\mathbf{S}_{ki,j} \qquad (4\text{-}15)$$

Similarly we obtain for $\mathbf{K}_{i,p}$ the relation:

$$\frac{1}{\rho}\left[\frac{1}{4}\left(\frac{\partial^2}{\partial x_j\,\partial x_j}\right)_{AB} + \left(\frac{\partial}{\partial x_j}\right)_{AB}\frac{\partial}{\partial \xi_j} + \frac{\partial^2}{\partial \xi_j\,\partial \xi_j}\right]\mathbf{K}_{i,p}$$
$$= -\left(\frac{\partial \bar{U}_j}{\partial x_k}\right)_B\left[\left(\frac{\partial}{\partial x_j}\right)_{AB} + 2\frac{\partial}{\partial \xi_j}\right]\mathbf{Q}_{i,k}$$
$$-\left[\frac{1}{4}\left(\frac{\partial^2}{\partial x_k\,\partial x_j}\right)_{AB} + \frac{\partial^2}{\partial \xi_k\,\partial \xi_j} + \frac{1}{2}\left\{\left(\frac{\partial}{\partial x_k}\right)_{AB}\frac{\partial}{\partial \xi_j} + \left(\frac{\partial}{\partial x_j}\right)_{AB}\frac{\partial}{\partial \xi_k}\right\}\right]\mathbf{S}_{i,kj} \qquad (4\text{-}16)$$

Homogeneous Turbulence

If the turbulence is homogeneous, all derivatives of the correlations with respect to $(x_k)_{AB}$ vanish, and Eq. (4-14) reduces to

$$\frac{\partial}{\partial t}\mathbf{Q}_{i,j} + \mathbf{Q}_{k,j}\left(\frac{\partial \bar{U}_i}{\partial x_k}\right)_A + \mathbf{Q}_{i,k}\left(\frac{\partial \bar{U}_j}{\partial x_k}\right)_B + [(\bar{U}_k)_B - (\bar{U}_k)_A]\frac{\partial}{\partial \xi_k}\mathbf{Q}_{i,j}$$
$$= -\frac{\partial}{\partial \xi_k}(\mathbf{S}_{i,kj} - \mathbf{S}_{ik,j}) - \frac{1}{\rho}\left(\frac{\partial}{\partial \xi_j}\mathbf{K}_{i,p} - \frac{\partial}{\partial \xi_i}\mathbf{K}_{p,j}\right) + 2\nu\frac{\partial^2}{\partial \xi_l\,\partial \xi_l}\mathbf{Q}_{i,j} \qquad (4\text{-}17)$$

The Eqs. (4-15) and (4-16) for $\mathbf{K}_{p,j}$ and $\mathbf{K}_{i,p}$ respectively, become

$$\frac{1}{\rho}\frac{\partial^2}{\partial \xi_i\,\partial \xi_i}\mathbf{K}_{p,j} = 2\left(\frac{\partial \bar{U}_i}{\partial x_k}\right)_A\frac{\partial}{\partial \xi_i}\mathbf{Q}_{k,j} - \frac{\partial^2}{\partial \xi_k\,\partial \xi_i}\mathbf{S}_{ki,j} \qquad (4\text{-}18)$$

and

$$\frac{1}{\rho}\frac{\partial^2}{\partial \xi_j\,\partial \xi_j}\mathbf{K}_{i,p} = -2\left(\frac{\partial \bar{U}_j}{\partial x_k}\right)_B\frac{\partial}{\partial \xi_j}\mathbf{Q}_{i,k} - \frac{\partial^2}{\partial \xi_k\,\partial \xi_j}\mathbf{S}_{i,kj} \qquad (4\text{-}19)$$

We have said in the introduction to this chapter that homogeneous shear turbulence exists only if the main motion has a constant velocity in a given direction and a constant lateral velocity gradient throughout the whole field.

Without loss of generality we may describe the turbulence by means of a coordinate system such that the main parallel flow is along one of the coordinate axes. Thus, with respect to this coordinate system, we have

$$\bar{U}_1 = f(x_2) \qquad \bar{U}_2 = \bar{U}_3 = 0 \qquad \frac{d\bar{U}_1}{dx_2} = \text{constant}$$

Equation (4-17) then reads

$$\frac{\partial}{\partial t}\mathbf{Q}_{i,j} + \left(\delta_{i1}\mathbf{Q}_{2,j} + \delta_{j1}\mathbf{Q}_{i,2} + \xi_2 \frac{\partial}{\partial \xi_1}\mathbf{Q}_{i,j}\right)\frac{d\bar{U}_1}{dx_2}$$
$$= -\frac{\partial}{\partial \xi_k}(\mathbf{S}_{i,kj} - \mathbf{S}_{ik,j}) - \frac{1}{\rho}\left(\frac{\partial}{\partial \xi_j}\mathbf{K}_{i,p} - \frac{\partial}{\partial \xi_i}\mathbf{K}_{p,j}\right) + 2\nu \frac{\partial^2}{\partial \xi_l\, \partial \xi_l}\mathbf{Q}_{i,j} \qquad (4\text{-}20)$$

While Eqs. (4-18) and (4-19) now read

$$\frac{1}{\rho}\frac{\partial^2}{\partial \xi_i\, \partial \xi_i}\mathbf{K}_{p,j} = 2\frac{d\bar{U}_1}{dx_2}\frac{\partial}{\partial \xi_1}\mathbf{Q}_{2,j} - \frac{\partial^2}{\partial \xi_2\, \partial \xi_1}\mathbf{S}_{21,j} \qquad (4\text{-}21)$$

and

$$\frac{1}{\rho}\frac{\partial^2}{\partial \xi_j\, \partial \xi_j}\mathbf{K}_{i,p} = -2\frac{d\bar{U}_1}{dx_2}\frac{\partial}{\partial \xi_1}\mathbf{Q}_{i,2} - \frac{\partial^2}{\partial \xi_2\, \partial \xi_1}\mathbf{S}_{i,21} \qquad (4\text{-}22)$$

With respect to this type of homogeneous turbulence and the special coordinate system considered, the statistical properties of the turbulence will not be changed if we change the signs of the coordinates simultaneously, thus replacing ξ_1 by $-\xi_1$, ξ_2 by $-\xi_2$, and ξ_3 by $-\xi_3$. That is, the point $\xi_i = 0$ is a center of symmetry for this type of turbulence.

In general, for any homogeneous flow field, we obtain from the condition of invariance under translation the relations:

$$\left[\overline{(u_i)_A(u_j)_B}\right](\xi_1,\xi_2,\xi_3) = \left[\overline{(u_j)_A(u_i)_B}\right](-\xi_1,-\xi_2,-\xi_3)$$
$$\left[\overline{(u_i)_A(u_k)_B(u_j)_B}\right](\xi_1,\xi_2,\xi_3) = \left[\overline{(u_i)_B(u_k)_A(u_j)_A}\right](-\xi_1,-\xi_2,-\xi_3)$$
$$\left[\overline{(u_i)_A(u_k)_A(u_j)_B}\right](\xi_1,\xi_2,\xi_3) = \left[\overline{(u_i)_B(u_k)_B(u_j)_A}\right](-\xi_1,-\xi_2,-\xi_3)$$
$$\left[\overline{p_A(u_j)_B}\right](\xi_1,\xi_2,\xi_3) = \left[\overline{p_B(u_j)_A}\right](-\xi_1,-\xi_2,-\xi_3)$$

In the case of the special coordinate system considered, from the condition of invariance under reflection we obtain the additional relations:

$$\left[\overline{(u_i)_A(u_j)_B}\right](\xi_1,\xi_2,\xi_3) = \left[\overline{(u_i)_A(u_j)_B}\right](-\xi_1,-\xi_2,-\xi_3)$$
$$\left[\overline{(u_i)_A(u_k)_B(u_j)_B}\right](\xi_1,\xi_2,\xi_3) = -\left[\overline{(u_i)_A(u_k)_B(u_j)_B}\right](-\xi_1,-\xi_2,-\xi_3)$$
$$\left[\overline{(u_i)_A(u_k)_A(u_j)_B}\right](\xi_1,\xi_2,\xi_3) = -\left[\overline{(u_i)_A(u_k)_A(u_j)_B}\right](-\xi_1,-\xi_2,-\xi_3)$$
$$\left[\overline{p_A(u_j)_B}\right](\xi_1,\xi_2,\xi_3) = -\left[\overline{p_A(u_j)_B}\right](-\xi_1,-\xi_2,-\xi_3)$$

Hence, if we combine the two sets of relations, we obtain

$$\mathbf{Q}_{i,j}(\xi_1,\xi_2,\xi_3) = \mathbf{Q}_{i,j}(-\xi_1,-\xi_2,-\xi_3) = \mathbf{Q}_{j,i}(\xi_1,\xi_2,\xi_3)$$

$$\mathbf{S}_{i,kj}(\xi_1,\xi_2,\xi_3) = -\mathbf{S}_{i,kj}(-\xi_1,-\xi_2,-\xi_3) = -\mathbf{S}_{kj,i}(\xi_1,\xi_2,\xi_3)$$

$$\mathbf{S}_{ik,j}(\xi_1,\xi_2,\xi_3) = -\mathbf{S}_{ik,j}(-\xi_1,-\xi_2,-\xi_3) = -\mathbf{S}_{j,ik}(\xi_1,\xi_2,\xi_3)$$

$$\mathbf{K}_{p,j}(\xi_1,\xi_2,\xi_3) = -\mathbf{K}_{p,j}(-\xi_1,-\xi_2,-\xi_3) = -\mathbf{K}_{j,p}(\xi_1,\xi_2,\xi_3)$$

We may further draw attention to the symmetry of the triple velocity correlations with respect to the indices referring to the same point.

Thus,

$$\mathbf{S}_{ik,j} = \mathbf{S}_{ki,j} \qquad \text{and} \qquad \mathbf{S}_{i,kj} = \mathbf{S}_{i,jk}$$

Hence, with the above relations for the triple velocity correlations and the pressure-velocity correlations, we may write

$$-\frac{\partial}{\partial \xi_k}(\mathbf{S}_{i,kj} - \mathbf{S}_{ik,j}) = \frac{\partial}{\partial \xi_k}(\mathbf{S}_{jk,i} + \mathbf{S}_{ik,j}) = \mathbf{S}_{i,j} \tag{4-23}$$

$$\frac{\partial}{\partial \xi_j}\mathbf{K}_{i,p} - \frac{\partial}{\partial \xi_i}\mathbf{K}_{p,j} = \frac{\partial}{\partial \xi_j}\mathbf{K}_{i,p} + \frac{\partial}{\partial \xi_i}\mathbf{K}_{j,p} = \mathbf{P}_{i,j} \tag{4-24}$$

and Eq. (4-20) becomes

$$\frac{\partial}{\partial t}\mathbf{Q}_{i,j} + \left(\delta_{i1}\mathbf{Q}_{2,j} + \delta_{j1}\mathbf{Q}_{i,2} + \xi_2 \frac{\partial}{\partial \xi_1}\mathbf{Q}_{i,j}\right)\frac{d\bar{U}_1}{dx_2}$$
$$= \mathbf{S}_{i,j} - \frac{1}{\rho}\mathbf{P}_{i,j} + 2\nu \frac{\partial^2}{\partial \xi_l\,\partial \xi_l}\mathbf{Q}_{i,j} \tag{4-25}$$

Since $\mathbf{S}_{i,kj}$ is an uneven function in ξ_i, we have $\mathbf{S}_{i,kj}(0) = 0$. But also $\mathbf{S}_{i,j}(0) = 0$, which follows from its defining equation (4-23). For from (4-12a) and (4-12b) follows

$$\frac{\partial}{\partial \xi_k} = \frac{1}{2}\left[\left(\frac{\partial}{\partial x_k}\right)_B - \left(\frac{\partial}{\partial x_k}\right)_A\right]$$

and so

$$\mathbf{S}_{i,j}(0) = -\left[\frac{\partial}{\partial \xi_k}(\mathbf{S}_{i,kj} - \mathbf{S}_{ik,j})\right]_{\xi_i=0}$$

$$= -\left\{\frac{1}{2}\left[\left(\frac{\partial}{\partial x_k}\right)_B - \left(\frac{\partial}{\partial x_k}\right)_A\right]\left[\overline{(u_i)_A(u_k)_B(u_j)_B} - \overline{(u_i)_A(u_k)_A(u_j)_B}\right]\right\}_{A=B} = 0 \tag{4-26}$$

This result is general and does not require homogeneity.

Furthermore, for homogeneous turbulence, there follows from the continuity equation

$$\mathbf{P}_{i,i}(\xi_1,\xi_2,\xi_3) = 0$$

because

$$\frac{\partial}{\partial \xi_i} \mathbf{K}_{i,p}(\xi_1,\xi_2,\xi_3) = 0$$

This, again, means that a redistribution of energy between the turbulent velocity components through the velocity-pressure correlations takes place without affecting the total turbulence kinetic energy. Thus, contraction of Eq. (4-25) yields

$$\frac{\partial}{\partial t}\mathbf{Q}_{i,i} + \left(2\mathbf{Q}_{1,2} + \xi_2 \frac{\partial}{\partial \xi_1}\mathbf{Q}_{i,i}\right)\frac{d\bar{U}_1}{dx_2} = \mathbf{S}_{i,i} + 2\nu \frac{\partial^2}{\partial \xi_l \, \partial \xi_l}\mathbf{Q}_{i,i} \tag{4-27}$$

For $\xi_i = 0$, this equation reduces to

$$\frac{\partial}{\partial t}\frac{\overline{q^2}}{2} + \overline{u_1 u_2}\frac{d\bar{U}_1}{dx_2} = \nu\left(\frac{\partial^2}{\partial \xi_l \, \partial \xi_l}\mathbf{Q}_{i,i}\right)_{\xi_i = 0} = -\nu \overline{\frac{\partial u_i}{\partial x_l}\frac{\partial u_i}{\partial x_l}} \tag{4-28}$$

that is, to Eq. (1-111) for the homogeneous case.

Because of the presence of the functions $\mathbf{S}_{i,j}$ and $\mathbf{S}_{i,i}$ in Eqs. (4-25) and (4-27), respectively, it is not possible fundamentally to obtain solutions of these equations without making an assumption concerning them.

Encouraged by the results obtained by making suitable assumptions concerning the Fourier transforms of these functions in the case of isotropic turbulence, Burgers and Mitchner,[4] and Tchen,[6,11] have tried to obtain solutions for the Fourier transform of $\mathbf{Q}_{i,i}$, making assumptions for the Fourier transform of $\mathbf{S}_{i,i}$ similar to those made in the corresponding theories of isotropic turbulence.

Deissler[13] obtained exact solutions by assuming a very weak turbulence, so that the dynamic effects given by the two-point triple velocity correlations $\mathbf{S}_{i,kj}$ and $\mathbf{S}_{ki,j}$ could be neglected. However, he retained the dynamic effects given by the pressure-velocity correlations. So in fact he assumed a low value of the Reynolds number of turbulence. First Deissler considered the case of a constant mean velocity, but a slightly inhomogeneous turbulence. With the inhomogeneity in the mean-flow direction he obtained an exponential decay of the turbulence in this direction. Of more importance is the other case considered, namely that of a steady, homogeneous turbulence superimposed on a steady mean flow with a constant velocity gradient $d\bar{U}_1/dx_2$. We shall come back to this case in the next section.

4-4 THE DYNAMIC EQUATION FOR THE ENERGY SPECTRUM

To derive this equation for a homogeneous turbulence we start from Eqs. (4-20), (4-21), and (4-22), and introduce the following Fourier transforms

$$\mathbf{Q}_{i,j}(\xi_1,\xi_2,\xi_3,t) = \iiint_{-\infty}^{+\infty} dk_1\, dk_2\, dk_3\, \mathbf{E}_{i,j}(k_1,k_2,k_3,t) \exp(\imath k_l \xi_l) \tag{4-29}$$

$$\mathbf{S}_{i,kj}(\xi_1,\xi_2,\xi_3,t) = \iiint_{-\infty}^{+\infty} dk_1\, dk_2\, dk_3\, \mathbf{F}_{i,kj}(k_1,k_2,k_3,t) \exp(\imath k_l \xi_l) \tag{4-30}$$

and similarly for $\mathbf{S}_{ik,j}$.

$$\mathbf{K}_{i,p}(\xi_1,\xi_2,\xi_3,t) = \iiint_{-\infty}^{+\infty} dk_1\, dk_2\, dk_3\, \mathbf{H}_{i,p}(k_1,k_2,k_3,t) \exp(\imath k_l \xi_l) \tag{4-31}$$

and similarly for $\mathbf{K}_{p,j}$.

In order to obtain the Fourier transform of

$$\xi_2 \frac{\partial}{\partial \xi_1} \mathbf{Q}_{i,j}$$

we proceed as follows.

From Eq. (4-29) we obtain

$$\xi_2 \frac{\partial}{\partial \xi_1} \mathbf{Q}_{i,j} = \imath \iiint_{-\infty}^{+\infty} dk_1\, dk_2\, dk_3\, \xi_2 k_1 \mathbf{E}_{i,j} \exp(\imath k_l \xi_l)$$

whence, after integration by parts with respect to k_2, assuming at the same time $\lim_{k_2 \to \infty} \mathbf{E}_{i,j} = 0$

$$\xi_2 \frac{\partial}{\partial \xi_1} \mathbf{Q}_{i,j} = -\iiint_{-\infty}^{+\infty} dk_1\, dk_2\, dk_3\, k_1 \frac{\partial \mathbf{E}_{i,j}}{\partial k_2} \exp(\imath k_l \xi_l) \tag{4-32}$$

With the expressions (4-29), (4-30), (4-31), and (4-32) we obtain from the dynamic equation (4-20), the dynamic equation for $\mathbf{E}_{i,j}$

$$\frac{\partial}{\partial t} \mathbf{E}_{i,j} + \left(\delta_{i1} \mathbf{E}_{2,j} + \delta_{j1} \mathbf{E}_{i,2} - k_1 \frac{\partial \mathbf{E}_{i,j}}{\partial k_2} \right) \frac{d\bar{U}_1}{dx_2}$$

$$= \imath k_k (\mathbf{F}_{jk,i} + \mathbf{F}_{ik,j}) - \frac{\imath}{\rho} (k_j \mathbf{H}_{i,p} + k_i \mathbf{H}_{j,p}) - 2\nu k^2 \mathbf{F}_{i,j} \tag{4-33}$$

While the Fourier transforms of Eqs. (4-21) and (4-22) become

$$\frac{1}{\rho} k^2 \mathbf{H}_{p,j} = -2\imath \frac{d\bar{U}_1}{dx_2} k_1 \mathbf{E}_{2,j} - k_2 k_1 \mathbf{F}_{21,j} \tag{4-34}$$

$$\frac{1}{\rho} k^2 \mathbf{H}_{i,p} = 2\imath \frac{d\bar{U}_1}{dx_2} k_1 \mathbf{E}_{i,2} - k_2 k_1 \mathbf{F}_{i,21} \tag{4-35}$$

With these expressions for $\mathbf{H}_{i,p}$ and $\mathbf{H}_{p,j}$, Eq. (4-33) can be presented in the following form, if we keep in mind that

$$\mathbf{H}_{p,j} = -\mathbf{H}_{j,p} \quad \text{and} \quad \mathbf{F}_{21,j} = -\mathbf{F}_{j,21}$$

$$\frac{\partial}{\partial t}\mathbf{E}_{i,j} + \left[\delta_{i1}\mathbf{E}_{2,j} + \delta_{j1}\mathbf{E}_{i,2} - k_1\frac{\partial \mathbf{E}_{i,j}}{\partial k_2} - 2\left(\frac{k_1 k_i}{k^2}\mathbf{E}_{2,j} + \frac{k_1 k_j}{k^2}\mathbf{E}_{i,2}\right)\right]\frac{d\bar{U}_1}{dx_2}$$

$$= \imath k_k(\mathbf{F}_{jk,i} + \mathbf{F}_{ik,j}) + \imath\frac{k_2 k_1}{k^2}(k_j\mathbf{F}_{i,21} + k_i\mathbf{F}_{j,21}) - 2\nu k^2\mathbf{E}_{i,j} \qquad (4\text{-}36)$$

When we introduce

$$\mathbf{F}_{i,j} = \imath k_k(\mathbf{F}_{ik,j} + \mathbf{F}_{jk,i}) \qquad (4\text{-}37)$$

and

$$\frac{1}{\rho}\mathbf{\Pi}_{i,j} = -2\left(\frac{k_1 k_j}{k^2}\mathbf{E}_{i,2} + \frac{k_1 k_i}{k^2}\mathbf{E}_{j,2}\right)\frac{d\bar{U}_1}{dx_2} - \imath\frac{k_2 k_1}{k^2}(k_j\mathbf{F}_{i,21} + k_i\mathbf{F}_{j,21}) \qquad (4\text{-}38)$$

the Fourier transforms of $\mathbf{S}_{i,j}$ and $\mathbf{P}_{i,j}$, respectively, Eq. (4-33) may also be given as follows

$$\frac{\partial}{\partial t}\mathbf{E}_{i,j} + \left(\delta_{i1}\mathbf{E}_{2,j} + \delta_{j1}\mathbf{E}_{i,2} - k_1\frac{\partial \mathbf{E}_{i,j}}{\partial k_2}\right)\frac{d\bar{U}_1}{dx_2} = \mathbf{F}_{i,j} - \frac{1}{\rho}\mathbf{\Pi}_{i,j} - 2\nu k^2\mathbf{E}_{i,j} \qquad (4\text{-}39)$$

When we apply a contraction with respect to i and j, the pressure term vanishes in the contracted equation. For, $\mathbf{\Pi}_{i,i} = 0$, since for an incompressible fluid $k_i\mathbf{E}_{i,j} = 0$ and $k_i\mathbf{F}_{i,jk} = 0$. So, from either Eq. (4-36) or Eq. (4-39) the dynamic equation for $\mathbf{E}_{i,i}$ is obtained and reads

$$\frac{\partial}{\partial t}\mathbf{E}_{i,i} + \left(\underset{\text{I}}{2\mathbf{E}_{1,2}} - \underset{\text{II}}{k_1\frac{\partial \mathbf{E}_{i,i}}{\partial k_2}}\right)\frac{d\bar{U}_1}{dx_2} = \underset{\text{III}}{\mathbf{F}_{i,i}} - \underset{\text{IV}}{2\nu k^2\mathbf{E}_{i,i}} \qquad (4\text{-}40)$$

The contracted spectral tensor $\mathbf{E}_{i,i}$ (and thus its trace) is related to the kinetic energy of the turbulence since from Eq. (4-29):

$$\mathbf{Q}_{i,i}(0,t) = \overline{u_i u_i} = \iiint_{-\infty}^{+\infty} dk_1\, dk_2\, dk_3\, \mathbf{E}_{i,i}$$

So we may interpret the terms I . . . IV in Eq. (4-40) respectively as: I, a "production" term; II, a term describing the transfer of energy between wavenumbers due to interaction with the mean flow through the mean strain rate (deformation of eddies by the mean flow); III, a term describing the transfer of energy between wavenumbers due to their mutual interaction and through the turbulent strain rates (deformation by the turbulence itself); and IV the viscous dissipation term. The transfer of energy between wavenumbers as described by the terms II and III is conservative. Integrated over all wavenumbers it yields a zero integral contribution. So

$$\iiint_{-\infty}^{+\infty} dk_1\, dk_2\, dk_3\, k_1\frac{\partial \mathbf{E}_{i,i}}{\partial k_2} = 0$$

since it is equal to

$$\lim_{\xi_2 \to 0} \xi_2 \frac{\partial \mathbf{Q}_{i,i}}{\partial \xi_1} = 0$$

Similarly

$$\iiint_{-\infty}^{+\infty} dk_1\, dk_2\, dk_3\, \mathbf{F}_{i,i} = 0$$

since according to Eq. (4-26) $\mathbf{S}_{i,i}(0) = 0$.

In isotropic turbulence it is possible to express the correlation functions and spectrum functions in terms of one single scalar, namely the distance r and the wavenumber k, respectively.

Batchelor[7] suggested doing the same in a nonisotropic but homogeneous turbulence by averaging the correlation and spectrum functions over all directions of r and k in the corresponding spaces, thus taking mean values of these functions over spherical surfaces $r =$ constant and $k =$ constant, respectively. For instance,

$$[\mathbf{E}_{i,i}(k_1,k_2,k_3,t)]_{\text{av}} = \frac{1}{4\pi k^2} \int dA(k) \mathbf{E}_{i,i}(k_1,k_2,k_3,t) \qquad (4\text{-}41)$$

$$[\mathbf{Q}_{i,i}(\xi_1,\xi_2,\xi_3,t)]_{\text{av}} = \frac{1}{4\pi r^2} \int dA(r) \mathbf{Q}_{i,i}(\xi_1,\xi_2,\xi_3,t) \qquad (4\text{-}42)$$

Between $(\mathbf{E}_{i,i})_{\text{av}}$ and $(\mathbf{Q}_{i,i})_{\text{av}}$, then, there is the relation [see Eq. (3-56)]

$$[\mathbf{Q}_{i,i}(\xi_1,\xi_2,\xi_3,t)]_{\text{av}} = 4\pi \int_0^\infty dk\, k^2 \frac{\sin kr}{kr} [\mathbf{E}_{i,i}(k_1,k_2,k_3,t)]_{\text{av}} \qquad (4\text{-}43)$$

As we did in the case of isotropic turbulence, we introduce the following new spectrum functions:

$$\mathbf{E}(k,t) = 2\pi k^2 [\mathbf{E}_{i,i}(k_1,k_2,k_3,t)]_{\text{av}} \qquad (4\text{-}44)$$

$$\mathbf{F}(k,t) = 2\pi k^2 [\mathbf{F}_{i,i}(k_1,k_2,k_3,t)]_{\text{av}} \qquad (4\text{-}45)$$

$$\mathscr{E}(k,t) = 2\pi k^2 \left[2(\mathbf{E}_{1,2})_{\text{av}} - \left(k_1 \frac{\partial \mathbf{E}_{i,i}}{\partial k_2} \right)_{\text{av}} \right] \qquad (4\text{-}46)$$

The physical meaning of $\mathbf{E}(k,t)$ may be deduced from (4-43) as follows:

$$[\mathbf{Q}_{i,i}(0,t)]_{\text{av}} = \mathbf{Q}_{i,i}(0,t) = (\overline{u_1{}^2} + \overline{u_2{}^2} + \overline{u_3{}^2})_{\text{av}} = \overline{q^2}$$

$$= 4\pi \int_0^\infty dk\, k^2 [\mathbf{E}_{i,i}(k_1,k_2,k_3,t)]_{\text{av}} = 2 \int_0^\infty dk\, \mathbf{E}(k,t)$$

Application to each term of Eq. (4-40) of the averaging procedure indicated by Eq. (4-41) and introduction of the spectrum functions according to Eqs. (4-44) to (4-46) yield

$$\frac{\partial}{\partial t}\mathbf{E}(k,t) + \mathscr{E}(k,t)\frac{d\bar{U}_1}{dx_2} = \mathbf{F}(k,t) - 2\nu k^2\mathbf{E}(k,t) \tag{4-47}$$

Formally a stationary condition is possible, in which case we have the following equation for the turbulence energy budget

$$\left\{4\pi k^2[\mathbf{E}_{1,2}]_{\text{av}} - 2\pi k^2\left[k_1\frac{\partial \mathbf{E}_{i,i}}{\partial k_2}\right]_{\text{av}}\right\}\frac{d\bar{U}_1}{dx_2} = \mathbf{F} - 2\nu k^2\mathbf{E} \tag{4-48}$$

While

$$4\pi\int_0^\infty dk\,k^2[\mathbf{E}_{1,2}]_{\text{av}}\frac{d\bar{U}_1}{dx_2} = -2\nu\int_0^\infty dk\,k^2\mathbf{E} \tag{4-49}$$

Earlier we have said that the term II in Eq. (4-40), which is the second term in Eq. (4-48), may be interpreted as describing the transfer of energy among wavenumbers through the deformation of eddies by the mean shear-flow.

Now this deformation is not the only effect of the mean shear $d\bar{U}_1/dx_2$. There is also a turning effect. We know that any flow of an incompressible fluid, whose velocity distribution is a linear function of the space coordinates can be decomposed into a pure rotation and a pure deformation. For the simple case $\bar{U}_i = ax_2\,\delta_{i1}$, the flow consists of a bodily rotation with angular speed $-\frac{1}{2}a$, and a pure deformation with a maximum strain rate $+\frac{1}{2}a$ along the principal axis in the direction $\pi/4$, and a minimum strain rate $-\frac{1}{2}a$ (i.e., a compression) in the direction $3\pi/4$. The turning effect, due to both rotation and deformation, is illustrated in Fig. 4-3, where an eddy flow pattern with one Fourier mode which initially has only one wavenumber component k_1 in the x_1-direction has been considered. Figure 4-3b shows the position of the wavenumber vector $k(t)$ at time t, at which instant the wavenumber k_2, which was initially zero, has increased to the value $k_2(t) \simeq k_1(0)a.t$. Figures 4-3c and 4-3d show the contributions respectively of the pure rotation and of the pure deformation.

When we consider two imaginary lines in the $\pi/4$ and the $3\pi/4$ directions respectively, we conclude that due to the rotation (Fig. 4-3c) the wavelength in the $\pi/4$ direction increases, whereas that in the $3\pi/4$ direction decreases. Hence the energy of the Fourier mode is shifted towards a smaller wavenumber in the $\pi/4$ direction, and towards a larger wavenumber in the $3\pi/4$ direction.

A similar conclusion can be made concerning the effect of the deformation shown in Fig. 4-3d.

This picture of the turning effect has been confirmed by a simple calculation given by Phillips,[14] starting from the linearized momentum-balance equation for the perturbation velocity u_i, and considering one Fourier mode as a solution of this equation. The effect decreases with decreasing wavenumber, and will be zero for the mode with $k_1 = 0$. Translated to a flow with many Fourier components, this means that such an eddy pattern evolves towards a preferred horizontal configuration of (big) eddies, so with a large-scale structure independent of x_1. At the same time these

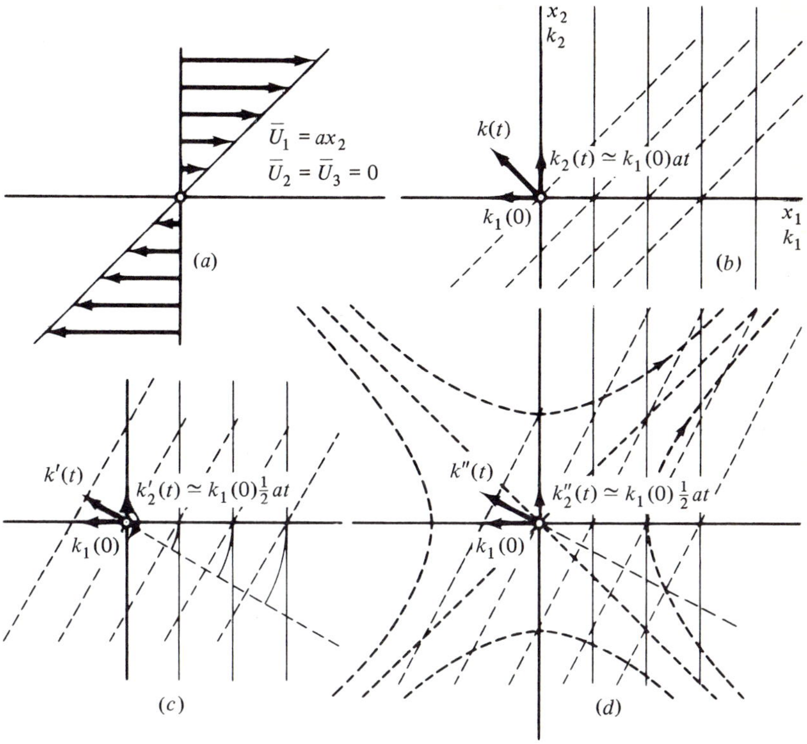

FIGURE 4-3
Effect of simple, uniform shear-flow on turning of wavenumber vector $k(t)$.

large eddies, elongated in the x_1-direction, have no further coupling with the mean shear-flow. Their energy can then only be maintained by nonlinear interaction with other Fourier components. The two important results from the above considerations are:[14] the spectral energy transfer due to the turning of wavenumber vectors, and the occurrence of persistent eddies, elongated in the main-flow direction with no dynamic coupling with the mean motion.

This spectral transfer of energy due to the interaction between eddies and the mean shear flow is described by the term

$$k^2\left[k_1 \frac{\partial \mathbf{E}_{i,i}}{\partial k_2}\right]_{\mathrm{av}}$$

in Eq. (4-40) or Eq. (4-48). It is indeed zero when $k_1 = 0$. In order to gain some insight in the character of this quantity as a function of k_1, we follow Lumley[15,16] in writing this quantity in a different form, namely:

$$\frac{\partial}{\partial k} k[k_1 k_2 \mathbf{E}_{i,i}]_{\mathrm{av}}$$

To show this introduce in the wavenumber space spherical polar coordinates k, Θ, and φ, where we take Θ as the colatitude, the angle between the k-vector and the k_2-axis

$$\left[k_1 \frac{\partial \mathbf{E}_{i,i}}{\partial k_2}\right]_{\text{av}} = \frac{1}{4\pi} \int_0^{2\pi} d\varphi \int_0^{\pi} d\Theta \sin \Theta \, k_1 \frac{\partial \mathbf{E}_{i,i}}{\partial k_2}$$

With $\Theta = \arccos (k_2/k)$, and $\varphi = \arctan (k_3/k_1)$ and

$$\frac{\partial \mathbf{E}_{i,i}}{\partial k_2} = \frac{\partial \mathbf{E}_{i,i}}{\partial k} \frac{\partial k}{\partial k_2} + \frac{\partial \mathbf{E}_{i,i}}{\partial \varphi} \frac{\partial \varphi}{\partial k_2} + \frac{\partial \mathbf{E}_{i,i}}{\partial \Theta} \frac{\partial \Theta}{\partial k_2}$$

the integral expression can be written as follows

$$\left[k_1 \frac{\partial \mathbf{E}_{i,i}}{\partial k_2}\right]_{\text{av}} = \frac{1}{4\pi} \int_0^{2\pi} d\varphi \int_0^{\pi} d\Theta \left[k \frac{\partial \mathbf{E}_{i,i}}{\partial k} \sin^2 \Theta \cos \Theta \cos \varphi - \frac{\partial \mathbf{E}_{i,i}}{\partial \Theta} \sin^3 \Theta \cos \varphi \right]$$

After applying an integration by parts to the second term on the right-hand side, we obtain

$$\left[k_1 \frac{\partial \mathbf{E}_{i,i}}{\partial k_2}\right]_{\text{av}} = \frac{1}{4\pi k^2} \int_0^{2\pi} d\varphi \int_0^{\pi} d\Theta \sin^2 \Theta \cos \Theta \cos \varphi \frac{\partial}{\partial k} (k^3 \mathbf{E}_{i,i})$$

Since $k \sin \Theta \cos \varphi = k_1$ and $k \cos \Theta = k_2$, the result is

$$\left[k_1 \frac{\partial \mathbf{E}_{i,i}}{\partial k_2}\right]_{\text{av}} = \frac{1}{4\pi k^2} \int_0^{2\pi} d\varphi \int_0^{\pi} d\Theta \sin \Theta \frac{\partial}{\partial k} (k k_1 k_2 \mathbf{E}_{i,i})$$

Hence

$$k^2 \left[k_1 \frac{\partial \mathbf{E}_{i,i}}{\partial k_2}\right]_{\text{av}} = \frac{\partial}{\partial k} k [k_1 k_2 \mathbf{E}_{i,i}]_{\text{av}}$$

This immediately shows that

$$\int_0^{\infty} dk \, k^2 \left[k_1 \frac{\partial \mathbf{E}_{i,i}}{\partial k_2}\right]_{\text{av}} = 0$$

so that $k^2[k_1 \, \partial \mathbf{E}_{i,i}/\partial k_2]_{\text{av}}$ indeed describes a transfer of energy within the spectral range. Let us consider further the function

$$\frac{\partial}{\partial k} k [k_1 k_2 \mathbf{E}_{i,i}]_{\text{av}}$$

This function must contain a negative and a positive part of equal size. Assume for simplicity a two-dimensional situation in the (x_1, x_2) plane. See Fig. 4-4. Let n_1 be the principal axis with maximum strain rate, n_2 that with minimum strain rate. We start from an initially "isotropic" condition, and waves in all directions with the same wavelength (Fig. 4-4a). Due to the stretching of the fluid element in the n_1-direction, the wavelength increases and thereby takes with it its kinetic energy. So this energy now belongs to a larger wavelength, or to a smaller wavenumber. The opposite takes place in the n_2-direction. Compare also the similar result in Fig. 4-3. Consider now contours of equal $\mathbf{E}_{i,i}(k^2)$ before and after deformation. In the isotropic condition they are concentric circles (Fig. 4-4a), in the deformed condition they form ellipses (Fig. 4-4b). Consequently for a given k we have after deformation that $\mathbf{E}_{i,i}(k^2)$ in the $\pi/4$-direction is smaller than $\mathbf{E}_{i,i}(k^2)$ in the $3\pi/4$-direction

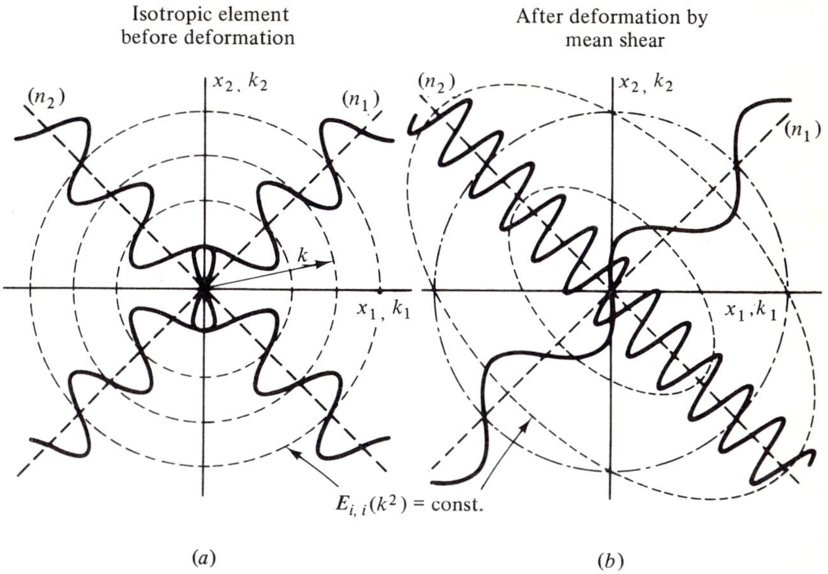

FIGURE 4-4
Deformation of an "isotropic" element by a mean shear. (After: *Lumley, J. L.*[15,16])

when $\mathbf{E}_{i,i}$ decreases with increasing k. Whereas the opposite is true when $\mathbf{E}_{i,i}$ increases with increasing k. This means that compared with the isotropic condition the $\mathbf{E}_{i,i}(k^2)$ distribution, and its peak value, is shifted to higher k-values in the $3\pi/4$-direction, and to lower k-values in the $\pi/4$-direction. Now since $k_1 k_2 = \frac{1}{2}k^2 \sin 2\Theta$, it is positive for $\Theta = \pi/4$ and negative for $\Theta = 3\pi/4$. So the spherical average $[k_1 k_2 \mathbf{E}_{i,i}]_{av}$ will be mainly negative as a function of k, except for small k. The same applies to $k[k_1 k_2 \mathbf{E}_{i,i}]_{av}$. Figure 4-5 shows this function and also its derivative. The latter distribution should be such that the integral of the negative part just equals in absolute value the integral of the positive part.

It is reasonable to assume that a similar mechanism is responsible for the transfer of energy from the larger eddies to the smaller eddies, described by the energy transfer function $\mathbf{F}(k)$. Namely, a stretching and turning effect on the smaller eddies caused by the local deformation rate of the larger eddies.

The energy balance for the stationary case, given by Eq. (4-48) is represented in Fig. 4-6. An interesting point to note is that interaction between turbulence and the mean shear-flow obviously takes place in the anisotropic, lower wavenumber, range of the spectrum. Also, direct interaction between the smaller eddies in the, universal, equilibrium range and the mean motion, resulting in direct energy transfer from the mean motion to these eddies is negligibly small. Furthermore, Fig. 4-6 shows the existence of a wavenumber range, where production and spectral transfer due to deformation by the mean shear-flow are both positive, and that for the stationary condition they have to be balanced by the negative transfer function $\mathbf{F}(k)$

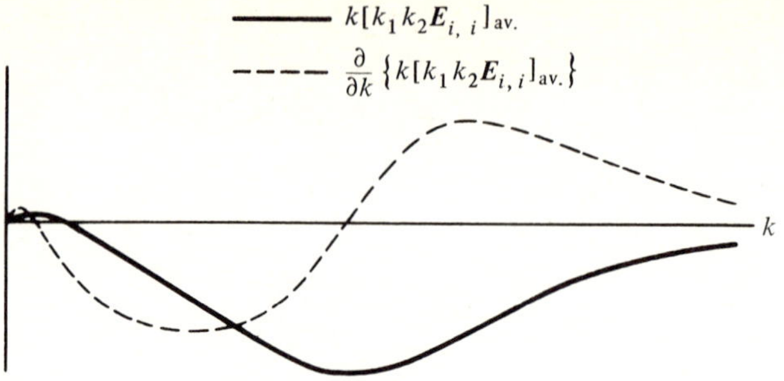

FIGURE 4-5
Spectral transfer due to deformation by a mean shear-flow.

in this range. So a stationary condition with negligible transfer function $\mathbf{F}(k)$ would be impossible on physical grounds.

In order to gain some insight into the quantitative effect of the pressure fluctuations, i.e., in the transfer of energy between velocity components, and in the transfer between wavenumbers due to the mean shear-flow, Deissler[13] considered Eqs. (4-39) and (4-38) for a *low Reynolds-number turbulence.* As mentioned earlier, Deissler neglected the inertial effects, represented by the triple velocity correlations. With the neglect of the term $\mathbf{F}_{i,j}$ in Eq. (4-39), and of the terms $\mathbf{F}_{i,21}$ and $\mathbf{F}_{j,21}$ in Eq. (4-38), the dynamic equation for $\mathbf{E}_{i,j}$ reads

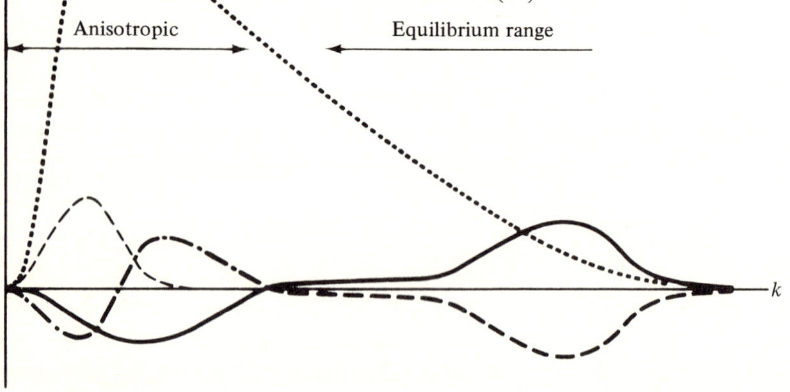

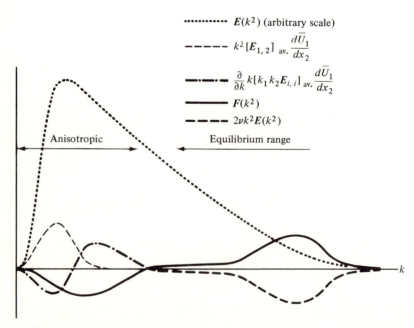

FIGURE 4-6
Energy balance in a stationary shear-flow. (After: *Lumley, J. L.*[15,16])

$$\frac{\partial}{\partial t}\mathbf{E}_{i,j} = \underbrace{-(\mathbf{E}_{2,j}\,\delta_{1i} + \mathbf{E}_{i,2}\,\delta_{1j})\frac{d\bar{U}_1}{dx_2}}_{\text{I}} + \underbrace{k_1\frac{\partial \mathbf{E}_{i,j}}{\partial k_2}\frac{d\bar{U}_1}{dx_2}}_{\text{II}}$$

$$+ \underbrace{2\left(\frac{k_1 k_i}{k^2}\mathbf{E}_{2,j} + \frac{k_1 k_j}{k^2}\mathbf{E}_{i,2}\right)\frac{d\bar{U}_1}{dx_2}}_{\text{III}} \underbrace{- 2\nu k^2 \mathbf{E}_{i,j}}_{\text{IV}} \tag{4-50}$$

Repeated once more, the term I describes the "production of turbulence," the term II the transfer between wavenumbers because of the turning caused by the mean shear-flow, the term III the transfer between velocity components due to pressure effects, and the term IV the viscous dissipation.

Deissler,[13] and Fox,[17] who supplemented Deissler's calculations, considered specifically the following equations:

$$\frac{\partial}{\partial t}\mathbf{E}_{1,1} = -2\mathbf{E}_{1,2}\frac{d\bar{U}_1}{dx_2} + k_1\frac{\partial \mathbf{E}_{1,1}}{\partial k_2}\frac{d\bar{U}_1}{dx_2} + 4\frac{k_1^2}{k^2}\mathbf{E}_{1,2}\frac{d\bar{U}_1}{dx_2} - 2\nu k^2\mathbf{E}_{1,1} \tag{4-51a}$$

$$\frac{\partial}{\partial t}\mathbf{E}_{2,2} = k_1\frac{\partial \mathbf{E}_{2,2}}{\partial k_2}\frac{d\bar{U}_1}{dx_2} + 4\frac{k_1 k_2}{k^2}\mathbf{E}_{2,2}\frac{d\bar{U}_1}{dx_2} - 2\nu k^2\mathbf{E}_{2,2} \tag{4-51b}$$

$$\frac{\partial}{\partial t}\mathbf{E}_{3,3} = k_1\frac{\partial \mathbf{E}_{3,3}}{\partial k_2}\frac{d\bar{U}_1}{dx_2} + 4\frac{k_1 k_3}{k^2}\mathbf{E}_{3,2}\frac{d\bar{U}_1}{dx_2} - 2\nu k^2\mathbf{E}_{3,3} \tag{4-51c}$$

$$\frac{\partial}{\partial t}\mathbf{E}_{i,i} = -2\mathbf{E}_{1,2}\frac{d\bar{U}_1}{dx_2} + k_1\frac{\partial \mathbf{E}_{i,i}}{\partial k_2}\frac{d\bar{U}_1}{dx_2} - 2\nu k^2\mathbf{E}_{i,i} \tag{4-51d}$$

$$\frac{\partial}{\partial t}\mathbf{E}_{1,2} = -\mathbf{E}_{2,2}\frac{d\bar{U}_1}{dx_2} + k_1\frac{\partial \mathbf{E}_{1,2}}{\partial k_2}\frac{d\bar{U}_1}{dx_2} + 2\left(\frac{k_1^2}{k^2}\mathbf{E}_{2,2} + \frac{k_1 k_2}{k^2}\mathbf{E}_{1,2}\right)\frac{d\bar{U}_1}{dx_2} - 2\nu k^2\mathbf{E}_{1,2} \tag{4-51e}$$

Note that the production term only occurs in the equation for $\mathbf{E}_{1,1}$ and neither in the equation for $\mathbf{E}_{2,2}$, nor in that for $\mathbf{E}_{3,3}$, and that $\mathbf{E}_{2,2} \neq \mathbf{E}_{3,3}$ because $k_1k_2\mathbf{E}_{2,2}/k^2 \neq k_1k_3\mathbf{E}_{3,2}/k^2$. Since Eq. (4-51b) contains only $\mathbf{E}_{2,2}$, this equation can be solved first. With the now known $\mathbf{E}_{2,2}$, Eq. (4-51e) for $\mathbf{E}_{1,2}$ can be solved, and then Eqs. (4-51a) for $\mathbf{E}_{1,1}$ and (4-51d) for $\mathbf{E}_{i,i}$. Deissler obtained exact solutions for $\mathbf{E}_{2,2}$, $\mathbf{E}_{1,2}$, and $\mathbf{E}_{i,i}$, while Fox supplemented these calculations by obtaining an exact solution of Eq. (4-51a) for $\mathbf{E}_{1,1}$, and so the solution for $\mathbf{E}_{3,3}$, since $\mathbf{E}_{3,3} = \mathbf{E}_{i,i} - \mathbf{E}_{1,1} - \mathbf{E}_{2,2}$. The initial conditions were assumed to be isotropic. Spherical averages of these magnitudes have then been obtained numerically. The main results of the calculations are the following. The term $[\mathbf{E}_{1,2}]_{\text{av}}$, which occurs also in the equations for $[\mathbf{E}_{1,1}]_{\text{av}}$ and $[\mathbf{E}_{i,i}]_{\text{av}}$ was positive in the whole k-range for any value of dU_1/dx_2. The term $[4k_1k_2\mathbf{E}_{2,2}]_{\text{av}}$ was positive when $d\bar{U}_1/dx_2$ was relatively small, but at relatively large $d\bar{U}_1/dx_2$ it was only positive at small wavenumbers, but negative at larger wavenumbers. So at a weak shear-rate $[\mathbf{E}_{2,2}]_{\text{av}}$ gains energy at the cost of $[\mathbf{E}_{1,1}]_{\text{av}}$. While with a strong shear-rate in the not small wavenumber range the pressure effects cause an energy loss of $[\mathbf{E}_{2,2}]_{\text{av}}$ in favor of

$[\mathbf{E}_{3,3}]_{av}$. For at this strong shear-rate the term $[4k_1k_3\mathbf{E}_{3,2}/k^2]_{av}$ in the equation for $[\mathbf{E}_{3,3}]_{av}$ was essentially positive. On the other hand the term $[k_1{}^2\mathbf{E}_{2,1}/k^2]_{av}$ occurring in the equation for $[\mathbf{E}_{1,1}]_{av}$ was essentially negative. Thus we conclude that production occurs only in favor of $[\mathbf{E}_{1,1}]_{av}$, but when the shear rate is strong enough energy is transferred through pressure effects to $[\mathbf{E}_{3,3}]_{av}$ and at small wavenumbers also to $[\mathbf{E}_{2,2}]_{av}$, which latter, however, loses energy in the higher wavenumber range again in favor of $[\mathbf{E}_{3,3}]_{av}$. So these pressure effects promote anisotropy, especially in the higher wavenumber range. Indeed $[\mathbf{E}_{2,2}]_{av}/[\mathbf{E}_{i,i}]_{av}$ dropped practically to zero in this range under the condition of strong shear-rate. Also, when $d\bar{U}_1/dx_2$ was increased $\overline{u_1{}^2}/\overline{u_iu_i}$ increased, $\overline{u_3{}^2}/\overline{u_iu_i}$ decreased slightly, while $\overline{u_2{}^2}/\overline{u_iu_i}$ decreased markedly. The correlation coefficient of the turbulent shear stress $\overline{u_2u_1}/u_1'u_2'$, remained practically constant as a function of $d\bar{U}_1/dx_2$, and equal to ~ 0.6.

The term $[k_1\,\partial\mathbf{E}_{i,i}/\partial k_2]_{av}$ which describes energy transfer between wavenumbers caused by rotation and deformation by the mean shear, was negative in the smaller wavenumber range, and positive in the large wavenumber range, thus supporting the qualitative picture of the spectral transfer shown in Fig. 4-5. As expected it was very small when the mean shear-rate was weak. It would be zero in the case of isotropy.

Let us return to Eq. (4-47) and consider its integral for the wavenumber range 0–k.

$$\frac{\partial}{\partial t}\int_0^k dk\,\mathbf{E}(k,t) + \frac{d\bar{U}_1}{dx_2}\int_0^k dk\,\mathscr{E}(k,t) = \int_0^k dk\,\mathbf{F}(k,t) - 2\nu\int_0^k dk\,k^2\mathbf{E}(k,t) \qquad (4\text{-}52)$$

If we make the upper limit of the integral $k = \infty$, we obtain an equation that is equivalent to Eq. (4-28). It then follows that

$$\int_0^\infty dk\,\mathscr{E}(k,t) = \overline{u_1u_2}$$

and

$$\int_0^\infty dk\,\mathbf{F}(k,t) = 0$$

These results can also be shown as follows.
From the Fourier transform relation

$$[\mathbf{S}_{i,i}(\xi_1,\xi_2,\xi_3,t)]_{av} = 4\pi\int_0^\infty dk\,k^2\,\frac{\sin kr}{kr}\,[\mathbf{F}_{i,i}(k_1,k_2,k_3,t)]_{av}$$

follows [see Eq. (4-26)]

$$[\mathbf{S}_{i,i}(0,t)]_{av} = 4\pi\int_0^\infty dk\,k^2[\mathbf{F}_{i,i}(k_1,k_2,k_3,t)]_{av} = 2\int_0^\infty dk\,\mathbf{F}(k,t) = 0$$

From

$$[\mathbf{Q}_{1,2}(\xi_1,\xi_2,\xi_3,t)]_{av} = 4\pi\int_0^\infty dk\,k^2\,\frac{\sin kr}{kr}\,[\mathbf{E}_{1,2}(k_1,k_2,k_3,t)]_{av}$$

follows

$$[\mathbf{Q}_{1,2}(0,t)]_{av} = \overline{u_1u_2} = 4\pi\int_0^\infty dk\,k^2[\mathbf{E}_{1,2}(k_1,k_2,k_3,t)]_{av}$$

Since, as shown earlier

$$\int_0^\infty dk\, k^2\left[k_1 \frac{\partial \mathbf{E}_{i,i}}{\partial k_2}\right]_{\mathrm{av}} = 0$$

we have

$$\int_0^\infty dk\, \mathscr{E}(k,t) = 2\pi \int_0^\infty dk\, k^2\left[2(\mathbf{E}_{1,2})_{\mathrm{av}} - \left(k_1 \frac{\partial \mathbf{E}_{i,i}}{\partial k_2}\right)_{\mathrm{av}}\right] = \overline{u_1 u_2}$$

From Eq. (4-52) we obtain

$$\frac{\partial}{\partial t}\int_0^\infty dk\, \mathbf{E}(k,t) + \overline{u_1 u_2}\frac{d\bar{U}_1}{dx_2} = -2\nu \int_0^\infty dk\, k^2 \mathbf{E}(k,t) = -\varepsilon \tag{4-53}$$

We may combine Eqs. (4-52) and (4-53) to

$$\begin{aligned}\varepsilon &= -\frac{\partial}{\partial t}\int_0^\infty dk\, \mathbf{E}(k,t) - \overline{u_1 u_2}\frac{d\bar{U}_1}{dx_2} \\ &= 2\nu \int_0^k dk\, k^2 \mathbf{E}(k,t) - \frac{\partial}{\partial t}\int_k^\infty dk\, \mathbf{E}(k,t) - \frac{d\bar{U}_1}{dx_2}\int_k^\infty dk\, \mathscr{E}(k,t) - \int_0^k dk\, \mathbf{F}(k,t)\end{aligned} \tag{4-54}$$

For the steady state this reduces to

$$\begin{aligned}\varepsilon &= -\overline{u_1 u_2}\frac{d\bar{U}_1}{dx_2} \\ &= 2\nu \int_0^k dk\, k^2 \mathbf{E}(k) - \frac{d\bar{U}_1}{dx_2}\int_k^\infty dk\, \mathscr{E}(k) - \int_0^k dk\, \mathbf{F}(k)\end{aligned} \tag{4-55}$$

The terms in this equation can be interpreted as follows:

$\varepsilon = -\overline{u_1 u_2}\dfrac{d\bar{U}_1}{dx_2}$ The total production of turbulence by the main motion, or the total "dissipation" by turbulence of the energy of the main motion.

$2\nu \displaystyle\int_0^k dk\, k^2 \mathbf{E}(k)$ The viscous dissipation of turbulence in the wavenumber range 0 to k.

$-\dfrac{d\bar{U}_1}{dx_2}\displaystyle\int_k^\infty dk\, \mathscr{E}(k)$ The production of turbulence in the range k to ∞, or the dissipation by turbulence of the main motion in this range.

$-\displaystyle\int_0^k dk\, \mathbf{F}(k)$ Transfer of turbulence energy in the range 0 to k to turbulence of higher wavenumbers.

This equation forms the starting point for the few solutions for $\mathbf{E}(k)$ obtained hitherto that are valid in the range of large wavenumbers.

In deriving these solutions we shall make use of suggestions and assumptions made by Tchen.[6]

In the first place, it is assumed that k is so large that the turbulence in this wavenumber range is not too remote from isotropy, so that use may be made of assumptions successfully applied in the theories of isotropic turbulence. Thus, Heisenberg's assumption (3-127) is taken to apply to the transfer-spectrum $\mathbf{F}(k)$:

$$\int_0^k dk\,\mathbf{F}(k) = -\int_k^\infty dk\,\mathbf{F}(k) = -2\alpha'\int_k^\infty dk\sqrt{\frac{\mathbf{E}(k)}{k^3}}\int_0^k dk\,k^2\mathbf{E}(k)$$

The other assumptions concern the term

$$\int_k^\infty dk\,\mathscr{E}(k)$$

in Eq. (4-55). Tchen[6] distinguishes two cases:

Case 1. The vorticity $d\bar{U}_1/dx_2$ of the main motion is small compared with the vorticity of the turbulence in the wavenumber range under consideration.

Case 2. The vorticity of the main motion is comparable to the vorticity of the turbulence in the wavenumber range under consideration.

When the vorticity of the main motion is small compared with that of the turbulent motion (*Case 1*), interaction between the two vorticities can only be slight. According to Tchen there is no resonance.

On the other hand, in *Case 2*, where the vorticities of the two motions are comparable, there may be violent interactions between them, and violent resonance may occur.

If we consider the full wavenumber range $0 < k < \infty$, we have

$$\int_0^\infty dk\,\mathscr{E}(k) = \overline{u_1u_2}$$

According to Boussinesq's concept, we introduce an eddy viscosity ϵ_m [see Eq. (1-21)], so that

$$\overline{u_1u_2} = -\epsilon_m\frac{d\bar{U}_1}{dx_2}$$

Tchen assumes that the same relation applies to *Case 1* above (relatively small vorticity of the main motion):

$$\int_k^\infty dk\,\mathscr{E}(k) = -\epsilon_m(k)\frac{d\bar{U}_1}{dx_2} \tag{4-56}$$

and for $\epsilon_m(k)$ he makes use of Heisenberg's concept:

$$\epsilon_m(k) = \alpha''\int_k^\infty dk\sqrt{\frac{\mathbf{E}(k)}{k^3}}$$

Hence

$$\int_k^\infty dk\,\mathscr{E}(k) = -\alpha''\int_k^\infty dk\sqrt{\frac{\mathbf{E}(k)}{k^3}}\frac{d\bar{U}_1}{dx_2}$$

With Eq. (4-56) we obtain

$$-\frac{d\bar{U}_1}{dx_2}\int_k^\infty dk\,\mathscr{E}(k) = \epsilon_m(k)\left(\frac{d\bar{U}_1}{dx_2}\right)^2 \tag{4-57}$$

which, according to Heisenberg's concept, may be interpreted as dissipation of energy of the main motion by eddy viscosity:

Dissipation = eddy viscosity × the square of the vorticity

In *Case 2,* that of violent interaction between the vorticities of the main and turbulent motions. Tchen suggests taking the product of the vorticities of the two motions in Eq. (4-57) instead of the square of the vorticity of the main motion.

The root-mean-square vorticity of the turbulence is $(\overline{\omega_k\omega_k})^{1/2}$, where

$$\omega_k = -\varepsilon_{ijk}\frac{\partial u_i}{\partial x_j}$$

Thus

$$\overline{\omega_k\omega_k} = \overline{\frac{\partial u_i}{\partial x_j}\frac{\partial u_p}{\partial x_q}}\varepsilon_{ijk}\varepsilon_{pqk} = \overline{\frac{\partial u_i}{\partial x_j}\frac{\partial u_p}{\partial x_q}}(\delta_{ip}\delta_{jq}-\delta_{iq}\delta_{jp})$$

$$= \left(\overline{\frac{\partial u_i}{\partial x_j}\frac{\partial u_i}{\partial x_j}} - \overline{\frac{\partial u_i}{\partial x_j}\frac{\partial u_j}{\partial x_i}}\right) = -\left\{\left[\frac{\partial^2 \mathbf{Q}_{i,i}}{\partial\xi_j\,\partial\xi_j}\right]_{r=0} - \left[\frac{\partial^2 \mathbf{Q}_{i,j}}{\partial\xi_j\,\partial\xi_i}\right]_{r=0}\right\}$$

In a homogeneous turbulence we have

$$\mathbf{Q}_{i,j}(\xi_1,\xi_2,\xi_3,t) = \mathbf{Q}_{j,i}(-\xi_1,-\xi_2,-\xi_3,t)$$

and, for incompressible fluid,

$$\left[\frac{\partial^2\mathbf{Q}_{i,j}(\xi_1,\xi_2,\xi_3,t)}{\partial\xi_j\,\partial\xi_i}\right]_{r=0} = \left[\frac{\partial^2\mathbf{Q}_{j,i}(-\xi_1,-\xi_2,-\xi_3,t)}{\partial\xi_j\,\partial\xi_i}\right]_{r=0} = 0$$

Hence

$$\overline{\omega_k\omega_k} = -\left[\frac{\partial^2\mathbf{Q}_{i,i}}{\partial\xi_j\,\partial\xi_j}\right]_{r=0} = \iiint_{-\infty}^{+\infty} dk_1\,dk_2\,dk_3\,k^2\mathbf{E}_{i,i}$$

and

$$(\overline{\omega_k\omega_k})_{\text{av}} = 4\pi\int_0^\infty dk\,k^4(\mathbf{E}_{i,i})_{\text{av}} = 2\int_0^\infty dk\,k^2\mathbf{E}(k,t)$$

We may remark in passing that, in a homogeneous turbulence, the dissipation per unit mass (1) becomes equal to the kinematic viscosity times the mean-square rate of strain (2) and to the kinematic viscosity times the mean-square vorticity (3). For all three the expression reads

$$\nu\overline{\frac{\partial u_i}{\partial x_j}\frac{\partial u_i}{\partial x_j}} = 2\nu\int_0^\infty dk\,k^2\mathbf{E}(k)$$

Thus, in *Case 2* instead of (4-56), we obtain

$$\int_k^\infty dk\, \mathscr{E}(k) = -\epsilon_m(k) \times (\text{vorticity of the turbulence in the range 0 to } k)$$

$$= -\epsilon_m(k)\left[2\int_0^k dk\, k^2 \mathbf{E}(k)\right]^{1/2}$$

With the assumptions made, the terms $\int_k^\infty dk\, \mathscr{E}(k)$ and $\int_0^k dk\, \mathbf{F}(k)$ are expressed in terms of the energy-spectrum function $\mathbf{E}(k)$. For the two cases considered, the equations for this function read:

Case 1:

$$\varepsilon = 2\nu \int_0^k dk\, k^2\mathbf{E}(k) + \alpha''\left(\frac{d\bar{U}_1}{dx_2}\right)^2 \int_k^\infty dk \sqrt{\frac{\mathbf{E}(k)}{k^3}} + 2\alpha' \int_k^\infty dk \sqrt{\frac{\mathbf{E}(k)}{k^3}} \int_0^k dk\, k^2\mathbf{E}(k) \tag{4-58}$$

Case 2:

$$\varepsilon = 2\nu \int_0^k dk\, k^2\mathbf{E}(k) + \alpha''\left(\frac{d\bar{U}_1}{dx_2}\right)\left[2\int_0^k dk\, k^2\mathbf{E}(k)\right]^{1/2} \int_k^\infty dk \sqrt{\frac{\mathbf{E}(k)}{k^3}} + 2\alpha' \int_k^\infty dk \sqrt{\frac{\mathbf{E}(k)}{k^3}} \int_0^k dk\, k^2\mathbf{E}(k) \tag{4-59}$$

Now in *Case 1* the mean shear $d\bar{U}_1/dx_2$ is assumed to be weak, so we may assume that $\nu(d\bar{U}_1/dx_2)^2 \ll \varepsilon$. This would mean that the second term in Eq. (4-58) will be small with respect to the third term. Indeed the solution obtained by Tchen of Eq. (4-58) turns out to be essentially the same as the solution, Eq. (3-12), except that $(k/k_d)^{-5/3}$ occurring in the numerator has to be multiplied by a correction factor f, while $(k/k_d)^4$ occurring in the denominator has to be divided by this factor f, where

$$f = 1 + \frac{\nu}{\varepsilon}\frac{\alpha''}{\alpha'}(d\bar{U}_1/dx_2)^2$$

which is almost equal to 1. We may recall that, according to Deissler's calculations, the term $[k_1\, \partial \mathbf{E}_{i,i}/\partial k_2]_{av}$ is very small when the mean shear $d\bar{U}_1/dx_2$ is weak.

Of more interest is *Case 2*. Since the vorticity $d\bar{U}_1/dx_2$ of the mean motion is large, the effect of its interaction with the turbulent motion predominates with respect to the viscous dissipation and eddy transfer. It is reasonable to expect that if this occurs, it will be in the non-viscous wavenumber range, where k is not large. Tchen consequently considered only the second term on the right-hand side of Eq. (4-59) and neglected the two others. He then obtained an exact solution of the approximate equation. However, this solution can also be obtained on dimensional grounds, if we conclude that the situation actually considered by Tchen refers to the wavenumber range, shown in Fig. 4-6, where the spectral transfer caused by interaction with the mean shear, and described by the second term on the left-hand side of Eq. (4-48) is practically balanced by the energy transfer by inertial effects between wavenumbers, and given by **F**.

$$-2\pi k^2\left[k_1\frac{\partial \mathbf{E}_{i,i}}{\partial k_2}\right]_{\text{av}}\frac{d\bar{U}_1}{dx_2}\simeq \mathbf{F} \qquad (4\text{-}60)$$

When we consider the spectral transfer by inertial interactions between eddies as described by **F**, we may assume this **F** to be determined solely by ε and k. Hence $\mathbf{F} = \text{const.}\ \varepsilon/k$, and **E** will be determined by $d\bar{U}_1/dx_2$ and ε/k.

Dimensional analysis then yields

$$\mathbf{E} = \text{const.}\ \frac{\varepsilon}{d\bar{U}_1/dx_2}\cdot k^{-1} \qquad (4\text{-}61)$$

The same result is obtained straightaway from Eq. (4-60) since the left-hand side is proportional with $\mathbf{E}\,d\bar{U}_1/dx_2$. Equation (4-61) is the solution obtained by Tchen, with the const. $= 1/\alpha''$. For the case of strong interaction we may expect the ratio $(d\bar{U}_1/dx_2)^2/(\varepsilon/\nu)$ not to be small, at least. A more direct appreciation of the degree of interaction between the vorticity of the main motion and that of the turbulence, will be obtained when for the latter the vorticity in the wavenumber range, where Eq. (4-60) is valid, is taken. For the vorticity corresponding with wavenumber k, we may take $u'_k k$, where u'_k is the intensity of the turbulence with wavenumber k. With Eq. (3-103), the expression for the local vorticity in wavenumber space, we then obtain $k\sqrt{k\mathbf{E}(k)}$. Hence for strong interaction we may expect

$$\frac{d\bar{U}_1/dx_2}{k\sqrt{k\mathbf{E}(k)}} \qquad (4\text{-}62)$$

to be at least of the $\mathcal{O}(1)$.

The role of the main motion in this *Case 2* is different from that in *Case 1*. Owing to its own strong vorticity, it not only serves as a constant source of energy but also interferes drastically with the turbulence. The result is a different energy spectrum for the wavenumber range in which such a violent interaction between main and turbulent motion occurs. Beyond that range where direct interaction between these two motions becomes weak and finally zero, the energy spectrum approaches that for isotropic turbulence. In an inertial subrange this means Kolmogoroff's $(-5/3)$ law, while for the still higher wavenumber range we may expect, e.g., Pao's expression, given by Eq. (3-146), to apply.

From Tchen's Eq. (4-59) for the case of strong interaction, it may be inferred that this interaction will be effective to higher wavenumbers when the mean shear-rate increases, as may be expected. For this effect will depend on the ratio

$$\frac{d\bar{U}_1}{dx_2}\Bigg/\left[\int_0^k dk\,k^2\mathbf{E}(k)\right]^{1/2}$$

This conclusion is confirmed by calculations carried out by Panchev.[18] He extended

Tchen's Eq. (4-59) by following a suggestion made by Stewart and Townsend[19] for a modified Heisenberg's expression for the eddy viscosity $\epsilon_m(k)$, namely

$$\epsilon_m(k) = \alpha''\left[\int_k^\infty dk\frac{1}{k}\left(\frac{\mathbf{E}(k)}{k}\right)^{1/(2c)}\right]^c \qquad (4\text{-}63)$$

For $c = 1$, Heisenberg's expression is obtained.

Panchev obtained exact solutions for $\mathbf{E}(k)$ in parametric form, with $c = 0, 0.5$, and 1. For the case $c = 0$, which yields $\epsilon_m(k) \propto (\mathbf{E}/k)^{1/2}$, curves are presented of $\mathbf{E}(k)$ for various values of

$$m = \frac{d\bar{U}_1}{dx_2}\bigg/\left(\frac{\varepsilon}{\nu}\right)^{1/2} \qquad (4\text{-}64)$$

These curves indeed clearly show an increasing (lower) wavenumber range with a slope (-1), with increasing m. With $m = 1$ there is already a definite (-1) slope at small wavenumbers. For this and higher values of m there is a continuous change of the slope from (-1) to (-7) when $k \to \infty$, which latter limiting value is inherent to Heisenberg's concept of the spectral energy transfer.

Experimental Results

Experimental results on homogeneous shear-flow turbulence have been obtained by Rose,[10] and by Champagne, Harris, and Corrsin,[12] mentioned earlier. They experimented in the same windtunnel of square cross-section $h^2 = 0.3 \times 0.3$ m^2. The mean shear-flow was generated by installing in the entry section by means of horizontal plates of 0.003 m thickness and 0.6 m length, 12 equal-width channels, spaced 0.025 m and provided with screens to obtain adjustable internal resistances. The homogeneous asymptotic condition was reached after $x_1 \simeq 8.5h$, leaving about 0.6 m length of nearly homogeneous turbulent shear flow in the working section. The measurements were made in this region $8.5 \leq x_1/h \leq 10.5$.

Some results obtained by Champagne, Harris, and Corrsin in the section $x_1/h = 10.5$ will be reported here. During the experiments the center-line mean-velocity $U_c \simeq 12.4$ m/s. The mean shear-rate obtained $d\bar{U}_1/dx_2 = 12.9$ s^{-1}. This is pretty weak, for with a viscous dissipation $\varepsilon = 0.23$ m^2/s^3 we obtain for m, according to Eq. (4-64) the value ~ 0.1.

So, clearly we have the *"weak-interaction,"* i.e., *Case 1* considered by Tchen. The correction factor f, introduced earlier for this case, and which accounts for departures from the homogeneous, shear-free, isotropic turbulence condition, then is almost equal to 1, being 1.01. Yet the intensity of the axial velocity component of the turbulence was appreciably greater than that of the other two components. The following values for the relative intensities were measured:

$$u_1'/\bar{U}_c = 0.018; \quad u_2'/\bar{U}_c = 0.013 \quad \text{and} \quad u_3'/\bar{U}_c = 0.014$$

So $u_1'/u_2' \simeq 1.4$. The turbulence shear stress amounted to:

$-\overline{u_1u_2}/\bar{U}_c^2 = 1.1 \times 10^{-4}$, with a correlation coefficient $\mathbf{R}_{1,2}(0) = -\overline{u_1u_2}/u_1'u_2' = 0.5$ Obviously the effect of pressure-fluctuations in transferring kinetic energy from the u_2'-component to the u_3'-component was negligible at this weak mean shear-rate. We recall that according to Deissler's and Fox's calculations such an energy transfer between velocity components produced an increased anisotropy only at relatively high mean shear-rates. One-dimensional energy spectra $\mathbf{E}_1(k_1)$, $\mathbf{E}_2(k_1)$ and $\mathbf{E}_3(k_1)$ have been obtained. They are shown in Fig. 4-7 and Fig. 4-8. Kolomogoroff's micro length-scale $\eta = (\nu^3/\varepsilon)^{1/4} \simeq 3.6 \times 10^{-4}$ m, corresponds with $k_d \simeq 2{,}800$ m^{-1}. The Reynolds number of turbulence was $\mathbf{Re}_\lambda \simeq 130$. Notwithstanding this rather low value the one-dimensional spectrum $\mathbf{E}_1(k_1)$ contains a distinct range $0.01 \gtrsim k_1/k_d \gtrsim 0.1$ where Kolomogoroff's $(-5/3)$ law applies. When Pao's expression, Eq. (3-146), is matched to the experimental curve at $k_1/k_d = 0.01$, the further agreement with the experimental data up to $k_1/k_d \simeq 0.5$ is satisfactory. Beyond this wavenumber the measured values drop below Pao's curve. A similar effect could be observed in Fig. 3-18 and Fig. 3-19. Though this difference may be attributed to the failure of Pao's equation at very high wavenumbers, in the present case the difference may also be a result of the poorer spatial resolution power of the hot-wire anemometer since the length of 3.5×10^{-4} m of the sensing portion of the hot-wire used in

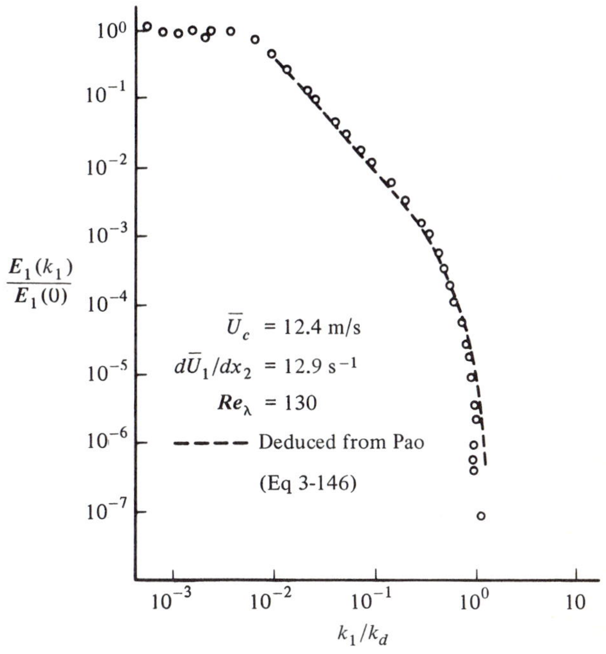

FIGURE 4-7
One-dimensional energy spectrum $\mathbf{E}_1(k_1)$ in a homogeneous shear-flow. (From: *Champagne, F. H., V. G. Harris, and S. Corrsin,*[12] *by permission of the Cambridge University Press.*)

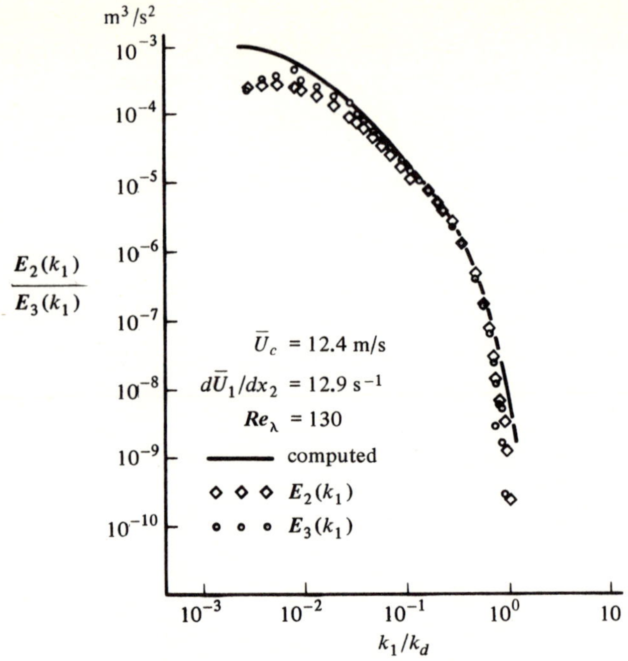

FIGURE 4-8
One-dimensional energy spectra $\mathbf{E}_2(k_1)$ and $\mathbf{E}_3(k_1)$ of u_2 and u_3 turbulence velocities. (From: *Champagne, F. H., V. G. Harris, and S. Corrsin,*[12] *by permission of the Cambridge University Press.*)

the spectral measurements was of the same magnitude as Kolmogoroff's dissipation scale η. No length correction had been applied. Figure 4-8 shows that the $\mathbf{E}_2(k_1)$ and $\mathbf{E}_3(k_1)$ spectra are almost identical, and do compare favorably in the range $0.04 < k_1/k_d < 0.5$ with the spectrum, computed according to the relation (3-53) and consistent with isotropy. These results of the spectra seem to point towards the existence of an inertial, locally isotropic subrange. On the other hand the co-spectrum of the turbulence shear correlation $\mathbf{R}_{12}$ showed an appreciable nonzero value still in this wavenumber range, thus indicating toward a significant nonisotropic condition. The spectrum of the shear correlation was only negligible for $k_1/k_d > 0.15$, i.e., in the viscous range.

Champagne, Harris, and Corrsin also determined spatial integral scales from the spatial correlations, and in a few cases also from the one-dimensional spectra. In the former case the integration was taken till the first zero-crossing of the correlation. The Λ_1, Λ_2, and Λ_3 so obtained from the correlations $\mathbf{R}_{1,1}(x_1)$, $\mathbf{R}_{2,2}(x_1)$ and $\mathbf{R}_{3,3}(x_1)$ respectively, had the values 0.05 m, 0.024 m, and 0.015 m, respectively. One concludes from these values that the large eddies are flattened in the x_3-direction, a conclusion supported by the measured spatial iso-correlation contours.

It will hardly be possible to achieve much higher mean shear-rates than obtained by Champagne, Harris, and Corrsin in similar kinds of experiments. For much higher shear-rates, so that the condition of strong interaction between the vorticity of the mean motion and turbulence vorticities as assumed by Tchen in his *Case 2* will be obtained, we have to look to turbulent flows very close to a wall, as in boundary layers and in pipe flow. Though discussions of this type of flow will be given in Chap. 7, it may be of interest to consider here results of measurements of the one-dimensional energy spectrum $\mathbf{E}_1(k_1)$ in the turbulent shear-flow made at a point very close to the wall. To this end we will consider the early measurements made by Klebanoff and Diehl,[8] and by Laufer.[9] These wall-turbulence flows are not homogeneous; moreover Tchen's considerations of *Case 2* pertain to the three-dimensional spectrum $\mathbf{E}(k)$.

Tchen[6] has shown that results for $\mathbf{E}(k)$ similar to those given above are obtained when the shear turbulence is not homogeneous.

As to the applicability of the results for the three-dimensional spectrum to the one-dimensional spectrum measured, we may remark that, if the turbulence in the high-wavenumber range is nearly isotropic, the relation between the three-dimensional and one-dimensional spectra given in Chap. 3 may apply approximately, so that similar expressions result. On the other hand, there is a difference between the dynamic equations for $\mathbf{E}_1(k_1)$ and for $\mathbf{E}(k)$ in that the $\mathbf{E}_1(k_1)$ equation contains the spectrum function of the pressure-velocity correlation. When discussing the corresponding equation (4-5) for the energy over all wavenumbers $\overline{u_1^2}$, we showed that the pressure-velocity-gradient correlation effects a transfer of energy to the other velocity components. Thus, although in the total-energy equation there is a transfer of energy only through the turbulence-transfer function between the various wavenumbers, the energy equation of one velocity component comprises in addition a transfer of energy among the three velocity components through the pressure-velocity-gradient correlations.

According to Eqs. (4-5) to (4-7), in a homogeneous parallel flow of the main motion the turbulence production by the main motion occurs only in the $\overline{u_1^2}$-component of the turbulence. Thus a strong interaction between the main motion and the turbulence such as is shown in *Case 2* can occur only for the $\overline{u_1^2}$-component. Hence a solution like Eq. (4-61) can be expected, if it does apply at all, only for the energy spectrum of the $\overline{u_1^2}$-component and not for the energy spectrum of the lateral components.

Figure 4-9 shows results of measurements made by Laufer[9] in a fully developed turbulent flow in a pipe of 25 cm diameter. The Reynolds number referred to the average flow velocity and the pipe diameter was 5×10^5. The Reynolds number of turbulence, $\mathbf{Re}_\lambda = u'_1\lambda_g/\nu$, was above 200, which is greater than any ever obtained in isotropic turbulent flows downstream of grids. This $\mathbf{Re}_\lambda$ is sufficiently large for one to expect a wavenumber range where an equilibrium occurs. As this figure shows, the spectrum curves for the points not too close to the wall even contain a rather wide range where $\mathbf{E}_1(k_1) \propto k_1^{-5/3}$ is closely followed. The curve for the

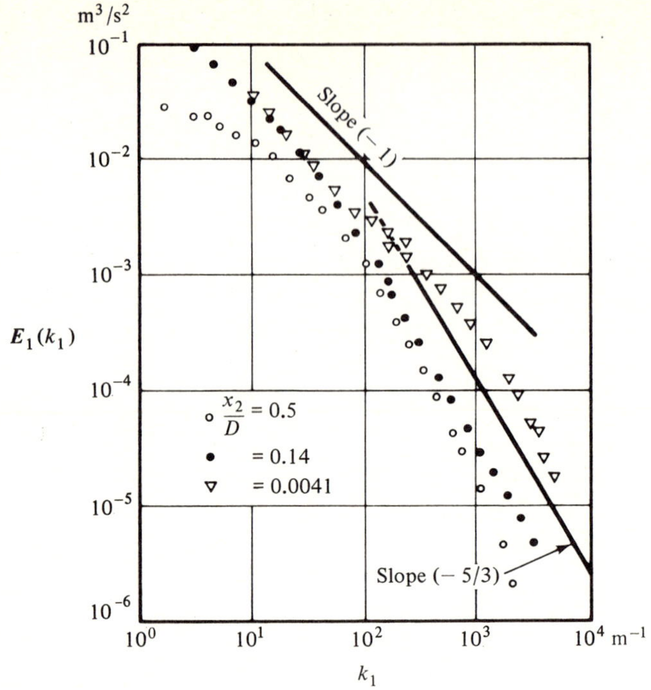

FIGURE 4-9
Energy spectra of u_1 in pipe flow with $\mathbf{Re}_D = \bar{U}_{av}D/\nu = 500{,}000$. (From: *Laufer, J.*,[9] *by permission of the National Advisory Committee for Aeronautics.*)

point very close to the wall contains a region where $\mathbf{E}_1(k_1) \propto k_1^{-1}$ is closely followed. At the distance $x_2 = 0.0041D$ from the wall the local dissipation was almost equal to the local production, so that the turbulence was in energy equilibrium. In that case

$$\varepsilon = -\overline{u_1 u_2}\frac{d\bar{U}_1}{dx_2}$$

When we introduce an eddy viscosity ϵ_m, so that $-\overline{u_1 u_2} = \epsilon_m\, d\bar{U}_1/dx_2$, we obtain for m of Eq. (4-64):

$$m = \frac{d\bar{U}_1}{dx_2}\bigg/\sqrt{\frac{\varepsilon}{\nu}} = \sqrt{\frac{\nu}{\epsilon_m}} \qquad (4\text{-}65)$$

So the value of m can be determined either from Eq. (4-64) or from $(\nu/\epsilon_m)^{1/2}$. At $x_2 = 0.0041D$ the vorticity of the mean motion, estimated from Laufer's mean-velocity distribution was $d\bar{U}_1/dx_2 \simeq 3{,}000\ \text{s}^{-1}$, a much higher value than obtained by Champagne, Harris, and Corrsin in their experiments. The value of $(\varepsilon/\nu)^{1/2}$ was estimated to be $\sim 10^4\ \text{s}^{-1}$, resulting in a value of $m \simeq 0.3$. This value agrees reasonably well with the value $(\nu/\epsilon_m)^{1/2} \simeq 0.25$.

In order to estimate the value of the ratio, given by Eq. (4-62), we may consider the wavenumber of the energy-containing eddies k_e. Since $\mathbf{E}(k_e) = \text{const.}\, u'^2/k_e$ and $k_e = \text{const.}\, u'^2/\mathbf{E}_1(0)$ [see Eq. (3-158)], we obtain

$$k_e\sqrt{k_e\mathbf{E}(k_e)} \simeq \text{const.}\, u'^3/\mathbf{E}_1(0)$$

So, Eq. (4-62) for the vorticity ratio becomes

$$\frac{d\bar{U}_1/dx_2}{k_e\sqrt{k_e\mathbf{E}(k_e)}} = A\frac{\mathbf{E}_1(0)\, d\bar{U}_1/dx_2}{u'^3} \tag{4-66}$$

When it is assumed that this ratio is small in the weak-interaction case, we conclude from Champagne, Harris, and Corrsin's experiments that the above constant A must be ~ 0.1 at the most. With a value of 0.1 for A we obtain for the above strong interaction case at the point $x_2 = 0.0041D$, the value of ~ 2 for the vorticity ratio, i.e. $\mathcal{O}(1)$. Use has been made of Fig. 7-57 and of Fig. 4-9 for estimating the values of u' and of $E_1(0)$ respectively.

Figure 4-10 shows energy spectra as measured by Klebanoff and Diehl.[8] Here too, the spectrum taken at $x_2 = 0.05\delta$ seems to contain a k^{-1} range, though not so pronounced as in Fig. 4-9. The local thickness of the boundary layer was ~ 0.075 m.

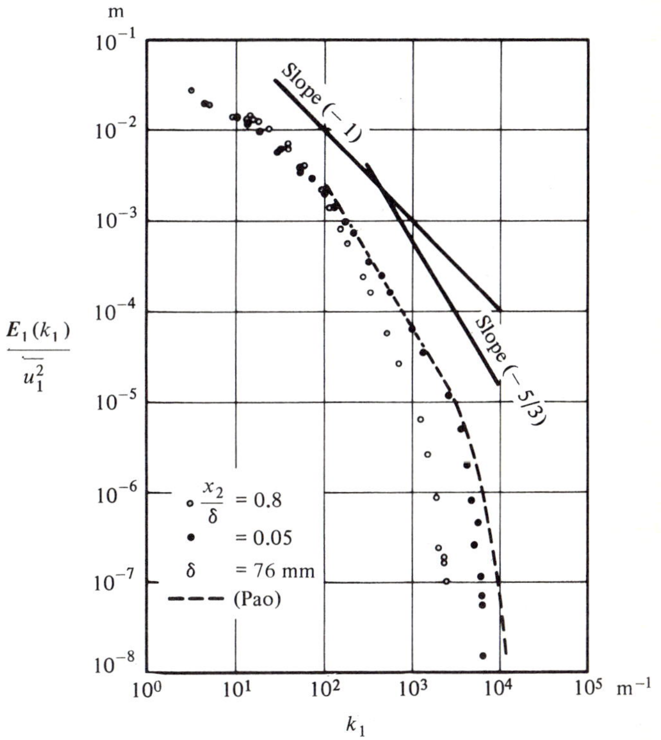

FIGURE 4-10
Energy spectra of u_1 in boundary-layer flow. $\mathbf{Re}_\delta = U_0\delta/\nu = 7.5 \times 10^4$. (After: *Klebanoff, P. S., and Z. W. Diehl,*[8] *by permission of the National Bureau of Standards, U.S. Department of Commerce.*)

At the point $x_2 = 0.05\delta$ the mean strain-rate or vorticity, estimated from the measured velocity distribution, was $\sim 380\ s^{-1}$. From the measured turbulence intensities and the viscous dissipation ε, one obtains with the relation (4-64), a value of $m \simeq 0.2$. While the relation (4-66), assuming again the value 0.1 for the numerical constant, would give the value ~ 0.6. In particular, this latter value is much smaller than the value of ~ 2 corresponding with the energy spectrum in Fig. 4-9, at $x_2 = 0.0041D$, in agreement with the much smaller k^{-1} range observed. In Fig. 4-10 also Pao's theoretical curve for $\mathbf{E}_1(k_1)$ has been given, by matching it to the experimental curve at the point $k_1 = 1{,}000\ m^{-1}$. There is hardly a $k_1^{-5/3}$ range, $(u'\lambda_g/\nu \simeq 200)$ which seems to support Panchev's calculations, mentioned earlier, for not very small values of m. Also the same departure of Pao's curve towards too high values in the viscous range can be observed, as shown earlier in Figs. 3-18, 3-19, and 4-7.

REFERENCES

1. Batchelor, G. K.: *Proc. Roy. Soc. London,* **186A,** 480 (1946)
2. Chandrasekhar, S.: *Phil. Trans. Roy. Soc. London,* **242A,** 557 (1950).
3. Chandrasekhar, S.: *Proc. Roy. Soc. London,* **203A,** 358 (1950).
4. Burgers, J. M., and M. Mitchner: *Proc. Koninkl. Ned. Akad. Wetenschap.,* **56,** 228, 343 (1953).
5. Rotta, J.: *Z. Physik,* **129,** 547 (1951).
6. Tchen, C. M.: *J. Research Natl. Bur. Standards,* **50,** 51 (1953).
7. Batchelor, G. K.: "The Theory of Homogeneous Turbulence," Cambridge University Press, New York, 1953.
8. Klebanoff, P. S., and Z. W. Diehl: *Natl. Advisory Comm. Aeronaut. Tech. Notes No.* 2475, 1951.
9. Laufer, J.: *Natl. Advisory Comm. Aeronaut. Tech. Repts. No.* 1174, 1954.
10. Rose, W. G.: *J. Fluid Mech.,* **25,** 97 (1966).
11. Tchen, C. M.: *Phys. Rev.,* **93,** 4 (1954).
12. Champagne, F. H., V. G. Harris, and S. Corrsin: *J. Fluid Mech.,* **41,** 81 (1970).
13. Deissler, R. G.: *Phys. Fluids,* **4,** 1187 (1961).
14. Phillips, O. M.: *Annual Review of Mechanics,* **1,** 245 (1969).
15. Lumley, J. L.: *Phys. Fluids,* **7,** 190 (1964).
16. Lumley, J. L., and H. A. Panofsky: "The Structure of Atmospheric Turbulence," p. 76, John Wiley & Sons, New York, 1964.
17. Fox, J.: *Phys. Fluids,* **7,** 562 (1964).
18. Panchev, S.: *Phys. Fluids,* **12,** 722 (1969).
19. Stewart, R. W., and A. A. Townsend: *Phil. Trans. Roy. Soc. London,* **243A,** 359 (1951).
20. Lumley, J. L.: *J. Fluid Mech.,* **41,** 413 (1970).

NOMENCLATURE FOR CHAPTER 4

$\mathbf{E}_{i,j}$ spectrum tensor of turbulence kinetic energy.

$\mathbf{E}(k,t)$ $2\pi k^2[\mathbf{E}_{i,i}(k_1,k_2,k_3,t)]_{av}$, three-dimensional energy-spectrum function averaged over a spherical surface in the wavenumber space.

$\mathbf{F}_{i,j}$	Fourier transform of $\mathbf{S}_{i,j}$.
$\mathbf{F}(k,t)$	$2\pi k^2[\mathbf{F}_{i,i}(k_1,k_2,k_3,t)]_{av}$, three-dimensional transfer-spectrum function averaged over a spherical surface in the wavenumber space.
$\mathbf{H}_i$	Fourier transform of $\mathbf{K}_i$
$\mathbf{K}_i$	first-order correlation tensor.
k	wavenumber.
P	static pressure; $\bar{P}$, time-mean value; p, turbulent fluctuation.
$\mathbf{P}_{i,j}$	$\dfrac{\partial}{\partial \xi_j}\mathbf{K}_{i,p} + \dfrac{\partial}{\partial \xi_i}\mathbf{K}_{j,p}$
$\mathbf{Q}_{i,j}$	second-order velocity-correlation tensor.
$\overline{q^2}$	twice the kinetic energy of turbulence.
$\mathbf{Re}_\lambda$	$u'\lambda_g/\nu$, Reynolds number of turbulence.
$\mathbf{S}_{i,kj}$ or $\mathbf{S}_{ik,j}$	third-order correlation tensor.
$\mathbf{S}_{i,j}$	$\dfrac{\partial}{\partial \xi_k}(\mathbf{S}_{jk,i} + \mathbf{S}_{ik,j})$.
t	time.
U_i	Eulerian velocity; $\bar{U}_i$, time-mean value; u_i, turbulence component; $u_i' = \sqrt{\overline{u_i^2}}$.
x_i	Eulerian coordinates.
α	Heisenberg constant in relation to the eddy viscosity ϵ_m.
δ_{ij}	Kronecker delta.
$\mathscr{E}(k,t)$	$2\pi k^2\left[2(\mathbf{E}_{1,2})_{av} - \left(k_1\dfrac{\partial \mathbf{E}_{i,i}}{\partial k_2}\right)_{av}\right]$.
ε_{ijk}	alternating unit tensor.
ϵ_m	eddy viscosity.
ε	dissipation by turbulence per unit of mass.
$\imath$	$\sqrt{-1}$.
Λ	spatial integral scale.
λ_g	lateral spatial dissipation scale.
ν	kinematic viscosity.
ω_k	vorticity of turbulence.
$\mathcal{O}$	order of magnitude.
$\mathbf{\Pi}_{i,j}$	Fourier transform of $\mathbf{P}_{i,j}$.
π	3.14159 . . . .
ρ	density.
ξ_i	Eulerian distance coordinates.

5

TRANSPORT PROCESSES IN TURBULENT FLOWS

5-1 INTRODUCTION

In Chap. 1 we showed that transport of a transferable quantity by random fluid motion is diffusive in nature. Since turbulent fluid motion is such a random fluid motion, a basic characteristic of turbulent flow is its diffusive action; accordingly, the exchange process is greater in turbulent than in nonturbulent flow. Any feature of turbulent fluid flow, eventually, finds its origin in processes of transport by turbulence; the study of such a feature then boils down to the study of the pertinent transport problem.

It seems useful to deal with the various turbulent-transport problems extensively and in detail in a chapter of their own. The general problem may be formulated as follows:[68] to express the turbulent-transport rate of a transferable quantity completely in terms of statistical functions of the turbulent velocity field and of boundary or initial conditions.

Thus a complete solution of the transport problem can be expected only if there is complete knowledge of the statistical functions describing the turbulent motion. As long as such complete knowledge is lacking, any solution of the transport problem must be incomplete and, at best, approximate. The more our knowledge of the turbulence increases, the greater will be the possibility of obtaining a correct and satisfactory solution.

It is obvious that, in the early days of turbulence investigations, when knowledge about and insight into the mechanism of turbulent fluid flow were rather poor, transport processes could be studied only in a rather rough way.

Yet these transport processes already formed a central subject in the earliest studies of turbulent flow. Among the theories arising from these studies, those based upon the concept of a "mixing length" (introduced by Taylor and independently by Prandtl) have proved to be most fruitful. This is not so much because they describe correctly the mechanisms underlying turbulent-transport processes as because they have resulted in useful and practical semiempirical relations.

These theories are purely phenomenological. Though later studies have shown that the physical picture based upon the concept of a mixing length cannot be correct in all details, the mixing-length theories still prove to be most useful to engineers. It is for this reason that so much space in the present chapter will be devoted to the treatment of these phenomenological theories.

More recent studies have considerably widened and deepened our insight into the actual mechanism of turbulent-transport processes. On the one hand they have shown the many serious shortcomings of the mixing-length theories but, on the other hand, they have confirmed some features of the mixing-length concept introduced by Prandtl on purely intuitive grounds.

In turbulent flow, parts of the fluid are moved in a random way. The nature of these movements depends on the size of the part of the fluid considered. In the previous chapters this dependence was expressed by introducing the various spectrum functions and scales (e.g., dissipation scale and integral scale). In the study of transport processes it will often be useful to distinguish between the motions of parts of the fluid of different size. Thus a distinction will be made between fluid particles and fluid lumps. By the *notion particle of fluid* we mean a very small volume of the fluid, small, for instance, with respect to the dissipation scale λ. The continuum forming such a fluid particle remains intact at least during a time interval sufficiently large compared with the interval to be considered during the transport process.[68] Any exchange with its direct surroundings is purely molecular in nature. The pressure and velocity distributions within such a particle are practically uniform.

In order to meet these conditions we have to consider particles, whose size are determined by the most fine-scale structure of the turbulence, which can be described solely in terms of ν and ε. So, such a fluid particle must be of the size of $\eta = (\nu^3/\varepsilon)^{1/4}$, with a velocity scale $(\nu\varepsilon)^{1/4}$ and a time scale $(\nu/\varepsilon)^{1/2}$ for its rate of deformation. Thus this time scale should be sufficiently large compared with the effective time available for the transport of a transferable property.

By the *notion lump of fluid* we understand a much larger part of the fluid continuum, consisting of a coherent conglomerate of fluid particles. The size of such a lump of fluid may be comparable to the larger eddies of the turbulence, i.e., of the order of magnitude of the dissipation scale or the integral scale. Exchange with its surroundings will be affected by the smaller scale turbulence, in addition to that by the molecular motion.

When we consider the combined effect of turbulent and molecular dispersion, it will be necessary to make a distinction between a *volume particle* and a *property or substance particle,* in particular in the case of short diffusion times. If there are no or negligible molecular effects, the volume and property particle remain identical during the dispersion process. The motion of the particle, e.g., represented by that of its centroid, is determined by the macroscopic velocity of the turbulent motion. However, consider now the case that there is a gradient in the property considered, and a non-negligible molecular diffusion of marked molecules describing the property. Then, as has been pointed out by Saffman,[9] due to the molecular migration down the property gradient, the motion of the centroid of these marked molecules, i.e., the property particle, will deviate from that of the original volume particle. As will be shown later in Sec. 5-5, the dispersion of a property due to molecular diffusion will not be statistically independent of that due to the macroscopic turbulent velocities.

It is customary to distinguish between two ways of describing turbulent-transport processes. The one may be referred to as the Eulerian description, the other as the Lagrangian description. In the Eulerian description the variation of some property with respect to a fixed coordinate system is considered; in the Lagrangian description the variation of the property connected with a given fluid particle or fluid lump is considered during the motion of this particle or lump through the flow field.

In Sec. 1-9, we mentioned that there is some similarity between dispersion by random molecular or Brownian motions and by the random turbulent motions. The earlier attempts to describe turbulence transport processes used procedures similar to those developed in the kinetic theory of gases. In these procedures two methods can be distinguished. In *method A* the rate of transport due to molecular motions through a control plane is considered, and expressed in terms of the local gradient of the mean property by means of a diffusion coefficient. In *method B* the diffusion of the property from a plane or point in space is considered. Here too the diffusion can be described with a diffusion coefficient. The latter method is often used in the case of a transient or nonstationary process. These two methods thus use a Lagrangian description.

Basically the two methods are the same in the sense that they lead to the same differential equation for the balance of the property concerned in a Eulerian description. Though, as mentioned, the description of the diffusion process according to the two methods is Lagrangian.

5-2 MIXING-LENGTH AND PHENOMENOLOGICAL THEORIES

These theories have been applied primarily in describing the distribution of mean values of a quantity (momentum, heat, mass, etc.) by the effect of turbulence.

In the earlier theories, among them the mixing-length theories, the diffusive action of turbulence is considered to result in an eddy viscosity or eddy heat-conductivity. This implies the assumption of a *gradient-type of diffusion*, so that the flux $\mathfrak{J}_{\mathscr{P}}$ of a transferable property $\mathscr{P}$ can be expressed in terms of the local gradient of the mean property.

We will start from the equations of motion and of transport given by Boussinesq [see Eq. (1-18) together with Eq. (1-21), and Eq. (1-33) together with Eq. (1-34)]. Boussinesq himself assumed scalar values for the coefficients of eddy diffusion ϵ_m and ϵ_γ. Prandtl[1] tried to connect these quantities with the characteristics of the turbulence. According to the kinetic theory of gases, the kinematic viscosity is equal to the product of the root-mean-square velocity of the molecules and the mean free path; analogously, Prandtl assumed that the coefficient of eddy viscosity was equal to the product of a "mixing" length and some suitable velocity (*mixing-length theory*). Prandtl's line of thought, analogous to that applied in the kinetic theory of gases to molecular-transport processes, is briefly as follows.

Consider for simplicity a two-dimensional flow, with a mean velocity $\bar{U}_1$ in the x_1-direction, and the average transport through a plane parallel to $\bar{U}_1$ of a transferable property $\mathscr{P}$ per unit of mass caused by the velocity u_2 of the turbulent motions. Prandtl made the assumption that each lump of fluid that is subjected to the turbulent motion may be considered as an individual entity and that the property $\mathscr{P}$ of this lump is conserved during a certain time, that is, over a certain distance. Entirely similar to molecular diffusion in the kinetic theory of gases we may put for the eddy diffusion coefficient $\epsilon_{\mathscr{P}}$ the expression

$$\epsilon_{\mathscr{P}} = \text{const.}\, u_2'\Lambda$$

where Λ is an integral length scale. In order to evaluate both u_2' and Λ, Prandtl made the following assumptions.

1 The distance $L_{\mathscr{P}}$ across which the fluid lumps transport a transferable property during their motions perpendicular to the control plane till adaptation with the new environment has taken place, is relatively small. So, consistent with the assumption of a gradient-type of diffusion, the gradient of the mean property is constant across the distance $L_{\mathscr{P}}$.

2 The axial and transverse components of the turbulence velocities are roughly of the same magnitude.

Let $\mathfrak{p}$ be the fluctuation of the property $\mathscr{P}$ with respect to the mean value $\bar{\mathscr{P}}$ at the control plane, which is assumed to be in $x_2 = 0$. Then the flux or average transport per unit of area and of time through the control plane is

$$(\mathfrak{J}_{\mathscr{P}})_2 = \rho\overline{u_2\mathfrak{p}} \tag{5-1}$$

where the average is taken over a large ensemble of fluid lumps passing the control plane. According to assumption (1) we may assume

$$\mathfrak{p} = L_{\mathscr{P}}\frac{d\bar{\mathscr{P}}}{dx_2} \tag{5-2}$$

Thus the fluctuation $\mathfrak{p}$ is assumed to be entirely due to the variation in $L_{\mathscr{P}}$ for the ensemble of fluid lumps, and not in addition by the instantaneous value of $\mathscr{P}$ at the point from which the fluid lump originates. We may subdivide this ensemble into subensembles, so that the lumps of each subensemble originate from the same distance $L_{\mathscr{P}}$. For this subensemble the average value of the difference in $\mathscr{P}$ with respect to $\mathscr{P}$ at $x_2 = 0$, then is given by the above expression for $\mathfrak{p}$, Eq. (5-2). For all subensembles together we thus obtain

$$(\mathfrak{J}_{\mathscr{P}})_2 = \rho\overline{u_2 L_{\mathscr{P}}} \cdot \frac{d\bar{\mathscr{P}}}{dx_2} \tag{5-3}$$

According to assumption (2) we put

$$u_2 = \text{const.}\, u_1 = -\text{const.}\, L_{\mathscr{P}} \left| \frac{d\overline{U}_1}{dx_2} \right|$$

The $(-)$ signs stems from the fact that the sign of u_2 is entirely determined by the sign of $L_{\mathscr{P}}$, which is assumed negative when the lump originates from the negative x_2-direction. Hence

$$(\mathfrak{J}_{\mathscr{P}})_2 = -\text{const.}\, \rho\overline{L_{\mathscr{P}}^2} \left| \frac{d\overline{U}_1}{dx_2} \right| \frac{d\bar{\mathscr{P}}}{dx_2}$$

The constant of proportionality may be included conveniently in a new length parameter

$$\mathfrak{l}_{\mathscr{P}}^2 = \text{const.}\, \overline{L_{\mathscr{P}}^2}$$

Since by definition

$$(\mathfrak{J}_{\mathscr{P}})_2 = -\rho\epsilon_{\mathscr{P}} \frac{d\bar{\mathscr{P}}}{dx_2} \tag{5-4}$$

we have

$$\epsilon_{\mathscr{P}} = -\overline{u_2 L_{\mathscr{P}}} = \mathfrak{l}_{\mathscr{P}}^2 \left| \frac{d\bar{U}_1}{dx_2} \right| \tag{5-5}$$

Note that $\overline{u_2 L_{\mathscr{P}}}$ is essentially negative. Also we may note that the procedure followed in Prandtl's mixing-length concept is according to method A, mentioned in the previous section.

Though originally Prandtl developed his mixing-length theory for the one-dimensional transport by turbulence, an *extension to the three-dimensional case* can easily be made. Following again method A, consider the transport in the x_i-direction through a control plane perpendicular to this direction. This transport of the property $\mathscr{P}$ is equal to

$$\overline{U_i \mathscr{P}} = \bar{U}_i \bar{\mathscr{P}} + \overline{u_i \mathfrak{p}}$$

Here, we are only interested in the transport $\overline{u_i \mathfrak{p}}$ by the turbulent motions. Now a fluid lump passing the control plane with velocity u_i may originate from a region whose center is a distance $(L_{\mathscr{P}})_j$ away from the point of the control plane passed by the center of the fluid lump. The assumption of a gradient-type of diffusion then yields

$$\mathfrak{p} = (L_{\mathscr{P}})_j \frac{\partial \bar{\mathscr{P}}}{\partial x_j}$$

and hence

$$(\mathfrak{J}_{\mathscr{P}})_i = \overline{\rho u_i (L_{\mathscr{P}})_j} \cdot \frac{\partial \bar{\mathscr{P}}}{\partial x_j}$$

If again we introduce a coefficient of eddy diffusion

$$(\epsilon_{\mathscr{P}})^*_{ij} = -\overline{u_i (L_{\mathscr{P}})_j}$$

we conclude that, in general, this coefficient is a tensor of the second order.

Now $\overline{u_i(L_{\mathscr{P}})_j}$ is not invariant under a change of i and j. Thus $(\epsilon_{\mathscr{P}})^*_{ij}$ is not symmetric in i and j. It has been provided with an asterisk to distinguish it from the eddy-diffusion tensor $(\epsilon_{\mathscr{P}})_{ij}$, which occurs in the Fickian diffusion equation, and which is symmetric in i and j. Now

$$\frac{\partial}{\partial x_i}(\mathfrak{J}_{\mathscr{P}})_i = -\frac{\partial}{\partial x_i}\left\{(\epsilon_{\mathscr{P}})^*_{ij} \frac{\partial \bar{\mathscr{P}}}{\partial x_j}\right\}$$

Bearing in mind that the index i in $(\epsilon_{\mathscr{P}})^*_{ij}$ corresponds with u_i and the index j with $(L_{\mathscr{P}})_j$, we may also write

$$\frac{\partial}{\partial x_j}(\mathfrak{J}_{\mathscr{P}})_j = -\frac{\partial}{\partial x_j}\left\{(\epsilon_{\mathscr{P}})^*_{ji} \frac{\partial \bar{\mathscr{P}}}{\partial x_i}\right\}$$

If we consider only the contribution by turbulence transport to the change with time of $\bar{\mathscr{P}}$ in a cubical volume, we obtain

$$\frac{\partial \bar{\mathscr{P}}}{\partial t} = \frac{\partial}{\partial x_i}(\mathfrak{J}_{\mathscr{P}})_i = -\frac{1}{2}\left[\frac{\partial}{\partial x_i}\left\{(\epsilon_{\mathscr{P}})^*_{ij} \frac{\partial \bar{\mathscr{P}}}{\partial x_j}\right\} + \frac{\partial}{\partial x_j}\left\{(\epsilon_{\mathscr{P}})^*_{ji} \frac{\partial \bar{\mathscr{P}}}{\partial x_i}\right\}\right]$$

For a *homogeneous turbulence*, we obtain the Fickian diffusion equation

$$\frac{\partial \bar{\mathscr{P}}}{\partial t} = -(\epsilon_{\mathscr{P}})_{ij} \frac{\partial^2 \bar{\mathscr{P}}}{\partial x_i \, \partial x_j} \qquad (5\text{-}6)$$

where

$$(\epsilon_{\mathscr{P}})_{ij} = \tfrac{1}{2}[(\epsilon_{\mathscr{P}})^*_{ij} + (\epsilon_{\mathscr{P}})^*_{ji}] \qquad (5\text{-}7)$$

and which is symmetric in i and j.

Let us now return to Prandtl's one-dimensional mixing-length theory, and consider the transport of axial momentum in the transverse, x_2-direction. (*Prandtl's momentum-transfer theory.*) The expression for the eddy viscosity then reads according to Eq. (5-5)

$$\epsilon_m = \mathrm{l}_m^2 \left| \frac{d\bar{U}_1}{dx_2} \right| \tag{5-8}$$

and, consequently,

$$\sigma_{21} = \rho \mathrm{l}_m^2 \left| \frac{d\bar{U}_1}{dx_2} \right| \frac{d\bar{U}_1}{dx_2} \tag{5-9}$$

This relation may also be obtained in a different way, namely from

$$\sigma_{21} = -\rho \overline{u_1 u_2}$$

Introduce the coefficient of the second-order correlation $\mathbf{R}_{21}$ between the velocity components u_1 and u_2 at the same point

$$\mathbf{R}_{21}(0) = \frac{\overline{u_1 u_2}}{u_1' u_2'}$$

Then the expression for σ_{21} reads

$$\sigma_{21} = -\rho \mathbf{R}_{21}(0) u_1' u_2'$$

If we again assume u_1' and u_2' both proportional to $|d\bar{U}_1/dx_2|$ and if we bear in mind that σ_{21} and $\mathbf{R}_{21}(0)$ have opposite signs, we obtain the relation (5-9); the constants of proportionality are put into the length parameter l_m.

These last considerations show that Prandtl's theory implies the assumptions of constant coefficient of correlation between u_1 and u_2 and constant proportionality between u_1' and u_2' across a section of the flow region if the length l_m is assumed constant across this region. In free turbulent flows Prandtl made the assumption that l_m was proportional to the width of the turbulent mixing zone and so a function only of the axial coordinate and not of the lateral coordinates (see also Chap. 6).

If $\mathscr{P}$ is a scalar quantity Γ, we may obtain, in analogy with Eq. (5-9),

$$(\mathfrak{J}_\gamma)_2 = \overline{\rho u_2 (L_\gamma)_2} \frac{d\bar{\Gamma}}{dx_2} = -\rho \mathrm{l}_\gamma^2 \left| \frac{d\bar{U}_1}{dx_2} \right| \frac{d\bar{\Gamma}}{dx_2} \tag{5-10}$$

And the expression for the eddy diffusivity reads

$$(\epsilon_\gamma)_2 = \mathrm{l}_\gamma^2 \left| \frac{d\bar{U}_1}{dx_2} \right| \tag{5-11}$$

It is, of course, not necessary a priori that the length l_γ be equal to l_m, though this has often been assumed.

The application of Prandtl's theory of momentum transport to the mean-velocity distribution in free jets and in the wake of cylinders and spheres has given reasonable agreement with measured velocity distributions (see Chap. 6). In these applications the value of l_m has been adapted in such a way that computed and measured distributions agree at the point where the mean velocity is half the maximum value across the section.

Deviations between computed and measured distributions are found especially at those points where $d\bar{U}_1/dx_2 = 0$. According to Prandtl's theory, $\epsilon_m = 0$ at those points; whereas actually it will have a finite value. Prandtl[2] himself has already become aware of this objection to his theory and has proposed the following more extended expression for the coefficient of eddy viscosity:

$$\epsilon_m = \mathrm{l}_m{}^2 \left[\left(\frac{d\bar{U}_1}{dx_2} \right)^2 + \mathrm{l}'^2_m \left(\frac{d^2\bar{U}_1}{dx_2{}^2} \right)^2 \right]^{1/2}$$

This assumption complicates the computations considerably. But it should, obviously, lead to better agreement with experiment, if only because the two parameters l_m and l'_m are available for adapting the theoretical to the measured distributions.

The inclusion of the second derivative $d^2\bar{U}_1/dx_2{}^2$ and the additional length l'_m may also be interpreted, in a series-expansion procedure, as permitting a variation of $d\bar{U}_1/dx_2$ over the mixing-length distance. Or, that the contribution by, relatively, larger-scale motions to the transport of momentum have to be taken into account, which becomes apparent for regions where $d\bar{U}_1/dx_2$ is very small, or zero.

Another more fundamental objection may be made to the momentum-transport theory. In this theory no account is taken of the effect of pressure fluctuations on momentum transport. As may be inferred from the equations of motion, momentum can be transferred by pressure differences alone without the lump of fluid itself being transported. Let us put it in another way. Because of the pressure fluctuations to which each lump of fluid is subjected during its path over the distance L, the momentum of the lump will not remain constant; so this momentum will not be preserved and, consequently, will not satisfy the requirement of being a transferable quantity in the sense of Prandtl.

According to one of the laws of Thompson concerning vortex motions, the vorticity must remain constant in a two-dimensional motion. Hence, in a two-dimensional flow field, the vorticity ω is a transferable quantity in the sense of Prandtl. Taylor[3] had developed, many years earlier, a theory similar to Prandtl's theory, but he had assumed that the vorticity might be considered a transferable quantity (*Taylor's vorticity-transport theory*).

If we consider again a two-dimensional flow uniform in the x_1-direction with velocity $\bar{U}_1$, so that

$$\frac{\partial}{\partial x_1} \overline{u_1{}^2} = \frac{\partial}{\partial x_1} \overline{u_2{}^2} = 0$$

we have

$$\frac{d\sigma_{21}}{dx_2} = -\rho \frac{d}{dx_2} \overline{u_1 u_2} = \rho \overline{u_2 \omega_3}$$

where

$$\omega_3 = \frac{\partial u_2}{\partial x_1} - \frac{\partial u_1}{\partial x_2}$$

Since ω_3 is assumed to be a transferable quantity, it follows that

$$\overline{u_2\omega_3} = +\overline{u_2 L_\omega}\frac{d\bar{\Omega}}{dx_2}$$

Because

$$\bar{\Omega} = -\frac{d\bar{U}_1}{dx_2}$$

we obtain

$$\frac{d\sigma_{21}}{dx_2} = -\rho\overline{u_2 L_\omega}\frac{d^2\bar{U}_1}{dx_2^{\,2}} = \rho \mathrm{l}_\omega^{\,2}\left|\frac{d\bar{U}_1}{dx_2}\right|\frac{d^2\bar{U}_1}{dx_2^{\,2}} \tag{5-12}$$

According to the momentum-transport theory we would obtain

$$\frac{d\sigma_{21}}{dx_2} = \rho\frac{d}{dx_2}\left[\mathrm{l}_m^{\,2}\left|\frac{d\bar{U}_1}{dx_2}\right|\frac{d\bar{U}_1}{dx_2}\right] \tag{5-13}$$

The expressions (5-12) and (5-13) are identical when l_m does not depend on x_2.

According to Prandtl this may be assumed in free turbulence with the mean velocity along the x_1-direction, so that in that case the two theories give identical solutions. The factor 2 by which the two formulas differ may be included in the value of l; l_ω will then be equal to $\mathrm{l}_m\sqrt{2}$.

The expression for the eddy viscosity reads

$$\epsilon_m = \epsilon_\omega = \frac{\mathrm{l}_\omega^{\,2}}{2}\left|\frac{d\bar{U}_1}{dx_2}\right| \tag{5-14}$$

If we compare the expressions (5-8), (5-11), and (5-14) for ϵ_m, ϵ_γ, and ϵ_ω, respectively, we find that

$$\frac{\epsilon_m}{\epsilon_\gamma} = \frac{\mathrm{l}_m^{\,2}}{\mathrm{l}_\gamma^{\,2}} \qquad \frac{\epsilon_m}{\epsilon_\omega} = \frac{2\mathrm{l}_m^{\,2}}{\mathrm{l}_\omega^{\,2}} = 1 \qquad \frac{\epsilon_\gamma}{\epsilon_\omega} = \frac{2\mathrm{l}_\gamma^{\,2}}{\mathrm{l}_\omega^{\,2}}$$

If we make $\mathrm{l}_\gamma = \mathrm{l}_\omega$, a greater spread of $\bar{\Gamma}$ than of $\bar{U}_1$ is obtained according to the vorticity-transport theory. This is not the case according to the momentum transport theory, if there we put $\mathrm{l}_\gamma = \mathrm{l}_m$.

If the turbulence is three-dimensional, the vorticity is no longer a transferable property in the sense mentioned above. Since actual turbulence is three-dimensional even in plane flows of the mean motion, the vorticity of lumps of fluid changes continuously because the vortex threads become longer during the deformations of fluid elements by the spatial motion of turbulence.

Taylor,[4] in taking account of this effect, extended his two-dimensional vorticity-transport theory to a three-dimensional theory (generalized vorticity-transport theory), but the expressions became practically intractable for further applications. Therefore, in order to gain simplicity, Taylor had to sacrifice exactness; he had to assume that in three-dimensional flow also the vorticity remains conserved. This simplified theory is now known as the *modified vorticity-transport theory*.

We assume a steady flow, and neglect the effect of viscosity. The equations of motion (1-18) then read

$$\bar{U}_j \frac{\partial \bar{U}_i}{\partial x_j} = -\frac{1}{\rho}\frac{\partial \bar{P}}{\partial x_i} - \frac{\partial}{\partial x_j}\overline{u_i u_j}$$

We transform the last term on the right-hand side as follows:

$$\frac{\partial}{\partial x_j}\overline{u_i u_j} = \overline{u_j \frac{\partial u_i}{\partial x_j}} - \overline{u_j \frac{\partial u_j}{\partial x_i}} + \overline{u_j \frac{\partial u_j}{\partial x_i}} = -\overline{u_j \omega_k \varepsilon_{ijk}} + \frac{\partial}{\partial x_i}\frac{\overline{u_j u_j}}{2}$$

where

$$\omega_k \varepsilon_{ijk} = \frac{\partial u_j}{\partial x_i} - \frac{\partial u_i}{\partial x_j}$$

(see Sec. 1-2). Hence

$$\bar{U}_j \frac{\partial \bar{U}_i}{\partial x_j} = -\frac{1}{\rho}\frac{\partial \bar{P}}{\partial x_i} - \frac{\partial}{\partial x_i}\frac{\overline{u_j u_j}}{2} + \overline{u_j \omega_k \varepsilon_{ijk}} \tag{5-15}$$

According to the mixing-length theory we write

$$\overline{u_j \omega_k} = \overline{u_j (\mathfrak{l}_\omega)_l}\frac{\partial \bar{\Omega}_k}{\partial x_l}$$

Taylor now further assumed the anisotropy of the turbulence sufficiently small, to take isotropic conditions for $\overline{u_j(\mathfrak{l}_\omega)_l}$. Thus, it is zero when $j \neq l$ (invariance under reflection) and equal to a scalar $-\epsilon_\omega$, when $j = l$. Hence

$$\overline{u_j \omega_k \varepsilon_{ijk}} = -\epsilon_\omega \varepsilon_{ijk}\frac{\partial \bar{\Omega}_k}{\partial x_l} = \epsilon_\omega \frac{\partial^2 \bar{U}_i}{\partial x_j \partial x_j}$$

Equation (5-15) then reads

$$\bar{U}_j \frac{\partial \bar{U}_i}{\partial x_j} = -\frac{1}{\rho}\frac{\partial \bar{P}}{\partial x_i} - \frac{\partial}{\partial x_i}\frac{\overline{u_j u_j}}{2} + \epsilon_\omega \frac{\partial^2 \bar{U}_i}{\partial x_j \partial x_j} \tag{5-16}$$

According to the momentum-transport theory we would have obtained

$$\bar{U}_j \frac{\partial \bar{U}_i}{\partial x_j} = -\frac{1}{\rho}\frac{\partial \bar{P}}{\partial x_i} + \frac{\partial}{\partial x_j}\left[(\epsilon_m)_{ik}\left(\frac{\partial \bar{U}_j}{\partial x_k} + \frac{\partial \bar{U}_k}{\partial x_j}\right)\right] \tag{5-17}$$

Thus we see that the two differential equations become identical only under certain conditions. The turbulence must be homogeneous so that the second term on the right-hand side of Eq. (5-16) may be neglected. An assumption concerning near isotropy of turbulence-transfer quantities similar to that made by Taylor with respect to the Eq. (5-15) has to be introduced, so that $(\epsilon_m)_{ik}$ will reduce to a scalar quantity, which, moreover, should be independent of x_j.

According to Prandtl's mixing-length hypothesis, the eddy viscosity ϵ_m may be written as the product of the square of the mixing length and the lateral derivative of the mean velocity. As already mentioned, Prandtl assumed (1) that this mixing length is constant in a cross section of the mixing zone in a free turbulent flow, and (2) that the mixing length is proportional to the width of the mixing zone. These assumptions imply that, in a cross section of the mixing zone, the scale of turbulence, which is related to the mixing length, must have on the average a constant value, determined by the total width of the mixing zone.

Von Kármán[5] made another assumption concerning the value of the mixing length, namely, that it is determined by the local flow conditions and that it may be described in terms of quantities determined by these local conditions. Such quantities are, for instance, the derivatives of the mean velocity at a point of the flow field. In addition, Von Kármán assumed that the flow pattern shows *geometrical similarity* everywhere throughout the flow field. According to the definition of complete similarity (see Sec. 3-1) a length scale and a velocity scale alone are sufficient to determine the structure of the turbulence. Since the derivatives of the mean velocity are assumed to be the quantities that are determined by the local conditions, the simplest way to obtain a characteristic length scale is from the ratio between two consecutive derivatives, for instance, the first and the second lateral derivatives of the mean velocity.

If we consider a plane flow with the mean velocity in the x_1-direction, the condition of geometrical similarity thus requires that

$$\mathrm{l}\frac{(\partial^2\bar{U}_1/\partial x_2{}^2)}{(\partial\bar{U}_1/\partial x_2)} = \text{const.} = K$$

or

$$\mathrm{l} = K\frac{(\partial\bar{U}_1/\partial x_2)}{(\partial^2\bar{U}_1/\partial x_2{}^2)} \tag{5-18}$$

For the coefficient of eddy viscosity ϵ_m we then obtain

$$\epsilon_m = \mathrm{l}_m{}^2\left|\frac{\partial\bar{U}_1}{\partial x_2}\right| = K^2\frac{(\partial\bar{U}_1/\partial x_2)^2}{(\partial^2\bar{U}_1/\partial x_2{}^2)^2}\left|\frac{\partial\bar{U}_1}{\partial x_2}\right| \tag{5-19}$$

This expression shows that ϵ_m becomes infinite at the points where $\partial^2\bar{U}_1/\partial x_2{}^2 = 0$ and $\partial\bar{U}_1/\partial x_2 \neq 0$. This result is not very reasonable. Formally we can get rid of this objection by taking for l^2 the ratio between $\partial\bar{U}_1/\partial x_2$ and $\partial^3\bar{U}_1/\partial x_2{}^3$, which, however, is not very satisfactory either. Also, it appears that Von Kármán's theory does not offer many advantages over Prandtl's much simpler assumption, that is, mixing length proportional to the width of the mixing zone (in the case of free turbulence) or proportional to the distance from a wall (in the case of wall turbulence; see Chap. 7).

Neither Prandtl nor Taylor discusses in detail the mechanism according to which the lumps of fluid adapt their transferable property to their new environments. It is simply assumed that such an adaptation ultimately takes place. Particularly in the case of the vorticity-transport theory where the vorticity remains conserved (two-dimensional case) or is assumed to be conserved (three-dimensional case), it is very difficult to see how the vorticity of the fluid lump adapts itself to the mean vorticity of the new environment unless the effect of viscosity is brought into the picture.

After these mixing-length theories, other theories have been developed. Von Kármán[6] and Prandtl[2] have proposed theories in which some direct relationship between the coefficient of eddy diffusion ϵ and the flow pattern is assumed. Other theories that will be discussed do not introduce a coefficient of eddy diffusion but follow an entirely different line of thoughts based upon phenomenology.

Von Kármán doubts whether the mean-velocity distribution can be described satisfactorily by mixing-length theories if (for instance, in the case of free turbulence) the mixing length is determined by the dimensions of the mixing zone, as assumed by Prandtl. Von Kármán still follows his earlier assumption that the turbulent-transport processes are determined by local flow conditions. Von Kármán writes instead of $\epsilon = \text{const.}\, u_2'\Lambda$

$$\epsilon = u_2'\mathfrak{l}^*$$

This length $\mathfrak{l}^*$ is not identical with Prandtl's mixing length. According to Von Kármán, the value of $\mathfrak{l}^*$ must be determined by quantities that have a local character, such as λ_g and the local Reynolds number or Reynolds number of turbulence $\mathbf{Re}_\lambda$. Thus

$$\mathfrak{l}^* = \lambda_g f(\mathbf{Re}_\lambda)$$

The simplest form is

$$\mathfrak{l}^* \propto \lambda_g \mathbf{Re}_\lambda$$

so that

$$\epsilon = \text{const.} \frac{u_2'^2 \lambda_g^2}{\nu} \tag{5-20}$$

However in a later publication Prandtl[113] made another suggestion, namely

$$\epsilon - \text{const.}\, (\overline{q^2})^{1/2} \cdot L \tag{5-20a}$$

where $\overline{q^2} = \overline{u_i u_i}$. The length-scale L may still be a function of the position in the flow. For free turbulent flows Prandtl assumed $L \propto B$, where B is the width of the mixing zone. In the case of the flow near the wall, L may be taken proportional with the distance to the wall.

In the preceding considerations we have shown a controversy between Prandtl's and Von Kármán's ideas: Prandtl tried to relate transport processes in mixing zones of free turbulent flows to their over-all conditions; Von Kármán tried to relate these processes to local conditions. This disparity is found again between the theory of Von Kármán just expounded and another theory of Prandtl pertinent to free turbulence. In this theory Prandtl assumes that the over-all flow character in the mixing zone is determined by the dimensions of this zone and the greatest velocity difference across this zone. According to the assumption that the eddy diffusivity is determined by these over-all conditions, he suggests for ϵ the relation

$$\epsilon = \text{const.} \times B(\bar{U}_{max} - \bar{U}_{min}) \qquad (5\text{-}21)$$

Thus ϵ must be constant (independent of x_2) across a section and a function of x_1 only, that is, of the direction of the main flow. A consequence of this is that, according to the relation (5-8), the mixing length l_m is not constant in a cross section of the mixing zone, contrary to the assumption made earlier by Prandtl.

According to Eq. (5-20), ϵ may also be a function of x_2. As we shall show in Chap. 6, experimental results on the velocity distribution across free jets and wakes of cylinders rather favor belief in a constant value of ϵ_m in a cross section of the mixing zone.

It may be useful to discuss the various theories in more detail in the light of more recent investigations on the structure of turbulence, because these clearly show the shortcomings of the theories. Since the investigations refer mainly to free turbulent flows (free jets and wake flow of cylinders) we shall restrict our considerations here to this type of turbulent flow. We have to make use of experimental results that will be considered later in Chap. 6, but it is felt that this will not be a serious objection.

If we write again $\epsilon = u_2'\mathrm{l}^*$, then a gradient type of transport implies that l^* must be small, at any rate so small that the gradient of the transported quantity is practically uniform through the distance l^*. Now it appears, from the relevant values of ϵ and u_2' in mixing zones, that the length l^* is usually not very small; so the above condition is not satisfied. According to Townsend's measurements in the wake flow of a cylinder, the empirical value of l^* is of the order of 0.1 times the half-width of the wake.

If we apply to u_2' and l^* Prandtl's mixing-length theory, so that the relation (5-8), for instance, applies to ϵ_m, then here, too, the mixing length l_m must be small compared with the dimensions of the mixing zone. As we shall see in Chap. 6, the values of l_m calculated from the experimental values of ϵ_m are anything but small compared with the width of the mixing zone.

We have mentioned that Prandtl's mixing-length theory also implies that u_1' and u_2' have a constant ratio across the mixing zone; so the turbulence structure must be similar in all parts of the mixing zone. But the same theory implies more, namely that the effect of diffusion and convection of turbulence energy must be negligibly small, so that the turbulence is in energy equilibrium, i.e. the turbulence energy generated locally by the work of the shear stresses is equal to the dissipation [see, e.g., Eq. (1-111)]. For in that case we have

$$-\overline{u_1 u_2}\frac{\partial \bar{U}_1}{\partial x_2} = \text{dissipation}$$

Now we use the assumption of similarity and take u_2' and λ as a characteristic velocity and length. We then have

$$\overline{u_1 u_2} \propto u_2'^2$$

and

$$\text{Dissipation} \propto \frac{\nu u_2'^2}{\lambda^2} \propto \frac{1}{\mathbf{Re}_\lambda} \frac{l_m}{\lambda} \frac{u_2'^3}{l_m}$$

so

$$u_2'^2 \frac{\partial \bar{U}_1}{\partial x_2} \propto \frac{1}{\mathbf{Re}_\lambda} \frac{l_m}{\lambda} \frac{u_2'^3}{l_m}$$

whence

$$u_2' \propto l_m \frac{\partial \bar{U}_1}{\partial x_2}$$

This is indeed one of the assumptions made by Prandtl. But, as we shall show in Chap. 6, diffusion and convection of turbulence energy are not at all negligible.

Experimentally it has been found for the wake flow as well as for the free jet that the coefficient of eddy viscosity ϵ_m is fairly constant over a large part of the mixing zone. If we follow Von Kármán and assume that the relation (5-20) applies to ϵ_m, this implies either that $u_2'^2$ and λ_g are both constant or that they vary in such a way that their product is constant. From Townsend's measurements in the wake flow of a cylinder it follows that λ_g is practically constant across the wake. And, because u_2' is found not to be constant across the wake, the relation (5-20) would lead to a nonconstant value of ϵ_m.

But the choice of u_2' as the characteristic velocity is rather arbitrary. In the Reynolds number $\mathbf{Re}_\lambda$ we should better take the root-mean-square of the three turbulence components and, consequently, assume instead of Eq. (5-20) a relation similar to Eq. (5-21a):

$$\epsilon_m = \text{const.} \times \frac{\overline{q^2} \lambda_g^2}{\nu} \tag{5-22}$$

According to Townsend's measurements, the turbulence kinetic energy $\overline{q^2}/2$ is nearly constant across the wake.

The constancy of the turbulence kinetic energy and of the dissipation scale across the mixing zone are experimental facts. It is possible to make these facts appear reasonable. To this end it will be necessary to consider more closely the transport process of the turbulence kinetic energy in the mixing zone. An analysis of the measurements concerning transport of this energy made by Townsend in the wake flow of a cylinder will be given in Chap. 6. It will suffice here to mention one of the results of this analysis, namely, that there appears to be a small region of the mixing zone where the diffusion of turbulence energy seems to occur up and not down the gradient, if we stick to the assumption that the total transport $\overline{u_2 q^2}$ is of the gradient type. In that case ϵ_q becomes negative in this region, which is impossible. The only way to solve this difficulty is to abandon the assumption that the total transport $\overline{u_2 q^2}$ is of the gradient type and to realize that a mechanism other than gradient-type diffusion must be responsible for the transport observed. For this mechanism it is postulated that the transport of $\overline{q^2}$ may be brought about by a gradient-type diffusion caused by small-scale turbulence plus convection by large-scale motions of eddies comparable in size to the width of the wake.

Hence

$$\overline{u_2 q^2} = -\epsilon_q \frac{\partial}{\partial x_2} \overline{q^2} + \overline{\mathscr{V} q^2} \tag{5-23}$$

where $\mathscr{V}$ denotes the velocity of the large-scale motions.

It is obvious that in principle the same complex of transport mechanisms will apply to every transferable property, momentum, heat, or matter. Townsend has found that, from measurements of the distribution of temperature $\bar{\Theta}$ and of $\overline{u_2 \theta}$, one must conclude that the total transport of heat must have been due not to gradient-type diffusion alone but also to transport by bulk movements.

The contribution to the total transport of either term on the right-hand side of Eq. (5-23) may depend on the character of the transferable property $\mathscr{P}$. From experimental results obtained by Townsend concerning the distributions of mean velocity, heat, and turbulence energy in the wake of a cylinder, it may be inferred that the transport of axial momentum is very probably due mainly to a gradient-type diffusion by small-scale turbulence, whereas transport of turbulence energy is caused primarily by the convective action of large-scale motion and transport of heat is possibly "mixed." This difference might be expressed in an exaggerated way as follows:

$$\overline{u_2 u_1} \simeq -\epsilon_m \frac{\partial \bar{U}_1}{\partial x_2} \tag{5-24}$$

$$\overline{u_2 q^2} \simeq \overline{\mathscr{V} q^2} \tag{5-25}$$

$$\overline{u_2 \theta} \simeq -\epsilon_\theta \frac{\partial \bar{\Theta}}{\partial x_2} + \overline{\mathscr{V} \theta} \tag{5-26}$$

but such a distinction would certainly be too extreme.

Since the transport of the turbulence energy $\overline{q^2}$ is caused mainly by the large-scale convective flows across the mixing zone, it is reasonable to accept the experimental result that the distribution of $\overline{q^2}$ is practically uniform across an essential part of the mixing region.

As already said, the relation (5-24) for the turbulent shear-stress is an approximate one, at last. And certainly so when larger scale turbulent motions do give a non-negligible contribution. We did mention already in Sec. 1-2 that this neglect may lead to negative values of eddy viscosity. Such regions where shear-stress and mean-velocity gradient have an opposite sign, have been observed in a number of experiments in turbulent flows with an asymmetric mean-velocity distribution, such as wall jets and channel flows with different roughness of the two opposite walls.[104,105,106,109,110] The regions concerned were localized around the maximum mean-velocity, where its gradient is zero or very small. In general the explanation is sought in the fact that the effect of larger-scale motions becomes apparent, due to the asymmetry of the velocity field around its maximum, and therefore may not be neglected.[107,108,109] The assumption of a gradient-type alone of transport of momentum is no longer valid.

The "*memory effect*" (see Sec. 1-2) just considered where it is still felt over distances not small compared with the length scale for the change of the gradient of the mean property has been restricted to the transport in transverse direction with respect to the main flow. The change in the main flow direction has been assumed to be very small indeed. However, this latter assumption is not always justified. It may be possible that the length scale for the change in the main flow direction of the gradient of the mean property is not relatively large enough in order that memory effects in this direction may be neglected. We will consider this point later in Sec. 5-4.

Before leaving this critical discussion of the transport theories considered, we want to mention a feature of free turbulent flows which certainly affects the results of theories that fail to take it into account, namely, the intermittent character of the turbulent flow in the boundary regions between the mixing zone and the undisturbed free stream outside it.

The description of transport processes by means of differential equations presupposes continuity of the turbulence in the entire region considered. But if we consider the turbulence of wake flow or of a free jet, we find such continuity only in the central part of the mixing zone; the continuity is interrupted more and more toward the outer boundaries of the mixing zone. Here, the turbulent state is only present during a fraction of the time, which fraction decreases to zero in outward direction. Consequently, solutions of differential equations based on continuous turbulence all the time with a constant coefficient of eddy diffusion must fail. Actually the effective value of such a coefficient decreases rapidly in these regions as the outer edge is approached; this is reasonable, because gradient-type diffusion at a point can occur only during the periods in which the flow at that point is turbulent. During these periods the instantaneous value of ϵ may be nearly equal to the value occurring in the continuous central part of the mixing region; that is, ϵ may be nearly constant during these periods.

In the theories advanced by Boussinesq, Prandtl, Taylor, and Von Kármán a *deductive method* is followed. A hypothesis is made concerning the turbulence shear stress or concerning the turbulence diffusion coefficient, from which, with the aid of the equations of motion and continuity and with assumed similarity conditions, velocity distributions are deduced. To this method Reichardt[7,8] objected, pointing out that relatively easily measurable quantities such as mean velocities are derived on the basis of more or less questionable hypotheses, and that the usual scatter of the experimental data used for comparison makes any decision in favor of one hypothesis or another difficult.

As against the deductive method Reichardt proposes an *inductive method.* Its basis is experimental data about directly measurable quantities. He has worked out an inductive theory for the case of free turbulent flows. Experiments have shown that the velocity and Γ-distributions across mixing zones follow Gaussian error or related functions rather closely. Such functions are also found in processes of leveling by molecular diffusion or heat conduction. Reichardt assumes that a turbulent-transport process is a statistical process exactly analogous to molecular-transport processes. If we consider, for instance, the distribution of the velocity U_1 in a free turbulent flow with the main velocity in the x_1-direction, it has been found that the total momentum flow $\rho\overline{U_1{}^2}$ (incorporating the mean and turbulent components) follows the Gaussian error curve. If this is true, the differential equations for $\rho\overline{U_1{}^2}$ must be identical with the equation for molecular diffusion or heat conduction, which yields a Gaussian solution. Hence we have to transform the equations of motion in such a way that the equation for heat conduction is obtained. This now requires a separate hypothesis.

Let us consider Reichardt's theory, which is purely phenomenological, in more detail for a simple two-dimensional case. If, as is usual in free-turbulence investigations, we neglect the pressure term and the terms containing the viscosity, the equation of motion for the instantaneous velocity U_1 reads

$$\frac{\partial U_1}{\partial t} + U_1\frac{\partial U_1}{\partial x_1} + U_2\frac{\partial U_1}{\partial x_2} = 0 \qquad (5\text{-}27)$$

With the aid of the equation of continuity this equation is transformed into

$$\frac{\partial U_1}{\partial t} + \frac{\partial U_1{}^2}{\partial x_1} + \frac{\partial}{\partial x_2}U_1U_2 = 0 \qquad (5\text{-}28)$$

And, since the mean motion is assumed steady,

$$\frac{\partial\overline{U_1{}^2}}{\partial x_1} + \frac{\partial}{\partial x_2}\overline{U_1U_2} = 0 \qquad (5\text{-}28a)$$

This is nothing but the equation for the conservation of the momentum component in the x_1-direction.

If $\overline{U_1{}^2}$ were to satisfy the parabolic differential equation

$$\frac{\partial\overline{U_1{}^2}}{\partial x_1} = \mathscr{L}\frac{\partial^2\overline{U_1{}^2}}{\partial x_2{}^2} \qquad (5\text{-}29)$$

its solution would be a Gaussian error function. The coefficient $\mathscr{L}$ may still be a function of x_1 determined by the width of the mixing zone.

In order to transform Eq. (5-28) into Eq. (5-29) we have simply to postulate that

$$\overline{U_1U_2} = -\mathscr{L}\frac{\partial\overline{U_1{}^2}}{\partial x_2} \tag{5-30}$$

This relation, which Reichardt calls the momentum-transfer law, may be interpreted as follows. The rate of transfer of U_1-momentum with the velocity U_2 in the lateral x_2-direction is proportional to the change of the momentum flux $\overline{U_1{}^2}$ in that lateral direction.

This "law" does not sound unreasonable. Yet it is not satisfactory, and is in fact found to be incorrect if it is considered in all its consequences. As Prandtl[2] has pointed out, Eq. (5-29) favors the x_1-direction as against the x_2-direction in an unjustified way—unless the justification is sought in the fact that the x_1-direction is that of the main flow.

Whereas Eq. (5-28) is invariant under a translation at constant velocity of the coordinate system, this appears not to be true of the momentum-transfer equation (5-30). So this conflicts with Newton's relativity, which states that forces in a mechanical system must be independent of the addition of a constant velocity.

Reichardt's counter argument that by selecting a fixed coordinate system in a wake flow or jet flow, Newton's relativity principle may no longer be applied, is not convincing.

In this light one is inclined not to accept Reichardt's theory. The only argument, admittedly a strong one, that Reichardt adduces in favor of his theory is its good agreement with experimental results. But this is not really very surprising, for the equations of motion have been transformed artificially so as to agree with the experimental evidence.

On the other hand, Eq. (5-29) has the advantage, not only mathematically but also from a practical engineering point of view, of being linear in $\overline{U^2}$, so that the law of superposition of elementary solutions holds. Alexander, Baron, and Comings[11] have shown that this principle of superposition can be applied with success to predict the momentum distribution across two parallel free jets, issuing with the same velocity from orifices in the same plane perpendicular to the jet axes.

Reichardt's theory may be extended to the case in which a scalar quantity Γ is considered instead of the momentum component ρU_1. For plane steady motion, if molecular effects are neglected, the equation for the transport of Γ reads

$$\frac{\partial\overline{U_1\Gamma}}{\partial x_1} + \frac{\partial}{\partial x_2}\overline{U_2\Gamma} = 0$$

We then have to postulate the "law"

$$\overline{U_2\Gamma} = -\mathscr{L}_\gamma\frac{\partial}{\partial x_2}\overline{U_1\Gamma} \tag{5-31}$$

in order to arrive at

$$\frac{\partial}{\partial x_1}\overline{U_1\Gamma} = \mathscr{L}_\gamma\frac{\partial^2}{\partial x_2{}^2}\overline{U_1\Gamma} \tag{5-32}$$

which yields the Gaussian solution for $\overline{U_1\Gamma}$.

Reichardt's theory does not give a true picture of the turbulence-flow mechanism or advance our insight into this mechanism. The only merit of this theory is the practical one already mentioned: that solutions for distributions of a transferable quantity in free turbulence can be constructed easily and that these solutions do agree very satisfactorily with experimental data.

Finally we may remark that the Gaussian solution according to Eq. (5-29) refers to the total momentum flux $\overline{U_1{}^2}$, which contains the contributions of both the mean and the turbulent motions, namely,

$$\overline{U_1{}^2} = \bar{U}_1{}^2 + \overline{u_1{}^2}$$

We shall show in Chap. 6 that in free turbulent flows it is possible under certain assumptions to obtain from the equations of motion Gaussian solutions which then refer to the mean velocity $\bar{U}_1$.

The experimental data usually show such scatter that it is difficult to decide whether the Gaussian solution fits the total momentum flux $\overline{U_1{}^2}$ or the mean velocity $\bar{U}$ more satisfactorily.

Investigations by Townsend and Batchelor on the wake flow behind a cylinder have shown that the Gaussian solution seems to be correct for the mean motion, at least for the central fully turbulent part of the mixing zone. It does not apply to the boundary region of the mixing zone. We shall discuss these investigations in more detail in Chap. 6.

Finally, we may note that Reichardt's inductive theory also assumes a gradient-type of turbulence transport. Basically it does not account for possible effects of transport by large-scale turbulent motions (big eddies), as discussed earlier.

5-3 ANALOGIES IN TURBULENT TRANSPORT

In studying transport processes in turbulent flows, it is reasonable to ask whether the transport processes of different quantities such as momentum, heat, matter, and turbulence energy can be analogous. If such an analogy existed between any pair of quantities, similar relations would be obtained for these quantities, and it might even be possible to express the parameters determining the transport of one quantity in terms of the parameters determining the transport of the other quantity. The question of the existence of analogies was, therefore, raised very early and has been answered more or less satisfactorily on the basis of existing theories.

If we want to discuss the problem of analogies, we have first of all to agree on what exactly is understood by an analogy.

We assume that the transport of a transferable quantity (say, momentum, heat, matter, turbulence energy) in a turbulent flow field may be described by means of differential equations. In that case a transport analogy holds if the corresponding differential equations and initial and boundary conditions are identical.

The differential equation describing transport in a turbulent flow field consists of various terms that give the respective contributions made to the total change in the quantity considered by: local change with time; convection by mean motion; turbulence and molecular diffusion; external effects; and, in the case of momentum transport, the mean-pressure gradient.

It is suitable and possible to distinguish between a transport process taken in a broad sense and in a narrow sense. By a *transport process in the broad sense* will be understood the entire process that causes the spread or dispersion in space of the quantity concerned and that determines it as a function of time. In this spread may be included the effects of external influences and sources and sinks which, together with the effect of the flow as such, cause a change in the quantity. These sources and sinks may have their origin in the flow itself. The generation and absorption of heat by a chemical reaction form a source and a sink, respectively, which do not originate directly in the flow itself. On the other hand, heat generated by dissipation of kinetic energy has its origin directly in the flow itself; it acts on the turbulence energy as a continuously distributed sink of energy and on the distribution of heat as a continuously distributed source of heat.

A *transport process in the narrow sense* will be understood to be confined to the *diffusion process* proper of the quantity under consideration.

In Chap. 1 and in the preceding section we have drawn a parallel between turbulent- and molecular-transport processes, both then considered in the narrow sense. It may be useful to consider first the analogies among molecular-transport processes of momentum, heat, and matter. As is well known, the assumption that the molecules behave like rigid spheres leads to the same quantitative results for the coefficient of kinematic viscosity, the coefficient of heat conductivity, and the coefficient of diffusion. But various corrections are necessary to take into account the effects of elasticity of the spheres, repellent forces, persistency of velocities after collisions, etc., effects which are different in the cases of momentum, heat (=molecular energy), and matter, and the three coefficients become unequal. For instance, the effect on the coefficient of diffusion of velocities that persist after collisions is different from the effect of these velocities on viscosity and on heat conductivity. This is because mass is not affected by collisions, but momentum and energy are. The differences between pairs of coefficients find expression in their ratios, for instance, the Prandtl and Schmidt numbers.

In turbulent motions, where large complexes of molecules are moved more or less "bodily," the kinetic energy associated with the motions of these lumps of fluid does not contribute to the internal energy of the fluid, that is, to the heat. Hence it is logical to assume that neither heat nor mass is affected by interaction between these lumps in turbulent motion otherwise than by exchanges of molecular nature alone during the time of contact. If these molecular-exchange effects are neglected, the eddy exchange of heat and matter must be the same. This has been expressed in earlier considerations by applying the same symbol Γ for the two quantities. Collision between lumps of fluid will influence their momentum and turbulence energy only. So, in turbulence-exchange processes, neglecting molecular-exchange effects, we may distinguish among three quantities: momentum, heat or matter, and turbulence energy.

However, it is not always permissible to neglect the molecular-exchange effects—for instance, in the case of transport processes through boundary layers along rigid walls. Even in free turbulence molecular effects may not always be neglected. As we shall discuss later on, the turbulent motions cause a stretching and deformation of the lumps of fluid, thus increasing the concentration and temperature gradients and, hence, the molecular exchange also.

We shall consider first the transport processes in free turbulence. Notwithstanding the above, we shall assume that the transport processes in the narrow sense are mainly turbulent in nature and that the effect of molecular transport is negligible. We shall then consider the analogies among transport of momentum, heat or matter, and turbulence energy in the light of the theories expounded in the preceding section.

For an *incompressible fluid,* the change of velocity is described by Eq. (1-18), which, if we neglect the viscous term, reads

$$\frac{\partial \bar{U}_i}{\partial t} + \bar{U}_j \frac{\partial \bar{U}_i}{\partial x_j} = -\frac{\partial}{\partial x_j}\overline{u_i u_j} + \bar{F}_i - \frac{1}{\rho}\frac{\partial \bar{P}}{\partial x_i} \tag{5-33}$$

Here $\bar{F}_i - (1/\rho)(\partial \bar{P}/\partial x_i)$ forms the total driving force per unit of mass. The equation for the change of Γ according to Eq. (1-33), again neglecting molecular effects, reads

$$\frac{\partial \bar{\Gamma}}{\partial t} + \bar{U}_j \frac{\partial \bar{\Gamma}}{\partial x_j} = -\frac{\partial}{\partial x_j}\overline{u_j \gamma} + \bar{F}_\gamma \tag{5-34}$$

where $\bar{F}_\gamma$ denotes the "driving force" for Γ.

From Eq. (1-110), for the change of the turbulence kinetic energy plus the term p/ρ we obtain

$$\frac{\partial}{\partial t}\overline{\left(\frac{p}{\rho} + \frac{q^2}{2}\right)} + \bar{U}_j \frac{\partial}{\partial x_j}\overline{\left(\frac{p}{\rho} + \frac{q^2}{2}\right)} = -\frac{\partial}{\partial x_j}\overline{u_j\left(\frac{p}{\rho} + \frac{q^2}{2}\right)} - \overline{u_i u_j}\frac{\partial \bar{U}_i}{\partial x_j}$$
$$- \nu \overline{\left(\frac{\partial u_i}{\partial x_j} + \frac{\partial u_j}{\partial x_i}\right)\frac{\partial u_j}{\partial x_i}} \tag{5-35}$$

In accord with the above, we have neglected the term representing the work done per unit of mass and time by the viscous shear stresses, but not the dissipation term.

If we compare Eqs. (5-33) and (5-34) we notice that they are similar except for the pressure term $(1/\rho)(\partial \bar{P}/\partial x_i)$ in Eq. (5-33). This difference expresses the fact that momentum transport, unlike the transport of Γ, is influenced by pressure differences.

In free turbulence, however, the pressure gradient $\partial \bar{P}/\partial x_i$ is usually so small that this term may be neglected; in this case the equations are identical if $\bar{F}_i$ and $\bar{F}_\gamma$ are identical. We may note that, if heat is the quantity transported, $\bar{F}_\gamma$ consists at least of the heat generated by dissipation of turbulence kinetic energy.

In Eq. (5-35) the last two terms on the right-hand side, denoting the production and dissipation of turbulence energy, may be regarded as a continuously distributed source and sink, respectively. It is obvious that no complete analogy between the transport of turbulence energy on the one side, and momentum or heat or mass on the other side is possible, since it would hardly be feasible to obtain identical functions for the driving forces $-\overline{u_i u_j}\,\partial\bar{U}_i/\partial x_j$, $-\overline{\nu(\partial u_i/\partial x_j + \partial u_j/\partial x_i)\,\partial u_j/\partial x_i}$ and $\bar{F}_i$ or $\bar{F}_\gamma$.

It may thus be concluded that, in free turbulent flows, no complete analogy in transport in the broad sense among momentum, heat or matter, and turbulence energy is possible. Only in the absence of the "driving forces" $\bar{F}_i$ and $\bar{F}_\gamma$ (thus neglecting heat generation by dissipation) is analogy possible between the transport of momentum and heat or matter.

If we consider only the transport in the narrow sense, that is, the transport by eddy diffusion alone, then there may be analogy among transports of momentum, Γ, and turbulence energy, if the mechanisms of transport that determine the form of the terms:

$$\frac{\partial}{\partial x_j}\overline{u_j u_i} \qquad \frac{\partial}{\partial x_j}\overline{u_j \gamma} \quad \text{and} \quad \frac{\partial}{\partial x_j}\overline{u_j\left(\frac{p}{\rho}+\frac{q^2}{2}\right)}$$

are identical in all three cases.

Let us now consider these three diffusion terms in the light of the theories expounded in the previous section.

According to Boussinesq, we may introduce the coefficients of eddy diffusion ϵ_m, ϵ_γ, and ϵ_q, such that

$$\overline{u_j u_i} = -\epsilon_m\left(\frac{\partial \bar{U}_i}{\partial x_j}+\frac{\partial \bar{U}_j}{\partial x_i}\right)$$

$$\overline{u_j \gamma} = -\epsilon_\gamma \frac{\partial \Gamma}{\partial x_j}$$

$$\overline{u_j\left(\frac{p}{\rho}+\frac{q^2}{2}\right)} = -\epsilon_q \frac{\partial}{\partial x_j}\overline{\left(\frac{p}{\rho}+\frac{q^2}{2}\right)} = -\epsilon_q \frac{\partial}{\partial x_j}\frac{\overline{q^2}}{2}$$

For an incompressible fluid and for constant values of the coefficients of eddy diffusion, the corresponding diffusion terms become

$$\frac{\partial}{\partial x_j}\overline{u_j u_i} = -\epsilon_m\left(\frac{\partial^2 \bar{U}_i}{\partial x_j\,\partial x_j}+\frac{\partial^2 \bar{U}_j}{\partial x_j\,\partial x_i}\right) = -\epsilon_m \frac{\partial^2 \bar{U}_i}{\partial x_j\,\partial x_j}$$

$$\frac{\partial}{\partial x_j}\overline{u_j \gamma} = -\epsilon_\gamma \frac{\partial^2 \Gamma}{\partial x_j\,\partial x_j}$$

$$\frac{\partial}{\partial x_j}\overline{u_j\left(\frac{p}{\rho}+\frac{q^2}{2}\right)} = -\epsilon_q \frac{\partial^2}{\partial x_j\,\partial x_j}\frac{\overline{q^2}}{2}$$

So we conclude that there is complete analogy. Of course ϵ_m, ϵ_γ, and ϵ_q may be different, resulting in a quantitative difference in the rate of transport among momentum, Γ, and turbulence energy. One may introduce an eddy Prandtl number, an eddy Schmidt number, and a similar number for the ratio ϵ_m/ϵ_q.

If we consider the mixing-length theories according to Prandtl and Taylor, we have to distinguish between two-dimensional and three-dimensional flows.

For plane flow we find that, according to Prandtl's momentum-transport theory and according to Taylor's vorticity-transport theory, complete analogy exists [see, e.g., Eqs. (5-11) and (5-14)] if the various mixing lengths are independent of the lateral direction x_2, except for quantitative differences due to different values for the mixing lengths $\mathfrak{l}_m$ or $\mathfrak{l}_\omega$, $\mathfrak{l}_\gamma$, and $\mathfrak{l}_q$.

In the three-dimensional case the eddy-diffusion terms are identical for momentum, Γ, and turbulence energy according to Prandtl's momentum-transport theory. But not according to Taylor's modified vorticity-transport theory. This can easily be shown for the axi-symmetric case. For this case we may derive for the diffusion term

$$\frac{1}{r}\mathfrak{l}_\omega^2\left|\frac{d\bar{U}_x}{dr}\right|\frac{d}{dr}\left(r\frac{d\bar{U}_x}{dr}\right)$$

whereas, for the quantities Γ and turbulence energy, the turbulence-diffusion terms read

$$\frac{1}{r}\frac{d}{dr}\left(r\mathfrak{l}_\gamma^2\left|\frac{d\bar{U}_x}{dr}\right|\frac{d\bar{\Gamma}}{dr}\right)$$

and

$$\frac{1}{r}\frac{d}{dr}\left(r\mathfrak{l}_q^2\left|\frac{d\bar{U}_x}{dr}\right|\frac{d}{dr}\frac{\overline{q^2}}{2}\right)$$

Thus, according to Taylor's modified vorticity-transport theory, there is no analogy in transport in the narrow sense between momentum on the one side and Γ or turbulence energy on the other; between Γ and turbulence energy there is analogy.

The experiments concerning the velocity, Γ, and turbulence-energy distributions in free jets and in wakes (see Chap. 6) have shown that the analogies according to Boussinesq's and Prandtl's momentum-transfer theory agree with experiment only within the continuous completely turbulent regions, and then only approximately. For the transport of turbulence energy the agreement is only very approximate. According to these two theories the ratios $\epsilon_\gamma/\epsilon_m$ and ϵ_q/ϵ_m should be constant. The empirical value of $\epsilon_\gamma/\epsilon_m$, however, is only approximately constant. The ratio ϵ_q/ϵ_m appears not to be constant at all.

If we have to do with flow regions such as the boundary regions of free jets and of wakes, where the turbulence has an intermittent character, it will be difficult to draw conclusions about the existence of analogies on the basis of differential equations that assume continuity of the turbulence in the region considered. If we insist, nevertheless, on describing the processes in these boundary regions by means of differential equations, we must at least introduce some kind of intermittency factor to correct for the intermittent character of the turbulence. In that case the above results are not affected, since the intermittency factor occurs in all three coefficients ϵ_m, ϵ_γ, and ϵ_q and, consequently, drops out of the ratios of these coefficients. There may be a difference in the transport of momentum and that of heat or matter in this intermittent region, due to a subtle difference in transport mechanism, though (see Sec. 6-12).

If we consider more closely the mechanism of transport in free turbulence, in particular if we consider such factors as gradient-type diffusion and large-scale-turbulence convection transport, the analogy in the narrow sense among the transport processes of momentum, Γ, and turbulence energy no longer holds. This is clearly demonstrated by the expressions (5-24), (5-25), and (5-26) for these three quantities, respectively.

As already mentioned, it is no longer justifiable to neglect the effect of molecular transport in transport processes through boundary layers along fixed walls. These molecular effects on transport of heat and on transport of matter are different. Therefore, it is not possible to treat these processes as if they were identical, even if the transport mechanisms of the two quantities are nearly the same. Especially very close to the wall where the flow is almost viscous, differences may become appreciable.

We shall consider first the transport of heat compared with the transport of momentum. For transport in the broad sense we consider the relevant differential equations:

$$\frac{\partial U_i}{\partial t} + U_j \frac{\partial U_i}{\partial x_j} = -\frac{1}{\rho}\frac{\partial P}{\partial x_i} + \nu \frac{\partial^2 U_i}{\partial x_j \partial x_j} + F_i \tag{5-36}$$

and

$$\frac{\partial \Theta}{\partial t} + U_j \frac{\partial \Theta}{\partial x_j} = \mathfrak{k}_\theta \frac{\partial^2 \Theta}{\partial x_j \partial x_j} + F_\theta \tag{5-37}$$

The "driving force" F_θ includes the heat generated by the dissipation. $\mathfrak{k}_\theta = \kappa/\rho c_p$, where κ is the coefficient of heat conduction.

There is analogy in the broad sense if the two equations (5-36) and (5-37) become identical for $U_i = \alpha\Theta$, where α is a constant of proportionality. Thus the equations are identical if

$$-\frac{1}{\rho}\frac{\partial P}{\partial x_i} + F_i = \alpha F_\theta$$

and if

$$\mathbf{Pr} = \frac{\nu}{\mathfrak{k}_\theta} = 1$$

In general it is not possible to satisfy these conditions. Because the dissipation alone cannot account for the pressure term $(1/\rho)(\partial P/\partial x_i)$, analogy is possible only if it is imagined that F_θ also includes heat sources or sinks distributed over the field of flow in such a way that the first condition is satisfied. Since, owing to the turbulence, the pressure P fluctuates, F_θ must also be a fluctuating quantity.

The constant of proportionality α depends on the kind of flow. For flow through a pipe we may obtain α from the condition that the difference between the average temperature of the fluid and the average temperature of the wall must be proportional to the average velocity of the fluid through the pipe and with the same constant of proportionality. Thus

$$\bar{U}_{\text{av}} = \alpha(\bar{\Theta}_{\text{av}} - \bar{\Theta}_w)$$

whence follows the value of α.

For the flow through a pipe we then have analogy between transport of heat and of momentum if

$$F_\theta = -\frac{\bar{\Theta}_{\text{av}} - \bar{\Theta}_w}{\bar{U}_{\text{av}}}\frac{\partial P}{\partial x_i}$$

since $F_i = 0$. This means that heat sources and sinks, the value and distribution of which are determined by $\partial P/\partial x_i$, must be present in the flow field. Prandtl[14] has drawn attention to this fact.

In the boundary layer along a flat plate, the average pressure gradient is zero if the free stream outside the boundary layer has a constant velocity. Schultz[10] has concluded that in this case complete analogy is possible if $\mathbf{Pr} = 1$. However, the dissipation must then be zero and F_θ must be just equal to the fluctuating part of $\partial P/\partial x_i$. If, on the other hand, we apply the analogy to the time-mean value, then analogy is possible only if the turbulence transports of momentum and of heat in the narrow sense are identical ($\mathbf{Pr}_{\text{turb}} = \epsilon_m/\epsilon_\theta = 1$) and if the dissipation can be neglected. In that case no heat sources are required to make the analogy complete.

For transport of mass we have an equation similar to Eq. (5-37). For a constant density ρ and a constant molecular-diffusion coefficient $\mathfrak{k}_C = D$ (or, more precisely, for a constant value of ρD) this equation reads

$$\frac{\partial C}{\partial t} + U_j\frac{\partial C}{\partial x_j} = D\frac{\partial^2 C}{\partial x_j \partial x_j} + F_C \qquad (5\text{-}38)$$

where C is the concentration per unit of mass. The "driving force" F_C may originate, for instance, from chemical reactions.

If we compare Eqs. (5-37) and (5-38), we find that there is complete analogy between the transports of heat and mass in the broad sense only if the driving forces F_θ (including heat production by dissipation and by chemical reactions) and F_C are negligibly small. The points made concerning the analogy between transport of heat and of momentum then apply to the analogy between transport of mass and of momentum.

We have not yet considered the boundary conditions at the wall. Perfect analogy in transport processes through boundary layers requires also that the boundary conditions at the wall be identical. If we consider transport of matter compared with transport of heat, the boundary conditions may be assumed identical only if the partial pressure of the diffusing matter is small. Assume, for instance, the evaporation of a liquid from a surface in a gas flowing along this surface. The transport of vapor by molecular diffusion reads

$$\mathfrak{I}_{\text{vap}} = -\rho D_{\text{vap}} \frac{\partial C_{\text{vap}}}{\partial x_2}$$

where C_{vap} is the concentration of the vapor, D_{vap} is the molecular diffusion coefficient, and ρ is the density of the mixture of vapor and gas. For the transport of the gas by molecular diffusion we have a similar relation

$$\mathfrak{I}_{\text{gas}} = -\rho D_{\text{gas}} \frac{\partial C_{\text{gas}}}{\partial x_2} = \rho D_{\text{gas}} \frac{\partial C_{\text{vap}}}{\partial x_2}$$

since in a binary system $C_{\text{gas}} + C_{\text{vap}} = 1$. Moreover, for a binary system we have $D_{\text{gas}} = D_{\text{vap}}$; so we may write D for short for the two diffusion coefficients.

Thus at the wall there would be a transport of gas $\mathfrak{I}_{\text{gas}}$ by molecular diffusion. If the wall is not porous and so cannot transmit the gas, the transport of gas at the wall must be zero. This requires that a convective transport by a current perpendicular to the wall takes place, that just compensates for the transport of gas by molecular diffusion. Consequently the total transport of gas

$$\rho D \frac{\partial C_{\text{vap}}}{\partial x_2} + \rho C_{\text{gas}} U_2 = 0$$

at the wall. Whence follows for the convective current

$$U_2 = -\frac{D}{C_{\text{gas}}} \frac{\partial C_{\text{vap}}}{\partial x_2}$$

This current U_2 also causes a convective transport of vapor; so the total transport of vapor reads

$$-\rho D \frac{\partial C_{\text{vap}}}{\partial x_2} + \rho C_{\text{vap}} U_2 = -\rho D \left(1 + \frac{C_{\text{vap}}}{C_{\text{gas}}}\right) \frac{\partial C_{\text{vap}}}{\partial x_2}$$

We may express C_{vap} and C_{gas} in terms of the partial pressures P_{vap} and $P_{\text{gas}} = P - P_{\text{vap}}$, respectively, by means of the gas law

$$C_{\text{vap}} = \frac{M_{\text{vap}}}{\rho \mathfrak{R} \Theta} P_{\text{vap}} \qquad C_{\text{gas}} = \frac{M_{\text{gas}}}{\rho \mathfrak{R} \Theta} P_{\text{gas}}$$

where M_{vap} and M_{gas} are the molar masses of the vapor and the gas, and where $\mathfrak{R}$ is the absolute gas constant. The expression for the total transport of vapor then becomes

$$\text{Transport of vapor} = -\rho D\left(1 + \frac{M_{\text{vap}}}{M_{\text{gas}}}\frac{P_{\text{vap}}}{P - P_{\text{vap}}}\right)\frac{\partial C_{\text{vap}}}{\partial x_2} \tag{5-39}$$

If the partial pressure of the vapor becomes appreciable, the second term on the right-hand side of Eq. (5-39) also becomes important and may no longer be neglected.

In view of this difference between transport of heat and of matter, there is no perfect analogy possible between the two.

We may also express $\partial C_{\text{vap}}/\partial x_2$ in terms of $\partial P_{\text{vap}}/\partial x_2$, by means of the above gas law. In carrying out the differentiation we must keep in mind that $\rho\Re\Theta = M_{\text{vap}}P_{\text{vap}} + M_{\text{gas}}(P - P_{\text{vap}})$ so that ρ is still a function of x_2. Thus we obtain

$$\frac{\partial C_{\text{vap}}}{\partial x_2} = \frac{M_{\text{gas}}M_{\text{vap}}P}{(\rho\Re\Theta)^2}\frac{\partial P_{\text{vap}}}{\partial x_2}$$

and substitution into Eq. (5-39) yields

$$\text{Transport of vapor} = -\frac{DM_{\text{vap}}}{\rho\Re\Theta}\frac{P}{P - P_{\text{vap}}}\frac{\partial P_{\text{vap}}}{\partial x_2} \tag{5-39a}$$

In passing it may be remarked that, since in the boundary layer P_{vap} varies with x_2, ρD is no longer an exact constant, and so the differential equation (5-38) no longer holds in an exact way.

Nusselt[12] was, perhaps, the first to focus attention on this effect of large partial pressures on the phenomena at the wall boundary. Eckert and Lieblein[13] have made approximate calculations of the evaporation for the transport through a laminar boundary layer taking this effect into account; their work has been followed by more exact calculations made by Schuh,[63] by Spalding,[62] and by Mickley, Ross, Squyers and Stewart.[64]

5-4 TRANSPORT BY TURBULENT DIFFUSION

In the mixing-length theories the transport of a transferable quantity by lumps of fluid crossing a control plane in a turbulent flow field is considered. The transport rate through such a plane perpendicular to the x_2-direction is given by the values of the one-point Eulerian correlations $\overline{u_2u_1}$ and $\overline{u_2\gamma}$ for the transport of momentum and for a scalar quantity, respectively. We know that it has become customary to express this rate of transport in terms of a coefficient of eddy diffusion and of the local gradient of the quantity under consideration and that the coefficient of eddy diffusion may be assumed to be equal to the product of a mixing length and the local gradient of the mean velocity. For instance [see Eqs. (5-10) and (5-11)],

$$\epsilon_\gamma = -\overline{u_2L_\gamma} = u_2'\mathrm{l}_\gamma^* = \mathrm{l}_\gamma{}^2\left|\frac{d\bar{U}_1}{dx_2}\right|$$

For reasons of simplicity we disregard the subscript 2 in $(\epsilon_\gamma)_2$, as given in Eq. (5-11).

The lumps of fluid that cross the control plane have transported the transferable quantity over distances L_γ over which the quantity is assumed to be conserved.

We may consider the wandering of the lumps through the distances L_γ more closely by applying to it Taylor's theory of diffusion of fluid particles, expounded in Sec. 1-9. It should then be possible to relate the Eulerian mixing length l_γ to the Lagrangian integral scale $(\Lambda_L)_\gamma$, since we have found in Sec. 1-9 that

$$\epsilon_\gamma = u_2'^2 \int_0^\infty d\tau \, f(\alpha,\tau)\mathbf{R}_L(\tau) = u_2'(\Lambda_L)_\gamma$$

We recall that $f(\alpha,\tau)$ accounts for the effect of exchange of a fluid lump with its surrounding during its path.

From comparison with the former expression for ϵ_γ we conclude that $\mathrm{l}_\gamma^* = (\Lambda_L)_\gamma$ and that l_γ is of the same order as $(\Lambda_L)_\gamma$.

We have also the following relation:

$$\overline{u_2\gamma} = u_2'\gamma'\mathbf{R}_{2,\gamma}(0) = -\epsilon_\gamma \frac{d\bar{\Gamma}}{dx_2}$$

whence follows the formal relation between the Eulerian one-point correlation coefficient $\mathbf{R}_{2,\gamma}(0)$ and the Lagrangian correlation coefficient $\mathbf{R}_L(\tau)$:

$$\mathbf{R}_{2,\gamma}(0) = \frac{u_2'}{\gamma'}\frac{d\bar{\Gamma}}{dx_2}\int_0^\infty d\tau \, f(\alpha,\tau)\mathbf{R}_L(\tau) \qquad (5\text{-}40)$$

We may obtain a connection between the mixing-length theory and Taylor's theory. To this end consider a control plane at x_2 perpendicular to the x_2-direction. We assume that fluid particles travel in the x_2-direction, transporting with them the property $\mathscr{P}$. They produce a transport of this property through the control plane equal to

$$\overline{v_2\mathfrak{p}(t)} = \frac{1}{T}\int_0^T dt_0 \mathfrak{p}(t_0,t)v_2(t_0) \qquad (5\text{-}41)$$

where t_0 is the instant that a fluid particle passes the plane at x_2, and t is the time of traveling, i.e., the diffusion time. According to the mixing-length theory, this fluid particle has carried its property from a point at a distance y_2 to the plane at x_2. Assuming a linear variation of $\bar{\mathscr{P}}$ with x_2, according to this theory, and disregarding any exchange of the fluid particle with its surrounding during its path, we have

$$\mathfrak{p}(t_0,t) = -y_2(t_0,t)\frac{d\bar{\mathscr{P}}}{dx_2}$$

Since methods A and B are essentially the same, we may put

$$v_2(t_0) = \frac{d}{dt}y_2(t_0,t)$$

and obtain

$$\overline{v_2 \mathfrak{p}} = -\frac{d\bar{\mathscr{P}}}{dx_2}\frac{1}{T}\int_0^T dt_0\, y_2(t_0,t)\frac{d}{dt}\,y_2(t_0,t) = -\frac{d\bar{\mathscr{P}}}{dx_2}\frac{1}{2}\frac{d}{dt}\,\overline{y_2^{\,2}}(t)$$

If again we put by definition

$$\overline{v_2 \mathfrak{p}} = -\epsilon_{\mathscr{P}}\frac{d\bar{\mathscr{P}}}{dx_2}$$

we obtain

$$\epsilon_{\mathscr{P}}(t) = \epsilon(t) = \frac{1}{2}\frac{d}{dt}\,\overline{y_2^{\,2}}(t) = \overline{y_2(t)v_2(t)} \tag{5-42}$$

Hence, with the expression (1-81) for $\overline{y_2^{\,2}}(t)$,

$$\epsilon_{\mathscr{P}}(t) = \epsilon(t) = \overline{v_2^{\,2}}\int_0^t d\tau\, \mathbf{R}_L(\tau) \tag{5-43}$$

The expressions (5-42) and (5-43) show that, for short diffusion times, the coefficient of eddy diffusion is still a function of time. In the mixing-length theory we have to consider long diffusion times for the fluid particles, and Eq. (5-43) then reduces to

$$[\epsilon_{\mathscr{P}}]_{t=\infty} = \epsilon = \overline{v_2^{\,2}}\int_0^\infty d\tau\, \mathbf{R}_L(\tau) = v_2'\Lambda_L \tag{1-87}$$

Here we have obtained for the diffusion coefficient $\epsilon_{\mathscr{P}}$ the same value as for the fluid particles ϵ, since we have not yet considered any loss of the property $\mathscr{P}$ during the travel of the fluid particle over the distance y_2.

Burgers[15] has tried to estimate the effect of this exchange with the environment, for such small fluid particles that a linear relationship for the rate of exchange may be assumed. Thus,

$$\frac{d\mathscr{P}(t')}{dt'} = \alpha_{\mathscr{P}}[\bar{\mathscr{P}}(t') - \mathscr{P}(t')] \tag{5-44}$$

Here $\bar{\mathscr{P}}(t')$ is the average value of $\mathscr{P}$ in the direct environment of the particle at the instant t' (see Fig. 5-1), and not the value of $\mathscr{P}$ in this environment at this instant. This is in accord with the mixing-length theory as explained in Sec. 5-2. The coefficient of exchange $\alpha_{\mathscr{P}}$ is assumed to be constant.

Now we consider the difference $\mathfrak{p}(t')$ between the instantaneous value of the particle property $\mathscr{P}(t')$ and the average value $\bar{\mathscr{P}}(x_2 = 0)$ i.e.,

$$\mathfrak{p}(t') = \mathscr{P}(t') - \bar{\mathscr{P}}(x_2 = 0)$$

Equation (5-44) then reads

$$\frac{d\mathfrak{p}(t')}{dt'} = \alpha_{\mathscr{P}}[\bar{\mathscr{P}}(t') - \bar{\mathscr{P}}(x_2 = 0) - \mathfrak{p}(t')]$$

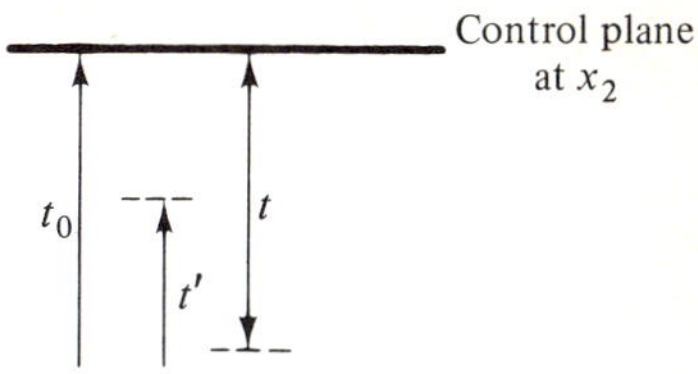

FIGURE 5-1
Time coordinates of a fluid particle which passes the control plane at time t_0 and has traveled during a time t.

In accord with the assumption of a gradient-type of diffusion we put

$$\bar{\mathscr{P}}(t') = \bar{\mathscr{P}}[-x_2(t_0 - t')] = \bar{\mathscr{P}}(x_2 = 0) - \left(\frac{d\bar{\mathscr{P}}}{dx_2}\right)_{x_2=0} \cdot y_2(t_0 - t')$$

$$= \bar{\mathscr{P}}(x_2 = 0) - \left(\frac{d\bar{\mathscr{P}}}{dx_2}\right)_{x_2=0} \cdot \int_{t'}^{t_0} dt''\, v_2(t'')$$

The differential equation for $\mathfrak{p}(t')$ thus becomes

$$\frac{d\mathfrak{p}(t')}{dt'} = \alpha_{\mathscr{P}}\left[-\left(\frac{d\bar{\mathscr{P}}}{dx_2}\right)_{x_2=0} \cdot \int_{t'}^{t_0} dt''\, v_2(t'') - \mathfrak{p}(t')\right]$$

Its solution reads

$$\mathfrak{p}(t_0) = -\alpha_{\mathscr{P}}\left(\frac{d\bar{\mathscr{P}}}{dx_2}\right)_{x_2=0} \cdot \int_{t_0-t}^{t_0} dt' \exp\left[-\alpha_{\mathscr{P}}(t_0 - t')\right] \cdot \int_{t'}^{t_0} dt''\, v_2(t'')$$
$$+ \mathfrak{p}(t_0 - t)\exp(-\alpha_{\mathscr{P}} t)$$

Or, after integration by parts,

$$\mathfrak{p}(t_0) = -\left(\frac{d\bar{\mathscr{P}}}{dx_2}\right)_{x_2=0} \cdot \left[\int_{t_0-t}^{t_0} dt' \exp\left[-\alpha_{\mathscr{P}}(t_0 - t')\right] v_2(t')\right.$$
$$\left. - \exp(-\alpha_{\mathscr{P}} t)\int_{t_0-t}^{t_0} dt'\, v_2(t')\right] + \mathfrak{p}(t_0 - t)\exp(-\alpha_{\mathscr{P}} t)$$

Here t is the total diffusion time.

If, again in accord with the mixing-length theory, we put for the subensemble of particles starting from the position $x_2 = -y_2(t_0 - t)$,

$$\mathfrak{p}(t_0 - t) = \bar{\mathscr{P}}(t_0 - t) - \bar{\mathscr{P}}(x_2 = 0) = -\left(\frac{d\bar{\mathscr{P}}}{dx_2}\right)_{x_2=0} \cdot \int_{t_0-t}^{t_0} dt'\, v_2(t')$$

we obtain

$$\mathfrak{p}(t_0) = -\left(\frac{d\bar{\mathscr{P}}}{dx_2}\right)_{x_2=0} \cdot \int_{t_0-t}^{t_0} dt' \exp\left[-\alpha_{\mathscr{P}}(t_0 - t')\right] v_2(t')$$

Equation (5-41) then becomes

$$(\mathfrak{I}_{\mathscr{P}})_2 = \rho\overline{v_2 \mathfrak{p}}(t) = -\rho\left(\frac{d\bar{\mathscr{P}}}{dx_2}\right)_{x_2=0} \cdot \frac{1}{T}\int_0^T dt_0 \int_{t_0-t}^{t_0} dt' \exp\left[-\alpha_{\mathscr{P}}(t_0 - t')\right] v_2(t') v_2(t_0)$$

With $t_0 - t' = \tau$, and with the Lagrangian auto-correlation coefficient

$$\mathbf{R}_L(\tau) = \frac{\overline{v_2(t_0) v_2(t_0 - \tau)}}{\overline{v_2^2}}$$

the expression for $(\mathfrak{I}_{\mathscr{P}})_2(t)$ becomes

$$(\mathfrak{I}_{\mathscr{P}})_2(t) = -\rho\left(\frac{d\bar{\mathscr{P}}}{dx_2}\right)_{x_2=0} \cdot \overline{v_2^2}\int_0^t d\tau \exp(-\alpha_{\mathscr{P}}\tau)\mathbf{R}_L(\tau) \qquad (5\text{-}45)$$

Consequently

$$\epsilon_{\mathscr{P}}(t) = \overline{v_2^2}\int_0^t d\tau \exp(-\alpha_{\mathscr{P}}\tau)\mathbf{R}_L(\tau) \qquad (5\text{-}46)$$

The expressions (5-45) and (5-46) refer to the contribution of particles with finite diffusion times only. When we consider the total average transport by turbulence through the control plane, we have to include all possible diffusion times. Actually we may suffice with a total diffusion time large compared with the Lagrangian integral time scale. Hence

$$\epsilon_{\mathscr{P}} = \overline{v_2^2}\int_0^\infty d\tau \exp(-\alpha_{\mathscr{P}}\tau)\mathbf{R}_L(\tau) \qquad (5\text{-}47)$$

When we introduce an effective Lagrangian integral time scale for the property $\mathscr{P}$, defined by

$$(\Lambda_L)_{\mathscr{P}} = v_2'\int_0^\infty d\tau \exp(-\alpha_{\mathscr{P}}\tau)\mathbf{R}_L(\tau) \qquad (5\text{-}48)$$

we may again write

$$\epsilon_{\mathscr{P}} = v_2'(\Lambda_L)_{\mathscr{P}}$$

For $\alpha_{\mathscr{P}} \to 0$, the expressions (5-47) and (5-48) reduce to Eqs. (1-86) and (1-88), respectively.

It may be emphasized that the relation (5-47) is based on the assumption that $\bar{\mathscr{P}}$ is a linear function of x_2 over distances comparable to the Lagrangian integral scale.

From Eq. (5-47) it may be inferred that, if the exchange coefficient α is very large, $\exp(-\alpha_{\mathscr{P}}\tau) \simeq 0$ even for moderate values of τ. In that case the Lagrangian coefficient $\mathbf{R}_L(\tau)$ may have deviated only slightly from $\mathbf{R}_L(0) = 1$. In the extreme case we would have

$$\epsilon_{\mathscr{P}} \simeq \overline{v_2^2}\mathbf{R}_L(0)\int_0^\infty d\tau \exp(-\alpha_{\mathscr{P}}\tau) \simeq \frac{\overline{v_2^2}}{\alpha_{\mathscr{P}}}$$

Hence all frequencies of the turbulence contribute to the eddy-diffusion coefficient. This is, of course, also well demonstrated if we introduce the Lagrangian energy-spectrum function $\mathbf{E}_L(n)$:

$$\mathbf{R}_L(\tau) = \frac{1}{\overline{v_2^2}} \int_0^\infty dn\, \mathbf{E}_L(n) \cos 2\pi n\tau \tag{5-49}$$

Introducing this expression for $\mathbf{R}_L(\tau)$ into Eq. (5-47), performing the integration with respect to t, and keeping in mind that $\exp(-\alpha_{\mathscr{P}} t) \simeq 0$, we obtain

$$\epsilon_{\mathscr{P}} = \int_0^\infty dn\, \mathbf{E}_L(n) \frac{\alpha_{\mathscr{P}}}{\alpha_{\mathscr{P}}^2 + 4\pi^2 n^2} \tag{5-50}$$

This result indeed shows that, for α very large,

$$\epsilon_{\mathscr{P}} \simeq \frac{1}{\alpha_{\mathscr{P}}} \int_0^\infty dn\, \mathbf{E}_L(n) = \frac{\overline{v_2^2}}{\alpha_{\mathscr{P}}}$$

so that all frequencies contribute.

Let us now assume that $\alpha_{\mathscr{P}}$ is very small. This implies that we have to consider very large diffusion times. Assume $2\pi n/\alpha_{\mathscr{P}} = \beta$. From Eq. (5-50) we obtain

$$\epsilon_{\mathscr{P}} = \frac{1}{2\pi} \int_0^\infty d\beta\, \mathbf{E}_L\left(\frac{\alpha_{\mathscr{P}}\beta}{2\pi}\right) \frac{1}{1+\beta^2}$$

For large β, the factor $(1+\beta^2)^{-1}$ is very small. Hence we may expect a contribution to the integral only for moderate values of β. But for $\alpha_{\mathscr{P}} \to 0$ a moderate value of β can be obtained only when n is small and, consequently, so is $\alpha_{\mathscr{P}}\beta/2\pi$.

Hence, for the extreme case $\alpha_{\mathscr{P}} \to 0$, we may put

$$\epsilon_{\mathscr{P}} \simeq \mathbf{E}_L(0) \frac{1}{2\pi} \int_0^\infty d\beta \frac{1}{1+\beta^2} = \tfrac{1}{4}\mathbf{E}_L(0)$$

This shows that, for large diffusion times, only the small frequencies contribute to the value of the eddy-diffusion coefficient.

If we compare the expression (5-47) for $\epsilon_{\mathscr{P}}$, which is based upon the linear exchange rate (5-44), with the more general expression

$$\epsilon_{\mathscr{P}} = u_2'^2 \int_0^\infty d\tau\, f(\alpha_{\mathscr{P}},\tau)\mathbf{R}_L(\tau)$$

apparently we have $f(\alpha_{\mathscr{P}},\tau) = \exp(-\alpha_{\mathscr{P}}\tau)$. It is reasonable to expect a linear exchange rate when the exchange with the surroundings is of molecular nature. This implies that the fluid particle considered must be sufficiently small.

An essential difficulty in applying the above Burgers's model to actual turbulence diffusion processes is the evaluation of the exchange coefficient $\alpha_{\mathscr{P}}$. Burgers assumed $\alpha_{\mathscr{P}}$ to be determined solely by molecular effects, which implies a Reynolds number based on the relative velocity of the particle with respect to its new environment, much smaller than one. When we assume a spherical fluid-particle, or an equivalent sphere with diameter d which has the same exchange rate as the actual, nonspherical, fluid particle, the equation for the exchange process in the case of a scalar property may be written

$$\frac{\pi}{6} d^3 \frac{d\Gamma}{dt} = h_\gamma \pi d^2 (\bar{\Gamma} - \Gamma)$$

whence follows

$$\alpha_\gamma = 6 \frac{h_\gamma}{d}$$

For vanishing value of the particle Reynolds number molecular diffusion alone would yield

$$\frac{h_\gamma d}{\mathfrak{k}_\gamma} = 2$$

so that

$$\alpha_\gamma = 12 \frac{\mathfrak{k}_\gamma}{d^2} \tag{5-51}$$

Here $\mathfrak{k}_\gamma$ is the molecular transport or diffusion coefficient. However, when $\mathscr{P}$ is momentum the simple molecular diffusion process just given for a scalar property can hardly be accepted, mainly because the velocity and momentum of the fluid particle form part of the flow field itself. Consequently an initial, almost discontinuous change of momentum between particle and environment will not occur. If the particle were a rigid sphere, Stokes's viscous resistance equal to $3\pi\mu d(\bar{v} - v)$ could be applied. The viscous, pressure forces at the interface contribute by $\frac{1}{3}$ to this resistance, the other $\frac{2}{3}$ is due to the viscous shear-stress. They are a result of the conditions of zero normal and tangential velocity at the rigid interface. This condition can not be imposed on a fluid particle. Yet if we assume for the initial condition a large velocity gradient at the "interface" of a "spherical" fluid particle, we may estimate the value of α_m for momentum by the expression

$$\alpha_m = A \cdot \frac{\nu}{d^2} \tag{5-52}$$

For a rigid sphere $A = 18$.

For large diffusion times, so that the behavior of $\mathbf{R}_L(\tau)$ for small τ is not of importance, we may approximate $\mathbf{R}_L(\tau)$ by a simple exponential function $\exp(-\tau/\mathfrak{I}_L)$. Substitution of this function for $\mathbf{R}_L(\tau)$ in Eq. (5-47) yields for a scalar and momentum respectively:

$$\epsilon_\gamma = \frac{\overline{v_2{}^2}\mathfrak{I}_L}{1 + \alpha_\gamma \mathfrak{I}_L} = \frac{\epsilon}{1 + (12/d^2)\mathfrak{k}_\gamma \mathfrak{I}_L} \tag{5-53}$$

and

$$\epsilon_m = \frac{\overline{v_2{}^2}\mathfrak{I}_L}{1 + \alpha_m \mathfrak{I}_L} = \frac{\epsilon}{1 + (A/d^2)\nu \mathfrak{I}_L} \tag{5-54}$$

For the ratio $\epsilon_m/\epsilon_\gamma$, which is a turbulence Schmidt number $(\mathbf{Sc})_{\text{turb}}$, we thus obtain

$$(\mathbf{Sc})_{\text{turb}} = \frac{\epsilon_m}{\epsilon_\gamma} = \frac{1 + \dfrac{12}{A}\dfrac{1}{\mathbf{Sc}}\dfrac{A}{d^2}\nu\mathfrak{I}_L}{1 + \dfrac{A}{d^2}\nu\mathfrak{I}_L} \tag{5-55}$$

where $\mathbf{Sc} = \nu/\mathfrak{k}_\gamma$ is the molecular Schmidt number. Hence

$$(\mathbf{Sc})_{\text{turb}} \lesseqgtr 1 \quad \text{when} \quad \mathbf{Sc} \gtreqless \frac{12}{A}$$

Since we may expect the value of A to be of $\mathcal{O}(10)$, we conclude that according to Burgers's model the turbulence Schmidt number becomes smaller than one for fluids with a molecular Schmidt number greater than $\mathcal{O}(1)$, and vice versa.

With the above assumption concerning an equivalent spherical particle to account for the exchange of a property with the particle's environment, we still have not solved the problem of how to apply Burgers's model in a quantitative way. For, the magnitude of the equivalent diameter d is still not known. If we follow the idea behind the definition of a fluid particle, given earlier in Sec. 5-1, we may assume d to be equal to the Kolmogoroff micro length-scale $\eta = (\nu^3/\varepsilon)^{1/4}$. However, it turns out that too large an effect is obtained in this way. The effective path for the transfer of a property attains unacceptably small values even for moderate values of the coefficient of molecular diffusion.

Another possibility might be to take the equivalent diameter proportional to the dissipation length-scale. Thus $d = B \cdot u'(\nu/\varepsilon)^{1/2}$, where B is a numerical constant roughly of the magnitude 5. As we will show later, this estimate of d may yield more realistic results.

In the mixing-length theories we have to do with lumps of fluid comparable in size to the mixing length and, consequently, also to $(\Lambda_L)_{\mathscr{P}}$. In this case it is more reasonable to expect an exchange of a nonlinear character between the wandering lumps of fluid and their surroundings, since this exchange is affected strongly by small-scale turbulent motions. As a matter of fact, Prandtl's estimate of the size of the lumps of fluid responsible for the transport processes was based upon the assumption that the momentum transfer during collisions of the lumps of fluid with their surroundings follows a quadratic relation.

We may extend the considerations for the turbulence transport transverse to the main flow direction to the case that changes of the gradient of the mean property are no longer negligibly small. In the expression (5-41) we may no longer assume that

$$\mathfrak{p}(t_0,t) = -y_2(t_0,t)\frac{d\bar{\mathscr{P}}}{dx_2}$$

but we have to extend the series expansion of $\bar{\mathscr{P}}$, namely

$$\mathfrak{p}(t_0,t) = -y_1(t_0,t)\frac{\partial\bar{\mathscr{P}}}{\partial x_1} - y_2(t_0,t)\frac{\partial\bar{\mathscr{P}}}{\partial x_2}$$

$$+ \tfrac{1}{2}y_1{}^2(t_0,t)\frac{\partial^2\bar{\mathscr{P}}}{\partial x_1{}^2} + \tfrac{1}{2}y_2{}^2(t_0,t)\frac{\partial^2\bar{\mathscr{P}}}{\partial x_2{}^2} + y_1y_2(t_0,t)\frac{\partial^2\bar{\mathscr{P}}}{\partial x_1\,\partial x_2} + \cdots$$

When discussing in Sec. 5-2 the possibility of situations where shear-stress and mean-velocity gradient are of opposite sign, in the theory referred to, the term $y_2{}^2(t_0,t)\,\partial^2\bar{U}_1/\partial x_2{}^2$ has been included.

Here we will assume that the change of $\partial\bar{\mathscr{P}}/\partial x_2$ in main-flow direction, i.e. $\partial^2\bar{\mathscr{P}}/\partial x_1\,\partial x_2$, has to be taken into account in the turbulence transport in transverse direction. Physically this means that this transport is not only determined by the local value of $\partial\bar{\mathscr{P}}/\partial x_2$. Because the change of this gradient is relatively rapid compared with the memory time of the turbulence, the fluid particle crossing the control plane still "remembers" the value of $\partial\bar{\mathscr{P}}/\partial x_2$ upstream of the control plane. Thus, retaining in the series expansion only the terms with $\partial\bar{\mathscr{P}}/\partial x_2$ and $\partial^2\bar{\mathscr{P}}/\partial x_1\,\partial x_2$, and putting

$$y_2(t_0,t) = \int_0^t dt'\, v_2(t_0 - t')$$

we obtain

$$-\overline{v_2\mathfrak{p}} = \int_0^t dt'\, \overline{v_2(t_0)v_2(t_0 - t')\left[\frac{\partial\bar{\mathscr{P}}}{\partial x_2} - y_1(t_0,t)\frac{\partial^2\bar{\mathscr{P}}}{\partial x_1\,\partial x_2}\right]}$$

We may introduce a memory function $M(t)$, such that $\int_0^\infty dt'\, M(t') = 1$. For instance $M(t) = 1/T_m \exp(-t/T_m)$, with T_m the memory or relaxation time.

When t is sufficiently large we may introduce an eddy-diffusion coefficient $\epsilon_{\mathscr{P}}^*$, defined by

$$-\overline{v_2 \mathrm{p}} = \epsilon_{\mathscr{P}}^* \int_0^\infty dt' \frac{\partial \bar{\mathscr{P}}}{\partial x_2}(-t') M(t')$$

If we may apply Taylor's hypothesis of a "frozen" turbulence, $y_1(t_0,t) \simeq \bar{U}_1 T_m$. Furthermore, if at the same time we use a constant coefficient of exchange $\alpha_{\mathscr{P}}$ as in Burgers's model, discussed earlier, we obtain

$$-\overline{v_2 \mathrm{p}} = \left(\frac{\partial \bar{\mathscr{P}}}{\partial x_2} - \bar{U}_1 T_m \frac{\partial^2 \bar{\mathscr{P}}}{\partial x_1\, \partial x_2}\right) \int_0^\infty dt' \exp\left[-\alpha_{\mathscr{P}}(t_0 - t')\right] \overline{v_2(t_0) v_2(t')}$$

$$= \epsilon_{\mathscr{P}}^* \left(\frac{\partial \bar{\mathscr{P}}}{\partial x_2} - \bar{U}_1 T_m \frac{\partial^2 \bar{\mathscr{P}}}{\partial x_1\, \partial x_2}\right)$$

It is reasonable to assume the relaxation time T_m to be approximately equal to the Lagrangian integral time scale $(\mathfrak{I}_L)_{\mathscr{P}} = \int_0^t dt'\, \mathbf{R}_L(t') \exp(-\alpha_{\mathscr{P}} t')$.

Hinze et al.[114] have applied the above relation to the shear-stress $-\overline{u_2 u_1}$ measured in a turbulent boundary-layer where a relatively rapid change in streamwise direction of $\partial \bar{U}_1/\partial x_2$ occurred caused by the wake of a submerged hemispherical cup attached to the wall. The eddy-viscosity ϵ_m^* obtained in this way showed a distribution almost identical to that for the undisturbed boundary-layer.

From Eq. (1-87) it follows that for long periods of time the diffusion of a marked fluid particle is determined by the Lagrangian integral scale Λ_L. Intuitively one would conclude that the diffusion for long periods of time is then determined by the large-scale motions with slow fluctuations. This idea is strongly supported by the result obtained previously, namely that, in the case of a slight exchange rate with the surroundings, the diffusion coefficient for *long diffusion times* is determined mainly by the motions with small frequencies.

Let us consider the effect on its diffusion of the various frequencies in the turbulent motion of a fluid particle. To this end we introduce the relation (5-49) between $\mathbf{R}_L(\tau)$ and $\mathbf{E}_L(n)$ into Eq. (1-82), where in its derivation we had followed method B.

$$\overline{y_2^2} = 2\overline{v_2^2} \int_0^t d\tau (t-\tau) \mathbf{R}_L(\tau) = 2 \int_0^\infty dn\, \mathbf{E}_L(n) \int_0^t d\tau\, (t-\tau) \cos 2\pi n \tau$$
$$= 2 \int_0^\infty dn\, \mathbf{E}_L(n) \frac{1 - \cos 2\pi n t}{4\pi^2 n^2} \qquad \text{(5-56)}$$

For small values of t we obtain

$$\overline{y_2^2} \simeq t^2 \int_0^\infty dn\, \mathbf{E}_L(n)$$

Hence, for small diffusion times, all frequency components of the motion of the fluid particle contribute to $\overline{y_2^2}$.

For large values of t we rewrite the expression for $\overline{y_2^2}$ as follows:[17]

$$\overline{y_2^2} = \frac{t}{\pi} \int_0^\infty d\theta \frac{1 - \cos \theta}{\theta^2} \mathbf{E}_L\left(\frac{\theta}{2\pi t}\right)$$

where $\theta = 2\pi n t$.

The value of $(1 - \cos\theta)/\theta^2$ is very small for large values of θ. Hence there is an appreciable contribution to the integral only from values of the integrand corresponding to moderate values of θ. Consequently we may put $\mathbf{E}_L(\theta/2\pi t) \simeq \mathbf{E}_L(0)$, and then obtain

$$\overline{y_2{}^2} \simeq \frac{t}{\pi}\mathbf{E}_L(0)\int_0^\infty d\theta\,\frac{1-\cos\theta}{\theta^2} = \frac{t}{2}\mathbf{E}_L(0) = 2\overline{v_2{}^2}\mathfrak{I}_L t$$

since from the Fourier transform of Eq. (5-49) it follows that

$$\mathbf{E}_L(0) = 4\overline{v_2{}^2}\int_0^\infty d\tau\,\mathbf{R}_L(\tau) = 4\overline{v_2{}^2}\mathfrak{I}_L$$

The above result is exactly the same as that given by Eq. (1-85), but we have now shown that only the small-frequency components of the motion of the fluid particle contribute to $\overline{y_2{}^2}$.

Thus, for large t the diffusion of a single fluid particle and, consequently, of the center of mass of a lump of fluid is determined mainly by the large-scale motions of the turbulence, and the Eulerian analysis appears to be most suitable for studying the spread and distribution of marked fluid particles in space by introducing a coefficient of diffusion $\epsilon = \overline{y_2{}^2}/2t = \overline{v_2'^2}\mathfrak{I}_L$. Indeed, in the mixing-length theories, for instance, the analysis of the problem is Eulerian. However, we want to emphasize again that in these Eulerian analyses we need the Lagrangian description of the motion of a single fluid particle.

The case of *not large diffusion times* is also of interest, in particular when we consider cases corresponding with *method B*. Even for a stationary, homogeneous, turbulence, the coefficient of eddy diffusion $\epsilon_{\mathscr{P}}$ then is a function of time, as shown by Eq. (5-46). When $\mathbf{R}_L(t)$ is known, the value of $\epsilon_{\mathscr{P}}(t)$ at all diffusion times can be determined with this equation. For *small diffusion times,* we may approximate $\mathbf{R}_L(t)$ by the parabolic relation [compare Eq. (1-66)].

$$\mathbf{R}_L(t) \simeq 1 - \frac{t^2}{\tau_L{}^2} \qquad (5\text{-}57)$$

where τ_L is the Lagrangian dissipation time-scale. For such small diffusion times that the relation (5-57) holds true, we obtain from Eq. (5-46)

$$\epsilon_{\mathscr{P}}(t) = \overline{v_2{}^2}t\left(1 - \frac{t^2}{3\tau_L{}^2}\right) \qquad (5\text{-}58)$$

The value of τ_L will be determined by the more fine-scale structure of the turbulence, i.e., by the quantities ν and ε [see also, e.g., ref. (71)]. These quantities also determine the changes of $v_2(t)$ as caused by the small-eddy structure. Hence we may put

$$\overline{[v_2(t) - v_2(0)]^2} = f(\nu,\varepsilon,t)$$

where an ensemble average is taken. Dimensional analysis results in

$$2\overline{v_2^2}\left[1 - \frac{\overline{v_2(t)v_2(0)}}{\overline{v_2^2}}\right] = \sqrt{\nu\varepsilon}\cdot f\left(t\sqrt{\frac{\varepsilon}{\nu}}\right)$$

Whence follows

$$\mathbf{R}_L(t) = 1 - \frac{\sqrt{\nu\varepsilon}}{2\overline{v_2^2}} f\left(t\sqrt{\frac{\varepsilon}{\nu}}\right)$$

Note that $f(0) = 0$.

Since in a *stationary, homogeneous,* turbulence $\mathbf{R}_L(t)$ is an even function of t, a series expansion of $f(t\sqrt{\varepsilon/\nu})$ yields

$$\mathbf{R}_L(t) = 1 - \frac{f''(0)}{4}\cdot\frac{\varepsilon}{\overline{v_2^2}}\sqrt{\frac{\varepsilon}{\nu}}\cdot t^2 + \cdots$$

Hence

$$\frac{1}{\tau_L^2} = \frac{f''(0)}{4}\frac{\varepsilon}{\overline{v_2^2}}\sqrt{\frac{\varepsilon}{\nu}} = A\cdot\frac{\varepsilon}{\overline{v_2^2}}\sqrt{\frac{\varepsilon}{\nu}} \tag{5-59}$$

However, when $\mathbf{Re}_\lambda$ is sufficiently large we may assume the existence of an inertial subrange where the fine structure is solely determined by ε. For diffusion times where changes of $v_2(t)$ are determined by this inertial subrange, we have

$$\overline{[v_2(t) - v_2(0)]^2} = f_1(\varepsilon,t)$$

Whence, by dimensional analysis $f_1(\varepsilon,t) = C\varepsilon t$. The numerical constant C then must be of $\mathcal{O}(1)$. So, we obtain for $\mathbf{R}_L(t)$ the expression, for values of t corresponding to the inertial subrange

$$\mathbf{R}_L(t) = 1 - C\frac{\varepsilon t}{2\overline{v_2^2}} \tag{5-60}$$

This linear relationship may be considered as the first approximation of an exponential function, which excludes the parabolic behavior of $\mathbf{R}_L(t)$ when $t \to 0$. So, for large $\mathbf{Re}_\lambda$, we assume

$$\mathbf{R}_L(t) = \exp(-t/\mathfrak{I}_L) = 1 - \frac{t}{\mathfrak{I}_L} + \cdots$$

Whence

$$\mathfrak{I}_L = \frac{1}{C}\frac{2\overline{v_2^2}}{\varepsilon} \tag{5-61}$$

When we include a parabolic part of $\mathbf{R}_L(t)$ for $t \to 0$, the two relations Eq. (5-57) and (5-60) will intersect at

$$t^* = C\frac{\varepsilon\tau_L^2}{2\overline{v_2^2}} = \frac{C}{2A}\sqrt{\frac{\nu}{\varepsilon}} \tag{5-62}$$

For $t < t^*$, $\mathbf{R}_L(t)$ follows Eq. (5-57), for $t > t^*$ and large $\mathbf{Re}_\lambda$ the Eq. (5-60) as the first part of an exponential function. The constants A and C are still unknown, except that at least C must be of $\mathcal{O}(1)$. An estimate of A can be made by following a different procedure for evaluating τ_L. For by definition we have

$$\frac{1}{\tau_L{}^2} = \frac{1}{2\overline{v_2{}^2}} \overline{\left(\frac{dv_2}{dt}\right)^2} \tag{5-63}$$

We follow Uberoi and Corrsin,[28] assume isotropy, and start from the momentum-balance equation for the velocity component v_2.

$$\frac{dv_2}{dt} = -\frac{1}{\rho}\frac{\partial p}{\partial x_2} + \nu \frac{\partial^2 u_2}{\partial x_k \partial x_k} \tag{5-64}$$

Since in isotropic turbulence

$$\overline{\frac{\partial p}{\partial x_2}\frac{\partial^2 u_2}{\partial x_k \partial x_k}} = 0$$

from Eq. (5-64) we obtain

$$\overline{\left(\frac{dv_2}{dt}\right)^2} = \frac{1}{\rho^2}\overline{\left(\frac{\partial p}{\partial x_2}\right)^2} + \nu^2 \overline{\frac{\partial^2 u_2}{\partial x_k \partial x_k}\frac{\partial^2 u_2}{\partial x_l \partial x_l}} \tag{5-65}$$

Each term on the right-hand side of Eq. (5-65) can be expressed in terms of the longitudinal-double-correlation coefficient $\mathbf{f}(r)$, the first term by the relation (3-297) and the second term by a relation which is obtained as follows.

Introduce the double-velocity-correlation tensor $(\mathbf{Q}_{i,j})_{A,B} = \overline{(u_i)_A(u_j)_B}$. Then we have

$$\overline{\left(\frac{\partial^2 u_i}{\partial x_k \partial x_k}\right)_A \left(\frac{\partial^2 u_j}{\partial x_l \partial x_l}\right)_B} = \left(\frac{\partial^2}{\partial x_k \partial x_k}\right)_A \left(\frac{\partial^2}{\partial x_l \partial x_l}\right)_B (\mathbf{Q}_{i,j})_{A,B}$$

$$= \frac{\partial^4}{\partial \xi_k \partial \xi_k \partial \xi_l \partial \xi_l}(\mathbf{Q}_{i,j})_{A,B}$$

Consequently,

$$\overline{\frac{\partial^2 u_i}{\partial x_k \partial x_k}\frac{\partial^2 u_j}{\partial x_l \partial x_l}} = \left[\frac{\partial^4}{\partial \xi_k \partial \xi_k \partial \xi_l \partial \xi_l}\mathbf{Q}_{i,j}\right]_{r=0}$$

From Eq. (3-19) we obtain

$$\left[\frac{\partial^4}{\partial \xi_k \partial \xi_k \partial \xi_l \partial \xi_l}\mathbf{Q}_{i,j}\right]_{r=0} = \tfrac{35}{3}\delta_{ij}u'^2 \left[\frac{\partial^4 \mathbf{f}}{\partial r^4}\right]_{r=0} \tag{5-66}$$

whence

$$\overline{\frac{\partial^2 u_2}{\partial x_k \partial x_k}\frac{\partial^2 u_2}{\partial x_l \partial x_l}} = \tfrac{35}{3}u'^2 \left[\frac{\partial^4 \mathbf{f}}{\partial r^4}\right]_{r=0} \tag{5-67}$$

Substitution for the two terms on the right-hand side of Eq. (5-65) by Eqs. (3-297) and (5-67), respectively, gives

$$\overline{\left(\frac{dv_2}{dt}\right)^2} = 4u'^4 \int_0^\infty dr \frac{1}{r}\left(\frac{\partial \mathbf{f}}{\partial r}\right)^2 + \tfrac{35}{3}\nu^2 u'^2 \left[\frac{\partial^4 \mathbf{f}}{\partial r^4}\right]_{r=0} \tag{5-68}$$

In order to evaluate the right-hand side of this equation it is necessary to know the relation for $\mathbf{f}(r)$ over a wide range of values of r. In Chap. 3 we have obtained an exact relation for $\mathbf{f}(r)$ in the complete viscous stage of the turbulence, that is, at very low values of the Reynolds number of turbulence $\mathbf{Re}_\lambda$. And, for very large values of $\mathbf{Re}_\lambda$, when there is an appreciable wavenumber range in which the Kolmogoroff spectrum applies, the value of $\mathbf{f}(r)$ is also known theoretically over a wide range of values of r.

Let us consider first the case of very large values of $\mathbf{Re}_\lambda$. We assume such large values that the expression (3-142) for the energy spectrum $\mathbf{E}(k)$ applies. The corresponding value of $\mathbf{f}(r)$ can be calculated with the aid of Eq. (3-68). Thus,

$$4\int_0^\infty dr \frac{1}{r}\left(\frac{\partial u'^2\mathbf{f}}{\partial r}\right)^2 = 16\int_0^\infty dr\frac{1}{r}\left[\int_0^\infty dk\,\mathbf{E}(k)\frac{\partial}{\partial r}\left(\frac{\sin kr}{k^3r^3} - \frac{\cos kr}{k^2r^2}\right)\right]^2$$

With the expression (3-142) for $\mathbf{E}(k)$, at the same time introducing $k_d = (\varepsilon/\nu^3)^{1/4}$ and $\varepsilon = 15\nu u'^2/\lambda_g{}^2$, after some calculation we obtain

$$4\int_0^\infty dr\frac{1}{r}\left(\frac{\partial u'^2\mathbf{f}}{\partial r}\right)^2 = \frac{2^{15/2}5^{3/2}}{3}A\frac{\nu u'^3}{\alpha\lambda_g{}^3}$$

where

$$A = \int_0^\infty ds\, s\left[\int_0^\infty dw \frac{s^4 w^{-2/3}}{(s^4+w^4)^{4/3}}\frac{d}{dw}\left(\frac{\sin w}{w^3} - \frac{\cos w}{w^2}\right)\right]^2 = 0.03776$$

The value of the integral was determined by numerical and graphical methods.

For the second term on the right-hand side of Eq. (5-68) we obtain

$$\begin{aligned}\tfrac{35}{3}\nu^2u'^2\left[\frac{\partial^4\mathbf{f}}{\partial r^4}\right]_{r=0} &= \tfrac{70}{3}\nu^2\int_0^\infty dk\,\mathbf{E}(k)\left[\frac{\partial^4}{\partial r^4}\left(\frac{\sin kr}{k^3r^3} - \frac{\cos kr}{k^2r^2}\right)\right]_{r=0}\\ &= \tfrac{2}{3}\nu^2\int_0^\infty dk\,k^4\mathbf{E}(k)\end{aligned}$$

and, with Eq. (3-142)

$$\begin{aligned}\tfrac{35}{3}\nu^2u'^2\left[\frac{\partial^4\mathbf{f}}{\partial r^4}\right]_{r=0} &= \nu^2k_d{}^5\left(\frac{8}{9\alpha}\right)^{2/3}(\varepsilon\nu^5)^{1/4}\left(\frac{8}{3\alpha^2}\right)^{-5/6}\frac{\pi\sqrt{\pi}}{\Gamma(\frac{1}{3})\Gamma(\frac{1}{6})}\\ &= \alpha\frac{\nu u'^3}{\lambda_g{}^3}\frac{15^{3/2}\pi\sqrt{\pi}}{\sqrt{6}\,\Gamma(\frac{1}{3})\Gamma(\frac{1}{6})}\end{aligned}$$

With Eq. (5-63) the relation (5-68) then becomes

$$\frac{1}{\tau_L{}^2} = 12.75 \frac{\nu u'}{\alpha \lambda_g{}^3} + \alpha \frac{3(5\pi/2)^{3/2}}{\Gamma(\frac{1}{3})\Gamma(\frac{1}{6})} \frac{\nu u'}{\lambda_g{}^3}$$

$$= \left(\frac{12.75}{\alpha} + 4.46\alpha\right)\frac{\nu u'}{\lambda_g{}^3}$$

So we obtain a relation between τ_L and λ_g for large values of $\mathbf{Re}_\lambda$, which reads

$$\frac{\lambda_g{}^2}{u'^2 \tau_L{}^2} = \left(\frac{12.75}{\alpha} + 4.46\alpha\right)\frac{1}{\mathbf{Re}_\lambda} \tag{5-69}$$

Since for an isotropic turbulence Eq. (3-99) applies, we can eliminate λ_g from this equation and Eq. (5-69), with the result

$$\frac{1}{\tau_L{}^2} = \left(\frac{0.22}{\alpha} + 0.077\alpha\right)\frac{\varepsilon}{\overline{u^2}}\sqrt{\frac{\varepsilon}{\nu}} \tag{5-70}$$

As discussed in Sec. 3-5, from measurements at high values of $\mathbf{Re}_\lambda$ in an almost isotropic turbulence Gibson[72] obtained for the Heisenberg constant α the value 0.45, while Grant, Stewart, and Moilliet[73] obtained from their measurements in a tidal channel a value of 0.52. On the other hand a reconsideration of the latter experiments by Pao resulted in a value of 0.405, which is close to the limiting value of 0.38 for $\mathbf{Re}_\lambda \to \infty$ obtained theoretically by Kraichnan. So taking $\alpha = 0.40$ the following value will be obtained for A:

$$A = \frac{0.22}{\alpha} + 0.077\alpha = 0.58$$

Since Heisenberg's expression Eq. (3-142) for $\mathbf{E}(k)$ is not acceptable in the viscous, high wavenumber range, the actual fall-off being much faster than according to this expression, the value of A will probably be slightly smaller. A value of $A = 0.55$ would not be a bad estimate. Another estimate has been offered by Tennekes,[121] making use of experimental results by Shlien and Corrsin.[122] A much smaller value is obtained, namely $A \simeq 0.1$.

The constant C occurring in Eq. (5-61) should be determined from suitable diffusion experiments. Lacking such direct experimental evidence an estimate can be made from measurements of $\beta = \Lambda_L/\Lambda_f$ in an isotropic turbulence.

In Sec. 3-5 with Von Kármán's interpolation formula we obtained the relation

$$k_e = \frac{0.51}{\alpha} \cdot \frac{\varepsilon}{u'^3}$$

From Eq. (5-61) we obtain for the Lagrangian integral time-scale the relation

$$\Lambda_L = \mathfrak{I}_L u' = \frac{2}{C}\frac{u'^3}{\varepsilon}$$

From these two relations, and Eq. (3-158), we then obtain

$$C = \frac{1.02}{0.75\alpha}\frac{\Lambda_f}{\Lambda_L} = \frac{1.36}{\alpha \cdot \beta}$$

As will be shown later, experimental determination of β by a number of experimenters, resulted in a large variation of β-values, ranging roughly from 1 down to 0.3, where the lower values refer in general to higher Reynolds numbers. Assuming for the Heisenberg constant $\alpha \simeq 0.4$, we so obtain for C values between 3 and 10, increasing with increasing Reynolds' number.

Interesting additional information is obtained when we consider Lagrangian acceleration correlations, as done by Lin.[74] Consider again a one-dimensional turbulent motion in the x_2-direction only, and a *homogeneous stationary turbulence.* In this case $\overline{v_2^2}$ is a dynamical invariant, independent of time. But also $\overline{(dv_2/dt)^2}$ will be a dynamical invariant [see Eq. (5-68)], which is to be expected since in a homogeneous, stationary turbulence, the forces, i.e., the accelerations of fluid particles will be statistically independent of the position of the flow field. So $a_2(t)$ being a stationary random function of time, we may introduce the acceleration correlation coefficient:

$$\mathbf{A}_L(t) = \frac{\overline{a_2(t')a_2(t'-t)}}{\overline{a_2^2}} \tag{5-71}$$

Since $v_2(t) = v_2(0) + \int_0^t dt'\, a_2(t')$, we obtain the following relation between $\mathbf{R}_L(t)$ and $\mathbf{A}_L(t)$

$$\mathbf{R}_L(t) = 1 - \frac{\overline{a_2^2}}{\overline{v_2^2}}\int_0^t dt' \int_0^{t'} d\tau\, \mathbf{A}_L(\tau) \tag{5-72}$$

Whence follows

$$\frac{d\mathbf{R}_L}{dt} = -\frac{\overline{a_2^2}}{\overline{v_2^2}}\int_0^t d\tau\, \mathbf{A}_L(\tau) \tag{5-73}$$

Since $\lim_{t\to\infty} d\mathbf{R}_L/dt = 0$, we must have

$$\int_0^\infty d\tau\, \mathbf{A}_L(\tau) = 0 \tag{5-74}$$

So there does not exist an integral time-scale for the accelerations. A finite integral time-scale would be inconsistent with a homogeneous, stationary random velocity field. Differentiation of Eq. (5-73) yields

$$\mathbf{A}_L(t) = -\frac{\overline{v_2^2}}{\overline{a_2^2}}\frac{d^2\mathbf{R}_L}{dt^2} \tag{5-75}$$

For small t, $\mathbf{R}_L(t)$ has a parabolic behavior.

For these small values of t we thus have

$$\mathbf{A}_L(t) \simeq +\frac{2\overline{v_2^2}}{\overline{a_2^2}\tau_L^2} = \text{constant} \tag{5-76}$$

Since, by definition $\mathbf{A}_L(0) = 1$, the constant $= 1$ and

$$\frac{\overline{a_2^2}}{\overline{v_2^2}} = \frac{2}{\tau_L^2}$$

With Eq. (5-59) we obtain the relation

$$\overline{a_2^2} = 2A\varepsilon\sqrt{\frac{\varepsilon}{\nu}} \simeq \varepsilon^{3/2}/\nu^{1/2} = \eta^2/\tau_k^4$$

where $\eta = (\nu^3/\varepsilon)^{1/4}$ and $\tau_k = (\nu/\varepsilon)^{1/2}$ are the Kolmogoroff micro length-scale and time-scale, respectively. There is not a similar relation for v_2^2, since $v_2^2 \neq \eta^2/\tau_k^2$. The conclusion drawn from the relation (5-76) is that for small times the acceleration $a_2(t)$ persists longer in its initial value than the corresponding velocity $v_2(t)$.

Other interesting inferences that can be made from the relations (5-74) and (5-75) are the following.

The acceleration correlation $\mathbf{A}_L(t)$ contains at least one negative part and one positive part, with equal areas between the curves and the abscissa. An exponential behavior $\mathbf{R}_L(t) = \exp(-t/\mathfrak{J}_L)$ at all times $t \gg \tau_L$ would conflict with the condition Eq. (5-74).

Hitherto we have considered the diffusion of fluid particles in one direction, viz., x_2, only. A generalization to the *diffusion of fluid particles in three-dimensional space* may be obtained by considering instead of $\overline{y_2^2}$ the "diffusion" tensor $\overline{y_i y_j}$ (Batchelor[16,17]). In homogeneous turbulence this diffusion tensor is a function of t alone. It may be expressed in terms of the Lagrangian correlation-tensor coefficient

$$[\mathbf{R}_{i,j}(\tau)]_L = \frac{\overline{v_i(t)v_j(t+\tau)}}{v_i' v_j'} \qquad (5\text{-}77)$$

Here, and in the following, we drop the summation convention for repeated indices.

We follow a procedure similar to that used in Sec. 1-9, i.e. we follow *method B* and consider an ensemble average over a large number of fluid particles.

$$\overline{y_i(t)y_j(t)} = \overline{y_i(0)y_j(0)} + \int_0^t dt' \int_0^{t'} dt'' \left[\overline{v_i(t')v_j(t'')} + \overline{v_i(t'')v_j(t')}\right]$$

In homogeneous steady turbulence $\overline{v_i(t')v_j(t'')}$ is a function of $t'' - t' = \tau$ only. Hence, with Eq. (5-77) we obtain

$$\overline{y_i(t)y_j(t)} = \overline{y_i(0)y_j(0)} + v_i' v_j' \int_0^t dt' \int_0^{t'} d\tau \left\{[\mathbf{R}_{i,j}(\tau)]_L + [\mathbf{R}_{j,i}(\tau)]_L\right\}$$

or, after integration by parts,

$$\overline{y_i(t)y_j(t)} = \overline{y_i(0)y_j(0)} + v_i' v_j' \int_0^t d\tau\,(t-\tau)\left\{[\mathbf{R}_{i,j}(\tau)]_L + [\mathbf{R}_{j,i}(\tau)]_L\right\} \qquad (5\text{-}78)$$

We may remark that $[\mathbf{R}_{i,j}(\tau)]_L + [\mathbf{R}_{j,i}(\tau)]_L$ is an even function of τ, since

$$[\mathbf{R}_{i,j}(\tau)]_L = [\mathbf{R}_{j,i}(-\tau)]_L \qquad (5\text{-}79)$$

The expression (5-78) is the three-dimensional generalization of the expression (1-82). Just as we did with the latter expression, we may consider the two limiting cases where t is very small and very large, respectively.

For very small values of the time, $[\mathbf{R}_{i,j}(\tau)]_L \simeq [\mathbf{R}_{i,j}(0)]_L$, and Eq. (5-78) reduces to

$$\overline{y_i(t)y_j(t)} \simeq \overline{y_i(0)y_j(0)} + v_i' v_j' [\mathbf{R}_{i,j}(0)]_L t^2$$

or

$$\overline{y_i(t)y_j(t)} \simeq \overline{y_i(0)y_j(0)} + \overline{v_i(t')v_j(t')}t^2 \qquad (5\text{-}80)$$

We may note that in a flow with zero mean-shear these relations have sense only when $i = j$, since for $i \neq j$, $[\mathbf{R}_{i,j}(0)]_L = 0$.

In the case of very large t we assume that $[\mathbf{R}_{i,j}(\tau)]_L \simeq 0$ for $\tau > t_0$, so that, for $t \gg t_0$, from Eq. (5-78) we obtain

$$\overline{y_i(t)y_j(t)} \simeq \overline{y_i(0)y_j(0)} + 2v_i'v_j'(\mathfrak{I}_{i,j})_L t \qquad (5\text{-}81)$$

where

$$(\mathfrak{I}_{i,j})_L = \frac{1}{2}\int_0^{t_0} d\tau\, \{[\mathbf{R}_{i,j}(\tau)]_L + [\mathbf{R}_{j,i}(\tau)]_L\} = \frac{1}{2}\int_0^{\infty} d\tau\, \{[\mathbf{R}_{i,j}(\tau)]_L + [\mathbf{R}_{j,i}(\tau)]_L\}$$

Let us return to the *one-dimensional case*. We have shown in Sec. 1-9 that, for large diffusion times, the diffusion in a homogeneous turbulence can be described by means of a constant coefficient of diffusion ϵ. For diffusion from a fixed plane, for instance, we then shall obtain a Gaussian expression like Eq. (1-73).

Batchelor[16] has shown that the description of the diffusion of a marked fluid particle by means of a coefficient of diffusion need not be restricted to large diffusion times provided that, in the one-dimensional case, the probability-density distribution of y_2 is Gaussian. Now experimental evidence shows that, in homogeneous turbulence without average shear, the probability-density distribution of y_2 is Gaussian for small and for large diffusion times.

Assume that at time $t = t_0$ the marked fluid particles are concentrated in the region $-a \leq x_2 \leq +a$, and let us again apply *method B*. Let $\mathfrak{P}(x_2,t)$ be the probability that a marked fluid particle is at the point x_2 at time t. $\mathfrak{P}(x_2,t)$ is then identical with the concentration of marked fluid particles, and in the absence of molecular diffusion also of any scalar property transferred by the fluid particles, at time t. For $t = t_0$, we have the initial condition

$$\mathfrak{P}(x_2,t_0) = 1 \quad \text{if } -a \leq x_2 \leq +a$$

$$\mathfrak{P}(x_2,t_0) = 0 \quad \text{outside this region}$$

and, at all times,

$$\int_{-\infty}^{+\infty} dx_2\, \mathfrak{P}(x_2,t) = \int_{-\infty}^{+\infty} dx_2\, \mathfrak{P}(x_2,t_0) = 2a$$

Let $\mathfrak{Q}(y_2,t;t_0)$ be the probability that the fluid particle has advanced over a distance y_2 during the time $t - t_0$. We then have the relation

$$\mathfrak{P}(x_2,t) = \int_{-\infty}^{+\infty} dx_2'\, \mathfrak{Q}(x_2 - x_2',t;t_0)\mathfrak{P}(x_2',t_0) \qquad (5\text{-}82)$$

since $y_2(t - t_0) = x_2 - x_2'$ and since there is a contribution to the integral only from marked fluid particles that were at $-a \leq x_2' \leq +a$ at $t = t_0$. Let us now assume that $\mathfrak{P}(x_2,t)$ satisfies the Fickian diffusion equation

$$\frac{\partial \mathfrak{P}}{\partial t} = \epsilon_{22}(t) \cdot \frac{\partial^2 \mathfrak{P}}{\partial {x_2}^2} \qquad (5\text{-}83)$$

The reader can easily verify that the expression (5-82) for $\mathfrak{P}(x_2,t)$ satisfies Eq. (5-83), when also $\mathfrak{Q}(y_2,t\,;t_0)$ satisfies an identical equation

$$\frac{\partial \mathfrak{Q}}{\partial t} = \epsilon_{22}(t) \cdot \frac{\partial^2 \mathfrak{Q}}{\partial {y_2}^2}$$

since

$$\frac{\partial}{\partial x_2} = \frac{\partial}{\partial y_2}$$

Again, the reader may easily verify that this equation is satisfied by the following solution

$$\mathfrak{Q}(y_2,t\,;t_0) = \frac{1}{\sqrt{2\pi \overline{{y_2}^2}}} \exp -({y_2}^2/2\overline{{y_2}^2}) \qquad (5\text{-}84)$$

provided

$$\epsilon_{22}(t) = \frac{1}{2}\frac{d}{dt}\overline{{y_2}^2} \qquad (5\text{-}85)$$

Earlier we mentioned that from diffusion experiments in a homogeneous, shear-free, turbulence a Gaussian behavior of $\mathfrak{P}(x_2,t)$ could only be concluded for small and large diffusion times. Often it has been assumed that such a Gaussian behavior is true also during the intermediate diffusion times. This, of course, need not necessarily be so. Earlier we showed that for short diffusion times all frequencies contribute to $\overline{{y_2}^2}$, whereas for large diffusion times only the low frequencies. In a homogeneous, shear-free turbulence the probability-density distribution of v_2 is Gaussian, and since for very short diffusion times $\overline{{y_2}^2} = \overline{{v_2}^2}t^2$, the distribution of y_2 too must be Gaussian. The same conclusion can also be drawn in the case of large diffusion times, when $\overline{{y_2}^2} = 2\overline{{v_2}^2}\mathfrak{I}_L t$. In this case $\overline{{y_2}^2}$ is determined by that part of the velocity field which has no further correlation with the Lagrangian velocity $v_2(0)$ at $t = 0$, whereas in the first case it is entirely determined by $v_2(0)$.

For *intermediate diffusion times* $v_2(t)$, and hence $\overline{{y_2}^2}$, will only be partly determined by $v_2(0)$. So we may introduce a memory function to describe the diminishing effect of $v_2(0)$ when time proceeds, for which memory function we take the Lagrangian correlation-coefficient $\mathbf{R}_L(t)$. Philip[75] has shown that for the intermediate diffusion times the distribution of y_2 need not be Gaussian, even if that of v_2 is Gaussian.

We thus assume $v_2(t)$ to consist of two parts. The first part is what the fluid particle remembers from its velocity $v_2(0)$ at $t = 0$, the second part is what it picks up from its new environment.

$$v_2(t) = v_{2_d}(t) + v_2^*(t) = v_2(0)\mathbf{R}_L(t) + v_2^*(t) \qquad (5\text{-}87)$$

Philip calls $v_{2_d}(t) = v_2(0)R_L(t)$ the *drift velocity*. $v_2^*(t)$ is not correlated with $v_2(0)$. It satisfies the conditions $v_2^*(0) = 0$, and $v_2^* = v_2(t)$ when $t \to \infty$.

Since in a stationary, homogeneous turbulence $\overline{[v_2(t)]^2} = \overline{[v_2(0)]^2} =$ constant, we have

$$\overline{v_2^{*2}}(t) = \overline{v_2^2}[1 - \mathbf{R}_L^2(t)]$$

Thus $\overline{v_2^{*2}}$ is not a dynamic invariant, and v_2^* therefore not a stationary, random function of time

Consistent with Eq. (5-87) we may write for the displacement $y_2(t)$:

$$y_2(t) = y_{2_d}(t) + y_2^*(t) \tag{5-88}$$

Here, $y_{2_d}(t)$ is the part due to the drift velocity $v_{2_d} = v_2(0)\mathbf{R}_L(t)$.

Since $v_2(0)$ and $v_2^*(t)$ are uncorrelated, and so are y_{2_d} and y_2^*, we obtain for the ensemble average.

$$\overline{y_2^2}(t) = \overline{y_{2_d}^2}(t) + \overline{y_2^{*2}}(t) \tag{5-89}$$

Now $\overline{y_2^2}(t)$ can be expressed in two ways in terms of $\mathbf{R}_L(t)$.

$$\overline{y_2^2}(t) = \overline{\left[\int_0^t dt'\, v_2(t')\right]^2} = 2\overline{v_2^2}\int_0^t dt' \int_0^{t'} d\tau\, \mathbf{R}_L(\tau) \tag{5-90}$$

And, with Eq. (5-87):

$$\overline{y_2^2}(t) = \overline{v_2^2}\left[\int_0^t d\tau\, \mathbf{R}_L(\tau)\right]^2 + \overline{\left[\int_0^t dt'\, v_2^*(t')\right]^2} \tag{5-91}$$

From Eqs. (5-89), (5-90), and (5-91) we thus obtain

$$\overline{y_2^{*2}}(t) = \overline{v_2^2}\left[2\int_0^t dt' \int_0^{t'} d\tau\, \mathbf{R}_L(\tau) - \left(\int_0^t d\tau\, \mathbf{R}_L(\tau)\right)^2\right] \tag{5-92}$$

Consider now such *short diffusion times* that

$$\mathbf{R}_L(t) \simeq 1 - t^2/\tau_L^2$$

This results in

$$\overline{y_2^2} = \overline{v_2^2}t^2\left(1 - \frac{t^2}{6\tau_L^2}\right)$$

$$\overline{y_{2_d}^2} = \overline{v_2^2}t^2\left(1 - \frac{2t^2}{3\tau_L^2}\right)$$

and

$$\overline{y_2^{*2}} = \overline{v_2^2}\frac{t^4}{2\tau_L^2}$$

On the other hand *when* $t \to \infty$, we again obtain

$$\overline{y_2^2} = 2\overline{v_2^2}t\int_0^\infty d\tau\, \mathbf{R}_L(\tau) = 2\overline{v_2^2}\mathfrak{J}_L t$$

while $\overline{y_{2_d}^2} = \overline{v_2^2}\mathfrak{J}_L^2 =$ constant, and becomes small with respect to

$$\overline{y_2^{*2}} = \overline{v_2^2}[2\mathfrak{J}_L t - \mathfrak{J}_L^2] \simeq 2\overline{v_2^2}\mathfrak{J}_L t = \overline{y_2^2}$$

From Eq. (5-85) we conclude that, since $\overline{y_2^2}$ is a nonlinear function of time, the coefficient of diffusion ϵ_{22} is still a function of time.

With the expression (1-82) for $\overline{y_2^2}$, from (5-85), we obtain

$$\epsilon_{22} = \overline{v_2'^2}\int_0^t d\tau\, \mathbf{R}_L(\tau) \tag{5-93}$$

This result reduces to $\epsilon_{22} = \overline{v_2'^2}t$ and to $\epsilon_{22} = v_2'\Lambda_L$ for small and large values of t, respectively.

Thus, plotting $\epsilon_{22}/v_2'\Lambda_L$ against $v_2't/\Lambda_L$, a curve is obtained which initially increases with a slope of 45°. For larger values of $v_2't/\Lambda_L$ the curve shows a departure from this linear increase toward an asymptotic approach to the constant value of 1 for large values of $v_2't/\Lambda_L$.

The *generalization to the three-dimensional case* given by Batchelor[16,18] is based on the assumption that the probability-density distributions of the three components y_i are not only separately but also jointly Gaussian. It is then shown that the probability $\mathfrak{P}(x_1,x_2,x_3,t)$ for a marked fluid particle to be at (x_1,x_2,x_3) at time t satisfies the differential equation

$$\frac{\partial}{\partial t}\mathfrak{P}(x_1,x_2,x_3,t) = \epsilon_{ij}\frac{\partial^2}{\partial x_i\,\partial x_j}\mathfrak{P}(x_1,x_2,x_3,t) \tag{5-94}$$

where

$$\epsilon_{ij} = \frac{1}{2}\frac{d}{dt}\overline{y_i y_j} \tag{5-95}$$

The coefficient of diffusion here is a second-order tensor and is related to the diffusion tensor $\overline{y_i y_j}$ according to Eq. (5-95). With respect to its principal axes, the tensor $\epsilon_{ij} = 0$ for $i \neq j$, and the right-hand side of Eq. (5-94) reduces to three terms. In the case of isotropic turbulence we have spherical symmetry and $\epsilon_{11} = \epsilon_{22} = \epsilon_{33} = \epsilon$; so the right-hand side of Eq. (5-94) reduces to the Laplacian form

$$\epsilon\frac{\partial^2}{\partial x_i\,\partial x_i}\mathfrak{P}(x_1,x_2,x_3,t)$$

From the expressions (5-80) and (5-81) for $\overline{y_i y_j}$ we obtain, respectively,

$$\text{For small } t \quad \epsilon_{ij} = \overline{v_i(t')v_j(t')}t \tag{5-96}$$

$$\text{For large } t \quad \epsilon_{ij} = v_i'v_j'(\mathfrak{I}_{i,j})_L \tag{5-97}$$

Again, we drop the summation convention in Eq. (5-97). For intermediate times ϵ_{ij} will change in a way similar to ϵ_{22}, as described earlier for the one-dimensional diffusion in the x_2-direction.

At present these generalizations to the three-dimensional diffusion are of theoretical interest only and have no direct practical value, since neither theoretically nor experimentally is anything known concerning the Lagrangian correlation-tensor coefficient $[\mathbf{R}_{i,j}(\tau)]_L$. As we shall see in the next section, all experimental results concerning turbulent diffusion that are accessible to a theoretical approach can be described sufficiently with the one-dimensional diffusion theories.

Hitherto we have assumed a stationary, homogeneous turbulence. It may be worth while to consider, though briefly, the consequences when we drop the condition of homogeneity. Since the mean flow field is still assumed to be stationary, we may use time-averages for the Eulerian flow field. For the diffusion of the fluid particles, however, we need in the Lagrangian description ensemble averages. In order to indicate what kind of average is taken, we will use the notation introduced in Sec. 1-1.

Let the Eulerian flow field be described by

$$U_i(x_j,t) = \overset{t}{\bar{U}}_i(x_j) + u_i(x_j,t)$$

Let $V_i(x_{ko},t)$ be the Lagrangian velocity of a fluid particle at time t, which particle was at x_{ko} at time $t = 0$. Since at time t this fluid particle is at x_k, $V_i(x_{ko},t)$ satisfies the condition

$$V_i(x_{ko},t) = U_i(x_k,t)$$

while $x_k - x_{ko}$ is equal to the distance covered by the fluid particle during the time t:

$$x_k - x_{ko} = Y_k(x_{ko},t) = \int_0^t dt'\, V_k(x_{ko},t') \tag{5-98}$$

Hence

$$\begin{aligned} V_i(x_{ko},t) &= U_i(x_k,t) = U_i\left[x_{ko} + \int_0^t dt'\, V_k(x_{ko},t'),t\right] \\ &= \overset{t}{\bar{U}}_i\left[x_{ko} + \int_0^t dt'\, V_k(x_{ko},t')\right] + u_i\left[x_{ko} + \int_0^t dt'\, V_k(x_{ko},t'),t\right] \end{aligned} \tag{5-99}$$

We define

$$v_i(x_{ko},t) = V_i(x_{ko},t) - \overset{e}{\bar{V}}_i(x_{ko},t) \tag{5-100}$$

where an ensemble is taken of many particles with the same starting point x_{ko}, and the same travelling time t. Integration of Eq. (5-100) with respect to time yields

$$y_i(x_{ko},t) = Y_i(x_{ko},t) - \overset{e}{\bar{Y}}_i(x_{ko},t) \tag{5-101}$$

Hcncc

$$y_i(x_{ko},t) = \int_0^t dt'\, v_i(x_{ko},t') = \int_0^t dt'\, V_i(x_{ko},t') - \int_0^t dt'\, \overset{e}{\bar{V}}_i(x_{ko},t') \tag{5-102}$$

where according to Eq. (5-99)

$$\overset{e}{\bar{V}}_i(x_{ko},t) = \overline{\overset{t}{\bar{U}}_i\left[x_{ko} + \int_0^t dt'\, V_k(x_{ko},t')\right]}^{e} + \overline{u_i\left[x_{ko} + \int_0^t dt'\, V_k(x_{ko},t'),t\right]}^{e} \tag{5-103}$$

We may note that

$$\overline{u_i\left[x_{ko} + \int_0^t dt'\, V_k(x_{ko},t'),t\right]}^{e} \neq 0$$

and that

$$\overline{\overset{t}{\bar{U}}_i\left[x_{ko} + \int_0^t dt'\, V_k(x_{ko},t')\right]}^{e} \neq \overset{t}{\bar{U}}_i\left[x_{ko} + \overline{\int_0^t dt'\, V_k(x_{ko},t')}^{e}\right]$$

Since $\overset{e}{\bar{V}}_i(x_{ko},t)$ is still a function of time, the Lagrangian velocity $v_i(x_{ko},t)$ of a fluid particle is not a stationary, random function of time.

There are two possibilities for the time variation of $v_i(x_{ko},t)$. One is because at each point the

Eulerian velocity $u_i(x_k,t)$ varies with time. The other is because $\overset{t}{\bar{U}}_i$ as well as u_i varies due to the inhomogeneity of the time-averaged flow field. Also we have to keep in mind that the relation (5-102) refers to fluid particles with the starting point x_{k0}. For any other starting point the quantitative result will be different.

Consider now the diffusion tensor $\overset{e}{\overline{(y_i y_j)}}(x_{k0},t)$ of particles with the same starting point x_{k0} and the same traveling time. Following a now familiar procedure we obtain

$$\overset{e}{\overline{y_i y_j}}(x_{k0},t) = \int_0^t dt' \int_0^t dt''\, \overset{e}{\overline{v_i(x_{k0},t')v_j(x_{k0},t'')}}$$

$$= \frac{1}{2}\int_0^t dt' \int_0^t dt'' \left[\overset{e}{\overline{v_i(x_{k0},t')v_j(x_{k0},t'')}} + \overset{e}{\overline{v_i(x_{k0},t'')v_j(x_{k0},t')}}\right]$$

The expression between brackets is symmetric in t' and t''. Consequently we may transform the integration procedure

$$\overset{e}{\overline{y_i y_j}}(x_{k0},t) = \int_0^t dt' \int_0^{t'} dt'' \left[\overset{e}{\overline{v_i(x_{k0},t')v_j(x_{k0},t'')}} + \overset{e}{\overline{v_i(x_{k0},t'')v_j(x_{k0},t')}}\right]$$

With $t' - t'' = \tau$, and with the introduction of the following Lagrangian-velocity correlations

$$[\mathbf{Q}_{i,j}(x_{k0},t';\tau)]_L = \overset{e}{\overline{v_i(x_{k0},t')v_j(x_{k0},t'-\tau)}}$$

$$[\mathbf{Q}_{j,i}(x_{k0},t';\tau)]_L = \overset{e}{\overline{v_j(x_{k0},t')v_i(x_{k0},t'-\tau)}}$$

we obtain

$$\overset{e}{\overline{y_i y_j}}(x_{k0},t) = \int_0^t dt' \int_0^{t'} d\tau \left\{[\mathbf{Q}_{i,j}(x_{k0},t';\tau)]_L + [\mathbf{Q}_{j,i}(x_{k0},t';\tau)]_L\right\} \tag{5-104}$$

We note that because of the inhomogeneity of the flow field, $\mathbf{Q}_{i,j}$ and $\mathbf{Q}_{j,i}$ not only depend on τ, but also on t'. Also we note that the second subscript refers to the velocity taken a time interval τ earlier. They are not symmetric in τ, and so not symmetric in i and j. However, their sum is symmetric in i and j, and so, according to Eq. (5-104), is $\overset{e}{\overline{(y_i y_j)}}$. From this equation we obtain after differentiation with respect to time.

$$\frac{\partial}{\partial t}\overset{e}{\overline{y_i y_j}}(x_{k0},t) = \int_0^t d\tau \left\{[\mathbf{Q}_{i,j}(x_{k0},t;\tau)]_L + [\mathbf{Q}_{j,i}(x_{k0},t;\tau)]_L\right\} \tag{5-105}$$

We may put

$$\frac{1}{2}\frac{\partial}{\partial t}\overset{e}{\overline{y_i y_j}}$$

equal to an eddy diffusion coefficient $\epsilon_{i,j}(x_{k0},t)$. However, such a coefficient would not satisfy a Fickian diffusion equation with a Gaussian probability density distribution for the tensor $\overset{e}{\overline{(y_i y_j)}}$. Also, the derivative of such an eddy diffusion coefficient would not yield the Lagrangian velocity correlation, but in addition to it an extra term, as shown by the following Eq. (5-106).

$$\frac{\partial^2}{\partial t^2}\overset{e}{\overline{y_i y_j}}(x_{k0},t) = \left\{[\mathbf{Q}_{i,j}(x_{k0},t;t)]_L + [\mathbf{Q}_{j,i}(x_{k0},t;t)]_L\right\}$$

$$+ \int_0^t d\tau \frac{\partial}{\partial t}\left\{[\mathbf{Q}_{i,j}(x_{k0},t;\tau)]_L + [\mathbf{Q}_{j,i}(x_{k0},t;\tau)]_L\right\} \tag{5-106}$$

If the flow field is not only inhomogeneous but also *non-stationary*, the same results will be obtained as given by Eqs. (5-104), (5-105), and (5-106). The only difference with the stationary case is, that $[\mathbf{Q}_{i,j}]_L$ depends on both t' and τ, not only because of the inhomogeneity of the flow field but also because of its non-stationarity.[119]

The interaction of a lump of fluid with its surroundings will cause a strong deformation of the lump. Now it is a well-known theorem in mechanics that the motion of a system of particles may be resolved into the motion of the center of gravity of this system and the motion of the individual particles relative to this center of gravity. It may be useful to apply this theorem to the study of the history of a lump of fluid in a turbulent flow, by regarding the lump as consisting of a large number of fluid particles. We may then study separately the movement of the center of mass of the lumps of fluid and the relative movement of the fluid particles of the lump with respect to this center of mass. The relative movement of the fluid particles determines the deformation and change in shape of the lump and the interaction and exchange of a transferable property with the surroundings. As Batchelor[16,17] has pointed out the study of the movement of the center of mass boils down to a Eulerian analysis of the diffusion of a single marked fluid particle. For the study of the relative movements of the fluid particles it is sufficient to consider a Lagrangian analysis of the relative motion of two marked fluid particles. The motion of the center of mass of the cluster of fluid particles and the mean concentration of the fluid particles are independent of the original shape of the lump of fluid. They are determined only by the diffusion of the individual fluid particles, whose diffusion is unaffected by the diffusion of the other particles.

The main contribution to the study of the relative diffusion of two fluid particles is due to Batchelor,[17] although many years earlier some basic considerations were given by Taylor,[19] and also by Richardson[20] in his study of atmospheric diffusion.

The Relative Diffusion of Two Fluid Particles

This of course, depends on their initial relative position and distance. If their initial distance is large compared with the integral scale of the turbulence, there will be no connection between the motions of the two fluid particles and they will wander independently of each other. On the other hand, if the two fluid particles are close to each other initially, their motions will be closely connected, at least during the first stages of their wandering, and will be determined by the turbulence motions of many if not all frequencies. These include those of the most fine-scale structure of the turbulence, which is determined by the two parameters ν and ε. According to its definition the fluid particle is of the magnitude of the Kolmogoroff micro length-scale η. But since we want to include in our considerations initial distances smaller than η, we will, contrary to the above, assume in what follows infinitely small fluid particles. The two fluid particles will be marked α and β. We introduce their relative velocity

$$w_i = (v_i)_\beta - (v_i)_\alpha \qquad (5\text{-}107)$$

Their relative displacement is

$$z_i = (y_i)_\beta - (y_i)_\alpha = z_i(0) + \int_0^t dt'\, w_i(t') \qquad (5\text{-}108)$$

We now consider in the *homogeneous, stationary turbulence* a large number of particle pairs, with the same initial distance $z_i(0)$. In what follows we assume an ensemble average for this large number of particle pairs in order to define the following quantities:

(a) the *relative diffusion*

$$s^2 = \overline{z_i z_i(t)} \tag{5-109}$$

with the initial separation ${s_0}^2 = z_i(0) \cdot z_i(0)$.

(b) The *relative diffusion coefficient*

$$\epsilon_r = \frac{1}{2}\frac{ds^2}{dt} \tag{5-110}$$

(c) the *separation velocity*

$$\frac{d}{dt}\sqrt{s^2 - {s_0}^2} \tag{5-111}$$

Just as in a study of single particle diffusion, we may express ϵ_r in terms of the correlation between the relative velocities at two instants. Since the behavior of a particle pair depends on both time and initial separation s_0, we have

$$\epsilon_r(s_0,t) = \frac{1}{2}\frac{ds^2}{dt} = \overline{z_i(s_0,t)\frac{d}{dt}z_i(s_0,t)} = \overline{z_i(s_0,t)w_i(s_0,t)}$$

Or, with Eq. (5-108), and noting that $\overline{w_i}(s_0,t) = 0$,

$$\epsilon_r(s_0,t) = \int_0^t d\tau\, \overline{w_i(s_0,t)w_i(s_0,t-\tau)} \tag{5-112}$$

So the velocity-difference correlation is not only a function of s_0 and τ, but also of t. Hence, $w_i(s_0,t)$ is not a stationary random function of time. Indeed, as we will show later, $\overline{w_i w_i}$ is not a dynamical invariant.

From Eqs. (5-111) and (5-112), we obtain

$$s^2(s_0,t) = {s_0}^2 + 2\int_0^t dt' \int_0^{t'} d\tau\, \overline{w_i(s_0,t')w_i(s_0,t'-\tau)} \tag{5-113}$$

and

$$\frac{d}{dt}\sqrt{s^2 - {s_0}^2} = \frac{d}{dt}\left\{2\int_0^t dt' \int_0^{t'} d\tau\, \overline{w_i(s_0,t')w_i(s_0,t-\tau)}\right\}^{1/2} \tag{5-114}$$

Now we will first assume such *small diffusion times* that the Lagrangian velocities remain essentially constant, and so the relative velocity $w_i(s_0,t) \simeq w_i(s_0,0)$.

Consequently the velocity-difference correlation occurring in Eqs. (5-112), (5-113), and (5-114) is also constant and equal to $\overline{w_i(s_0,0)w_i(s_0,0)}$. These equations then yield

$$\epsilon_r(s_0,t) = \overline{w_i(s_0,0)w_i(s_0,0)} \cdot t \qquad (5\text{-}115)$$

$$s^2(s_0,t) = s_0{}^2 + \overline{w_i(s_0,0)w_i(s_0,0)} \cdot t^2 \qquad (5\text{-}116)$$

and

$$\frac{d}{dt}\sqrt{s^2 - s_0{}^2} = \left[\overline{w_i(s_0,0)w_i(s_0,0)}\right]^{1/2} = \text{constant} \qquad (5\text{-}117)$$

With respect to s_0 we may distinguish between different cases (see also refs. 17 and 71).

Let $s_0 < \eta = (v^3/\varepsilon)^{1/4}$. The turbulence at this scale is determined by s_0, v, and ε. From dimensional analysis there follows

$$\overline{w_i(s_0,0)w_i(s_0,0)} = (v\varepsilon)^{1/2}F(s_0/\eta)$$

The function $F(s_0/\eta)$ can be evaluated, when we assume *local isotropy*. For the very small separation distances considered we may put

$$v_i(z_j) = v_i(0) + \frac{\partial u_i}{\partial x_j} z_j(0)$$

and

$$w_i(s_0,0) = \frac{\partial u_i}{\partial x_j} z_j(0)$$

Hence

$$\overline{w_i(s_0,0)w_i(s_0,0)} = \overline{\frac{\partial u_i}{\partial x_j}\frac{\partial u_i}{\partial x_k}} z_j(0)z_k(0)$$

Since we assume local isotropy, we obtain from the isotropic relations (3-20) and (3-99)

$$\overline{\frac{\partial u_i}{\partial x_j}\frac{\partial u_i}{\partial x_k}} = -5\overline{u^2}\,\delta_{jk}\left(\frac{\partial^2 f}{\partial r^2}\right)_0 = \frac{1}{3}\frac{\varepsilon}{v}\delta_{jk}$$

Thus

$$\overline{w_i(s_0,0)w_i(s_0,0)} = \frac{1}{3}\frac{\varepsilon}{v}\delta_{jk}z_j(0)z_k(0) = \frac{1}{3}\frac{\varepsilon}{v}s_0{}^2 \qquad (5\text{-}118)$$

and consequently

$$\epsilon_r(s_0,t) = \tfrac{1}{3}s_0{}^2\frac{\varepsilon}{v}t$$

$$s^2(s_0,t) = s_0{}^2 + \tfrac{1}{3}s_0{}^2\frac{\varepsilon}{v}t^2$$

and

$$\frac{d}{dt}\sqrt{s^2 - s_0^{\,2}} = \frac{s_0}{\sqrt{3}}\sqrt{\frac{\varepsilon}{\nu}}$$

Let now $s_0 \gg \eta$, and let $\mathbf{Re}_\lambda$ be sufficiently large for the existence of an *inertial subrange,* where the turbulence is solely determined by ε. Dimensional analysis yields $\overline{w_i(s_0,0)w_i(s_0,0)} = A(s_0\varepsilon)^{2/3}$. Corresponding relations then may be obtained for ϵ_r, $\overline{s^2}$, and $d/dt(\overline{s^2} - s_0^{\,2})^{1/2}$.

When we again assume local isotropy, then the constant A can be further evaluated. Since for a homogeneous turbulence $\overline{v_i(z_j,0)v_i(z_j,0)} = \overline{v_i(0,0)v_i(0,0)}$, we obtain with $w_i(s_0,0) = v_i(z_j,0) - v_i(0,0)$,

$$\overline{w_i(s_0,0)w_i(s_0,0)} = 2[\overline{v_i(0,0)v_i(0,0)} - \overline{v_i(z_j,0)v_i(0,0)}] = 2[3\overline{u^2} - \mathbf{Q}_{i,i}(z_j,0)]$$

According to Eq. (3-12) we have the following relation for an isotropic turbulence

$$\mathbf{Q}_{i,i}(z_j,0) = \frac{\overline{u^2}}{s_0^{\,2}}\frac{\partial}{\partial s_0}[s_0^{\,3}\mathbf{f}(s_0)]$$

Also for values of s_0 falling in the inertial subrange, we obtain from Eq. (3-153):

$$\mathbf{f}(s_0) = 1 - \frac{C}{\overline{u^2}}(\varepsilon s_0)^{2/3}$$

where

$$C = \tfrac{18}{55}\Gamma(\tfrac{4}{3})\left(\frac{3}{\alpha}\right)^{2/3}$$

and α is Heisenberg's constant. Hence

$$\overline{w_i(s_0,0)w_i(s_0,0)} = \tfrac{22}{3}C(\varepsilon s_0)^{2/3} = \tfrac{12}{5}\Gamma(\tfrac{4}{3})\left(\frac{3}{\alpha}\right)^{2/3}(\varepsilon s_0)^{2/3} \simeq 8.25(\varepsilon s_0)^{2/3} \qquad (5\text{-}119)$$

when we assume $\alpha \simeq 0.4$. With $\alpha = 0.5$ we would have obtained a value 7.1 for the above numerical constant, which latter value is in agreement with the value of 2.32 obtained by Kraichnan[76] for $\overline{w_1^{\,2}(s_0,0)}$, according to his Lagrangian-History-Direct-Interaction theory.

When we consider now *not-small diffusion times,* $w_i(s_0,t)$ will no longer be independent of time. We then may assume $s \gg s_0$, so that the effect of the initial separation s_0 can be neglected. When we further assume that the relative diffusion is still determined by the inertial subrange, dimensional analysis yields:

$$\overline{w_i(t)w_i(t)} = \text{constant } \varepsilon t \qquad (5\text{-}120)$$

This, incidentally, clearly shows that $\overline{w_iw_i}$ is not a dynamical invariant.

The coefficient of relative diffusion ϵ_r will be determined by ε and t, and so must be equal to

$$\epsilon_r(t) = A_1\varepsilon t^2 \qquad (5\text{-}121)$$

Whence follows, with $s_0 \ll s$

$$s^2(t) = 2\int_0^t dt' \,\epsilon_r(t') = \tfrac{2}{3}A_1 \varepsilon t^3 \qquad (5\text{-}122)$$

and

$$\frac{ds}{dt} = \sqrt{\tfrac{3}{2}A_1\varepsilon}\, t^{1/2} \qquad (5\text{-}123)$$

When we eliminate the time t from the expressions for $\epsilon_r(t)$ and $s^2(t)$, we obtain *Richardson's*[20] *relative-diffusion law,*

$$\epsilon_r(s) = (\tfrac{9}{4}A_1\varepsilon)^{1/3}(s^2)^{2/3} \qquad (5\text{-}124)$$

Sullivan[117] made large-scale diffusion experiments in Lake Huron. From his results one obtains $A_1 = 0.368$ ($\varepsilon = 2.1 \times 10^{-7}$ m^2/s^3). From the above it follows that this law holds true for intermediate diffusion times and for a very large value of $\mathbf{Re}_\lambda$, so that the behavior of the particle pair is wholly determined by the inertial subrange.

For *very large diffusion times* the separation distance of the two particles becomes greater than the large eddies, so that the relative behavior becomes determined by the large-scale motions, and so uncorrelated. Equation (5-109) for s^2 may then be expressed in terms of $\overline{y_i y_i}$, since

$$s^2(t) = \overline{z_i z_i}(t) = \overline{[(y_i)_\beta - (y_i)_\alpha][(y_i)_\beta - (y_i)_\alpha]} = 2\overline{y_i y_i} \qquad (5\text{-}125)$$

The coefficient of relative diffusion becomes constant

$$\epsilon_r = \frac{1}{2}\frac{ds^2}{dt} = 2\overline{y_i v_i} = 2\int_0^\infty d\tau\, [\mathbf{Q}_{i,i}(\tau)]_L \qquad (5\text{-}126)$$

Hence $s^2 \to 2\epsilon_r t$ for very large t, and

$$\frac{ds}{dt} = \sqrt{\frac{\epsilon_r}{2}}\, t^{-1/2} \qquad (5\text{-}127)$$

In summarizing we see that the separation velocity ds/dt, reckoned from $t = 0$, first remains constant, then increases according to $t^{1/2}$ during the intermediate diffusion times, and finally after passing through a maximum decreases according to $t^{-1/2}$. In the intermediate, $t^{1/2}$-range we thus have an accelerated separation s, though the acceleration decreases with time. This accelerated behavior of s may be expected from a physical point of view, since the separation distance is determined only by those eddies whose size is smaller than the distance s. The bigger eddies only cause a meandering of the particle pair. As diffusion proceeds, more and bigger eddies affect the separation distance.

A separation distance accelerating with time would also occur between two fluid particles in a flow field of uniform rate of strain.

Now, in a turbulent flow field, the small regions within the smallest scale of turbulence may be regarded as temporary regions with uniform rate of strain. We may consider the accelerating separation of two fluid particles in a turbulent flow field during the initial instants of diffusion by assuming, again, that at all these instants the distance between the fluid particles remains smaller than the smallest scale of turbulence. Thus not only at $t = 0$ do we have $s_0 \simeq 0$, but at all times considered $s < \eta = (\nu^3/\varepsilon)^{1/4}$. We must keep this in mind, even if for convenience in the following analysis the time t is assumed to become very large. Then the rate of strain in the fluid is the same at all points of the material line joining the two marked fluid particles.

A more general description of the relative diffusion of two fluid particles is given by means of a *relative-diffusion tensor* $\overline{z_i z_j}$. Expressed in terms of $(y_i)_\alpha$ and $(y_i)_\beta$, we obtain

$$\begin{aligned}\overline{z_i z_j} &= \overline{(y_i y_j)_\beta} + \overline{(y_i y_j)_\alpha} - \overline{(y_i)_\beta (y_j)_\alpha} - \overline{(y_j)_\beta (y_i)_\alpha} \\ &= 2\overline{y_i y_j} - \overline{(y_i)_\beta (y_j)_\alpha} - \overline{(y_j)_\beta (y_i)_\alpha}\end{aligned}$$

since in a homogeneous turbulence the diffusion tensors for the two fluid particles will be the same. The tensor $\overline{(y_i)_\alpha (y_j)_\beta}$ can be expressed in terms of Lagrangian double-correlation tensors $[(\mathbf{Q}_{i,j})_L]_{\alpha,\beta} = \overline{(v_i)_\alpha (v_j)_\beta}$, but nothing is known experimentally about such two-particle Lagrangian correlation tensors. But they too will become zero at large diffusion times, so that the two fluid particles move independently, and

$$\overline{z_i z_j} \simeq 2\overline{y_i y_j} = 4v_i' v_j' (\mathfrak{I}_{i,j})_L t \qquad \text{(no summation)}$$

For small diffusion times similar considerations can be given as those for $s^2 = \overline{z_i z_i}(t)$. For instance, when the relative distance is still within the inertial subrange, where

$$\mathbf{f}(r) \simeq 1 - \frac{A}{u'^2} (\varepsilon r)^{2/3}$$

and Eq. (3-11) for $\mathbf{Q}_{i,j}(\xi_1, \xi_2, \xi_3, 0)$ may be used.

For, $\overline{(y_i)_\alpha (y_j)_\beta}$ can be expressed in terms of this Eulerian space correlation since

$$(y_i)_\alpha \simeq [v_i(0)]_\alpha t = [u_i(x_1, x_2, x_3, 0)]_\alpha t$$

when t is small.

The result then is

$$\overline{z_i z_j} \simeq \tfrac{2}{3} A \left(4\delta_{ij} - \frac{\xi_i \xi_j}{r^2} \right) (\varepsilon r)^{2/3} t^2 + \xi_i \xi_j$$

where $\xi_i = (x_i)_\beta - (x_i)_\alpha$.

Deardorff and Peskin[77] studied the relative diffusion of two fluid particles with a three-dimensional numerical model of two-dimensional channel flow at large Reynolds numbers. They performed a direct numerical integration of the nonlinear equations of motion for an inviscid incompressible fluid. They obtained the interesting result that the Lagrangian correlation coefficients

$$[(\mathbf{R}_{1,1})_L]_{\alpha,\beta}(t) = \overline{(v_1)_\alpha(v_1)_\beta(t)}^{e}/\overline{v_1}^2$$

and similar for the other two directions, decay much slower than the single particle Lagrangian correlation. Which is not entirely surprising when we consider two particles which at the time of release have been at a short distance only, so that their velocities were highly correlated. As time proceeds they tend to remain in each other's neighborhood, and their velocities remain more correlated than the velocity of each particle separately after the same time interval.

We may note that $[(\mathbf{R}_{1,1})_L]_{\alpha,\beta}$ of the velocities $(v_1)_\alpha$ and $(v_1)_\beta$ taken at the same time instant, is still a function of the diffusion time t, since it can not be stationary because $z_1 = (y_1)_\beta - (y_1)_\alpha$ depends on the time t.

Deardorff and Peskin extended their numerical calculations to the case of a homogeneous, stationary turbulence with a constant mean shear $d\bar{U}_1/dx_2$. They found that the leading term in the expression for the mean downstream dispersion

$$\overline{[(\bar{y}_1)_\beta - (\bar{y}_1)_\alpha]^2}^{e} \propto \left(\frac{d\bar{U}_1}{dx_2}\right)^2 t^3$$

Here

$$(\bar{y}_1)_\alpha = (\bar{y}_1)_\alpha(0) + \int_0^t dt'\, \bar{U}_1[(y_2')_\alpha]$$

a result obtained earlier for long diffusion times by Smith.[78] This dependence of the relative dispersion proportional with t^3 is similar to the result obtained for the inertial subrange and given by Eq. (5-122), and from which Richardson's relative-diffusion law Eq. (5-124) follows. Deardorff and Peskin concluded that the experimental t^3-dependency corresponding with this law, might perhaps be due to the presence of a mean shear rather than to the particles still being in the inertial subrange.

Deformation of a Fluid Element

Consider two particles situated on the instantaneous principal axis of maximum rate of strain. Assume a coordinate system with the x_1-axis along this principal axis. When z_1 is sufficiently small

$$w_1 = \frac{dz_1}{dt} = \frac{\partial u_1}{\partial x_1} z_1 = \zeta_1 z_1$$

Whence follows, when during the time considered $\zeta_1 = \partial u_1/\partial x_1$ may be assumed to remain constant,

$$z_1(t) = z_1(0)\exp(\zeta_1 t) \qquad (5\text{-}128)$$

Similar expressions are obtained for particle pairs situated on the other two principal axes, subject to the condition that for an incompressible fluid

$$\zeta_1 + \zeta_2 + \zeta_3 = 0 \qquad (5\text{-}129)$$

For a particle pair situated on an arbitrary material line with the instantaneous direction l, we have similarly

$$\frac{1}{|z|}\frac{d|z|}{dt} = \mathbf{e}_{li}\mathbf{e}_{lj}\frac{\partial u_i}{\partial x_j} \qquad (5\text{-}130)$$

where $|z| = (z_i z_i)^{1/2}$ and $\mathbf{e}_{li}$ and $\mathbf{e}_{lj}$ are the direction cosines of l.

Consider now a large number of particle pairs with the same initial distance s_0 and orientation. Since for all those particle pairs initially $\mathbf{e}_{li}\mathbf{e}_{lj}$ is the same, the ensemble average of Eq. (5-130) will be zero, since $\partial u_i/\partial x_j$ has a zero mean. After some time, however, through the random, meandering, action of the bigger eddies the orientation of the particle pairs will become randomly distributed, so that the average value of the right-hand side of Eq. (5-130) will no longer be zero. Since experience shows that in a turbulent flow neighboring fluid particles tend to separate, Batchelor[21] made the assumption that after a sufficiently long time

$$\overline{\frac{d \ln |z|}{dt}} = \overline{\mathbf{e}_{li}\mathbf{e}_{lj}\frac{\partial u_i}{\partial x_j}} = \zeta$$

where ζ is independent of time, and positive.

Batchelor further conjectured that for these relatively long diffusion times where the effect of the initial separation s_0 becomes negligible, $|z|/\overline{|z|}$ will be a stationary random function of time. He then showed that

$$\overline{\frac{d \ln |z|}{dt}} = \frac{d}{dt} \ln \overline{|z|} = \zeta$$

whence follows

$$\overline{|z(t)|} = \overline{|z(0)|}\, e^{\zeta t} \qquad (5\text{-}131)$$

with $\zeta > 0$.

A more rigorous proof of Batchelor's conjecture has been given by Cocke,[79] who showed that for an isotropic, homogeneous turbulence the expectation value of the logarithmic change of the length of a particular line element subjected to turbulent distortion is greater than zero. For a completely random orientation of $z(0)$, however, $\langle d \ln |z|/dt \rangle = 0$ when averaged over all angles of orientation. The rate of strain ζ will be proportional to $(\varepsilon/\nu)^{1/2}$, in accord with a result obtained earlier, namely that for $s < \eta$, and $s \gg s_0$, $ds/dt \simeq s_0(\varepsilon/3\nu)^{1/2}$.

Batchelor[21] and Reid[22] have studied what the effect of this stretching process of material lines will be on the deformation of lumps of fluid and on the local changes with time of scalar and vector quantities carried along by the fluid, such as heat and turbulence vorticities. They conclude that any small spherical element is ultimately drawn out into a long thin ribbon, such that

Length increases in proportion to $\exp(\zeta_1 t)$ (5-132a)

Width increases in proportion to $\exp(\zeta_2 t)$ (5-132b)

Thickness increases in proportion to $\exp(\zeta_3 t)$ (5-132c)

where ζ_1 and ζ_2 are positive, and $\zeta_1 \gg \zeta_2$. Since for an incompressible fluid we have the condition Eq. (5-129) we must have

$$\zeta_3 = -(\zeta_1 + \zeta_2) \simeq -\zeta_1 < 0$$

It must be emphasized that the spherical element must be sufficiently small so that the relation (5-131) and the underlying assumption apply. The spherical element is then subjected to a rate of strain that is nearly uniform. Observations by Townsend[23] on the diffusion of heat spots in a homogeneous turbulence suggest that there is little rotation, relative to the fluid, of the principal axes of the rate of strain to which a fluid element is subjected, and that these axes remain practically coincident with the principal axes of the ellipsoid into which a spherical element is initially deformed. Townsend's observations suggest further that the ellipsoid ultimately has two principal axes that are large compared with the third and also large compared with the diameter of the original sphere.

Thus all material lines and surfaces tend to become parallel to the direction of the maximum rate of stretching. The manner of deformation of any fluid element has an important bearing on the changes of the scalar and vector quantities entrained by the fluid element.

If there is a nonuniform distribution of a scalar quantity, for instance, heat, the gradients of this scalar quantity will change during the deformation of the element. The gradient in the direction where the rate of stretching is ζ_i varies inversely with $\exp(\zeta_i t)$. Thus the gradient of the scalar quantity increases strongly in the direction of the shortest principal axis of the ellipsoid into which an original spherical element is deformed, since two neighboring surfaces of constant concentration of this scalar quantity tend to come closer to each other.

So if Γ is the scalar quantity and if we consider a coordinate system x_i^* moving with the fluid element and with the axes coincident with the principal axes of the elongating ellipsoid, we have

$$\frac{\partial \Gamma}{\partial x_i^*} \propto \exp(-\zeta_i t) \qquad (5\text{-}133)$$

On the other hand, a vector quantity, for instance, the vorticity ω_k, will increase in proportion to the increase of the line element that coincides with the vector quantity, since the vector magnitude is inversely proportional to the area of the cross section of the vector tube. Thus, with respect to the coordinate system x_i^*,

$$\omega_1 \propto \exp[-(\zeta_2 + \zeta_3)t] = \exp(\zeta_1 t)$$

and, generalized,

$$\omega_k \propto \exp(\zeta_k t) \qquad (5\text{-}134)$$

From Eqs. (5-133) and (5-134) the following interesting result is obtained:

$$\delta_{ik}\omega_k \frac{\partial \Gamma}{\partial x_i^*} = \omega_i \frac{\partial \Gamma}{\partial x_i^*} \propto \exp(-\zeta_i t)\exp(\zeta_i t) = 1$$

or

$$\omega_i \frac{\partial \Gamma}{\partial x_i^*} = \text{const.} \qquad (5\text{-}135)$$

This result, given for the general case of scalar and vector quantities by Batchelor and Townsend,[25] had been obtained earlier by Ertel,[24] in his meteorological studies, for the temperature gradient and vorticity.

The results (5-133) to (5-135) are valid only if there is no molecular diffusion. In reality, however, there is molecular diffusion, and its effect is increased strongly by the stretching process of fluid elements. This means that, even if the molecular diffusion is very small compared with the eddy diffusivity, its effect need by no means be negligibly small, because of the strongly increased local gradients due to the action of turbulence. Consequently, the increased molecular diffusion and the eddy diffusion are not independent processes but are strongly related. A further consequence is that in most cases it will be very difficult, if not impossible, to correct diffusion processes in turbulent flows for the effect of molecular diffusion.

An idea of the quantitative effect of deformation of a fluid element on the transport of a scalar quantity Γ by molecular diffusion can be obtained from the following simple model. Consider an initially cubical volume-element, with the edges in the direction of the principal axes of strain rates. The Fickian diffusion equation reads

$$\frac{\partial \Gamma}{\partial t} = \mathfrak{k}_\gamma \frac{\partial^2 \Gamma}{\partial x_i \, \partial x_i} \tag{5-136}$$

Assume for simplicity that initially

$$\left(\frac{\partial^2 \Gamma}{\partial x_1^2}\right)_0 = \left(\frac{\partial^2 \Gamma}{\partial x_2^2}\right)_0 = \left(\frac{\partial^2 \Gamma}{\partial x_3^2}\right)_0 < 0 \tag{5-137}$$

We may express the right-hand side of Eq. (5-136) at time t in terms of the original length of material lines, according to Eq. (5-131). Since

$$\frac{\partial}{\partial x_1} = \left(\frac{\partial}{\partial x_1}\right)_0 \frac{\partial x_1(0)}{\partial x_1} = \exp(-\zeta_1 t)\left(\frac{\partial}{\partial x_1}\right)_0$$

$$\frac{\partial^2}{\partial x_1^2} = \exp(-2\zeta_1 t)\left(\frac{\partial^2}{\partial x_1^2}\right)_0$$

and similar relations for the two other directions, Eq. (5-136) becomes, using Eq. (5-137):

$$\frac{\partial \Gamma}{\partial t} = \mathfrak{k}_\gamma \left(\frac{\partial^2 \Gamma}{\partial x_1^2}\right)_0 [\exp(-2\zeta_1 t) + \exp(-2\zeta_2 t) + \exp(-2\zeta_3 t)]$$

For small times, this reduces to

$$\frac{\partial \Gamma}{\partial t} \simeq \mathfrak{k}_\gamma \left(\frac{\partial^2 \Gamma}{\partial x_1^2}\right)_0 [3 + 2(\zeta_1^2 + \zeta_2^2 + \zeta_3^2)t^2 + \cdots]$$

Or, since $|\zeta_1| \simeq |\zeta_3| \gg |\zeta_2|$.

$$\frac{\partial \Gamma}{\partial t} \simeq 3\mathfrak{k}_\gamma \left(\frac{\partial^2 \Gamma}{\partial x_1^2}\right)_0 [1 + \tfrac{4}{3}\zeta_1^2 t^2 + \cdots]$$

and

$$\Gamma(t) \simeq \Gamma(0) + 3\mathfrak{k}_\gamma \left(\frac{\partial^2 \Gamma}{\partial x_1^2}\right)_0 [t + \tfrac{4}{9}\zeta_1^2 t^3 + \cdots] \tag{5-138}$$

This clearly shows an accelerating effect of the deformation on the diffusion of Γ.

Let us return to the mixing-length theories and see how the above considerations affect these theories. The lumps of fluid that are assumed to conserve a transferable property over a distance equal to the mixing length are of the same order of magnitude as this length and as the Lagrangian integral scale of diffusion for a single fluid particle. On its way through the mixing length the fluid lump will be deformed by the small-scale motions of the turbulence. The deformation and stretching of fluid elements of this lump cause an increased exchange of the transferable property with the ambient fluid owing to small-scale eddy diffusion and accelerated molecular diffusion, as expounded above. The complexity of this process of diffusion and, hence, of exchange of the transferable property between the lump and its environment makes it most improbable that the exchange will follow a simple law like Eq. (5-44). As noted, Burgers assumed that the linear relation (5-44) applied only for very small fluid particles, but, even in that case, the accelerated molecular diffusion caused by stretching would affect appreciably the value of the coefficient of exchange α that occurs in Eq. (5-44).

Relationship between Eulerian and Lagrangian correlations

Because the two correlations are determined by the same turbulent field, it is reasonable to expect a more or less close relationship between the two. Yet the problem of obtaining theoretically this relationship has not yet been solved, except for very short diffusion times. Indeed, as we shall show, relations can be obtained between Lagrangian and Eulerian dissipation scales, though these relations generally involve higher-order velocity correlations, about which nothing or very little is known.

For longer diffusion times such a relationship will be very complicated, even for homogeneous and isotropic turbulence. Let us consider, for instance, the Lagrangian correlation between the velocities v_2 in the x_2-direction of a marked fluid particle at two instants:

$$[\mathbf{Q}_{2,2}(\tau)]_L = \overline{v_2(t)v_2(t+\tau)}$$

The marked fluid particle is assumed to be at the point with coordinates y_k at time t. It has traveled a distance η_k during the time τ. The average will be taken as an ensemble average.

The velocity component v_2 of the marked fluid particle will be equal to the velocity u_2 at the point that is just being passed by the fluid particle. Consequently,

$$v_2(t) = u_2(y_k,t) \qquad \text{(5-139a)}$$

and

$$v_2(t+\tau) = u_2(y_k + \eta_k, t+\tau) \qquad \text{(5-139b)}$$

where η_k and τ are connected by

$$\eta_k = \int_t^{t+\tau} v_k(t')\,dt' \qquad \text{(5-140)}$$

Hence

$$\overline{v_2(t)v_2(t+\tau)} = \overline{u_2(y_k,t)u_2(y_k+\eta_k,t+\tau)}$$

or

$$[\mathbf{Q}_{2,2}(\tau)]_L = [\mathbf{Q}_{2,2}(\eta_k,\tau)]_E \qquad (5\text{-}141)$$

The Eulerian correlation on the right-hand side is neither the usual spatial double velocity correlation between two points at distance η_k taken at the same instant nor the double velocity correlation at one point at a time interval τ, but is a space and time correlation. We may note that in this Eulerian space-time correlation η_k and τ are not independent, but related by Eq. (5-140). Also, since $v_k(t)$ is a random function, the same will be so for $\eta_k(t)$.

Chandrasekhar[26] has studied, for a steady isotropic turbulence, the kinematics and dynamics of a double velocity correlation with respect to both space and time. It appears possible to express this double velocity correlation in terms of one defining scalar, e.g., $\mathbf{f}(r,\tau)$, the longitudinal-double-velocity-correlation coefficient between two points at a spatial distance r and with a time interval τ:

$$\mathbf{f}(r,\tau) = \frac{\overline{u_r(0,0)u_r(r,\tau)}}{u'^2}$$

It is reasonable to expect a relation between $\mathbf{f}(r,\tau)$ and the longitudinal-spatial-correlation coefficient $\mathbf{f}(r)$ and the Eulerian-time-correlation coefficient $\mathbf{R}_E(\tau) = \overline{u(t)u(t+\tau)}/u'^2$. But as yet nothing is known about such a relation for the general case of arbitrary values of r and τ.

It is possible to obtain a relation by expanding $\mathbf{f}(r,\tau)$ into a series with respect to r and τ, with coefficients that are the derivatives of $\mathbf{f}(r)$ and $\mathbf{R}_E(\tau)$ for $r = 0$ and $\tau = 0$. The same procedure may, of course, be applied directly to the relation (5-141). Thus

$$[\mathbf{Q}_{2,2}(\tau)]_L = [\mathbf{Q}_{2,2}(0)]_L + \frac{\tau^2}{2}\left[\frac{d^2(\mathbf{Q}_{2,2})_L}{d\tau^2}\right]_{\tau=0} + \frac{\tau^4}{4!}\left[\frac{d^4(\mathbf{Q}_{2,2})_L}{d\tau^4}\right]_{\tau=0} + \cdots$$

$$= [\overline{v_2{}^2}]_{\tau=0} + \frac{\tau^2}{2}\left[\overline{v_2\frac{d^2v_2}{d\tau^2}}\right]_{\tau=0} + \frac{\tau^4}{4!}\left[\overline{v_2\frac{d^4v_2}{d\tau^4}}\right]_{\tau=0} + \cdots \qquad (5\text{-}142)$$

since in a homogeneous steady turbulence

$$\left[\overline{v_2\frac{dv_2}{d\tau}}\right]_{\tau=0} = 0 \qquad \left[\overline{v_2\frac{d^3v_2}{d\tau^3}}\right]_{\tau=0} = 0 \qquad \cdots$$

Similarly for $[\mathbf{Q}_{2,2}(\eta_k,\tau)]_E$ we obtain

$$[\mathbf{Q}_{2,2}(\eta_k,\tau)]_E = \overline{u_2{}^2} + \frac{\eta_k\eta_l}{2}\left[\overline{u_2\frac{\partial^2u_2}{\partial\eta_k\,\partial\eta_l}}\right]_{\substack{\tau=0\\ \eta_k=0}} + \frac{\tau^2}{2}\left[\overline{u_2\frac{\partial^2u_2}{\partial\tau^2}}\right]_{\substack{\tau=0\\ \eta_k=0}}$$

$$+ \eta_k\tau\left[\overline{u_2\frac{\partial^2u_2}{\partial\eta_k\,\partial\tau}}\right]_{\substack{\tau=0\\ \eta_k=0}} + \cdots \qquad (5\text{-}143)$$

The relation (5-140) may also be expanded in a series

$$\eta_k = [v_k]_{\tau=0}\tau + \left[\frac{dv_k}{d\tau}\right]_{\tau=0}\frac{\tau^2}{2} + \cdots \tag{5-144}$$

This procedure of series expansion to obtain a relation between Lagrangian and Eulerian correlations has practical value only if applied to short time intervals τ. The above considerations, however, illustrate well how involved the problem is and what difficulties are encountered in finding the relation considered.

Let us consider further only very short time intervals τ. From Eqs. (5-142) to (5-144) we may obtain a relation between $\overline{[v_2\, d^2v_2/d\tau^2]}_{\tau=0}$ and the corresponding expressions for the Eulerian velocity $u_2(y_k,t)$. However, this relation can be obtained readily also by starting from the relation

$$\frac{dv_2}{d\tau} = \frac{\partial u_2}{d\tau} + u_i\frac{\partial u_2}{\partial x_i} \tag{5-145}$$

Since in a homogeneous steady turbulence

$$\overline{\left[v_2\frac{d^2v_2}{d\tau^2}\right]_{\tau=0}} = -\overline{\left[\frac{dv_2}{d\tau}\right]^2_{\tau=0}}$$

we obtain the relation required by squaring both sides of the relation (5-145). Thus

$$-\overline{\left[v_2\frac{d^2v_2}{d\tau^2}\right]_{\tau=0}} = \overline{\left(\frac{dv_2}{d\tau}\right)^2} = \overline{\left(\frac{\partial u_2}{\partial \tau}\right)^2} + 2\overline{u_i\frac{\partial u_2}{\partial \tau}\frac{\partial u_2}{\partial x_i}} + \overline{u_iu_j\frac{\partial u_2}{\partial x_i}\frac{\partial u_2}{\partial x_j}} \tag{5-146}$$

Now we have $\overline{(dv_2/d\tau)^2} = 2\overline{u_2^2}/\tau_L^2$ and $\overline{(\partial u_2/\partial\tau)^2} = 2\overline{u_2^2}/\tau_E^2$. Furthermore we may follow Frenkiel[27] in expressing the last two terms on the right-hand side of Eq. (5-146) in terms of the corresponding correlation coefficients and the dissipation time scale τ_E and the dissipation spatial scales λ_{2i}. If, moreover, we assume *isotropy*, so that all λ_{2i} can be expressed in terms of λ_g, eventually we obtain the relation

$$\frac{1}{\tau_L^2} = \frac{1}{\tau_E^2} + A\frac{u'}{\tau_E\lambda_g} + B\frac{u'^2}{\lambda_g^2} \tag{5-147}$$

where A and B are numerical constants, involving among other things the values of the correlation coefficients for $\tau = 0$ and $x_i = 0$. Since the values of these correlation coefficients and, therefore, those of A and B are unknown, the relation (5-147) between Lagrangian and Eulerian dissipation scales is of no practical value for calculating the Lagrangian dissipation scale if the Eulerian dissipation scales are known, for instance, from direct measurements.

Another method, which leads to a relation between Lagrangian and Eulerian dissipation scales that does not contain unknown coefficients, is due to Uberoi and Corrsin[28] and given by Eq. (5-69). Since $\tau_E^2 = \text{const.}\, u^2/\lambda_g^2$, one concludes from Eq. (5-69) that $\tau_L/\tau_E = \text{const.}\ \mathbf{Re}_\lambda^{1/2}$, where the const. $= \mathcal{O}(1)$. By assuming Taylor's hypothesis of a frozen turbulence, and a statistical independency between the small and large eddies, Tennekes[121] obtained the same relationship, from which a value of the const. $= (3/4)(3/5)^{1/4} \simeq 0.66$ results.

For very small values of $\mathbf{Re}_\lambda$, we may use the expression (3-45) for $\mathbf{f}(r)$, which in terms of λ_g reads

$$\mathbf{f}(r) = \exp(-r^2/2\lambda_g^2)$$

From Eq. (5-68) we then obtain

$$\overline{\left(\frac{dv_2}{dt}\right)^2} = \frac{u'^4}{\lambda_g^2}\left(2 + 35\frac{1}{\mathbf{Re}_\lambda^2}\right) \simeq 35\frac{u'^4}{\lambda_g^2}\frac{1}{\mathbf{Re}_\lambda^2}$$

for very small values of $\mathbf{Re}_\lambda$; which means that the effect of the pressure term in Eq. (5-69) may be neglected.

With such a large viscosity effect, we cannot disregard the decay as we have done hitherto, assuming a steady turbulence: Uberoi and Corrsin showed that, in a decaying turbulence, we have

$$\frac{2u'^2}{\tau_L^2} = \overline{\left(\frac{dv_2}{dt}\right)^2} - \left(\frac{dv_2'}{dt}\right)^2$$

With $dv_2'/dt = du'/dt = -5\nu u'/\lambda_g^2$, we obtain finally

$$\frac{2u'^2}{\tau_L^2} \simeq 35\frac{u'^4}{\lambda_g^2}\frac{1}{\mathbf{Re}_\lambda^2} - 25\nu^2\frac{u'^2}{\lambda_g^4} = 10\frac{u'^4}{\lambda_g^2}\frac{1}{\mathbf{Re}_\lambda^2}$$

or

$$\frac{\lambda_g^2}{u'^2\tau_L^2} \simeq \frac{5}{\mathbf{Re}_\lambda^2} \tag{5-148}$$

There is some experimental evidence concerning the ratio $\lambda_g/u'\tau_L$, obtained by Uberoi and Corrsin[28] in a grid-generated isotropic turbulence, but the data showed so much scatter that it was not possible to obtain reliable confirmation of the theoretical relation (5-69), except for the order of magnitude.

With Eqs. (5-69) and (5-148) we have obtained direct relations between the Lagrangian dissipation scale τ_L and the Eulerian spatial dissipation scale λ_g. It is possible and probable that relations also exist between the Lagrangian and Eulerian spatial integral scales, but for these integral scales we would have to consider long diffusion times. We do know that the relations in that case become very involved and that the problem of finding them is at present insoluble.

On the other hand, experimental evidence has shown that, at least at large Reynolds numbers, the over-all shapes of the Lagrangian-correlation coefficient $\mathbf{R}_L(\tau)$ and the longitudinal-spatial-correlation coefficient $\mathbf{f}(r)$ can be approximated by the exponential functions $\exp(-\tau/\mathfrak{I}_L)$ and $\exp(-r/\Lambda_f)$, respectively, except very close to the vertex. Thus the correlation coefficients have the same functional form. We may now assume that these Lagrangian and Eulerian correlation coefficients in general have the same functional form; in this case the Lagrangian and Eulerian integral scales must be proportional. Such an assumption has been made by many investigators, who made diffusion measurements. We may mention Uberoi and

Corrsin,[28] Mickelsen,[30] Hay and Pasquil,[80] Baldwin and Mickelsen.[81] Thus if $\mathbf{f}(r) = \varphi(r/\Lambda_f)$ and $\mathbf{R}_L(\tau) = \varphi(\tau/\mathfrak{I}_L)$, we must have

$$\tau = \frac{r\mathfrak{I}_L}{\Lambda_f} = \frac{r}{u'}\frac{\Lambda_L}{\Lambda_f} = \beta\frac{r}{u'} \tag{5-149}$$

where $\beta = \Lambda_L/\Lambda_f$.

Unfortunately not all experimenters mentioned used the same definition of β, while also different integral scales have been considered depending on the experimental technique used. Apart from the fact that β is not a universal constant but depends on the kind of turbulent flow, so being at least a function of $\mathbf{Re}_\lambda$, a difference in the value of β obtained by two experimenters may also be attributed to a difference in technique used.

The Lagrangian integral length-scale Λ_L is usually deduced from lateral-diffusion measurements, i.e., from $[\mathbf{Q}_{2,2}(t)]_L$, whence follows $(\mathfrak{I}_{2,2})_L$ and so, according to Eq. (1-88), $\Lambda_L = v_2'(\mathfrak{I}_{2,2})_L$.

Now the Eulerian integral length-scale Λ_f should be determined from direct measurements of the two-point, one-time, velocity correlation $[\mathbf{Q}_{1,1}(x_1,0)]_E = \overline{u_1^2}f(x_1)$, with two probes. However, in many cases Λ_f has been determined from measurements with one probe of the one-point Eulerian time-correlation $[\mathbf{Q}_{1,1}(0,t)]_E$. This correlation should refer to the turbulence proper, with zero mean velocity. Since actually there is a mean velocity, e.g., $\bar{U}_1$, this correlation can only be measured properly with a probe moving with this velocity $\bar{U}_1$. If, as is usually done, a time correlation is measured with a stationary probe in such a moving turbulence, the correlation measured is not the above $[\mathbf{Q}_{1,1}(0,t)]_E$. In the way in which such measurements with a stationary probe has been evaluated we may distinguish between two methods. In the one (usual) method Taylor's hypothesis of a "frozen" turbulence is assumed to apply. For a "frozen" turbulence with $\bar{U}_1 = 0$.

$$\frac{1}{\overline{u_1^2}}[\mathbf{Q}_{1,1}(0,t)]_E = 1$$

Actually the turbulence is not frozen, and $[\mathbf{Q}_{1,1}(0,t)]_E \leq 1$. The assumption made is

$$\{[\mathbf{Q}_{1,1}(0,t)]_E\}_{\text{meas}} \simeq \{[\mathbf{Q}_{1,1}(0,t)]_E\}_{\text{froz}} \simeq [\mathbf{Q}_{1,1}(x_1 = \bar{U}_1 t,0)]_E, \tag{5-150}$$

where the second term refers to a stationary system, with respect to which the turbulence moves in a frozen condition with $\bar{U}_1 \neq 0$.

In the other, more exact method for evaluating the one-point correlation measurements one has to appreciate that

$$\{[\mathbf{Q}_{1,1}(0,t)]_E\}_{\text{meas}} \neq \{[\mathbf{Q}_{1,1}(0,t)]_E\}_{\text{froz}} \neq [\mathbf{Q}_{1,1}(\bar{U}_1 t,0)]_E$$

being actually smaller, since

$$\{[\mathbf{Q}_{1,1}(0,t)]_E\}_{\text{meas}} = [\mathbf{Q}_{1,1}(-\bar{U}_1 t,t)]_E \tag{5-151}$$

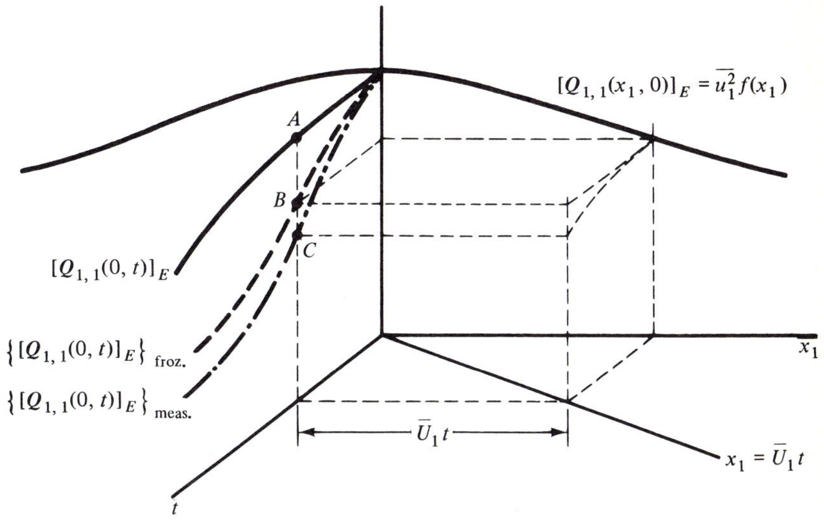

FIGURE 5-2a
Relation between Eulerian correlations.

i.e., equal to the Eulerian space-time correlation. Figure 5-2a shows the three time-correlations considered: $[\mathbf{Q}_{1,1}(0,t)]_E$, $\{[\mathbf{Q}_{1,1}(0,t)]_E\}_{\text{froz}}$, and $\{[\mathbf{Q}_{1,1}(0,t)]_E\}_{\text{meas}}$, the latter two as derived from the actual Eulerian space correlation $[\mathbf{Q}_{1,1}(x_1,0)]_E$. Corresponding with the different Eulerian time-correlations we have to distinguish between $\mathfrak{I}_E$, $(\mathfrak{I}_E)_{\text{froz}}$ and $(\mathfrak{I}_E)_{\text{meas}}$. Figures 5-2b and 5-2c may be useful in elucidating the behavior of Eulerian space-time correlations in a homogeneous stationary turbulence moving with a constant mean-velocity $\bar{U}_1$. Two probes, e.g., two hot-wire anemometers, are placed such that one is at the origin $x_1 = 0$ of the stationary coordinate system, the other at $x_1 = x_{1m}$. The correlations between the velocities measured by the probes can be determined by either keeping the time-delay constant, or the distance x_{1m} constant, or by varying both.

Figure 5-2b refers to the hypothetical "frozen" condition. The correlation curves in the $(\mathbf{Q}_{1,1} - x_1)$-plane are the spatial correlation with constant time-delay $t_m = x_{1m}/\bar{U}_1$.

The correlation curves in the $(\mathbf{Q}_{1,1} - t)$-plane are the time-correlations measured with the second probe at constant distance x_{1m} from the origin. Because of the "frozen" condition the correlations are moved unchanged with time in space, and so with constant peak-values. In the $(\bar{U}_1 t - x_1)$ mapping shown in the figure, the iso-correlation curves are straight parallel lines with a 45°-slope. Figure 5-2c shows the change with time and in space of the correlations for a "nonfrozen" turbulence. The change with time is expressed by the purely time-correlation $[\mathbf{Q}_{1,1}(0,t)]_E$ when the mean velocity $\bar{U}_1$ would be zero. The result is a decrease of the correlation-curves with time or distance as shown in Fig. 5-2c. The absolute maximum of the

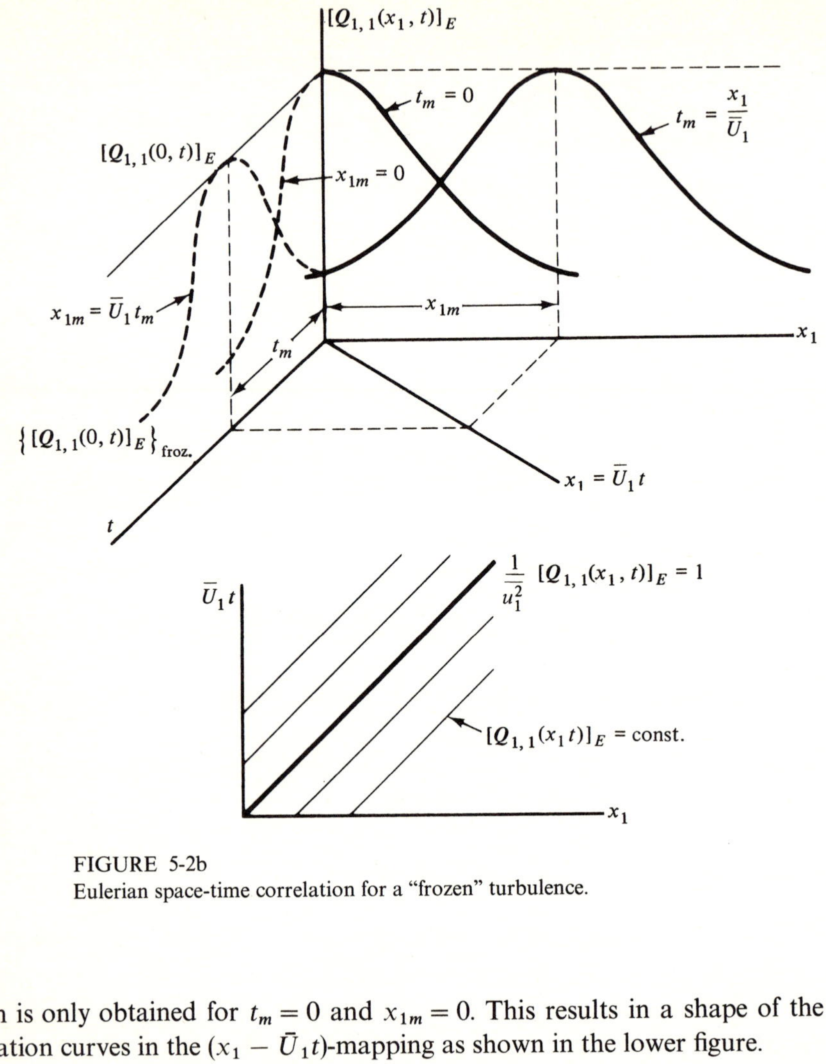

FIGURE 5-2b
Eulerian space-time correlation for a "frozen" turbulence.

correlation is only obtained for $t_m = 0$ and $x_{1m} = 0$. This results in a shape of the iso-correlation curves in the $(x_1 - \bar{U}_1 t)$-mapping as shown in the lower figure.

It is of interest to know what the cause may be for a "nonfrozen" condition, that is, for a not-constant value of $[\mathbf{Q}_{1,1}(0,t)]_E$. Various factors may be effective, namely, viscous dissipation, transfer of energy or diffusion of correlations by turbulence interactions, local production of turbulence, and distortion of turbulence by a mean shear. In a homogeneous turbulence the last two factors are nonexistent, while one may expect that viscous dissipation effects are negligible during the correlation times considered. So there remains the possibility of loss of correlation due to turbulent diffusion. Kovasznay[82] has worked out this possibility. We will come back to this later, at a more appropriate place, in Sec. 5-5.

The earliest *measurements* that will be considered here in determining values of β, are those made by Corrsin and Uberoi.[28] They measured the Lagrangian time correlation and the Eulerian spatial, longitudinal correlation in a grid-produced

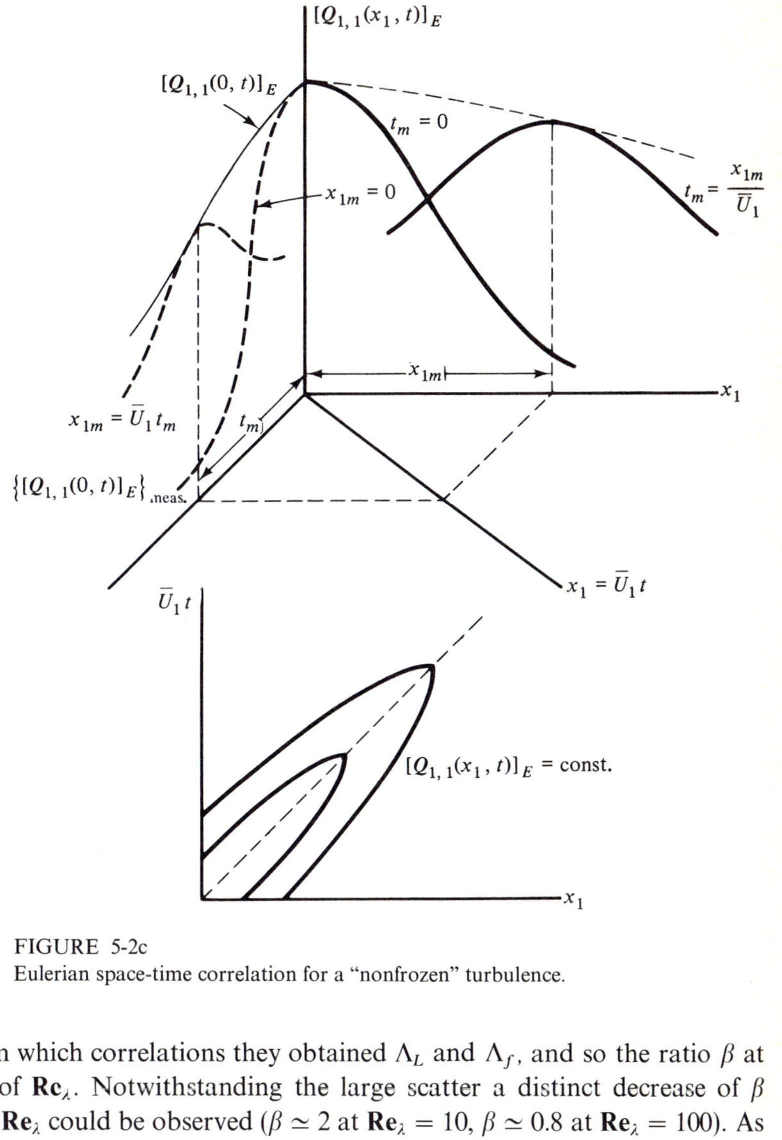

FIGURE 5-2c
Eulerian space-time correlation for a "nonfrozen" turbulence.

turbulence, from which correlations they obtained Λ_L and Λ_f, and so the ratio β at various values of $\mathbf{Rc}_\lambda$. Notwithstanding the large scatter a distinct decrease of β with increasing $\mathbf{Re}_\lambda$ could be observed ($\beta \simeq 2$ at $\mathbf{Re}_\lambda = 10$, $\beta \simeq 0.8$ at $\mathbf{Re}_\lambda = 100$). As shown in Chap. 3, $\mathbf{Re}_\lambda$ decreases with time in a decaying turbulence, except during the initial period, because u' decreases at a faster rate than λ_g increases. So Corrsin's and Uberoi's results may indicate an increase of β with decreasing u'. Also Mickelsen's experiments[30] in the turbulent core-region of pipe flow ($D = 0.2$ m; $\mathbf{Re}_D \simeq 5 \times 10^5$) seem to show the same trend. Mickelsen measured separately $[\mathbf{Q}_{1,1}(x_1,0)]_E$, and $[\mathbf{Q}_{2,2}(t)]_L$ from the diffusion of helium downstream of a fixed injection point. The average value of β obtained from these measurements was rather high, namely ~ 1.6. Baldwin and Walsh,[83] however, reported that after applying a correction a much lower value is obtained, varying from $\beta = 0.5$ at $\bar{U}_1 = 15$ m/s to $\beta \simeq 0.2$ at $\bar{U}_1 = 48$ m/s. However, experiments by Baldwin and Walsh themselves under the same flow conditions, with a heat-linesource placed across the pipe cross-section yielded values of β ranging from 0.29 to 0.15 in the speed range $\bar{U}_1 = 20$ to 50 m/s. In accord with these values Gifford[84] suggested values between 0.2 and 0.4. Much greater variations are obtained from the atmospheric diffusion experiments by Hay and Pasquil.[80] They measured the one-point Eulerian time-correlation $\{[\mathbf{Q}_{1,1}(t)]_E\}_{\text{meas}}$, and so determined the ratio $\mathfrak{I}_L/(\mathfrak{I}_E)_{\text{meas}}$. The value of β has been obtained from $\Lambda_L = u'\mathfrak{I}_L$, and by assuming $\Lambda_f \simeq \bar{U}_1(\mathfrak{I}_E)_{\text{meas}}$, which is not entirely correct, as discussed earlier. When one extreme value of $\beta = 1.05$ is disregarded, the variations of β are between 0.15 and 0.6. The experiments by Mickelsen in turbulent pipe-flow have also been repeated and extended later by Baldwin and Mickelsen.[81] In addition to the experiments with helium, they now also made experiments with heat as a tracer. The mean-velocity in the axis of the pipe was varied between $\bar{U}_1 = 15$ m/s and $\bar{U}_1 = 49$ m/s. The relative intensity of the turbulence was almost constant $u'_1/\bar{U}_1 \simeq 0.03$. Again a decrease of β with increasing u'_1 is obtained, with β varying between 0.2 and 0.4 for helium, and between 0.15 and 0.3 for heat, the same values as reported by Baldwin and Walsh, and referred to earlier. These values are in reasonable agreement with those obtained from Hay's and Pasquil's experiments.

When we exclude the values obtained from Uberoi's and Corrsin's experiments, the above results then show $\beta < 1$. However, this result does not imply $\mathfrak{I}_L/\mathfrak{I}_E < 1$, since $\Lambda_f \neq \bar{U}_1\mathfrak{I}_E$, but $\Lambda_f \simeq \bar{U}_1(\mathfrak{I}_E)_{\text{meas}}$.

Calculations by Corrsin and Lumley on a one-dimensional random walk with both Lagrangian and Eulerian statistics resulted in values $\mathfrak{I}_L/\mathfrak{I}_E \lesseqgtr 1$, depending on the statistics applied to the random walk. On the other hand Kraichnan[86] showed that it is easy to devise a model of turbulent flow for which definitely $\mathfrak{I}_L/\mathfrak{I}_E < 1$. Assume a time-independent, but random spatial, Eulerian velocity yield generated by randomly distributed pumps in the fluid. Because of the stationarity of the flow $\mathbf{R}_E(t) = 1$ with an infinite value of $\mathfrak{I}_E$, whereas the Lagrangian integral time-scale is finite. More recent *computer experiments on random walks* with both Eulerian and Lagrangian statistics by Patterson and Corrsin[87] showed that no single Eulerian two-point correlation function can be a good approximation of the Lagrangian correlation. For a certain "experiment" the Lagrangian function was below the Eulerian one, if the time separation only was taken, though the integral time scales were close to each other. However, when inertia effects were introduced an increase of the Lagrangian integral time-scale was obtained.

Also attempts have been made to solve the problem by starting from the formal relations given by Eq. (5-141). The main problem is that $\eta_k(t)$ is a statistical quantity. For long diffusion times we may assume some statistical average of $\eta_k(t)$, since then the individual particle displacements will be small compared with the total displacement and practically uncorrelated with it. We may, for instance, consider $(\overline{\eta_k\eta_k})^{1/2}$, whose value, however, requires again the Lagrangian correlation. A useful suggestion has been made by Corrsin,[88] namely to weight η_k with a probability function. For diffusion in the x_2-direction, the formal expression for $[\mathbf{Q}_{2,2}(t)]_L$ then reads

$$[\mathbf{Q}_{2,2}(t)]_L = \iiint_{-\infty}^{+\infty} dy_1\, dy_2\, dy_3\, [\mathbf{Q}_{2,2}(y_k,t)]_E \mathfrak{Q}(y_k,t) \qquad (5\text{-}152)$$

where $\mathfrak{Q}(y_k,t)$ is a probability-density function for a displacement y_k in time t. This expression presumes sufficiently long diffusion times, so that the velocity of a fluid particle has no correlation any more with the total distance covered. Saffman[90] has pursued Corrsin's suggestion. At the same time he assumed a *stationary, homogeneous and isotropic turbulence,* with zero mean-velocity, and Gaussian distributions for the Fourier transforms of $[\mathbf{Q}_{2,2}(y_k,t)]_E$ and $\mathfrak{Q}(y_k,t)$. The corresponding expressions for the Eulerian space-time correlation and for the displacement probability-density then are also Gaussian, and turn out to be:†

$$[\mathbf{Q}_{2,2}(y_k,t)]_E = \frac{\overline{u^2}}{a^{5/2}}\left(1 - \frac{k_0{}^2}{4}\frac{y_1{}^2 + y_3{}^2}{a}\right) \exp\left[-(k_0{}^2 y_i y_i/4a\right] \qquad (5\text{-}153)$$

where $a = 1 + \frac{1}{2}k_0{}^2\overline{u^2}t^2$, and

$$k_0 = \sqrt{\pi}/\Lambda_f. \qquad (5\text{-}155)$$

$$\mathfrak{Q}(y_k,t) = \frac{1}{[2\pi\overline{y_2{}^2}(t)]^{3/2}} \exp\left[-\frac{y_i y_i}{2\overline{y_2{}^2}(t)}\right] \qquad (5\text{-}156)$$

Substitution of Eqs. (5-153) and (5-156) in Eq. (5-152), yields after carrying out the integration.

$$[\mathbf{Q}_{2,2}(t)]_L = \overline{u^2}[a + \tfrac{1}{2}k_0{}^2\overline{y_2{}^2}(t)]^{-5/2} = \overline{u^2}[1 + \tfrac{1}{2}k_0{}^2\overline{u^2}t^2 + \tfrac{1}{2}k_0{}^2\overline{y_2{}^2}(t)]^{-5/2} \qquad (5\text{-}157)$$

Since $[\mathbf{Q}_{2,2}(t)]_L = \frac{1}{2}\, d^2/dt^2\, \overline{y_2{}^2}$, Eq. (5-157) gives a nonlinear differential equation for $\overline{y_2{}^2}$. When $t \to \infty$, the equation yields $d^2\overline{y_2{}^2}/dt^2 = 0$, or $\overline{y_2{}^2} = \text{const.}\, t$, as it should. We may then also conclude that for large times the second term on the right-hand side of Eq. (5-157) becomes the dominating term, so that the asymptotic behavior of the Lagrangian correlation is algebraic according to t^{-5}. Consequently, not an exponential behavior, as has been usually assumed. Saffman has solved the differential equation for $\overline{y_2{}^2}(t)$, numerically, thereby dropping Corrsin's assumption that Eq. (5-152) may be approximately true for long diffusion times only. With the solution for $\overline{y_2{}^2}(t)$, the Lagrangian correlation can be calculated. The Lagrangian integral time-scale is found to be

$$\mathfrak{J}_L = \frac{1}{\overline{u^2}}\int_0^\infty dt\, [\mathbf{Q}_{2,2}(t)]_L = \frac{0.71}{u'k_0} = \frac{0.40}{u'}\Lambda_f \qquad (5\text{-}158)$$

† The Fourier transform of $\mathbf{Q}_{2,2}(y_k,t)_E$ contains the term $\exp[-\overline{u^2}k^2t^2/2]$ for the correlation with time. This form has been suggested by Kraichnan[89] who showed that when the probability-density of u is Gaussian, as a consequence we must have

$$[\mathbf{E}(k,t)]_E = \mathbf{E}(k)\exp[-\overline{u^2}k^2t^2/2]$$

Whence follows $\beta = u'_L/\Lambda_f = 0.40$. This value agrees nicely with the average value obtained experimentally, though it must be remarked that the same experiments reveal a rather weak, but distinct, dependence on the intensity of the turbulence, as discussed earlier.

Earlier Corrsin[115] suggested, making a number of rather crude assumptions, the relation $\mathfrak{I}_L \simeq \frac{2}{3}\Lambda_f/u'$, whence would follow $\beta \simeq \frac{2}{3}$, which is of the same order as the experimental values. More recently Shlien and Corrsin[122] obtained $\beta \simeq 1$ from direct measurements of Λ_f and of $\mathfrak{I}_L$ in a grid-generated turbulence, using heat as a tracer. In contrast to this, experiments by Snyder and Lumley[112] yielded $\beta \simeq 0.4$.

Saffman's calculations also yield an asymptotic value for the eddy-diffusion coefficient $\epsilon_{2,2}$. With $\beta = 0.40$, this corresponds to $t \simeq 2.8\mathfrak{I}_L$, which again agrees satisfactorily with the value $t \simeq 3\mathfrak{I}_L$, usually taken for an exponential behavior of the Lagrangian correlation coefficient (see Sec. 5.5). From the result given by Eq. (5-158) we may calculate the value of the constant C occurring in Eq. (5-61), when at the same time we make use of the relations (3-158) and $k_e = 0.51\varepsilon/\alpha u'^3$ for an isotropic turbulence. Assuming the value 0.4 for the Heisenberg constant α, we obtain $C \simeq 8.5$, which is probably on the high side. This would yield

$$\mathfrak{I}_L = \frac{2}{C}\frac{\overline{u^2}}{\varepsilon} \simeq \frac{2}{8.5}\frac{\overline{u^2}}{\varepsilon} \simeq 0.08\overline{q^2}/\varepsilon$$

the numerical constant being rather close to the value of $\frac{1}{8}$, proposed by Lumley.[116]

With the expression (5-153) for $[\mathbf{Q}_{2,2}(y_k,t)]_E$ we may calculate

$$(\mathfrak{I}_E)_{\text{meas}} = \frac{1}{\overline{u^2}}\int_0^\infty dt\,[\mathbf{Q}_{2,2}(-\bar{U}_1 t,0,0;t)]_E$$

and

$$\mathfrak{I}_E = \frac{1}{\overline{u^2}}\int_0^\infty dt\,[\mathbf{Q}_{2,2}(0,0,0;t)]_E$$

For $(\mathfrak{I}_E)_{\text{meas}}$ Saffman obtained a relation, which for small relative intensity $u'/\bar{U}_1 \ll 1$ reduces to

$$(\mathfrak{I}_E)_{\text{meas}} = \frac{\sqrt{\pi}}{2\bar{U}_1 k_0} = 1.25\frac{u'}{\bar{U}_1}\mathfrak{I}_L$$

For $\mathfrak{I}_E$ there is obtained

$$\mathfrak{I}_E = \frac{\sqrt{2}}{3k_0 u'}$$

Hence

$$\frac{(\mathfrak{I}_E)_{\text{meas}}}{\mathfrak{I}_E} = \frac{3\sqrt{\pi}}{2\sqrt{2}}\frac{u'}{\overline{\overline{U_1}}} = 1.88\frac{u'}{\overline{\overline{U_1}}}$$

which shows that the "frozen" condition is obtained with $\mathfrak{I}_E \gg (\mathfrak{I}_E)_{\text{meas}}$ when $\overline{u'/U_1} \to 0$.

A similar procedure for obtaining a relation between Lagrangian and Eulerian statistics has been followed by Philip.[91] He too started from Eq. (5-152) and assumed Gaussian distributions for $[\mathbf{Q}_{1,1}(y_k,t)]_E$ and $\mathfrak{Q}(y_k,t)$, but different from the expressions (5-153) and (5-156), namely

$$[\mathbf{Q}_{1,1}(x,r;t)]_E = \overline{u^2} \exp\left[-\frac{\pi}{4}\left(\frac{x^2 + 3.138r^2}{\Lambda_f^2} + \frac{t^2}{\mathfrak{I}_E^2}\right)\right]$$

and

$$\mathfrak{Q}(x,r;t) = (2\pi\overline{y_1^2})^{-3/2} \exp\left[-\frac{x^2 + r^2}{2\overline{y_1^2}}\right]$$

Though Philip assumed isotropy, he chose a cylindrical coordinate system with the x-axis in the mean-flow direction and considered diffusion in the x-direction, determined by $\overline{y_1^2}$. He obtained numerically a solution for $\mathfrak{I}_L/\mathfrak{I}_E$ as a function of $u'\mathfrak{I}_E/\Lambda_f$. Comparison with experimental data obtained by Mickelsen, and by Hay and Pasquil showed a satisfactory agreement when $u'\mathfrak{I}_E/\Lambda_f \simeq 1$. The corresponding value of $\mathfrak{I}_L/\mathfrak{I}_E$ was 0.36, and consequently $\beta = \Lambda_L/u'\mathfrak{I}_E = \Lambda_L/\Lambda_f = 0.36$. This value is close to the value of 0.4 obtained by Saffman. Apparently the experiments mentioned suggest $\Lambda_f = u'\mathfrak{I}_E$.

Philip also calculated $(\mathfrak{I}_E)_{\text{meas}}$. Again with $\Lambda_f = u\mathfrak{I}_E$ he obtained

$$\frac{(\mathfrak{I}_E)_{\text{meas}}}{\mathfrak{I}_E} = \frac{\overline{u'/U_1}}{\sqrt{1 + \overline{u^2}/\bar{U}_1{}^2}} \simeq \overline{u'/U_1}$$

when $\overline{u'/U_1} \ll 1$.

In their numerical experiments, mentioned earlier, Deardorff and Peskin[77] also calculated the single-particle diffusion in the three x_i-directions in the turbulent channel flow, with the point of release at 25 per cent height from one of the walls. They found that the Lagrangian correlation in axial and spanwise direction could fairly well be approximated by an exponential function. Comparison with the corresponding one-time spatial Eulerian correlations as calculated by Deardorff[92] in similar numerical experiments revealed a value of $\beta = 0.14$, which agrees nicely with the value of 0.15 obtained by Baldwin and Walsh,[83] referred to earlier, and by Hay and Pasquil in the experiments with the highest Reynolds number, but still much smaller than the above value of 0.4. In the foregoing, both Saffman and Philip assumed Eulerian space-time correlations of a Gaussian form. Such a Gaussian form is also included partly in the expression for the Eulerian space-time correlation suggested by Kovaznay,[82] to which reference has been made earlier, and which will be discussed later at the end of Sec. 5-5.

This expression simplifies in the case of isotropy, since then $\overline{y_1^2} = \overline{y_2^2} = \overline{y_3^2}$. In that case Eq. (3-11) may be taken for $[\mathbf{Q}_{1,1}(\xi_k,0)]_E$. In principle it will be possible to determine with the above expression the space-time correlation by assuming a suitable function for $\mathbf{f}(r)$, and so to calculate the Lagrangian correlation. An attempt to compare Lagrangian correlations as obtained from diffusion experiments with those calculated from space-time correlations has been made by Rodriguez and Patterson,[93] though for very small diffusion times.

5-5 DIFFUSION FROM A FIXED SOURCE IN A UNIFORM FLOW

We shall consider here the turbulent diffusion of matter or heat emitted continuously from a fixed source, either a point or infinite line, placed in a turbulent flow. It is assumed that the amount of matter or heat released per unit of time and volume is so small that any effect on the turbulence may be neglected. The mean velocity of the turbulent flow will be assumed to be uniform, with a constant velocity $\bar{U}_1$ in the x_1-direction. In the next section we shall consider the case of turbulent shear flow where the velocity $\bar{U}_1$ shows a gradient in the x_2-direction only.

Most experiments on turbulent diffusion have been concerned with diffusion from a fixed source, and the available data may be used to check pertinent theories. Important theoretical contributions have been made in a series of publications by Frenkiel,[32-35] where the distribution of fluid particles is studied for small and large distances from the source, i.e., for short and long diffusion times, and for arbitrarily large values of the diffusion coefficients.

We shall consider the theoretical aspects of diffusion from a fixed source before giving the results of experimental investigations.

To begin with, we consider the diffusion and distribution in space of marked fluid particles introduced at a constant volume rate at a fixed source. The simplest case is where it is assumed that the diffusion of these marked fluid particles can be defined by a diffusion constant ϵ. This would be the case if the turbulence were homogeneous and isotropic and if long diffusion times were considered. The problem is then identical with those of molecular diffusion and of heat conduction in an isotropic medium. Solutions for the concentration or temperature distribution downstream from a point source or a line source are given, for instance, by Carslaw and Jaeger.[31] The mean concentration of marked fluid particles at a point (x_1,x_2,x_3) is identical with the probability $\mathfrak{P}(x_1,x_2,x_3)$ of finding a marked fluid particle at that point. The differential equation for this mean concentration or for $\mathfrak{P}$ reads [compare Eq. (5-94)]

$$\bar{U}_1 \frac{\partial \mathfrak{P}}{\partial x_1} = \epsilon \frac{\partial^2}{\partial x_i \, \partial x_i} \mathfrak{P} \tag{5-159}$$

The solutions are:

1 *Point source* of strength S at the origin:

$$\mathfrak{P}(x_1,x_2,x_3) = \frac{S}{4\pi r\epsilon} \exp\left[-\bar{U}_1(r - x_1)/2\epsilon\right] \tag{5-160}$$

where $r^2 = x_i x_i$.

This solution can be simplified if we consider points close to the x_1-axis, so that $x_2{}^2 + x_3{}^2 \ll x_1{}^2$ or $r/x_1 \simeq 1$:

$$\mathfrak{P}(x_1,x_2,x_3) \simeq \frac{S}{4\pi\epsilon |x_1|} \exp\left[-\frac{\bar{U}_1(x_2{}^2 + x_3{}^2)}{4\epsilon |x_1|}\right] \tag{5-160a}$$

2 *Line source* of strength S^* per unit of length and along the x_3-axis:

$$\mathfrak{P}(x_1,x_2) = \frac{S^*}{2\pi\epsilon} K_0\left(\frac{\bar{U}_1\sqrt{x_1{}^2 + x_2{}^2}}{2\epsilon}\right) \exp(\bar{U}_1 x_1/2\epsilon) \tag{5-161}$$

where K_0 is the modified Bessel function of the second kind and of zero order.

Since $K_0(z) \to (\pi/2z)^{1/2} \exp(-z)$ as $z \to \infty$, the solution (5-161) simplifies, for small $\epsilon/\bar{U}_1 x_1$ and not large values of x_2/x_1, into

$$\mathfrak{P}(x_1,x_2) \simeq \frac{S^*}{2\sqrt{\pi\epsilon\bar{U}_1|x_1|}} \exp\left(-\frac{\bar{U}_1 x_2^2}{4\epsilon|x_1|}\right) \quad (5\text{-}161a)$$

Note that the solutions (5-160) and (5-161) yield finite concentrations at points upstream from the source, decreasing rapidly with increasing upstream distance.

We have mentioned in the previous section Batchelor's proof that the probability $\mathfrak{P}(x_1,x_2,x_3,t)$ satisfies a differential equation similar to Eq. (5-159) if it is assumed that the probability-densities of the three components y_i of a diffusing marked fluid particle are separately as well as jointly Gaussian. As we shall show later on, experimental evidence has shown that, for short and long diffusion times, the probability-densities of y_i are Gaussian in a homogeneous turbulence with uniform mean velocities. Fleishman and Frenkiel[36] have shown that it is then possible to obtain solutions for the concentration distribution for long and short diffusion times directly without considering the pertinent differential equations. For short diffusion times the diffusion coefficients are functions of time. Moreover, it is not necessary to restrict ourselves to isotropic turbulence, that is, to turbulence with a diffusion coefficient independent of direction.

Let $\mathfrak{Q}(x_i - x_i',t;t_0)$ be the probability density that a marked fluid particle has arrived at the point (x_i) at time t from the point (x_i') at time t_0. We then have [see Eq. (5-82)]

$$\mathfrak{P}(x_1,x_2,x_3,t) = \iiint_{-\infty}^{+\infty} dx_1'\,dx_2'\,dx_3'\,\mathfrak{Q}(x_1 - x_1',x_2 - x_2',x_3 - x_3',t;t_0)\mathfrak{P}(x_1',x_2',x_3',t_0) \quad (5\text{-}162)$$

We assume that the principal axes of the turbulence-stress tensor coincide with the coordinate axes. The expression for the probability density $\mathfrak{Q}$ then reads

$$\mathfrak{Q} = \frac{1}{(2\pi)^{3/2}(\overline{y_1^2}\,\overline{y_2^2}\,\overline{y_3^2})^{1/2}} \exp\left\{-\frac{1}{2}\left[\frac{(x_1 - x_1')^2}{\overline{y_1^2}} + \frac{(x_2 - x_2')^2}{\overline{y_2^2}} + \frac{(x_3 - x_3')^2}{\overline{y_3^2}}\right]\right\} \quad (5\text{-}163)$$

where $\overline{y_i^2}$ are still functions of time.

If at the instant $t = 0$ a volume s of marked fluid is emitted at the origin $(x_i' = 0)$ in a flow with zero mean velocity, the probability $\mathfrak{P}(x_i,t)$ obtained from Eqs. (5-162) and (5-163) reads

$$\mathfrak{P}(x_1,x_2,x_3,t) = \frac{s}{(2\pi)^{3/2}(\overline{y_1^2}\,\overline{y_2^2}\,\overline{y_3^2})^{1/2}} \exp\left[-\frac{1}{2}\left(\frac{x_1^2}{\overline{y_1^2}} + \frac{x_2^2}{\overline{y_2^2}} + \frac{x_3^2}{\overline{y_3^2}}\right)\right]$$

Here it is assumed that $\overline{y_1^2}(t) = \overline{y_1^2}(t_0) + \overline{y_1^2}(t - t_0)$, and similarly for $\overline{y_2^2}$ and $\overline{y_3^2}$.

If at time t_0 during a short interval dt_0, marked fluid is emitted at the origin at a volume rate S, while the whole flow field has a constant mean velocity $\bar{U}_1$ in the x_1-direction, the above expression then applies to a coordinate system x_i^* moving with the flow field with the velocity $\bar{U}_1$. At time t we have, between x_1 and x_1^*, the relation $x_1^* = x_1 - \bar{U}_1(t - t_0)$, but $x_2^* = x_2$ and $x_3^* = x_3$. Hence the probability of finding marked fluid emitted during the interval dt_0 at the point (x_i) at time t reads

$$\mathfrak{P}'(x_1,x_2,x_3,t - t_0)\, dt_0$$
$$= \frac{S\, dt_0}{(2\pi)^{3/2}(\overline{y_1^2}\,\overline{y_2^2}\,\overline{y_3^2})^{1/2}} \exp\left(-\frac{1}{2}\left\{\frac{[x_1 - \bar{U}_1(t - t_0)]^2}{\overline{y_1^2}} + \frac{x_2^2}{\overline{y_2^2}} + \frac{x_3^2}{\overline{y_3^2}}\right\}\right)$$

whence, for a continuous injection with a volume rate S, we obtain

$$\mathfrak{P}(x_1,x_2,x_3)$$
$$= \int_{-\infty}^{t} dt_0 \frac{S}{(2\pi)^{3/2}(\overline{y_1^2}\,\overline{y_2^2}\,\overline{y_3^2})^{1/2}} \exp\left(-\frac{1}{2}\left\{\frac{[x_1 - \bar{U}_1(t - t_0)]^2}{\overline{y_1^2}} + \frac{x_2^2}{\overline{y_2^2}} + \frac{x_3^2}{\overline{y_3^2}}\right\}\right)$$

Or, with $\tau = t - t_0$,

$$\mathfrak{P}(x_1,x_2,x_3) = \int_0^{\infty} d\tau \frac{S}{(2\pi)^{3/2}(\overline{y_1^2}\,\overline{y_2^2}\,\overline{y_3^2})^{1/2}} \exp\left\{-\frac{1}{2}\left[\frac{(x_1 - \bar{U}_1\tau)^2}{\overline{y_1^2}} + \frac{x_2^2}{\overline{y_2^2}} + \frac{x_3^2}{\overline{y_3^2}}\right]\right\} \tag{5-164}$$

This is thus the general solution for the diffusion from a continuous *point source* in a uniform flow and homogeneous turbulence.

Let us now first consider short diffusion times, that is, small distances from the point source. In this case we may assume [see Eq. (1-83)]

$$\overline{y_i^2} = \overline{v_i^2}(t - t_0)^2 = \overline{u_i^2}\tau^2 \qquad i = 1,2,3$$

The integral (5-164) then reads

$$\mathfrak{P}(x_1,x_2,x_3)$$
$$= \frac{S}{(2\pi)^{3/2}(\overline{u_1^2}\,\overline{u_2^2}\,\overline{u_3^2})^{1/2}} \int_0^{\infty} d\tau \frac{1}{\tau^3} \exp\left\{-\frac{1}{2}\left[\frac{(x_1 - \bar{U}_1\tau)^2}{\overline{u_1^2}\tau^2} + \frac{x_2^2}{\overline{u_2^2}\tau^2} + \frac{x_3^2}{\overline{u_3^2}\tau^2}\right]\right\}$$

After performing the integration we obtain

$$\mathfrak{P}(x_1,x_2,x_3) = \frac{S \exp(-\bar{U}_1^2/2\overline{u_1^2})}{(2\pi)^{3/2}(\overline{u_1^2}\,\overline{u_2^2}\,\overline{u_3^2})^{1/2}\left(\frac{x_1^2}{\overline{u_1^2}} + \frac{x_2^2}{\overline{u_2^2}} + \frac{x_3^2}{\overline{u_3^2}}\right)}$$
$$\times \left\{1 + \sqrt{\frac{\pi}{2}} \frac{x_1\bar{U}_1 \exp\left[\frac{x_1^2\bar{U}_1^2}{2(\overline{u_1^2})^2\left(\frac{x_1^2}{\overline{u_1^2}} + \frac{x_2^2}{\overline{u_2^2}} + \frac{x_3^2}{\overline{u_3^2}}\right)}\right]}{\overline{u_1^2}\left(\frac{x_1^2}{\overline{u_1^2}} + \frac{x_2^2}{\overline{u_2^2}} + \frac{x_3^2}{\overline{u_3^2}}\right)^{1/2}}\right.$$

$$\times \operatorname{erfc}\left[\frac{-x_1\bar{U}_1}{\sqrt{2\overline{u_1^2}}\left(\dfrac{x_1^2}{\overline{u_1^2}}+\dfrac{x_2^2}{\overline{u_2^2}}+\dfrac{x_3^2}{\overline{u_3^2}}\right)^{1/2}}\right]\Bigg\} \tag{5-165}$$

where

$$\operatorname{erfc}(z) = 1 - \operatorname{erf}(z) = 1 - \frac{2}{\sqrt{\pi}}\int_0^z d\alpha \exp(-\alpha^2)$$

The solution (5-165) has only restricted validity if it is to remain consistent with the assumption of short diffusion times. According to this assumption, the diffusion time τ should remain small, say, with respect to the Lagrangian integral scale. Thus $\tau \ll \mathfrak{I}_L$. Now in the integration τ covers the range from zero to infinity. Hence the integral can be consistent with the assumption $\tau \ll \mathfrak{I}_L$ only if the contribution to the integral for greater values of τ is negligibly small. That is, the value of the integrand must become vanishingly small for values of τ comparable to $\mathfrak{I}_L$. It can be shown that this condition is fulfilled if for $x_1 > 0$

$$\frac{x_1\bar{U}_1}{\overline{u_1^2}\left(\dfrac{x_1^2}{\overline{u_1^2}}+\dfrac{x_2^2}{\overline{u_2^2}}+\dfrac{x_3^2}{\overline{u_3^2}}\right)^{1/2}} \gg 1$$

In this case the solution (5-165) reduces to

$$\mathfrak{P}(x_1,x_2,x_3) \simeq \frac{S}{2\pi(\overline{u_1^2}\,\overline{u_2^2}\,\overline{u_3^2})^{1/2}\left(\dfrac{x_1^2}{\overline{u_1^2}}+\dfrac{x_2^2}{\overline{u_2^2}}+\dfrac{x_3^2}{\overline{u_3^2}}\right)^{3/2}}$$

$$\times \frac{x_1\bar{U}_1}{\overline{u_1^2}} \exp\left[-\frac{\dfrac{\bar{U}_1^2}{2\overline{u_1^2}}\left(\dfrac{x_2^2}{\overline{u_2^2}}+\dfrac{x_3^2}{\overline{u_3^2}}\right)}{\dfrac{x_1^2}{\overline{u_1^2}}+\dfrac{x_2^2}{\overline{u_2^2}}+\dfrac{x_3^2}{\overline{u_3^2}}}\right] \tag{5-165a}$$

A much more drastic simplification is obtained on the assumption that the relative intensity of turbulence is small, $\overline{u_1^2}/\bar{U}_1^2 \ll 1$. This at the same time implies that $(x_2^2 + x_3^2)/x_1^2 \ll 1$, because $\overline{u_2^2}$ and $\overline{u_3^2}$ are of the same order as $\overline{u_1^2}$. With this condition the solution (5-165) reduces to

$$\mathfrak{P}(x_1,x_2,x_3) \simeq \frac{S\bar{U}_1}{2\pi(\overline{u_2^2}\,\overline{u_3^2})^{1/2}x_1^2}\exp\left[-\frac{\bar{U}_1^2}{2\overline{u_1^2}}\left(\frac{x_2^2\,\overline{u_1^2}}{x_1^2\,\overline{u_2^2}}+\frac{x_3^2\,\overline{u_1^2}}{x_1^2\,\overline{u_3^2}}\right)\right] \tag{5-165b}$$

For long diffusion times, that is, for large distances from the point source, we may introduce the asymptotic constant values of the diffusion coefficients:

$$\epsilon_{11} = \frac{\overline{y_1^2}}{2(t-t_0)} \quad \text{or} \quad \overline{y_1^2} = 2\epsilon_{11}\tau$$

and, similarly,

$$\overline{y_2^2} = 2\epsilon_{22}\tau \qquad \overline{y_3^2} = 2\epsilon_{33}\tau$$

From Eq. (5-164) we then obtain

$$\mathfrak{P}(x_1,x_2,x_3) = \frac{S}{(4\pi)^{3/2}(\epsilon_{11}\epsilon_{22}\epsilon_{33})^{1/2}} \int_0^\infty d\tau \frac{1}{\tau^{3/2}}$$

$$\times \exp\left\{-\frac{1}{4}\left[\frac{(x_1 - \bar{U}_1\tau)^2}{\epsilon_{11}\tau} + \frac{x_2^2}{\epsilon_{22}\tau} + \frac{x_3^2}{\epsilon_{33}\tau}\right]\right\}$$

or, after integration,

$$\mathfrak{P}(x_1,x_2,x_3) = \frac{S}{4\pi(\epsilon_{11}\epsilon_{22}\epsilon_{33})^{1/2}\left(\dfrac{x_1^2}{\epsilon_{11}} + \dfrac{x_2^2}{\epsilon_{22}} + \dfrac{x_3^2}{\epsilon_{33}}\right)^{1/2}}$$

$$\times \exp\left\{-\frac{\bar{U}_1}{2\epsilon_{11}}\left[\epsilon_{11}^{1/2}\left(\frac{x_1^2}{\epsilon_{11}} + \frac{x_2^2}{\epsilon_{22}} + \frac{x_3^2}{\epsilon_{33}}\right)^{1/2} - x_1\right]\right\} \tag{5-166}$$

The reader may easily verify that this result is identical with Eq. (5-160) if we assume isotropy, so that $\epsilon_{11} = \epsilon_{22} = \epsilon_{33} = \epsilon$.

For points close to the x_1-axis we obtain the approximation that corresponds to Eq. (5-160a).

$$\mathfrak{P}(x_1,x_2,x_3) \simeq \frac{S}{4\pi(\epsilon_{22}\epsilon_{33})^{1/2}|x_1|} \exp\left[-\frac{\bar{U}_1}{4\epsilon_{11}|x_1|}\left(x_2^2\frac{\epsilon_{11}}{\epsilon_{22}} + x_3^2\frac{\epsilon_{11}}{\epsilon_{33}}\right)\right] \tag{5-166a}$$

The problem of diffusion in a uniform flow from a continuous *line source* along the x_3-axis can be solved by integration of Eq. (5-164) with respect to x_3 from $-\infty$ to $+\infty$. There is then obtained

$$\mathfrak{P}(x_1,x_2) = \frac{S^*}{2\pi}\int_0^\infty d\tau \frac{1}{(\overline{y_1^2}\,\overline{y_2^2})^{1/2}} \exp\left\{-\frac{1}{2}\left[\frac{(x_1 - \bar{U}_1\tau)^2}{\overline{y_1^2}} + \frac{x_2^2}{\overline{y_2^2}}\right]\right\} \tag{5-167}$$

where S^* is the source strength per unit of length. For short distances we may again assume

$$\overline{y_1^2} = \overline{u_1^2}\tau^2 \quad \text{and} \quad \overline{y_2^2} = \overline{u_2^2}\tau^2$$

Hence the solution of Eq. (5-167) becomes

$$\mathfrak{P}(x_1,x_2) = \frac{S^*}{2\sqrt{2\pi}\,(\overline{u_2^2}x_1^2 + \overline{u_1^2}x_2^2)^{1/2}} \operatorname{erfc}\left[\frac{-u_2'\bar{U}_1x_1}{u_1'(2\overline{u_2^2}x_1^2 + 2\overline{u_1^2}x_2^2)^{1/2}}\right]$$

$$\times \exp\left[-\frac{\bar{U}_1^2x_2^2}{2(\overline{u_2^2}x_1^2 + \overline{u_1^2}x_2^2)}\right] \tag{5-168}$$

Of course, here, too, we have to reckon with the restriction imposed by the condition of short diffusion times, $\tau \ll \mathfrak{I}_L$. This leads to the condition

$$\frac{x_1\bar{U}_1}{\overline{u_1^2}\left(\frac{x_1^2}{\overline{u_1^2}}+\frac{x_2^2}{\overline{u_2^2}}\right)^{1/2}} \gg 1$$

The solution (5-168) then reduces to

$$\mathfrak{P}(x_1,x_2) \simeq \frac{S^*}{\sqrt{2\pi}\,(\overline{u_2^2}x_1^2+\overline{u_1^2}x_2^2)^{1/2}} \exp\left[-\frac{\bar{U}_1^2x_2^2}{2(\overline{u_2^2}x_1^2+\overline{u_1^2}x_2^2)}\right] \tag{5-168a}$$

If, again, we take the condition $\overline{u_1^2}/\bar{U}_1^2 \ll 1$ and, consequently, $x_2^2/x_1^2 \ll 1$, we obtain the further simplified solution

$$\mathfrak{P}(x_1,x_2) \simeq \frac{S^*}{\sqrt{2\pi}\,u_2'\,|x_1|} \exp\left(-\bar{U}_1^2x_2^2/2\overline{u_2^2}x_1^2\right) \tag{5-168b}$$

For long diffusion times, after the introduction of the constant diffusion coefficients ϵ_{11} and ϵ_{22}, we obtain from Eq. (5-167)

$$\mathfrak{P}(x_1,x_2) = \frac{S^*}{2\pi(\epsilon_{11}\epsilon_{22})^{1/2}} K_0\left[\frac{1}{2}\bar{U}_1\left(\frac{x_1^2}{\epsilon_{11}^2}+\frac{\epsilon_{22}}{\epsilon_{11}}\frac{x_2^2}{\epsilon_{22}^2}\right)^{1/2}\right] \exp\left(\bar{U}_1x_1/2\epsilon_{11}\right) \tag{5-169}$$

where $K_0(z)$ denotes the modified Bessel function of the second kind.

This solution may be simplified by assuming

$$\frac{\epsilon_{11}}{\bar{U}_1x_1} \ll 1 \quad \text{and} \quad \frac{x_2^2}{x_1^2} \ll 1$$

The result is

$$\mathfrak{P}(x_1,x_2) \simeq \frac{S^*}{2(\pi\epsilon_{22}\bar{U}_1\,|x_1|)^{1/2}} \exp\left(-\frac{\bar{U}_1x_2^2}{4\epsilon_{22}\,|x_1|}\right) \tag{5-169a}$$

Compare with Eqs. (5-169) and (5-169a) the corresponding expressions (5-161) and (5-161a). From the expressions (5-165), (5-166), (5-168), and (5-169) it may be inferred that, for small relative turbulence intensities, the diffusion is essentially transverse only, whereas, if the relative turbulence intensities are not small, the diffusion in the upstream direction may become appreciable. In that case, however, we may soon come into conflict with the condition imposed by the assumption of short diffusion times. Nevertheless, in order to show the effect of upstream diffusion in an exaggerated way, we give Fig. 5-3, taken from Fleishman and Frenkiel's paper.[36] In this figure the isoconcentration curves have been computed according to the short-distance formula (5-168), with relative intensity $u_2'/\bar{U}_1 = 0.5$. Dimensionless coordinate parameters x_1/L and $u_1'x_2/u_2'L$, where L is a reference length, have been used, and also the dimensionless group $\mathfrak{P}L\bar{U}_1u_2'/S^*u_1'$.

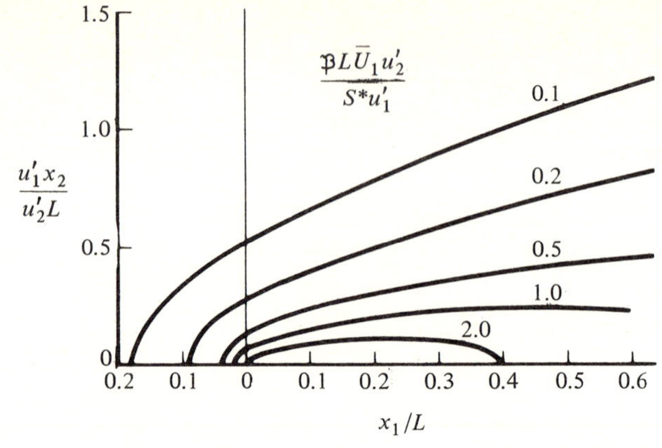

FIGURE 5-3
Isoconcentration curves according to the short-distance formula (5-168) with $u'_2/\bar{U}_1 = 0.5$. (After: *Fleishman, B. A., and F. N. Frenkiel.*[36])

In all the above considerations we have studied the diffusion of marked fluid particles as it is determined by turbulent motion alone, without taking account of molecular-diffusion effects. So the solutions given for the concentration distribution $\mathfrak{P}(x_1,x_2,x_3)$ would apply only if it were possible to introduce into the fluid a tracer that was not subject to molecular diffusion—foreign particles, for instance, with density equal to that of the fluid. These would have to be small enough not to affect the turbulent motions of the fluid, but not too small or we would obtain Brownian movements. Let us assume that it is possible to work with such foreign particles. In that case it would be possible in principle to determine the values of $\overline{y_i^2}$ from the distribution of these particles, introduced continuously into the fluid at a fixed source, and to determine, consequently, the turbulence intensities, Lagrangian integral scales, eddy-diffusion coefficients, and even the Lagrangian correlations.

Let us consider, for instance, the case of diffusion from a line source in a homogeneous isotropic turbulence with low relative turbulence intensity. The mean square $\overline{y_2^2}$ of the lateral turbulent diffusion may then be obtained from the concentration distribution $\mathfrak{P}(x_1,x_2)$ downstream from the source, according to the relation

$$\overline{y_2^2}(x_1) = \frac{\int_{-\infty}^{+\infty} dx_2\, x_2^2 \mathfrak{P}(x_1,x_2)}{\int_{-\infty}^{+\infty} dx_2\, \mathfrak{P}(x_1,x_2)} \tag{5-170}$$

that is, by taking the zero moment and the second moment of the concentration distribution in a section at distance x_1 from the source.

Substituting the solutions (5-168b) and (5-169a) into (5-170) and carrying out the integrations, we obtain respectively,

$$\overline{y_2^2} = \frac{\overline{u_2^2} x_1^2}{\bar{U}_1^2} = \overline{u_2^2}(t - t_0)^2 \quad \text{for short diffusion times}$$

and

$$\overline{y_2^2} = \frac{2\epsilon x_1}{\bar{U}_1} = 2\epsilon(t - t_0) \quad \text{for long diffusion times}$$

as we should. Thus the intensity of turbulence and the coefficient of eddy diffusion may be obtained directly from the values of $\overline{y_2^2}$ at small and at large distances from the source.

From the values of $\overline{y_2^2}$ at intermediate distances, as determined by Eq. (5-170), and from a relation similar to Eq. (1-82) we may obtain the Lagrangian correlation coefficient $\mathbf{R}_L$.

$$\overline{y_2^2}(x_1) = 2\frac{\overline{u_2^2}}{\bar{U}_1^2}\int_0^{x_1} d\xi_1 \, (x_1 - \xi_1)\mathbf{R}_L\left(\frac{\xi_1}{\bar{U}_1}\right) \tag{1-82a}$$

Differentiation with respect to x_1 gives

$$\frac{d}{dx_1}\overline{y_2^2}(x_1) = 2\frac{\overline{u_2^2}}{\bar{U}_1^2}\int_0^{x_1} d\xi_1 \, \mathbf{R}_L\left(\frac{\xi_1}{\bar{U}_1}\right) \tag{5-171}$$

whence

$$\lim_{x_1 \to \infty} \frac{d}{dx_1}\overline{y_2^2}(x_1) = 2\frac{\overline{u_2^2}}{\bar{U}_1}\mathfrak{I}_L = 2\frac{u_2'}{\bar{U}_1}\Lambda_L = \frac{2\epsilon}{\bar{U}_1} \tag{5-171a}$$

Two differentiations of Eq. (1-82a) give

$$\frac{d^2}{dx_1^2}\overline{y_2^2}(x_1) = 2\frac{\overline{u_2^2}}{\bar{U}_1^2}\mathbf{R}_L\left(\frac{x_1}{\bar{U}_1}\right) \tag{5-172}$$

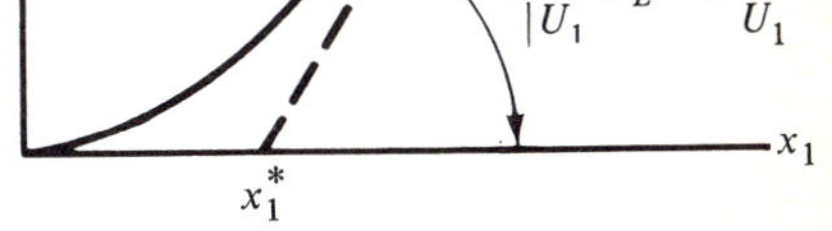

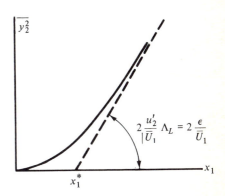

FIGURE 5-4
Mean square $\overline{y_2^2}$ of lateral particle diffusion in isotropic turbulence as function of distance x_1 from fixed point source.

If we plot $\overline{y_2^2}$ as a function of x_1, we obtain a curve shaped as shown in Fig. 5-4; $\overline{y_2^2}$ increases first proportionally with x_1^2 and finally proportionally with x_1 with a slope equal to $2u_2'\Lambda_L/\bar{U}_1$. The intersection x_1^* of the asymptote to the curve with the x_1-axis, according to the relation (1-82a) equals

$$x_1^* = \frac{u_2'}{\bar{U}_1\Lambda_L}\int_0^\infty d\xi_1\,\xi_1\mathbf{R}_L\left(\frac{\xi_1}{\bar{U}_1}\right) \tag{5-173}$$

The equation for the asymptote reads

$$\overline{y_2^2} = 2\frac{u_2'}{\bar{U}_1}\Lambda_L(x_1 - x_1^*) = 2\frac{\epsilon}{\bar{U}_1}(x_1 - x_1^*) \tag{5-174}$$

Various experimenters have tried to obtain the characteristics of a homogeneous almost isotropic turbulence by following the method just expounded. Most of them used as a tracer either a dye miscible with the fluid or heat. Since the diffusion of a dye or of heat is certainly affected by molecular diffusion, these diffusion measurements cannot be used to determine turbulence characteristics, especially not the Lagrangian correlation. Kampé de Fériet[37] reports diffusion experiments by Dupuit in a windtunnel, where soap bubbles about 3 mm in diameter were used and introduced at a point in the windtunnel stream. The concentration of the bubbles passing a plane perpendicular to the mean flow showed a Gaussian distribution.

Kalinske and Pien[38] have performed diffusion experiments in water flowing through a channel. At a certain distance from the bottom and the side walls the turbulent flow was nearly isotropic. By injecting a mixture of carbon tetrachloride and benzene of the same density as water, which formed small droplets in the water, the value of $\overline{y_2^2}$ was determined from photographic records. The concentration distribution of the droplets passing a plane perpendicular to the flow at a distance x_1 from the point of injection was very close to Gaussian:

$$\mathfrak{P}(x_1,x_2) = \mathfrak{P}(x_1,0)\exp\left(-x_2^2/2\overline{y_2^2}\right)$$

If we assume that the droplets are small enough with respect to the smallest turbulence scale, then their diffusion may be considered to be representative of the diffusion of the fluid particles caused by turbulence alone, since diffusion of the droplets is not influence by molecular effects.

After $\overline{y_2^2}$ has been evaluated from the distribution in planes at different distances from the source, $\overline{y_2^2}$ as a function of x_1 could be determined. From the curve for small values of x_1 the turbulence intensity $\overline{u_2^2}/\bar{U}_1^2 = \overline{y_2^2}/x_1^2$ was calculated, and from the curve for large values of x_1, the value of $\epsilon = \overline{y_2^2}\bar{U}_1/2x_1$. Furthermore, the Lagrangian correlation $\mathbf{R}_L(x_1/\bar{U}_1)$ was determined by applying the relation (5-172). Figure 5-5 shows a plot of $\overline{y_2^2}\bar{U}_1/2\epsilon$ against x_1.

It appeared that the rough shape of the Lagrangian-correlation coefficient could be well approximated by an exponential function

$$\mathbf{R}_L = \exp(-x_1/a)$$

where

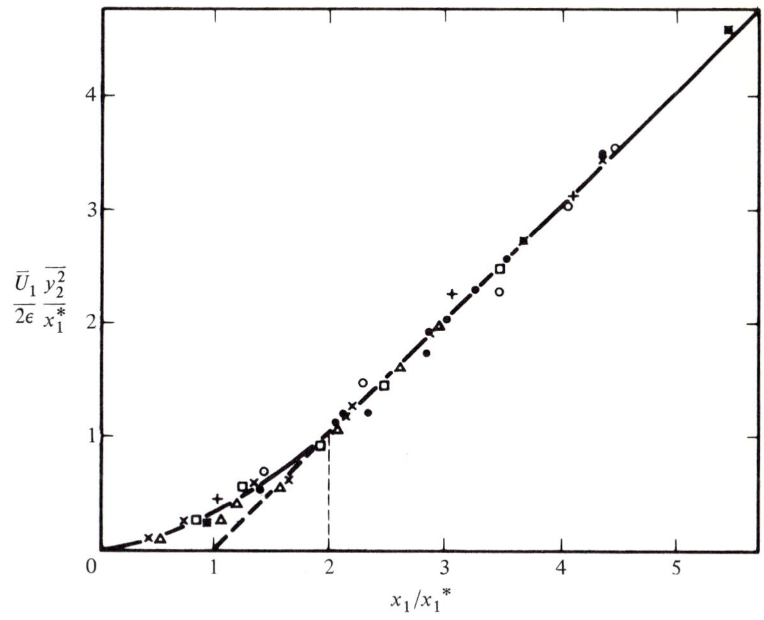

FIGURE 5-5
Lateral diffusion of benzene–carbon tetrachloride droplets in a turbulent flow of water downstream of the injection point. (From: *Kalinske, A. A., and C. L. Pien,*[38] *by permission of the American Chemical Society.*)

$$a = \int_0^\infty \mathbf{R}_L(x_1)\,dx_1 = \bar{U}_1\mathfrak{J}_L = \frac{\bar{U}_1}{u_2'}\Lambda_L = \frac{\bar{U}_1}{\overline{u_2^2}}\epsilon$$

Substitution of the exponential function for $\mathbf{R}_L$ in Eq. (5-173) yields

$$x_1^* = \frac{u_2'}{\bar{U}_1\Lambda_L}a^2 = a = \bar{U}_1\mathfrak{J}_L$$

Hence, for an exponential function for the Lagrangian correlation, we have

$$\mathbf{R}_L = \exp(-x_1/x_1^*)$$

whence

$$\overline{y_2^2} = \frac{2\epsilon x_1^*}{\bar{U}_1}\left\{\frac{x_1}{x_1^*} - [1 - \exp(-x_1/x_1^*)]\right\}$$

Since $\exp(-3) = 0.05$, we see that, for this shape of the Lagrangian correlation, $\overline{y_2^2}$ approximates the value determined by the asymptotic relation (5-174) for $x_1 > 3x_1^*$, to which value reference has been made earlier in Sec. 5.4. Figure 5-5 shows that, considering the usual spread of the experimental data, even an approximation with $x_1 > 2x_1^*$ is justified.

Let us now consider the effect of molecular diffusion on the spread of, say, heat, downstream from a line source placed in a homogeneous turbulence with uniform mean velocity $\bar{U}_1$.

According to the investigations by Townsend[39] and by Batchelor and Townsend,[25] we may distinguish three stages in the development of the hot wake downsteam from the heat source. During the first two stages the wake consists of a continuous connected layer that becomes more and more distorted as the distance from the heat source increases. During the third stage it is no longer possible to speak of a continuous, connected sheet of hot fluid. The distortion has become so strong that the dispersion consists of a number of practically discrete, marked fluid particles clustered more or less around the original plane of the undisturbed sheet (plane of symmetry).

Let us consider these three stages separately in more detail. During the first stage the fluid particles passing the heat source and picking up heat from it form a thin sheet whose thickness may be considered small with respect to the dissipation scale of the turbulence. This sheet is waved to and fro by the larger-scale motions and is at the same time increasingly deformed by the small-scale motions. The thickness of the sheet increases because of molecular diffusion. The mean-square width increases with time according to $2\mathfrak{k}_\theta(t - t_0)$, where t_0 denotes the instant when the fluid particles pass the heat source. The total spread of the heat is determined by this molecular diffusion and by the action of turbulence, whose large-scale motions cause the lateral movements of the hot sheet. The mean square of the temperature distribution is determined by

$$\overline{(y_2)_\theta{}^2} = \frac{\displaystyle\int_{-\infty}^{+\infty} dx_2\, x_2{}^2\bar{\Theta}(t,x_2)}{\displaystyle\int_{-\infty}^{+\infty} dx_2\, \bar{\Theta}(t,x_2)} = \overline{y_2{}^2} + 2\mathfrak{k}_\theta(t - t_0)$$

since at this stage the diffusion by molecular effects and the spread by turbulence are statistically independent of each other, so that the two contributions to the dispersion are additive. This, however, holds only as long as the deformations of the sheet caused by the small-scale motions do not affect appreciably the accelerated molecular diffusion due to the stretching processes of the fluid elements. This condition is satisfied as long as the diffusion time $t - t_0$ is small compared with the time scale of the smallest eddies, which in the case of large Reynolds numbers of turbulence is determined by $(\nu/\varepsilon)^{1/2}$.

Or, we may also put the condition that the spread by molecular diffusion must be small compared with the dimensions of the smallest eddies, that is, with $\nu^{3/4}/\varepsilon^{1/4}$. Hence

$$\mathfrak{k}_\theta(t - t_0) \ll \nu^{3/2}\varepsilon^{-1/2}$$

or

$$(t - t_0)\left(\frac{\varepsilon}{\nu}\right)^{1/2} \ll \frac{\nu}{\mathfrak{k}_\theta} = \mathbf{Pr}$$

During the second stage the distortion of the hot sheet by the smallest eddies becomes important. The effect of the turbulence is not only to displace the fluid particles but also to stretch and rotate the fluid elements. Only the last two processes affect the molecular diffusion; they accelerate it.

Saffman[9,94] has evaluated the effect of molecular diffusion due to distortion of fluid particles on the total diffusion of a scalar property. He assumed such small diffusion times that the Lagrangian behavior of the marked fluid particles passing the same starting point is entirely determined by the Eulerian turbulence field at that point. Since molecular diffusion takes place simultaneously with the turbulent diffusion, we must distinguish between *volume particle* and *property particle* (see Sec. 5-1) formed by the marked molecules. We then are interested in the diffusion of these marked molecules due to the macroscopic (convective) turbulent motions and to the molecular motion. The total diffusion of the marked molecules can be determined by considering first the displacement of the centroid of the marked molecules, and then the molecular diffusion with respect to this centroid. Due to the molecular diffusion the displacement of the centroid of the marked molecules, i.e., of the property particle will not be the same as that of the centroid of the volume particle. The diffusion of molecules is a stochastic process and since one property particle contains a very large number of marked molecules, in fact we have to carry out two averaging procedures. One with respect to the ensemble of marked molecules for each property particle, and one with respect to a large ensemble of particles passing the starting point.

Let us first consider the *displacement of the centroid of one property particle.* Assume a coordinate system with the origin at the starting point. After time t the centroid of the volume particle that passes the origin at $t = 0$ is at a position $(x_k)_\upsilon$, which is equal to the displacement $(y_k)_\upsilon$. At the same instant the centroid of the property particle will be at $(x_k)_\mathscr{P} = (y_k)_\mathscr{P}$. The Lagrangian velocities of the two centroids are respectively $(v_k)_\upsilon = d(y_k)_\upsilon/dt$ and $(v_k)_\mathscr{P} = d(y_k)_\mathscr{P}/dt$. Due to the effect of molecular migration $(v_k)_\upsilon \neq (v_k)_\mathscr{P}$. These velocities are macroscopic velocities, and ensemble averages of the macroscopic velocities of the large number of molecules forming the volume particle and the property particle, respectively. These macroscopic velocities are determined by the local Eulerian velocity of the fluid at the point of the instantaneous position of the molecules.

Let $\mathfrak{P}(x_k,t)\,dx_1\,dx_2\,dx_3$ be the probability that a marked molecule is at (x_k,dx_k) at time t. At that instant it has a macroscopic velocity $u_i(x_k,t)$ of the fluid at x_k. Hence

$$(v_i)_\mathscr{P}(t) = \iiint_{-\infty}^{+\infty} dx_1\,dx_2\,dx_3\,u_i[(x_k)_\mathscr{P},t]\,\mathfrak{P}[(x_k)_\mathscr{P},t]$$

Since the diffusion times t are assumed to be very small, the displacements are small too. We may then describe the Eulerian velocity field in the small region covered by the displacements in terms of the Eulerian velocity and its spatial derivatives at the origin $x_k = 0$:

$$u_i[(x_k)_\mathscr{P},t] = u_i[(y_k)_\upsilon,t] + \xi_k\left(\frac{\partial u_i}{\partial x_k}\right)_0 + \tfrac{1}{2}\xi_k\xi_l\left(\frac{\partial^2 u_i}{\partial x_k\,\partial x_l}\right)_0 + \cdots$$

where $\xi_k = (y_k)_\mathscr{P} - (y_k)_\upsilon$.

Since $u_i[(y_k)_v,t] = (v_i)_v(t)$, we thus have

$$(v_i)_{\mathscr{P}}(t) = \iiint_{-\infty}^{+\infty} dx_1\, dx_2\, dx_3 \left[(v_i)_v + \xi_k \left(\frac{\partial u_i}{\partial x_k} \right)_0 + \tfrac{1}{2}\xi_k \xi_l \left(\frac{\partial^2 u_i}{\partial x_k\, \partial x_l} \right)_0 + \cdots \right] \mathfrak{P}[(x_k)_{\mathscr{P}},t] \qquad (5\text{-}175)$$

In order to simplify the expressions we assume a new coordinate system x_i^*, with the same origin as the x_i-system, but with the coordinate-axes coinciding with the principal axes of the local deformation tensor, which for the small times considered may be assumed to be sufficiently stationary. This assumption holds true only when the local rotation is small relative to the local rate of deformation. However, as shown by Lumley[120] in the case of large rotation the fluid particle will have turned to a new orientation before the deformation becomes effective.

The position $(y_k^*)_{\mathscr{P}}(t)$ with respect to $(y_k^*)_v(t)$, i.e., $\xi_k^*(t)$, is determined by a molecular diffusion process superimposed on a Eulerian convection field. The mass-balance equation for the marked molecules including the convection terms, reads

$$\frac{\partial \mathfrak{P}}{\partial t} + [(v_k^*)_{\mathscr{P}}(t) - (v_k^*)_v(t)] \frac{\partial \mathfrak{P}}{\partial \xi_k^*} = \mathfrak{k}_\gamma \frac{\partial^2 \mathfrak{P}}{\partial \xi_k^*\, \partial \xi_k^*}$$

since the concentration of marked molecules is equal to the probability density $\mathfrak{P}$.

For small t we may assume

$$(v_1^*)_{\mathscr{P}} - (v_1^*)_v = \xi_1^* \left(\frac{\partial u_1^*}{\partial x_1^*} \right)_0 = \xi_1^* \zeta_1^*$$

and similarly for the x_2^* and x_3^* direction. So

$$\frac{\partial \mathfrak{P}}{\partial t} + \zeta_1^* \xi_1^* \frac{\partial \mathfrak{P}}{\partial \xi_1^*} + \zeta_2^* \xi_2^* \frac{\partial \mathfrak{P}}{\partial \xi_2^*} + \zeta_3^* \xi_3^* \frac{\partial \mathfrak{P}}{\partial \xi_3^*} = \mathfrak{k}_\gamma \frac{\partial^2 \mathfrak{P}}{\partial \xi_k^*\, \partial \xi_k^*} \qquad (5\text{-}176)$$

The solution of this equation is

$$\mathfrak{P}(\xi_k^*,t) = \frac{1}{(2\pi \mathfrak{k}_\gamma)^{3/2} \beta_1 \beta_2 \beta_3} \exp \left[-\frac{1}{2\mathfrak{k}_\gamma} \left(\frac{\xi_1^{*2}}{\beta_1^2} + \frac{\xi_2^{*2}}{\beta_2^2} + \frac{\xi_3^{*2}}{\beta_3^2} \right) \right] \qquad (5\text{-}177)$$

where, again for the small times considered

$$\beta_1^2 \simeq 2t + 2\zeta_1^* t^2 + \cdots = 2t + \mathscr{O}(t^2); \text{etc.} \qquad (5\text{-}178)$$

After substitution of this Gaussian distribution for $\mathfrak{P}(\xi_k^*,t)$ in Eq. (5-175) we may carry out the integration in space with respect to a ξ_k^*-coordinate system instead of with respect to the x_k-coordinate system. However, we then have also to transform the expression within square brackets in Eq. (5-175) from the ξ_k-coordinate system to the ξ_k^*-coordinate system, with the transformation formulae [see Eq. (A-6)]:

$$\xi_k = \xi_m^* \cos(\xi_m^*, \xi_k) = \xi_m^* \mathbf{e}_{m^*k}$$

Because $\mathfrak{P}(\xi_k^*,t)$ is symmetric in ξ_k^*, all uneven moments are zero, and only the even moments contribute to the integral. The result of the integration reads for the v_2 component

$$(v_2)_{\mathscr{P}} = (v_2)_v + \frac{1}{2} \left(\frac{\partial^2 u_2}{\partial x_k\, \partial x_l} \right)_0 \mathfrak{k}_\gamma [\beta_1^2 \mathbf{e}_{1^*k} \mathbf{e}_{1^*l} + \beta_2^2 \mathbf{e}_{2^*k} \mathbf{e}_{2^*l} + \beta_3^2 \mathbf{e}_{3^*k} \mathbf{e}_{3^*l}] + \cdots$$

With

$$\beta_1^2 = 2t + \mathscr{O}(t^2), \text{etc.}; \quad \mathbf{e}_{n^*k} \mathbf{e}_{n^*l} = \delta_{kl}$$

and

$$\left(\frac{\partial^2 u_2}{\partial x_k\, \partial x_l} \right)_0 \delta_{kl} = \left(\frac{\partial^2 u_2}{\partial x_k\, \partial x_k} \right)_0$$

we finally obtain

$$(v_2)_{\mathscr{P}} = (v_2)_v + \mathfrak{k}_\gamma \left(\frac{\partial^2 u_2}{\partial x_k \, \partial x_k} \right)_0 t + \mathcal{O}(t^2)$$

Whence follows:

$$(y_2)_{\mathscr{P}} = (y_2)_v + \tfrac{1}{2}\mathfrak{k}_\gamma \left(\frac{\partial^2 u_2}{\partial x_k \, \partial x_k} \right)_0 t^2 + \mathcal{O}(t^3)$$

Taking an ensemble average of a large number of particles, the mean-square of the displacements of the centroids of the property particles, expressed in those of the volume particles becomes

$$\overline{(y_2)_{\mathscr{P}}^2(t)} = \overline{(y_2)_v^2(t)} + \mathfrak{k}_\gamma \overline{(y_2)_v \left(\frac{\partial^2 u_2}{\partial x_k \, \partial x_k} \right)_0} t^2 + \overline{2(y_2)_v \mathcal{O}(t^3)}$$

Since for small t, $(y_2)_v(t) \simeq (v_2)_v(0)t = u_2(0)t$, the expression reads, up to t^3:

$$\overline{(y_2)_{\mathscr{P}}^2(t)} = \overline{(y_2)_v^2(t)} + \mathfrak{k}_\gamma \overline{\left(u_2 \frac{\partial^2 u_2}{\partial x_k \, \partial x_k} \right)_0} t^3 + \mathcal{O}(t^4) \qquad (5\text{-}179)$$

Let us now consider the *molecular diffusion relative to the centroid* of the property particle. Since the particle is in a deforming Eulerian velocity-field, as given by ζ_1, ζ_2, and ζ_3, we have also to include the effect of deformation on the molecular diffusion. For the small times considered we may use the result obtained earlier in Sec. (5-4) for the case of the effect of deformation in a uniform, straining field. For an ensemble average of many particles this result reads

$$\frac{\partial \bar{\Gamma}}{\partial t} = \mathfrak{k}_\gamma \left(\frac{\partial^2 \bar{\Gamma}}{\partial x_2^2} \right)_0 [3 + 2\overline{\zeta_k^* \zeta_k^*} t^2 + \cdots]$$

But the diffusion of marked molecules is also given by $\langle (\eta_1^*)^2 \rangle$, $\langle (\eta_2^*)^2 \rangle$, and $\langle (\eta_3^*)^2 \rangle$, where η_i^* is the displacement of a marked molecule relative to the centroid of the property particle, and $\langle \ \rangle$ denotes an ensemble average of the marked molecules. For an ensemble average of many property particles we may also write

$$\frac{\partial \bar{\Gamma}}{\partial t} = 3 \left(\frac{\partial^2 \bar{\Gamma}}{\partial x_2^2} \right)_0 \frac{1}{2} \frac{\partial}{\partial t} \tfrac{1}{3} \overline{\langle \eta_k^* \eta_k^* \rangle}$$

where a spherical average is taken of $\langle (\eta_1^*)^2 \rangle$, $\langle (\eta_2^*)^2 \rangle$, and $\langle (\eta_3^*)^2 \rangle$. Whence follows

$$\frac{1}{2} \frac{\partial}{\partial t} \tfrac{1}{3} \overline{\langle \eta_k^* \eta_k^* \rangle} = \mathfrak{k}_\gamma (1 + \tfrac{2}{3} \overline{\zeta_k^* \zeta_k^*} t^2 + \cdots)$$

or

$$\tfrac{1}{3} \overline{\langle \eta_k^* \eta_k^* \rangle} = 2\mathfrak{k}_\gamma t (1 + \tfrac{2}{9} \overline{\zeta_k^* \zeta_k^*} t^2 + \cdots) \qquad (5\text{-}180)$$

The total diffusion in the x_2-direction of marked molecules thus becomes

$$[\overline{(y_2)_\gamma^2}]_{\text{total}} = [\overline{(y_2)_{\mathscr{P}}^2}]_{\text{centroid}} + \tfrac{1}{3} \overline{\langle \eta_k^* \eta_k^* \rangle}$$

With Eqs. (5-179) and (5-180) we thus obtain

$$[\overline{(y_2)_\gamma^2}]_{\text{total}} = \underbrace{\overline{(y_2)_v^2} + 2\mathfrak{k}_\gamma t}_{\text{(a)}} + \underbrace{\left\{ \overline{\left(u_2 \frac{\partial^2 u_2}{\partial x_k \, \partial x_k} \right)} + \tfrac{4}{9} \overline{\zeta_k^* \zeta_k^*} \right\} \mathfrak{k}_\gamma t^3}_{\text{(b)}} + \mathcal{O}(t^4) \qquad (5\text{-}181)$$

The two terms, indicated by (a), on the right-hand side describe the diffusion due to the displacement of the centroid of the property particle, the two terms indicated by (b), the diffusion relative to the centroid.

The terms describing the effect of the Eulerian velocity field can be evaluated in terms of ε and

ν when we assume local isotropy. They have the dimension $[T^{-2}]$, so they must be proportional to ε/ν. For a *homogeneous, isotropic turbulence* we have

$$\overline{u_2 \frac{\partial^2 u_2}{\partial x_k \, \partial x_k}} = -\overline{\frac{\partial u_2}{\partial x_k}\frac{\partial u_2}{\partial x_k}} = -\frac{1}{3}\frac{\varepsilon}{\nu}$$

Also with respect to the x_k^*-coordinate system, where $(\partial u_i^*/\partial x_j^* + \partial u_j^*/\partial x_i^*) = 0$ when $i \neq j$:

$$\varepsilon = \nu \overline{\frac{\partial u_i^*}{\partial x_j^*}\left(\frac{\partial u_i^*}{\partial x_j^*} + \frac{\partial u_j^*}{\partial x_i^*}\right)} = 2\nu\overline{\zeta_k^*\zeta_k^*}; \quad \text{or} \quad \overline{\zeta_k^*\zeta_k^*} = \frac{1}{2}\frac{\varepsilon}{\nu}$$

The result obtained by Saffman for a locally homogeneous, isotropic, turbulence is

$$\overline{[(y_2)_\gamma{}^2]}_{\text{total}} = \overline{(y_2)_v{}^2} + 2\mathfrak{k}_\gamma t - \frac{1}{9}\frac{\varepsilon}{\nu}\mathfrak{k}_\gamma t^3 + \mathcal{O}(t^4) \qquad (5\text{-}182)$$

With

$$(\epsilon_\gamma)_{2,2} + \mathfrak{k}_\gamma = \frac{1}{2}\frac{d}{dt}\overline{[(y_2)_\gamma{}^2]}_{\text{total}} \quad \text{and} \quad \epsilon_{2,2} = \frac{1}{2}\frac{d}{dt}[(y_2)_v{}^2) \simeq \overline{u_2{}^2}t$$

$$(\epsilon_\gamma)_{2,2} = \epsilon_{2,2}\left(1 - \frac{\varepsilon}{6\overline{u_2{}^2}}\frac{\mathfrak{k}_\gamma}{\nu}t + \cdots\right) \qquad (5\text{-}183)$$

So returning to the case of diffusion from a fixed source, initially we have $(\epsilon_\gamma)_{2,2} = \mathfrak{k}_\gamma + \epsilon_{2,2}$. But during the later stage, where we have the combined effect of molecular diffusion and the turbulence, giving a negative contribution due to the diffusion of the centroid of the property particle and a positive contribution due to the accelerated relative diffusion through the deformation of the fluid particle, there is a decrease in the total diffusion. According to Saffman's theory the transition from the first to the second stage is determined by

$$2\mathfrak{k}_\gamma t = \frac{1}{9}\frac{\varepsilon}{\nu}\mathfrak{k}_\gamma t^3$$

whence follows

$$t_{\text{I}-\text{II}} = \sqrt{18\frac{\nu}{\varepsilon}}$$

This seems to be on the high side, since one would expect a value much smaller than the characteristic deformation time of the Kolmogoroff micro-eddy η. Still for these short times it will be hardly possible to obtain reliable experimental results for the effect of molecular diffusion during the first stage (see Fig. 5-6).

We may compare Eq. (5-183) with the expression for ϵ_γ according to Burgers's model and for very short diffusion times. For these short times we may assume $\mathbf{R}_L(t) \simeq 1$. Equation (5-46) then yields

$$(\epsilon_\gamma)_{2,2} = \overline{v_2{}^2}t(1 - \tfrac{1}{2}\alpha_\gamma t) = (\epsilon)_{2,2}(1 - \tfrac{1}{2}\alpha_\gamma t)$$

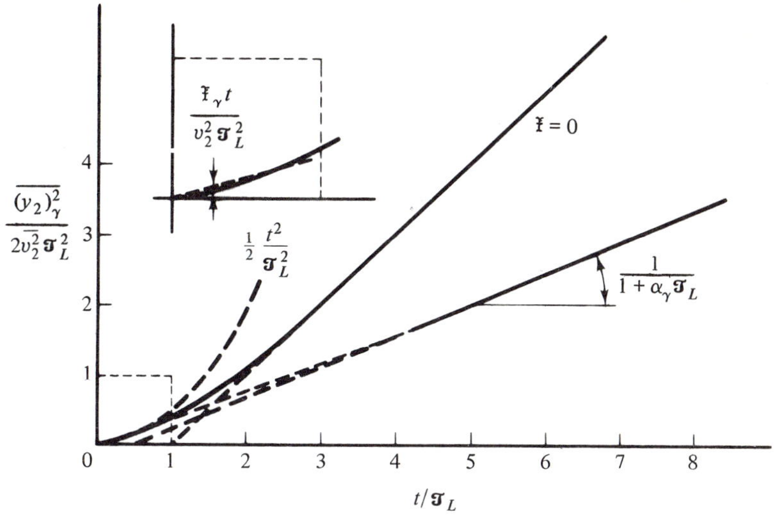

FIGURE 5-6
Comparison between lateral diffusion of a scalar property and that of volume particles downstream of a fixed source.

With Eq. (5-51) and putting $d = Bu'(\nu/\varepsilon)^{1/2}$, we obtain

$$(\epsilon_\gamma)_{2,2} = \epsilon_{2,2}\left(1 - \frac{6}{B^2}\frac{\mathfrak{k}_\gamma \varepsilon}{\overline{u^2}\nu}t\right)$$

Comparison with Eq. (5-183) shows that with $B = 6$ the same result is obtained.

One of the conditions in Burgers's model is that the fluid particle should not lose its property in too short a time. According to Burgers $\alpha_\gamma \gtrsim \mathfrak{I}_L^{-1}$. This yields with Eq. (5-51) $d^2 > 12\mathfrak{k}_\gamma\mathfrak{I}_L$, or with $d = Bu'(\nu/\varepsilon)^{1/2}$ and Eq. (5-61) the condition is satisfied for fluids with $\mathbf{Sc} > 24/B^2C$. Earlier C has been estimated to have values between 3 and 10. Thus with $B = 6$ we may expect Burgers's simple model to be applicable when $\mathbf{Sc} > \mathcal{O}(0.1)$.

The experiments by Townsend[39] and similar experiments by Uberoi and Corrsin[28] with heat as the diffusing property have shown that during the third stage $\overline{(y_2)_\theta^2}$ increases linearly with the diffusion time. This would be the case if the spread of heat were caused by a diffusion process with a constant diffusion coefficient. The separate hot fluid particles will continue to be distorted by the turbulent motions and will, therefore, be subjected still to an accelerated molecular-diffusion process, which thus results in a decrease in the effective diffusion coefficient $(\epsilon_\gamma)_{2,2}$ relative to that for the volume particles $\epsilon_{2,2}$. See for instance Eq. (5-47). When we assume an exponential function for the Lagrangian correlation coefficient, we obtain Eq. (5-53), while for the mean-square $\overline{(y_2)_\gamma^2}$ there is obtained

$$\overline{(y_2)_\gamma^2} = \frac{2\overline{v_2^2}\mathfrak{I}_L t}{1 + \alpha_\gamma \mathfrak{I}_L} \qquad (5\text{-}184)$$

Figure 5-6 shows $\overline{(y_2)_\gamma{}^2}$ compared with $\overline{y_2{}^2}$ during the three stages considered.

We have shown that, if the Lagrangian correlation coefficient for the diffusion of fluid particles can be represented by an exponential function, the distance where $\overline{y_2{}^2}$ is almost equal to the distance given by the asymptote to the $(\overline{y_2{}^2},x_1)$ curve is roughly equal to $3x_1^*$. This no longer holds for the $[\overline{(y_2)_\gamma{}^2},x_1]$ curve. The corresponding distance becomes much greater than $3(x_1^*)_\gamma$ since the curve is flatter.

These effects of molecular diffusion show clearly once again that it is not possible to obtain reliable information concerning the Lagrangian correlation coefficient $\mathbf{R}_L$ and the diffusion coefficient ϵ on the fluid particles from diffusion experiments in which the diffusing agent is subject to molecular-diffusion effects. Since, as is also shown by the relation (5-132), these molecular effects are proportional to $\mathbf{Sc}^{-1}$, large values of **Sc** make the error smaller.

Of course it is possible to determine ϵ_γ from such diffusion experiments.

It is also possible to determine $\overline{y_2{}^2}$ and, hence, $\overline{v_2{}^2}$ from such diffusion experiments provided we consider only very short diffusion times, so that (1) $t \ll \mathfrak{I}_L$ and there exists perfect correlation ($\mathbf{R}_L \simeq 1$), and (2) we are still in the first stage of the diffusion process. Though in this first stage it is possible to correct for molecular-diffusion effects, a large value of **Sc** is recommended in order to minimize the correction needed.

Many of the experiments on diffusion in a homogeneous turbulence have been carried out in the approximate isotropic turbulent flow field downstream from a grid. Now we know that this turbulent flow field decays with distance from the grid. This decay is a relatively slow process compared with the time scale of the turbulence fluctuation; so the assumption of homogeneity of the field appears to be tolerable, at least in regions whose extent does not exceed many times the integral scale of the turbulence. Properties of various quantities, for instance, velocity correlations, based upon the assumption of spatial homogeneity still hold.

Yet, in making diffusion experiments in such a decaying isotropic turbulence, particularly in considering long diffusion times, we cannot entirely neglect the effect of the decay process, since the time and length scales vary with distance from the grid and, consequently, vary also during the diffusion process downstream from a fixed source.

Let this source be placed at a distance x_0 from the grid. If we measure the time from the instant the fluid passes the grid, the fluid will reach the fixed source at the time t_0, where $t_0 = x_0/\bar{U}_1$. The lateral diffusion of fluid particles passing the source is then determined [see Eq. (1-79)] by

$$\overline{y_2{}^2}(t) = 2\int_{t_0}^{t} dt' \int_0^{t'-t_0} d\tau\, \overline{v_2(t')v_2(t'-\tau)}$$

If we introduce the Lagrangian correlation coefficient

$$\mathbf{R}_L(\tau) = \frac{\overline{v_2(t)v_2(t-\tau)}}{v_2'(t)v_2'(t-\tau)}$$

we have to keep in mind that now $v_2'(t) \neq v_2'(t - \tau)$, because of the decay process.

We thus obtain

$$\overline{y_2^2}(t) = 2 \int_{t_0}^{t} dt'\, v_2'(t') \int_0^{t'-t_0} d\tau\, v_2'(t' - \tau)\mathbf{R}_L(\tau)$$

whence

$$\frac{1}{2}\frac{d}{dt}\overline{y_2^2}(t) = v_2'(t) \int_0^{t-t_0} d\tau\, v_2'(t - \tau)\mathbf{R}_L(\tau)$$

We may express this relation in terms of the distance x_1, which is rendered dimensionless in combination with the mesh length M of the grid.

$$\frac{1}{2M}\frac{d\overline{y_2^2}}{dx_1} = \frac{1}{\bar{U}_1^2 M} v_2'\left(\frac{x_1 - x_0}{M}\right) \int_0^{(x_1 - x_0)/M} d\xi_1\, v_2'\left(\frac{x_1 - \xi_1}{M}\right)\mathbf{R}_L\left(\frac{\xi_1}{M}\right) \tag{5-185}$$

Experimental evidence has shown that, within the experimental accuracy, $v_2'/\bar{U}_1$ is approximately independent of $\bar{U}_1$ and is a function only of x_1/M. Also, the width of the thermal wake downstream from a heat source is a function only of the distance from the heat source and is independent of $\bar{U}_1$. This result led Taylor[40] to assume that the diffusion process is dependent only on a variable η given by

$$d\eta = \frac{v_2'}{\bar{U}_1 M} dx_1$$

We then obtain

$$\frac{1}{2M^2}\frac{d\overline{y_2^2}}{d\eta} = \int_0^{\eta} d\eta'\, \mathbf{R}_L(\eta')$$

and

$$\frac{1}{2M^2}\frac{d^2\overline{y_2^2}}{d\eta^2} = \mathbf{R}_L(\eta)$$

We may make use of another empirical fact, namely that, during the initial period of the decay process, the length scale of the turbulence increases approximately as $t^{1/2}$ and the velocity scale decreases approximately as $t^{-1/2}$. Hence the time scale increases as t. Accordingly, Batchelor and Townsend[25] suggest that the diffusion from a fixed source in a decaying grid turbulence should be a stationary random function of a new variable t^*, such that

$$dt^* = \frac{dt}{t} \quad \text{or} \quad t^* = \ln\frac{t}{t_1}$$

where t_1 is arbitrary.

Expressed in terms of this new variable, the diffusion equations for the fluid particles become

$$\frac{\overline{y_2{}^2}}{M^2} = 2\frac{v_2'^2(t_0)t_0t_1}{M^2}\int_{\ln(t_0/t_1)}^{\ln(t/t_1)} dt'^* \exp(t'^*)\int_0^{t'^*+\ln(t_0/t_1)} d\tau^* \exp(-\tfrac{1}{2}\tau^*)\mathbf{R}_L(\tau^*) \qquad (5\text{-}186)$$

and

$$\frac{1}{2M^2}\frac{d\overline{y_2{}^2}}{d(t/t_0)} = \frac{\overline{v_2'^2(t_0)t_0{}^2}}{M^2}\int_0^{\ln(t/t_0)} d\tau^* \exp(-\tfrac{1}{2}\tau^*)\mathbf{R}_L(\tau^*)$$

Since $\overline{v_2'^2}(t_0)t_0$ is constant for a given grid, the right-hand sides of the equations appear to be a function of t/t_0 and of $t_0 f(t/t_0)$, respectively. The equations show further that, for long diffusion times, the spread $\overline{y_2{}^2}$ increases linearly with time, notwithstanding the decay of the turbulence. For diffusion in a grid-generated turbulence the equation (5-182) for $\overline{(y_2)_\gamma{}^2}$ may be written as follows

$$\frac{\overline{(y_2)^2}}{M^2} = \frac{\overline{y_2{}^2}}{M^2} + \frac{2}{\mathbf{Sc}}\frac{\nu}{M\bar{U}_1}\frac{\bar{U}_1(t-t_0)}{M} - \frac{1}{9}\frac{1}{\mathbf{Sc}}\frac{\varepsilon M}{\bar{U}_1{}^3}\frac{\bar{U}_1{}^3(t-t_0)^3}{M^3} \qquad (5\text{-}187)$$

From this equation and Eq. (5-186) we conclude that the spread of a property is also determined by a function of the form $t_0 f(t/t_0)$.

From Townsend's experiments[39] on the spread of heat as shown in Fig. 5-7 it may be concluded that this appears true also for all values of $(t - t_0)/t_0$, just as for the spread $\overline{y_2{}^2}$ of the volume particles. Townsend used a line source placed in a windtunnel downstream of a square-mesh grid. For heat $\mathfrak{k}_\theta = 2 \times 10^{-5}$ m²/s

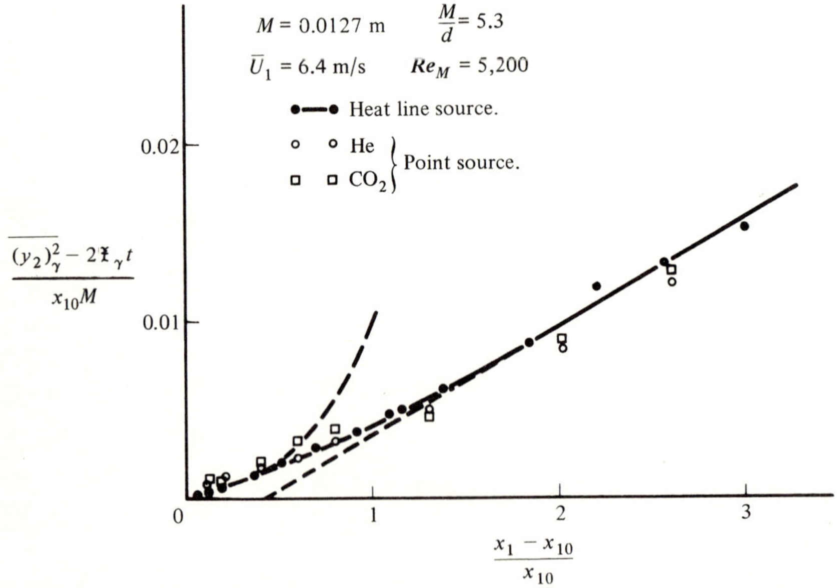

FIGURE 5-7
Lateral diffusion of heat downstream of a line source (*Townsend*[39]) and of helium and carbon dioxide downstream of a point source (*Mickelsen*[95]).

(**Pr** = 0.74). In the same Fig. 5-7 are also shown results obtained by Mickelsen[95] under the same experimental conditions, however, using a point source and both helium and carbon dioxide as tracers. For helium, $\mathfrak{k}_\gamma = 7.25 \times 10^{-5}$ m^2/s, and for carbon dioxide, $\mathfrak{k}_\gamma = 1.67 \times 10^{-5}$ m^2/s. Taking into account the possible difference between the lateral diffusion from a line-source and a point-source respectively, and the relatively large scatter of the experimental data, no conclusion can be drawn concerning the effect of molecular diffusion notwithstanding the marked difference in Schmidt number. At the most one may conclude a smaller lateral diffusion of helium for the larger diffusion times, as it should be.

As mentioned earlier, diffusion experiments with helium and heat have also been made by Baldwin and Mickelsen.[81] The experiments were conducted in the turbulent airflow of a circular pipe of 0.2 m internal diameter, and along the centerline of the pipe. For the helium experiments a thin, hollow, and streamlined tube was used, and for the experiments with heat, electrically heated wires of decreasing diameters from 1.65 mm to 0.25 mm were used as line sources. Figure 5-8 shows a set of results obtained with helium. The drawn line is the result which is obtained by extrapolation to zero source-diameter. The result of measurements with heat did not show a significant difference from those obtained with helium. Also shown in this figure is a curve calculated according to Eq. (5-182). The value of the dissipation per unit of mass ε, has been estimated by making use of experimental data obtained by Laufer in pipeflow at almost the same Reynolds number (see Sec. 7-13). The part of the theoretical curve shown in Fig. 5-8 is well above the experimental curve. Since the time scale $(\nu/\varepsilon)^{1/2}$ amounts to less than 10^{-3} s, obviously the dotted curve falls

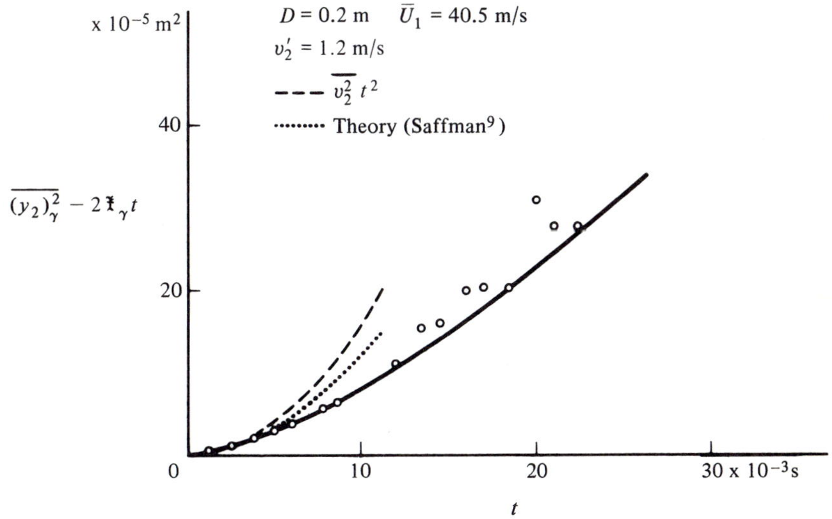

FIGURE 5-8
Lateral diffusion of helium downstream of a line source in the center of a pipeflow. (Adapted from: *Baldwin, L. V., and W., R. Mickelsen.*[81])

in a region where the theory may no longer be applied. The conclusion is that it will be practically impossible to check experimentally diffusion theories for the first two stages of the diffusion process. At present it seems only possible to obtain reliable experimental information concerning the third stage, for instance in order to determine values of Burgers's exchange coefficient α_y (see Eq. (5-184). But for that the Lagrangian integral time-scale $\mathfrak{I}_L$ must be known, which then has to be determined from experiments with tracers for which the molecular Schmidt number is very large, and under the same experimental conditions.

Before concluding this section with a further consideration of Eulerian space-time correlation as has been promised in the preceeding section, we will give a brief discussion of a number of diffusion experiments which may be of practical value.

In 1939 Towle and Sherwood[41] published results of experiments made in a round duct of 0.3 m diameter. They introduced carbon dioxide and hydrogen through a fine tube placed in the duct axis and obtained concentration traverses at great distances from the source by continuous sampling with a rag of fine tubes. They studied the effect of turbulence produced by a grid ($M = 12.7$ mm, $M/d = 6$) installed at various distances upstream from the source. From the concentration distributions, which were close to Gaussian error curves, the coefficient of eddy diffusion was determined according to the method outlined in Sec. 2-10 based upon Eq. (5-160a). No corrections were made for the effect of molecular diffusion or for the effect of the decay of the turbulence produced by the grids. The results are interesting in that they showed clearly a marked increase in the coefficient of eddy diffusion when the distance between source and grid exceeded 25 times the duct diameter, achieving a constant value close to that obtained without grids beyond about 50 duct diameters. Obviously, the scale of the turbulence of the free duct flow without grid was much greater than the scale of the turbulence produced by the grid, making the coefficient of eddy diffusion appreciably smaller in the flow with a grid.

A similar result was obtained by McCarter, Stutzman, and Koch,[42] who made heat-diffusion experiments by introducing hot air in the centerline of a vertical pipe of 0.2 m diameter, through which cold air was blown.

Just as in the experiments by Towle and Sherwood, the eddy diffusion coefficients were much smaller with grids installed than without. With a 40-mesh grid, ϵ_θ was roughly half the value found for the free duct flow. It was further concluded that the values of ϵ_θ for the free duct flow were roughly equal to the values obtained from the empirical formula given by Sherwood and Woertz:[43]

$$\epsilon_y = 0.02\bar{U}_{\mathrm{av}} D\sqrt{f}$$

where $\bar{U}_{\mathrm{av}}$ is the average flow velocity through the duct, D is its diameter, and f is the friction factor, defined by the pressure-drop relation

$$-\frac{d\bar{P}}{dx} = \frac{f}{D}\tfrac{1}{2}\rho\bar{U}_{\mathrm{av}}^2$$

From these experiments and those of Towle and Sherwood it may be concluded that introducing grids for the purpose of increasing the over-all rate of mixing will not always serve this purpose. Such a grid must not be too fine, because the reduction in the coefficient of eddy diffusion by the decrease in the scale of turbulence is often greater than the increase in value caused by the augmented intensity of turbulence.

A relation to the above empirical formula for ϵ_γ can be obtained for the eddy viscosity ϵ_m from measurements by Laufer carried out in the core region of a turbulent pipe-flow, as will be discussed in Sec. 7-13. Laufer obtained the relation $\epsilon_m \simeq 0.035u^*D$, where $u^* = (\sigma_w/\rho)^{1/2}$ is the wall-friction velocity. From the above expression for $(-d\bar{P}/dx)$ and $-d\bar{P}/dx = 4\sigma_w/D$, we obtain $u^* = \bar{U}_{\text{av}}(\lambda/8)^{1/2}$, and consequently

$$\epsilon_m \simeq 0.0125\bar{U}_{\text{av}}D\sqrt{\lambda}$$

With the above expression for ϵ_γ, we would obtain a turbulent Schmidt number $\epsilon_m/\epsilon_\gamma \simeq 0.625$.

We still have to consider Eulerian space-time correlations in the light of a change of correlation due to a turbulence diffusion process, as suggested by Kovasznay.[82] In a *homogeneous turbulence* moving with a constant mean velocity $\bar{U}_1$, we consider the correlation $[\mathbf{Q}_{1,1}(x_1,0,0;t)]_E$ between the velocity u_1 at the origin of a stationary coordinate system, and the same velocity-component at a point $(x_1,0,0)$ a time interval t later. According to the diffusion idea the loss of correlation is due to the fact that at the point $(x_1,0,0)$ at time t a fluid particle has arrived that was in another point (ξ_1,ξ_2,ξ_3) at time $t = 0$. Using Eq. (5-163) we may consider first a scalar $\Gamma\, d\xi_1\, d\xi_2\, d\xi_3$ introduced at time $t = 0$ at the point (ξ_1,ξ_2,ξ_3). The probability of finding it at the point $(x_1,0,0)$ at time t, taking into account the effect of the mean velocity, then is

$$\Gamma'(x_1,0,0;t) = \frac{\Gamma(\xi_1,\xi_2,\xi_3;0)\, d\xi_1\, d\xi_2\, d\xi_3}{(2\pi)^{3/2}(\overline{y_1^2}\,\overline{y_2^2}\,\overline{y_3^2})^{1/2}} \times \exp\left[-\frac{1}{2}\left\{\frac{(x_1-\xi_1-\bar{U}_1t)^2}{\overline{y_1^2}} + \frac{\xi_2^2}{\overline{y_2^2}} + \frac{\xi_3^2}{\overline{y_3^2}}\right\}\right]$$

Hence, considering $[\mathbf{Q}_{1,1}]_E$ as a diffusing scalar, we may write the relation

$$[\mathbf{Q}_{1,1}(x_1,0,0;t)]_E = \iiint_{-\infty}^{+\infty} d\xi_1\, d\xi_2\, d\xi_3 \frac{[\mathbf{Q}_{1,1}(\xi_1,\xi_2,\xi_3;0)]_E}{(2\pi)^{3/2}(\overline{y_1^2}\,\overline{y_2^2}\,\overline{y_3^2})^{1/2}} \times \exp\left[-\frac{1}{2}\left\{\frac{(x_1-\xi_1-\bar{U}_1t)^2}{\overline{y_1^2}} + \frac{\xi_2^2}{\overline{y_2^2}} + \frac{\xi_3^2}{\overline{y_3^2}}\right\}\right]$$

Assuming that only a real contribution is obtained by particles with a diffusion time short with respect to the Lagrangian correlation time, and further assuming isotropy so that

$$\overline{y_1^2} = \overline{y_2^2} = \overline{y_3^2} = \overline{u^2}t^2$$

the equation for $[\mathbf{Q}_{1,1}(x_1,0,0,t)]_E$ reads

$$[\mathbf{Q}_{1,1}(x_1,0,0;t)]_E = \iiint_{-\infty}^{+\infty} d\xi_1\, d\xi_2\, d\xi_3 \frac{[\mathbf{Q}_{1,1}(\xi_1,\xi_2,\xi_3;0)]_E}{(2\pi\overline{u^2}t^2)^{3/2}}$$

$$\times \exp\left[-\frac{1}{2\overline{u^2}t^2}\{(x_1-\xi_1-\bar{U}_1 t)^2+\xi_2{}^2+\xi_3{}^2\}\right] \tag{5-188}$$

Since we assumed isotropy we may use Eq. (3-8) for a further evaluation of the integral. The purely Eulerian time-correlation $[\mathbf{Q}_{1,1}(0,0,0;t)]_E$ is obtained with $x_1 = 0$.

Favre[96] has applied Eq. (5-188) to Eulerian space-time correlations measured in a grid-generated turbulence. Figure 5-9 shows the comparison between computed and measured correlations. When considered as a rough approximation to the actual process determining the time effects on Eulerian correlations, the agreement between the two seems to be satisfactory. However, one has to bear in mind that Eq. (5-188) has been obtained on the assumption of very short diffusion times, while in Fig. 5-9 time intervals are considered which are not very small compared with the Lagrangian integral time-scale. So one would expect a slower drop of $[\mathbf{Q}_{1,1}(0,0,0;t)]_E$ with time than according to Eq. (5-188), though the difference may not be large, since from a rough estimate based on relations between $(\mathfrak{I}_E)_{\text{meas}}$ and $\mathfrak{I}_L$, one may conclude that at $\bar{U}_1 t/M = 10$, and for values of the relative turbulence-intensity of a few percent $t/\mathfrak{I}_L$ is still much smaller than 1. It is also clear that $[\mathbf{R}_{1,1}(0,0,0;t)]_E$ cannot be approximated by the simple function $\exp(-t/\mathfrak{I}_E)$.

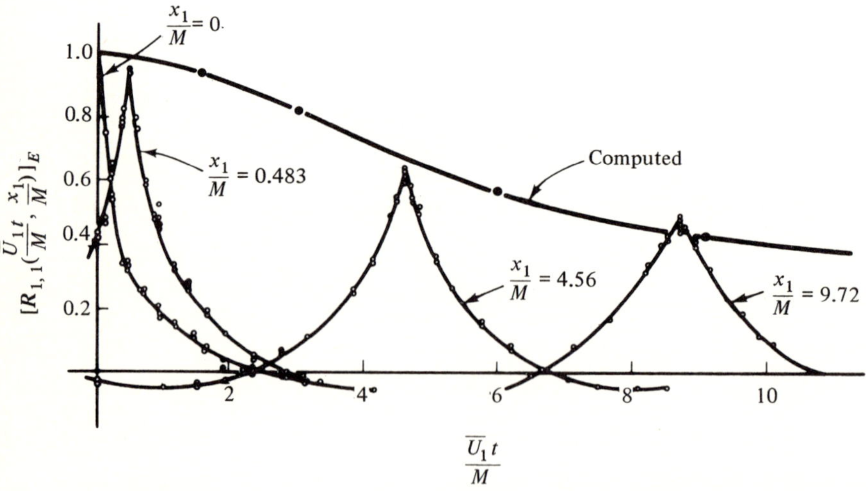

FIGURE 5-9
Measured and computed space-time correlations with optimum delay. (From: *Favre, A.,*[96] *by permission of the American Society of Mechanical Engineers.*)

5-6 DIFFUSION FROM A FIXED SOURCE IN A TURBULENT SHEAR FLOW

In the previous section it was assumed that the turbulent flow had a uniform mean velocity and that the turbulence was homogeneous. The concentration-distribution curves then show a symmetry with respect to an axis (in the case of a point source) or with respect to a plane (in the case of a line source). If the mean velocity is not uniform but is some function of the space coordinates, it may be expected that there will no longer be symmetry in the concentration-distribution curves. In general the shape of these distribution curves will be determined by the characteristics of the turbulence and by the spatial distribution of the mean velocity. If, however, we consider only short diffusion times, that is, diffusion within short distances from the source, the shape of the concentration-distribution curve may be determined mainly by the local distribution of the mean velocity and the local turbulence at the source.

The simplest case of a turbulent shear flow is where the shear flow is steady and homogeneous, that is, parallel flow with a constant gradient of the mean velocity in a direction perpendicular to this velocity:

$$\bar{U}_1 = \text{const.} \times x_2 \qquad \bar{U}_2 = \bar{U}_3 = 0$$

We may assume a line source placed perpendicular to the mean velocity and to the direction of its gradient. The distribution of the diffused matter downstream from the source will be skew. Such a skewness has indeed been shown in experiments carried out by Skramstad[44] in a turbulent boundary layer, by Corrsin and Uberoi[45] in the turbulent mixing zone of a round free jet, and by Hinze and Van der Hegge Zijnen[46] in the turbulent mixing zone of a plane free jet. All investigators used a thin heated wire as a line source of heat. The skewness was such that the greater lateral spread occurred at the side of the greater value of the mean velocity.

If the spread could be described by means of a constant coefficient of eddy diffusion, the effect of the gradient in the mean velocity would be skewness just the opposite of that actually observed by the experimenters. The effect may be explained as follows. In general, the lateral spread at a given distance from a source is determined by the time the flow needs to cover this distance. Now this time is shorter at the side where the flow velocity is greater, and vice versa. Hence, the lateral spread will be narrower at the side of greater mean velocity and wider at the side of smaller mean velocity.

Mathematically this problem has been solved by Lauwerier.[47] Lauwerier started from the diffusion equation with convection

$$\bar{U}_{1,0}(1 + ax_2)\frac{\partial \bar{\Gamma}}{\partial x_1} = \epsilon_\gamma \frac{\partial^2 \bar{\Gamma}}{\partial {x_2}^2}$$

and assumed $\bar{\Gamma} = 0$ at the plane $x_2 = -1/a$. Lauwerier's solution reads

$$\bar{\Gamma} = \frac{S_0}{3ax_1}(1 + ax_2)^{1/2} I_{1/3}\left[\tfrac{2}{9}\bar{U}_{1,0}\frac{(1 + ax_2)^{3/2}}{a^2 \epsilon_\gamma x_1}\right]$$

$$\times \exp\left[-\tfrac{1}{9}\bar{U}_{1,0}\frac{1+(1+ax_2)^3}{a^2\epsilon_y x_1}\right] \tag{5-189}$$

where S_0 is the strength of the source and $I_{1/3}$ is the Bessel function of the order $\frac{1}{3}$. The distribution of $\bar{\Gamma}$ given by Eq. (5-189) is skew in the way described just above.

In practical cases, for instance, where the temperature distribution behind a heated wire is measured, the dimension of the wire must generally be small compared with the dimensions of the turbulent flow region; so the lateral diffusion is restricted to distances in which the mean velocity $\bar{U}_1$ varies only slightly. Consequently, the skewness of the temperature distribution will be very small.

The assumption of a constant coefficient of eddy diffusion in the theoretical case has the formal objection that it is not possible to associate a finite integral scale with a homogeneous shear-flow turbulence, as we have explained in Chap. 4. Hence it is also impossible to associate with this turbulence a constant coefficient of eddy diffusion, which would then be expressible in terms of an integral scale and an intensity of turbulence. However, in actual experiments the turbulence has a finite scale due to the device generating the shear flow (see Chap. 4).

But, apart from this formal objection, there are other more stringent arguments which make the solution (5-189) of very little practical value for diffusion processes in turbulent flows.

In the first place, most if not all turbulent flows show only approximately a constant gradient in the mean-velocity distribution, and then only in regions that are certainly not large with respect to the integral scale of turbulence. Hence there is no actual case where diffusion times large compared with the Lagrangian integral scale can be considered without an associated change in the mean-velocity gradient.

In the second place, the relative intensity of turbulence in shear flow is generally not small. The turbulence spread of, say, heat from a fine hot wire within the integral scale of turbulence is already so great that, beyond this scale, the temperature rise is only a very small fraction of the initial excess temperature just around the heated wire, scarcely measurable with reasonable accuracy. Hence, measurements must be restricted to the region of the integral scale. This applies to the measurements made by Skramstad, by Corrsin and Uberoi, and by Hinze and Van der Hegge Zijnen, who all observed a spread increasing linearly with distance.

The picture of the diffusion process close behind the source and within the integral scale of turbulence is different from that seen at greater distance. Hinze[48] has shown that we may obtain an idea of this picture if we consider such short distances that perfect correlation exists between the velocities at subsequent times of a fluid particle passing the source. Thus we assume the Lagrangian correlation coefficient $\mathbf{R}_L(t) \simeq 1$ during the time of diffusion t considered. We then have, according to Eq. (1-83),

$$\overline{y_2{}^2(t)} = v_2'^2 t^2$$

The same result would be obtained if we assumed for the displacement of a fluid particle

$$y_2(t) = \int_0^t d\tau\, v_2(\tau) \simeq v_2(0)t = u_2(0)t$$

thus neglecting the higher-order terms in the development of $v_2(t)$

$$v_2(t) = v_2(0) + \left[\frac{dv_2}{dt}\right]_{t=0} t + \frac{1}{2!}\left[\frac{d^2v_2}{dt^2}\right]_{t=0} t^2 + \cdots$$

To the same approximation we may put for the displacement of a fluid particle in the axial direction

$$y_1(t) \simeq [\bar{U}_1 + u_1(0)]t$$

Here $\bar{U}_1$ is the mean velocity at the point of the source.

Now we want to consider the distribution of the marked fluid particles that have passed the source over a plane at a distance x_1 from the source. This distribution is the same as that of $y_2(x_1)$ for a given x_1, and we have to find the probability density of $y_2(x_1)$. According to the above, $x_1 = [\bar{U}_1 + u_1(0)]t$ or $t = x_1/[\bar{U}_1 + u_1(0)]$; so

$$y_2(x_1) = \frac{u_2(0)}{\bar{U}_1 + u_1(0)} x_1 \tag{5-190}$$

If $u_1(0)/\bar{U}_1 \ll 1$ the probability-density distribution of $y_2(x_1)$ is identical with that of $u_2(0)$, that is, of $v_2(0)$. But, if the relative intensity of turbulence is not that small, this identity no longer holds. As a matter of fact, it has often been observed that the skewness of the $\bar{\Gamma}$-distribution—that is, neglecting molecular effects, of the y_2-distribution over a plane at distance x_1—not only is much greater than that of the v_2-distribution but may also be of opposite sign. This will be the case, for instance, in the examples to be given later.

From Eq. (5-190) we may obtain various important relations. Let us first take the average value of y_2 taken over a large number of fluid particles that have passed the source at subsequent times t_0.

$$\bar{y}_2(x_1) = \frac{x_1}{T}\int_0^T dt_0 \frac{u_2(t_0)}{\bar{U}_1 + u_1(t_0)}$$

We assume that $u_1(t_0)/\bar{U}_1 < 1$, so that we may develop $(1 + u_1/\bar{U}_1)^{-1}$ into a Taylor series.

$$\frac{\bar{y}_2(x_1)}{x_1} = \frac{1}{T}\int_0^T dt_0 \frac{u_2(t_0)}{\bar{U}_1}\left[1 - \frac{u_1(t_0)}{\bar{U}_1} + \frac{u_1^2(t_0)}{\bar{U}_1^2} - \cdots\right]$$

$$= -\frac{\overline{u_1u_2}}{\bar{U}_1^2} + \frac{\overline{u_1^2u_2}}{\bar{U}_1^3} - \frac{\overline{u_1^3u_2}}{\bar{U}_1^4} + \cdots \tag{5-191}$$

In the same way the average value of the squared and cubed y_2's of a large number of fluid particles is found to be

$$\frac{\overline{y_2^2}}{x_1^2} = \frac{\overline{u_2^2}}{\bar{U}_1^2} - 2\frac{\overline{u_1u_2^2}}{\bar{U}_1^3} + 3\frac{\overline{u_1^2u_2^2}}{\bar{U}_1^4} - \cdots \tag{5-192}$$

and

$$\frac{\overline{y_2^3}}{x_1^3} = \frac{\overline{u_2^3}}{\bar{U}_1^3} - 3\frac{\overline{u_1 u_2^3}}{\bar{U}_1^4} + 6\frac{\overline{u_1^2 u_2^3}}{\bar{U}_1^5} - \cdots \tag{5-193}$$

From the last two relations we may easily determine the skewness factor S_{y_2} of the distribution of y_2. The skewness factor is defined by

$$S_{y_2} = \frac{\overline{y_2^3}}{(\overline{y_2^2})^{3/2}}$$

From Eqs. (5-192) and (5-193) we obtain

$$S_{y_2} = \frac{\overline{u_2^3}}{(\overline{u_2^2})^{3/2}} - 3\frac{\overline{u_1 u_2^3}}{(\overline{u_2^2})^{3/2}\bar{U}_1} + 3\frac{\overline{u_2^3}\,\overline{u_1 u_2^2}}{(\overline{u_2^2})^{5/2}\bar{U}_1} + 6\frac{\overline{u_1^2 u_2^3}}{(\overline{u_2^2})^{3/2}\bar{U}_1^2}$$
$$-4.5\frac{\overline{u_2^3}\,\overline{u_1^2 u_2^2}}{(\overline{u_2^2})^{5/2}\bar{U}_1^2} + 7.5\frac{\overline{u_2^3}(\overline{u_1 u_2^2})^2}{(\overline{u_2^2})^{7/2}\bar{U}_1^2} - 9\frac{\overline{u_1 u_2^3}\,\overline{u_1 u_2^2}}{(\overline{u_2^2})^{5/2}\bar{U}_1^2} + \cdots \tag{5-194}$$

If, in the expression (5-191) for $\bar{y}_2/x_1$, the resultant effect of the triple and higher-order correlations is negligibly small, we obtain the interesting result that the value of $\bar{y}_2/x_1$ must be proportional to the shearing stress in the flow at the point of the source.

$$\frac{\bar{y}_2}{x_1} \simeq -\frac{\overline{u_1 u_2}}{\bar{U}_1^2} = \frac{\sigma_{12}}{\rho \bar{U}_1^2} \tag{5-195}$$

$\bar{y}_2/x_1$ is identical with the mean, that is, the centroid of the $\bar{\Gamma}$-distribution surface, since, by definition,

$$\frac{\bar{y}_2}{x_1} = \int_{-\infty}^{+\infty} d\left(\frac{y_2}{x_1}\right)\frac{\bar{\Gamma}(y_2/x_1)}{\bar{\Gamma}(0)}\left(\frac{y_2}{x_1}\right)$$

Thus the ordinate of this centroid is, in first approximation, proportional to the local turbulent shearing stress.

Figure 5-10 shows a skewed temperature distribution downstream from a fine heated wire placed in a horizontal-plane free jet of air and measured by Van der Hegge Zijnen.[46] The discharge orifice of the jet had a width of $d = 5$ mm. The jet had an issuing velocity of 40 m/s. In the cross section of the jet at $20d$ distance from the orifice (the maximum velocity amounted to 22 m/s and the jet width was about 40 mm), a platinum-iridium wire 0.05 mm in diameter, 10 mm long, and heated to about 400°C was installed at 6 mm below the axis of the jet. The temperature distribution is given with respect to a coordinate system with the origin at the location of the heat source and with the x_1-axis along the streamline of the mean velocity through the heat source.

The skewness is such that the spread of $\bar{\Gamma}$ is largest toward the region of the greater mean velocity $\bar{U}_1$. This can easily be explained. If, for instance, $d\bar{U}_1/dx_2 > 0$ and, consequently, $\sigma_{12} > 0$, then $\overline{u_1 u_2} < 0$. This means that, for positive values of

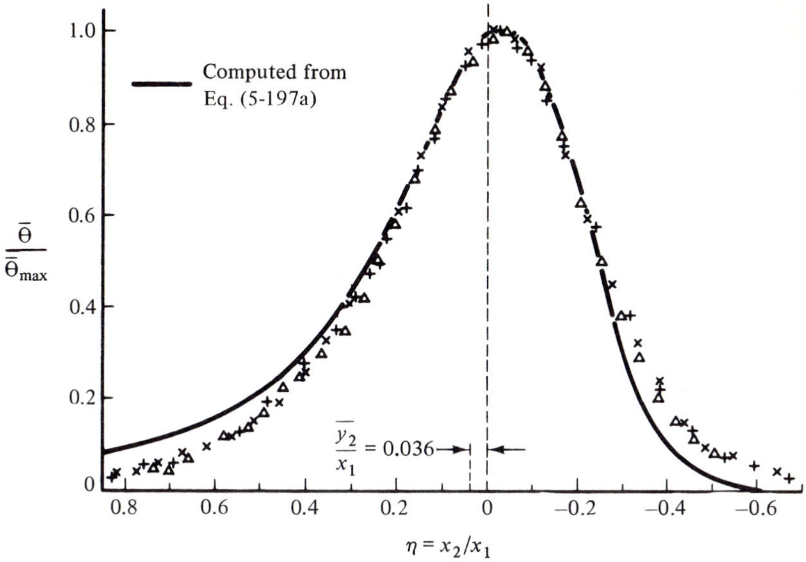

FIGURE 5-10
Skewed temperature distribution downstream of a heated wire in a plane free air jet. Location 6 mm below axis at distance 100 mm from an orifice of 5 mm width. Issuing velocity 40 m/s. (From: *Hinze, J. O., and B. G. van der Hegge Zijnen.*[46])

u_2, on the average more fluid particles with negative u_1 will be transported through a control plane, and vice versa. To positive values of u_2 of fluid particles passing the control plane correspond also positive values of x_2. Hence, for the same value of x_1, on the average $t > x_1/\bar{U}_1$ for positive x_2. Similarly, for negative values of x_2, on the average $t < x_1/\bar{U}_1$. Consequently, the spread of fluid particles for positive values of x_2 must be greater than the spread of fluid particles for negative values of x_2.

The ordinate of the centroid of the temperature-distribution surface is $\bar{y}_2/x_1 = 0.036$, which agrees with the value $\sigma_{12}/\rho\bar{U}_1{}^2 = 0.036$ calculated from the mean-velocity distribution.

For the same temperature distribution we calculate $\overline{y_2{}^2}/x_1{}^2 = 0.074$ and $\overline{y_2{}^3}/x_1{}^3 = 0.01705$. Hence we obtain a skewness factor

$$S_{y_2} = 0.85$$

No measurements have been made of $\overline{u_2{}^3}$ or of the various third- and higher-order velocity correlations occurring in Eq. (5-194). However, we may use this relation to calculate the skewness factor of u_2 if we follow a suggestion made by Batchelor and Townsend,[25] namely, to consider only the first three terms on the right-hand side of Eq. (5-194) and to assume (1) that the fourth-order velocity correlation $\overline{u_1u_2{}^3}$ is related to $\overline{u_1u_2}$ as if the joint-probability-density distribution of u_1 and u_2 were normal, and (2) that $\overline{u_1u_2{}^2}$ is less than $\overline{u_2{}^3}$ in the ratio $\overline{u_1u_2}/\overline{u_2{}^2}$.

If $\mathfrak{P}(u_1,u_2)$ is the joint-probability-density distribution of u_1 and u_2, the expression when it is normal reads (see, e.g., Ref. 49)

$$\mathfrak{P}\left(\frac{u_1}{\bar{U}_1},\frac{u_2}{\bar{U}_1}\right)=\frac{\bar{U}_1^{\,2}}{2\pi u_1' u_2'[1-\mathbf{R}_{12}^{\,2}(0)]^{1/2}}$$

$$\times \exp\left\{\frac{-1}{2[1-\mathbf{R}_{12}^{\,2}(0)]}\left[\frac{u_1^{\,2}}{u_1'^2}-\frac{2u_1u_2\mathbf{R}_{12}(0)}{u_1'u_2'}+\frac{u_2^{\,2}}{u_2'^2}\right]\right\} \tag{5-196}$$

where $\mathbf{R}_{12}(0)=\overline{u_1u_2}/u_1'u_2'$.

Such a normal joint-probability-density distribution has the following features:

1 The marginal distributions of u_1 and u_2 are normal Gaussian distributions.

2 All uneven moments are zero:

$$\overline{u_1^{\,m}u_2^{\,n}}=\frac{1}{\bar{U}_1^{\,2}}\iint_{-\infty}^{+\infty} du_1\, du_2\, \mathfrak{P}(u_1,u_2)u_1^{\,m}u_2^{\,n}=0$$

for m and n integers and $m+n=$ uneven

The fourth-order moment $\overline{u_1u_2^{\,3}}$ is related to $\overline{u_1u_2}$ according to

$$\overline{u_1u_2^{\,3}}=\frac{1}{\bar{U}_1^{\,2}}\iint_{-\infty}^{+\infty} du_1\, du_2\, \mathfrak{P}(u_1,u_2)u_1u_2^{\,3}=3\mathbf{R}_{12}(0)u_1'u_2'^3$$

Thus, with Batchelor and Townsend's suggestion, we obtain from Eq. (5-194)

$$S_{y_2}\simeq S_{u_2}-9\frac{u_1'}{\bar{U}_1}\mathbf{R}_{12}(0)+3S_{u_2}^{\;2}\frac{u_1'}{\bar{U}_1}\mathbf{R}_{12}(0)$$

where S_{u_2} is the skewness factor of u_2.

As will be shown shortly, it is possible to calculate the skewed temperature distribution if the joint-probability-density function $\mathfrak{P}(u_1,u_2)$ is known. From this skewed temperature distribution it is then possible to estimate the values of u_1' and u_2'. For the temperature distribution given in Fig. 5-10, Hinze and Van der Hegge Zijnen obtained $u_1'/\bar{U}_1=0.28$ and $u_2'/\bar{U}_1=0.215$. Hence, with $\overline{u_1u_2}/\bar{U}_1^{\,2}=-\sigma_{12}/\rho\bar{U}_1^{\,2}=-0.036$, there is obtained $\mathbf{R}_{12}(0)=-0.60$. Thus, with $S_{y_2}=0.85$, a value of -0.48 for S_{u_2} results. A negative value of the skewness factor of u_2 could be expected, since the turbulence intensity u_2' showed a positive gradient $du_2'/dx_2>0$ at the point of the heat source. This value of S_{u_2} is of the same magnitude as the values obtained by Townsend[50] in the wake of a circular cylinder near the point of maximum shear stress, namely, -0.4 to -0.5.

If we compare the value of $\overline{u_2^{\,2}}/\bar{U}_1^{\,2}=0.046$ with the value $\overline{y_2^{\,2}}/x_1^{\,2}=0.074$, we conclude from the relation (5-192) that the contribution of the higher-order velocity correlations cannot be neglected.

Also, from the skewness factor $S_{u_2} = -0.48$ we obtain

$$\frac{\overline{u_2^3}}{\bar{U}_1^3} = -0.0048$$

whereas $\overline{y_2^3}/x_1^3 = 0.01705$. Hence, here, too, the contribution of the higher-order velocity correlations to the relation (5-193) cannot be neglected. Apparently these results for $\overline{y_2^2}/x_1^2$ and $\overline{y_2^3}/x_1^3$ are contrary to the result obtained for $\bar{y}_2/x_1$. This latter result, namely, that the resultant contribution of the third- and higher-order correlations in the relation (5-191) can be neglected, appeared also to apply to other points in the same cross section of the jet. Figure 5-11 shows the distribution

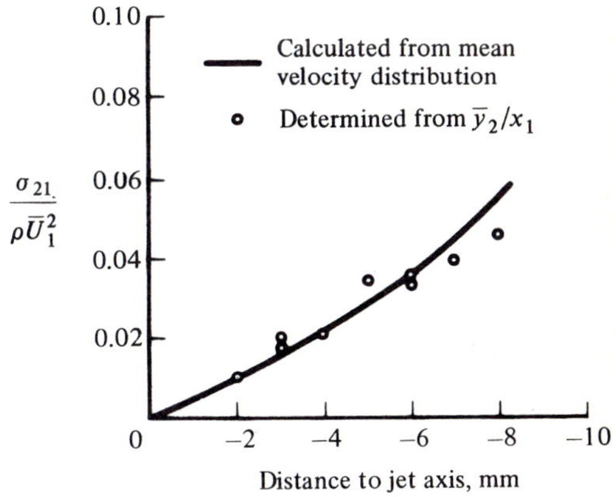

FIGURE 5-11
Shear-stress distribution across a plane free jet in a section 100 mm from the orifice of 5 mm width. (From: *Hinze, J. O., and B. G. van der Hegge Zijnen.*[46])

of $\sigma_{21}/\rho\bar{U}_1^2$ over the cross section, first determined from the values of $\bar{y}_2/x_1$ and, second, calculated from the mean-velocity distribution according to a procedure to be given in Chap. 6.

Let us now show the calculation of the temperature distribution if it is assumed that the joint-probability-density function $\mathfrak{P}(u_1,u_2)$ is known. We follow a procedure similar to that expounded in the previous section.

Consider a point with coordinates (x_1,x_2) downstream from the heat source. For a fluid particle which passes the heat source at t_0 with velocities $\bar{U}_1 + u_1$ and u_2 and which arrives at the point under consideration at time t, the velocity components and the time interval $t - t_0$ are related by

$$x_1 = (\bar{U}_1 + u_1)(t - t_0)$$

$$x_2 = u_2(t - t_0)$$

Let $\mathfrak{P}'(x_1,x_2,t-t_0)\,dx_1\,dx_2$ be the probability that the fluid particle arrives in the area element $dx_1\,dx_2$ at the point under consideration after the time interval $t-t_0$. We want to express this probability in terms of the joint-probability-density function $\mathfrak{P}(u_1,u_2)$. Now $\mathfrak{P}'(x_1,x_2,t-t_0)\,dx_1\,dx_2$ must be just equal to $\mathfrak{P}(u_1,u_2)\,du_1\,du_2$ provided that u_1, u_2, x_1, x_2, and $t-t_0$ are connected by the above relations. Thus, if again S^* is the source strength, the probability that fluid passing the heat source during dt_0 will arrive in the area element $dx_1\,dx_2$ is given by

$$\mathfrak{P}'(x_1,x_2,t-t_0)\,dt_0\,dx_1\,dx_2 = S^*\,dt_0\,\mathfrak{P}(u_1,u_2)\,du_1\,du_2/\bar{U}_1{}^2$$

For a continuous source, the probability of finding in $dx_1\,dx_2$ a fluid particle originating from the source becomes

$$\mathfrak{P}(x_1,x_2)\,dx_1\,dx_2 = \frac{1}{\bar{U}_1{}^2}\int_{-\infty}^{t} dt_0\,S^*\mathfrak{P}(u_1,u_2)\,du_1\,du_2$$

Since, for a given $t-t_0$, we have

$$du_1 = \frac{dx_1}{t-t_0} \qquad du_2 = \frac{dx_2}{t-t_0}$$

the above relation becomes

$$\mathfrak{P}(x_1,x_2) = \frac{S^*}{\bar{U}_1{}^2}\int_{-\infty}^{t} dt_0\,\mathfrak{P}(u_1,u_2)\frac{1}{(t-t_0)^2}$$

For a given value of x_1, we may write

$$\frac{dt_0}{(t-t_0)^2} = d\frac{1}{t-t_0} = d\frac{\bar{U}_1+u_1}{x_1} = \frac{du_1}{x_1}$$

Consequently,

$$\mathfrak{P}(x_1,x_2) = \frac{S^*}{x_1}\frac{1}{\bar{U}_1{}^2}\int_{-\bar{U}_1}^{\infty} du_1\,\mathfrak{P}(u_1,u_2)$$

or

$$\mathfrak{P}(x_1,x_2) = \frac{S^*}{x_1\bar{U}_1}\int_{-1}^{\infty} d\frac{u_1}{\bar{U}_1}\,\mathfrak{P}\left[\frac{u_1}{\bar{U}_1},\left(1+\frac{u_1}{\bar{U}_1}\right)\frac{x_2}{x_1}\right]$$

Substituting the expression (5-196) for a normal joint-probability-density function and carrying out the integration yield

$$\mathfrak{P}(x_1,x_2) = \frac{S^*}{2\sqrt{2\pi}\,[\overline{u_2{}^2}x_1{}^2 - 2u'_1u'_2\mathbf{R}_{12}(0)x_2x_1 + \overline{u_1{}^2}x_2{}^2]^{1/2}}$$
$$\times\,\text{erfc}\left\{-\frac{u'_2\bar{U}_1x_1\left[1-\dfrac{u'_1}{u'_2}\mathbf{R}_{12}(0)\right]}{u'_1\{2[1-\mathbf{R}_{12}(0)][\overline{u_2{}^2}x_1{}^2 - 2u'_1u'_2R_{12}(0)x_2x_1 + \overline{u_1{}^2}x_2{}^2]\}^{1/2}}\right\}$$
$$\times\,\exp\left\{\frac{-\bar{U}_1{}^2x_2{}^2}{2[\overline{u_2{}^2}x_1{}^2 - 2u'_1u'_2\mathbf{R}_{12}(0)x_2x_1 + \overline{u_1{}^2}x_2{}^2]}\right\} \tag{5-197a}$$

The reader may easily verify that this expression reduces to Eq. (5-168) if $\mathbf{R}_{12}(0) = 0$, that is, if the mean flow is uniform.

If we assume that the effect of molecular diffusion is negligible, the $\bar{\Gamma}$-distribution is identical with $\mathfrak{P}(x_1, x_2)$. The distribution is a function only of $\eta = x_2/x_1$; that is, the temperature distributions in subsequent cross sections of the wake of the heat source are similar.

If we want to apply Eq. (5-197a) to actual $\bar{\Gamma}$-distributions, we have to keep in mind (1) that the assumption of a normal joint-probability distribution of u_1 and u_2 is perhaps not justified; (2) that the expressions are valid only for short diffusion times $t - t_0$, that is, small values of $x_1/(\bar{U}_1 + u_1)$ and x_2/u_2. If $u_1'/\bar{U}_1$ is much smaller than 1, the condition of short diffusion times is equivalent to requiring small values of $\eta = x_2/x_1$. An estimate of the value of η beyond which, for small values of $u_1'/\bar{U}_1$, the assumption of short diffusion times is no longer valid may be obtained from the value of $u_2'/\bar{U}_1$. Thus we have to take η not very much greater than $u_2'/\bar{U}_1$, that is, roughly 0.2 if the diffusion is studied in a free jet.

For short diffusion times the value of the erfc occurring in the expression (5-197a) is practically equal to 2. For the $\bar{\Gamma}$-distribution in a cross section, expressed in terms of $\bar{\Gamma}(0)$, we then obtain

$$\frac{\bar{\Gamma}(\eta)}{\bar{\Gamma}(0)} = \frac{1}{[1 - 2(u_1'/u_2')\mathbf{R}_{12}(0)\eta + (u_1'^2/u_2'^2)\eta^2]^{1/2}} \times \exp\frac{-\bar{U}_1{}^2\eta^2}{2u_2'^2[1 - 2(u_1'/u_2')\mathbf{R}_{12}(0)\eta + (u_1'^2/u_2'^2)\eta^2]}$$

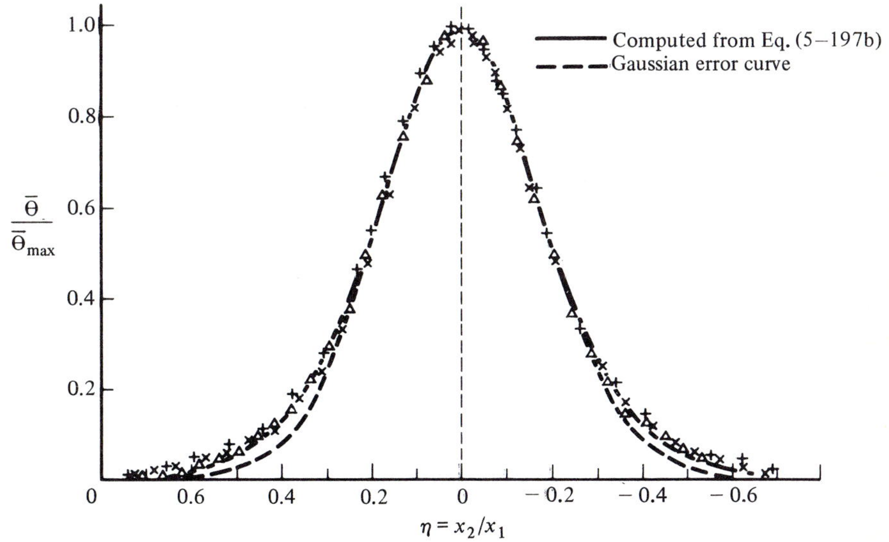

FIGURE 5-12
Temperature distribution downstream of a heated wire in the axis of a plane free air jet at 100 mm from the orifice of 5 mm width. (From: *Hinze, J. O., and B. G. van der Hegge Zijnen.*[46])

For the case of isotropic turbulence, where $u'_1 = u'_2$ and $\mathbf{R}_{12}(0) = 0$, we would obtain

$$\frac{\bar{\Gamma}(\eta)}{\bar{\Gamma}(0)} = \frac{1}{(1+\eta^2)^{1/2}} \exp\left[\frac{-\bar{U}_1{}^2\eta^2}{2u_2'^2(1+\eta^2)}\right] \qquad (5\text{-}197b)$$

According to this result, the spread of $\bar{\Gamma}$ for greater values of η would be somewhat greater than according to a Gaussian distribution.

Figure 5-10 also shows the temperature-distribution curve calculated according to Eq. (5-197a) using the value $\mathbf{R}_{12}(0)$ obtained from $\bar{y}_2/x_1$ and values of u'_1 and u'_2 obtained by trial and error. In this way the values of 0.28 and 0.215 for $u'_1/\bar{U}_1$ and $u'_2/\bar{U}_1$, respectively, given earlier, were obtained. The agreement with experiment in the region $-0.2 < x_2/x_1 < 0.2$ is satisfactory. For greater values of x_2/x_1 the computed curve is too skew.

Figure 5-12 shows a temperature distribution downstream from a heated wire located at a point in the axis of a plane free jet and measured by Van der Hegge Zijnen.[46] Also given are the curves computed according to Eq. (5-197b) and according to a Gaussian error function.

The difference between the two computed curves is noticeable only for values of x_2/x_1 much greater than 0.2. Though the curve computed according to Eq. (5-197b) shows an improvement over the Gaussian curve in the right sense, yet it must be regarded with some reserve, for reasons explained earlier.

5-7 DIFFUSION OF DISCRETE PARTICLES IN A HOMOGENEOUS TURBULENCE

Knowledge of the behavior of discrete particles in a turbulent flow is of great interest to many branches of technology, particularly if there is a substantial difference in density between the particle and the fluid. The combined flow of solids and fluids or of atomized liquids and gases (flow of a mist) is encountered, for instance, in one or more of the following technical applications: gas and liquid cleaners (e.g., cyclone separators), pneumatic conveying, coal washing and mineral dressing, chemical reactors based on the fluidized-solids system, combustion of pulverized or atomized fuels, etc.

The behavior of a swarm of discrete particles in a turbulent fluid depends largely (1) on the concentration of the particles, and (2) on the size of the particles with respect to the scale of turbulence of the fluid.

At great concentration there is interaction between the particles through collisions and through effects on the flow of the fluid in the neighborhood of the particles. The relative motion of the particles with respect to the fluid may cause, via the creation of additional turbulence, extra dissipation of the kinetic energy of the turbulent fluid, thus damping the turbulence in the fluid. At extremely high concentrations, near that of maximum packing of the particles, the turbulence may even be "frozen"—a term introduced by Bagnold[51] to denote a condition of almost entirely damped turbulent motion. Through collisions and saltation phenomena, such a dense concentration of particles may still exhibit appreciable normal and shear stresses, as shown by Bagnold.[51,52] If the particles are very fine, effects other than the purely hydrodynamic may play a role, causing, for instance, agglomeration of particles.

In the other extreme case, namely, when the concentration of particles is very low, we may neglect the interference of the particles and regard each particle as being alone in the turbulent flow field. If the particles are big compared with the scale of turbulence, the main effect of the turbulence on the particle will be to increase its flow resistance, and the particle will, at most, more or less follow the slow large-scale turbulent motions of the fluid. If, on the other hand, the particles are very small compared with the smallest scale of the turbulence, they will tend to follow all turbulence components of the fluid. Usually the flow resistance of the particle with respect to the ambient fluid will be entirely viscous in nature. Here, too, the presence of the particles will increase the dissipation in the turbulent flow. Since this additional dissipation is caused by the lag in turbulent motion between particle and fluid and since the lag increases with increasing wavenumber of the turbulence, it is reasonable to expect that the presence of the particles will influence the energy spectrum of the turbulence mainly in the high-wavenumber range.

Theoretically very little is known about the fluid dynamics of a mixture of discrete particles of arbitrary size and concentration with a fluid, where both are in turbulent motion. For high concentrations of very fine particles, the mixture may be considered a homogeneous suspension and may be treated as a non-Newtonian fluid with certain over-all flow features. For bigger particles, such as grains and sand, Bagnold has studied stress-strain relationships of the mixture in nearly turbulent motion.

The first extensive theoretical study has been made by Tchen[53] on the motion of a very small particle suspended in a turbulent fluid. Some ten years later this study was taken up again by Corrsin and Lumley,[55,69] and by Friedlander.[70]

In what follows we shall mainly consider Tchen's studies, but treated in a different way mathematically. Although the case considered by Tchen is extreme because of the simplifying assumptions made, some of the results obtained are of general interest.

In Tchen's theory the following assumptions are made:

1 The turbulence of the fluid is homogeneous and steady.
2 The domain of turbulence is infinite in extent.
3 The particle is spherical and so small that its motion relative to the ambient fluid follows Stokes' law of resistance.
4 The particle is small compared with the smallest wavelength present in the turbulence, i.e, with the Kolmogoroff micro-scale η, say.
5 During the motion of the particle the neighborhood will be formed by the same fluid.
6 Any external force acting on the particle originates from a potential field, such as a gravity field.

All assumptions, except assumption (5), are rather reasonable and may actually be satisfied. As to assumption (5), the mechanism of a real turbulence is such that it is hardly possible for this to be satisfied. If the element of fluid containing a small discrete particle could be considered an undeformable entity, say, a spherical one, then it could, possibly, satisfy this assumption, provided that its size was larger than the amplitude of the motion of the discrete particle relative to the fluid (no overshooting). The motions of discrete particle and fluid element might then actually be as illustrated in Fig. 5-13. However, we know that in turbulent motion the fluid elements are distorted and stretched into long thin ribbons and it seems unreasonable that the fluid element should continue to contain the same discrete particles during this stretching process.

The condition of no-overshooting implies not only that the discrete particle cannot be too heavy relative to the fluid, but also that the Lagrangian velocity of the discrete particle relative to that of the fluid must be sufficiently small. The situation pictured in Fig. 5-13 can only be realized if the displacement of the particle relative to that of the centroid of the fluid element during the available time is smaller than Kolmogoroff's microscale $\eta = (\nu^3/\varepsilon)^{1/4}$. The available time is determined by the deformation time of the fluid element, i.e., by its characteristic time $(\nu/\varepsilon)^{1/2}$. Hence, the relative velocity must be smaller than the characteristic velocity $(\varepsilon\nu)^{1/4}$. These severe conditions, on which Tchen's theory is based make the results of this theory of limited value only. Still the results obtained concerning the relative diffusion of discrete particles are of sufficient interest to warrant a discussion of the theory. Though the results may not be applied quantitatively to cases where the above conditions are not satisfied, they may be used to estimate some trends in the diffusive behavior of discrete particles with respect to that of the carrier fluid.

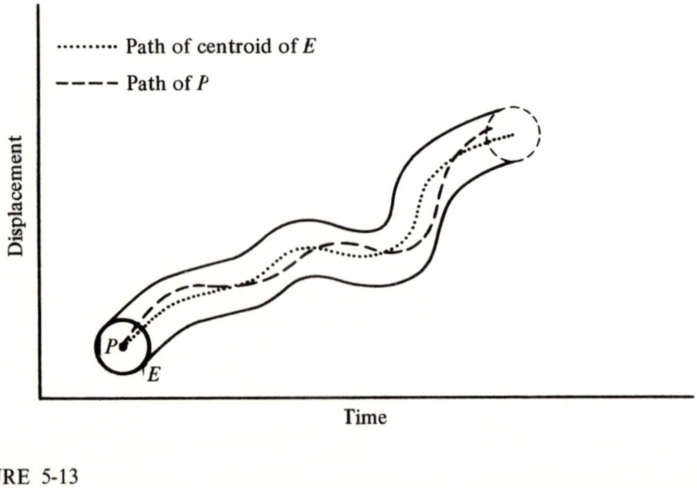

FIGURE 5-13
Displacement of a discrete particle P and that of a spherical fluid element E.

We start from the equation for the slow motion of a spherical particle, derived by Basset,[65] Boussinesq,[66] and Oseen[67] for a fluid at rest but extended by Tchen[53] to the case of a fluid moving with variable velocity.

$$\frac{\pi}{6} d^3 \rho_p \frac{dv_p}{dt} = 3\pi\mu d(v_f - v_p) + \frac{\pi}{6} d^3 \rho_f \frac{dv_f}{dt} + \frac{1}{2}\frac{\pi}{6} d^3 \rho_f \left(\frac{dv_f}{dt} - \frac{dv_p}{dt}\right)$$

$$+ \tfrac{3}{2} d^2 \sqrt{\pi \rho_f \mu} \int_{t_0}^{t} dt' \frac{\dfrac{dv_f}{dt'} - \dfrac{dv_p}{dt'}}{\sqrt{t - t'}} + F_e \qquad (5\text{-}198)$$

where t_0 is the starting time; the index f refers to the fluid and the index p, to the particle; and the fluid velocity v_f is that of the fluid in the neighborhood of the discrete particle but sufficiently remote not to be disturbed by the relative motion of the particle.

The meaning of the various terms is as follows.

The term on the left-hand side is the force required to accelerate the particle. The first term on the right-hand side is the viscous resistance force according to Stokes' law. The second term is due to the pressure gradient in the fluid surrounding the particle, caused by acceleration of the fluid. The third term is the force to accelerate the virtual, "added" mass of the particle relative to the ambient fluid. The fourth term is the so-called "Basset" term, which takes account of the effect of the deviation in flow pattern from steady state. The last term F_e is an external potential force.

We may remark here that the second, third, and fourth terms on the right-hand side become important only if the density of the fluid becomes comparable to or higher than the density of the particle. The Basset term increases the instantaneous flow resistance. At time $t' = t$ the integrand becomes infinite, but, since it is according to $\lim_{\tau=0} \tau^{-1/2}$, the integral remains finite. When the particle is accelerated at a high rate by a strong external force, the Basset term may become substantial, increasing the instantaneous coefficient of resistance to many times its value at steady state. Calculations to show this increase have been made by Hughes and Gilliland.[54]

Since we are interested mainly in the motion of a discrete particle caused by the fluid motion, from now on we shall no longer consider effects of external forces. According to the above condition (5), the relative velocity between discrete particle and fluid is sufficiently small when $|v_f - v_p|/(\varepsilon\nu)^{1/4} \lesssim 1$. This implies that

$$\left|\frac{v_f - v_p}{v_f}\right| \ll 1$$

This restriction makes it possible to consider the particle motion in a turbulent flow relative to that of the fluid as being described by a linear equation. For, formally we have to make a distinction between

$$\left(\frac{D}{Dt}\right)_f = \frac{\partial}{\partial t} + (u_f)_k \frac{\partial}{\partial x_k}$$

and

$$\left(\frac{D}{Dt}\right)_p = \frac{\partial}{\partial t} + (u_p)_k \frac{\partial}{\partial x_k}$$

But owing to the above condition for $(v_f - v_p)$, the difference between the two Lagrangian derivatives becomes negligibly small to the degree of approximation considered here.

Another point which may have to be considered has been mentioned by Corrsin and Lumley,[55] namely that the local pressure gradient in the fluid surrounding the particle is not only caused by the acceleration of the fluid, but also by viscous effects. According to the Navier-Stokes equation, we have for the component $(v_f)_i$

$$-\frac{\partial p}{\partial x_i} = \rho_f \left[\frac{\partial}{\partial t}(v_f)_i + (v_f)_k \frac{\partial}{\partial x_k}(v_f)_i \right] - \mu \frac{\partial^2 (v_f)_i}{\partial x_j \partial x_j}$$

However, since we assume $d \ll \eta$, and time intervals only smaller than the deformation time, the fluid element containing the particle may be considered to have a sufficiently uniform velocity that the viscous effects within this fluid element on $\partial p/\partial x_i$ may be neglected. While the viscous term may still be neglected in larger regions beyond the fluid element when $\mathbf{Re}_\lambda$ is not small. When all these conditions are met, we may apply the linear, first-order, differential equation (5-198) being time-dependent only to the motion of a discrete particle in a turbulent flow. Without loss of generality we may consider only one and the same component of v_p and v_f, respectively. Dropping the term F_e, Eq. (5-198) may be rewritten as follows

$$\frac{dv_p}{dt} + av_p = av_f + b\frac{dv_f}{dt} + c\int_{t_0}^{t} dt' \frac{dv_f/dt' - dv_p/dt'}{\sqrt{t - t'}} \tag{5-199}$$

where

$$a = \frac{36\mu}{(2\rho_p + \rho_f)d^2} \qquad b = \frac{3\rho_f}{2\rho_p + \rho_f} \qquad c = \frac{18}{(2\rho_p + \rho_f)d}\sqrt{\frac{\rho_f \mu}{\pi}}$$

We shall consider first the *stationary case*, the case also studied by Tchen. The starting time t_0 may then be taken at minus infinity. Tchen showed in a general way that, for *long diffusion times*, the diffusion coefficients for the discrete particles and for the fluid particles are the same. This interesting result can be obtained directly from Eq. (5-199) if we assume that both v_p and v_f may be represented by a Fourier integral [compare Eq. (1-192)]:

$$v_f = \int_0^\infty d\omega\,(\alpha \cos \omega t + \beta \sin \omega t)$$

and

$$v_p = \int_0^\infty d\omega\,(\gamma \cos \omega t + \delta \sin \omega t) \tag{5-200}$$

where $\omega = 2\pi n$, the angular frequency.

If we introduce the expressions (5-200) for v_f and v_p into Eq. (5-199), and change the order of integration, we have to solve the following integrals:

$$\int_{-\infty}^{t} dt' \frac{\sin \omega t'}{\sqrt{t-t'}} \quad \text{and} \quad \int_{-\infty}^{t} dt' \frac{\cos \omega t'}{\sqrt{t-t'}}$$

These integrals can be solved by substituting τ for $t - t'$ and by using the known definite integrals

$$\int_0^\infty d\tau \frac{\sin \omega\tau}{\sqrt{\tau}} = \int_0^\infty d\tau \frac{\cos \omega\tau}{\sqrt{\tau}} = \sqrt{\frac{\pi}{2\omega}}$$

Thus, introducing Eq. (5-200) into Eq. (5-199) and carrying out the differentiations and integrations, we find that Eq. (5-199) is identically satisfied if we put the sum of the sine terms equal to zero and the sum of the cosine terms equal to zero. This yields the following relations between γ and δ, and α and β:

$$\begin{aligned} \gamma &= [1 + f_1(\omega)]\alpha + f_2(\omega)\beta \\ \delta &= -f_2(\omega)\alpha + [1 + f_1(\omega)]\beta \end{aligned} \qquad (5\text{-}201)$$

where

$$f_1(\omega) = \frac{\omega(\omega + c\sqrt{\pi\omega/2})(b-1)}{(a + c\sqrt{\pi\omega/2})^2 + (\omega + c\sqrt{\pi\omega/2})^2}$$

$$f_2(\omega) = \frac{\omega(a + c\sqrt{\pi\omega/2})(b-1)}{(a + c\sqrt{\pi\omega/2})^2 + (\omega + c\sqrt{\pi\omega/2})^2}$$

We now introduce the Lagrangian energy-spectrum functions $\mathbf{E}_{f_L}(n)$ for the fluid and $\mathbf{E}_{p_L}(n)$ for the discrete particles. We can then show, as in Sec. 1-12, that

$$\begin{aligned} \overline{v_f^2} &= \pi^2 \int_0^\infty dn \frac{\alpha^2 + \beta^2}{T} = \int_0^\infty dn\, \mathbf{E}_{f_L}(n) \\ \overline{v_p^2} &= \pi^2 \int_0^\infty dn \frac{\gamma^2 + \delta^2}{T} = \int_0^\infty dn\, \mathbf{E}_{p_L}(n) \end{aligned} \qquad (5\text{-}202)$$

where T is a sufficiently long period of time. Furthermore,

$$\mathbf{R}_{f_L}(t) = \frac{\overline{v_f(t_0)v_f(t_0+t)}}{\overline{v_f'^2}} = \frac{1}{\overline{v_f'^2}} \int_0^\infty dn\, \mathbf{E}_{f_L}(n) \cos 2\pi n t$$

and

$$\mathbf{R}_{p_L}(t) = \frac{\overline{v_p(t_0)v_p(t_0+t)}}{\overline{v_p'^2}} = \frac{1}{\overline{v_p'^2}} \int_0^\infty dn\, \mathbf{E}_{p_L}(n) \cos 2\pi n t \qquad (5\text{-}203)$$

Equation (5-56) gives the diffusion of particles as a function of time, expressed in terms of the Lagrangian energy-spectrum function. For the present case we have

$$\overline{y_f^2}(t) = 2\int_0^\infty dn\, \mathbf{E}_{f_L}(n) \frac{1 - \cos 2\pi n t}{4\pi^2 n^2}$$

and

$$\overline{y_p^2}(t) = 2\int_0^\infty dn\, \mathbf{E}_{p_L}(n) \frac{1 - \cos 2\pi n t}{4\pi^2 n^2} \qquad (5\text{-}204)$$

whence, with the relation (5-42), we obtain for the diffusion coefficients

$$\epsilon_f = \frac{1}{2}\frac{d}{dt}\overline{y_f^2}(t) = \int_0^\infty dn\, \mathbf{E}_{f_L}(n)\frac{\sin 2\pi nt}{2\pi n}$$

and

$$\epsilon_p = \int_0^\infty dn\, \mathbf{E}_{p_L}(n)\frac{\sin 2\pi nt}{2\pi n} \tag{5-205}$$

For *short diffusion times,*

$$\epsilon_f = t\int_0^\infty dn\, \mathbf{E}_{f_L}(n) = \overline{v_f^2}t$$

and

$$\epsilon_p = t\int_0^\infty dn\, \mathbf{E}_{p_L}(n) = \overline{v_p^2}t$$

Hence

$$\frac{\epsilon_p}{\epsilon_f} = \frac{\overline{v_p^2}}{\overline{v_f^2}} \tag{5-206}$$

For *long diffusion times,* the main contribution is again from the low-frequency components of the motion.

$$\epsilon_f = \lim_{t\to\infty}\frac{1}{2\pi}\int_0^\infty d\theta\,\frac{\sin\theta}{\theta}\mathbf{E}_{f_L}\left(\frac{\theta}{2\pi t}\right) = \tfrac{1}{4}\mathbf{E}_{f_L}(0)$$

and, similarly, $\epsilon_p = \frac{1}{4}\mathbf{E}_{p_L}(0)$.

These relations can also be obtained from the Fourier cosine transforms of Eq. (5-203) by putting $n = 0$.

Hence,

$$\frac{\epsilon_p}{\epsilon_f} = \frac{\mathbf{E}_{p_L}(0)}{\mathbf{E}_{f_L}(0)} = 1 \tag{5-207}$$

which follows from the relation

$$\frac{\mathbf{E}_{p_L}(n)}{\mathbf{E}_{f_L}(n)} = \frac{\gamma^2 + \delta^2}{\alpha^2 + \beta^2} = [1 + f_1(\omega)]^2 + f_2^2(\omega) \tag{5-208}$$

since $f_1(0) = f_2(0) = 0$.

This is the result obtained by Tchen.

In order to get an idea of the difference between ϵ_p and ϵ_f when the diffusion time is not long, we assume that the Lagrangian correlation coefficient for the fluid motion may be represented by an exponential function

$$\mathbf{R}_{f_L}(t) = \exp(-t/\mathfrak{J}_{f_L})$$

The corresponding Lagrangian energy spectrum then reads

$$\mathbf{E}_{f_L}(n) = 4\overline{v_f{}^2}\frac{\mathfrak{I}_{f_L}}{1 + \omega^2\mathfrak{I}_{f_L}{}^2}$$

We further assume that the Basset term in Eq. (5-199) is negligibly small, so that we may disregard the c in the expression for $f_1(\omega)$ and $f_2(\omega)$. We then have from Eq. (5-208)

$$\mathbf{E}_{p_L}(n) = \frac{a^2 + b^2\omega^2}{a^2 + \omega^2}\mathbf{E}_{f_L}(n) = 4\overline{v_f{}^2}\frac{a^2 + b^2\omega^2}{a^2 + \omega^2}\frac{\mathfrak{I}_{f_L}}{1 + \omega^2\mathfrak{I}_{f_L}{}^2}$$

and

$$\overline{v_p{}^2} = \int_0^\infty dn\,\mathbf{E}_{p_L}(n) = \frac{2}{\pi}\overline{v_f{}^2}\mathfrak{I}_{f_L}\int_0^\infty d\omega\frac{a^2 + b^2\omega^2}{(a^2 + \omega^2)(1 + \omega^2\mathfrak{I}_{f_L}{}^2)} = \frac{a\mathfrak{I}_{f_L} + b^2}{a\mathfrak{I}_{f_L} + 1}\overline{v_f{}^2} \tag{5-209}$$

According to this relation between $\overline{v_p{}^2}$ and $\overline{v_f{}^2}$, we would obtain $\overline{v_p{}^2} = \overline{v_f{}^2}$ only if $b = 1$, that is, if $\rho_p = \rho_f$—an obvious result.

$$\mathbf{R}_{p_L}(t) = \frac{1}{\overline{v_p{}^2}}\int_0^\infty dn\,\mathbf{E}_{p_L}(n)\cos\omega t = \frac{2}{\pi}\frac{\overline{v_f{}^2}}{\overline{v_p{}^2}}\mathfrak{I}_{f_L}\int_0^\infty d\omega\frac{(a^2 + b^2\omega^2)\cos\omega t}{(a^2 + \omega^2)(1 + \omega^2\mathfrak{I}_{f_L}{}^2)}$$

$$= \frac{\overline{v_f{}^2}}{\overline{v_p{}^2}}\frac{1}{a^2\mathfrak{I}_{f_L}{}^2 - 1}[(a^2\mathfrak{I}_{f_L}{}^2 - b^2)\exp(-t/\mathfrak{I}_{f_L}) - a\mathfrak{I}_{f_L}(1 - b^2)\exp(-at)]$$

$$= \frac{1}{a\mathfrak{I}_{f_L} - 1}\left[\frac{(a^2\mathfrak{I}_{f_L}{}^2 - b^2)\exp(-t/\mathfrak{I}_{f_L})}{a\mathfrak{I}_{f_L} + b^2} - \frac{a\mathfrak{I}_{f_L}(1 - b^2)}{a\mathfrak{I}_{f_L} + b^2}\exp(-at)\right] \tag{5-210}$$

So we conclude that the Lagrangian correlation coefficient for the discrete-particle motion is no longer a single exponential function.

From the Lagrangian correlation coefficient we obtain the diffusion coefficient according to Eq. (5-43):

$$\epsilon_p = \overline{v_p{}^2}\int_0^1 d\tau\,\mathbf{R}_{p_L}(\tau)$$

$$= \overline{v_f{}^2}\mathfrak{I}_{f_L}\left[1 - \frac{(a^2\mathfrak{I}_{f_L}{}^2 - b^2)\exp(-t/\mathfrak{I}_{f_L}) - (1 - b^2)\exp(-at)}{a^2\mathfrak{I}_{f_L}{}^2 - 1}\right]$$

With $\epsilon_f = \overline{v_f{}^2}\mathfrak{I}_{f_L}[1 - \exp(-t/\mathfrak{I}_{f_L})]$ after some calculation we then obtain

$$\frac{\epsilon_p}{\epsilon_f} = 1 + \frac{1 - b^2}{a^2\mathfrak{I}_{f_L}{}^2 - 1}\frac{\exp(-at) - \exp(-t/\mathfrak{I}_{f_L})}{1 - \exp(-t/\mathfrak{I}_{f_L})} \tag{5-211}$$

For long diffusion times, the diffusion coefficient may be written like $\epsilon_f = v'_f\Lambda_{f_L}$. Since $v'_p \neq v'_f$, apparently $\Lambda_{p_L} \neq \Lambda_{f_L}$ in such a manner that $v'_p\Lambda_{p_L}$ just equals $v'_f\Lambda_{f_L}$.

The result (5-207) $\epsilon_p = \epsilon_f$ is in sharp contrast with what follows from the relation (5-206) for short diffusion times, namely that $\epsilon_p \neq \epsilon_f$ since $\overline{v_p{}^2} \neq \overline{v_f{}^2}$.

According to the equation $\epsilon_f = \frac{1}{4}\mathbf{E}_{fL}(0)$, the diffusion coefficient is proportional to the fractional kinetic energy contributed by the turbulent motion with zero frequency. Now, for zero frequency, there is no difference between the motion of the particle and the motion of the fluid. Hence, from a physical point of view, it is quite logical that, given this equation, the diffusion coefficients for the discrete particle and for the fluid particles should be the same. Hitherto we have assumed a linear resistance law for the discrete particle. But at zero frequency the nature of the resistance law plays no role, since there is no lag in motion between discrete particle and fluid. The conclusion is obvious that the result (5-207) should hold regardless of the nature of the resistance law.

Of course, this conclusion does not hold when $n \neq 0$, since the relation (5-208) is obtained specifically for a linear resistance law and would be different for a nonlinear resistance law. Furthermore, the result (5-207) only holds for an infinitely long diffusion time. Consequently, if one wants to apply this result to the turbulent flow in an apparatus of finite dimensions, one has to make sure that the turbulence domain is not only sufficiently homogeneous, but also sufficiently large compared with, for instance, the Lagrangian spatial integral scale. In most actual flow cases, this will be hardly the case, so there might be not only a difference between ϵ_p and ϵ_f, albeit slight, but also a resistance-law effect on ϵ_p and, hence, on ϵ_p/ϵ_f.

The neglect of the Basset term calls for a justification. According to the expressions for $f_1(\omega)$ and $f_2(\omega)$, such a neglect would only be justified when the two following conditions are satisfied simultaneously

$$c\sqrt{\frac{\pi\omega}{2}} \ll \omega \quad \text{and} \quad c\sqrt{\frac{\pi\omega}{2}} \ll a$$

Or, since

$$c = \frac{6b}{d}\sqrt{\frac{\nu}{\pi}}$$

when

$$\frac{b}{d}\sqrt{\frac{\nu}{\omega}} \ll 1 \quad \text{and} \quad \frac{1}{d}\sqrt{\frac{\nu}{\omega}} = \mathcal{O}(1) \text{ or larger}$$

For a finite value of b, these conditions are conflicting. In general this would be so in the case of gas bubbles in a liquid, while the two conditions can be satisfied for solid particles in a gas, where $b = \mathcal{O}(0.001)$.

According to calculations made by Hjelmfelt and Mockros[97] the effect of the neglect of the Basset term becomes of importance to the value of the amplitude of the oscillating motion of a particle when

$$\frac{1}{d}\sqrt{\frac{\nu}{\omega}} \lesssim 0.6$$

While the effect on the phase shift becomes already appreciable at much larger values of $1/d\sqrt{\nu/\omega}$. Hjelmfelt and Mockros considered also the effect of a neglect of the virtual "added" mass, so $b = \rho_f/\rho_p$ in addition to the neglect of the Basset term. Differences in the amplitudes because of the neglect of the Basset term alone, were only obtained when, again,

$$\frac{1}{d}\sqrt{\frac{\nu}{\omega}} \lesssim 0.6$$

When considering the theoretical results given by Eqs. (5-209), (5-210), and (5-211), we still have to keep in mind that in order to satisfy the condition (5), the value of b may not differ too much from unity. Which means that both $\overline{v_p^2}/\overline{v_f^2}$ and ϵ_p/ϵ_f are close to one.

In order to get some idea of the difference between discrete particle behavior and that of the fluid also when b is either small or much greater than one, we may apply the above equations to a hypothetical case where a very large volume of the fluid surrounding the discrete particle moves "bodily" in a random way. For the case of a solid particle in air under atmospheric conditions $\rho_f/\rho_p \simeq 1/2{,}000$; so we may put $b = 0$, and also $c = 0$. We then obtain for the ratio of the energy spectra

$$\frac{E_{p_L}(n)}{E_{f_L}(n)} = \frac{a^2}{a^2 + 4\pi^2 n^2}$$

When $2\pi n/a = 2$, the ratio has decreased to 0.2. According to Eq. (5-209), $\overline{v_p^2}/\overline{v_f^2}$ becomes smaller than one, depending on the value of $a\mathfrak{I}_L$. Also the ratio ϵ_p/ϵ_f will approach unity in a shorter time when $a\mathfrak{I}_L$ increases. Another interesting extreme case is when the density of the discrete particle is very small compared with that of the fluid, as in the case of a tiny gas bubble in a liquid. Then $\rho_p/\rho_f \to 0$ and $b \to 3$. Usually the Basset term may not be neglected. For the limiting case that the viscosity of the liquid tends to zero, however, both a and c become zero. Then from Eq. (5-199) we conclude that $v_p = bv_f = 3v_f$, so $\overline{v_p^2} = 9\overline{v_f^2}$. While it follows from Eq. (5-203) that $\mathbf{R}_{f_{L(t)}} = \mathbf{R}_{p_L}(t)$ and so $\mathfrak{I}_{p_L} = \mathfrak{I}_{f_L}$.

We will conclude this part by considering the effect on the *induction period* of a density difference between discrete particle and fluid.[70] This effect will be maximal when $\rho_p/\rho_f \to \infty$, or when $b \simeq 0$ and $c \simeq 0$. Equation (5-199) reduces to

$$\frac{dv_p}{dt} + av_p = av_f$$

Let $v_{p_\infty}(t)$ be the solution for the stationary state. The solution that satisfies the initial condition $v_p = 0$ at $t = 0$, reads

$$v_p(t) = v_{p_\infty}(t) - v_{p_\infty}(0)\exp(-at)$$

Considering an ensemble-average, one easily obtains

$$\overline{v_p^2}(t) = \overline{v_{p_\infty}^2}[1 - 2\exp(-at)\mathbf{R}_{p_L}(t) + \exp(-2at)] \qquad (5\text{-}212)$$

The induction period is determined by $\exp(-at)$. For a solid particle of 100 μm diameter in air, $a \simeq 10\ \text{s}^{-1}$. If we assume $\exp(-at) = 0.05$, we obtain an induction period of 0.3 s. Though this is a rather short time, in a high-speed flow the particle may have covered a considerable distance during this time.

As has been pointed out more than once in the foregoing, Tchen's theory may be considered as a limiting case due to the drastic simplifications made, the theory being consequently of very limited practical value. Or, stated in a different way, in most actual cases of the diffusion of discrete particles in a turbulent flow the simplifying assumptions made by Tchen, in particular the assumptions (3), (4), and (5), do not apply. When we drop these assumptions the problem becomes very complicated indeed, certainly not allowing a simple solution. In the first place, when dropping assumption (5), Eq. (5-198) actually is a nonlinear, stochastic integro-differential equation, the Lagrangian fluid velocity v_f occurring on the right-hand side of the equation being the velocity of the "surrounding" fluid encountered by the discrete particle on its way through the fluid, and which "surrounding" fluid does not remain the same at least when the diffusion time exceeds the deformation time. It originates from elsewhere subject to its own Lagrangian correlation. When the discrete particle leaves its originally "surrounding" fluid when the time exceeds the deformation time, or earlier in the case of a large overshooting effect, its Lagrangian auto-correlation will drop at a faster rate than it would do if it could have followed its originally "surrounding" fluid. In first approximation we may consider the auto-correlation of the discrete particle as the result of the combined effect of the Lagrangian auto-correlation of the originally "surrounding" fluid and of the spatial Eulerian correlation corresponding to the distance at the instant considered between the centroids of the discrete particle and of the originally "surrounding" fluid, provided the distance is still small enough for a high degree of spatial correlation. Peskin[111] has studied this problem. He assumed a joint Gaussian distribution of the fluid velocities, in which case the expected velocity of the fluid encountered by the discrete particle may be put equal to the product of the Lagrangian velocity and the spatial Eulerian correlation. For small distances between the discrete particle and the "originally surrounding fluid," and thus a high degree of the latter correlation Peskin indeed obtained $\epsilon_p/\epsilon_f < 1$, though the deviation from unity was still small.

Experiments by Snyder and Lumley[112] in a well-defined, vertical, grid-generated turbulent flow and carried out with discrete particles of different size and different density ratio ρ_p/ρ_f, but all greater than one, resulted in $v'_p/v'_f < 1$, $\mathfrak{I}_{p_L} < \mathfrak{I}_{f_L}$ and $\epsilon_p/\epsilon_f < 1$, with values as low as ~ 0.1. Also from the experiments by Abramovich[118] in free jets with very small discrete particles but $\rho_p/\rho_f \gg 1$, one obtains a value of $\epsilon_p/\epsilon_f < 1$. On the other hand Ahmadi and Goldschmidt[98] and Goldschmidt *et al.*[99] obtained from experiments with small oil droplets in a plane air jet and with small gas bubbles in a round water jet values of $\epsilon_p/\epsilon_f > 1$. However, it may be suspected that in the highly nonhomogeneous turbulent flow field like free jets, at least in the transverse direction, "overshoot" effects may become important. This is so when the response time of the discrete particle is no longer small compared with the time-scale for the rate of change of the mean velocity of the fluid experienced by the discrete particle when crossing the region concerned.[103]

Yet values of $\epsilon_p/\epsilon_f > 1$ may be possible. For instance when the size of the discrete particles becomes greater than the small eddies of the turbulence. On the basis of a simple model Hinze[103] showed that for a neutrally buoyant discrete particle this "filtering" effect may yield a value of ϵ_p/ϵ_f slightly greater than one, with a maximum for a discrete particle with a size about half the size of the energy-containing eddies of the turbulence.

We mentioned already possible overshoot effects due to the inhomogeneity of the mean flow field. In many investigations the response of a discrete particle not only to spatial changes of the mean flow, but also to the turbulence has been taken into account insufficiently or not at all. For, of great importance for the behavior of a discrete particle is the ratio between its response time to a change of environmental conditions, and the many characteristic times of the turbulence as given by the spectral distribution of its energy [Hinze;[103] Snyder and Lumley[112]].

The "overshooting" and "filtering" effects just considered are not the only effects that influence the eddy diffusion of a discrete particle. When one considers the turbulent flow of a suspension, one has to reckon with direct and indirect interference effects, unless the concentration by volume of the discrete particles is very small indeed. The indirect interference effects include those due to a decreased fractional volume of the continuous, fluid phase, which may cause an increase in the turbulence intensity of the fluid[100] with increasing concentration, at least as long as the suspension is still rather dilute. For particles not neutrally buoyant, relative velocity and drag effects also may change the turbulence, in particular the highest wavenumber range of the turbulence. Apart from the fact that they are far from being solved in a satisfactory way, a further consideration of the problems associated with the above effects, would go far beyond the scope of the present book. For more general information concerning turbulent flows of suspensions references 102, 103 may be consulted.

5-8 EFFECT OF COMPRESSIBILITY

The relations between the diffusion of fluid particles and the Lagrangian velocity correlation, given in the preceding sections, are general and unaffected by compressibility effects, notwithstanding the diffusion itself and the correlations are so influenced, since the fluid-particle movement is subject to these effects. Consider, for instance, the fluid-particle displacement in the x_2-direction since the time $t = 0$.

$$y_2(t) = \int_0^t dt'\, v_2(t') = v_2(0)t + \frac{1}{2!}\left[\frac{dv_2}{dt}\right]_{t=0} t^2 + \frac{1}{3!}\left[\frac{d^2v_2}{dt^2}\right]_{t=0} t^3 + \cdots$$

where, according to the Navier-Stokes equation of motion [see Eq. (1-10) or (1-12)], the quantities $dv_2/dt, d^2v_2/dt^2, \ldots$ each contain a compressibility term.

Thus, the rates of diffusion of marked fluid particles and of transferable quantities depend on the compressibility of the fluid. If we introduce a coefficient of diffusion to describe the diffusion process, we have to reckon with a value different from the value for an incompressible fluid.

The spread of a transferable quantity is then still determined by the gradient of the mean value and occurs in a direction corresponding to this gradient.

However, there is a case where the spread of heat may be in a direction opposite to that determined by the mean-temperature gradient, that is, up and not down the local gradient; namely, when the mean-pressure distribution is not uniform and when the mean-temperature distribution deviates from the adiabatic. This adiabatic mean-temperature distribution is determined by the relation

$$\frac{\Theta}{\Theta_0} = \left(\frac{P}{P_0}\right)^{(\varkappa-1)/\varkappa}$$

whence, for instance,

$$\frac{d\Theta}{dt} = \frac{\varkappa - 1}{\varkappa}\frac{\Theta}{P}\frac{dP}{dt}$$

where $\varkappa = c_p/c_v$.

It has long been known that, in a stable stratified atmosphere, there may be transport of heat directed up the temperature gradient, from the lower to the higher temperature.[56] If the temperature distribution with the height x_2 is polytropic with a coefficient n, the stratification is stable if

$$n < \varkappa$$

so that

$$\frac{d}{dx_2}(\Theta_{\text{ad}} - \Theta_{\text{pol}}) > 0$$

In the range $1 < n < \varkappa$, transport "against" the temperature gradient occurs if there is a process of transport by turbulence in the atmosphere.

The turbulence causes a mixing of the upper with the lower layers of the atmosphere, tending to make the temperature distribution approach the adiabatic ($n \to \varkappa$). Air particles moved by turbulence to and fro from one layer to another are compressed or expanded adiabatically. The temperature rise or fall due to adiabatic compression or expansion causes an air particle to have a higher temperature than its surroundings when moving down and a lower temperature when moving up. As a consequence of this temperature difference there is heat exchange with the surroundings, and the net effect is transport of heat from the upper to the lower layers of the atmosphere. As Schultz-Grunow[57] has pointed out, if we assume that the same air particle is moved up and down by the turbulent motion, it will be subjected to a cyclical thermodynamic process. In a pressure-volume diagram of this process it can be seen that the air particle picks up heat at low pressure and delivers heat at high pressure. This is just what occurs in the process of refrigeration.

The same process of heat transport may occur in the wake flow of bodies in compressible flow at high Mach numbers,[58] where the stagnation temperatures (for instance, at the wall of the body) have been observed to be much lower than expected. Usual values for the recovery factor, that is, the ratio of the actual stagnation temperature to the adiabatic stagnation temperature, in a laminar boundary layer are not far from the value $\mathbf{Pr}^{0.5}$, that is, ~ 0.85 in the case of air. Here, however, much lower values, even negative values, have been measured.

An interesting apparatus, which has already found technical application, is based upon the phenomenon just described. It is the Ranque-Hilsch vortex tube for the "separation" of a gas into a hot and a cold fraction.[59-61] In essence the apparatus consists of a cylindrical tube with one end containing a central orifice; the other end can be closed more or less by means of a throttle valve. The gas is introduced at sonic or supersonic velocity through a tangential nozzle at the periphery of the tube, at a location preferably close to the end containing the central orifice. The gas escaping through this central orifice has a much lower stagnation temperature than the incoming gas, and the gas escaping through the throttle valve at the other end has a much higher stagnation temperature. The fraction ratio and the corresponding temperature ratio of the gases escaping at the two ends of the tube depend on the degree of throttling at the "hot" end.

Within the cylindrical tube we have a stratified turbulent gas flow, the static pressure P decreasing in the radial direction toward the tube axis. Apparently the radial temperature distribution is not adiabatic, and the turbulence exchange among the gas layers results in a flow of heat from the inner regions to the tube wall, that is, directed up the temperature gradient.

This effect of turbulence exchange in a compressible gas with non-uniform mean-temperature and mean-pressure distributions can be shown by means of the transport theory expounded in Sec. 5-4.

Though the mean-pressure distribution and possibly the mean-velocity distribution are not uniform, we still assume that the turbulence flow field is sufficiently homogeneous so that we may make use of the features of a homogeneous turbulence.

We consider first the equilibrium temperature distribution in the steady state, obtained from the balance between the total flux of heat through a unit volume of the space and the heat production by compression and viscous dissipation.

Writing $\mathscr{E}_{\text{int}} = \rho c_v \Theta$, Eq. (1-107) reads for the stationary case

$$\rho c_v U_i \frac{\partial \Theta}{\partial x_i} - \frac{\partial}{\partial x_i}\left(\kappa \frac{\partial \Theta}{\partial x_i}\right) = -P \frac{\partial U_j}{\partial x_j} + \mu\left[\left(\frac{\partial U_i}{\partial x_j} + \frac{\partial U_j}{\partial x_i}\right)\frac{\partial U_i}{\partial x_j} - \frac{2}{3}\left(\frac{\partial U_j}{\partial x_j}\right)^2\right]$$

Or, with $P = \rho(c_p - c_v)\Theta$ for a perfect gas, and with Eq. (1-2)

$$-P\frac{\partial U_j}{\partial x_j} = U_j \frac{\partial P}{\partial x_j} - \frac{\partial}{\partial x_j} PU_j = U_j \frac{\partial P}{\partial x_j} + (c_v - c_p)\rho U_j \frac{\partial \Theta}{\partial x_j}$$

The equation giving the heat balance becomes

$$\rho c_p U_i \frac{\partial \Theta}{\partial x_i} - \frac{\partial}{\partial x_i}\left(\kappa \frac{\partial \Theta}{\partial x_i}\right) = U_i \frac{\partial P}{\partial x_i} + \mu \frac{\partial U_j}{\partial x_i}\left(\frac{\partial U_j}{\partial x_i} + \frac{\partial U_i}{\partial x_j}\right) - \tfrac{2}{3}\mu\left(\frac{\partial U_j}{\partial x_j}\right)^2$$

Now put again

$$\rho = \bar{\rho} + \tilde{\rho} \quad \mu = \bar{\mu} + \tilde{\mu} \quad U_i = \bar{U}_i + u_i \quad \Theta = \bar{\Theta} + \theta \quad P = \bar{P} + p$$

If we neglect the effect of molecular conduction and the viscous dissipation by the mean motion, the time-mean condition of the heat balance reads

$$\bar{\rho} c_p \bar{U}_i \frac{\partial \bar{\Theta}}{\partial x_i} = -\overline{\tilde{\rho} u_i} c_p \frac{\partial \bar{\Theta}}{\partial x_i} - \frac{\partial}{\partial x_i}(\bar{\rho} c_p \overline{\theta u_i} + \overline{\tilde{\rho}\theta} c_p \bar{U}_i + c_p \overline{\tilde{\rho}\theta u_i})$$

$$+ \left[\bar{U}_i \frac{\partial \bar{P}}{\partial x_i} + \overline{u_i \frac{\partial p}{\partial x_i}} + \overline{\tilde{\mu}\frac{\partial u_j}{\partial x_i}}\left(\frac{\partial \bar{U}_j}{\partial x_i} + \frac{\partial \bar{U}_i}{\partial x_j}\right) + \overline{\tilde{\mu}\left(\frac{\partial u_j}{\partial x_i} + \frac{\partial u_i}{\partial x_j}\right)}\frac{\partial \bar{U}_j}{\partial x_i} \right.$$

$$\left. + \overline{\mu \frac{\partial u_j}{\partial x_i}\left(\frac{\partial u_j}{\partial x_i} + \frac{\partial u_i}{\partial x_j}\right)} - \tfrac{2}{3}\overline{\tilde{\mu}\frac{\partial u_j}{\partial x_j}}\frac{\partial \bar{U}_j}{\partial x_j} - \tfrac{2}{3}\overline{\mu\left(\frac{\partial u_j}{\partial x_j}\right)^2} \right]$$

This equation contains many turbulence terms, whose orders of magnitude are probably not the same. In general we may assume that the magnitude of $\tilde{\rho}/\bar{\rho}$, $\tilde{\mu}/\bar{\mu}$, and $p/\bar{P}$ is one order smaller than that of $u_i/\bar{U}_i$. If this assumption is correct, we may simplify the above equation by neglecting the terms containing fluctuations in density, viscosity and pressure.

$$\bar{\rho} c_p \bar{U}_i \frac{\partial \bar{\Theta}}{\partial x_i} = -\frac{\partial}{\partial x_i}(\bar{\rho} c_p \overline{\theta u_i}) + \left[\bar{U}_i \frac{\partial \bar{P}}{\partial x_i} + \bar{\mu}\overline{\frac{\partial u_j}{\partial x_i}\left(\frac{\partial u_j}{\partial x_i} + \frac{\partial u_i}{\partial x_j}\right)} - \tfrac{2}{3}\bar{\mu}\overline{\left(\frac{\partial u_j}{\partial x_j}\right)^2} \right] \tag{5-213}$$

The turbulence term $\overline{\theta u_i} = \overline{\Theta u_i}$ then determines the transport of heat by the turbulence velocity u_i through a plane perpendicular to the x_i-direction:

$$\overline{\Theta u_i} = \frac{1}{T}\int_0^T dt_0\, \Theta(t_0) v_i(t_0)$$

We now assume:

1 That the temperature $\Theta(t_0)$ of the fluid particle that passes the plane x_i at time t_0 is determined by the mean-temperature and mean-pressure distributions in the region through which it travels, and that the fluid particle is subjected at the same time to an adiabatic compression or expansion and to an exchange with its surroundings at a rate directly proportional to the temperature difference.
2 That on the one hand the Lagrangian-correlation distance is relatively small, so that over this distance the variations in mean temperature and in mean pressure may be approximated by linear laws.
3 That on the other hand the fluid particle has been under way for such a long time that we may put this time infinitely long.

The differential equation for the temperature of the fluid particle, resulting from a heat balance, reads

$$\frac{d\Theta}{dt} = \frac{\varkappa - 1}{\varkappa}\frac{\Theta}{\bar{P}}\frac{d\bar{P}}{dt} + \alpha(\bar{\Theta} - \Theta) + \frac{\varepsilon}{c_p} \qquad (5\text{-}214)$$

The first term on the right-hand side represents the contribution by the adiabatic compression of the fluid particle when it is subjected to a pressure change $d\bar{P}/dt$. The third term on the right-hand side represents the contribution by viscous dissipation. In the following this term will be neglected. However, if one wants to maintain this term the only difference in the final result will be that the mean temperature $\bar{\Theta}$ in the expression for $\overline{\Theta u_i}$ has to be replaced by $\bar{\Theta} + \varepsilon/\alpha c_p$.

Let t be the time of traveling. Then, with the initial condition $\Theta = \Theta_0$ at $t = 0$, the solution of the differential equation reads

$$\Theta = \alpha \exp(-\alpha t)\int_0^t dt'\,\bar{\Theta}(t')\left[\frac{\bar{P}(t)}{\bar{P}(t')}\right]^{(\varkappa-1)/\varkappa}\exp(\alpha t') + \Theta_0\left[\frac{\bar{P}(t)}{\bar{P}(0)}\right]^{(\varkappa-1)/\varkappa}\exp(-\alpha t)$$

The expression for the temperature of a fluid particle that passes the plane x_i at time t_0, in view of assumption 3, becomes

$$\Theta(t_0) = \alpha\int_{-\infty}^{t_0} dt'\,\bar{\Theta}(t')\left[\frac{\bar{P}(t_0)}{\bar{P}(t')}\right]^{(\varkappa-1)/\varkappa}\exp\left[-\alpha(t_0 - t')\right]$$

Now we have approximately

$$\bar{\Theta}(t') = \bar{\Theta}(t_0) - \frac{d\bar{\Theta}}{dt'}(t_0 - t') = \bar{\Theta}(x_j) - \frac{\partial\bar{\Theta}}{\partial x_j}\int_{t'}^{t_0} dt''\,v_j(t'')$$

and

$$\bar{P}(t') = \bar{P}(x_j) - \frac{\partial\bar{P}}{\partial x_j}\int_{t'}^{t_0} dt''\,v_j(t'')$$

So, to the same degree of approximation,

$$\left[\frac{\bar{P}(t_0)}{\bar{P}(t')}\right]^{(\varkappa-1)/\varkappa} = 1 + \frac{\varkappa - 1}{\varkappa}\frac{1}{\bar{P}}\frac{\partial\bar{P}}{\partial x_j}\int_{t'}^{t_0} dt''\,v_j(t'')$$

Consequently,

$$\bar{\Theta}(t')\left[\frac{\bar{P}(t_0)}{\bar{P}(t')}\right]^{(\varkappa-1)/\varkappa} = \bar{\Theta}(x_j) - \left(\frac{\partial\bar{\Theta}}{\partial x_j} - \frac{\varkappa - 1}{\varkappa}\frac{\bar{\Theta}}{\bar{P}}\frac{\partial\bar{P}}{\partial x_j}\right)\int_{t'}^{t_0} dt''\,v_j(t'')$$

hence,

$$\Theta(t_0) = \bar{\Theta}(x_j) - \alpha\left(\frac{\partial\bar{\Theta}}{\partial x_j} - \frac{\varkappa - 1}{\varkappa}\frac{\bar{\Theta}}{\bar{P}}\frac{\partial\bar{P}}{\partial x_j}\right)\int_{-\infty}^{t_0} dt'\exp\left[-\alpha(t_0 - t')\right]\int_{t'}^{t_0} dt''\,v_j(t'')$$

The expression for the correlation $\overline{\Theta u_i}$ then becomes

$$\overline{\Theta u_i} = -\left(\frac{\partial\bar{\Theta}}{\partial x_j} - \frac{\varkappa - 1}{\varkappa}\frac{\bar{\Theta}}{\bar{P}}\frac{\partial\bar{P}}{\partial x_j}\right)\frac{\alpha}{T}\int_0^T dt_0\int_{-\infty}^{t_0} dt'\exp\left[-\alpha(t_0 - t')\right]\int_{t'}^{t_0} dt''\,v_i(t_0)v_j(t'')$$

By carrying out the procedure followed in Sec. 5-4, this expression is transformed into

$$\overline{\Theta u_i} = -\left(\frac{\partial \bar{\Theta}}{\partial x_j} - \frac{\varkappa - 1}{\varkappa}\frac{\bar{\Theta}}{\bar{P}}\frac{\partial \bar{P}}{\partial x_j}\right)\int_0^\infty d\tau\, \overline{v_i(t_0)v_j(t_0+\tau)}\exp(-\alpha\tau)$$

For a polytropic "atmosphere" we have

$$\bar{P} \propto (\bar{\Theta})^{n/n-1}$$

Hence,

$$\overline{\Theta u_i} = -\left[1 - \frac{\varkappa - 1}{\varkappa}\frac{n}{n-1}\right]\frac{\partial \bar{\Theta}}{\partial x_j}\int_0^\infty d\tau\, \overline{v_i(t_0)v_j(t_0+\tau)}\exp(-\alpha\tau)$$

This shows that the transport of heat is down the temperature gradient when $n > \varkappa$, and up the temperature gradient when $n < \varkappa$, while there is no transport of heat for an adiabatic mean-temperature distribution.

With the introduction of a symmetrical eddy-diffusion tensor $(\epsilon_\theta)_{ij}$, defined by

$$(\epsilon_\theta)_{ij} = \frac{1}{2}\int_0^\infty d\tau\,\{[\mathbf{Q}_{i,j}(\tau)]_L + [\mathbf{Q}_{j,i}(\tau)]_L\}\exp(-\alpha\tau) \qquad (5\text{-}215)$$

the equation for the heat balance reads:

$$\bar{\rho}c_p\bar{U}_i\frac{\partial \bar{\Theta}}{\partial x_i} - \frac{\partial}{\partial x_i}\bar{\rho}c_p(\epsilon_\theta)_{ij}\left[\frac{\partial \bar{\Theta}}{\partial x_j} - \frac{\varkappa - 1}{\varkappa}\frac{\bar{\Theta}}{\bar{P}}\frac{\partial \bar{P}}{\partial x_j}\right]$$
$$= \left[\bar{U}_i\frac{\partial \bar{P}}{\partial x_i} + \bar{\mu}\overline{\frac{\partial u_j}{\partial x_i}\left(\frac{\partial u_j}{\partial x_i} + \frac{\partial u_i}{\partial x_j}\right)} - \tfrac{2}{3}\bar{\mu}\overline{\left(\frac{\partial u_j}{\partial x_j}\right)^2}\right] \qquad (5\text{-}216)$$

Equation (5-216) in its general form is entirely intractable for many applications. But in specific cases considerable simplification is possible. If this equation is applied to the heat-transport processes in a vortex tube, it is usual to consider only the rotational-velocity component U_φ, since this velocity is much greater than the radial and axial components. Furthermore, changes in radial direction are much greater than those in the other directions. As a consequence we may put $\bar{U}_i\partial/\partial x_i \simeq 0$.

Another simplification is the assumption that, in stratified flow with transport mainly in the radial direction, it is possible to introduce a momentum-transport coefficient ϵ_m and to assume a constant value for it. The dissipation can then be calculated from

$$\epsilon_m\left(\frac{\partial \bar{U}_\varphi}{\partial r} - \frac{\bar{U}_\varphi}{r}\right)^2$$

Hence, the heat balance equation becomes

$$\frac{1}{r}\frac{\partial}{\partial r}\bar{\rho}c_p\epsilon_\theta\left(r\frac{\partial \bar{\Theta}}{\partial r} - \frac{\varkappa - 1}{\varkappa}\frac{\bar{\Theta}}{\bar{P}}r\frac{\partial \bar{P}}{\partial r}\right) + \epsilon_m\bar{\rho}\left(\frac{\partial \bar{U}_\varphi}{\partial r} - \frac{\bar{U}_\varphi}{r}\right)^2 = 0$$

or, since

$$\frac{\partial P}{\partial r} = \bar{\rho}\frac{\bar{U}_\varphi^2}{r}$$

and

$$\bar{\rho}\frac{\bar{\Theta}}{\bar{P}} = \frac{1}{c_p - c_v} = \frac{\varkappa}{c_p(\varkappa - 1)}$$

$$\frac{1}{r}\frac{\partial}{\partial r}\bar{\rho}c_p\epsilon_\theta r\left(\frac{\partial\bar{\Theta}}{\partial r} - \frac{1}{c_p}\frac{\bar{U}_\varphi^2}{r}\right) + \epsilon_m\bar{\rho}\left(\frac{\partial\bar{U}_\varphi}{\partial r} - \frac{\bar{U}_\varphi}{r}\right)^2 = 0$$

The radial temperature distribution can be calculated if the velocity distribution $\bar{U}_\varphi(r)$ is known.

REFERENCES

1. Prandtl, L.: *Z. angew. Math. u. Mech.*, **5,** 136 (1925). Also, Goldstein, S.: "Modern Developments in Fluid Dynamics," vol. 1, p. 205, Oxford University Press, New York, 1938.
2. Prandtl, L.: *Z. angew. Math. u. Mech.*, **22,** 241 (1942).
3. Taylor, G. I.: *Phil. Trans. Roy. Soc. London,* **215A,** 1 (1915).
4. Taylor, G. I.: *Proc. Roy. Soc. London,* **135A,** 685 (1932).
5. Kármán, Th. von: *Nachr. Akad. Wiss. Göttingen Math.-phys. Kl.*, 58 (1930).
6. Kármán, Th. von: *J. Aeronaut. Sci.*, **4,** 131 (1937).
7. Reichardt, H.: *Z. angew. Math. u. Mech.*, **21,** 257 (1941).
8. Reichardt, H.: *Forsch. Gebiete Ingenieurw.*, no. 414, 1951. Also, Oudart, A.: *Publ. sci. et tech. ministè air No.* 234, 1949.
9. Saffman, P. G.: *J. Fluid Mech.*, **8,** 273 (1960).
10. Schultz, B. H.: *Appl. Sci. Research,* **1A,** 387, 400 (1949).
11. Alexander, L. G., T. Baron, and E. W. Comings: *Univ. Illinois Bull. No.* 413, 1953.
12. Nusselt, W.: *Z. angew. Math. u. Mech.*, **10,** 105 (1930).
13. Eckert, E., and V. Lieblein: *Forsch. Gebiete Ingenieurw.*, **16B,** 33 (1949).
14. Prandtl, L.: *Physik. A.*, **29,** 487 (1928).
15. Burgers, J. M.: lecture notes, California Institute of Technology, Pasadena, 1951.
16. Batchelor, G. K.: *Australian J. Sci. Research,* **2,** 437 (1949).
17. Batchelor, G. K.: *Proc. Cambridge Phil. Soc.*, **48,** 345 (1952).
18. Batchelor, G. K.: *Inst. Fluid Dynamics and Appl. Math., Lecture Ser. No.* 4, University of Maryland, College Park, 1951.
19. Taylor, G. I.: *Proc. London Math. Soc.*, **20,** 196 (1921).
20. Richardson, L. F.: *Proc. Roy. Soc. London,* **110A,** 709 (1926).
21. Batchelor, G. K.: *Proc. Roy. Soc. London,* **213A,** 349 (1952).
22. Reid, W. H.: *Proc. Cambridge Phil. Soc.*, **51,** 350 (1955).
23. Townsend, A. A.: *Proc. Roy. Soc. London,* **209A,** 418 (1951).
24. Ertel, H.: *Z. Meteorol.*, **59,** 277 (1942).
25. Batchelor, G. K., and A. A. Townsend, in G. K. Batchelor and R. M. Daires (eds.): "Surveys in Mechanics," p. 352, Cambridge University Press, New York, 1956.
26. Chandrasekhar, S.: *Proc. Roy. Soc. London,* **229A,** 1 (1955).

27. Frenkiel, F. N.: *Proc. 7th Intern. Congr. Appl. Mech.,* London, vol. 2, p. 112, 1948.
28. Uberoi, M. S., and S. Corrsin: *Natl. Advisory Comm. Aeronaut. Tech. Repts. No.* 1142, 1953.
29. Batchelor, G. K., and A. A. Townsend: *Proc. Roy. Soc. London,* **190A,** 534 (1947).
30. Mickelsen, W. R.: *Natl. Advisory Comm. Aeronaut. Tech. Notes No.* 3570, 1955.
31. Carslaw, H. S., and J. C. Jaeger: "Conduction of Heat in Solids," p. 223, Clarendon Press, Oxford, 1947.
32. Frenkiel, F. N.: Symposium on Turbulence, *Naval Ord. Lab. Rept. No.* 1136, p. 67, July, 1949.
33. Frenkiel, F. N.: *Proc. 1st U.S. Natl. Congr. Appl. Mech.,* p. 837, 1951.
34. Frenkiel, F. N.: *Proc. Natl. Acad. Sci. U.S.,* **38,** 509 (1952).
35. Frenkiel, F. N.: "Advances in Applied Mechanics," vol. 3, p. 61, Academic Press, New York, 1953.
36. Fleishman, B. A., and F. N. Frenkiel: *J. Meteorol.,* **12,** 141 (1955).
37. Kampé de Fériet, J.: *Proc. 5th Intern. Congr. Appl. Mech.,* Cambridge, Mass., 1938, p. 352.
38. Kalinske, A. A., and C. L. Pien: *Ind. Eng. Chem.,* **36,** 220 (1944).
39. Townsend, A. A.: *Proc. Roy. Soc. London,* **224A,** 487 (1954).
40. Taylor, G. I.: *Proc. Roy. Soc. London,* **151A,** 421 (1935).
41. Towle, W. L., and T. K. Sherwood: *Ind. Eng. Chem.,* **31,** 457 (1939).
42. McCarter, R. J., L. F. Stutzman, and H. A. Koch, Jr.: *Ind. Eng. Chem.,* **41,** 1290 (1949).
43. Sherwood, T. K., and B. B. Woertz: *Ind. Eng. Chem.,* **31,** 1034 (1939).
44. Dryden, H. L.: *Ind. Eng. Chem.,* **31,** 416 (1939).
45. Corrsin, S., and M. S. Uberoi: *Natl. Advisory Comm. Aeronaut. Tech. Repts. No.* 1040, 1951.
46. Hinze, J. O., and B. G. van der Hegge Zijnen: "General Discussion on Heat Transfer," p. 188, Institute of Mechanical Engineering, London, 1951.
47. Lauwerier, H. A.: *Appl. Sci. Research,* **4A,** 153 (1954).
48. Hinze, J. O.: *J. Aeronaut. Sci.,* **18,** 565 (1951).
49. Kendall, M. G.: "The Advanced Theory of Statistics," vol. 1, Griffin & Co., London, 1947.
50. Townsend, A. A.: *Australian J. Sci. Research,* **2A,** 451 (1949).
51. Bagnold, R. A.: *Proc. Roy. Soc. London,* **225A,** 49 (1954).
52. Bagnold, R. A.: *Phil. Trans. Roy. Soc. London,* **249A,** 235 (1956).
53. Tchen, C. M.: Mean Value and Correlation Problems Connected with the Motion of Small Particles Suspended in a Turbulent Fluid, Ph.D. thesis, Delft, 1947.
54. Hughes, R. R., and E. R. Gilliland: *Chem. Eng. Progr.,* **48,** 497 (1952).
55. Corrsin, S., and J. Lumley: *Appl. Sci. Research,* **6A,** 114 (1956).
56. Schmidt, W.: "Der Massenaustausch in freier Luft," p. 18, Henri Grand Verlag, Hamburg, 1925.
57. Schultz-Grunow, F.: *Forsch. Gebiete Ingenieurw.,* **17,** 65 (1951).
58. Ryan, L. F.: Experiments on Aerodynamic Cooling, *Mitt. Inst. Aerodyn. E.T.H. Zürich, No.* 18, 1951.
59. Ranque, M. G.: *J. phys. radium,* **4,** 112 (1933).
60. Hilsch, R.: *Z. Naturforsch.,* **1,** 208 (1946).
61. Deemter, J. J. van: *Appl. Sci. Research,* **3A,** 174 (1953).
62. Spalding, D. B.: *Proc. Roy. Soc. London,* **221A,** 78, 100 (1954).
63. Schuh, H.: *Z. angew. Math. u. Mech.,* **25–27,** 54 (1947).

64. Mickley, H. S., R. C. Ross, A. L. Squyers, and W. E. Stewart: *Natl. Advisory Comm. Aeronaut. Tech. Notes No.* 3208, 1954.
65. Basset, A. B.: "A Treatise on Hydrodynamics," vol. 2, ch. 5, Deighton, Bell and Co., Cambridge, England, 1888.
66. Boussinesq, J.: "Théorie analytique de la chaleur," vol. 2, p. 224, Gauthier-Villars, Paris, 1903.
67. Oseen, C. W.: "Hydrodynamik," p. 132, Leipzig, 1927.
68. Corrsin, S.: lecture notes, APL (JHU-TG-63-29), p. 49, Johns Hopkins University, Baltimore, Md., 1953.
69. Lumley, J. L.: Some Problems Connected with the Motion of Small Particles in Turbulent Fluid, Ph.D. thesis, Johns Hopkins University, Baltimore, 1957.
70. Friedlander, S. K.: *A.I.Ch.E. Journal,* **3,** 381 (1957).
71. Corrsin, S.: Symposium "Mécanique de la Turbulence," p. 27, Marseilles, 1961.
72. Gibson, M. M.: *Nature, Lond.,* **195,** 1281 (1962).
73. Grant, H. L., R. W. Stewart, and A. Moilliet: *J. Fluid Mech.,* **12,** 241 (1962).
74. Lin, C. C.: *Proc. Nat. Acad. Sci.,* **46,** 566 (1960).
75. Philip, J. R.: *Phys. Fluids,* **11,** 38 (1968).
76. Kraichnan, R. H.: *Phys. Fluids,* **9,** 1937 (1966).
77. Deardorff, J. W., and R. L. Peskin: *Phys. Fluids,* **13,** 584 (1970).
78. Smith, F. B.: *Quart. J. Roy. Meteor. Soc.,* **91,** 318 (1965).
79. Cocke, W. J.: *Phys. Fluids,* **12,** 2488 (1969).
80. Hay, J. S., and F. Pasquil: "Advances in Geophysics," vol. 6, p. 345 (1959).
81. Baldwin, L. V., and W. R. Mickelsen: *Proc. A.S. Civ. Eng.,* **88** EM2, 37 (1962).
82. Kovasznay, L. S. G.: *J. Aeronaut. Sci.,* **20,** 657 (1953).
83. Baldwin, L. V., and Th. J. Walsh: *A.I.Ch.E. Jl.,* **7,** 53 (1961).
84. Gifford, F.: *U.S. Weather Bureau, Monthly Weather Review,* **83,** 293 (1955).
85. Corrsin, S., and J. L. Lumley: "Advances in Geophysics," vol. 6, p. 179 (1959).
86. Kraichnan, R. H.: *Phys. Fluids,* **7,** 1717 (1964).
87. Patterson, G. S., and S. Corrsin: "Dynamics of Fluids and Plasmas," p. 275, Academic Press Inc., New York, 1966.
88. Corrsin, S.: "Advances in Geophysics," vol. 6, p. 161 (1959).
89. Kraichnan, R. H.: *J. Fluid Mech.,* **5,** 497 (1959).
90. Saffman, P. G.: *Appl. Sci. Res.,* **11A,** 245 (1963).
91. Philip, J. R.: *Phys. Fluids,* **10,** Suppl. 9, S 69 (1967).
92. Deardorff, J. W.: *J. Fluid Mech.,* **41,** 453 (1970).
93. Rodriguez, J. M., and G. K. Patterson: *A.I.Ch.E. Jl.,* **15,** 790 (1969).
94. Saffman, P. G.: Symposium "Mécanique de la Turbulence," p. 53, Marseilles (1961).
95. Mickelsen, W. R.: *J. Fluid Mech.,* **7,** 397 (1960).
96. Favre, A.: *J. Appl. Mech., Trans. A.S.M.E.,* **32E,** 241 (1965).
97. Hjelmfelt, A. T., and L. F. Mockros: *Appl. Sci. Res.,* **16,** 149 (1966).
98. Ahmadi, G., and V. W. Goldschmidt: *Techn. Rep. F.M.T.R.* 70-3, Purdue Univ., 1970.
99. Goldschmidt, V. W., M. K. Householder, G. Ahmadi, and S. C. Chuang: "Progress in Heat and Mass Transfer," vol. 6, p. 487, Pergamon Press, New York, 1972.
100. Elata, C., and A. T. Ippen: *Techn. Rep. 45. Hydrodynamics Laboratory,* M.I.T., Cambridge, Mass., 1961.
101. Hino, M.: *J. Hydro. Div., Proc. Am. Soc. Civil Engrs.,* **89,** 161 (1963).

102. Soo, S. L.: "Fluid Dynamics of Multiphase Systems," Blaisdell Publishing Co., Toronto, 1967.
103. Hinze, J. O.: "Progress in Heat and Mass Transfer," vol. **6,** p. 433, Pergamon Press, New York, 1972.
104. Eskinazi, S., and H. Yeh: *J. Aeronaut. Sci.,* **23,** 23 (1956).
105. Tailland, A., and J. Mathieu: *J. Mécanique,* **6,** 103 (1967).
106. Kjellström, B., and S. Hedberg: *Aktiebolaget Atomenergie Stockholm. Rep. AE-243,* 1966.
107. Eskinazi, S., and F. F. Erian: *Phys. Fluids,* **12,** 1988 (1969).
108. Hinze, J. O.: *Appl. Sci. Res.,* **22,** 163 (1970).
109. Béguier, C.: "Etude du jet plan dissymétrique en régime turbulent incompressible." Ph.D. Thesis Université d'Aix-Marseille, 1971.
110. Hanjalic, K., and B. E. Launder: *J. Fluid Mech.,* **52,** 609 (1972).
111. Peskin, R. L.: "The diffusivity of small suspended particles in turbulent fluids," National Meeting A.J.Ch.E., Baltimore, 1962.
112. Snyder, W. H., and J. L. Lumley: *J. Fluid Mech.,* **48,** 41 (1971).
113. Prandtl, L.: *Nachr. Akad. Wiss. Göttingen, Math.-Phys. Klasse,* 6–19, 1945.
114. Hinze, J. O., R. E. Sonnenberg, and P. J. H. Builtjes: "XIII Intern. Congress of Theor. and Appl. Mech.," Moscow, 1972. *Appl. Sci. Res.,* **29,** 1 (1974).
115. Corrsin, S.: *J. Atm. Sciences,* **20,** 115 (1963).
116. Lumley, J. L.: I.A.H.R. Int. Symp. on Stratified Flows, Novosibirsk, September 1972.
117. Sullivan, P. L.: *J. Fluid Mech.,* **47,** 601 (1971).
118. Abramovich, G. N.: *Int. J. Heat Mass Transfer,* **14,** 1039 (1971).
119. Hinze, J. O.: General Lecture XVth Congress of the Int. Association for Hydraulic Research, Istanbul, 2–7 September 1973.
120. Lumley, J. L.: *Instituto nazionale di alta matematica Symposia Matematica,* **9,** 315 (1972).
121. Tennekes, H.: *J. Fluid Mech.,* **67,** 561 (1975) (in press).
122. Shlien, D. J., and S. Corrsin: *J. Fluid Mech.,* **62,** 255 (1974).

NOMENCLATURE FOR CHAPTER 5

C concentration.
c_p specific heat at constant pressure.
D diameter of pipe; coefficient of molecular diffusion.
d diameter of cylinder or rod.
$\mathbf{E}_L$ Lagrangian energy-spectrum function.
e_{ij} direction cosines.
F driving force.
f friction factor in pipe flow.
$\mathbf{f}$ coefficient of spatial, longitudinal velocity correlation.
L distance through which a transferable property of a fluid lump is conserved; L_m, for momentum; L_γ, for a scalar quantity.
M mesh length of grid; molar mass.
n frequency.
P static pressure; $\bar{P}$, time-mean value; p, turbulent fluctuation.

Pr Prandtl number $\nu/\mathfrak{k}_\theta$; $\mathbf{Pr}_{turb}$, $\epsilon_m/\epsilon_\theta$.
$\overline{q^2}$ $\overline{u_i u_i}$, twice kinetic energy of turbulence.
$\mathbf{R}_{ij}$ coefficient of spatial second-order velocity-correlation tensor.
$(\mathbf{R}_{ij})_L$ coefficient of Lagrangian second-order velocity-correlation tensor.
$\mathbf{R}_L$ coefficient of Lagrangian time correlation.
Re Reynolds number; $\mathbf{Re}_\lambda$, $u'\lambda_g/\nu$; $\mathbf{Re}_M$, $\bar{U}_1 M/\nu$; $\mathbf{Re}_{\Lambda_f}$, $u'\Lambda_f/\nu$.
r distance between two points.
S strength of point source.
S^* strength of unit of length of line source.
Sc Schmidt number $\nu/\mathfrak{k}_\gamma$, ν/D.
s $(\overline{z_i z_i})^{1/2}$
t time.
U_i Eulerian velocity; $\bar{U}_i$, time-mean value; u_i, turbulence component; u_i', $\sqrt{\overline{u_i^2}}$.
V_i Lagrangian velocity; v_i, turbulence component.
w_i relative velocity of two fluid particles.
x_i Eulerian coordinates.
y_i Lagrangian coordinates of fluid particle.
z_i relative displacement of two fluid particles.
α exchange coefficient.
β $2\pi n/\alpha$; Λ_L/Λ_f.
δ_{ij} Kronecker delta.
ε_{ijk} alternating unit tensor.
ϵ coefficient of eddy diffusion; ϵ_m, for momentum; ϵ_ω, for vorticity; ϵ_γ, for scalar; ϵ_q, for turbulence kinetic energy; ϵ_θ, for heat.
ϵ_{ij} coefficient of eddy-diffusion tensor.
ε dissipation by turbulence per unit of mass.
η $(\nu^3/\varepsilon)^{1/4}$, Kolmogoroff micro length-scale.
η_i Lagrangian distance traveled by a fluid particle.
Γ scalar quantity; $\bar{\Gamma}$, time-mean value; γ, turbulent fluctuation.
Γ gamma function.
$\mathfrak{J}$ flux per unit of area; $\mathfrak{J}_{\mathscr{P}}$, for transferable property; $\mathfrak{J}_m$, for momentum; $\mathfrak{J}_\gamma$, for scalar.
$\varkappa$ c_p/c_v, ratio of specific heats.
κ heat conductivity.
$\mathfrak{k}_\theta$ $\kappa/c_p\rho$; $\mathfrak{k}_c$, D, coefficient of molecular diffusion.
Λ_L Lagrangian integral scale $v'\mathscr{I}_L$.
λ_g spatial lateral dissipation scale.
$\mathscr{L}$ coefficient in Reichardt's "diffusion" equation; $\mathscr{L}_\gamma$, for scalar.
$\mathfrak{l}$ Prandtl's mixing length; $\mathfrak{l}_m$, for momentum; $\mathfrak{l}_\gamma$, for scalar.
μ dynamic viscosity; $\bar{\mu}$, mean value; $\tilde{\mu}$, turbulent fluctuation.
ν kinematic viscosity.
Ω_k vorticity; $\bar{\Omega}_k$, time-mean value; ω_k, turbulent fluctuation.

ω	angular frequency.
$\mathcal{O}$	order of magnitude.
$\mathscr{P}$	transferable property; $\bar{\mathscr{P}}$, time-mean value; $\mathfrak{p}$, turbulent fluctuation.
$\mathfrak{P}$	probability.
$\mathfrak{P}(u_1,u_2)$	joint-probability-density distribution.
π	3.14159....
$\mathfrak{Q}$	displacement probability.
$\mathfrak{R}$	gas constant.
ρ	density; $\bar{\rho}$, mean value; $\tilde{\rho}$, turbulent fluctuation.
σ_{ij}	stress tensor.
$\mathfrak{I}_L$	Lagrangian integral time scale.
$(\mathfrak{I}_{i,j})_L$	tensor of Lagrangian integral time scale.
τ	time.
τ_k	$(\nu/\varepsilon)^{1/2}$, Kolmogoroff micro time-scale.
τ_E	Eulerian dissipation time-scale.
τ_L	Lagrangian dissipation time-scale.
Θ	dilatation.
Θ	temperature; $\bar{\Theta}$, time-mean value, θ, turbulent fluctuation.
$\mathscr{V}$	velocity of large-scale turbulent motions.
ξ_i	Eulerian coordinates.
ζ	rate of strain, $\overline{e_{li}e_{lj}\,\partial u_i/\partial x_j}$ for $t \to \infty$.

6
FREE TURBULENT SHEAR FLOWS

6-1 INTRODUCTION

Free turbulence is encountered in a flow field where there is no direct effect of any fixed boundary on the turbulence in the flow. There may be indirect influence, because a fixed boundary may create the conditions for the existence of a free turbulence; for example, an obstacle placed in a general stream may cause a flow pattern in which free turbulence occurs. In nonisotropic shear flows whole regions in the flow field differ—usually substantially—in mean-flow velocity. The gradients in the mean velocity, and the associated mean shear-stresses generate turbulence (see, e.g., Sec. 1-13, Eq. 1-110). This local generation of turbulence makes possible a steady-flow condition when it just compensates the local losses by viscous dissipation and diffusion.

The free turbulent flow fields studied may be classified into two main groups: (1) jet flows, and (2) wake flows.

The flow patterns investigated to date both theoretically and experimentally are: the boundary flow between two parallel flows with different mean velocities (often called the half-jet); the plane and the round jet; the plane wake behind a cylindrical rod; the round wake behind an axi-symmetric body; the plane wake flow close behind a row of cylindrical rods.

In the flow patterns considered it is possible to distinguish one main flow direction with velocity substantially greater than in any other. This fact is very welcome, because it simplifies considerably the theoretical investigation of these flows. Simplification is based on the following assumptions:

1 The mean flow velocity transverse to the main flow is very small compared with the main flow velocity and may even be neglected in some cases.

2 Changes of quantities in the direction of the main flow are correspondingly slow with respect to those in the transverse direction.

This assumption is consistent with the general observation that the free-turbulent-flow region is very much longer in the main-flow direction than in a direction transverse to it.

3 As will be shown in Sec. 6-3, the mean-pressure variation across the flow region in the transverse direction is mainly determined by the variation in turbulent-velocity intensities. In the main-flow direction it varies mainly according to the pressure distribution in the undisturbed flow. Usually this pressure distribution in the main-flow direction is uniform; so for calculating the mean-velocity distribution the assumption is made that the mean pressure is uniform throughout the whole turbulence region.

In Chap. 4 we have already discussed the incompleteness of theories about turbulent shear flow. There is no statistical theory equivalent to that developed for isotropic turbulence, which enables us to attack turbulent shear flow and to predict, say, mean-velocity distributions; although our knowledge based upon experimental evidence of the details of the turbulent-flow mechanism is already very helpful in understanding this mechanism and in interpreting experimental data.

The various theories which have been applied to free turbulent flows and which have gradually come to be considered more or less classical are purely phenomenological. We have considered these theories in Chap. 5: those based upon the assumption that the mean-velocity (or temperature or concentration) distribution can be calculated with the aid of a suitably chosen coefficient of eddy diffusion (Boussinesq's and the mixing-length theories) and Reichardt's inductive theory of free turbulence. Since these theories, though not satisfactory from a fundamental point of view, are still generally accepted for predicting the distribution of mean properties in free turbulent flows, we will consider them in what follows in their application to jet and wake flows.

In the application of these phenomenological theories to free turbulent flows, much use has been made of the assumption of *similarity*, or, better, of *self-preservation*, in the sense given in the introduction to Chap. 3. We recall that self-preservation is used to indicate that the turbulence maintains its structure during the development of the turbulent region in the downstream direction of the main flow.

There is now sufficient experimental evidence about similarity of mean-velocity profiles in successive sections of jet and wake flows in the downstream direction. For each particular free turbulent flow, the more or less bell-shaped curves are seen to be geometrically similar when the velocities at any section are made dimensionless with the maximum mean-velocity difference as a velocity scale and when the lateral distances are made dimensionless with the local width of the turbulent region as a scale—or, better, in the case of symmetrical mean-velocity profiles, using the so-called *"half-value" distance*. This is the distance from the axis of symmetry, at which the mean-velocity difference is half the maximum value. These velocity and length scales are sole functions of the distance from some apparent origin where the free turbulent flow (jet or wake) seems to originate.

This similarity in flow patterns in various sections and self-preservation of turbulence patterns moving downstream with the main flow are made probable by the empirical fact, mentioned earlier, that free-turbulent-flow regions are relatively narrow, with a main-flow velocity much greater than the transverse velocities, whereas spatial changes in the main-flow direction are much smaller than the corresponding changes in the transverse direction. Hence, a turbulence flow pattern moving downstream is strongly dependent on its history. This is particularly true of wake flows and of those jet flows in a secondary uniform stream whose velocity is large compared with the difference between absolute-jet velocity and secondary-stream velocity.

If there were no production of turbulence by mean shear stresses, the turbulence would decay in the downstream direction, and self-preservation would hold for the main part of the wavenumber range of the turbulence, just as it does in a decaying isotropic turbulence. In free turbulent shear flow, however, there is production of turbulence. But this production is determined by the gradient of the mean-velocity distribution, which depends in turn on the turbulence generated upstream and transported downstream by turbulence diffusion and by convection. This close connection between the turbulence and the mean-velocity distribution at any section and those upstream makes it reasonable to expect similarity of the total patterns, even though turbulence is continuously being produced by the main motion through the turbulence shear stresses.

However, the flow requires some time or distance downstream of the above apparent origin before the self-preserving condition is reached. As will be discussed later, in the case of wake flows the distance expressed in terms of the dimension of the body may be appreciable, while this distance may be much greater for the fine-scale structure of the turbulence than for the gross structure and for the mean-velocities.

The above mentioned special features of jet flows and wake flows, namely being relatively very narrow in directions transverse to the main-flow direction, and their tendency to become self-preserving, form the basis for two important theoretical tools, by which mathematically the governing differential equations may be considerably simplified. The assumption of *self-preservation* of the flow makes it possible to reduce the number of independent variables by one. So in the case of steady plane and axi-symmetric flows with two independent variables, the partial differential equations can be reduced to a set of ordinary differential equations. Also, the relative narrowness of the turbulent regions makes it possible to apply so-called *"boundary-layer approximations,"* by which it is possible to reduce the number of terms in the equations by neglecting, in first approximation, those terms which appear to be an order of magnitude smaller than the other terms.

In what follows we shall not treat all investigations into free turbulent flows. We shall confine the discussion mainly to two special cases: (1) the wake flow behind a cylinder, representing the group of wake flows, and (2) the round free jet, representing the group of jet flows. These two cases also represent a plane flow and an axi-symmetrical flow, respectively. Before treating these two special cases it may be useful to consider in separate sections in a more general sense the conditions for similarity and the "boundary-layer approximations."

6-2 SIMILARITY CONSIDERATIONS

Referring to Sec. 3-1 we recall that two flow patterns are defined as completely similar when it is possible to reduce all velocities by one *velocity scale* $\mathcal{U}$, and all dimensions by one *length scale* $\mathcal{L}$, so that the two flow patterns expressed in the reduced quantities become identical. In the case considered here of self-preservation the velocity scale $\mathcal{U}$ and the length scale $\mathcal{L}$ are functions of the coordinate in the main-flow direction. Since the flows to be considered are either steady plane flows or steady axi-symmetric flows, it will be sufficient in the following general treatment to consider only one of the two, for which we take the two-dimensional situation.

Let x_1 be the coordinate in the main-flow direction, and x_2 the coordinate in the transverse direction. The mass and momentum balance equations to be considered are

$$\frac{\partial \bar{U}_i}{\partial x_i} = 0 \qquad i = 1,2 \qquad j = 1,2 \tag{6-1}$$

$$\bar{U}_j \frac{\partial \bar{U}_i}{\partial x_j} = -\frac{1}{\rho}\frac{\partial \bar{P}}{\partial x_i} + \nu \frac{\partial^2 \bar{U}_i}{\partial x_j \partial x_j} - \frac{\partial}{\partial x_j}\overline{u_j u_i} \tag{6-2}$$

Let $\mathcal{U}(x_1)$ and $\mathcal{L}(x_1)$ be the velocity and length scale. In case of *complete similarity* we must have

$$\frac{\bar{U}_i}{\mathcal{U}} = f_i\left(\frac{x_2}{\mathcal{L}}\right)$$

$$-\frac{\overline{u_j u_i}}{\mathcal{U}^2} = f_{ji}\left(\frac{x_2}{\mathcal{L}}\right)$$

and

$$\frac{\bar{P}}{\rho \mathscr{U}^2} = \Pi\left(\frac{x_2}{\mathscr{L}}\right)$$

By substituting these reduced quantities in the above mass and momentum balance equations, the conditions can be obtained which have to be satisfied for complete similarity. We will not do this, because the procedure is the same as the one which will be carried out in the following for the case of incomplete similarity. But the results of the procedure applied to the above equations are that complete similarity requires that the length scale $\mathscr{L}$ must vary in proportion to the distance x_1 to the apparent origin of similarity, and the velocity scale $\mathscr{U}$ inversely proportional to x_1, so that the Reynolds number $\mathscr{U}\mathscr{L}/v$ remains constant, while the pressure must either vary with x_1^{-2}, or be constant. Since these conditions are very restrictive we have to make allowance for departures from complete similarity, and consider incomplete similarity. For the jet flows and wake flows it appears possible to maintain one length scale $\mathscr{L}(x_1)$, but to consider different velocity scales for the different velocity components, all velocity scales being functions of x_1 alone.

The above procedure for complete similarity also reveals that the velocity scale and length scale are power functions of x_1. Therefore, in the case of *incomplete similarity* we put

$$\bar{U}_1 = \mathscr{U}_1 f(\xi_2) = \mathscr{U}_0\left(\frac{x_1}{\mathscr{L}_0}\right)^p f(\xi_2) \tag{6-3a}$$

$$\bar{U}_2 = \mathscr{U}_2 g(\xi_2) = \mathscr{U}_0\left(\frac{x_1}{\mathscr{L}_0}\right)^r g(\xi_2) \tag{6-3b}$$

$$-\bar{P} = \rho\mathscr{U}_3{}^2\Pi(\xi_2) = \rho\mathscr{U}_0{}^2\left(\frac{x_1}{\mathscr{L}_0}\right)^t \Pi(\xi_2) \tag{6-3c}$$

$$-\overline{u_1{}^2} = \mathrm{u}_1{}^2 l(\xi_2) = \mathscr{U}_0{}^2\left(\frac{x_1}{\mathscr{L}_0}\right)^{s_1} l(\xi_2) \tag{6-3d}$$

$$-\overline{u_2 u_1} = \mathrm{u}_{21}{}^2 h(\xi_2) = \mathscr{U}_0{}^2\left(\frac{x_1}{\mathscr{L}_0}\right)^{s_{21}} h(\xi_2) \tag{6-3e}$$

$$-\overline{u_2{}^2} = \mathrm{u}_2{}^2 m(\xi_2) = \mathscr{U}_0{}^2\left(\frac{x_1}{\mathscr{L}_0}\right)^{s_2} m(\xi_2) \tag{6-3f}$$

where

$$\xi_2 = \frac{x_2/\mathscr{L}_0}{(x_1/\mathscr{L}_0)^q} \tag{6-3g}$$

$\mathscr{U}_0$ and $\mathscr{L}_0$ are a *characteristic velocity* and a *characteristic length* respectively for which as a suitable quantity may be taken a velocity and dimension primarily determining the condition of the particular flow (e.g., the issuing velocity and nozzle diameter in case of a free jet).

The reader may note that the *condition of complete similarity* is included in the above relations (6-3), for which condition we must have

$$2p = 2r = t = s_1 = s_{21} = s_2 \tag{6-4}$$

Let us first consider the *mass-balance equation* (6-1)

$$\frac{\partial \bar{U}_1}{\partial x_1} = \frac{\partial}{\partial x_1}\left[\mathscr{U}_0\left(\frac{x_1}{\mathscr{L}_0}\right)^p f(\xi_2)\right]$$

$$= p\frac{\mathscr{U}_0}{\mathscr{L}_0}\left(\frac{x_1}{\mathscr{L}_0}\right)^{p-1} f(\xi_2) + \mathscr{U}_0\left(\frac{x_1}{\mathscr{L}_0}\right)^p \frac{d}{d\xi_2} f(\xi_2)\frac{\partial \xi_2}{\partial x_1}$$

$$= p\frac{\mathscr{U}_0}{\mathscr{L}_0}\left(\frac{x_1}{\mathscr{L}_0}\right)^{p-1} f(\xi_2) - q\frac{\mathscr{U}_0}{\mathscr{L}_0}\left(\frac{x_1}{\mathscr{L}_0}\right)^{p-1} \xi_2\frac{df}{d\xi_2}$$

Similarly

$$\frac{\partial \bar{U}_2}{\partial x_2} = \mathscr{U}_0\left(\frac{x_1}{\mathscr{L}_0}\right)^r \frac{d}{d\xi_2} g(\xi_2)\frac{\partial \xi_2}{\partial x_2} = \frac{\mathscr{U}_0}{\mathscr{L}_0}\left(\frac{x_1}{\mathscr{L}_0}\right)^{r-q} \frac{dg}{d\xi_2}$$

Substitution of these expressions in Eq. (6-1) results, after dividing all terms by $\mathscr{U}_0/\mathscr{L}_0 \cdot (x_1/\mathscr{L}_0)^{p-1}$, in the equation

$$pf(\xi_2) - q\xi_2\frac{df}{d\xi_2} + \left(\frac{x_1}{\mathscr{L}_0}\right)^{1-p+r-q}\frac{dg}{d\xi_2} = 0$$

The first two terms are not a function of x_1, and since the solutions of $f(\xi_2)$ and $g(\xi_2)$ must hold true for any value of x_1, the part of the last term containing x_1 must be a constant, equal to 1, which requires

$$p + q - r = 1 \tag{6-5a}$$

The relation between $f(\xi_2)$ and $g(\xi_2)$ then becomes

$$pf(\xi_2) - q\xi_2\frac{df}{d\xi_2} + \frac{dg}{d\xi_2} = 0 \tag{6-6}$$

The *momentum-balance equations* can be treated in a similar way. We will do this for the momentum-balance equation in the x_1-direction. Equation (6-2) rewritten for this direction reads

$$\bar{U}_1\frac{\partial \bar{U}_1}{\partial x_1} + \bar{U}_2\frac{\partial \bar{U}_1}{\partial x_2} = -\frac{1}{\rho}\frac{\partial \bar{P}}{\partial x_1} + \nu\left(\frac{\partial^2 \bar{U}_1}{\partial {x_1}^2} + \frac{\partial^2 \bar{U}_1}{\partial {x_2}^2}\right) - \frac{\partial}{\partial x_1}\overline{{u_1}^2} - \frac{\partial}{\partial x_2}\overline{u_2 u_1} \tag{6-7}$$

Again substitution of the relations (6-3) into this equation yields after division of all terms by ${\mathscr{U}_0}^2/\mathscr{L}_0(x_1/\mathscr{L}_0)^{2p-1}$:

$$pf^2(\xi_2) - q\xi_2 f(\xi_2)\frac{df}{d\xi_2} + \left(\frac{x_1}{\mathscr{L}_0}\right)^{1-p+r-q} g(\xi_2)\frac{df}{d\xi_2}$$

$$= \left(\frac{x_1}{\mathscr{L}_0}\right)^{t-2p}\left[t\Pi(\xi_2) - q\xi_2\frac{d\Pi}{d\xi_2}\right] + \frac{\nu}{\mathscr{U}_0\mathscr{L}_0}\left(\frac{x_1}{\mathscr{L}_0}\right)^{1-p-2q}\frac{d^2 f}{d{\xi_2}^2}$$

$$+ \frac{\nu}{\mathscr{U}_0 \mathscr{L}_0}\left(\frac{x_1}{\mathscr{L}_0}\right)^{-p-1}\left[p(p-1)f(\xi_2) - q(2p-q-1)\xi_2 \frac{df}{d\xi_2}\right.$$

$$\left. + q^2 \xi_2{}^2 \frac{d^2 f}{d\xi_2{}^2}\right] + \left(\frac{x_1}{\mathscr{L}_0}\right)^{s_1 - 2p}\left[s_1 l(\xi_2) - q\xi_2 \frac{dl}{d\xi_2}\right]$$

$$+ \left(\frac{x_1}{\mathscr{L}_0}\right)^{s_{21} - 2p - q + 1} \frac{dh}{d\xi_2}$$

Since here too the terms cannot contain functions of x_1, the assumption of similarity requires

$$p + q - r = 1 \tag{6-5a}$$

$$2p - t = 0 \tag{6-5b}$$

$$p = -1 \tag{6-5c}$$

$$p + 2q = 1 \tag{6-5d}$$

$$s_1 - 2p = 0 \tag{6-5e}$$

$$2p + q - s_{21} = 1 \tag{6-5f}$$

These relations are satisfied when

$$p = -1; \quad q = 1; \quad r = -1; \quad s_1 = -2; \quad s_{21} = -2; \quad t = -2 \tag{6-8}$$

The same results are obtained from the momentum-balance equation in the x_2-direction, except of course the value $s_1 = -2$, but $s_2 = -2$ is obtained instead.

The values of $p, q \ldots$ given in Eq. (6-8) satisfy the conditions (6-4) for complete similarity. This, of course, is not surprising since we have already indicated earlier that complete similarity is possible. However, as said, in that case the lateral extent of the turbulent flow region must increase linearly with x_1, all velocities must decrease inversely proportional with x_1, the Reynolds-number remains constant and the pressure $\bar{P} \propto x_1{}^{-2}$, or be constant.

In the next section we will consider the "boundary-layer approximations" applied to the momentum-balance equations, by which method through order of magnitude estimates some terms may be neglected. However, by this procedure at the same time we may lose one or more of the conditions (6-5), making the set of algebraical equations incomplete. By means of suitable assumptions additional relations may be obtained, making the set of equations complete again. Solutions may then be obtained which, however, no longer satisfy the condition of complete similarity but are incompletely similar.

6-3 APPROXIMATIONS APPLIED TO THE EQUATIONS OF MOTION

The approximations considered here are usually, and have already been in the previous sections, called *boundary-layer approximations.* This name stems from the historical fact that Prandtl in his first boundary-layer studies applied these approximations to the Navier-Stokes equations.

As mentioned earlier, the application of "boundary-layer approximations" is possible because the regions of free turbulent shear flows considered here are extremely elongated in the main-flow direction. Assume the x_1-axis of the coordinate system and the corresponding mean velocity $\bar{U}_1$ to coincide with the main-flow direction. We introduce length scales L_1, L_2, and L_3 for the x_1-, x_2-, and x_3-directions, respectively. Since the turbulent region is relatively narrow in the x_2- and x_3-directions the length scales are chosen appropriately if

$$\frac{L_2}{L_1} \ll 1 \qquad \frac{L_3}{L_1} \ll 1 \qquad \frac{L_2}{L_3} \sim 1 \tag{6-9}$$

Furthermore when we reduce the coordinates x_i by these length scales L_i, the reduced coordinates ξ_i are such that $d\xi_i$ become of the same order of magnitude for $i = 1$, 2, and 3. Let $\mathscr{V}_i$ be a suitable velocity scale for the corresponding velocity $\bar{U}_i$. In many cases, and certainly so in the shear flows considered, the changes in reduced velocities may become of the same order of magnitude, i.e.,

$$d\left(\frac{\bar{U}_i}{\mathscr{V}_1}\right) \sim d\left(\frac{\bar{U}_2}{\mathscr{V}_2}\right) \sim d\left(\frac{\bar{U}_3}{\mathscr{V}_3}\right)$$

Of course, this need not be true in general, since a velocity scale for the velocity itself need not be the same as that for the change of the velocity. Also there may be regions in the flow field where locally $\partial\bar{U}_i/\partial x_j$ differs in order of magnitude from $\mathscr{V}_i/L_j$. However, again for the shear flows considered, we may estimate the average value of $\partial\bar{U}_i/\partial x_j$ across the whole flow region as of the order of magnitude $\mathscr{V}_i/L_j$. Provided we take for $\bar{U}_i$ the difference in velocity with the velocity U_0 outside the shear-flow region, if there is such a velocity U_0.

The magnitudes of the velocity scales $\mathscr{V}_2$ and $\mathscr{V}_3$ for the velocity components $\bar{U}_2$ and $\bar{U}_3$, respectively, are no longer independent of the velocity scale $\mathscr{V}_1$, because changes in the velocities in various directions are not independent, on the ground of the equation of continuity

$$\begin{array}{ccccccc} \dfrac{\partial\bar{U}_1}{\partial x_1} & + & \dfrac{\partial\bar{U}_2}{\partial x_2} & + & \dfrac{\partial\bar{U}_3}{\partial x_3} & = & 0 \\[2ex] \dfrac{\mathscr{V}_1}{L_1} & & \dfrac{\mathscr{V}_2}{L_2} & & \dfrac{\mathscr{V}_3}{L_3} & & \end{array}$$

The order of magnitude of the various terms is written below the corresponding terms of the equation of continuity. They must be the same.

Hence

$$\frac{\mathscr{V}_1}{L_1} \sim \frac{\mathscr{V}_2}{L_2} \sim \frac{\mathscr{V}_3}{L_3}$$

or

$$\frac{\mathscr{V}_2}{\mathscr{V}_1} = \frac{L_2}{L_1} \qquad \frac{\mathscr{V}_3}{\mathscr{V}_1} = \frac{L_3}{L_1} \qquad \frac{\mathscr{V}_2}{\mathscr{V}_3} \sim 1 \tag{6-10}$$

Consequently, also,

$$\frac{\bar{U}_2}{\bar{U}_1} = \frac{L_2}{L_1} \qquad \frac{\bar{U}_3}{\bar{U}_1} = \frac{L_3}{L_1} \qquad \frac{\bar{U}_2}{\bar{U}_3} \sim 1$$

The Reynolds equations of motion contain the turbulence Reynolds stresses $\rho\overline{u_i u_j}$, for which, too, we have to introduce a suitable scale. We know from experience that, in anisotropic shear flows, the intensities $\overline{u_1^2}, \overline{u_2^2}$, and $\overline{u_3^2}$ are roughly of the same magnitude. So we may introduce the same scale $\mathfrak{v}$ for u'_1, u'_2, and u'_3. The magnitudes of the shear stresses $\rho\overline{u_i u_j}$ for $i \neq j$, compared with $\rho\overline{u_1^2}$, $\rho\overline{u_2^2}$, and $\rho\overline{u_3^2}$, depend on the magnitudes of the correlation coefficients $\mathbf{R}_{ij}$ for $i \neq j$. Though in free turbulent shear-flows the turbulence intensities may be of the same order of magnitude as the mean velocities, yet it may be convenient to assume a separate velocity scale $\mathfrak{v}$ for the turbulence velocities:

$$\overline{u_1^2} \sim \overline{u_2^2} \sim \overline{u_3^2} \sim \mathfrak{v}^2$$

and (6-11)

$$\overline{u_i u_j} \sim \mathbf{R}_{ij}\mathfrak{v}^2 \qquad \text{for } i \neq j$$

It is convenient to distinguish between two cases of flow types.

1 There is a constant velocity U_0 superimposed on the main velocity $\bar{U}_1$, and, furthermore,

$$\frac{U_0}{\bar{U}_1} \gg 1$$

We shall see later that the order of magnitude of this ratio is not independent of the order of magnitude of L_1/L_2.

This case is encountered in wake flows and in free-jet flows in a general stream with constant velocity U_0 large compared with the relative issuing velocity.

2 There is no superimposed velocity U_0 or, if there is such a velocity, it is of the same order as $\bar{U}_1$. This is the case of free jets issuing in a quiescent or low-velocity ambient medium.

Let us consider first the Reynolds equation of motion for *case 1*. Since, in the comparison of the phenomena in the various directions, there is no essential difference between the x_2- and the x_3-direction, it is sufficient to consider only the equations for the x_1- and x_2-directions. Any conclusion reached with respect to the x_2-direction applies also to the x_3-direction.

We write down the equation of motion and, below each term, its relative order of magnitude with respect to the other terms, expressed in terms of the velocity and length scales.

We assume a *steady state* for the mean motion, and such *large Reynolds numbers* that molecular effects are negligible with respect to the turbulence effects, a situation which is often true in the shear flows of technological or geophysical interest.

For the x_1-direction we obtain

$$
\begin{array}{ccccccc}
(U_0+\bar{U}_1)\dfrac{\partial \bar{U}_1}{\partial x_1} & +\,\bar{U}_2\dfrac{\partial \bar{U}_1}{\partial x_2} & +\,\bar{U}_3\dfrac{\partial \bar{U}_1}{\partial x_3} & = -\dfrac{1}{\rho}\dfrac{\partial \bar{P}}{\partial x_1} & -\dfrac{\partial}{\partial x_1}\overline{u_1^2} & -\dfrac{\partial}{\partial x_2}\overline{u_1u_2} & -\dfrac{\partial}{\partial x_3}\overline{u_1u_3} \\[2ex]
\dfrac{U_0}{\bar{U}_1}\dfrac{\mathscr{V}_1^2}{L_1} & \dfrac{\mathscr{V}_1^2}{L_1} & \dfrac{\mathscr{V}_1^2}{L_1} & \dfrac{\Delta P_1}{\rho L_1} & \dfrac{\mathfrak{v}^2}{L_1} & \mathbf{R}_{12}\dfrac{\mathfrak{v}^2}{L_2} & \mathbf{R}_{13}\dfrac{\mathfrak{v}^2}{L_3}
\end{array}
$$

or

$$
\begin{array}{ccccccc}
\dfrac{U_0}{\bar{U}_1} & 1 & 1 & \dfrac{\Delta P_1}{\rho \mathscr{V}_1^2} & \dfrac{\mathfrak{v}^2}{\mathscr{V}_1^2} & \mathbf{R}_{12}\dfrac{\mathfrak{v}^2}{\mathscr{V}_1^2}\dfrac{L_1}{L_2} & \mathbf{R}_{13}\dfrac{\mathfrak{v}^2}{\mathscr{V}_1^2}\dfrac{L_1}{L_3}
\end{array}
\tag{6-12}
$$

By ΔP_1 we mean a change in the x_1-direction. We see that $\bar{U}_2\,\partial\bar{U}_1/\partial x_2$ and $\bar{U}_3\,\partial\bar{U}_1/\partial x_3$ are of the same order of magnitude as $\bar{U}_1\,\partial\bar{U}_1/\partial x_1$. This is shown as follows:

$\bar{U}_2\,\partial\bar{U}_1/\partial x_2$ is of the order of magnitude $\mathscr{V}_2\mathscr{V}_1/L_2 \sim \mathscr{V}_1^2/L_1$ because $\mathscr{V}_2 = \mathscr{V}_1 L_2/L_1$.

The same result is obtained for the term $\bar{U}_3\,\partial\bar{U}_1/\partial x_3$.

For the equation of motion in the x_2-direction we obtain

$$
\begin{array}{ccccccc}
(U_0+\bar{U}_1)\dfrac{\partial \bar{U}_2}{\partial x_1} & +\,\bar{U}_2\dfrac{\partial \bar{U}_2}{\partial x_2} & +\,\bar{U}_3\dfrac{\partial \bar{U}_2}{\partial x_3} & = -\dfrac{1}{\rho}\dfrac{\partial \bar{P}}{\partial x_2} & -\dfrac{\partial}{\partial x_1}\overline{u_2u_1} & -\dfrac{\partial}{\partial x_2}\overline{u_2^2} & -\dfrac{\partial}{\partial x_3}\overline{u_2u_3} \\[2ex]
\dfrac{U_0}{\bar{U}_1}\dfrac{\mathscr{V}_1^2L_2}{L_1^2} & \dfrac{\mathscr{V}_1^2L_2}{L_1^2} & \dfrac{\mathscr{V}_1^2L_2}{L_1^2} & \dfrac{\Delta P_2}{\rho L_2} & \mathbf{R}_{12}\dfrac{\mathfrak{v}^2}{L_1} & \dfrac{\mathfrak{v}^2}{L_2} & \mathbf{R}_{23}\dfrac{\mathfrak{v}^2}{L_3}
\end{array}
$$

or

$$
\begin{array}{ccccccc}
\dfrac{U_0}{\bar{U}_1} & 1 & 1 & \dfrac{L_1^2}{L_2^2}\dfrac{\Delta P_2}{\rho \mathscr{V}_1^2} & \mathbf{R}_{12}\dfrac{\mathfrak{v}^2}{\mathscr{V}_1^2}\dfrac{L_1}{L_2} & \dfrac{\mathfrak{v}^2}{\mathscr{V}_1^2}\dfrac{L_1^2}{L_2^2} & \mathbf{R}_{23}\dfrac{\mathfrak{v}^2}{\mathscr{V}_1^2}\dfrac{L_1^2}{L_2L_3}
\end{array}
\tag{6-13}
$$

On the ground of the assumption that $U_0/\bar{U}_1 \gg 1$ (*case 1*), the second and third terms on the left-hand side of Eqs. (6-12) and (6-13) are an order of magnitude smaller. If it is assumed that the correlation coefficients $\mathbf{R}_{12}$, $\mathbf{R}_{13}$, and $\mathbf{R}_{23}$ are not small, $\partial\overline{u_1^2}/\partial x_1$ in Eq. (6-12) is small compared with $\partial\overline{u_1u_2}/\partial x_2$ and $\partial\overline{u_1u_3}/\partial x_3$, and $\partial\overline{u_2u_1}/\partial x_1$ in Eq. (6-13) is small compared with the other turbulence terms.

Since at the most $\mathfrak{v}^2/\mathscr{V}_1^2$ may be of the order 1, it follows from Eq. (6-12) that $U_0/\bar{U}_1$ cannot be of a higher order of magnitude than L_1/L_2; otherwise all turbulence terms would be small compared with the first term on the left-hand side of Eq. (6-12). In other words, $U_0/\bar{U}_1$ must be of the same order of magnitude as L_1/L_2. But, if this is so, the first term on the left-hand side of Eq. (6-13) also remains small compared with the turbulence terms $\partial\overline{u_2^2}/\partial x_2$ and $\partial\overline{u_2u_3}/\partial x_3$. Hence the only other term that can balance these turbulence terms is the pressure term $\partial\bar{P}/\rho\,\partial x_2$. Thus $\Delta P_2/\rho\mathscr{V}_1^2$ is of the same order of magnitude as $\mathfrak{v}^2/\mathscr{V}_1^2$.

Consequently, if we retain only the terms of the highest order of magnitude, the equation of motion (6-13) may be reduced to

$$\frac{\partial\bar{P}}{\partial x_2}+\rho\frac{\partial}{\partial x_2}\overline{u_2^2}+\rho\frac{\partial}{\partial x_3}\overline{u_2u_3}=0 \tag{6-14}$$

or, after integration,

$$\bar{P}+\rho\overline{u_2^2}+\rho\int dx_2\frac{\partial}{\partial x_3}\overline{u_2u_3}=P_0 \tag{6-14a}$$

where P_0 is the pressure at the same section but outside the turbulence region.

From Eq. (6-14a) we obtain

$$\frac{1}{\rho}\frac{\partial\bar{P}}{\partial x_1}=\frac{1}{\rho}\frac{dP_0}{dx_1}-\frac{\partial}{\partial x_1}\overline{u_2^2}-\frac{\partial}{\partial x_1}\int dx_2\frac{\partial}{\partial x_3}\overline{u_2u_3} \tag{6-15}$$

But, since the velocity outside the turbulence region is equal to the constant velocity U_0, the pressure gradient $dP_0/dx_1=0$.

Furthermore, the turbulence terms are of the same order of magnitude as the turbulence term $\partial\overline{u_1^2}/\partial x_1$ in Eq. (6-12). Hence, if, in this equation also we retain only the terms of the highest order of magnitude, we obtain for the reduced Eq. (6-12)

$$U_0\frac{\partial\bar{U}_1}{\partial x_1}=-\frac{\partial}{\partial x_2}\overline{u_1u_2}-\frac{\partial}{\partial x_3}\overline{u_1u_3} \tag{6-16}$$

This equation can be solved if the functions $\overline{u_1u_2}$ and $\overline{u_1u_3}$ are known, or their relations to the distribution of $\bar{U}_1$. From Eq. (6-15) we can then calculate the pressure distribution.

The turbulence terms can be expressed in terms of the distribution of $\bar{U}_1$ by introducing an eddy-viscosity. With reference to Sec. 1-2, we may first consider the modified Boussinesq concept

$$-\overline{u_iu_j}=-\tfrac{1}{3}\overline{u_ku_k}\,\delta_{ij}+\epsilon_m\left(\frac{\partial\bar{U}_i}{\partial x_j}+\frac{\partial\bar{U}_j}{\partial x_i}\right)$$

Equation (6-16) then becomes

$$U_0\frac{\partial\bar{U}_1}{\partial x_1}=-\frac{\partial}{\partial x_2}\overline{u_1u_2}-\frac{\partial}{\partial x_3}\overline{u_1u_3}=\frac{\partial}{\partial x_2}\left(\epsilon_m\frac{\partial\bar{U}_1}{\partial x_2}\right)+\frac{\partial}{\partial x_3}\left(\epsilon_m\frac{\partial\bar{U}_1}{\partial x_3}\right) \tag{6-17}$$

If we assume a second order eddy-viscosity tensor $(\epsilon_m)_{il}$, the right-hand-side of Eq. (6-16) becomes identical to that of Eq. (6-17), namely

$$\frac{\partial}{\partial x_2}\left[(\epsilon_m)_{11}\frac{\partial\bar{U}_1}{\partial x_2}\right]+\frac{\partial}{\partial x_3}\left[(\epsilon_m)_{11}\frac{\partial\bar{U}_1}{\partial x_3}\right]$$

On the other hand if we were to assume a fourth-order tensor for the eddy-viscosity we would obtain

$$-\overline{u_1u_2} = [(\epsilon_m)_{1212} + (\epsilon_m)_{1221}]\frac{\partial \bar{U}_1}{\partial x_2} + [(\epsilon_m)_{1213} + (\epsilon_m)_{1231}]\frac{\partial \bar{U}_1}{\partial x_3}$$

$$-\overline{u_1u_3} = [(\epsilon_m)_{1312} + (\epsilon_m)_{1321}]\frac{\partial \bar{U}_1}{\partial x_2} + [(\epsilon_m)_{1313} + (\epsilon_m)_{1331}]\frac{\partial \bar{U}_1}{\partial x_3}$$

This leads to the same result as the other two assumptions in the case of a two-dimensional flow, so that one coefficient of the eddy-viscosity will then suffice. For a constant eddy-viscosity the Eq. (6-16) yields a Gaussian solution.

If the flow is two-dimensional, so that $\bar{U}_1$ is a function only of x_1 and x_2, another solution can be obtained if Prandtl's mixing-length theory is applied. The equation then reads

$$U_0\frac{\partial \bar{U}_1}{\partial x_1} = \mathfrak{l}_m{}^2\frac{\partial}{\partial x_2}\left(\frac{\partial \bar{U}_1}{\partial x_2}\right)^2 \tag{6-18}$$

A solution of this equation for the wake flow behind a circular cylinder will be given in the next section.

In the *second case* to be considered we shall not show a possible superimposed constant velocity U_0 explicitly in the equations of motion, since it is assumed that such a velocity, if it occurs, is small or of the same order of magnitude as $\bar{U}_1$, so that it does not affect the order-of-magnitude considerations.

$$\begin{array}{cccccccc}
\bar{U}_1\dfrac{\partial \bar{U}_1}{\partial x_1} & +\ \bar{U}_2\dfrac{\partial \bar{U}_1}{\partial x_2} & +\ \bar{U}_3\dfrac{\partial \bar{U}_1}{\partial x_3} & = -\dfrac{1}{\rho}\dfrac{\partial \bar{P}}{\partial x_1} & -\ \dfrac{\partial}{\partial x_1}\overline{u_1{}^2} & -\ \dfrac{\partial}{\partial x_2}\overline{u_1u_2} & -\ \dfrac{\partial}{\partial x_3}\overline{u_1u_3} & \\[2ex]
\dfrac{\mathscr{V}_1{}^2}{L_1} & \dfrac{\mathscr{V}_1{}^2}{L_1} & \dfrac{\mathscr{V}_1{}^2}{L_1} & \dfrac{\Delta P_1}{\rho L_1} & \dfrac{\mathfrak{v}^2}{L_1} & \mathbf{R}_{12}\dfrac{\mathfrak{v}^2}{L_2} & \mathbf{R}_{13}\dfrac{\mathfrak{v}^2}{L_3} & (6\text{-}19) \\[2ex]
\text{or}\quad 1 & 1 & 1 & \dfrac{\Delta P_1}{\rho\mathscr{V}_1{}^2} & \dfrac{\mathfrak{v}^2}{\mathscr{V}_1{}^2} & \mathbf{R}_{12}\dfrac{\mathfrak{v}^2}{\mathscr{V}_1{}^2}\dfrac{L_1}{L_2} & \mathbf{R}_{13}\dfrac{\mathfrak{v}^2}{\mathscr{V}_1{}^2}\dfrac{L_1}{L_3} &
\end{array}$$

From this relation we conclude immediately that $\mathfrak{v}^2/\mathscr{V}_1{}^2$ can be at the most of the order of magnitude L_2/L_1, that is, one order of magnitude smaller than in the previous case.

The second equation of motion shows the following:

$$\begin{array}{cccccccc}
\bar{U}_1\dfrac{\partial \bar{U}_2}{\partial x_1} & +\ \bar{U}_2\dfrac{\partial \bar{U}_2}{\partial x_2} & +\ \bar{U}_3\dfrac{\partial \bar{U}_2}{\partial x_3} & = -\dfrac{1}{\rho}\dfrac{\partial \bar{P}}{\partial x_2} & -\ \dfrac{\partial}{\partial x_1}\overline{u_2u_1} & -\ \dfrac{\partial}{\partial x_2}\overline{u_2{}^2} & -\ \dfrac{\partial}{\partial x_3}\overline{u_2u_3} & \\[2ex]
\dfrac{\mathscr{V}_1{}^2L_2}{L_1{}^2} & \dfrac{\mathscr{V}_1{}^2L_2}{L_1{}^2} & \dfrac{\mathscr{V}_1{}^2L_2}{L_1{}^2} & \dfrac{\Delta P_2}{\rho L_2} & \mathbf{R}_{12}\dfrac{\mathfrak{v}^2}{L_1} & \dfrac{\mathfrak{v}^2}{L_2} & \mathbf{R}_{23}\dfrac{\mathfrak{v}^2}{L_3} & (6\text{-}20) \\[2ex]
\text{or}\quad 1 & 1 & 1 & \dfrac{\Delta P_2}{\rho\mathscr{V}_1{}^2}\dfrac{L_1{}^2}{L_2{}^2} & \mathbf{R}_{12}\dfrac{\mathfrak{v}^2}{\mathscr{V}_1{}^2}\dfrac{L_1}{L_2} & \dfrac{\mathfrak{v}^2}{\mathscr{V}_1{}^2}\dfrac{L_1{}^2}{L_2{}^2} & \mathbf{R}_{23}\dfrac{\mathfrak{v}^2}{\mathscr{V}_1{}^2}\dfrac{L_1{}^2}{L_3{}^2} &
\end{array}$$

Since by Eq. (6-19) $\overline{\mathfrak{v}^2}/\mathscr{V}_1^2$ is of the order of magnitude L_2/L_1, the turbulence terms $\partial\overline{u_2^2}/\partial x_2$ and $\partial\overline{u_2 u_3}/\partial x_3$ in Eq. (6-20) must be large compared with the velocity terms on the left-hand side. And the pressure term $\Delta P_2/\rho\mathscr{V}_1^2$ must be of the same order of magnitude.

Thus Eq. (6-20) reduces to Eq. (6-14), and Eq. (6-18) to

$$\bar{U}_1\frac{\partial\bar{U}_1}{\partial x_1}+\bar{U}_2\frac{\partial\bar{U}_1}{\partial x_2}+\bar{U}_3\frac{\partial\bar{U}_1}{\partial x_3}=-\frac{1}{\rho}\frac{dP_0}{dx_1}-\frac{\partial}{\partial x_2}\overline{u_1u_2}-\frac{\partial}{\partial x_3}\overline{u_1u_3} \tag{6-21}$$

Here, too, we may assume a transport of momentum due to a gradient type of diffusion and introduce an eddy-diffusion coefficient. The equation then reads

$$\bar{U}_1\frac{\partial\bar{U}_1}{\partial x_1}+\bar{U}_2\frac{\partial\bar{U}_1}{\partial x_2}+\bar{U}_3\frac{\partial\bar{U}_1}{\partial x_3}=-\frac{1}{\rho}\frac{dP_0}{dx_1}+\frac{\partial}{\partial x_2}\left[\epsilon_m\frac{\partial\bar{U}_1}{\partial x_2}\right]+\frac{\partial}{\partial x_3}\left[\epsilon_m\frac{\partial\bar{U}_1}{\partial x_3}\right] \tag{6-22}$$

For the two-dimensional case solutions may be obtained for

$$\epsilon_m=\text{const.}$$

and for

$$\epsilon_m=\mathrm{l}_m^2\,\partial\bar{U}_1/\partial x_2$$

The same applies to an axi-symmetric flow in the z-direction. The equation for the main-flow-velocity component $\bar{U}_z$ in this case reads

$$\bar{U}_z\frac{\partial\bar{U}_z}{\partial z}+\bar{U}_r\frac{\partial\bar{U}_z}{\partial r}=-\frac{1}{\rho}\frac{dP_0}{dz}+\frac{1}{r}\frac{\partial}{\partial r}\left[\epsilon_m r\frac{\partial\bar{U}_z}{\partial r}\right] \tag{6-23}$$

and the pressure distribution across the turbulent-flow region follows from

$$\frac{1}{\rho}\frac{\partial\bar{P}}{\partial r}+\frac{\partial}{\partial r}\overline{u_r^2}+\frac{\overline{u_r^2}-\overline{u_\varphi^2}}{r}=0 \tag{6-24}$$

or, after integration,

$$\bar{P}+\rho\overline{u_r^2}+\rho\int dr\frac{\overline{u_r^2}-\overline{u_\varphi^2}}{r}=P_0 \tag{6-24a}$$

Finally we note that the various terms in Eqs. (6-19) and (6-20) differ at the most by one order of magnitude. Hence, if the approximations underlying the Eq. (6-21) are too rough, we have to consider the full equation (6-19), this being the next possible step. Since in Eq. (6-19) we now retain the turbulence term $\partial\overline{u_1^2}/\partial x_1$, we cannot neglect the turbulence terms in the relation (6-15). Hence, instead of Eq. (6-21) we then obtain

$$\bar{U}_1\frac{\partial\bar{U}_1}{\partial x_1}+\bar{U}_2\frac{\partial\bar{U}_1}{\partial x_2}+\bar{U}_3\frac{\partial\bar{U}_1}{\partial x_3}=-\frac{1}{\rho}\frac{dP_0}{dx_1}$$

$$-\frac{\partial}{\partial x_1}\left(\overline{u_1^2}-\overline{u_2^2}-\int dx_2\frac{\partial}{\partial x_3}\overline{u_2u_3}\right)-\frac{\partial}{\partial x_2}\overline{u_1u_2}-\frac{\partial}{\partial x_3}\overline{u_1u_3} \tag{6-25}$$

6-4 VELOCITY DISTRIBUTION BEHIND A CYLINDER ACCORDING TO THE CLASSICAL THEORIES

In this section we shall consider the distributions of the mean velocity obtained according to the classical theories of Boussinesq, Prandtl, and Taylor. In these theories use is made of the assumption of similarity between velocity profiles in subsequent sections beyond a certain distance from the cylinder.

Let U_0 be the velocity of the undisturbed oncoming stream.

The origin of the coordinate system is set at the center of the cylinder, the x_1-axis in the direction of the undisturbed flow U_0. The x_2-axis is assumed perpendicular to the axis of the cylinder (see Fig. 6-1).

The deviation from the velocity U_0 caused by the cylinder is U_1, and $\bar{U}_1$ is its mean value, while we put $U_1 = \bar{U}_1 - u_1$.

With the approximation shown in the previous section, the equation of motion in the x_1-direction reads

$$U_0 \frac{\partial \bar{U}_1}{\partial x_1} = -\frac{\partial}{\partial x_2}(-\overline{u_2 u_1}) = -\frac{\partial}{\partial x_2}\frac{\sigma_{21}}{\rho}$$

or (6-26)

$$\frac{\partial}{\partial x_1}\frac{\bar{U}_1}{U_0} = -\frac{\partial}{\partial x_2}\frac{-\overline{u_2 u_1}}{U_0^2} = -\frac{\partial}{\partial x_2}\frac{\sigma_{21}}{\rho U_0^2}$$

On the ground of the assumed similarity we will make use of the Eqs. (6-3a), (6-3e), and (6-3g). We choose the free-stream velocity U_0 as the characteristic velocity, and the diameter d as the characteristic length. Since the origin of similarity will, in general, not coincide with the center of the cylinder (i.e. with the origin of the coordinate system) in the Eqs. (6-3a), (6-3e), and (6-3g) we have to read $x_1 + a$ for x_1 (see Fig. 6-1).

After substitution of the latter relations in Eq. (6-26) we obtain

$$\frac{1}{d}\left(\frac{x_1 + a}{d}\right)^{p-1}\left[pf(\xi_2) - q\xi_2 \frac{df}{d\xi_2}\right] = -\left(\frac{x_1 + a}{d}\right)^{s_{21}-q} \frac{1}{d}\frac{dh}{d\xi_2}$$

Whence follows the relation

$$p + q - s_{21} = 1 \qquad (6\text{-}27)$$

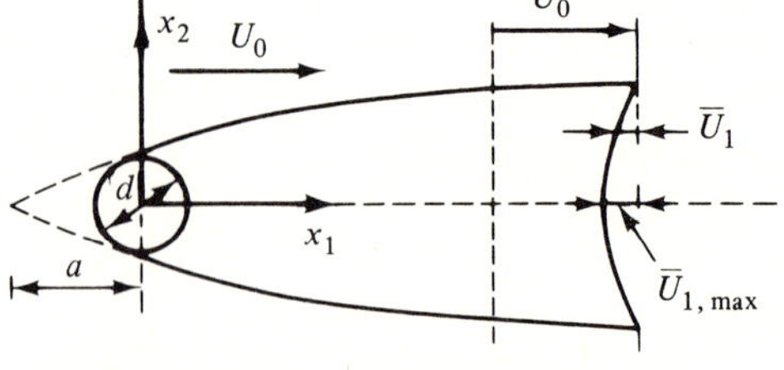

FIGURE 6-1
Scheme of wake flow behind a cylinder.

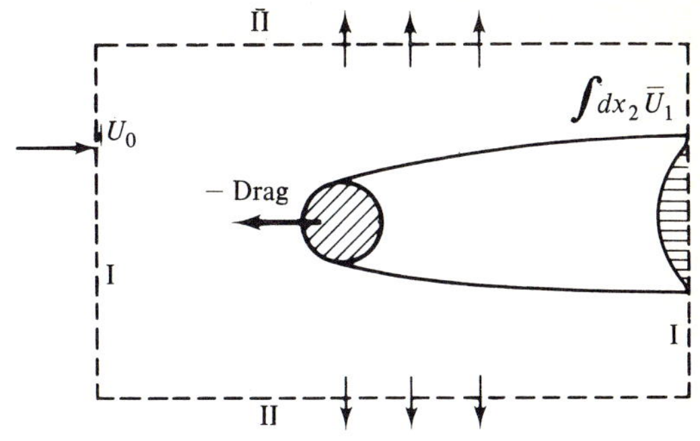

FIGURE 6-2
Imaginary control planes for determining the balance of momentum flux.

Consequently we have only one relation for the three unknowns p, q, and s_{21}. This is the penalty for having accepted the boundary-layer approximation, because by dropping a number of terms we lost at the same time a number of relations between p, q, and s_{21}. So we have to look for other conditions. One possibility is to consider boundary conditions. Since $\sigma_{21} = 0$ at $x_2 = +\infty$ and $x_2 = -\infty$, integration of Eq. (6-26) yields

$$-\frac{d}{dx_1}\int_{-\infty}^{+\infty} dx_2 \frac{\bar{U}_1}{U_0} = \left.\frac{\sigma_{12}}{\rho U_0^2}\right|_{-\infty}^{+\infty} = 0$$

whence

$$\int_{-\infty}^{+\infty} dx_2 \frac{\bar{U}_1}{U_0} = \text{const.}$$

This constant value appears to be directly related to the drag of the cylinder. This follows from a momentum-flux balance for the flow around the cylinder.

In order to determine this balance we assume a set of imaginary control planes at large distances from the cylinder (see Fig. 6-2). Through the set of planes I there is a flux of momentum in the x_1-direction.

$$\int_{-\infty}^{+\infty} dx_2\, \rho[U_0^2 - \overline{(U_0 - U_1)^2}] = \int_{-\infty}^{+\infty} dx_2\, \rho[2U_0\bar{U}_1 - \overline{U_1^2}]$$

$$\simeq 2\rho U_0 \int_{-\infty}^{+\infty} dx_2\, \bar{U}_1$$

Through the control planes I there is a defect in mass flow equal to $\rho \int_{-\infty}^{+\infty} dx_2\, \bar{U}_1$. This is balanced by an efflux to an equal amount through the set of control planes II, causing a loss in momentum flux in the x_1-direction

$$-\rho U_0 \int_{-\infty}^{+\infty} dx_2\, \bar{U}_1$$

The sum of these momentum fluxes must be equal to the drag of the cylinder.

Hence, since at sufficiently large distances $\overline{U_1^2}/\bar{U}_1 U_0 \ll 1$,

$$\frac{\text{Drag}}{\rho U_0{}^2 d} = \frac{1}{d}\int_{-\infty}^{+\infty} dx_2 \frac{\bar{U}_1}{U_0}$$

Again, with the relation (6-3a) and (6-3g), we obtain

$$\frac{\text{Drag}}{\rho U_0{}^2 d} = \left(\frac{x_1 + a}{d}\right)^{p+q} \int_{-\infty}^{+\infty} d\xi_2\, f(\xi_2)$$

Whence follows the relation

$$p + q = 0 \qquad (6\text{-}28)$$

Hence, with Eq. (6-27), we have $s_{21} = -1$.

Still we need another relation. Other assumptions which have been shown to yield useful relations are: (1) the assumption of similarity in downstream sections between the turbulence and the mean-velocity field, so that they have the same dependence on x_1, and (2) the assumption of an eddy-viscosity ϵ_m independent of x_1.

Let us first consider the *first assumption*, according to which $-\overline{u_2 u_1}/\bar{U}_1{}^2$ is independent of $(x_1 + a)$. Since

$$-\frac{\overline{u_2 u_1}}{\bar{U}_1{}^2} = \left(\frac{x_1 + a}{d}\right)^{s_{21} - 2p} \frac{h(\xi_2)}{f^2(\xi_2)}$$

there follows the relation

$$s_{21} - 2p = 0 \qquad (6\text{-}29)$$

From Eqs. (6-27), (6-28), and (6-29) we thus obtain

$$p = -\tfrac{1}{2} \qquad q = \tfrac{1}{2} \qquad s_{21} = -1 \qquad (6\text{-}30)$$

Comparison with the values of Eq. (6-8), shows that there is no complete similarity, though according to the assumption the turbulence shear-stress and the mean-velocity component $\bar{U}_1$ have the same velocity scale. However, as will be shown later, the mean-velocity component $\bar{U}_2$ has a different velocity scale.

With $p = -\frac{1}{2}$, $q = \frac{1}{2}$ and $s_{21} = -1$, we obtain

$$\xi_2 = \frac{x_2}{d}\sqrt{\frac{d}{x_1 + a}} = \frac{x_2}{\sqrt{d(x_1 + a)}} \qquad (6\text{-}31)$$

$$\frac{\bar{U}_1}{U_0} = \left(\frac{x_1 + a}{d}\right)^{-1/2} f(\xi_2) \qquad (6\text{-}32)$$

while the momentum-balance equation becomes

$$f(\xi_2) + \xi_2 \frac{df}{d\xi_2} = 2\frac{dh}{d\xi_2}$$

Since the left-hand-side can be written $d/d\xi_2(\xi_2 f)$, the equation can be readily integrated to give

$$h(\xi_2) = \tfrac{1}{2}\xi_2 f(\xi_2)$$

The constant of integration is zero, since the shear stress is zero at $\xi_2 = 0$, because of symmetry. The shear-stress distribution expressed in terms of the distribution of $\bar{U}_1$ reads

$$\frac{\sigma_{21}}{\rho U_0^2} = \left(\frac{x_1 + a}{d}\right)^{-1} h(\xi_2) = \left(\frac{x_1 + a}{d}\right)^{-1} \tfrac{1}{2}\xi_2 f(\xi_2) = \frac{1}{2}\frac{x_2}{x_1 + a}\frac{\bar{U}_1}{U_0} \tag{6-33}$$

From this we may deduce an expression for the eddy-viscosity ϵ_m, if we want to express the shear stress in terms of the mean-velocity gradient. Since according to Eqs. (6-17) and (6-26)

$$\frac{\sigma_{21}}{\rho U_0^2} = -\frac{\epsilon_m}{U_0^2}\frac{\partial \bar{U}_1}{\partial x_2} = -\frac{\epsilon_m}{U_0 d}\left(\frac{x_1 + a}{d}\right)^{-1}\frac{df}{d\xi_2}$$

we obtain with Eq. (6-33):

$$\frac{\epsilon_m}{U_0 d} = -\frac{1}{2}\frac{\xi_2 f(\xi_2)}{df/d\xi_2} = -\frac{1}{2}\frac{\xi_2 \bar{U}_1/U_0}{d/d\xi_2(\bar{U}_1/U_0)} \tag{6-34}$$

This result shows that the eddy-viscosity is independent of x_1. Consequently the *second, alternative assumption* of an eddy viscosity independent of x_1 is for the present case already included in the first assumption. With the aid of the continuity equation the mean-velocity component $\bar{U}_2$ can be obtained. Since $\bar{U}_2 = 0$ at $x_2 = 0$, $\bar{U}_2$ is determined by

$$\frac{\bar{U}_2}{U_0} = -\int_0^{x_2} dx_2 \frac{\partial}{\partial x_1}\frac{\bar{U}_1}{U_0}$$

With Eqs. (6-31) and (6-32):

$$\frac{\bar{U}_2}{U_0} = -\frac{1}{2}\left(\frac{x_1 + a}{d}\right)^{-1}\int_0^{\xi_2} d\xi_2\left[f(\xi_2) + \xi_2\frac{df}{d\xi_2}\right] = -\frac{1}{2}\left(\frac{x_1 + a}{d}\right)^{-1}\xi_2 f(\xi_2) \tag{6-35}$$

From this result we conclude that (1) the distribution of the $\bar{U}_2$-component is the same as the shear-stress distribution [see Eq. (6-33)]; (2) $\bar{U}_2$ has a velocity scale proportional to $(x_1 + a)^{-1}$, and therefore different from that for $\bar{U}_1$.

Integration of Eq. (6-34) yields

$$\frac{\bar{U}_1}{\bar{U}_{1,\max}} = \exp\left[-\frac{U_0 d}{2}\int_0^{\xi_2}\frac{\xi_2 d\xi_2}{\epsilon_m}\right] \tag{6-36}$$

The integral in Eq. (6-36) can be solved and the velocity distribution obtained if some assumption is made concerning ϵ_m. Let us first consider the simplest one, namely that it is constant. The integration then yields

$$\frac{\bar{U}_1}{\bar{U}_{1,\max}} = \exp\left[-\frac{U_0 d}{4\epsilon_m}\xi_2^2\right] \tag{6-36a}$$

The maximum velocity $\bar{U}_{1,\max}$ is still a function of $(x_1 + a)$. It varies, according to its velocity scale, proportionally to $(x_1 + a)^{-1/2}$. Thus from Eq. (6-32), writing A for $f(0)$:

$$\frac{\bar{U}_{1,\max}}{U_0} = A\left(\frac{x_1 + a}{d}\right)^{-1/2} \tag{6-37}$$

In the *mixing-length theories* we may apply either Prandtl's hypothesis that momentum is the transferable quantity to be considered or Taylor's hypothesis that vorticity is the transferable quantity. Now we have shown in Chap. 5 that in a two-dimensional flow the two hypotheses lead to the same result, provided the mixing length for the transfer of vorticity l_ω is $\sqrt{2}$ times the mixing length for momentum transfer.

Let us assume here that momentum is the transferable quantity. According to the mixing-length theory, we then have [see Eq. (5-14)]

$$\epsilon_m = l_m{}^2 \left|\frac{\partial \bar{U}_1}{\partial x_2}\right|$$

whence

$$\frac{\epsilon_m}{U_0 d} = \frac{l_m{}^2}{d(x_1 + a)} \frac{df}{d\xi_2}$$

Since ϵ_m is independent of x_1 and the mixing length l_m is assumed to be a function of x_1 alone, it follows that $l_m = c_m\sqrt{d(x_1 + a)}$.

Substitution of the latter expression for $\epsilon_m/U_0 d$ in Eq. (6-34) yields a differential equation for $f(\xi)$ which can be readily integrated. The result reads

$$f(\xi_2) = \frac{(\xi_2)_0{}^3}{18c_m{}^2}\left\{1 - \left[\frac{|\xi_2|}{(\xi_2)_0}\right]^{3/2}\right\}^2$$

Here $(\xi_2)_0$ is the value of ξ_2 at which $\bar{U}_1$, and consequently $f(\xi_2)$, becomes zero. Hence this solution shows a finite value for the width of the wake in contrast to the Gaussian distribution given by Eq. (6-36a). For $\bar{U}_1/\bar{U}_{1,\max}$ we obtain, again with $f(0) = A$,

$$\frac{\bar{U}_1}{\bar{U}_{1,\max}} = \frac{(\xi_2)_0{}^3}{18c_m{}^2 A}\left\{1 - \left[\frac{|\xi_2|}{(\xi_2)_0}\right]^{3/2}\right\}^2 \tag{6-38}$$

Since $\bar{U}_1/\bar{U}_{1,\max} = 1$ for $\xi_2 = 0$, the following relation among $(\xi_2)_0$, A, and c_m must hold:

$$(\xi_2)_0{}^3 = 18c_m{}^2 A$$

Thus the final expression for the velocity distribution becomes

$$\frac{\bar{U}_1}{\bar{U}_{1,\max}} = \left\{1 - \left[\frac{|\xi_2|}{(\xi_2)_0}\right]^{3/2}\right\}^2 \tag{6-39}$$

For the axial distribution of $\bar{U}_{1,\max}$ we obtain

$$\frac{\bar{U}_{1,\max}}{U_0} = A\sqrt{\frac{d}{x_1 + a}} = \frac{(\xi_2)_0{}^3}{18c_m{}^2}\sqrt{\frac{d}{x_1 + a}} \tag{6-40}$$

In principle it is possible to determine the value of $(\xi_2)_0$ from measured velocity distributions, namely, from the points where $\bar{U}_1 = 0$. From the velocity distribution in the axial direction we may find the value of c_m by matching the theoretical and measured velocity distributions along the axis of the wake. It is also possible to determine the value of c_m from the drag of the cylinder, because

$$\frac{\text{Drag}}{\rho U_0{}^2 d} = \frac{(\xi_2)_0{}^3}{18c_m{}^2}\int_{-(\xi_2)_0}^{+(\xi_2)_0} d\xi_2 \left\{1 - \left[\frac{|\xi_2|}{(\xi_2)_0}\right]^{3/2}\right\}^2 = \frac{(\xi_2)_0{}^4}{20c_m{}^2}$$

Because of the low accuracy of the velocity measurements near the boundaries of the wake, the experimental data show great scatter in this region. It is, therefore, very difficult to determine the value of $(\xi_2)_0$ accurately. For that reason one prefers to determine this value from the half-value width $(\xi_2)_{1/2}$, that is, the value of ξ_2 at which $\bar{U}_1/\bar{U}_{1,\max} = 0.5$. For this half-value width Eq. (6-39) gives

$$\frac{(\xi_2)_{1/2}}{(\xi_2)_0} = 0.441$$

Though *Reichardt's inductive theory* is not yet considered a classical theory, yet we will discuss briefly the results obtained with this theory for the wake flow. According to this theory, the momentum flux $\rho\overline{(U_0 + U_1)^2}$ satisfies the diffusion equation (5-29). Since

$$\overline{(U_0 + U_1)^2} = U_0{}^2 + 2U_0\bar{U}_1 + \bar{U}_1{}^2 + \overline{u_1{}^2}$$

and $U_0{}^2$ has a constant value, Reichardt[1] considers instead of $\overline{(U_0 + U_1)^2}$ the quantity

$$\overline{\mathfrak{B}^2} = 2U_0\bar{U}_1 + \bar{U}_1{}^2 + \overline{u_1{}^2} \tag{6-41}$$

The diffusion equation thus reads

$$\frac{\partial}{\partial x_1}\overline{\mathfrak{B}^2} = \mathscr{L}_m \frac{\partial^2}{\partial x_2{}^2}\overline{\mathfrak{B}^2} \tag{6-42}$$

Since earlier we introduced with success the assumption of self-preservation of the turbulence and the $\bar{U}_1$-distribution in downstream direction, we assume here that the $\overline{\mathfrak{B}^2}$ profiles remain similar in subsequent sections. Now the quantity $\overline{\mathfrak{B}^2}$ contains the velocity $\bar{U}_1$, the square of it, and $\overline{u_1{}^2}$. According to the earlier results $\bar{U}_1$ varies with $(x_1 + a)^{-1/2}$, however $\bar{U}_1{}^2$ and $\overline{u_1{}^2}$ vary with $(x_1 + a)^{-1}$. According to the assumption made that $\bar{U}_1/U_0 \ll 1$, the term $2U_0\bar{U}_1$ is the leading term in Eq. (6-41). Therefore we put

$$\frac{\overline{\mathfrak{B}^2}}{U_0{}^2} = \left(\frac{x_1 + a}{d}\right)^{-1/2} f_1(\xi_2) \tag{6-43}$$

where ξ_2 is defined by Eq. (6-31).

Substitution in Eq. (6-42) yields:

$$\frac{\mathscr{L}_m}{d}\frac{d^2f_1}{d\xi_2^2}+\frac{\xi_2}{2}\frac{df_1}{d\xi_2}+\tfrac{1}{2}f_1(\xi_2)=0 \qquad (6\text{-}44)$$

Thus similarity requires that the length parameter $\mathscr{L}_m$ be independent of $(x_1 + a)$, i.e., constant in the whole turbulence region.

The reader can easily check that

$$f_1 = \exp\left(-\frac{d}{4\mathscr{L}_m}\xi_2^2\right)$$

is a solution of the differential equation (6-44). Hence,

$$\frac{\overline{\mathfrak{B}^2}}{U_0^2} = A_1\sqrt{\frac{d}{x_1+a}}\exp\left(-\frac{d}{4\mathscr{L}_m}\xi_2^2\right) \qquad (6\text{-}45)$$

If we compare this solution with the other Gaussian solution (6-36a), there seems to be a discrepancy: in the solution (6-36a) it is $\bar{U}_{1,\max}/U_0$ that is proportional to $\sqrt{d/(x_1+a)}$ but here it is $\overline{\mathfrak{B}_{\max}^2}/U_0^2$. But

$$\frac{\overline{\mathfrak{B}^2}}{U_0^2} = 2\frac{\bar{U}_1}{U_0}+\frac{\bar{U}_1^2}{U_0^2}+\frac{\overline{u_1^2}}{U_0^2}$$

and, since it has been assumed that $\bar{U}_1/U_0 \ll 1$, the quadratic terms on the right-hand side may be neglected. Thus in fact the same solution is obtained, namely,

$$\frac{\bar{U}_1}{U_0} = \tfrac{1}{2}A_1\sqrt{\frac{d}{x_1+a}}\exp\left(-\frac{d}{4\mathscr{L}_m}\xi_2^2\right) \qquad (6\text{-}45a)$$

Comparison with the expressions (6-36a) and (6-37) yields the formal relations

$$A_1 = 2A \qquad \text{and} \qquad U_0\mathscr{L}_m = \epsilon_m$$

Axi-symmetric Wake

It may be of interest, if only for the sake of comparison, to give here the results of the same similarity considerations for the axi-symmetric case. Again on the assumption of large Reynolds number and small values of the velocity defect $\bar{U}_z \ll U_0$, the momentum balance equations may be simplified accordingly. We assume a cylindrical coordinate system (z,r). The equation for the axial-momentum balance yields again one equation, viz. Eq. (6-27) for the parameters p, q, and s_{21}. On the other hand the axial momentum integral which must be equal to the total drag, yields a relation between p and q different from Eq. (6-28), namely

$$p + 2q = 0$$

The assumption of independence of $-\overline{u_r u_z}/\bar{U}_z^2$ of $(z+a)$, yields $s_{21} = 2p$. With this assumption we obtain

$$p = -\tfrac{2}{3} \qquad q = \tfrac{1}{3} \qquad s_{21} = -\tfrac{4}{3}$$

while the equation of continuity yields $r = -4/3$ for the dependence of the radial mean velocity component on $(z + a)$. There is only *incomplete similarity*. The eddy viscosity does not remain independent of $(z + a)$, as in the plane wake, but varies according to $(z + a)^{-1/3}$. Hence the alternative assumption of an eddy viscosity independent of $(z + a)$ will result in different values of p, q, and s_{21}, namely

$$p = -1 \qquad q = \tfrac{1}{2} \qquad s_{21} = -\tfrac{3}{2}$$

While for the dependence of the radial mean-velocity component on $(z + a)$ we obtain $r = -\frac{3}{2}$. Consequently again there is *incomplete similarity*. At this stage of our considerations of wake flows it cannot yet be decided which of the two assumptions is to be preferred. We will come back to this subject in Sec. 6-6.

6-5 THE TRANSPORT OF A SCALAR QUANTITY IN THE WAKE FLOW OF A CYLINDER

If we apply the same approximations to Eq. (1-33) for the transport of the scalar quantity Γ, we obtain

$$U_0 \frac{\partial \bar{\Gamma}}{\partial x_1} = -\frac{\partial}{\partial x_2} \overline{u_2 \gamma} \qquad \text{(6-46)}$$

Here too we will assume that the turbulence transport of γ can be described by means of an eddy-diffusion coefficient. We may either take the relation (1-34), or the relation (1-35). However, since to the same approximation we need only one component $(\epsilon_\gamma)_{22}$ of the diffusion tensor Eq. (1-35), we may for reasons of simplicity just write ϵ_γ. So we put

$$-\overline{u_2 \gamma} = \epsilon_\gamma \frac{\partial \bar{\Gamma}}{\partial x_2}$$

and, consequently,

$$U_0 \frac{\partial \bar{\Gamma}}{\partial x_1} = \frac{\partial}{\partial x_2} \left[\epsilon_\gamma \frac{\partial \bar{\Gamma}}{\partial x_2} \right] \qquad \text{(6-47)}$$

A similar equation would have been obtained for $\bar{U}_1$ if in Eq. (6-26) we had assumed

$$\sigma_{21} = \rho \epsilon_m \frac{\partial \bar{U}_1}{\partial x_2}$$

Consequently, if the $\bar{\Gamma}$-profiles are also assumed to be similar, a relation similar to the relation (6-34) is obtained for $\epsilon_\gamma / U_0 d$, namely,

$$\frac{\epsilon_\gamma}{U_0 d} = -\frac{1}{2} \frac{\xi_2 \bar{\Gamma}/\bar{\Gamma}_{\max}}{d/d\xi_2 (\bar{\Gamma}/\bar{\Gamma}_{\max})}$$

The integration of this equation yields

$$\frac{\bar{\Gamma}}{\bar{\Gamma}_{\max}} = \exp\left[-\frac{U_0 d}{2} \int_0^{\xi_2} \frac{\xi_2 \, d\xi_2}{\epsilon_\gamma} \right] \qquad \text{(6-49)}$$

For the simplest case, namely a constant value of ϵ_γ, the Gaussian solution is again obtained:

$$\frac{\bar{\Gamma}}{\bar{\Gamma}_{max}} = \exp\left[-\frac{U_0 d}{4\epsilon_\gamma}\xi_2{}^2\right] \tag{6-49a}$$

From the solutions (6-36a) and (6-49a) it is concluded that

$$\frac{\bar{\Gamma}}{\bar{\Gamma}_{max}} = \left(\frac{\bar{U}_1}{\bar{U}_{1,max}}\right)^{\epsilon_m/\epsilon_\gamma} \tag{6-50}$$

However, this result appears to be more general. It is also obtained from the relations (6-36) and (6-49) if it is assumed that the ratio $\epsilon_m/\epsilon_\gamma$ is constant throughout the wake-flow region, so that ϵ_m and ϵ_γ separately need not be constant.

If we apply to ϵ_γ *Prandtl's mixing-length hypothesis,* we obtain

$$\epsilon_\gamma = \mathrm{l}_\gamma{}^2\left|\frac{\partial \bar{U}_1}{\partial x_2}\right| \tag{6-51}$$

With the velocity distribution (6-39), according to the momentum-transport theory and with $\mathrm{l}_\gamma = c_\gamma\sqrt{d(x_1 + a)}$, the expression for ϵ_γ becomes

$$\frac{\epsilon_\gamma}{U_0 d} = \frac{1}{6}\frac{c_\gamma{}^2}{c_m{}^2}(\xi_2)_0{}^2\left\{1 - \left[\frac{|\xi_2|}{(\xi_2)_0}\right]^{3/2}\right\}\left[\frac{|\xi_2|}{(\xi_2)_0}\right]^{1/2} \tag{6-52}$$

Substituting in the relation (6-49) and carrying out the integration, we obtain

$$\frac{\bar{\Gamma}}{\bar{\Gamma}_{max}} = \left\{1 - \left[\frac{|\xi_2|}{(\xi_2)_0}\right]^{3/2}\right\}^{2c_m{}^2/c_\gamma{}^2} \tag{6-53}$$

but, since $\epsilon_m/\epsilon_\gamma = \mathrm{l}_m{}^2/\mathrm{l}_\gamma{}^2 = c_m{}^2/c_\gamma{}^2 = \text{const.}$, this result had been found immediately from the relations (6-50) and (6-39).

In the earlier literature (see, e.g., Ref. 2) it has often been assumed that $\mathrm{l}_\gamma = \mathrm{l}_m$, which, however, is not true a priori; if it were the $\bar{\Gamma}$-distribution would become identical with the velocity distribution. Because the experiments all showed a greater spread of the temperature than of the velocity in the warm wake of a heated cylinder, the momentum-transport theory was rejected. This rejection is not justified by these arguments for the experimental evidence showed only that the assumption $\mathrm{l}_\gamma = \mathrm{l}_m$ must be incorrect.

In *Reichardt's inductive theory* we have here to consider the transport of $(U_0 + U_1)\Gamma$. But, with the same degree of approximation assumed in the other theories, we have only to take account of $U_0\bar{\Gamma}$, so that the diffusion equation (5-32) becomes

$$\frac{\partial \bar{\Gamma}}{\partial x_1} = \mathscr{L}_\gamma \frac{\partial^2 \bar{\Gamma}}{\partial x_2{}^2} \tag{6-54}$$

The solution of this equation, based upon the assumption of similarity of $\bar{\Gamma}$-profiles, reads

$$\frac{\bar{\Gamma}}{\bar{\Gamma}_{max}} = \exp\left(-\frac{d}{4\mathscr{L}_\gamma}\xi_2^2\right) \tag{6-55}$$

Comparison with the solution (6-49a) yields the formal relation

$$U_0\mathscr{L}_\gamma = \epsilon_\gamma$$

6-6 MEASUREMENTS OF MEAN-VELOCITY AND MEAN-TEMPERATURE DISTRIBUTION IN THE WAKE FLOW OF A CYLINDER

In the theories concerning the distribution of the mean velocity discussed in the previous section, similarity of the distribution profiles in consecutive sections of the wake has been assumed. Also it has been assumed that the Reynolds number in the whole flow region is sufficiently large to allow the neglect of the direct viscous effects. Hence, when comparing the measured distributions with the theoretical distributions, we have to investigate first whether the actual wake-flow considered experimentally does meet with these assumptions. It is known that at sufficiently high Reynolds numbers the drag coefficient of a cylinder becomes independent of this number over a wide range of values. Since at some distance from the cylinder the large-scale structure of the turbulent flow is nearly wholly determined by the work done by the drag force, it is reasonable to expect that at these high Reynolds numbers this structure becomes independent of the Reynolds number, so that to this structure the so-called *Reynolds-number-similarity* applies.

Early measurements on the distribution of mean velocity have been made by Schlichting,[3] by Fage and Falkner,[4] and by Townsend[5] at various distances from the cylinder. Townsend has found practically no effect of the Reynolds number on the mean-velocity distribution for $\mathbf{Re}_d > 800$.

As to the similarity of the flow pattern in consecutive sections, it has been found that self-preservation of the over-all structure, as shown by the mean-velocity distribution and by the increase of the width of the wake with distance from the cylinder, occurs at distances beyond

$$\xi_1 = \frac{x_1 + a}{d} > 90$$

but that the detail turbulence structure requires much longer distances $\xi_1 = 500$ to 1000 to become self-preserving. In particular this applies to the most fine-scale structure of the turbulence, i.e., to the viscous wavenumber range of the energy spectrum. At these large distances the wake may have entered the final stage of decay, depending on the Reynolds number.

We can apply the similarity considerations also to the turbulence-energy balance equation after having subjected it to the same boundary-layer approximations as for the momentum-balance equation. Writing the dissipation term $\varepsilon =$

const. $\nu u'^2/\lambda^2$, and assuming similarity between u' and $\bar{U}_1$ it can be shown that similarity may also exist for the dissipation length-scale, so that $\lambda \propto [d(x_1 + a)]^{1/2}$. Indeed, measurements by Uberoi and Freymuth[11] of energy spectra of the three velocity components have shown that at sufficiently high Reynolds numbers the spectra scale satisfactorily well with $[d(x_1 + a)]^{1/2}$ even up to the dissipation range. We will come back to their measurements in Sec. 6-7.

If we compare Townsend's[5,6] experimental results for values of $\xi_1 > 500$ with the theoretical distributions according to the "classical" theories, it turns out that the momentum-transport theory (and, consequently, also, the vorticity-transport theory, which gives the same solution) gives a solution in reasonable agreement with the measured distribution, except in the neighborhood of the axis and near the wake boundary. Figure 6-3 shows the measured and computed velocity distributions. The constant $(\xi_2)_0$ occurring in the expression (6-39) is equal to 0.48.

For reasons to be given later, the disagreement near the wake boundary is not serious. Of more importance for judging the theories is the fact that, near the axis of the wake, the calculated distribution gives values lower than the measured velocities; this means that the theoretical transport coefficient is too small (even zero at the axis).

A very good agreement is obtained with the Gaussian distribution (6-36a), which in the present case reads

$$\frac{\bar{U}_1}{\bar{U}_{1,\max}} = \exp\left[-\left(\frac{\xi_2}{0.256}\right)^2\right]$$

From this expression we obtain a half-value width given by $(\xi_2)_{1/2} = 0.215$.

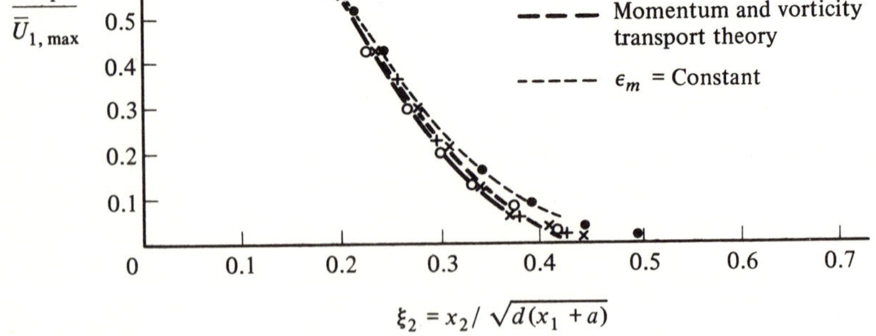

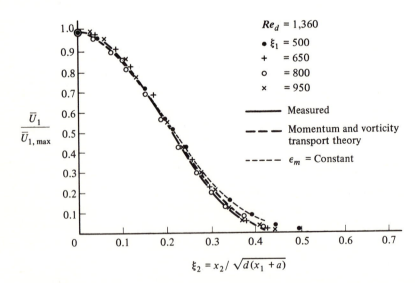

FIGURE 6-3
Mean-velocity distribution in the wake of a circular cylinder. (Adapted from: *Townsend, A. A.*[6])

The above Gaussian distribution shows only a deviation from the measured distribution close to the wake boundary. There the actual value of the apparent coefficient of eddy diffusion ϵ_m is much smaller than the assumed constant value, which applies to the main (central) part of the velocity-distribution curve. The constant value according to the experimental data is found to be

$$\epsilon_m = \tfrac{1}{4}(0.256)^2 U_0 d = 0.0164 U_0 d$$

If one calculates the value of ϵ_m directly from the measured velocity distribution, using the expression (6-34), one finds that ϵ_m is practically constant over the main (central) part of the wake.

If one calculates the value of the mixing length $\mathfrak{l}_m$ from the values of ϵ_m, obtained directly from the measured velocity distribution, one finds that $\mathfrak{l}_m$ is not constant across the wake; it is infinite at the axis and zero at the wake boundary.

It is possible to calculate an average value of $\mathfrak{l}_m$ from the constant c_m occurring in the expression (6-40). From Townsend's data one would obtain the value $\mathfrak{l}_m/(x_2)_{1/2} \simeq 0.4$, where $(x_2)_{1/2}$ is the half-value width. From Schlichting's experimental data a value of 0.46 would be obtained. These values are anything but small, contrary to what has been assumed in the mixing-length theories.

These facts, namely, a nonconstant value of the length $\mathfrak{l}_m$ and an average value too large compared with the wake width, are strong arguments against the mixing-length theories.

With the Eqs. (6-36a) and (6-37) we can obtain a relation between the constant A and the drag coefficient of the cylinder per unit of length. This coefficient is defined by

$$\text{Drag} = C_D d \tfrac{1}{2} \rho U_0^2$$

By calculating with Eqs. (6-36a) and (6-37) the drag of the cylinder, we obtain

$$A = \frac{1}{4}\sqrt{\frac{U_0 d}{\pi \epsilon_m}}\, C_D$$

With $\epsilon_m/U_0 d = 0.0164$ according to Townsend's measurements, $A \simeq 1.1 C_D$.

From the measured velocity distribution, the distribution of the shear stress may be calculated with the aid of the relation (6-33). If we assume, for instance, that the Gaussian error curve gives an adequate approximation to the actual velocity distribution, calculation shows that the maximum shear stress is equal to

$$\left(\frac{\sigma_{21}}{\rho U_0^2}\right)_{\max} = -\frac{A}{2}\frac{d}{x_1 + a}\sqrt{\frac{2\epsilon_m}{e U_0 d}}$$

(here e denotes the base of the natural logarithm).

This maximum value occurs at the point

$$(\xi_2)_{\text{opt}} = \sqrt{\frac{2\epsilon_m}{U_0 d}}$$

Since

$$(\xi_2)_{1/2} = \sqrt{\frac{4\epsilon_m}{U_0 d} \ln 2}$$

we see that $(\xi_2)_{opt}$ is smaller than the half-value width $(\xi_2)_{1/2}$. As we shall see later, maximum production of turbulence occurs beyond $(\xi_2)_{1/2}$, that is, not at $(\xi_2)_{opt}$.

Temperature measurements in the warm wake of heated cylinder have been carried out by Fage and Falkner,[4] by Townsend,[6] and more recently by Freymuth and Uberoi.[84] Those by Townsend show that similarity between temperature profiles in consecutive sections is obtained beyond $\xi_1 = 500$. Also, the heat-flux profiles $\overline{u_2\theta}$ then show similarity.

In contrast to this Freymuth and Uberoi obtained self-preservation of the mean-velocity and the mean-temperature distribution already when $\xi_1 \gtrsim 150$ (the origin of similarity was at $a = -50d$). They experimented with an internally-heated cylinder with $d = 2.8$ mm, placed vertically in a windtunnel. Most of the measurements were done with $U_0 = 61$ m/s ($\mathbf{Re}_d = 960$).

The experimental data obtained by these experimenters, which are in fair agreement with each other, show that the spread of the heat must have been appreciably larger than the spread of momentum; at least the width of the temperature wake was much greater than that of the velocity wake.

Figure 6-4 shows the results of the measurements by Fage and Falkner together with the curves according to the various theories, namely:

1 Prandtl's mixing-length theory, with $c_\gamma = c_m$ [see the solution (6-53)]:

$$\frac{\bar{\Gamma}}{\bar{\Gamma}_{max}} = \left\{1 - \left[\frac{|\xi_2|}{(\xi_2)_0}\right]^{3/2}\right\}^2$$

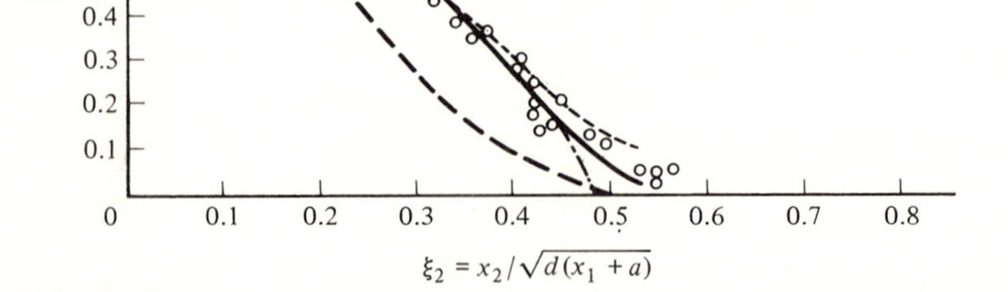

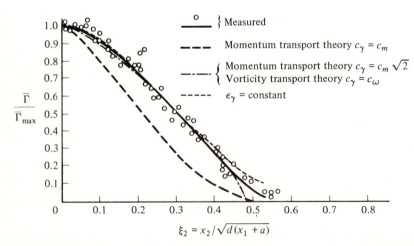

FIGURE 6-4
Mean-temperature distribution in the wake of a circular cylinder. (Adapted from: *Fage, A., and V. M. Falkner.*[4])

2 Prandtl's mixing-length theory, with $c_\gamma = c_m\sqrt{2}$, or Taylor's vorticity-transport theory with $c_\gamma = c_\omega$

$$\frac{\bar{\Gamma}}{\bar{\Gamma}_{max}} = 1 - \left[\frac{|\xi_2|}{(\xi_2)_0}\right]^{3/2}$$

3 Boussinesq's theory, with $\epsilon_\gamma = \text{const.}$ [see the solution (6-49a)].

The number of experimental data was not very great, and since, moreover, these data show rather wide scatter, it is difficult to decide which theory to prefer, except for Prandtl's momentum-transport theory with $c_\gamma = c_m$, which gives identical temperature and velocity distributions. In the above expressions for $\bar{\Gamma}/\bar{\Gamma}_{max}$, again $(\xi_2)_0 = 0.48$.

If, however, in Prandtl's mixing-length theory we take

$$c_\gamma = c_m\sqrt{2}$$

or, in Taylor's vorticity-transport theory, $c_\gamma = c_\omega$, the agreement with the experimental data is reasonable over a large part of the wake, with the exception of the most central part and the wake boundary region.

The coefficient of eddy diffusion ϵ_γ may be calculated from the measured temperature distribution with the aid of the relation (6-48). If this is done, one obtains a nearly constant value of ϵ_γ across the wake, with an average value $\epsilon_\gamma \simeq 0.03\, U_0 d$. This is almost twice the value of ϵ_m, namely $1.85\epsilon_m$. According to Taylor's vorticity-transport theory with $c_\gamma = c_\omega$, one would obtain $\epsilon_\gamma = 2\epsilon_m$.

The Gaussian error curve again gives the best agreement, outside the boundary region. The corresponding value for $\epsilon_\gamma = 0.027 U_0 d$, which agrees quite well with the above value of $0.03\, U_0 d$.

The results obtained by Freymuth and Uberoi[84] in the self-preserving region were in reasonable agreement with those shown in Fig. 6-4.

Axi-symmetric Wake

The theoretical similarity considerations given in Sec. 6-4 have shown a different dependence of the flow pattern on $(z + a)$, depending on the assumptions made. Measurements[13,42] in the wakes of axi-symmetric bodies have revealed a dependency on $(z + a)$ closer to the one following from the similarity assumption for $-\overline{u_r u_z}/\bar{U}_z^2$. We will refer here only to the measurements by Carmody[13] in a windtunnel with round disks of 0.05 m and 0.15 m diameter respectively, producing the wakes. The experiments were made at two Reynolds numbers, 3.2×10^4 and 7×10^4, based on disk diameter and free-stream velocity. Similarity according to the theory mentioned was obtained for $z + a > 20d$. The diameter of the wake then increased as $(z + a)^{1/3}$, while the maximum velocity defect $\bar{U}_{z,max}$ decreased as $(z + a)^{-2/3}$. When we assume an eddy-viscosity ϵ_m varying as $(z + a)^{-1/3}$, but constant in a cross-section the solution of the axial-momentum balance equation is again a Gaussian distribution.

$$\frac{\bar{U}_z}{\bar{U}_{z,max}} = \exp\left[-\frac{\xi_r^2}{\beta}\right]$$

where

$$\xi_r = \frac{r/d}{[(z+a)/d]^{-1/3}}, \quad \text{and} \quad \beta = \frac{6\epsilon_m}{U_0 d}\left(\frac{z+a}{d}\right)^{1/3}$$

Consequently

$$\frac{\epsilon_m}{U_0 d} = \frac{\beta}{6}\left(\frac{z+a}{d}\right)^{-1/3}$$

When

$$\bar{U}_{z,\max}/U_0 = A\left(\frac{z+a}{d}\right)^{-2/3}$$

the relation between A and the drag coefficient C_D of the disk reads

$$A = \frac{U_0 d}{48\epsilon_m}\left(\frac{z+a}{d}\right)^{-1/3} C_D$$

From Carmody's measurements we obtain $a/d \simeq -0.5$; $(\xi_r)_{1/2} \simeq 0.64$; $\beta \simeq 0.58$. This yields

$$\frac{\epsilon_m}{U_0 d} = 0.097\left(\frac{z+a}{d}\right)^{-1/3} \quad \text{and} \quad A \simeq 0.2C_D$$

which is a much smaller value than obtained from Townsend's measurements in the plane wake of a cylinder.

6-7 MEASUREMENTS OF TURBULENCE QUANTITIES IN THE WAKE FLOW OF A CYLINDER

Extended measurements of these quantities have already been made more than twenty years ago by Townsend.[5-8] Though since then a number of investigators have made measurements of a similar nature, we will in the following restrict ourselves mainly to Townsend's experimental results, except that we will also include

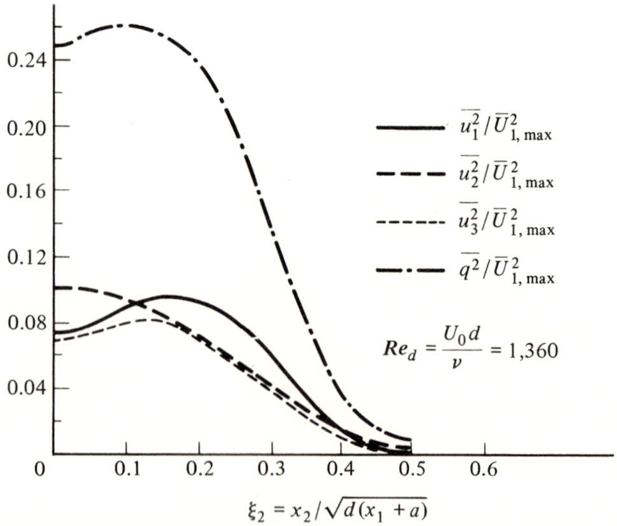

FIGURE 6-5
Relative turbulence intensities at the sections $\xi_1 = 500$, 650, 800, and 950 in the wake of a circular cylinder. (After: *Townsend, A. A.*[6])

results obtained by Uberoi and Freymuth[11] because they give additional information of importance on the wake flow of a cylinder. The measurements by Townsend were made at various sections in the wake at relative distances from $\xi_1 = 80$ up to 950. Most of the measurements refer to a free-stream velocity $U_0 = 12.8$ m/s and to a cylinder 1.59 mm in diameter; some to a cylinder 9.53 mm in diameter.

Most of his results considered here refer to measurements in sections beyond $\xi_1 = 500$, so that they certainly belong to the region of close self-preservation.

Figure 6-5 shows the values of the squared relative turbulence intensities $\overline{u_1{}^2}/\bar{U}^2_{1,\max}$, $\overline{u_2{}^2}/\bar{U}^2_{1,\max}$, and $\overline{u_3{}^2}/\bar{U}^2_{1,\max}$. The curves represent the average values for the four sections at $\xi_1 = 500$, 650, 800, and 950, respectively. The differences between the quantities at these sections are small and are within experimental scatter; so similarity among the various curves is good indeed.

Similar results for the distribution of $\overline{u_1{}^2}/\bar{U}^2_{1,\max}$ have been obtained by Uberoi and Freymuth, except that their values were more than 10 per cent higher; namely 0.1 at $\xi_2 = 0$ and 0.125 for the maximum value, also roughly at $\xi_2 \simeq 0.15$, which is smaller than $(\xi_2)_{1/2}$. Considering the still limited accuracy of turbulence measurements a difference of 10 per cent, say, may not be considered serious. Uberoi and Freymuth carried out their experiments in a low-turbulence windtunnel. Cylinders of 3, 12.5 and 50 mm diameter respectively were used, while the air velocity could be varied so as to cover a range of Reynolds numbers $\mathbf{Re}_d$ between 320 and 95,000. Hot-wire anemometry was used for making the turbulence measurements in cross-sections of the wake from $(x_1 + a)/d = 25$ up to 800. At the center of the wake,

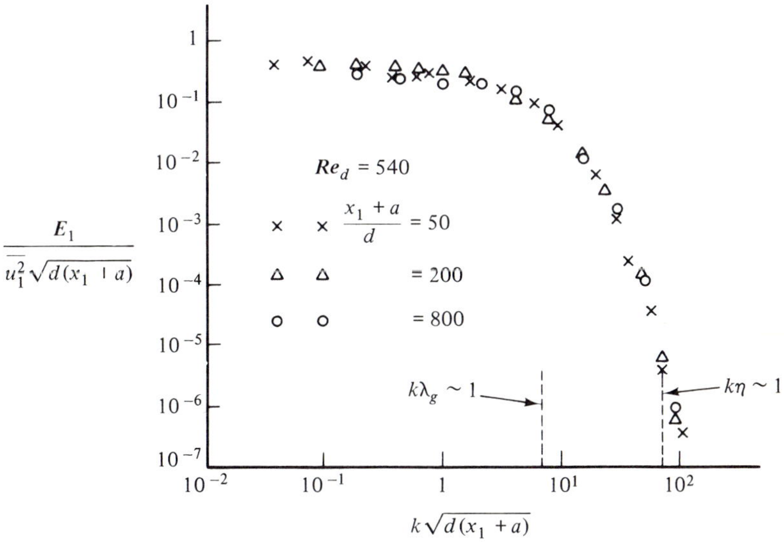

FIGURE 6-6
Energy spectra of the u_1-velocity taken at the center of the wake at three different x_1-stations. (After: *Uberoi, M. S., and P. Freymuth,*[11] *by permission of the American Institute of Physics.*)

$\xi_2 = 0$, $\overline{u_1{}^2}/\bar{U}^2_{1,\max}$ turned out to be independent of $(x_1 + a)/d$ and equal to 0.1 for $(x_1 + a)/d > 100$. The same similarity was observed in the whole central part of the wake up to $\xi_2 \simeq 0.25$, which is just beyond the half-value width $(\xi_2)_{1/2} = 0.215$. For $\xi_2 > 0.25$ the $\overline{u_1{}^2}/\bar{U}^2_{1,\max}$ distributions became similar only when $(x_1 + a)/d \geq 400$. The energy spectra of the u_1, u_2 and u_3 turbulence velocities taken at the center of the wake showed almost isotropic conditions in the wavenumber range $k\sqrt{d(x_1 + a)} > 10$ with anisotropy only in the large-scale structure.

The similarity of almost the whole turbulence structure, i.e., over almost the whole wavenumber range is shown in Fig. 6-6, where $\sqrt{d(x_1 + a)}$ is used as the length-scale. No measurable differences between the spectra taken at the three x_1-stations can be observed up to $k\eta = 1$. This suggests that also $\eta \propto \sqrt{d(x_1 + a)}$, which would agree with a universal high-wavenumber spectrum with the Kolmogoroff micro-length scale η as a length-scale. In that case the relation following from the expressions for the energy dissipation, namely $u'^2/\lambda_g{}^2 \propto \int_0^\infty dk\, k^2\mathbf{E}$, yields $\eta \propto \lambda_g \propto \sqrt{d(x_1 + a)}$. The spectra taken at two x_1-stations and various Reynolds numbers $\mathbf{Re}_\lambda = 18$ to 308, and rendered dimensionless with η and ν (see Table 3-1) showed clearly a universal curve for $k\eta > 0.1$. For $\mathbf{Re}_\lambda > 1{,}000$ a distinct inertial subrange over at least one decade of wavenumbers was present.

In Fig. 6-7 are shown the courses of the shear stress $\overline{u_1 u_2}/\bar{U}^2_{1,\max}$ and of the eddy viscosity $\epsilon_m/U_0 d$, in Fig. 6-8 those of the lateral transport quantities $\overline{u_1{}^2 u_2}/\bar{U}^3_{1,\max}$, $\overline{u_2{}^3}/\bar{U}^3_{1,\max}$, and $\overline{u_3{}^2 u_2}/\bar{U}^3_{1,\max}$.

The distribution curves of $\overline{u_1{}^2 u_2}/\bar{U}^3_{1,\max}$ and $\overline{u_3{}^2 u_2}/\bar{U}^3_{1,\max}$ show negative parts in the center region of the wake, which means an inward transport to the axis of the wake. This picture accords with the course of the intensity gradients shown in Fig. 6-5, though closer observation of the respective curves reveals that the regions of inward transport and of positive intensity gradients in the center of the wake do not entirely

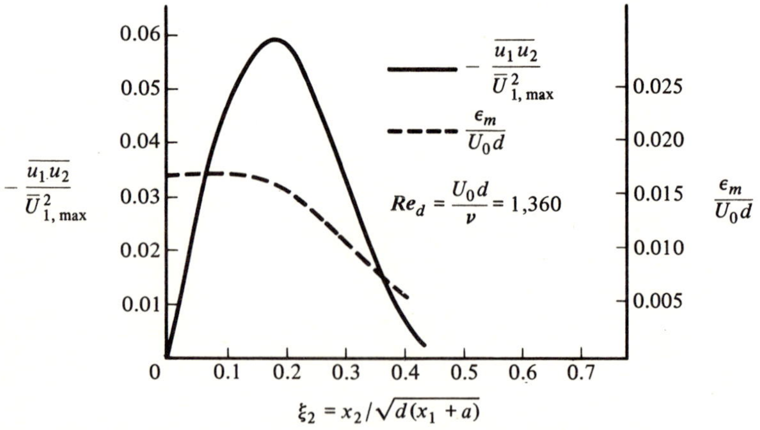

FIGURE 6-7
Distribution of shear stress and eddy viscosity at the sections $\xi_1 = 500$, 650, and 800 in the wake of a circular cylinder. (After *Townsend, A. A.*[6])

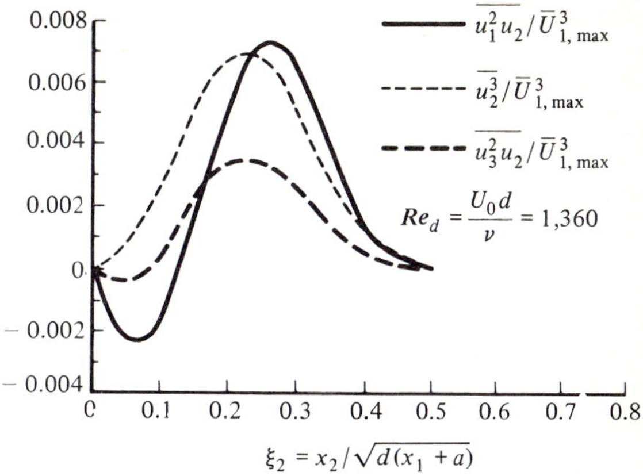

FIGURE 6-8
Lateral transport rate by turbulence of turbulence-intensity components at the sections $\xi_1 = 500$, 650, 800, and 950 in the wake of a circular cylinder. (After: *Townsend, A. A.*[6])

overlap, so that there exists a small part of the region where transport seems to be in the direction opposite to that given by the intensity gradient, namely in the region $0.13 \gtrsim \xi_2 \gtrsim 0.15$. We shall come back to this point later on.

The dissipation of turbulence per unit of mass in a nonisotropic turbulent flow is given by the expression

$$\varepsilon = \nu \overline{\left(\frac{\partial u_i}{\partial x_j} + \frac{\partial u_j}{\partial x_i} \right) \frac{\partial u_j}{\partial x_i}}$$

An exact determination of the dissipation would require the measurement of all quantities $\overline{(\partial u_i/\partial x_j)(\partial u_k/\partial x_l)}$. We know that, for isotropy, the complicated expression for ε reduces to

$$\varepsilon = -15\nu u'^2 \left[\frac{\partial^2 \mathbf{f}}{\partial r^2} \right]_{r=0} = 15\nu \overline{\left(\frac{\partial u_1}{\partial x_1} \right)^2}$$

It would be very gratifying to be able to use this simple expression for ε. Fortunately the turbulence in the wake shows a sufficient degree of local isotropy to justify this procedure. Townsend[8] has measured the following skewness and flattening factors in the wake of the cylinder:

$$S = \overline{\left(\frac{\partial u_1}{\partial x_1} \right)^3} \Bigg/ \left[\overline{\left(\frac{\partial u_1}{\partial x_1} \right)^2} \right]^{3/2} \qquad T = \overline{\left(\frac{\partial u_1}{\partial x_1} \right)^4} \Bigg/ \left[\overline{\left(\frac{\partial u_1}{\partial x_1} \right)^2} \right]^2$$

He observed that these factors were nearly constant in the central region of the wake (up to $\xi_2 \simeq 0.15$). He found, moreover, that the following relations, valid for isotropic turbulence, were closely satisfied in the central regions:

$$\overline{\left(\frac{\partial u_1}{\partial x_1} \right)^2} = \frac{1}{2} \overline{\left(\frac{\partial u_2}{\partial x_1} \right)^2} = \frac{1}{2} \overline{\left(\frac{\partial u_3}{\partial x_1} \right)^2}$$

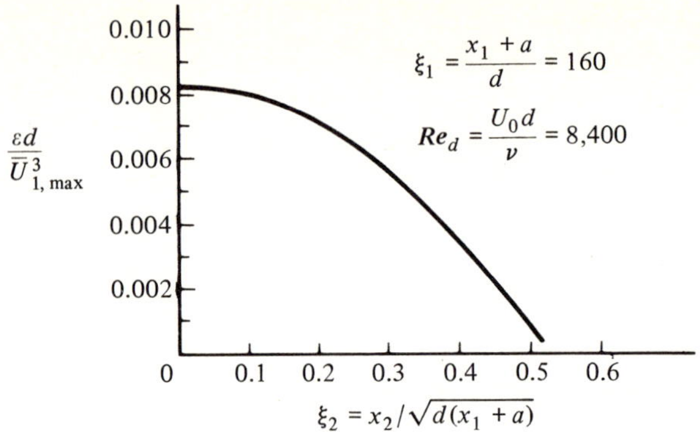

FIGURE 6-9
Energy dissipation at the section $\xi_1 = 160$ in the wake of a circular cylinder (From: *Townsend, A. A.,*[9] *by permission of the Cambridge University Press.*)

This near-isotropy of the small-scale structure of the turbulence has also been observed in later experiments by Uberoi and Freymuth, as already mentioned. Consequently, Townsend measured the energy dissipation in the wake of a cylinder using the simple "isotropic" relation for ε. The results of the measurements at the section $\xi_1 = 160$ are shown in Fig. 6-9.

As discussed in Sec. 3-5, for a normal distribution of $\partial u_1/\partial x_1$, the value of the flattening factor would be $T = 3$. A higher value of T might also indicate an intermittency factor $\Omega < 1$, corresponding to some degree of intermittency in the smaller-scale structure of the turbulence. Townsend obtained $T = 3.5$.

With the data of all the turbulence quantities given above, Townsend was able to develop a turbulence-energy balance. This turbulence-energy balance is given by Eq. (1-110), which for the wake flow considered may be reduced to the approximate relation

$$\frac{\partial}{\partial \xi_1'} \frac{1}{2} \frac{\overline{q^2}}{U_0^2} + \frac{\partial}{\partial \xi_2'} \frac{1}{U_0^3} \overline{u_2\left(\frac{p}{\rho} + \frac{q^2}{2}\right)} + \frac{\overline{u_1 u_2}}{U_0^2} \frac{\partial}{\partial \xi_2'} \frac{\bar{U}_1}{U_0} + 15 \frac{\nu}{U_0 d} \overline{\left(\frac{\partial}{\partial \xi_1'} \frac{u_1}{U_0}\right)^2} = 0$$

where $\xi_1' = x_1/d$ and $\xi_2' = x_2/d$.

In this equation all terms were measured except the pressure-transport term $\overline{u_2 p}$. This term was obtained by difference. This is a weak point in the procedure, since, if this pressure term had been measured too, the equation could have been nicely used to check the reliability of the measurement of the individual terms.

An attempt to measure this pressure term has been made by Kobashi.[44] He used a pressure pickup with a condenser microphone as a transducer for measuring the static-pressure fluctuations, in combination with a single rotary-type hot-wire anemometer placed close to the static holes of the pressure pickup tube. However, it must be appreciated that even at present the correct measurement of the static-pressure fluctuations is still an unsolved problem. This means that we have to consider the results of the measurements of $\overline{u_2 p}$ by Kobashi with reserve, though it must be admitted that the values obtained by him of the pressure term appear to be roughly of the same magnitude as the values obtained by Townsend from the closing entry of the energy-balance equation.

The various terms of the energy balance, obtained by Townsend[7] with respect to the section $\xi_1 = 160$, are shown in Fig. 6-10. If we consider a volume element in the wake flow, the contribution of viscous dissipation to the turbulence energy in that element is always negative.

In Fig. 6-10 the dissipation term is given positive. Hence a positive value in the diagram of Fig. 6-10 means a negative contribution, or a loss of energy. Conversely a negative value in the diagram means a gain in energy.

From this Fig. 6-10 we observe that the contributions made by the various terms are different in the various parts of the wake. Near the center, production is negligible and the main gain is by convection in the axial direction through the main flow. This gain is balanced by the dissipation and by outward turbulence transport (diffusion). Production is maximum at $\xi_2 \simeq 0.25$. Maximum shear stress occurs at $(\xi_2)_{opt} \simeq 0.18$ (see Fig. 6-7), which is smaller than $(\xi_2)_{1/2}$, but slightly greater than the location of maximum u_1'.

At the point $\xi_2 \simeq 0.325$ diffusion and convection are negligible, and production equals the local dissipation. This is the only point in the whole wake region where, strictly speaking, Prandtl's hypothesis $u_2' \propto l_m\, \partial\bar{U}_1/\partial x_2$ would apply (see Sec. 5-2).

In the outer boundary region, diffusion of kinetic energy, production, and dissipation are small. Convection in the axial direction withdraws energy, which is supplied by the lateral transport of pressure energy.

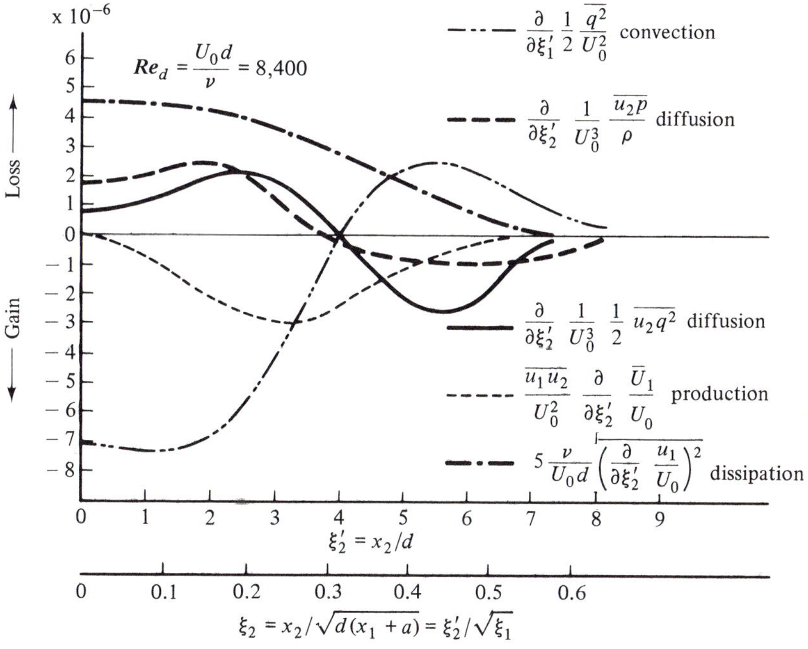

FIGURE 6-10
Energy balance in the wake of a circular cylinder at the section $\xi_1 = 160$ (From: *Townsend, A. A.,*[7] *by permission of the Royal Society.*)

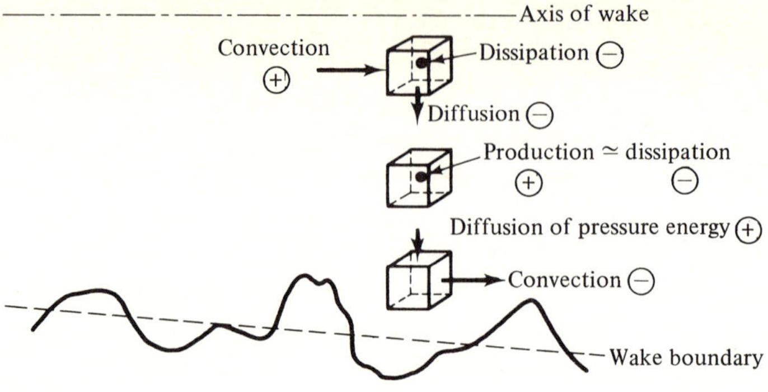

FIGURE 6-11
Contributions to turbulence energy in various parts of the wake of a cylinder.

These differences in the contributions to the turbulence energy at the various parts of the wake are illustrated in Fig. 6-11. As will be shown shortly, the flow in the outer region $\xi_2 > 0.3$ is highly intermittent, so that for this region it will be difficult to draw definite conclusions concerning the energy-budget for the turbulence, without taking account of this intermittency. The results obtained for the total time averaged situation shown in Figs. 6-10 and 6-11, must be considered in this light.

Freymuth and Uberoi[84] made extensive measurements of the turbulent temperature field in the wake of the heated cylinder, mentioned in Sec. 6-6. At $\mathbf{Re}_d = 960$ also this field was self-preserving when $\xi_1 \gtrsim 150$. The main results were the following.

In the cross-section $\xi_1 \simeq 1200$, $\overline{u_2\gamma}$ was zero at the axis, as it should be, and showed a maximum at $\xi_2 \simeq 0.21$. The intensity of the temperature fluctuations γ' had its maximum at $\xi_2 \simeq 0.26$. These two values are smaller than $(\xi_2)_{1/2} \simeq 0.3$ for the mean-temperature distribution. In this respect there is some difference with the relative locations of the optimum ξ_2-values for $\overline{u_2u_1}$, u'_1, and $(\xi_2)_{1/2}$ for the mean-velocity distribution. However, the turbulence heat-balance showed marked differences with the turbulence energy-balance (see Fig. 6-10). The dominating terms were the "production" $2\overline{u_2\gamma}\,\partial\bar{\Gamma}/\partial x_2$ and the "dissipation" $2\mathfrak{k}\overline{(\partial\gamma/\partial x_i)(\partial\gamma/\partial x_i)} = 6\mathfrak{k}\,\overline{(\partial\gamma/\partial x_1)^2}$. [The measurements showed $\overline{(\partial\gamma/\partial x_1)^2} \simeq \overline{(\partial\gamma/\partial x_2)^2} \simeq \overline{(\partial\gamma/\partial x_3)^2}$.] The mean-flow convection and the turbulent diffusion showed only moderate contributions. The contribution by molecular diffusion was almost negligible. The "production" term had its maximum at $\xi_2 \simeq 0.25$, which is almost equal to that for $(\gamma')_{\max}$. The "dissipation" had its maximum at $\xi_2 \simeq 0.22$, that is practically the same location where $u_2\gamma$ showed its maximum. Compare these with the locations for the maxima of production and dissipation in the turbulence energy balance (Fig. 6-10).

The one-dimensional spectrum $\mathbf{E}_{\gamma,1}$ (see Eq. 3-220) showed an almost similar distribution as that for $\mathbf{E}_1$ shown in Fig. 6-6, though Freymuth and Uberoi found that $\mathbf{E}_{\gamma,1}$ was almost equal to $\mathbf{E}_2$ computed from $\mathbf{E}_1$ according to the isotropic relation (3-71). No Kolmogoroff ($-5/3$) subrange could be observed at this Reynolds number. This subrange was only observed at $\mathbf{Re}_d \gtrsim 15{,}000$. As may be expected, the spectrum of $\overline{u_2\gamma}$ dropped at a faster rate in the high wavenumber range than that of $\overline{\gamma^2}$.

If we consider the courses of various quantities across the wake, we may notice that some of them, e.g., the turbulence kinetic energy, the dissipation, and the eddy viscosity, vary only slightly in the central region of the wake and decrease more or less sharply toward the outer boundary region. This behavior of the quantities would indicate a certain homogeneity of the turbulence in the central part of the wake. This idea is strengthened by the observation that the dissipation and integral scales also show only slight variation across the wake.

Now, in the central part of the wake, the turbulence is practically wholly continuous with respect to time, whereas, near the boundary of the wake, the turbulent flow becomes more and more intermittent. This is illustrated by Fig. 6-12; reference may also be made to the oscillograms shown in Fig. 1-9. The actual continuous turbulent region is much narrower than shown in Fig. 6-12.

Townsend has suggested that the turbulence might be nearly homogeneous in a much greater region than that corresponding to the continuous central core region. This would mean that the decrease of various turbulence quantities toward the outer region is apparent only and due to the intermittent character of the flow. A study of velocity oscillograms taken at various points in the wake (such as shown by Fig. 1-9) reveals that the fluctuations in the turbulent regions of the boundary have the same character as those occurring in the fully turbulent core region. If this is true, the velocity correlation $\overline{u(t)u(t-\tau)}$ must have the same value in the turbulent core region and in the turbulent regions within the protuberances at the boundary of the wake. This appears to be virtually true.

As a measure of the degree of intermittence of turbulence, one may take the ratio Ω between the time during which turbulence occurs and the total time. (See Sec. 3-5.)

Townsend[8] has used two methods for determining the intermittency factor in the wake of a cylinder. The one method is indirect, based on the measurement of time-mean values of powers of velocity derivatives. For we have

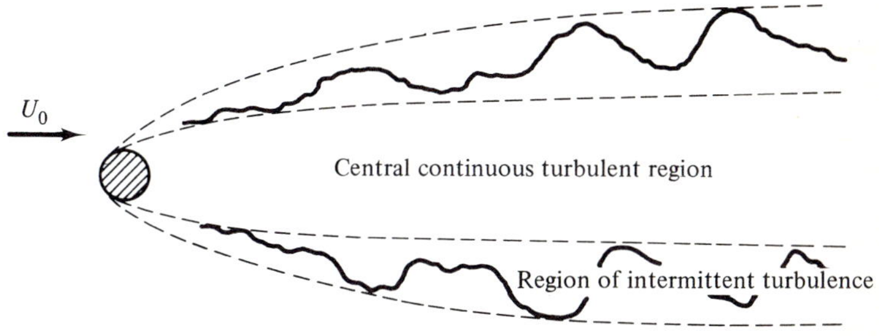

FIGURE 6-12
Instantaneous picture of the wake flow behind a cylinder.

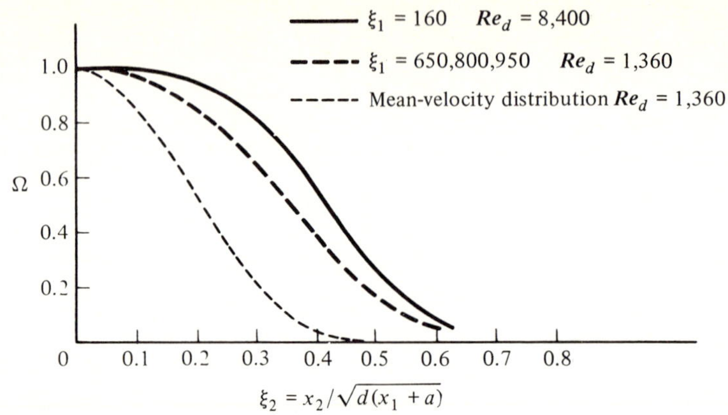

FIGURE 6-13
Intermittency factor Ω in the wake flow of a circular cylinder. (From: *Townsend, A. A.,*[9] *by permission of the Cambridge University Press.)*

$$\overline{\left(\frac{\partial u_1}{\partial x_1}\right)^2_{\text{intermit.}}} = \Omega\overline{\left(\frac{\partial u_1}{\partial x_1}\right)^2_{\text{cont.}}}$$

$$\overline{\left(\frac{\partial u_1}{\partial x_1}\right)^4_{\text{intermit.}}} = \Omega\overline{\left(\frac{\partial u_1}{\partial x_1}\right)^4_{\text{cont.}}}$$

Thus the flattening factor T_1 of $\partial u_1/\partial x_1$ becomes

$$T\left(\frac{\partial u_1}{\partial x_1}\right) = \frac{\overline{(\partial u_1/\partial x_1)^4_{\text{intermit.}}}}{[\overline{(\partial u_1/\partial x_1)^2_{\text{intermit.}}}]^2} = \frac{1}{\Omega}\frac{\overline{(\partial u_1/\partial x_1)^4_{\text{cont.}}}}{[\overline{(\partial u_1/\partial x_1)^2_{\text{cont.}}}]^2}$$

Hence the ratio $T_{1,\text{cont.}}/T_{1,\text{intermit.}} = \Omega$ is a direct measure of the intermittency factor. This method can be used if the turbulence in the continuous and intermittent parts of the wake is sufficiently homogeneous in its fine structure, and if either $T_{1,\text{cont.}}$ is known, or $\Omega = 1$ at $\xi_2 = 0$.

The other method is based on the construction of an electrical circuit that produces, from the output of a hot wire, a signal which is zero when $(\partial u_1/\partial x_1)^2$ is zero and which is unity when $(\partial u_1/\partial x_1)^2$ exceeds a suitably chosen threshold value. The average value of this output signal is then equal to the intermittency factor. This method can be applied satisfactorily only if a large number of complete velocity fluctuations occur during a period of turbulent motion. This condition is usually satisfied if the Reynolds number of the turbulent flow is large. Simple as this method seems to be, there is still a problem of a practical nature in the proper choice of the threshold value, i.e., by the criterion what part of the fluctuations has to be considered turbulent still.

Figure 6-13 shows the results of measurements of the intermittency factor Ω. It may be noticed that this intermittency factor deviates appreciably from unity already at relatively short distances from the wake axis, that is, in regions that are usually observed as being fully turbulent.

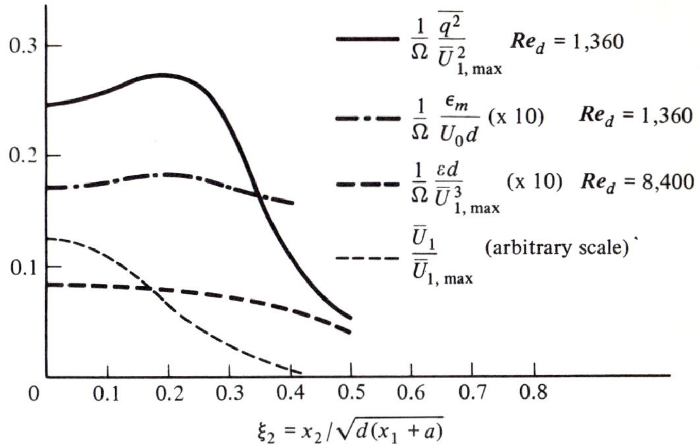

FIGURE 6-14
Turbulence quantities in the turbulent regions across the whole wake of a cylinder.

If the turbulence is almost homogeneous in the turbulent regions, including the regions within the protuberances, the pertinent turbulence quantities would obtain practically constant values across the wake when corrected for the intermittency factor. Figure 6-14 shows that this is nearly so. The turbulence kinetic energy, the dissipation, and the eddy viscosity given in Figs. 6-5, 6-9, and 6-7, respectively, when divided by the intermittency factor given in Fig. 6-13, are practically constant across a large part of the wake width. In particular, the constancy of the eddy viscosity and of the dissipation across the wake is noteworthy.

Strictly speaking the corrections made by simply dividing the measured mean values by the intermittency factor are not entirely correct. In particular the correction applied to the eddy-viscosity made in this way may still be slightly in error in the outer region. For, in the first place it has been assumed that the flow between the turbulent periods is entirely devoid of fluctuations, whereas actually this is not the case. (Consequently $\overline{q^2}$ will be overcorrected however little it may be.) Since in the outer region of the wake the mean-velocity is practically equal to U_0, and $\bar{U}_1/U_0 \ll 1$, it is taken for granted that a more correct correction-procedure will not make much difference in the results shown in Fig. 6-14. The same is true for the "corrected" dissipation. The correct procedure will be discussed in Sec. 6-11, while the nature of the flow in the intermittency region will be discussed in more detail in Sec. 6-12.

Axi-symmetric Wake

We will conclude this section with the discussion of a number of experimental results on turbulence measurements in axi-symmetric wakes, mainly because these experiments reveal differences concerning the similarity behavior of the turbulence as compared with the plane wake.

Hwang and Baldwin[45] performed measurements in a windtunnel on the decay of the turbulence in the wakes produced by disks of 20 mm, 45 mm, and 75 mm diameter at free-stream velocities ranging from 3 to 22 m/s. We recall that just like the plane wake only incomplete similarity can be obtained. Hwang and Baldwin observed this similarity for $z + a > 100d$, though a satisfactory degree of similarity was observed for the turbulence intensity u_z', the dissipation length-scale and the half-value radius $r_{1/2}$ of the wake in the region $z + a > 40d$. For $z + a > 400d$ the turbulence approached the final stage of decay, where viscous effects become dominant. In the region $40 < (z + a)/d < 300$, $u_z' \propto (z + a)^{-2/3}$, $\lambda_g \propto (z + a)^{1/2}$ and $r_{1/2} \propto (z + a)^{1/3}$. The behavior of u_z' and $r_{1/2}$ are according to the similarity laws discussed earlier for the axi-symmetric case. In the region $100 < (z + a)/d < 400$ the decay of turbulence energy shows a similarity when $\lambda_g \propto (z + a)^{1/2}$. This agrees with the decay relation

$$U_0 \frac{d\overline{q^2}}{dz} \simeq -\text{const.}\, \nu \frac{\overline{q^2}}{\lambda_g^2}$$

provided the effects of local production and turbulence transport of turbulence energy are negligible.

With $u_z' \propto (z + a)^{-2/3}$ and $\lambda_g \propto (z + a)^{1/2}$, we would obtain for the viscous dissipation $\varepsilon \propto \overline{u_z^2}/\lambda_g^2 \propto (z + a)^{-7/3}$. Since we may also put $\varepsilon \propto (u_z')^3/\Lambda$, where Λ is an integral length-scale, there would follow $\Lambda \propto (z + a)^{1/3} \propto r_{1/2}$. This agrees with the behavior of the eddy viscosity, discussed earlier, namely $\epsilon_m \propto (z + a)^{-1/3}$, if we assume $\epsilon_m \propto u_r'\Lambda$, and $u_r' \propto u_z'$ which was found to be approximately true according to the measurements. In contrast to the plane wake where λ_g and Λ both vary with $(x_1 + a)^{1/2}$, here there is no similarity between λ_g and Λ. This is also demonstrated by the behavior of the Kolmogoroff micro-length-scale as a function of $(z + a)$. Again from

$$\overline{u_z'^2}/\lambda_g^2 \propto \int_0^\infty dk\, k^2\mathbf{E} \propto 1/\eta^4$$

we may deduce $\eta \propto (z + a)^{7/12}$, which is neither similar to the behavior of λ_g, nor to that of Λ.

Similar decay behaviors have been observed by Gibson, Chen, and Lin[46] in the wake flow of a sphere placed in a windtunnel. They restricted their measurements to $z + a < 80d$, and observed a decay of ε very close to the asymptotic value $(z + a)^{-7/3}$. The decay of u_z' in this region, however, was faster than $(z + a)^{-2/3}$. These investigators also observed some intermittency of the turbulence at the axis of the wake. This intermittency has been studied in more detail by Baldwin and Sandborn[85] in windtunnel experiments with the three circular disks used by Hwang and Baldwin.[45] In their measurements in a region up to $z + a = 600d$, they observed an almost linear decrease of Ω with $(z + a)/d$ to 0.86 at the end of the region considered. A Gaussian distribution of Ω from the wake axis outward was found. As may be expected they also observed a faster decay of the smaller eddies, resulting in an increasing intermittency (i.e. a smaller Ω) of the small-scale turbulence. The decrease of Ω with $(z + a)/d$, mentioned above may be explained through this progressive decay of smaller eddies, since measuring the velocity derivative favors the effect of these eddies.

6-8 VELOCITY DISTRIBUTION IN A ROUND FREE JET ACCORDING TO THE CLASSICAL THEORIES

We shall consider here the application of the theories of Boussinesq, Prandtl, and Reichardt to the distribution of mean velocity in a round free jet. Our procedure will be similar to that of Sec. 6-4.

The general case will be assumed, where a jet issues from a circular orifice into a stream of uniform velocity. Figure 6-15 shows this general case. The jet issues with a velocity U_p from an orifice with diameter d into a stream with uniform velocity U_s. The jet stream proper and the ambient stream are usually called primary and secondary flow, respectively.

We assume axi-symmetry, and cylindrical coordinates (z,r) with the axial coordinate z along the jet axis. The origin is taken at the plane of the orifice.

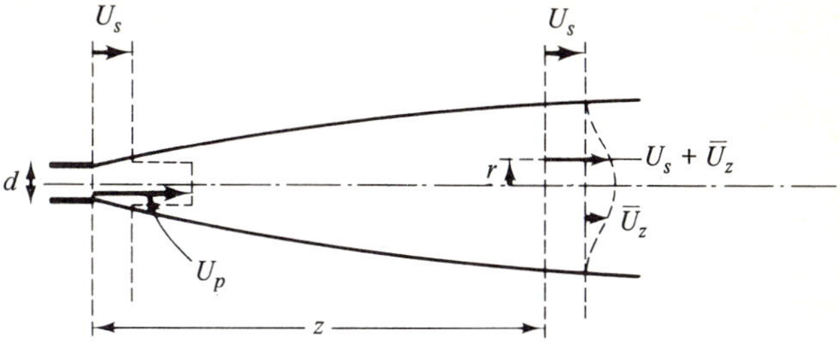

FIGURE 6-15
Round free jet in a general stream of constant velocity U_s.

In the jet region, the axial velocity component is assumed to be composed of the velocity U_s of the ambient stream and the velocity $\bar{U}_z$, which is the deviation from the velocity U_s caused by the jet. Outside the jet region the velocity U_s, and hence the pressure, are assumed to the uniform, i.e., $dP_0/dz = 0$. We will follow the same procedure as applied to the theoretical study of the wake flow, thus restricting ourselves to those axial distances from the orifice, where the flow patterns in subsequent sections may be expected to be similar, either complete or incomplete. We take the issuing velocity U_p for the characteristic velocity $\mathscr{U}_0$, and the orifice diameter d for the characteristic length $\mathscr{L}_0$.

Since we also apply boundary-layer approximations, it is possible to consider only the similarity relations corresponding with Eqs. (6-3a), (6-3b), (6-3e), and (6-3g), namely:

$$\bar{U}_z = U_p\left(\frac{z+a}{d}\right)^p f(\xi_2) \qquad \text{(6-56a)}$$

$$\bar{U}_r = U_p\left(\frac{z+a}{d}\right)^r g(\xi_2) \qquad \text{(6-56b)}$$

$$-\overline{u_r u_z} = U_p^2\left(\frac{z+a}{d}\right)^{s_{21}} h(\xi_2) \qquad \text{(6-56c)}$$

where

$$\xi_2 = \frac{r/d}{[(z+a)/d]^q}$$

Here $z = -a$ is the position of the *"origin of similarity."* The mass-balance equation for the mean-motion reads

$$\frac{\partial}{\partial z}(U_s + \bar{U}_z) + \frac{1}{r}\frac{\partial}{\partial r}(r\bar{U}_r) = 0 \qquad \text{(6-57)}$$

Substitution of the above similarity relations, yields

$$pf(\xi_2) - q\xi_2\frac{df}{d\xi_2} + \frac{1}{\xi_2}\frac{d}{d\xi_2}[\xi_2 g(\xi_2)] = 0 \qquad \text{(6-58)}$$

independent of z, provided p, q and r satisfy the condition (6-5a), i.e.,

$$p + q - r = 1 \qquad \text{(6-5a)}$$

The momentum-balance equation for the axial direction reads, after applying the boundary-layer approximations and neglecting viscous effects

$$\frac{\partial}{\partial z}(U_s + \bar{U}_z)^2 + \frac{1}{r}\frac{\partial}{\partial r} r\bar{U}_r(U_s + \bar{U}_z) = \frac{1}{r}\frac{\partial}{\partial r} r(-\overline{u_r u_z}) \qquad \text{(6-59)}$$

Or, with Eq. (6-57)

$$(U_s + \bar{U}_z)\frac{\partial \bar{U}_z}{\partial z} + \bar{U}_r\frac{\partial \bar{U}_z}{\partial r} = \frac{1}{r}\frac{\partial}{\partial r} r(-\overline{u_r u_z}) \qquad \text{(6-60)}$$

Substitution of the similarity expressions yields

$$\left[\left(\frac{z+a}{d}\right)^p f(\xi_2) + \frac{U_s}{U_p}\right]\left[pf(\xi_2) - q\xi_2\frac{df}{d\xi_2}\right] + \left(\frac{z+a}{d}\right)^{r+1-q} g(\xi_2)\frac{df}{d\xi_2}$$
$$= \left(\frac{z+a}{d}\right)^{s_{21}-p-q+1}\frac{1}{\xi_2}\frac{d}{d\xi_2}[\xi_2 h(\xi_2)] \qquad \text{(6-61)}$$

Hence, similarity is only possible when

$$p = 0; \quad q - r = 1 \quad \text{and} \quad q - s_{21} = 1$$

The first condition, $p = 0$, would mean that $\bar{U}_z$ remains independent of z. This is not a very realistic situation, since it would indicate no interaction and mixing of the jet with its environment. Indeed, as we will show, from the conservation of the total axial momentum follows $q = 0$, implying no spread of the jet at all.

Integration of the mean-balance equation (6-57) with respect to r yields

$$\frac{d}{dz}\int_0^\infty dr\, r(U_s + \bar{U}_z) + \int_0^\infty dr \frac{\partial}{\partial r} r\bar{U}_r = 0$$

Whence follows

$$\lim_{r\to\infty}(r\bar{U}_r) = -\frac{d}{dz}\int_0^\infty dr\, r(U_s + \bar{U}_z)$$

Applying the same integration to the momentum-balance equation (6-59) yields

$$\frac{d}{dz}\int_0^\infty dr\, r(U_s + \bar{U}_z)^2 + [r\bar{U}_r(U_s + \bar{U}_z)]_0^\infty = [r(-\overline{u_r u_z})]_0^\infty$$

Since $\lim_{r\to\infty}(-\overline{u_r u_z}) = 0$ and $\lim_{r\to\infty}\bar{U}_z = 0$, we obtain with the above relation for $\lim_{r\to\infty}(r\bar{U}_r)$,

$$\frac{d}{dz}\int_0^\infty dr\, r[(U_s + \bar{U}_z)^2 - U_s(U_s + \bar{U}_z)] = 0$$

Whence,

$$\int_0^{\infty} dr\, r\bar{U}_z(U_s + \bar{U}_z) = \text{constant}$$

independent of z. From the initial condition at $z = 0$ we obtain for the constant the value

$$\tfrac{1}{8}d^2 U_p(U_p - U_s)$$

With the similarity relations (6-56a) and (6-56d) the integral momentum-balance equation becomes

$$\int_0^{\infty} d\xi_2 \xi_2 \left(\frac{z+a}{d}\right)^{2q+p} f(\xi_2)\left[\left(\frac{z+a}{d}\right)^p f(\xi_2) + \frac{U_s}{U_p}\right] = \frac{1}{8}\left(1 - \frac{U_s}{U_p}\right) \tag{6-62}$$

Thus, with $p = 0$, this equation yields the condition $q = 0$. Hence, as concluded earlier, the jet does not spread, and $f(\xi_2)$ would be equal to the distribution at $z = 0$. Since, with $q = 0$, we would obtain $r = s_{21} = -1$, a possible solution would only be with $g(\xi_2) = 0$ and $h(\xi_2) = 0$, implying no turbulence. It is impossible therefore, to maintain the assumption of similarity in a fully developed turbulent jet in a co-flowing outer stream, when U_s/U_p is neither small, nor large compared with $\bar{U}_z/U_p$. Only for two extreme cases can similarity be maintained, namely, when either

$$\frac{\bar{U}_z}{U_s} \gg 1 \quad \text{or} \quad \frac{\bar{U}_z}{U_s} \ll 1$$

Let us consider these cases separately.

Case: $\bar{U}_z/U_s \gg 1$. This case then resembles the case of a free jet in a quiescent environment. In Eqs. (6-61) and (6-62) we may then neglect the term U_s/U_p with respect to $\bar{U}_z/U_p$. The Eq. (6-61) then yields the conditions for similarity: $p + q - r = 1$, which is the same as Eq. (6-5a), and $2p + q - s_{21} = 1$ which is the same as Eq. (6-5f). The integral momentum balance equation (6-62) yields the condition

$$p + q = 0 \tag{6-63}$$

So we have three relations, Eqs. (6-5a), (6-5f), and (6-63) for the four unknowns p, q, r, and s_{21}. Thus we lack an additional relation. Again we will consider the same alternative assumptions as we made in the case of wake flows, namely the one where $-\overline{u_r u_z}/\bar{U}_z^2$ is independent of z, and the other where the eddy viscosity ϵ_m is independent of z.

The assumption $-\overline{u_r u_z}/\bar{U}_z^2$ independent of z, which means similarity between turbulence and mean-flow pattern, yields the condition

$$s_{21} - 2p = 0 \tag{6-64}$$

So from Eqs. (6-5a), (6-5f), (6-63), and (6-64) we obtain

$$p = -1 \qquad q = 1 \qquad r = -1 \quad \text{and} \quad s_{21} = -2$$

i.e. *complete similarity*. The expressions (6-56) become

$$\frac{\bar{U}_z}{U_p} = \left(\frac{z+a}{d}\right)^{-1} f(\xi_2) \qquad \text{(6-65a)}$$

$$\frac{\bar{U}_r}{U_p} = \left(\frac{z+a}{d}\right)^{-1} g(\xi_2) \qquad \text{(6-65b)}$$

$$\frac{\sigma_{rz}}{\rho U_p^2} = \frac{-\overline{u_r u_z}}{U_p^2} = \left(\frac{z+a}{d}\right)^{-2} h(\xi_2) \qquad \text{(6-65c)}$$

and

$$\xi_2 = \frac{r/d}{(z+a)/d} = \frac{r}{z+a} \qquad \text{(6-65d)}$$

When we do express the turbulence shear-stress in terms of an eddy viscosity ϵ_m and the mean-velocity gradient $\partial \bar{U}_z/\partial r$, we obtain from

$$-\overline{u_r u_z} = \epsilon_m \, \partial \bar{U}_z/\partial r$$

and the relations (6-65a), (6-65c), and (6-65d), a relation for ϵ_m that is independent of z. So the first alternative assumption includes the second one.

Case: $\bar{U}_z/U_s \ll 1$. This case resembles that of the wake flow. Indeed neglecting in Eqs. (6-61) and (6-62) the term $\bar{U}_z/U_p$ with respect to U_s/U_p, and since $\bar{U}_r$ is not larger than $\bar{U}_z$, so that we may also neglect the term with $\bar{U}_r/U_p$, the same relations for p, q, and s_{21} are obtained as for the axi-symmetric wake flow, namely

$$p + q - s_{21} = 1 \qquad \text{(6-27)}$$

$$p + 2q = 0 \qquad \text{(6-66)}$$

The assumption $-\overline{u_r u_z}/\bar{U}_z^2$ independent of z yields again the relation (6-64). Thus from Eqs. (6-27), (6-64), and (6-66) we obtain

$$p = -\tfrac{2}{3} \qquad q = \tfrac{1}{3} \qquad s_{21} = -\tfrac{4}{3}$$

The mass-balance equation yields

$$r = p + q - 1 = -\tfrac{4}{3}$$

Thus we have

$$\frac{\bar{U}_z}{U_p} = \left(\frac{z+a}{d}\right)^{-2/3} f(\xi_2) \qquad \text{(6-67a)}$$

$$\frac{\bar{U}_r}{U_p} = \left(\frac{z+a}{d}\right)^{-4/3} g(\xi_2) \qquad \text{(6-67b)}$$

$$\frac{\sigma_{rz}}{\rho U_p^2} = \frac{-\overline{u_r u_z}}{U_p^2} = \left(\frac{z+a}{d}\right)^{-4/3} h(\xi_2) \qquad \text{(6-67c)}$$

and

$$\xi_2 = \frac{r/d}{[(z+a)/d]^{1/3}} \qquad \text{(6-67d)}$$

Hence there is only *incomplete similarity.*

For the dependence of the eddy viscosity ϵ_m on z we then obtain

$$\frac{\epsilon_m}{U_p d} = \left(\frac{z+a}{d}\right)^{-1/3} \frac{h(\xi_2)}{df/d\xi_2} \qquad \text{(6-68)}$$

On the other hand, when we consider the other alternative assumption, namely ϵ_m to be independent of z, we obtain instead of the relation (6-64)

$$s_{21} = p - q \qquad \text{(6-69)}$$

So from Eqs. (6-27), (6-66), and (6-69) there is obtained

$$p = -1 \qquad q = \tfrac{1}{2} \qquad \text{and} \qquad s_{21} = -\tfrac{3}{2}$$

while Eq. (6-5a) then yields $r = -\frac{3}{2}$.

Consequently again only *incomplete similarity* can be obtained. In so far as experimental evidence is available it seems to be in favor of the first assumption, i.e., similarity of $-\overline{u_r u_z}/\bar{U}_z^2$, a result which may be expected on the ground of experimental results obtained for axi-symmetric wake flows. Therefore, in the following we will no longer consider the second alternative.

We then conclude that, in the case $\bar{U}_z/U_s \gg 1$, which occurs when U_s is small (or zero, as the case may be) with respect to $\bar{U}_z$, the spread of the jet is linear with $z + a$, and the jet velocity decreases in the axial direction hyperbolically with increasing distance $z + a$.

If, however, $\bar{U}_z/U_s \ll 1$, that is, if U_s is large compared with $\bar{U}_z$, the spread of the jet is proportional to $(z + a)^{1/3}$, and the velocity decreases according to $\bar{U}_{z,\max}/U_s \propto (z + a)^{-2/3}$.

If a jet issues in a region with a constant velocity U_s in the axial direction, so that U_s is small with respect to the issuing velocity U_p of the jet, the spread of the jet will be linear with z at short distances from the orifice, and the jet velocity will decrease in inverse proportion to z. But, since U_s is constant, whereas the jet velocity decreases with increasing z, a condition will be reached in which the jet velocity not only is no longer large compared with U_s but may have become small with respect to U_s. Then the spread in this region will be proportional to $z^{1/3}$. Between the two regions there is a transitional region in which one may expect a gradual decrease in the jet velocity from the z^{-1} law into the $z^{-2/3}$ law.

Let us now try to obtain solutions for $f(\xi_2)$, $g(\xi_2)$, and $h(\xi_2)$ for the two extreme cases. For the case $\bar{U}_z/U_s \gg 1$ we may conveniently assume $U_s = 0$, that is, the jet issues in a quiescent medium.

Case: $U_s = 0$. With the relations (6-65) and with $U_s = 0$, the equation of motion (6-60) becomes

$$-\left(f^2 + \xi_2 f \frac{df}{d\xi_2}\right) + g\frac{df}{d\xi_2} = \frac{1}{\xi_2}\frac{d}{d\xi_2}(\xi_2 h) \tag{6-70}$$

We may eliminate $g(\xi_2)$ with the aid of the equation of continuity. With $p = -1$, $q = 1$, Eq. (6-58) becomes

$$-\left(f + \xi_2\frac{df}{d\xi_2}\right) + \frac{1}{\xi_2}\frac{d}{d\xi_2}(\xi_2 g) = 0$$

or

$$\frac{d}{d\xi_2}(\xi_2 f) = \frac{1}{\xi_2}\frac{d}{d\xi_2}(\xi_2 g)$$

Whence, after integration,

$$\xi_2 g = \xi_2{}^2 f - \int_0^{\xi_2} d\xi_2\, \xi_2 f$$

Now put

$$\int_0^{\xi_2} d\xi_2\, \xi_2 f = F(\xi_2)$$

so that

$$f = \frac{1}{\xi_2}\frac{dF}{d\xi_2} \qquad \text{and} \qquad g = \frac{dF}{d\xi_2} - \frac{1}{\xi_2}F \tag{6-71}$$

Substitution in Eq. (6-70) yields

$$-\frac{F}{\xi_2}\frac{d^2F}{d\xi_2{}^2} - \frac{1}{\xi_2}\left(\frac{dF}{d\xi_2}\right)^2 + \frac{F}{\xi_2{}^2}\frac{dF}{d\xi_2} = \frac{d}{d\xi_2}(\xi_2 h)$$

or

$$-\frac{d}{d\xi_2}\left(\frac{F}{\xi_2}\frac{dF}{d\xi_2}\right) = \frac{d}{d\xi_2}(\xi_2 h)$$

Integration of this differential equation yields

$$\xi_2 h = -\frac{F}{\xi_2}\frac{dF}{d\xi_2} = -f\int_0^{\xi_2} d\xi_2\, \xi_2 f \tag{6-72}$$

where the constant of integration must be zero, since h is finite at $\xi_2 = 0$. From this relation, with the expressions (6-65a) and (6-65c), we obtain

$$\frac{\sigma_{rz}}{\rho \bar{U}^2_{z,\max}} = -\frac{1}{\xi_2}\frac{\bar{U}_z}{\bar{U}_{z,\max}}\int_0^{\xi_2} d\xi_2\, \xi_2 \frac{\bar{U}_z}{\bar{U}_{z,\max}} \tag{6-73}$$

We had already concluded earlier that for this case of complete similarity the eddy viscosity ϵ_m is independent of z, and must be a function of ξ_2 alone. For this function we will first assume a constant, thus

$$\epsilon_m = \text{const.}$$

Since

$$\epsilon_m = \frac{-\overline{u_r u_z}}{\partial \bar{U}_z/\partial r}$$

we obtain with Eqs. (6-71) and (6-72):

$$\frac{\epsilon_m}{\bar{U}_{z,\max}(z+a)} = -\frac{(F/\xi_2)(dF/d\xi_2)}{\xi_2(d/d\xi_2)[(1/\xi_2)(dF/d\xi_2)]}\frac{1}{f(0)} \tag{6-74}$$

or

$$-F\frac{dF}{d\xi_2} = \frac{\epsilon_m f(0)}{\bar{U}_{z,\max}(z+a)}\xi_2^{\,2}\frac{d}{d\xi_2}\left(\frac{1}{\xi_2}\frac{dF}{d\xi_2}\right)$$

This equation can be integrated readily to

$$\frac{\epsilon_m f(0)}{\bar{U}_{z,\max}(z+a)}\left(\xi_2\frac{dF}{d\xi_2} - 2F\right) = -\tfrac{1}{2}F^2$$

The constant of integration must be zero, because at $\xi_2 = 0$ also $F(\xi_2)$ must be zero.
The next integration yields

$$F(\xi_2) = \frac{2C\xi_2^{\,2}}{1 + [\bar{U}_{z,\max}(z+a)/2\epsilon_m f(0)]C\xi_2^{\,2}}$$

whence

$$f(\xi_2) = \frac{1}{\xi_2}\frac{dF}{d\xi_2} = \frac{4C}{\{1 + C[\bar{U}_{z,\max}(z+a)/2\epsilon_m f(0)]\xi_2^{\,2}\}^2}$$

The constant of integration C follows from the condition at $\xi_2 = 0$. This yields $C = \frac{1}{4}f(0)$. Consequently,

$$\frac{\bar{U}_z}{\bar{U}_{z,\max}} = \left[1 + \frac{\bar{U}_{z,\max}(z+a)}{8\epsilon_m}\xi_2^{\,2}\right]^{-2} \tag{6-75}$$

Prandtl's Mixing-Length Theory

Here we put

$$\frac{\epsilon_m}{\bar{U}_{z,\max}(z+a)} = \frac{\mathrm{l}_m^{\,2}}{(z+a)^2}\left|\frac{d}{d\xi_2}\frac{\bar{U}_z}{\bar{U}_{z,\max}}\right|$$

We have already mentioned that the term on the left-hand side is a function of ξ_2 only. The mixing-length l_m is assumed to be a function of $z + a$ only. Consequently,

$$\mathrm{l}_m = c_m(z+a)$$

Substituting these expressions for $\epsilon_m/\bar{U}_{z,\max}(z+a)$ and l_m, respectively, in Eq. (6-74) and with Eq. (6-71), we obtain

$$c_m^{\,2}\left(\frac{d^2F}{d\xi_2^{\,2}} - \frac{1}{\xi_2}\frac{dF}{d\xi_2}\right)^2 = F\frac{dF}{d\xi_2}$$

or, with $\xi_2^* = \xi_2/c_m^{\,2/3}$,

$$\left(\frac{d^2F}{d\xi_2^{*2}} - \frac{1}{\xi_2^*}\frac{dF}{d\xi_2^*}\right)^2 = F\frac{dF}{d\xi_2^*}$$

Tollmien[10] has solved this equation. He put

$$F = \exp\left(\int Z\, d\xi_2^*\right)$$

with which the differential equation becomes

$$\frac{dZ}{d\xi_2^*} = \frac{Z}{\xi_2^*} - Z^2 - \sqrt{Z}$$

This equation he solved by means of series expansion. Thus Tollmien obtained

$$Z = \frac{2}{\xi_2^*} - \frac{2\sqrt{2}}{7}(\xi_2^*)^{1/2} - \frac{1}{245}(\xi_2^*)^2 + \frac{\sqrt{2}}{1{,}715}(\xi_2^*)^{7/2} + \cdots$$

The velocity distribution in terms of ξ_2^* is given by

$$\frac{\bar{U}_1}{\bar{U}_{1,\max}} = f(\xi_2^*) = \frac{1}{\xi_2^*}\frac{dF}{d\xi_2^*}$$

$$= \left[1 - \frac{\sqrt{2}}{7}(\xi_2^*)^{3/2} - \cdots\right]\exp\left[-\frac{4\sqrt{2}}{21}(\xi_2^*)^{3/2} - \cdots\right] \tag{6-76}$$

From this expression it follows that, for small ξ_2^*, that is, in the neighborhood of $\xi_2^* = 0$, the function $f(\xi_2^*)$ behaves like

$$f(\xi_2^*) \simeq 1 - \frac{\sqrt{2}}{3}(\xi_2^*)^{3/2} - \cdots \tag{6-76a}$$

This result shows that the curve has an infinitely great curvature at the top $\xi_2^* = 0$.

Case: $\bar{U}_z/U_s \ll 1$ and $\bar{U}_r/U_s \ll 1$, so $U_s/U_p \to 1$. In this case we had obtained

$$p = -\tfrac{2}{3} \qquad q = \tfrac{1}{3} \qquad s_{21} = -\tfrac{4}{3}$$

Eq. (6-61) then reduces to

$$2f(\xi_2) + \xi_2\frac{df}{d\xi_2} = -\frac{3}{\xi_2}\frac{d}{d\xi_2}(\xi_2 h)$$

This equation can readily be integrated to

$$\xi_2 f(\xi_2) = -3h(\xi_2) \tag{6-77}$$

The constant of integration must be zero, since $h(0) = 0$. The shear-stress distribution can then be expressed in terms of the mean-velocity distribution with the aid of Eq. (6-67c). Since according to Eq. (6-67a)

$$\frac{\bar{U}_{z,\max}}{U_p} = \left(\frac{z+a}{d}\right)^{-2/3} f(0)$$

from Eq. (6-67c) we obtain

$$\frac{\sigma_{rz}}{\rho \bar{U}_{z,\max}^2} = -\frac{1}{3f(0)} \xi_2 \frac{\bar{U}_z}{\bar{U}_{z,\max}} \qquad (6\text{-}78)$$

From Eq. (6-68) we conclude that

$$\frac{\epsilon_m}{U_p d}\left(\frac{z+a}{d}\right)^{+1/3} = \frac{h(\xi_2)}{df/d\xi_2} \qquad (6\text{-}79)$$

Just as we did in the case $U_s = 0$, we again assume that the eddy viscosity is independent of ξ_2. The distribution of $\bar{U}_z$ follows from Eqs. (6-77) and (6-79), namely

$$\frac{df}{d\xi_2} \cdot \left[\frac{\epsilon_m}{U_p d}\left(\frac{z+a}{d}\right)^{1/3}\right] + \tfrac{1}{3}\xi_2 f(\xi_2) = 0$$

Integration then leads to the Gaussian error law for the distribution of $\bar{U}_z$ in a cross-section (see also ref. 12).

$$\frac{\bar{U}_z}{\bar{U}_{z,\max}} = \frac{f(\xi_2)}{f(0)} = \exp\left[-\frac{U_p d^{4/3}}{6\epsilon_m (z+a)^{1/3}} \xi_2{}^2\right] \qquad (6\text{-}80)$$

Reichardt's Inductive Theory

In order to apply Reichardt's inductive theory, we have to start from the momentum equation, which, neglecting the pressure and viscous terms, reads

$$\frac{\partial}{\partial z}\overline{(U_s + U_z)^2} + \frac{1}{r}\frac{\partial}{\partial r} r\overline{(U_s + U_z)U_r} = 0 \qquad (6\text{-}81)$$

This equation transforms into the diffusion equation, if we assume Reichardt's hypothesis

$$\overline{(U_s + U_z)U_r} = -\mathscr{L}_m \frac{\partial}{\partial r}\overline{(U_s + U_z)^2}$$

Again,

$$\overline{(U_s + U_z)^2} = U_s{}^2 + \overline{\mathfrak{V}^2}$$

where

$$\overline{\mathfrak{V}^2} = 2U_s\bar{U}_z + \bar{U}_z{}^2 + \overline{u_z{}^2}$$

and the diffusion equation becomes

$$\frac{\partial}{\partial z}\overline{\mathfrak{V}^2} = \mathscr{L}\frac{1}{r}\frac{\partial}{\partial r}\left(r\frac{\partial \overline{\mathfrak{V}^2}}{\partial r}\right) \qquad (6\text{-}82)$$

We have seen that the assumption of similarity of velocity profiles can be applied only to two distinct cases: $\bar{U}_z/U_s \gg 1$ (for example, $U_s = 0$) and $\bar{U}_z/U_s \ll 1$. In the case $U_s = 0$ we put, in accordance with the relation (6-65a):

$$\frac{\overline{\mathfrak{B}^2}}{U_p^{\,2}} = \left(\frac{z+a}{d}\right)^{-2} f_1(\xi_2)$$

Similarity also requires

$$\frac{\mathscr{L}_m}{z+a} = \text{const.} \tag{6-83}$$

Then $f_1(\xi_2)$ satisfies the differential equation

$$\frac{\mathscr{L}_m}{z+a}\frac{1}{\xi_2}\frac{d}{d\xi_2}\left(\xi_2\frac{df_1}{d\xi_2}\right) + \xi_2\frac{df_1}{d\xi_2} + 2f_1 = 0 \tag{6-84}$$

The solution of this differential equation, which satisfies the condition $f_1 = f_1(0)$ for $\xi_2 = 0$, is the Gaussian error law

$$f_1 = f_1(0)\exp\left(-\frac{z+a}{2\mathscr{L}_m}\xi_2^{\,2}\right)$$

Hence

$$\frac{\overline{\mathfrak{B}^2}}{U_p^{\,2}} = f_1(0)\left(\frac{d}{z+a}\right)^2 \exp\left(-\frac{z+a}{2\mathscr{L}_m}\xi_2^{\,2}\right) \tag{6-85}$$

In case $\bar{U}_z/U_s \ll 1$, the expression for $\overline{\mathfrak{B}^2}/U_s^{\,2}$ reduces to

$$\frac{\overline{\mathfrak{B}^2}}{U_s^{\,2}} \simeq 2\frac{\bar{U}_z}{U_s}$$

so that Eq. (6-82) reads

$$\frac{\partial}{\partial z}\frac{\bar{U}_z}{U_s} = \mathscr{L}_m\frac{1}{r}\frac{\partial}{\partial r}\left(r\frac{\partial \bar{U}_z/U_s}{\partial r}\right)$$

Since $U_s \simeq U_p$, we may use for $\bar{U}_z/U_s$ the same relation (6-67a) as for $\bar{U}_z/U_p$. If we put $\mathscr{L}_m(z+a)^{1/3}/d^{4/3} = \text{const.}$, exactly the same differential equation is obtained for $f(\xi_2)$ as obtained earlier on the assumption of an eddy viscosity independent of ξ_2 [see Eq. (6-79)]. It led to the Gaussian solution Eq. (6-80). In the present case the Gaussian solution reads

$$\frac{\overline{\mathfrak{B}^2}}{U_s^{\,2}} = 2\frac{\bar{U}_z}{U_s} = 2f(0)\left(\frac{d}{z+a}\right)^{2/3} \exp\left[-\frac{d^{4/3}}{6\mathscr{L}_m(z+a)^{1/3}}\xi_2^{\,2}\right]$$

where $\mathscr{L}_m(z+a)^{1/3}/d^{4/3} = \text{const.}$ As we have found earlier for the wake flow behind a cylinder, the relation between $\mathscr{L}_m/d$ and $\epsilon_m/U_s d$ is given by $\mathscr{L}_m U_s = \epsilon_m$.

6-9 THE TRANSPORT OF A SCALAR QUANTITY IN A ROUND FREE JET

The equation of transport for $\bar{\Gamma}$ in cylindrical coordinates, to which the approximations considered in free turbulent flow have been applied, reads

$$(U_s + \bar{U}_z)\frac{\partial\bar{\Gamma}}{\partial z} + \bar{U}_r\frac{\partial\bar{\Gamma}}{\partial r} = -\frac{1}{r}\frac{\partial}{\partial r}(r\overline{u_r\gamma}) \tag{6-86}$$

We assume again that the turbulence transport of Γ can be described by means of a diffusion tensor $(\epsilon_\gamma)_{ij}$. Then, to the same degree of approximation, the expression for $\overline{u_r\gamma}$ reads

$$-\overline{u_r\gamma} = \epsilon_\gamma\frac{\partial\bar{\Gamma}}{\partial r} \tag{6-87}$$

where ϵ_γ is actually the component $(\epsilon_\gamma)_{rr}$ of the diffusion tensor. With

$$\bar{\Gamma} = \Gamma_s + \bar{\Gamma}_1$$

$$U_p = \text{issuing velocity}$$

$$\Gamma_p = \text{value of } \Gamma \text{ for the issuing fluid}$$

the equation of transport becomes

$$\left(\frac{U_s}{U_p} + \frac{\bar{U}_z}{U_p}\right)\frac{\partial}{\partial z}\frac{\bar{\Gamma}_1}{\Gamma_p} + \frac{\bar{U}_r}{U_p}\frac{\partial}{\partial r}\frac{\bar{\Gamma}_1}{\Gamma_p} = \frac{1}{r}\frac{\partial}{\partial r}\left[r\frac{\epsilon_\gamma}{U_p}\frac{\partial}{\partial r}\frac{\bar{\Gamma}_1}{\Gamma_p}\right] \tag{6-88}$$

If we assume similarity of the $\bar{\Gamma}_1$ and $\bar{U}_z$ profiles in consecutive sections, the same difficulties will be encountered in solving this equation as in solving the equation for the velocity distribution. Only the cases $\bar{U}_z/U_s \gg 1$ and $\bar{U}_z/U_p \ll 1$, $U_s/U_p \simeq 1$ can give solutions based upon complete similarity.

In the case $\bar{U}_z/U_p \ll 1$, where $\bar{U}_r/U_p \ll 1$ also, a constant value of the coefficient of eddy diffusion ϵ_γ will result in a Gaussian solution, analogous to the solution (6-80).

Thus we shall consider in what follows only the case $\bar{U}_z/U_s \gg 1$, for instance, the limiting case $U_s = 0$. From the requirements of similarity it follows that we may put

$$\frac{\bar{\Gamma}_1}{\Gamma_p} = \frac{d}{z+a}k(\xi_2) \tag{6-89}$$

where $\xi_2 = r/(z+a)$.

We assume that the origin a of the apparent source for Γ_1 is the same as that for the velocity U_z. Together with the corresponding relations (6-65) for $\bar{U}_z/U_p$ and $\bar{U}_r/U_p$, Eq. (6-88), reduced for $U_s = 0$, becomes

$$-f\frac{d}{d\xi_2}(\xi_2 k) + g\frac{dk}{d\xi_2} = \frac{1}{\xi_2}\frac{d}{d\xi_2}\left[\xi_2\frac{\epsilon_\gamma}{U_p d}\frac{dk}{d\xi_2}\right] \tag{6-90}$$

Now substitute for $f(\xi_2)$ and $g(\xi_2)$ the relations (6-71) involving the function $F(\xi_2)$. Equation (6-90) then reads

$$-\frac{1}{\xi_2}\frac{d}{d\xi_2}(Fk) = \frac{1}{\xi_2}\frac{d}{d\xi_2}\left[\xi_2\frac{\epsilon_\gamma}{U_p d}\frac{dk}{d\xi_2}\right]$$

This equation can be integrated. Since the constant of integration must be zero because of the boundary conditions, the solution reads

$$F(\xi_2)k(\xi_2) = -\frac{\epsilon_\gamma}{U_p d}\xi_2\frac{dk}{d\xi_2} \tag{6-91}$$

The next integration yields

$$\frac{k(\xi_2)}{k(0)} = \frac{\bar{\Gamma}_1}{\bar{\Gamma}_{1,\max}} = \exp\left[-\int_0^{\xi_2}\frac{U_p d}{\epsilon_\gamma}F(\xi_2)\frac{d\xi_2}{\xi_2}\right] \tag{6-92}$$

Thus the $\bar{\Gamma}_1$ distribution is expressed directly in terms of the velocity distribution and can be calculated if the velocity distribution and the coefficient of eddy diffusion ϵ_γ are known.

The expression (6-92) can be worked out further by expressing the velocity distribution in terms of the coefficient of eddy diffusion ϵ_m. For, from Eqs. (6-65c), (6-72), and the relation $-\overline{u_r u_z} = \epsilon_m \,\partial\bar{U}_z/\partial r$, a relation similar to Eq. (6-91) can be obtained, namely

$$F(\xi_2)f(\xi_2) = -\frac{\epsilon_m}{U_p d}\xi_2\frac{df}{d\xi_2} \tag{6-93}$$

Elimination of $F(\xi_2)$ from Eqs. (6-92) and (6-93) yields

$$\frac{1}{k(\xi_2)}\frac{dk}{d\xi_2} = \frac{\epsilon_m}{\epsilon_\gamma}\frac{1}{f(\xi_2)}\frac{df}{d\xi_2}$$

Whence follows for $k(\xi_2)/k(0) = \bar{\Gamma}_1/\bar{\Gamma}_{1,\max}$

$$\frac{\bar{\Gamma}_1}{\bar{\Gamma}_{1,\max}} = \exp\left[\int_0^{\xi_2} d\xi_2\frac{\epsilon_m}{\epsilon_\gamma}\frac{d}{d\xi_2}\left(\ln\frac{\bar{U}_z}{\bar{U}_{z,\max}}\right)\right] \tag{6-94}$$

If the ratio $\epsilon_m/\epsilon_\gamma$ is independent of ξ_2, an expression identical with the expression (6-50) for the wake flow behind a cylinder is obtained.

This constancy of the ratio is assumed, for instance, if *Prandtl's mixing-length hypothesis* is applied. For we then have

$$\frac{\epsilon_m}{U_p d} = c_m^2\left|\frac{d}{d\xi_2}\frac{\bar{U}_z}{\bar{U}_{z,\max}}\right|$$

$$\frac{\epsilon_\gamma}{U_p d} = c_\gamma^2\left|\frac{d}{d\xi_2}\frac{\bar{U}_z}{\bar{U}_{z,\max}}\right|$$

so that $\epsilon_m/\epsilon_\gamma = c_m^2/c_\gamma^2$.

The relation (6-94) then becomes

$$\frac{\bar{\Gamma}_1}{\bar{\Gamma}_{1,\max}} = \left(\frac{\bar{U}_z}{\bar{U}_{z,\max}}\right)^{c_m{}^2/c_\gamma{}^2} \tag{6-95}$$

Thus, for $c_m = c_\gamma$, the $\bar{\Gamma}_1$ distribution in the same as the velocity distribution.

Finally, let us consider *Reichardt's inductive theory.* If we neglect molecular effects, we obtain the following simplified equation of Γ-transport:

$$\frac{\partial}{\partial z}\overline{(U_s + U_z)\Gamma} + \frac{1}{r}\frac{\partial}{\partial r}(r\overline{U_r\Gamma}) = 0$$

Following Reichardt, we introduce the hypothesis [see the relation (5-31)]

$$\overline{U_r\Gamma} = -\mathscr{L}_\gamma \frac{\partial}{\partial r}\overline{(U_s + U_z)\Gamma} \tag{6-96}$$

so that we obtain

$$\frac{\partial}{\partial z}\overline{(U_s + U_z)\Gamma} = \mathscr{L}_\gamma \frac{1}{r}\frac{\partial}{\partial r}\left[r\frac{\partial}{\partial r}\overline{(U_s + U_z)\Gamma}\right] \tag{6-97}$$

With

$$U_z = \bar{U}_z + u_z$$

$$\Gamma = \Gamma_s + \bar{\Gamma}_1 + \gamma$$

we have

$$\overline{(U_s + U_z)\Gamma} = U_s\Gamma_s + U_s\bar{\Gamma}_1 + \Gamma_s\bar{U}_z + \bar{U}_z\bar{\Gamma}_1 + \overline{u_z\gamma} \tag{6-98}$$

Since $U_s\Gamma_s$ is constant, we need consider only the sum of the last four terms on the right-hand side of Eq. (6-98). The introduction of the similarity assumption concerning the distributions of momentum and Γ in consecutive sections of the jet leads, for the two cases for which this assumption is applicable, to a Gaussian solution for the quantity

$$\bar{U}_z\bar{\Gamma}_1 + \overline{u_z\gamma} \quad \text{if} \quad U_s = \Gamma_s = 0$$

whereby

$$\frac{\mathscr{L}_\gamma}{z + a} = \text{const.}$$

and for the quantity

$$U_s\bar{\Gamma}_1 + \Gamma_s\bar{U}_z \quad \text{if} \quad \bar{U}_z/U_s \ll 1 \quad \text{and} \quad \bar{\Gamma}_1/\Gamma_s \ll 1$$

whereby

$$\mathscr{L}_\gamma \frac{(z + a)^{1/3}}{d^{4/3}} = \text{const.}$$

Hence, it must be emphasized here again that only in the second case $\bar{U}_z/U_s \ll 1$; $\bar{\Gamma}_1/\Gamma_s \ll 1$ does the Reichardt theory lead to a Gaussian distribution for the mean velocity $\bar{U}_z$ and for $\bar{\Gamma}_1$. But in the other case this Gaussian distribution applies only to the momentum flux $\rho\overline{\mathfrak{V}^2}$ and to $(U_s\bar{\Gamma}_1 + \Gamma_s\bar{U}_z + \bar{U}_z\bar{\Gamma}_1 + \overline{u_z\gamma})$.

6-10 MEASUREMENTS OF MEAN-VELOCITY AND MEAN-TEMPERATURE DISTRIBUTION IN A ROUND FREE JET

The round free jet has been the subject of many experimental investigations. Most of them have been concerned with a jet issuing in a still ambient fluid, and only a few of them with the more general case of a jet in a secondary stream. Perhaps the first to make an extensive investigation of a jet in a secondary stream were Forstall and Shapiro.[14–16] Later Acharya,[17] and Alexander, Kivnick, Comings, and Henze[34] made similar experiments.

More recently also measurements have been made in coaxial-jet configurations (see, e.g., references 47, 48, 49) which resemble the case of a jet (here inner jet) in a coaxial free stream (here outer, annular, jet). However, most of the configurations studied have a too narrow outer, annular, jet so that the inner and outer mixing regions already meet before the inner jet has been fully developed.

Forstall and Shapiro obtained empirical relations for the spread of the jet as a function of the distance from the orifice and for the decrease in the jet velocity, both for different ratios between the velocity U_s of the secondary flow and the issuing velocity U_p of the jet.

When the issuing velocity U_p is assumed to be uniform at the orifice, as well as the ambient velocity U_s (no boundary-layer effects), then close behind the orifice the turbulent mixing zone has zero width; this width increases with increasing distance from the orifice, up to a distance z_c such that the mixing zone covers the entire jet (see Fig. 6-16). The velocity at the axis maintains its constant value U_p up to this distance z_c. It is often said that, up to z_c, the jet contains a central potential core region. Farther downstream from the orifice the velocity at the axis of the jet decreases with increasing distance. It will be some distance farther before the jet flow becomes fully developed and the flow patterns in consecutive sections become similar. To this fully developed part of the jet the theories considered up to

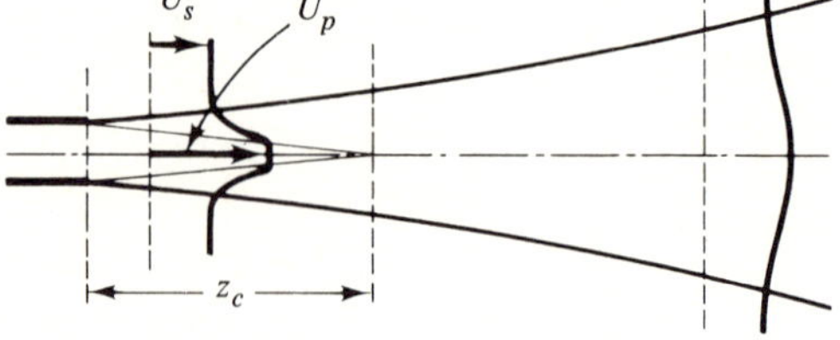

FIGURE 6-16
Mean-velocity distribution in the first, potential-core region and farther downstream of the orifice.

now may be applied. Usually the transition region between core region and fully developed turbulent region is quite short.

The length of the core region z_c is determined mainly by the value $\mu = U_s/U_p$; it increases with increasing μ. According to Forstall,[14] the following empirical relation holds

$$\frac{z_c}{d} = 4 + 12\mu \qquad (6\text{-}99)$$

Forstall carried out his measurements in the range $\mu \simeq 0.2$ to 0.5. Obviously, the relation (6-99) cannot be applied to cases differing too much from those covered by the experiments. From (6-99) it would follow that $x_c = 4d$ for $U_s/U_p = 0$, but experiments on free jets issuing in still ambient fluid have disclosed much greater values of z_c (see hereinafter).

According to Forstall's measurements the relative velocity of the jet at the axis decreases hyperbolically with increasing z

$$\frac{\bar{U}_{z,\max}}{U_p - U_s} = \frac{z_c}{z} \quad \text{for} \quad z > z_c \qquad (6\text{-}100)$$

For the change in the jet width (half-value radius $r_{1/2}$) he obtained

$$\frac{2r_{1/2}}{d} = \left(\frac{z}{z_c}\right)^{1-\mu} \quad \text{for} \quad z > z_c \qquad (6\text{-}101)$$

These two results are not entirely consistent with each other, if they are considered in the light of the theory expounded in the previous sections. The empirical hyperbolic relation (6-100) between mean velocity and distance seems to apply even at large distances from the orifice (for $U_s/U_p = 0.5$ Forstall's measurements covered the range from $z = 0$ to $z = 120d$). But, if this result were correct, one would expect a linear increase of the jet width with increasing z, not the relation (6-101). For, on the basis of the momentum-integral equation, Eq. (6-101) does not agree with Eq. (6-100); at any rate this is so if the velocity profiles in consecutive sections are similar. And, according to Forstall's measurements, the profiles are practically similar if at each section the velocities are expressed in terms of $\bar{U}_{z,\max}$ and the radial distances in terms of the half-value radius.

Forstall compared the measured velocity profiles with the Gaussian error curve. From this comparison he concluded that, for the central part of the jet, the Gaussian error curve gives a satisfactory agreement with the measured distribution.

According to the empirical relation (6-101), the increase in the jet width becomes zero if μ approaches the value 1. This seems logical, provided that both the primary and secondary flows are free from turbulence. On the other hand, according to the perturbation theory, the increase of jet width becomes proportional to $(z + a)^{1/3}$ in the limiting case $U_s/U_p \to 1$ [see the relation (6-67d)].

This seeming discrepancy may be explained as follows. Forstall's result holds only for this limiting case if the primary and secondary flows are indeed entirely free from turbulence, whereas the result of the perturbation theory holds only if any small difference between secondary and primary flow is sufficient to create a turbulence (causing the turbulent spread of the primary flow) such as is assumed in the corresponding theory.

Forstall and Shapiro[15] carried out measurements with a jet of a mixture of air and helium issuing in air. They found that the concentration distribution across a jet section could be satisfactorily given by a cosine or by a Gaussian error curve. The spread of the concentration turned out to be greater than that of the axial momentum. From their measurements one obtains a ratio between the coefficients of eddy diffusion

$$\frac{\epsilon_\gamma}{\epsilon_m} \simeq 1.4$$

This ratio is equal to the reciprocal of the eddy Schmidt number.

Landis and Shapiro[18] have shown by similar experiments with a heated jet issuing in a secondary flow that the same quantitative results are obtained for the spread of heat. Alexander, Kivnick, Comings, and Henze applied Reichardt's hypothesis to the results of their measurements on the spread of a heated air jet in a secondary airflow. They found a ratio $\mathscr{L}_\gamma/\mathscr{L}_m \simeq \frac{4}{3}$; this ratio is equivalent to the ratio $\epsilon_\gamma/\epsilon_m$. Acharya[17] obtained the same result in general as Landis and Shapiro. He observed that the ratio $\epsilon_\gamma/\epsilon_m$ seems to depend on μ, but no distinct law could be obtained, for the values obtained for ϵ_γ and ϵ_m as a function of μ were not very consistent.

More numerous and more extensive investigations have been made concerning a free jet issuing in still ambient fluid.

Most of the experiments on the fully developed part of round free jets are already relatively old, but many of the results obtained are still useful. We may mention those on gas jets issuing in air carried out by Trüpel,[19] Ruden,[20] Reichardt,[1] Corrsin,[21] Corrsin and Uberoi,[22,23] Albertson, Dai, Jenson, and Rouse,[25] Alexander, Baron, and Comings,[28] and by Hinze and Van der Hegge Zijnen.[24] Experiments on submerged water jets have been made by Forstall and Gaylord.[26] Of the more recent investigations we may mention those by Sami[50] and by Wygnanski and Fiedler.[51]

In general there is satisfactory agreement among the experimental results obtained by the various investigators. At large distances from the orifice a linear spread of the jet is observed and a hyperbolic decrease of the jet velocity with distance from some apparent source, and the radial velocity distributions in consecutive sections show good similarity. When z_c is defined as the distance from the orifice of the point of intersection of the constant, issuing, velocity in the jet axis and the curve of the hyperbolic decrease of $\bar{U}_{z,\max}$, then according to the measurements by Van der Hegge Zijnen, $z_c \simeq 6$ to $8d$. However, we may expect that this value depends on the actual, initial velocity distribution at the orifice, mainly because of a more or less thick boundary-layer at the inner wall of the orifice. Consequently

z_c should depend on the shape of the nozzle (beveled or sharp edge of the orifice, round tube, etc.) and on the Reynolds number. The latter effect decreases with increasing Reynolds number, and appears to become practically negligible when $\mathbf{Re}_d = U_p d/\nu > 10^5$–$10^6$. In general at distances much larger than z_c (i.e. in the self-preserving region) one observes that the velocity distribution is practically unaffected by the initial distribution, and determined only by the total jet momentum.

Comparison of the theoretical radial distribution of the axial velocity component, calculated according to the momentum-transport and modified vorticity-transport theories, with the measured distribution, gives almost the same picture as that found for the wake flow behind a cylinder. Both theories give too sharp an apex of the velocity distribution on the axis of the jet. The momentum-transport theory gives a slightly smaller deviation from the measured distribution than the modified vorticity-transport theory. Near the jet boundary the deviation, particularly that obtained with the modified vorticity-transport theory, becomes appreciable.

If the computed and the measured velocity distributions coincide at the half-value point $(\xi_2)_{1/2}$ [according to the measurements made by Van der Hegge Zijnen,[24] $(\xi_2)_{1/2} = 0.08$], the value of the mixing length according to the momentum-transport theory is

$$\mathrm{l}_m = c_m(z + a) = 0.017(z + a)$$

If we compare this value with the value of the half-value radius $r_{1/2} = 0.08(z + a)$, we obtain $\mathrm{l}_m = 0.21 r_{1/2}$. Just as in wake flow, these values of the mixing length are anything but small compared with the width of the mixing zone.

The Gaussian error curve gives very good agreement practically across the entire jet width. Values slightly too high are obtained near the apex of the velocity-distribution curve; at the boundary region the values are too low. Figure 6-17 shows the Gaussian error curve and the experimental curve obtained by Van der

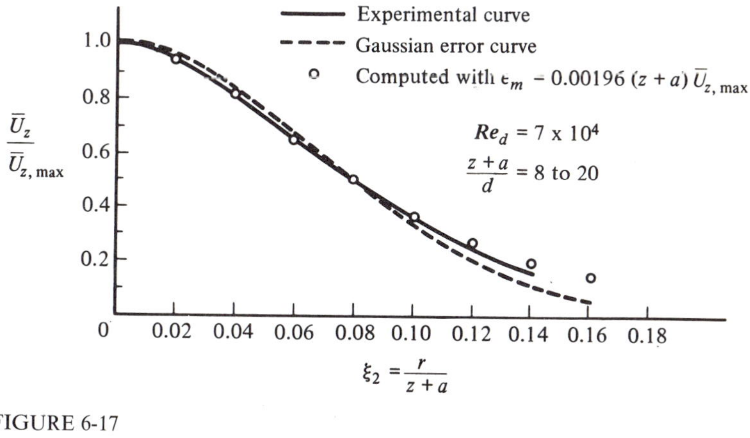

FIGURE 6-17
Mean-velocity distribution in the self-preserving part of a round free jet.

Hegge Zijnen. Van der Hegge Zijnen experimented with a jet of air issuing from an orifice of 25 mm diameter with an initial velocity of 40 m/s. The axial velocity component was measured with a thin total-head tube. The readings were corrected for the effect of turbulence, on the assumption that only the axial turbulence component contributes to this effect—an assumption which is certainly not free from doubt (see Sec. 2-10).

The Gaussian distribution matching the experimental curve at $(\xi_2)_{1/2} = 0.08$ reads

$$\frac{\bar{U}_z}{\bar{U}_{z,\max}} = \exp(-108\xi_2^2)$$

As mentioned earlier, beyond $z_c = 6$ to $8d$ the velocity distributions were similar and the velocity at the axis decreased in inverse proportion to the distance $z + a$; that is,

$$\frac{\bar{U}_{z,\max}}{U_p} = A\frac{d}{z + a}$$

From Van der Hegge Zijnen's measurements one obtains $A = 5.9$ and $a = -0.5d$. However, these values may depend on the initial conditions, so other values have been obtained by other investigators. For instance, Wygnanski and Fiedler[51] obtained $A = 5.9$ and $a = -3d$ when only the experimental data in the region $10 < (z + a)/d < 50$ are considered, the range covered by Van der Hegge Zijnen. However, Wygnanski and Fiedler observed a change in A and a, at larger distances. In the region $25 < (z + a)/d < 90$ they obtained a better agreement with $A = 5.4$ and $a = -7d$. The measurements by Wygnanski and Fiedler have been carried out under almost the same conditions as those by Van der Hegge Zijnen, namely with an air jet issuing from an orifice of 26 mm diameter, while most of their measurements have been made at an issuing velocity of 51 m/s. For $z/d > 40$ the mean-velocity profiles were similar in consecutive sections, and agreed nicely with the experimental curve shown in Fig. 6-17. When the mean radial velocity $\bar{U}_r$ is calculated from the $\bar{U}_z$-distribution using the mass-balance equation, the $\bar{U}_r$-distribution shows a maximum equal to $0.017\bar{U}_{z,\max}$ at $\xi_2 = 0.06$. It becomes zero at $\xi_2 = 0.127$ and negative (i.e., directed inward toward the jet axis) beyond this value of ξ_2, as it should be. For, this inward radial velocity at larger distances from the axis should account for the entrainment of ambient fluid by the jet. We must notice that the radial velocity is very small indeed, being at least one order of magnitude smaller than the turbulence velocities (see the next section).

The *rate of the entrainment* mentioned above can be easily calculated for the self-preserving part of the jet from its linear increase in diameter with distance from the origin of similarity. For it is equal to the increase of volume flow rate in consecutive sections. Since

$$\frac{\Phi}{\Phi_0} = \frac{8}{d^2}\int_0^\infty dr\, r\bar{U}_z/U_p$$

where $\Phi_0 = (\pi/4)d^2 U_p$, it can easily be shown that for the self-preserving region $\Phi/\Phi_0 = A(z + a)/d$, where

$$A = 8 \int_0^\infty d\xi_2 \, \xi_2 f(\xi_2)$$

Since theoretical distributions of $f(\xi_2)$ do not account for the intermittency of the turbulent flow in the outer part of the jet, and therefore do not agree with the experimental distribution in this part of the jet, the calculation of A from these theoretical distributions does not yield a reliable result. For instance, the theoretical distribution given by Eq. (6-75) for a constant ϵ_m and shown in Fig. 6-17 would yield too large a value. Indeed with $f(0) = 5.9$ one would obtain $A = 0.37$, whereas direct measurement of Φ in a special experimental setup by Ricou and Spalding[52] yielded a value of $A = 0.32$. It should be possible to calculate the value of A directly from the measured mean-velocity distribution. However, since the measurement of this velocity at the edge of the jet is rather inaccurate, and an extrapolation of $f(\xi_2)$ to zero has to be made, this calculation procedure yields an unreliable value of A also. The best figure calculated in this way has been obtained by Crow and Champagne[57] who obtained a value of $A = 0.292$ from their measured mean-velocity distributions in the range $6 < z/d < 10$. This value of A is still lower than the above value given by Ricou and Spalding.

By a method similar to Van der Hegge Zijnen's, Alexander, Baron, and Comings[28] measured the momentum flux $\rho\overline{U_z^2}$ with a total-head tube. For the radial distribution of this quantity they obtained satisfactory agreement with the Gaussian error curve across the entire jet width.

Also shown in Fig. 6-17 are the values computed on the basis of a constant coefficient of eddy diffusion ϵ_m [see Eq. (6-75)]. These theoretical values show the best agreement, particularly in the central region of the jet. Near the boundaries the theoretical values are somewhat too high. The value of the coefficient of eddy diffusion used to obtain this result amounts to

$$\epsilon_m = 0.00196\bar{U}_{z,\max}(z + a)$$

or, if expressed in terms of $U_p d$,

$$\epsilon_m = 0.0116 U_p d$$

(compare the corresponding value given for the wake flow behind a cylinder). For an issuing velocity $U_p = 40$ m/s and $d = 25$ mm we obtain $\epsilon_m = 0.0116$ m^2/s, which is roughly 1,000 times the value of the kinematic viscosity of air under atmospheric conditions.

It may be remarked that, since ϵ_m is proportional to the square root of the jet momentum flux, the whole mixing process in the jet is determined by the value of this momentum flux.

If we calculate the value of ϵ_m directly from the radial velocity distribution

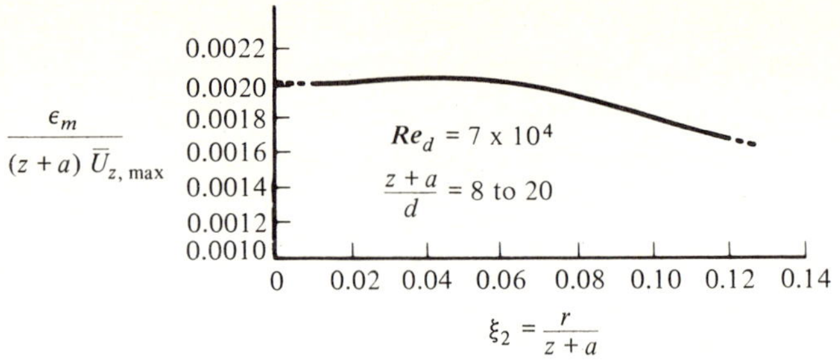

FIGURE 6-18
Coefficient of eddy viscosity computed from the mean-velocity distribution given in Fig. 6-17.

given in Fig. 6-17, we obtain an almost constant value of ϵ_m for the central region of the jet. This is shown in Fig. 6-18.

Accepting a constant value of ϵ_m across the main part of the jet and, thus, a velocity distribution according to Eq. (6-75), we obtain the shear-stress distribution [see Eq. (6-73)]

$$\frac{\sigma_{rz}}{\rho \bar{U}^2_{z,\max}} = -\tfrac{1}{2}\xi_2 \left[1 + \frac{\bar{U}_{z,\max}(z+a)}{8\epsilon_m}\xi_2{}^2\right]^{-3} \tag{6-102}$$

The shear stress has a maximum value at

$$(\xi_2)_{\text{opt}} = \left[\frac{8\epsilon_m}{5\bar{U}_{z,\max}(z+a)}\right]^{1/2}$$

Since

$$(\xi_2)_{1/2} = \left[\frac{8(\sqrt{2}-1)\epsilon_m}{\bar{U}_{z,\max}(z+a)}\right]^{1/2}$$

$(\xi_2)_{\text{opt}}$ is smaller than $(\xi_2)_{1/2}$, just as in the case of wake flow behind a cylinder. With the empirical value according to Van der Hegge Zijnen's measurements $\epsilon_m/\bar{U}_{z,\max}(z+a) = 0.00196$, there is obtained $(\xi_2)_{\text{opt}} = 0.056$ and $(\xi_2)_{1/2} = 0.08$.

Experiments on heated jets have been made by Ruden,[20] Corrsin,[21] Corrsin and Uberoi,[22,23] and by Hinze and Van der Hegge Zijnen,[24] who also investigated the spread of tracer gas (town gas, hydrogen, carbon dioxide, and methane) in an isothermal jet. In all their experiments the temperature differentials and concentrations were sufficiently small to exclude appreciable density effects. Hinze and Van der Hegge Zijnen found that in that case there is no difference in the spread of heat and matter; if the spread is expressed as a coefficient of eddy diffusion ϵ_γ, this coefficient has the same value for heat and matter.

The spread of heat and matter is much greater than that of momentum. Van der Hegge Zijnen obtained an averaged value of $\epsilon_\gamma/\epsilon_m \simeq 1.36$ across the jet. Ruden obtained for his heated jet a ratio of 1.32. Corrsin, in his experiments with heated jets, obtained a ratio of 1.43 independent of the magnitude of the temperature differential, up to an initial excess temperature of 300°C. All these values of $\epsilon_\gamma/\epsilon_m$ are in rough agreement with the value of 1.38 found by Keagy and Weller[27] for an isothermal jet of pure nitrogen issuing in still air and with the value 1.4 mentioned earlier in connection with Forstall's measurements on the jet issuing in a moving fluid.

It may be remarked that the value $\epsilon_\gamma/\epsilon_m \simeq 1.4$, which has been found to apply to the axi-symmetric air jet, is close to the reciprocal molecular Prandtl number for air under atmospheric conditions. This agreement, however, is only fortuitous. For in the case of the plane wake flow behind a cylinder a much higher value is obtained, namely 1.85. This value of about 1.85 was also found by Reichardt[1] in a plane free jet issuing in still air. Similarly, Van der Hegge Zijnen[53] obtained from his measurements in a *plane free jet* an average value of the same order of magnitude; the local value varied from 1.7 to 2.4 across the jet. More recent experiments by Wygnanski and Fiedler[54] in a two-dimensional mixing region, where the lateral turbulent diffusion of axial momentum and that of turbulence kinetic energy have been determined, showed an average value across the mixing region of $\epsilon_\gamma/\epsilon_m \simeq 2.2$.

Consider in what follows only the experimental results obtained by Van der Hegge Zijnen. Beyond z_c the *concentration* distribution in the axial direction along the axis of the jet followed closely the hyperbolic relation $\bar{\Gamma}_{1,\max}/\Gamma_p = 5.27d/(z+a)$. The value of a was different from that for the velocity distribution, namely $0.8d$ as compared with $-0.5d$.

The axial *temperature* distribution, however, showed noticeable deviations from this hyperbolic relationship. Apparently the differences in density caused by the cooling process during mixing with the cold ambient air did not noticeably affect the radial temperature distribution, but did so affect the axial temperature distribution. As was to be expected, the temperature in the jet decreased in the axial direction faster than it would according to a true hyperbolical relationship. At an initial temperature of the issuing jet amounting to 30°C above ambient air temperature, the deviation from the hyperbolic law was only a few per cent.

If a comparison is made between the measured radial temperature or concentration distribution and the theoretical $\bar{\Gamma}_1$ distribution according to the mixing-length theory, deviations in the same sense as those obtained for the wake flow of a cylinder are found.

From Van der Hegge Zijnen's data we obtain for the mixing length according to Prandtl's theory

$$l_\gamma = c_\gamma(z+a) = 0.02(z+a)$$

The $\bar{\Gamma}_1$-distribution calculated on the assumption of a constant coefficient of eddy diffusion ϵ_γ, together with a constant coefficient of eddy viscosity ϵ_m for the velocity distribution, reads

$$\frac{\bar{\Gamma}_1}{\bar{\Gamma}_{1,\max}} = \left(\frac{\bar{U}_z}{\bar{U}_{z,\max}}\right)^{\epsilon_m/\epsilon_\gamma} = \left[1 + \frac{\bar{U}_{z,\max}(z+a)}{8\epsilon_m}\xi_2{}^2\right]^{-2\epsilon_m/\epsilon_\gamma} \tag{6-103}$$

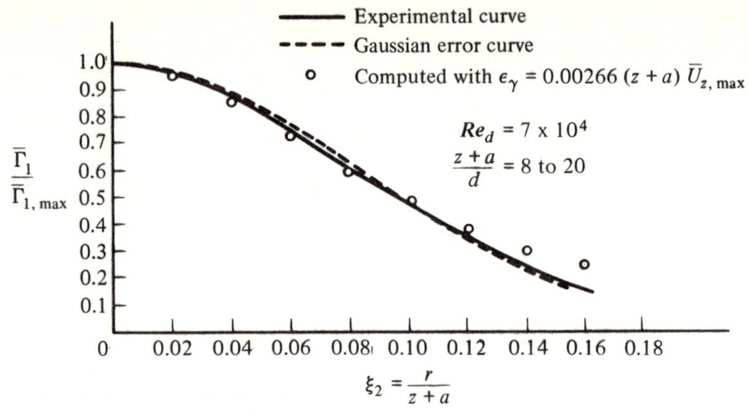

FIGURE 6-19
Radial distribution of $\bar{\Gamma}$ in a round free jet.

This yields much better agreement with the measured distribution with $\epsilon_\gamma/\epsilon_m = 1.36$. The corresponding value of ϵ_γ then reads $\epsilon_\gamma = 0.00266\bar{U}_{z,\max}(z + a)$. Yet, as may be concluded from Fig. 6-19, this agreement is not so good as that for the velocity distribution based on a constant value of ϵ_m. The Gaussian error curve is also shown in Fig. 6-19; it represents very satisfactorily the actual experimental curve across the whole jet.

This Gaussian distribution matching the experimental curve at $(\xi_2)_{1/2} = 0.098$ reads

$$\frac{\bar{\Gamma}_1}{\bar{\Gamma}_{1,\max}} = \exp(-72\xi_2^2)$$

According to Eq. (6-103) the $\bar{\Gamma}_1$-distribution can be obtained from the $\bar{U}_z$-distribution by a single transformation. The same holds true in the case of Gaussian distributions, where here the transformation factor is equal to $\frac{72}{108} = 0.66$.

From the fact that a constant value of the coefficient of eddy diffusion ϵ_γ for the $\bar{\Gamma}_1$ distribution does not yield the same result as a constant value of the coefficient of eddy viscosity ϵ_m for the velocity distribution, one may infer that the ratio $\epsilon_\gamma/\epsilon_m$ is certainly not constant across the jet, even in the central part of it. In Fig. 6-20 this ratio as computed from the experimental $\bar{\Gamma}_1$ and $\bar{U}_z$ distributions is shown; it decreases monotonously from the center of the jet toward its boundary.

The data obtained by Forstall and Gaylord[26] for the axial and radial velocity distributions in a submerged water jet agree nicely with those obtained by Hinze and Van der Hegge Zijnen. By introducing sodium chloride as a tracer into the jet liquid, the spread of matter in the jet could be studied. The data on the axial distribution agree with those obtained by Hinze and Van der Hegge Zijnen with gas jets, but there is some difference between the sets of data on radial distribution. The ratio $\epsilon_\gamma/\epsilon_m$ deduced from Forstall and Gaylord's data indicates a value in the range 1.2 to 1.3.

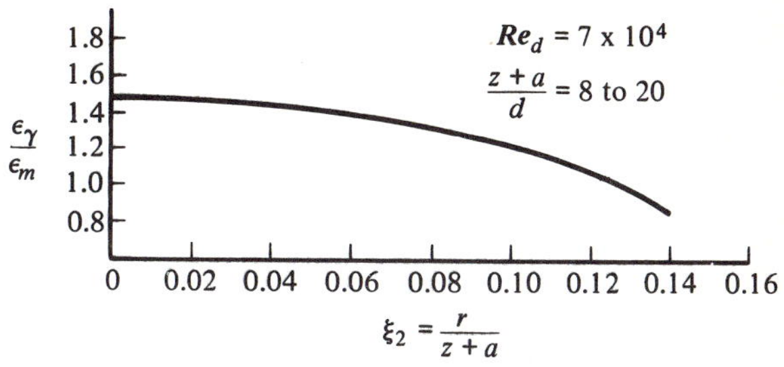

FIGURE 6-20
Radial distribution of $\epsilon_\gamma/\epsilon_m$ in a round free jet.

In the theories on free turbulent flow it has been assumed that the variation in static pressure across the mixing zone is negligibly small. This assumption was based on the fact that radial mean velocities are small compared with the axial mean velocities. Further support for this assumption is given by the calculations made by Tollmien.[10] In the central part of a jet the direction of the mean velocity is slightly outward, whereas in the outer region of the jet there occurs an inward flow. Thus a minimum in static pressure may be expected in the outer region of the jet. For the round free jet Tollmien obtained from his calculations the following relations:

$$\bar{P}_c - \bar{P}_{r^*} = +0.0012(\tfrac{1}{2}\rho\bar{U}^2_{z,\max})$$
$$\bar{P}_c - P_0 = +0.00075(\tfrac{1}{2}\rho\bar{U}^2_{z,\max})$$

where $\bar{P}_c$, $\bar{P}_{r^*}$, and P_0 are the static pressures at the center of the jet, at its average boundary, and in the undisturbed outside region, respectively.

These relations show that the static pressure differences are very small indeed.

However, Tollmien did not consider the effect of the turbulence on the static-pressure variation across the jet. The estimates of the orders of magnitude of the various terms in the equation of motion for the radial directions, given in Sec. 6-3, show that these static-pressure variations must be an order of magnitude larger than the variations caused by the mean velocities. The static-pressure variation due to turbulence is given by Eq. (6-24) or by its integrated form (6-24a). As will be discussed in the next section Corrsin[21] measured the radial distributions of the turbulence intensities u'_z and u'_r at $(z + a)/d = 20$, while Wygnanski and Fiedler[51] measured in addition also the intensity of the third component u'_φ in the region $30 < z/d < 97.5$. Wygnanski and Fiedler obtained $u'_z > u'_r \simeq u'_\varphi$. If $u'_r \simeq u'_\varphi$, then we may neglect the integral term in Eq. (6-24a). From Corrsin's measurements of u'_r at $(z + a)/d = 20$ (see Fig. 6-24) a static-pressure difference

$$\bar{P}_c - P_0 \simeq -0.15(\tfrac{1}{2}\rho\bar{U}^2_{z,\max})$$

is obtained. However, Wygnanski and Fiedler measured much smaller values of u'_r. At $z/d = 30$ they found $u'_r/\bar{U}_{z,\max} = 0.185$, increasing to ~ 0.24 for $z/d > 60$. With these values we would obtain for the numerical constant in the expression for $\bar{P}_c - P_0$ the values 0.07 and 0.116 respectively, as compared to the value 0.15 deduced from Corrsin's measurements.

A direct experimental determination of the static-pressure differences would be very welcome. But, owing to the high relative-turbulence intensities in free jets, it is extremely difficult to measure static pressures within a jet in a reliable way. Yet such measurements have been made. Barat[37] performed the measurements in a round free jet ($\mathbf{Re}_d = U_p d/\nu = 400{,}000$) up to a maximum distance $z/d = 10$. At this distance a minimum static pressure at the axis of the jet was found equal to

$$\bar{P}_c - P_0 \simeq -0.075(\tfrac{1}{2}\rho \bar{U}^2_{z,\max})$$

The absolute pressure difference decreased monotonously to zero toward the edge of the jet. Barat did not mention the method he used to measure the static pressure in the highly turbulent region.

Similar measurements, but in a plane free jet ($\mathbf{Re} = U_p a/\nu = 18{,}000$, $a =$ slot-width) have been made by Miller and Comings.[43] The static pressure in the jet was measured through a 0.5 mm hole in a 12.5 mm-diameter disk parallel to the main flow. Static-pressure profiles similar to Barat's were obtained at various cross sections of the jet. The relative negative pressure difference increased in the downstream direction without reaching a constant level at the farthest cross section $x_1/a = 40$. At $x_1/a = 10$, Miller and Comings obtained for the ratio $(\bar{P}_c - P_0)/\frac{1}{2}\rho \bar{U}^2_{1,\max}$ the value ~ -0.052, at $x_1/a = 40$ the value ~ -0.116.

The experimental results by Barat and by Miller and Comings show that the static pressure differences in a jet are of the same order of magnitude (at least) as those expected from the relative turbulence intensities on the ground of the Eq. (6-24a). The lateral static-pressure differences are well in balance with the lateral variation of the turbulence stresses.

In all the foregoing considerations we have deliberately neglected the *effect of density variations*. In general these effects have proved to be small indeed, even at moderate density differences. Experiments on free jets issuing in still air and having substantial initial density differences showed only small differences in the general shape of the jet. The velocity distribution is practically the same as that in jets at constant density. However, one may expect an effect of a density difference, when this is not small, on the penetration rate of the jet, i.e., on the width of the jet. A number of investigators have found, indeed, that the value of $(\xi_2)_{1/2}$ changes with this density difference. The value becomes smaller (larger) if the density of the jet fluid is greater (smaller) than that of the ambient fluid.

At equal density the half-value radius of the jet amounts to about $(\xi_2)_{1/2} \simeq 0.08$. According to Corrsin and Uberoi,[22] this value increased to 0.094 when the initial temperature of the jet was 300°C above ambient temperature (initial density ratio about 0.5). They found, moreover, that the length of the "potential" core region decreased substantially, and that the velocity at the axis decreased at a higher rate with increasing distance from the orifice than it did in the case of a constant-density jet. Szablewski[38] made calculations for the density ratios $\rho_p/\rho_0 = 0.5$, 1, and 2.5. He calculated that the length of the potential core increased but the width of the jet decreased with increasing ρ_p/ρ_0.

Similar results have been obtained more recently from experiments with cold air-jets, flowing upward from a 6 mm diameter orifice, by Uberoi and Garby.[55] The lowest initial temperature of the jet was -173°C. At $z/d = 10$, $(\xi_2)_{1/2}$ increases linearly with the absolute initial temperature of the jet. The value of z_c decreased and the fall of $\bar{U}_{z,\max}$ with z/d increased with increasing initial temperature.

Keagy and Weller[39] experimented with vertical jets of helium and of carbon dioxide issuing in still air. They obtained for the helium jet ($\rho_p/\rho_0 = 0.14$) a half-value radius $(\xi_2)_{1/2} \simeq 0.1$ and for the carbon dioxide jet ($\rho_p/\rho_0 = 1.5$) a value of $\simeq 0.076$. It must be remarked that no corrections were made for buoyancy effects. Sunavala, Hulse, and Thring[40] experimented with air jets of different initial temperatures (up to 320°C above ambient temperature) issuing from orifices ranging from 4.7 to 9.5 mm in diameter. They, too, found that the rate of decrease of the velocity at the jet axis with increasing distance was greater at higher initial jet temperatures. The curves could be made to coincide with that for an isothermal constant-density jet if the orifice diameter was replaced by an equivalent diameter:

$$d_e = d\left(\frac{\Theta_0}{\Theta_p}\right)^{1/2} = d\left(\frac{\rho_p}{\rho_0}\right)^{1/2}$$

This concept of an *equivalent orifice diameter* was introduced by Thring and Newby.[41] It is based on the assumption that the whole development of a jet is determined by the total jet momentum. The equation for the total momentum flux reads ($U_s = 0$)

$$2\pi \int_0^\infty r\,dr\,\overline{\rho U_z^2} = \frac{\pi}{4} d^2 \rho_p U_p^2 = \frac{\pi}{4} d_e^2 \rho_0 U_p^2$$

whence follows

$$d_e = d\sqrt{\frac{\rho_p}{\rho_0}} \tag{6-104}$$

At sufficiently large distances from the orifice, for instance, at $x_1/d > 10$, the density within the jet will have approached that of the ambient fluid. If for such large distances we assume $\rho \simeq \rho_0$, we obtain

$$2\pi \int_0^\infty r\,dr\,\overline{U_z^2} = \frac{\pi}{4} d_e^2 U_p^2$$

The procedure applied to the constant-density jet may be followed. The assumption of similarity of velocity profiles again yields [see Eq. (6-65a)]

$$\frac{\bar{U}_z}{U_p} = \frac{d_e}{z+a} f(\xi_2)$$

If a is unaffected by the density ratio ρ_p/ρ_0, this relation shows that the velocity has decreased to a lower value with increasing $(z + a)$ if $d_e < d$, that is, if $\rho_p/\rho_0 < 1$. But the distribution function $f(\xi_2)$ may remain the same.

Ricou and Spalding[52] also found that the equivalent diameter d_e can be used to express the increase of mass-flow rate of a jet due to its entrainment action, namely $\Phi_m/\Phi_{m,0} = A(z + a)/d_e$, where again $A \simeq 0.32$.

Density differences may also occur from pressure differences, for example, in the case of high-speed compressible-flow jets. Pai[12] calculated the spread of a jet in a high-speed secondary stream where $U_p/U_s \simeq 1$, applying a small-perturbation method and assuming a constant eddy viscosity across any section of the jet. The differential equation for the velocity in that case becomes identical with the corresponding equation for a laminar jet, so that an identical solution applies.

Since at some distance from the orifice the density differences with the ambient fluid become small, here, too, the concept of an equivalent orifice diameter can be usefully applied to describe the over-all decay of the jet velocity with distance from the orifice.

6-11 MEASUREMENTS OF TURBULENCE QUANTITIES IN A ROUND FREE JET

For the fully developed turbulent jet, experimental results are available from work by Corrsin,[21] Corrsin and Uberoi,[22,23] and Corrsin and Kistler,[29] while extensive turbulence measurements have been made by Wygnanski and Fiedler.[51] Earlier similar measurements were carried out by Liepmann and Laufer[30] in the mixing zone between the potential core region and the ambient air of the first part of an isothermal air jet. These measurements were later extended by Laurence.[31]

All these experimenters used the hot-wire anemometer essentially for measuring turbulence characteristics. An entirely different technique: use of photographic records to study the movements of lint particles in a low-speed turbulent air jet, was applied by Chin and Rib[32] for determining energy spectra and the rate of energy dissipation. However, their data showed such large scatter that the results obtained are not very convincing.

In this section we shall deal mainly with the results obtained by Wygnanski and Fiedler.

As mentioned earlier they used an orifice of 26 mm diameter, and experimented with air jets with an issuing velocity of 51 m/s, while some measurements were made at 72 m/s. Constant-temperature anemometers and linearizers were used, and hot wires of 5 μm diameter and 1.2 mm length. An important remark should be made concerning the electronic equipment. Wygnanski and Fiedler observed that an important fraction of the kinetic energy was present at frequencies as low as 1 Hz. Consequently, for measurements of the turbulence, a.c. capacitive couplings were used, so as to obtain a flat response down to frequencies of 0.05 Hz. Hitherto

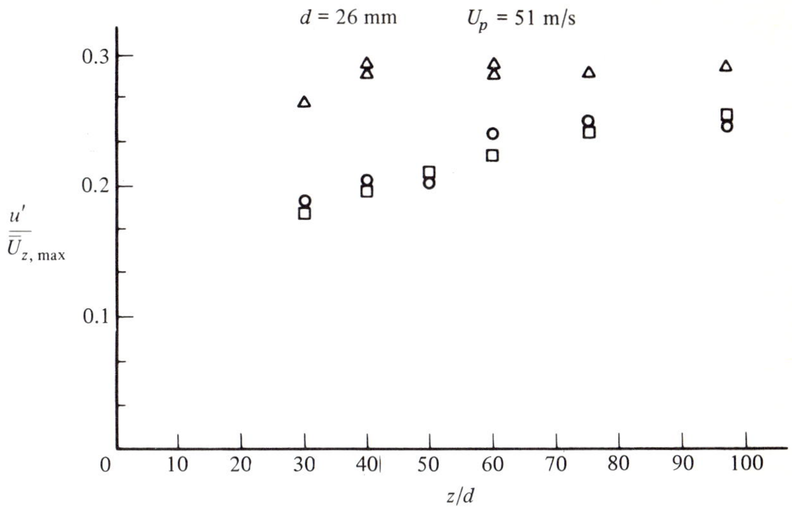

FIGURE 6-21
Axial distribution of $u'_z/\bar{U}_{z,\max}$ (△), $u'_r/\bar{U}_{z,\max}$ (□) and $u'_\varphi/\bar{U}_{z,\max}$ (○) on the jet axis. (From: *Wygnanski, I., and H. Fiedler,*[51] *by permission of the Cambridge University Press.*)

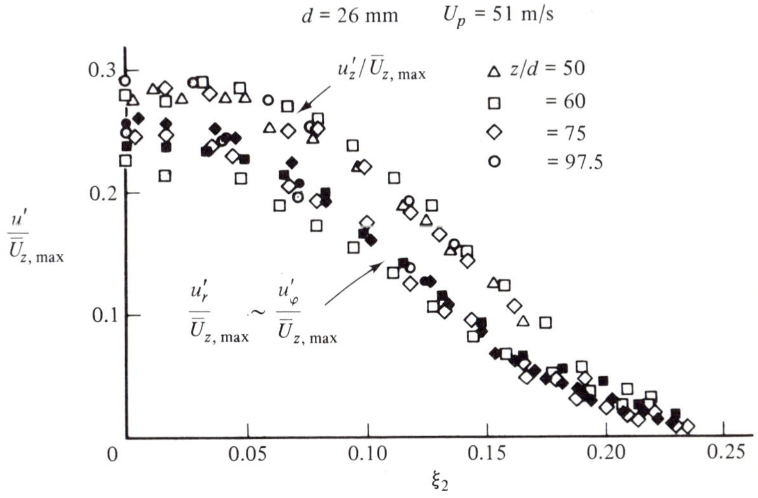

FIGURE 6-22
Radial distribution of $u'_z/\bar{U}_{z,\max}$, $u'_r/\bar{U}_{z,\max}$ (open marks) and $u'_\varphi/\bar{U}_{z,\max}$ (black marks). (After: *Wygnanski, I., and H. Fiedler,*[51] *by permission of the Cambridge University Press.*)

this point has been overlooked by many investigators, or insufficiently taken into account. For instance, it explains why earlier investigators measured too low values for the turbulence shear-stress in the outer parts of the jet. Figure 6-21 shows the variation of the relative intensities $u_z'/\bar{U}_{z,\max}$, $u_r'/\bar{U}_{z,\max}$ and $u_\varphi'/\bar{U}_{z,\max}$ along the jet-axis. We may recall that up to $z/d = 60$ the mean-velocity profiles were similar with an apparent origin of similarity at $a/d = -3$, while in the region $30 < z/d < 100$ this origin appeared to be at $a/d = -7$. We note from this figure that $u_r' \simeq u_\varphi'$, but that the two are much smaller than the axial component u_z'. This in contrast to results obtained by earlier investigators as Corrsin[21] and Gibson,[56] who observed almost isotropy of the turbulence at the jet-axis, a result which may be expected since at the jet-axis there is no production of turbulence, and the turbulence present is only due to transport by the mean and the turbulent motion. Figure 6-22 shows the radial distribution of the three turbulence intensities in the self-preserving region. In Fig. 6-23 the shear-stress distribution measured and computed from the mean-velocity distribution is shown. The agreement between the two is satisfactory indeed, in contrast to earlier measurements and referred to above. The maximum value of the shear stress is reported to occur at $(\xi_2)_{\text{opt}} = 0.058$, which is in satisfactory agreement with the value of 0.056 mentioned earlier, and obtained from Eq. (6-102). For comparison Fig. 6-24 shows the radial distribution of the relative intensity of the axial and radial turbulence velocities as measured by Corrsin[21] at $(z + a)/d = 20$ in an isothermal air jet, while Fig. 6-25 shows the radial distribution of the relative intensity of the axial turbulence velocity and of the turbulent temperature fluctuation, also measured at the same cross-section in a heated air jet by Corrsin and Uberoi.[22] Comparison of the two figures reveals that in the hot jet the relative intensity of the turbulence velocity is smaller than that of the isothermal jet.

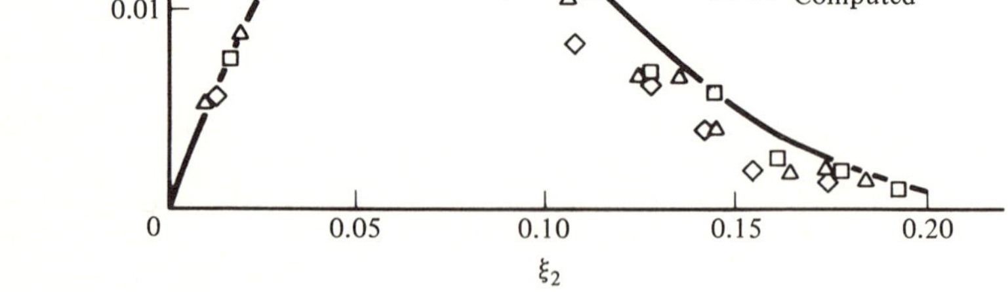

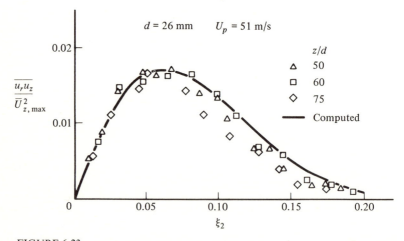

FIGURE 6-23
Shear-stress distribution, measured, and computed from mean-velocity distribution. (From: *Wygnanski, I., and H. Fiedler,*[51] *by permission of the Cambridge University Press.*)

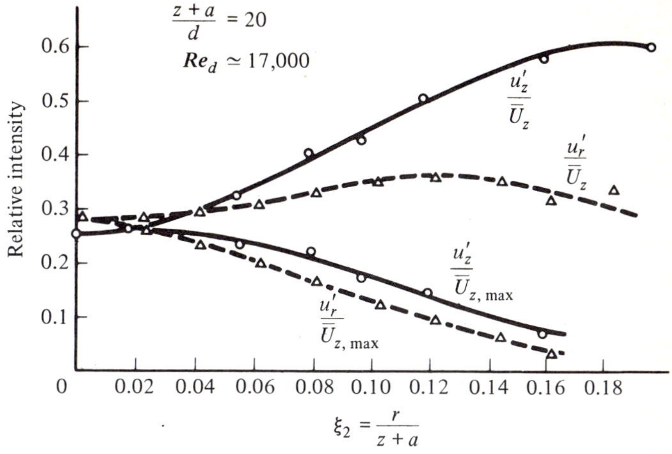

FIGURE 6-24
Distribution of relative turbulence intensities in an isothermal round free jet. (From: *Corrsin, S.,*[21] *by permission of the National Advisory Committee for Aeronautics.*)

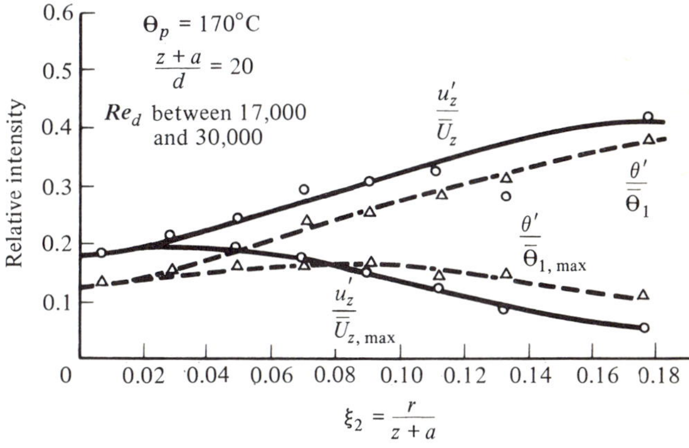

FIGURE 6-25
Distribution of relative intensity of turbulent velocity and temperature fluctuations in a hot round free jet. (From: *Corrsin, S. and M. S. Uberoi,*[22] *by permission of the National Advisory Committee for Aeronautics.*)

If we compare both the course and the values of the relative turbulence intensities with those obtained in the wake flow of a cylinder (see Fig. 6-5, where the distributions of the squares of the relative turbulence intensities are given), there is great similarity. With respect to the maximum velocity difference at the axis, the relative intensities of the turbulence components are of the same magnitude in the two flows, those of the wake flow being in general slightly higher. This difference may be real, but it may also be due to a greater measuring accuracy of the hot-wire anemometer in the wake flow than in the jet flow considered, even when the constant-temperature method with a linearization circuit is used. For, in the case of a free jet in a quiescent environment the value of $u'/\bar{U}_{z,\max}$ is decisive for the applicability of the hot-wire anemometer as a reliable instrument, when the usual method based upon small fluctuations is used. Since in the wake-flow measurements by Townsend $\bar{U}_1 \ll U_0$, the value of $u'/(U_0 - \bar{U}_1)$, which is decisive here, must have been very small, in contrast to the relatively high values of $u'/\bar{U}_{z,\max}$ in the jet flow.

A comparison between the variation in axial direction at the jet axis of mean-velocity $\bar{U}_{z,\max}$, turbulence intensity u'_z on the one side, and mean temperature $\bar{\Theta}_{1,\max}$ and turbulence intensity θ' on the other side according to the measurements by Corrsin and Uberoi[22] is shown in Fig. 6-26. In a self-preserving turbulent flow the relative intensities $u'_z/\bar{U}_{z,\max}$ and $\theta'/\bar{\Theta}_{1,\max}$ should be independent of z. Figure 6-26 shows that at $(z + a)/d \simeq 20$ the condition of self-preservation is hardly obtained, though, as discussed earlier, the radial distributions of mean-velocity and mean-temperature are already practically similar beyond $(z + a)/d = 10$. The departures from self-preservation for the turbulence intensities even increase toward the jet boundary. Other measurements by Corrsin and Uberoi on the radial distributions of these turbulence intensities show that self-preservation was not yet attained at $(z + a)/d = 40$, which was the maximum distance from the orifice where measurements were made. As shown in Fig. 6-21, according to the measurements by Wygnanski and Fiedler, self-preservation of the turbulence occurs only beyond $z/d = 40$ to 60.

Further measurements were made by Corrsin and Uberoi on the one-dimensional spectra $\mathbf{E}_1(k_1)$ and $\mathbf{E}_{\gamma 1}(k_1)$ of the turbulent-velocity and temperature

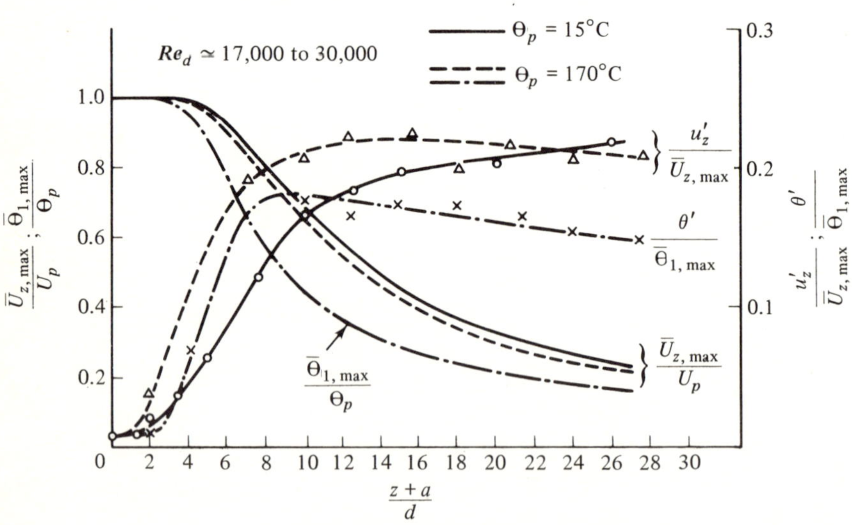

FIGURE 6-26
Distributions of mean velocity, mean temperature, and relative intensities of axial velocity and temperature fluctuations along the axis of a round free jet. (From: *Corrsin, S., and M. S. Uberoi,*[22] *by permission of the National Advisory Committee for Aeronautics.*)

fluctuations, respectively. The measurements were made at two locations in the jet: in a cross section $(x + a)/d = 20$ at the point on the jet axis, and at the point of maximum shear stress.

In the u_1-spectrum measurements the jet was cold; in the θ-spectrum measurements the jet was hot, having an issuing temperature $\Theta_p = 170°C$. The average curves through the experimental data are presented in Fig. 6-27. They show that (1) the energy spectra of the turbulence velocity u_1 were practically the same at the two points; (2) for the θ-spectra a difference occurred in the higher-wavenumber range: the spectral distribution at the point of maximum shear stress decreased at a higher rate than the distribution at the point on the jet axis; (3) no appreciable difference was observed between the u_z- and θ-spectra, particularly at the point of maximum shear stress; (4) the u_z-spectra showed no appreciable range where the $-\frac{5}{3}$ law applied. Such a Kolmogoroff range might have been expected, since the Reynolds number of turbulence was rather high, roughly $\mathbf{Re}_\lambda \simeq 500$. On the other hand, Corrsin and Uberoi found that the two u_1-energy spectra could be approximated well by the Von Kármán interpolation formula (3-155)

$$\mathbf{E}_1(k_1) = \frac{\text{const.}}{[1 + (k_1/k_e)^2]^{5/6}}$$

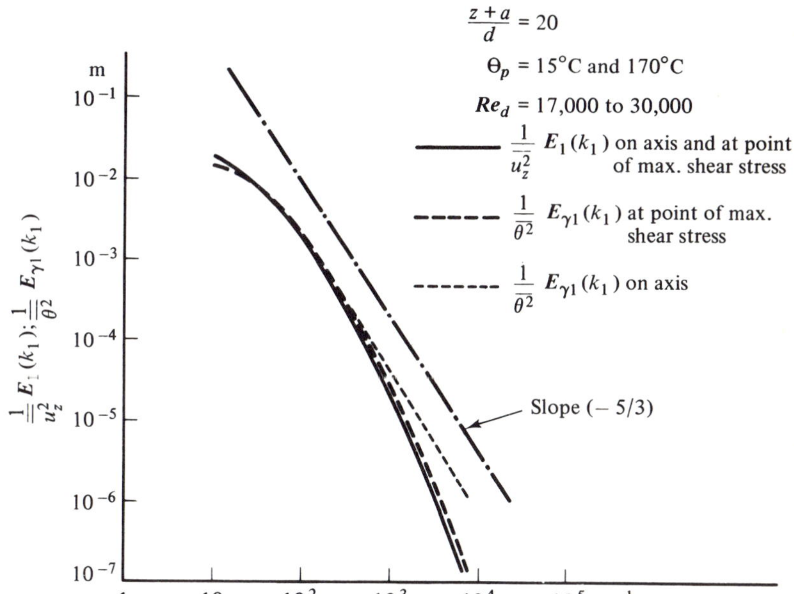

FIGURE 6-27
One-dimensional spectra of u_z and θ fluctuations on the axis and at the point of maximum shear stress in a round free jet. (From: *Corrsin, S., and M. S. Uberoi,*[23] *by permission of the National Advisory Committee for Aeronautics.*)

though in a restricted region only: $0 < k_1 < 125\ \text{m}^{-1}$. During the experiments the velocity at the jet axis at $(z + a)/d = 20$ amounted to about 12 m/s. With this value there follows from $k_1 = 2\pi n_1/\bar{U}_1 = 1.25$ a frequency $n_1 \simeq 250\ \text{s}^{-1}$. Above this frequency $\mathbf{E}_1(k_1)$ decreased at a higher rate than according to the $-\frac{5}{3}$ law.

When discussing wake flows we considered the distribution of the intermittency of the turbulent flow across the wake. For the free jet in a quiescent environment Corrsin and Kistler[29] measured this distribution. Figure 6-28 shows that the distribution of the intermittency factor Ω at the various cross sections investigated in the range $(z + a)/d = 20$ to 76 are pretty well similar. Exactly the same result has been obtained by Wygnanski and Fiedler,[51] using in essence the same experimental technique, namely by recording directly the turbulent periods of the flow. Wygnanski and Fiedler used $\partial u_z/\partial t$ and $\partial^2 u_z/\partial t^2$ as basic signals, thus eliminating undesired contributions of the relatively low frequencies in the non-turbulence periods. The triggering level was checked against traces of oscilloscope photographs. They also determined Ω from the flattening factor of the u_z-fluctuations. Slightly higher values were so obtained, mainly because the flattening factor also includes the effect of fluctuations in the non-turbulence periods. If we compare Fig. 6-28 with the corresponding Fig. 6-13 for the wake flow of a cylinder, we note that the lateral extent of the region where $\Omega \simeq 1$ is much larger for the round free jet than for the plane wake flow. At $(\xi_2)_{1/2}$, $\Omega \simeq 1$ for the jet flow, whereas it is much less than unity for the wake flow. Wygnanski and Fiedler[54] observed in the plane half-jet also a value of $\Omega \simeq 1$ at the "half-value" point of the mixing region.

As will be discussed in the next section, the flow during the "non-turbulence" periods is essentially irrotational, i.e. of a potential-flow nature, though random. In contrast, according to the description given in Sec. 1-1, turbulence is rotational. Yet we will indicate time-mean values referring solely to the turbulence periods

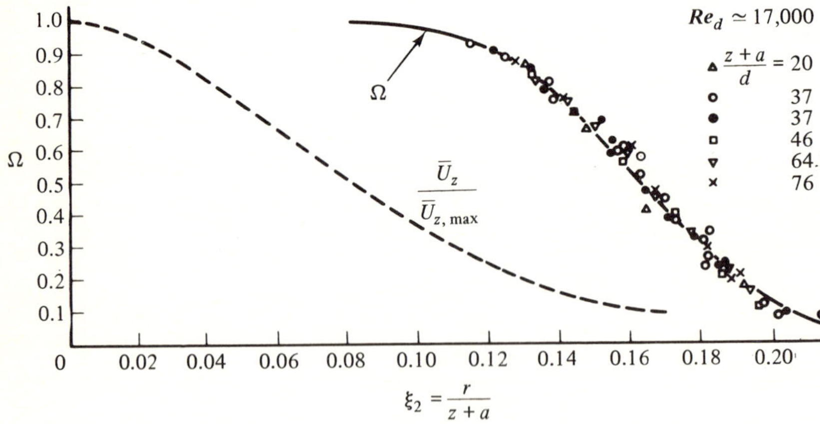

FIGURE 6-28
Distribution of intermittency factor Ω in a round free jet. (From: *Corrsin, S., and A. L. Kistler,*[29] *by permission of the National Advisory Committee for Aeronautics.*)

by the label "turb," while time-mean values referring to the irrotational periods are distinguished by the label "irrot." The total time-mean value of the mean-velocity measured by an anemometer from a continuous recording may be written

$$\bar{U}_z = \Omega(\bar{U}_z)_{\text{turb}} + (1 - \Omega)(\bar{U}_z)_{\text{irrot}} \qquad (6\text{-}105)$$

Similarly we may write for the sum of the mean-squares of the velocity fluctuations $\overline{q^2} = \overline{u_i u_i}$

$$\overline{q^2} = \Omega(\overline{q^2})_{\text{turb}} + (1 - \Omega)(\overline{q^2})_{\text{irrot}} \qquad (6\text{-}106)$$

and for the turbulence shear-stress

$$\overline{u_r u_z} = \Omega(\overline{u_r u_z})_{\text{turb}} + (1 - \Omega)(\overline{u_r u_z})_{\text{irrot}} \qquad (6\text{-}107)$$

A non-viscous irrotational flow cannot produce a shear-stress. So we may expect $(\overline{u_r u_z})_{\text{irrot}} = 0$. Indeed, Wygnanski and Fiedler found this correlation to be practically zero. Figure 6-29, which shows oscillograms of U_z, $u_z{}^2$, and $u_r u_z$, taken by Wygnanski and Fiedler in a highly intermittent region of the jet flow, clearly demonstrates that $(\overline{u_r u_z})_{\text{irrot}}$ is vanishingly small during the nonturbulence periods. Also $(\overline{q^2})_{\text{irrot}}$ was two orders of magnitude smaller than $(\overline{q^2})_{\text{turb}}$. At $\xi_2 = 0.22$, where $\Omega \simeq 0$, they measured $(q')_{\text{irrot}}/\bar{U}_{z,\text{max}} \simeq 3.9 \times 10^{-3}$, which value has to be compared with the value of 3.95×10^{-2} for $q'/\bar{U}_{z,\text{max}}$, also at $\xi_2 = 0.22$. In Eq. (6-105) $(\bar{U}_z)_{\text{irrot}}$ is not zero because of the accelerating effect of the faster moving turbulent bulges of the jet. Though smaller, it is not an order of magnitude smaller than $(\bar{U}_z)_{\text{turb}}$ (see also Fig. 6-29). In a plane half-jet, Wygnanski and Fiedler[54] measured both $(\bar{U}_1)_{\text{turb}}$ and $(\bar{U}_1)_{\text{irrot}}$ directly. At the point where $\Omega = 0.1$, $(\bar{U}_1)_{\text{irrot}}$ was roughly 30 per cent of $(\bar{U}_1)_{\text{turb}}$.

This *intermittency of the turbulence* in the outer region of the jet makes it difficult to introduce in a satisfactory way the use of an eddy diffusivity, even if the transport could be assumed to be of the gradient-type of diffusion. When considering the total time-mean condition, we could write

$$-\overline{u_r u_z} = \epsilon_m \frac{\partial \bar{U}_z}{\partial r}$$

We then found earlier that a constant value of the eddy viscosity across the jet, resulted in too high values of $\bar{U}_z/\bar{U}_{z,\text{max}}$ in the outer region. If we consider only the turbulence periods, then we should write

$$-(\overline{u_r u_z})_{\text{turb}} = (\epsilon_m)_{\text{turb}} \frac{\partial}{\partial r}(\bar{U}_z)_{\text{turb}} \qquad (6\text{-}108)$$

The former relation can be written, according to Eqs. (6-105) and (6-107).

$$-[\Omega(\overline{u_r u_z})_{\text{turb}} + (1 - \Omega)(\overline{u_r u_z})_{\text{irrot}}] = \epsilon_m \frac{\partial}{\partial r}[\Omega(\bar{U}_z)_{\text{turb}} + (1 - \Omega)(\bar{U}_z)_{\text{irrot}}]$$

As mentioned earlier, the term $(\overline{u_r u_z})_{\text{irrot}}$ may be neglected. However, on the right-

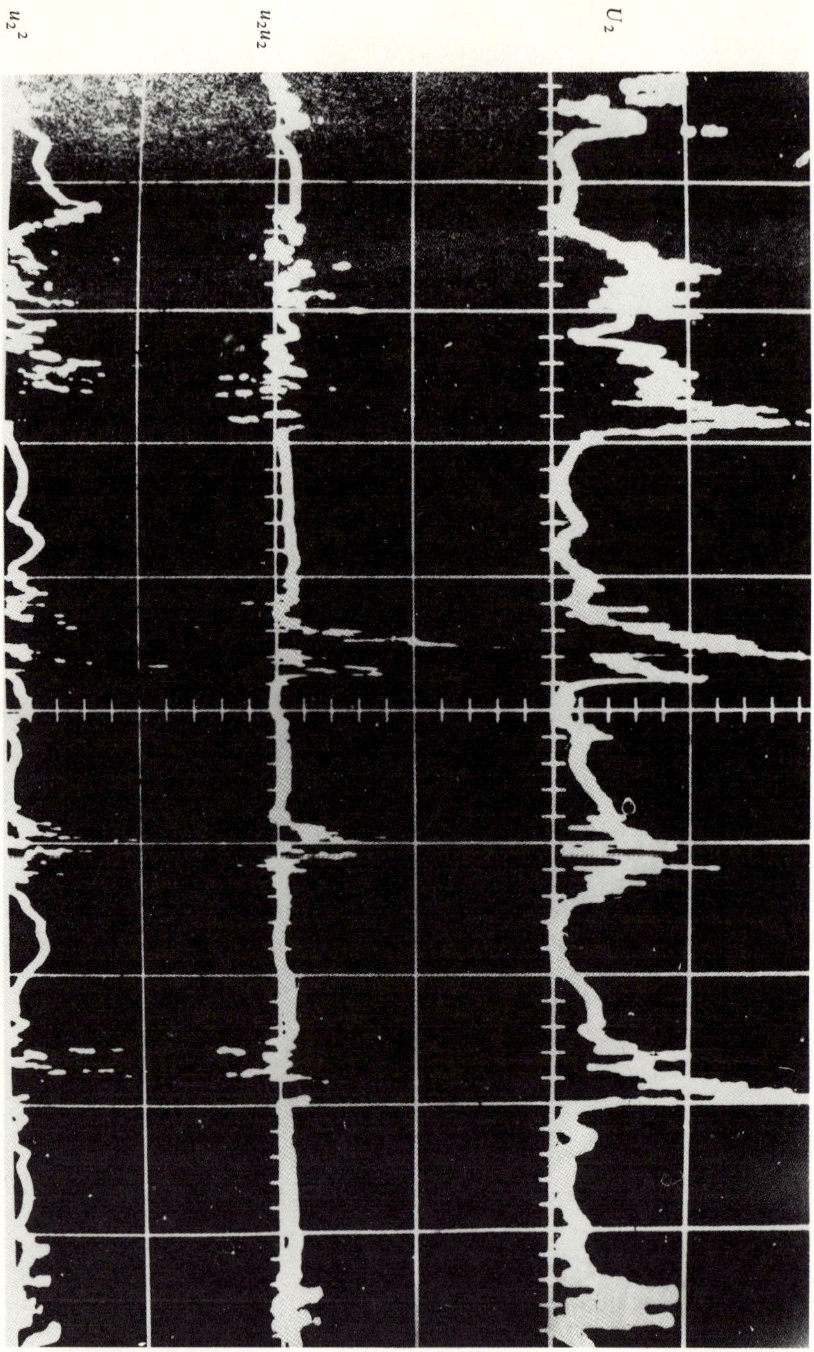

FIGURE 6-29
Oscillograms of U_z, $u_r u_z$, and u_z^2 taken in an intermittent region of a jet flow. (From: *Wygnanski, I., and H. Fiedler,*[51] *by permission of the Cambridge University Press.*)

hand side the term with $(\bar{U}_z)_{\text{irrot}}$ may not always be neglected with respect to the term with $(\bar{U}_z)_{\text{turb}}$. Yet, this is often done, in which case one uses the relation (6-108) with $(\bar{U}_z)_{\text{turb}} \simeq \bar{U}_z/\Omega$.

$$-\overline{u_r u_z} \simeq -\Omega(\overline{u_r u_z})_{\text{turb}} \simeq \Omega(\epsilon_m)_{\text{turb}} \frac{\partial}{\partial r} \frac{\bar{U}_z}{\Omega} \tag{6-109}$$

from which relation $(\epsilon_m)_{\text{turb}}$ can be calculated, when all other quantities have been determined experimentally. This relation shows, that $(\epsilon_m)_{\text{turb}}$ is not equal to ϵ_m/Ω. So obtaining $(\epsilon_m)_{\text{turb}}$ from ϵ_m simply by dividing it by the intermittency factor is not correct, though in some cases the error made may be acceptable. We referred to this, when discussing the plane wake flow at the end of Sec. 6-7 (Fig. 6-14). Wygnanski and Fiedler[51] determined $(\epsilon_m)_{\text{turb}}$ according to Eq. (6-109) and found it to be practically constant across the jet. Since their mean-velocity measurements agree nicely with those by Van der Hegge Zijnen, a value of $(\epsilon_m)_{\text{turb}}$ equal to $\sim 0.00196\bar{U}_{z,\max}(z + a)$ mentioned earlier may be expected. Wygnanski and Fiedler[51] also determined $(\epsilon_{\overline{q^2}})_{\text{turb}}$ from the measured distributions of $\overline{q^2}/\Omega$ and of $\overline{u_r q^2}$. Up to $\xi_2 = 0.10$ the eddy-diffusion coefficient was constant, and decreased beyond this value of ξ_2. From the value of $(\epsilon_{\overline{q^2}})_{\text{turb}}$ obtained at $z/d = 90$ and valid for the central part of the jet, one deduces

$$(\epsilon_{\overline{q^2}})_{\text{turb}} \simeq 0.00255\bar{U}_{z,\max}(z + a) = 0.0138U_p d$$

This would lead to $(\epsilon_{\overline{q^2}})_{\text{turb}}/(\epsilon_m)_{\text{turb}} \simeq 1.3$, which is roughly of the same magnitude as obtained for the ratio $\epsilon_\gamma/\epsilon_m$, and considered earlier.

Other measurements made by Wygnanski and Fiedler[51] concern *two-point one-time correlations,* Eulerian time-correlations, and the turbulence energy-balance. The longitudinal Eulerian correlation coefficient $\mathbf{f}(\Delta z)$ could fairly well be represented by an exponential function $\exp(-\Delta z/\Lambda_f)$, for not too small values of Δz. The lateral correlation coefficient $\mathbf{g}(\Delta r)$ becomes negative at large Δr, as may be expected. The relation between the two correlations deviated markedly from that for an isotropic turbulence. In the self-preserving region of the jet the integral scales Λ_f and Λ_g varied linearly with z. At the jet axis Wygnanski and Fiedler obtained $\Lambda_f = 0.0385z$ and $\Lambda_g = 0.0157z$.

Both scales increased with increasing distance to the jet axis, but their ratio remained essentially constant up to $\xi_2 = 0.15$, and equal to $\Lambda_f/\Lambda_g \simeq 2.35$ (for the isotropic case this ratio would be 2). At $\xi_2 = 0.058$, the point of maximum shear-stress, $\Lambda_f = 0.0525z$ and $\Lambda_g = 0.0227z$.

The dissipation length-scales λ_f and λ_g as determined from their defining equations [see, e.g., Eq. (1-60)] also increased linearly with z. At the jet axis for $z/d > 50$, $\lambda_f = 0.0048z$ and $\lambda_g = 0.0037z$.

Whence follows $\lambda_f/\lambda_g = 1.3$ (the isotropic value is $\sqrt{2}$). Consequently, the ratio λ_f/Λ_f and λ_g/Λ_g are independent of z. The ratio $\Lambda_g/r_{1/2}$ varied from 0.2 at the jet axis to 0.38 at $\xi_2 = 0.15$, and is comparable with the Prandtl mixing-length l_m as obtained by Van der Hegge Zijnen.

Measurements at $z/d = 90$ revealed that λ_f increased, like Λ_f, with increasing ξ_2, though in a more pronounced way, whereas in contrast with it λ_g decreased slightly with increasing ξ_2. At $\xi_2 = 0.15$ the value of 5.5 was obtained for λ_f/λ_g.

An interesting result is that the *Eulerian time-correlation* $\mathbf{R}_E(t)$ could not be obtained from the spatial correlation $\mathbf{f}(\Delta z)$ by using the Taylor hypothesis of a "frozen" turbulence, resulting in the relation (1-70). So $\Lambda_f \neq \bar{U}_z \mathfrak{J}_E$ [see Eq. (1-71)]. This is not surprising, since one of the conditions for this hypothesis is that $u'/\bar{U}_z \ll 1$, which condition is definitely not satisfied here. In a frozen turbulence, the flow pattern is convected with a convection velocity U_c equal to the mean velocity $\bar{U}_z$, but the flow pattern remains unchanged with respect to an observer traveling with this mean velocity. See Fig. 5-2b. In the case of a "nonfrozen" turbulence in a homogeneous mean flow field, the whole turbulence flow pattern may be convected again with $U_c = \bar{U}_z$, however, with respect to the moving system the flow pattern changes with time. This has been illustrated in Fig. 5-2c. The convection velocity can be obtained from the iso-correlation curves in the $(\bar{U}_1 t, x_1)$-mapping either from $x_1/\bar{U}_1 t$ at the point where $[\partial \mathbf{Q}_{1,1}/\partial t]_{x_1 = \text{const}} = 0$, or from $x_1/\bar{U}_1 t$ at the point where $[\partial \mathbf{Q}_{1,1}/\partial x_1]_{t = \text{const}} = 0$. In Fig. 5-2c these two points coincide at the dashed straight line, yielding the same value of U_c. In a nonhomogeneous flow the line connecting the points where $[\partial \mathbf{Q}_{1,1}/\partial x_1]_{t = \text{const}} = 0$ is not a straight line in general, and does not coincide with the line connecting the points where $[\partial \mathbf{Q}_{1,1}/\partial t]_{x_1 = \text{const}} = 0$. Hence different values of the convection velocity will be obtained depending on its definition. We will come back to this subject in the next chapter, in the discussion of the structure of boundary-layer flow. Here we will content ourselves with mentioning some results obtained by Wygnanski and Fiedler.[51] From space-time correlation measurements they calculated the *convection velocity* in two ways, one from $\partial \Delta z/\partial \tau$ at the points where

$$\frac{\partial}{\partial \Delta z}[\mathbf{R}_{1,1}(\Delta z, 0, 0; \tau)]_{\tau = \text{const}} = 0$$

and the other from $\partial \Delta z/\partial \tau$ at the points where

$$\frac{\partial}{\partial \tau}[\mathbf{R}_{1,1}(\Delta z, 0, 0; \tau)]_{\Delta z = \text{const}} = 0$$

Let us denote the first convection velocity by $(U_c)_\tau$, and the second by $(U_c)_{\Delta z}$.

In general the first method for $(U_c)_\tau$, yielded slightly smaller values. The ratio $(U_c)_\tau/\bar{U}_z$ increased with increasing ξ_2, being smaller than 1 at and near the jet axis, equal to 1 at $\xi_2 \simeq 0.0615$, where $\bar{U}_z/\bar{U}_{z,\max} = 0.65$, and greater than 1 beyond this point. They also found from measurements at the axis, that within the turbulence the convection velocity $(U_c)_\tau$ increased with increasing wavenumber.

When considering Townsend's measurements of the turbulence energy balance in the wake of a cylinder, the general equation (1-110) could be drastically simplified, mainly because all relative velocities in the wake were very small with respect to U_0. In the case of the jet flow, Wygnanski and Fiedler found out that no such drastic simplifications could be permitted. In this case they started from Eq. (1-111), neglecting at the same time the viscous term

$$\frac{1}{2}\nu\frac{\partial^2}{\partial x_i\,\partial x_i}\overline{q^2}$$

For the axi-symmetric case the equation considered reads

$$\underbrace{\frac{\partial}{\partial z}\left(\bar{U}_z\frac{\overline{q^2}}{2}\right)+\frac{1}{r}\frac{\partial}{\partial r}r\left(\bar{U}_r\frac{\overline{q^2}}{2}\right)}_{\text{convection}}+\underbrace{\frac{\partial}{\partial z}\overline{u_z\frac{q^2}{2}}+\frac{1}{r}\frac{\partial}{\partial r}\overline{r\,u_r\frac{q^2}{2}}}_{\text{diffusion}}$$

$$\underbrace{+\overline{u_r u_z}\frac{\partial\bar{U}_z}{\partial r}+\overline{u_z^2}\frac{\partial\bar{U}_z}{\partial z}+\overline{u_r^2}\frac{\partial\bar{U}_r}{\partial r}}_{\text{production}}+\frac{1}{\rho}\underbrace{\left(\frac{\partial}{\partial z}\overline{u_z p}+\frac{1}{r}\frac{\partial}{\partial r}r\,\overline{u_r p}\right)}_{\text{pressure diffusion}}$$

$$+\,\nu\underbrace{\overline{\frac{\partial u_i}{\partial x_j}\frac{\partial u_i}{\partial x_j}}}_{\text{dissipation }\varepsilon'}=0 \qquad \text{(6-110)}$$

Note that the dissipation term is written with velocity derivatives in a local Cartesian coordinate system.

Because there is still no method known for measuring in a reliable way the pressure-velocity correlation, the usual assumption has been made that the pressure-diffusion term can be considered as the closing entry. Of the dissipation term ε' five out of the nine terms could be measured, namely

$$\overline{(\partial u_z/\partial z)^2},\ \overline{(\partial u_r/\partial z)^2},\ \overline{(\partial u_\varphi/\partial z)^2},\ \overline{(\partial u_z/\partial r)^2},\quad\text{and}\quad \overline{(\partial u_z/r\partial\varphi)^2}$$

considered with respect to a local Cartesian coordinate system. Since the turbulence away from the axis was not isotropic, even in the smaller eddy range, no isotropic relations could be used for determining the remaining four terms. Wygnanski and Fiedler, however, observed that the relation

$$\overline{\left(\frac{\partial u_r}{\partial z}\right)^2}=\overline{\left(\frac{\partial u_\varphi}{\partial z}\right)^2}\simeq k\overline{\left(\frac{\partial u_z}{\partial z}\right)^2}$$

where $k=[1+\exp(-200\xi_2{}^2)]$, held true approximately. So they assumed the following similar relations

$$\overline{\left(\frac{\partial u_\varphi}{\partial r}\right)^2}=k\overline{\left(\frac{\partial u_r}{\partial r}\right)^2}=\overline{\left(\frac{\partial u_z}{\partial r}\right)^2}$$

$$\overline{\left(\frac{\partial u_\varphi}{r\partial\varphi}\right)^2}=k\overline{\left(\frac{\partial u_\varphi}{r\partial\varphi}\right)^2}=\overline{\left(\frac{\partial u_z}{r\partial\varphi}\right)^2}$$

Figure 6-30 shows the result obtained by Wygnanski and Fiedler for the energy balance. In this presentation the terms in Eq. (6-110) has been rendered dimensionless

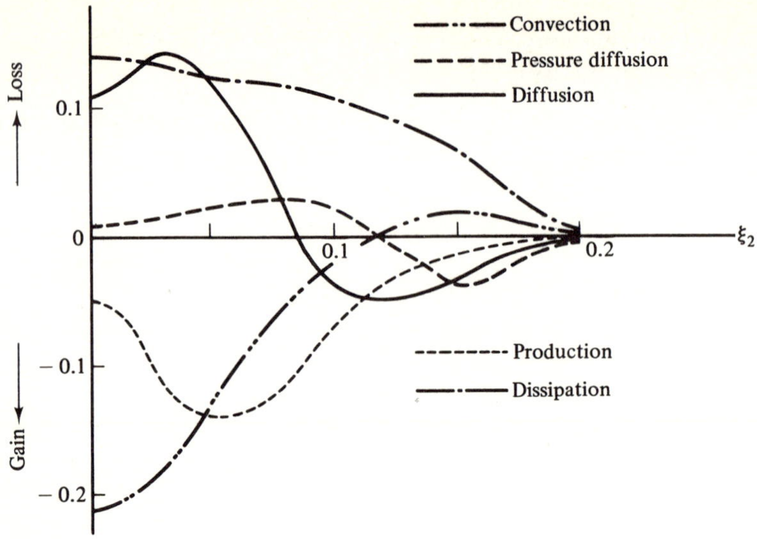

FIGURE 6-30
Energy balance of turbulence in a round free jet at $z = 90d$. $\mathbf{Re}_d \simeq 8 \times 10^4$. (From: *Wygnanski, I., and H. Fiedler,*[51] *by permission of the Cambridge University Press.*)

by expressing all velocities in terms of $\bar{U}_{z,\max}$, taking into account that $\bar{U}_{z,\max} \simeq 5.4.\ U_p d/z$, and expressing the derivatives $\partial/\partial z$ and $\partial/\partial r$ in $d/d\xi_2$, except for the dissipation term which has been written

$$\frac{z}{\bar{U}_{z,\max}}\varepsilon' = \frac{z^2}{5.4U_p d\bar{U}_{z,\max}^2}\varepsilon'$$

As Eq. (6-110) shows, the various terms of the balance equation consist separately of more than one subterm. In contrast to the wake flow case the contribution of convection by the lateral mean velocity is not negligible. In the center region it amounts to 30 per cent of the convection by the axial mean velocity. On the other hand the contribution by the axial turbulence diffusion is small relative to that by the radial turbulence diffusion. Also the contributions to the production by $\overline{u_z^2}$ and $\overline{u_r^2}$ are only of importance at and near the jet axis, where the shear-stress $\overline{u_r u_z}$ is practically zero. When we compare Fig. 6-30 with the corresponding Fig. 6-10 for the wake flow, we notice a great deal of similarity. Except that in the jet flow the contributions by diffusion and by production in the center region is much greater than in the wake flow. Also, in contrast with the wake flow, there is no subregion, however small, in the jet flow where the turbulence is locally in equilibrium, so that there production just equals dissipation.

6-12 THE STRUCTURE OF FREE TURBULENT SHEAR FLOW AND TRANSPORT PROCESSES

From the shape of free turbulent shear flows (wake flows and jets) and from the flow phenomena observed and the distributions of various turbulence quantities, a certain picture may be gained about turbulence flow patterns and processes. Summarized, the empirical facts are:

1 The lateral spread of a turbulent region with respect to the main-flow direction is a relatively slow process.
2 The turbulent region is separated from the irrotational region by an irregularly distorted boundary surface.
3 Turbulent flow has an intermittent character, in particular toward the "average" boundary of the turbulent region.
4 Many properties, such as turbulence intensities and dissipation, show great uniformity across the main body of the turbulent flow, decreasing appreciably to zero only in the intermittent boundary region.
5 This uniformity is increased over a larger region if the distribution function of the quantity considered is corrected for the intermittency factor.
6 The central part of a turbulent flow may show small regions where the lateral transport of turbulence energy appears to be directed up the gradient of its distribution and not down the gradient.

When considering the structure of free turbulent shear flow it appears useful to consider first the *developing region* of the flow before discussing the fully developed and self-preserving region far downstream from the origin. In doing this it also appears useful to make a distinction between plane and axisymmetric flows, and in addition for the two between wake flows and jet flows. The reason for considering first the developing region is that many features of the turbulence in the self-preserving region are already present or find their origin in the developing region. So some knowledge of the structure of the turbulence in this region may be helpful in better understanding the corresponding features in the self-preserving region. Experimental evidence, as discussed in the foregoing sections, has shown that in general wake flows require a much longer distance before attaining the self-preserving condition. Just for the sake of convenience, and also because of the available experimental material, we may think of a cylinder producing a plane wake, and of a disk or sphere producing the axi-symmetric wake. We then have observed that at least a hundred diameters downstream of the body are required for obtaining self-preservation of the mean flow, whereas in the case of jet flows in a quiescent environment 20 to 50 orifice diameters are already sufficient. Fortunately, when using, e.g., the hot-wire anemometer, measurements with sufficient reliability can be performed up to much greater downstream distances (1,000 diameters, say) in the case of wake flows than in the case of jet flows. This is possible because in wake flows the free stream velocity U_0 is decisive for the response of the sensing probe, and small deviations from U_0 can so still be measured. In case of a free jet in a quiescent environment the jet velocity itself is decisive, and most of the measurements have therefore to be restricted to distances less than, say, 100 orifice diameters.

The developing region usually starts with two, more or less thin shear layers of high vorticity in the plane flow case, or with a thin annular shear-layer in the axi-symmetric case. The thin shear layers are inherently unstable. We may refer to the well-known development of the two shear layers in the plane flow case to the Von-Kármán vortex street. The more or less discrete vortices resulting from the two original shear layers may affect each other. The question is whether this is always the case, so that there is for instance a corresponding behavior of the two shear layers at two opposite points. There are three possibilities: (1) the two shear layers move in an alternating fashion in the same lateral direction simultaneously; (2) they move in an opposite direction; and (3) each shear layer moves and develops independently of the other. For the axi-symmetric jet Crow and Champagne[57] have observed the situation (2). On the other hand Wygnanski and Gutmark[58] concluded from their experiments on a plane jet, that here the situation (3) is present. They measured the correlation between the intermittency factors at two corresponding points of the two shear layers with two hot-wire anemometers, and found this correlation to be zero.

In the *developing region of wake flows* the shear layers may originate from the boundary layer of the wake-producing body. In the plane wake of a cylinder there is a mutual effect between the boundary layer and the "discrete" vortices which in the two rows of the Von Kármán vortex street have the stable, alternating, position. The mutual effect is that boundary-layer separation at the two sides of the cylinder occurs in an alternating fashion, with a frequency expressed by the Strouhal number nd/U_0. So, close behind the cylinder the flow is highly intermittent. This intermittency persists in the downstream direction. As Fig. 6-13 shows, even far downstream there is still some intermittency near the axis of the wake. The originally big eddies break down, through nonlinear effects, into ever smaller eddies, but remains of the big eddies are still noticeable far downstream. One may imagine that after separation a straight vortex parallel to the cylinder is formed. This straight

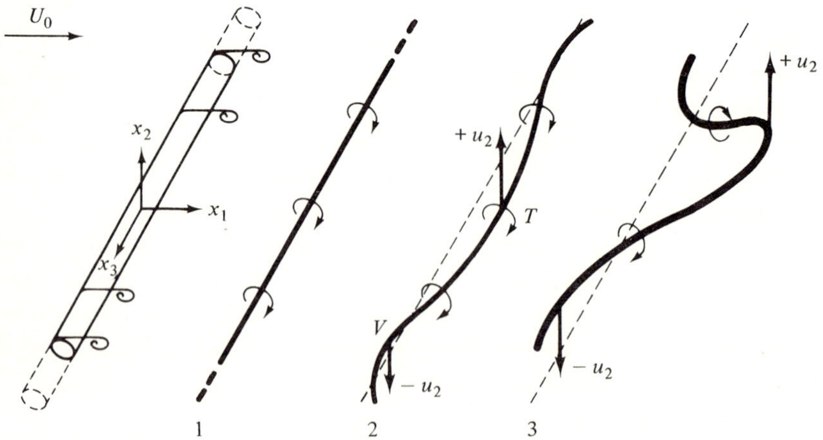

FIGURE 6-31
U-loop deformation of a vortex line due to a disturbance in the (x_1, x_3) plane. u_2 is induced velocity caused by the deformed vortex.

vortex appears not to be stable against certain disturbances. If such a disturbance slightly deforms the vortex, the combination of self-induction effect and nonuniform mean-velocity distribution further deforms the vortex farther downstream in such a way that it consists of more or less regularly spaced U-shaped loops. These loops obtain an orientation with a downstream component. Figure 6-31 illustrates the development of such a U-loop out of a disturbance in the (x_1,x_3) plane. In the first stage 1 a straight vortex-line is assumed that has just been separated. In stage 2 the slight deformation in the (x_1,x_3) plane of the vortex causes a velocity component u_2 as indicated in the points T (top) and V (valley) of the vortex. The top-part of the vortex is transported into a region of high velocity, the valley-part is transported into a low-velocity region. This movement further produces a stretching effect in the x_1-direction of the U-loop as shown in stage 3. We thus see that stream-wise components of the vortex develop, so that farther downstream eddies with axes inclined in the x_1-direction occur, forming pairs with opposite sign of vorticity. As will be discussed later, the large-eddy structure in the cylinder wake does show such a configuration. It may be left to the reader to show that a disturbance in the (x_2,x_3)-plane will be turned by the nonuniform mean-velocity distribution towards the (x_1,x_3)-plane, so that the former situation occurs again.

A corresponding eddy behavior in the axi-symmetric wake is less clear. The point of separation of the boundary-layer of a sphere may wander in circumferential direction, either irregularly or with a preferred rotational direction, resulting in a helicoidal vortex line downstream of the sphere.

When we now consider jet flows, we will restrict ourselves to the *axi-symmetric jet*. Mainly because this type of jet has been investigated most intensively, in particular more recently in connection with noise-production by a jet.

Of course, the initial, issuing, velocity distribution is of importance, since it determines the thickness of the initial shear layer. In the case of an almost uniform distribution at the orifice of the issuing velocity, the initial shear layer is very thin indeed. Experiments by Crow and Champagne,[57] showed a preferential wavelength for maximum spatial growth of $\lambda = 7$ to 8δ, where δ is the thickness of the shearlayer. The phase velocity was $\sim 0.5U_p$. The layer was very thin in their experiments, namely $\delta \simeq 0.025d = 0.63$ mm. So $\lambda/d \simeq 0.2$. The velocity distribution agreed well with the hyperbolic-tangent form assumed in a theoretical study by Michalke.[59,60] He obtained an optimum wavelength for spatial growth equal to 7.8δ and a phase velocity $0.513U_p$. These waves show up as surface ripples which scale on δ. They become shorter with increasing Reynolds number. The instability of the shear layer causes a steepening of the waves, a combination of the annular vortices so formed to longer waves and resulting in bigger eddies. The further spatial development of these eddies gives the developing region of the jet a typical shape. After 2 to 4 orifice diameters temporary puffs of spade-like shape are formed. They can be observed in the Schlieren photograph of Fig. 6-32 taken by Bradshaw, Ferriss, and Johnson[61] of an air jet from a 51 mm diameter orifice. The Reynolds number at the orifice was 300,000. Thus, as Crow and Champagne concluded from their experiments

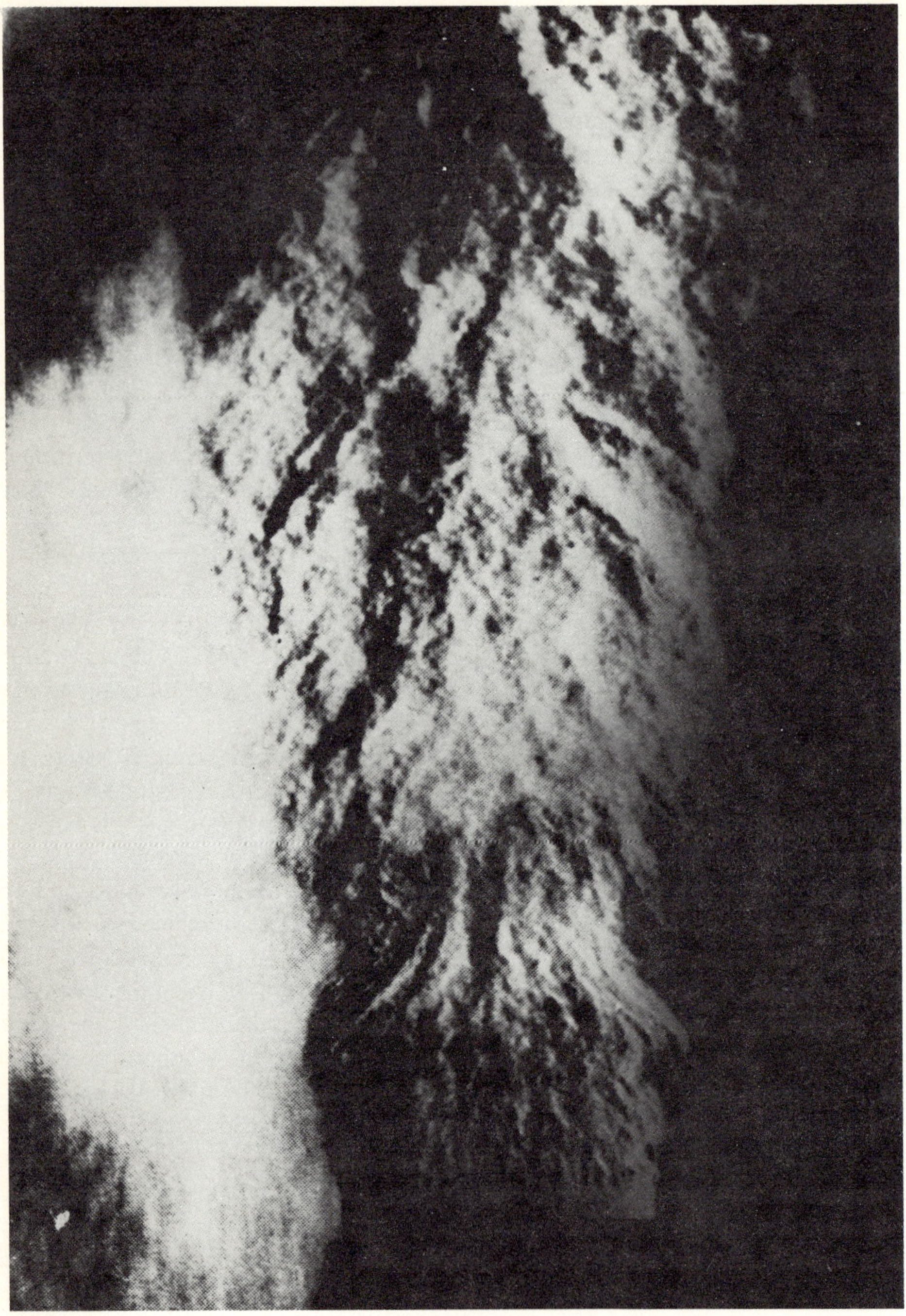

FIGURE 6-32
Schlieren photograph of the developing region of a round free air-jet. $d = 51$ mm. $U_p = 84$ m/s. (From: *Bradshaw, P., D. H. Ferriss, and R. J. Johnson,*[61] *by permission of the Cambridge University Press.*)

where flash-photographs of a jet using fog as a flow-visualization medium showed much more clearly the train of puffs, two kinds of eddies can be discerned: the earlier mentioned surface ripples, and a train of large-scale puffs, which scale on the jet diameter. The puffs show a distinct frequency, which according to Crow and Champagne follows the relation given by the value of the Strouhal number $nd/U_p = 0.3$. The part of the energy spectrum of the turbulence in the developing region, corresponding to this Strouhal number appears to be most sensitive to external disturbances, including background turbulence and turbulence of the jet itself from which it selects the above frequency.

The annular vortices form a rather stable configuration. No clear deformations of the annular vortices, like those discussed for the straight vortices of a plane wake, have been observed in the developing region. However, if similar disturbances on the annular vortices would occur, they would lead to formation of U-loops whose tops penetrate into the jet, again towards the high-velocity region. The superposition of small-scale and large-scale eddies is still present in the self-preserving region. Figure 6-33 shows a photograph taken by Davies and Fisher[62] of a free jet covering the region up to $z/d \simeq 20$. [See also Fig. 1-7.] The large-scale structure corresponding to the initial puffs is more or less masked by smaller-scale eddies, but with some imagination a spade-like large-scale structure is still discernible far downstream. We thus conclude that in the developing region of a round jet two subregions may be discerned. A wedge-shaped mixing region widening linearly with distance from the orifice, forms the first subregion. The mean flow pattern which develops out of the instability of the initial shear layer shows up as a similarity mixing region. According to Nayar, Siddon, and Chu[73] the region extends to about $z \simeq 4.5d$. This distance agrees reasonably well with the distance where the above temporary puffs of spade-like shape are being formed. The second subregion extending to about $z = 10d$, is a transition region to the fully developed jet, where the mean flow pattern becomes more and more self-preserving. Detailed investigations of the turbulence in the wedge-shaped mixing layer and the transition region of the developing region have been made by a number of investigators (see, e.g., references 61 to 66, and 73). In some respects results agree, in other respects they do not. For instance Bradshaw *et al.*[61] found the intensities of the three turbulence velocity components roughly of the same magnitude, while Sami, Carmody, and Rouse[65] observed distinct axial symmetry, the intensity of the axial turbulence velocity being 1.4 times that of the two other components. An important result is that the intensity u_z' decreases towards the high-velocity region from its maximum at $\xi_2' = (r - d/2)/z = 0$ at a faster rate than in the region $\xi_2' > 0$. The maximum shear-stress too occurs at $\xi_2' = 0$. This skewed radial distribution of u_z' may explain the observed skewness in the probability density $\mathfrak{P}(u_z)$ of u_z. For $\xi_2' < 0$, $\mathfrak{P}(u_z)$ has a peak at $u_z < 0$, and so a skewness factor $S > 0$, whereas for $\xi_2' < 0$, $S < 0$. In the region $d < z < 6d$, Davies[64] observed a linear increase of the longitudinal integral scale Λ_f with z, namely $\Lambda_f \simeq 0.13z$. Since the width B of the mixing layer increases linearly, roughly according to $0.32z$, we conclude that $\Lambda_f \simeq 0.4B$. The lateral integral scale of u_z, Λ_g was found

FIGURE 6-33
Flash photograph of a subsonic, round, free air jet. $d = 25$ mm. $U_p = 60$ m/s. (From: *Davies, P. O. A. L., and M. J. Fisher,*[62] *by permission of the University of Southampton.*)

to be roughly $\Lambda_g \simeq \frac{1}{3}\Lambda_f \simeq 0.11B$. This latter value is of the same magnitude as measured by Wygnanski and Fiedler in the self-preserving region of the jet, namely $\Lambda_g/B \simeq 0.1$ at the point of maximum shear stress ($\xi_2 = 0.058$). For B we assume half the diameter of the jet, which is roughly determined by $0.22z$. On the other hand, according to Wygnanski's and Fiedler's measurements $\Lambda_f/B \simeq 0.24$.

Davies[64] and Bradshaw *et al.*[61] also determined the *convection velocity* of turbulence pattern $(U_c)'_\tau$, defined by $\Delta z/\tau$ at the point where $\partial/\partial\Delta z[\mathbf{R}_{1,1}(\Delta z,0,0;\tau)]_{\tau=\text{const}} = 0$. As will be discussed in the next chapter $(U_c)'_\tau \neq (U_c)_\tau$ (see Sec. 6-11), though the difference may be small. They found $(U_c)'_\tau \simeq \bar{U}_z$ at $\xi'_2 = 0$, where $\bar{U}_z \simeq 0.6U_p$. For $\xi'_2 < 0$, $(U_c)'_\tau < \bar{U}_z$, and for $\xi'_2 > 0$, $(U_c)'_\tau > \bar{U}_z$. For the mixing layer of the first subregion Nayar, Siddon, and Chu[73] observed the same behavior. In the transition region the convection velocity tends to match up with $\bar{U}_z$. On the other hand $(U_c)'_\tau \gtrless \bar{U}_z$ for $\xi'_2 \gtrless 0$ has also been measured in the developed jet by Wygnanski and Fiedler, and discussed earlier. Also in agreement with the observations of these investigators is that the convection velocity as measured by Davies at $\xi'_2 = 0$ increases with the wavenumber. A clear-cut explanation cannot be offered. In the case of the mixing layer a possible explanation might be sought in the effect of the large-scale eddies, having their center at $\xi'_2 = 0$. We will come back to this point in the next chapter. Just as in the self-preserving region, the turbulence is not in energy-equilibrium at any point of the mixing layer, the contributions by convection and diffusion to the energy balance not being small.

Before closing this discussion of the developing region, some interesting results obtained by Wygnanski and Fiedler[54] from their experiments with a *two-dimensional mixing region* between a high-velocity airstream with undisturbed velocity U_0, and a quiescent environment will be considered. They observed that the spread of the turbulent mixing region into the quiescent environment occurs at a faster rate than into the high-velocity stream. Taking the origin of the lateral coordinate at the half-value point ($\bar{U}_1/U_0 = 0.5$), and rendering this coordinate dimensionless with the distance to the origin of similarity, the average position of the turbulent/irrotational interface was at $\xi''_2 = -0.11$ on the high-velocity side, and at $\xi''_2 = +0.07$ on the low-velocity side. The intermittency factor Ω was nowhere equal to one, but had a maximum value of ~ 0.96 at $\xi''_2 = -0.03$, i.e., towards the high-velocity region. As mentioned earlier, Wygnanski and Fielder measured separately $\bar{U}_1$, $(\bar{U}_1)_{\text{irrot}}$, and $(\bar{U}_1)_{\text{turb}}$. The three distribution curves intersect at this point $\xi''_2 = -0.03$. On the high-velocity side $(\bar{U}_1)_{\text{turb}} < \bar{U}_1 < (\bar{U}_1)_{\text{irrot}}$, while on the low-velocity side the situation changes, in that there $(\bar{U}_1)_{\text{turb}} > \bar{U}_1 > (\bar{U}_1)_{\text{irrot}}$. They also measured the average of the U_1-distribution across the mixing region at the instant that the interface passes a detector located at a pre-determined position ξ''_2. At $\xi''_2 = 0$, this average velocity $\langle U_1 \rangle$ differed only slightly from $0.5U_0$ for the four different locations of the detector, corresponding to $\Omega = 0.1; 0.3; 0.5$; and 0.7. The leading-edge and trailing-edge of the bulges of the turbulent front had identical average velocities. For values of $|\xi''_2|$ greater than the value of the position of the detector, the velocity $\langle U_1 \rangle$ was almost equal to U_0. However, for smaller values of $|\xi''_2|$ (for positions of the anemometer "inside" the position of the detector), the $\langle U_1 \rangle$-distribution was linear. Consequently $d\langle U_1 \rangle / d\xi_2''$ changed over a very short distance from zero to the constant value inside the turbulent region when crossing the interface. This change was more marked on the high-speed side than on the low-speed side. The value of the constant velocity-gradient varied with the position of the detector, increasing with increasing Ω. This means that at locations of the detector deep in the "valleys" (large value of Ω) the turbulence is sheared at a higher rate than at locations of the detector near the "top" of the turbulent bulges (small value of Ω). Hence, for a stationary observer the instantaneous linear distributions of $\langle U_1 \rangle$ fluctuate as the turbulence pattern passes, the value of the gradient depending on whether a "valley" or a "top" is being observed. The total time-mean value of the fluctuating linear distributions gives the S-shape distribution of $\bar{U}_1$. This linear distribution of $\langle U_1 \rangle$ implies that the turbulent fluid is subjected to an almost constant rate of strain, with a larger strain rate in the "valley" parts than in the "top" parts of the turbulent mixing zone.

When turning now our attention to the *fully-developed self-preserving region* of wake flows and jet flows, it is of interest to keep in mind the various features of the turbulence structure in the developing region, as mentioned earlier. A number of the phenomena just described, and which are also present in the self-preserving regions have already been observed and studied many years ago by Townsend[9] in his early, extensive, studies of the wake flow behind a circular cylinder. Since many of the inferences made and the conclusions arrived at by Townsend are still of importance, it is worth while to quote him in the description of a model structure of free turbulent shear flow.[9]

The description reads:†

"(1) The fully turbulent fluid (the part having appreciable vorticity fluctuations) is bounded by contorted surfaces which may in some places approach the central plane of the flow. These bounding surfaces are moved by the convective action of a system of large eddies whose dimensions are comparable with the width of that part of the flow with mean shear everywhere of the same sign, as well as by the process of entrainment of undisturbed fluid.

(2) The undisturbed fluid outside the bounding surfaces is not turbulent and its motion is irrotational, composed of the mean velocity of translation (if any) and potential flow due to the movement at the boundary.

(3) Except near the boundaries where the process of entrainment of quiescent fluid is proceeding, the turbulent intensity is nearly uniform. This is possible, since the scale of the turbulence is small and its time scales for production and dissipation are comparable with the time for appreciable flow development.

(4) The large eddies, which distort the bounding surface, are simple eddies with central vorticity along the principal axis of positive mean rate of strain, elongated in the direction of flow and centered near the plane of maximum rate of shear. Their life is comparable with the time for appreciable development of the

† By permission of the Cambridge University Press.

flow, but they are not permanent structures, new ones arising as old ones disappear."

Thus wake flow, for instance, consists of a core of fully developed continuous turbulence. From this core bulges are formed moving outward by large-eddy motion, with a scale comparable in size to the width of the wake. These large eddies and bulges contain small-scale eddies at least one order of magnitude smaller than the large eddies. These small-scale eddies, which originate mostly in the core region, are entrained with the large-scale motions of the bulges.

This might explain immediately why the distributions of turbulence energy and dissipation are nearly constant in the core region and within the bulges of the boundary region; it points toward a uniform internal small-scale structure of turbulence throughout the whole turbulent region. Indeed, the spatial integral scales Λ_g [determined by the correlation coefficients $\mathbf{R}_{2,2}(x_1,0,0)$ and $\mathbf{R}_{3,3}(x_1,0,0)$] are nearly constant across the wake and have values about one-fourth the half-width value of the wake.

The large-eddy motions, responsible for the general, large-scale bulgy shape of the boundary, increase in size in downstream direction, and so do the amplitudes of the "waves" of the interface, causing an increase of the average width of the turbulent region. In the turbulence zone the motion is essentially rotational, whereas outside this zone the motion is irrotational. How does this increase of the turbulence zone take place? The slow large-scale motions in the turbulence zone induce via pressure forces equivalent motions in the irrotational region; thereby the bulges increase in amplitude. But during this process the total volume of turbulent fluid must remain the same unless other effects, not yet considered, occur which cause at the same time an increase of the volume of turbulent fluid. Now, the only way in which irrotational fluid can become rotational is by the action of viscous shearing forces, that is, by "direct contact" between rotational and irrotational fluid. At sufficiently high Reynolds numbers the effect of viscosity in the turbulent fluid is restricted to the smallest-scale motions, i.e., to the smallest eddies. Hence the propagation of the turbulence "front" by viscous shearing forces can be caused only by these smallest eddies, and the real boundary between rotational and irrotational fluid must be of the nature of an interface; that is, the viscous actions must be restricted to a very thin layer. A further consequence is that the turbulence front must be continuous and that there can be no isolated regions of turbulence in the irrotational field outside the turbulence field (Corrsin and Kistler[29]).

The thickness of the interface where the viscous shearing forces are concentrated should be of the order of magnitude of the smallest eddies. Corrsin and Kistler have shown that the thickness is likely to be of the same order of magnitude as Kolmogoroff's micro length-scale η [see Eq. (3-106a)]. They call this thin layer, very appropriately, the *viscous superlayer,* because of its similarity to the viscous sublayer along a fixed wall, where, too, viscous shear stresses are concentrated. (See Sec. 7-1.) There is, however, a dissimilarity in that the viscous sublayer consists mainly of the same fluid particles except for exchange with the fluid in the fully turbulent region, whereas new fluid particles are introduced continuously into the viscous superlayer from the irrotational field through the diffusive propagation of the turbulence into this field. Corrsin and Kistler[29] have estimated this propagation velocity. By dimensional reasoning it is taken proportional to $(\nu\omega')^{1/2}$, where ω' is the root-mean-square vorticity of the turbulence. We will come back to this point in Sec. 7-9. The sharp boundary between turbulent and irrotational fluid, consisting of a viscous superlayer, is nicely shown by the shadowgraph given in Fig. 6-34, and made by Schapker.[67]

Thus the propagation of the turbulence front causes an increase in the volume of turbulent fluid. At the same time the larger eddies cause larger-scale deformation of the interface, and the high bulges are eventually caused by the slow largest eddies. These largest eddies control the rate of spread of the whole turbulence region (Townsend[9]). As stated, outside the turbulence zone potential flows are set up. These flows are of the same magnitude as the large-scale motion at the other side of the interface; this interface produces no discontinuity in velocities of small or of large scale (Townsend,[9] Phillips,[35] Stewart[36]). Phillips and Stewart have studied especially the motion in the irrotational field set up by a randomly distributed velocity field at an interface. They found that in the neighborhood of the interface the velocity fluctuations must be essentially anisotropic and follow the relation

$$\overline{u_2^2} = \overline{u_1^2} + \overline{u_3^2} \tag{6-111}$$

where u_2 is normal to the interface. Stewart has shown that this relation is an immediate consequence of the requirement of irrotational flow when the fluctuations are stationary in the x_1- and x_3-directions.

Let us consider $(\partial/\partial x_j)\overline{u_j u_i}$. For an incompressible fluid this derivative may be written as follows

$$\frac{\partial}{\partial x_j}\overline{u_j u_i} = \overline{u_j \frac{\partial u_i}{\partial x_j}} = \overline{u_j\left(\frac{\partial u_i}{\partial x_j} - \frac{\partial u_j}{\partial x_i}\right)} + \frac{\partial}{\partial x_i}\frac{\overline{u_j u_j}}{2}$$

Now

$$\varepsilon_{ijk}\omega_k = -\varepsilon_{ijk}\varepsilon_{lmk}\frac{\partial u_l}{\partial x_m} = -[\delta_{il}\delta_{jm} - \delta_{im}\delta_{jl}]\frac{\partial u_l}{\partial x_m}$$

$$= -\left(\frac{\partial u_i}{\partial x_j} - \frac{\partial u_j}{\partial x_i}\right)$$

Hence

$$\frac{\partial}{\partial x_j}\overline{u_j u_i} = -\varepsilon_{ijk}\overline{u_j\omega_k} + \frac{\partial}{\partial x_i}\frac{\overline{u_j u_j}}{2} \tag{6-112}$$

For an irrotational flow the relation reduces to

$$\frac{\partial}{\partial x_j}\overline{u_j u_i} = \frac{\partial}{\partial x_i}\frac{\overline{u_j u_j}}{2} \tag{6-112a}$$

When $i = 1$, there is obtained

$$\frac{\partial}{\partial x_1}\overline{u_1^2} + \frac{\partial}{\partial x_2}\overline{u_2 u_1} + \frac{\partial}{\partial x_3}\overline{u_3 u_1} = \frac{\partial}{\partial x_1}\frac{\overline{u_j u_j}}{2}$$

or

$$\frac{\partial}{\partial x_2}\overline{u_2 u_1} = -\frac{\partial}{\partial x_3}\overline{u_3 u_1} - \frac{\partial}{\partial x_1}\frac{\overline{u_1^2} - \overline{u_2^2} - \overline{u_3^2}}{2} \tag{6-113}$$

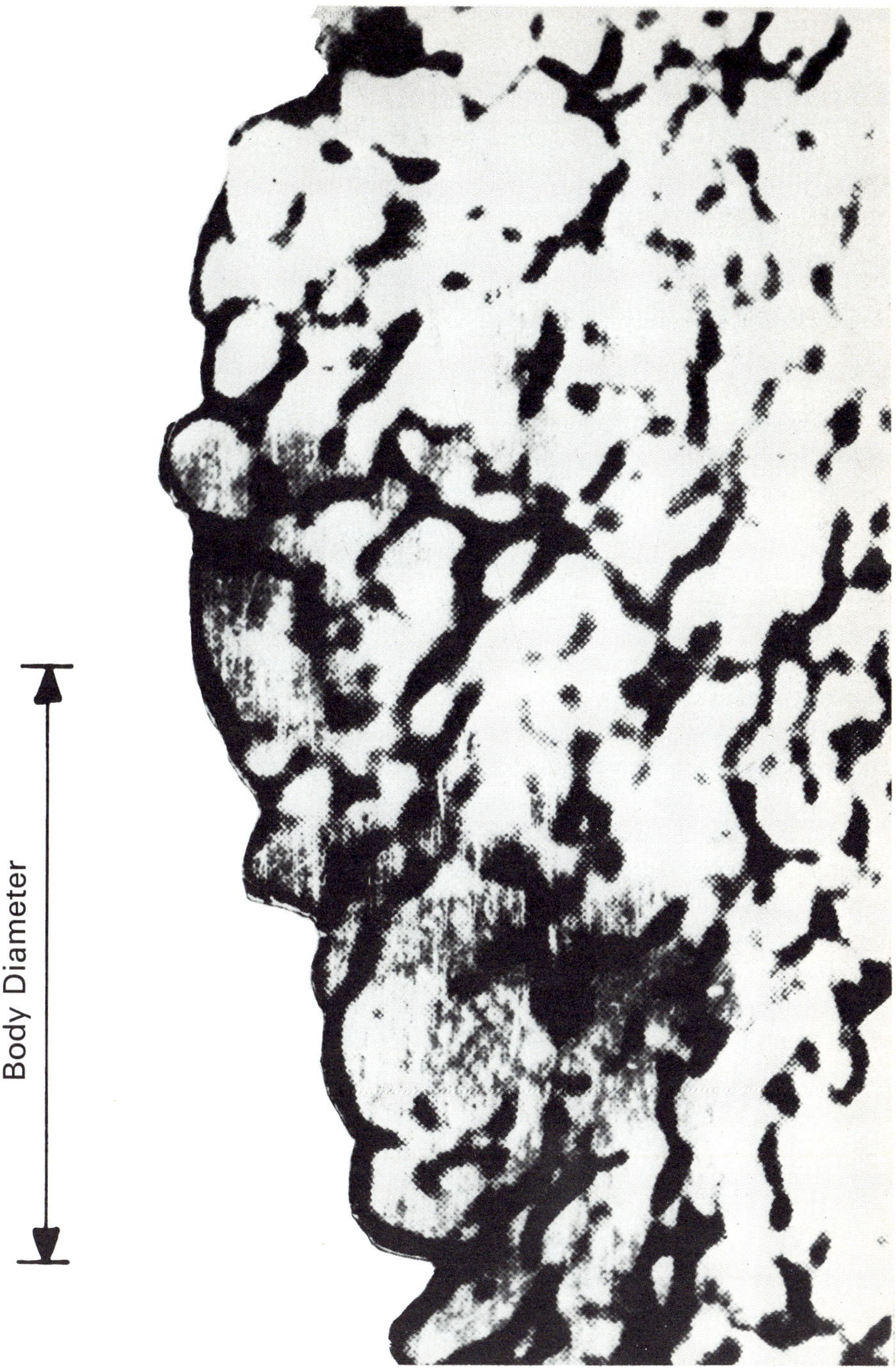

FIGURE 6-34
Blow-up of a shadowgraph of wake-boundary region downstream of a sphere of 6.4 mm diameter. $\mathbf{Re}_d = 71 \times 10^5$. $\mathbf{Ma} = 4.7$. $z/d = 180$. (From: *Schapker, R. L.*[67])

When we assume homogeneity of the flow in the x_1, and in the x_3-direction, $(\partial/\partial x_2)\overline{u_2 u_1} = 0$. Since at least at $x_2 = \infty$, $\overline{u_2 u_1} = 0$, it is zero everywhere.

When we take $i = 2$, Eq. (6-112a) becomes

$$\frac{\partial}{\partial x_2}\overline{u_2^2} = -\frac{\partial}{\partial x_1}\overline{u_1 u_2} - \frac{\partial}{\partial x_3}\overline{u_3 u_2} + \frac{\partial}{\partial x_2}\frac{\overline{u_j u_j}}{2}$$

Again, when we assume homogeneity in the x_1, and in the x_3-direction

$$\frac{\partial}{\partial x_2}\frac{\overline{u_2^2} - \overline{u_1^2} - \overline{u_3^2}}{2} = 0$$

whence follows the relation (6-111), since the constant of integration is zero because all velocities u_i vanish at $x_2 = \infty$.

Townsend's measurements of $\overline{u_1^2}$, $\overline{u_2^2}$, and $\overline{u_3^2}$ in the wake flow of a cylinder appear to satisfy the above relation in the region of small intermittency factors, namely $\Omega < 0.1$. Also the measurements of $(\overline{u_z^2})_{\text{irrot}}$, $(\overline{u_r^2})_{\text{irrot}}$, and $(\overline{u_\varphi^2})_{\text{irrot}}$ by Wygnanski and Fiedler[51] in the axi-symmetric jet at $\xi_2 = 0.22$ confirmed this relation satisfactorily, taking into account the limited accuracy of the measurements of these velocities.

Phillips[35] assumed these irrotational motions at x_2-distances large relative to the size of the energy-containing eddies as if they are generated by randomly distributed point sources with zero total strength, located in a x_2-plane somewhere in the intermittent region. Since velocities decrease in inverse proportion to the square of the distance from the point source, $(u_i')_{\text{irrot}}$ must decrease with x_2^{-2} at large x_2. We will come back to this in Sec. 7-9.

Stewart has obtained the interesting result that, for plane wake flow, in the irrotational region there is no mean flow in the x_2-direction into the wake if the flow conditions within the wake are self-preserving. For a round free jet, however, there is a mean irrotational flow inward into the jet, essentially perpendicular to the jet axis.

The irrotational flow between the bulges of turbulent fluid tends to have the same mean direction as the bulges themselves, owing to the action of the pressure forces. From the fact that the equations governing the behavior of the fluctuation velocities u_1, u_2, and u_3 have exactly the same form as the equations for the flow of gravity water waves according to the classical theory, Stewart draws the conclusion that the flow pattern in the irrotational flow region close to the turbulence boundary may have a close similarity to the flow pattern in gravity water waves.

From the above considerations concerning the phenomena in the intermittent boundary region of free turbulent flows we may form a picture of the flow pattern, which is illustrated by Fig. 6-35 for the case of plane wake flow.

A similar picture applies to a jet flow at large distances from the origin, where the jet velocity has become small. At shorter distances to the origin, when the jet velocity is still large, the bulges will show a more skewed shape, as can be observed in Fig. 6-33.

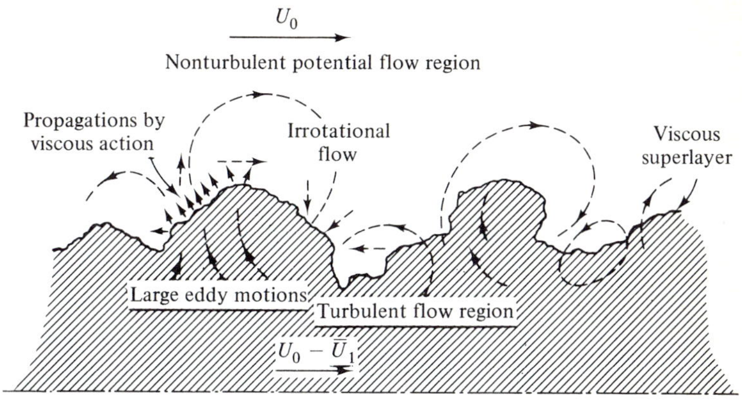

FIGURE 6-35
Possible flow pattern in the intermittent boundary region of a plane wake flow.

Since the broadening of the turbulent region in the downstream direction results mainly from the larger-scale motions in the turbulent region, it is reasonable to expect that this spreading is a statistical process determined by statistical laws for the turbulence in the turbulent region. Corrsin and Kistler[29] have shown that it is possible to describe this spreading process as a Lagrangian diffusion process, using the statistical properties of the turbulence in the fully turbulent region. The height of the bulges shows a statistical variation; on the average it increases with downstream distance, as does the average position of the interface between turbulent and irrotational flow. Corrsin and Kistler have determined the increase of the root-mean-square of the height of the bulges and the average position of the interface (i.e., the average width of the turbulence region) with increasing distance from the orifice, for the round free jet. Figure 6-36 shows $[\overline{(r^* - \bar{r}^*)^2}]^{1/2}/d$ and $\bar{r}^*/d$ together with $r_{1/2}/d$ as functions of z/d; r^* is the radial position of the interface at a given

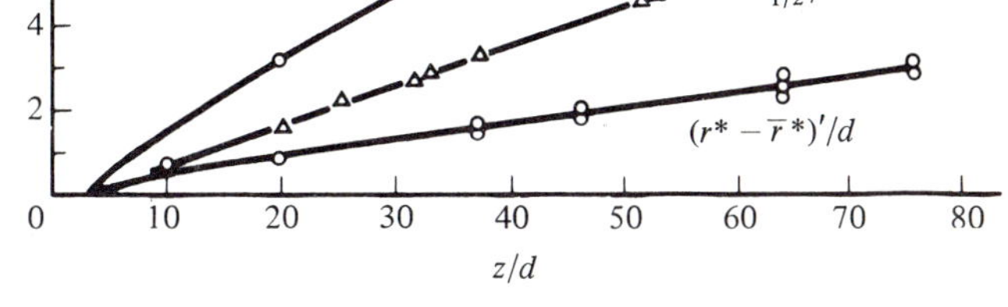

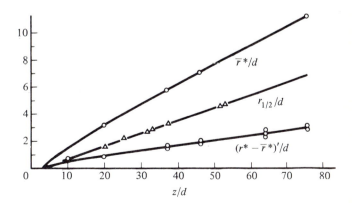

FIGURE 6-36
Average position of turbulent front and root-mean-square height of bulges in a round free jet as functions of z/d. (From: *Corrsin, S., and A. L. Kistler,*[29] *by permission of the National Advisory Committee for Aeronautics.*)

instant. Whereas $r_{1/2}/d$ shows an exact linear variation with the distance $(z/d - 3)$ to an apparent origin, the data on $\bar{r}^*/d$ and $(r^* - \bar{r}^*)/d$ seem to indicate a different variation if the same apparent origin is chosen. However, in view of the experimental inaccuracy, these ratios too should be considered roughly linear functions of z/d.

For the plane wake flow these quantities should vary proportionally with $[(z + a)/d]^{1/2}$ in a self-preserving turbulence. The few data available from Townsend's measurements are insufficient to conclude whether this was actually the case in his experiments, but at any rate the data do not contradict this prediction.

The statistical variations in the height of the bulges make it reasonable that the intermittency factor Ω should follow a Gaussian distribution. According to Corrsin and Kistler,[29] the data on Ω for a round free jet and a plane wake flow follow such a Gaussian distribution closely. The curves in Figs. 6-13 and 6-28 are such distributions

$$\Omega = \frac{1}{\sqrt{2\pi}} \int_{\alpha\xi_2+\beta}^{\infty} dz \exp\left(-\tfrac{1}{2}z^2\right)$$

where $\alpha\xi_2 + \beta = (r - \bar{r}^*)/(r^* - \bar{r}^*)'$, and α and β are constants.

We have mentioned several times the close agreement between measured and predicted mean-velocity distributions, if the predicted distributions are based on the assumptions of gradient-type diffusion of momentum and constant coefficient of eddy viscosity and if, moreover, they are corrected for intermittency.

The calculation of the velocity distribution from Eq. (6-36), where instead of ϵ_m we read $\Omega\epsilon_m$, is not an easy matter if the above Gaussian distribution is taken for Ω. To simplify the calculation Townsend[9] approximated the Gaussian distribution of Ω for plane wake flow by the function

$$\Omega \simeq [1 + \text{const.} \times \xi_2{}^4]^{-1}$$

Integration of Eq. (6-36) was then easily performed. The result reads

$$\frac{\bar{U}_1}{\bar{U}_{1,\max}} = \exp\left[-\frac{U_0 d\xi_2{}^2}{4\epsilon_m}\left(1 + \tfrac{1}{3}\,\text{const.} \times \xi_2{}^4\right)\right]$$

Figure 6-37 shows the improvement made over the simple Gaussian solution (6-29a) if the correction for intermittency is made and if at the same time a slightly different value of ϵ_m is taken. The corrected curve in Fig. 6-37 reads

$$\frac{\bar{U}_1}{\bar{U}_{1,\max}} = \exp\left\{-\left(\frac{\xi_2}{0.264}\right)^2\left[1 + \frac{1}{3}\left(\frac{\xi_2}{0.35}\right)^4\right]\right\}$$

The value of the constant occurring in the above approximate relation for Ω was taken equal to 67. A reconsideration of the experimental data including those obtained at $\mathbf{Re}_d = 8{,}400$ led Yen[68] to assume a value of 39 for said constant. This, however, hardly modifies the corrected curve for $\bar{U}_1/\bar{U}_{1,\max}$. Instead of 0.264 in the equation for $\bar{U}_1/\bar{U}_{1,\max}$ one then has to take 0.255, and instead of 0.35 the value 0.40.

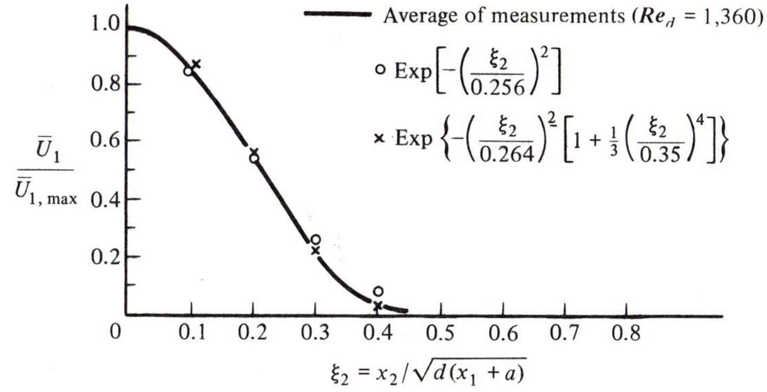

FIGURE 6-37
Comparison between measured mean-velocity distribution in the wake flow of a cylinder and distributions computed using a constant eddy viscosity not corrected and corrected for the effect of the intermittency factor. (From: *Townsend, A. A.*,[9] *by permission of the Cambridge University Press.*)

From the general description of the structure of wake flows and jet flows one concludes that the random flow consists of the following three kinds of motions:

1 relatively small-scale turbulent motions;
2 relatively large-scale turbulent motions;
3 relatively large-scale irrotational motions.

Only the irrotational motions are observed outside the turbulent region, but they will continue to exist also inside the turbulent region. As explained earlier, they do not contribute to the shear-stress if the flow is purely irrotational, though they may contribute to the transport of a scalar quantity as heat and mass. Inside the turbulent region we may split each turbulence velocity component into a deformation mode and a vorticity mode (the dilatation mode is zero for an incompressible fluid). As is known, the deformation mode is irrotational and solenoidal, while the vorticity mode represents the vorticity of the total velocity and is solenoidal too. So, if we consider a practically one-dimensional flow in the x_1-direction, the turbulence shear-stress $-\overline{u_2 u_1}$ can be expressed in the above irrotational and rotational modes. The cross-correlation of the irrotational modes is zero, so that the contribution to the shear-stress is determined mainly by the cross-correlation of the rotational modes, and possibly also by the two cross-correlations between the rotational mode of the one velocity component and the irrotational mode of the other velocity component. It may be surmised that indeed the cross-correlation of the rotational modes gives the main contribution to the shear-stress. In that case the irrotational mode of the turbulence velocity is "inactive" in the sense of not contributing to the shear-stress[69] (see also Sec. 7-8). If these larger-scale turbulent motions find their origin in the deformation and break-down of the initial Von-Kármán vortices (in the case of a plane wake flow) as illustrated in Fig. 6-31, the large-scale structure will be three-dimensional with streamwise vorticity components, which in the turbulent region occur in pairs. It is well known that the instantaneous flow pattern around a straight cylinder is not purely two-dimensional because of back-coupling effects of the wake. For instance, finite, large-scale spanwise correlations of the pressure-fluctuations on the wall of the cylinder have been observed. According to measurements by Prendergast[70] of two-point pressure correlations along a line on the cylinder at 90° from the stagnation point, the corresponding integral length-scale varies between 3 and $4d$ in the range $\mathbf{Re}_d = 2 \times 10^4$ to 10^5. Beyond the latter value of $\mathbf{Re}_d$, so approaching the critical region where the drag coefficient drops markedly, the correlation distance decreases to very low values.

The existence of *large-scale eddies* in the wake flow of a cylinder has already been suggested by Townsend,[9] on the ground of two-point one-time velocity-correlation measurements. Similar, but more extended measurements have been made by Grant[71] in a plane at $x_1 = 533d$, with the fixed hot-wire anemometer at the positions $x_2 = 0$; $x_2 = 0.173\sqrt{x_1 d}$ (point of maximum shear-stress) and $x_2 = 0.33\sqrt{x_1 d}$. Nine correlations have been determined, namely $\mathbf{R}_{1,1}(\xi_1,0,0)$; $\mathbf{R}_{1,1}(0,\xi_2,0)$; $\mathbf{R}_{1,1}(0,0,\xi_3)$ and similar three for $\mathbf{R}_{2,2}$ and $\mathbf{R}_{3,3}$. Here ξ_1, ξ_2, and ξ_3, respectively, denote the distance between the second, movable, hot wire and the fixed one. By considering only the results of $\mathbf{R}_{1,1}$ and $\mathbf{R}_{3,3}$, Grant concluded that their behavior could reasonably well be described by a model consisting of a pair of large eddies as illustrated in Fig. 6-38. The upper parts have been drawn with dashed curves, since the two vortices must continue in a closed loop though viscous diffusion causing a spatial spread will weaken the vorticity there. The model is not complete in that the u_2-component has been assumed to be zero. The actual picture of the flow pattern due to the u_2-component is not clear, but on the basis of the $\mathbf{R}_{2,2}$ measurements Grant concluded that the behavior of the u_2-component could be explained by assuming the existence of local and temporary, rather concentrated "jets" in outward direction, alternating with weaker inward motions or re-entrant

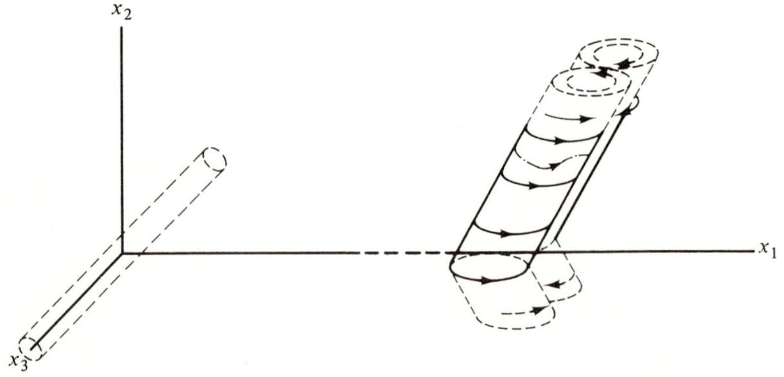

FIGURE 6-38
Schematic model of a pair of large eddies. Solid streamlines according to Grant. Dash-dot streamlines according to Payne-Lumley, and to Townsend. (Adapted from: *Grant, H. L.,*[71] *and Townsend, A. A.*[75])

jets. The picture obtained from this model and the behavior of the u_2-component shows some similarity with the picture which can be deduced from Fig. 6-31, if indeed at such a large distance from the cylinder the large eddies still have some connection with the Von-Kármán vortex street in the developing part of the wake. Since in Grant's model the size of the large eddies is not well defined, Payne and Lumley[72] have re-analyzed Grant's experimental results by expressing the u_i-distributions in a series of orthogonal functions (see, e.g., reference 82). By assuming homogeneity of the main field in the x_1 and x_3-directions, a Fourier transformation could be applied, yielding the energy-spectrum. The large eddies then are defined as the eddy with the wavenumber corresponding with the dominant eigenfunction. The picture obtained by Payne and Lumley of the large-eddy pattern corresponds quite well with that suggested by Grant, though with a slightly different shape of the counter-rotating eddy pair. The planes of circulation are roughly normal to the direction of the maximum positive rate of strain. Thus the streamlines of the eddy pair are not parallel with the (x_1, x_3)-plane, as shown in Fig. 6-38 by the fully drawn lines, but are inclined with respect to this plane. As follows from this analysis, and emphasized by Townsend,[75] the circulation corresponding with the streamlines of the eddy pair cannot be exactly in a plane, but must have a shape as shown by the dash-dot curve in Fig. 6-38. This picture suggests that the outward and inward motions which Grant assumed are the u_2-velocities between the oblique "cylinders," as already indicated by Paine and Lumley's analysis.

An important aspect of free turbulent shear-flow, closely connected with its large-scale structure, is the *entrainment of ambient fluid*. Earlier, in Sec. 6-10, we showed that for the self-preserving region of an axi-symmetric jet in a quiescent environment the volume flowrate of the jet Φ increases proportionally with $(z + a)$. Hence, the rate of increase $d\Phi/dz$ is a constant, i.e., independent of z. We may identify $d\Phi/dz$ with the effective rate of entrainment of ambient fluid by the jet. In the case of a plane jet, however, in the self-preserving region the mean-velocity $\bar{U}_1$ varies as $(x_1 + a)^{-1/2}$, while the width of the jet increases as $(x_1 + a)$. Hence, here Φ increases with $(x_1 + a)^{1/2}$, and the effective rate of entrainment $d\Phi/dx_1$, is not independent of $(x_1 + a)$, but varies as $(x_1 + a)^{-1/2}$. When we consider the wake flows in the same way, wc conclude that for the plane wake, as well as for the axi-symmetric wake the volume flow rate Φ corresponding with the velocity defect with respect to the free-stream flow is a constant in the self-preserving region. So the effective rates of entrainment $d\Phi/dx_1$ and $d\Phi/dz$ are zero. Here we have to bear in mind that such large distances downstream of the wake-producing object have been assumed that $\bar{U}_{1,\max} \ll U_0$, so that the momentum balance equation reduces to Eq. (6-26).

It is reasonable to connect the entrainment with the bulgy nature of the interface, as shown, e.g., in Fig. 6-35. And also to expect an increase of the entrainment with the depth or amplitude of the large-scale indentations. Since these indentations are stronger for wake flows than for jet flows [compare, e.g., Figs. 1-2 and 6-33], apparently in the case of wake flows the entrainment in the self-preserving region is a local affair resulting in a zero net effective rate of entrainment.

Therefore $d\Phi/dx_1$ is not a suitable quantity to describe local entrainment effects due to the pulsating nature of the bulgy interface. Townsend[74,75] concluded from the momentum-balance and energy-balance equations that the following non-dimensional quantity would be more suitable to characterize the entrainment action of the interface, which he defines as the *entrainment constant* β.

$$\beta = \frac{2U_0 + \bar{U}_{1,\max}}{\bar{U}_{1,\max}} \frac{d\overline{x_2^*}}{dx_1} \tag{6-114}$$

Here $\overline{x_2^*}$ is the mean lateral position of the interface, and the same quantity as $\overline{r^*}$ shown in Fig. 6-34 for the axi-symmetric jet. Because for the regions of wake flows considered $\bar{U}_{1,\max} \ll U_0$, here Eq. (6-114) reduces to

$$\beta = 2 \frac{U_0}{\bar{U}_{1,\max}} \frac{d\overline{x_2^*}}{dx_1} \tag{6-114a}$$

On the other hand for free jets in a quiescent environment $U_0 = 0$, so there Eq. (6-114) becomes

$$\beta = \frac{d\overline{x_2^*}}{dx_1} \tag{6-114b}$$

The reader may convince himself that for all wake flows and jet flows considered here β becomes a constant in the self-preserving region, but with a different value depending on the specific flow considered. Townsend[74] mentions the value $\beta = 0.40$ for the plane wake flow and $\beta = 0.18$ for the axi-symmetric jet. From Corrsin and Kistler's measurements on such a jet, shown in Fig. 6-36 one would obtain $\beta = 0.15$.

The R.M.S.-value $(r^* - \overline{r^*})'$ of the indentations shown in this figure is roughly $0.5r_{1/2}$, while we deduce from the same figure $(r^* - \overline{r^*})'/\overline{r^*} \simeq 0.23$. Townsend mentions a value of 0.2 for the jet, and a value of 0.4 for the plane wake.

The values of β and of $(r^* - \overline{r^*})'/\overline{r^*}$ show in a quantitative way that indeed in wake flows the large-scale indentations are relatively stronger than in jet flows. Also in agreement with it is the fact that the position of the intermittency-curve is relatively much more outwards for the jet (Fig. 6-28) than for the wake (Fig. 6-13). The mechanism of the entrainment process, or that of the movements of the interface resulting in the lateral spread of the turbulence regions is still not completely understood. There are, however, a number of empirical facts which throw some light on this mechanism. The main facts are the following:

1 Grant,[71] and Keffer[76] observed in wakes, in the dynamic processes taking place in the boundary regions, and associated with the movements of the interface, periods of high turbulence activity alternating with periods of relative quiescence. The surface of the interface seems to distort into groups of periodic waves which develop in a way similar to those resulting from an instability of a vortex sheet. During the periods of high activity the entrainment is high, while practically no entrainment occurs during the periods of quiescence.

2 The distortion of the interface is three-dimensional. So in plane flows the bulges have finite dimensions in the direction transverse to the plane of the mean flow.

3 We recall the observations made by Wygnanski and Fiedler[51] in an axi-symmetric jet concerning the convection velocity of the turbulence pattern. This velocity becomes greater than the local mean-velocity in the jet, when approaching the outer boundary regions. For an observer moving with the mean velocity in this region, the interface is subjected to distortions in time like traveling waves.

4 We also recall the measurements made by Wygnanski and Fiedler[54] in a plane half-jet, of $(\bar{U}_1)_{\text{turb}}$ and $(\bar{U}_1)_{\text{irrot}}$ in the region of low values of the intermittency factor Ω. There $(\bar{U}_1)_{\text{irrot}}$ was much smaller than $(\bar{U}_1)_{\text{turb}}$. So the irrotational parts are overtaken by the turbulent parts in the bulges of the interface. This means that relative to the bulges there is an "overflow" of irrotational fluid in a direction opposite to that of the mean-velocity of the jet. This overflowing motion around the bulges has been studied in more detail by Kovasznay, Kibens, and Blackwelder[77] for the case of a plane boundary-layer, where the overflow was in the direction of the free-stream velocity. We shall come back to this study in Sec. 7-9.

5 When a turbulence is subjected to a high rate of shear, and so to a high strain rate, the turbulence will be intensified, becomes more anisotropic and develops high turbulence-stresses. This has been investigated experimentally by Keffer,[76] Tucker and Reynolds,[83] and by Mobbs[81] in a distorting constant-area duct. In such a duct $\bar{U}_1$ remains constant, so the axial strain-rate $\bar{\zeta}_1 = \partial \bar{U}_1/\partial x_1 = 0$. The other two components of the strain-rate are equal, but opposite in sign for an incompressible fluid, i.e., $\bar{\zeta}_2 = -\bar{\zeta}_3$. After some time a fluid particle has undergone a deformation by the plane straining, such that

$$e_1 = \exp \int_0^t dt' \, \zeta_1(t') = 1; \qquad e_2 = e_3^{-1} = \exp \int_0^t dt' \, \zeta_2(t')$$

The response of an initially isotropic turbulence to such a plane straining has been studied already many years ago by Townsend,[9] in particular the case of such a rapid distortion rate, that only small values of t need to be considered (*rapid-distortion theory*). The findings obtained by Tucker and Reynolds, and by Mobbs were generally in agreement with the rapid-distortion theory. Mobbs also observed that downstream of the distorting section of the duct, where the distortion ends, the turbulence subsides due to the absence of any straining action, with a tendency to return to a more isotropic state and with a decay accordingly. An interesting point observed by Mobbs of the turbulent motions near the interface of a turbulent/irrotational region with uniform mean axial velocity generated by a specially constructed device installed in the distorting duct, was a return to a laminar state due to viscous dissipation. So the volume of turbulent fluid seemed to decrease at the absence of a straining action. Mobbs concluded from his experiments that the irrotational/turbulent flow interface will advance into the irrotational fluid at a rate that depends on the local mean strain-rate and the local degree of anisotropy of the turbulence near the interface. A reduction in the volume of turbulent fluid may occur when the interface is in a region of zero mean strain-rate, or when the degree of anisotropy is low.

Based on the above facts Townsend[74,75] proposed the following picture of the processes occurring in the boundary region of a free shear-flow. Initially during a period of quiescence the turbulence has a fine structure of high intensity, which decays due to viscous dissipation. The interface becomes unstable according to a Helmholtz instability theory. Relatively large modes or eddies are selected from the unorganized turbulence by the general shearing motion according to the rapid-distortion theory. Calculations of the correlation coefficients $\mathbf{R}_{1,1}$, $\mathbf{R}_{2,2}$ and $\mathbf{R}_{3,3}$ according to this theory appear to be of the same order of magnitude as those measured by Grant in a wake flow.[71] During the growth of the large eddies there is a rapid entrainment due to folding of the interface and engulfing of irrotational fluid. This interaction and deformation of the large eddies cause the production of small eddies with increasing intensity. The extraction of energy from the large eddies prevent a further growth of these large eddies. They become more stable. Due to the continued extraction of energy by the small eddies the large eddies decay, and the associated large-scale motions of the interface subside. A period of relative quiescence with again initially a high-intensity fine-scale structure sets in. Then the next cycle begins.

According to this picture, the combination of shear flow and Helmholtz instability, there is a self-excitation power in the free-shear flow, which may be entirely independent of the processes in the developing region. However, it is quite possible that the latter processes and the turbulence in the developing region may act as a trigger, in that they produce the conditions as deformation of the interface and formation of bulges, suitable for the self-maintenance of the cyclic processes farther downstream. The gradual increase of the bulges or amplitude of the indentations of the interface farther downstream, leading to the observed broadening of the turbulent region, is in accord with the fact that notwithstanding the continuous production of turbulence, there is a decrease in turbulence intensity in the downstream direction. This results in less initial damping by the fine-scale turbulence during a period of quiescence, so that a different mode of the Helmholtz instability will be preferably selected and may grow to a higher amplitude.

Let us now consider the consequences of the above picture of the turbulence structure on the transport of various quantities. In this respect one of the most important features to recognize is the existence of smaller-scale high-intensity turbulent motions together with large-scale slow motions, part of which are of an irrotational nature.

The quantities that we shall consider are axial momentum, heat, and turbulence energy. The turbulent transport of these quantities will be caused by the two types of turbulent motion just mentioned. If the small-scale turbulence has a scale small enough, so that the mean value of the quantity considered is practically uniform over distances equal to this scale, it is reasonable to assume that the transport caused by the small-scale turbulence may be described by means of a gradient-type diffusion. On the other hand, the transport by large-scale turbulence may be assumed to be purely convective.

Let $\mathscr{P}$ be the transferable property considered. We may write

$$\overline{u_j\mathscr{P}} = \overline{u_j^*\mathscr{P}} + \overline{\mathscr{V}_j\mathscr{P}} = -\Omega(\epsilon_{\mathscr{P}})_{ji}\frac{\partial(\bar{\mathscr{P}}/\Omega)}{\partial x_i} + \overline{\mathscr{V}_j\mathscr{P}} \tag{6-115}$$

where u_j^* and $\mathscr{V}_j$ represent the small-scale and large-scale parts of the turbulence velocity u_j.

The contributions by gradient-type diffusion and by convective transport will not be independent of the nature of the property transported. Indeed, they are different for momentum, heat, and turbulence energy; this can be inferred from the differences in distribution of their mean values. To what extent they are different is still very difficult to say. But what does the shape of the distribution curves tell us?

If, on the one hand, the transport were due entirely to the bulk convective motion $\mathscr{V}_j$, one might expect that the value of this quantity would vary only slightly in the core region and would decrease rapidly to zero in the boundary region, provided that the effects that cause local changes of the property transported were not large, so that the changes were not large compared with the absolute value of the property. Thus, where turbulence energy is the quantity transported, the changes due to production and dissipation would have to be small compared with the energy level. Also, in the case of momentum, if the large-scale motions were important in the transport process, it is only the rotational part that would contribute.

If, on the other hand, the transport occurred in consequence solely of gradient-type diffusion due to small-scale turbulence, one might expect a gradual decrease of the quantity from the center of the turbulence region toward its outer boundary.

Consider in this light the distribution curves of mean axial velocity (which is the distribution of mean axial momentum), heat, and turbulence kinetic energy, which are given, respectively in Figs. 6-3, 6-4, and 6-5.

From these figures it appears that the mean axial momentum is distributed as if by gradient-type diffusion, and the turbulence kinetic energy as if by convective transport; the distribution of heat follows a pattern in between. Of course, we would go too far if we concluded that the distribution of momentum occurs by gradient-type diffusion alone and distribution of turbulence energy by convection alone—though a pretty good description of the distribution of these quantities is obtained if this is done.

There are arguments, however, against the concept of a dominating transport of momentum by gradient-type diffusion. One of these arguments is based on the observation that, in the region of maximum shear, the ratio $\overline{u_1u_2}/\overline{q^2}$ has an almost uniform distribution. This seems to indicate that the transport of momentum does not depend on the gradient of the mean momentum (mean-velocity gradient). Also, the fact that momentum transport is affected by the action of pressure gradients makes the possibility of a pure gradient-type diffusion not very great.

To show that transport of momentum cannot be described by a gradient-type of diffusion alone, we may refer to experiments in flows where the distribution of the main-flow velocity in the lateral direction is markedly asymmetric. For instance, in the cases of wall-jets, or flow through a channel of unequal roughness of the two parallel walls. There, local regions have been observed in the vicinity of the maximum velocity, where the shear-stress $-\overline{u_1u_2}$ and the mean-velocity gradient $\partial\bar{U}_1/\partial x_2$ have opposite sign. So a gradient-type of diffusion would imply a transport up the gradient, instead of down the gradient. See, e.g., references 78, 79, and 80. The explanations offered by Eskinazi[78] and by Hinze[80] are based on the assumption that larger-scale motions have an effect that becomes apparent in said regions, where $\partial\bar{U}_1/\partial x_2$ is small or zero. Moreover, Hinze[80] showed that the sign of the gradient $\partial u_2'/\partial x_2$ is of importance in these regions.

On the other hand, if we exclude these small regions, or if we have to do with flows having a symmetric mean-velocity distribution, experimental evidence has shown that the distribution of mean momentum is surprisingly well described if the concept of a constant coefficient of eddy viscosity is introduced.

Hence, with the situation as it is, the only conclusion that can safely be drawn at present is that the transport processes in free turbulent flow are closely related to the above specific structure of turbulence consisting of motions of large scale and those of a much smaller scale. The transport processes may be described by means of relations like (6-115), where the contributions of the two mechanisms [thus, for instance, the value of $(\epsilon_{\mathscr{P}})_{ij}$] are different for the various transferable quantities.

Although the concept of transport by means of gradient-type diffusion, together with a constant coefficient of eddy viscosity, is useful and although it may yield a successful description of mean-momentum distribution, yet it may have little bearing on the real physical processes (Townsend[9]). It may also be remarked that the same concept applied to the distribution of heat and turbulence energy does not lead to equivalent results. It is very likely that the contribution by convective transport of these quantities is by no means small, particularly in the case of turbulence energy. In Fig. 6-8, we may observe, as mentioned in Sec. 6-7, a small region, namely $0.13 \gtrsim \xi_2 \gtrsim 0.15$ where $\overline{u_1{}^2u_2} > 0$ whereas according to Fig. 6-5 in the same region $d\overline{u_1{}^2}/d\xi_2 > 0$. Thus again a transport of $u_1{}^2$ up the gradient. Since also according to Fig. 6-5, in this small region $du_2'/d\xi_2 < 0$, the explanation suggested by Hinze[80] may be applied here.

In the preceding sections we have tried to describe the transport of heat by means of a diffusion alone, with a coefficient of eddy diffusion ϵ_γ. This coefficient should then include the effect of convection by large-scale motions. It has been found that $\epsilon_\gamma/\epsilon_m > 1$ and that this ratio is larger for plane than for axi-symmetrical mixing zones. If ϵ_γ does include an important contribution by convective transport, then this difference may be explained as follows.

Townsend has estimated the velocity of the large-scale motion from the distribution of turbulence kinetic energy in the wake flow of a cylinder, by assuming

$$\overline{q^2\mathscr{V}} = \overline{q^2}\mathscr{V}^*$$

He found that this apparent velocity $\mathscr{V}^*$ was directed outward, increased rapidly toward the boundary of the wake up to a maximum value at about $\xi_2 = 0.4$, then decreased very rapidly to zero in the boundary region. The maximum value amounted to roughly $\mathscr{V}^*/\bar{U}_{1,\max} \simeq 0.15$. So this large-scale outward flow is of the same order of magnitude as the relative intensity of turbulence, though smaller, which seems reasonable.

It may be expected that the behavior of this outward motion will be different in a mixing zone of axi-symmetry (e.g., round free jet) than in a plane mixing zone; in the axi-symmetric mixing zone the velocity $\mathscr{V}$ will change at a lower rate in the outward direction, owing to the divergent character of the outward flow in a zone of axi-symmetry. Consequently the effect of the convection flow $\mathscr{V}$ on the total transport in the outward direction will be smaller in a round jet or wake than in a plane jet or wake. Indeed the measurements by Wygnanski and Fiedler[51] in an *axi-symmetric jet* of the distribution of $\overline{u_z^2}$ and $\overline{u_z^2u_r}$ do not show regions, which would lead to a transport of u_z^2 up the gradient, in contrast to the plane wake. Also, as referred to earlier, they showed that the turbulence transport of q^2, namely as given by $\overline{q^2u_r}$, may be reasonably well described with a constant eddy diffusivity $\epsilon_{\overline{q^2}}$ across the main part of the jet cross-section when corrected for the intermittency factor Ω, which implies a negligible effect here of $\overline{q^2\mathscr{V}}$.

REFERENCES

1. Reichardt, H.: *Forsch. Gebiete Ingenieurw.*, no. 414, 1951.
2. Goldstein, S.: "Modern Developments in Fluid Dynamics," vol. 2, Clarendon Press, Oxford, 1938.
3. Schlichting, H.: *Ingr. Arch.*, **1,** 533 (1930).
4. Fage, A., and V. M. Falkner: *Proc. Roy. Soc. London,* **135A,** 702 (1935).
5. Townsend, A. A.: *Proc. Roy. Soc. London,* **190A,** 551 (1947).
6. Townsend, A. A.: *Australian J. Research,* **2A,** 451 (1949).
7. Townsend, A. A.: *Proc. Roy. Soc. London,* **197A,** 124 (1949).
8. Townsend, A. A.: *Australian J. Research,* **1A,** 161 (1948).
9. Townsend, A. A.: "The Structure of Turbulent Shear Flow," Cambridge University Press, New York, 1956.
10. Tollmien, W.: *Z. angew. Math. u. Mech.*, **6,** 468 (1926).
11. Uberoi, M. S., and P. Freymuth: *Phys. Fluids,* **12,** 1359 (1969).
12. Pai, S. I.: *Quart. Appl. Math.*, **10,** 141 (1952).

13. Carmody, Th.: *J. Basic Engng., Trans. A.S.M.E.* **86D,** 689 (1964).
14. Forstall, W., Jr.: Material and Momentum Transfer in Coaxial Gas Streams, Ph.D. thesis, Massachusetts Institute of Technology, Cambridge, 1949.
15. Forstall, W., Jr., and A. H. Shapiro: *J. Appl. Mechanics,* **17,** 399 (1950).
16. Forstall, W., Jr., and A. H. Shapiro: *J. Appl. Mechanics,* **18,** 219 (1951).
17. Acharya, Y. V. G.: Momentum Transfer and Heat Diffusion in the Mixing of Coaxial Turbulent Jets Surrounded by a Pipe, Ph.D. thesis, Delft, 1954.
18. Landis, F., and A. H. Shapiro: *Proc. Heat Transfer and Fluid Mechanics Inst.,* Palo Alto, Calif., 1951, p. 133.
19. Trüpel, Th.: *Z. ges. Turbinenwesen,* **12,** 52 (1915).
20. Ruden, P.: *Naturwissenschaften,* **21,** 375 (1933).
21. Corrsin, S.: *Natl. Advisory Comm. Aeronaut. Wartime Repts. No.* 94, 1943.
22. Corrsin, S., and M. S. Uberoi: *Natl. Advisory Comm. Aeronaut. Tech. Notes No.* 1865, 1949.
23. Corrsin, S., and M. S. Uberoi: *Natl. Advisory Comm. Aeronaut. Tech. Notes No.* 2124, 1950.
24. Hinze, J. O., and B. G. van der Hegge Zijnen: *Appl. Sci. Research,* **1A,** 435 (1949).
25. Albertson, M. L., Y. B. Dai, R. A. Jenson, and H. Rouse: *Proc. Am. Soc. Civil Engrs.,* **74,** 1571 (1948).
26. Forstall, W., and E. W. Gaylord: *J. Appl. Mechanics,* **22,** 161 (1955).
27. Keagy, W. R., and A. E. Weller: *Proc. Heat Transfer and Fluid Mechanics Inst.,* Berkeley, Calif., 1949, p. 89.
28. Alexander, L. G., T. Baron, and E. W. Comings: *Univ. Illinois Bull. No.* 413, 1953.
29. Corrsin, S., and A. L. Kistler: *Natl. Advisory Comm. Aeronaut. Tech. Notes No.* 3133, 1954.
30. Liepmann, H. W., and J. Laufer: *Natl. Advisory Comm. Aeronaut. Tech. Notes No.* 1257, 1947.
31. Laurence, J. C.: *Natl. Advisory Comm. Aeronaut. Tech. Notes No.* 3561, 1955.
32. Chin, W. C., and L. N. Rib: *Trans. Am. Geophys. Union,* **37,** 13 (1956).
33. Van der Hegge Zijnen, B. G.: *Appl. Sci. Research,* **7A,** 293 (1958).
34. Alexander, L. G., A. Kivnick, E. W. Comings, and E. D. Henze: *A.I.Ch.E. Journal,* **1,** 55 (1955).
35. Phillips, O. M.: *Proc. Cambridge Phil. Soc.,* **51,** 220 (1955).
36. Stewart, R. W.: *J. Fluid Mech.,* **1,** 593 (1956).
37. Barat, M.: *Compt. rend.,* **238,** 445 (1954).
38. Szablewski, W.: *Ingr. Arch.,* **20,** 567, 573 (1952).
39. Keagy, W. R., and A. E. Weller: *Proc. Heat Transfer and Fluid Mechanics Inst.,* Berkeley, Calif., 1949, p. 89.
40. Sunavala, P. D., C. Hulse, and M. W. Thring: *Combustion and Flame,* **1,** 179 (1957).
41. Thring, M. W., and M. P. Newby: *4th Symposium on Combustion,* p. 789, Cambridge, Mass., 1952.
42. Chevray, R.: *J. Basic Engng., Trans. A.S.M.E.,* **90D,** 275 (1968).
43. Miller, D. R., and E. W. Comings: *J. Fluid Mech.,* **3,** 1 (1957).
44. Kobashi, Y.: *J. Phys. Soc. Japan,* **12,** 533 (1957).
45. Hwang, N. H. C., and L. V. Baldwin: *J. Basic Engng., Trans. A.S.M.E.,* **88D,** 261 (1966).
46. Gibson, C. H., C. C. Chen, and S. C. Lin: *A.I.A.A. Jl.,* **6,** 642 (1968).
47. Chigier, N. A., and J. M. Beér: *J. Basic Engng., Trans. A.S.M.E.,* **86D,** 797 (1964).
48. Morton, B. R.: *Int. J. Heat Mass Transf.,* **5,** 955 (1962).
49. Champagne, F. H., and I. Wygnanski: *Boeing Scientific Research Laboratories, Rep. D 1-82-0958,* 1970.

50. Sami, S.: *J. Fluid Mech.,* **29,** 81 (1967).
51. Wygnanski, I., and H. Fiedler: *J. Fluid Mech.,* **38,** 577 (1969).
52. Ricou, F. P., and D. B. Spalding: *J. Fluid Mech.,* **11,** 21 (1961).
53. Van der Hegge Zijnen, B. G.: *Appl. Sci. Res.,* **7A,** 277 (1958).
54. Wygnanski, I., and H. E. Fiedler: *J. Fluid Mech.,* **41,** 327 (1970).
55. Uberoi, M. S., and L. C. Garby: *Phys. Fluids,* Supplement **10,** No. 9, S. 200 (1967).
56. Gibson, M. M.: *J. Fluid Mech.,* **15,** 161 (1963).
57. Crow, S. C., and F. H. Champagne: *J. Fluid Mech.,* **48,** 547 (1971).
58. Wygnanski, I., and E. Gutmark: *Technion, Israel Inst. Technology, TAE Rep. 103,* Haifa, 1969.
59. Michalke, A.: *J. Fluid Mech.,* **19,** 543 (1964).
60. Michalke, A.: *J. Fluid Mech.,* **23,** 521 (1965).
61. Bradshaw, P., D. H. Ferriss, and R. F. Johnson: *J. Fluid Mech.,* **19,** 591 (1964).
62. Davies, P. O. A. L., and M. J. Fisher: *University of Southampton A.A.S.U. Rep.* 233, 1963.
63. Bradshaw, P., and D. H. Ferriss: *Aero. Res. Council, Current Papers:* C.P. 899, 1965.
64. Davies, P. O. A. L.: *A.I.A.A. Jl.,* **4,** 1971 (1966).
65. Sami, S., Th. Carmody, and H. Rouse: *J. Fluid Mech.,* **27,** 231 (1967).
66. Sami, S.: *J. Fluid Mech.,* **29,** 81 (1967).
67. Schapker, R. L.: *A.I.A.A. Jl.,* **4,** 1979 (1966).
68. Yen, K. T.: *Space Sci. Lab., Techn. Inf. Series, Report R 67 SD 3* (1967).
69. Townsend, A. A.: *J. Fluid Mech.,* **11,** 97 (1961).
70. Prendergast, V.: *University of Toronto, UTIA Techn. Note 23,* 1958.
71. Grant, H. L.: *J. Fluid Mech.,* **4,** 149 (1958).
72. Payne, F. R., and J. L. Lumley: *Phys. Fluids,* Supplement, **10,** No. 9, S 194 (1967).
73. Nayar, B. M., T. E. Siddon, and W. T. Chu: *University of Toronto, UTIA Techn. Note 131,* 1969.
74. Townsend, A. A.: *J. Fluid Mech.,* **26,** 689 (1966).
75. Townsend, A. A.: *J. Fluid Mech.,* **41,** 13 (1970).
76. Keffer, J. F.: *J. Fluid Mech.,* **22,** 135 (1965).
77. Kovasznay, L. S. G., V. Kibens, and R. F. Blackwelder: *J. Fluid Mech.,* **41,** 283 (1970).
78. Eskinazi, S., and F. F. Erian: *Phys. Fluids,* **12,** 1988 (1969).
79. Beguier, C.: *Compt. rend. acad. sci.,* **268A,** 69 (1969).
80. Hinze, J. O.: *Appl. Sci. Res.,* **22,** 163 (1970).
81. Mobbs, F. R.: *J. Fluid Mech.,* **33,** 227 (1968).
82. Lumley, J. L.: "Stochastic Tools in Turbulence," Academic Press, New York, 1972.
83. Tucker, H. J., and A. R. Reynolds: *J. Fluid Mech.,* **32,** 657 (1968).
84. Freymuth, P., and M. S. Uberoi: *Phys. Fluids,* **14,** 2574 (1971).
85. Baldwin, L. V., and V. A. Sandborn: *A.I.A.A. Jl.,* **6,** 1163 (1968).

NOMENCLATURE FOR CHAPTER 6

a distance of virtual source of wake or jet from the origin.

c $1/\sqrt{d(x_1 + a)}$ in wake flow of cylinder; $1/(x + a)$. in jet flow; c_m, for momentum; c_γ, for scalar; c_ω, for vorticity.

d diameter of cylinder or orifice.

$\mathbf{E}_1(k_1)$	one-dimensional energy spectrum.
k_1	$2\pi n/\bar{U}_1$ wavenumber.
L_i	length scale in direction x_i.
n	frequency.
P	static pressure; $\bar{P}$, time-mean value; p, turbulent fluctuation; P_0, ambient pressure; P_s, pressure in secondary stream.
$\mathbf{Q}_{i,j}$	spatial second-order velocity-scale.
$\overline{q^2}$	$\overline{u_i u_i}$, twice kinetic energy of turbulence.
$\mathbf{R}_{ij}$	coefficient of spatial second-order velocity-correlation tensor.
$\mathbf{R}_E$	coefficient of Eulerian time correlation.
$\mathbf{Re}_d$	$U_p d/\nu$.
r	cylindrical polar coordinate in radial direction.
t	time.
U_i	Eulerian velocity; $\bar{U}_i$, time-mean value; u_i, turbulence component; u_i', $\sqrt{\overline{u_i^2}}$; U_p, velocity of primary stream; U_s, velocity of secondary stream; subscripts r, φ, x refer to cylindrical polar coordinates; U_c, convection velocity.
x	cylindrical polar coordinate in axial direction; x_i, Cartesian coordinates.
ε	dissipation by turbulence per unit of mass.
ϵ_{ij}	eddy-diffusion tensor; $(\epsilon_{\mathscr{P}})_{ij}$, for any transferable property; $(\epsilon_m)_{ij}$, for momentum; $(\epsilon_\gamma)_{ij}$, for scalar.
η	$(\nu^3/\varepsilon)^{1/4}$.
φ	cylindrical polar coordinate in angular direction.
Γ	scalar quantity; $\bar{\Gamma}$, time-mean value; γ, turbulent fluctuation.
Λ_g	lateral spatial integral scale.
$\mathscr{L}$	coefficient in Reichardt's "diffusion" equation; $\mathscr{L}_\gamma$ for scalar; similarity length-scale.
$\mathfrak{l}$	Prandtl's mixing length; $\mathfrak{l}_m$, for momentum; $\mathfrak{l}_\gamma$, for scalar; $\mathfrak{l}_\omega$, for vorticity.
ν	kinematic viscosity.
Ω	intermittency factor.
$\mathscr{P}$	transferable property.
$\mathfrak{P}$	probability density.
ρ	density.
σ_{ij}	stress tensor.
$\mathfrak{I}$	integral time scale.
τ	time.
Θ	temperature; $\bar{\Theta}$, time-mean value; θ, turbulent fluctuation.
$\mathscr{U}_i$	similarity velocity scale in direction x_i.
$\mathfrak{u}_i$	similarity scale for turbulence velocity in direction x_i.
$\mathscr{V}_i$	velocity scale in direction x_i.
$\mathscr{V}$	velocity of large-scale turbulent motions.
$\mathscr{V}^*$	apparent outward velocity in wake flow.

$\overline{\mathfrak{V}^2}$	$2U_0\bar{U}_1 + \bar{U}_1{}^2 + \overline{u_1{}^2}$.
$\mathfrak{v}$	scale for turbulence velocity.
ξ_1	$(x_1 + a)/d$; ξ_2, $x_2/\sqrt{d(x_1 + a)}$, for wake flow of cylinders.
ξ_2	$r/(x + a)$, for jet flow; ξ'_1, x_1/d; ξ'_2, x_2/d.

7
"WALL"-TURBULENT SHEAR FLOWS

7-1 INTRODUCTION

"Wall" turbulence may be described as turbulence whose structure is directly influenced by the presence of a solid boundary. There is great variety in the possibilities of wall turbulence, depending on the nature and configuration of the boundary.

In what follows we shall confine ourselves in the first place to the case of impervious, rigid walls. Even along such walls many types of turbulent flow are still possible, but they may be classified into two main groups. One group comprises the flows around rigid bodies; the other group comprises the flows within a space bounded by rigid walls.

An essential difference between the two groups is that, in the first, the domain where wall turbulence occurs increases along the body in the downstream direction, whereas, in the second, this domain of wall turbulence remains restricted to the space bounded by the rigid walls.

The first group of flows is usually designated more specifically by the name "boundary-layer flows," because the domain of wall turbulence remains confined to a relatively thin layer along the surface of the body, though this layer usually increases in absolute thickness in the downstream direction; outside this boundary layer we have the undisturbed free stream.

The simplest types of flow belonging to these two groups are, respectively, the two-dimensional boundary-layer flow along a plane surface with zero pressure-gradient in the free stream, and the fully developed flow through a straight two-dimensional channel or a round tube of uniform cross-section. In the latter case the flow is homogeneous in its time-mean structure in the downstream direction.

We shall restrict ourselves in the second place to considering only these two simplest flow types: the "constant-pressure" flow along a flat plate, and the flow through a straight tube of uniform circular cross section. This is done not only because of their relative simplicity, but also because most of the investigations concerning the details of the turbulence pertain to these two types of flow.

Finally we shall consider mainly the flow of a fluid with constant properties.

As elsewhere, it is the mechanism of the turbulence and of its transport processes that we want to consider in detail, in so far as this is made possible by the available results of experimental investigations. At present our knowledge of this mechanism for wall turbulence is still inadequate as a basis for a theory that would be sound and complete. Present theories about time-mean-velocity and scalar-quantity distributions and about the corresponding resistance coefficients are still semiempirical in nature. In the early days much use was made of the various mixing-length theories. We shall not consider these extensively. The reader can find an extensive treatment of the application of Prandtl's momentum-transport and Taylor's vorticity-transport theories to boundary-layer and tube flow in Ref. 1 and elsewhere.

In the turbulent flow along a rigid wall the turbulence is directly affected by the wall, at least in a region close to it. For a smooth wall this effect occurs through the action of viscous stresses, for a rough wall through the action of forces resulting from the flow around the roughness elements.

In free turbulence, at sufficiently large Reynolds numbers the effect of viscosity on the over-all behavior and gross structure of the flow can be neglected; in wall turbulence, at any Reynolds number, there is a region close to the wall where the behavior of the flow is determined by the fluid viscosity, and so by the Reynolds number if the wall is smooth. At a distance from the wall, the direct effect of fluid viscosity on the gross structure of the turbulence may become small and negligible; so some similarity with free turbulent flow may be expected there.

Indeed, this similarity may become very close, particularly in boundary-layer flow. In the outer region of this flow, near the boundary with the undisturbed free stream, there is the same interaction between turbulent-flow region and free-stream region that occurs in free turbulence with a restricted turbulence region.

Consequently, in the turbulent flow past a rigid boundary we may distinguish two regions. A region adjacent to the wall, in which the flow is directly affected by the condition at the wall. This condition is expressed by the wall shear stress, and in the case of a smooth wall by the fluid viscosity, or by the wall roughness if this roughness is sufficiently pronounced. We will refer to this region as the *"wall" region.*

This region is also referred to as the *"inner" region*. Beyond this wall region we have another region, where the flow is only indirectly affected by the wall through its wall shear stress. This second region is usually referred to as the *"outer" region*, but in the case of tube flow or channel flow we will prefer to refer to this region as the *"core" region*.

When the wall is perfectly smooth there is an extremely thin layer at the wall where the flow is predominantly viscous. Because of this fact we will call it the *viscous sublayer*. An earlier name used to be laminar sublayer, but this name appears not to be appropriate since, as will be shown later, the flow in this sublayer is not laminar in its strict sense. The average thickness of this sublayer is usually denoted by δ_l. The subscript l still refers to the earlier name, laminar sublayer.

Outside this thin viscous sublayer inertial effects become more and more of importance with respect to the viscous effects, until at some distance from the wall the flow becomes fully turbulent with dominating inertial effects and negligible direct viscous effects on the main flow.

Let δ_t be the average distance from the wall beyond which the flow is completely turbulent. The range of the *transition or "buffer" region* is then specified by $\delta_l < x_2 < \delta_t$.

We conclude that in the case of a smooth wall three subregions may be distinguished in the wall region. A viscous sublayer at the wall, a fully turbulent region, separated from the sublayer by a buffer region where viscous and turbulence inertial effects on the main flow are of the same order of magnitude. Of course, the changes in the flow, going through these three subregions are gradual. The flow in the outer region is completely turbulent, and the distinction with the turbulent flow in the wall region stems from the fact that the turbulence in the latter region is directly influenced by the wall, whereas it is not in the outer region, though there is still some interaction between the two regions. As will be shown later the viscous sublayer and the buffer region are both very thin compared with the local thickness of the boundary layer or with the radius of the tube. The wall region covers roughly 15 per cent of the whole boundary layer.

In the following sections we shall consider first the boundary-layer flow and then the flow through a straight tube of circular cross-section.

7-2 APPROXIMATIONS TO THE EQUATIONS OF MOTION, AND THEIR INTEGRAL RELATIONS FOR A PLANE BOUNDARY LAYER

Before considering approximations and integral relations we will first make a few remarks concerning similarity. In Sec. 6-2 we introduced the notions of *complete* and *incomplete similarity*. We showed that as a consequence of complete similarity the length scale must increase linearly with the axial distance from the origin of similarity. In the case of a boundary layer this would mean a linear increase of the boundary-layer thickness in the free-stream direction. At the same time all velocities, including the free-stream velocity, should decrease inversely proportional with x_1, the stream-wise coordinate, while the pressure P should either be constant or vary according to x_1^{-2}. Since usually it may be assumed that the free stream is irrotational, so that Bernoulli's theorem holds true, the first possibility would then be ruled out for complete similarity. For the boundary layer with zero pressure gradient (and hence constant free-stream velocity) considered mainly in this chapter, the conclusion is that no complete similarity is possible. This result on similarity is general, and so applies to a laminar boundary layer as well. It is known that a laminar boundary layer with zero-pressure gradient increases in thickness proportional to the square root of the axial distance, which implies incomplete similarity. Also, the velocity scales for the axial and transverse components of the velocity are different.

We choose a coordinate system with the origin at the plane rigid boundary, with the x_1-axis in the free-stream direction and the x_2-axis perpendicular to the surface. The position of the origin in the x_1-direction is considered arbitrary for the time being. The corresponding components of the velocity are U_1 and U_2.

Boundary-Layer Approximations

When making so-called boundary-layer approximations it is useful to make a distinction between the wall region and the outer region.

Let us first consider the *wall region*. Let L_1 and L_2 be length scales for the variations in the x_1 and x_2 directions respectively. The small relative thickness of the boundary-layer region leads to the conclusion that

$$\frac{L_2}{L_1} \ll 1$$

Let $\mathscr{V}_1$ and $\mathscr{V}_2$ be the velocity scales for the U_1 and U_2 components, then subject to the restrictions put forward in Sec. 6-3, we may conclude by virtue of the mass-balance equation for an incompressible fluid and a two-dimensional flow

$$\frac{\partial U_1}{\partial x_1} + \frac{\partial U_2}{\partial x_2} = 0 \tag{7-1}$$

that

$$\frac{\mathscr{V}_2}{\mathscr{V}_1} \sim \frac{L_2}{L_1} \ll 1$$

Although close to the wall the degree of anisotropy of the turbulence-velocity components is much greater than in free turbulence (see Sec. 7-7), the turbulence intensities in the various directions are still of the same order of magnitude; accordingly, it is certainly permissible to consider one velocity scale $\mathfrak{v}$ for u_1', u_2', and u_3'.

For the x_1-direction the equation of motion for an incompressible fluid reads

$$\begin{array}{ccccccc}
\bar{U}_1 \dfrac{\partial \bar{U}_1}{\partial x_1} & + \bar{U}_2 \dfrac{\partial \bar{U}_1}{\partial x_2} & = -\dfrac{1}{\rho}\dfrac{\partial \bar{P}}{\partial x_1} & - \dfrac{\partial}{\partial x_1}\overline{u_1^2} & - \dfrac{\partial}{\partial x_2}\overline{u_1 u_2} + & \nu & \left(\dfrac{\partial^2 \bar{U}_1}{\partial x_1^2} + \dfrac{\partial^2 \bar{U}_1}{\partial x_2^2}\right) \\[2ex]
1 & 1 & \dfrac{\Delta P_1}{\rho \mathscr{V}_1^2} & \dfrac{\mathfrak{v}^2}{\mathscr{V}_1^2} & \mathbf{R}_{12}\dfrac{\mathfrak{v}^2}{\mathscr{V}_1^2}\dfrac{L_1}{L_2} & \dfrac{L_1}{L_2}\dfrac{\nu}{\mathscr{V}_1 L_2} & \left(\dfrac{L_2^2}{L_1^2} \quad 1\right)
\end{array}$$

The relative orders of magnitude of the terms are shown below the corresponding terms.

We conclude that the viscous terms are important only if the Reynolds number $\mathscr{V}_1 L_2/\nu$ is at most of the order L_1/L_2. Furthermore, the order of magnitude of $\mathfrak{v}^2/\mathscr{V}_1^2$ can at most be equal to L_2/L_1.

For the x_2-direction the equation of motion reads

$$\begin{array}{ccccccc}
\bar{U}_1 \dfrac{\partial \bar{U}_2}{\partial x_1} & + \bar{U}_2 \dfrac{\partial \bar{U}_2}{\partial x_2} & = -\dfrac{1}{\rho}\dfrac{\partial \bar{P}}{\partial x_2} & - \dfrac{\partial}{\partial x_1}\overline{u_2 u_1} & - \dfrac{\partial}{\partial x_2}\overline{u_2^2} + & \nu & \left(\dfrac{\partial^2 \bar{U}_2}{\partial x_1^2} + \dfrac{\partial^2 \bar{U}_2}{\partial x_2^2}\right) \\[2ex]
\dfrac{L_2}{L_1} & \dfrac{L_2}{L_1} & \dfrac{\Delta P_2}{\rho \mathscr{V}_1^2}\dfrac{L_1}{L_2} & R_{12}\dfrac{\mathfrak{v}^2}{\mathscr{V}_1^2} & \dfrac{\mathfrak{v}^2}{\mathscr{V}_1^2}\dfrac{L_1}{L_2} & \dfrac{L_2}{L_1}\dfrac{\nu}{\mathscr{V}_1 L_2} & \left(\dfrac{L_2}{L_1} \quad \dfrac{L_1}{L_2}\right)
\end{array}$$

The orders of magnitude written below each term are taken relative to the order of magnitude of the terms on the left-hand side of the equation for the x_1-direction.

We have found that $\mathfrak{v}^2/\mathscr{V}_1^2$ is at most of the order L_2/L_1 and that $\nu/\mathscr{V}_1 L_2$ must be at least of the order L_2/L_1 if the viscous term in the equation of motion in the x_1-direction is to play a part. Then, if $\nu/\mathscr{V}_1 L_2$ is of the order L_2/L_1, the viscous term in the second equation of motion may be neglected, being one order of magnitude smaller than the turbulence term $\partial \overline{u_2^2}/\partial x_2$. If, however, $\mathscr{V}_1 L_2/\nu$ becomes of the order 1, the viscous term should be taken into account. But it then follows from the first equation of motion that the viscous term there becomes the dominating term and the whole flow becomes viscous. This is what may be expected to occur in the viscous layer adjoining the wall.

In the adjacent buffer region where viscosity still plays a part, the Reynolds number $\mathscr{V}_1 L_2/\nu$ should be of the order L_1/L_2. The second equation of motion, retaining only the terms of the highest order of magnitude, then reduces to

$$\frac{\partial \bar{P}}{\partial x_2} + \rho \frac{\partial}{\partial x_2}\overline{u_2^2} = 0 \qquad (7\text{-}2)$$

Let us now consider the *outer region*. Let $\mathscr{V}_1$ now be the velocity scale for the maximum difference of $\bar{U}_1$ in this region. From Eq. (7-1) we obtain

$$\bar{U}_2 = -\int_0^{x_2} dx_2 \frac{\partial \bar{U}_1}{\partial x_1}$$

This relation suggests that $\bar{U}_2$, and so its velocity scale $\mathscr{V}_2$ is at the most of the order of magnitude $\mathscr{V}_1 L_2/L_1$.

In the momentum-balance equation for the x_1-direction the term $\bar{U}_2\,\partial \bar{U}_1/\partial x_2$ becomes of the order of magnitude $(\mathscr{V}_1 L_2/L_1)(\mathscr{V}_1/L_2) = \mathscr{V}_1^2/L_1$, while the term

$\bar{U}_1\, \partial \bar{U}_1/\partial x_1$ becomes of the order of magnitude $\bar{U}_1 \mathscr{V}_1/L_1$. Since in the outer region $\mathscr{V}_1$ is much smaller than $\bar{U}_1$, but not an order of magnitude smaller, the same results are obtained as for the wall region for the relative order of magnitude of the various terms. Also, when considering the momentum-balance equation in the x_2-direction, the same result as Eq. (7-2) is obtained.

We so may integrate this equation to yield

$$\bar{P} + \rho \overline{u_2^2} = P_0 \tag{7-2a}$$

where P_0 is the pressure at the same section, outside the turbulence region.

The first equation of motion reduces to

$$\bar{U}_1 \frac{\partial \bar{U}_1}{\partial x_1} + \bar{U}_2 \frac{\partial \bar{U}_1}{\partial x_2} = -\frac{1}{\rho}\frac{dP_0}{dx_1} - \frac{\partial}{\partial x_2}\overline{u_1 u_2} + \nu \frac{\partial^2 \bar{U}_1}{\partial x_2^2} \tag{7-3}$$

since $\partial \bar{P}/\partial x_1 = dP_0/dx_1 - \rho\, \partial \overline{u_2^2}/\partial x_1$ and since $\partial \overline{u_2^2}/\partial x_1$ is of the same order of magnitude as $\partial \overline{u_1^2}/\partial x_1$ and, accordingly, negligible.

For the outer region the Reynolds number $\mathscr{V}_1 L_2/\nu$ may become of greater order of magnitude than L_1/L_2, and the viscous term in Eq. (7-3) may be neglected; the equation then becomes identical with Eq. (6-21) which was obtained for free turbulent flow.

Similar approximations may be applied to the energy equation. Let us consider first the energy equation for the mean motion. This equation contains the term $\bar{P}/\rho + \frac{1}{2}(\bar{U}_1^2 + \bar{U}_2^2) \simeq \bar{P}/\rho + \frac{1}{2}\bar{U}_1^2$, since $\bar{U}_2/\bar{U}_1 \ll 1$. The equation then reads (see Sec. 1-13)

$$\begin{array}{llll}
\dfrac{\partial}{\partial x_1}\bar{U}_1\left(\dfrac{\bar{P}}{\rho} + \dfrac{\bar{U}_1^2}{2}\right) & + \dfrac{\partial}{\partial x_2}\bar{U}_2\left(\dfrac{\bar{P}}{\rho} + \dfrac{\bar{U}_1^2}{2}\right) & & \\
\dfrac{\Delta P_1}{\rho \mathscr{V}_1^2} \quad 1 & \dfrac{\Delta P_2}{\rho \mathscr{V}_1^2} \quad 1 & &
\end{array}$$

$$\begin{array}{llll}
= -\bar{U}_1\left(\dfrac{\partial}{\partial x_1}\overline{u_1^2} \right. & \left. + \dfrac{\partial}{\partial x_2}\overline{u_1 u_2}\right) & - \bar{U}_2\left(\dfrac{\partial}{\partial x_1}\overline{u_2 u_1} \right. & \left. + \dfrac{\partial}{\partial x_2}\overline{u_2^2}\right) \\
\quad \dfrac{\mathfrak{v}^2}{\mathscr{V}_1^2} & \quad \dfrac{L_1}{L_2}\dfrac{\mathfrak{v}^2}{\mathscr{V}_1^2}\mathbf{R}_{12} & \quad \dfrac{L_2}{L_1}\dfrac{\mathfrak{v}^2}{\mathscr{V}_1^2}\mathbf{R}_{12} & \quad \dfrac{\mathfrak{v}^2}{\mathscr{V}_1^2}
\end{array}$$

$$\begin{array}{llllll}
+ \nu\left\{\bar{U}_1\left[2\dfrac{\partial^2 \bar{U}_1}{\partial x_1^2} \right.\right. & + \dfrac{\partial}{\partial x_2}\left(\dfrac{\partial \bar{U}_1}{\partial x_2} \right. & \left.\left. + \dfrac{\partial \bar{U}_2}{\partial x_1}\right)\right] & + \bar{U}_2\left[\dfrac{\partial}{\partial x_1}\left(\dfrac{\partial \bar{U}_2}{\partial x_1} \right.\right. & \left. + \dfrac{\partial \bar{U}_1}{\partial x_2}\right) & \left.\left. + 2\dfrac{\partial^2 \bar{U}_2}{\partial x_2^2}\right]\right\} \\
\dfrac{\nu}{\mathscr{V}_1 L_2}\left\{ \quad \dfrac{L_2}{L_1} \right. & \quad \dfrac{L_1}{L_2} & \quad \dfrac{L_2}{L_1} & \quad \dfrac{L_2^3}{L_1^3} & \quad \dfrac{L_2}{L_1} & \left. \quad \dfrac{L_2}{L_1} \right\}
\end{array}$$

We have already shown that $\Delta P_2/\rho \mathscr{V}_1^2$ is of the order $\mathfrak{v}^2/\mathscr{V}_1^2$ and that this order of magnitude is equal to L_2/L_1. If the Reynolds number $\mathscr{V}_1 L_2/\nu$ is of the order L_1/L_2, in which case the viscous term in the equation of motion (7-3) is retained, and if only the terms of the highest order are considered, the following result is obtained:

$$\frac{\partial}{\partial x_1}\bar{U}_1\left(\frac{P_0}{\rho}+\frac{\bar{U}_1^{\,2}}{2}\right)+\frac{\partial}{\partial x_2}\bar{U}_2\left(\frac{P_0}{\rho}+\frac{\bar{U}_1^{\,2}}{2}\right)=-\bar{U}_1\frac{\partial}{\partial x_2}\overline{u_1u_2}+\nu\bar{U}_1\frac{\partial^2\bar{U}_1}{\partial x_2^{\,2}} \tag{7-4}$$

For the outer region, where $\mathscr{V}_1L_2/\nu$ may become much larger than L_1/L_2, the viscous term in Eq. (7-4) may again be neglected.

A first approximation applied to the turbulence-energy equation (1-110) yields the following result:

$$\begin{aligned}
&\bar{U}_1\frac{\partial}{\partial x_1}\frac{\overline{q^2}}{2}+\bar{U}_2\frac{\partial}{\partial x_2}\frac{\overline{q^2}}{2}=-\overline{u_1u_2}\frac{\partial\bar{U}_1}{\partial x_2}-\frac{\partial}{\partial x_2}\overline{u_2\left(\frac{p}{\rho}+\frac{q^2}{2}\right)}\\
&\qquad 1 \qquad\qquad\quad 1 \qquad\qquad\qquad \mathbf{R}_{12}\frac{L_1}{L_2} \qquad\qquad\quad \frac{L_1}{L_2}\frac{\mathfrak{v}}{\mathscr{V}_1}\\
&\qquad\qquad +\nu\frac{\partial}{\partial x_2}\overline{u_j\left(\frac{\partial u_2}{\partial x_j}+\frac{\partial u_j}{\partial x_2}\right)}-\nu\overline{\left(\frac{\partial u_i}{\partial x_j}+\frac{\partial u_j}{\partial x_i}\right)\frac{\partial u_j}{\partial x_i}}\\
&\qquad\qquad\qquad \frac{\nu}{\mathscr{V}_1L_2}\frac{L_1}{l} \qquad\qquad\qquad\quad \frac{\nu}{\mathscr{V}_1L_2}\frac{L_1L_2}{l^2}
\end{aligned} \tag{7-5}$$

where the length l is a suitable scale for the spatial variations of the turbulence velocities. It may be expected that, except very close to the wall, this scale will be much smaller than the length scale L_2 and, therefore, that the term giving the work done per unit of mass and time by the viscous shear stresses of the turbulent motion will be small compared with the viscous-dissipation term.

It may also be expected that, depending on the subregion considered of the boundary layer, not all terms of Eq. (7-5) will be of equal importance. The shear-correlation coefficient $\mathbf{R}_{21}$ is of the $\mathcal{O}(1)$ throughout the greater part of the boundary layer (in general it is roughly equal to 0.5). As long as $L_1/L_2 \gg 1$, the convective terms on the left-hand side of the balance equation may be neglected with respect to the production term. In the subregion close to the wall, including the viscous sublayer and the buffer region, where the direct viscous effects are of importance, $\mathscr{V}_1L_2/\nu$ is not large and may be smaller than $\mathcal{O}(L_1/L_2)$. Also $\mathfrak{v}/\mathscr{V}_1$ is not small, while L_2 and l are of the same order of magnitude.

So we may expect that for this subregion close to the wall, Eq. (7-5) reduces to

$$\overline{u_1u_2}\frac{\partial\bar{U}_1}{\partial x_2}+\frac{\partial}{\partial x_2}\overline{u_2\left(\frac{p}{\rho}+\frac{q^2}{2}\right)}=\nu\frac{\partial}{\partial x_2}\overline{u_j\left(\frac{\partial u_2}{\partial x_j}+\frac{\partial u_j}{\partial x_2}\right)}-\nu\overline{\left(\frac{\partial u_j}{\partial x_i}+\frac{\partial u_i}{\partial x_j}\right)\frac{\partial u_j}{\partial x_i}} \tag{7-5a}$$

The right-hand side may be transformed to make the terms more suitable for measuring [see Eq. (1-111)]; accordingly,

$$\overline{u_1u_2}\frac{\partial\bar{U}_1}{\partial x_2}+\frac{\partial}{\partial x_2}\overline{u_2\left(\frac{p}{\rho}+\frac{q^2}{2}\right)}=\nu\frac{\partial^2}{\partial x_2^{\,2}}\frac{\overline{q^2}}{2}-\nu\overline{\frac{\partial u_j}{\partial x_i}\frac{\partial u_j}{\partial x_i}} \tag{7-5b}$$

Farther from the wall, when approaching the fully turbulent region, $\mathfrak{v}^2/\mathscr{V}_1^{\,2}$ may become of the $\mathcal{O}(L_2/L_1)$, so that the diffusion term may become much smaller than the production term and may so be neglected. The production term then must balance the viscous terms. Of these two terms, however, the dissipation term will dominate the term describing the work by the viscous stresses, since $L_2/l \gg 1$. It is sufficient that L_2/l be of $\mathcal{O}(L_1/L_2)^{1/2}$. In order that the dissipation balances the production, $\mathscr{V}_1 L_2/\nu$ must be of the $\mathcal{O}(L_2^{\,2}/l^2)$ or $\mathcal{O}(L_1/L_2)$, so sufficiently large. In this subregion we thus have the condition of energy equilibrium of the turbulence. The energy equation further simplifies to

$$\overline{u_1 u_2}\frac{\partial \bar{U}_1}{\partial x_2} + \varepsilon = 0 \qquad (7\text{-}6)$$

where, to the same degree of approximation,

$$\varepsilon \simeq \varepsilon' = \nu\overline{\frac{\partial u_j}{\partial x_i}\frac{\partial u_j}{\partial x_i}}$$

In the outer region the velocity scale $\mathscr{V}_1$ becomes smaller than that for the wall region, while also the length scale ratio L_1/L_2 becomes relatively smaller. So the transport of turbulence energy both by turbulent diffusion and by convection by the mean motion have to be considered. It is only the term describing the work by the viscous stresses that may be neglected, and we then have to consider the equation

$$\bar{U}_1\frac{\partial}{\partial x_1}\tfrac{1}{2}\overline{q^2} + \bar{U}_2\frac{\partial}{\partial x_2}\tfrac{1}{2}\overline{q^2} = -\overline{u_1 u_2}\frac{\partial \bar{U}_1}{\partial x_2} - \frac{\partial}{\partial x_2}\overline{u_2\left(\frac{p}{\rho} + \frac{q^2}{2}\right)} - \varepsilon' \qquad (7\text{-}7)$$

Finally when approaching the outer boundary of the boundary layer the correlation coefficient $\mathbf{R}_{2,1}$ appears to decrease, while at the same time $\partial\bar{U}_1/\partial x_2$ becomes small. Thus the production becomes negligible, and since the dissipation becomes small too, we may expect only a balance between convective transport by the mean motion and diffusion by the turbulent motion of turbulence energy.

Integral Relations

Interesting and simple relations can be obtained by integrating the equation of continuity (7-1), the equation of motion (7-3), and the energy equations (7-4) and (7-5) with respect to x_2 across the whole boundary layer.

Let us first integrate the equation of continuity with respect to x_2 from 0 to ∞. Accordingly, we change the order of integration with respect to x_2 and differentiation with respect to x_1. We then obtain

$$\frac{d}{dx_1}\int_0^\infty dx_2\,\bar{U}_1 + \bar{U}_2\Big|_0^\infty = 0$$

whence

$$(\bar{U}_2)_\infty = -\frac{d}{dx_1}\int_0^\infty dx_2\,\bar{U}_1 \tag{7-8}$$

Before carrying out the integration of the momentum-balance equation (7-3), we write it in the following form

$$\frac{\partial}{\partial x_1}\bar{U}_1{}^2 + \frac{\partial}{\partial x_2}\bar{U}_2\bar{U}_1 = U_0\frac{dU_0}{dx_1} - \frac{\partial}{\partial x_2}\overline{u_2u_1} + \nu\frac{\partial^2\bar{U}_1}{\partial x_2{}^2}$$

assuming potential flow in the free stream, so that Bernoulli's equation, i.e., $P_0 + \frac{1}{2}\rho U_0{}^2 =$ constant, applies.

The integration with respect to x_2 from 0 to ∞ yields

$$\frac{d}{dx_1}\int_0^\infty dx_2\,\bar{U}_1{}^2 - \int_0^\infty dx_2\,U_0\frac{dU_0}{dx_1} + \bar{U}_2\bar{U}_1\Big|_0^\infty = -\overline{u_2u_1}\Big|_0^\infty + \nu\frac{\partial\bar{U}_1}{\partial x_2}\Big|_0^\infty$$

With the boundary conditions we obtain:

$$\bar{U}_2\bar{U}_1\Big|_0^\infty = U_0(\bar{U}_2)_\infty = -U_0\frac{d}{dx_1}\int_0^\infty dx_2\,\bar{U}_1$$

$$-\overline{u_2u_1}\Big|_0^\infty = 0$$

and

$$+\nu\frac{\partial\bar{U}_1}{\partial x_2}\Big|_0^\infty = -\nu\left(\frac{\partial\bar{U}_1}{\partial x_2}\right)_0 = -\frac{\sigma_w}{\rho}$$

where σ_w is the viscous shear stress at the wall. Here we have assumed a smooth wall. In the case of a wall so rough that the direct viscous effects at the wall are negligible, instead of $\nu(\partial\bar{U}_1/\partial x_2)_0$, we then have to take $(\overline{u_2u_1})_0 = -\sigma_w/\rho$.

Now we rewrite the following two terms

$$-\int_0^\infty dx_2\,U_0\frac{dU_0}{dx_1} + \bar{U}_2\bar{U}_1\Big|_0^\infty = -\int_0^\infty dx_2\,U_0\frac{dU_0}{dx_1} - U_0\frac{d}{dx_1}\int_0^\infty dx_2\,\bar{U}_1$$

$$= -\frac{d}{dx_1}\int_0^\infty dx_2\,U_0\bar{U}_1 + \frac{dU_0}{dx_1}\int_0^\infty dx_2(\bar{U}_1 - U_0)$$

Hence the integrated momentum balance equation becomes

$$\frac{d}{dx_1}\int_0^\infty dx_2\,\bar{U}_1(U_0 - \bar{U}_1) + \frac{dU_0}{dx_1}\int_0^\infty dx_2\,(U_0 - \bar{U}_1) = \frac{\sigma_w}{\rho} \tag{7-9}$$

This equation is often referred to as the Von-Kármán integral momentum-balance equation.

Let δ be the thickness of the boundary layer, that is, the value of x_2 where $\bar{U}_1$ is practically equal to the free-stream velocity U_0 outside the boundary layer.

Then the integral

$$\int_0^\delta dx_2\,\bar{U}_1$$

is the total flux of fluid in the x_1-direction through the boundary layer. We may introduce a length δ' such that

$$U_0\delta' = \int_0^{\delta} dx_2\, \bar{U}_1$$

It is obvious that $\delta' < \delta$, and so, because of the retardation of the flow in the boundary layer, the streamlines in the free stream outside the boundary layer are displaced over a distance $\delta_d = \delta - \delta'$ in the positive x_2-direction. Since the velocity $\bar{U}_1$ approaches the free-stream velocity U_0 more or less asymptotically with increasing x_2 it is difficult to determine the value of δ accurately. In this respect δ_d is a much more useful parameter, since its value can be obtained exactly. From

$$\delta_d U_0 = \delta U_0 - \delta' U_0 = \delta U_0 - \int_0^{\delta} dx_2\, \bar{U}_1$$

$$= \int_0^{\delta} dx\,(U_0 - \bar{U}_1)$$

follows

$$\delta_d = \int_0^{\delta} dx_2\left(1 - \frac{\bar{U}_1}{U_0}\right) = \int_0^{\infty} dx_2\left(1 - \frac{\bar{U}_1}{U_0}\right) \tag{7-10}$$

The quantity δ_d is usually called the *displacement thickness* of the boundary layer, for obvious reasons.

Similarly we may introduce another length δ_m defined by

$$\delta_m = \int_0^{\infty} dx_2\, \frac{\bar{U}_1}{U_0}\left(1 - \frac{\bar{U}_1}{U_0}\right) \tag{7-11}$$

This length δ_m will be called the *momentum-loss thickness*† of the boundary layer, since $\rho U_0^2 \delta_m$ is the total loss in momentum of the fluid in the boundary layer at the section x_1. Equation (7-9) can now be written in a different way, including the two length parameters δ_d and δ_m according to Eqs. (7-10) and (7-11). To this end the first term of Eq. (7-9) is first transformed as follows:

$$\frac{d}{dx_1}\int_0^{\infty} dx_2\, \bar{U}_1(U_0 - \bar{U}_1) - \frac{d}{dx_1}\left[U_0^2 \int_0^{\infty} dx_2\, \frac{\bar{U}_1}{U_0}\left(1 \quad \frac{\bar{U}_1}{U_0}\right)\right]$$

$$= U_0^2 \frac{d}{dx_1}\int_0^{\infty} dx_2\, \frac{\bar{U}_1}{U_0}\left(1 - \frac{\bar{U}_1}{U_0}\right) + 2U_0 \frac{dU_0}{dx_1}\int_0^{\infty} dx_2\, \frac{\bar{U}_1}{U_0}\left(1 - \frac{\bar{U}_1}{U_0}\right)$$

$$= U_0^2 \frac{d\delta_m}{dx_1} + 2U_0 \frac{dU_0}{dx_1}\delta_m$$

Equation (7-9) then becomes, after dividing all terms by U_0^2

$$\frac{d\delta_m}{dx_1} + \frac{2\delta_m + \delta_d}{U_0}\frac{dU_0}{dx_1} = \frac{\sigma_w}{\rho U_0^2} \tag{7-9a}$$

† This length is usually referred to as momentum thickness, which is, however, less appropriate.

Von Doenhoff and Tetervin[68] introduced the ratio $H = \delta_d/\delta_m$ as a *shape factor*, since this parameter characterizes the mean-velocity distribution, i.e., the shape of the mean-velocity profile, in the boundary layer

$$\frac{d\delta_m}{dx_1} + \frac{H+2}{U_0}\frac{dU_0}{dx_1}\delta_m = \frac{\sigma_w}{\rho U_0{}^2} \tag{7-9b}$$

In the case of a constant-pressure boundary layer, where $dU_0/dx_1 = 0$, Eq. (7-9b) reduces to the simple relation between the rate of change of the momentum-loss thickness and the wall shear stress

$$\frac{d\delta_m}{dx_1} = \frac{\sigma_w}{\rho U_0{}^2} \tag{7-9c}$$

When we integrate the energy balance equation (7-4) for the mean motion with respect to x_2 from 0 to ∞, the second, third, and fourth terms can be transformed as follows

$$\left[\bar{U}_2\left(\frac{P_0}{\rho} + \frac{\bar{U}_1{}^2}{2}\right)\right]\Bigg|_0^\infty = -\left(\frac{P_0}{\rho} + \frac{U_0{}^2}{2}\right)\frac{d}{dx_1}\int_0^\infty dx_2\,\bar{U}_1$$
$$= -\frac{d}{dx_1}\int_0^\infty dx_2\,\bar{U}_1\left(\frac{P_0}{\rho} + \frac{U_0{}^2}{2}\right)$$

since $P_0/\rho + U_0{}^2/2 = \text{constant}$. Also,

$$\int_0^\infty dx_2\,\bar{U}_1\frac{\partial}{\partial x_2}\overline{u_2u_1} = U_0\,\overline{u_2u_1}\Big|_0^\infty - \int_0^\infty dx_2\,\overline{u_2u_1}\frac{\partial\bar{U}_1}{\partial x_2} = -\int_0^\infty dx_2\,\overline{u_2u_1}\frac{\partial\bar{U}_1}{\partial x_2}$$

$$\int_0^\infty dx_2\,\bar{U}_1\frac{\partial^2\bar{U}_1}{\partial x_2{}^2} = \bar{U}_1\frac{\partial\bar{U}_1}{\partial x_2}\Big|_0^\infty - \int_0^\infty dx_2\left(\frac{\partial\bar{U}_1}{\partial x_2}\right)^2 = -\int_0^\infty dx_2\left(\frac{\partial\bar{U}_1}{\partial x_2}\right)^2$$

When we introduce these terms in the integrated Eq. (7-4), the result is that the pressure term drops out. We thus obtain

$$\frac{d}{dx_1}\int_0^\infty dx_2\,\bar{U}_1\left(\frac{U_0{}^2}{2} - \frac{\bar{U}_1{}^2}{2}\right) = \int_0^\infty dx_2\,(-\overline{u_2u_1})\frac{\partial\bar{U}_1}{\partial x_2} + \nu\int_0^\infty dx_2\left(\frac{\partial\bar{U}_1}{\partial x_2}\right)^2 \tag{7-12}$$

Or, treating this equation in a similar way to the momentum-balance equation

$$\frac{1}{2}\frac{d}{dx_1}\int_0^\infty dx_2\,\frac{\bar{U}_1}{U_0}\left(1 - \frac{\bar{U}_1{}^2}{U_0{}^2}\right) + \frac{3}{2}\frac{1}{U_0}\frac{dU_0}{dx_1}\int_0^\infty dx_2\,\frac{\bar{U}_1}{U_0}\left(1 - \frac{\bar{U}_1{}^2}{U_0{}^2}\right)$$
$$= \frac{1}{U_0{}^3}\int_0^\infty dx_2\,(-\overline{u_2u_1})\frac{\partial\bar{U}_1}{\partial x_2} + \frac{\nu}{U_0}\int_0^\infty dx_2\left(\frac{\partial}{\partial x_2}\frac{\bar{U}_1}{U_0}\right)^2$$

We may again introduce suitable length-parameters which also characterize the boundary layer, namely

the *energy-loss thickness*
$$\delta_e = \int_0^\infty dx_2\,\frac{\bar{U}_1}{U_0}\left(1 - \frac{\bar{U}_1{}^2}{U_0{}^2}\right) \tag{7-13}$$

and

the *dissipation thickness*
$$\Delta = \left[\int_0^\infty dx_2 \left(\frac{\partial}{\partial x_2}\frac{\bar{U}_1}{U_0}\right)^2\right]^{-1} \tag{7-14}$$

Equation (7-12) may then be written

$$\frac{1}{2}\frac{d\delta_e}{dx_1} + \frac{3}{2}\frac{dU_0}{U_0\, dx_1}\delta_e = \frac{1}{U_0^3}\int_0^\infty dx_2\,(-\overline{u_2u_1})\frac{\partial \bar{U}_1}{\partial x_2} + \frac{\nu}{U_0\Delta} \tag{7-12a}$$

For a constant-pressure boundary layer this equation reduces to

$$\frac{1}{2}\frac{d\delta_e}{dx_1} = \frac{1}{U_0^3}\int_0^\infty dx_2\,(-\overline{u_2u_1})\frac{\partial \bar{U}_1}{\partial x_2} + \frac{\nu}{U_0\Delta} \tag{7-12b}$$

Wieghardt[2] first introduced the energy-loss thickness δ_e and the dissipation thickness Δ for the case of a laminar boundary layer, where the turbulence term on the right-hand side of Eq. (7-12) is zero. On the other hand, at large Reynolds numbers, the last term on the right-hand side of Eq. (7-12) becomes negligibly small, and the energy-loss thickness is determined only by the turbulence term; the turbulence energy is produced at the cost of the kinetic energy of the mean motion.

We may relate this turbulence-energy production to the viscous dissipation by the turbulence and to the convection of turbulence energy by the mean motion through the integrated form of Eq. (7-5). Again using the equation of continuity, and integrating Eq. (7-5), we obtain

$$\frac{d}{dx_1}\int_0^\infty dx_2 \frac{\overline{q^2}}{2}\bar{U}_1 = -\int_0^\infty dx_2\,\overline{u_1u_2}\frac{\partial \bar{U}_1}{\partial x_2} - \int_0^\infty dx_2\,\varepsilon \tag{7-15}$$

The combination of Eqs. (7-12a) and (7-15) yields an equation that gives the total energy balance of the turbulent boundary layer:

$$\frac{1}{2}\frac{d}{dx_1}\left[\delta_e - \int_0^\infty dx_2 \frac{\overline{q^2}\bar{U}_1}{U_0^3}\right] + \frac{3}{2}\frac{1}{U_0}\frac{dU_0}{dx_1}\left[\delta_e - \int_0^\infty dx_2 \frac{\overline{q^2}\bar{U}_1}{U_0^3}\right] = \frac{\nu}{U_0\Delta} + \int_0^\infty dx_2 \frac{\varepsilon}{U_0^3} \tag{7-16}$$

At large Reynolds numbers the term $\nu/U_0\Delta$, giving the viscous dissipation by the mean motion, may be neglected.

7-3 THE LAMINAR BOUNDARY LAYER AND TRANSITION

Before considering the turbulent boundary layer along a flat plate, it may be useful to make some general remarks about boundary layers and to give the velocity distribution in the laminar boundary layer along a flat plate with zero pressure gradient.

If one considers the boundary layer along a rigid body, several regions may be distinguished in the downstream direction. From the stagnation point on, the boundary-layer thickness increases in the downstream direction. The flow in the boundary layer is at first laminar but, from a certain point on, the flow becomes unstable, and transition into turbulent flow may occur if disturbances are present. The location of this point along the body depends on the flow velocity, on the curvature and roughness of the surface, and on the degree of turbulence in the free stream outside the boundary layer. After transition into turbulent flow, the increase in thickness of the boundary layer occurs at a higher rate than before.

If the surface of the body is convex to the free stream or if there is an adverse, i.e. positive, pressure-gradient in the flow direction, there is a possibility of "separation" of the boundary layer from the surface.

If the body is a straight flat plate placed in a uniform parallel flow, the pressure gradient in the flow direction is zero, and no separation of the boundary layer occurs; the thickness of the boundary layer increases indefinitely. We then have only a laminar and a turbulent region, separated by a relatively short transitional region.

Let us consider briefly the solution for the streamwise velocity component as obtained by Blasius[5,14] for a laminar boundary layer.

According to the boundary-layer approximations, the equation of motion in the x_1-direction reads

$$U_1\frac{\partial U_1}{\partial x_1} + U_2\frac{\partial U_1}{\partial x_2} = \nu\frac{\partial^2 U_1}{\partial {x_2}^2}$$

Blasius's solution, which is expressed in terms of

$$\eta = \tfrac{1}{2}x_2\left(\frac{U_0}{\nu x_1}\right)^{1/2}$$

consists of two parts: a solution in the form of a power series for small values of η, and an asymptotic solution for large η.

The power series for small η reads

$$U_1 = \frac{U_0}{2}\left(\frac{\alpha\eta}{1!} - \frac{\alpha^2\eta^4}{4!} + 11\frac{\alpha^3\eta^7}{7!} - 375\frac{\alpha^4\eta^{10}}{10!} + \cdots\right) \qquad \text{(7-17a)}$$

where $\alpha = 1.328$.

The asymptotic solution for large η reads

$$U_1 = U_0\left\{1 + \gamma\int_{\infty}^{\eta} d\eta \exp\left[-(\eta - \beta)^2\right]\right\} \qquad \text{(7-17b)}$$

With $\beta = 0.865$ and $\gamma = 0.461$ the two solutions give the same values of U_1 and $\partial U_1/\partial x_2$ at a suitable point ($\eta = 1.5$, $U_1/U_0 = 0.85$), where both solutions are still applicable.

According to Blasius's solution, the velocity U_0 is reached asymptotically. Hence, in determining the thickness δ, one has to agree on the degree of deviation from the constant free-stream velocity that is allowable if one wishes to obtain a finite value for this thickness.

For $\eta = 2.75$ Blasius obtained the value $U_1 = 0.997U_0$. Thus, if a deviation of 0.3 per cent is allowed, we obtain

$$\delta = 5.5\sqrt{\frac{\nu x_1}{U_0}} \tag{7-18}$$

Instead of the ill-defined thickness δ we do better to take the displacement thickness δ_d or the momentum-loss thickness δ_m. For these boundary-layer thicknesses with the Blasius velocity distribution we obtain

$$\delta_d = 1.731\sqrt{\frac{\nu x_1}{U_0}} \tag{7-19}$$

and

$$\delta_m = 0.664\sqrt{\frac{\nu x_1}{U_0}} \tag{7-20}$$

whence follow the relations:

$$\delta_d = 0.31\delta \qquad \delta_m = 0.12\delta \qquad \delta_d = 2.6\delta_m$$

Thus, Von Doenhoff's and Tetervin's shape factor $H = \delta_d/\delta_m$ is equal to 2.6 according to the above Blasius solution.

Any one of the formulas (7-18) to (7-20) shows how the thickness increases parabolically with increasing x_1.

Two definitions of the Reynolds number, which characterizes the flow in the boundary layer are common: (1) one with reference to any of the lengths determining the thickness of the boundary layer, that is

$$\mathbf{Re}_\delta = \frac{U_0\delta}{\nu}, \qquad \mathbf{Re}_{\delta_d} = \frac{U_0\delta_d}{\nu}, \qquad \mathbf{Re}_{\delta_m} = \frac{U_0\delta_m}{\nu}$$

and so forth; and (2) one with reference to the distance x_1 from the stagnation point, that is

$$\mathbf{Re}_{x_1} = \frac{U_0 x_1}{\nu}$$

According to Blasius's solution for the plane laminar boundary layer the relation between $\mathbf{Re}_{x_1}$ and, for instance, $\mathbf{Re}_\delta$ reads

$$\mathbf{Re}_\delta = 5.5(\mathbf{Re}_{x_1})^{1/2} \tag{7-21}$$

Transition may occur if the Reynolds number of the boundary layer flow exceeds a critical value. Schubauer and Skramstad[3] determined the critical value of $\mathbf{Re}_{x_1}$ for the onset of turbulence in the laminar boundary layer of a flat plate installed in a windtunnel. They did this for different turbulence conditions of the free stream. The turbulence condition was characterized only by the relative intensity of the turbulence, here defined as the ratio $(\frac{1}{3}\overline{q^2})^{1/2}/U_0$, where $\overline{q^2} = \overline{u_i u_i}$.

Figure 7-1 shows the results. Apparently the critical value of $\mathbf{Re}_{x_1}$ for transition was no longer affected by the relative intensity of turbulence of the free stream if this intensity decreased below roughly 0.1 per cent. The data in Fig. 7-1 suggest that, if free-stream turbulence is to promote transition into turbulent flow in a boundary layer, the relative intensity of the free-stream turbulence must be at least 0.2 per cent.

When the relative intensity of the free-stream turbulence was 0.5 per cent, the critical value of $\mathbf{Re}_{x_1}$ decreased from 3×10^6 to $\sim 10^6$. According to Dryden,[4] this value decreased further to $\sim 10^5$ when the relative intensity of the turbulence was increased to 3 per cent.

From this critical value of $\mathbf{Re}_{x_1}$ the critical distance x_{crit} can be evaluated

$$(x_1)_{\text{crit}} = \frac{\nu}{U_0}(\mathbf{Re}_{x_1})_{\text{crit}}$$

With the above values for $(\mathbf{Re}_{x_1})_{\text{crit}}$ obtained by Dryden, Eq. (7-21) yields for $\mathbf{Re}_\delta$ the critical values for transition to turbulence between 11,000 and 1,700 when the relative intensity of the free-stream turbulence varies from zero to 3 per cent.

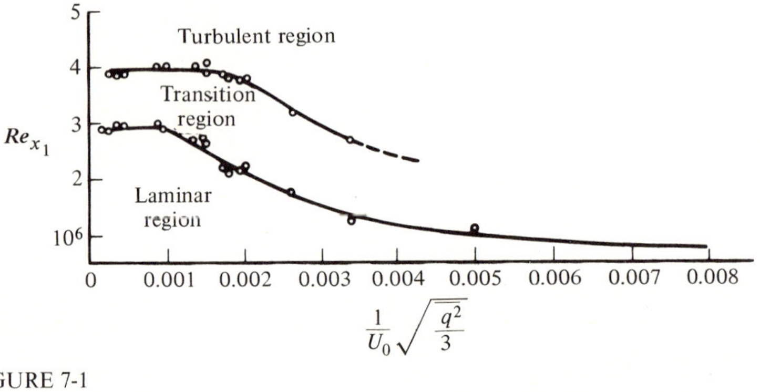

FIGURE 7-1
Effect of free-stream turbulence on the transition into turbulence of a laminar boundary layer along a flat plate with zero pressure gradient. (From: *Schubauer, G. B., and H. K. Skramstad,*[3] *by permission of the Institute of the Aeronautical Sciences.*)

7-4 FURTHER CONSIDERATIONS ON TRANSITION TO TURBULENCE OF A LAMINAR BOUNDARY LAYER

The many experimental and theoretical studies that have been made during the last two decades of the transition to turbulence of a laminar boundary layer, and of the phenomena in the viscous sublayer and the buffer layer in fully developed wall turbulence, have led to the conclusion that there is a great similarity in many aspects of the mechanisms leading to transition and of the continuous generation of turbulence in wall-bounded flows, respectively. It is for this reason that we will consider here the transition studies, since some knowledge of the phenomena during transition may also lead to a better understanding of the mechanism of fully developed wall turbulence.

Experimental evidence has shown that the transition process in a boundary layer starts locally at a number of points which are more or less randomly distributed in both streamwise and spanwise direction near the wall in a region of finite streamwise extent at some distance from the leading edge. These transitions probably occur as a result of local instabilities of the mean basic flow to disturbances. This evidence suggests application of a "local perturbation" method to the equations of motion. Let us consider as the simplest case a basic flow field $U_1(x_2)$ on which a perturbation $u_1(x_1,x_2,t)$; $u_2(x_1,x_2,t)$ is superimposed. With $\omega_3 = \partial u_2/\partial x_1 - \partial u_1/\partial x_2$, the dynamic equation for ω_3 reads

$$\frac{\partial \omega_3}{\partial t} + U_1 \frac{\partial \omega_3}{\partial x_1} - u_2 \frac{d^2 U_1}{d{x_2}^2} = \nu \nabla^2 \omega_3$$

where $\nabla^2 = \partial^2/\partial {x_1}^2 + \partial^2/\partial {x_2}^2$.

Introduce a stream function ψ, so that $u_2 = -\partial\psi/\partial x_1$ and $\omega_3 = -\nabla^2\psi$. With all quantities rendered dimensionless with the free-stream velocity U_0, and the boundary-layer thickness δ, the nondimensional equation reads

$$\frac{\partial}{\partial t}\nabla^2\psi + U_1 \frac{\partial}{\partial x_1}\nabla^2\psi - \frac{\partial \psi}{\partial x_1}\frac{d^2 U_1}{d{x_2}^2} = \frac{1}{\mathbf{Re}_\delta}\nabla^4\psi$$

The variation of δ with x_1 is assumed to be negligibly small in the x_1-region considered.

Assume a perturbation

$$\psi = \varphi(x_2)\exp\left[\imath\alpha(x_1 - ct)\right]$$

where $\alpha = 2\pi\delta/\lambda$ and $x = c_r + \imath c_\imath$; λ is the wavelength in the x_1-direction, c_r the phase velocity or the wave propagation velocity, while amplification occurs when $c_\imath > 0$, and damping when $c_\imath < 0$.

Substitution of the expression for ψ in the dynamic equation yields the *Orr-Sommerfeld* equation[14]

$$(U_1 - c)\left(\frac{d^2\varphi}{d{x_2}^2} - \alpha^2\varphi\right) - \varphi\frac{d^2 U_1}{d{x_2}^2} = -\frac{\imath}{\alpha\,\mathbf{Re}_\delta}\left[\frac{d^4\varphi}{d{x_2}^4} - 2\alpha^2\frac{d^2\varphi}{d{x_2}^2} + \alpha^4\varphi\right] \qquad (7\text{-}22)$$

The equation contains four parameters: c_r, $c_\imath$, α, and $\mathbf{Re}_\delta$. The problem is to determine, for given values of α and $\mathbf{Re}_\delta$, the eigenfunctions of φ and the corresponding eigenvalues of c_r and $c_\imath$, subject to the boundary conditions:

$$x_2 = 0 \quad \partial\psi/\partial x_1 = 0 \quad \partial\psi/\partial x_2 = 0; \qquad x_2 = \infty \quad \partial\psi/\partial x_2 = U_0$$

Because of the complexity of Eq. (7-22), in the first attempts to solve the problem, it has been assumed that the right-hand side of the equation could be disregarded, because of the large value of $\alpha\,\mathbf{Re}_\delta$ occurring at transition. Two important theorems, first given by Lord Rayleigh, are obtained from the "frictionless" solution [see, e.g., ref. (14), also for a further introduction to the instability and transition to turbulence of a boundary layer].

I. When d^2U_1/dx_2^2 does not change sign (e.g., $d^2U_1/dx_2^2 < 0$, i.e., a convex velocity profile), there is at least one neutral disturbance ($c_i = 0$) for which $U_1 = c_r$ at some point x_2 in the boundary layer. At this point $u_1 \to \infty$. For the boundary layer considered this "critical" point $(x_2)_{crit} \simeq 0.2\delta$.

II. A necessary and sufficient condition for instability ($c_i > 0$) is the existence of an inflexion point in the velocity profile ($d^2U_1/dx_2^2 = 0$).

Consequently in the non-viscous case only inflexional instability can occur. Since, in contrast to the free turbulence shear flows considered in Chap. 6, the laminar boundary layer has a convex velocity profile, the nonviscous approximation cannot be considered to explain the observed instability and transition. It has been concluded that apparently the viscous term in Eq. (7-22) cannot be neglected. Besides the two "nonviscous" solutions φ_1 and φ_2, for the boundary layer a third, "viscous" solution φ_3 has to be considered. The complete Orr-Sommerfeld equation has been solved first, and in an approximate way, by Tollmien.[69] He showed that the effect of viscosity, necessary to keep the perturbation velocity u_1 finite in the critical layer, changes the conditions in and around this layer considerably. Indeed, depending on the values of α and $\mathbf{Re}_\delta$, instability may set in. In the course of years following Tollmien's publication, Tollmien and Schlichting,[14] and many others have made improvements and refinements in the calculations, however, without changing basically the general nature of the instability conditions as a function of α and

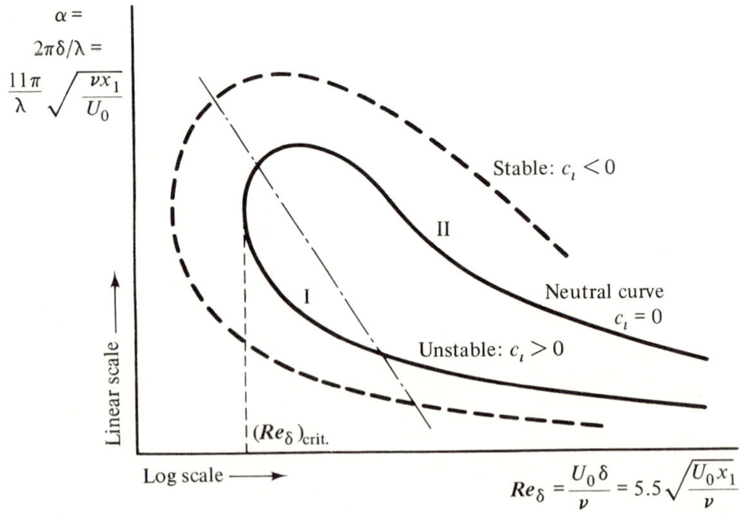

FIGURE 7-2
Neutral curve (solid line) separating stable and unstable regions in the $(\alpha - \mathbf{Re}_\delta)$-plane.
— — — — — neutral curve for inviscid instability with inflexion point in the velocity profile.
— · — · — line followed when ν is varied with U_0, x_1, and λ kept constant.

$\mathbf{Re}_\delta$. Figure 7-2 shows the general shape of the neutral curve ($c_i = 0$), separating the unstable and stable regions in the (α, $\mathbf{Re}_\delta$)-plane.

The effect of viscosity in making possible the existence of unstable solutions may be surprising, since it is usually supposed that viscosity has a damping effect on disturbances. Of course there is a damping effect in that viscosity prevents the disturbance velocity becoming infinite in the critical layer. However, in addition to it there may also be an amplifying effect. These opposite effects can be concluded from the theoretical results shown in Fig. 7-2. Keeping x_1 and U_0 constant, and varying only the viscosity, the dash-dot line is followed. Starting from a point on the lower part of this line in the stable region, by increasing the viscosity the unstable region is entered when crossing branch I of the neutral curve. When passing through the unstable region the amplification factor, i.e., c_i, first increases up to a maximum value and then decreases to zero again at the neutral curve.

Apart from the zero-condition for the disturbance velocity at the wall due to viscosity, according to Schlichting[14] the theoretical solution also shows a jump in phase, i.e., a 180 degree phase shift, at the critical layer when the basic velocity profile is convex. Under the unstable conditions a disturbance can grow because viscosity then appears to produce an extraction of energy from the basic motion in favor of the disturbance. It may also be remarked that, other conditions remaining the same, a higher viscosity results in a less convex, that means a less stable, velocity profile. Once the disturbance is no longer infinitesimal, so that nonlinear effects become of importance, the zero condition at the wall results in a phase shift between the u_1 and the u_2 component such that at the wall $(-\overline{u_2 u_1})$ becomes nonzero and positive. Since at the wall also $dU_1/dx_2 > 0$, there is a positive "production" term in the balance equation for $\overline{u_1^2}$.

Lin[70] has shown this effect, to which reference has already been made in Chapter 1, using a simple model. For very small values of x_2 both U_1 and d^2U_1/dx_2^2 are roughly zero. The equation for the vorticity component $\omega_3 = -\nabla^2\psi$ then reduces to

$$\frac{\partial}{\partial t}\nabla^2\psi = \nu\frac{\partial^2}{\partial x_2^2}\nabla^2\psi$$

Consider a neutral disturbance

$$\psi = \varphi(x_2)\exp\left[\iota\alpha(x_1 - c_r t)\right]$$

and a series expansion of $\varphi(x_2)$. The solution that satisfies the boundary condition at the wall yields for the average value of $u_2 u_1$ during one period:

$$-\overline{u_2 u_1} = A^2\alpha\sqrt{\frac{\alpha c_r}{2\nu}}\,x_2^4 + \cdots > 0$$

where A is real. This result also shows, that though the viscosity produces the destabilizing or "production" effect, the magnitude of this effect decreases with increasing viscosity. Also note that the shear-stress $-\rho\overline{u_2 u_1}$ starts with x_2^4 when $x_2 \to 0$.

The dual role of the wall through the fluid viscosity, namely both a damping effect and a "turbulence production" effect, has been clearly demonstrated experimentally by Uzkan and Reynolds.[71] They studied a boundary layer along a streamwise movable wall in a water-tunnel, where the free stream outside the boundary layer was made turbulent with a grid. By making the flow pattern near the wall visible with a dye tracer, they observed a disappearance of the pattern typical for the "production" of turbulence at the wall, as soon as the wall had the same velocity as the free stream. The only effect of the wall then was a damping, i.e., an increased rate of decay of the grid-produced turbulence near the wall.

According to Fig. 7-2 there is a minimum value of the Reynolds number below which all infinitesimal disturbances are damped. A value, generally accepted, for the *critical Reynolds number for instability* is the one calculated by Shen:[72]

$$(\mathbf{Re}_{\delta_d})_{\text{crit}} \simeq 420 \quad \text{or} \quad (\mathbf{Re}_{\delta})_{\text{crit}} \simeq 1{,}220$$

whence follows $(\mathbf{Re}_{x_1})_{\text{crit}} \simeq 6.10^4$.

The corresponding value for the propagation velocity is $c_r = 0.42U_0$, and the wavelength following from $\alpha = 1.05$ is $\lambda \simeq 6\delta$. This wavelength increases with the Reynolds number. For a point on branch I of Fig. 7-2, at $\mathbf{Re}_{\delta_d} = 893$, the calculation results in $c_r = 0.35U_0$ and $\lambda \simeq 14\delta$. We thus conclude that the wavelengths of unstable disturbances are large compared with the boundary-layer thickness.

The above value of $(\mathbf{Re}_{\delta_d})_{\text{crit}} = 420$ agrees nicely with present experimental data, which show a value of ~ 400. As Ross *et al.*[73] have shown, the value of 500 obtained by Barry *et al.*[74] from a numerical integration of the Orr-Sommerfeld equation can be reduced to the value obtained by Shen, and consequently also to the experimental value, if due account is taken of the additional amplification because of the streamwise variation of $|d\varphi/dx_2|_{\max}$, occurring in the expression for the maximum value of u_1'.

In the unstable region the linearized theory results in an exponential growth with time of the disturbance. Such a disturbance, however, cannot grow indefinitely since soon the nonlinear effects become of importance. In a general discussion of nonlinear stability theory, Stuart[75] showed that there are three main nonlinear processes that modify the linearized theory: (1) a distortion of the mean flow because of the generation of Reynolds' stresses through the withdrawal of energy from the mean flow; (2) the fundamental oscillation is modified in its distribution in the x_2-direction; (3) higher harmonics are generated. From a simple mathematical model, already used by Landau and Lifschitz[76] for the same purpose, Stuart showed that when $\alpha c_i < 0$, the squared, time-dependent, amplitude reaches a threshold value asymptotically with time.

Another interesting result of the nonlinear theory is (according to Benney and Bergeron[77]) that for sufficiently large values of $\mathbf{Re}^{2/3}u_1/U_0$, the nonlinear effects dominate the viscous effects in the critical layer resulting in a suppression of the aforementioned 180 degree phase shift.

Any disturbance introduced in one way or another in the unstable region of a plane boundary layer along a smooth wall, will grow due to the unstable nature of the flow. In the case of natural transition the disturbances develop from points in the boundary layer which are distributed randomly in time and in space (in streamwise and spanwise directions). Each disturbance develops in streamwise and spanwise direction into a turbulence patch, or *turbulence spot* according to Emmons.[78] In a plan view such a turbulence spot has a kidney shape, elongated in streamwise direction. Since the turbulence spots developing from the randomly distributed disturbances grow in spanwise direction, ultimately they will overlap and breakdown of the whole boundary layer to turbulence soon follows. In order to make possible the experimental study in a more controlled way of transition to turbulence, Schubauer and Skramstad[79] developed their *"vibrating ribbon"* technique. A thin metal ribbon is stretched parallel to the wall in spanwise direction, at a short distance from the wall, preferably below the critical layer. The ribbon is vibrated electromagnetically with a frequency which corresponds with a point, preferably, on or near branch I of the neutral stability curve. In this way a two-dimensional disturbance in the form of a spanwise vortex varying harmonically in intensity with time is introduced in the unstable region of the boundary layer. However, Schubauer and Skramstad observed that the disturbances so introduced did not remain two-dimensional, but soon showed a more or less regularly spaced spanwise variation. This, notwithstanding that very careful precautions were taken to maintain a two-dimensional situation. Accepting such a three-dimensional development of originally two-dimensional disturbances, Klebanoff *et al.*[80] modified the vibrating-ribbon technique in that they tried to control the spanwise variation in wave amplitude, or vortex intensity, of the disturbance by glueing strips of Cellophane tape onto the wall beneath the vibrating ribbon. These strips of 0.076 mm thickness were regularly spaced, leaving gaps between the strips equal to the width of the strips, namely 12.5 mm.

The result is that the intensity of the vortex is stronger downstream of the gaps than downstream of the strips. Consequently a regular three-dimensional disturbance due to the trailing streamwise vortices caused by the spanwise changes of disturbance vortex is produced. A similar technique has been used by Hinze *et al.*[81] They studied the development of a more localized disturbance produced by one single gap left open between two Cellophane strips beneath the vibrating ribbon and parallel with it. The gap had a width of 18 mm.

From these experiments, and also from those by a number of other experimenters, the following general picture can be obtained of the processes taking place between the introduction of a disturbance and the breakdown to turbulence. It is mainly based on measurements in an axial plane of symmetry through the center of a gap.

1 The initial amplification of a disturbance is in accordance with the so-called *"Tollmien-Schlichting waves"* of the linear instability theory. The amplification rate appears to depend on the nature of the disturbance. According to experimental evidence it was higher in the "multi-gap" case than in the "single-gap" case, while in the case of a freely vibrating ribbon it had the lowest value. After some distance from the vibrating ribbon the actual amplification becomes greater than the amplification of the Tollmien-Schlichting waves.

2 The periodically formed vortices parallel with the vibrating ribbon are deformed into U-shaped loops. This results in the creation of a u_3-velocity component of the disturbance. Due to self-induction effect the downstream top of the vortex loop lifts away from the wall. Since it then arrives in a higher mean-velocity region, the top of the loop becomes more peaked due to a stretching effect. At the same time the local intensity of the u_1 and u_2 components around the top is strongly increased.

3 Crossing the critical layer ($x_2 \simeq 0.2\delta$) seems to have no noticable effect on the disturbance vortex. Also the predicted phase-shift of the Tollmien-Schlichting waves at the critical layer is hardly noticeable. Such a 180-degree phase shift is observed in a much more pronounced way at $x_2 = 0.6$ to 0.8δ.

4 The peak of the loop seems not to move farther away from the wall than the region $x_2 = 0.6$ to 0.8δ.

5 Subsequently, second and third harmonics are observed. They might already be present, but unnoticeably weak, right from the beginning during the generation of the disturbance by the vibrating ribbon. Anyway when they are measurable, nonlinear effects must have taken place.

6 In the u_1-oscillograms taken in the region $x_2 = 0.6$ to 0.8δ sudden dips in the velocity are observed. Because of their sharply peaked shape they are referred to as *"spikes."* First single "spikes" are observed, but soon followed at a later stage by "double-spikes" roughly at the same x_2-distance from the wall. Also a second "single-spike" may be observed in the u_1-oscillograms, but at a different x_2-location. The occurrence of these "spikes" is accompanied with an increase of the energy of the second and third harmonics. Moreover, they are in phase at a "spike," and out of phase elsewhere.

7 The "spikes" condition, and in particular the "double-spikes" condition is followed by a *turbulence burst* at roughly the same x_2-location. Thereby the energy of the three harmonics decreases, in particular that of the first harmonic, suggesting the generation of many more and higher harmonics.

8 Measured as a function of the x_1-position the total energy of the u_1-component flattens in the x_1-region where "spikes" are observed. After the occurrence of turbulence bursts it decreases to the level of fully developed turbulent flow. At the same time the energy of the two other velocity components increases. In particular the u_2-component may become much greater than the u_1 and u_3-component at a "spike" stage.

9 By the time that the second and third harmonics become appreciable, at the position of their maximum value (roughly at $x_2 = 0.6\delta$), the mean-velocity distribution shows a kink with two inflexion-points. This introduces the possibility of inflexional instability. The sequence in the x_1-direction of mean-velocity profiles suggests a relatively slower moving fluid in the region $x_2 < 0.6$ to 0.8δ, overtaken by a faster moving fluid layer $x_2 > 0.6$ to 0.8δ.

10 Breakdown to complete turbulence soon follows the stage where an increasing number of turbulence bursts is observed.

Let us now consider in more detail a few of the above phenomena, which may be of interest in connection with the phenomena taking place and which have been observed in the fully developed turbulent flow in the region close to the wall.

Under (2) we mentioned the deformation of the periodically formed vortices into U-shaped loops, regularly spaced in the x_3-direction in the case of the experiments by Klebanoff *et al.* Due to self-induction the U-shaped loop moves away from its original x_2-plane, in a fashion similar to the description given in Sec. 6-12, and illustrated by Fig. 6-31. In the present case, however, the vortex drifts away from the wall, with a velocity that is higher for the parts with greater curvature. Consequently the tip of the U-shaped loop has the highest velocity away from the wall. This is illustrated in Fig. 7-3a. Arriving in a region with increasing $\bar{U}_1$-velocity, due to the accompanying stretching of the U-shaped loop, the vorticity obtains a higher intensity. As shown by Hama,[82] and illustrated in Fig. 7-3b in both the plan view and the side view, the tip of the stretched U-shaped loop obtains a typical bottle-neck, or Ω-shape, again due to self-induction effects.

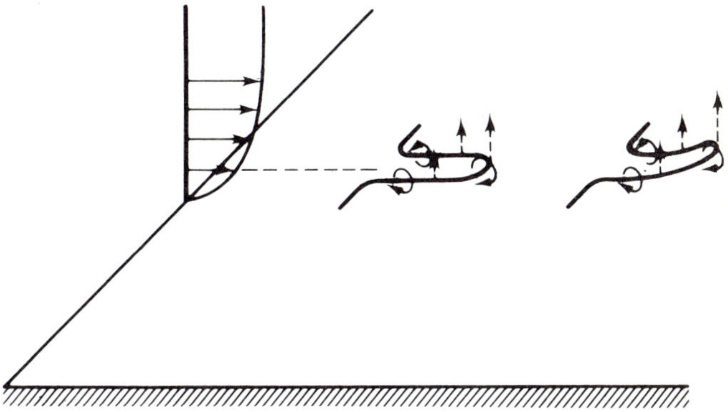

FIGURE 7-3a
Progressive movement and deformation of a U-shaped vortex loop due to self-induction.

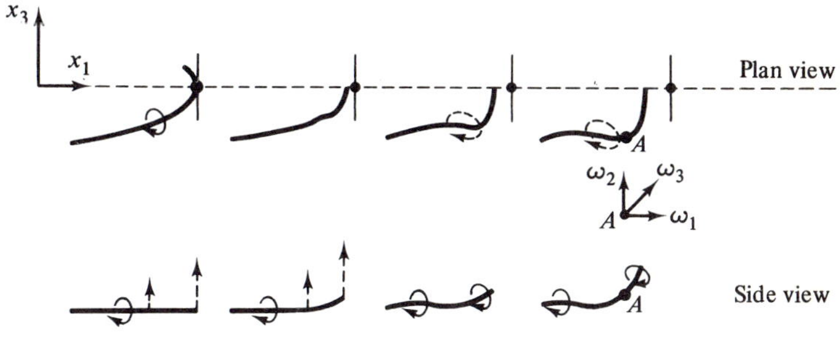

FIGURE 7-3b
Plan view and side view of the progressive deformation of a U-shaped vortex loop, resulting in a Ω-shape of the peak. At point *A* the vorticity vector has components in the three directions.

This is caused primarily by the ω_2-component of the vortex, which arises when the vortex tip drifts away from the wall. It induces a dent in the U-shape vortex near its tip, producing the Ω-shape.

At point A of the Ω-shaped part, the three components ω_1, ω_2, and ω_3 of the vortex are shown in Fig. 7-3b. This shape has been observed experimentally by Hama and Nutant,[83] and is also shown in Fig. 7-4, taken from Knapp and Roache.[84]

Because of the intensification of the vorticity, in particular at the Ω-shaped peak, the u_1 and u_2 components of the disturbance strongly increase in the vicinity of the peak, resulting in a local and temporary deformation of the velocity profile of the boundary layer, and consequently in an intense shear-layer. Figure 7-5 shows the deformation of this velocity profile during one cycle, as measured by Van

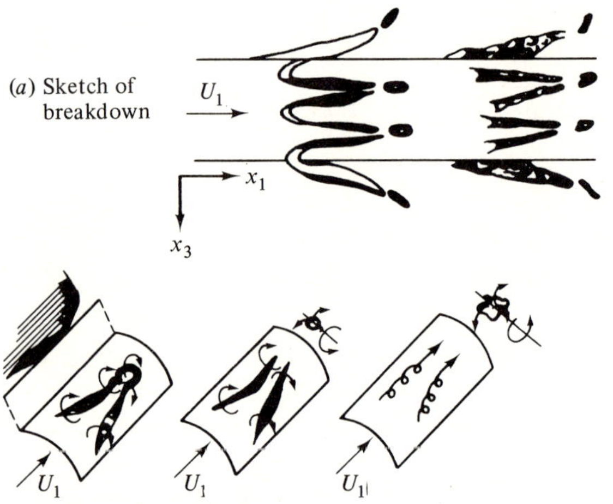

(b) Time history of breakdown

FIGURE 7-4
Observed distortion of a U-shaped vortex loop. (From: *Knapp, C. F., and P. J. Roache,*[84] *by permission of the American Institute of Aeronautics and Astronautics.*)

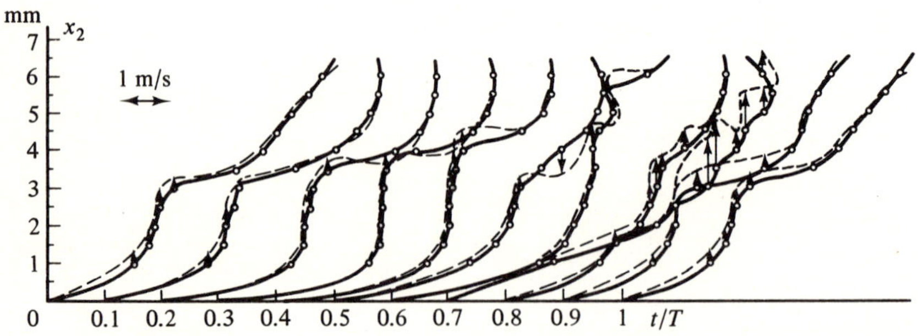

FIGURE 7-5
Measured instantaneous velocity profiles during one cycle T. $U_0 = 7.7$ m/s; $\delta \simeq 6$ mm; $T = 0.015$ s.

Muiswinkel[85] with a hot-wire anemometer in the plane of symmetry through the tip of the U-shaped vortex ("single-gap" experiments), at a station 0.28 m downstream of the vibrating ribbon. The local thickness of the undisturbed boundary layer was $\delta \simeq 6$ mm, the free-stream velocity $U_0 = 7.7$ m/s. At this station the u_1-oscillograms showed "double-spikes" at $x_2 \simeq 4$ mm, and a "single-spike" at $x_2 \simeq 5$ mm. The boundary layer became much thicker than the undisturbed boundary during the passage of the disturbance wave. With some imagination the occurrence of spikes in the u_1-oscillograms can be deduced from the instantaneous velocity profiles shown in Fig. 7-5. Figure 7-6 shows six cinematographic frames obtained in a water flow by Hama and Nutant[83] with the hydrogen-bubble technique at a situation of the disturbed boundary layer comparable with that shown in Fig. 7-5. The bubble-producing wire is to the left of the pictures, and placed in the x_2-direction. The pulse frequency was 13.6 s^{-1} and the time between two frames $\frac{1}{3}$ s. The free stream velocity was 0.088 m/s, the boundary-layer thickness roughly 25 mm. Of course the curves in Figs. 7-5 and 7-6 respectively should not be compared quantitatively. In Fig. 7-5 the solid lines (open-circle data) show the end points of the instantaneous U_1-component, the dashed lines the end points of the total velocity vector, including the U_1-component and the u_2-component. Note that at a location where the U_1-component shows a kink in its distribution, the u_2-component becomes appreciable, up to some 30 per cent of the local U_1 value, while in Fig. 7-6 each curve indicates the position of bubbles generated at the same time at the bubble-producing wire. From these pictures Hama and Nutant obtained the corresponding instantaneous U_1-profiles, which showed the same kink formations as in Fig. 7-5. The intense shear-layers associated with these kink formations are highly unstable (inflexional instability), and through the breakdown process develop into the local turbulence bursts mentioned earlier under item (7). These turbulence bursts may be identified with the turbulence spots observed by Emmons. In plan view these turbulence spots are kidney-shaped, but since they are much more elongated in the main-stream direction than in spanwise direction, a rounded-off elongated delta may be a more proper description of its plan-view shape. Figure 7-7 shows a nice picture of a growing turbulence spot in plan view, taken by Elder.[86] Note that the flow condition upstream of the turbulence spot, after its passage, is laminar again. Earlier, other investigators had observed the existence of turbulence spots during the transition process (amongst others Mitchner[87] and Schubauer and Klebanoff[88]). Figure 7-8 shows a schematic drawing, both in plan view and in side view, of the shape of a turbulence spot generated upstream by an electric spark, located at 0.7 m from the leading edge of the plate. The spot as shown refers to the situation, again at 0.7 m downstream of the location of the electric spark. The solid-line figure in the plan view is the shape at the wall, the dashed-line figure that in the x_2-plane where the spot has its maximum extent. The velocity figures given are the propagation velocities of the boundary near the wall. Since there the velocity of the undisturbed laminar boundary layer is smaller than these propagation velocities, apparently the fluid downstream of the spot is overtaken by the spot, becomes turbulent, and after the

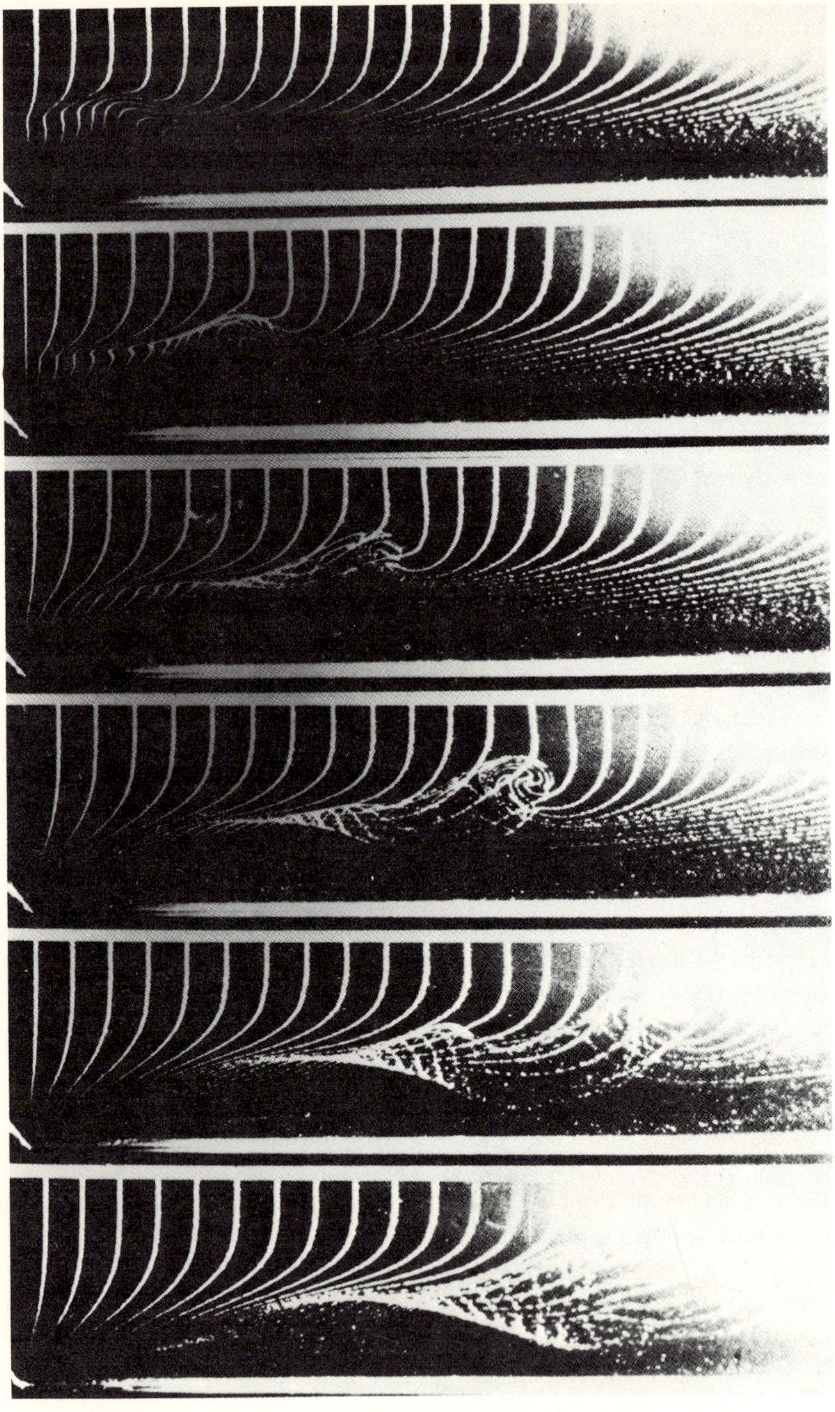

FIGURE 7-6
Secondary instability prior to breakdown. (From: *Hama, F. R., and J. Nutant,*[83] *by permission of the Heat Transfer and Fluid Mechanics Institute.*)

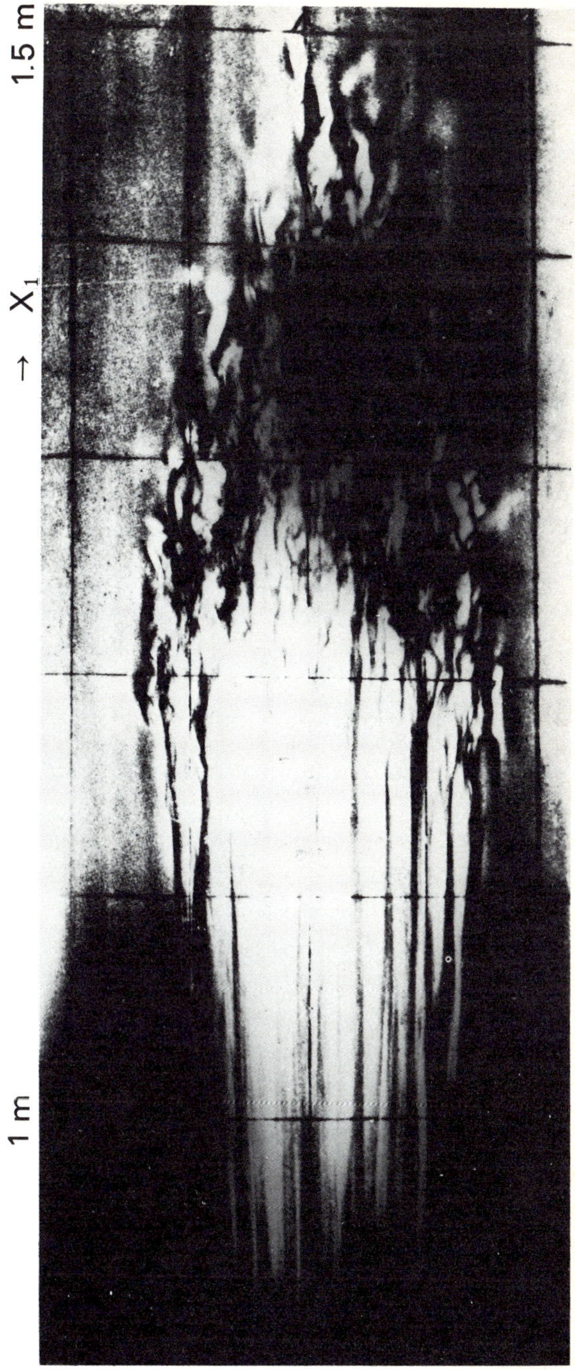

FIGURE 7-7
Photograph of a turbulence spot 3.9 s after a 0.1 s disturbance, introduced at $x_1 = 0.5$ m on a flat plate. $U_0 = 0.23$ m/s; $\delta = 12$ mm at $x_1 = 1.3$ m. (From: *Elder, J. W.,*[86] *by permission of the Cambridge University Press.*)

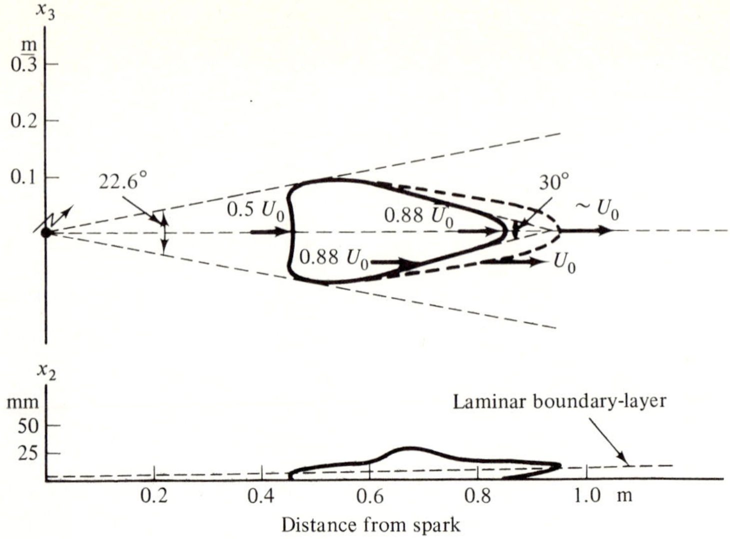

FIGURE 7-8
Plan and side views of a turbulence spot generated by an electric spark. $U_0 = 9$ m/s. (After: *Schubauer, G. B., and P. S. Klebanoff.*[88])

passage of the trailing boundary resumes the laminar flow condition. This seems to suggest the possibility of relaminarization, or reverse-transition from the turbulent to the laminar condition. We will come back to this point later in Sec. 7-11, when discussing similar phenomena during the transition process laminar-turbulent in pipe flow.

In the case of the spark-induced disturbance, and also for natural transition it can be shown that a local disturbance (a point-disturbance, say) will develop into a kidney-shaped patch, according to the linear instability theory. This has been done by Criminale and Kovasznay.[89] The theory is based on Squire's theorem,[90] which says that if the wavenumber vector of a disturbance makes an angle φ with the flow velocity U_1, the component $U_1 \cos\varphi$ which is parallel with the wavenumber vector is decisive for the development of an unstable disturbance. Starting from a circular disturbance in the $(x_1, 0, x_3)$-plane the propagation and amplification are different for the different directions. When α again is the nondimensional wavenumber (see Fig. 7-2), the wavenumbers β and γ in the x_1 and x_3 directions respectively are: $\beta = \alpha\cos\varphi$ and $\gamma = \alpha\sin\varphi$. Let $\mathbf{Re}_\delta(\varphi = 0)$ be the condition for such a local disturbance on branch I of the neutral curve in Fig. (7-2), and $\mathbf{Re}_\delta(\varphi = 0) > (\mathbf{Re}_\delta)_{\text{crit}}$. Then $\mathbf{Re}_\delta(\varphi) = \mathbf{Re}_\delta(\varphi = 0)\cos\varphi$, where $0 < \varphi < \varphi_{\max}$, and $\varphi_{\max}$ corresponds with $(\mathbf{Re}_\delta)_{\text{crit}}$. By considering the curves with constant c_i in Fig. (7-2) the reader may convince himself that for the above range of φ-values and a constant value of $\mathbf{Re}_\delta(\varphi = 0)$, the corresponding curves in a (β,γ)-plane are kidney-shaped, symmetric with respect to the β-axis, i.e., the x_1-axis. The kidney-shaped wave-packets then develop into the elongated delta-shaped turbulence spots by non-linear effects. An experimental confirmation of the theoretical results obtained by Criminale and Kovasznay has been given by Vasudeva.[91]

Let us again return to the case of transition produced artificially with the vibrating-ribbon technique, and in particular when (with Cellophane spacers) a regular spanwise variation in disturbance intensity is obtained ("multi-gap" situation). Downstream, the intensity of the growing disturbance has a maximum at x_3-values corresponding with the center of a gap, and a minimum at x_3-values corresponding with the center of a spacer. Klebanoff *et al.*[80] refer to this u'_1-variation as the peaks and valleys in the spanwise distribution. During the development of the disturbance vortex to U-shaped loops, the "trailing" parts, or the "legs" of the U-shaped loops, because of their streamwise orientation, induce u_2-flows away from the wall in the peak-regions and towards the wall in the valley-regions. Consequently, the thickness of the boundary layer becomes greater in the peak regions, and smaller in the valley regions. As we have seen, in the peak regions the U_1-velocity profiles become less convex, and begin to show inflexion points as the disturbance grows, whereas in the valley regions the U_1-velocity profiles become more convex. Close to the wall in a peak region of the u'_1-component the $\bar{U}_1$ is smaller than in a valley region of u'_1. Hence, peak and valley regions of u'_1 correspond with valley and peak regions of the $\bar{U}_1$ velocity, respectively.

When the amplitude of the disturbance becomes large, *vertical shear-layers* are created at the boundaries between peak and valley regions, which shear-layers become more and more intense in downstream direction. These vertical shear-layers arise because near the wall the U_1-velocity in a u'_1-peak region is smaller than in a u'_1-valley region at the same value of x_2, thus generating at the boundaries between peak and valley regions a ω_2-vortex component growing in intensity. Its rotation is clockwise looking in the positive x_2-direction when we cross the boundary from a peak region towards a valley region of the U_1 velocity.

Tani and Komoda,[92] and Komoda[93] have shown experimentally the existence of these vertical shear-layers by controlling the spanwise variation in boundary-layer thickness, and the associated variation in U_1-velocity profiles, by means of the trailing, i.e., streamwise, vortices of a row of wings placed outside the boundary layer and upstream of the vibrating ribbon. It appeared possible to obtain breakdown of these vertical shear-layers in the case of a large variation in the thickness of the boundary layer. However, in contrast to the breakdown of the horizontal shear-layers, no spikes in the u_1-oscillograms are observed prior to the breakdown of the vertical shear-layers. This may seem surprising, but we have to bear in mind that, though the spanwise distribution of the U_1-velocity at $x_2 = \text{constant}$ may show locally large changes, the variation in the streamwise direction, which is the one recorded in the u_1-oscillogram, may still be relatively gradual.

Mathematical Models

A number of the experimental facts described above, and of primary importance for the transition process, are consistent with results of theoretical analyses based on simplified *mathematical models*. Benney and Lin,[94] and Benney[95,96] studied mainly the effect of streamwise vortices, generated by a spanwise periodicity in the perturbation, and after amplification, by nonlinear second-order interactions between these two perturbations. As may be expected, these secondary streamwise vortices modify the mean flow by transporting axial momentum of the mean flow at uniformly spaced x_3-positions where the secondary u_2-velocity is away from, or towards the wall. In the positions with the u_2-velocity away from the wall the mean-velocity distribution will obtain an inflexion point at an x_2-location above the critical layer. Also Stuart[97] considered the deformation of the mean-velocity distribution due to longitudinal vortices, however, by assuming a flow pattern with steady u_2- and u_3-components, varying periodically in the x_3-direction, and a u_1-component which at $t = 0$ has the form $u_1 = U_0[1 - \exp(-\beta x_2)]$. When $t > 0$, the u_1-profile deforms, becoming more and more S-shaped at the "peak" regions. As a result a concentration of vorticity grows at this "peak" region at $\beta x_2 \simeq 2.7$, while at the "valley" region a concentration of vorticity grows near $x_2 = 0$. The position of the concentrated shear-layer at $\beta x_2 \simeq 2.7$, corresponds roughly with $x_2 \simeq 0.6$ to 0.8δ.

The secondary, inflexional instability has been considered by Greenspan and Benney.[98] They restricted themselves to a linearized theory, and assumed the non-viscous Orr-Sommerfeld equation. The basic flow was approximated by a $U_1(x_2,t)$ distribution, consisting of straight portions, and which resembled the observed U_1-distribution at $t/T \simeq 0.4$ in Fig. 7-5. At constant value of t, the amplified disturbance showed a maximum at an optimum wavenumber. This maximum value increased with time, while at the same time the optimum wavenumber shifted towards a higher value. This implies a shift of the maximum kinetic energy towards higher wavenumbers during the development of the shear-layer, and growth of the unstable disturbance. Other interesting results consistent with experimental observations are that after one half cycle the optimum wavenumber is about five times that of the primary disturbance, and that the increase in kinetic energy during this half cycle was more than hundred-fold. Furthermore, the propagation velocity of the secondary disturbance was twice that of the primary disturbance. We may recall that the Tollmien-Schlichting waves had a propagation velocity of 0.3 to $0.4U_0$, while the front of a turbulence spot propagates with a velocity of $\sim 0.8U_0$.

7-5 TURBULENT BOUNDARY LAYER ALONG A FLAT PLATE. CLASSICAL THEORIES

In Sec. 7-1 we made a distinction between a number of regions in a boundary layer. This fact was recognized very early by Prandtl, Von Kármán, and others. Investigations have been directed accordingly by many experimenters, some of the most important by Nikuradse.[6]

For the flow along a smooth wall, we introduced the notion of the viscous sublayer with an average thickness δ_l. In the case of the flow along a rough wall the flow, and hence its velocity distribution may become affected by this wall roughness, particularly when the average height, k, of the roughness elements becomes greater than the thickness of the viscous sublayer.

The flow in the *wall region* is, according to its definition, determined by the condition at the wall as expressed by the wall shear stress and its roughness. Let us assume that the roughness can be characterized by one length parameter k, for instance the average height of the roughness elements. The mean-velocity component $\bar{U}_1$ in this region then will be a function of the wall shear-stress σ_w, the roughness parameter k, the distance x_2 to the wall, and the fluid properties ρ and μ. Dimensional analysis would yield three dimensionless groups. From the equations of motion for an incompressible fluid it becomes evident that the ratio σ_w/ρ is the quantity that determines the velocity distribution. Since this ratio has the dimension of a velocity squared, instead of σ_w we will consider the characteristic velocity

$$u^* = \sqrt{\frac{\sigma_w}{\rho}} \tag{7-23}$$

This velocity is usually referred to as the *wall-friction velocity*, or wall shear stress velocity.

Also from the equations of motion it follows that the kinematic viscosity $\nu = \mu/\rho$ is the fluid property of primary importance. Hence, for the wall region, $\bar{U}_1$ is a function of u^*, k, ν, and x_2. Whence follows, from dimensional analysis

$$\frac{\bar{U}_1}{u^*} = f\left(\frac{u^* x_2}{\nu}, \frac{u^* k}{\nu}\right) \tag{7-24}$$

This should be a universal relationship, subject, of course, to the conditions assumed in the foregoing. Equation (7-24) is often called the *law of the wall*.

For a *smooth boundary* ($k = 0$), the relation reduces to

$$\frac{\bar{U}_1}{u^*} = f\left(\frac{u^* x_2}{\nu}\right) \tag{7-25}$$

Since $\bar{U}_1 = 0$ at the wall, a linear relationship is obtained when $x_2 \to 0$. It can be easily shown that the constant of proportionality is unity. For, in the viscous sublayer the velocity distribution is in first approximation determined by

$$\mu \frac{\partial U_1}{\partial x_2} = \sigma_{21} \simeq \sigma_\omega$$

whence follows

$$U_1 = \frac{\sigma_\omega}{\mu} x_2$$

or

$$\frac{\bar{U}_1}{u^*} = \frac{u^* x_2}{\nu} \tag{7-26}$$

If we assume that across the wall region the local shear stress deviates only little from the wall shear stress, for the turbulent part of the wall region we may write

$$\epsilon_m \frac{\partial \bar{U}_1}{\partial x_2} = \frac{\sigma_w}{\rho}$$

It is obvious that a constant value of the eddy viscosity ϵ_m, which would give again a linear velocity distribution, is not a satisfactory assumption, since such a distribution would conflict with experimental evidence.

Now the eddy viscosity ϵ_m is connected with the transport of momentum from one layer to the other. This transport is determined by the diffusion of fluid particles due to turbulence. We have shown in Chap. 5 that, for short diffusion distances, that is, for diffusion times short with respect to the Lagrangian integral time scale, the diffusion proceeds linearly with time and that, if a coefficient of eddy diffusion is introduced, it varies linearly with distance [see, for example, Eq. (5-96)]. Since we are considering here a region at short distance from the wall, so that diffusion of fluid particles within that region is restricted to short distances, we may assume an eddy viscosity proportional to the distance x_2 and expect a better result than would be obtained assuming a constant eddy viscosity.

An equivalent result is obtained if it is assumed that, in this region close to the wall, the size of the eddies, which are responsible for the eddy viscosity, varies in proportion to the distance from the wall; though the mechanism upon which this latter assumption is based differs from that corresponding to the first assumption.

Thus we assume

$$\epsilon_m = \varkappa u^* x_2 \tag{7-27}$$

The equation for the velocity distribution then reads

$$\varkappa u^* x_2 \frac{\partial \bar{U}_1}{\partial x_2} = \frac{\sigma_w}{\rho} = u^{*2}$$

or, written in dimensionless quantities,

$$\varkappa x_2^+ \frac{\partial \bar{U}_1^+}{\partial x_2^+} = 1 \tag{7-28}$$

where

$$\bar{U}_1^+ = \frac{\bar{U}_1}{u^*} \quad \text{and} \quad x_2^+ = \frac{u^* x_2}{\nu} \tag{7-29}$$

The solution of this differential equation reads

$$\bar{U}_1^+ = \frac{1}{\varkappa} \ln x_2^+ + \text{const.} \qquad \text{for } x_2 > \delta_t \tag{7-30}$$

Prandtl obtained the same solution, but he applied his momentum-transport theory.

According to this theory we have

$$\sigma_{21} = \rho l_m^2 \left|\frac{\partial \bar{U}_1}{\partial x_2}\right| \frac{\partial \bar{U}_1}{\partial x_2}$$

Prandtl now assumed intuitively that the mixing length l_m at such short distances from the wall must be proportional to the distance. Substitution of $l_m = \varkappa x_2$ and of $\sigma_{21} \simeq \sigma_w = \rho u^{*2}$ leads to the same differential equation (7-28) and, consequently, to the same logarithmic velocity distribution (7-30). If, instead of Prandtl's assumption on the mixing length l_m we use Von Kármán's expression (5-18) for this mixing length, we again obtain the logarithmic velocity distribution.

Prandtl's assumption is essentially the same as the assumption (7-27). For we have

$$\epsilon_m = l_m^2 \left|\frac{\partial \bar{U}_1}{\partial x_2}\right| = \varkappa^2 x_2^2 \left|\frac{\partial \bar{U}_1}{\partial x_2}\right|$$

With the velocity distribution (7-30) we obtain the relation (7-27).

For the turbulent boundary layer along a *rough wall,* the same assumption concerning the eddy viscosity ϵ_m or the mixing length l_m may still be made. If the average height of the roughnesses is greater than the thickness of the buffer-layer, that is, if $k > \delta_t$, it is reasonable to expect no viscosity effect; so the expression for the velocity distribution should not involve the viscosity of the fluid, but should include the roughness parameter k.

Elimination of the kinematic viscosity in the relation (7-24) results in the following one for the law of the wall valid for a completely rough wall

$$\frac{\bar{U}_1}{u^*} = f\left(\frac{x_2}{k}\right) \tag{7-31}$$

The above assumption concerning the eddy viscosity, or the mixing length again yields a logarithmic distribution

$$\bar{U}_1^{\,+} = \frac{1}{\varkappa} \ln \frac{x_2}{k} + \text{const.} \tag{7-32}$$

There is uncertainty in defining the position of the plane $x_2 = 0$ if the wall is rough. If not otherwise stated, it will be assumed that the roughness elements are "attached" to an imaginary smooth wall whose surface is at $x_2 = 0$. This assumption will not give rise to difficulties as long as the roughness elements are not too irregular or too large. We will come back to this point later.

According to the solutions (7-30) and (7-32), the $\bar{U}_1$-velocity distributions in consecutive sections of the boundary layer are similar, at least in the region close to the wall where the shear stress is nearly constant. The wall-region is, therefore, also frequently called the *constant-stress layer.*

For the outer region the logarithmic distributions according to Eqs. (7-30) and (7-32) may no longer apply, since the conditions on which they are based are no longer valid. No equivalent theories have been given by Prandtl for this outer region. However, experimental evidence has shown that when we consider the velocity difference $(U_0 - \bar{U}_1)$, similarity is obtained for its distribution in consecutive sections, if this velocity difference is reduced by u^*, and the distance x_2 reduced by the local boundary-layer thickness δ. Namely

$$\frac{U_0 - \bar{U}_1}{u^*} = \psi\left(\frac{x_2}{\delta}\right) = \psi(\xi_2) \qquad (7\text{-}33)$$

where $\xi_2 = x_2/\delta$.

This similarity relation turns out to apply to the whole turbulent part of the boundary layer, including that of the wall region. Von Kármán made the suggestion to consider it as a postulate, called the *velocity-defect law*. In passing it may be remarked that this relation follows also from either Eqs. (7-30) or (7-32), though these equations should not strictly apply to the outer region of the boundary layer.

We have mentioned in the introduction and in Sec. (7-2) the close similarity between flow conditions in this outer region of the boundary layer and in free turbulent flow. We may, therefore, expect a satisfactory solution for the velocity distribution if we apply the assumption of a constant eddy viscosity. Equation (7-3), neglecting the pressure and viscous terms, should then apply, but it would be well to consider $(U_0 - \bar{U}_1)/U_0$ rather than $\bar{U}_1/U_0$ in this equation. Considerations based on this concept will be given later, in Sec. (7-10) on newer theories about the turbulent boundary layer.

As mentioned, until recently no theories equivalent to those for the wall region have been given for the velocity distribution in the outer region. Perhaps the need has not been very urgent, for it appears that, from a practical engineering point of view, the logarithmic velocity distribution gives sufficiently satisfactory results even when applied to this outer region. As we shall show in the next section, the deviations from the actual velocity distribution are small, and the resulting approximate average velocity over the boundary-layer cross section is sufficiently close to the actual value for engineering purposes.

The logarithmic velocity distributions (7-30) and (7-32) are universal in character, determined only by the wall conditions and the distance from the wall. The constants occurring in the expressions (7-30) and (7-32) are numerical constants. Indeed, these expressions can be obtained in a general way by dimensional reasoning on the basis of some assumptions of a general nature, without entering into details of flow structure. Such reasoning has been adduced by Millikan.[7]

As mentioned earlier the velocity-defect law (7-33) is supposed to apply to the whole turbulent part of the boundary layer, including that of the wall region. For the turbulent part of the wall region also the relation (7-24) should apply. Consequently the velocity distribution for the turbulent part of the wall region has to satisfy the law of the wall as well as the velocity-defect law. As we shall show this condition leads to a logarithmic distribution. We shall give this proof in a different way than Millikan did (see, e.g., ref. 99). Consider first the case of a *smooth wall*. If the expression for the distribution $\bar{U}_1^+$ in this "overlapping" region should satisfy simultaneously the relations (7-25) and (7-33), then this should also hold true for the velocity gradient $\partial \bar{U}_1^+/\partial x_2^+$.

From the two relations then follows

$$\frac{\partial \bar{U}_1^+}{\partial x_2^+} = \frac{df}{dx_2^+} = -\frac{d\psi}{d\xi_2}\frac{d\xi_2}{dx_2^+}$$

Since $d\xi_2/dx_2^+ = \nu/\delta u^* = \xi_2/x_2^+$, we obtain

$$x_2^+ \frac{df}{dx_2^+} = -\xi_2 \frac{d\psi}{d\xi_2} = \text{constant} = A$$

Whence follows

$$\bar{U}_1^+ = f(x_2^+) = A \ln x_2^+ + B \tag{7-30a}$$

and

$$U_0^+ - U_1^+ = \psi(\xi_2) = -A \ln \xi_2 + B' \tag{7-34}$$

A similar reasoning applied to the relation (7-31) for a *rough wall* and the relation (7-33) yields

$$\bar{U}_1^+ = A \ln \frac{x_2}{k} + B_1 \tag{7-32a}$$

while, as it should, for the velocity defect the same expression (7-34) as for a smooth wall is obtained. As follows from the corresponding expressions (7-30) and (7-32) obtained earlier, the constant A in Eq. (7-30a) is the same as that in Eq. (7-32a), and equal to the reciprocal of Von Kármán's univeral constant $\varkappa$.

It should be stressed that the universal logarithmic velocity-distribution applies only to the turbulent part of the wall region, where the shear stress is practically constant.

For the region $x_2 < \delta_t$ many assumptions about the velocity distribution have been made in the past. In calculating the heat transfer through the boundary layer along a smooth wall, Von Kármán[8] approached the velocity distribution in the buffer region $\delta_l < x_2 < \delta_t$ by a straight line in the $(\bar{U}_1^+, \ln x_2^+)$ diagram, thus assuming simply that $\bar{U}_1^+$ must be proportional to $\ln x_2^+$. Hofmann[9] and Reichardt[10] assumed other functions for $\bar{U}_1^+$ in this buffer region: a third-degree polynomial in $(x_2^+ - \delta_l^+)$ and an exponential function, respectively, but made no essential progress.

Rotta[11] assumed that in this buffer region the total shear stress is determined by viscous and turbulence effects. For the turbulence part he applied Prandtl's mixing-length hypothesis, and put

$$\sigma_{21} = \rho\left[\nu + \mathrm{l}_m^2\left|\frac{\partial \bar{U}_1}{\partial x_2}\right|\right]\frac{\partial \bar{U}_1}{\partial x_2} \simeq \sigma_w \tag{7-35}$$

where $l_m = \varkappa(x_2 - \delta_l)$. Thus for $x_2 \leq \delta_l$ the flow is completely viscous. The solution for $\bar{U}_1$ then reads

$$\bar{U}_1^{+} = \frac{1}{2\varkappa l_m^{+}}(1 - \sqrt{1 + 4l_m^{+2}}) + \frac{1}{\varkappa}\ln(2l_m^{+} + \sqrt{1 + 4l_m^{+2}}) + \delta_l^{+} \tag{7-36}$$

where $l_m^{+} = u^* l_m/\nu$ and $\delta_l^{+} = u^*\delta_l/\nu$. An identical solution, based upon the same assumption that Rotta made, was given later by Miles.[24]

Hitherto a certain value of the thickness δ_l of the viscous sublayer has been assumed, and that there should be no turbulent motion at all. It has long been known, however, that this is not a true picture of the actual flow near a wall but that turbulence fluctuations continue within this layer even when they are damped strongly by viscous effects. It would, therefore, be more reasonable to make no abrupt distinction between the viscous sublayer and the buffer region with respect to the function describing the mean-velocity distribution.

In order to study the effect on the transport of heat by turbulence vanishing toward the wall, Hofmann[9] assumed another function for the velocity distribution, namely, a third-degree polynomial in x_2^{+}. In later publications Reichardt[12] and Elrod[25] showed that, since the turbulence-velocity components have to satisfy the equation of continuity, the eddy viscosity ϵ_m must increase with $(x_2^{+})^3$ for $x_2^{+} \to 0$ if there is still a streamwise variation $\partial/\partial x_1$ in the time-mean values of turbulence quantities, whereas ϵ_m must increase with $(x_2^{+})^4$ for $x_2^{+} \to 0$ if there is no longer any streamwise variation.

However, as we will show, the latter result is only approximately true. A series expansion of u_1, u_2, and u_3 with respect to x_2 reads

$$u_1 = \left(\frac{\partial u_1}{\partial x_2}\right)_0 x_2 + \frac{1}{2}\left(\frac{\partial^2 u_1}{\partial x_2^2}\right)_0 x_2^2 + \cdots$$

$$u_2 = \frac{1}{2}\left(\frac{\partial^2 u_2}{\partial x_2^2}\right)_0 x_2^2 + \frac{1}{6}\left(\frac{\partial^3 u_2}{\partial x_2^3}\right)_0 x_2^3 + \cdots$$

$$u_3 = \left(\frac{\partial u_3}{\partial x_2}\right)_0 x_2 + \frac{1}{2}\left(\frac{\partial^2 u_3}{\partial x_2^2}\right)_0 x_2^2 + \cdots$$

Use has been made of the condition $\partial u_i/\partial x_i = 0$ for an incompressible fluid.

Consequently we obtain for the R.M.S. values

$$u_1' = \left(\frac{\partial u_1}{\partial x_2}\right)_0' x_2 + \cdots$$

$$u_2' = \frac{1}{2}\left(\frac{\partial^2 u_2}{\partial x_2^2}\right)_0' x_2^2 + \cdots$$

$$u_3' = \left(\frac{\partial u_3}{\partial x_2}\right)_0' x_2 + \cdots$$

As will be shown in Sec. 7-8, the value of $(\partial u_3/\partial x_2)_0'$ is about $\frac{1}{3}(\partial u_1/\partial x_2)_0'$ according to experimental evidence. The series expansion for the turbulence shear stress $-\overline{u_2 u_1}$ reads after some calculation and using the condition $\partial u_i/\partial x_i = 0$:

$$\frac{-\overline{u_2u_1}}{u^{*2}} = Ax_2^{+3} + Bx_2^{+4} + \cdots$$

where

$$A = \frac{1}{2}\frac{\partial}{\partial x_1^+}\overline{\left(\frac{\partial u_1^+}{\partial x_2^+}\right)_0^2} + \overline{\left(\frac{\partial u_1^+}{\partial x_2^+}\right)_0 \frac{\partial}{\partial x_3^+}\left(\frac{\partial u_3^+}{\partial x_2^+}\right)_0}$$

$$B = \frac{1}{4}\overline{\left(\frac{\partial^2 u_1^+}{\partial x_2^{+2}}\right)_0\left(\frac{\partial^2 u_2^+}{\partial x_2^{+2}}\right)_0} + \frac{1}{6}\overline{\left(\frac{\partial u_1^+}{\partial x_2^+}\right)_0\left(\frac{\partial^3 u_2^+}{\partial x_2^{+3}}\right)_0}$$

Since in the viscous sublayer $\bar{U}_1^+$ varies linearly with x_2^+, the same behavior as that of $-\overline{u_2u_1}/u^{*2}$ is obtained for the eddy viscosity ϵ_m/ν.

We conclude from the above that even if the flow is homogeneous in the streamwise direction, the coefficient A does not become zero, unless the second term on the right-hand side of the expression for A becomex zero or negligibly small. An estimate can be made of the value of A if it is allowed to use the value of $\lambda_3^+ \simeq 100$ for the variation in the x_3-direction of the instantaneous turbulence pattern as observed experimentally, and to assume $(\partial u_1^+/\partial x_2^+)'_0 \simeq 0.3$ and $(\partial u_3^+/\partial x_2^+)'_0 \simeq 0.1$ (see Sec. 7-8). This estimate yields $A = \mathcal{O}(10^{-4})$. However, the value of B is not known either. Further it should be kept in mind that the above behavior of $-\overline{u_2u_1}/u^{*2}$ only holds true for the limiting situation $x_2 \to 0$. Therefore an experimental verification seems almost impossible. Yet usually it is assumed that in a flow that is homogeneous in the streamwise direction $A \simeq 0$. Since in a constant-pressure boundary layer the changes in streamwise direction are relatively very slow, a x_2^{+4} behavior of ϵ_m/ν might then apply as well. Finally we may remark that the correlation coefficient $\mathbf{R}_{21} \to 0$ for $x_2 \to 0$ only when $A \simeq 0$, but remains finite if $A \neq 0$. If there is a decoupling effect of viscosity on the u_1 and u_2 velocities (see Sec. 7-11) then $\mathbf{R}_{21} \to 0$ and consequently $A \simeq 0$ may be expected.

For larger values of x_2^+, ϵ_m should change monotonously into a linear function of x_2^+ as the region beyond the buffer region is approached. Thus Reichardt assumed the following function

$$\frac{\epsilon_m}{\nu} = \varkappa\left(x_2^+ - \delta_l^+ \tanh\frac{x_2^+}{\delta_l^+}\right) \tag{7-37}$$

The velocity distribution follows from

$$\sigma_{12} = [\mu + \rho\epsilon_m]\frac{\partial \bar{U}_1}{\partial x_2} \simeq \sigma_w \tag{7-38}$$

The integral occurring in the solution of this differential equation cannot be solved in closed form. Reichardt approximated the integrand of the integral by a function containing exponential functions of x_2^+, which can be easily integrated. He finally obtained the velocity distribution

$$\bar{U}_1^+ = \frac{1}{\varkappa}\ln(1 + \varkappa x_2^+) + c\left[1 - \exp(-x_2^+/\delta_l^+) - \frac{x_2^+}{\delta_l^+}\exp(-0.33x_2^+)\right] \tag{7-39}$$

where c is a numerical constant.

Reichardt used the following values for the numerical constants: $\varkappa = 0.4$, $\delta_l^+ = 11$, and $c = 7.4$.

Another formula for the eddy viscosity ϵ_m is suggested by Deissler,[13] who also tried to take account of a turbulence diminishing toward the wall; Deissler put

$$\frac{\epsilon_m}{\nu} = a\bar{U}_1^+x_2^+[1 - \exp(-a\bar{U}_1^+x_2^+)] \tag{7-40}$$

where a is a numerical constant. This expression shows that, for $x_2^+ \to 0$, ϵ_m/ν varies proportionally with $(x_2^+)^4$, in contrast with the expression (7-37) proposed by Reichardt, according to which ϵ_m/ν varies proportionally with $(x_2^+)^3$. The velocity distribution is obtained from Eq. (7-38) by numerical integration.

Still another solution has been suggested by Van Driest,[26] who assumed the following modified expression for the mixing length l_m in Prandtl's mixing-length theory:

$$l_m = \varkappa x_2[1 - \exp(-x_2^+/A)]$$

so that

$$\epsilon_m = \varkappa^2 x_2^2[1 - \exp(-x_2^+/A)]^2 \left|\frac{\partial \bar{U}_1}{\partial x_2}\right| \tag{7-41}$$

In this case too, $\epsilon_m \propto x_2^4$ as $x_2 \to 0$. The solution obtained with Van Driest's assumption reads

$$\bar{U}_1^+ = 2\int_0^{x_2^+} \frac{dx_2^+}{1 + \{1 + 4\varkappa^2 x_2^{+2}[1 - \exp(-x_2^+/A)]^2\}^{1/2}}$$

Van Driest assumed $\varkappa = 0.4$ and $A = 2.7$.

A more recent relation for calculating the mean-velocity distribution covering the whole wall region, and which has proven to be useful for engineering purposes, has been suggested by Spalding.[100] Here the law of the wall, Eq. (7-25), is given in its inverse form, namely

$$x_2^+ = \bar{U}_1^+ + C\left[\exp(\varkappa\bar{U}_1^+) - 1 - \varkappa\bar{U}_1^+ - \frac{(\varkappa\bar{U}_1^+)^2}{2!} - \frac{(\varkappa\bar{U}_1^+)^3}{3!} - \frac{(\varkappa\bar{U}_1^+)^4}{4!}\right] \tag{7-42}$$

where $\varkappa$ is Von Kármán's universal constant, and $C = \exp(-\varkappa B)$ with B the constant occurring in Eq. (7-30a). This relation reduces to the linear distribution for small $\bar{U}_1^+$, i.e., small x_2^+, and to the logarithmic distribution at large $\bar{U}_1^+$ and x_2^+. The relation for ϵ_m/ν then is linear with x_2^+, while it reduces to x_2^{+4} when $x_2^+ \to 0$. Spalding suggested $\varkappa = 0.4$ and $C = 0.1108$ which corresponds with $B = 5.5$.

The foregoing expressions for ϵ_m/ν all show a behavior either with x_2^{+3} or x_2^{+4} when $x_2^+ \to 0$. However, it has to be borne in mind that this behavior applies only to the limit mentioned. As will be shown later, when considering the behavior of ϵ_m/ν for finite values of x_2^+, the behavior is more like x_2^{+2}. With this in mind the following expression for ϵ_m/ν suggested by Rannie[50] for the viscous sublayer and buffer region, is acceptable still, since it varies with x_2^{+2} for small x_2^+.

$$\frac{\epsilon_m}{\nu} = \sinh^2(\varkappa_1 x_2^+) \tag{7-43}$$

The mean-velocity distribution then reads

$$\bar{U}_1^+ = \frac{1}{\varkappa_1}\tanh(\varkappa_1 x_2^+) \tag{7-44}$$

It joints the logarithmic distribution at $\delta_t^+ = 27.5$ if $\varkappa_1 = 0.0688$.

The case of the velocity distribution in the buffer region of a boundary layer along a *rough wall* has been considered by Rotta.[11] Rotta assumed that, because of the roughness, the mixing length occurring in the expression (7-35) at $x_2 = 0$ is not zero but has some finite value l_0. He assumed

$$l_m = l_0 + \varkappa x_2$$

The solution of Eq. (7-35) when $\bar{U}_1^+ = 0$ at $x_2 = 0$ then reads

$$\bar{U}_1^+ = \frac{1}{2\varkappa}\left(\frac{1}{l_m^+} - \frac{1}{l_0^+}\right) - \frac{1}{2\varkappa}\left[\frac{(1 + 4l_m^{+2})^{1/2}}{l_m^+} - \frac{(1 + 4l_0^{+2})^{1/2}}{l_0^+}\right]$$
$$+ \frac{1}{\varkappa}\ln\frac{2l_m^+ + (1 + 4l_m^{+2})^{1/2}}{2l_0^+ + (1 + 4l_0^{+2})^{1/2}} \qquad (7\text{-}45)$$

For further information concerning the mean-velocity distribution in the vicinity of a rough boundary the reader is referred to the excellent review article by Rotta.[101]

A comparison of the logarithmic velocity distribution and of the various velocity distributions given for the buffer region with the measured velocity distribution will be given in the next section.

From the law of the wall (7-25) for a *smooth boundary* a relation for the wall-resistance can be obtained by putting $x_2 = \delta$:

$$U_0^+ = f(\delta^+)$$

Hence, if we assume the logarithmic distribution still valid approximately in the outer region, with Eq. (7-30a) the relation determining the wall resistance or wall shear stress $\sigma_w = \rho u^{*2}$ reads

$$\frac{U_0}{u^*} = A\ln(\delta^+) + B \qquad (7\text{-}46)$$

It is common to express the wall resistance in terms of a *friction coefficient* c_f, defined by

$$\sigma_w = c_f \tfrac{1}{2}\rho U_0^2 \qquad (7\text{-}47)$$

Between c_f and u^*/U_0 then exists the relation

$$c_f = 2\left(\frac{u^*}{U_0}\right)^2 \qquad (7\text{-}48)$$

Since

$$\delta^+ = \frac{u^*}{U_0}\frac{U_0\delta}{\nu} = \frac{u^*}{U_0}\mathbf{Re}_\delta$$

it follows from the relations (7-48) and (7-46) that

$$\frac{c_f}{2} = \left[\frac{A}{2}\ln\left(\frac{c_f}{2}\right) + A\ln\mathbf{Re}_\delta + B\right]^{-2} \qquad (7\text{-}49)$$

For the boundary layer along a *rough wall*, the relations corresponding to (7-46) and (7-49) read, respectively,

$$\frac{U_0}{u^*} = A \ln\left(\frac{\delta}{k}\right) + B_1 \tag{7-50}$$

and

$$\frac{c_f}{2} = \left[A \ln\left(\frac{\delta}{k}\right) + B_1\right]^{-2} \tag{7-51}$$

Furthermore, we can obtain expressions for the average mean velocity across the boundary layer $\bar{U}_{1,\text{av}}$, the displacement thickness δ_d, the momentum-loss thickness δ_m, and so for the shape factor H. They read

Smooth wall:
$$\frac{\bar{U}_{1,\text{av}}}{u^*} = A \ln \delta^+ + B - A = \frac{U_0}{u^*} - A \tag{7-52a}$$

Rough wall:
$$\frac{\bar{U}_{1,\text{av}}}{u^*} = A \ln \frac{\delta}{k} + B_1 - A = \frac{U_0}{u^*} - A \tag{7-52b}$$

Hence it follows that for a smooth as well as for a rough wall

$$\frac{\bar{U}_{1,\text{av}}}{U_0} = 1 - A\frac{u^*}{U_0} = 1 - A\sqrt{\frac{c_f}{2}} \tag{7-53}$$

Another consequence of the velocity-defect law is that the expressions for the displacement thickness and the momentum-loss thickness are the same for smooth and rough walls:

$$\frac{\delta_d}{\delta} = \int_0^1 d\xi_2\left(1 - \frac{\bar{U}_1}{U_0}\right) = A\frac{u^*}{U_0} \tag{7-54}$$

$$\frac{\delta_m}{\delta} = \int_0^1 d\xi_2 \frac{\bar{U}_1}{U_0}\left(1 - \frac{\bar{U}_1}{U_0}\right) = A\frac{u^*}{U_0}\left(1 - 2A\frac{u^*}{U_0}\right) = \frac{\delta_d}{\delta}\left(1 - 2\frac{\delta_d}{\delta}\right) \tag{7-55}$$

The shape factor H becomes

$$H = \frac{\delta_d}{\delta_m} = \left(1 - 2A\frac{u^*}{U_0}\right)^{-1} = \left(1 - 2\frac{\delta_d}{\delta}\right)^{-1} \tag{7-56}$$

For a smooth wall u^*/U_0 is a function of $\mathbf{Re}_\delta$, and for a rough wall it is a function of δ/k. Consequently, the quantities δ_d/δ, δ_m/δ, and H are functions of $\mathbf{Re}_\delta$ and of δ/k for a smooth wall and a rough wall, respectively.

According to the relation (7-55) the wall-friction velocity u^*/U_0 can be expressed in terms of δ_m/δ. With Eq. (7-9c) and Eq. (7-46) it is then possible to obtain a differential equation for δ, which might yield a solution for δ as a function of x_1. However, no solution in closed form can be obtained. The logarithmic velocity distribution does not yield a simple relation between δ and x_1.

Such a simple relation can be obtained if use is made of an old assumption concerning the velocity distribution, namely, that it can be represented by a simple power law (see, e.g., Ref. 14)

$$\frac{\bar{U}_1}{u^*} = C\left(\frac{u^* x_2}{\nu}\right)^{1/n} = C(\delta^+ \xi_2)^{1/n} \tag{7-57}$$

The exponent $1/n$ may still be a function of the Reynolds number. It appears that the logarithmic velocity distribution can be approximated by such a power law in a large fraction of the boundary-layer cross section. The velocity distribution (7-57), however, does not yield a velocity-defect law that is independent of wall conditions.

From Eq. (7-57) we obtain

$$\frac{\bar{U}_1}{U_0} = \xi_2^{1/n} \tag{7-58}$$

and

$$\frac{\bar{U}_{1,\text{av}}}{U_0} = \frac{n}{n+1} \tag{7-59}$$

Further,

$$\frac{U_0}{u^*} = C_1 \mathbf{Re}_\delta^{1/(n+1)}$$

so the relation for the wall-friction coefficient becomes

$$\frac{c_f}{2} = C_1^{-2} \mathbf{Re}_\delta^{-2/(n+1)} \tag{7-60}$$

For the displacement thickness δ_d and the momentum-loss thickness δ_m we obtain

$$\frac{\delta_d}{\delta} = \frac{1}{n+1} \quad \text{and} \quad \frac{\delta_m}{\delta} = \frac{n}{(n+1)(n+2)} \tag{7-61}$$

so

$$H = \frac{\delta_d}{\delta_m} = \frac{n+2}{n}$$

The increase in δ_d and δ_m with increasing x_1 can be approximately calculated from Eq. (7-9c):

$$\frac{d\delta_m}{dx_1} = C_1^{-2} \mathbf{Re}_\delta^{-2/(n+1)} = C_1^{-2}\left[\frac{n}{(n+1)(n+2)}\right]^{2/(n+1)} \left(\frac{U_0 \delta_m}{\nu}\right)^{-2/(n+1)}$$

In this equation n is still a function of the Reynolds number, and consequently of x_1. This dependence is assumed so small that the equation may be integrated as if n were constant. The integration then yields

$$\left(\frac{U_0 \delta_m}{\nu}\right)^{(n+3)/(n+1)} = \left(\frac{U_0 \delta_{m0}}{\nu}\right)^{(n+3)/(n+1)} + C_2 \frac{U_0}{\nu}(x_1 - x_{10}) \tag{7-62}$$

where

$$C_2 = C_1^{-2} \frac{n+3}{n+1} \left[\frac{n}{(n+1)(n+3)} \right]^{2/(n+1)}$$

If $\delta_{m0} = 0$ at $x_{10} = 0$, Eq. (7-62) may also be written

$$\frac{U_0 \delta_m}{\nu} = \left(C_2 \frac{U_0}{\nu} x_1 \right)^{(n+1)/(n+3)} \tag{7-62a}$$

Exactly the same expressions, but with different values for the constant C_2, are obtained for δ and δ_d.

Here again the reader may be referred to Rotta's review article[101] for further information, and to the more recent survey given in Ref. 102.

7-6 MEASUREMENTS ON MEAN-VELOCITY DISTRIBUTION

In the course of years an abundant amount of data on the mean-velocity distribution across a turbulent boundary layer has been collected. For the boundary layer along a *smooth flat plate* with zero pressure gradient, we have available the experimental data obtained by Ludwieg and Tillmann,[15] Schubauer and Klebanoff,[16] Klebanoff and Diehl,[17] Schultz-Grunow,[18] Allan and Cutland,[19] Hama,[20] and Smith and Walker.[103] These are all rather early measurements. Since then, of course, incidental measurements have been made by many other experimenters, in most cases for other purposes than for the measured velocity distribution itself. Since, in general, their results do not differ much from those obtained by the above experimenters, for our purpose we may suffice with showing a number of the earlier results only.

Except for Ludwieg and Tillman, and for Smith and Walker, all the experimenters made their measurements only in the fully turbulent region of the boundary layer.

The results of the measurements made by Ludwieg and Tillmann, Klebanoff and Diehl, and Schultz-Grunow with reference to the "constant-stress" wall region are shown in Fig. 7-9. In the fully turbulent region the data are closely grouped along the universal logarithmic velocity distribution (7-30) or (7-30a), at least for not too large values of $u^* x_2/\nu$. The measurements have been made at different values of the Reynolds number $\mathbf{Re}_\delta$. Yet in the range of x_2^+ up to roughly 1,000, they follow the same straight line in the semi-logarithmic plot, so confirming the underlying similarity assumption. This similarity of the flow in the wall region has been clearly shown also by the measurements of Smith and Walker,[103] who varied the Reynolds number $\mathbf{Re}_{\delta_d}$ between 5,300 and 61,500.

The corresponding values of u^*/U_0 varied from 0.040 to 0.032. However, the value of x_2^+ beyond which the experimental data diverge from the straight line depends on the Reynolds number. They all deviate in a similar fashion as shown in Fig. 7-9, indicating that the logarithmic distribution, when extrapolated to the outer

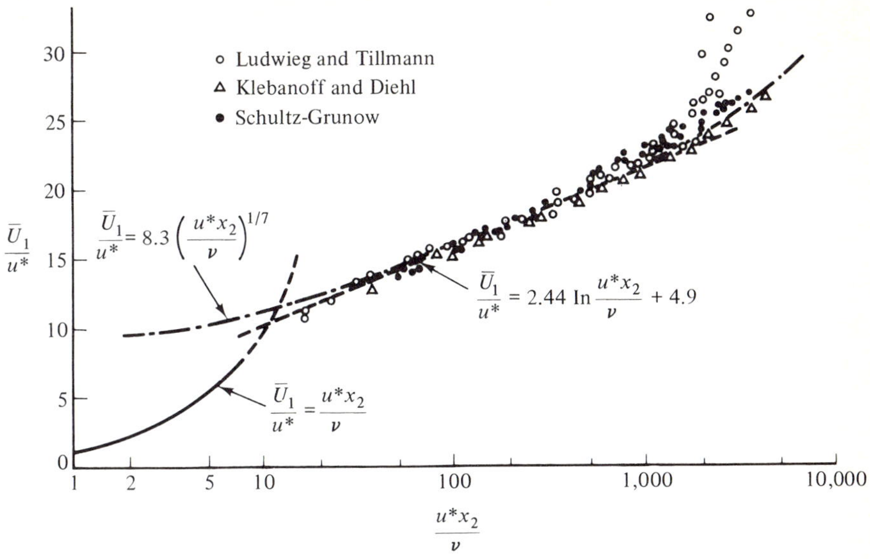

FIGURE 7-9
Mean-velocity distribution near smooth walls.

region, would yield too low values for the mean velocity. In Fig. 7-9 the values of the constants A and B in Eq. (7-30a) are those suggested by Clauser;[21] they are a good average when adapted to the experimental data obtained by the various investigators. Equation (7-30a) then reads

$$\frac{\bar{U}_1}{u^*} = 2.44 \ln \frac{u^*x_2}{\nu} + 4.9 \qquad (7\text{-}63)$$

Other investigators, however, prefer other values for these constants, which give better representations of their own experimental data. So there is appreciable variation in the values of A and B. Townsend[31] remarked that many of the observed data seem to indicate a value of B nearer to 7 than to the above value 4.9. Opinions about the value of the constant A or the "universal" constant $\varkappa$ of Eq. (7-30) are less divergent. In the earlier investigations a value of $A = 2.5$, or $\varkappa = 0.4$, based upon the experiments by Nikuradse[6] on pipe flow, was usually accepted.

The value of B seems to be connected with the thickness of the viscous sublayer. This becomes especially apparent from Rotta's theory on the velocity distribution close to the wall. For large values of u^*x_2/ν the formula (7-36) reduces to

$$\frac{\bar{U}_1}{u^*} = \frac{1}{\varkappa} \ln \frac{u^*x_2}{\nu} + \frac{1}{\varkappa}(\ln 4\varkappa - 1) + \frac{u^*\delta_l}{\nu}$$

from which we conclude that the constant B in the identical equation (7-30a) must read

$$B = \frac{1}{\varkappa}(\ln 4\varkappa - 1) + \frac{u^*\delta_l}{\nu}$$

Rotta assumed values of $\varkappa = 0.4$ and of $u^*\delta_l/\nu = 6.7$ on the basis of Nikuradse's experimental data. We then obtain a value of $B = 5.37$. On the other hand Nikuradse himself suggested a value $u^*\delta_l/\nu = 5$ to characterize the average thickness of the viscous sublayer. In that case, again with $\varkappa = 0.4$, a value of $B = 3.68$ would be obtained. With $\varkappa = 0.41$ and $B = 4.9$ suggested by Clauser would correspond $u^*\delta_l/\nu = 6.2$ according to Rotta's theory.

It should be appreciated that slight variations in the value of A, determining the slope of the straight line in the semilogarithmic plot, correspond with much greater variations in the value of B. Seeing the relatively large scatter of the experimental data, it is understandable that many present-day investigators again prefer the value of $A = 2.5$ instead of the value 2.44 given in Eq. (7-63). For the constant B then values between 5.2 and 5.5 are taken, with a slight preference for the latter value. However, a more recent re-evaluation of the available experimental evidence has led Huffman and Bradshaw[164] to conclude that A is a universal constant even at low Reynolds numbers ($\mathbf{Re}_\delta \lesssim 35{,}000$ say). For A they suggested again the value 2.44, but for a constant-pressure boundary-layer the value 5 for the constant B, which is very close to the value 4.9 given in Eq. (7-63). Universal values of these constants are in agreement with the existence of *Reynolds-number similarity* even at the low Reynolds numbers. This may indeed be correct for the mean-velocity distribution and for that part of the turbulence structure which is not directly affected by viscosity. Consequently one may expect that Reynolds number similarity will no longer hold true in the fine-structure of the turbulence and in the region adjacent the wall. Indeed, Huffman and Bradshaw found a non-negligible effect at the low Reynolds numbers on the viscous sublayer. Also at low Reynolds numbers the velocity-defect law becomes affected, which they attributed to the effect of the viscous superlayer (see Sec. 6-12). The experimental data shown in Fig. 7-9 refer to different values of $\mathbf{Re}_\delta$. No separate effect of the Reynolds number in the wall region, apart from its effect on u^*, can be observed.

In Fig. 7-10 the velocity distribution in the viscous sublayer and the adjacent buffer region is shown. Since no experimental data are available for the boundary-layer flow along a flat plate, data for the flow through a smooth two-dimensional channel and a smooth circular pipe obtained by Reichardt[12] and Laufer,[41] respectively, are inserted instead. According to present-day views, this should make no difference.

It may be concluded from this figure that, because of the scatter in the experimental data, any value of $u^*\delta_l/\nu$ between 5 and 6.7 may be assumed, though there seems to be a preference for the value of 5, or even slightly smaller values. Depending on the Reynolds number δ_l is of the order of 0.001 to 0.01δ.

Because of the relatively large scatter of the experimental data, no definite conclusion can be drawn concerning a preference for any of the semi-empirical relations as suggested by Reichardt, Rotta, Deissler, Van Driest, Rannie, and Spalding. None of these relations appear to be in strong contradiction with the experimental data.

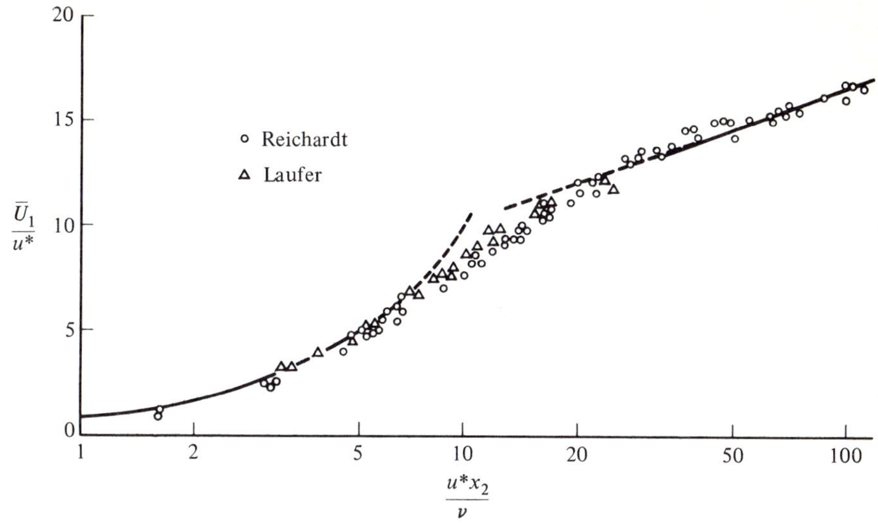

FIGURE 7-10
Velocity distribution in the buffer region of a turbulent boundary layer, according to experimental data by Reichardt for channel flow and by Laufer for pipe flow. (*Reichardt, H.*,[12] *Laufer, J.*[41])

The thickness of the buffer region as determined by the value of δ_t^+, is difficult to estimate from the experimental data. Figure 7-10 seems to suggest a value of $\delta_t^+ \simeq 30$. However, it might well be that still higher values are required for the fully turbulent flow condition with negligible direct viscosity effect. For instance, as will be shown later the eddy viscosity ϵ_m assumes a linear variation with x_2 only at values of x_2^+ much greater than 30; nearer $x_2^+ = 50$ to 60.

In the case of a wall of extreme roughness, where the height of the roughness elements is not very small compared with δ, a logarithmic velocity-distribution is only obtained at much higher values of x_2^+.

As Fig. 7-9 shows, beyond $u^*x_2/\nu = 500$ to 1,000, the experimental data diverge from the logarithmic velocity distribution; the actual velocities become greater than they would be according to the logarithmic velocity distribution. In the same figure a curve according to the simple power law (7-57) is also given, namely

$$\frac{\bar{U}_1}{u^*} = 8.3\left(\frac{u^*x_2}{\nu}\right)^{1/7} \tag{7-64}$$

The value $C = 8.3$ is obtained if this "power-law" velocity distribution is made to fit the logarithmic velocity distribution in the overlapping region between $u^*x_2/\nu = 100$ and 1,000. The value $n = 7$ is obtained if Blasius's resistance law for the flow along a smooth plate

$$c_f \propto \left(\frac{U_0\delta}{\nu}\right)^{-1/4}$$

[see Eq. (7-60)] is assumed.

It may be inferred from Fig. 7-9 that the above power law for the velocity distribution gives not only the same satisfactory result as the logarithmic velocity distribution in the region $u^*x_2/\nu = 100$ to 1,000 but even an improvement in the region beyond $u^*x_2/\nu = 1{,}000$. There it seems to follow reasonably closely the experimental data obtained by Klebanoff and Diehl.

The values given in the literature for the constants C and n in Eq. (7-57) again differ widely. On the assumption that Blasius's resistance law for smooth pipes applies to the flow along a smooth plate, Schlichting[14] obtained a value $C = 8.74$; Klebanoff and Diehl propose the value $C = 8.16$. The constants C and n depend on the Reynolds number. Blasius's resistance law is only approximately independent of the Reynolds number and then only in a certain Reynolds-number range. According to Wieghardt[22] there is a substantial increase in the values of both C and n with increasing Reynolds number. Yet Wieghardt suggests the same value of $n = 7.7$ for the flow along a smooth plate with positive, negative, and zero pressure gradients for not large values of u^*x_2/ν. Clauser[21] came to the conclusion that no universal value can be assigned to C and n, since n for instance may vary from 3 to 10 for the various velocity distributions considered.

For large values of u^*x_2/ν the logarithmic velocity distribution deviates from the actual velocity distribution; apparently it no longer holds in the outer region of the boundary layer. For that region it may be more appropriate to consider the velocity defect $(U_0 - \bar{U}_1)/u^*$ rather than $\bar{U}_1/u^*$. For the overlapping region, where the logarithmic velocity distribution applies, we also have from the relation (7-30a)

$$\frac{U_0 - \bar{U}_1}{u^*} = -A \ln \frac{x_2}{\delta} \tag{7-65}$$

But, since the value of U_0/u^* obtained from (7-30a) for $x_2 = \delta$ does not give the correct actual value of U_0/u^*, we cannot expect a good representation of the measured velocity distribution by the logarithmic distribution if we put in Eq. (7-65) the actual values of U_0/u^*. In order to take account of the difference between the actual value of U_0/u^* and the value according to Eq. (7-65), we have to complete the latter equation with an additional correction term. Thus instead of Eq. (7-65) we write

$$\frac{U_0 - \bar{U}_1}{u^*} = -A \ln \frac{x_2}{\delta} + B^* \tag{7-65a}$$

Figure 7-11 shows a logarithmic plot of universal velocity distributions which was obtained by Clauser[21] from data determined by Klebanoff and Diehl, Schultz-Grunow, Hama, and others, for smooth plates as well as for rough plates.

For $x_2/\delta < 0.15$ the data are nicely grouped (represented by the shaded strip //////) around the logarithmic relation.

$$\frac{U_0 - \bar{U}_1}{u^*} = -2.44 \ln \frac{x_2}{\delta} + 2.5 \tag{7-66}$$

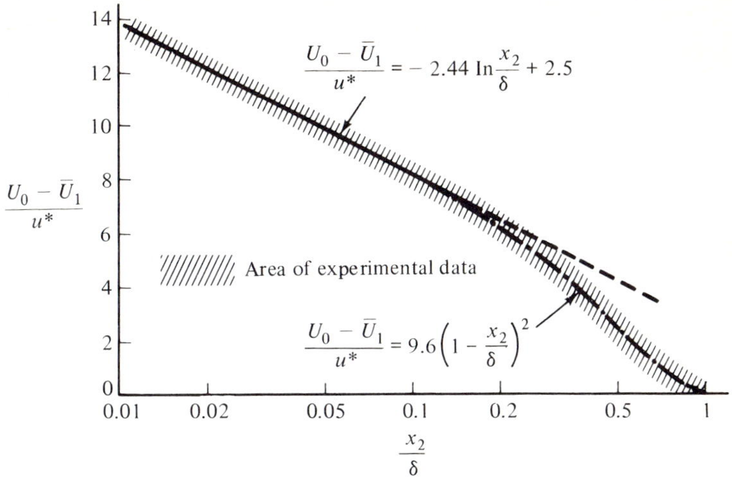

FIGURE 7-11
Logarithmic plot of velocity distributions in the outer part of turbulent boundary layers. (After: *Clauser, F. H.*[21])

However, if one would prefer the value $A = 2.5$, a value $B^* = 3.7$ should be taken.

According to Fig. 7-9 the value of $x_2/\delta \simeq 0.15$ beyond which the logarithmic distribution deviates from the actual velocity distribution, seems to correspond roughly with the value $u^*x_2/\nu = 750$. Actually, however, the value of u^*x_2/ν marking the above deviation should depend on the Reynolds number, in that it increases with increasing Reynolds number, roughly proportional with $(\mathbf{Re}_\delta)^{n/(n+1)}$. The dependency on the Reynolds number has already been shown by Clauser,[21] and more recently clearly demonstrated by the measurements by Smith and Walker.[103]

Hama[20] has proposed the following simple empirical formula for the mean-velocity distribution in the region $x_2/\delta > 0.15$:

$$\frac{U_0 - \bar{U}_1}{u^*} = 9.6\left(1 - \frac{x_2}{\delta}\right)^2 \tag{7-67}$$

This velocity distribution is also shown in Fig. 7-11, from which it may be concluded that a very satisfactory agreement indeed with experimental data is obtained, at any rate for practical purposes.

If Hama's formula is assumed to be universal, then it follows from Eqs. (7-65a) and (7-67), that the value of x_2/δ marking the boundary between the wall region and the outer region is independent of the Reynolds number.

If we want to apply the power-law velocity distribution (7-58) to the velocity-defect distribution, the effect of the Reynolds number, which in the other relations is included in the wall-friction velocity u^*, should be found in the value of the exponent n. In Fig. 7-9 the curve with $n = 7$ is shown to agree satisfactorily with the experimental data of Klebanoff and Diehl. These data refer to a Reynolds number

$\mathbf{Re}_\delta = 152{,}000$, which corresponds roughly to $\mathbf{Re}_{x_1} = 10^7$. In Fig. 7-12, logarithmic plots similar to that of Fig. 7-11 are given for the experimental data obtained by Schultz-Grunow with reference to $\mathbf{Re}_{x_1} = 0.7 \times 10^6$ and to $\mathbf{Re}_{x_1} = 7 \times 10^6$. There is agreement with the power-law velocity distribution with $n = 7$ only for the data corresponding to $\mathbf{Re}_{x_1} = 7 \times 10^6$; for the data corresponding to $\mathbf{Re}_{x_1} = 0.7 \times 10^6$ a value of $n = 5.3$ should be taken. This result clearly demonstrates the effect of the Reynolds number on the value of the exponent n. It is indeed impossible to regard the power-law velocity distribution as a universal distribution.

On the other hand, since the experimental data of Fig. 7-11 refer to different values of the Reynolds number and to smooth and rough walls, the velocity-defect ratio $(U_0 - \bar{U}_1)/u^*$ is, apparently, a universal function of $\xi_2 = x_2/\delta$:

$$\frac{U_0 - \bar{U}_1}{u^*} = f(\xi_2) \tag{7-68}$$

But

$$\frac{U_0 - \bar{U}_1}{U_0} = \frac{u^*}{U_0} f(\xi_2) \tag{7-69}$$

is not a universal function of ξ_2.

For the displacement thickness δ_d and the momentum-loss thickness δ_m we then obtain [compare Eqs. (7-54) to (7-56)]

$$\frac{\delta_d}{\delta} = \frac{u^*}{U_0} \int_0^1 d\xi_2\, f(\xi_2) = I_1 \frac{u^*}{U_0} \tag{7-70}$$

and

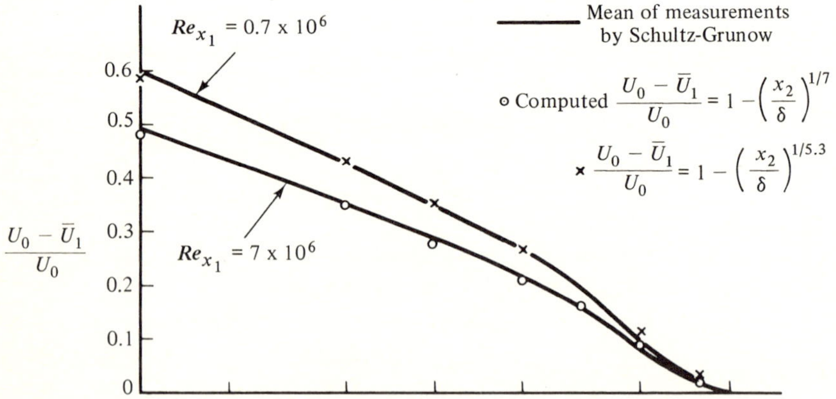

FIGURE 7-12
Comparison of power-law velocity distributions with measurements by Schultz-Grunow.[18]

$$\frac{\delta_m}{\delta} = \frac{u^*}{U_0}\int_0^1 d\xi_2\, f(\xi_2)\left[1 - \frac{u^*}{U_0} f(\xi_2)\right]$$

$$= I_1 \frac{u^*}{U_0} - I_2 \frac{u^{*2}}{U_0^{\,2}} \tag{7-71}$$

Hence

$$\frac{U_0 - \bar{U}_1}{U_0}\frac{\delta}{\delta_d} = \frac{1}{I_1} f(\xi_2) \tag{7-72}$$

must be a universal function of ξ_2, whereas the shape factor

$$H = \frac{\delta_d}{\delta_m} = \left(1 - \frac{u^*}{U_0}\frac{I_2}{I_1}\right)^{-1} \tag{7-73}$$

is still a function of the conditions at the wall. Since I_1 and I_2 are universal constants, the ratio I_2/I_1 is also a universal constant for flows without a pressure gradient.

Goethals[23] determined the function $(1 - \bar{U}_1/U_0)(\delta/\delta_d)$ for different types of flow, even including flows with positive and negative pressure gradients. The differences among the functions of the various types of flow were within the experimental scatter of the data belonging to each type of flow.

Hama[20] plotted the shape factor H as a function of u^*/U_0, using the data for a smooth and rough plate. This plot is shown in Fig. 7-13. The value $I_2/I_1 = 6.1$ was used for the curve in this figure, though a value of $I_2/I_1 = 6.3$ would better fit the experimental data.

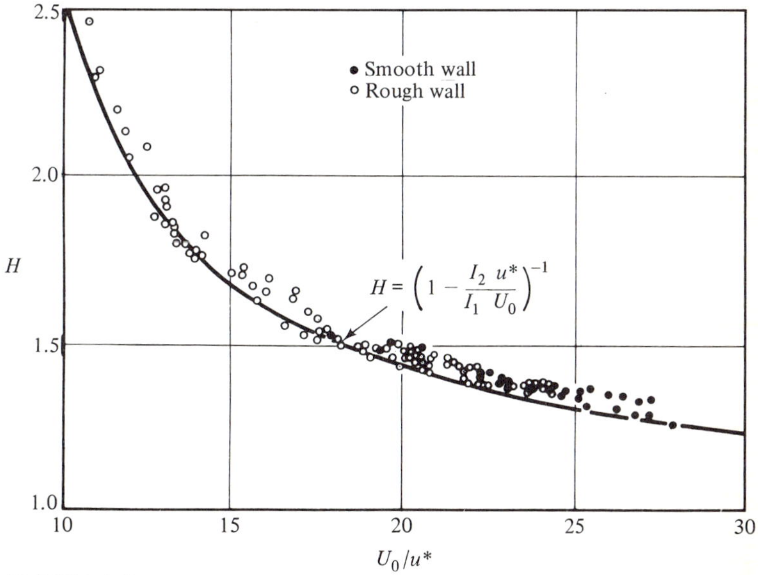

FIGURE 7-13
Shape factor H as a function of u^*/U_0. (After: *Hama, F. R.*[20])

It may be remarked that, if the logarithmic velocity distribution (7-65) is used, the value $I_2/I_1 = 2A = 4.88$ is obtained [compare Eq. (7-56)] and, if the logarithmic velocity distribution (7-65a) is used, we obtain

$$\frac{I_2}{I_1} = \frac{2A^2 + 2AB^* + B^{*2}}{A + B^*} = 6.15$$

We have noticed that for $x_2/\delta > 0.15$ the logarithmic velocity distribution (7-66) deviates from the actual velocity distribution. Following a suggestion by Millikan,[7] we may introduce a correction function $h(\xi_2)$ such that

$$\frac{U_0 - \bar{U}_1}{u^*} = -A \ln \xi_2 + B^* + h(\xi_2) \qquad (7\text{-}74)$$

is equal to the actual velocity distribution $f(\xi_2)$. This correction function $h(\xi_2)$ is shown in Fig. 7-14.

We have shown that the velocity-defect law applies to smooth and rough walls. The only effect of the wall condition is found in the value of u^*. But the velocity distribution close to the wall is directly related to the kind and magnitude of the wall roughness. Now, there is an uncertainty in the estimation of the value of x_2 where the velocity $\bar{U}_1$ is on the average zero along the wall. Though it is usually assumed that $\bar{U}_1 = 0$ at $x_2 = 0$, the actual point where $\bar{U}_1 = 0$ may be somewhere between $x_2 = 0$ and $x_2 = k$, the average height of the roughnesses. As long as the thickness of the viscous sublayer $\delta_l \gg k$, the deviation from the velocity distribution valid for a smooth wall will be small. There is an effect of the roughnesses, which is found in a decrease in the value of δ_l with increasing roughness k, since these roughnesses cause disturbances in the flow which result in an extension of the buffer region toward smaller values of x_2.

From Nikuradse's experiments on flow through circular pipes with walls of uniform and sand-grain roughness, Rotta[11] deduced a relationship between δ_l and k,

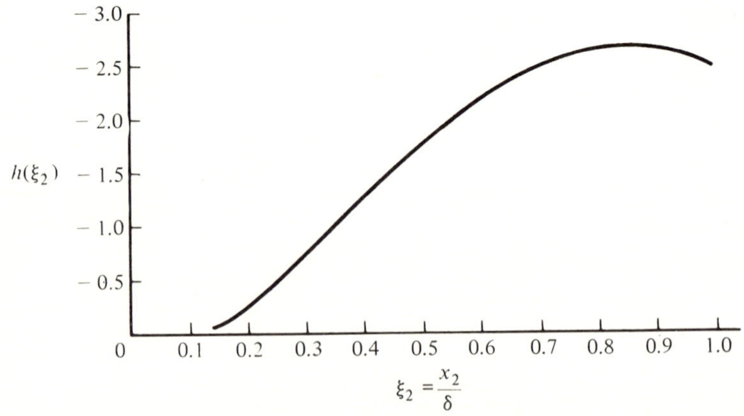

FIGURE 7-14
Correction function $h(\xi_2)$ for the logarithmic velocity-defect distribution.

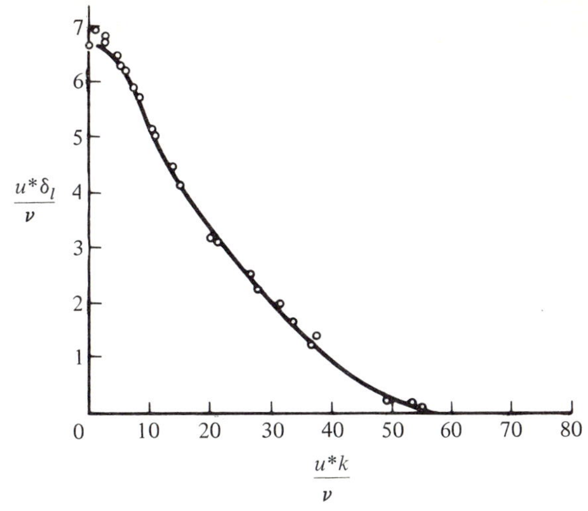

FIGURE 7-15
Effect of wall roughness on δ_l according to Nikuradse's experiments with uniform sand-grain roughness. (From: *Rotta, J.,*[11] *by permission of the Springer Verlag.*)

as shown in Fig. 7-15. From this figure it may be concluded that for $u^*k/\nu < 5$ the effect on $u^*\delta_l/\nu$ is small, but that for $u^*k/\nu > 55$ we have a fully rough-wall condition, with no effective viscous sublayer. These values of 5 and 55, respectively, refer to Nikuradse's experiments. They may differ, and even appreciably, for other types of roughness. But at any rate it may be expected that there will be a lower value of u^*k/ν below which the wall is fluid-dynamically smooth and no deviation from the perfect smooth-wall condition is found.†

This fact led Hama[20] to rewrite Eq. (7-32a) as follows:

$$\frac{\bar{U}_1}{u^*} = A \ln \frac{u^* x_2}{\nu} + B - \left(A \ln \frac{u^* k}{\nu} + B - B_1 \right)$$

This shows that the effect of the wall roughness is expressed as a vertical shift in the velocity distribution for a smooth wall with a value

$$\left(\frac{\bar{U}_1}{u^*}\right)_s - \left(\frac{\bar{U}_1}{u^*}\right)_r = \frac{\Delta \bar{U}_1}{u^*} = A \ln \frac{u^* k}{\nu} + B - B_1 \qquad (7\text{-}75)$$

depending on the kind and magnitude of the roughnesses. For large values of u^*k/ν, the first term on the right-hand side becomes predominant, and $\Delta\bar{U}_1/u^*$ becomes proportional to $\ln(u^*k/\nu)$ with a constant of proportionality equal to A. The fully rough-wall condition is then reached.

† It may be of interest to remark that the value of $u^*k/\nu \simeq 5$ might correspond with the generation of eddies shed by the roughness elements. For a cylinder or sphere discrete vortices are shed when the Reynolds number exceeds the value 40 to 50.

When we apply this value to a roughness element with height k, taking the velocity $\bar{U}_1^+ = u^*k/\nu$ at the top of the roughness element, then from $\bar{U}_1 k/\nu = 20$ to 25, there would follow $u^*k/\nu = (20)^{1/2}$ to $(25)^{1/2} = 4.5$ to 5.

Clauser[21] and Hama[20] have determined the value of $\Delta\bar{U}_1/u^*$ for quite different types of roughnesses. The results are shown in Fig. 7-16. It may be noticed that the above-mentioned values $u^*k/\nu = 5$ and 55 are valid only for uniform sand-grain roughness and that $\Delta\bar{U}_1/u^*$ is by no means a universal function of u^*k/ν. For large values of u^*k/ν the slope of the straight lines is equal to $A = 2.44$, but the values of $B - B_1$ occurring in $\Delta\bar{U}_1/u^*$ may still differ for the various types of roughness. In the region of intermediate roughness, where the condition at the wall is neither fluid-dynamically smooth nor fully rough, the course of the $\Delta\bar{U}_1/u^*$ curve is entirely different depending on the roughness pattern.

This is not very surprising since the roughness effect is not only determined by some average height k of the roughness elements, but also by the size distribution and shape of the roughness elements, and their density distribution over the wall surface. It is customary in technical applications to introduce an *equivalent sand roughness* k_s. By definition it is the size of uniform sand-grains that produces the same wall shear stress as the actual roughness under the same flow conditions. When the roughness elements are not small, there is a difficulty in fixing the origin where $\bar{U}_1 = 0$, as mentioned earlier. If we assume $x_2 = 0$ at a smooth surface to which the rough elements are supposed to be attached, this origin will be situated somewhere between $x_2 = 0$ and $x_2 = k$. Einstein and El-Samni[104] already observed

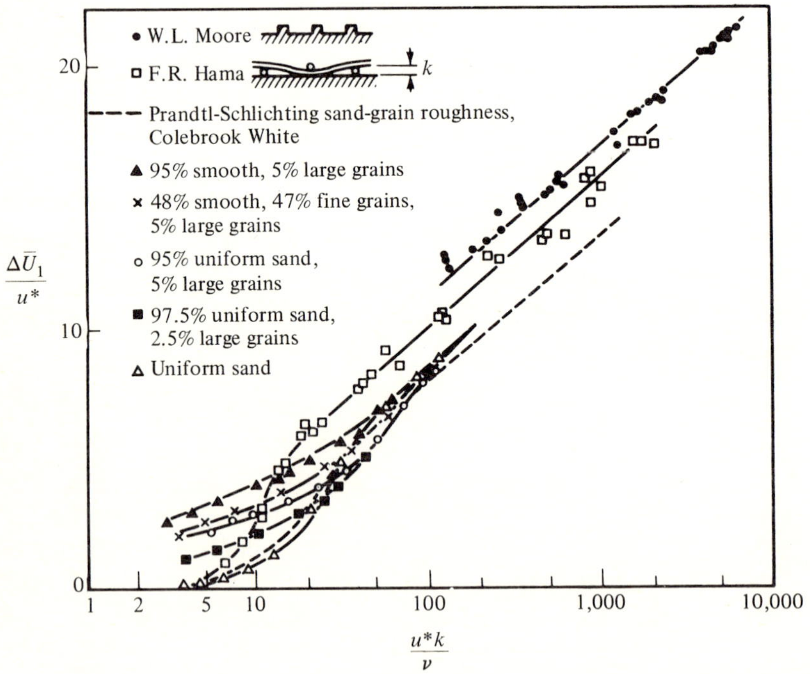

FIGURE 7-16
Effect of wall roughness on the shift $\Delta\bar{U}_1/u^*$ of the velocity-distribution profile. (From: *Clauser, F. H.,*[21] *by permission of Academic Press, Inc.*)

that the "origin" of the logarithmic velocity distribution was situated below the top of the roughness elements consisting of hemispherical caps, and of pebbles, with which they carried out their experiments. This has been confirmed by others. For a more recent investigation, see, e.g., ref. 165. The location of such an apparent origin depends on the geometry of the roughness elements. For spheres, pebbles and sand roughness of a uniform nature the origin is roughly located $0.25k_s$ below the tops of the roughness elements. For a given roughness pattern this apparent origin may be obtained by shifting the ordinate of this origin in such a way that the best straight line in a semi-logarithmic plot of the $\bar{U}_1$-velocity data is obtained. In Fig. 7-16, k is the actual height of the roughness elements. For the roughness pattern investigated by Moore,[105] like Einstein and El-Samni, an apparent origin is found below the tops of the rectangular roughness elements. Recent experiments by Perry, Schofield, and Joubert[106] on rough-wall boundary layers with a similar roughness pattern to that used by Moore, but with much more closely packed elements, showed that the location k_e of the apparent origin below the crests, decreases when the ratio: width to height of the gap decreases. They obtained for $\Delta\bar{U}_1/u^*$ a relation similar to Eq. (7-75), namely

$$\frac{\Delta\bar{U}_1}{u^*} = 2.5 \ln \frac{u^* k_e}{\nu} + C$$

where C still depends on the geometry, with an average value of -0.4. When compared with Moore's data in Fig. 7-16, for which roughness configurations according to Perry *et al.* $k_e \simeq k$, the data by Perry *et al.* are slightly below Moore's data. For sand roughness a value of $B - B_1 \simeq -3.5$ may be obtained from Fig. 7-16, a value close to that to be found from measurements by Blake.[145]

As has been remarked earlier the distance beyond which the logarithmic velocity-distribution applies may become considerably greater than that for a smooth boundary layer. This may also be concluded from Fig. 7-16.

It is possible to follow a different procedure based on the use of the equivalent sand roughness concept. From Eq. (7-50) or Eq. (7-51) the value of k_s can be determined for a given roughness. This value of k_s can then be used, for the fully rough-wall condition, in Eq. (7-32a), written as follows

$$\bar{U}_1^+ = A \ln \frac{x_2 + k_e}{k_s} + B_1 \qquad (7\text{-}76)$$

where x_2 is measured from the crests of the roughness elements, and k_e is determined from the best straight-line representation in a semi-logarithmic plot. It may then be expected, when writing

$$\bar{U}_1^+ = A \ln \frac{u^*(x_2 + k_e)}{\nu} + B - \Delta\bar{U}_1^+ \qquad (7\text{-}77)$$

that the differences between $\Delta\bar{U}_1^+$ as a function of k_s^+ will be much smaller for the various types of roughness than shown in Fig. 7-16.

We have shown in the previous section that the increase in the boundary-layer thickness with increasing x_1 may be obtained if u^*/U_0 or c_f is known. Ludwieg and Tillmann[15] have succeeded in obtaining, from accurate measurements of the local wall shear stress σ_w for the flow along a smooth wall, the following empirical relation:

$$c_f = 0.246 \times 10^{-0.678H}\left(\frac{U_0\delta_m}{\nu}\right)^{-0.268} \tag{7-78}$$

valid in the range $U_0\delta_m/\nu = 10^3$ to 10^4.

Since the shape factor H is a function of u^*/U_0, that is, of c_f, the relation is still too involved to be used for the calculation of δ_m as a function of x_1 in an explicit form. However, in the range of $U_0\delta_m/\nu$ considered, the shape factor H varies only slightly and may be approximated by an average value. For this average we obtain $H \simeq 1.36$; so the expression for c_f becomes

$$c_f \simeq 0.03\left(\frac{U_0\delta_m}{\nu}\right)^{-0.268} \tag{7-79}$$

This corresponds to a value of $n = 6.45$ if a power-law function is assumed for the velocity distribution [see Eq. (7-60)]. On the other hand, Eq. (7-61) would yield $n = 5.56$.

With $n = 6.45$ we obtain for the simplified relation (7-62a)

$$\frac{U_0\delta_m}{\nu} = 0.045\left(\frac{U_0x_1}{\nu}\right)^{0.79} \tag{7-80}$$

It must be emphasized that this relation is valid only in the range $U_0\delta_m/\nu = 10^3$ to 10^4, and then only approximately. But it may be used to obtain an estimate of the increase in boundary-layer thickness with downstream distance. Since the value of n varies with the Reynolds number, so will the constant C_2 and the exponent of U_0x_1/ν of the relation (7-62a). For $n = 5$ the exponent becomes 0.75 and for $n = 7$ it becomes 0.80; so the effect of the Reynolds number on the exponent of U_0x_1/ν is only slight.

7-7 MEASUREMENTS ON TURBULENCE QUANTITIES IN A BOUNDARY LAYER

Extensive measurements on turbulence quantities have been made by Townsend, Schubauer and his collaborators, and by Corrsin and Kistler.

Townsend[28] used one of the walls of a windtunnel with a $0.38 \times 0.38\text{-m}^2$ section, the boundary layer being slightly thickened and stabilized by introducing a projection, 2 mm high and 50 mm long in the downstream direction, at the entrance of the working section. The wall surface was provided with a smooth sheet of Bakelite. The measurements were made at sufficient distance from this entrance to ensure equilibrium conditions in the boundary layer. The free-stream velocity U_0 was 12.80 m/s, and the free-stream turbulence had a relative intensity of $u_1'/U_0 = 0.06$ per cent, with $u_2'/U_0 = u_3'/U_0 = 0.15$ per cent at the entrance section. The Reynolds number of the boundary layer, based upon the displacement thickness δ_d, varied between 3,630 and 5,080, depending on the distance from the entrance.

Schubauer and collaborators[16,17,27] made most of their measurements in a boundary layer along a smooth flat aluminum plate 6 mm thick with a symmetrical and pointed leading edge, mounted vertically and centrally in the octagonal test section with 1.36 m distance between the walls. Here, too, the boundary layer was artificially thickened by covering the first 0.6 m of the plate with sandpaper. The free stream had a relative turbulence intensity of roughly 0.03 per cent. Most of the measurements discussed below were made at a section where the boundary layer had a thickness of 75 mm, with a free-stream velocity of 15 m/s. It was thought that at that section the boundary layer had reached equilibrium conditions.

Corrsin and Kistler,[29] like Townsend, used a wall of a windtunnel with 0.60×0.60-m working section, but roughened by corrugated paper along its entire length. The corrugations, set perpendicular to the flow, were nearly sinusoidal, with about 8 mm wavelength and 2 mm amplitude. The free stream had a velocity of 11.2 m/s and a relative turbulence intensity of roughly 0.06 per cent. The flow in the boundary layer corresponded to a "fully rough" wall condition. At the end of the test section it had a thickness of about 90 mm.

Figure 7-17 shows the intensities of the three turbulence-velocity components

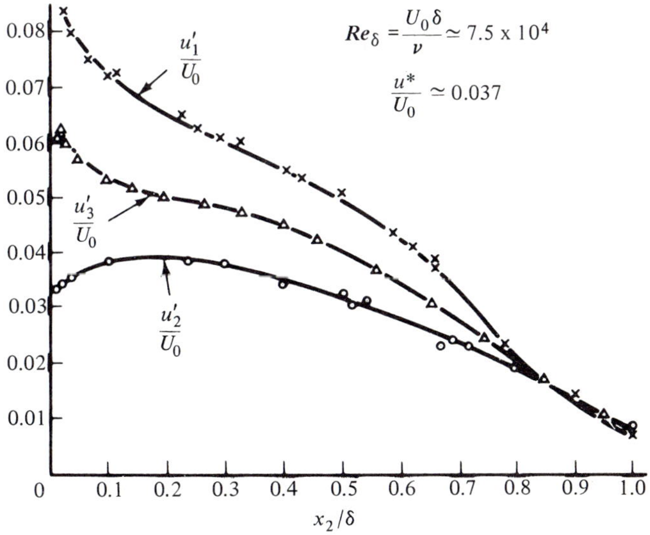

FIGURE 7-17
Relative turbulence intensities in a boundary layer along a smooth wall with zero pressure gradient. (From: *Klebanoff, P. S.,*[27] *by permission of the National Advisory Committee for Aeronautics.*)

relative to the free-stream velocity, according to measurements by Klebanoff[27] in a boundary layer along a smooth wall. It may be noticed that the intensities of the three turbulence velocities differ appreciably from one another over the main inner part of the boundary layer. The degree of anisotropy increases toward the wall. The intensity of the axial turbulence-velocity component has the highest value; that of the transverse component perpendicular to the wall has the smallest value.

As Fig. 7-18 shows, the axial component does reach a maximum value $u_1' \simeq 3u^*$ in a region very close to the wall, roughly at $x_2^+ = 25$. The two transverse components could not be measured so close to the wall. However, as will be shown later, the spanwise component will decrease to zero at the wall in a fashion similar to that of the axial component, i.e. linearly with the distance. On the other hand the u_2-component must decrease at least quadratically with the distance x_2, because of the condition imposed by the equation of continuity. Since such a quadratic behavior can be shown only for the limiting condition $x_2 \to 0$, an experimental verification seems hardly possible. Anyway no measurements of the u_2-component so close to the wall are known that clearly show said quadratic behavior. Since in turbulent pipe flow or channel flow measurements much closer to the wall could be made, so that corresponding data on turbulence are available, we will come back to this subject later when considering the wall region in turbulent pipe flow.

In Fig. 7-18 we have also shown the intensity of the axial turbulence-velocity component relative to the local mean velocity. This relative intensity reaches a

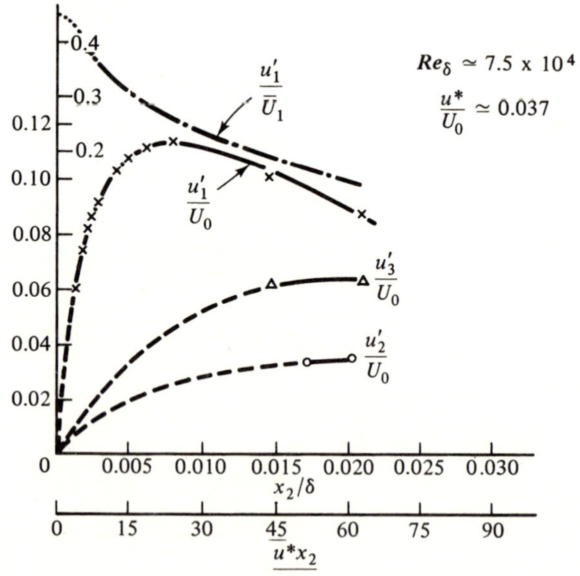

FIGURE 7-18
Relative turbulence intensities in the constant-stress layer of a boundary layer along a smooth wall. (From: *Klebanoff, P. S.,*[27] *by permission of the National Advisory Committee for Aeronautics.*)

constant value in the viscous sublayer; from which it may be inferred that the flow in this layer is still turbulent-fluctuating in character, though viscous, and that the velocities vary linearly with distance from the wall. When extrapolated to the wall, the curve for $u'_1/\bar{U}_1$ seems to suggest a limiting value of roughly 0.4.

As will be discussed later, the limiting value appears to be smaller than 0.4.

A similar distribution of turbulence intensities across the boundary layer was obtained by Corrsin and Kistler[29] in the case of a *rough wall.* The measurements, however, were performed outside the wall region. A comparison of their measurements, which are shown in Fig. 7-19, with those given in Fig. 7-17, shows that the relative turbulence intensities measured by Corrsin and Kistler are much higher than those measured by Klebanoff. But the ratio between the relative turbulence intensities for the rough-wall boundary layer and those for the smooth-wall boundary layer is roughly equal to the ratio between the corresponding wall shear stresses. Consequently, if the relative turbulence intensities are reduced to the same wall shear stress, practically the same values are obtained for the smooth-wall and rough-wall boundary layers, as they should be.

From Klebanoff's data on the relative turbulence intensities, the distribution of the turbulence kinetic energy $\overline{q^2}/U_0{}^2$ across the boundary layer has been calculated.

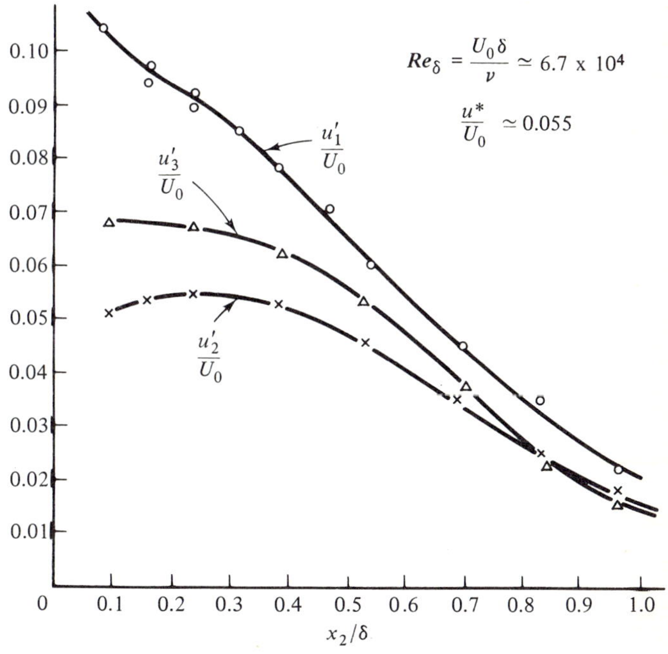

FIGURE 7-19
Relative turbulence intensities in a boundary layer along a rough wall with zero pressure gradient. (From: *Corrsin, S., and A. L. Kistler,*[29] *by permission of the National Advisory Committee for Aeronautics.*)

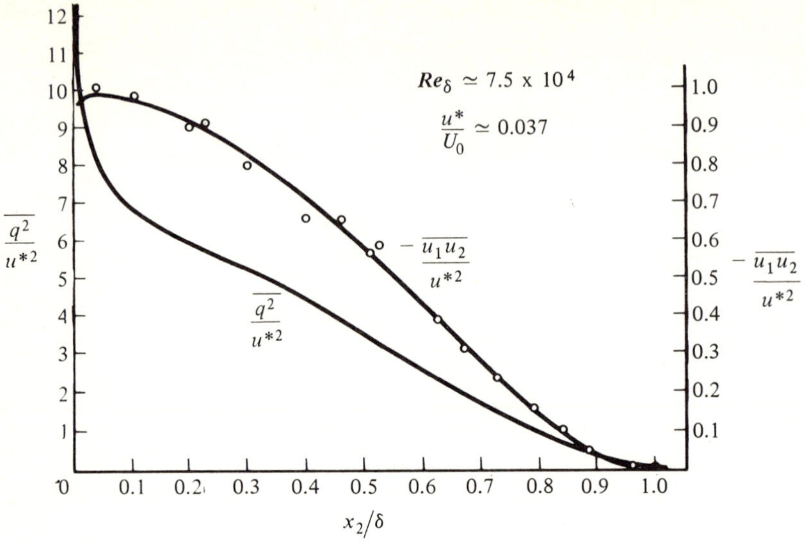

FIGURE 7-20
Distribution of turbulence kinetic energy and turbulence shear stress across a boundary layer. (From: *Klebanoff, P. S.,*[27] *by permission of the National Advisory Committee for Aeronautics.*)

The result is shown in Fig. 7-20, together with the distribution of the turbulence shear stress measured directly by Klebanoff.[27] It may be noticed that, in the region $x_2/\delta < 0.1$, the shear stress is practically constant (see also Fig. 7-21, where the distribution of $-\overline{u_1 u_2}/u^{*2}$ in this region is given). Hence, the assumption of constant shear stress in the wall region seems justified.

If we consider only the region not too close to the wall, a similarity between the turbulence-shear-stress distribution and the turbulence-kinetic-energy distribu-

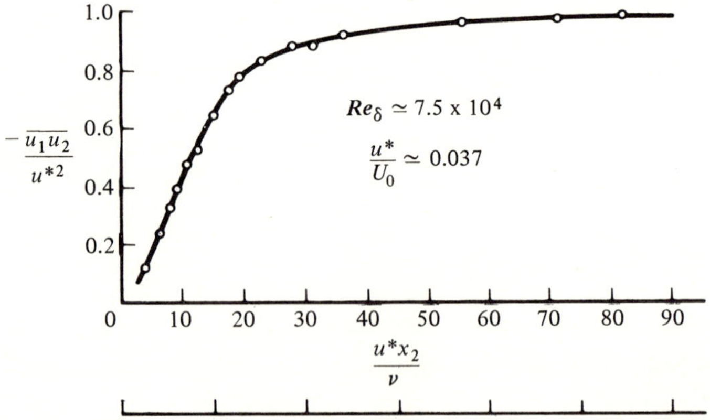

FIGURE 7-21
Distribution of turbulence shear stress in the wall region of a boundary layer. (From: *Schubauer, G. B.,*[30] *by permission of the American Institute of Physics.*)

tion may be observed. Complete similarity between the two distributions, that is, a constant ratio between the two quantities, would be obtained if Von Kármán's similarity hypothesis concerning the structure of turbulence were true. The distribution of the ratio $-\overline{u_1u_2}/\overline{q^2}$ is given in Fig. 7-22; it is nearly constant across a large portion of the boundary layer.

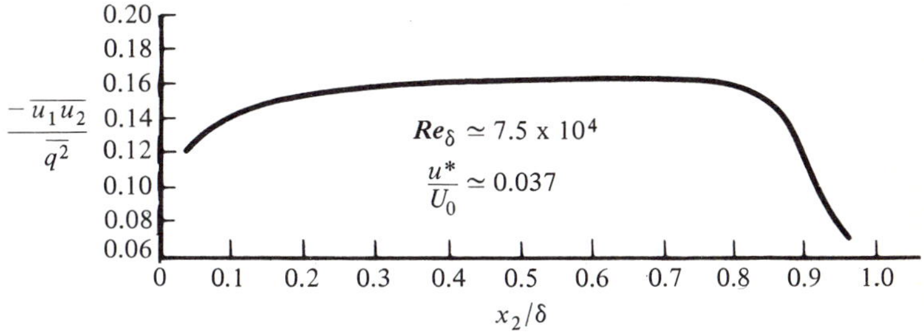

FIGURE 7-22
Ratio between turbulence shear stress and turbulence kinetic energy across a boundary layer.

Some of Prandtl's assumptions also yield results in agreement with the above. If we introduce the coefficient of correlation $\mathbf{R}_{1,2} = \overline{u_1u_2}/u_1'u_2'$, the ratio $\overline{u_1u_2}/\overline{q^2}$ may be written $\mathbf{R}_{1,2}u_1'u_2'/\overline{q^2}$. Since Prandtl assumed $\mathbf{R}_{1,2} = \text{const.}$ and $u_1' \propto u_2' \propto q' \propto \mathrm{l}_m \partial\bar{U}_1/\partial x_2$, the ratio $\overline{u_1u_2}/\overline{q^2}$ must be constant. However, the converse conclusion, namely, that from the constant ratio would follow the correctness of Prandtl's assumption, is not correct. For instance, the distribution curves of the relative turbulence intensities given in Figs. 7-17 and 7-19 show that the ratio u_1'/u_2' is not constant across the boundary layer, contrary to Prandtl's assumption.

The distribution of an eddy viscosity ϵ_m may be obtained from the turbulence-shear-stress distribution and the mean-velocity distribution, according to the relation

$$-\frac{\overline{u_1u_2}}{U_0{}^2} = \frac{\epsilon_m}{U_0\delta}\frac{\partial\bar{U}_1/U_0}{\partial\xi_2}$$

$$(7\text{-}81)$$

or

$$-\frac{\overline{u_1u_2}}{u^{*2}} = \frac{\epsilon_m}{u^*\delta}\frac{\partial\bar{U}_1/u^*}{\partial\xi_2}$$

In Fig. 7-23 the distribution of $\epsilon_m/u^*\delta$, calculated from data published by Schubauer,[30] is given for the constant-stress layer. The course of this distribution near the wall seems to indicate that there $\epsilon_m/u^*\delta$ varies proportionally with $(x_2{}^+)^n$, where $n \simeq 2$, though it should be noted that this value of n is not very reliable, since it is based on too little data.

With increasing distance from the wall and in the fully turbulent region, the distribution of the eddy viscosity should approach a linear variation with distance

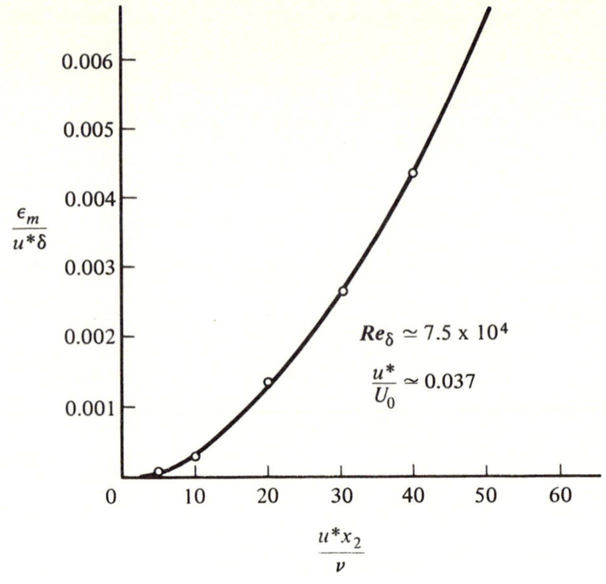

FIGURE 7-23
Distribution of eddy viscosity in the wall region of a boundary layer, calculated from Schubauer's data.[30]

from the wall. The course of the $\epsilon_m/u^*\delta$ curve, given in Fig. 7-24 and obtained from Klebanoff's[27] data, seems to support such a linear variation in the wall region. The eddy viscosity appears to attain a maximum value roughly at $x_2/\delta = 0.3$. In the same figure are also included values of $\epsilon_m/u^*\delta$ calculated from data obtained by Townsend.[28] The agreement with the values obtained from Klebanoff's data may be considered reasonable.

We mentioned in Sec. 7-1 the similarity between the flow conditions in the outer region of the boundary layer and in free turbulent flow; there is also similar interaction between turbulent flow and free stream at the outer boundary of the boundary layer. Such interaction is reflected, for instance, in the intermittent character of the flow. Intermittency factors have been obtained by Klebanoff[27] and by Corrsin and Kistler.[29]

Klebanoff obtained the intermittency factors from measured flattening factors (see Chap. 6). Corrsin and Kistler applied a different method: measuring the mean-square output of an on off signal that was triggered by passing the intermittent signal from a hot-wire anemometer through a gate.

The distribution of the intermittency factor Ω showed no difference from that found in free turbulent flows. This distribution could very well be described by the Gaussian error function

$$\Omega = \frac{1}{\sqrt{2\pi}} \int_{\alpha\xi_2+\beta}^{\infty} dz \exp\left(-\tfrac{1}{2}z^2\right) \qquad \text{(7-82)}$$

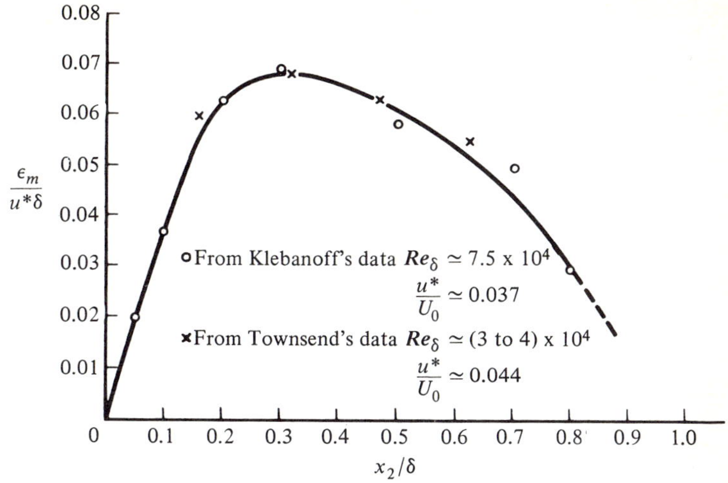

FIGURE 7-24
Distribution of eddy viscosity across a boundary layer, calculated from Klebanoff's[27] and Townsend's[28] data.

where

$$\alpha\xi_2 + \beta = (x_2 - \bar{x}_2^*)/(x_2^* - \bar{x}_2^*)'$$

$\bar{x}_2^*$ = average position of the interface between turbulent and non-turbulent fluid

$(x_2^* - \bar{x}_2^*)'$ = root-mean-square value of the difference between the instantaneous and average positions of the interface

The curves in Fig. 7-25 are such Gaussian error functions, adapted to Corrsin and Kistler's and to Klebanoff's data, respectively. There is a slight shift between the two curves. Corrsin and Kistler obtained a value of $\bar{x}_2^*/\delta = 0.8$, Klebanoff a value of 0.78. The difference is rather small and may be explained by the uncertainty in determining the value of the boundary-layer thickness δ. Corrsin and Kistler as well as Klebanoff found a standard deviation of roughly $(x_2^* - \bar{x}_2^*)' = 0.14\delta$. Since Corrsin and Kistler made the experiments at various sections in the boundary layer, the different points in Fig. 7-25 refer to these various stations, and the values $\bar{x}_2^*/\delta = 0.8$ and $(x_2^* - \bar{x}_2^*)'/\delta = 0.14$ obtained are average values. Actually the values changed slightly in the downstream direction.

Later measurements by other experimenters have, in general, confirmed the shape and position of the intermittency distribution curve as shown in Fig. 7-25.

For instance, a recent investigation by Kovasznay, Kibens, and Blackwelder[107] yielded also a Gaussian error function for the Ω-distribution, with $\bar{x}_2^*/\delta = 0.8$ and $(x_2^* - \bar{x}_2^*)'/\delta = 0.15$.

Just as we did in the case of free turbulent flows, we may correct the distributions of various turbulence quantities for the effect of intermittency, so as to get an idea

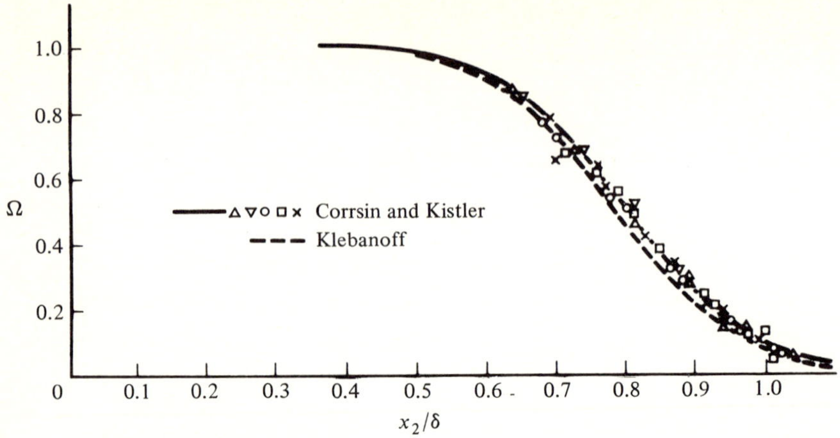

FIGURE 7-25
Intermittency distribution across a boundary layer as measured by Klebanoff,[27] and by Corrsin and Kistler.[29] (*By permission of the National Advisory Committee for Aeronautics.*)

of the distribution of these quantities in the fully turbulent fluid portions. For calculating the eddy viscosity, corrected for the intermittency, we may start from equations for $\bar{U}_1$ and $\overline{u_2 u_1}$ similar to Eqs. (6-105) and (6-107) respectively. In boundary-layer flow the rotational part of $\bar{U}_1$ is almost equal to $\bar{U}_1$, except near the crests of "wavy" interface. See, e.g., Kovasznay *et al.*[107]

So no large error will be made if, in calculating ϵ_m, we use $\partial \bar{U}_1/\partial x_2$. On the other hand $(\overline{u_2 u_1})_{\text{irrot}} = 0$, hence we may take for the corrected eddy viscosity the expression ϵ_m/Ω. Also the same remark may be made as in Sec. 6-7, when

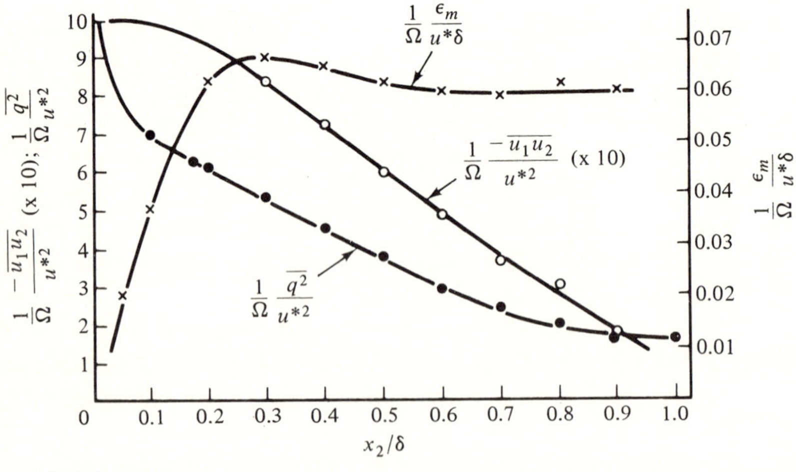

FIGURE 7-26
Distribution of turbulence kinetic energy, turbulence shear stress, and eddy viscosity in a boundary layer after correction for intermittency.

considering the effect of intermittency in the wake flow of a cylinder, namely that the turbulence kinetic energy might be overcorrected, since this energy is not negligible in the irrotational parts.

In Fig. 7-26 the turbulence kinetic energy, the turbulence shear stress, and the eddy viscosity, each multiplied by Ω^{-1}, are shown. The striking results are obtained that $-\overline{u_1u_2}/\Omega u^{*2}$ shows a nearly linear variation with x_2/δ and that $\epsilon_m/\Omega u^*\delta$ is nearly constant across the main outer part (the velocity-defect region) of the boundary layer. The latter result is completely similar to that obtained for the wake flow of a cylinder (see Fig. 6-14), which seems to point again toward similarity between the turbulence flow patterns of wake flow and of the outer portion of the boundary layer. On the other hand, the same result is not obtained with respect to the turbulence-energy dissipation. Whereas in the wake flow the distribution after correction for intermittency is nearly uniform (see Fig. 6-14), in the boundary-layer flow this is by no means true. Figure 7-27 shows the distribution of the turbulence-energy dissipation as measured by Klebanoff.[27] There is a gradual decrease in the dissipation toward the outer edge of the boundary layer, even after correction for intermittency.

In order to determine the turbulence-energy dissipation one should measure all terms of the dissipation function, which in nondimensional form reads

$$\frac{\varepsilon\delta}{u^{*3}} = \frac{\nu}{u^*\delta}\overline{\left(\frac{\partial u_j/u^*}{\partial \xi_i} + \frac{\partial u_i/u^*}{\partial \xi_j}\right)\frac{\partial u_j/u^*}{\partial \xi_i}}$$

However, since not all terms on the right-hand side are suitable for direct measure-

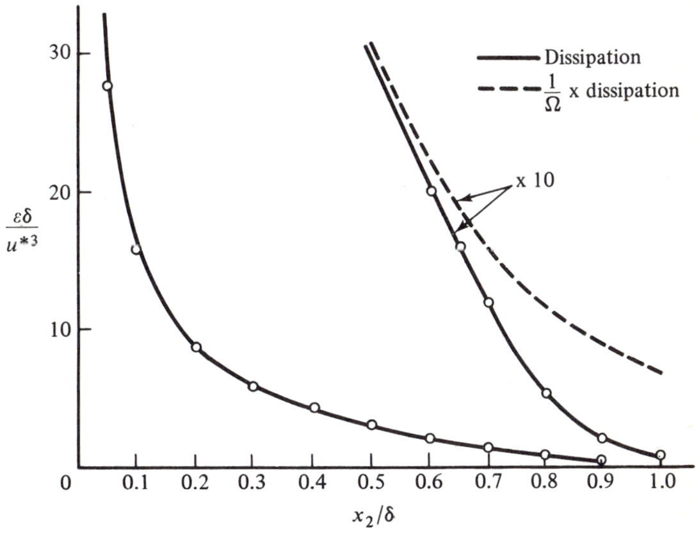

FIGURE 7-27
Distribution of turbulence dissipation in a boundary layer. (From: *Klebanoff, P. S.*,[27] *by permission of the National Advisory Committee for Aeronautics.*)

ment, Klebanoff considered instead the whole viscous term in the turbulence-energy equation written in the form (7-5b), that is,

$$\frac{\nu}{u^*\delta}\left(\frac{\partial^2}{\partial \xi_2{}^2}\frac{\overline{q^2}}{2u^{*2}} - \overline{\frac{\partial u_j/u^*}{\partial \xi_i}\frac{\partial u_j/u^*}{\partial \xi_i}}\right)$$

thus including the work done by the viscous shear stresses. If the departure from homogeneity is not too strong, which may be assumed to be true in the boundary layer except for the constant-stress layer and the region near it, the above whole viscous term is nearly equal to the dissipation for a homogeneous turbulence.

Of the nine terms $\overline{(\partial u_j/\partial x_i)(\partial u_j/\partial x_i)}$, Klebanoff measured only $\overline{(\partial u_1/\partial x_1)^2}$, $\overline{(\partial u_2/\partial x_1)^2}$, $\overline{(\partial u_3/\partial x_1)^2}$, $\overline{(\partial u_1/\partial x_2)^2}$, and $\overline{(\partial u_1/\partial x_3)^2}$, and assumed for the remaining terms the isotropic relations

$$\overline{\left(\frac{\partial u_3}{\partial x_2}\right)^2} = \overline{\left(\frac{\partial u_1}{\partial x_2}\right)^2} \qquad \overline{\left(\frac{\partial u_2}{\partial x_2}\right)^2} = \frac{1}{2}\overline{\left(\frac{\partial u_1}{\partial x_2}\right)^2}$$

$$\overline{\left(\frac{\partial u_2}{\partial x_3}\right)^2} = \overline{\left(\frac{\partial u_1}{\partial x_3}\right)^2} \qquad \overline{\left(\frac{\partial u_3}{\partial x_3}\right)^2} = \frac{1}{2}\overline{\left(\frac{\partial u_1}{\partial x_3}\right)^2}$$

Thus we obtain for ε'/ν:

$$\varepsilon'/\nu = \overline{\frac{\partial u_i}{\partial x_j}\frac{\partial u_i}{\partial x_j}} = \overline{\left(\frac{\partial u_1}{\partial x_1}\right)^2} + \overline{\left(\frac{\partial u_2}{\partial x_1}\right)^2} + \overline{\left(\frac{\partial u_3}{\partial x_1}\right)^2} + \frac{5}{2}\left[\overline{\left(\frac{\partial u_1}{\partial x_2}\right)^2} + \overline{\left(\frac{\partial u_1}{\partial x_3}\right)^2}\right] \qquad (7\text{-}83)$$

Another possibility is to assume isotropic relations between the four terms that cannot be measured directly and the term $\overline{(\partial u_1/\partial x_1)^2}$. With this assumption we obtain

$$\varepsilon'/\nu = 7\overline{\left(\frac{\partial u_1}{\partial x_1}\right)^2} + \overline{\left(\frac{\partial u_2}{\partial x_1}\right)^2} + \overline{\left(\frac{\partial u_3}{\partial x_1}\right)^2} + \overline{\left(\frac{\partial u_1}{\partial x_2}\right)^2} + \overline{\left(\frac{\partial u_1}{\partial x_3}\right)^2} \qquad (7\text{-}84)$$

However, in general not much difference is obtained between the values of ε'/ν determined according to the relations (7-83) and (7-84), and even when calculated according to the isotropic relation $\varepsilon'/\nu = 15\overline{(\partial u_1/\partial x_1)^2}$, at least for the outer region of the boundary layer. This notwithstanding the fact that Klebanoff obtained the isotropic relation $\overline{(\partial u_2/\partial x_1)^2} = 2\overline{(\partial u_1/\partial x_1)^2}$ only approximately in the region $x_2/\delta > 0.7$. Very close to the wall, in particular when approaching the buffer region, strong deviations from isotropy become apparent, and unacceptable errors may be obtained when the assumption of isotropy is made for calculating the dissipation.

From the measured distributions of turbulence kinetic energy, turbulence shear stress, and dissipation, it is possible to obtain a turbulence-energy balance, according to either Eq. (7-5) or (7-7), with the turbulence-diffusion term as the closing entry. Klebanoff made such a balance. Since Townsend,[28] in his similar experiments, also measured separately the distribution of $\overline{u_2 q^2}$, Townsend was able to make a more accurate balance, with the diffusion of pressure energy $\overline{u_2 p}/\rho$ as the closing entry. Figure 7-28 shows the energy balance as given by Townsend.[31] The energy balance obtained from Klebanoff's measurements shows the same general trends of the contributions by the various terms of the energy-balance equation, though there are quantitative differences.

If we compare this energy balance with the corresponding one for the wake flow behind a cylinder (see Fig. 6-10), we notice striking differences as well as similarities. The main contribution to the energy balance is given by the production and dissipation terms, if we exclude the region near the outer edge of the boundary layer. Production and dissipation are nearly in balance with each other, the more so as the wall is approached. The contribution of convection by the mean motion is practically negligible, except near the outer edge of the boundary layer. This term is essentially positive, which means a loss of energy. Near the outer edge the dissipation is greater than the production. The difference plus the loss by convection by the mean motion is counterbalanced by a gain by turbulence diffusion. Farther inward this turbulence diffusion changes sign, reaches a maximum (loss) just outside the wall region, and seems to change sign again when approaching the wall, producing there a gain in turbulence energy. Apparently there is a transfer of energy by turbulence diffusion from the inner part of the boundary layer toward its outer part.

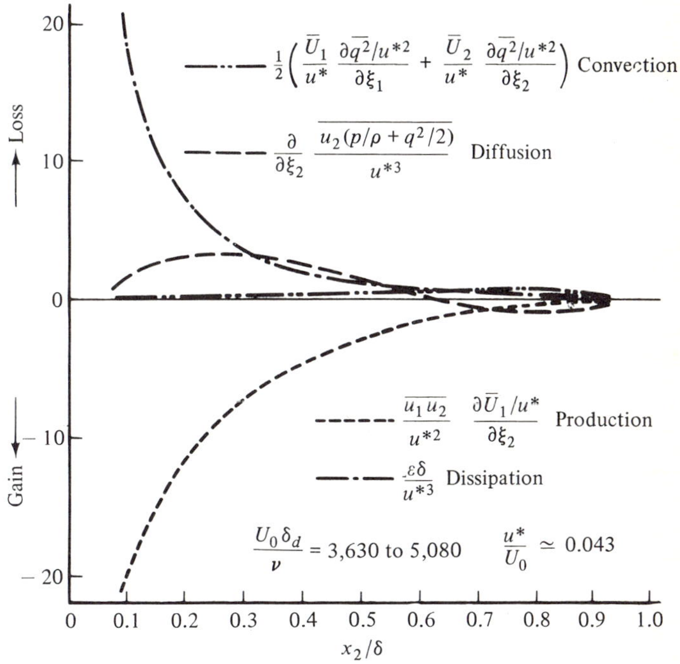

FIGURE 7-28
Energy balance in the boundary layer along a smooth wall with zero pressure gradient. (From: *Townsend, A. A.,*[31] *by permission of the Cambridge University Press.*)

Townsend[31] also determined the energy balance for the mean motion, as given by Eq. (7-4). Neglecting the viscous term in this equation and writing the equation in nondimensional form, we have

$$\frac{1}{2}\left[\frac{\partial}{\partial \xi_1}\frac{\bar{U}_1{}^3}{u^{*3}}+\frac{\partial}{\partial \xi_2}\frac{\bar{U}_2\bar{U}_1{}^2}{u^{*3}}\right]+\frac{\bar{U}_1}{u^*}\frac{\partial}{\partial \xi_2}\frac{\overline{u_1u_2}}{u^{*2}}=0$$

The last term may be written as follows:

$$\frac{\bar{U}_1}{u^*}\frac{\partial}{\partial \xi_2}\frac{\overline{u_1u_2}}{u^{*2}}=\frac{\partial}{\partial \xi_2}\frac{\overline{u_1u_2}\bar{U}_1}{u^{*3}}-\frac{\overline{u_1u_2}}{u^{*2}}\frac{\partial \bar{U}_1/u^*}{\partial \xi_2}$$

The first term on the right-hand side denotes the total work done by the turbulence shear stress. The second term may be interpreted either as the work of deformation done by the turbulence shear stress, as the production of turbulence energy, or as the "dissipation" of kinetic energy of the mean motion, since we may write $-\overline{u_1u_2}=\epsilon_m\,\partial\bar{U}_1/\partial x_2$.

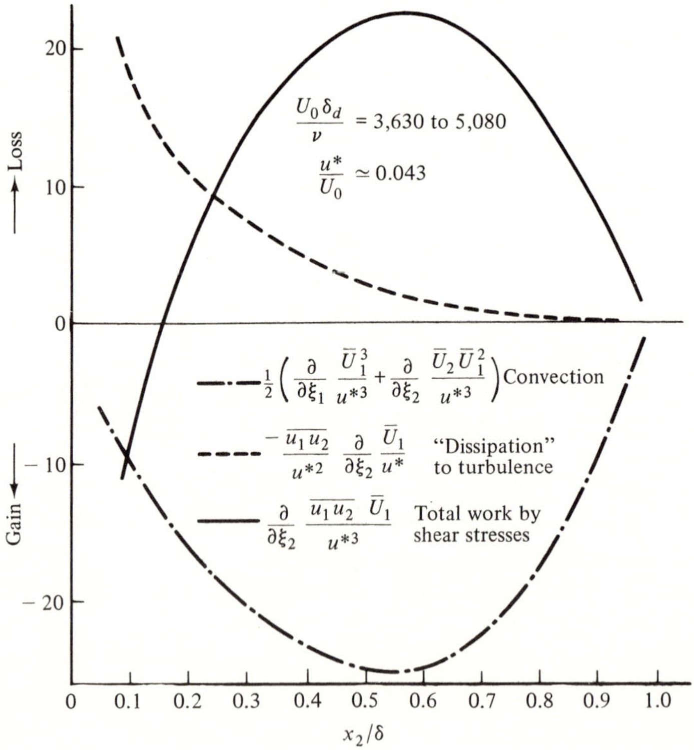

FIGURE 7-29
Energy balance of the mean motion in a boundary layer along a smooth wall with zero pressure gradient. (From: *Townsend, A. A.,*[31] *by permission of the Cambridge University Press.*)

Thus the energy-balance equation for the mean motion becomes

$$\frac{1}{2}\left[\frac{\partial}{\partial \xi_1}\frac{\bar{U}_1^3}{u^{*3}}+\frac{\partial}{\partial \xi_2}\frac{\bar{U}_2\bar{U}_1^2}{u^{*3}}\right]-\frac{\overline{u_1u_2}}{u^{*2}}\frac{\partial \bar{U}_1/u^*}{\partial \xi_2}+\frac{\partial}{\partial \xi_2}\frac{\overline{u_1u_2}\bar{U}_1}{u^{*3}}=0 \tag{7-85}$$

Figure 7-29 shows the distribution of the various terms of this equation across the boundary layer. The loss of kinetic energy of the mean motion due to the "dissipation" to turbulence and, thus, to production of turbulence energy increases strongly toward the wall. In the outer region the gain due to convection by the mean motion is practically equal to the loss by the work of the turbulence shear stresses. In the wall region the production of turbulence energy exceeds the supply directly through convection by the mean motion. The difference is obtained through the total work by the turbulence shear stresses. There seems to be a flux of energy toward the inner wall region, where it is "converted" into the energy of the turbulent motion.

If we combine the results obtained from the energy balances of the mean motion and of the turbulent motion, we come to the interesting and important conclusion that there must be strong interaction between the flows in the wall region of the boundary layer and the outer region. The two regions have different turbulent-flow mechanisms, but there is a strong energy exchange between them.

In the outer region there is a supply of mean-flow kinetic energy from upstream parts of the boundary layer. This energy is converted through the work done by the turbulence shear stresses into production of turbulence energy in the inner part of the boundary layer. Or, put in a different way, the mean flow is retarded by the action of the turbulence shear stresses, thereby losing some of its kinetic energy.

In the wall region of the boundary layer most of the turbulence energy produced is converted directly into heat by turbulence dissipation. But part of this turbulence energy is transported by turbulent diffusion toward the outer region of the boundary layer, where it is dissipated. Thus there is an influx toward the wall of energy originating from the mean flow, converted into turbulence energy, which is in part directly dissipated but in part diffused back by turbulence into the outer region.

Since no measurements have yet been made in the innermost part of the boundary layer, that is, close to the wall, not very much more can be said about what really occurs there. Measurements have been made, however, in the case of flow through circular pipes and channels; they will be considered later, in Sec. 7-13.

Some further information can be obtained from measurements on energy-spectra and spatial-velocity correlations in the boundary layer. Measurements of the spectra of the axial and transverse turbulence velocities $\overline{u_1^2}$ and $\overline{u_2^2}$ and of the turbulence shear stress $-\rho\overline{u_1u_2}$ have been made by Klebanoff.[27] Figure 7-30, which is similar to Fig. 4-10, gives the results for the axial turbulence component obtained at the relative distances $x_2/\delta = 0.0011$, 0.05, and 0.80 from the wall. From these results we may note that the spectrum taken at the point $x_2/\delta = 0.80$, that is, in the outer region of the boundary layer, follows the $-\frac{5}{3}$ law rather closely in a

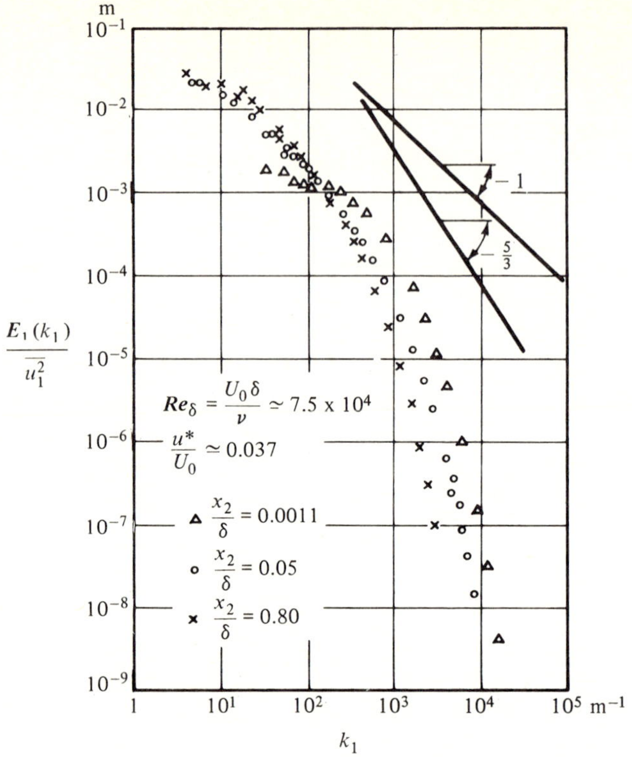

FIGURE 7-30
Spectra of u_1 in a boundary layer along a smooth flat plate with zero pressure gradient. (From: *Klebanoff, P. S.*,[27] *by permission of the National Advisory Committee for Aeronautics.*)

small, but discernible wavenumber range $k_1 \simeq 50$ to $500\ m^{-1}$. The spectrum taken at the point $x_2/\delta = 0.05$ shows a range where $\mathbf{E}_1(k_1)$ varies nearly according to k_1^{-1}, thus indicating strong interaction between mean and turbulent flow. At this point a strong production of turbulence energy takes place (see Fig. 7-28). The spectra in Fig. 7-30 also reveal that the contribution to the turbulence energy in the low-wavenumber range, that is, made by the larger eddies, decreases as the wall is approached, but that the contribution in the high-wavenumber range is increased.

The same conclusion about the high-wavenumber range may be drawn from Fig. 7-31, which shows the spectra of the turbulence shear stress $-\rho\overline{u_2u_1}$ at the points $x_2/\delta = 0.05$ and 0.80. A comparison with Fig. 7-30 reveals that the spectrum $\mathbf{E}_{2,1}(k_1)$ in the high-wavenumber range decreases with increasing wavenumber at a steeper rate than the spectrum $\mathbf{E}_1(k_1)$; which means that, beyond a certain wavenumber, the contribution to the shear stress becomes negligible. But there is still a contribution to the energy spectrum of $\overline{u_1^2}$. This fact points toward some local isotropy in the highest-wavenumber range.

The spectrum of the lateral turbulence component $\overline{u_2^2}$ behaved differently with respect to the spectrum of $\overline{u_1^2}$ than it would if the turbulence were isotropic.

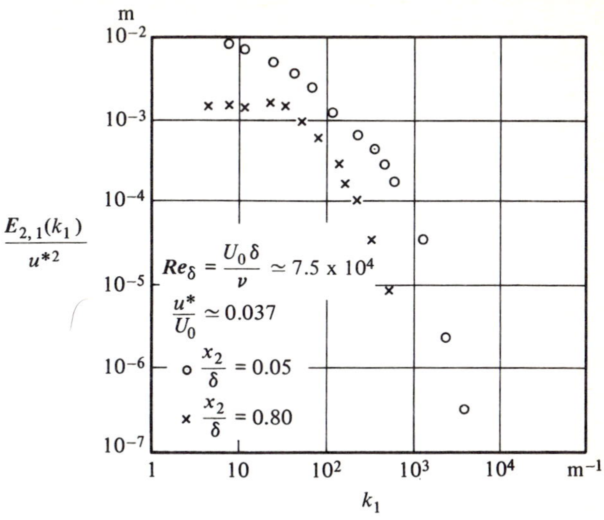

FIGURE 7-31
Spectra of the turbulence shear stress u_2u_1 in a boundary layer along a smooth flat plate with zero pressure gradient. (From: *Klebanoff, P. S.*,[27] *by permission of the National Advisory Committee for Aeronautics.*)

The difference is marked in the low-wavenumber range, in particular at the point $x_2/\delta = 0.05$ very close to the wall. In the high-wavenumber range some local isotropy may be observed.

This anisotropy is also found in the spatial correlations $\mathbf{f}(r)$ and $\mathbf{g}(r)$ and in the corresponding values of the integral scales Λ_f and Λ_g. For the longitudinal correlation Schubauer and Klebanoff[16] measured the correlation between the axial turbulence component u_1 at two points a distance x_1 apart but at the same plane x_2; and for the lateral correlation they measured the correlation between u_1 at two points a distance x_2 apart but at the same cross section x_1. For isotropic turbulence we would have $\Lambda_f = 2\Lambda_g$. Schubauer and Klebanoff's measurements in the outer region of the boundary layer ($x_2/\delta > 0.15$) show that Λ_f/Λ_g took values of 3 and more. Moreover, it is found that Λ_f may become of the same magnitude as δ. Thus it may be concluded that the turbulence in the outer region of the boundary layer contains eddies of the same size as the boundary-layer thickness. These eddies are very much elongated in a downstream direction, and correlations exist over large downstream distances. This conclusion is supported by measurements of Favre, Gaviglio, and Dumas[33] on space-time correlations in a boundary layer. This and later investigations by Favre *et al.* will be considered in more detail in the following sections.

Effect of Fluid Compressibility

Little information is available on the effect of compressibility of the fluid on the mechanism of turbulent flow in a boundary layer or in a pipe. This effect was too small to be accounted for in the low-speed flows studied before, but it can no longer be ignored in high-speed (supersonic) flow. In a boundary layer the retardation of the free-stream flow of high velocity causes a conversion of kinetic energy, through compression, into heat and, hence, an increase in temperature. Since the generation of heat of compression is not uniform in the boundary layer, a transport of heat occurs by molecular and turbulence diffusion. As a result the temperature at the wall may become different from the stagnation temperature. The difference depends on the molecular Prandtl number and on the degree of insulation of the wall against heat exchange.

A major effect of compressibility in a high-speed boundary layer may be expected from the changes of fluid properties due to temperature variations.

Some details about this effect on the mean-velocity distribution in a turbulent boundary layer at high free-stream Mach numbers have been obtained by Lobb, Winkler, and Persh[65] from total-head and temperature traverses. Free-stream Mach numbers $\mathbf{Ma}_0$ of 5.0, 6.8, and 7.7 were achieved.

The velocity distributions in the viscous sublayer were linear and followed the relation $\bar{U}_1^+ = x_2^+$, with the viscosity taken at the wall temperature. This distribution, however, applied in a much larger region than for incompressible flow, namely up to $\delta_l^+ \simeq 12$ at $\mathbf{Ma}_0 = 6.8$, as compared with $\delta_l^+ \simeq 5$ for incompressible flow. The transition to the fully turbulent region was much less gradual, and, in the fully turbulent part of the wall region, deviations from the logarithmic velocity distribution occurred. These deviations depended on the degree of heat exchange with the wall.

In the outer region of the turbulent boundary layer the velocity distribution could reasonably well be approximated by a power law

$$\frac{\bar{U}_1}{U_0} = \xi_2^{1/n}$$

The value of n decreased with increasing Mach number. For $\mathbf{Ma}_0 = 5.0$ the value of n was approximately 7 and the Reynolds number $\mathbf{Re}_\delta$ varied slightly from 1.5×10^5 to 1.65×10^5. For $\mathbf{Ma}_0 = 6.8$ ($\mathbf{Re}_\delta = 1.81$ to 2.94×10^5), $n = 6$; and for $\mathbf{Ma}_0 = 7.7$ ($\mathbf{Re}_\delta = 2 \times 10^5$), $n \simeq 5.5$. Since the exponent n also depends on $\mathbf{Re}_\delta$ (it increases with increasing $\mathbf{Re}_\delta$) some effect of $\mathbf{Re}_\delta$ is included in the above values of n.

From the velocity distribution in the viscous sublayer Lobb, Winkler, and Persh also determined the wall shear stress σ_w and from σ_w the friction coefficient $c_f = \sigma_w/\frac{1}{2}\rho_0 U_0^2$. Kuethe[32] collected these data and data obtained by others in a graph which we have shown in Fig. 7-32. There the ratio $c_f/c_{f,i}$, where $c_{f,i}$ refers to incompressible flow, is plotted against the free-stream Mach number.

Actually there is also some effect of the Reynolds number, in that the ratio $c_f/c_{f,i}$ decreases slightly with increasing Reynolds number. In the range $2{,}000 < \mathbf{Re}_{\delta_m} < 7{,}000$ the data agree satisfactorily well with the flat-plate theory by Wilson.[108] Also an appreciable effect of a favorable pressure gradient and/or heat transfer does not show up in the data by Lobb *et al.* More recent data by

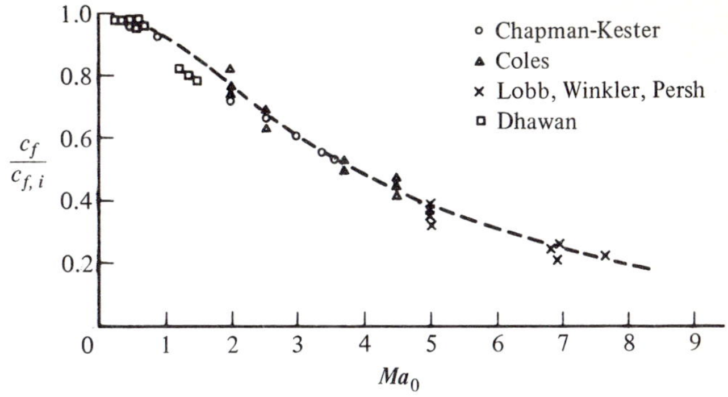

FIGURE 7-32
Wall-friction coefficient c_f for compressible flow along a flat plate at various free-stream Mach numbers compared with the coefficient $c_{f,i}$ for incompressible flow. (From: *Kuethe, A. M.,*[32] *by permission of the Institute of the Aeronautical Sciences.*)

Nagamatsu, Graber, and Sheer[109] from experiments with small-angle sharp cones at high Mach numbers and including heat transfer are in agreement with Fig. 7-32. Though some effect of the Reynolds number is present, the major variation is with the Mach number. The decrease of the ratio with increasing Mach number stems from the fact that this ratio is roughly proportional to some positive power of Θ_0/Θ_w, which decreases with increasing Mach number, that is, with increasing stagnation temperature. For an incompressible isothermal flow we have approximately [see Eq. (7-60)]

$$c_{f,i} = \text{const.} \times \left(\frac{\rho_0 U_0 \delta}{\mu_0}\right)^{-m}$$

If we assume that the same relation holds for a compressible boundary layer, provided that ρ and μ are taken at the wall conditions, we obtain

$$\frac{\sigma_w}{\frac{1}{2}\rho_w U_0^2} = \text{const.} \times \left(\frac{\rho_w U_0 \delta}{\mu_w}\right)^{-m}$$

whence

$$c_f = \frac{\sigma_w}{\frac{1}{2}\rho_0 U_0^2} = \text{const.} \times \frac{\rho_w}{\rho_0}\left(\frac{\rho_w U_0 \delta}{\mu_w}\right)^{-m}$$

$$= c_{f,i}\left(\frac{\rho_w}{\rho_0}\right)^{1-m}\left(\frac{\mu_w}{\mu_0}\right)^{m} \simeq c_{f,i}\left(\frac{\Theta_0}{\Theta_w}\right)^{1-(n+1)m}$$

if it is assumed that $\mu_w/\mu_0 \simeq (\Theta_w/\Theta_0)^n$. For the relevant values of m and n, the value of $(n+1)m$ is less than 1.

Goddard[67] has investigated the effect of wall roughness on a compressible boundary layer. Up to **Ma** $= 5$ the effect of compressibility is only indirect if the full rough condition is reached. The friction coefficient c_f is only a function of u^*k/ν, just as in the incompressible case. Since for the full rough condition there is no direct viscosity effect, we have $c_f/c_{f,i} = \rho_w/\rho_0$, so the whole effect of compressibility turns out to be a reduction of the density of the fluid at the surface. The shift in the velocity profile $\Delta\bar{U}_1/u^*$ appears to be a function of only u^*k/ν (compare Fig. 7-16 pertaining to the incompressible case).

7-8 MORE DETAILED INFORMATION ON THE STRUCTURE OF THE TURBULENT BOUNDARY LAYER

During the past ten to fifteen years a lot of experimental material concerning the structure and the mechanism of wall turbulence, not yet covered in the previous sections, has become available. The progress in experimental techniques, including also new methods, has been of great help in this respect.

In Sec. 7-1 we made a distinction between an inner or wall region, and an outer region. For the gross description of the turbulent flow it appeared useful to further subdivide the wall region into a viscous sublayer adjacent to the wall, a buffer region, and a fully turbulent region. Though at the same time we stressed the fact that this subdivision should not be considered too strictly, since the transition between these subregions is gradual. There appears to be an appreciable interaction between these subregions. Also, the turbulence in the entire wall region cannot be considered as being entirely independent of the turbulence in the outer region, as already mentioned earlier.

As may become clear from Fig. 7-28, the buffer region near the wall ($x_2^+ \gtrsim 70$) is the most active part as far as production and dissipation of turbulence energy is concerned, with pronounced maxima of these quantities at $x_2^+ \simeq 10$. See also Fig. 7-68, referring to the wall region in turbulent pipe flow. For this reason we will consider the turbulent flow processes, starting from the wall. The early experiments by Fage and Townend,[110] where the trajectories of suspended small particles were followed by means of an ultramicroscope, already revealed that the flow in the viscous sublayer is not laminar in its strict sense. The particle trajectories showed a spanwise, u_3-velocity component varying irregularly with time. The pictures of these trajectories remained quite sharp during a relatively long traveling distance. Because the depth of focus of the ultra-microscope was very small, it may be concluded that the flow adjacent the wall must have been almost two-dimensional in planes parallel to the wall. This conclusion is supported by the fact that, due to continuity conditions, the normal, u_2-velocity component varies at least proportional with x_2^2. Consequently the flow becomes more and more "two-dimensional" in a plane parallel to the wall as the wall is approached, if at the same time it is assumed that the other two turbulence velocity components u_1 and u_3 decrease to zero in a similar fashion as $\bar{U}_1$, i.e., linearly with x_2. Indeed, Fig. 7-18 seems to suggest a limiting value of $u'_1/\bar{U}_1 \simeq 0.4$ at $x_2 = 0$. However, these measurements made closest to the wall were not very reliable. The more accurate measurements very close to the wall carried out in pipe flow by Laufer seem to yield a lower limiting-value, namely 0.3 (see Fig. 7-59), a value which also may be obtained from the measurements in pipe flow by Coantic,[111] and has been obtained by Bakewell and Lumley[112] in a glycerine tunnel ($\delta_l \simeq 3$ mm). Mitchell and Hanratty[113,114] obtained with the use of an electrochemical technique the following limiting value

$$[\overline{(\partial u_1/\partial x_2)^2}]^{1/2}/(d\bar{U}_1/dx_2) = 0.32$$

whence follows $u_1'/\bar{U}_1 = 0.32$ for a linear variation of the total velocity component U_1 with x_2. A lower value still, namely $(u_1'/\bar{U}_1)_{x_2=0} = 0.24$ has been obtained by Eckelmann,[115] who made measurements with both a hot-film anemometer and a hot-film wall-surface sensor in an oil channel. The measurements were made at two values of the Reynolds number based on the maximum, center, velocity and the width of the channel, namely 11,200 and 16,400. The thickness of the viscous sublayer corresponding with $\delta_l^+ = 5$, was about 5 mm, thus enabling pretty accurate measurements deep into the sublayer. The distributions of $u_1'/\bar{U}_1$ and u_1'/u^* in the region $x_2^+ < 20$, are shown in Fig. 7-33. We note that the ratio $u_1'/\bar{U}_1$ obtains a maximum value of ~ 0.38 at $x_2^+ \simeq 4$. Because of this sharp increase near the wall it is possible that measurements in thinner sublayers become inaccurate. For instance, the measurements with the electrochemical technique where the electric current measured is determined by the flow in a fluid layer of very small but still finite thickness adjacent to the electrode surface, yielded a higher value than the limiting value obtained by Eckelmann with the surface sensor. On the other hand it may also be remarked that in Fig. 7-33 a relation $u_1'/u^* \simeq 0.3x_2^+$ is not in contradiction with the data for $x_2^+ < 2.5$, which might explain the value obtained with the electro-

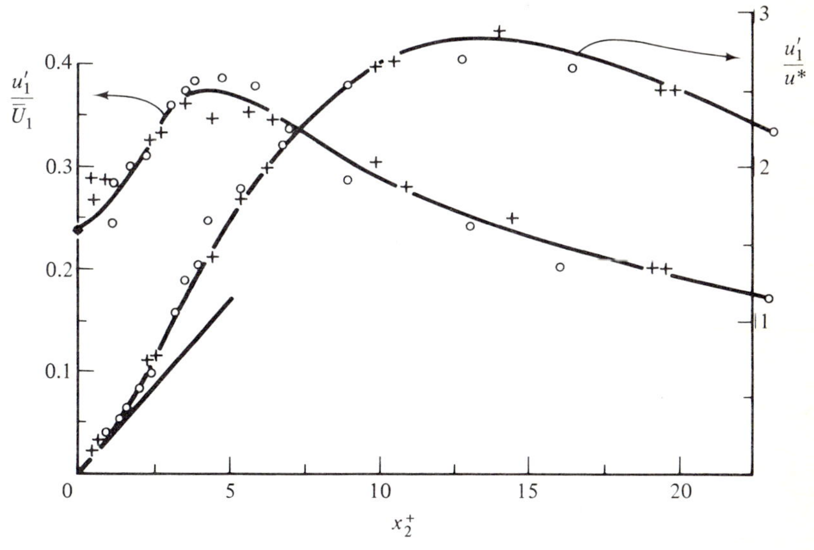

FIGURE 7-33
Distributions of $u_1'/\bar{U}_1$ and u_1'/u^* across the viscous sublayer. (After: *Eckelmann, H.*,[115] *by permission of the Max Planck Institut für Strömungsforschung.*)

chemical method. An uncertainty is still found in the fact that the turbulence fluctuations measured do not include all frequencies, the values depending on the cut-off frequency of sensor and electronic circuit, in particular at the low frequency side.

Figure 7-33 also shows that u_1'/u^* reaches a maximum of about 2.8 between $x_2^+ = 10$ and 15. The measurements of u_2'/u^*, on the other hand, show a maximum value of about 1.0 at a larger distance from the wall, namely at $x_2^+ \simeq 50$ (see also Fig. 7-18). The behavior of u_2' when $x_2 \to 0$ could not be determined from these measurements. The shear stress $-\overline{u_2 u_1}$ showed the linear distribution for steady channel flow only for $x_2^+ \gtrsim 50$. The center of the channel corresponded with $x_2^+ = 142$ at $\mathbf{Re} = 11{,}200$, and with $x_2^+ = 210$ at $\mathbf{Re} = 16{,}400$. The shear-stress correlation coefficient $-\overline{u_2 u_1}/u_2' u_1'$ was constant and equal to 0.4 over roughly 60 per cent of the half-width of the channel, approaching the zero value at the axis.

Sirkar and Hanratty[116] also determined with the electrochemical method the behavior of the spanwise turbulence velocity u_3 near the wall. The measurements confirmed the above-mentioned expectation of a linear variation with x_2. For the ratio

$$[\overline{(\partial u_3/\partial x_2)^2}]^{1/2}/(d\bar{U}_1/dx_2)$$

they obtained the limiting value of 0.085 to 0.09 when $x_2 \to 0$. This value is in agreement with that which can be deduced from measurements by Fowles *et al.*,[117] namely $u_3'/\bar{U}_1 \simeq 0.08$–$0.09$ for $x_2 \to 0$. Here too, for the same reasons mentioned above, the actual limiting value might be somewhat lower. We may note that the limiting value of $u_1'/u_3' \simeq 3$.

The turbulent fluctuation of $\partial u_1/\partial x_2$ at the wall results in a turbulent fluctuation of the wall shear-stress σ_w. Popovich and Hummel[118] studied the region adjacent to the wall using a colored tracer in a photosensitive liquid. The instantaneous x_2-distributions of U_1, as deduced from the dye traces, were linear across x_2-distances which varied irregularly with time. The statistical distribution of these distances was skew with a maximum at $x_2^+ \simeq 4.3$, which is equal to $0.7\delta_l^+$ since the average value was $\delta_l^+ \simeq 6.2$. The experiments also showed the existence of a very thin layer $x_2^+ \simeq 1.6 \pm 0.4$, where the instantaneous velocity distribution was linear all the time, but with a slope changing irregularly with time. The statistical distribution of $(\partial U_1/\partial x_2)_{x_2=0}$, and hence that of σ_w was slightly skew, with a maximum at $0.9\bar{\sigma}_w$. In agreement with Popovich and Hummel's results are the findings by Armistead and Keyes.[119] They measured with a wall hot-film sensor the probability density of the heat-transfer fluctuations in pipe flow. A high degree of correlation between the heat-transfer fluctuations and the simultaneously measured u_1-fluctuations in a region up to $x_2^+ \simeq 3$ was obtained. The correlation decreased to zero at $x_2^+ > 25$ ($\mathbf{Re}_D \simeq 28{,}000$). On the other hand Eckelmann's measurements showed that the correlation $u_1(\partial u_1/\partial x_2)_{x_2=0}$ did not decrease to zero earlier than $x_2^+ \simeq 100$. At $x_2^+ = 25$ the correlation coefficient was still about 0.25.

The measurements with a hot-film anemometer carried out by Bakewell and

Lumley[112] in a glycerine tunnel at a $\mathbf{Re}_D = 8{,}700$ showed that the turbulence near the wall could reasonably well be described using ν/u^* and ν/u^{*2} as scaling parameters for length and time, respectively. Within the viscous sublayer ($x_2^+ \gtrsim 5$) the spectral distribution $\mathbf{E}_1(n)$ of the u_1-velocity component was universal when $\mathbf{E}_1(n)/x_2^2 n$ was plotted versus $n\nu/u^{*2}$. Normalizing $\mathbf{E}_1(n)$ with $x_2^2 n$ instead of with $\overline{u_1^2}/n$ is permitted when in the viscous sublayer $u_1' \propto x_2$. The nondimensional spectral distribution showed a distinct (-1) power law in the nondimensional frequency range $0.0005 \gtrsim n\nu/u^{*2} \gtrsim 0.005$ [see Eq. (4-61)], but the nonexistence of a distinct $(-\frac{5}{3})$ power law in the higher frequency range, as may be expected from the low values of $\mathbf{Re}_\lambda$. Measurements of $\mathbf{E}_1(n)$ at $x_2^+ = 10$, 20, and 40 also showed, but much less distinctly, a (-1) slope in the curve in a frequency range that shifted towards higher frequencies with increasing x_2^+, i.e., with increasing $\mathbf{Re}_\lambda$. With a novel orthogonal decomposition of the u_1-velocity component proposed by Lumley,[129] Bakewell and Lumley[112] made an attempt to obtain some information concerning the "large-eddy" structure close to the wall. These "large eddies," whose existence has already been suggested by Townsend,[31] have to be considered in a relative sense, because of the relatively thin region near the wall where these eddies occur. Lumley[129] identified these "large eddies" with the most energetic eigenfunctions in the decomposition. The size of these "large eddies" thus corresponds roughly with the size of the energy-containing eddies.† The dominant eigenfunction appears to increase linearly with x_2^+ in the viscous sublayer, i.e., similar to the x_2^+-variation of $\bar{U}_1^+$. Based on a mixing-length type of assumption Bakewell and Lumley made an estimate of the flow pattern of a large eddy. They concluded that the structure must consist of a pair of contra-rotating streamwise vortices with a strongly concentrated ejection from the wall, creating in this way a defect in the U_1-distribution at some distance from the wall. The structure is similar to the picture given by Townsend[31] of "attached" eddies elongated in the streamwise direction and with the axis making an average angle of 45° with the mean flow. The position of the centers of these eddies was estimated to be roughly at $x_2^+ \simeq 50$, while the spanwise spacing was roughly $\lambda_3^+ \simeq 80$, which is of the same order of magnitude as has been concluded from direct visual-observation studies, to be discussed shortly.

Mechanism of the Turbulence in the Wall Region

Interesting results have been obtained concerning the mechanism of the turbulence in the wall region, and in particular in the turbulence production part of it, from a number of experimental investigations in which either visualization techniques, or hot-wire and hot-film anemometry have been used. Kline *et al.*,[120,121,122] Clark and Markland,[128] and Grass[124] used the hydrogen-bubble technique (Sec. 2-9) in a waterchannel, Corino and Brodkey,[126] and Nychas *et al.*[131] made high-speed

† In a later publication,[130] Lumley proposed a slightly different definition, namely that the large eddy corresponds with the motion which can most efficiently extract energy from the mean motion, and loses as little as possible energy through dissipation.

motion pictures of the turbulent flow of a liquid using colloidal-sized particles as a tracer, Laufer *et al.*,[125] and Willmarth and Lu[132] applied hot-wire anemometry for measurements in the boundary layer along one of the walls of a windtunnel, while Morrison *et al.*[123] did the same in a circular pipe.

Earlier we focused attention on the fact that many aspects of the laminar-turbulent transition process may also be observed in the wall-adjacent region of fully developed turbulent flow. In order to understand better the results of the experiments in fully developed turbulent flow in the light of this similarity concerning many aspects of the mechanism, it may be useful to recapitulate a number of characteristic and salient points of the transition process. These are: the formation of U-shaped vortex loops, with the top drifting away from the wall due to self-induction (Figs. 7-3 and 7-4); the occurrence of a dent in the instantaneous U_1-velocity profile, with inflexion points, in the region between the trailing streamwise vortices of the U-shaped loop and in the vicinity of its top (Figs. 7-5 and 7-6); associated with it the formation of an intense "horizontal" shear-layer, which breaks down and which then shows up as a turbulence burst in the velocity oscillogram (secondary, inflexional, instability); the temporal occurrence of adverse values of u_2u_1 with respect to the local mean-velocity gradient (Fig. 7-5). In the case of many parallel U-shaped loops, "peak" regions of u_1'-intensity between the legs of these loops alternate with "valley" regions of u_1'-intensity in spanwise direction. There is a strong u_2-component away from the wall in the "peak" region of u_1', and a u_2-component towards the wall in the "valley" region of u_1'. Consequently these u_2-components transport low axial-momentum fluid outward (resulting in a low U_1 or "valley" region of U_1) and high axial-momentum inward (resulting in a high U_1 or "peak" region of U_1), respectively. "Vertical" shear-layers may thus be formed, temporarily and locally with respect to the spanwise direction, which, however, do not break down. In a three-dimensional picture a turbulence spot is formed causing a local thickening of the boundary layer. Since the average convection velocity of the turbulence spot is smaller than the free-stream velocity the moving "bubble" of the boundary layer is overtaken by the free-stream fluid.

Let us now consider the results of the experiments in the fully developed turbulent flow. Figure 7-34 shows an instantaneous picture of streaklines formed by hydrogen bubbles produced by a spanwise wire at $x_2 = 1.3$ mm ($x_2^+ = 8$) from the wall.[120] The marked streaks that show up in the picture are a result of concentration of hydrogen bubbles in the "valley" region of U_1, where the u_2-component is away from the wall. The streaks show a wavy shape, though also sometimes a streamwise spiral motion may be observed locally. The wavy motion becomes more distinct in x_2-planes closer to the wall, while the spiral, vortex, motion is more distinct at greater distances from the wall. This may also be concluded from similar hydrogen-bubble pictures obtained by Clark and Markland,[128] shown in Fig. 7-35. These experiments were made in a watertunnel of 65 mm depth. The average velocity was $\bar{U}_{1,\mathrm{aver}} = 0.083$ m/s and the wall-friction velocity $u^* = 0.0043$ m/s. Four spanwise wires were used with a streamwise spacing $l^+ = 100$. Kline *et al.* and Clark and

FIGURE 7-34
Streaklines formed by hydrogen bubbles showing an instaneous picture of the turbulent boundary-layer structure at $x_2^+ = 8$; $U_0 = 0.13$ m/s. (From: *Rundstadler, P. W., et al.,*[120] *by permission of the authors.*)

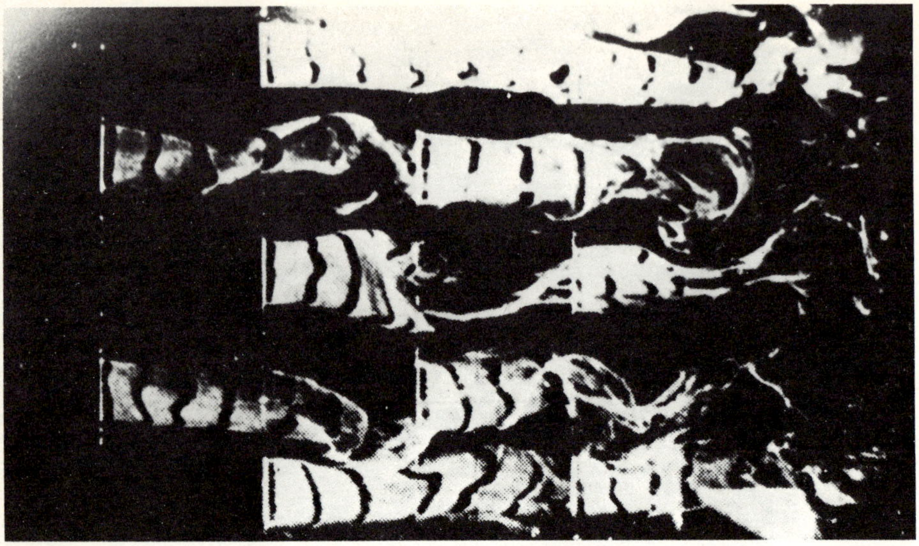

$X_2^+ = 40$ 8 c/s

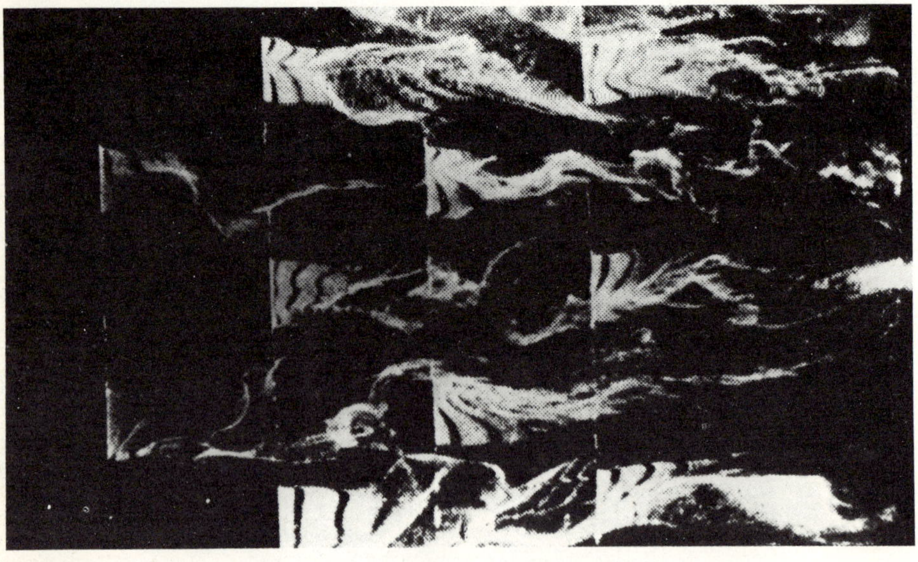

$X_2^+ = 10$ 8 c/s

FIGURE 7-35
Vortex structure in a turbulent boundary layer. (From: *Clark, J. A., and E. Markland,*[128] *by permission of the Royal Aeronautical Society.*)

Markland observed occasionally U-shaped vortices, and sometimes pairs of contra-rotating streamwise vortices. The streaks revealed a most violent motion in the region ${x_2}^+ = 8 - 12$. Due to the u_2-velocities in the U_1 valley regions the streaks are ejected away from the wall to break up mainly at distances ${x_2}^+ > 30$ in a random and intermittent way. Clark and Markland also observed locally, short spanwise vortices, probably corresponding with the top of U-shaped loops, which persisted over distances ${x_1}^+ \simeq 300$ at ${x_2}^+ = 12$, but over much greater distances ${x_1}^+ \simeq 1{,}000$ at ${x_2}^+ = 100$. An average spanwise spacing of these U-shaped vortices of ${\lambda_3}^+ \simeq 100$ was found, and a streamwise spacing of ${\lambda_1}^+ \simeq 440$. On the average, the axes of the streamwise vortices were inclined 5° away from the wall. They could be identified up to ${x_2}^+ \simeq 70$. The same value of ${\lambda_3}^+ \simeq 100$ for the spanwise spacing had already been reported by Kline *et al.* The spanwise extent of the low-speed streaks amounted to $10 < {x_3}^+ < 30$. From spanwise, spatial correlations of the u_1-velocities measured in the boundary layer along one of the walls of a windtunnel, Gupta *et al.*[125] found the spanwise spacing to increase from ${\lambda_3}^+ \simeq 100$ to 150, when the Reynolds number was increased from $\mathbf{Re}_{\delta_m} = 2{,}200$ to 6,500. Also a Reynolds dependency of ${\lambda_1}^+$ may be expected. As will become clear from considerations to be given later, it is more reasonable to expect the streamwise spacing to scale with the depth of the watertunnel rather than with ν/u^*. Expressed in terms of the half-depth of the watertunnel, under the experimental conditions ${\lambda_1}^+ = 440$ corresponds roughly with λ_1 equal to 3–4 times this half-depth. Morrison *et al.*[123] expect even more pronounced effects of the Reynolds number on the structure of the turbulence near the wall at large values of this number. They studied the possibility of a wavelike behavior in the viscous sublayer and buffer region by measuring the two-dimensional spectra $\mathbf{E}_1(k_1,n)$ and $\mathbf{E}_1(k_3,n)$ in the fully developed turbulent airflow through a tube of 134 mm I.D., employing constant-temperature hot-wire anemometry. The wavenumber k_1 and the frequency n occur as two independent variables because of the failure of the Taylor hypothesis $k_1 = 2\pi n/\bar{U}_1$ in the region close to the wall. From the measurements farther away from the wall they obtained an increase of the fraction of low-frequency, low-wavenumber ${k_3}^+$ energy with increasing Reynolds number. Because, most probably, this additional energy will also appear in the wall-adjacent layers, it is possible that the universal nature of the turbulence structure with u^* and ν as characteristic quantities will diminish, the flow becoming also influenced by quantities characteristic for the outer layers when the Reynolds number becomes high enough. This additional low-frequency, low-wavenumber energy which becomes significant at higher Reynolds numbers, results from disturbances which are convected with velocities much greater than the convection velocity characteristic for the wall-adjacent layers. Thus, if these layers obtain a broader range of wavenumbers as the Reynolds number increases, a change in the structure may be expected. Morrison *et al.* expect this change to become quite drastic at Reynolds numbers much greater than $\mathbf{Re}_D = 60{,}000$. The streaky nature, observed by Kline *et al.* and by others, may decrease in importance, and may even vanish eventually at very high Reynolds numbers.

In the following paragraphs a rather detailed description will be given of experimental results concerning the processes occurring in the wall region, as obtained by the aforementioned investigators using either visualization techniques or hot-wire anemometry. These results not only agree qualitatively but also in many aspects quantitatively in a reasonable way. They all point towards a mechanism consisting of processes which appear to be repetitive in nature, statistically speaking, both in time and in space. The observations show local and temporary *ejections* of low-momentum fluid from the wall outward, *inrush movements* towards the wall accompanied by *sweep movements* almost parallel to the wall, and alternating with the generation locally of highly unstable instantaneous velocity-distributions, resulting in turbulence bursts, which show up in the velocity oscillograms as intensive high-frequency fluctuations. These bursts thus occur twice during one "cycle" of the ejection and inrush events. It may be expected that the first, i.e., the ejection burst will be more intensive than the second, i.e., inrush burst. However, the actual oscillograms do not always clearly show the difference. Associated with it, the turbulence shear stress, the production, and dissipation are intermittent. During short instants negative values of $-u_2u_1$ are observed. Again, phenomena which show a great similarity with those observed during transition.

The important role that the u_2-velocity component, fluctuating between positive and negative values, plays in the turbulence phenomena close to the wall, is also an important result from the experiments by Corino and Brodkey.[126] The high-speed motion pictures they made, were of the flow of trichloroethylene through a circular pipe of 50 mm I.D., using magnesium oxide of 0.6 μm as the colloidal-sized particles. The experiments were carried out at Reynolds numbers in the range $\mathbf{Re}_D = 20{,}000$ to 50,000, but the clearest pictures of the events were obtained at $\mathbf{Re}_D = 20{,}000$. They too found the region $0 < x_2^+ \gtrsim 30$ to be most important, with the characteristic feature of the ejection of more or less discrete fluid elements from the wall. These ejections are local in nature, random in space and time. They originate from positions in the region $5 < x_2^+ < 15$, with the majority from positions in the region of maximum turbulence production, though occasionally they may originate from positions as near as $x_2^+ = 2.5$ to the wall. The sublayer $x_2^+ < 2.5$ seems to be predominantly passive in this respect. This value corresponds with the value mentioned by Popovich *et al.*,[118] and by Armistead *et al.*,[119] referred to earlier, for the thin layer adjacent to the wall where the instantaneous velocity-distribution is linear all the time, though with randomly varying slope. The ejection angles with the x_1-direction observed by Corino and Brodkey varied between 1.5° and 21°, with an average angle of 8.5°. The outward penetration depth of the ejected fluid elements was random, the fluid elements being disrupted eventually, accelerated and swept downstream in axial direction at the expense of the mean flow. With respect to a stationary observer the ejection phase ended with the entry from upstream of fluid which was directly mainly axially, and with a U_1-velocity roughly equal to the mean value. These ejections contribute to the production of turbulence through the value of $-u_2u_1$. The instantaneous values were predominantly positive, and the contribution of the ejections to the shear stress is intermittent. The ejections occurred about 20 percent of the time, yet at $\mathbf{Re}_D = 20{,}000$ they contributed to 70 per cent of the shear stress. Associated with the ejection events a deceleration of the axial velocity in local regions near the wall was observed, with deficiencies up to 50 per cent of the local mean velocity. No negative U_1-values were ever observed, in accord with the findings by Popovich and Hummel. The instantaneous acceleration of ejected fluid elements due to the entering of high-momentum fluid from upstream causes, as mentioned, a sweep almost in axial direction. Though Corino and Brodkey also observed that the entire sweeping motion are on the average directed weakly towards the wall, the velocity vector making an angle varying from 5° to 15° with the wall. If the attention is to be focused mainly on this wallward motion of the sweep phase, following Grass[124] it may be convenient to refer to it as the inrush phase.

In spanwise direction the above events do not take place simultaneously. In this direction regions of higher and lower axial velocities occur in a similar fashion as the spanwise variation of the u_1-velocity observed by Kline *et al.* The ratio between the axial velocities of the faster and slower moving fluid varied between 1.2 and 2.9, with an average of 1.5. This variation of axial velocity in spanwise direction corresponds with the existence of "vertical" shear-layers. They do not contribute to the shear stress, but may form layers of higher viscous dissipation. On the other hand, when during the ejection and sweep phases entering fluid streams accelerate and displace fluid above a particular x_2-position, the underlying fluid remains more or less in a retarded situation. The interface between the accelerated and retarded fluid is rather sharp, which implies an intensive "horizontal" shear-layer. This intensive shear-layer varies its position between $x_2^+ = 4$ and 32.

A confirmation of the observations by Kline *et al.* and by Corino and Brodkey has been obtained by Grass[124] with his experiments in a waterchannel (depth 50 mm, average velocity $\bar{U}_{1,\text{aver}} = 0.245$ m/s). As mentioned earlier, Grass also applied the hydrogen-bubble technique. Figure 7-36 shows typical instantaneous velocity profiles of the U_1-component and the u_2-component during an ejection phase, and an inrush phase when high momentum fluid enters the region closer to the wall in a wallward motion. The strong defect of the instantaneous U_1-velocity during the ejection phase, as well as the marked increase of the same velocity component during the inrush phase are clearly demonstrated. This picture of the course of instantaneous U_1-velocity profiles obtained from the above experiments has been confirmed by the experimental results obtained by Blackwelder and Kaplan.[154] In the turbulent boundary layer of a windtunnel ($\delta = 75$ mm, $\mathbf{Re}_{\delta_m} = 2{,}550$) they measured the instantaneous U_1-velocity profile using a probe containing ten hot-wires placed in a row in the x_2-direction. The hot-wire nearest to the wall was at $x_2^+ = 5$, the outermost hot-wire at $x_2^+ = 100$. From the u_1-oscillograms taken simultaneously a high degree of correlation between velocities is noticed in the region up to $x_2^+ \simeq 20$. Bursts have been observed. In agreement with the above picture a momentum defect occurred just prior to the first burst event, for $x_2^+ \gtrsim 40$, and a momentum excess for $x_2^+ > 40$. This outer excess velocity gradually covers the whole region $x_2^+ < 100$. At about the same time as this burst and shortly after it there is a steep rise of the u_1-velocity from a negative to a positive value. Thus shortly after the ejection burst event there is a velocity excess everywhere (inrush). Prior to this burst event there is first a strong deceleration of the fluid in the region $x_2 \gtrsim 20$ before it is accelerated during and after the burst.

Noticeable in Fig. 7-36 is a negative value of $-u_2u_1$ for $x_2 = 10$–15 mm, both u_1 and u_2 being positive simultaneously during the ejection and inrush phases. On the other hand, the local and temporary negative values of $-u_2u_1$ were observed by Kim *et al.*[122] to occur much closer to the wall, namely at $x_2^+ = 10$ to 20. As will be discussed later Wallace *et al.*[127] obtained the same result. The inflexion point in the U_1-profile during the ejection phase is favorable for a local instability, similar to the secondary instability during transition. Grass concluded from the instantaneous velocity distributions like those shown in Fig. 7-36, that the contributions to the shear stress were equally strong during the ejection and sweep-inrush phases.

Later experiments by Nychas *et al.*[131] in a turbulent boundary layer along a flat plate placed centrally in a watertunnel, using particles of polyvinyltoluene butadiene (density 1,026 kg/m^3) in the size range of 44–74 μm as tracers, confirmed many aspects of the events in the wall region discussed above. The events responsible for the production of turbulence were the ejections and following sweeps when the ejected fluid elements are caught by the oncoming fluid and swept downstream. The instantaneous contributions during the ejections and sweeps, being very short in time may attain very high peak-values. Nychas *et al.* reported instantaneous values of $-u_2u_1$ as high as 40 times the local mean value $-\overline{u_2u_1}$.

Even higher instantaneous values of $-u_2u_1$ have been reported by Willmarth and Lu,[153] applying a conditional sampling hot-wire anemometer technique for obtaining these instantaneous values. Namely, up to 60 times the local mean value, occurring just prior and during an ejection "burst" event. The u_1-velocity near the wall was negative and decreasing with time. Also important contributions to $-u_2u_1$ were observed during the sweep-inrush period following the first burst, but the $-u_2u_1$ value was not as peaked as it was during bursting and roughly equal to $7(-\overline{u_2u_1})$. In general they observed the largest contributions when $u_1 < 0$ and $u_2 > 0$, less large contributions when $u_1 > 0$ and $u_2 < 0$ (sweep-inrush periods), and relatively small negative contributions of about equal value for the two cases $u_1 < 0$, $u_2 < 0$ and $u_1 > 0$, $u_2 > 0$.

The experiments by Eckelmann[115] in the oil channel were continued by Wallace *et al.*,[127] by making simultaneous recordings of the instantaneous u_1, u_2, and u_2u_1 signals. These recordings too showed

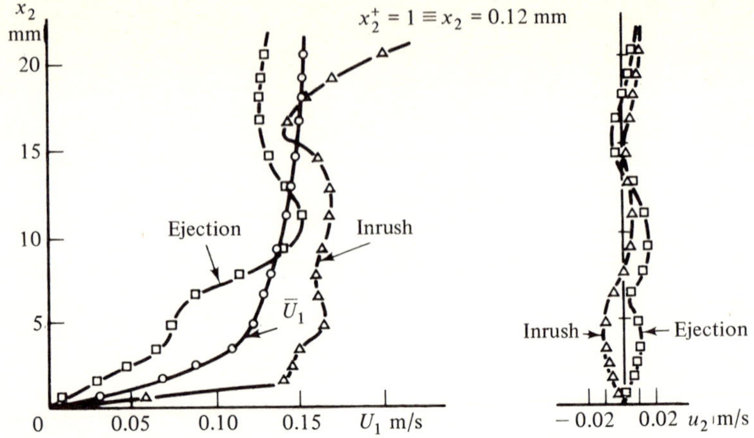

FIGURE 7-36
Instantaneous lateral distributions of U_1 and u_2 during an ejection and inrush phase. (After: *Grass, A. J.*,[124] *by permission of the Cambridge University Press.*)

the events of ejection and sweep-inrush movements of fluid elements, both usually coupled with period of high activity, alternated with periods of low activity. As may be expected the main contribution to the shear stress during the sweep-inrush phases was by the inrush part with much higher values of the u_2-velocity component. This followed clearly from a comparison of the simultaneous traces of the above three signals. The largest contributions were due, either to low-speed fluid moving outward during an ejection, or to high-speed fluid moving wallward during an inrush, in accord with the descriptions given by Corino and Brodkey,[126] Nychas *et al.*,[131] and also for a part by Kim *et al.*[122] The total contributions during the ejection periods were roughly equal to those during the inrush periods, though, as discussed earlier, according to Willmarth and Lu[153] much more peaked, each being about $\frac{2}{3}$ of the net shear stress at $x_2^+ = 15$, which is close to the region of maximum production. This would make in total $\frac{4}{3}$ times the net shear stress. However, as mentioned earlier, occasionally during the outward and wallward motions, interactions resulted in values of the u_1-velocity and u_2-velocity components of equal sign, and consequently in a negative value of $-u_2u_1$. Hence, a situation similar to the one mentioned in the description of the laminar-turbulent transition process. These negative contributions were about equal during the outward and wallward motions, each contribution being about $\frac{1}{6}$ of the net shear stress.

The R.M.S. of the difference between the instantaneous value of $-u_2u_1$ and the mean value $-\overline{u_2u_1}$, i.e.

$$\left[\overline{\left(u_2u_1 - \overline{u_2u_1}\right)^2}\right]^{1/2} = \left[\overline{(u_2u_1)^2} - \left(\overline{u_2u_1}\right)^2\right]^{1/2}$$

has been measured by Antonia[233] in a turbulent boundary layer for both the smooth-wall and rough-wall condition. Across the main part of the boundary layer the standard deviation was roughly equal to three times the local mean shear stress, and about twice the square of the wall-friction velocity.

The occurrence of temporal negative contributions to the shear stress has also been reported by Kim *et al.*[122] They further observed that during the periods of high activity, which they referred to as the "bursting" periods, the u_1-traces showed large amplitude oscillations downstream of the zone where the instantaneous U_1-velocity profile showed a dent with inflexion points. These oscillations were quite regular and organized during a number, up to 10, cycles. These "bursting" periods alternate with periods of relative quietness with smaller-amplitude random motions. During these latter periods there was practically no contribution to the turbulence production. By introducing a "bursting" intermittency factor, they obtained an average value greater than one at $x_2^+ \simeq 10$. This led them to the above conclusion of the presence of periods of negative turbulence production. The longest time-scale present in the turbulence at this distance from the wall was that of the entire process including the period of random chaotic motions following the "burst" and the relatively quiet period of little activity. The average time $\bar{T}_B$ between "bursts," when compared with the auto-correlation of the u_1-velocity component, coincides roughly with the first re-rise maximum value of the correlation coefficient $[\mathbf{R}_{1,1}(t)]_E$. When scaled with the wall-region variables u^* and ν, the period $\bar{T}_B$ satisfies the relation

$$\frac{u^{*2}\bar{T}_B}{\nu} = 0.65\,\mathbf{Re}_{\delta_m}^{\ 0.73} \tag{7-86}$$

i.e., dependent on the Reynolds number. However, when scaled with the outer-region variables, the non-dimensional period $U_0\bar{T}_B/\delta$ turned out to be independent of the Reynolds number. The above relation has been confirmed by the results obtained by Laufer and Badri Narayanan[133] from hot-wire anemometer measurements in the boundary layer along one of the walls of a windtunnel. These measurements too showed the turbulence production to occur in a roughly periodic way through the formation and straining of vortices and their breakdown ("bursts"), with a probabilistic mean period that includes both the ejection burst and inrush burst satisfying the above relation given by Kim *et al.* If the $\frac{1}{7}$ power-law for the mean-velocity distribution is assumed and the corresponding Blasius friction law, taking at the same time the above non-dimensional period $\bar{T}_B$ proportional with $\mathbf{Re}_{\delta_m}^{\ 0.75}$, it can easily be shown that $U_0\bar{T}_B/\delta \simeq 5$. As Kim *et al.* pointed out, this result which they obtained from the hydrogen-bubble data suggests that the low-speed streak-lifting seems to be triggered by large-scale disturbances already present in the flow, so that $\bar{T}_B$ should scale with the outer-region variables. The primary energy transfer occurs as a result of local intermittent instability, which has a preferred range of frequencies of occurrence and of oscillation. It is of interest to note that as a consequence of these results Kim *et al.* suggested that an appropriate description of the turbulent boundary-layer flow should be based on a two-part model, one part considering an unorganized eddy structure, the other part being a wave or wavepacket model. A distinction more or less similar to Townsend's[31] model of "attached" large eddies and purely randomly orientated smaller eddies.

The interrelation between the processes taking place in the outer and inner region is confirmed by the observation made by Laufer and Badri Narayanan[133] that the value of $\bar{T}_B$ following from $U_0\bar{T}_B/\delta \simeq 5$ is of the same order as the convection time for the average wavelength of the "wavy" outer interface of the boundary layer. Or, put in other words, the characteristic wavelength corresponding with $\bar{T}_B$ is roughly equal to the average wavelength of the "wavy" interface. For, putting $\bar{T}_B \simeq \lambda_1/U_c$, where λ_1 is a streamwise wavelength, we obtain from $U_0\bar{T}_B/\delta = 5$, $\lambda_1/\delta = 5U_c/U_0$. If we assume $U_c/U_0 = 0.8$, we obtain $\lambda_1/\delta = 4$ which would mean that the spatial periodicity of the "bursts" is of the magnitude of the large eddies in the outer region. Rao *et al.*[134] made a special study of the "burst" phenomena. They observed that the "bursts" occur throughout the whole turbulent boundary layer with practically the same value of $\bar{T}_B$ for the two bursts even at distances from the wall beyond the inner region proper, up to $x_2^+ \simeq 2{,}000$ at $\mathbf{Re}_{\delta_m} = 6{,}550$. They obtained a value $U_0\bar{T}_B/\delta_d = 32$, i.e., somewhat lower than the value 40 deduced from the experiments by Kim *et al.* Assuming again a $\frac{1}{7}$-power law for the mean-velocity distributions, one would obtain values for $U_0\bar{T}_B/\delta$ of 4 and 5, respectively. On the other hand Blackwelder and Kaplan[154] reported at a slightly lower Reynolds number a value of 10, the duration of the burst event proper being 25 to 30 per cent of the total period.

In the fully turbulent region the "burst" phenomena are obscured by the general "background" turbulence, and consequently not directly noticeable in the velocity oscillograms. However, by filtering out this "background" turbulence by means of a narrow bandwidth wave analyzer, Rao *et al.*[134] found that the "bursts" clearly showed up in the filtered traces. The probability distribution of the "burst" frequency T_B^{-1}, was almost log-normal. They suggested that these "burst" phenomena in the fully turbulent region could be related to the smaller-scale intermittency observed by Batchelor and Townsend in grid-generated turbulence and in a turbulent wakeflow (see Sec. 3-5, and ref. 30).

We mentioned that both the ejection and inrush events are accompanied by a period of relatively high activity or a burst, the inrush burst being in general weaker than the ejection burst. Both are observed in the fully turbulent part of the wall region, at least.

However, when approaching the wall, and in particular the viscous sublayer, the effect of the ejection burst will diminish so that finally only the inrush burst will be observed. Indeed, the results obtained by Kim *et al.*[122] in the layers just beyond the viscous sublayer (at $x_2^+ \simeq 10$) giving the value of $U_0\bar{T}_B/\delta \simeq 5$ refer to one burst per cycle only, while Rao *et al.*[134] determined the same relation for $\bar{T}_B$ referring to a cycle including both the ejection and inrush burst.

Also Ueda and Hinze[234] concluded from narrow-band pass signals of $(\partial u_1/\partial t)^3$ taken in a constant-pressure turbulent boundary layer ($U_0 = 4.1$ m/s and 13.5 m/s; $\mathbf{Re}_\delta = 11{,}450$ and 35,500 respectively) in a windtunnel to a value of $U_0\bar{T}_B/\delta \simeq 5$ for $x_2^+ \gtrsim 10$, which value decreased to $\simeq 2.5$ for $x_2^+ \gtrless 40$. The single burst during one cycle in the region $x_2^+ \gtrsim 10$ referred most probably to the effect of the inrush event since the skewness factor of $\partial u_1/\partial t$ was positive and quite large in this region.

Wall-pressure Fluctuations

Additional information concerning the turbulence structure near the wall can be gained from measurements of the pressure fluctuations at the wall, in particular when these pressure fluctuations are correlated with velocity measurements near the wall. Extensive measurements have been made of space-time correlations between wall-pressure fluctuations p_w in two points of the wall, and between p_w in a point of the wall and u_1 and u_2, respectively, in another point near the wall, by Willmarth and Wooldridge.[135,136] Later Willmarth and Tu[137,138] extended these correlation measurements by including the third velocity component u_3. The measurements were carried out in a windtunnel. The boundary-layer thickness at the location of the measurements was $\delta \simeq 125$ mm $[\delta_d = 12.5$ mm; $\delta_m = 9.6$ mm$]$. The free-stream velocity was $U_0 = 62$ m/s. $[\mathbf{Re}_{\delta_m} = 38{,}000.]$ The pressure-transducer used consisted of a lead-zirconate-titanate disk of 1.5 mm diameter and 0.5 mm thickness, mounted flush with the wall. The turbulence velocities u_1, u_2, and u_3 were measured with a hot-wire anemometer with the wires in X-array, while another hot wire was placed at a fixed point at 0.05 mm distance from the wall ($\delta_l \simeq 0.04$ mm) for measuring the wall shear stress σ_w. With the origin of the coordinate system at the pressure transducer, the following space-time correlations were measured

$$\mathbf{R}_{p,p}(\xi_1,\xi_3;\tau) = \frac{\overline{p_w(0,0;t)p_w(\xi_1,\xi_3;t+\tau)}^{\,t}}{\left\{\overline{[p_w(0,0;t)]^2}^{\,t}\;\overline{[p_w(\xi_1,\xi_3;t+\tau)]^2}^{\,t}\right\}^{1/2}} \tag{7-87a}$$

and

$$\mathbf{R}_{p,i}(\xi_k;\tau) = \frac{\overline{p_w(0,0;t)u_i(\xi_k;t+\tau)}^{\,t}}{\left\{\overline{[p_w(0,0;t)]^2}^{\,t}\;\overline{[u_i(\xi_k;t+\tau)]^2}^{\,t}\right\}^{1/2}} \tag{7-87b}$$

Also the two-point velocity-correlations $\mathbf{R}_{1,1}(\xi_1,x_2,0;0)$ and $\mathbf{R}_{1,1}(0,x_2,\xi_3;0)$ were measured. In the plane $x_2 = 0.004\delta_d$ $(x_2^+ \simeq 6)$ the correlations were confined to a narrow strip of about $12\delta_d$ length and $2\delta_d$ width. With increasing ξ_1 the correlations changed sign at roughly $\xi_1 = \pm 2\delta_d$. One of the important results of the pressure-velocity correlation measurements was that the correlation $\overline{p_w u_2(\xi_k;t+\tau)}$ appeared to correspond with a turbulence convected approximately with the local mean-velocity $\bar{U}_1(\xi_2)$ at the position where u_2 was measured. The zero-time delay correlation $\overline{p_w u_2(\xi_k)}$ turned out to be an odd function of ξ_1, being positive for $\xi_1 > 0$, negative for $\xi_1 < 0$, and zero for $\xi_1 = 0$. The iso-correlation contours for zero-time delay, in planes parallel to the wall showed the same behavior for planes $\xi_2 > 0.5\delta_d$, the contours being almost symmetrical with respect to $\xi_1 = 0$, the correlations changing sign when $\xi_1 \gtrless 0$. However, for planes closer to the wall this symmetry in shape no longer showed up. The contours had a swept-back nature in that, though for $\xi_3 = 0$ the correlation was still negative for $\xi_1 < 0$, it attained positive values for $\xi_3 > 0$ and for negative, but small, values of ξ_1. The degree of sweep-back increased with decreasing ξ_2. With $\xi_1 = 0$ the $\overline{p_w u_2}$ correlation was practically zero in spanwise direction for $\xi_2/\delta_d > 0.5$. Because of the swept-back nature for $\xi_2/\delta_d < 0.5$ it was positive but decreased rapidly to zero with increasing ξ_3. It became zero and changed sign for $\xi_2/\delta_d = 0.1$ and 0.2 at $\xi_3/\delta_d \simeq 0.5$ and

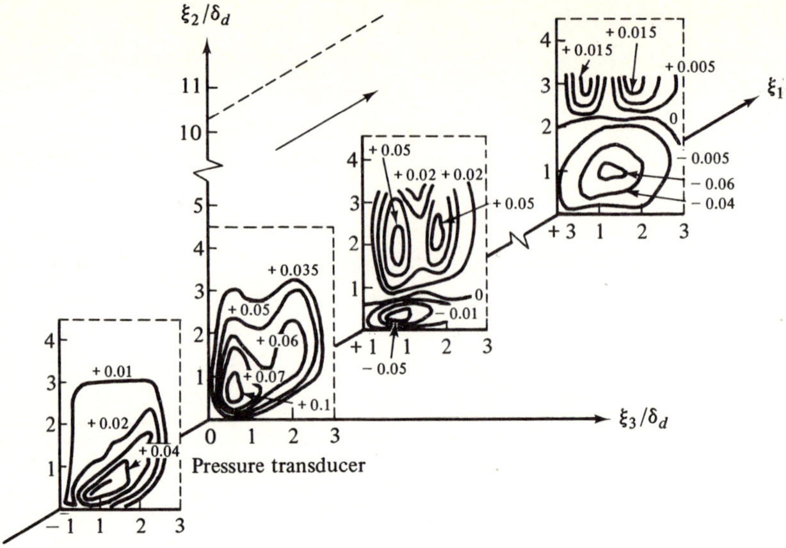

FIGURE 7-37
Three-dimensional diagram of isocorrelation contours of $\mathbf{R}_{p,3}(\xi_k)$. (From: *Willmarth, W. W., and B. J. Tu,*[138] *by permission of the American Institute of Physics.*)

$\simeq 1$, respectively. The isocorrelation contours were strongly elongated in the streamwise direction.

Of particular importance in connection with the three-dimensional structure of the turbulence near the wall are the measurements of the correlations $\overline{p_w u_3(\xi_k)}$ with zero-time delay. Figure 7-37 shows iso-correlation contours $\mathbf{R}_{p,3}(\xi_2,\xi_3) =$ constant in (x_2,x_3) planes for different ξ_1 values.

The pressure-velocity correlations are rather weak. The maximum values observed for the correlation coefficients are ~ 0.3 for $\mathbf{R}_{p,1}$ and ~ 0.1 for $\mathbf{R}_{p,2}$ and $\mathbf{R}_{p,3}$.

The results of the pressure-velocity correlations just considered are consistent with the following picture, based upon the behavior of U-shaped vortices as described by Hama.[82] A slightly different explanation has been offered by Willmarth *et al.*[138] The U-shaped vortices pass at random the pressure transducer. It is reasonable to expect that an optimum effect of these vortices on the measured p_w will occur at the situation when the lower top of the U-shaped loop is near or just above the pressure transducer. This situation is illustrated in Fig. 7-38a. At this position of the U-vortex the pressure measured by the transducer will be negative $p_w < 0$. Figure 7-38b shows the regions in (x_2,x_3)-planes where the induced velocity u_3 is either positive or negative, resulting in an either negative or positive value of $p_w u_3$. For $\xi_1 \leq 0$ and $\xi_3 > 0$ (up to points A and A'), $u_3 < 0$, while for $\xi_1 > 0$ there is a region $O''A''BC$ where $u_3 > 0$. This figure also shows that $\overline{p_w u_3} = 0$ at $\xi_3 = 0$, and that $\overline{p_w u_2} \gtrless 0$ when $\xi_1 \gtrless 0$. It may be noticed from Fig. 7-37 that very close to the wall, the $p_w u_3$ correlation extends in the x_3-direction up to $\xi_3/\delta_d \simeq 1$ to 2, before

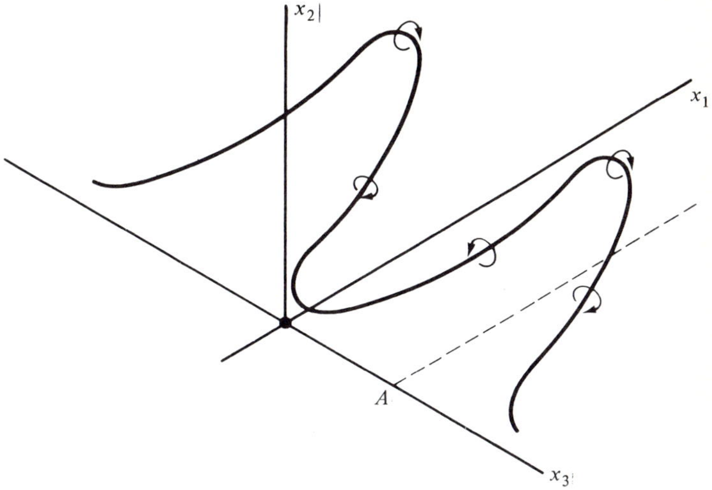

FIGURE 7-38a
Situation of maximum effect of vortex loop on p_w, measured with a transducer at 0.

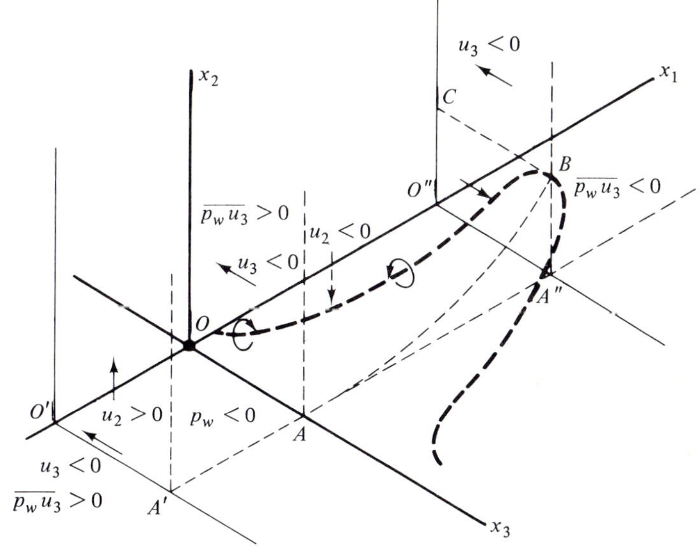

FIGURE 7-38b
Effect of the lower part of the U-vortex when passing the transducer at O, on the induced velocities u_2 and u_3, and on the correlation $\overline{p_w u_3}$ (Adapted from: *Willmarth, W. W., and B. J. Tu.*[138])

it changes sign. This value of ξ_3/δ_d is roughly of the same magnitude as that at which the correlation $\overline{p_w u_2}$ changes sign, and mentioned earlier. At the high Reynolds number $\mathbf{Re}_{\delta_m} = 38{,}000$ at which Willmarth *et al.* carried out their experiments, the value of $u^*/\nu \simeq 1.28 \times 10^5\ \mathrm{m}^{-1}$, whence follows a value of $\delta_d^+ \simeq 1{,}600$. Consequently the value of ξ_3^+ at which the pressure-velocity correlations $\overline{p_w u_2}$ and $\overline{p_w u_3}$ change sign is an order of magnitude larger than the spanwise spacing $\lambda_3^+ \simeq 100$ mentioned earlier for the streaky structure of the turbulence in and near the sublayer, and observed at much lower values of the Reynolds number. This may be due partly to the Reynolds number effect suggested by Morrison *et al.*,[123] but also to the fact that because of the random nature of the phenomena concerned the correlations may extend over greater distances than might be concluded from regularly spaced U-shaped vortices as shown, for instance, in Fig. 7-38. Such a picture should only be considered in a qualitative way.

The isocorrelation curves for the two-point velocity correlations and pressure-velocity correlations respectively with zero-time-delay are elongated in the streamwise direction. In contrast to this the *two-point pressure correlations* with zero-time-delay measured by Willmarth and Wooldridge[135,136] showed a greater persistence in the x_3-direction than in the streamwise direction. At $\mathbf{Re}_{\delta_m} = 38{,}000$ Willmarth and Wooldridge measured a decrease of $\mathbf{R}_{p,p}$ to 0.1 when $\xi_1/\delta_d \simeq 2$ in the streamwise direction, whereas the same correlation value was measured not earlier than at a spanwise separation distance of $\xi_3 \simeq 3\delta_d$.

A similar difference in streamwise and spanwise direction of the wall-pressure correlation has been observed by Emmerling,[169] who applied a novel method for obtaining pictures of the instantaneous wall-pressure distribution under a turbulent boundary-layer flow. The method is an optical one and is based on the Mickelson-interferometer. One mirror of the interferometer consisted of a reflecting flexible part of the wall. Positive and negative pressure peaks up to $6p'_w$ were measured. In spanwise direction the pressure peaks extended mostly over a distance $\xi_3 \simeq 0.7\delta_d$, while in streamwise direction they extended over a distance $\xi_1 \simeq 0.5\delta_d$. In spanwise direction the instantaneous structure showed a correlation often exceeding $\xi_3 = 6.5\delta_d$ (the width of the flexible part of the wall), being greater than that in the streamwise direction. Apart from the fact that the pressure fluctuations at the wall are also affected by the larger-scale motions at a distance from the wall, no clear-cut explanation can be given for this difference in behavior between the pressure correlations, and the velocity or pressure-velocity correlations.

The measurements of wall-pressure fluctuations also yield a *relation between the intensity* $p'_w = (\overline{p_w^2})^{1/2}$ *and the mean wall shear stress* σ_w.

Willmarth and Roos[139] obtained the relation

$$p'_w = 2.66\sigma_w$$

The numerical constant has been obtained after extrapolation to zero diameter of the pressure transducer. The constant varied practically linearly with the diameter d of the transducer in the range $d/\delta_d = 0.122$ to 0.442. It decreased with increasing d.

With the smallest transducer used the numerical constant was 2.54.

In general one may expect a relation of the form

$$p' = C_1(\mathbf{Re}_\lambda)\rho\overline{u^2} \qquad (7\text{-}88)$$

where $\overline{u^2} = \frac{1}{3}\overline{u_i u_i}$. The factor C_1 is a weak function of $\mathbf{Re}_\lambda$. In grid-generated turbulence Uberoi obtained values between 0.6 and 0.8 for this factor (see Fig. 3-32). If we assume a similar relationship for p'_w, and further assume p'_w to be determined mainly by the maximum values of the turbulence intensities in the wall region, namely

$$(u'_1/u^*)_{\max} \simeq 3, \quad (u'_2/u^*)_{\max} \simeq 1.1 \quad \text{and} \quad (u'_3/u^*)_{\max} \simeq 1.7$$

[See Fig. 7-77], we would obtain

$$\tfrac{1}{3}\rho\overline{u_i u_i} \simeq 4.3\rho u^{*2} = 4.3\sigma_w$$

Consequently, with the above values for C_1:

$$p'_w = C\sigma_w = (2.58 \text{ to } 3.44)\ \sigma_w$$

A number of other investigators have also tried to determine the value of C.[140-146,169] Kraichnan[140] predicted possible values ranging from 2 to 12. A value close to the lower limit, namely $C = 2.3$ has been obtained by Lilley[141] from theoretical considerations. The values obtained experimentally were in general higher, up to the value $C = 5$, suggested by Harrison.[143] Emmerling[169] measured p'_w with wall-microphones of different diameters. From his results one would obtain after extrapolation to zero diameter a value of $C \simeq 3.75$. Blake[145] obtained a value of 3.6 for a smooth wall but an average value of 3.4 for different rough walls, actual values varying between 2.9 and 3.8 depending on the kind and degree of roughness. As a reasonable average from the experimental data the following relation may be suggested

$$p'_w \simeq 3\sigma_w \qquad (7\text{-}89)$$

Convection Velocities

The spectral density of the pressure fluctuations at the wall is determined by the turbulence pressure field in the whole boundary layer passing the pressure transducer. Since the turbulence pattern is convected with a convection velocity depending on the size of the eddy and the distance from the wall, and since the pressure transducer measures the pressure fluctuations as a function of the frequency n, it cannot discriminate between n corresponding to a small wavenumber k_1 and large convection velocity (larger eddies farther away from the wall) and n corresponding to a large wavenumber and small convection velocity (small eddies closer to the wall).

Consequently Wills[146] pointed out that, when determining convection velocities from wall-pressure space-time correlations, one has to distinguish between a convection velocity $(U_c)_{k_1}$ at constant wavenumber, and a convection velocity $(U_c)_n$

at constant frequency, $c_r = -2\pi n/k_1$ being the phase velocity. From the space-time wall-pressure correlations $\mathbf{R}_{p,p}(\xi_1,\tau)$ we may obtain the Fourier transform[146]

$$\mathbf{E}_p(k_1,n) = \frac{\overline{p_w^2}}{4\pi^2} \int\!\!\int_{-\infty}^{+\infty} d\xi_1\, d\tau\, \mathbf{R}_{p,p}(\xi_1,\tau) \exp\left[\imath(k_1\xi_1 + 2\pi n\tau)\right] \qquad \text{(7-90)}$$

With one pressure transducer we measure

$$\mathbf{E}_p(0,n) = \frac{\overline{p_w^2}}{2\pi} \int_{-\infty}^{+\infty} d\tau\, \mathbf{R}_{p,p}(0,\tau) \exp(\imath 2\pi n\tau) \qquad \text{(7-90a)}$$

Figure 7-39 shows the nondimensional spectrum $U_0\mathbf{E}_p(0,n)/\overline{p_w^2}\,\delta_d$ as a function of $2\pi n\,\delta_d/U_0$, measured by Willmarth and Young,[147] and after correction for the finite size of the pressure transducer.

From Eq. (7-90) the function $\mathbf{E}_p(k_1,c_r/k_1)$ is obtained by substituting the phase velocity $c_r = -2\pi n/k_1$. For a given wavenumber k_1 it shows the power density as a

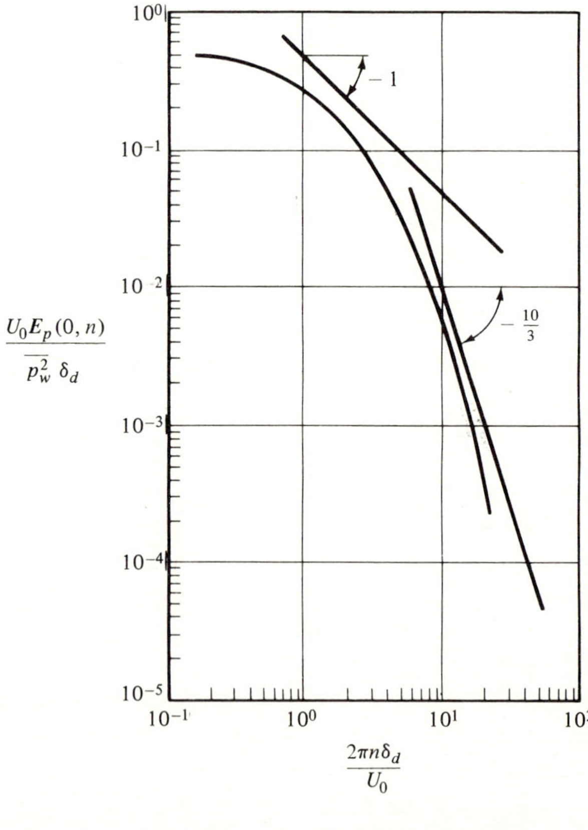

FIGURE 7-39
Spectral density of wall-pressure fluctuations. (From: *Willmarth, W. W., and C. S. Young,*[147] *by permission of the Cambridge University Press.*)

function of the phase velocity c_r. From this function Wills[146] obtained the convection velocity $(U_c)_{k_1}$, defined as the value of the phase velocity for which $\mathbf{E}_p(k_1,c_r/k_1)$ has a maximum, i.e.:

$$\left[\frac{\partial}{\partial c_r}\mathbf{E}_p\left(k_1,\frac{c_r}{k_1}\right)\right]=0 \quad \text{for} \quad c_r=(U_c)_{k_1} \tag{7-91}$$

Similarly $(U_c)_n$ is defined by

$$\left[\frac{\partial}{\partial c_r}\mathbf{E}_p\left(\frac{n}{c_r},n\right)\right]=0 \quad \text{for} \quad c_r=(U_c)_n \tag{7-92}$$

Wills carried out measurements in a windtunnel ($U_0 = 38$ m/s) with two pressure transducers (piezo-electric type of 3.2 mm diameter, and orifice-hot-wire probe of 0.92 mm diameter used for small separation distances). At the upstream transducer the boundary-layer thickness $\delta_{.995} = 3.3$ mm ($\mathbf{Re}_\delta \simeq 70{,}000$). At this Reynolds number the difference between $(U_c)_{k_1}$ and $(U_c)_n$ was small. In the frequency range $2 \gtrsim 2\pi n\delta_d/U_0 \gtrsim 5$ they were practically constant and equal to $(U_c)_{k_1} = 0.575\ U_0$ and $(U_c)_n = 0.59\ U_0$, respectively. For $2\pi n\delta_d/U_0 < 2$ they increased and became equal to each other at $2\pi n\delta_d/U_0 \simeq 0.5$. They attained a maximum of $0.87\ U_0$ at $2\pi n\delta_d/U_0 \simeq 0.25$, which corresponds roughly with a wavenumber $k_1\delta = 1.2$. These results are in agreement with earlier results on convection velocities deduced from space-time wall-pressure correlation measurements. Willmarth and Wooldridge[135] found that the convection velocity increased with increasing streamwise separation distance of two pressure transducers. The convection velocity amounted to $0.56\ U_0$ for small spatial separations, and increased to $0.83\ U_0$ for large separation distances. The lower values of the convection velocity are probably due to the contribution to the wall-pressure fluctuations by the smaller eddies of the wall region, which do decay more rapidly. This idea is supported by results obtained by Bull.[148] He distinguishes between two groups of convected wavenumber components. The group of high-wavenumber components which originate from the turbulence in the wall region, and another group with wavenumber $k_1\delta \gtrsim 0.5$ which is associated with the larger eddies of the outer region. The "broad-band" convection velocity showed initially a strong contribution from the turbulence in the wall region, but when the spatial separation increased beyond a few δ_d, the broad-band convection velocity increased rapidly to $0.825\ U_0$, which is roughly equal to the average mean-velocity across the boundary layer.

Extensive information on convection velocities in the boundary-layer have been gathered by Favre, Gaviglio, and Dumas.[149–152] In the above, Wills[146] made a distinction between two convection velocities. Also in Sec. 6-11 we discussed the possibility of defining different kinds of convection velocities. However, many more definitions are possible. Favre *et al.*[152] showed that when iso-space-time velocity-correlation curves $\mathbf{R}_{1,1}(\xi_1,0,0;\tau) =$ constant are drawn in a (τ,ξ_1)-plane, the following convection velocities, at least, may be distinguished:

$$(U_c)_\tau = \partial\xi_1/\partial\tau \quad \text{at the point where} \quad \frac{\partial}{\partial\xi_1}[\mathbf{R}_{1,1}(\xi_1;\tau)]_{\tau=\text{const.}} = 0 \qquad (7\text{-}93\text{a})$$

$$(U_c)'_\tau = \xi_1/\tau \quad \text{at the same point} \qquad (7\text{-}93\text{b})$$

$$(U_c)_{\xi_1} = \partial\xi_1/\partial\tau \quad \text{at the point where} \quad \frac{\partial}{\partial\tau}[\mathbf{R}_{1,1}(\xi_1;\tau)]_{\xi_1=\text{const.}} = 0 \qquad (7\text{-}94\text{a})$$

$$(U_c)'_{\xi_1} = \xi_1/\tau \quad \text{at the same point} \qquad (7\text{-}94\text{b})$$

Figure 7-40 shows how isocorrelation curves in a (ξ_1,τ)-plane would look for an inhomogeneous turbulent flow, like that in a boundary layer. In this figure the various ways of obtaining a convection velocity according to the above definitions are indicated. In addition to the above four possibilities one may also define a convection velocity

$$U_c = \xi_1/\tau \quad \text{at point } A \qquad (7\text{-}95)$$

As has been pointed out in Sec. 5-4, in the case of a homogeneous turbulent flow, the isocorrelation curves are more like those shown in Fig. 5-2c, as obtained, amongst others, by Favre *et al.*[152] in a grid-generated turbulence. In that case

$$(U_c)_\tau = (U_c)'_\tau = (U_c)_{\xi_1} = (U_c)'_{\xi_1} = (U_c)_A$$

We may conclude from Fig. 7-40 that for large separation distances the convection velocities according to the above definitions Eqs. (7-93) to (7-95) tend to the same value.

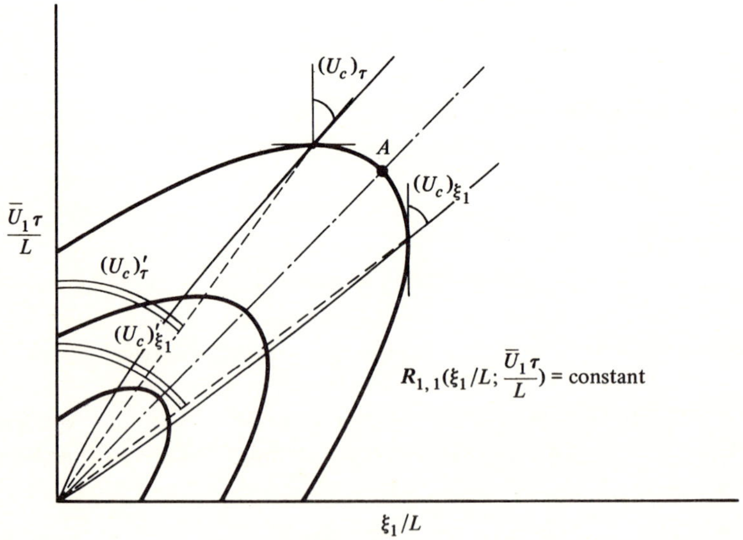

FIGURE 7-40
Convection velocities obtained from isocorrelation curves according to different definitions. L is a suitable length-scale.

Since for the general, inhomogeneous case $(U_c)_\tau \neq (U_c)_{\xi_1}$, Favre *et al.*[152] suggested yet another convection velocity, namely

$$U_c = [(U_c)_{\xi_1}(U_c)_\tau]^{1/2} \tag{7-96}$$

Favre also suggested, in particular in the case of relatively strong inhomogeneity, taking, in the dimensionless time $\bar{U}_1\tau/L$, an average value of $\bar{U}_1$ across the distance ξ_1, i.e.

$$\bar{\bar{U}}_1 = \frac{1}{\xi_1}\int_0^{\xi_1} d\xi_1' \, \bar{U}_1(x_1 + \xi_1') \simeq \tfrac{1}{2}[\bar{U}_1(x_1) + \bar{U}_1(x_1 + \xi_1)]$$

We may also conclude from Fig. 7-40, that the convection velocity $(U_c)_{\xi_1}$ appears to be almost independent of the separation distance, and equal to the limiting values for large separations of all the above convection velocities.

By passing the hot-wire anemometer signal through a low-pass and a high-pass filter, respectively, Favre *et al.*[152] found the lower wavenumber part of the space-time correlations to be much more persistent than the high wavenumber part, as may be expected. Also, since it is reasonable to assume that the smaller eddies will be less distorted by the large-scale motions and the mean shear than the bigger eddies, we may expect the smaller scale motions to behave more according to the Taylor hypothesis, and thus to be convected with the local mean-velocity. Indeed, the results obtained by Favre *et al.* showed this to be the case. Up to $\tau = \xi_1/\bar{U}_1 \simeq 1$ s the correlations $\mathbf{R}_{1,1}(\xi_1;0)$ and $\mathbf{R}_{1,1}(0;\tau)$ were the same across the whole boundary layer ($\mathbf{Re}_\delta = 27{,}900$; $\delta = 33$ mm). For larger separations, however, $\mathbf{R}_{1,1}(\xi_1;0) \gtrless \mathbf{R}_{1,1}(0;\tau)$ when $x_2/\delta \lessgtr 0.24$, while at $x_2/\delta = 0.24$ they were equal, independent of the separation. This leads to the conclusion that $U_c/\bar{U}_1 \gtrless 1$ when $x_2/\delta \lessgtr 0.24$. Since at the Reynolds number concerned, $\bar{U}_1/U_0 \simeq 0.8$ at $x_2/\delta = 0.24$, the conclusion obtained earlier from the space-time wall-pressure correlation that the larger scale motions are convected with roughly the average mean-velocity across the boundary layer finds support from the measurements by Favre *et al.* Additional support is obtained from the measurements of $(U_c)_{\xi_1}$ by Favre *et al.* of a "filtered" turbulent field, now at different distances from the wall. The value of $(U_c)_{\xi_1}$ approached the value of $\bar{U}_1$ when $(U_c)_{\xi_1}/2\pi n\delta$ decreased to zero. Again, for $x_2/\delta > 0.24$ the ratio $(U_c)_{\xi_1}/\bar{U}_1$ became smaller than 1 as $(U_c)_{\xi_1}/2\pi n\delta$ increased, whereas for $x_2/\delta < 0.24$ it increased with increasing $(U_c)_{\xi_1}/2\pi n\delta$, and the more so as the distance to the wall decreased.

Favre *et al.* extended their measurements in the same boundary layer and at the same $\mathbf{Re}_\delta = 27{,}900$ to the more general *space-time two-point velocity correlations* $\mathbf{R}_{1,1}(\xi_1,\xi_2,\xi_3;\tau)$. From these measurements they calculated *isocorrelation curves for optimum delay* τ_m (see Fig. 5-9 and ref. 151) and with different positions (x_2/δ) of the fixed hot wire. Figures 7-41a and 7-41b show typical results obtained with $x_2/\delta = 0.03$ and 0.152, respectively. Figure 7-42 shows $\mathbf{R}_{1,1}(\xi_1;\tau_m)$-curves, again with $x_2/\delta = 0.03$ and 0.152, respectively. The correlations $\mathbf{R}_{1,1}(\xi_1;0)$, also shown in Fig. 7-42, were determined along the average streamline through the point x_2/δ of the fixed hot wire,

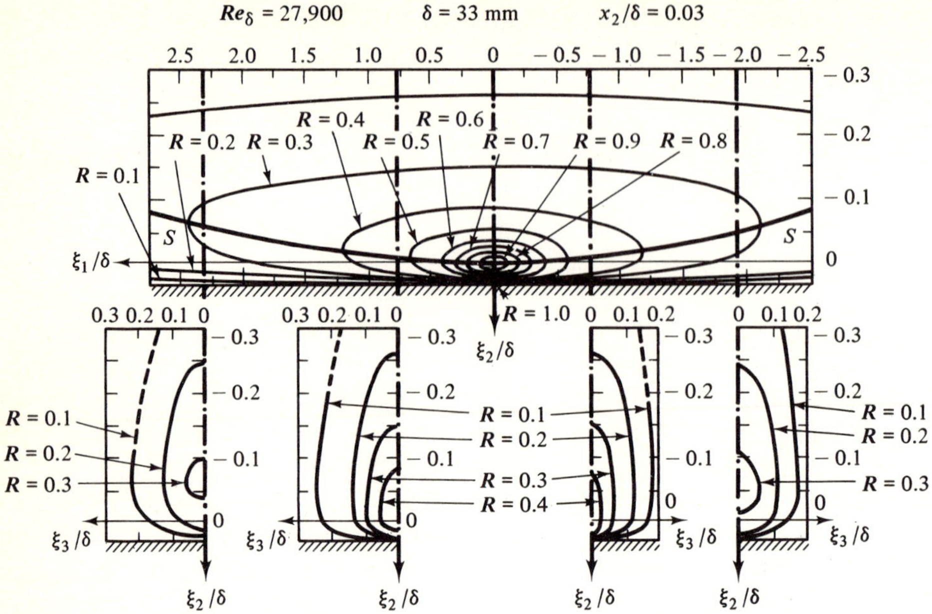

FIGURE 7-41a
Iso-space-time correlation curves $\mathbf{R}_{1,1}(\xi_1, \xi_2, \xi_3; \tau_m)$ = constant for optimum delay τ_m. Note the different scales for ξ_1/δ, ξ_2/δ, and ξ_3/δ. (After: *Favre, A., et al.*[151])

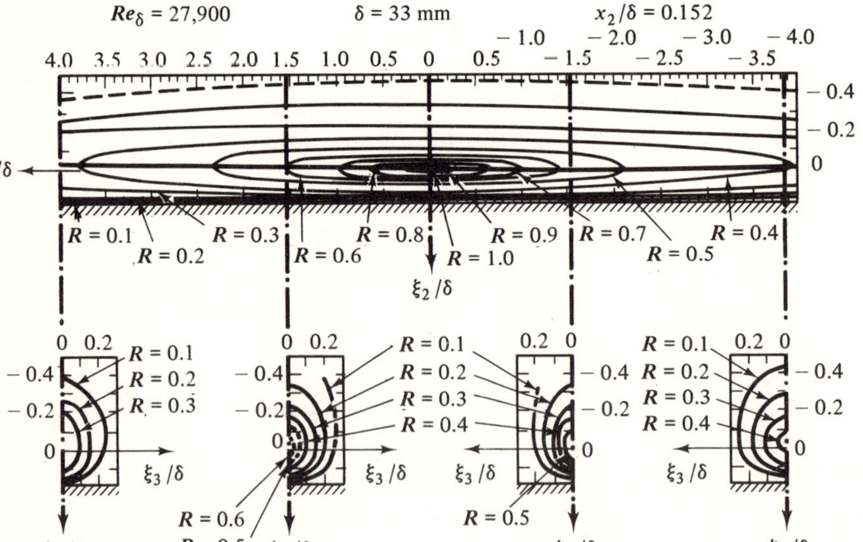

FIGURE 7-41b
Iso-space-time correlation curves $\mathbf{R}_{1,1}(\xi_1, \xi_2, \xi_3; \tau_m)$ = constant for optimum delay τ_m. Note the different scales for ξ_1/δ, ξ_2/δ, and ξ_3/δ. (After: *Favre, A. et al.*[151])

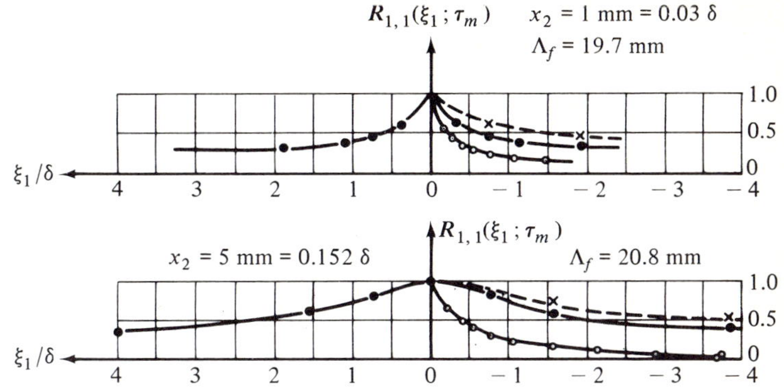

FIGURE 7-42
Space-time correlations $\mathbf{R}_{1,1}(\xi_1;\tau_m)$. $\mathbf{Re}_\delta = 27{,}900$; $\delta = 33$ mm.

• • • • along line of maximum correlation and optimum delay with wide-band filter (1 to 2,500 s^{-1}).

× × × × the same, but with low-pass filter (1 to 275 s^{-1}).

○○○○ along mean streamline through the point of the fixed hot wire, and for $\tau = 0$.

(From: *Favre, A., et al.,*[151] *by permission of the American Society of Mechanical Engineers.*)

the other correlations were determined along a line of maximum correlation and optimum delay τ_m. We note that the correlations measured with a low-pass filter ($n = 1$ to 275 s^{-1}) persist longer than those measured with a wide-band filter ($n = 1$ to 2500 s^{-1}).

The isocorrelation curves shown in Figs. 7-41a and 7-41b are very elongated in the streamwise direction, and not symmetric with respect to $\xi_1 = 0$. The line S, marking the points of *maximum maximorum* of the correlations, is slightly curved and convex towards the wall. However, for the isocorrelation curves taken at x_2/δ values corresponding with the outer region, the S-lines are slightly concave towards the wall. Sternberg[155] has offered a possible explanation of a qualitative nature for the observed shapes of the S-lines. Namely, that this may be caused by the mean shear, deforming the turbulence flow-pattern as determined by its vorticity distribution. The convection-velocity of this turbulence flow-pattern equals the mean-velocity $\bar{U}_1$ roughly in the plane $x_2/\delta = 0.24$ where this velocity is practically equal to its average value across the boundary layer. We may imagine that the maximum correlation will be determined by a point of the vorticity pattern that passes the fixed hot wire at $\tau = 0$, and is observed some time τ later. The bigger eddies which have a longer correlation-distance, will "rotate," due to the distortion by the mean shear, with an angular velocity proportional with $\partial\bar{U}_1/\partial x_2$ at $x_2/\delta = 0.24$ in clockwise sense, and with a "center of rotation" in this plane. When the fixed hot wire is in a plane $x_2/\delta < 0.24$, a point of the flow pattern passing the fixed hot wire at $\tau = 0$ will be displaced on the average from the wall by this rotation for $\tau > 0$, while for $\tau < 0$ this point has arrived from an upstream position also at a greater distance from the wall. This rough picture would indeed lead to an S-line convex towards the wall when $x_2/\delta < 0.24$. According to the same picture it will remain roughly parallel to the wall when $x_2/\delta = 0.24$, whereas it will be curved concave to the wall when $x_2/\delta > 0.24$, as has been observed, indeed, by Favre *et al.* For, in this case a point of the flow pattern passing the fixed hot wire at $x_2/\delta > 0.24$ will be moved towards the wall downstream of the hot wire due to the above "rotation."

Active and Inactive Motions

A study of the details of the structure of the turbulent boundary layer flow, in particular that in the wall region, led Townsend[156] to the conclusion that a distinction can be made in the turbulent motions between an *active* part and an *inactive* or *passive* part. The active part is responsible for the shear stress, while the inactive part does not contribute essentially to this shear stress. From considerations as given in Sec. 6-12, here too we may expect the inactive part to be associated with the irrotational mode of the turbulence. We may assume it possible to split the turbulence stress $\overline{u_2u_1}$ into a "rotational" and an "irrotational" mode. From Eq. (6-112) we obtain

$$\frac{\partial}{\partial x_2}\overline{u_2u_1} = -\overline{u_2\omega_3} + \overline{u_3\omega_2} - \frac{\partial}{\partial x_1}\frac{\overline{u_1^2} - \overline{u_2^2} - \overline{u_3^2}}{2}$$

With the above assumption we put

$$\frac{\partial}{\partial x_2}\overline{(u_2u_1)}_{\text{rot}} = -\overline{u_2\omega_3} + \overline{u_3\omega_2}$$

and

$$\frac{\partial}{\partial x_2}\overline{(u_2u_1)}_{\text{irrot}} = -\frac{\partial}{\partial x_1}\frac{\overline{u_1^2} - \overline{u_2^2} - \overline{u_3^2}}{2}$$

However, since $\overline{(u_2u_1)}_{\text{rot}}$ is not equal to the correlation between the vorticity modes of the pertinent velocity components, the assumption can only be approximately correct. But if the assumption is accepted, the irrotational mode of $\overline{u_2u_1}$ will not contribute much to the shear stress, the contribution being zero in a turbulence which is homogeneous in the x_1-direction. In that case $\overline{(u_2u_1)}_{\text{irrot}} = \text{constant}$, and since it is zero at the wall, it is zero everywhere.

The inactive part is observed mainly in the lower wavenumber region of the energy spectra, so it is of a relatively large scale. Hence, it is reasonable to associate the irrotational part of the turbulence with the pressure fluctuations and larger-scale motions of the outer region. However, according to Bradshaw,[157] who made a detailed study of the active and inactive part of the turbulence, the latter part contains not only the irrotational mode, but in addition also the large-scale vorticity field of the outer region, which acts on the turbulence of the wall region as if an unsteady external stream were present. An estimate has been made by Bradshaw on the basis of a simple mathematical model. To this end he assumed a large-scale periodic motion with a time scale much greater than that of the smaller-scale motions. By considering both a long-time average and a short-time average, he showed that the contribution of the large-scale motion to the relative turbulence intensity may be quite strong, whereas at the same time the effect on the mean-velocity was negligibly small.

Earlier we mentioned that the turbulence in the wall region is more or less in energy equilibrium, which means that the local production of turbulence is roughly equal to the local viscous dissipation. This suggests the existence of local similarity and a universal character of the smaller-scale part of the turbulence structure. According to Townsend[156] and Bradshaw[157] it is the *active shear-stress producing part* of the turbulence that appears to be similar and universal. It scales with $(\sigma_{21}/\rho)^{1/2}$ and the distance x_2 to the wall. Apparently the remaining part of the turbulence is not determined by the local conditions. Being *inactive,* it does not transfer momentum, though it contributes to the transfer of a scalar quantity. Also it does not interact with the universal "component," because of a difference in scale.

In the case of a constant-pressure boundary layer the shear stress σ_{21} varies very little across the major part of the wall region, thus supporting the assumption made earlier that the wall-friction velocity u^* may be taken as characterizing the turbulence in this region. For this reason the region close to the wall has been frequently referred to as the *constant-stress layer.* It suggests a direct effect on and control of the turbulence by the local wall conditions. On the other hand, when we have to deal with a non-constant-pressure boundary layer, where σ_{21} is no longer nearly constant close to the wall, we may expect that the assumption of the turbulence being directly determined by the local wall shear-stress σ_w may become incorrect. Taking the local value of the shear-stress velocity $(\sigma_{21}/\rho)^{1/2}$ instead of u^* in local similarity considerations concerning the turbulence velocity field would then be a better approach. Support for this idea is obtained from the measurements by Bradshaw[157] of the spectral distributions of $\overline{u_1^2}$, $\overline{u_2^2}$, $\overline{u_3^2}$, and $\overline{u_2u_1}$ in boundary layers with zero and adverse pressure-gradients. Since the larger inactive motions do not contribute to the turbulence shear stress, but do contribute to the kinetic energy, we may expect a departure from similarity below an upper limit of the wavenumber, which limit is at a higher wavenumber for the turbulence kinetic energy than for the turbulence shear stress, when taken at the same x_2. Indeed, Bradshaw's measurements did reveal similarity for the spectral distributions of $\overline{u_1^2}$, $\overline{u_2^2}$, and $\overline{u_3^2}$ for the different boundary layers only when $k_1x_2 \gtrsim 2$, whereas those for $\overline{u_2u_1}$ already showed similarity when $k_1x_2 \gtrsim 0.1$.

Dependency between the Turbulence in the Outer Region and that in the Wall Region

The earlier idea of making a distinction between a wall region and an outer region practically independent of each other, has proven to be successful in describing the mean-velocity distribution in the two regions, at least for a constant-pressure boundary layer. However, it is now clear from the considerations just given above, that newer experimental evidence points towards an interdependency or a non-negligible interaction between the two regions. Main experimental facts are: (1) the distinction between smaller-scale active motions and large-scale inactive motions in the wall region, which latter motions scale with outer variables; (2) the convection velocity of the larger-scale motions in the wall region, being roughly equal to 0.8 U_0, is larger than the local mean-velocity; (3) the average "bursting" period $\bar{T}_B$ is almost constant across the whole boundary layer; it scales with the outer variables ($U_0 \bar{T}_B/\delta \simeq 5$) and it is of the same order of magnitude as the average periodicity of the "wavy" outer interface of the boundary layer. Laufer[158] deduced from his own experiments an average wavelength between large-amplitude bulges of the outer interface roughly equal to 4δ. Assuming a propagation velocity equal to the free-stream velocity U_0 (actually it is slightly lower), one would obtain for the average periodicity $U_0 T/\delta \simeq 4$.

The direct effect of the outer region through the large-scale "inactive" motions on the wall region may be relatively small. However, the indirect effect through creating local regions of intense shear in the wall region by interaction with the flow in this region, and by that triggering local instabilities, is of major importance for the energy production. Vice versa, these secondary instabilities and energy production in the inner layers of the wall region may, through the generation of pressure fluctuations including large periodicities equal to the bursting periods, react on the flow of the outer region. Consequently, the wall-adjacent layers will be neither passive, nor only acting as a turbulence-generating zone, but they interact strongly with the outer region.[134]

Putting together all facts obtained from the experimental investigations, we may arrive at the following qualitative description of the turbulence mechanism in the wall region of a boundary layer. This description and the picture obtained from it are very general, the actual processes being much more complex. Also, the picture given may not be considered to be definite, but suggestive only.

It is striking that, notwithstanding the random nature of the turbulence, a repetition of similar processes may be distinguished, with a distinct and recognizable average spacing in both spanwise and streamwise directions. In time it corresponds with, on the average, some cyclic process, with many features similar to the laminar-turbulent transition process. When trying to give a description of this "cyclic" process in the fully developed turbulent flow, it is immaterial where the beginning of the "cycle" is fixed. Because of the similarity mentioned with the transition process we will begin the "cycle" with the situation where, owing to a large-scale disturbance already present in the outer region and outer part of the wall region, a horseshoe-shaped vortex is beginning to be formed locally at the wall. This vortex is deformed by the flow into a more and more elongated U-shaped loop in streamwise direction. Because of self-induction processes the tip of the loop moves away from the wall thereby coming into regions of ever-increasing velocities. Consequently the vorticity increases due to stretching processes. At the same time it gives rise to an outward flow between the legs of the U-loop, with a strong u_2-component near the tip. Between the vortex moving away from the wall and the wall a local deceleration of the fluid is effected. This ejection process transports low-momentum fluid away from the wall, thus producing a positive and marked contribution to the shear stress. Moreover, at distances $x_2^+ = 5$ to 30, an intense horizontal shear-layer is formed, showing up in the instantaneous U_1-velocity profile as a dent with inflection points. The resultant local inflexional instability and breakdown of the flow surrounding the original tip of the vortex produces a turbulence burst, similar to that observed during the laminar-turbulent transition process. The pressure waves associated with the turbulence burst are propagated through the whole boundary layer. At the same time the blob of fluid of high turbulence intensity produced during the burst is convected downstream and moves farther away from the wall, thereby increasing in scale, amongst others, by turbulent diffusion. Since at the same time high-momentum fluid is entering from upstream, the above blob of fluid is convected in an accelerated way or swept in downstream direction. The above pressure waves may add to the movement of fluid towards the wall, resulting in a sweep-inrush flow. The inrush process has already been preceded and initiated by a negative u_2-component downstream of the tip of the U-looped vortex before its breakdown. The sweep-inrush flow makes a very small angle (5° to 15°) with the wall, which at the wall also is observed as the entry of higher momentum fluid in almost horizontal direction. Both the ejection burst process, as well as the sweep-inrush flow and burst contribute to the shear-stress, and consequently are responsible for the turbulence production, mainly in the region $x_2^+ = 10$ to 15 from the wall.

The horizontal movement during the sweep-inrush period will be strongly retarded near the wall. It may eventually, in conjunction with the action of overtaking faster moving fluid at a greater distance from the wall develop into another horseshoe-type vortex.

In Fig. 7-43 an attempt is made to show this "cyclic" process in a rather crude schematic drawing showing only the ejection burst. Of course, this conceptual model is highly idealized. Anyway, it does show, through the instantaneous U_1-velocity profiles given, that the region up to $x_2^+ \simeq 40$ is most important. Marked deviations from the mean-velocity profile (dashed curves) are shown up to $x_2^+ \simeq 80$. It may be of interest to note that the dimensionless thickness δ^+ of a laminar boundary layer when approaching the station of breakdown to turbulence is roughly of the same magnitude. Also, from a number of incidental cases[80,81] which could be considered, it was possible to calculate the value of $[U_0 T_{\text{instab}}/\delta]$ at the station of breakdown. Here T_{instab} is the period of the unstable disturbance introduced to produce the artificial transition during the experiments concerned. When the boundary-layer thickness of δ just after breakdown to turbulence is taken values close to ~ 10 are obtained.

In the foregoing we considered mainly the processes in a (x_1,x_2)-plane through the top of a U-shaped vortex (see Fig. 7-43). Prior to the burst there is an ejection of fluid from the wall. However, in another (x_1,x_2)-plane, at $\lambda_3^+ = 50$, say, this vortex induces a flow towards the wall, leading to a

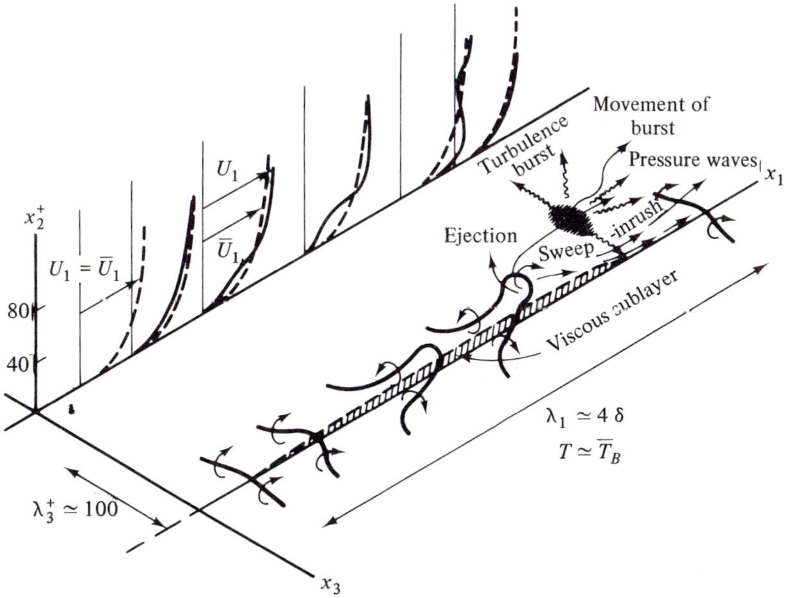

FIGURE 7-43
Conceptual model of the turbulence near the wall during a "cyclic" process, with average spacings λ_1 and λ_3.

velocity excess near the wall with respect to the mean-velocity $\bar{U}_1$-distribution, i.e., a peak region of the U_1-velocity. We may recall the existence of vertical shear-layers between valley and peak regions of the U_1-velocity. This may suggest some phase-shift in spanwise direction of the above "cyclic" processes. Or, put in a different way, at the same x_1-station ejection and inrush-sweep phenomena may alternate in spanwise direction. Again, of course, all these "cyclic" processes actually take place in a random way.

7-9 THE STRUCTURE OF THE OUTER REGION

The flow in this region, which for a constant-pressure boundary layer extends from $x_2 \gtrsim 0.15\delta$, shows in many aspects a great similarity with free-turbulent shear-flows, in particular with plane wake flows. In Sec. 6-4, dealing with the plane wake flow, we also considered the velocity defect with respect to the free stream, expressed in a relation similar to Eq. (7-33). Another similarity is shown by the intermittent nature of the turbulent flow near the outer boundary, though the amplitudes of the "wavy" interface are much smaller than in the case of the plane wake flow. This is quite clear from the corresponding distributions of the intermittency factor Ω, as shown in Fig. 7-25 and Fig. 6-13, respectively.

The turbulence in the outer region is characterized by large eddies, elongated in the main flow direction. The turbulence obtains its energy mainly by convection, and by turbulent transport from the inner region of the upstream part of the boundary layer. Consequently the turbulence is also determined by the conditions in the inner region, i.e., by the distribution of the wall shear stress farther upstream. Accordingly $-\overline{u_2 u_1}/\overline{q^2}$ is nearly constant in the outer region (see Fig. 7-22). Apparently the turbulent fluid has been sheared sufficiently long to attain its equilibrium structure (Townsend[31]).

In the foregoing section we already focused attention on the interaction processes between the inner and outer region. The above dependence of the turbulence on the conditions farther upstream in the inner region is through the convective contributions. But there is also a connection between the turbulence in outer and inner regions through pressure effects (inactive motion), which connection is much less extended in the main flow direction. In other words these pressure effects are relatively more local in streamwise direction, though they are convected mainly with the convection velocities of the larger eddies, i.e., roughly with the average mean-velocity across the boundary layer.

The dependency of part of the structure of the turbulence in the outer region on upstream condition points towards a *long "memory"* of the flow in this region. In sharp contrast, the inner region should have a much shorter "memory." Consequently the recovery from any disturbance will be much quicker when this disturbance is introduced in the inner region, than it will be when introduced in the outer region. This difference in rate of recovery is clearly demonstrated by results obtained experimentally by Clauser.[21] He measured the decay of a disturbance produced by a rod of 12.5 mm diameter placed at distances 38 and 140 mm from the wall in a boundary layer whose thickness at that station was 235 mm. The decay of the maximum deviations $\Delta\bar{U}_1$ of the distorted mean-velocity profile from the equilibrium profile is shown in Fig. 7-44.

Though, as Fig. 7-25 shows, the intermittency of the turbulent flow becomes distinct only for $x_2/\delta \gtrsim 0.4$, and consequently does not cover the whole outer region of the boundary layer, it is such an important feature of the outer region that we will devote the rest of this section to the intermittent part and the resultant consequences for the whole flow behavior of this region. Since, as already mentioned, the phenomena associated with the *intermittent turbulent/irrotational flow* are essentially the same as for the free-turbulent shear flow discussed in Sec. 6-12, it may be useful to reread the pertinent parts of that section.

In Sec. 6-12 reference has been made to an extensive investigation by Kovasznay *et al.*[107] on the structure of the intermittent part of the outer region of a boundary layer. This boundary layer was the one formed at one of the vertical walls of a windtunnel. The free-stream velocity was $U_0 = 4.3$ m/s, the thickness of the boundary layer at the station of the measurements was $\delta_{.99} = 100$ mm, which with the value of the free-stream velocity gives a $\mathbf{Re}_\delta \simeq 27{,}500$. With one hot-wire anemometer the instant was determined when the anemometer entered a turbulence zone, as well as the instant when it left the same turbulence zone. These instants thus correspond respectively with those when the front (downstream part) or the back (upstream part) of a bulge of the interface passes this detector probe. An instantaneous intermittency function $\tilde{\Omega}$ was defined, such that $\tilde{\Omega} = 1$ at the instant of entering, and $\tilde{\Omega} = 0$ at the instant of leaving the turbulence zone. With this hot-wire anemometer at different x_2-positions, corresponding with different values of the time-average intermittency factor Ω, "zone average" values of the U_1 and U_2 velocity components were measured with another hot-wire anemometer which could be traversed in the x_2-direction. This was achieved by making the measurements with the second probe either during the time that the detector probe was in a turbulence

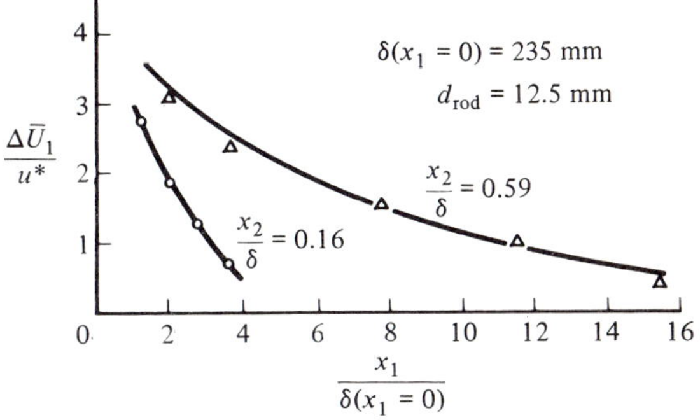

FIGURE 7-44
Decay of mean-velocity distortion in a turbulent boundary layer. (From: *Clauser, F. H.*,[21] *by permission of Academic Press, Inc.*)

zone, or during the time that it was outside this zone, i.e., in the irrotational zone. The turbulence-zone average $(\bar{U}_1)_{\text{turb}}$ was defined by

$$(\bar{U}_1)_{\text{turb}} = \frac{\overline{\tilde{\Omega} U_1}}{\Omega} \qquad \text{(7-97a)}$$

while the irrotational-zone average $(\bar{U}_1)_{\text{irrot}}$ was defined by

$$(\bar{U}_1)_{\text{irrot}} = \frac{\overline{(1-\tilde{\Omega})U_1}}{1-\Omega} \qquad \text{(7-97b)}$$

Distributions of these average velocities are shown in Fig. 7-45, together with the usual mean-velocity $\bar{U}_1$. Note that deep in the valley region of the interface $(\Omega \to 1)$, the irrotational mean-velocity is higher than $\bar{U}_1$, while in the crest region of the interface $(\Omega \to 0)$ in the turbulence zone the mean-velocity $(\bar{U}_1)_{\text{turb}}$ is lower than $\bar{U}_1$.

Kovasznay *et al.* also determined the average $\langle U_1 \rangle$-distribution in the turbulence zone at the instant that the front of this zone passed the detector probe, and this for different x_2-positions of this probe. A similar result was obtained for these distributions as found by Wygnanski and Fiedler for the $\langle U_1 \rangle$-distribution inside the turbulence zone of a two-dimensional mixing region, and discussed in Sec. 6-12. Namely that the distributions were almost linear with x_2/δ across a substantial part of the intermittency region. With $(x_2)_D$ denoting the position of the detector probe,

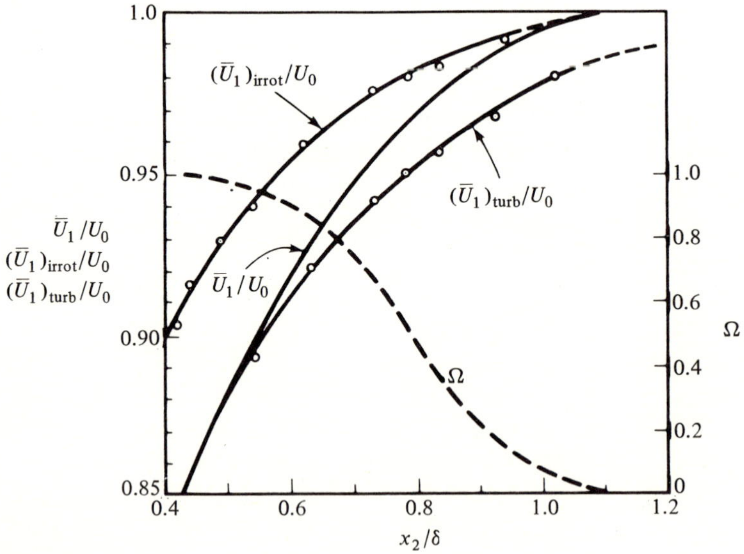

FIGURE 7-45
Distributions of the mean velocity $\bar{U}_1/U_0$, the turbulence-zone average $(\bar{U}_1)_{\text{turb}}/U_0$, and the irrotational-zone average $(\bar{U}_1)_{\text{irrot}}/U_0$ in the intermittent region of a boundary layer. (After: *Kovasznay, L. S. G., et al.*,[107] *by permission of the Cambridge University Press.*)

and with $\langle U_1 \rangle_D$ the average velocity at $(x_2)_D$, the distribution of $[\langle U_1 \rangle - \langle U_1 \rangle_D]/U_0$ was linear in the region $-0.15 < [x_2 - (x_2)_D]/\delta < 0$. The nondimensional velocity-gradient in this region had a value of

$$\frac{\delta}{u^*} \frac{d\langle U_1 \rangle}{dx_2} \simeq 8.3 \qquad (7\text{-}98)$$

Consequently in the turbulence regions of the bulges the fluid is subjected to an almost constant rate of strain. Compare in this respect the experiments by Wygnanski and Fiedler discussed in Sec. 6-12 (ref. 54 of Chap. 6) and the investigations by Townsend, Keffer, and Mobbs considered in the same section (references 9, 76, and 81 respectively of Chap. 6) on the effect of this straining action on the intensification of the turbulence and associated increase of viscous dissipation.

A very small positive value of $\bar{U}_2$ of the order of 0.001 U_0, i.e., a slight outward expansion of the mean boundary, was measured in the intermittent region. The turbulence zone-average $(\bar{U}_2)_{\text{turb}}$ was positive, $[(\bar{U}_2)_{\text{turb}} - \bar{U}_2]/U_0$ increasing to $\sim$0.012 in the crests of the turbulence bulges. On the other hand the irrotational zone average $(\bar{U}_2)_{\text{irrot}}$ was negative, with a maximum value of $[(\bar{U}_2)_{\text{irrot}} - U_2]/U_0$ equal to ~ -0.025 deep in the valley regions. Note that the absolute value of the inward irrotational motion is larger than that of the outward motion in the turbulence zone.

The intensity of the random irrotational fluctuations outside the turbulence zone showed a relation with the distance x_2 in nice agreement with Phillips' theory.[159] From the measurements the following relation may be obtained

$$\left[\frac{\overline{(u_1)_{\text{irrot}}^2}}{U_0^2}\right]^{-1/4} \simeq 13.5\left(\frac{x_2}{\delta} - 0.26\right) \qquad (7\text{-}99)$$

Thus this irrotational motion seems to have its effective sources according to Phillips' theory, located within the boundary layer, roughly at $x_2/\delta = 0.26$, which is just outside the wall region.

On the other hand both Bradbury[160] and Bradshaw[161] located the effective source at the average position of the interface, i.e., at $x_2/\delta \simeq 0.8$, with a correspondingly larger value of the constant of proportionality in the above relation. Since Phillips' theory applies only for large values of x_2/δ, the difference in the numerical factors is immaterial. Also from a physical point of view the difference in location of the effective source is immaterial if it is agreed upon that the motion of the interface and the occurrence of turbulence bursts within the boundary layer are not independent of each other.

By making a large number of *space-time two-point correlation* measurements of the u_1 and u_2 velocity components, with the fixed hot wire at different positions in the boundary layer and the movable hot wire at different ξ_2 and ξ_3 distances from the fixed hot wire, maps of isocorrelation curves in the (ξ_2,τ) and (ξ_3,τ)-planes could be obtained. The conclusions obtained therefrom by Kovasznay *et al.*[107] may be briefly summarized. The interface is highly corrugated with a root-mean-square slope in the (x_1,x_2)-plane roughly equal to 0.5. The individual bulges are three-dimensional and elongated in the streamwise direction with an aspect ratio of 2:1. They have a characteristic dimension varying between 0.5δ and δ. They appear at random and are roughly similar to each other. It is surmised that they originate from inside the boundary layer as a result of a strong outward motion of a blob of turbulent fluid with a typical diameter of $\sim 0.5\delta$. Probably they may be identified with the turbulence bursts mentioned in the previous section. Since they originate from a lower-velocity region they have a velocity defect with respect to $(\bar{U}_1)_{\text{turb}}$, but maintaining their angular momentum proportional with $\partial\bar{U}_1/\partial x_2$ at their original x_2 position. When this blob of turbulent fluid arrives in the new, higher-velocity, environment it is overtaken and entrained by the surrounding fluid, causing at the same time a wake downstream of this blob of fluid. Consequently it loses its momentum defect, spreading its angular momentum over a larger region, which may result in the almost constant value of $\partial\bar{U}_1/\partial x_2$ within the turbulence zone of the bulge observed in the interface region. The straining actions cause locally concentrated regions of turbulence and so may contribute to the Reynolds stress in these concentrated regions. As is clear from the foregoing the major contributions to the Reynolds stress originate from the short and active initial stages of the bursts. The resulting bulges themselves at the interface are rather passive. It is conjectured that these large-scale motions in the outer region and in the interface region may be connected with the large-scale wall-pressure fluctuations.

From the space-time correlation measurements, Kovasznay *et al.* deduced a *convection velocity* of the front of the interface near the crest of a bulge equal to $\simeq 0.97\ U_0$, and of the back of the interface down in the valley equal to $\simeq 0.92\ U_0$. Consequently the length of the turbulence zone at a given x_2-position increases continuously in streamwise direction roughly at a rate of $0.05\ U_0$. The average of the two convection velocities is $\sim 0.94\ U_0$, which means that the propagation velocity of the bulges of the interface relative to the free stream is $-0.06\ U_0$.

When we compare the "wavy" shape of the turbulent/irrotational interface of the boundary layer with the shapes observed in the free-turbulent shear flows discussed in Chap. 6, the agreement is qualitatively very close, but quantitatively there are differences, in particular concerning the depth of the indentations, or amplitude of the "waves," as mentioned earlier. Quantitatively this difference may be expressed by the values of $\overline{(x_2^* - \overline{x_2^*})'}/\overline{x_2^*}$ and the *entrainment constant* β. If for the boundary layer we take the values $\overline{x_2^*}/\delta = 0.8$ and $\overline{(x_2^* - \overline{x_2^*})'}/\delta = 0.15$ as given by Kovasznay *et al.*,[107] and mentioned earlier, we obtain $\overline{(x_2^* - \overline{x_2^*})'}/\overline{x_2^*} = 0.175$.

For the plane wake the value is roughly 0.4, and for the axisymmetric jet 0.23 (see Sec. 6-12). Accordingly, the entrainment constant β, as defined by Eq. (6-114), is largest for the plane wake and smallest for the boundary layer. In Sec. 6-12, we mentioned the values $\beta = 0.4$ and 0.18 for the plane wake and axi-symmetric jet respectively, while Townsend[162] estimated a value of $\beta \simeq 0.05$ from the results of the experiments on a boundary layer by Corrsin and Kistler.[29]

In Sec. 6-12 we mentioned that, though the *entrainment process* is still not yet completely understood, it is closely connected with the lateral spreading of the turbulence regions as determined mainly by the increase of amplitudes of the "wavy" interface. But the actual increase of the volume of turbulent fluid is due to an advancing velocity V normal to the interface. Figure 7-46 shows a sketch of the instantaneous shape of the interface, roughly one "wavelength" long. Phillips[163] estimated the value of this advancing velocity V from the measurements by Kovasznay *et al.*[107] to be $V \simeq 2$ to $3\ u^*$. Since spread of vorticity into a velocity-free region can only occur through viscous diffusion, Corrsin and Kistler[29] estimated this advancing velocity to be of the order $(\nu\omega')^{1/2}$ or $(\nu\varepsilon)^{1/4}$. However, by estimating the viscous dissipation ε from the measurements by Kovasznay *et al.*, Phillips obtained $(\nu\varepsilon)^{1/4} \simeq 0.17\ u^*$, and consequently $V \simeq 12$ to $18\ (\nu\varepsilon)^{1/4}$, which is one order of magnitude larger. Phillips himself offered an explanation for this disagreement. Because of the occurrence of very small eddies in the turbulence at the turbulence-side of the interface, this interface contains a large number of micro-convolutions superimposed on larger scale corrugations. It might well be that the probes used in the experiments were unable to discriminate between these micro-convolutions, and that in this way more integral effects were observed. The effective advancing velocity V of the interface as deduced from the measurements may quite well be much greater than the actual advancing velocity of the micro convolutions due to viscous diffusion, owing to an effect of these micro convolutions similar to that of the large scale "waves" of the interface on the global lateral spread of the boundary layer.

In Fig. 7-46 the closed dashed curves indicate the large-scale irrotational motions, associated with pressure fluctuations. Since, as mentioned earlier, and shown in Figs. 7-45 and 7-46, the free stream overflows the "wavy" interface, Laufer[158] suggested that at the crests and fronts of the interface local mixing-zones are generated which on the scale of the bulges may effectively contribute to the increase of rotational fluid.

It is possible to suggest the following, *conceptual model* of the processes occurring in the whole boundary layer (see also ref. 158). Induced by effects originating from regions away from the wall, at the wall local U-shaped vortices are generated which while moving away from the wall cause, through

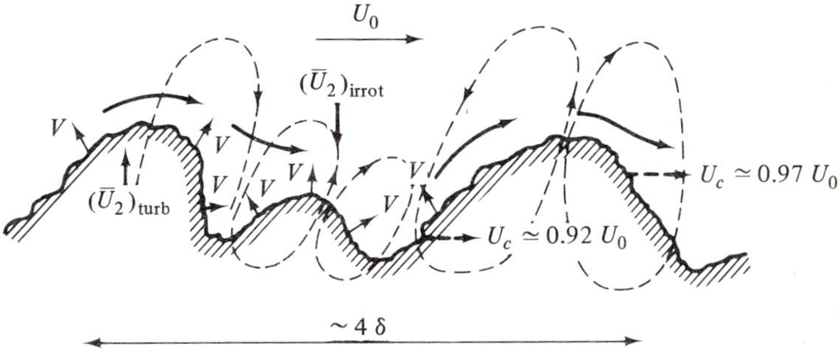

FIGURE 7-46
"Wavy" shape of the interface between turbulent and irrotational parts.

local intensive "horizontal" shear-layers, instabilities leading to turbulence bursts. The corresponding highly turbulent blob of fluid further moves away from the wall, is entrained by the higher-velocity fluid of the outer region and through turbulence diffusion increases in size. At the same time large-scale motions of irrotational nature are induced by the pressure-fluctuations associated with the turbulence bursts. Near the interface the blob of turbulent fluid causes a local bulge at the interface. The interaction with the free-stream causes pressure gradients through momentum exchange which pressure gradients induce higher velocity flow towards the wall (sweep-inrush phase) during a "valley" stage of the interface. Due to the retarding effect of the wall by viscous shear again locally a U-shaped vortex will be formed at the wall. While the small-scale processes in the wall region may be suitably described in terms of the local wall parameters u^* and ν/u^*, the large-scale processes of a repetitive nature, observed as time and spatial "periodicities" of wall-pressure fluctuations, of the turbulence bursts and of the large-amplitude bulges of the interface, are determined by the parameters U_0 and δ, characteristic for the outer region of the boundary layer.

In Sec. 6-12 when discussing Townsend's concept of the "cyclic" processes in the boundary region of a free shear-flow, we directed attention to the self-maintenance of these "cyclic" processes through a combination of shear flow and *Helmholtz instability effects,* and to the possibility that processes in the developing region may act as a trigger. If one agrees with the idea of similarity between processes in free-turbulent shear-flows, and in the outer region of a boundary-layer flow, one may expect a similar self-maintenance action in the boundary layer. From the turbulence-bursts periodicities and the "wavelength" of the large-amplitude bulges of the interface we obtained as a rough estimate $U_0T/\delta \simeq 5$, for the "periodicity" of the "cycles." An estimate of the value of U_0T/δ, where T is the period of the unstable disturbance introduced artificially and causing transition of the laminar boundary layer to turbulence, and taking for δ the value of the boundary-layer thickness just after breakdown, results not in the same value of 5, but at least in the same order of magnitude. One may speculate that this original periodicity initiates through pressure and velocity changes at the wall the generation of the above U-shaped vortices, and consequently the beginning of "cyclic" processes in both the inner and outer region of the boundary layer. More experimental evidence, however, will be required to support this idea and to clarify the interplay between the cyclic processes in the whole boundary layer, determined quantitatively roughly by $U_0T/\delta \simeq 5$, and the Helmholtz instability of the interface, if such an instability does play an important role.

7-10 NONCLASSICAL THEORIES ON THE DISTRIBUTION OF MEAN PROPERTIES IN A PLANE BOUNDARY LAYER

We have seen that in the case of a plane boundary layer with zero pressure-gradient the logarithmic mean-velocity distribution gives excellent agreement with measured distributions in the fully turbulent part of the inner or wall region. In the viscous sublayer the mean-velocity distribution is linear. As already mentioned, no complete and satisfactory theory is yet available for the adjacent buffer region, though many attempts have been made, yielding results more or less in agreement with experimental results. We will discuss in what follows a number of these theories.

One of the earlier theories has been proposed, independently, by Einstein and Li,[34] and by Hanratty.[35] These investigators considered a "*renewal*" or "*discontinuous-film*" *model of the viscous sublayer*, where the exchange of mass and momentum in the viscous sublayer occur in a repetitive way. Due to the disturbances from the outer layers the viscous sublayer is momentarily completely disrupted, and built up again by viscous diffusion. Thus, the fluid particles in the viscous sublayer are assumed to have only a finite contact time with the wall. The disrupting or breakdown period of the viscous sublayer is assumed to be very small compared with the building-up or recovery period.

During this period the exchange of axial momentum occurs according to a Fickian equation of unsteady viscous diffusion:

$$\frac{\partial U_1}{\partial t} = \nu \frac{\partial^2 U_1}{\partial {x_2}^2}$$

Note that the pressure-gradient term has been neglected. With the initial and boundary-conditions: $t = 0$, $U_1 = U_{1,\infty}$; $x_2 = 0$, $U_1 = 0$, and $x_2 = \infty$, $U_1 = U_{1,\infty}$, the solution reads

$$U_1 = U_{1,\infty} \operatorname{erf}\left(\frac{x_2}{2\sqrt{\nu t}}\right) \qquad (7\text{-}100)$$

Assume an average contact-time $\bar{t}_c$ and an average value $\bar{U}_{1,\infty}$, then one obtains

$$\bar{U}_1{}^+ = \bar{U}_{1,\infty}{}^+ \int_0^1 d\tau \operatorname{erf}\left(\frac{\sqrt{\pi}\, x_2{}^+}{4\bar{U}_{1,\infty}{}^+ \sqrt{\tau}}\right) \qquad (7\text{-}101)$$

where $\tau = t/\bar{t}_c$.

From Eq. (7-101) it follows that $u^{*2} = 2\bar{U}_{1,\infty}\sqrt{\nu/\pi\bar{t}_c}$, or

$$n_c = \frac{1}{\bar{t}_c} = \frac{\pi u^{*4}}{4\nu \bar{U}_{1,\infty}^2} \qquad (7\text{-}102)$$

With $\bar{U}_{1,\infty}{}^+ = 13.5$ Hanratty obtained a mean-velocity distribution deviating slightly from the linear distribution $\bar{U}_1{}^+ = x_2{}^+$ already at $x_2{}^+ = 3$, and following Reichardt's data (see Fig. 7-10) rather closely in the first part of the buffer region. The value of $\bar{U}_{1,\infty}{}^+ = 13.5$ corresponds with $x_2{}^+ = 30$, that is well within the buffer region, where the linearized diffusion equation should no longer apply. We may note that the above value of $x_2{}^+ = 3$ is close to the values mentioned earlier in Sec. 7-8, up to which values according to Popovich and Hummel the instantaneous velocity distributions are linear all the time, and according to Armistead and Keyes a high degree of correlation exists between heat-transfer fluctuations and the U_1-fluctuations.

By calculating the root-mean-square of the difference between instantaneous and mean velocity as given by Eqs. (7-100) and (7-101) respectively, Einstein and Li determined the distribution of u_1'/u^* as a function of $x_2{}^+$. It shows first a linear increase with $x_2{}^+$, with a slope equal to 0.9, which is a factor of 3 greater than the experimental value obtained by Laufer. It attains a maximum value of about 2.5

between $x_2^+ = 10$ and 20, when $\bar{U}_{1,\infty}^+ = 13.5$ is taken. This behavior is roughly in agreement with the experimental evidence. However, beyond its maximum value it drops rather steeply, in contrast to measured values.

With $\bar{U}_{1,\infty}^+ = 13.5$ Eq. (7-102) yields $\bar{t}_c = 230\nu/u^{*2}$, which is of the same order of magnitude as observed by Kline *et al.*[121] at the relatively low Reynolds numbers of their experiments. If we identify $\bar{t}_c$ with the average "burst" period $\bar{T}_B$, according to Eq. (7-86), the above numerical constant 230 would correspond with $\mathbf{Re}_{\delta_m} \simeq 2500$.

We conclude that the simple model on the one hand describes a number of aspects rather satisfactorily, even quantitatively, on the other hand, however, it yields values at variance with actual experimental data. Another shortcoming of the model is, that it yields the following expression: $(\epsilon_m/\nu)_{x_2\to 0} = \pi x_2^+/4\bar{U}_{1,\infty}^+ \simeq 0.074 x_2^+$, that is a linear increase of the eddy viscosity, which should increase in proportion to $\propto x_2^{+3}$, at least. One of the reasons for the failure of this model, apart from the restriction to a linear dynamic equation, is the neglect of the pressure term.

This pressure term has been taken into account by, amongst others, Schubert and Corcos,[166] and by Sternberg.[167] These authors also use linearized dynamic equations. Sternberg considers wave-type solutions of the equation

$$\frac{\partial U_i}{\partial t} = -\frac{1}{\rho}\frac{\partial p}{\partial x_i} + \nu \frac{\partial^2 U_i}{\partial x_2{}^2} \tag{7-103}$$

Since the normal velocity component u_2 drops to zero at a much greater rate than the other two components, Sternberg essentially considers waves with oblique positions in the (x_1,x_3) plane. Though Sternberg extends the calculation to $x_2^+ \simeq 100$, agreement between theory and experiment of the behavior of, e.g., u_1' is only obtained in and just beyond the viscous sublayer, as may be expected because of the applied linearization. The same remark can be made concerning Schubert and Corcos's theory. They considered the set of equations:

$$\begin{gathered}\frac{\partial u_1}{\partial t} + \bar{U}_1 \frac{\partial u_1}{\partial x_1} + u_2 \frac{\partial \bar{U}_1}{\partial x_2} = -\frac{1}{\rho}\frac{\partial p}{\partial x_1} + \nu \frac{\partial^2 u_1}{\partial x_2{}^2} \\ \frac{\partial u_3}{\partial t} + \bar{U}_1 \frac{\partial u_3}{\partial x_1} = -\frac{1}{\rho}\frac{\partial p}{\partial x_3} + \nu \frac{\partial^2 u_3}{\partial x_2{}^2}; \qquad \frac{\partial u_i}{\partial x_i} = 0\end{gathered} \tag{7-104}$$

and also wave-type solutions. Also in this case the theory appears to be useful only to explain qualitatively certain aspects of the turbulence. For instance, why u_1' shows a peak at some distance from the wall, and why the structure of u_1 and u_2 is much more elongated in the x_1-direction than that of u_3. The quantitative results, however, even differ by an order of magnitude from actual values. In general, for reasons mentioned above, one should consider results of linearized theories with reserve, when they are applied far beyond the viscous sublayer. A *wave-guide model* for turbulent shear-flow of a more general nature has been proposed by Landahl.[168] His aim was to show the importance of a wave propagation mechanism due to a shear flow and the effect on the turbulence. For, slightly damped waves tend to dominate over other kinds of disturbances because waves can correlate over relatively large distances, of the order of the inverse of their damping ratio. We may recall that the existence of wave-type motions in the layers close to the wall has been shown experimentally by Kline *et al.*, Morrison *et al.*, and others (see Sec. 7-8). Landahl considered a parallel wall shear-flow in the x_1-direction with one mean-velocity component $\bar{U}_1(x_2)$.

From Eq. (4-1) we obtain for the u_2-component

$$\left(\frac{\partial}{\partial t} + \bar{U}_1 \frac{\partial}{\partial x_1}\right) u_2 = -\frac{1}{\rho}\frac{\partial p}{\partial x_2} + \nu \frac{\partial^2 u_2}{\partial x_k\,\partial x_k} + \frac{\partial}{\partial x_k}(\overline{u_k u_2} - u_k u_2)$$

Take also the divergence $\partial/\partial x_i$ of Eq. (4-1). For an incompressible fluid we then obtain

$$\frac{1}{\rho}\frac{\partial^2 p}{\partial x_i\,\partial x_i} = -2\frac{\partial u_2}{\partial x_1}\frac{d\bar{U}_1}{dx_2} + \frac{\partial^2}{\partial x_k\,\partial x_i}(\overline{u_k u_i} - u_k u_i)$$

By differentiating this equation with respect to x_2, and subtracting it from the first equation after the Laplacian $\nabla^2 = \partial^2/\partial x_l\,\partial x_l$ has been taken from this equation, we obtain

$$\left(\frac{\partial}{\partial t} + \bar{U}_1 \frac{\partial}{\partial x_1}\right)\nabla^2 u_2 - \frac{\partial u_2}{\partial x_1}\frac{d^2\bar{U}_1}{dx_2{}^2} - \nu\nabla^4 u_2$$
$$= \nabla^2\left[\frac{\partial}{\partial x_k}(\overline{u_k u_2} - u_k u_2)\right] - \frac{\partial}{\partial x_2}\left[\frac{\partial^2}{\partial x_k\,\partial x_i}(\overline{u_k u_i} - u_k u_i)\right] \qquad (7\text{-}105)$$

The term

$$-2\frac{\partial^2 u_2}{\partial x_2\,\partial x_1}\frac{d\bar{U}_1}{dx_2}$$

has been neglected since $\partial u_2/\partial x_2$ is very small close to the wall.

Waves in the (x_1,x_3) plane are considered. Consequently, a Fourier transformation with respect to t, x_1, and x_3 only is taken of the latter equation. Following further a procedure similar to the one for obtaining the Orr-Sommerfeld equation in an instability analysis (which may yield wave-type solutions), results in a nonhomogeneous Orr-Sommerfeld equation, the nonlinear terms of the right-hand side being considered as a known forcing function. Landahl then considers the possibility of wave propagation through an expansion in terms of the eigenfunctions of the homogeneous part of the Orr-Sommerfeld equation. A formal solution can be obtained in this way.[174] Landahl obtained a numerical solution for the eigenfunction with the smallest damping. It is then shown that due to this damping each wave component practically loses its identity after a distance 6 times its wavelength, in agreement with the results obtained experimentally by Willmarth and Wooldridge.[135] Thus the only suitable length-scale to describe the spatial behavior of an eddy is its own size.

As a complement to Landahl's wave theory, Lighthill[189] considered the behavior of a "*wave-packet*" in a shear flow, with a group velocity different from the local mean-velocity. The interaction with the mean shear causes a decrease of the wavenumber, i.e., an increase in size while moving in the x_2-direction, behavior that is similar to that of a "turbulence-burst." Since the propagation velocity of "wave-packet" and that of "turbulence-burst" differ from the local mean-velocity, they are not connected with the same material of the fluid. The energy of a wave-packet can change because of viscous dissipation but also due to an interaction between wave-packets which may overlap. When the spatial extent of the wave-packet increases, the theory appears to approach that given by Landahl.

Black[170,171,172] proposed a theory which should explain also quantitatively the main features as mean-velocity and shear-stress distributions in the region $x_2 \gtrsim \delta_t$. Mainly on the basis of the experimental results as obtained by Kline *et al.*,[121] Black assumed a periodic process taking place in this region. This process includes the formation and breakdown of horseshoe-type main eddies. We may refer to the discussion at the end of Sec. 7-8. Black made a distinction between an organized periodic motion, the primary motion, and a smaller-scale stochastic motion, the secondary motion. The cycle starts with a completely disturbed viscous sublayer while the flow at that instant is turbulent up to the wall, showing a logarithmic velocity distribution for the primary motion. This logarithmic distribution is gradually broken up by purely viscous effects, with a motion as generated by a developing horseshoe eddy. A viscous sublayer builds up in which a Blasius laminar profile is being restored, till it attains a thickness corresponding to the instability of the Blasius profile. For a laminar boundary layer the nondimensional thickness $u^*\delta/\nu$ for instability, and corresponding with $(\mathbf{Re}_\delta)_{\text{crit}} = 1220$ (see Sec. 7-4), is practically equal to 50. This value has been taken by Black for the maximum value of δ_t^+, when the Blasius profile breaks down due to a turbulence burst at the end of the cycle.

The damping of the primary motion is assumed to be determined by the diffusion equation

$$\frac{\partial}{\partial t}(U_1)_p = \nu \frac{\partial^2}{\partial {x_2}^2}(U_1)_p - \frac{1}{\rho}\frac{\partial P}{\partial x_1}$$

The secondary motion is assumed not to contribute to the shear stress. This is produced by the periodically changing viscous shear stress during the building up of the Blasius profile.

The theory results in a frequency of the cycle $n = 0.009u^{*2}/\nu$, which compares nicely with the value of $0.01u^{*2}/\nu$ suggested by Kline *et al.* According to Eq. (7-86) this would correspond with $\mathbf{Re}_{\delta_m} \simeq 1000$, a pretty low value.

The average of the instantaneous $(U_1)_p$-distributions over one cycle, appeared to agree very closely with the mean-velocity profile according to Coles's experimental law of the wall.[173] According to Black the primary motion should scale on the wall variables u^* and ν, whereas the random secondary motion generated during the breakdown farther away from the wall should scale on the outer variables U_0 and δ. Accordingly the low-frequency part of the wall-pressure fluctuations should scale on u^* and ν. Black obtained the relation $\mathbf{E}_p(n) = \text{const. } \rho^2 u^{*2}\nu(n\nu/u^{*2})^{-1}$ for $n > 0.009u^{*2}/\nu$, independent of the Reynolds number. This relation should hold up to a Reynolds-number-dependent upper limit n^*, beyond which Black obtained for the high-frequency part the relation $\mathbf{E}_p(n) = \text{const. } \rho^2 U_0{}^3 \delta_d(n\delta_d/U_0)^{-10/3}$. In Fig. 7-39 two lines with slope (-1) and $(-10/3)$ respectively are shown. The experimental curve obtained by Willmarth and Young does not confirm these relations in a convincing way. The shear stress $(-\overline{u_2 u_1})$ calculated according to Black's theory showed a nice agreement with Laufer's data, except close to the wall, where for $x_2{}^+ < 10$ the theory yielded an almost linear increase with $x_2{}^+$, in contrast with the lower experimental values.

The merit of Black's theory is that it includes the observed repetitive character of the flow. On the other hand, the quantitative agreement might be fortuitous, since the assumptions made are not all in agreement with present experimental evidence. Amongst others, the assumptions that the primary motion and the basis cyclic motion should scale with the inner variables, whereas the smaller-scale random motions should scale with the outer variables. Also a purely viscous damping according to the linear equation used for the primary motion may be questioned.

More recently a straightforward linear theory has been given by Reichardt.[175] Correctly, this theory was intended to apply only to the region adjacent to the wall within the viscous sublayer. Since there $\partial u_2/\partial x_2 \simeq 0$, a two-dimensional motion in the (x_1, x_3) plane is considered, governed by the linear equations.

$$\frac{\partial u_1}{\partial t} = -\frac{1}{\rho}\frac{\partial p}{\partial x_1} + \nu \frac{\partial^2 u_1}{\partial x_2{}^2} \qquad \text{(7-106a)}$$

$$\frac{\partial u_3}{\partial t} = -\frac{1}{\rho}\frac{\partial p}{\partial x_3} + \nu \frac{\partial^2 u_3}{\partial x_2{}^2} \qquad \text{(7-106b)}$$

In the wall-adjacent layer $\partial p/\partial x_2 \simeq 0$, since there $u_2 \sim 0$. Hence, the pressure term may be eliminated by differentiating Eqs. (7-106a) and (7-106b) with respect to x_2. With $(\sigma_{21})_{\text{visc}} = \mu\, \partial u_1/\partial x_2$, Eq. (7-106a) then becomes

$$\frac{\partial}{\partial t}(\sigma_{21})_{\text{visc}} = \nu \frac{\partial^2}{\partial x_2{}^2}(\sigma_{21})_{\text{visc}} \qquad \text{(7-107)}$$

An identical equation is obtained for $(\sigma_{32})_{\text{visc}} = \mu\, \partial u_3/\partial x_2$. Damped, wave-type, solutions may be obtained. Of particular interest is the behavior as a function of x_2. To this end Reichardt considered this function for a fixed value of x_1 and x_3, so that the wave propagation in the x_1 and x_3 directions need not be taken into account. A solution of Eq. (7-107) is

$$(\sigma_{21})_{\text{visc}} = \mu \frac{\partial u_1}{\partial x_2} = A \exp(kx_2) \cos(\omega t + kx_2) \qquad \text{(7-108)}$$

where $k = (\omega/2\nu)^{1/2}$.

For small values of x_2 Reichardt gives the following approximate solution for u_1:

$$u_1 = \left(\frac{\partial u_1}{\partial x_2}\right)_0 x_2 \exp\left(\tfrac{1}{2}kx_2\right) \cos(\omega t + \phi)$$

with

$$\phi = \tfrac{1}{2}kx_2\left(1 + \tfrac{1}{6}kx_2\right)$$

For the relative intensity u_1'/u^* the following relation is then obtained

$$\frac{u_1'}{u^*} = \beta x_2^+ \exp(\alpha x_2^+/2) \qquad (7\text{-}109)$$

Adapting this relation to the experimental results obtained by Eckelmann (see Fig. 7-33) yielded $\alpha = 0.2$ and $\beta = 0.24$. Figure 7-33 indeed shows a satisfactory agreement, if with these values of α and β the relation (7-109) is introduced in this figure. We may then also conclude that the tangent $u_1'/u^* = 0.24x_2^+$ fits the data well up to $x_2^+ \simeq 2$.

By considering a wave solution for the wall shear stress

$$\sigma_w = A\cos(\omega t - 2\pi x_1/\lambda_1)$$

and using the relation $(\partial p/\partial x_1)_0 = (\partial\sigma_{21}/\partial x_2)_0$, Reichardt calculated the amplitude of the wall pressure fluctuations. Making use again of Eckelmann's experimental results and assuming $\lambda_1^+ \simeq 500$, there is obtained $p_w' \simeq 5\sigma_w$, which is within the range of experimental data (see Sec. 7-8 and Harrison[143]).

For the *outer region* Hama's empirical formula (7-67) for the $\bar{U}_1$-distribution has been suggested, useful for practical applications.

Struck by the similarity between the flow in this region and wake flow, for both show large-scale mixing processes controlled primarily by inertia rather than by viscosity effects, Coles[36] proposed a purely empirical correction function $w(\xi_2)$ similar to Millikan's correction function $h_1(\xi_2)$. Coles writes the relation for $\bar{U}_1^+$ which includes both the turbulent part of the wall region and the outer region as follows

$$\frac{\bar{U}_1}{u^*} = \frac{1}{\varkappa}\ln\left(\frac{u^* x_2}{\nu}\right) + B + \frac{\Pi}{\varkappa} w(\xi_2) \qquad (7\text{-}110)$$

The correction function $w(\xi_2)$, which unlike $h(\xi_2)$, is a positive function, has the same general form as $h(\xi_2)$, as shown in Fig. 7-14. Coles called the equation for this function $w(\xi_2)$ the *law of the wake*, thus expressing the similarity with wake flow that it reflects. Experimental evidence suggests that this law is of universal character.

The quantity Π is a profile parameter and does not depend on x_2. It is related to the local-friction coefficient $c_f = 2u^{*2}/U_0^2$ by

$$\frac{U_0}{u^*} = \frac{1}{\varkappa}\ln\left(\frac{u^*\delta}{\nu}\right) + B + \frac{\Pi}{\varkappa} w(1)$$

Or,

$$\frac{\Pi}{\varkappa} w(1) = \frac{U_0}{u^*} + \frac{1}{\varkappa} \ln \frac{U_0}{u^*} - \frac{1}{\varkappa} \ln \mathbf{Re}_\delta - B \qquad (7\text{-}111)$$

Consequently Π is, through u^*/U_0 and $\mathbf{Re}_\delta$, a weak function of x_1. Yet, for a boundary layer with zero pressure-gradient Coles assumed Π independent of x_1. He obtained $\Pi = 0.55$ with $\varkappa = 0.40$ and $B = 5.1$.

The expression for the velocity defect becomes

$$U_0^+ - \bar{U}_1^+ = -\frac{1}{\varkappa} \ln \xi_2 + \frac{\Pi}{\varkappa}[2 - w(\xi_2)] \qquad (7\text{-}112)$$

With $\Pi = 0.55$ and $\varkappa = 0.40$, $\Pi/\varkappa = 1.38$.

The function $w(\xi_2)$ has a physical meaning, in that it turns out to be equal to twice the $\bar{U}_1$-distribution at a point of separation. Separation may occur when $dP_0/dx_1 > 0$. The parameter Π in a plane boundary layer with positive pressure gradient may no longer be assumed independent of x_1. We obtain $\Pi = \infty$ at the point of separation where $u^* = 0$. This follows directly from Eq. (7-111).

From Eq. (7-110) we obtain

$$\frac{\bar{U}_1}{U_0} = \frac{1}{\varkappa} \frac{u^*}{U_0} \ln \frac{u^* x_2}{\nu} + \frac{u^*}{U_0} B + \frac{u^* \Pi}{U_0 \varkappa} w(\xi_2)$$

whence follows for $x_2 = \delta$:

$$1 = \frac{1}{\varkappa} \frac{u^*}{U_0} \ln \frac{u^* \delta}{\nu} + \frac{u^*}{U_0} B + \frac{u^* \Pi}{U_0 \varkappa} w(1)$$

Since

$$\lim_{u^* \to 0} \frac{u^*}{U_0} \ln \frac{u^* \delta}{\nu} = 0$$

and $w(1) = 2$, a value chosen by Coles as a normalizing condition, we obtain

$$\lim_{u^* \to 0} \frac{u^* \Pi}{U_0 \varkappa} = \frac{1}{2}$$

Consequently at a point of separation

$$\left(\frac{\bar{U}_1}{U_0}\right)_{u^*=0} = \tfrac{1}{2} w(\xi_2) \qquad (7\text{-}113)$$

At the point of separation then obviously the logarithmic part of the $\bar{U}_1/U_0$ distribution is zero.

A tentative determination of the universal wake function $w(\xi_2)$ is shown in the accompanying table and also in Fig. 7-47.

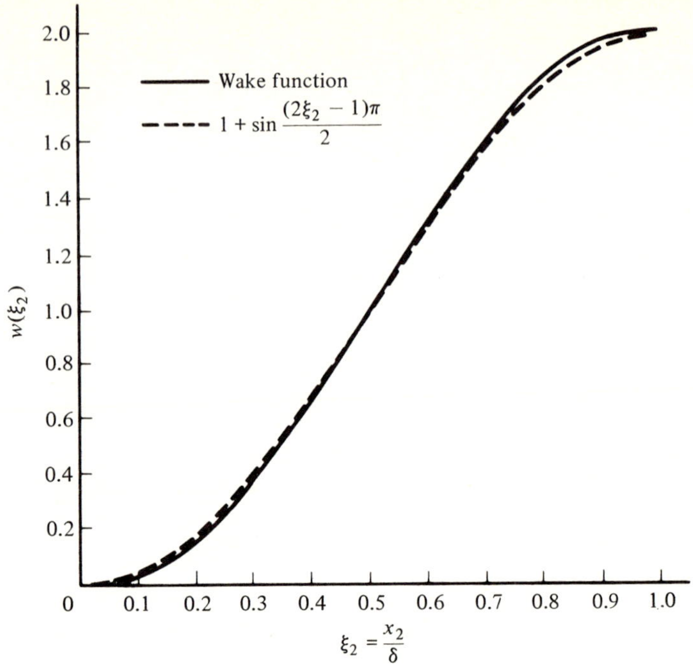

FIGURE 7-47
The universal wake function $w(\xi_2)$ according to Coles.[36]

THE WAKE FUNCTION $w(\xi_2)$

ξ_2	0	0.05	0.10	0.15	0.20	0.25	0.30	0.35	0.40	0.45	0.50
$w(\xi_2)$	0	0.004	0.029	0.084	0.168	0.272	0.396	0.535	0.685	0.838	0.994

ξ_2	0.55	0.60	0.65	0.70	0.75	0.80	0.85	0.90	0.95	1.00
$w(\xi_2)$	1.152	1.307	1.458	1.600	1.729	1.840	1.926	1.980	1.999	2.000

As a second normalizing condition Coles proposed

$$\int_0^2 \xi_2 \, dw = 1 \quad \text{or} \quad \int_0^1 w \, d\xi_2 = 1 \tag{7-114}$$

Inspection of the empirical curve for $w(\xi_2)$ reveals that this function is not exactly antisymmetrical with respect to $\xi_2 = 0.5$ but that the deviations are small. The actual wake function $w(\xi_2)$ may be approximated by the antisymmetrical function

$$w(\xi_2) = 1 + \sin \frac{(2\xi_2 - 1)\pi}{2}$$

which is a sufficiently close approximation for many practical applications. As Fig. 7-47 shows, the difference between the sine curve and the empirical $w(\xi_2)$ curve is of importance only near $\xi_2 = 0.1$ and $\xi_2 = 0.9$. This mean-velocity distribution has

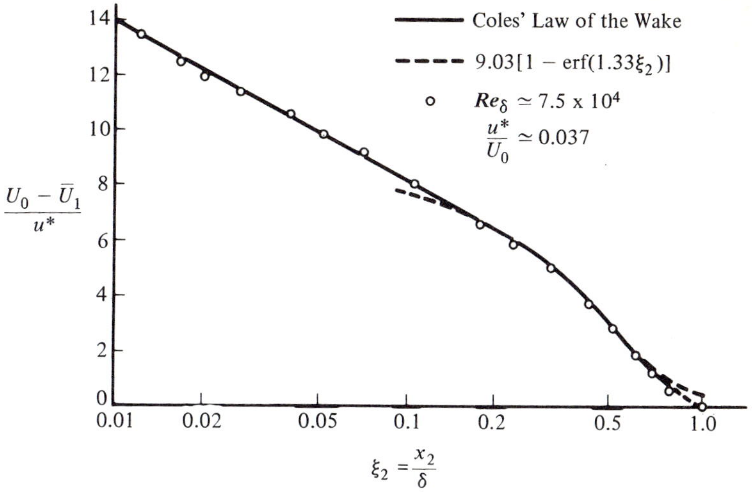

FIGURE 7-48
Comparison between computed velocity defect and experimental data by Klebanoff.[27]

been compared with the distribution by Klebanoff[27] and shown in Fig. 7-48. The drawn curve represents the empirical $w(\xi_2)$ function, suggested by Coles. Practically the same curve is obtained if the sine function is taken for $w(\xi_2)$, except near $\xi_2 = 0.9$ where the sine function yields slightly higher values for $(U_0 - \bar{U}_1)/u^*$.

By combining the expression for $\bar{U}_1^+$ valid for the wall region, e.g., Eq. (7-39) with the correction function $(\Pi/\varkappa)w(\xi_2)$ as follows

$$\bar{U}_1^+ = (\bar{U}_1^+)_{\text{wall region}} + \frac{\Pi}{\varkappa} w(\xi_2)$$

an acceptable approximation for the $\bar{U}_1$-distribution across the whole boundary layer is obtained.

With Eq. (7-110) it is possible to calculate the distribution of the $\bar{U}_2$-component. Since

$$\frac{\partial \bar{U}_1}{\partial x_1} = \left(\frac{\bar{U}_1}{u^*} + \frac{1}{\varkappa}\right)\frac{du^*}{dx_1} + \frac{u^*}{\varkappa}\frac{d}{dx_1}[\Pi w(\xi_2)]$$

the equation of continuity yields

$$\bar{U}_2 = -\int_0^{x_2} dx_2 \frac{\partial \bar{U}_1}{\partial x_1} = -\frac{x_2}{\varkappa}\frac{du^*}{dx_1} - \frac{1}{u^*}\frac{du^*}{dx_1}\int_0^{x_2} dx_2\, \bar{U}_1 - \frac{u^*}{\varkappa}\int_0^{x_2} dx_2 \frac{d}{dx_1}(\Pi w) \quad (7\text{-}115)$$

The $\bar{U}_1$ and $\bar{U}_2$ distributions being known, Eq. (7-3) enables one to calculate the shear-stress distribution.

For a constant-pressure boundary layer Eq. (7-3) yields for the fully turbulent part:

$$-\frac{\overline{u_1u_2}}{u^{*2}} = 1 + \frac{1}{u^{*2}}\int_0^{x_2} dx_2\left(\bar{U}_1\frac{\partial\bar{U}_1}{\partial x_1} + \bar{U}_2\frac{\partial\bar{U}_1}{\partial x_2}\right) \qquad (7\text{-}116)$$

Coles made such a calculation for the boundary layer investigated by Klebanoff, whose experimental data are shown in Fig. 7-49. The agreement between measured and computed values is very satisfactory. This agreement is not very surprising, since the results are a consequence only of the equations of continuity and of motion, with a good representation of the mean-velocity distribution. It may rather be considered a good demonstration of the reliability of Klebanoff's measurements. The experimental data, which are the same as those shown in Fig. 7-20, show a slight shift with respect to the ξ_2 position, because Coles estimated a smaller value for the local width of the boundary layer, namely, $\delta = 72$ mm, instead of the $\delta = 76$ mm assumed by Klebanoff.

In Sec. 7-2 we showed that for a plane boundary layer with zero pressure-gradient *no complete similarity* is possible. If the $\bar{U}_1/u^*$ distributions in subsequent cross sections are similar, it is no longer true of the distributions of $\bar{U}_2$ using the same velocity scale as for the $\bar{U}_1$-distribution. Also the shear-stress distributions can no longer be similar, as follows from the equation

$$\bar{U}_1\frac{\partial\bar{U}_1}{\partial x_1} + \bar{U}_2\frac{\partial\bar{U}_1}{\partial x_2} = \frac{1}{\rho}\frac{\partial}{\partial x_2}\sigma_{21} \qquad (7\text{-}117)$$

if we apply to it the same procedure as applied in Sec. 6-2.

Clauser[21] made calculations on velocity profiles corresponding to various values of u^*, assuming a universal shear-stress distribution. Also, he made calculations on the shear-stress distributions assuming similarity of the velocity profiles. He came to the conclusion that, though turbulent boundary layers are never

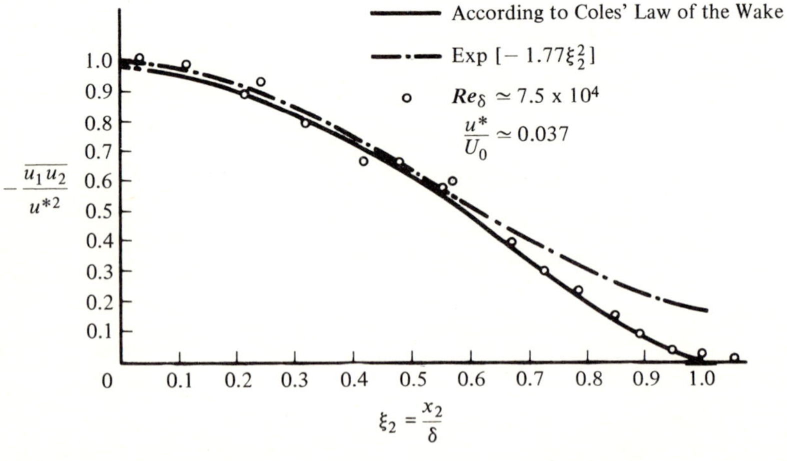

FIGURE 7-49
Computed shear-stress distributions and measured values by Klebanoff.[27]

strictly universal with respect to mean-velocity, shear-stress, or turbulence-energy distributions simultaneously, the departures from universality are small and may even be neglected for many practical purposes.

Accordingly, if we may assume similarity simultaneously among mean-velocity-distribution profiles and among shear-stress-distribution profiles, we may go one step further and assume the flow in the outer region to be self-preserving, the turbulence flow pattern being entrained by a mean motion with a relatively large velocity, deviating not much from the free-stream velocity. At the same time we may make use of two empirical facts: (1) the similarity in flow pattern between the outer region of the turbulent boundary layer and wake flow, and (2) the near-constancy of the eddy viscosity in this region. We may then expect a solution for the mean-velocity distribution similar to the solution for wake flow. This is possible if the differential equations in the two cases are similar. To this end we write Eq. (7-117) as follows:

$$U_0\left(1-\frac{U_0-\bar{U}_1}{U_0}\right)\frac{\partial}{\partial x_1}(U_0-\bar{U}_1)+\bar{U}_2\frac{\partial}{\partial x_2}(U_0-\bar{U}_1)=-\frac{1}{\rho}\frac{\partial}{\partial x_2}\sigma_{21}$$

If

$$\frac{U_0-\bar{U}_1}{U_0}\ll 1$$

and if we then apply the same order-of-magnitude procedure we used in Sec. 6-3, we arrive at

$$U_0\frac{\partial}{\partial x_1}(U_0-\bar{U}_1)\simeq-\frac{1}{\rho}\frac{\partial}{\partial x_2}\sigma_{21}$$

This approximate equation, which is the same as the one used in the case of the wake flow behind a cylinder, was used by Townsend[31] to calculate the mean-velocity distribution in the outer region, assuming at the same time self-preservation of the flow pattern (at least of the large-eddy motion) in this region. Townsend further assumed

$$\sigma_{21}=-\rho\epsilon_m\frac{\partial}{\partial x_2}(U_0-\bar{U}_1) \qquad (7\text{-}118)$$

where ϵ_m may still be a function of x_1.

Now put

$$U_0-\bar{U}_1=u^*f(\xi_2) \qquad (7\text{-}69)$$

If we keep in mind that both u^* and δ depend on x_1 we obtain for the equation of motion

$$U_0\left[\frac{du^*}{dx_1}f(\xi_2)-\frac{u^*}{\delta}\frac{d\delta}{dx_1}\xi_2\frac{d}{d\xi_2}f(\xi_2)\right]=\epsilon_m\frac{u^*}{\delta^2}\frac{d^2}{d\xi_2{}^2}f(\xi_2)$$

or

$$-\frac{d\delta}{dx_1}\xi_2\frac{d}{d\xi_2}f(\xi_2)+\delta\frac{U_0}{u^*}\frac{d}{dx_1}\left(\frac{u^*}{U_0}\right)f(\xi_2)=\frac{\epsilon_m}{U_0\delta}\frac{d^2}{d\xi_2{}^2}f(\xi_2) \qquad (7\text{-}119)$$

Except near $\xi_2 = 1$, $f(\xi_2)$ and $\xi_2(d/d\xi_2)f(\xi_2)$ are of the same order of magnitude.

Now at large Reynolds numbers the second term on the left-hand side becomes small with respect to the first term, since

$$\frac{U_0}{u^*}\frac{d}{dx_1}\left(\frac{u^*}{U_0}\right) \ll \frac{1}{\delta}\frac{d\delta}{dx_1}$$

This may be shown, using, for instance, the relation (7-46). The friction coefficient c_f and the relative wall-friction velocity u^*/U_0 become very small at large Reynolds numbers. Differentiation of Eq. (7-46) with respect to x_1 yields

$$-\frac{1}{\delta}\frac{d\delta}{dx_1}=\left[\frac{1}{A}\left(\frac{U_0}{u^*}\right)^2+\frac{U_0}{u^*}\right]\frac{d}{dx_1}\left(\frac{u^*}{U_0}\right)$$

from which it may be inferred that

$$\frac{U_0}{u^*}\frac{d}{dx_1}\left(\frac{u^*}{U_0}\right)$$

does become small with respect to $(1/\delta)(d\delta/dx_1)$ at large Reynolds numbers. Hence it is permissible to neglect the second term on the left-hand side of Eq. (7-119), which then becomes

$$\frac{d^2}{d\xi_2{}^2}f(\xi_2)+\frac{U_0\delta}{\epsilon_m}\frac{d\delta}{dx_1}\xi_2\frac{d}{d\xi_2}f(\xi_2)=0 \qquad (7\text{-}120)$$

Since the first term is a function of ξ_2 alone, the equation can be correct only if

$$\frac{U_0\delta}{\epsilon_m}\frac{d\delta}{dx_1}=a^2 \qquad (7\text{-}121)$$

that is, a constant, independent of x_1.

The solution of the differential equation (7-120) reads

$$\frac{U_0-\bar{U}_1}{u^*}=f(\xi_2)=C\int_{\xi_2}^{\infty}d\xi_2\exp\left(-\tfrac{1}{2}a^2\xi_2{}^2\right)$$

or

$$\frac{U_0-\bar{U}_1}{u^*}=\frac{C}{a}\frac{\sqrt{\pi}}{2}\left[1-\operatorname{erf}\left(\frac{a}{\sqrt{2}}\xi_2\right)\right] \qquad (7\text{-}122)$$

where

$$\operatorname{erf}(\alpha)=\frac{2}{\sqrt{\pi}}\int_0^{\alpha}du\exp(-u^2)$$

For the shear-stress distribution we then obtain from the relation (7-118)

$$-\frac{\overline{u_1 u_2}}{u^{*2}} = \exp\left(-\tfrac{1}{2}a^2\xi_2{}^2\right) \qquad (7\text{-}123)$$

and

$$C\frac{\epsilon_m}{u^*\delta} = 1$$

These solutions for the velocity-defect and the shear-stress distributions, respectively, have been applied to the measurements by Klebanoff. Figure 7-48 shows the result for the velocity distribution, obtained with $a^2 = 3.54$ and $C = 13.5$. The agreement is satisfactory only in a restricted region. Appreciable deviations from the measured values occur if ξ_2 becomes much smaller than 0.2 and if $\xi_2 > 0.7$.

Figure 7-49 shows the shear-stress distribution according to Eq. (7-123). The agreement with the measured values is satisfactory in the region $\xi_2 < 0.6$, but beyond $\xi_2 = 0.7$ the calculated values are much too high, just as in the case of the velocity distribution. These too-high values in the region $\xi_2 > 0.7$ were to be expected since, in assuming a constant value of the eddy viscosity ϵ_m across a section of the boundary layer no account was taken of the effect of intermittency in this region. As Fig. 7-25 shows, the intermittency factor Ω becomes much smaller than unity beyond $\xi_2 = 0.7$. Thus better agreement might have been expected if we had assumed $\epsilon_m\Omega$ in the relation (7-118) instead of ϵ_m alone.

On the other hand it is surprising that for the not very high value of the Reynolds number at which the experimental results of Figs. 7-48 and 7-49 have been obtained, there still is satisfactory agreement with the theory in the region $\xi_2 < 0.5$ where the assumption $(U_0 - \bar{U}_1)/U_0 \ll 1$ no longer applies.

The theories considered in the foregoing, some of them being of a semi-empirical nature, are satisfactory in describing the distributions of mean velocity and shear stress in a constant-pressure boundary layer. They may fail for non-constant-pressure boundary layers, and are inadequate in predicting the distribution of other turbulence quantities, such as turbulence kinetic energy and dissipation.

As shown, in a constant-pressure boundary layer the shear stress in the wall region is almost constant across this region, and the flow in it is directly determined by the local conditions at the wall. The change in the main-flow direction is slow. These facts may no longer be true in relatively rapidly developing flows and in non-self-preserving boundary layers with arbitrary pressure-gradients, in particular when the gradient is positive. For these reasons newer theories, based on various models, have been developed.

"Model" theories

The situations that have to be faced may be placed in two categories. In the *first category,* to which practically all the "model" theories belong, the flow is still assumed to be determined by local conditions, at least for a number of relations occurring in the theory. The primary, governing equations are, of course, the mass-balance and momentum-balance or Reynolds equations. For the unknown second-order turbulence quantities occurring in the latter equations, additional transport equations are introduced. These equations, however, again contain new unknowns, since we are still faced with the closure problem. To overcome this difficulty, certain relations with the mean flow or ad-hoc assumptions of a heuristic nature are introduced. The models are "local" because in these relations often use is made of the idea of a gradient type of transport, with the introduction of a diffusion coefficient. Though this gradient type of transport does not exclude some degree of memory behavior of the turbulence, yet the situation may still be considered as "local" in character. Additional assumptions that are often made, are those concerning local similarity and the turbulence being in energy equilibrium.

The *second category* comprises cases where memory effects are strong, with relaxation times and relaxation distances during or along which the changes of gradients of the mean-properties in the main-flow direction are no longer small. Any transport equation used then should exclude relations of a "local" character, such as gradient-type transport relations.

As stated, the "model" theories presented hitherto fall in the first category, though not all of them are completely "local" in all aspects. In what follows, we shall briefly consider the salient points of the most important theories. It must be mentioned that none of the theories considered is final. The whole situation of obtaining satisfactory theoretical solutions for boundary layers under general, arbitrary conditions, is still in a developing state. Further, it should be put to the fore, that the theories mainly apply to the more inner region of the boundary layer, since they do not account for the intermittent character of the flow in the outer region (except that corrections for the distributions of mean properties may be made, by assuming a normal probability distribution for the position of the turbulent/irrotational interface). Also, it should be mentioned that, in principle, these "model" theories may be applied to other shear-flows, such as free-turbulent flows and pipe-flows.

With reference to Sec. 5-2, we may recall that one of the suggestions made by Prandtl[188] was to relate the eddy-viscosity with the local value of the turbulence kinetic energy $\overline{q^2}/2 = \overline{u_i u_i}/2$, thus assuming local similarity, namely

$$\epsilon_m = A\left(\frac{\overline{q^2}}{2}\right)^{1/2}\Lambda \qquad (7\text{-}124)$$

where Λ is an integral length-scale.

With the relation (7-124), the relation $-\overline{u_2 u_1} = \epsilon_m\, \partial \bar{U}_1/\partial x_2$ is introduced in the Reynolds' equation for $\bar{U}_1$, after applying the usual boundary-layer approximation to it. Both $\overline{q^2}$ and Λ are unknown still. For $\overline{q^2}$ the balance equation (1-110) is used, which for the stationary, two-dimensional case reads

$$\left(\bar{U}_1\frac{\partial}{\partial x_1} + \bar{U}_2\frac{\partial}{\partial x_2}\right)\frac{\overline{q^2}}{2} = -\overline{u_2 u_1}\frac{\partial \bar{U}_1}{\partial x_2} - (\overline{u_1^2} - \overline{u_2^2})\frac{\partial \bar{U}_1}{\partial x_1} - \frac{\partial}{\partial x_2}\overline{u_2\left(\frac{p}{\rho} + \frac{q^2}{2}\right)}$$

$$-\frac{\partial}{\partial x_1}\overline{u_1\left(\frac{p}{\rho} + \frac{q^2}{2}\right)} - \varepsilon \qquad (1\text{-}110a)$$

The viscous term IV of Eq. (1-110) has been neglected.

The length-scale Λ can be determined for instance by following Prandtl's suggestion $\Lambda \propto B$, where B is the width of the mixing region (which in the case of free turbulence is still a good assumption), or by considering an additional transport equation for $\overline{q^2}\Lambda$, as suggested by Ng and Spalding.[181] The right-hand side of such an equation contains a production term, a turbulence diffusion term of the gradient type and a dissipation term.

Since in most cases of the first category the terms

$$(\overline{u_1^2} - \overline{u_2^2})\,\partial \bar{U}_1/dx_2 \quad \text{and} \quad \frac{\partial}{\partial x_1}\overline{u_1\left(\frac{p}{\rho} + \frac{q^2}{2}\right)}$$

are an order of magnitude smaller than the other terms in Eq. (1-110a), they too may be neglected. The transport term

$$\overline{u_2\left(\frac{p}{\rho} + \frac{q^2}{2}\right)}$$

is expressed in terms of the lateral gradient of $\overline{q^2}/2$ with a diffusion coefficient proportional with ϵ_m, for which Eq. (7-124) is taken. If for the dissipation ε the relation (3-109) is used, the equation for $\overline{q^2}$ as obtained from Eq. (1-110a) becomes

$$\frac{D}{Dt}\frac{\overline{q^2}}{2} = \epsilon_m\left(\frac{\partial \bar{U}_1}{\partial x_2}\right)^2 - \frac{\partial}{\partial x_2}\left(A_1\epsilon_m\frac{\partial}{\partial x_2}\frac{\overline{q^2}}{2}\right) - C\frac{1}{\Lambda}\left(\frac{\overline{q^2}}{2}\right)^{3/2} \tag{7-125}$$

A, A_1, C, and Λ are parameters which have to be adapted to the flow concerned.

Instead of Eq. (7-124), Jones and Launder[182] suggested

$$\epsilon_m = C_1\frac{(\overline{q^2})^2}{4\varepsilon} \tag{7-126}$$

With Eq. (3-109) for ε this is equivalent to Eq. (7-124). We may recall (see Sec. 5-4) that $\overline{q^2}/\varepsilon$ may be identified with the Lagrangian integral time-scale by a numerical factor, since for a homogeneous, isotropic turbulence $\mathfrak{I}_L \simeq 0.1A\overline{q^2}/\varepsilon \simeq B\overline{q^2}/\varepsilon$ where $B \simeq 0.1$.

Jones *et al.* and others[183–187] use a separate transport equation for the dissipation ε. Lumley[186,187] derived the following approximate equation for ε:

$$\frac{D\varepsilon}{Dt} = -\frac{\partial}{\partial x_k}\overline{\varepsilon u_k} - 2\nu\overline{\frac{\partial u_i}{\partial x_k}\frac{\partial u_i}{\partial x_j}\frac{\partial u_j}{\partial x_k}} - 2\nu^2\overline{\frac{\partial^2 u_i}{\partial x_k\,\partial x_j}\frac{\partial^2 u_i}{\partial x_k\,\partial x_j}} \tag{7-127}$$

where the second term on the right-hand side may be interpreted as a "production" of dissipation due to stretching effects of the fluctuating strain-rates, and the last term as a destruction of turbulence velocity-gradients by viscosity. Other forms of transport equations for ε have been suggested, but Eq. (7-127) can be obtained in a straightforward way from the dynamic equation for the R.M.S. vorticity $\overline{\omega_k\omega_k}$ of the turbulence[190] since, on the assumption of weak inhomogeneity of second-order quantities $\varepsilon/\nu \simeq \overline{\omega_k\omega_k} \simeq \overline{(\partial u_i/\partial x_j)(\partial u_i/\partial x_j)}$.

The transport term $\overline{\varepsilon u_k}$ is expressed in $\partial\varepsilon/\partial x_2$ with an eddy-diffusion coefficient either according to Eq. (7-124) or to Eq. (7-126), or in a more general fashion[185]

$$\frac{\partial}{\partial x_k}\left(\frac{\overline{q^2}}{\varepsilon}\overline{u_k u_l}\frac{\partial\varepsilon}{\partial x_l}\right)$$

The destruction term is written as a constant times $\varepsilon^2/\overline{q^2}$.[182,185–187] For the production term Jones *et al.* suggested a constant times

$$\epsilon_m\frac{\varepsilon}{\overline{q^2}}\left(\frac{\partial \bar{U}_1}{\partial x_2}\right)^2$$

while Launder *et al.*[185] suggested, more generally,

$$\overline{u_i u_j}\frac{\varepsilon}{\overline{q^2}}\partial\bar{U}_i/\partial x_j$$

Bradshaw, Ferriss, and Atwell[178] followed a different approach by transforming the energy equation (7-125) into one for the turbulence shear stress. To this end they assumed local similarity, and put

$$-\overline{u_2 u_1} = a\frac{\overline{q^2}}{2} = \sigma_{21}/\rho \tag{7-128}$$

The transport term $\overline{u_2(p + \rho q^2/2)}$ was approximated by

$$\overline{u_2\left(p + \rho\frac{q^2}{2}\right)} = g\left(\frac{x_2}{\delta}\right)\left(\frac{\sigma_{21}}{\rho}\right)^{3/2} \tag{7-129}$$

while the dissipation term was approximated by $[\sigma_{21}/a\rho]^{3/2}/\Lambda$.

Hence, with the neglect of the two small terms Eq. (1-110a) yields

$$\frac{D}{Dt}\sigma_{21} = a\sigma_{21}\frac{\partial \bar{U}_1}{\partial x_2} - a\frac{\partial}{\partial x_2}\left[g\left(\frac{\sigma_{21}{}^3}{\rho}\right)^{1/2}\right] - \frac{1}{\Lambda}\left(\frac{\sigma_{21}{}^3}{a\rho}\right)^{1/2} \tag{7-130}$$

Furthermore they put $\Lambda = \delta f(x_2/\delta)$. The functions $f(x_2/\delta)$ and $g(x_2/\delta)$ were determined on the basis of experimental data for flows of a similar nature.

A simpler method has been proposed by Nee and Kovasznay.[180] The turbulence shear stress occurring in the Reynolds' equation is put equal to $\epsilon_m\, \partial\bar{U}_1/\partial x_2$. A transport equation for ϵ_m is constructed, by assuming the equation to be analogous to the energy-balance equation. For the fully turbulent part of the flow the equation for ϵ_m reads

$$\frac{D}{Dt}\epsilon_m = \frac{\partial}{\partial x_2}\left(\epsilon_m \frac{\partial \epsilon_m}{\partial x_2}\right) + A\epsilon_m \frac{\partial \bar{U}_1}{\partial x_2} - B\frac{\epsilon_m{}^2}{\Lambda^2} - C\frac{\epsilon_m{}^2}{U_0{}^2}\frac{dU_0}{dx_1}\frac{\partial \bar{U}_1}{\partial x_2} \tag{7-131}$$

Note that the eddy-viscosity ϵ_m acts as its own eddy-diffusion coefficient. The other terms on the right-hand side are respectively: a "production" term, a "dissipation" term, and a term accounting for the effect of the streamwise pressure-gradient.

Instead of transforming the energy equation to an equation for the shear stress as done by Bradshaw *et al.*, it is possible to start from the basic equation (4-3) for the second-order velocity-correlation $\overline{u_i u_j}$. This has been done by Hanjalic and Launder[184] for two-dimensional flows. For the triple-velocity correlations they used the nonviscous dynamic equation, in which the pressure-double-velocity correlations are assumed to be related to the triple-velocity correlations, while a normal joint-probability distribution is assumed to express the quadruple-velocity correlations in the double-velocity correlations, i.e., a zero fourth-order cumulant tensor (see Sec. 3-3). As we do know, the latter assumption is basically incorrect, though the error made may be small. For the pressure term occurring in Eq. (4-3), they used the approximation suggested by Rotta[176] (see Sec. 4-2):

$$\overline{p\left(\frac{\partial u_i}{\partial x_j} + \frac{\partial u_j}{\partial x_i}\right)} = -A\rho\frac{\sqrt{\overline{q^2}}}{\Lambda}(\overline{u_i u_j} - \tfrac{1}{3}\delta_{ij}\overline{q^2})$$

Finally the viscous terms are connected with the viscous dissipation ε. Again additional equations are required for $\overline{q^2}/2$ and for ε.

The above "model" theories contain a number of parameters, part of which are just constants, others are functions of the space coordinates. These parameters have either to be guessed, or adjusted on the basis of available experimental evidence for certain types of flows. They are, in general, not universal, being therefore of restricted value only.

Now, on the one hand a large number of adjustable parameters (up to 8 in the Ng-Spalding "model"[181]) may be considered of advantage from a practical point of view because the theory concerned becomes more flexible. On the other hand, from a fundamental point of view, one may object against this kind of flexibility. For, the greater the number of adjustable parameters the better an otherwise not entirely correct, or even wrong, theory may be made to agree with experimental results.

Anyway, with the above "model" theories in a number of cases remarkably good agreement has been obtained between predicted and measured distributions of mean velocity, turbulence shear stress, and turbulence kinetic energy, once the parameters had been given the correct values.

We may also note that all the above "model" theories use transport equations containing turbulence transport terms being assumed of the gradient type, implying length and time scales small compared to those of the mean motion, which is not entirely true.

The objection made above concerning a large number of adjustable parameters no longer holds true when these parameters are universal, i.e., not limited to a certain type of flow only. Recently Lumley *et al.*[186,187] made an attempt to obtain solutions for fully turbulent flows with no direct viscous effects, of the set of equations consisting of the mass-balance equation, the three Reynolds' equations, the six dynamic equations for $\overline{u_i u_j}$, and the dissipation equation (7-127). Thus in total we have 11 equations for the 11 unknowns $\bar{P}$, $\bar{U}_i$, $\overline{u_i u_j}$ and ε. In the dynamic equations for $\overline{u_i u_j}$ the third-order terms $\overline{u_i u_j u_k}$ and $\overline{(u_j\, \partial p/\partial x_i + u_i\, \partial p/\partial x_j)}$ are connected with the dissipation ε, the kinetic energy $\overline{q^2}/2$ and the

Lagrangian integral time-scale $\mathfrak{I}_L$, on a rational basis. At the same time the assumptions of near-isotropy and of the applicability of weak-inhomogeneity approximations to these third-order terms are made. Similarly, the transport term $\overline{\tilde{\varepsilon}u_k}$ is modeled with $\overline{q^2}$, ε, and $\mathfrak{I}_L$, while the viscous terms in Eq. (7-127) are shown to be proportional with $\varepsilon^2/\overline{q^2}$, with a constant of proportionality of the $\mathcal{O}(1)$. Order of magnitude considerations are applied to the viscous terms of the equations, so that a further simplification can be obtained by the neglect of a number of relatively small terms. It appears that the 11 final equations contain 11 constants, out of which 8 could be evaluated on a rational basis using available experimental evidence, leaving 3 nondetermined constants still left. At the present juncture insufficient data are available to check the permissibility of the neglections made, the universal nature of the relations and parameters, and if they are universal, to obtain their values.

It is obvious that no analytical solutions of the set of equations may be expected to be obtainable, so that one has to rely on numerical methods. The presently available computation techniques still pose limitations in many cases because of the required computer time, or of convergence troubles, so that one has to deal with insufficient accuracy of the numerical solutions.

7-11 TRANSITION TO TURBULENCE OF A LAMINAR POISEUILLE FLOW

In Sec. 1-13 we already considered briefly a number of aspects of the transition of a laminar flow to the turbulent condition, including the case of pipe flow. We mentioned that according to general experience the fully developed flow in a straight pipe will be laminar when the Reynolds number $\mathbf{Re}_D = \bar{U}_{av}D/\nu \gtrsim 2{,}300$, and that it will be turbulent at greater values of $\mathbf{Re}_D$.

However, it is also known that the flow can maintain the laminar condition at far greater values if the necessary precautions are taken. These are: an entry of the fluid as free as possible from disturbances, the absence of any vibration in the whole pipe system, a pipe being straight, with uniform circular cross section and a smooth wall as perfect as can be achieved with present, most advanced production techniques. To the author's knowledge, the highest value of $\mathbf{Re}_D$ that has been obtained with still laminar Poiseuille flow is 100,000.[191] This suggests that the Poiseuille flow seems to be pretty stable against infinitesimal disturbances.

In Sec. 1-13 we made a distinction between integral energy-methods and methods based on local instability of the flow, for predicting the value of the transition Reynolds number. We mentioned that the classical integral energy-method, using Eq. (1-115) as the basic equation, yielded too low values in contrast to experimental evidence. An explanation might be that the initial disturbances, when strong enough, may modify the original flow through the, nonlinear, Reynolds' stresses. Also the disturbances can not grow indefinitely, because of the nonlinear effects, which effects result in the generation of higher harmonics which require a higher critical $\mathbf{Re}_D$ according to the energy method.

Much more recently the *integral energy method* has been taken up again, amongst others by Hopf,[192] Serrin,[193] and Velte,[194,196] at the same time employing a *variational principle*. We may remark that in general, when using a variational principle to establish the most likely flow configuration, a basic difficulty is presented in selecting the proper criterion. In the case of energy methods a reasonable criterion might be based on the viscous dissipation or entropy production. For purely viscous flow, for which the dynamic equations are linear, it can be shown that the viscous dissipation should be a minimum[195] (theorem of Helmholtz and Korteweg). However, in the case of the transition to turbulence, the nonlinear terms may not be neglected. Considering the fact that turbulence is a disordered phenomenon on a macroscopic scale, a maximum of entropy production might be a better criterion. Indeed, for a Poiseuille flow the average, nondimensional, dissipation $\varepsilon D^4/\nu^3$ is equal to $32\mathbf{Re}_D{}^2$, while for turbulent flow, using the Blasius resistance law for a smooth pipe, the nondimensional dissipation becomes equal to $0.158\mathbf{Re}_D{}^{11/4}$. In a log-log plot of $\varepsilon D^4/\nu^3$ versus $\mathbf{Re}_D$, the two relations intersect at $\mathbf{Re}_D \simeq 1000$. It does show that for the Poiseuille flow the dissipation is greater than it would be for the turbulent flow when $\mathbf{Re}_D \gtrsim 1{,}000$, whereas when $\mathbf{Re}_D \lesssim 2{,}000$ the dissipation is greater for the turbulent than for the laminar flow condition. But it should be emphasized that it has not yet been proven that fundamentally the maximum entropy-production is the correct criterion to be used.

In general, also these modern modifications of the integral energy-method, still yield too low values of the critical $\mathbf{Re}_D$. In a critical account of these energy methods Lumley[130] showed that the failure of the integral method is due to the fact that in this method the integral of the nonlinear transport effects is zero. Consequently these methods are basically insensitive to the strength of the disturbance. The disturbance that most efficiently extracts energy from the main flow need not necessarily be small.

Therefore Lumley suggested, as a generalization of Serrin's method, to consider the extremum of the ratio between $d\mathscr{E}_t/dt$ according to Eq. (1-115) and the integral kinetic energy $\mathscr{E}_t$. That is, he considered the disturbance that can most efficiently extract energy from the main flow. He applied this idea to the flow in the constant stress or wall region, in order to obtain the conditions for stable eddies in this region (see Sec. 7-8).

For pipe flows also the *local-instability method* has not been successful when small perturbations are assumed, which makes it possible to use linearized Orr-Sommerfeld type equations. This in contrast to the plane boundary layer and plane Poiseuille flow. Two-dimensional as well as three-dimensional (including rotational symmetric and helical type) disturbances have been assumed. The axi-symmetric Poiseuille flow proved to be stable for these infinitesimal disturbances. Only Shibuya[197] obtained instability with $(\mathbf{Re}_D)_{\text{crit}} \simeq 300$ for a three-dimensional disturbance that was not only periodic in time, but also in the spatial, axial, and tangential direction. When the disturbance was independent of the tangential coordinate φ, the Poiseuille flow became stable again at any $\mathbf{Re}_D$. A confirmation of Shibuya's results might be seen in the experimental results obtained by Fox *et al.*[198] They introduced a three-dimensional disturbance, whose flow pattern was periodic in time and a function of r and φ. To this end a metal strip, placed in the axis of the pipe in longitudinal direction, was vibrated in transverse direction. The amplitude of the velocity disturbance was 1 to 2 per cent of the maximum velocity (it may be questioned whether this disturbance was sufficiently small to be considered as an infinitesimal one for the Poiseuille flow). They observed instability with an $(\mathbf{Re}_D)_{crit}$ being frequency dependent. For the pipe flow concerned a minimum value of $(\mathbf{Re}_D)_{crit} = 2{,}130$ was obtained with a frequency of 5 s^{-1}. Note that this disturbance has been introduced in the center of the pipe.

If the entry of the flow in the pipe is uniform, a laminar boundary layer develops along the wall. This laminar boundary layer may become unstable against infinitesimal disturbances, thus triggering the transition to turbulence in the fully developed pipe flow. Tatsumi[199] considered this possibility and obtained a critical $\mathbf{Re}_D = 9{,}700$. A rough estimate based on this idea may be obtained as follows. Assume that, just as for the plane boundary layer, transition in the boundary layer along the pipe wall takes place when $(\bar{U}_x)_{av}x/\nu = 10^5$ to 10^6. Assume further that the transition distance x_{trans} be equal to the entrance length for fully developed Poiseuille flow, for which length we take $\simeq 0.06D\,\mathbf{Re}_D$. This then yields $(\mathbf{Re}_D)_{crit}^2 = 1.66\ (10^6$ to $10^7)$, or $(\mathbf{Re}_D)_{crit} \simeq 2{,}500$. However, the fully developed Poiseuille flow then should be unstable against the finite disturbances produced by the turbulent boundary layer in the entrance region of the pipe. We have also to account for the stabilizing effect on the boundary layer due to the accelerated flow in the entry-region. If use is made of this stabilizing effect with boundary layers in a favorable, i.e., negative, pressure gradient dP/dx, the above value of 2,500 should be increased at least by a factor 10. Wygnanski and Champagne[235] studied experimentally the transition to turbulence caused by unstable disturbances in the boundary layer in the entrance region of a smooth circular pipe. They indeed observed the occurrence of turbulent slugs due to this instability at values of $\mathbf{Re}_D$ as high as 5×10^4. We will come back to these experiments later.

The theories based on the assumption of instability of the Poiseuille flow against infinitesimal disturbances being not successful, theories have been presented considering *finite disturbances*. Here we will only briefly consider the results of a theory given by Davey and Drazin,[200] and by Davey and Nguyen.[201] Davey and Drazin considered the kinetic energy per unit of mass of the disturbance, integrated over the cross section of the pipe and per unit of length of the pipe. This quantity, rendered nondimensional with $(U_x)_{max}{}^2D^2$, namely $d\mathscr{E}/dx$ has been calculated for the critical disturbance as a function of the $\mathbf{Re}_D$ and the nondimensional wavenumber. An essential point is that they made a distinction between disturbances occurring in the core region, and those occurring in the wall region. For the core region the curvature of the velocity-profile at the center, $(d^2U_x/dr^2)_{r=0}$ and the kinematic viscosity are the basic parameters, while for the wall region u^* and ν are these parameters. With these parameters a length-scale and a velocity scale can be formed. For the core region their calculations yielded a minimum value for the critical, nondimensional quantity, $(d\mathscr{E}/dx)_{min} = 560\,\mathbf{Re}_D{}^{-2}$. For the wall region this value was $(d\mathscr{E}/dx)_{min} = 130\mathbf{Re}_D{}^{-3/2}$. It thus may be concluded that according to these calculations, $(d\mathscr{E}/dx)_{min}$ for a critical disturbance decreases with increasing $\mathbf{Re}_D$ at a faster rate in the core region than in the wall region. Hence, we may expect that transition from laminar to turbulent flow in a pipe, and caused by finite disturbances begins in the core region. A confirmation of this theoretical result may be seen already in the classical experiments by Reynolds. It is also in agreement with general experience. The above experiments by Fox *et al.*, may also be mentioned, while an indirect confirmation may be found in the experiments by Paterson and Abernathy.[202] They observed no difference in the phenomena during transition in the experiments with water and with a dilute solution of polyethylene oxide in water as used for drag reduction. See Fig. 7-50. It is in general accepted that the effect of drag-reducing solutions should be sought in the processes occurring in the wall-adjacent layers (increase of an effective viscous sublayer at the same Reynolds number). If transition indeed begins in the core-region then no effect of the drag-reducing solution on $(\mathbf{Re}_D)_{\text{crit}}$ has to be expected.

Another confirmation still is found in the generally accepted assumption that wall roughness has no effect on the value of $(\mathbf{Re}_D)_{\text{crit}}$. This assumption is partly based on the experiments by Nikuradse. However, the coarsest sand used by Nikuradse, namely with $D/k_s = 30$, corresponds at $(\mathbf{Re}_D)_{\text{crit}}$ with $k_s^+ \simeq 4$ which is smaller than $\delta_l^+ \simeq 5$. But also Lindgren[203] did not find an effect of wall roughness though in his experiments $k_s^+ \simeq 7$.

Finally, it must be noted that, even when transition starts in the core region, the wall still plays an essential role in the continuous generation of turbulence in order to maintain the turbulent condition.

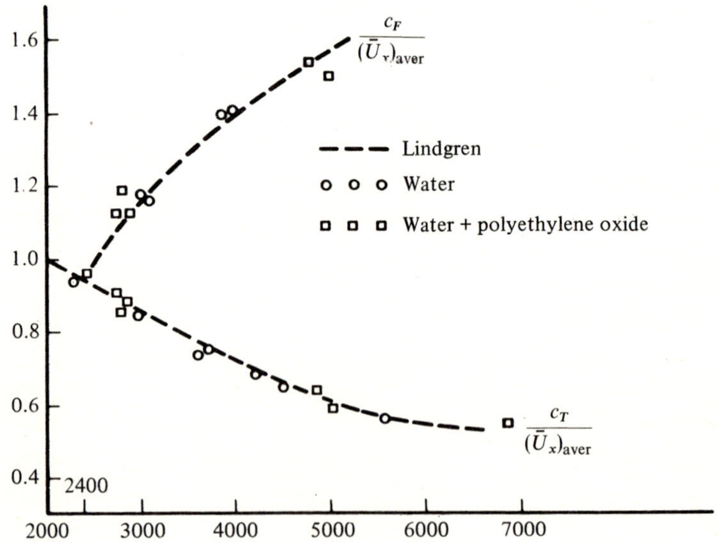

FIGURE 7-50
Propagation velocities of the front and the tail of a turbulent slug. (After: *Lindgren, E. R.;*[203] *Paterson, R. W., and F. H. Abernathy.*[202])

A series of interesting *experiments on the transition process* has been made by Lindgren.[203] In his experiments he used two straight pipes of transparent material, one with an I.D. of 6 mm, the other of 23.7 mm. The entry flow of the fluid could be strongly disturbed, for the smaller pipe by means of a circular disk placed perpendicular to the pipe axis at a variable distance from the inlet, for the other pipe by means of a variable slit in the otherwise closed entry section. An optical birefringence method was employed, using as the fluid water to which 0.1 to 1 per cent of bentonite was added.

Starting with a low value of $\mathbf{Re}_D$ and gradually increasing $\mathbf{Re}_D$ the following phenomena were observed.

$\mathbf{Re}_D < 200$. Any initial disturbance was damped in the entrance region.

$\mathbf{Re}_D > 200$. Initial disturbances developed to *turbulence slugs* over an "accumulation" distance of 20 to $70D$. These, downstream propagating, turbulence slugs showed high-frequency and high-intensity fluctuations, even if the initial disturbances were rather weak.

$\mathbf{Re}_D < (\underline{\mathbf{Re}}_D)_{\text{crit}} \simeq 2{,}000$. After 100–$130D$ distance eventually the slugs vanished due to damping. [It may be remarked that at $\mathbf{Re}_D = 1{,}000$, for instance, the entrance length for Poiseuille flow is $\sim 60D$. Since in the entrance section the wall shear-stress σ_w is greater than in the fully developed flow, it might well be that initial disturbances are first amplified by interaction with the wall, but are damped farther downstream where σ_w has a lower value.]

$\mathbf{Re}_D > (\underline{\mathbf{Re}}_D)_{\text{crit}}$. The lifetimes of the turbulence slugs increased. They became more and more self-sustaining. At the same time the frequency of the occurrence of slugs and the intermittency turbulent/nonturbulent increased.

$\mathbf{Re}_D > (\overline{\mathbf{Re}}_D)_{\text{crit}} \simeq 2{,}400$. The slugs grew in length while propagating downstream, till they overlapped. There then is a continuous turbulent-flow condition with practically no intermittency with nonturbulent periods.

Apparently Lindgren distinguished between a lower and an upper limit for the critical Reynolds number, $(\underline{\mathbf{Re}}_D)_{\text{crit}}$ and $(\overline{\mathbf{Re}}_D)_{\text{crit}}$, respectively. He suggested that transition is not a sudden phenomenon, and that consequently $2{,}000 < (\mathbf{Re}_D)_{\text{crit}} < 2{,}400$. The same is true for boundary-layer transition, where the actual transition and breakdown to complete turbulence takes place over a certain distance during which the Reynolds number $U_0\delta/\nu$ increases.

With a highly disturbed entry by means of a jet generated by a rather narrow slit, the flow in the first part of the pipe showed a high level of turbulence with a fine structure. This fine structure, however, decayed when flowing downstream till the coarser equilibrium structure belonging to the fully developed turbulent pipe flow was reached.

When for $\mathbf{Re}_D > 2{,}400$ the turbulent slugs increased in length, they had roughly an arrow shape, with a propagation velocity of the peaked front higher than that of the tail. Figure 7-50 shows the result of the measurements of the propagation velocities of the front, c_F, and of the tail, c_T, of the slugs as a function of $\mathbf{Re}_D$ obtained by Lindgren,[203] and later by Paterson and Abernathy.[202] As the latter experiments clearly show, there is no effect noticeable of the added polyethylene oxide, which at the concentration of 50 ppm already does reduce the flow resistance in the fully developed turbulence state considerably. Note that $c_T/(\bar{U}_x)_{av} < 1$, $c_F/(\bar{U}_x)_{av} > 1$, but that $c_T < c_F < (\bar{U}_x)_{max}$ for Poiseuille flow.

The turbulence within a slug is nonhomogeneous. Oscillograms of the velocity fluctuations taken at the axis of the pipe showed that during the passage of a turbulence slug along the probe, the fluctuations were first relatively slow and weak, but increased in frequency and intensity when the tail end approached. The drop off was quite sudden after the passage of the tail end. The finest structure of the turbulence was observed near the wall.

These observations lead to the following general picture of the phenomena occurring in a *turbulence slug*. Turbulence is generated at the wall. It is convected with a velocity greater than the local mean-velocity. At the same time a diffusive transport of turbulence elements takes place towards the center of the pipe, while at the same there is a decay of the smallest eddies. Arriving in the center region where the mean-velocity is greater than near the wall the turbulence elements are entrained downstream, thus overtaking the fluid near the wall. Probably due to pressure fluctuations of the remaining turbulence of the entrained turbulence element, disturbances are induced at the wall ahead of the main turbulence slug. These disturbances initiate the generation of new turbulence at the wall. Within the turbulence slug the mean-velocity distribution resembles much more that of fully developed turbulent flow. This difference in mean-velocity distribution between the parts upstream of, within, and downstream of the turbulence slug must be accompanied with secondary currents. At the tail end nonturbulent fluid enters the turbulence slug, whereas at the front turbulent fluid becomes nonturbulent again. This seems to point towards the possibility of retransition to laminar flow, or *relaminarization*. We shall come back to this point later.

Figure 7-51 shows a picture of a turbulence slug in which the basic elements of the foregoing descriptions are shown.

When we compare the phenomena in these *turbulence slugs* with those in *turbulence spots* occurring during transition of a laminar boundary layer, we notice similarities as well as dissimilarities. Consider first the similarities. Both turbulence

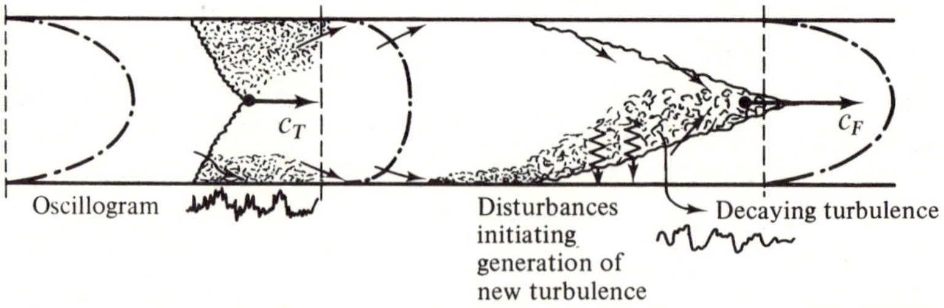

FIGURE 7-51
Conceptual sketch of a turbulent slug at $\mathbf{Re}_D > 2{,}400$.

slugs and turbulence spots are elongated in the streamwise direction and more or less arrow shaped. The tail has a smaller propagation velocity than the front. At $\mathbf{Re}_D = 3{,}600$, $c_T/\bar{U}_{x,\max} = 0.5$ and $c_F/\bar{U}_{x,\max} = 0.88$ when $\bar{U}_{x,\max}$ for turbulent flow at this $\mathbf{Re}_D$ is taken. When $\bar{U}_{x,\max}$ for Poiseuille flow is taken the values become 0.35 and 0.66 respectively. For a turbulence spot, near the wall $c_T/U_0 = 0.5$ and $c_F/U_0 = 0.88$. In a turbulence slug and in a turbulence spot secondary currents are present of a similar nature, due to the required redistribution of fluid because of the difference in $\bar{U}_x$-distribution during the laminar and turbulent state.

The dissimilarities are that in pipe flow there is no free boundary so that $(\bar{U}_x)_{\text{av}}$ must remain constant, while in boundary-layer flow there is a free boundary with the possibility for a local thickening of the boundary layer at a turbulence spot; the $(\bar{U}_x)_{\text{av}}$ does not remain constant. In contrast to fully developed pipe flow there is a growth in thickness of the boundary layer, albeit slight, in downstream direction.

The turbulence slugs generated by the growth of unstable small disturbances in the laminar boundary layer of the entrance region as studied by Wygnanski and Champagne[235] appear to be of a somewhat different nature to the slugs studied by Lindgren and caused by relatively strong entrance disturbances. Wygnanski and Champagne used a seamless aluminum pipe of 33 mm diameter and a total length of about 500 diameters. It consisted of five sections, and special care was taken to minimize surface discontinuities at the junctions and any degree of eccentricity and misalignment.

They observed transition of the same nature as studied by Lindgren for entrance disturbances in excess of 2 to 3 per cent, while transition due to instability of the entrance boundary layer occurred when the disturbances were smaller than ~ 0.3 per cent, at which slugs were formed for $\mathbf{Re}_D > 2{,}700$. Extensive measurements using conditional sampling and applying ensemble averaging were made to obtain distributions of mean-velocity, turbulence intensities, and turbulence energy balances within a slug. Most of the measurements were made at $\mathbf{Re}_D = 1.9 \times 10^4$ and 2.37×10^4. At these values of $\mathbf{Re}_D$ the entrance length for fully developed Poiseuille flow would be $x/D \simeq 0.06\mathbf{Re}_D \simeq 1{,}200$, i.e., much greater than the total length of the pipe used. The turbulence slugs studied by Wygnanski and Champagne had shapes of the interfaces at the front and tail different from those caused by relatively strong entrance disturbances at the much lower Reynolds numbers. In particular the frontal interface was much more blunt, and differed but little from the tail interface. The propagation velocities c_F and c_T (see Fig. 7-50), were determined up to $\mathbf{Re}_D \simeq 6 \times 10^4$. It turned out that $c_F/(\bar{U}_x)_{\text{av}}$ obtained a maximum of ~ 1.7 at $\mathbf{Re}_D \simeq 10^4$, and decreased monotonously to 1.05 at $\mathbf{Re}_D = 6 \times 10^4$. At this Reynolds number $c_T/(\bar{U}_x)_{\text{av}}$ had decreased to ~ 0.2. We may note that the maximum value of c_F is still smaller than the maximum velocity for fully developed Poiseuille flow. Those turbulence slugs grew in length exceeding in most cases the total length of the pipe used. Of the interesting results obtained we will mention the following. The turbulence in the interior of the slug appeared similar to that in the fully developed pipe flow. Near the interfaces the velocity profiles showed inflexions. Maxima in the Reynolds stresses occurred approximately at these locations. The turbulence intensities imparted to newly entrained fluid, i.e., close to the slug interface, exceeded the values in the interior of the slug by a factor 4 to 7, and had values up to 15 per cent of $\bar{U}_{x,\max}$. The interface at the front moves with a velocity equal to $\bar{U}_{x,\max}$ in the laminar region downstream. Wygnanski and Champagne concluded that a unique relation exists between the propagation velocity of an interface and the velocity of the fluid, thus preventing any turbulent fluid from leaving the slug, except very close to the wall within the viscous sublayer. However, we have to keep in mind that $c_F/(\bar{U}_x)_{av}$ was still smaller than 2 and that at the high Reynolds numbers of the experiments the pipe was not long enough to attain a fully developed Poiseuille flow.

Let us now consider the possibility of *relaminarization*. The above experiments, but also those carried out by others, showed that turbulence-free fluid enters a turbulent region, while at another boundary of a turbulent region fluid may leave this region in a laminar state. In pipe flow the fluid, when entering the turbulence slug in the core region of the pipe is decelerated, whereas in Lindgren's experiments, when it leaves the slug it is accelerated. Now, deceleration promotes turbulence, and acceleration has a damping effect. However, since deceleration and acceleration take place because the fluid becomes turbulent and laminar respectively, this leads to no acceptable explanation (chicken-egg problem). The experiments by Lindgren and similar experiments by Rotta[204] clearly showed that the transition laminar–turbulent takes place at a much faster rate than relaminarization which appears to be a relatively slow process. In this respect the experiments carried out by Coles and Van Atta[205] on spiral turbulence have been very illuminating. They studied, using hot-wire anemometry for the measurements of the air flow, the phenomena taking placc in the annular space between two concentric, but counter-rotating, straight cylinders. Under certain conditions the flow consists of alternate helical stripes in laminar and turbulent state, rotating with practically the mean angular velocity of the cylinders. Figure 7-52a shows a picture of the shape of the mean interface between laminar and turbulent flow, Fig. 7-52b the mean-velocity distribution during the laminar and turbulent flow condition.

Again we may note that near the outer wall turbulence-free fluid enters the turbulent region, at the same time decreasing in mean-velocity, while near the inner wall fluid leaves the turbulent region in a laminar flow condition thereby increasing in mean-velocity. Detailed measurements revealed that the entry of the fluid in the turbulent region occurred in a direction almost perpendicular to the interface, whereas the fluid leaves this region in a direction almost tangential to the interface. Consequently relaminarization takes place at a much slower rate, with much more time available for viscous damping. In this respect also mention may be made of the experiments by Badri Narayanan.[206] He studied the relaminarization of an initially turbulent airflow through a duct of rectangular cross section. The width of the duct was constant and equal to 12.5 mm, while the height was kept equal to 75 mm over the first 1.27 m of the duct, and then increased in a diffuser linearly

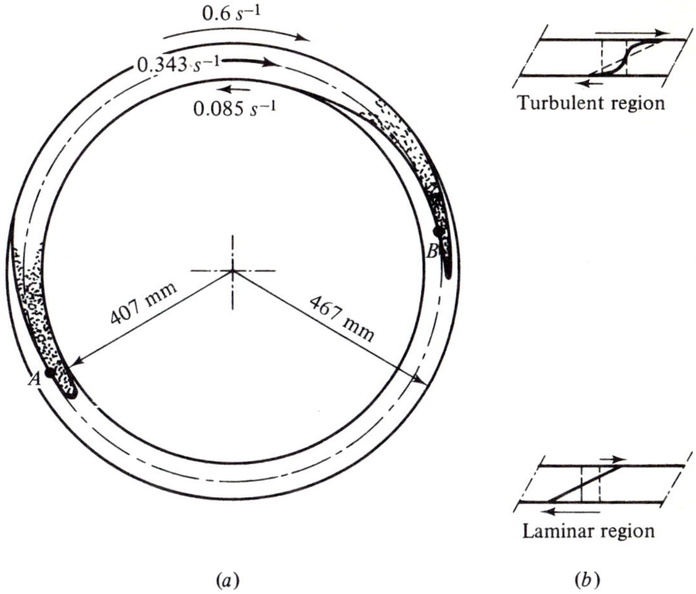

FIGURE 7-52
Alternate laminar–turbulent flow between two counter-rotating, concentric cylinders. (After: *Coles, D., and C. W. van Atta.*[205])

to 225 mm over a distance of 2.03 m. During the flow through the diffuser the Reynolds number, based on the hydraulic diameter, decreases. When at the end of the diffuser the Reynolds number was lower than $\simeq$2,000 laminar flow was obtained downstream. Measurements of $\overline{u_1^2}$, $\overline{u_2^2}$, $\overline{u_3^2}$, and $\overline{u_2 u_1}$ showed that they decayed to zero indeed, but in a different way and at a different rate. The decay started at the wall where viscous effects are strongest. The shear-stress $\overline{u_2 u_1}$ decreased practically inversely proportional with distance, and at a faster rate than u_1' and u_2'. Consequently, the correlation coefficient $\mathbf{R}_{21}$ decreased, which points towards a decoupling process between u_1 and u_2 velocity components. These experiments clearly showed that "retransition" is an entirely different process from transition, and that re-laminarization without 100 per cent viscous damping does not seem possible. Finally, we may refer to the experiments by Mobbs, considered in Sec. 6-12.

7-12 TURBULENT FLOW THROUGH A STRAIGHT CIRCULAR PIPE. MEAN-VELOCITY DISTRIBUTION

For the fully developed turbulent flow in the pipe the mean-flow conditions are independent of the axial coordinate x and axi-symmetric, assuming a uniform wall condition. Since $\bar{U}_\varphi = 0$ and $\bar{U}_x$ is independent of x the radial velocity component $\bar{U}_r = 0$. Hence, Eqs. (1-30) reduce to

$$\frac{1}{\rho}\frac{\partial \bar{P}}{\partial x} = -\frac{1}{r}\frac{d}{dr}(r\overline{u_r u_x}) + \nu\left(\frac{d^2\bar{U}_x}{dr^2} + \frac{1}{r}\frac{d\bar{U}_x}{dr}\right)$$

and

$$\frac{1}{\rho}\frac{\partial \bar{P}}{\partial r} = -\frac{1}{r}\frac{d}{dr}(r\overline{u_r^2}) + \frac{\overline{u_\varphi^2}}{r}$$

Integration of the second equation yields

$$\bar{P}(x,r) + \rho\overline{u_r^2} - \rho\int_r^{1/2D} dr\,\frac{\overline{u_r^2} - \overline{u_\varphi^2}}{r} = \bar{P}_w(x) \qquad \text{(7-132)}$$

where $\bar{P}_w$ is the static pressure at the wall and a function of x, and where D is the diameter of the cross section of the tube. The first equation may then be written

$$\frac{1}{\rho}\frac{d\bar{P}_w}{dx} = -\frac{1}{r}\frac{d}{dr}(r\overline{u_r u_x}) + \nu\left(\frac{d^2\bar{U}_x}{dr^2} + \frac{1}{r}\frac{d\bar{U}_x}{dr}\right)$$

or, after multiplication by r and integration with respect to r,

$$\frac{r}{2}\frac{d\bar{P}_w}{dx} = -\rho\overline{u_r u_x} + \mu\frac{d\bar{U}_x}{dr} \qquad \text{(7-133)}$$

Because of the assumed homogeneity of the mean conditions in the axial direction, the terms on the right-hand side of Eq. (7-133) are functions of r alone. Hence $\bar{P}_w(x)$ must be a linear function of x—an obvious result. Thus, if we put $r = \frac{1}{2}D$ and integrate with respect to x, we obtain

$$\bar{P}_w(x) - \bar{P}_w(0) = \frac{4}{D}\mu\left[\frac{d\bar{U}_x}{dr}\right]_{r=1/2D} x = -\frac{4}{D}\sigma_w x \qquad \text{(7-134)}$$

With this expression for $\bar{P}_w(x)$, Eq. (7-132) becomes

$$\bar{P}(x,r) - \bar{P}_w(0) = -\frac{4\sigma_w}{D}x - \rho\overline{u_r^2} + \rho\int_r^{1/2D} dr\,\frac{\overline{u_r^2} - \overline{u_\varphi^2}}{r} \qquad \text{(7-132a)}$$

We may approximate the turbulence-energy equation much as we did the corresponding equation for boundary-layer flow. In the region very close to the wall, the flow is practically two-dimensional and no different from boundary-layer flow. Hence, for this region, the same approximate energy equation (7-5a) may be applied. In the region farther away from the wall but still close enough for the flow to be nearly two-dimensional we may expect the same result but a lesser degree of similarity. To this last region in boundary-layer flow the reduced energy equation (7-6) applies.

For the main part of the turbulence region in tube flow, an equation similar to Eq. (7-7) should apply, namely,

$$\overline{u_x u_r}\frac{\partial \bar{U}_x}{\partial r} + \frac{1}{r}\frac{\partial}{\partial r}r\left[\overline{u_r\left(\frac{p}{\rho} + \frac{q^2}{2}\right)}\right] + \varepsilon' = 0 \qquad \text{(7-135)}$$

where

$$\varepsilon' = \nu \overline{\frac{\partial u_j}{\partial x_i} \frac{\partial u_j}{\partial x_i}}$$

$$= \nu \left[\overline{\left(\frac{\partial u_x}{\partial x}\right)^2} + \overline{\left(\frac{\partial u_x}{\partial r}\right)^2} + \overline{\left(\frac{\partial u_x}{r\,\partial \varphi}\right)^2} + \overline{\left(\frac{\partial u_r}{\partial x}\right)^2} + \overline{\left(\frac{\partial u_r}{\partial r}\right)^2} + \overline{\left(\frac{\partial u_r}{r\,\partial \varphi}\right)^2} \right.$$

$$\left. + \overline{\left(\frac{\partial u_\varphi}{\partial x}\right)^2} + \overline{\left(\frac{\partial u_\varphi}{\partial r}\right)^2} + \overline{\left(\frac{\partial u_\varphi}{r\,\partial \varphi}\right)^2} \right] \tag{7-136}$$

The uniform condition for fully developed turbulent flow considered above requires some distance from the entrance section of the pipe. In the case in which the flow of the entering fluid is already turbulent, Latzko[37] calculated the distance from the entrance at which the velocity distribution approached a distribution according to the $\frac{1}{7}$-power law across the entire cross section of the pipe. He obtained

$$\frac{x}{D} = 0.693\mathbf{Re}_D{}^{1/4}$$

This formula yields a much shorter entry length than ever found experimentally. Kirsten[38] measured entry lengths of 50 to 100 diameters, depending on the Reynolds number. On the other hand Nikuradse[39] obtained fully turbulent flow at $x/D = 25$ to 40 in the case of disturbed-entry flow, the 40-diameter value at $\mathbf{Re}_D = 9 \times 10^5$. Latzko's theoretical formula would have given an entry length of 21 diameters.

For practical purposes Nikuradse's value of 40 diameters may be recommended as a minimum value.

As in boundary-layer flow, in pipe flow also we may distinguish between an "inner" wall region and an "outer" wall-remote or "core" region, the latter covering most of the cross section of the pipe. The flow in the wall region is not affected by the flow conditions far from the wall and may, therefore, be considered to be the same for boundary-layer and for pipe flow. The flow in the outer region may be different in the two cases. As a matter of fact, in pipe flow there is no interaction with a turbulence-free stream, and no intermittency of turbulence; this constitutes a dissimilarity from boundary-layer flow. Another dissimilarity is found in the fact that in pipe flow the conditions are independent of x, with, as a consequence, no mean radial velocity and a uniform distribution across the pipe of the axial pressure-gradient. From the equilibrium between shear-stress forces and pressure forces it follows that the shear-stress distribution must be exactly linear

$$\frac{\sigma_{21}}{\sigma_w} = \frac{2r}{D} = 1 - \frac{2x_2}{D} = 1 - \xi_2 \tag{7-137}$$

where r is the distance from the axis of the pipe, and x_2 is the distance from the wall.

We have seen that, in the turbulent constant-pressure boundary layer, the shear-stress distribution is linear only in the outer region and then only approximately (see Fig. 7-26), whereas in the wall region the shear stress is nearly constant. In the wall region of the pipe flow the shear stress is essentially not constant, though the deviations from the wall shear stress σ_w may be only small, because the wall region is relatively thin. For this reason, the wall region in pipe flow is usually also referred to as the constant-stress layer, notwithstanding the incorrectness of this designation. At any rate, the similarity assumed between the flow in the wall regions for the two cases has not been contradicted by the experimental data obtained to date.

Thus in the wall region the mean-velocity distribution is described by the general relation (7-24), where $\bar{U}_x$ has to be read for $\bar{U}_1$. Further, the more specific relations (7-26) for the viscous sublayer and (7-30a) for the fully turbulent part hold true for a smooth wall and Eq. (7-32a) for a rough wall.

Let us refer again to Fig. 7-10 where the results of measurements by Laufer[41] and Reichardt,[12] made in the wall region, are shown. Comparison with Fig. 7-9 shows that, in the overlapping region around $u^* x_2/\nu = 100$, the agreement between pipe-flow measurements and boundary-layer measurements is satisfactory.

Thus, for the fully turbulent part of the wall region, where the universal logarithmic mean-velocity distribution (7-30a) holds, the same values for the constants, namely $A = 2.44$ and $B = 4.9$, should apply—though it must be noted that, just as for boundary-layer flow, various other values of these constants have been used by various experimenters. Townsend,[31] for instance, suggests the same value for A, but a value of 5.85 for B.

However, when compared with the plane, zero-pressure-gradient boundary layer, the spread in the empirical values of A and B appears to be greater in the case of the turbulent pipe flow. It has been suspected that a departure from the assumed Reynolds number similarity might be responsible for it. If this were the case, an effect of the Reynolds number on the two constants should be expected. For this reason Hinze[207] re-examined the old data obtained by Nikuradse. Notwithstanding the great scatter of the data a trend of the values of A and B as a function of $\mathbf{Re}_D$ appeared noticeable. The general trend for the parameter A is to increase with increasing $\mathbf{Re}_D$ both for smooth and rough pipes, with a tendency to level off for $\mathbf{Re}_D \gtrsim 10^6$. For smooth pipes the value of $A \simeq 2.8$ at $\mathbf{Re}_D = 10^6$. The general trend of the values of B and B^* (Eqs. 7-30a and 7-34) is to decrease with increasing $\mathbf{Re}_D$ in the case of a smooth wall, while no such trend can be found due to the large scatter of the data in the case of a rough wall. A similar effect of $\mathbf{Re}_D$ on the exponent n in the case of a power-law distribution for the velocity defect then can be expected. According to Nikuradse's measurements a power law is applicable only in the region $0.05 < 2r/D < 0.7$, in which region n has a trend to increase with increasing $\mathbf{Re}_D$.

Finally one may deduce from Nikuradse's experiments that the eddy viscosity $2\epsilon_m/u^*D$ in general decreases with increasing $\mathbf{Re}_D$, with a tendency to remain constant for $\mathbf{Re}_D > 5.10^5$ in the core region. At $2r/D = 0.98$, however, the eddy viscosity still decreases slightly with increasing $\mathbf{Re}_D$ up to the highest value of $\mathbf{Re}_D$, namely 3×10^6. Hence, for the eddy viscosity the Reynolds number similarity required a much higher Reynolds number close to the wall, which is not surprising considering the increasing viscosity effect, as the wall is approached.

An explanation which at the same time shows the increasing trend of the constant A with increasing Reynolds number has been offered by Tennekes.[208] He attributed the variation of the slope of the logarithmic velocity profile, i.e., the value of A, to a second-order effect caused by the streamwise pressure-gradient $d\bar{P}_w/dx$. Since it is convenient when studying the wall region to assume coordinates x_1 and x_2, with $x_2 = D/2 - r$, Eq. (7-133) may be rewritten with $u_2 = -u_r$:

$$-\overline{u_2 u_1} + \nu \frac{d\bar{U}_x}{dx_2} = \left(\frac{D}{4} - \frac{x_2}{2}\right)\frac{1}{\rho}\left(-\frac{d\bar{P}_w}{dx_1}\right) = u^{*2} - \frac{x_2}{2\rho}\left(-\frac{d\bar{P}_w}{dx_1}\right)$$

when

$$u^{*2} = \frac{D}{4\rho}\left(-\frac{d\bar{P}_w}{dx_1}\right)$$

For the additional effect of the pressure gradient we may introduce a velocity-scale u_1^*, so that this effect on $\bar{U}_x$ is determined by x_2, ν and u_1^*. Consequently we may extend the relation (7-25) to

$$\bar{U}_x/u^* = f\left(\frac{u^* x_2}{\nu}, \frac{u_1^*}{u^*}\right)$$

Since u_1^*/u^* yields a second-order effect we may assume

$$\bar{U}_x{}^+ = f_1\left(\frac{u_1^*}{u^*}\right) f_2(x_2{}^+) \simeq \text{const.}\left(1 + \alpha\frac{u_1^*}{u^*}\right) f_2(x_2{}^+)$$

With the velocity-defect law (7-33), with $\bar{U}_{x,\max}$ for U_0, and with the same procedure as applied to obtain Eq. (7-30a), the result is

$$\bar{U}_x{}^+ = A\left(1 - \alpha\frac{u_1^*}{u^*}\right) \ln x_2{}^+ + B$$

where B is also a function of u_1^*/u^*.

The velocity-scale u_1^* is determined by $(-d\bar{P}_w/dx_1)/\rho = 4u^{*2}/D$ and by ν, since this is the only relevant parameter to construct a velocity scale. Dimensional analysis then yields $u_1^{*3} = \nu 4u^{*2}/D$, whence follows $u_1^*/u^* = [4\nu/u^* D]^{1/3}$. Consequently, the second-order effect due to the pressure gradient yields for the logarithmic velocity-distribution instead of Eq. (7-30a), the relation

$$\bar{U}_x{}^+ = A(1 - \alpha_1 D^{+^{-1/3}}) \ln x_2{}^+ + B \qquad (7\text{-}138)$$

A then is the asymptotic value for $D^+ \to \infty$, or $\mathbf{Re}_D \to \infty$.

From the available data Tennekes obtained $A = 3$ and $\alpha_1 = 2.1$. When allowance is made for the large scatter in the empirical data, the slope of the logarithmic distribution according to Eq. (7-138) is in acceptable agreement with these data.

As we mentioned in discussing boundary-layer flow, the "thickness" of the viscous sublayer is determined by a value of $u^*\delta_l/\nu \simeq 5$ and the "thickness" of the buffer region beyond which the flow is fully turbulent, by a value of $u^*\delta_t/\nu \simeq 30$. In order to get an idea of the thickness δ_t, the quantity δ_t/D has been calculated for different values of $\mathbf{Re}_D = \bar{U}_{av}D/\nu$. The results are shown in the accompanying table.

$\mathbf{Re}_D$	5×10^3	10^4	10^5	10^6
$\dfrac{\delta_t}{D}$	0.1	0.05	0.006	0.0008

For the "core" region the mean velocity may again be conveniently described as a velocity defect from the maximum velocity at the center of the pipe. In accord with the corresponding expression (7-65a) for boundary-layer flow, we write the velocity distribution in the wall region

$$\frac{\bar{U}_{x,\max} - \bar{U}_x}{u^*} = -A \ln \xi_2 + B^* \qquad (7\text{-}139)$$

where $\xi_2 = 2x_2/D = 1 - 2r/D$.

The velocity defect (7-139) applies to smooth and to rough walls, as it should. Here again the slope of the logarithmic distribution is, through the second-order effect of the pressure gradient, a weak function of the Reynolds number.

Figure 7-53 shows the logarithmic plot of this velocity-defect distribution

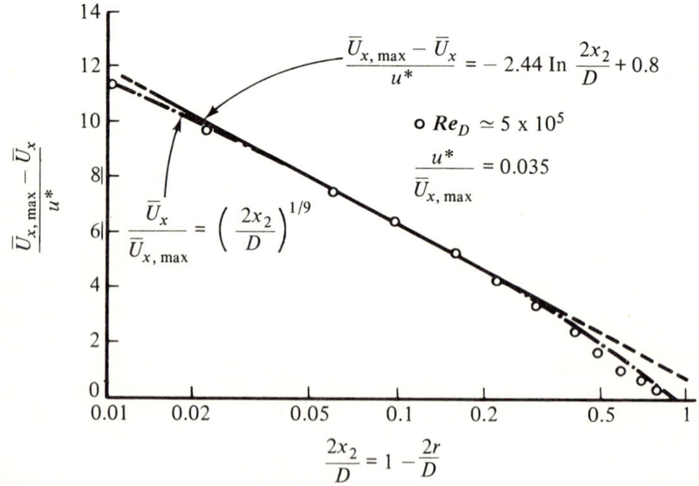

FIGURE 7-53
Logarithmic plot of mean velocity distribution in the core region of turbulent pipe flow. (After: *Laufer, J.*[41])

versus ξ_2 for the data measured by Laufer[41] ($\bar{U}_{x,\max}D/\nu = 5 \times 10^5$), together with the curve for the logarithmic distribution with $A = 2.44$ and $B^* = 0.8$.

If we compare these values with the corresponding values for boundary-layer flow (see Fig. 7-11), we see that B^* is much smaller for pipe flow. This means that the deviation of the actual velocity distribution from the logarithmic velocity distribution is smaller for pipe flow. It does not imply that the wall region to which the logarithmic velocity distribution applies is greater in extent for pipe flow. As may be inferred from Fig. 7-53, deviations from the logarithmic velocity distribution become noticeable beyond $\xi_2 \simeq 0.15$, say, just as they did for boundary-layer flow. But the magnitude of the deviations is much smaller.

Consequently, if we introduce here, too, a correction function $h(\xi_2)$, so that

$$\frac{\bar{U}_{x,\max} - \bar{U}_x}{u^*} = -2.44 \ln \xi_2 + 0.8 + h(\xi_2) \qquad (7\text{-}140)$$

this correction function $h(\xi_2)$ will be much smaller than the corresponding function for boundary-layer flow. Figure 7-54 shows the correction function $h(\xi_2)$ deduced from Laufer's data; this figure should be compared with Fig. 7-14.

These relatively small deviations of the actual velocity distribution from the logarithmic velocity distribution close to the center of the pipe make it understandable that, in the early days and even nowadays for practical applications, the logarithmic velocity distribution has been assumed to apply to the whole flow region of the pipe.

From his experiments on flow through circular pipes with smooth and with rough walls, Nikuradse[39,40] obtained the following logarithmic velocity distributions:

$$\text{Smooth pipe:} \qquad \frac{\bar{U}_x}{u^*} = 2.5 \ln \frac{u^* x_2}{\nu} + 5.5$$

$$\text{Rough pipe:} \qquad \frac{\bar{U}_x}{u^*} = 2.5 \ln \frac{x_2}{k} + 8.48 \qquad (7\text{-}141)$$

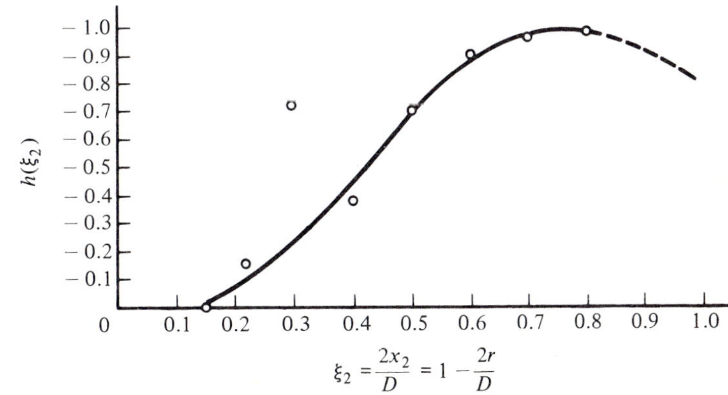

FIGURE 7-54
Correction function $h(\xi_2)$ for the logarithmic velocity-defect distribution in pipe flow.

For the average velocity $\bar{U}_{av}$ we then obtain

$$\text{Smooth pipe:} \qquad \frac{\bar{U}_{av}}{u^*} = 2.5 \ln \frac{u^*D}{2\nu} + 1.75$$

$$\text{Rough pipe:} \qquad \frac{\bar{U}_{av}}{u^*} = 2.5 \ln \frac{D}{2k} + 4.73 \qquad (7\text{-}142)$$

The wall-friction velocity u^* may be expressed in terms of the wall shear stress or the pressure gradient, according to the relations

$$\frac{d\bar{P}}{dx} = -\frac{4}{D}\sigma_w = -\frac{4}{D}\rho u^{*2}$$

Direct determination of $\bar{U}_{av}/u^*$ from Nikuradse's data on the pressure drop yields

$$\text{Smooth pipe:} \qquad \frac{\bar{U}_{av}}{u^*} = 2.44 \ln \frac{u^*D}{2\nu} + 2.0$$

$$\text{Rough pipe:} \qquad \frac{\bar{U}_{av}}{u^*} = 2.44 \ln \frac{D}{2k} + 4.9 \qquad (7\text{-}143)$$

The values of the constants are slightly different from the values in former expressions obtained from the velocity distributions; this difference may be ascribed to the slight deviations of the logarithmic from the actual velocity distributions.

It should be recalled that the numerical constants in the equations for rough pipe refer to more or less uniform sand roughness. For other kinds of wall roughnesses these constants will have different values depending on the way in which the roughness parameter k has been defined. It is possible and usual to define an equivalent sand roughness so that the same equations (7-142) or (7-143) apply.

It is customary in pipe-flow technique to express the flow pressure drop in terms of the friction factor f instead of the wall-friction velocity u^*. This coefficient f is defined by

$$\frac{d\bar{P}}{dx} = -\frac{f}{D}\tfrac{1}{2}\rho \bar{U}_{av}{}^2$$

Hence

$$f = 8\left(\frac{u^*}{\bar{U}_{av}}\right)^2 \qquad (7\text{-}144)$$

Accordingly, it is possible to express the velocity distributions in terms of f instead of u^*. The equations of (7-143) yield relations between f and $\mathbf{Re}_D$ and between f and k/D, respectively:

$$\sqrt{\frac{8}{f}} = 2.44 \ln\left(\sqrt{\frac{f}{32}}\mathbf{Re}_D\right) + 2.0$$

and

$$\sqrt{\frac{8}{f}} = 2.44 \ln \frac{D}{2k} + 4.9 \qquad (7\text{-}143a)$$

If we now express the velocity distribution in terms of f we arrive at the result that $\bar{U}_x/\bar{U}_{x,\max}$ is a function of $2x_2/D$ and f only:

$$\frac{\bar{U}_x}{\bar{U}_{x,\max}} = f\left(\frac{2x_2}{D}, f\right)$$

This applies to smooth as well as to rough pipes.

We have shown in Sec. 7-6 that the mean-velocity distribution in a turbulent boundary layer may well be approximated by a simple power law provided that the exponent of the power law is still a function of the Reynolds number.

The same approximation may be made to the velocity distribution in pipe flow. Thus we may apply Eq. (7-57) or instead of it

$$\frac{\bar{U}_x}{\bar{U}_{x,\max}} = \left(\frac{2x_2}{D}\right)^{1/n}$$

For smooth pipes n is still a function of $\mathbf{Re}_D$. According to the measurements by Nikuradse $n = 6$ at $\mathbf{Re}_D = 4 \times 10^3$, $n = 7$ at $\mathbf{Re}_D = 10^5$, $n \simeq 9$ at $\mathbf{Re}_D = 10^6$ and further increases to $n = 10$ at $\mathbf{Re}_D = 3 \times 10^6$. For rough pipes n attains smaller values, namely between 4 and 5.

Since this velocity distribution is a function only of the friction factor f, the conclusion is that n is a function of f only, for both smooth and rough pipes.

Nunner[42] determined n as a function of f from his own measurements on the velocity distribution in smooth and rough pipes and from Nikuradse's[39,40] measurements. As Fig. 7-55 shows, this function is approximately a simple one, namely,

$$\frac{1}{n} = \sqrt{f} \qquad \text{(7-145)}$$

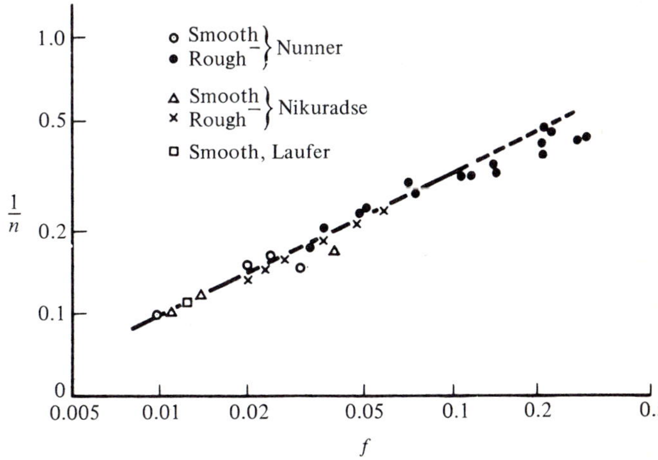

FIGURE 7-55
Relation between the exponent $1/n$ of the power-law velocity distribution and the friction factor f. (From: *Nunner, W.*,[42] *by permission of the Verlag des Vereins Deutscher Ingenieure.*)

valid in the range $f < 0.1$. For $f > 0.1$ the values of n are slightly larger.

However, a value of $f = 0.1$ applies to such a high degree of roughness that it is seldom encountered in practice.

According to the relation (7-145) the value of $n\sqrt{f} = 1$ is independent of the Reynolds number. Such a relation can be obtained if it is assumed that the relation (7-57) and the relation (7-149), to be considered later, both give a satisfactory description of the mean-velocity distribution in the core region in that at least the same value for the average mean-velocity in a cross section is obtained. Though the power-law distribution fails when $x_2 \to 0$ and $x_2 \simeq D/2$, and Eq. (7-149) is not valid in the wall region, we may roughly put

$$\int_0^1 d\xi_2 (1 - \xi_2)\xi_2^{1/n}\bar{U}_{x,\max} = \int_0^1 d\xi_2 (1 - \xi_2)\left[\bar{U}_{x,\max} - \frac{u^*}{2C}(1 - \xi_2)^2\right]$$

This yields

$$\frac{u^*}{\bar{U}_{x,\max}} = 4\frac{3n + 1}{(n + 1)(2n + 1)}C$$

With the relations (7-144) and

$$(\bar{U}_x)_{\text{av}} = \frac{2n^2}{(n + 1)(2n + 1)}\bar{U}_{x,\max}$$

we obtain

$$n\sqrt{f} = 4\sqrt{2}\frac{3n + 1}{n}C \simeq 12\sqrt{2}C$$

for $n > 5$, say.

If $C = 2\epsilon_m/u^*D$ is assumed independent of $\mathbf{Re}_D$, which is not strictly true, and if we assume $C \simeq 0.07$ (see Fig. 7-63), we would obtain $n\sqrt{f} \simeq 1.2$.

The power-law velocity distribution has also been applied to Laufer's data. This is shown in Fig. 7-53; with $n = 9$, corresponding to $f = 0.013$ for a smooth pipe and $\mathbf{Re}_D \simeq 5 \times 10^5$, the agreement between the computed and measured velocity distributions is satisfactory indeed.

7-13 MEASUREMENTS OF TURBULENCE QUANTITIES IN PIPE FLOW

An almost complete set of measurements of turbulence quantities were made as early as 1954 by Laufer.[41] Since then others have partly repeated, partly extended these measurements. Here mention will be made only of the investigations by Coantic[111] and by Lawn.[209] Since most of the results obtained by Laufer still agree satisfactorily well with the results of the later experiments, we will in the following mainly consider Laufer's experiments. Only when necessary will the other results be included.

Laufer's test tube was a straight, seamless brass tube, about 5 m long with an internal diameter of 247 mm. Two air speeds were applied, one corresponding to a maximum mean velocity $\bar{U}_{x,\max} \simeq 3$ m/s and the other corresponding to $\bar{U}_{x,\max} \simeq$ 30 m/s. The corresponding Reynolds numbers are $\bar{U}_{x,\max}D/\nu = 50{,}000$ and $500{,}000$.

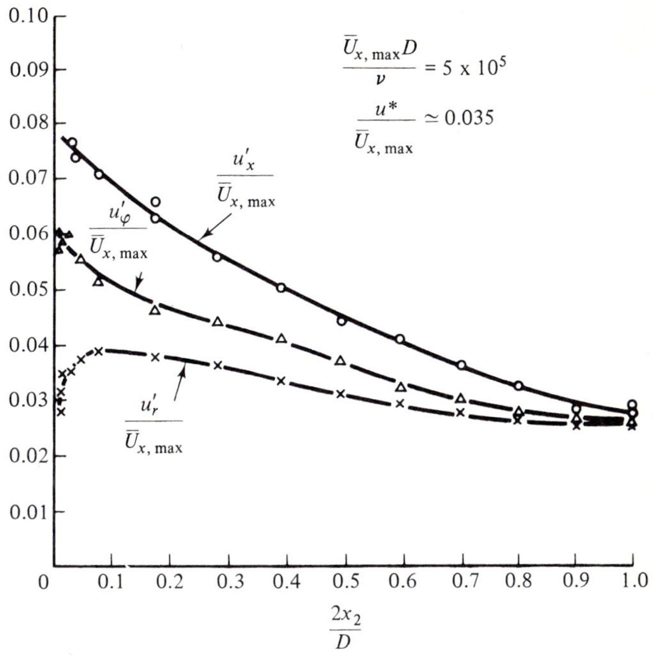

FIGURE 7-56
Relative turbulence intensities in pipe flow. (From: *Laufer, J.*,[41] *by permission of the National Advisory Committee for Aeronautics.*)

Mean-velocity distributions and turbulence quantities were measured even into the viscous sublayer. From the measured mean-velocity gradient at the wall and also from the measured axial pressure drop, the wall shear stress could be determined. For the two Reynolds numbers the wall-friction velocity amounted to $u^*/\bar{U}_{x,\max} \simeq 0.042$ and 0.035, respectively.

Distributions of the relative turbulence intensities $u'_x/\bar{U}_{x,\max}$, $u'_r/\bar{U}_{x,\max}$, and $u'_\varphi/\bar{U}_{x,\max}$ are shown in Fig. 7-56. Comparison with Fig. 7-17 shows that these quantities are practically of the same magnitude as the corresponding quantities in a boundary layer (the wall friction velocities are nearly equal). The condition of isotropy is approached in the remote region of the wall just as in the boundary layer. Near the wall, at $2x_2/D \simeq 0.1$, the intensity of the radial turbulence component u'_r is roughly equal to $1.1u^*$, while the intensity of the axial turbulence component u'_x is roughly equal to $2u^*$ at this radial position. With decreasing distance to the wall it continues to increase till a maximum value is reached. A similar behavior holds true for the tangential component u'_φ. The maxima of u'_x and u'_φ are reached so close to the wall that they cannot be shown in Fig. 7-56. For u'_x this maximum occurs at $2x_2/D \simeq 0.0017$. For the region close to the wall $u'_x/\bar{U}_{x,\max}$ is shown in Fig. 7-57, and the ratio u'_x/u^* together with u'_r/u^* and u'_φ/u^* are shown in Fig. 7-58. The maximum of u'_x/u^* occurs in the neighborhood of $x_2^+ = 15$, a value which roughly applies also to the boundary layer investigated by Klebanoff

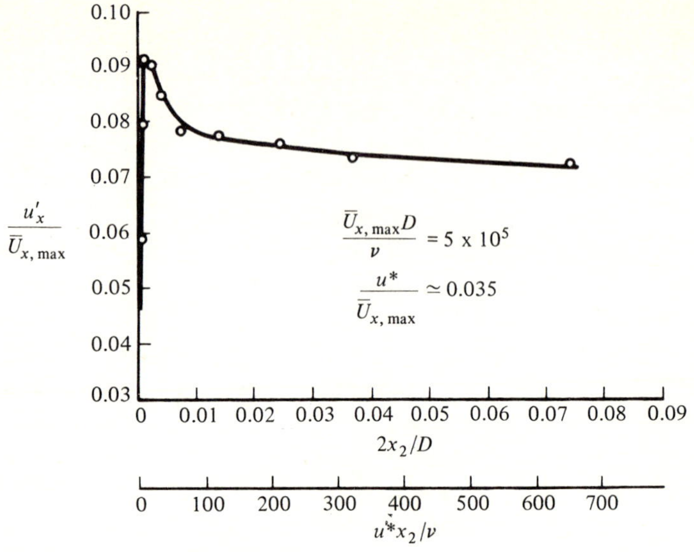

FIGURE 7-57
Relative turbulence intensity $u'_x/\bar{U}_{x,\max}$ near the wall in pipe flow. (From: *Laufer, J.,*[41] *by permission of the National Advisory Committee for Aeronautics.*)

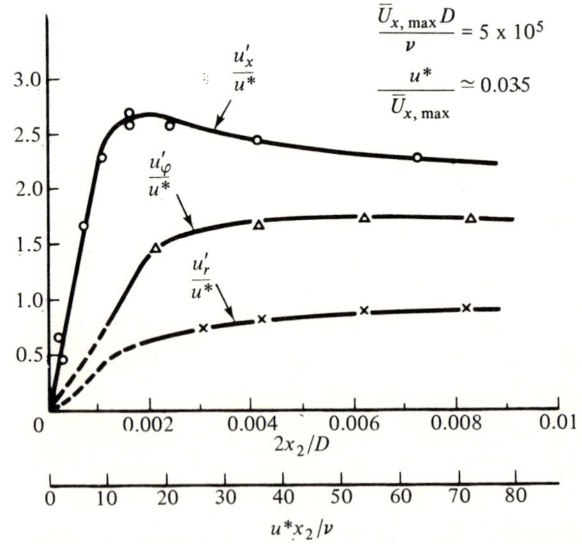

FIGURE 7-58
Relative turbulence intensities near the wall in pipe flow. (From: *Laufer, J.,*[41] *by permission of the National Advisory Committee for Aeronautics.*)

(see Fig. 7-18). Laufer's measurements at $\bar{U}_{x,\max}D/\nu = 5 \times 10^4$ yield the same result as Klebanoff's measurements. The measurements of u'_x, u'_r, and u'_φ by Coantic and by Lawn yielded results which agree satisfactorily well with those by Laufer. The maximum differences are less than 15 per cent. The main differences are found for the data taken in the axis of the pipe. Both Coantic and Lawn obtained $u'_\varphi \simeq u'_r$ there, but the two much smaller than u'_x in contrast to Laufer's results.

Coantic[111] experimented with an airflow in a pipe of 76.5 mm I.D., and the majority of the measurements were carried out at $\mathbf{Re}_D = 5 \times 10^4$. The pipe used by Lawn[209] had an I.D. of 144 mm, while measurements were made at different air velocities so that the value of $\mathbf{Re}_D$ could be varied from 4×10^4 to 25×10^4.

Figure 7-59 shows the distribution of turbulence intensities relative to the local mean velocity $\bar{U}_x$ in the wall region. The turbulence component $u'_x/\bar{U}_x$ seems to approach a constant value (at the same time a maximum) in the viscous sublayer. In contrast to this, the components $u'_\varphi/\bar{U}_x$ and $u'_r/\bar{U}_x$ seem to approach zero at the wall.

However, we now do know from the measurements by Eckelmann[115] (see Sec. 7-8) that $u'_x/\bar{U}_x$ should attain a maximum around the edge of the viscous sublayer, and decreases to a finite value at the wall smaller than 0.3. Also from the results obtained by Sirkar and Hanratty,[116] and by Fowles *et al.*[117] it has been concluded that $u'_\varphi/\bar{U}_x$ remains finite at the wall. The value of $\sim$0.09 obtained by them is, however, higher than the values close to the wall as measured by Laufer.

With known distributions of u'_x, u'_r, and u'_φ it is possible to calculate the distribution of the static pressure $\bar{P}$ according to Eq. (7-132). This has been done by Patterson *et al.*[232] They measured the distribution of the static pressure with a Prandtl static tube in an air flow through a pipe of 145 mm I.D. and at $\mathbf{Re}_D \simeq 80{,}000$

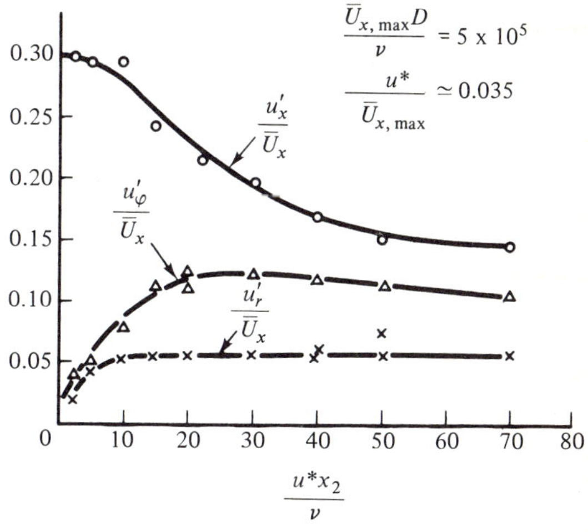

FIGURE 7-59
Distributions of turbulence intensities relative to local mean velocity near the wall in pipe flow. (After: *Laufer, J.*[41])

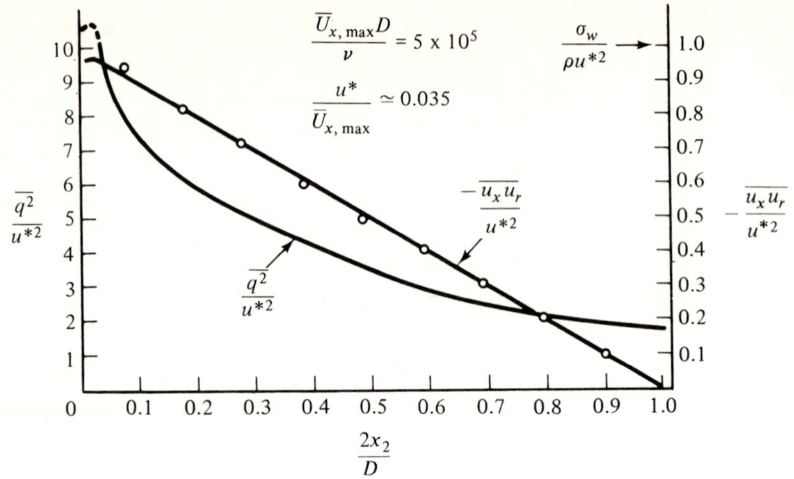

FIGURE 7-60
Distributions of turbulence kinetic energy and turbulence shear stress in pipe flow. (From: *Laufer, J.*,[41] *by permission of the National Advisory Committee for Aeronautics.*)

and 147,000. These distributions were compared with the calculated distributions using distributions of u_r' and u_φ' measured earlier by Sandborn in a 100 mm I.D. tube and at $\mathbf{Re}_D = 50{,}000$ and 100,000. Taking into account the uncertainty in measuring static pressures in a turbulent flow, the agreement appeared reasonable. The calculated and measured values were of the same order of magnitude, showed the same trend in the distribution and in the Reynolds-number effect. $(\bar{P}_w - \bar{P})/\frac{1}{2}\rho\bar{U}_{av}{}^2$ showed a maximum of roughly 0.003 around $2x_2/D \simeq 0.1$, where u_r' had its maximum (see Fig. 7-56).

From Laufer's data on $u_x'/\bar{U}_{x,\max}$, $u_r'/\bar{U}_{x,\max}$, and $u_\varphi'/\bar{U}_{x,\max}$, the turbulence kinetic energy $\overline{q^2}/u^{*2}$ has been calculated. Its distribution across a pipe section is shown in Fig. 7-60. Figure 7-60 also gives the distribution of the turbulence shear stress $-\overline{u_x u_r}/u^{*2}$, measured by Laufer. The attention of the reader may be drawn to the close similarity with the corresponding results obtained in boundary-layer flow, as shown in Fig. 7-26, where the curves are corrected for intermittency. Thus, if the effect of the intermittency of the boundary-layer flow near the outer boundary is taken into account, it becomes clear that there is close similarity between the larger-scale motions in pipe flow and in the fully turbulent regions of the boundary layer. These motions are mainly responsible for the turbulence kinetic energy and the turbulence shear stress. Consequently, a similar result will be obtained for the ratio $-\overline{u_x u_r}/\overline{q^2}$ which is shown for pipe flow in Fig. 7-61 and for boundary-layer flow in Fig. 7-22.

The turbulence shear-stress distribution in the wall region of pipe flow is shown in Fig. 7-62. This must be compared with Fig. 7-21 for boundary-layer flow: the qualitative and quantitative agreement between the two curves is worth noting.

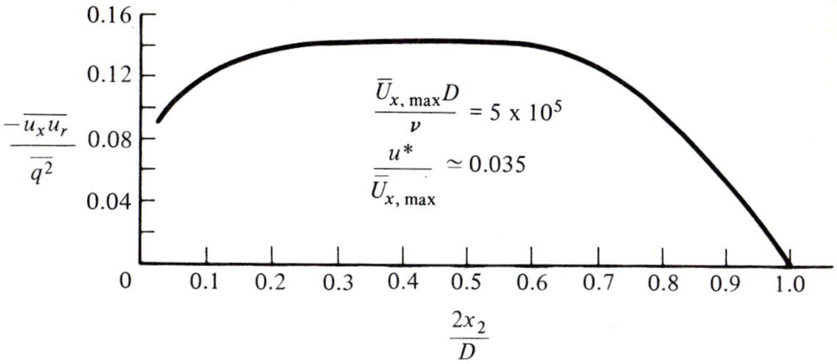

FIGURE 7-61
Ratio between turbulence shear stress and turbulence kinetic energy in pipe flow. (After: *Laufer, J.*[41])

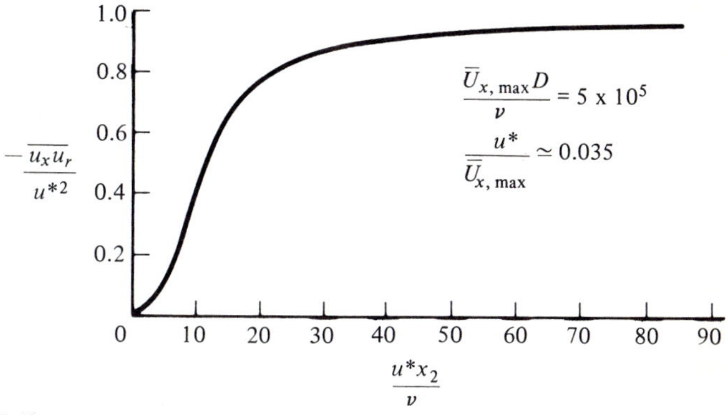

FIGURE 7-62
Distribution of turbulence shear stress near the wall in pipe flow. (From: *Laufer, J.*,[41] *by permission of the National Advisory Committee for Aeronautics.*)

From the mean-velocity distribution and the turbulence shear-stress distribution, the eddy viscosity ϵ_m has been calculated. Except for the region at the wall, where the viscous shear stress becomes important, the turbulence shear stress is practically equal to the total shear stress, whose distribution is linear according to Eq. (7-137). With this equation we obtain

$$\frac{2\epsilon_m}{\bar{U}_{x,\max} D} = \frac{u^{*2}}{\bar{U}^2_{x,\max}} \frac{1 - \xi_2}{(d/d\xi_2)(\bar{U}_x/\bar{U}_{x,\max})} \tag{7-146}$$

In the wall region, where close to the wall the turbulence shear stress is only a part of the total shear stress, the eddy viscosity may be obtained directly from the turbulence-shear-stress distribution. Instead of the relation (7-146) we must then consider

$$\frac{2\epsilon_m}{u^* D} = \frac{\bar{U}_{x,\max}}{u^*} \frac{2\nu}{\bar{U}_{x,\max} D} \frac{-\overline{u_x u_r}/u^{*2}}{(d/dx_2{}^+)(\bar{U}_x/u^*)} \tag{7-147}$$

where $x_2{}^+ = u^* x_2/\nu$.

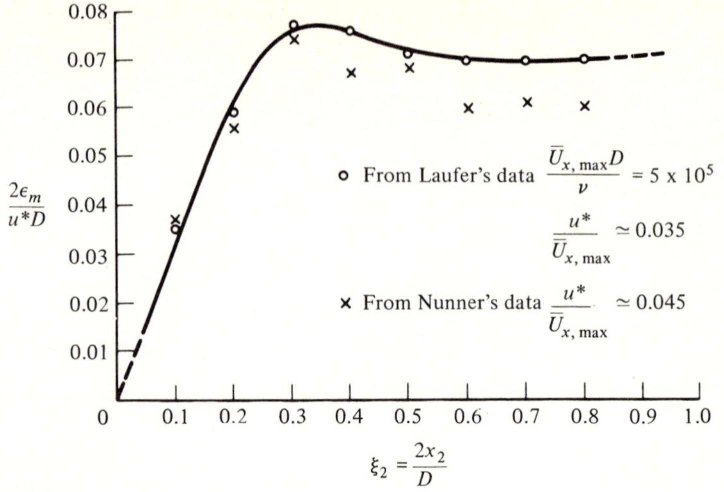

FIGURE 7-63
Distribution of eddy viscosity in pipe flow calculated from Laufer's[41] and Nunner's[42] data.

The eddy viscosities computed in this way are shown in Fig. 7-63 for the core region and in Fig. 7-64 for the wall region. For the core region not only Laufer's data for the mean-velocity distribution have been used but also the data obtained by Nunner.[42] In Nunner's experiments the wall-friction velocity amounted to $u^*/\bar{U}_{x,\max} \simeq 0.045$.

For the core region of the pipe flow the distribution of the eddy viscosity resembles both qualitatively and quantitatively the corresponding distribution for the outer region of the boundary-layer flow, again considering only the fully turbulent portions (see Fig. 7-26). The eddy viscosity first increases linearly with $\xi_2 = 2x_2/D$, then reaches a maximum at about $\xi_2 = 0.3$, then decreases slightly, and attains a nearly constant value beyond $\xi_2 = 0.5$. The same course of the distribution curve was obtained by Reichardt[12] in a two-dimensional channel flow, although the maximum there occurred at about $\xi_2 = 0.4$ to 0.5.

Accepting an almost constant value of $2\epsilon_m/u^*D$ in the core region, the value of ϵ_m/ν can easily be obtained from a relation with $\mathbf{Re}_D$ and the friction factor f. For, with $2\epsilon_m/u^*D = C$ we obtain, using Eq. (7-144):

$$\frac{\epsilon_m}{\nu} = C\frac{u^*D}{2\nu} = \frac{C}{2}\mathbf{Re}_D\frac{u^*}{\bar{U}_{\text{av}}} = \frac{C}{2}\mathbf{Re}_D\sqrt{\frac{f}{8}} \qquad (7\text{-}148)$$

As mentioned earlier, because of the departure from Reynolds number similarity, C is still a (decreasing) function of $\mathbf{Re}_D$ according to Nikuradse's measurements.

From the measurements by Laufer, shown in Fig. 7-63 we would obtain $C \simeq 0.07$.

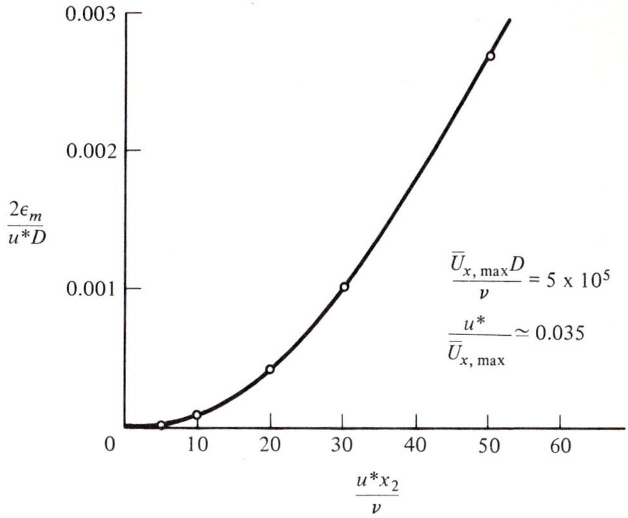

FIGURE 7-64
Distribution of eddy viscosity near the wall in pipe flow. (After: *Laufer, J.*[41])

For the wall region the values of $2\epsilon_m/u^*D$ for pipe flow are much smaller than those of $\epsilon_m/u^*\delta$ for the boundary layer. Now neither $D/2$ nor δ is a suitable length scale for the wall region. Since the flow there is determined solely by u^* and ν, the length scale to be used should be ν/u^*. If this length scale is taken, then we have to consider ϵ_m/ν instead of $2\epsilon_m/u^*D$ or $\epsilon_m/u^*\delta$.

In Fig. 7-65 the distribution of ϵ_m/ν is shown for pipe flow as obtained from the measurements by Laufer[41] and by Abbrecht and Churchill,[52] and for boundary-layer flow obtained from Schubauer's[30] measurements. The measurements by Abbrecht and Churchill were carried out at $\mathbf{Re}_D \simeq 65{,}000$. The data obtained from the latter measurements clearly show an S-shape behavior approaching the linear distribution for the fully turbulent part of the wall region. A quadratic variation $\epsilon_m/\nu = 0.007{x_2^+}^2$ describes these results reasonably well up to $x_2^+ \simeq 40$. On the other hand, Rannie's Eq. (7-43) with $\varkappa_1 = 0.0688$ already shows a departure at a lower value than $x_2^+ = \delta_t^+ = 27.5$ where the velocity-distribution according to his Eq. (7-44) should join the logarithmic distribution.

As mentioned earlier, a quadratic variation of ϵ_m/ν does not exclude a behavior proportional with ${x_2^+}^3$, or even ${x_2^+}^4$ when $x_2^+ \to 0$. Indeed, measurements on heat transfer very near to the wall (see Sec. 7-14) indicate a behavior of ϵ_θ/ν proportional with ${x_2^+}^3$ up to $x_2^+ \simeq 10$, and with ${x_2^+}^2$ beyond this distance up to $x_2^+ \simeq 35$. However, it should be kept in mind that ϵ_θ and ϵ_m need not necessarily decrease at the same rate when approaching the wall.

The linear increase of $2\epsilon_m/u^*D$ with the distance to the wall (see Fig. 7-63) in the turbulent part of the wall region is consistent with the logarithmic distribution and a constant shear stress, though this is only approximately true in pipe flow where this stress decreases linearly with the distance x_2.

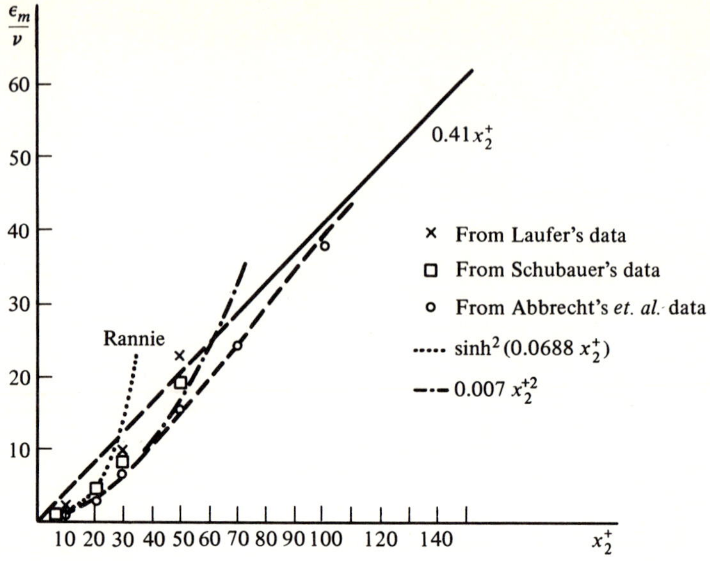

FIGURE 7-65
Distribution of eddy viscosity near the wall in pipe flow. (Adapted from: *Laufer J.*;[41] *Abbrecht, P. H., and S. W. Churchill*[52]), and in boundary-layer flow. (Adapted from: *Schubauer, G. B.*[30])

If for the core region a constant eddy-viscosity is accepted, with a linear variation of the shear stress a parabolic mean-velocity distribution must be obtained. The differential equation for this distribution reads

$$\epsilon_m \frac{d}{dx_2}\bar{U}_x = \frac{1}{\rho}\sigma_{21} = u^{*2}(1 - \xi_2)$$

or

$$\frac{d}{d\xi_2}\frac{\bar{U}_x}{\bar{U}_{x,\max}} = \frac{u^*D}{2\epsilon_m}\frac{u^*}{\bar{U}_{x,\max}}(1 - \xi_2) = \frac{1}{C}\frac{u^*}{\bar{U}_{x,\max}}(1 - \xi_2)$$

whence follows

$$\bar{U}^+_{x,\max} - \bar{U}_x{}^+ = \frac{1}{2C}(1 - \xi_2)^2 \qquad (7\text{-}149)$$

We may compare this solution with the velocity distribution measured by Laufer at $\bar{U}_{x,\max}D/\nu = 5 \times 10^5$ (see Fig. 7-53). With $C = 0.07$ Eq. (7-149) yields

$$\bar{U}^+_{x,\max} - \bar{U}_x{}^+ = 7.15(1 - \xi_2)^2$$

Figure 7-66 shows the result; the agreement between computed and measured distributions is very satisfactory indeed in the region $\xi_2 > 0.2$. Also it matches reasonably well with the logarithmic distribution at $\xi_2 = 0.2$. The logarithmic

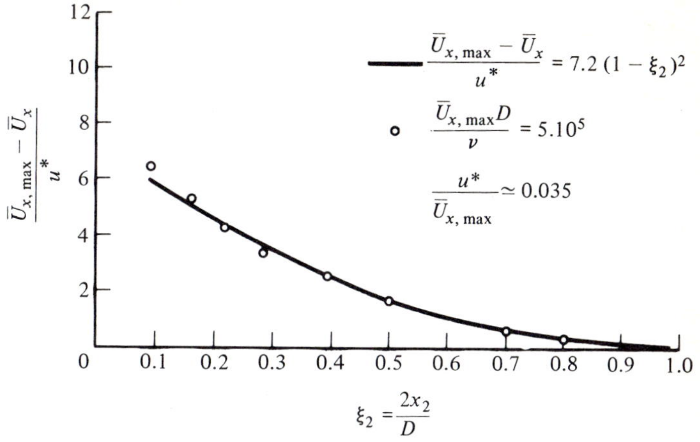

FIGURE 7-66
Comparison between computed velocity defect and Laufer's[41] experimental data.

distribution, with $A = 2.44$ and $B^* = 0.8$ yields at $\xi_2 = 0.2$, $\bar{U}^+_{x,\max} - \bar{U}_x^+ = 4.7$ and $d\bar{U}_x^+/d\xi_2 \simeq 12$, while the above parabolic distribution yields for these quantities the values 4.6 and 11.5 respectively.

Of course it is possible to approximate the actual distribution of the eddy viscosity in a better way and across the whole pipe section. Reichardt[12] approximated this distribution across the two-dimensional channel flow that he investigated, using an even fourth-degree polynomial in $1 - \xi_2$. The resulting velocity distribution showed excellent agreement with the measured distribution, but the improvement over that based on constant eddy viscosity was appreciable only in the region $\xi_2 < 0.2$, where the logarithmic velocity distribution also gives a satisfactory solution.

Laufer determined the dissipation by turbulence by starting from the approximate relation (7-136) for ε'. He measured the quantities $\overline{(\partial u_x/\partial x)^2}$, $\overline{(\partial u_r/\partial x)^2}$, and $\overline{(\partial u_\varphi/\partial x)^2}$ by assuming that $\partial/\partial x \simeq -\bar{U}_x^{-1}\,\partial/\partial t$, which is approximately true in shear flow only if at least $\overline{(u_x/\bar{U}_x)^2} \ll 1$ (see Sec. 1-8). It was not possible to measure $\overline{(\partial u_r/\partial r)^2}$, $\overline{(\partial u_\varphi/\partial r)^2}$, $\overline{(\partial u_r/r\,\partial\varphi)^2}$, and $\overline{(\partial u_\varphi/r\,\partial\varphi)^2}$; but, like Klebanoff in his boundary-layer investigations, Laufer assumed a sufficient degree of local isotropy to justify use of the isotropic relations

$$\overline{\left(\frac{\partial u_r}{\partial r}\right)^2} = \frac{1}{2}\overline{\left(\frac{\partial u_\varphi}{\partial r}\right)^2} = \frac{1}{2}\overline{\left(\frac{\partial u_x}{\partial r}\right)^2}$$

and

$$\overline{\left(\frac{\partial u_\varphi}{r\,\partial\varphi}\right)^2} = \frac{1}{2}\overline{\left(\frac{\partial u_r}{r\,\partial\varphi}\right)^2} = \frac{1}{2}\overline{\left(\frac{\partial u_x}{r\,\partial\varphi}\right)^2}$$

In order to avoid any misunderstanding in the above relations and in Eq. (7-136), at the point considered, for convenience a local Cartesian coordinate system is assumed, with the three axes coinciding with the x, the r, and the local φ direction respectively. For, even in an axi-symmetric situation there is a preferred, axial, direction which is contradictory with the idea of isotropy.

The measurable terms showed a fair degree of isotropy except in the region close to the wall. According to this procedure, the relation (7-136) becomes

$$\varepsilon' = \nu\left[\overline{\left(\frac{\partial u_x}{\partial x}\right)^2} + \overline{\left(\frac{\partial u_r}{\partial x}\right)^2} + \overline{\left(\frac{\partial u_\varphi}{\partial x}\right)^2} + \frac{5}{2}\overline{\left(\frac{\partial u_x}{\partial r}\right)^2} + \frac{5}{2}\overline{\left(\frac{\partial u_x}{r\,\partial\varphi}\right)^2}\right]$$

As discussed in Sec. 7-7, another possibility is to assume isotropic relations for the terms $\overline{(\partial u_r/\partial x_r)^2}$, $\overline{(\partial u_\varphi/r\,\partial\varphi)^2}$, $\overline{(\partial u_\varphi/\partial r)^2}$, $\overline{(\partial u_r/r\,\partial\varphi)^2}$, and $\overline{(\partial u_x/\partial x)^2}$.

The expression for ε' then reads [see Eq. (7-84)]

$$\frac{\varepsilon'}{\nu} = 7\overline{\left(\frac{\partial u_x}{\partial x}\right)^2} + \overline{\left(\frac{\partial u_r}{\partial x}\right)^2} + \overline{\left(\frac{\partial u_\varphi}{\partial x}\right)^2} + \overline{\left(\frac{\partial u_x}{\partial r}\right)^2} + \overline{\left(\frac{\partial u_x}{r\,\partial\varphi}\right)^2}$$

However, if the local dissipation and production give the major contributions to the energy balance, and the turbulence in the dissipation range of the energy spectrum is almost isotropic, the simple expression $\varepsilon = 15\nu\overline{(\partial u_x/\partial x)^2}$ for an isotropic turbulence may be used as well. Indeed, when excluding the region very close to the wall ($x_2^+ < 100$, say), experimental evidence shows but little difference in the determination of ε' or ε, according to the above three methods.[209,211]

The turbulence dissipation ε' determined according to the expression used by Laufer, and rendered dimensionless with u^* and $D/2$ is shown in Fig. 7-67.

At the higher Reynolds number it was extremely difficult to measure the turbulence-dissipation terms. The values obtained are believed to be too low, because

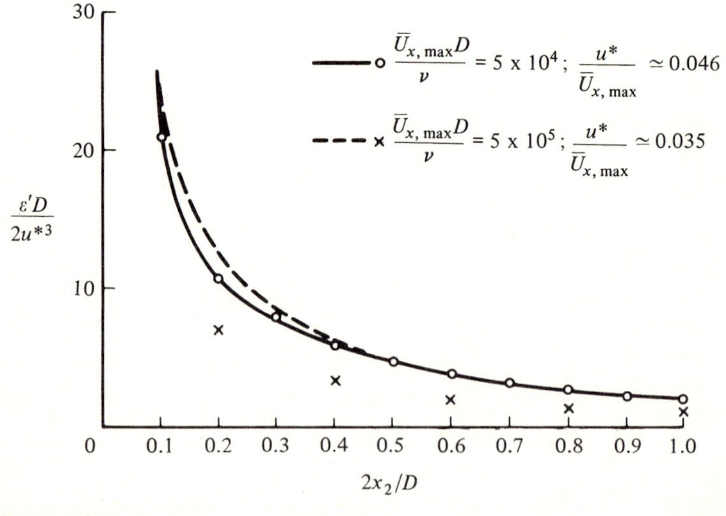

FIGURE 7-67
Turbulence dissipation in pipe flow. (From: *Laufer, J.*,[41] *by permission of the National Advisory Committee for Aeronautics.*)

the contribution of the high-wavenumber part of the turbulence spectrum could be measured only inaccurately and definitely too low. Therefore the dissipation was estimated from the results obtained at the lower Reynolds number by assuming for the region close to the wall (1) similarity, if u^* and ν/u^* are taken as velocity and length scales, and (2) the same errors in percentage across the pipe section. The dashed curve in Fig. 7-67 is the dissipation so estimated for the higher Reynolds number. The distribution of the dissipation greatly resembles the distribution of the dissipation in the boundary layer (Fig. 7-27) and, like that, is completely different from the distribution of the dissipation in free turbulent flow, where a high degree of uniformity across the turbulent regions occurs. This points to the fact that the small-scale structure of wall turbulence is much less uniform than the small-scale structure of free turbulence.

By applying a hot-wire-anemometer technique first suggested and applied by Townsend (see Sec. 2-5), Laufer measured the triple velocity correlations $\overline{u_r^3}$, $\overline{u_x^2 u_r}$, and $\overline{u_\varphi^2 u_r}$. Thus he was able to determine the distribution of $\overline{u_r q^2}$. Accordingly, with known distributions for the mean velocity, the turbulence shear stress, the turbulence kinetic energy, and the dissipation, it is now possible to make a turbulence-energy balance, since the only unknown term in this balance is the distribution of the diffusion of pressure energy $\overline{u_r p}$, which may be considered the closing entry.

As explained in Sec. 7-12, the complete equation for the turbulence-energy balance may be simplified more or less, depending on the region of the pipe flow considered. If we consider the part of the wall region closest to the wall, Eq. (7-5a) can be used. With u^* and ν/u^* as velocity and length scales, respectively, Eq. (7-5b) for the pipe flow may be written in the following nondimensional form:

$$\frac{\overline{u_x u_r}}{u^{*2}} \frac{d}{dx_2^{+}} \left(\frac{\bar{U}_x}{u^*} \right) + \frac{1}{u^{*3}} \frac{d}{dx_2^{+}} \overline{u_r \left(\frac{p}{\rho} + \frac{q^2}{2} \right)} = \frac{\nu \varepsilon'}{u^{*4}} - \frac{d^2}{dx_2^{+2}} \frac{\overline{q^2}}{2u^{*2}}$$

The signs of some terms differ from those in Eq. (7-5b) because the positive direction of u_r is opposite to that of x_2.

Figure 7-68 shows the energy balance in the wall region measured by Laufer and corrected by Townsend,[31] for Laufer did not take into account the effect on the computed dissipation of the large gradients close to the wall.

In the region closest to the wall all the terms that occur in the above equation are of importance. It may be noticed that the dissipation and production terms are nearly equal but opposite to each other, and so are the terms representing diffusion by turbulence of kinetic energy and of pressure energy.

It may be noted that the turbulence kinetic energy, its production, and its dissipation all show a sharp maximum in the buffer region ($x_2^+ \simeq 10$) near the wall.

In the region farther from the wall ($30 \gtrsim x_2^+ \gtrsim 80$) the main contributions to the energy balance are made by the production and the dissipation. Though there is some contribution from the diffusion terms, one may consider the turbulence to be in energy equilibrium, and the simpler Eq. (7-6) may suffice. In the fully turbulent region, and certainly so in the core region, Eq. (7-135) may be more appropriate.

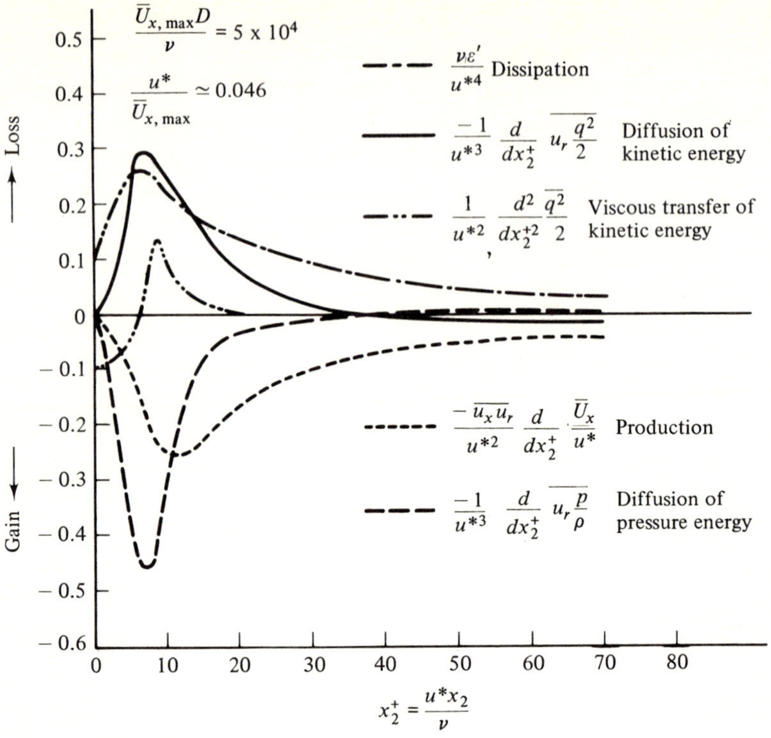

FIGURE 7-68
Energy balance in the wall region of pipe flow. (From: *Laufer, J.*,[41] *by permission of the National Advisory Committee for Aeronautics.*)

Using u^* and $D/2$ as suitable velocity and length scales, this equation can be written in nondimensional form:

$$\frac{\overline{u_x u_r}}{u^{*2}} \frac{d}{d\xi_2'}\left(\frac{\bar{U}_x}{u^*}\right) + \frac{1}{u^{*3}} \frac{1}{\xi_2'} \frac{d}{d\xi_2'}\left\{\xi_2'\left[\overline{u_r\left(\frac{p}{\rho} + \frac{q^2}{2}\right)}\right]\right\} + \frac{\varepsilon' D}{2u^{*3}} = 0$$

where $\xi_2' = 1 - \xi_2 = 2r/D$.

But as Fig. 7-69 shows, in the region $\xi_2 < 0.7$ the diffusion of pressure energy and kinetic energy together contribute but little to the energy balance, so that in this region too the turbulence may behave as if it were in energy equilibrium. Whereas the turbulent diffusion of pressure energy decreases and becomes negligible in the center region ($\xi_2 > 0.7$) the turbulent diffusion of kinetic energy becomes increasingly important. Close to the axis of the pipe the dissipation of energy is balanced by the diffusion of kinetic energy alone.

The results obtained by Laufer have been confirmed by the much more recent experiments carried out by Lawn.[209] The dissipation measured by Lawn at $\mathbf{Re}_D = 9 \times 10^4$ agrees very closely with Laufer's result shown in Fig. 7-69. There are some differences in the contribution by diffusion of pressure and kinetic energy.

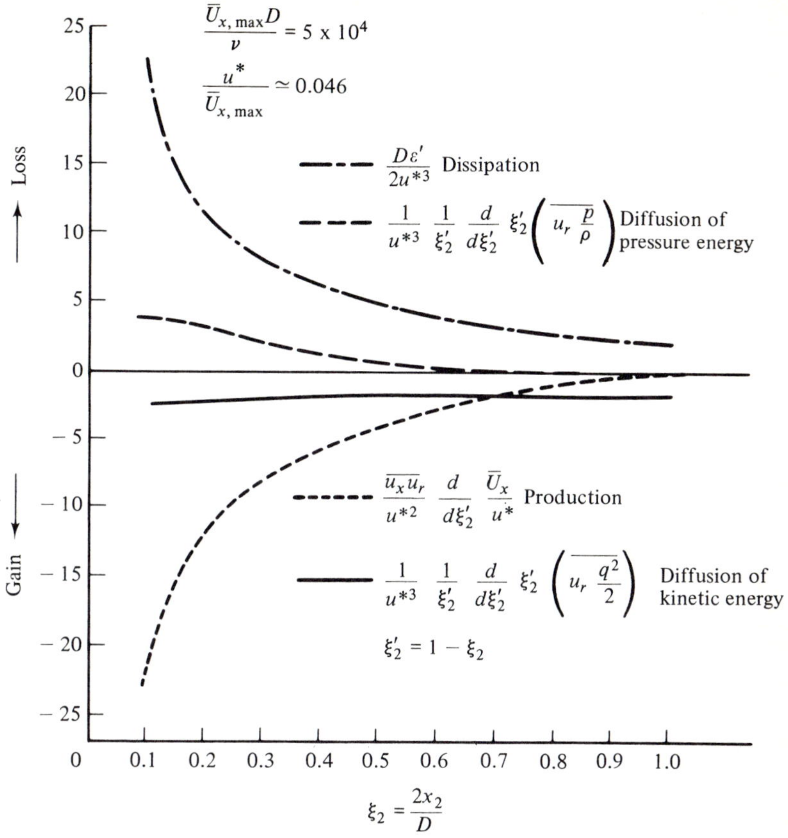

FIGURE 7-69
Energy balance in the core region of pipe flow. (From: *Laufer, J.*,[41] *by permission of the National Advisory Committee for Aeronautics.*)

Lawn estimated the contribution by diffusion of pressure energy, but did not measure all terms of the diffusion of kinetic energy.

Figures 7-68 and 7-69 seem to indicate a continuous flux of turbulence kinetic energy by turbulence diffusion from the wall toward the center of the pipe. In the buffer region ($x_2^+ < 30$) this results in a loss of kinetic energy; outside this region and in the core region it results in a gain. Similarly, there seems to be a continuous flux of pressure energy toward the wall in the region $\xi_2 < 0.5$, producing a gain in the buffer region and a loss in the region outside it. It may be noticed that this latter transfer of energy is against the gradient of the turbulence kinetic energy (compare Fig. 7-60).

Energy spectra of the u_x turbulence component measured by Laufer at various points of the pipe section have already been shown in Fig. 4-9. If the turbulence were isotropic the energy spectra of the u_r and u_φ turbulence components could be computed using the relation (3-53). These spectra then would be the same and would show higher values in the higher wavenumber range and lower values in

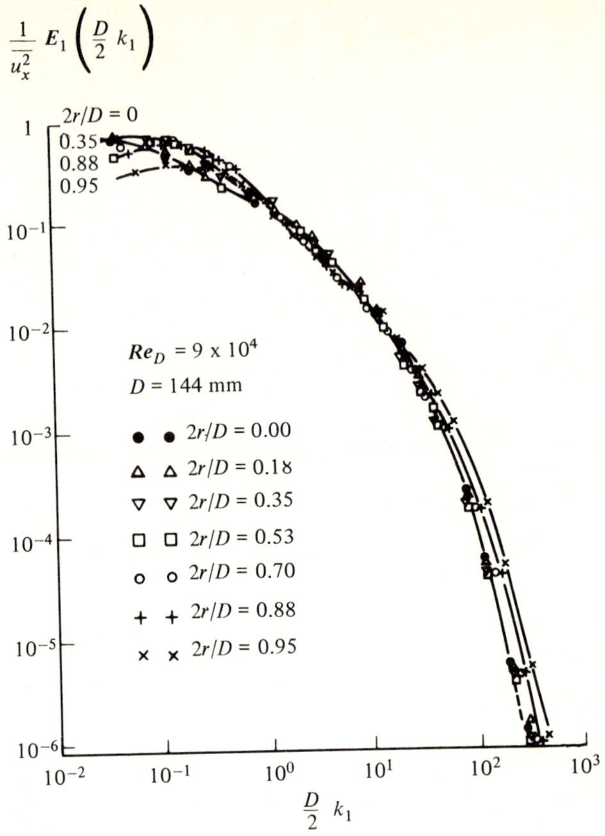

FIGURE 7-70
Energy-density spectra of u_x in pipe flow. (After: *Lawn, C. J.,*[209] *by permission of the Cambridge University Press.*)

the low wavenumber range than the spectrum of the u_x-component. However, pipe-flow turbulence is not isotropic and though the measured spectra $\mathbf{E}_2$ (of u_r) and $\mathbf{E}_3$ (of u_φ) do show such a difference with the measured $\mathbf{E}_1$ spectrum, the actual quantitative differences are much greater. Lawn[209] has made extensive measurements of the spectra $\mathbf{E}_1$, $\mathbf{E}_2$, and $\mathbf{E}_3$, and of $\mathbf{E}_{21}$ of the turbulence shear stress $(-u_r u_x)$. They are reproduced in Figs. 7-70, 7-71, 7-72, and 7-73 respectively. These spectra have been measured at $\mathbf{Re}_D = 9 \times 10^4$. Measurements of the $\mathbf{E}_1$-spectrum in the point $2r/D = 0.35$ and at different values of $\mathbf{Re}_D$ (varying from 3.67×10^4 to 25×10^4) did not reveal a marked difference, except in the higher wavenumber range $\frac{1}{2}Dk_1 \gtrsim 50$ where due to viscous effects the $\mathbf{E}_1$-spectrum decreases at a faster rate for the lower Reynolds numbers.

The wavenumber has been rendered dimensionless with $D/2$, and the spectra have been normalized such that

$$\frac{1}{\overline{u_x^2}} \int_0^\infty d\left(\frac{D}{2}k_1\right)\mathbf{E}_1\left(\frac{D}{2}k_1\right) = 1$$

and similarly for $\mathbf{E}_2$, $\mathbf{E}_3$, and $\mathbf{E}_{21}$.

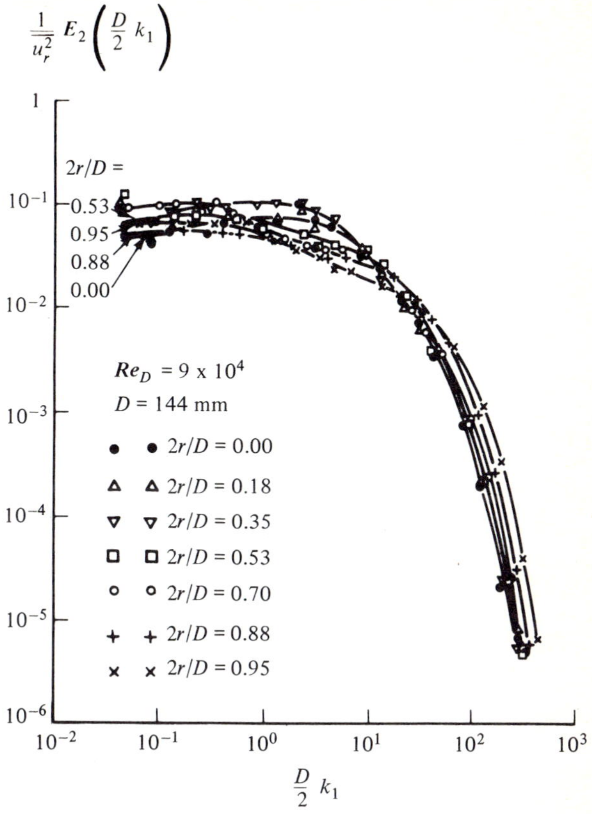

FIGURE 7-71
Energy-density spectra of u_r in pipe flow. (After: *Lawn, C. J.*,[209] *by permission of the Cambridge University Press.*)

The measurements had not been corrected for the effect of finite length of the hot-wire probe. Lawn estimated this effect to become appreciable when $\frac{1}{2}Dk_1 > 100$.

All these figures show that in the higher wavenumber range ($\frac{1}{2}Dk_1 \gtrsim 20$) the values of the spectra are higher in the points closer to the wall, the turbulence being finer grained there. The opposite is in general true in the low wavenumber range ($\frac{1}{2}Dk_1 \lesssim 1$). Furthermore, a comparison of Fig. 7-73 with the other three reveals that the shear-stress spectra fall off at a faster rate at wavenumbers approaching the viscous range, indicating an approach to isotropy there and a decoupling of the higher wavenumber components of u_r and u_φ owing to viscous effects.

A comparison of the non-normalized, dimensional spectra showed that $\mathbf{E}_2$ agreed with the computed one according to the relation (3-53) in the high wavenumber range $\frac{1}{2}Dk_1 > 20$. For $\frac{1}{2}Dk_1 < 2, \mathbf{E}_2 < \mathbf{E}_{21} < \mathbf{E}_1$, while for $\frac{1}{2}Dk_1 > 20, \mathbf{E}_{21} < \mathbf{E}_2 \simeq \mathbf{E}_1$. The correlation coefficient $\mathbf{R}_{21}$ was roughly constant and equal to 0.37 in the region $0.5 < 2r/D < 0.95$. It decreased to zero when $2r/D \to 0$. However, the local correlation coefficients in wavenumber space showed a different behavior. Locally much higher values than 0.37 were obtained up to ~ 0.9 at $\frac{1}{2}Dk_1 \simeq 0.1$ in the region $2r/D = 0.5$

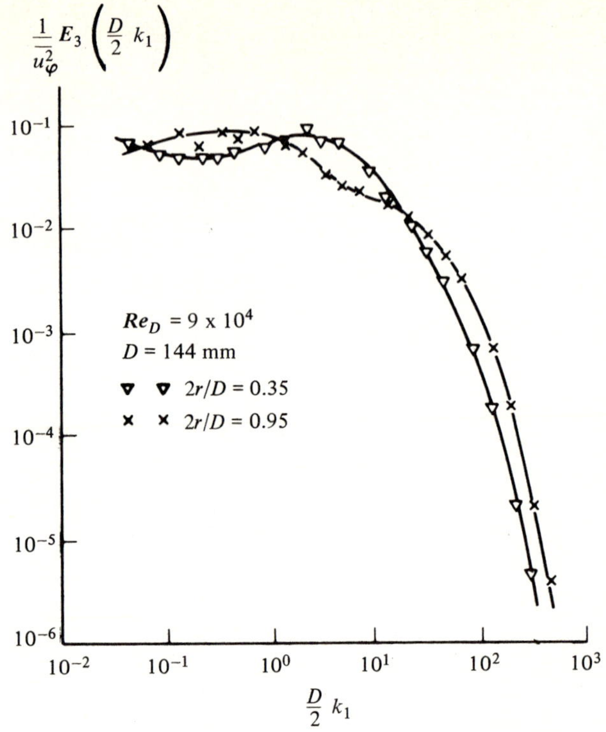

FIGURE 7-72
Energy-density spectra of u_φ in pipe flow. (After: *Lawn, C. J.*,[209] *by permission of the Cambridge University Press.*)

to 0.7, and to $\sim$0.75 at $\frac{1}{2}Dk_1 \simeq 5$ for $2r/D = 0.95$. At $\frac{1}{2}Dk_1 = 100$, the correlations were still finite. The value of the coefficient in the point $2r/D = 0.18$ was still $\sim$0.03.

Laufer also made measurements on spatial correlations of turbulence-velocity components for flow in a two-dimensional channel.[43] From these measurements it is concluded that the longitudinal correlation in the axial direction $\overline{u_1(\xi_1)u_1(\xi_1 + x_1)}$ extends over much longer distances than the transverse correlations $\overline{u_1(\xi_2)u_1(\xi_2 + x_2)}$ and $\overline{u_1(\xi_3)u_1(\xi_3 + x_3)}$ (the coordinate x_2 is perpendicular to the wall). The integral scale Λ_f in the x_1-direction was roughly equal to 0.8 times the half-width of the channel; the integral scales Λ_g in the x_2- and x_3-directions were very roughly equal to each other, and 0.2 to 0.3 times the half-width of the channel. The variation of these integral scales across the core region was not large. On the other hand the corresponding dissipation scales λ_f and λ_g showed an increase with distance from the wall, and their maximum values were of the order of one-tenth of the half-width of the channel. Apparently the turbulence in the core region consisted of large eddies elongated in the axial direction. The same results with respect to the magnitude of the integral and dissipation scales are obtained from Laufer's measurements on the kinetic energy and dissipation in pipe flow, if use is made of the relations (3-109) and (3-99) given in Chap. 3 (Townsend[31]; the length l_e is of the magnitude of the integral scale).

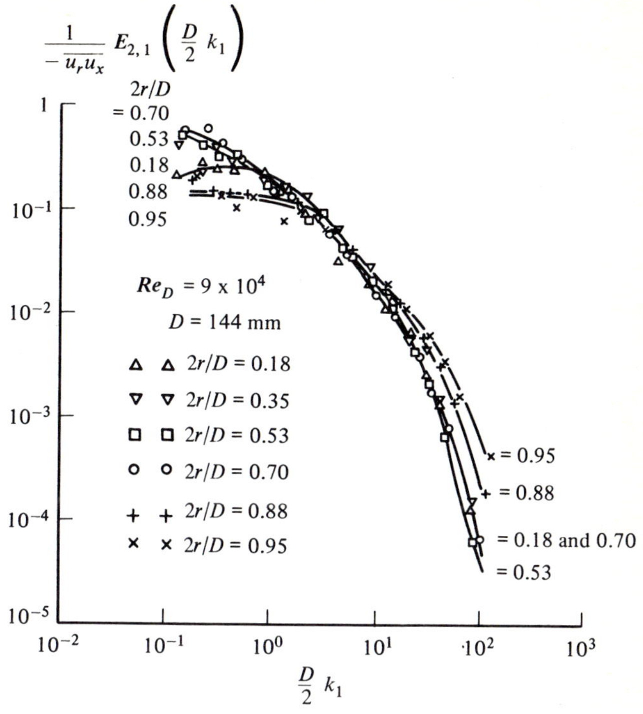

FIGURE 7-73
Power-density spectra of $-u_r u_x$ in pipe flow. (After: *Lawn, C. J.*,[209] *by permission of the Cambridge University Press.*)

Lawn[209] also determined the dissipation length-scales λ_f and λ_g (both in r and φ direction) from the spatial u_x-correlations in the three directions. They too showed a decrease when approaching the wall, as may be expected, and were roughly constant in the region $2r/D \gtrsim 0.5$, with a value of ~ 0.06 for $2\lambda_f/D$ and ~ 0.045 for the two $2\lambda_g/D$. Actually $2\lambda_f/D$ had a maximum equal to 0.061 at $2r/D = 0.35$. In that point $2\lambda_g/D \simeq 0.047$, so that $\lambda_f/\lambda_g \simeq 1.3$. The value of 1.4 for this ratio was obtained in the region $2r/D = 0.5$ to 0.9 (for istropic turbulence $\lambda_f/\lambda_g = \sqrt{2}$). These values refer to $\mathbf{Re}_D = 9 \times 10^4$. As may be expected, a definite effect of the Reynolds number on these dissipation length-scales was observed. Lawn obtained the following relation

$$2\lambda_f/D = A(\tfrac{1}{2}D^+)^{-1/2}$$

with $A = 2.8$ in the center of the pipe, $A = 2.9$ at $2r/D = 0.35$, and $A = 1.4$ at $2r/D = 0.95$.

We will close this section with a few remarks concerning the *structure of the turbulence* in pipe flow. In general there is a great similarity with the constant-pressure boundary-layer flow, in particular in the wall region. The degree of similarity decreases in the more wall-remote regions. One reason is that there is convective transport by the mean motion in the axial and transverse directions in boundary-layer flow in contrast to pipe flow. See the turbulence energy balances shown in Figs. 7-28 and 7-69. Another reason is the absence of intermittency of the whole turbulence in pipe flow as present in the outer region of boundary-layer flow.

There is intermittency in the fine structure of the turbulence in pipe flow as well, of the nature as discussed for boundary-layer flow in Sec. 7-8. We may recall that in the fine structure turbulence bursts take place with a period that is practically constant across the whole boundary-layer, ($U_0\bar{T}_B/\delta \simeq 5$). Measurements by Narasimha *et al.*[210] in a two-dimensional channel revealed not only the occurrence of these turbulence bursts, but yielded a burst-period satisfying the relation $\bar{U}_{x,\max}\bar{T}_B/h \simeq 5$ (h = half width), i.e., in agreement with the value for boundary-layer flow. In Sec. 7-9 when discussing the "cyclic" processes occurring both at the wall and in the outer region at the interface, with a "period" roughly satisfying the above relation, the possibility of a connection between these interface "periodic" processes and the similar wall-region processes through interaction between outer and inner region has been indicated. In pipe or channel flow no free interface is present, but one may expect that through the larger-scale pressure fluctuations across the pipe section there will be a connection between "cyclic" processes in the wall region and the core region. This expectation is supported by the near-constancy and the magnitude of the integral length-scale in the core region. The large eddies are elongated in the axial direction and have a diameter roughly equal to one-quarter to one half the radius of the pipe (see, e.g., Fig. 1-10). Incidentally, these large eddies in the core region, the consequent near-constancy of the integral scale, and the small variation of the turbulence intensity of the radial velocity component (Fig. 7-56) in the core region may explain the near-constancy of the eddy-viscosity in this region.

7-14 TRANSPORT OF A SCALAR QUANTITY IN WALL TURBULENCE

In this section we will restrict ourselves to considering only a number of salient aspects of the transport of a scalar quantity. For a more complete discussion of the transfer of heat and mass in turbulent pipe flow and through turbulent boundary layers the reader is referred to textbooks on this subject. Most of the experimental evidence available in the literature refers to data and semi-empirical relations for the coefficients of heat and mass-transfer between the wall and the fluid, while the information on distributions of mean-temperature and mean-concentrations, as well as of characteristics of turbulent temperature and concentration fluctuations is much more limited. In particular the information on such quantities as eddy diffusivity and cross-correlations between turbulent fluctuations of velocities and a scalar quantity is very limited. Moreover, the data presented in the literature do not always agree and are often conflicting, and therefore confusing. Appreciating the great difficulties encountered in making the appropriate measurements sufficiently reliable and the complexity of the problem as will become apparent from the following considerations, the above becomes less surprising though remaining disappointing still. It is therefore understandable that the earlier attempts to determine wall transfer-coefficients and mean-scalar distributions have been based on an assumed *analogy between the transport of momentum and a scalar quantity*. If we assume it possible to neglect the effect of possible driving forces F_θ and F_c and of the partial pressure of the diffusing matter near the wall as discussed in Sec. 5-3, then no distinction needs to be made between the transport of heat and mass. For simplicity, we will therefore in the following only consider the transport of heat and the analogy in broad and narrow sense with that of momentum. If such an analogy existed, it would be possible, in principle, to calculate the mean-velocity distribution and to obtain a relation between the coefficients of heat transfer and of wall friction.

Some support to this idea of analogy was given by Elias's measurements of the mean-velocity and mean-temperature distributions in the boundary layer along a heated flat smooth plate,[44] though the mean-temperature measurements were restricted to the outer region only. See Fig. 7-74, from which it is possible indeed to conclude that there is similarity between velocity and temperature distributions in the outer region, but that the similarity seems to decrease when approaching the wall.

Let us now see whether an analogy between momentum and heat transport, resulting in similar mean-velocity and mean-temperature distributions, is possible.

An approach to this problem is to express the equation for the temperature distribution in terms of the velocity distribution and then to consider whether the consequences of an assumed analogy, that is, similar distributions, are physically possible and also consistent with experimental evidence.

In deducing the equation for the temperature distribution expressed in terms of the velocity distributions, we assume that the turbulence transport of momentum

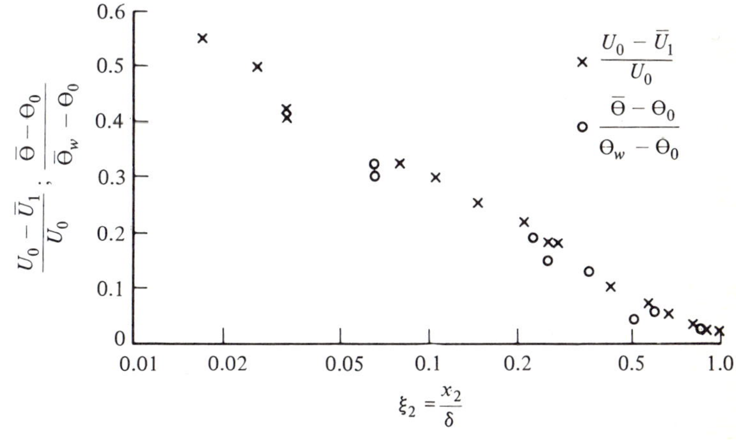

FIGURE 7-74
Velocity defect and temperature defect in the boundary layer along a heated flat smooth plate. (After: *Elias, F.*[44])

in the narrow sense and that of heat are of the gradient type of diffusion, so that we may introduce a transport coefficient.

The total shear stress, consisting of a part determined by the molecular viscosity and a part determined by the eddy viscosity, may be written

$$\sigma_{21} = (\sigma_{21})_{\text{mol}} + (\sigma_{21})_{\text{turb}} = \mu_{\text{eff}} \frac{\partial \bar{U}_1}{\partial x_2} = (\mu + \rho \epsilon_m) \frac{\partial \bar{U}_1}{\partial x_2} \tag{7-150}$$

Likewise, for the transport of heat $\mathfrak{J}_\theta$ per unit of area and of time through a plane parallel to the wall, we write

$$\mathfrak{J}_\theta = (\mathfrak{J}_\theta)_{\text{mol}} + (\mathfrak{J}_\theta)_{\text{turb}} = -\kappa_{\text{eff}} \frac{\partial \bar{\Theta}}{\partial x_2} = -(\kappa + \rho c_p \epsilon_\theta) \frac{\partial \bar{\Theta}}{\partial x_2} \tag{7-151}$$

Since the temperature distribution is not uniform in the region considered, neither are the physical properties of the flowing fluid. We shall now assume that the molecular viscosity μ, the conductivity κ, and the density ρ of the fluid vary with the temperature. In general the specific heat c_p of a gas varies only slightly with temperature; it will, therefore, be assumed to be constant.

From Eq. (7-151) after integration follows

$$\bar{\Theta}_w - \bar{\Theta} = \int_0^{x_2} dx_2 \frac{\mathfrak{J}_\theta}{\kappa_{\text{eff}}}$$

or, in dimensionless form,

$$\frac{\mu_w c_p(\bar{\Theta}_w - \bar{\Theta})}{(\mathfrak{J}_\theta)_w \delta} = \int_0^{\xi_2} d\xi_2 \frac{\mathfrak{J}_\theta}{(\mathfrak{J}_\theta)_w} \frac{\mu_w}{\mu_{\text{eff}}} \mathbf{Pr}_{\text{eff}} \tag{7-152}$$

where the index w refers to the wall condition and where $\mathbf{Pr}_{\text{eff}}$ is the effective Prandtl number

$$\mathbf{Pr}_{\text{eff}} = \frac{c_p \mu_{\text{eff}}}{\kappa_{\text{eff}}} \tag{7-153}$$

Equation (7-152) is written for boundary-layer flow, but the same expression applies to pipe flow if we read for the boundary-layer thickness δ the pipe radius $D/2$.

The effective Prandtl number and the temperature distribution can now be expressed in terms of the velocity distribution, using μ_{eff} and the relation (7-150). Thus

$$\begin{aligned} \mathbf{Pr}_{\text{eff}} &= \frac{c_p}{\kappa + \rho c_p \epsilon_\theta} \frac{\sigma_{21}}{\partial \bar{U}_1/\partial x_2} \\ &= \frac{c_p}{\kappa + \rho c_p \varepsilon_\theta} \frac{\sigma_{21}}{\sigma_w} \frac{\mu_w \delta^+}{(d/d\xi_2)(\bar{U}_1/u^*)} \end{aligned} \tag{7-154}$$

where $\delta^+ = \rho_w u^* \delta/\mu_w$.

This expression may be transformed still further by (1) introducing the turbulence Prandtl number

$$\mathbf{Pr}_{\text{turb}} = \frac{\epsilon_m}{\epsilon_\theta} \tag{7-155}$$

and (2) by expressing ϵ_m in terms of the velocity distribution using the relation (7-150). After some calculation we obtain

$$\mathbf{Pr}_{\text{eff}} = \frac{\mathbf{Pr}\dfrac{\sigma_{21}}{\sigma_w}\dfrac{\mu_w}{\mu}}{\dfrac{\mathbf{Pr}}{\mathbf{Pr}_{\text{turb}}}\dfrac{\sigma_{21}}{\sigma_w}\dfrac{\mu_w}{\mu} + \left(1 - \dfrac{\mathbf{Pr}}{\mathbf{Pr}_{\text{turb}}}\right)\dfrac{1}{\delta^+}\dfrac{d}{d\xi_2}\dfrac{\bar{U}_1}{u^*}} \tag{7-156}$$

Similarly the temperature distribution (7-152) can be expressed in terms of the velocity distribution. There is then obtained

$$\frac{\mu_w c_p(\bar{\Theta}_w - \bar{\Theta})}{(\mathfrak{J}_\theta)_w \delta} = \frac{1}{\delta^+}\int_0^{\xi_2} d\xi_2 \frac{\mathfrak{J}_\theta}{(\mathfrak{J}_\theta)_w}\frac{\sigma_w}{\sigma_{21}}\mathbf{Pr}_{\text{eff}}\frac{d}{d\xi_2}\frac{\bar{U}_1}{u^*} \tag{7-157}$$

We may in a similar way as for the velocity field, where velocities are rendered dimensionless with the wall-friction velocity u^*; introduce a *wall heat-transfer temperature*[214] θ^*, defined by

$$(\mathfrak{J}_\theta)_w = \rho_w c_p u^* \theta^* \tag{7-158}$$

Equation (7-157) may then be written

$$\frac{\bar{\Theta}_w - \bar{\Theta}}{\theta^*} = \int_0^{\xi_2} d\xi_2 \frac{\mathfrak{J}_\theta}{(\mathfrak{J}_\theta)_w}\frac{\sigma_w}{\sigma_{21}}\mathbf{Pr}_{\text{eff}}\frac{d}{d\xi_2}\bar{U}_1^{+} \tag{7-159}$$

From this equation for the temperature distribution the total heat transfer through the wall to the flowing fluid can be calculated. It is customary then to introduce either the Nusselt number or the coefficient c_h for the heat-flow resistance at the wall.

The Nusselt number is defined by

$$\mathbf{Nu} = \frac{\delta}{\kappa_w(\bar{\Theta}_w - \Theta_0)}(\mathfrak{J}_\theta)_w = \frac{h\delta}{\kappa_w} \tag{7-160}$$

where h is the coefficient of heat transfer.

From Eq. (7-157) we obtain for **Nu**

$$\frac{1}{\mathbf{Nu}} = \frac{1}{\delta^+}\int_0^1 d\xi_2 \frac{\mathfrak{J}_\theta}{(\mathfrak{J}_\theta)_w}\frac{\sigma_w}{\sigma_{21}}\frac{\mathbf{Pr}_{\text{eff}}}{\mathbf{Pr}_w}\frac{d}{d\xi_2}\frac{\bar{U}_1}{u^*} \tag{7-161}$$

The definition of the coefficient c_h for the heat-flow resistance is analogous to that of the friction coefficient c_f given by Eq. (7-47), namely,

$$c_h = \frac{(\mathfrak{J}_\theta)_w}{c_p\rho_w U_0(\bar{\Theta}_w - \Theta_0)} = \frac{h}{c_p\rho_w U_0} \tag{7-162}$$

This dimensionless number c_h is also known as the Stanton number.

From Eq. (7-157) we then obtain

$$\frac{1}{c_h} = \frac{U_0}{u^*}\int_0^1 d\xi_2 \frac{\mathfrak{J}_\theta}{(\mathfrak{J}_\theta)_w}\frac{\sigma_w}{\sigma_{21}}\mathbf{Pr}_{\text{eff}}\frac{d}{d\xi_2}\frac{\bar{U}_1}{u^*}$$

or, by the relation (7-48) between u^*/U_0 and c_f,

$$\frac{1}{c_h} = \sqrt{\frac{2}{c_f}} \int_0^1 d\xi_2 \frac{\mathfrak{I}_\theta}{(\mathfrak{I}_\theta)_w} \frac{\sigma_w}{\sigma_{21}} \mathbf{Pr}_{\text{eff}} \frac{d}{d\xi_2} \frac{\bar{U}_1}{u^*} \qquad (7\text{-}163)$$

From the definitions (7-160) and (7-162) follows the well-known relation

$$\mathbf{Nu} = \mathbf{Re}_\delta \mathbf{Pr}_w c_h \qquad (7\text{-}164)$$

where $\mathbf{Re}_\delta = \rho_w U_0 \delta / \mu_w$.

Equation (7-157) shows that the temperature distribution can be calculated from the velocity distribution only if $\mathfrak{I}_\theta/(\mathfrak{I}_\theta)_w$, σ_w/σ_{21}, and $\mathbf{Pr}_{\text{eff}}$ as a function of ξ_2 are known.

Furthermore similarity between temperature and velocity distributions can be obtained only if

$$\frac{\mathfrak{I}_\theta}{(\mathfrak{I}_\theta)_w} \frac{\sigma_w}{\sigma_{21}} \mathbf{Pr}_{\text{eff}} = \text{const.} \qquad (7\text{-}165)$$

It is clear that this will not hold in general, but in some cases this condition may be approximately fulfilled.

In the oldest theory concerning the transport of heat in turbulent flows, namely, that of Reynolds, it is simply assumed that there is complete analogy between transport of momentum and transport of heat. The consequence of this assumption is that the condition (7-165) must be satisfied. It is frequently assumed that Reynold's analogy implies that

$$\mathbf{Pr} = 1 \quad \mathbf{Pr}_{\text{turb}} = 1 \quad \frac{\mathfrak{I}_\theta}{(\mathfrak{I}_\theta)_w} = \frac{\sigma_{21}}{\sigma_w} \frac{\mu_w}{\mu} = 1 \quad \text{and} \quad \frac{\rho_w}{\rho} = 1$$

but we have shown that this need not necessarily be so. The constant occurring in Eq. (7-165) should be equal to 1. It then follows from the Eqs. (7-163) and (7-48) that, according to Reynolds's analogy,

$$c_h = \tfrac{1}{2} c_f \qquad (7\text{-}166)$$

The same result concerning the equality of $(\bar{\Theta} - \Theta_0)/(\bar{\Theta}_w - \Theta_0)$ and $(U_0 - \bar{U}_1)/U_0$ is obtained if it is assumed that

$$\frac{\mathfrak{I}_\theta}{(\mathfrak{I}_\theta)_w} = \frac{\sigma_{21}}{\sigma_w} \quad \text{and} \quad \mathbf{Pr} = \mathbf{Pr}_{\text{turb}} = \text{const.}$$

Then $\mathbf{Pr}_{\text{eff}} = \mathbf{Pr}_{\text{turb}} = \mathbf{Pr}$, and Eq. (7-157) reduces to

$$\frac{\mu_w c_p (\bar{\Theta}_w - \bar{\Theta})}{(\mathfrak{I}_\theta)_w \delta} = \frac{\mathbf{Pr}}{\delta^+} \frac{\bar{U}_1}{u^*}$$

which does give $(\bar{\Theta} - \Theta_0)/(\bar{\Theta}_w - \Theta_0) = (U_0 - \bar{U}_1)/U_0$. From Eq. (7-163) however, instead of the relation (7-166) there is obtained

$$c_h = \frac{1}{2\mathbf{Pr}} c_f \qquad (7\text{-}167)$$

Now for air $\mathbf{Pr} \simeq 0.71$. Various measurements on the distribution of heat and matter in pipe flow seem to indicate an average $\mathbf{Pr}_{\text{turb}} \simeq 0.65$ to 0.7. Woertz and Sherwood[45] obtained the value 0.72 from diffusion experiments with helium, town gas, and carbon dioxide in the turbulent flow through a two-dimensional channel. From experiments made by Friedrich and reported by Lorenz[46] with air flowing through a heated pipe ($\mathbf{Re}_D \simeq 10^5$), Reichardt[10] deduced a value of 0.65.

McCarter, Stutzman, and Koch[47] found that the diffusion of heat from a source in the airflow through a vertical pipe ($\mathbf{Re}_D = 7{,}000$ to 26,000) could be well described by (see Sec. 5-5)

$$\epsilon_\theta = 0.02\bar{U}_{\text{av}} D\sqrt{f}$$

If we assume for the core region of pipe flow $\epsilon_m \simeq 0.035u^*D$ (see Fig. 7-63), we obtain with $u^* = \bar{U}_{\text{av}}(f/8)^{1/2}$

$$\epsilon_m = 0.0125\bar{U}_{\text{av}} D\sqrt{f}$$

Consequently, here, too, $\mathbf{Pr}_{\text{turb}} = \epsilon_m/\epsilon_\theta \simeq 0.63$.

Hence, if we assume $\mathbf{Pr}_{\text{turb}} = 0.65$ to 0.70, with $\mathbf{Pr} \simeq 0.71$ for air we have $\mathbf{Pr}_{\text{turb}} \simeq \mathbf{Pr}$. This might explain why Reynolds's analogy seems to apply to the outer region in boundary-layer flow (see Fig. 7-74), and according to measurements by Nunner[42] to the core region in pipe flow. However, for the wall region, $\mathbf{Pr}_{\text{turb}}$ may quite well differ from the core-region value, so that Reynolds's analogy will no longer apply, even approximately. This will be reflected in a dissimilarity between temperature and velocity distributions in the wall region and in a departure from the simple relation (7-166) between c_h and c_f. Equation (7-166) does not contain the Prandtl number, although an effect of this number should be expected because of the existence of a sublayer where molecular transport processes dominate.

The various modifications of Reynolds's analogy proposed in the course of time are concerned mainly with the transport in the wall region. Accordingly, assumptions had to be made concerning $\mathbf{Pr}_{\text{eff}}$, $\mathfrak{J}_\theta/(\mathfrak{J}_\theta)_w$, and σ_w/σ_{21}.

We have shown that in the wall region the velocity distribution may well be described by formulas based upon the assumption $\sigma_w/\sigma_{21} \simeq 1$ (constant-stress layer). An obvious assumption to make concerning the ratio $\mathfrak{J}_\theta/(\mathfrak{J}_\theta)_w$ is that likewise $\mathfrak{J}_\theta/(\mathfrak{J}_\theta)_w \simeq 1$ in the wall region. Moreover, it is assumed that $\mu_w/\mu \simeq 1$ and $\rho_w/\rho \simeq 1$. Equation (7-163) then becomes

$$\frac{1}{c_h} = \sqrt{\frac{2}{c_f}} \int_0^1 d\xi_2\, \mathbf{Pr}_{\text{eff}} \frac{d}{d\xi_2} \frac{\bar{U}_1}{u^*} \tag{7-168}$$

and from Eq. (7-156) it follows that

$$\mathbf{Pr}_{\text{eff}} = \left[\frac{1}{\mathbf{Pr}_{\text{turb}}} + \left(\frac{1}{\mathbf{Pr}} - \frac{1}{\mathbf{Pr}_{\text{turb}}}\right)\frac{1}{\delta^+}\frac{d}{d\xi_2}\frac{\bar{U}_1}{u^*}\right]^{-1} \tag{7-169}$$

As mentioned above, in the course of time various modifications of the Reynolds's analogy have been proposed. Taylor, Prandtl, and Colburn independently developed modifications that agree in broad outline. They only made a distinction between a wall-adjacent layer where the transport processes are of molecular nature (i.e., the viscous sublayer) and a fully turbulent region, where Reynolds's analogy is assumed to apply ($\mathbf{Pr}_{\text{eff}} = 1$). It resulted in a (c_h, c_f) relation including the molecular Prandtl number, which yields reasonable agreement with experiment only if **Pr** does not differ much from unity. Von Kármán appreciated the importance of the buffer region where the effect of **Pr** may not be neglected, though in this region he still assumed $\mathbf{Pr}_{\text{turb}} = 1$. Outside this region again Reynolds's analogy was used. Thus for the buffer region $\mathbf{Pr}_{\text{turb}} = 1$ is put in Eq. (7-167). For the mean-velocity distribution he assumed simply a relation which in a semi-log plot gives a straight line in the region $\delta_l^+ = 5 < x_2^+ < \delta_t^+ = 30$. The relation between c_h and c_f obtained in this way gives an improvement over the Taylor-Prandtl-Colburn relation though deviations from actual values of c_h still occur beyond $\mathbf{Pr} = 10$; the deviations seem to depend on the Reynolds number. Further modifications by Hofmann[9] differ mainly in the assumptions made about the velocity distribution in the buffer region and about the values of δ_l^+ and δ_t^+. Like the others he too assumed $\mathbf{Pr}_{\text{turb}} = 1$.

Hofmann calculated **Nu** as a function of **Pr**, and compared his computed results with the empirical relations given by Kraussold,[49] namely

$$\mathbf{Nu} = 0.024\mathbf{Re}_D{}^{0.8}\mathbf{Pr}^{0.37} \quad \text{for heating}$$

and

$$\mathbf{Nu} = 0.024\mathbf{Re}_D{}^{0.8}\mathbf{Pr}^{0.3} \quad \text{for cooling}$$

Hofmann's calculations gave results in fair agreement with the empirical relation only for the "heating" experiments up to $\mathbf{Pr} \simeq 50$ with ~ 10 per cent difference and being on the high side. For higher Prandtl numbers the computed values became increasingly optimistic. The "cooling" experiments yielded according to the above empirical relation lower values. Since Kraussold's experiments were done with liquids whose viscosity increases with decreasing temperature, the lower values may be explained by assuming a reduced turbulence near the wall as viscosity increases and consequently also viscous damping.

Apparently the temperature sensitivity of the fluid viscosity is of importance. This is expressed explicitly, amongst others by the well-known relation given by Sieder and Tate[212]

$$\mathbf{Nu} = 0.023\left(\frac{\mu_b}{\mu_w}\right)^{0.14}\mathbf{Re}_D{}^{0.8}\mathbf{Pr}^{1/3} \qquad (7\text{-}170)$$

where all the properties of the fluid are taken at the bulk temperature, except the viscosity μ_w occurring in the correction term $(\mu_b/\mu_w)^{0.14}$, which refers to the wall temperature.

For large values of **Pr** the thickness of the thermal sublayer becomes small with respect to δ_l, which means that the thermal resistance becomes more concentrated in a very thin layer at the wall. An increase of the Reynolds number results in a decrease of δ_l, and consequently in a deeper penetration of turbulence towards the wall, thereby enhancing the heat transport. Hence one may expect that at large **Pr**, the effect of the Reynolds number becomes stronger than given by $\mathbf{Re}_D{}^{0.8}$. Indeed, experiments by Dorresteyn[213] with viscous liquids and $\mathbf{Pr} > 50$ resulted in the following empirical relation

$$\mathbf{Nu} = 0.0123 \left(\frac{\mu_b}{\mu_w}\right)^{0.14} \mathbf{Re}_D{}^{0.9} \mathbf{Pr}^{1/3} \qquad (7\text{-}171)$$

Reichardt[10] improved on the previous theories in two important respects. (1) He was the first to take account of the variability in the physical properties of the fluid caused by the temperature field across the boundary layer. (2) He did not make any assumption concerning possible analogies between transport of momentum and heat in any region of the boundary layer. Thus, Reichardt used the complete equations (7-157) and (7-163). Concerning the ratios $\mathfrak{J}_\theta/(\mathfrak{J}_\theta)_w$ and σ_{21}/σ_w, he assumed that $[\mathfrak{J}_\theta/(\mathfrak{J}_\theta)_w](\sigma_w/\sigma_{21})$ remains of the order of magnitude 1. This assumption is supported, among other evidence, by Reichardt's own calculations of the course of $\mathfrak{J}_\theta/(\mathfrak{J}_\theta)_w$ (see below). Reichardt took account of the variation of the physical properties of the fluid by taking average values across the viscous sublayer and other average values across the buffer region. He approximated the velocity distribution in the buffer region by an exponential function in order to obtain integrable expressions for the temperature distributions and for c_h. Yet the result is quite involved.

We have shown that Reichardt[12] later proposed another velocity distribution [see Eq. (7-39)] which follows the actual velocity distribution in a better way than the exponential function. Of course the improved velocity distribution (7-39) might better be used, but then the integrals occurring in Eqs. (7-157) and (7-163) become too involved and permit only a numerical solution

However, if a numerical solution is accepted, an exact solution can be obtained by applying an iteration method. In Eqs. (7-157) and (7-163) data for the actual velocity distribution are used. In first approximation $[\mathfrak{J}_\theta/(\mathfrak{J}_\theta)_w](\sigma_w/\sigma_{21})$ may be taken equal to 1, and average values for μ_w/μ, **Pr**, and $\mathbf{Pr}_{\text{turb}}$ may be assumed. From the resulting temperature distribution $\mathfrak{J}_\theta/(\mathfrak{J}_\theta)_w$ can be calculated according to a method to be discussed later [see Eq. (7-173)]. Also μ_w/μ and **Pr** can be corrected for their temperature dependence. We so obtain $\mathfrak{J}_\theta/(\mathfrak{J}_\theta)_w$, μ_w/μ, and **Pr** as new functions of ξ_2. The Eqs. (7-157) and (7-163) can be solved again numerically using these new functions.

With the corrected temperature distribution a second approximation may be obtained for the variable quantities considered; with this the integrals can be solved again. And so on, and so forth.

But instead of following this elaborate procedure, one always tries to obtain solutions in closed form. The only possible way to accomplish this is that followed by the previously mentioned investigators, with the result that the value of the results obtained depends entirely on the suitability of the assumptions.

Deissler[13] too assumed $\mathbf{Pr}_{\text{turb}} = 1$ and a negligible effect on the velocity and temperature distributions of the variations of σ_{21} and $\mathfrak{I}_\theta$ across the pipe or boundary layer. He showed that, as long as **Pr** does not differ much from unity, the assumption that σ_{21} and $\mathfrak{I}_\theta$ vary linearly yields practically the same velocity and temperature distributions as the assumption that they are uniformly distributed, at any rate for values of u^*D/v ranging from 1,000 to 10,000. Deissler started from Eq. (7-151), in which he thus assumed $\mathfrak{I}_\theta = (\mathfrak{I}_\theta)_w$ and in which he used for ϵ_θ the same expression as for ϵ_m, namely, Eq. (7-40). Like the velocity distribution, the temperature distribution could be solved only numerically.

Deissler also studied the effect of variation of fluid viscosity with temperature. He obtained the interesting result that, even with a temperature-dependent viscosity, the calculations may be performed with a constant value for the viscosity, provided that this value is taken at a suitable reference temperature. This reference temperature is a function of **Pr**, decreasing with increasing **Pr**, and the function is different for heat transfer and for momentum transfer (friction) and also different for heating and for cooling. Figure 7-75 shows these functions for liquids. For **Pr** > 10 the reference temperature for heat transfer may be roughly taken equal to $\Theta_{ref} - \bar{\Theta}_{av} \simeq 0.5(\bar{\Theta}_w - \bar{\Theta}_{av})$, which is the so-called *film temperature* that is usually adopted.

Deissler's analytical results for c_h give remarkably good agreement with measured data obtained by many investigators with various liquids and values of the Prandtl number up to 3,000. The computed curve for c_h as a function of **Pr** is shown in Fig. 7-76. Most of the experimental data deviated by less than 10 per cent from the computed values, and maximum deviations were of the order of 20 per cent. These experimental data include mass-transfer data also. In the case of mass transfer

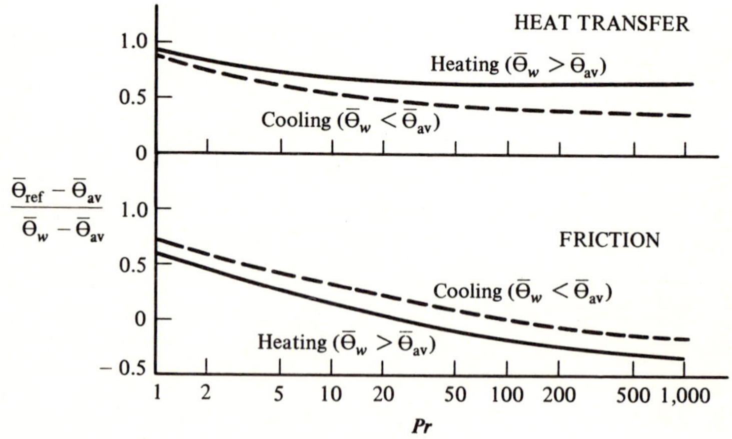

FIGURE 7-75
Reference temperature as a function of **Pr** for evaluating the apparent constant fluid properties. (From: *Deissler, R. G.,*[13] *by permission of the National Advisory Committee for Aeronautics.*)

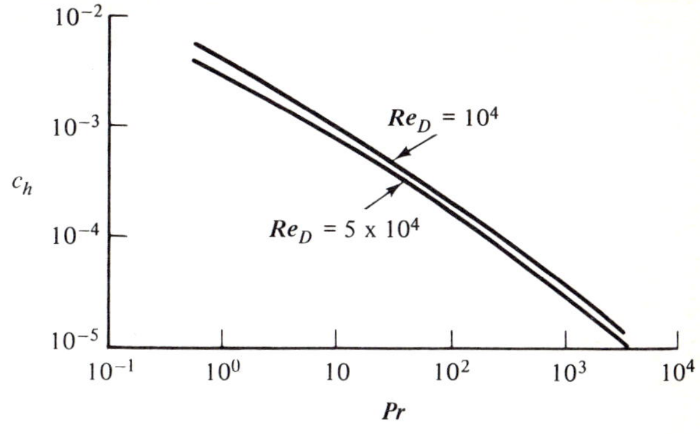

FIGURE 7-76
Computed coefficient of heat-flow resistance as a function of Prandtl number. (From: *Deissler, R. G.,*[13] *by permission of the National Advisory Committee for Aeronautics.*)

the same analysis applies but we must read for **Pr** the Schmidt number **Sc** and for c_h the corresponding coefficient for mass-flow resistance.

Deissler's investigations showed that the coefficient c_h for heat-flow resistance is quite insensitive to the assumptions made concerning the course of the various quantities outside the wall region, and that it is determined mainly by conditions in the wall region, provided that the Reynolds number of the flow is large. This conclusion is supported by a comparison of the results obtained by various investigators who made different assumptions.

Hence we may safely assume for the region outside the wall region that Reynolds's analogy applies, and for the wall region we may not only assume $\sigma_{21}/\sigma_w \simeq 1$ but also $\mathfrak{I}_\theta/(\mathfrak{I}_\theta)_w \simeq 1$. Equation (7-163) may then be transformed into

$$\frac{1}{c_h} = \frac{2}{c_f} + \sqrt{\frac{2}{c_f}} \int_0^{\xi_2''} d\xi_2 \, (\mathbf{Pr}_{\text{eff}} - 1) \frac{d}{d\xi_2} \frac{\bar{U}_1}{u^*}$$

$$= \frac{2}{c_f} + \sqrt{\frac{2}{c_f}} F(\mathbf{Pr}) \tag{7-172}$$

since the integral is mainly a function of **Pr**. The value ξ_2'' refers to the limit of the wall region.

An analytical expression for the velocity distribution that applies reasonably well to the entire wall region is Reichardt's expression (7-39). But, as mentioned, this leads to integrals that cannot be solved in closed form.

For the fully turbulent part of the wall region the logarithmic velocity distribution holds. Such a logarithmic velocity distribution is consistent with the assumption that the eddy viscosity varies linearly with x_2^+. If it is assumed that ϵ_θ is also a linear function of x_2^+, there then follows a logarithmic temperature distribution, as we may conclude directly from the Eq. (7-151), if $\mathfrak{I}_\theta = (\mathfrak{I}_\theta)_w =$ const. and κ is neglected with respect to $\rho c_p \epsilon_\theta$.

Besides the relations for c_h and for **Nu** as functions of **Re** and **Pr**, which are based upon some more or less realistic concepts of the transport processes, of course there are many purely empirical correlations. For such correlations, however, the reader may refer to textbooks on heat and mass transfer (see, for instance, Ref. 51).

Returning to the more theoretical solutions, we have seen that the value of these solutions is based upon underlying concepts about the transport processes, in particular in the wall region, and the assumptions that follow from these concepts.

We may recall that these assumptions concern the velocity distribution and the distributions of **Pr**, μ_w/μ, σ_{21}/σ_w, $\mathfrak{I}_\theta/(\mathfrak{I}_\theta)_w$, and $\mathbf{Pr}_{\text{turb}}$.

Deissler has shown that the variation of **Pr** and of μ_w/μ can be taken into account by choosing a correct reference temperature. The variation of σ_{21}/σ_w in pipe flow is exactly linear with distance from the wall, but the assumption $\sigma_{21}/\sigma_w \simeq 1$ in the wall region has proved to be permissible both in boundary-layer and in pipe flow.

We thus have to consider essentially the variation of $\mathfrak{I}_\theta/(\mathfrak{I}_\theta)_w$ and $\mathbf{Pr}_{\text{turb}}$.

For pipe flow Reichardt[10] calculated the course of $\mathfrak{I}_\theta/(\mathfrak{I}_\theta)_w$ across the pipe for various Reynolds and Prandtl numbers from a heat balance. In so doing he made the usual assumption that the effect of diffusion of heat in the flow direction is negligible with respect to the convective transport by the mean motion, an assumption which may be true except perhaps very close to the wall.

Consider an elementary annular space of length dx_1 enclosed by cylindrical envelopes of radius $D/2 - x_2$ and $D/2 - (x_2 + dx_2)$, respectively.

The heat balance for this volume element reads

$$\frac{d}{dx_2}\left[2\pi\left(\frac{D}{2} - x_2\right)dx_1\,\mathfrak{I}_\theta\right]dx_2 = -2\pi\left(\frac{D}{2} - x_2\right)dx_2\,\frac{\partial}{\partial x_1}(\rho\bar{U}_1 c_p\bar{\Theta})\,dx_1$$

whence

$$\left(\frac{D}{2} - x_2\right)\mathfrak{I}_\theta = -c_p\int_0^{x_2} dx_2\left(\frac{D}{2} - x_2\right)\rho\bar{U}_1\frac{\partial\bar{\Theta}}{\partial x_1} + \frac{D}{2}(\mathfrak{I}_\theta)_w$$

or

$$(1 - \xi_2)\frac{\mathfrak{I}_\theta}{(\mathfrak{I}_\theta)_w} = 1 - \frac{\displaystyle\int_0^{\xi_2} d\xi_2\,(1 - \xi_2)\rho\bar{U}_1\frac{\partial\bar{\Theta}}{\partial x_1}}{\displaystyle\int_0^{1} d\xi_2\,(1 - \xi_2)\rho\bar{U}_1\frac{\partial\bar{\Theta}}{\partial x_1}}$$

Reichardt further assumed a fully established temperature distribution and, hence,

$$\frac{\partial}{\partial x_1}(\bar{\Theta}_w - \bar{\Theta}) = -\text{const.}\,(\bar{\Theta}_w - \bar{\Theta})$$

The above equation then becomes

$$(1-\xi_2)\frac{\mathfrak{J}_\theta}{(\mathfrak{J}_\theta)_w} = 1 - \frac{\displaystyle\int_0^{\xi_2} d\xi_2\,(1-\xi_2)\bar{U}_1\rho(\bar{\Theta}_w - \bar{\Theta})}{\displaystyle\int_0^{1} d\xi_2\,(1-\xi_2)\bar{U}_1\rho(\bar{\Theta}_w - \bar{\Theta})} \qquad (7\text{-}173)$$

In this way Reichardt calculated $\mathfrak{J}_\theta/(\mathfrak{J}_\theta)_w$ in first approximation, using the temperature distribution calculated according to his own method of approximation, discussed earlier. Figure 7-77 shows the results for **Pr** = 0.72 and 200 and for $\mathbf{Re}_D = 5 \times 10^3$ and 4×10^5.

From this figure it may be concluded that $\mathfrak{J}_\theta/(\mathfrak{J}_\theta)_w > \sigma_{21}/\sigma_w$ and that their ratio approaches unity for high values of **Pr** and $\mathbf{Re}_D$. Hence the assumption that $\mathfrak{J}_\theta \simeq (\mathfrak{J}_\theta)_w$ in the wall region is certainly less disputable than the assumption that $\sigma_{21} \simeq \sigma_w$.

Let us consider the turbulence Prandtl number $\mathbf{Pr}_{\text{turb}}$, which is the ratio of the eddy diffusivities for momentum and heat.

Various investigators have determined this turbulence Prandtl number in pipe flow and in two-dimensional channel flow. Over-all values as well as local values of $\mathbf{Pr}_{\text{turb}}$ have been reported. We mentioned earlier in the present section that over-all Prandtl numbers varying between 0.6 and 0.75 have been observed for air as the flowing fluid.

However, $\mathbf{Pr}_{\text{turb}}$ is not only a function of the molecular Prandtl number, but should also depend on the local turbulence conditions, thus on the mean-velocity distribution and the Reynolds number. For the region away from the wall and at sufficiently high Reynolds number $[\mathfrak{J}_\theta/(\mathfrak{J}_\theta)_w]\cdot(\sigma_w/\sigma_{21}) \simeq 1$ (see Fig. 7-77). With $\mu_w/\mu \simeq 1$, and **Pr** not too much different from unity, according to Eq. (7-159) the

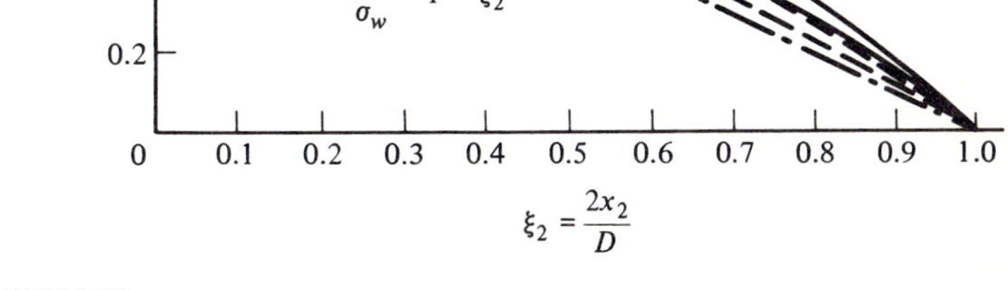

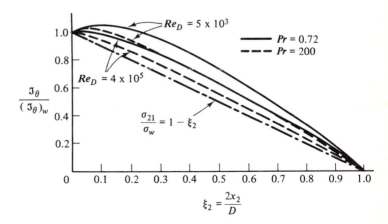

FIGURE 7-77
Effect of $\mathbf{Re}_D$ and **Pr** on the lateral heat-flux distribution in pipe flow. (From: *Reichardt, H.,*[10] *by permission of the Akademie-Verlag.*)

nondimensional temperature defect becomes equal to the nondimensional velocity defect, when $\mathbf{Pr}_{\text{turb}} = 1$:

$$\frac{\bar{\Theta} - \bar{\Theta}_{\text{min}}}{\theta^*} = \frac{\bar{U}_{x,\text{max}} - \bar{U}}{u^*}$$

But this is no longer true when $\mathbf{Pr}_{\text{turb}} \neq 1$. Theoretically the effect of $\mathbf{Pr}_{\text{turb}}$ on the temperature distribution across a turbulent boundary-layer has been calculated by Rotta[63] for air ($\mathbf{Pr} = 0.72$) and $\mathbf{Pr}_{\text{turb}} < 1$, but decreasing in different ways with increasing distance to the wall. For the four cases considered ($\mathbf{Pr}_{\text{turb}} = 0.9 =$ constant, a convex, a linear, and a concave distribution starting with $\mathbf{Pr}_{\text{turb}} = 0.9$ at $x_2/\delta = 0$ and decreasing to 0.5 at $x_2/\delta = 1$), there was no effect on the value of $(\Theta - \Theta_0)/\theta^*$ at about $x_2/\delta = 0.5$. For smaller distances to the wall the $\mathbf{Pr}_{\text{turb}} =$ constant gave the highest temperature defect, while this temperature defect had the lowest value for the concave distribution. For $x_2/\delta > 0.5$ it was just the opposite. The differences, however, were not large.

According to the simple model considered by Burgers, and discussed in Sec. 5-4, the $\mathbf{Pr}_{\text{turb}}$ would be $\lesssim 1$, when $\mathbf{Pr} \gtrless 12/A$, where A is of $\mathcal{O}(10)$. [See Eq. (5-55).] If it is allowed to use this model with the cases considered by Rotta, $\mathbf{Pr}_{\text{turb}} < 0.9$ would require, with $\mathbf{Pr} = 0.72$ a value of $A > 18$ which is not a reasonable value. But in Sec. 5-4, we pointed out already that Burgers's model should be used in a qualitative way only. Anyway it may be expected from this model and from Eq. (5-55) that $\mathbf{Pr}_{\text{turb}}$ will depend on the molecular Prandtl number, the Reynolds number of turbulence and ϵ_m/ν:

$$\mathbf{Pr}_{\text{turb}} = f\left(\mathbf{Pr}, \mathbf{Re}, \frac{\epsilon_m}{\nu}\right)$$

The experimental determination of $\mathbf{Pr}_{\text{turb}}$ appears to be a very difficult procedure. For, in the procedures used, a differentiation is carried out on the distributions of mean velocity and mean temperature, to obtain the local spatial gradients. This is a rather inaccurate procedure, in particular in the wall-remote region where these gradients are small. Close to the wall, there is the difficulty in measuring very accurately the mean-velocity and mean-temperature. The second major difficulty is presented by the measurement of the cross-correlation $\overline{u_2\theta}$ where, in addition, one has to take into account corrections for finite probe dimensions.

A survey of $\mathbf{Pr}_{\text{turb}}$ obtained experimentally by a number of investigators, and its distribution across the wall region of turbulent boundary-layers and pipe or duct flows has been given by Blom.[215] They include his own measurements in air flow ($\mathbf{Pr} \simeq 0.7$) and measurements by others in air and liquid flows ($\mathbf{Pr} \simeq 0.02$ up to 14.3). They show $\mathbf{Pr}_{\text{turb}}$ as a function of x_2^+.[53,216–222] The large scatter and the sometimes contradictory behavior of the values of $\mathbf{Pr}_{\text{turb}}$ are clearly demonstrated in this survey. There is a general trend noticeable, however. For air ($\mathbf{Pr} \simeq 0.7$) $\mathbf{Pr}_{\text{turb}}$ appears to decrease in the region $x_2^+ \gtrsim 10$. In the turbulent part of the wall region $\mathbf{Pr}_{\text{turb}}$ is roughly constant and of the magnitude $\sim$0.7 to 0.8.

In contrast with this behavior of $\mathbf{Pr}_{\text{turb}}$ very close to the wall the measurements by Meier and Rotta[224] in supersonic air boundary layers showed a sharp increase of $\mathbf{Pr}_{\text{turb}}$ with decreasing x_2. Extrapolation to $x_2 = 0$ yields a value between 1.2 and 1.5. Here too, across the main part of the boundary layer $\mathbf{Pr}_{\text{turb}} \simeq 0.8$, but decreasing for $x_2/\delta \gtrsim 0.8$.

In Blom's survey the above values of 0.7 to 0.8 have also been obtained in the same region for liquids with $\mathbf{Pr} \simeq 6$ and $\simeq 14$. On the other hand the results with low $\mathbf{Pr}$-number liquids show a strong increase of $\mathbf{Pr}_{\text{turb}}$ when approaching the wall, which thus would indicate a strong decrease of ϵ_θ near the wall. Values of $\mathbf{Pr}_{\text{turb}}$ for air, water, 30 per cent ethylene glycol, and sodium-potassium alloy, i.e., with $\mathbf{Pr}$ values varying between 0.03 and 14, and shown by Hughmark[226] are scattered around the value of 1, across the main part of turbulence region in the whole cross section of the pipe. This seems to justify the still rather crude assumption of $\mathbf{Pr}_{\text{turb}} \simeq 1$ in this region and explains the reasonable results of theories where this assumption is used. Hughmark further found that close to the wall for all the above fluids ϵ_θ/ν satisfied the relations

$$\frac{\epsilon_\theta}{\nu} \simeq 0.001 {x_2^+}^{+3} \quad \text{for} \quad 4 < x_2^+ < 12 \tag{7-174a}$$

$$\frac{\epsilon_\theta}{\nu} \simeq 0.0155 {x_2^+}^{+1.9} \quad \text{for} \quad 12 < x_2^+ < 35 \tag{7-174b}$$

In Sec. 7-13 we mentioned a relation $\epsilon_m/\nu = c{x_2^+}^{+2}$ describing the general behavior in the transition region. Abbrecht and Churchill's results yield $c = 0.007$, while according to Rannie's relation $c \simeq 0.0047$. From a comparison with the above relations given by Hughmark one would conclude that $\mathbf{Pr}_{\text{turb}} \simeq (4.7 \text{ to } 7){x_2^+}^{-1}$ for $4 < x_2^+ < 12$, and $\mathbf{Pr}_{\text{turb}} \simeq 0.3$ to 0.58 for $12 < x_2^+ < 35$. Hence, indeed an increase of $\mathbf{Pr}_{\text{turb}}$ when $x_2^+ \to 0$. It is surprising that no effect of $\mathbf{Pr}$ has been found by Hughmark in the above relations. For one would expect a strong effect of $\mathbf{Pr}$, at least for $\mathbf{Pr} < 1$, in the viscous sublayer and adjacent buffer region. Namely that the "wall value" of $\mathbf{Pr}_{\text{turb}}$ would increase with decreasing $\mathbf{Pr}$, as mentioned already earlier. This behavior is also included in a number of theoretical models.

A recent model is, for instance, that proposed by Cebeci.[225] In essence he applied the modified mixing-length theory proposed by Van Driest [see Eq. (7-41)] also to the transport of heat or mass. Cebeci thus obtained the following relation for $\mathbf{Pr}_{\text{turb}}$

$$\mathbf{Pr}_{\text{turb}} = \frac{\varkappa_m[1 - \exp(-x_2/A_m)]}{\varkappa_\gamma[1 - \exp(-x_2/A_\gamma)]} \tag{7-175}$$

At the wall this reduces to $\mathbf{Pr}_{\text{turb}} = \varkappa_m A_\gamma/\varkappa_\gamma A_m$, while at large distance from the wall it reduces to the constant value $\mathbf{Pr}_{\text{turb}} = \varkappa_m/\varkappa_\gamma$.

The "constants" A_m and A_γ act as a "damping" parameter for the turbulent fluctuations of momentum and a scalar quantity. A_γ is a rather complicated function of $\mathbf{Pr}$. It is assumed that the total effect of $\mathbf{Pr}$ is included in A_γ, so that $\varkappa_\gamma$ is not affected by **Pr**. This need not necessarily be so, and indeed we concluded already that also at distance from the wall Hughmark's results still show an effect of **Pr** and $\mathbf{Pr}_{turb}$. From a study of available experimental data Cebeci concluded that $\varkappa_m = 0.40$ and $\varkappa_\gamma = 0.44$, hence $\mathbf{Pr}_{turb} = 0.9$ for the wall-remote region. Further he suggested $A^+ = u^+A/\nu = 26$. For the "wall value" of $\mathbf{Pr}_{turb}$ he thus obtained $\mathbf{Pr}_{turb} = 1.22$ for air at sufficiently high Reynolds numbers, which is in satisfactory agreement with the extrapolated value of the results obtained by Meier and Rotta.[224] For a liquid metal with $\mathbf{Pr} = 0.02$ the predicted "wall value" was 10.9.

Interesting results have been obtained by Hughmark concerning the mean-temperature distribution in a cross-section in a vertical airflow through a pipe with 78 mm I.D. The wall was heated, and $\mathbf{Re}_D$ was varied from 18,000 to 71,000. For the turbulent part of the wall region the distribution was distinctly logarithmic

$$\bar{\Theta}/\theta^* = A_\theta \ln x_2^{\,+} + B_\theta \qquad (7\text{-}175)$$

A_θ decreased from 2.4 at $\mathbf{Re}_D = 18{,}000$ to 2.15 at $\mathbf{Re}_D = 71{,}000$, while B_θ increased from 3.06 to 3.72 in the same $\mathbf{Re}_D$ range.

For the core region the distribution was parabolic

$$\frac{\bar{\Theta} - \bar{\Theta}_{min}}{\theta^*} = 7.2\left(\frac{2r}{D}\right)^2 \qquad (7\text{-}176)$$

i.e. equal to that for the velocity defect.

For $x_2^{\,+} > 30$, also a power-law could be assumed

$$\bar{\Theta}/\theta^* = C_\theta x_2^{\,+1/n_\theta} \qquad (7\text{-}177)$$

where C_θ increased from 6.03 at $\mathbf{Re}_D = 18{,}000$ to 6.49 at $\mathbf{Re}_D = 71{,}000$, while n_θ too increased from 5.4 to 6.25.

The eddy diffusivity $2\epsilon_\theta/u^*D$ increased almost linearly with $x_2^{\,+}$ in the region $2x_2/D \gtrsim 0.2$, showed a maximum roughly halfway along the pipe radius, and after a slight decrease remained almost constant in the central region with a value $2\epsilon_\theta/u^*D \simeq 0.8$. Comparing this value with the values of $2\epsilon_m/u^*D$ obtained by Laufer and Nunner, and shown in Fig. 7-63, one would conclude for air and for the core-region that $\mathbf{Pr}_{turb} \simeq 0.75$ to 1.

In general the distribution curves $(\bar{\Theta} - \bar{\Theta}_{min})/(\bar{\Theta}_w - \bar{\Theta}_{min})$ were slightly lower than the corresponding curves for $\bar{U}_x/\bar{U}_{x,max}$. The same was found by Bremhorst and Bullock.[223] They made their measurements again in an airflow through a horizontal pipe with 135 mm I.D. The Reynolds number was $\bar{U}_{x,max}D/\nu \simeq 7 \times 10^4$. They measured not only the mean-temperature and mean-velocity distribution, but also the distributions of $u_1'/\bar{U}_x$ and $\theta'/(\bar{\Theta} - \bar{\Theta}_{min})$. These latter two distributions were almost identical in the region $0.05 < 2x_2/D \leqslant 1$. But since in their experiments $u^*/\bar{U}_{x,max} = 0.04$ and $\theta^*/(\bar{\Theta}_w - \bar{\Theta}_{min}) = 0.05$, there follows $\theta'/\theta^* < u'/u^*$, with a difference of roughly 20 per cent.

The spectral distribution $\mathbf{E}_\theta(k_1)$ of θ showed very close to the wall ($2x_2/D = 0.0374$) a distinct $k_1^{\,-1}$ portion in the low wavenumber range. Neither a distinct $(-5/3)$-law, nor a distinct $(-17/3)$-law (see Sec. 3-7) was obtained in the higher wavenumber range. In general $\mathbf{E}_\theta(k_1) < \mathbf{E}_1(k_1)$ for $k_1D \gtrsim 5$, and $\mathbf{E}_\theta(k_1) > \mathbf{E}_1(k_1)$ for $k_1D \lesssim 5$. The differences, however, were small and very small at $2x_2/D = 0.0374$. Not much difference has to be expected in the high wavenumber range since for air $\mathbf{Pr} = 0.72$.

In the above considerations we have deliberately neglected any effect of non-gradient-type diffusion on the transport, i.e., the effect of very large-scale motions. Such an effect may be expected in the outer region of a boundary layer and in the core region of pipe flow. This effect may obscure results of turbulence-transport measurements that are based on the concept of gradient-type diffusion. Furthermore, as far as the $\mathbf{Pr}_{\text{turb}}$ is concerned the following aspect may be worth considering. In Sec. 7-8 we made a distinction between active and inactive turbulent motion in the transport of momentum. At the same time we made the remark that the inactive part of the turbulent motion may contribute to the transport of a scalar quantity. In the case of a turbulent boundary layer the irrotational, inactive motion continues in the free stream outside the turbulent/irrotational interface, and consequently may transport heat outside the interface even if $\mathbf{Pr} > 1$. This would mean that then the boundary-layer thickness for heat δ_θ may become greater than that for momentum. An indication of the effect of irrotational motions on the transport of a scalar quantity may be found, perhaps, in the decrease of $\mathbf{Pr}_{\text{turb}}$ obtained by Meier and Rotta in the boundary-layer flow for $x_2/\delta \lesssim 0.8$.

Effect of Wall Roughness on Heat Transfer

Experimental evidence has shown that the effect of wall roughness on the turbulence flow pattern is negligible if the relative roughness is sufficiently low. In the case of sand roughness it may be deduced from Nikuradse's experiments that this holds, for $u^*k/\nu < 5$. Consequently one may expect that for not large values of $\mathbf{Pr}$ at least, the transport of heat in a flow past a rough wall with low relative roughness will be like the transport of heat in a flow past a smooth wall, except perhaps for a slight effect associated with the increased surface area of the rough wall.

With increasing roughness, such that (for sand roughness) $5 < u^*k/\nu < 55$, there is an increasing effect on the flow pattern; among other things, the effective thickness of the viscous sublayer decreases. This decrease affects the transport of heat across the boundary layer, for the roughnesses cause disturbances in the viscous sublayer which promote turbulence transport. In other words, the roughnesses increase the "active" action of the viscous sublayer, already present in the case of a smooth wall. This beneficial effect on the transport of heat will be greater at higher values of the Prandtl number.

If the roughness is so great that $u^*k/\nu > 55$ (the numerical value again refers to sand roughness) the effective thickness of the viscous sublayer is zero (see Figs. 7-15 and 7-16). The flow resistance then becomes independent of the Reynolds number. It seems reasonable to expect that the transport of heat will also be unaffected by molecular transport, and so will be independent not only of the Reynolds number but also of the Prandtl number for not small values of it.

This seems logical, but not unassailable. In the fully turbulent part of the flow the transport of heat is determined by ϵ_θ, and this coefficient of eddy diffusion is a function of the Prandtl number. So at least this effect of the Prandtl number on the transport of heat remains.

Let us now consider what occurs at the wall. We assume a fully rough-wall condition of the flow, where the flow resistance is independent of the Reynolds number. The flow resistance shown by the roughness elements consists of the viscous wall resistance and the shape resistance. At the fully rough-wall condition the roughness elements protrude so far into the fully turbulent part of the wall region that the shape resistance becomes predominant and a quadratic resistance law results, independent of the Reynolds number.

However, the transport of heat need not follow the same pattern as the momentum transport. It is known, from heat-transfer measurements on cylinders and spheres exposed to uniform flow, that the Nusselt number remains a function of the Reynolds number even when the resistance coefficient is already independent of the Reynolds number. This is also true if the free stream itself is turbulent (see below). Hence, if we consider the roughness elements to be small bodies attached to a smooth wall and exposed to a turbulent stream, in analogy to the above we may expect an effect of the Reynolds number on the heat transfer from the wall to the fluid at any value of the Reynolds number—and, obviously, an effect of the Prandtl number also, although these effects may perhaps be smaller than they would be for a smooth wall.

Unfortunately, the early experiment by Pohl,[54] and by Smith and Epstein[55] set up to study the effect of wall roughness on heat transfer were inconclusive since even at the highest Reynolds number obtained in the experiments the fully rough-wall condition was hardly attained.

More conclusive are the results of experiments by Nunner.[42] The artificial roughnesses Nunner applied to his copper tube ($D = 50$ mm) by means of rings had equivalent sand roughnesses up to $k = 22.8$ mm! The Reynolds numbers were sufficiently high to obtain a large range where the fully rough-wall condition (friction factor f independent of $\mathbf{Re}_D$) occurred. However, the experiments were done with air only, and so the Prandtl number was not varied. Nunner was able to obtain a ($\mathbf{Nu}$,$\mathbf{Re}_D$) relation for each roughness condition. In a double logarithmic representation of $\mathbf{Nu}$ versus $\mathbf{Re}_D$ these relations showed practically straight lines in the range $\mathbf{Re}_D > 3{,}000$, just as they would for a smooth wall. These lines are almost parallel to the smooth-wall line and show a vertical shift with respect to it that increases with increased effective roughness. Hence, no marked departure from the Reynolds-number dependence for a smooth wall was obtained. Figure 7-78 shows two curves for the extreme case: a smooth pipe, and a pipe with the highest equivalent sand roughness of $k = 22.8$ mm. For the rough pipe a mean diameter D_m was used to correlate the experimental data. This mean diameter was defined as the equivalent diameter of a smooth pipe having the same volume per unit of length.

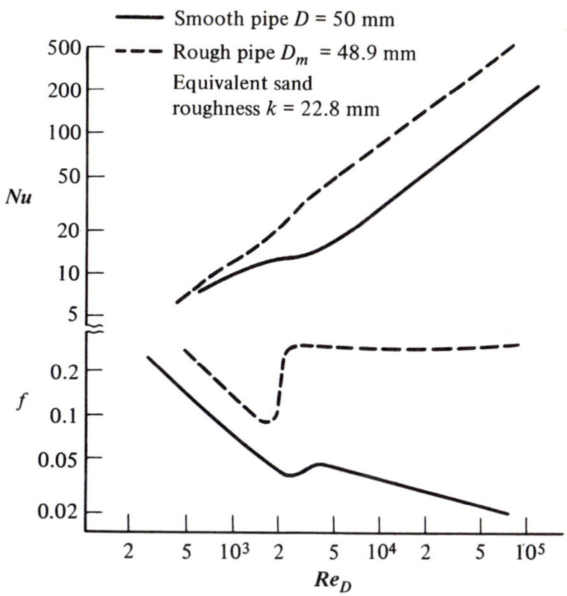

FIGURE 7-78
Effect of wall roughness on heat transfer and friction for air flow in a pipe. (From: *Nunner, W.,*[42] *by permission of the Verlag des Vereins Deutscher Ingenieure.*)

It is interesting to note that for the rough pipe no marked distinction is observable between the $(\mathbf{Nu}, \mathbf{Re}_D)$ relations for the "laminar"-flow ($\mathbf{Re}_D < 2{,}000$) and turbulent-flow regions, whereas the $(f, \mathbf{Re}_D)$ relation shows the usual transition range around $\mathbf{Re}_D = 2{,}000$, even though much higher friction coefficients are obtained in the laminar-flow region than for the smooth pipe. The relative increase in the smooth-pipe Nusselt number with decreasing Reynolds number in the transition and laminar-flow ranges may be caused in Nunner's experiments with the horizontal pipe by secondary flows set up by buoyancy effects, which increase with decreasing mean-flow velocity, i.e., here, with decreasing Reynolds number. The occurrence of these secondary flows could be inferred from the peculiar nonsymmetrical temperature-distribution profiles measured by Nunner. Apparently, the same effects are responsible for the "straight" continuation of the correlation curve for the rough pipe with the laminar-flow region, where it approaches the curve for the smooth pipe.

Nunner concluded from his investigations that the effect of wall roughness on heat transfer is similar to the effect that may be expected from a variation in the Prandtl number. He extended the $(\mathbf{Nu}, \mathbf{Re}_D, \mathbf{Pr})$ relation which may be deduced from the Taylor-Prandtl-Colburn model for the transport process near the wall, mentioned earlier, to the following, which includes the effect of wall roughness.

$$\mathbf{Nu} = \frac{f}{8} \frac{\mathbf{Re}_D \, \mathbf{Pr}}{1 + 1.5\mathbf{Re}_D^{-1/8} \, \mathbf{Pr}^{-1/6}(\mathbf{Pr} \, f/f_0 - 1)} \tag{7-178}$$

where f_0 denotes the friction coefficient for a smooth wall at the same $\mathbf{Re}_D$. This relation predicts Nunner's own data within 20 per cent.

The occurrence of the term $\mathbf{Pr}\, f/f_0$ reflects the equivalence of Prandtl number and roughness effect on the heat transfer. But, since Nunner's experiments were done with air only, the effect of the Prandtl number requires still further experimental investigation.

Since both the heat transfer and the flow resistance increase with increasing roughness of the pipe wall, an interesting question is whether it would be advantageous from an economic point of view to increase the roughness by artificial means, as has been done for instance by Nunner[42] and by Koch,[66] who extended Nunner's experiments. For the same value of **Nu** or c_h can be obtained in a smooth pipe of the same diameter as the rough pipe by increasing the mass flow rate. If the product of mass flow rate and pressure drop is higher for the smooth pipe than for the rough pipe, it is economical to apply the rough pipe. Because of the complexity of the heat-transfer processes in turbulent flow through rough pipes, it is not possible to predict which of the two cases would be more advantageous. In general it turns out that the rough pipe is less economical, which might perhaps be attributed to a relatively large contribution of the shape resistance of the roughness elements to the total flow resistance. But cases where the rough pipe becomes more economical are also possible. This may be deduced for instance from Nunner's experimental results, and has been shown specifically by Koch.

Effect of Free-Stream Turbulence on Boundary-Layer Transport

If the free stream outside the boundary layer of a body is itself turbulent, it will affect the turbulence transport processes through the boundary layer.

The free-stream turbulence, depending on its intensity, will affect the transition from the laminar state to the turbulent state of a boundary layer, promoting the transition as the intensity of the free-stream turbulence increases. As long as this turbulence intensity is small relative to the intensity of the turbulence in a fully developed turbulent boundary layer, the effect may be expected to remain restricted to the outermost regions of the boundary layer. One also may expect an effect of the integral length-scale of the free-stream turbulence, in particular if it is of the magnitude of the boundary-layer thickness and the "wavelength" and periodicity of the outer interface of the boundary layer. With increasing intensity of the free-stream turbulence, its effect may penetrate deeper into the boundary layer and may enhance the turbulence heat-transport process. But since, except for very small **Pr**-values, the main resistance to heat flow is concentrated near the wall where the turbulence intensity is high, not much influence on the heat transfer will occur for free-stream turbulence of not extremely high relative intensity.

For the flow around a non-slender body the effect of free-stream turbulence may be threefold.

1 Usually the upstream part of the boundary layer of the body is not yet turbulent, and transition into turbulence occurs at some point downstream from the forward stagnation point. Turbulence in the free stream may then not only influence the location of the transition point but may also cause disturbances in the laminar part of the boundary layer and so affect the transport processes through it.

2 The free-stream turbulence may affect the turbulent part of the boundary layer along the body, as noted above.

3 If the body has a wake, the free-stream turbulence may interact with the wake flow.

It is known that for spheres and cylinders in a turbulence-free cross flow the boundary layer starting from the forward stagnation point is laminar first. Up to a critical Reynolds number which for a cylinder is around $\mathbf{Re}_D = U_0D/\nu = 10^5$ the boundary layer remains laminar till the point of separation which is located $\sim 85°$ from the forward stagnation point. At higher Reynolds numbers the boundary layer becomes turbulent which results in a shift downstream of the point of separation to a location at $\sim 140°$ from the forward stagnation point. With increasing $\mathbf{Re}_D$ the location of transition from laminar to turbulent flow in the boundary layer shifts in upstream direction. The effect of free-stream turbulence on the boundary layer before and the wake flow after separation will in principle be determined by the relative intensity u'/U_0 and an integral length scale (e.g., Λ_f) of this turbulence. It is mainly the relative intensity which is of influence. But one may expect an influence of Λ_f in particular when some resonance occurs between the energy-containing eddies of the free stream and the periodic separation of the boundary layer. Up to the critical $\mathbf{Re}_D$ the shedding frequency of eddies from the cylinder is roughly equal to $0.2U_0/D$, and about $0.4U_0/D$ at $\mathbf{Re}_D \gtrsim 10^6$. We thus may expect the above resonance to occur when the "frequency" of the energy-containing eddies which is proportional with U_0/Λ_f is equal to the shedding frequency. This results in a possible maximum effect when $\Lambda_f/D \simeq \mathcal{O}(1)$.

When considering the effect of free-stream turbulence on heat transfer from a cylinder, one has to take into account that the distribution of the local heat-transfer coefficient h depends markedly on $\mathbf{Re}_D$. At rather low values ($\mathbf{Re}_D \gtrsim 10{,}000$) h shows a maximum at the forward stagnation point, decreases to a minimum at $\sim 90°$ from the forward stagnation point and increases again to another maximum at the rear stagnation point, but with a lower value than at the forward stagnation point. With increasing $\mathbf{Re}_D$ the distribution remains essentially the same, but depending on the Prandtl number of the fluid the above minimum may increase relatively. At super-critical $\mathbf{Re}_D$, however, near the location of separation after a first decrease from the maximum at the forward stagnation point, a pronounced and very localized maximum is observed at the points of separation, which maximum is even higher than that at the forward stagnation point.

Thus we may expect that at lower subcritical $\mathbf{Re}_D$ the effect of free-stream turbulence on the $\mathbf{Nu}_D$ will be mainly through the effect of its relative intensity on the local heat transfer around the forward stagnation point. According to Sutera *et al.*[227] a possible explanation is that the turbulence vorticity component transverse to the axis of the cylinder is strongly increased through a stretching process when such a vortex is deformed by the mainstream approaching the forward stagnation point.

Of importance to the effect is the average separation of these transverse vortices in the direction of the cylinder axis. According to Kestin and Wood[228] a theoretical analysis by Weiss showed that the most amplified wavelength is given by

$$\lambda/D = 33.1\mathbf{Re}_D^{-1/2} \qquad (7\text{-}179)$$

When the $\mathbf{Re}_D$ is increased and approaches the critical value the effect of the intensity of the free-stream turbulence becomes also more and more pronounced at the location just after separation of the flow. This is clearly shown by experiments by Kestin and Wood[229] on mass transfer to air with two cylinders coated with a layer of paradichloro-benzene ($\mathbf{Sc} = 2.4$). The measurements were done at $\mathbf{Re}_D = 75{,}000$, 100,000, and 125,000. The results were in agreement with a correlation also valid for heat transfer processes, namely

$$\mathbf{Nu}_D/\mathbf{Re}_D^{1/2} = 0.945 + 3.48\left(\frac{u'}{\bar{U}_1}\frac{\mathbf{Re}_D^{1/2}}{100}\right) - 3.99\left(\frac{u'}{\bar{U}_1}\frac{\mathbf{Re}_D^{1/2}}{100}\right)^2 \qquad (7\text{-}180)$$

in the range $0 < (u'/\bar{U}_1)\mathbf{Re}_D^{1/2} < 40$.

One may expect that the effect on the local increase of the transfer coefficient in the region of separation will become stronger when the molecular **Pr** or **Sc** becomes large. An indication in this direction may be found in similar mass-transfer experiments by Mizushina *et al.*[230]

As mentioned above an effect of the integral length scale, mainly through some resonance effect, may only be expected to occur at values of $\mathbf{Re}_D$ near and beyond the critical $\mathbf{Re}_D$. No clear support for this idea can be found in published material, since either the $\mathbf{Re}_D$ was too low, or no information concerning the values of Λ_f/D occurring in the experiments have been given. Results of earlier investigations[56–62] have been reported in an incomplete way in this respect. For more recent information on the effect of free-stream turbulence on heat-transfer rates the reader may be referred to the survey article by Kestin.[231]

REFERENCES

1. Goldstein, S.: "Modern Developments in Fluid Dynamics," vol. 2, p. 331, Oxford University Press, New York, 1938.
2. Wieghardt, K.: Ph.D. thesis, University of Göttingen, 1945.
3. Schubauer, G. B., and H. K. Skramstad: *J. Aeronaut. Sci.*, **14**, 69 (1947).

4. Dryden, H. L.: *Natl. Advisory Comm. Aeronaut. Tech. Notes No.* 1168, 1947.
5. Blasius, H.: *Z. Math. u. Physik,* **56,** 4 (1908).
6. Nikuradse, J.: *VDI-Forschungsheft No.* 361, 1933.
7. Millikan, C. B.: *Proc. 5th Intern. Congr. Appl. Mech.,* Cambridge, Mass., 1938, p. 386.
8. Kármán, Th. von: *Trans. ASME,* **61,** 705 (1939).
9. Hofmann, E.: *Forsch. Gebiete Ingenieurw.,* **11A,** 159 (1940).
10. Reichardt, H.: *Z. angew. Math. u. Mech.,* **20,** 297 (1940).
11. Rotta, J.: *Ingr. Arch.,* **18,** 277 (1950).
12. Reichardt, H.: *Z. angew. Math. u. Mech.,* **31,** 208 (1951).
13. Deissler, R. G.: *Natl. Advisory Comm. Aeronaut. Tech. Notes No.* 3145, 1954.
14. Schlichting, H.: "Boundary Layer Theory," Pergamon Press Ltd., London, 1968.
15. Ludwieg, H., and W. Tillmann: *Ingr. Arch.,* **17,** 288 (1949).
16. Schubauer, G. B., and P. S. Klebanoff: *Natl. Advisory Comm. Aeronaut. Tech. Repts. No.* 1030, 1951.
17. Klebanoff, P. S., and F. W. Diehl: *Natl. Advisory Comm. Aeronaut. Tech. Notes No.* 2475, 1951.
18. Schultz-Grunow, F.: *Luftfahrt-Forsch.,* **17,** 239 (1940).
19. Allan, J. F., and R. S. Cutland: *Trans. North East Coast Inst. Engrs. & Shipbuilders,* **69,** 245 (1953).
20. Hama, F. R.: *Soc. Naval Architects Marine Engrs. Trans.,* **62,** 333 (1954).
21. Clauser, F. H.: *Advances in Appl. Mechanics,* **4,** 1 (1956).
22. Wieghardt, K.: *Z. angew. Math. u. Mech.,* **25–27,** 146 (1947).
23. Goethals, R.: *Compt. rend.,* **226,** 1073 (1948).
24. Miles, J. W.: *J. Aeronaut. Sci.,* **24,** 704 (1957).
25. Elrod, H. G.: *J. Aeronaut. Sci.,* **24,** 468 (1957).
26. Driest, E. R. van: *J. Aeronaut. Sci.,* **23,** 1007 (1956).
27. Klebanoff, P. S.: *Natl. Advisory Comm. Aeronaut. Tech. Notes No.* 3178, 1954.
28. Townsend, A. A.: *Proc. Cambridge Phil. Soc.,* **47,** 375 (1951).
29. Corrsin, S., and A. L. Kistler: *Natl. Advisory Comm. Aeronaut. Tech. Notes* 3133, 1954.
30. Schubauer, G. B.: *J. Appl. Phys.,* **25,** 188 (1954).
31. Townsend, A. A.: "The Structure of Turbulent Shear Flow," Cambridge University Press, New York, 1956.
32. Kuethe, A. M.: *J. Aeronaut. Sci.,* **23,** 444 (1956).
33. Favre, A., J. Gaviglio, and R. Dumas: *9th Intern. Congr. Appl. Mechanics,* Brussels, 1956; also, *J. Fluid Mech.,* **3,** 329 (1958).
34. Einstein, H. A., and H. Li: *Proc. Am. Soc. Civil Engrs. Paper No.* 945, 1956.
35. Hanratty, T. J.: *A.I.Ch.E. Journal,* **2,** 359 (1956).
36. Coles, D.: *J. Fluid Mech.,* **1,** 191 (1956).
37. Latzko, H.: *Z. angew. Math. u. Mech.,* **1,** 277 (1921).
38. Kirsten, H.: Experimental Investigation on the Development of the Velocity Distribution in Turbulent Pipe Flow, Ph.D. thesis, University of Leipzig, 1927.
39. Nikuradse, J.: *VDI-Forschungsheft No.* 356, 1932.
40. Nikuradse, J.: *VDI-Forschungsheft No.* 361, 1933.
41. Laufer, J.: *Natl. Advisory Comm. Aeronaut. Tech. Repts. No.* 1174, 1954.
42. Nunner, W.: *VDI-Forschungsheft No.* 455, 1956.
43. Laufer, J.: *Natl. Advisory Comm. Aeronaut. Tech. Notes No.* 2123, 1950.
44. Elias, F.: *Z. angew. Math. u. Mech.,* **9,** 434 (1929); **10,** 1 (1930).

45. Woertz, B. B., and T. K. Sherwood: *Trans. AIChE,* **35,** 517 (1939).
46. Lorenz, H.: *Z. tech. Phys.,* **15,** 376 (1934).
47. McCarter, R. J., L. F. Stutzman, and H. A. Koch, Jr.: *Ind. Eng. Chem.,* **41,** 1290 (1949).
48. Kármán, Th. von: *Trans. ASME,* **61,** 705 (1939).
49. Kraussold, H.: *VDI-Forschungsheft No.* 351, 1931.
50. Rannie, W. D.: *J. Aeronaut. Sci.,* **23,** 485 (1956).
51. McAdams, W. H.: "Heat Transmission," McGraw-Hill Book Company, Inc., New York, 1954.
52. Abbrecht, P. H., and S. W. Churchill: *A.I.Ch.E. Jl.,* **6,** 268 (1960).
53. Isakoff, S. E., and T. B. Drew: *Proc. General Discussion on Heat Transfer,* pp. 405, 479, New York, 1951.
54. Pohl, W.: *Forsch. Gebiet. Ingenieurw.,* **4A,** 230 (1933).
55. Smith, J. W., and N. Epstein: *A.I.Ch.E. Jl.,* **3,** 242 (1957).
56. Reiher, H.: *Mitt. Forsch.,* **269,** 1 (1925).
57. Loitsianskii, L. G., and B. A. Schwab: *Central Aerodynamics Hydrodynamics Inst. U.S.S.R. Rept. No.* 329, 1935.
58. Comings, E. W., J. T. Clapp, and J. F. Taylor: *Ind. Eng. Chem.,* **40,** 1096 (1948).
59. De Haas van Dorsser, A. H., H. A. Leniger, and D. A. van Meel: *De Ingenieur,* **61,** ch. 25 (1949).
60. Maisel, D. S., and T. K. Sherwood: *Chem. Eng. Progr.,* **46,** 172 (1950).
61. Sherwood, T. K., and J. M. Petrie: *Ind. Eng. Chem.,* **24,** 736 (1932).
62. Sato, K., and B. H. Sage: *ASME Paper No.* 57-A-20.
63. Rotta, J. C.: *Int. J. Heat Mass Transf.,* **7,** 215 (1964).
64. Edwards, A., and B. N. Furber: *Proc. Inst. Mech. Engrs. London,* **170,** 941 (1956).
65. Lobb, R. K., E. M. Winkler, and J. Persh: *J. Aeronaut. Sci.,* **22** (1955).
66. Koch, R.: *VDI-Forschungsheft No.* 469, 1958.
67. Goddard, F. E.: *J. Aero. Space Sci.,* **26,** 1 (1959).
68. Von Doenhoff, A. E., and N. Tetervin: *Natl. Advisory Comm. Aeronaut., Techn. Repts. No.* 772, 1943.
69. Tollmien, W.: *Nachr. Ges. Wiss. Göttingen, Math. Phys. Klasse,* **21** (1929); also *N.A.C.A. Techn. Mem. Nr.* 609 (1931).
70. Lin, C. C.: "The Theory of Hydrodynamic Stability," p. 61, Cambridge University Press, New York, 1955.
71. Uzkan, T., and W. C. Reynolds: *J. Fluid Mech.,* **28,** 803 (1967).
72. Shen, S. F.: *J. Aeronaut. Sci.,* **21,** 62 (1954).
73. Ross, J. A., F. H. Barnes, J. G. Burns, and M. A. S. Ross: *J. Fluid Mech.,* **43,** 819 (1970).
74. Barry, M. D. J., and M. A. S. Ross: *J. Fluid Mech.,* **43,** 813 (1970).
75. Stuart, J. T.: *Annual Review of Fluid Mechanics,* **3,** 347 (1971).
76. Landau, L. D., and E. M. Lifschitz: "Fluid Mechanics," p. 103, Pergamon Press, New York, 1959.
77. Benney, D. J., and R. F. Bergeron: *Stud. Appl. Math.,* **48,** 181 (1969).
78. Emmons, H. W.: *J. Aeronaut. Sci.,* **18,** 490 (1951).
79. Schubauer, G. B., and H. K. Skramstad: *N.A.C.A. Techn. Repts.* No. 909, 1948.
80. Klebanoff, P. S., K. D. Tidstrom, and L. M. Sargent: *J. Fluid Mech.,* **12,** 1 (1962).
81. Hinze, J. O., H. Leijdens, J. B. van den Burg, and D. Kleiweg: Proc. "8 Journées de l'Hydraulique de la Soc. Hydrotechnique de France," p. 43, Lille, 1964.

82. Hama, F. R.: *Phys. Fluids,* **5,** 1156 (1962).
83. Hama, F. R., and J. Nutant: "Heat Transfer and Fluid Mechanics Institute," p. 77, Stanford, 1963.
84. Knapp, C. F., and P. J. Roache: *A.I.A.A. J.,* **6,** 29 (1968).
85. Van Muiswinkel, J. C.: Unpublished results, Fluid Mechanics Laboratory, Delft University of Technology, 1970.
86. Elder, J. W.: *J. Fluid Mech.,* **9,** 235 (1960).
87. Mitchner, M. J.: *J. Aeronaut. Sci.,* **21,** 350 (1954).
88. Schubauer, G. B., and P. S. Klebanoff: *N.A.C.A. Techn. Note* 3489 (1955).
89. Criminale Jr, W. O., and L. S. G. Kovasznay: *J. Fluid Mech.,* **14,** 59 (1962).
90. Squire, H. B.: *Proc. Roy. Soc. London,* A **142,** 621 (1933).
91. Vasudeva, B. R.: *J. Fluid Mech.,* **29,** 745 (1967).
92. Tani, I., and H. Komoda: *J. Aeronaut. Sci.,* **29,** 440 (1962).
93. Komoda, H.: *Phys. Fluids,* **10,** Suppl., 87 (1967).
94. Benney, D. J., and C. C. Lin: *Phys. Fluids,* **3,** 656 (1960).
95. Benney, D. J.: *J. Fluid Mech.,* **10,** 209 (1961).
96. Benney, D. J.: *Phys. Fluids,* **7,** 319 (1964).
97. Stuart, J. T.: *AGARD Rep.* 514 (1965).
98. Greenspan, H. R., and D. J. Benney: *J. Fluid Mech.,* **15,** 133 (1963).
99. Tennekes, H.: "Similarity Laws for Turbulent Boundary Layers with Suction or Injection." Ph.D. Thesis, Delft University of Technology, 1964.
100. Spalding, D. B.: *Trans. A.S.M.E., J. Appl. Mech.,* **28E,** 455 (1961).
101. Rotta, J. C.: "Turbulent Boundary Layers in Incompressible Flows," *Progress in Aeronaut. Sci.,* **2,** p. 1, Pergamon Press, 1962.
102. Proc. "Computation of Turbulent Boundary Layers," vol. **1,** edited by Kline, S. J., M. V. Morkovin, G. Sovran, and D. J. Cockrell, Stanford University, California, August 18–25, 1968.
103. Smith, D. W., and J. H. Walker: *N.A.S.A. Rep.* R-26 (1959).
104. Einstein, H. A., and E. A. El-Samni: *Rev. Modern Phys.* **21,** 520 (1949).
105. Moore, W. L.: Ph.D. Thesis, State University of Iowa, 1951.
106. Perry, A. E., W. H. Schofield, and P. N. Joubert: *J. Fluid Mech.,* **37,** 383 (1969).
107. Kovasznay, L. S. G., V. Kibens, and R. F. Blackwelder: *J. Fluid Mech.,* **41,** 283 (1970).
108. Wilson, R. E.: *J. Aeronaut. Sci.,* **17,** 585 (1950).
109. Nagamatsu, H. T., B. C. Graber, and R. E. Sheer: *J. Fluid Mech.,* **24,** 1 (1966).
110. Fage, A., and H. C. H. Townend: *Proc. Roy. Soc. London,* **135,** 656 (1934).
111. Coantic, M.: *Publ. Sci. et Tech. du Ministère de l'Air.* No. N.T. 113, 1962.
112. Bakewell Jr., H. P., and J. L. Lumley: *Phys. Fluids,* **10,** 1880 (1967).
113. Mitchell, J. E., and T. J. Hanratty: *J. Fluid Mech.,* **26,** 199 (1966).
114. Hanratty, T. J.: *Phys. Fluids,* Suppl. **10,** Part II, S 126 (1967).
115. Eckelmann, H.: *Mitteilungen Max-Planck-Inst. für Strömungsforschung, Göttingen,* No. 48, 1970.
116. Sirkar, K. K., and T. J. Hanratty: *J. Fluid Mech.,* **44,** 605 (1970).
117. Fowles, P. E., K. A. Smith, and T. K. Sherwood: *Chem. Eng. Sci.,* **23,** 1225 (1968).
118. Popovich, A. T., and R. L. Hummel: *A.I.Ch.E. Jl.,* **13,** 854 (1967).
119. Armistead, R. A., and J. J. Keyes: *J. Heat Transf., Trans. A.S.M.E.,* **90** C, 13 (1968).
120. Rundstadler, P. W., S. J. Kline, and W. C. Reynolds: *Dept. Mech. Engng., Thermoscience Div., Stanford University, Rep.* MD-8, 1963.

121. Kline, S. J., W. C. Reynolds, F. A. Schraub, and P. W. Rundstadler: *J. Fluid Mech.,* **30,** 741 (1967).
122. Kim, H. T., S. J. Kline, and W. C. Reynolds: *J. Fluid Mech.,* **50,** 133 (1971).
123. Morrison, W. R. B., K. J. Bullock, and R. E. Kronauer: *J. Fluid Mech.,* **47,** 639 (1971).
124. Grass, A. J.: *J. Fluid Mech.,* **50,** 233 (1971).
125. Gupta, A. K., J. Laufer, and R. E. Kaplan: *J. Fluid Mech.,* **50,** 493 (1971).
126. Corino, E. R., and R. S. Brodkey: *J. Fluid Mech.,* **37,** 1 (1969).
127. Wallace, J. M., H. Ecklemann, and R. S. Brodkey: *J. Fluid Mech.,* **54,** 39 (1972).
128. Clark, J. A., and E. Markland: *Aeronaut. J.,* **74,** 243 (1970).
129. Lumley, J. L.: "Proc. Intern. Colloquium on the Fine Scale Structure of the Atmosphere and its Influence on Radio Wave Propagation," p. 166, Moscow, 1967.
130. Lumley, J. L.: "Developments in Mechanics," **6,** Proc. 12th Mid-western Mech. Conference, p. 63, 1971.
131. Nychas, S. G., H. C. Hershey, and R. S. Brodkey: *J. Fluid Mech.,* **61,** 513 (1973).
132. Willmarth, W. W., and S. S. Lu: *J. Fluid Mech.,* **55,** 65 (1972).
133. Laufer, J., and M. A. Badri Narayanan: *Phys. Fluids,* **14,** 182 (1971).
134. Rao, K., R. Narasimha, and M. A. Badri Narayanan: *J. Fluid Mech.,* **48,** 339 (1971).
135. Willmarth, W. W., and C. E. Wooldridge: *J. Fluid Mech.,* **14,** 187 (1962).
136. Willmarth, W. W., and C. E. Wooldridge: *NATO AGARD REP,* 456, 1963.
137. Tu, B. J., and W. W. Willmarth: *Univ. Michigan Techn. Rep. ORA* 0290-3-T, 1966.
138. Willmarth, W. W., and B. J. Tu: *Phys. Fluids,* Suppl. **10,** Part II, S 134 (1967).
139. Willmarth, W. W., and F. W. Roos: *J. Fluid Mech.,* **22,** 81 (1965).
140. Kraichnan, R. H.: *J. Acoust. Soc. Amer.,* **28,** 64 (1965).
141. Lilley, G. M.: *College Aeron. Cranfield, Co A Rep.* 133 (1960).
142. Lilley, G. M., and T. H. Hodgson: *AGARD Rep.* 276 (1960).
143. Harrison, M.: *David Taylor Model Basin, Hydromech. Lab. Rep.* 1260 (1958).
144. Bull, M. K., and J. L. Willis: *Univ. Southampton, A.A.S.U. Rep.* 199 (1961).
145. Blake, W. K.: *J. Fluid Mech.,* **44,** 637 (1970).
146. Wills, J. A. B.: *J. Fluid Mech.,* **45,** 65 (1971).
147. Willmarth, W. W., and C. S. Young: *J. Fluid Mech.,* **41,** 47 (1970).
148. Bull, M. K.: *J. Fluid Mech.,* **28,** 719 (1967).
149. Favre, A., J. Gaviglio, and R. Dumas: "Mécanique de la Turbulence," p. 419, Paris, 1962.
150. Favre, A., J. Gaviglio, and J. P. Fohr: Proc. 11th Int. Congress Appl. Mech., p. 878, Munich, 1964.
151. Favre, A.: *J. Appl. Mech.,* **32E,** 241 (1965).
152. Favre, A., J. Gaviglio, and R. Dumas: *Phys. Fluids,* Suppl. **10,** Part II, S 138 (1967).
153. Willmarth, W. W., and S. S. Lu: *J. Fluid Mech.,* **55,** 65 (1972).
154. Blackwelder, R. F., and R. E. Kaplan: "The Intermittent Structure of the Wall Region of the Turbulent Boundary-Layer," Univ. Southern California Rep. USCAE 1-22, 1972.
155. Sternberg, J.: *Phys. Fluids,* Suppl. **10,** Part II, S 146 (1967).
156. Townsend, A. A.: *J. Fluid Mech.,* **11,** 97 (1961).
157. Bradshaw, P.: *J. Fluid Mech.,* **30,** 241 (1967).
158. Laufer, J.: *Instituto Nazionale di Alta Matematica, Symposia Matematica,* **9,** 299 (1972).
159. Phillips, O. M.: *Proc. Camb. Phil. Soc.,* **51,** 220 (1955).
160. Bradbury, L. J. S.: *J. Fluid Mech.,* **23,** 31 (1965).
161. Bradshaw, P.: *J. Fluid Mech.,* **27,** 209 (1967).

162. Townsend, A. A.: *J. Fluid Mech.,* **26,** 689 (1966).
163. Phillips, O. M.: *J. Fluid Mech.,* **51,** 97 (1972).
164. Huffman, G. D., and P. Bradshaw: *J. Fluid Mech.,* **53,** 45 (1972).
165. Blinco, P. H., and E. Partheniades: *J. Hydr. Res.,* **9,** 43 (1971).
166. Schubert, G., and G. M. Corcos: *J. Fluid Mech.,* **29,** 113 (1967).
167. Sternberg, J.: *J. Fluid Mech.,* **13,** 241 (1962).
168. Landahl, M. T.: *J. Fluid Mech.,* **29,** 441 (1967).
169. Emmerling, R.: *Mitteilungen Max-Planck Inst. für Strömungsforschung, Göttingen,* No. 56, 1973.
170. Black, T. J.: *Proc. Heat Transfer and Fluid Mech. Inst.,* p. 366, 1966.
171. Black, T. J.: *N.A.S.A.* CR 888, 1968.
172. Black, T. J.: "Viscous Drag Reduction," p. 383, Plenum Press, New York, 1969.
173. Coles, D.: *Z. angew. Math. u. Physik,* **5,** 3 (1954).
174. Eckhaus, W.: "Studies in Non-linear Stability Theory," Springer-Verlag, Berlin, 1965.
175. Reichardt, H.: *Max Planck Inst. für Strömungsforschung, Göttingen, Bericht* No. 6, 1971.
176. Rotta, J.: *Z. für Physik,* **129,** 547 (1951).
177. Townsend, A. A.: *J. Fluid Mech.,* **12,** 536 (1961).
178. Bradshaw, P., D. H. Ferriss, and N. P. Atwell: *J. Fluid Mech.,* **28,** 593 (1967).
179. Bradshaw, P.: *J. Fluid Mech.,* **29,** 625 (1967).
180. Nee, V. W., and L. S. G. Kovasznay: *Phys. Fluids,* **12,** 473 (1969).
181. Ng, K. H., and D. B. Spalding: *Phys. Fluids,* **15,** 20 (1972).
182. Jones, W. P., and B. E. Launder: *Int. J. Heat Mass Transf.,* **15,** 301 (1972).
183. Launder, B. E., and D. B. Spalding: "Lectures in Mathematical Models of Turbulence," Academic Press, London, 1972.
184. Hanjalic, K., and B. E. Launder: *J. Fluid Mech.,* **52,** 689 (1972).
185. Launder, B. E., A. Morse, W. Rodi, and D. B. Spalding: Nasa Conference on Free Shear Flows, Hampton, Virginia, July, 1972.
186. Lumley, J. L.: I.A.H.R. Int. Symp. on Stratified Flows, Novosibirsk, September, 1972.
187. Lumley, J. L., and B. Khajeh-Nouri: IUTAM-IUGG "Symposium on Turbulent Diffusion in Environmental Pollution," Charlottesville, Virginia, April, 1973.
188. Prandtl, L.: *Nachr. Akad. Wiss. Göttingen, Math.-Phys. Klasse,* 6–19, 1945.
189. Lighthill, M. J.: Proceedings Conference on Computation of Turbulent Boundary-Layers, Part I, AFOSR-IFP, p. 55, Stanford, 1968.
190. Tennekes, H., and J. L. Lumley: "A First Course in Turbulence," p. 86, The M.I.T.-Press, London, 1972.
191. Pfenniger, W.: "Boundary Layer and Flow Control," p. 970, Pergamon Press, Oxford, 1961.
192. Hopf, E.: Lecture Series Symp. on Partial Differential Equations, Univ. California Press, Berkeley, p. 7, 1955.
193. Serrin, J.: *Archiv for Rational Mechanics and Analysis,* **3,** 1 (1959).
194. Velte, W.: *Archiv for Rational Mechanics and Analysis,* **9,** 9 (1962).
195. Lamb, H.: "Hydrodynamics," 6th Ed., p. 619, Cambridge University Press, New York, 1932.
196. Görtler, H., and W. Velte: *Phys. Fluids,* Suppl.. Part II, **10,** S 3 (1967).
197. Shibuya, I.: *Rep. Inst. High-Speed Mech. Tohoku Univ.,* **1,** 37 (1951).
198. Fox, J. A., M. Lessen, and N. V. Bhat: *Phys. Fluids,* **11,** 1 (1968).

199. Tatsumi, T.: *J. Phys. Soc. Japan,* **7,** 489 (1952).
200. Davey, A., and P. G. Drazin: *J. Fluid Mech.,* **36,** 209 (1969).
201. Davey, A., and H. P. F. Nguyen: *J. Fluid Mech.,* **45,** 701 (1971).
202. Paterson, R. W., and F. H. Abernathy: *J. Fluid Mech.,* **51,** 177 (1972).
203. Lindgren, E. R.: *Archiv för Fysik,* **12,** 1 (1957); **15,** 97 and 503 (1959); **16,** 101 (1959); **18,** 449 (1960); **23,** 403 (1962); **24,** 269 (1963).
204. Rotta, J.: *Ing. Archiv,* **24,** 258 (1956).
205. Coles, D., and C. W. Van Atta: *Phys. Fluids,* Suppl., Part II, **10,** S 120 (1967).
206. Badri Narayanan, M. A.: *J. Fluid Mech.,* **31,** 609 (1968).
207. Hinze, J. O.: "Mécanique de la Turbulence," p. 129, Paris, 1962.
208. Tennekes, H.: *A.I.A.A. Jl.,* **6,** 1735 (1968).
209. Lawn, C. J.: *J. Fluid Mech.,* **48,** 477 (1971).
210. Narasimha, R., K. N. Rao, and M. A. Badri Narayanan, IUTAM-IUGG, "Symposium on Turbulent Diffusion in Environmental Pollution," April 8–14, Charlottesville, 1973. Advances in Geophysics, vol. **18B,** 372 (1974).
211. Hinze, J. O.: *Appl. Sci. Res.,* **28,** 453 (1973).
212. Sieder, E. N., and G. E. Tate: *Ind. Eng. Chem.,* **28,** 1429 (1936).
213. Dorresteyn, W. R.: *De Ingenieur,* **84,** Ch. 86 (1972).
214. Squire, H. B.: "General Discussion on Heat Transfer," p. 11, Institution of Mechanical Engineers, London, 1951.
215. Blom, J.: "An Experimental Determination of the Turbulent Prandtl Number in a Developing Temperature Boundary-Layer," Ph.D. Thesis, Eindhoven University of Technology, 1970.
216. Sleicher, C. A.: *Trans. A.S.M.E.,* **80,** 693 (1958).
217. Sage, B. H., and E. Venezian: *A.I.Ch.E. Jl.,* **7,** 688 (1961).
218. Ludwieg, H.: *Z. Flugwissenschaften,* **4,** 73 (1956).
219. Johnk, R. E., and T. J. Hanratty: *Chem. Eng. Sci.,* **17,** 867 (1962).
220. Johnk, R. E., and T. J. Hanratty: *Chem. Eng. Sci.,* **17,** 881 (1962).
221. Sesonski, A., *et al.*: *Chem. Eng. Progress,* **61,** 101 (1965).
222. Gowen, R. A., and J. W. Smith: *Int. J. Heat Mass Transf.,* **11,** 1657 (1968).
223. Bremhorst, K., and K. J. Bullock: *Int. J. Heat Mass Transf.,* **13,** 1313 (1970).
224. Meier, H. U., and J. C. Rotta: "A.I.A.A. Third Fluid and Plasma Dynamics Conference," A.I.A.A. Paper No. 70-744, Los Angeles, 1970.
225. Cebeci, T.: *Trans. A.S.M.E., J. Heat Transfer,* **95** C, 227 (1973).
226. Hughmark, G. A.: *A.I.Ch.E. Jl.,* **17,** 902 (1971).
227. Sutera, S. P., J. Kestin, and P. F. Maeder: *J. Fluid Mech.,* **16,** 497 (1963).
228. Kestin, J., and R. T. Wood: "Progress in Heat and Mass Transfer," vol. 2, p. 249, Pergamon Press, London, 1969.
229. Kestin, J., and R. T. Wood: *Trans. A.S.M.E., J. Heat Transfer,* **93** C, 321 (1971).
230. Mizushina, T., H. Ueda, and N. Umemiya: *Int. J. Heat Mass Transf.,* **15,** 769 (1972).
231. Kestin, J.: "Advances in Heat Transfer," vol. 3, p. 1, Academic Press, New York, 1966.
232. Patterson, G. K., W. J. Ewbank, and V. A. Sandborn: *Phys. Fluids,* **10,** 2082 (1967).
233. Antonia, R. A.: *Phys. Fluids,* **15,** 1669 (1972).
234. Ueda, H., and J. O. Hinze: To be published.
235. Wygnanski, I., and F. H. Champagne: *J. Fluid Mech.,* **59,** 281 (1973).

NOMENCLATURE FOR CHAPTER 7

c — $= c_r + ic_i$.
c_r — phase velocity.
c_i — amplification factor.
c_p — specific heat at constant pressure.
c_f — $\sigma_w/\frac{1}{2}\rho U_0{}^2$, coefficient of wall friction.
c_h — coefficient of heat flow resistance at the wall.
D — pipe diameter.
$\mathbf{E}_2(k_1)$, $\mathbf{E}_2(k_2)$ — one-dimensional spectra of turbulence kinetic energy.
f — friction factor.
f — coefficient of spatial, longitudinal velocity correlation.
g — coefficient of spatial, lateral velocity correlation.
H — δ_d/δ_m, shape factor of boundary layer
$h(\xi_2)$ — correction function for logarithmic velocity-defect law.
$h_1(\xi_2)$ — Millikan's correction function for logarithmic velocity-distribution law.
k — roughness parameter, average height of roughness elements; k_s, equivalent sand roughness.
L_i — length scales in directions x_i.
l — length scale for spatial variations of turbulent velocities.
$\mathbf{Ma}_0$ — free-stream Mach number.
Nu — Nusselt number.
P — static pressure; $\bar{P}$, time-mean value; p, turbulent fluctuation; P_0, static pressure in free stream.
Pr — ν/κ_θ, Prandtl number; $\mathbf{Pr}_{\text{turb}} = \epsilon_m/\epsilon_\theta$; $\mathbf{Pr}_{\text{eff}} = c_p\mu_{\text{eff}}/\kappa_{\text{eff}}$.
$\overline{q^2}$ — $\overline{u_i u_i}$, twice kinetic energy of turbulence.
$\mathbf{R}_E$ — coefficient of Eulerian time correlation.
Re — Reynolds number; $\mathbf{Re}_D$, $\bar{U}_{\text{av}}D/\nu$; $\mathbf{Re}_{x_1}$, $U_0 x_1/\nu$; $\mathbf{Re}_\delta$, $U_0\delta/\nu$; $\mathbf{Re}_{\delta_d}$, $U_0\delta_d/\nu$; $\mathbf{Re}_{\delta_m}$, $U_0\delta_m/\nu$.
r — cylindrical polar coordinate in radial direction; radial distance to pipe axis.
Sc — Schmidt number.
T_B — "burst" period.
U_i — Eulerian velocity; $\bar{U}_i$, time-mean value; u_i, turbulence component; u_i', $\sqrt{\overline{u_i{}^2}}$; U_0, free-stream velocity; $\bar{U}_{x,\max}$, maximum velocity at center of pipe flow; $\bar{U}_{\text{av}}$, average velocity across a section of pipe flow; u^*, $(\sigma_w/\rho)^{1/2}$ wall-friction velocity; $\bar{U}_i{}^+$, $\bar{U}_i/u^*$; subscripts r, φ, and x refer to cylindrical polar coordinates; U_c, convection velocity.
$w(\xi_2)$ — Coles's correction function for logarithmic velocity-distribution law.
x — cylindrical polar coordinate in axial direction; x_i, Cartesian coordinates; $x_2{}^+$, $x_2 u^*/\nu$.

α — $2\pi\delta/\lambda$.
β — entrainment constant.

δ	boundary-layer thickness; δ_d, displacement thickness of boundary layer; δ_m, momentum-loss thickness of boundary layer; δ_e, energy-loss thickness of boundary layer; δ^+, $\delta u^*/\nu$; δ_l, thickness of viscous sublayer, $\delta_t - \delta_l$, thickness of buffer region.
Δ	dissipation thickness of boundary layer.
ε	dissipation by turbulence per unit of mass.
ε'	$\nu\overline{(\partial u_j/\partial x_j)(\partial u_j/\partial x_i)}$.
ϵ_m	eddy viscosity; ϵ_θ, eddy diffusivity for heat.
φ	cylindrical polar coordinate in angular direction.
$\mathfrak{J}$	flux per unit of area; $\mathfrak{J}_\theta$, for heat.
$\varkappa$	Von Kármán's universal constant.
κ	heat conductivity; κ_{eff}, $\kappa + \rho c_p \epsilon_\theta$.
$\mathfrak{k}_\theta$	$\kappa/c_p\rho$.
$\mathfrak{l}$	Prandtl's mixing length; $\mathfrak{l}_m$, for momentum.
Λ_f	spatial longitudinal integral scale.
Λ_g	spatial lateral integral scale.
λ	wavelength.
μ	dynamic viscosity; μ_{eff}, $\mu + \rho\epsilon_m$.
ν	kinematic viscosity.
$\mathcal{O}$	order of magnitude.
Ω	intermittency factor.
ω_k	k-component of turbulence vorticity.
Π	Coles's profile parameter of boundary layer.
π	3.14159....
ρ	density.
σ_{ij}	stress tensor.
σ_w	wall shear-stress.
Θ	temperature; $\bar{\Theta}$, time-mean value; θ, turbulent fluctuation; $\bar{\Theta}_w$, wall temperature; Θ_0, free-stream temperature; $\bar{\Theta}_{\min}$, temperature at center of pipe flow.
ψ	stream function.
$\mathscr{V}_i$	velocity scale in direction x_i.
$\mathfrak{v}$	scale for turbulence velocity.
ξ_i	x_i/δ; ξ_2, $2x_2/D$; ξ'_2, $1 - \xi_2$.

APPENDIX

ELEMENTS OF CARTESIAN TENSORS

For those readers who are not familiar with tensor calculus, we shall consider briefly the elements that are frequently applied in this book.

A tensor is an object that remains invariant and transforms into itself under rotation of the coordinate system.

Depending on the number of independent parameters or components that determine them, we distinguish tensors of different order, increasing in complexity with increasing order, that is, with an increasing number of components.

The simplest tensor is determined by one component only. It is a tensor of zero order, or a scalar, and as such it is invariant under rotation of the coordinate system.

Next in simplicity is a tensor of the first order, or a vector, which is determined by three components.

However, before considering further these tensors and their invariance under rotation of the coordinate system, we shall consider the transformation of the coordinates of a point in the three-dimensional space under such rotation.

Let P be a point whose coordinates are x_1, x_2, and x_3 with respect to an (x_1,x_2,x_3) coordinate system (see Fig. A-1). Let (x_1^*,x_2^*,x_3^*) be a new coordinate

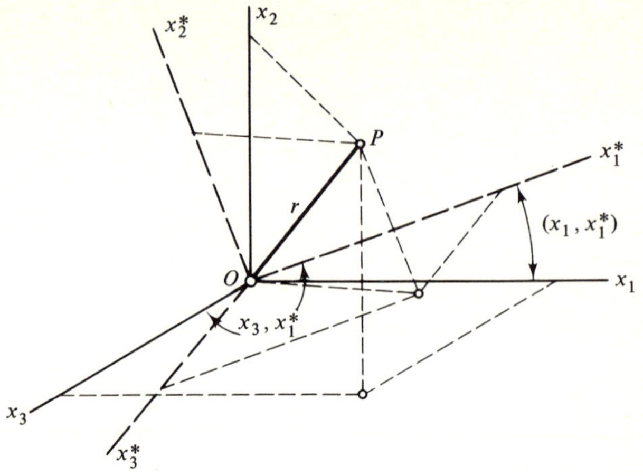

FIGURE A-1
Transformation of coordinates.

system and let x_1^*, x_2^*, and x_3^* be the coordinates of P with respect to this new system.

We then have the identity

$$r^2 = {x_1}^2 + {x_2}^2 + {x_3}^2 = x_1^{*2} + x_2^{*2} + x_3^{*2} \tag{A-1}$$

The coordinates x_1, x_2, and x_3 can be expressed in terms of the coordinates x_1^*, x_2^*, and x_3^* of the new system, and vice versa, by means of the angles between the respective coordinate axes of the old and the new system.

Let (x_1, x_1^*) be the angle through which the positive-x_1-axis has to turn to make it coincide with the positive-x_1^*-axis.

Similarly (x_1, x_3^*) is the angle between the positive-x_1-axis and the positive-x_3^*-axis.

In general (x_i, x_j^*) is the angle between the positive-x_i-axis of the old coordinate system and the positive-x_j^*-axis of the new coordinate system ($i = 1,2,3$; $j = 1,2,3$).

Then we have:

$$x_1^* = x_1 \cos(x_1, x_1^*) + x_2 \cos(x_2, x_1^*) + x_3 \cos(x_3, x_1^*)$$

$$x_2^* = x_1 \cos(x_1, x_2^*) + x_2 \cos(x_2, x_2^*) + x_3 \cos(x_3, x_2^*)$$

$$x_3^* = x_1 \cos(x_1, x_3^*) + x_2 \cos(x_2, x_3^*) + x_3 \cos(x_3, x_3^*)$$

If for the various direction cosines we write for short:

$$e_{11} = \cos(x_1, x_1^*) \qquad e_{21} = \cos(x_2, x_1^*)$$

and

$$e_{ij} = \cos(x_i, x_j^*) \quad \text{in general}$$

the above set of three equations reads:

$$x_1^* = x_1e_{11} + x_2e_{21} + x_3e_{31}$$
$$x_2^* = x_1e_{12} + x_2e_{22} + x_3e_{32} \qquad \text{(A-2)}$$
$$x_3^* = x_1e_{13} + x_2e_{23} + x_3e_{33}$$

Note (1) that the first index in e_{ij} refers to the old coordinate system, the second index to the new coordinate system, and (2) that in general $e_{ij} \neq e_{ji}$, because $\cos(x_i,x_j^*) \neq \cos(x_j,x_i^*)$.

Conversely, we can express the x_i of the old system in terms of x_j^* of the new system and we can obtain another set of three equations similar to Eq. (A-2):

$$x_1 = x_1^*e_{11} + x_2^*e_{12} + x_3^*e_{13}$$
$$x_2 = x_1^*e_{21} + x_2^*e_{22} + x_3^*e_{23} \qquad \text{(A-3)}$$
$$x_3 = x_1^*e_{31} + x_2^*e_{32} + x_3^*e_{33}$$

The direction cosines e_{ij} form a normalized orthogonal system. Their orthogonal relations can be derived as follows, by making use of the identity (A-1). For instance, introduce the expressions (A-2) for x_i^* into the right-hand side of Eq. (A-1),

$$\begin{aligned}
r^2 &= x_1^{*2} + x_2^{*2} + x_3^{*2} \\
&= (x_1e_{11} + x_2e_{21} + x_3e_{31})^2 + (x_1e_{12} + x_2e_{22} + x_3e_{32})^2 \\
&\quad + (x_1e_{13} + x_2e_{23} + x_3e_{33})^2 \\
&= x_1^2(e_{11}^2 + e_{12}^2 + e_{13}^2) + x_2^2(e_{21}^2 + e_{22}^2 + e_{23}^2) + x_3^2(e_{31}^2 + e_{32}^2 + e_{33}^2) \\
&\quad + 2x_1x_2(e_{11}e_{21} + e_{12}e_{22} + e_{13}e_{23}) + 2x_1x_3(e_{11}e_{31} + e_{12}e_{32} + e_{13}e_{33}) \\
&\quad + 2x_2x_3(e_{21}e_{31} + e_{22}e_{32} + e_{23}e_{33})
\end{aligned}$$

Since the identity (A-1) holds for any value of x_1, x_2, and x_3, the only possibility is that the coefficients of x_1^2, x_2^2, x_3^2 be equal to unity, and those of x_1x_2, x_1x_3, and x_2x_3 be equal to zero.

Hence,

$$\begin{aligned}
e_{11}^2 + e_{12}^2 + e_{13}^2 &= 1 \\
e_{21}^2 + e_{22}^2 + e_{23}^2 &= 1 \\
e_{31}^2 + e_{32}^2 + e_{33}^2 &= 1 \\
e_{11}e_{21} + e_{12}e_{22} + e_{13}e_{23} &= 0 \\
e_{11}e_{31} + e_{12}e_{32} + e_{13}e_{33} &= 0 \\
e_{21}e_{31} + e_{22}e_{32} + e_{23}e_{33} &= 0
\end{aligned} \qquad \text{(A-4)}$$

This set of six equations comprises the orthogonal relations among the direction cosines e_{ij} $(i,j = 1,2,3)$.

In order to simplify the writing further, we abbreviate the set of equations (A-2), for instance, to

$$x_i^* = x_je_{ji} \qquad \text{(A-5)}$$

where the indices i and j can take the values 1, 2, and 3. We adopt Einstein's summation convention, according to which a summation should be made over the three values of a repeated index occurring in the expression for the general term. For instance, for every value 1, 2, or 3 of i, the Eq. (A-5) should be read

$$x_i^* = x_1 e_{1i} + x_2 e_{2i} + x_3 e_{3i} \qquad \text{for } i = 1, 2, \text{ or } 3$$

Hence Eq. (A-5) really means a set of three equations, each with three terms in its right-hand side.

Similarly, the set of equations (A-3) may be written

$$x_j = x_i^* e_{ji} \tag{A-6}$$

The orthogonal relations (A-4) can now be written very briefly, namely,

$$e_{ij} e_{kj} = 1 \qquad \text{for } i = k$$

$$e_{ij} e_{kj} = 0 \qquad \text{for } i \neq k$$

or, in one equation,

$$e_{ij} e_{kj} = \delta_{ik} \tag{A-7a}$$

where δ_{ik} is the Kronecker delta, defined by

$$\delta_{ik} = 1 \qquad \text{for } i = k$$

$$\delta_{ik} = 0 \qquad \text{for } i \neq k \tag{A-8}$$

In a similar way it can be shown that

$$e_{ij} e_{ik} = \delta_{jk} \tag{A-7b}$$

The quantities x_i form the coordinates of the line segment r, which has a given direction. Thus such a directed line segment is determined by three components, namely the three coordinates x_i.

Such a quantity is called a vector, or a tensor of the first order. The components of this vector, or tensor of the first order, transform according to Eq. (A-5) under rotation of the coordinate system.

Any object that is determined by three components A_i and whose components transform like

$$A_i^* = A_j e_{ji} \tag{A-9}$$

is called a tensor of the first order. It is written symbolically A_i, that is, it is indicated by its three components; and Eq. (A-9) is its transformation rule.

Its reciprocal relation reads

$$A_j = A_i^* e_{ji} \tag{A-10}$$

If each component of a tensor A_j is multiplied by a scalar B, the result is a new tensor of the first order $BA_j = C_j$.

$$BA_j = BA_i^* e_{ji} = (BA_i)^* e_{ji} \tag{A-11}$$

whence

$$C_j = C_i^* e_{ji}$$

The sum of two first-order tensors A_i and B_i is again a first-order tensor C_i; that is,

$$A_j + B_j = A_i^* e_{ji} + B_i^* e_{ji} = (A_i + B_i)^* e_{ji} \qquad \text{(A-12)}$$

or

$$C_j = C_i^* e_{ji}$$

so C_j obeys the transformation rule.

The product of two first-order tensors A_i and B_j is a second-order tensor C_{ij}. By definition it is obtained by multiplying each component of B_j by each component of A_i, or each component of A_i by each component of B_j. Symbolically written,

$$A_i B_j = C_{ij} \qquad \text{(A-13)}$$

By Eq. (A-10),

$$A_i B_j = A_k^* e_{ik} B_l^* e_{jl} = (A_k B_l)^* e_{ik} e_{jl}$$

or

$$C_{ij} = C_{kl}^* e_{ik} e_{jl} \qquad \text{(A-14)}$$

Any object A_{ij} that has nine components and that obeys the transformation rule (A-14) is defined as a tensor of the second order.

In general a tensor of the order n has 3^n components, and its transformation rule reads

$$\underset{n \text{ indices}}{A_{ijk\ldots lm}^*} = \underset{n \text{ indices}}{A_{pqr\ldots st}} e_{pi} e_{qj} e_{rk} \cdots e_{sl} e_{tm} \qquad \text{(A-15)}$$

There are two important tensors of a special type, one a second-order tensor, the other a third-order tensor. The second-order tensor has already been introduced; this is the Kronecker delta, defined by Eq. (A-8). It is the unit tensor of the second order and it is symmetrical. That it is a tensor can be shown as follows. Assume for the moment that it is a second-order tensor, so that it obeys the rule

$$\delta_{ij}^* = \delta_{kl} e_{ki} e_{lj}$$

Then δ_{ij}^* should follow Eq. (A-8) if δ_{kl} follows Eq. (A-8).

.For $k \neq l$, $\delta_{kl} = 0$ and the corresponding terms are zero.

For $k = l$, $\delta_{kl} = 1$; so

$$\delta_{ij}^* = e_{ki} e_{kj}$$

But, according to the orthogonal relations, $e_{ki} e_{kj} = 0$ if $i \neq j$ and $e_{ki} e_{kl} = 1$ if $i = j$. Hence δ_{ij}^* follows Eq. (A-8).

The other special tensor of importance is the antisymmetrical or alternating tensor ε_{ijk}.

By definition,

$\varepsilon_{ijk} = 1$ for all different indices and even numbers of permutations of the indices

$\varepsilon_{ijk} = -1$ for all different indices and uneven numbers of permutations of the indices

$\varepsilon_{ijk} = 0$ if two arbitrary indices are equal

Different methods have been proposed for memorizing these properties of the antisymmetrical tensor. Here the following simple rule is suggested:

$$\varepsilon_{123} = \varepsilon_{231} = \varepsilon_{312} = +1$$

$$\varepsilon_{321} = \varepsilon_{213} = \varepsilon_{132} = -1$$

Note the cyclic permutation.

The multiplication rule given above for the product of two tensors of the first order can be extended to higher-order tensors.

Thus, by the general product of two tensors of order n and m, respectively, we understand a tensor of the order $n + m$ formed by multiplying each component of the first tensor by each component of the second tensor, or each component of the second tensor by each component of the first tensor. For example,

$$A_{ij}B_{klm} = C_{ijklm} \tag{A-16}$$

According to this multiplication rule, multiplication of a tensor by δ_{ij} yields a tensor two orders higher

$$\delta_{ij}A_{pq} = B_{ijpq}$$

If $j = p$, we obtain

$$\delta_{ij}A_{jq} = B_{ijjq}$$

But, if a tensor has two equal indices, the order of the tensor is reduced by 2, because a pair of equal indices means a summation with respect to them, and the tensor now has two less free indices. Hence,

$$B_{ijjq} = B_{i11q} + B_{i22q} + B_{i33q} = C_{iq}$$

We shall now show that $C_{iq} = A_{iq}$.

By Eqs. (A-7) and (A-14), we have

$$C_{iq} = \delta_{ij}A_{jq} = e_{ik}e_{jk}e_{js}e_{qr}A^*_{sr}$$

But $e_{jk}e_{js} = 0$ for $k \neq s$ and 1 for $k = s$; so we may take $e_{jk}e_{js} = 1$ and $k = s$.

$$\delta_{ij}A_{jq} = e_{is}e_{qr}A^*_{sr} = A_{iq}$$

Thus multiplication of A_{jq} by δ_{ij} means changing the index j of A_{jq} to i.

Generalized,

$$\delta_{ij}A_{kjrs} = A_{kirs} \tag{A-17}$$

This process is usually referred to as a substitution.

If, in addition, $i = k$, a tensor of order less by 2 is obtained:

$$\delta_{ij}A_{ijrs} = A_{iirs} = B_{rs} \tag{A-18}$$

Thus, multiplying A_{kjrs} by δ_{kj} means making the indices k and j of A_{kjrs} equal, that is, repeating the index and consequently lowering the order by 2. This latter process is usually referred to as a contraction.

If two tensors A_i and B_j of order 1 are multiplied and if their product A_iB_j

is multiplied by δ_{ij}, so that we apply a contraction, the result is a tensor of zero order, or a scalar; that is,

$$\delta_{ij}A_iB_j = A_iB_i = A_1B_1 + A_2B_2 + A_3B_3 \qquad \text{(A-19)}$$

This result is simply the "scalar" or "dot" product of the vectors A_i and B_j, denoted in vector algebra by $\mathbf{A} \cdot \mathbf{B}$.

We may apply this rule generally to tensors of higher order. Thus the scalar product of the tensors A_{ij} and B_{pqr} with respect to the indices i and p is

$$\delta_{ip}A_{ij}B_{pqr} = A_{ij}B_{iqr} = C_{jqr} \qquad \text{(A-20)}$$

Since multiplication by δ_{ip} means a contraction with respect to the indices i and p in $A_{ij}B_{pqr}$, we may identify the scalar product with respect to certain indices with a contraction with respect to these indices.

A further generalization is:

The scalar multiplication of the tensors $A_{ij\ldots}$ and $B_{pqr\ldots}$ with respect to the indices i,p and j,q

$$\delta_{ip}\delta_{jq}A_{ij\ldots}B_{pqr\ldots} = A_{ij\ldots}B_{ijr\ldots}$$

is the same as a contraction with respect to the indices i,p and j,q.

The antisymmetrical character of the tensor ε_{ijk} may be illustrated as follows:

Multiply the first-order tensor A_l by ε_{ijk}. A fourth-order tensor is obtained. If we contract with respect to the indices l,k, a second-order tensor is obtained:

$$\varepsilon_{ijk}A_k = B_{ij} \qquad \text{(A-21)}$$

This second-order tensor B_{ij} will be antisymmetrical. For we have

$$B_{ij} = \varepsilon_{ij1}A_1 + \varepsilon_{ij2}A_2 + \varepsilon_{ij3}A_3$$

With the values of ε_{ijk} for different i and j we obtain

$$\begin{array}{lll} i=1 & j=2 & B_{12} = A_3 \\ & j=3 & B_{13} = -A_2 \\ i=2 & j=1 & B_{21} = -A_3 \\ & j=3 & B_{23} = +A_1 \\ i=3 & j=1 & B_{31} = A_2 \\ & j=2 & B_{22} = -A_1 \end{array}$$

The components of B_{ij} are thus

$$B_{ij} = \begin{pmatrix} 0 & A_3 & -A_2 \\ -A_3 & 0 & A_1 \\ A_2 & -A_1 & 0 \end{pmatrix} \qquad \text{(A-22)}$$

a scheme which shows clearly the antisymmetrical character of B_{ij}.

The vector A_k is usually called the vector of the antisymmetrical tensor $B_{ij} = \varepsilon_{ijk}A_k$. And, conversely, the antisymmetrical tensor associated with a vector is obtained by multiplication by ε_{ijk} and a contraction.

Let us now multiply the product of two first-order tensors A_iB_j by ε_{ijk}. The result then appears to be simply the "vector" or "cross" product of two vectors $\mathbf{A} \times \mathbf{B}$, well-known in vector algebra.

$$\varepsilon_{ijk}A_iB_j = C_{iijjk} = D_k \qquad \text{(A-23)}$$

and

$$D_1 = \varepsilon_{ij1}A_iB_j = A_2B_3 - A_3B_2$$
$$D_2 = \varepsilon_{ij2}A_iB_j = A_3B_1 - A_1B_3$$
$$\cdots\cdots\cdots\cdots\cdots\cdots$$

In the case of three vectors A_i, B_j, and C_k the product $\varepsilon_{ijk}A_iB_jC_k$ is a scalar:

$$\begin{aligned}\varepsilon_{ijk}A_iB_jC_k &= A_1B_2C_3 + A_2B_3C_1 + A_3B_1C_2 - A_3B_2C_1 - A_2B_1C_3 - A_1B_3C_2 \\ &= A_1(B_2C_3 - B_3C_2) + A_2(B_3C_1 - B_1C_3) + A_3(B_1C_2 - B_2C_1) \\ &= A_iD_i\end{aligned}$$

where

$$D_i = \varepsilon_{ijk}B_jC_k$$

In vector algebra this product of three vectors is usually called the scalar triple product $\mathbf{A} \cdot (\mathbf{B} \times \mathbf{C})$.

We conclude this algebraic section about tensors by mentioning the following properties of the unit tensor δ_{ij} and of ε_{ijk}:

$$\delta_{ii} = \delta_{11} + \delta_{22} + \delta_{33} = 3$$
$$\delta_{ij}\delta_{jl} = \delta_{il}$$
$$\delta_{ik}\varepsilon_{ikm} = 0$$
$$\varepsilon_{ijk}\varepsilon_{lmn} = \begin{vmatrix} \delta_{il}\delta_{im}\delta_{in} \\ \delta_{jl}\delta_{jm}\delta_{jn} \\ \delta_{kl}\delta_{km}\delta_{kn} \end{vmatrix}$$

or, written out,

$$\varepsilon_{ijk}\varepsilon_{lmn} = \delta_{il}\delta_{jm}\delta_{kn} + \delta_{im}\delta_{jn}\delta_{kl} + \delta_{in}\delta_{jl}\delta_{km} - (\delta_{il}\delta_{jn}\delta_{km} + \delta_{im}\delta_{jl}\delta_{kn} + \delta_{in}\delta_{jm}\delta_{kl})$$

A single contraction, e.g., with respect to k and n, yields

$$\varepsilon_{ijk}\varepsilon_{lmk} = \delta_{il}\delta_{jm} - \delta_{im}\delta_{jl}$$

Twice a contraction yields

$$\varepsilon_{ijk}\varepsilon_{ljk} = 2\delta_{il}$$

and eventually,

$$\varepsilon_{ijk}\varepsilon_{ijk} = 6$$

The *derivative* of a scalar with respect to a coordinate is a tensor of the first order. Let A be the scalar; then the derivative with respect to x_i reads $\partial A/\partial x_i$. Since

$$\frac{\partial A}{\partial x_i} = \frac{\partial A}{\partial x_j^*}\frac{\partial x_j^*}{\partial x_i} = \frac{\partial A}{\partial x_j^*} e_{ij} \tag{A-24}$$

the derivative does indeed transform according to the transformation rule (A-6) for a tensor of the first order.

Similarly, the second derivative of a scalar is a tensor of the second order, etc.

$$\frac{\partial^2 A}{\partial x_i\,\partial x_k} = \frac{\partial}{\partial x_l^*}\left(\frac{\partial A}{\partial x_j^*}\frac{\partial x_j^*}{\partial x_i}\right)\frac{\partial x_l^*}{\partial x_k} = \frac{\partial^2 A}{\partial x_l^*\,\partial x_j^*} e_{ij}e_{kl} \tag{A-25}$$

The derivative of a tensor of the first order with respect to a coordinate is a tensor of the second order. Let A_i be that first-order tensor; then the derivative with respect to x_j reads $\partial A_i/\partial x_j$.

Since

$$A_i = A_k^* e_{ik} \quad \text{and} \quad \frac{\partial}{\partial x_j} = \frac{\partial}{\partial x_l^*}\frac{\partial x_l^*}{\partial x_j} = \frac{\partial}{\partial x_l^*} e_{jl}$$

we have

$$\frac{\partial A_i}{\partial x_j} = e_{jl}\frac{\partial}{\partial x_l^*}(A_k^* e_{ik}) = e_{jl}e_{ik}\frac{\partial A_k^*}{\partial x_l^*} \tag{A-26}$$

which is the transformation rule (A-14) for a second-order tensor.

Generalized, the mth derivative of a tensor of the nth order is a tensor of the order $m + n$.

A special case is presented if we take for the first-order tensor one of the coordinates x_i. The derivative with respect to the coordinate x_j, namely,

$$\frac{\partial x_i}{\partial x_j}$$

is a second-order tensor. But, since $\partial x_i/\partial x_j = 1$ if $i = j$ and $\partial x_i/\partial x_j = 0$ if $i \neq j$, we have

$$\frac{\partial x_i}{\partial x_j} = \delta_{ij} \tag{A-27}$$

Note that $\partial x_i/\partial x_i = \delta_{ii} = 3$.

Let A be a function of x_j; determine the derivative of A with respect to x_i:

$$\frac{\partial}{\partial x_i} A(x_j) = \frac{\partial A}{\partial x_j}\frac{\partial x_j}{\partial x_i} = \delta_{ij}\frac{\partial A(x_j)}{\partial x_j}$$

Let A be a function of u and let u be a function of x_j. Application of the known rule for the derivation of a function of a function yields

$$\frac{\partial A}{\partial x_i} = \frac{\partial A}{\partial u}\frac{\partial u}{\partial x_j}\frac{\partial x_j}{\partial x_i} = \delta_{ij}\frac{\partial A(u)}{\partial u}\frac{\partial u(x_j)}{\partial x_j}$$

In this book we have frequently applied these rules to the radius vector $r = \sqrt{x_j x_j}$. Accordingly,

$$\frac{\partial r}{\partial x_i} = \frac{1}{2\sqrt{x_j x_j}} 2x_i = \frac{x_i}{r}$$

If $A = A(r)$; then we have

$$\frac{\partial A}{\partial x_j} = \frac{\partial A}{\partial r}\frac{\partial r}{\partial x_j} = \frac{x_j}{r}\frac{\partial A}{\partial r}$$

Similarly, for the higher derivatives

$$\frac{\partial^2 A}{\partial x_k \partial x_j} = \frac{\partial}{\partial x_k}\left(\frac{x_j}{r}\frac{\partial A}{\partial r}\right) = \frac{1}{r}\frac{\partial A}{\partial r}\frac{\partial x_j}{\partial x_k} + x_j\frac{\partial A}{\partial r}\frac{\partial}{\partial x_k}\frac{1}{r} + \frac{x_j}{r}\frac{\partial}{\partial x_k}\frac{\partial A}{\partial r}$$

$$= \delta_{jk}\frac{1}{r}\frac{\partial A}{\partial r} - \frac{x_j}{r^2}\frac{\partial A}{\partial r}\frac{\partial r}{\partial x_k} + \frac{x_j}{r}\frac{\partial^2 A}{\partial r^2}\frac{\partial r}{\partial x_k}$$

$$= \delta_{jk}\frac{1}{r}\frac{\partial A}{\partial r} - \frac{x_j x_k}{r^3}\frac{\partial A}{\partial r} + \frac{x_j x_k}{r^2}\frac{\partial^2 A}{\partial r^2}$$

$$= \delta_{jk}\frac{1}{r}\frac{\partial A}{\partial r} + \left(\frac{\partial^2 A}{\partial r^2} - \frac{1}{r}\frac{\partial A}{\partial r}\right)\frac{x_j x_k}{r^2}$$

Consequently,

$$\frac{\partial^2 A}{\partial x_k \partial x_j} = \left(\frac{\partial^2 A}{\partial r^2} - \frac{1}{r}\frac{\partial A}{\partial r}\right)\frac{x_j x_k}{r^2} \qquad \text{for } j \neq k$$

$$\frac{\partial^2 A}{\partial x_j \partial x_j} = \frac{3}{r}\frac{\partial A}{\partial r} + \frac{\partial^2 A}{\partial r^2} - \frac{1}{r}\frac{\partial A}{\partial r} = \frac{2}{r}\frac{\partial A}{\partial r} + \frac{\partial^2 A}{\partial r^2}$$

but

$$\frac{\partial^2 A}{\partial x_1 \partial x_1} = \frac{1}{r}\frac{\partial A}{\partial r} + \left(\frac{\partial^2 A}{\partial r^2} - \frac{1}{r}\frac{\partial A}{\partial r}\right)\frac{{x_1}^2}{r^2}$$

NAME INDEX

Abbrecht, P. H., 731, 755
Abernathy, F. H., 710, 711
Abramovich, G. N., 470
Acharya, Y. V. G., 534, 536
Agostini, L., 150, 200
Ahmadi, G., 470
Ahmed, A. M., 90
Albertson, M. L., 536
Alexander, L. G., 166, 375, 534, 536, 539
Allan, J. F., 626
Antonia, R. A., 666
Armistead, R. A., Jr., 144, 153, 658, 664, 691
Atwell, N. P., 705

Badri Narayanan, M. A., 667, 714
Bagnold, R. A., 460, 461
Baines, W. D., 269, 270
Bakewell, H. P., Jr., 657, 658
Baldwin, L. V., 420, 423, 424, 427, 447, 519, 520
Barat, M., 544
Baron, T., 166, 375, 536, 539
Barry, M. D. J., 604
Bass, J., 200, 237
Basset, A. B., 463
Batchelor, G. K., 58, 177, 200, 202, 214, 216, 221, 232, 234, 235, 242, 259, 264, 268–270, 273, 275, 278, 293, 294, 306, 309, 321, 337, 376, 399, 400, 403, 406, 413, 415, 438, 445, 668
Becker, H. A., 160, 162, 167
Bellhouse, B. J., 144
Benney, D. J., 604, 613, 614
Bergeron, R. F., 604
Betchov, R., 98, 119
Binder, G., 153
Birkhoff, G., 217
Black, T. J., 694, 695
Blackwelder, R. F., 577, 645, 665, 668
Blake, W. K., 637, 673
Blasius, H., 598
Blom, J., 754, 755
Bourot, J. M., 156
Boussinesq, J., 23, 361, 463
Bradbury, L. J. S., 90, 687
Bradshaw, P., 269, 561, 563, 565, 628, 680, 681, 687, 705, 706
Brazier, J. G., 155
Bremshorst, K., 756
Brodkey, R. S., 278, 297, 659, 664, 666
Broer, L. J. F., 145
Brown, A. P. G., 167
Bull, M. K., 675
Bullock, K. J., 756
Burgers, J. M., 78, 104, 233, 239, 322, 328, 334, 386, 389, 754

Carmody, T., 509, 563
Carslaw, H. S., 428
Castro, J. P., 90
Cebeci, T., 755, 756
Cermak, J. E., 152
Champagne, F. H., 124, 125, 322, 350, 352, 354, 539, 560, 561, 709, 713
Chandrasekhar, S., 237, 311, 321, 417
Chen, W. Y., 184, 191, 243, 253, 254, 257, 520
Chin, W. C., 546
Chou, P. Y., 197
Chu, B. T., 314, 563
Chu, W. T., 565
Chuang, H., 152
Churchill, S. W., 731, 755
Clark, J. A., 659, 660
Clauser, F. H., 627, 628, 630, 631, 636, 684, 700
Clutter, D. W., 155, 157
Coantic, M., 657, 724, 727

Cocke, W. J., 413
Cole, J. D., 89
Coles, D., 694, 696, 700, 714
Collis, D. C., 89, 90, 94, 141, 143
Comings, E. W., 166, 375, 534, 536, 539, 544
Comte-Bellot, G., 117, 253, 257, 268–270, 272, 273
Corcos, G. M., 692
Corino, E. R., 659, 664, 666
Corrsin, S., 46, 105, 107, 138, 139, 141, 239, 243, 253, 257, 268–270, 272, 273, 278, 292, 303, 322, 350, 352, 354, 395, 397, 418, 422, 424, 426, 443, 451, 452, 461, 464, 536, 540, 543, 544, 546, 548, 550, 552, 567, 568, 571, 639, 641, 644, 688, 689
Criminale, W. O., Jr., 612
Crow, S. C., 539, 560, 561
Csallner, K., 148
Cummins, H., 162
Cutland, R. S., 626

Dai, Y. B., 536
Das, H. K., 148
Davey, A., 709
Davies, P. O. A. L., 86, 87, 90, 100, 536, 565
Davis, M. R., 86, 87
Deardorff, J. W., 411, 427
De Haan, R. E., 145, 147
Deissler, R. G., 197, 334, 342, 343, 351, 621, 749
Delleur, J. W., 89, 144
Dewey, C. F., Jr., 93
Diehl, Z. W., 353, 355, 626, 630
Doob, J. L., 59
Dorresteyn, W. R., 749
Drake, R. M., 87
Drazin, P. G., 709
Dryden, H. L., 4, 58, 114, 269, 600
Dugstadt, I., 239
Dumas, R., 186, 653, 675

Eckelmann, H., 657, 658, 665, 696, 727
Eckert, E., 384
Eckert, E. R. G., 92
Einstein, H. A., 636, 691
Elder, J. W., 148, 609
Elias, F., 743
Ellington, E., 142
Ellison, T. H., 235
Elrod, H. G., 620
El-Samni, E. A., 636
Emmerling, R., 672, 673
Emmons, H. W., 605
Epstein, N., 758
Ertel, H., 415
Eschenroeder, A. Q., 236
Eskinazi, S., 152, 580

Fabula, A. G., 143, 144
Fage, A., 155, 505, 508, 656
Falkner, V. M., 505, 508
Farell, C., 273, 274
Favre, A., 21, 47, 71, 128, 186, 450, 653, 675–677, 680
Ferriss, D. H., 561, 705
Ffowcs Williams, J. E., 314
Fiedler, H., 536, 538, 541, 543, 546, 552, 555, 556, 565, 570, 577, 581, 686, 687
Fingerson, L. M., 141, 142, 144
Fisher, M. J., 90, 100, 161, 563
Fleishman, B. A., 429, 433
Flügel, G., 12
Forstall, W., Jr., 534–536, 542
Fowles, P. E., 658, 727
Fox, J., 343, 351
Fox, J. A., 708, 710
Franzen, B., 150
Frenkiel, F. N., 60, 122, 131, 158, 184, 418, 428, 429, 433
Freymuth, P., 506, 508, 509, 511, 514, 516
Friedlander, S. K., 461
Fuchs, W., 150
Fujita, H., 126

Garby, L. C., 545
Garside, J. E., 13
Gaviglio, E. W., 186, 653, 675
Gaylord, E. W., 536, 542
Gibson, C. H., 152, 278, 292, 297, 299, 303, 520, 548
Gibson, M. M., 255, 397
Gifford, F., 424
Gilliland, E. R., 463
Goddard, F. E., 655
Goethals, R., 633
Goldschmidt, V. W., 470
Goldstein, R. J., 162, 164
Goldstein, S., 166
Görtler, H., 78
Graber, B. C., 655
Gracey, W., 166
Grant, H. L., 224, 255, 257, 269, 278, 297, 397, 574, 576
Grass, A. J., 659, 664, 665
Greenspan, H. R., 614
Grossman, L. M., 152
Gupta, A. K., 663
Gurvich, A. S., 242, 243
Gutmark, E., 560

Hall, A. A., 269
Hama, F. R., 157, 607–609, 626, 630, 631, 633, 635, 670, 696
Hanjalic, K., 25, 706
Hanratty, T. J., 153, 657, 658, 691, 727
Harris, V. G., 322, 350, 352, 354
Harrison, M., 673, 696
Hartung, F., 148
Hay, J. S., 420, 424, 427
Heisenberg, W., 230, 232, 262, 306
Henze, E. D., 534, 536
Hill, J. C., 124
Hilsch, R., 473
Hinze, J. O., 26, 158, 166, 167, 243, 392, 451, 452, 456, 471, 536, 540, 542, 580, 605, 668, 718
Hjelmfelt, A. T., 468
Hofmann, E., 619, 620, 748
Hopf, E., 707
Hottel, H. C., 160, 162
Howarth, L., 178, 185, 196
Howels, I. D., 293
Hubbard, J. C., 160
Hubbard, P. G., 143
Huffman, G. D., 628
Hughes, B. A., 278, 297
Hughes, R. R., 463
Hughmark, G. A., 755
Hulse, C., 545
Hummel, R. L., 658, 664, 691
Hwang, N. H. C., 519, 520

Irmay, S., 23
Iwasa, Y., 149

Jaeger, J. C., 428
Jeans, J. H., 48
Jenson, R. A., 536
Johnson, R. F., 561
Jones, W. P., 705
Joubert, P. N., 637

Kalinske, A. A., 59, 436
Kampé de Fériet, J., 53, 68, 436
Kaplan, R. E., 665, 668
Karlson, S. K. F., 153
Keagy, W. R., 541, 545
Keffer, J. F., 576, 577, 687
Kennedy, D. A., 242
Kesich, D., 236, 241, 295
Kestin, J., 762
Keyes, J. J., Jr., 144, 153, 658, 691
Kibens, V., 577, 645
Kim, H. T., 665–668
Kirsten, H., 717
Kistler, A. L., 254, 257, 269, 270, 272, 273, 278, 292, 303, 546, 552, 567, 568, 571, 639, 641, 644, 688, 689
Kivnick, E. W., 534, 536
Klebanoff, P. S., 131, 184, 353, 355, 605, 609, 613, 626, 630, 640, 642, 644, 647, 651, 653, 699, 700, 703, 725
Kline, S. J., 155, 659, 660, 663, 665, 692–694
Knapp, C. F., 608
Kobashi, Y., 514
Koch, H. A., Jr., 448, 747, 760
Kolin, A., 152
Kolmogoroff, A. N., 221, 223, 226, 228, 242, 267
Komoda, H., 613
Kovasznay, L. S. G., 91, 92, 112, 126, 139, 159, 160, 230, 237, 314, 422, 427, 449, 577, 612, 645, 685, 687–689, 706
Kraichnan, R. H., 55, 198, 224, 241, 249, 256, 397, 409, 424, 425, 673
Kramers, H., 88, 94, 118
Krause, F. R., 161
Kraussold, H., 748
Krzywoblocki, M. Z. E., 311

Kuethe, A. M., 4, 144, 654
Kuo, A. Y. S., 242, 243

Lambert, R. B., 153
Landahl, M. T., 692, 693
Landau, L. D., 604
Landis, F., 536
Latzko, H., 717
Laufer, J., 91, 92, 133, 134, 353, 354, 449, 546, 628, 656, 660, 667, 682, 689, 718, 721, 724, 727, 728, 731, 733, 737, 740
Launder, B. E., 705, 706
Laurence, J. C., 546
Lauwerier, H. A., 451
Lawn, C. J., 724, 727, 736, 738, 741
Lee, D. A., 277
Lee, J., 197, 198
Leijdens, H., 125
Li, H., 691
Lieblein, V., 384
Liepmann, H. W., 133–135, 251, 546
Lifschitz, E. M., 604
Lighthill, M. J., 140, 314, 315, 693
Lilley, G. M., 673
Limber, D. N., 306
Lin, C. C., 46, 78, 221, 227, 264, 398, 613
Lin, S. C., 520
Lindgren, E. R., 78, 710, 711, 714
Lindvall, F. C., 149
Ling, S. C., 143, 270, 275–277
Liu, C. L., 89, 144
Lobb, R. K., 654
Loeffler, A. L., 197
Loitsianski, L. G., 199, 200
Lorentz, H. A., 76, 77
Lorenz, H., 747
Lowell, H. H., 91, 100, 105
Lu, S. S., 660, 665, 666
Ludwieg, H., 144, 626, 638
Lumley, J. L., 25, 218, 237, 339, 424, 426, 440, 461, 464, 470, 471, 575, 657, 659, 705, 706, 708

McAdams, W. H., 88
McCarter, R. J., 448, 747
McClellan, R., 91, 92
Markland, E., 659, 660
Meier, H. U., 755–757
Michalke, A., 561
Mickelsen, W. R., 420, 423, 424, 427, 447
Mickley, H. S., 384
Middlebrook, G. B., 143
Miles, J. W., 620
Miller, D. R., 544
Millikan, C. B., 618, 634, 696
Mills, R. R., 257, 273, 303
Mintzer, D., 258
Mitchell, J. E., 153, 657
Mitchner, M., 322, 328, 334, 609
Mizushina, T., 762
Mobbs, F. R., 577, 687, 715
Mockros, L. F., 468
Moilliet, A., 224, 255, 278, 297, 397

Moore, W. L., 637
Morgan, P. G., 269
Morkovin, M. V., 112
Morrison, W. R. B., 660, 663, 672, 693
Munk, M., 78

Nagamatsu, H. T., 655
Narasimha, R., 742
Naudascher, E., 273, 274
Nayar, B. M., 563, 565
Nee, V. W., 706
Newby, M. P., 545
Ng, R. H., 704
Nguyen, H. P. F., 709
Nieuwenhuizen, J. K., 154
Nikuradse, J., 614, 627, 628, 717, 721, 723
Nisbet, I. C. T., 269
Nunner, W., 723, 730, 747, 758, 760
Nusselt, W., 384
Nutant, J., 157, 608, 609
Nychas, S. G., 659, 665, 666
Nye, J. O., 278, 297

O'Brien, V., 273, 278, 292, 303
Obukhoff, A. M., 230, 242, 306
Ogura, Y., 198, 305
Onsager, L., 228
Orr, W. M. F., 76
Orszag, S. A., 243, 256
Oseen, C. W., 463

Pai, S. I., 546
Panchev, S., 236, 237, 241, 295
Pao, Y. H., 239, 240, 255, 256, 278, 292, 299, 397
Parthasarathy, S. P., 111
Pasquil, F., 420, 424, 427
Patel, V. C., 147
Paterson, R. W., 710, 711
Patterson, G. S., 424, 427, 727
Payne, F. R., 575
Perry, A. E., 637
Persh, J., 654
Peskin, R. L., 411, 427, 470
Peterson, E. G., 269, 270
Philip, J. R., 401, 427
Phillips, O. M., 568, 570, 687, 689
Pien, C. L., 59, 436
Piercy, N. A. V., 142
Pigott, M. T., 145
Piret, E. L., 143
Plate, E. J., 149
Pohl, W., 758
Pond, S., 243
Popovich, A. T., 664, 691
Prandtl, L., 359, 361, 362, 365, 368, 371, 375, 380, 391, 490, 614, 616, 643, 704, 747
Prendergast, V., 574
Preston, J. H., 147
Proudman, I., 200, 214–216, 250, 309, 315

Quarmby, A., 148

Rannie, W. D., 622, 755
Ranque, M. G., 473
Rao, K., 668
Rasmussen, R. A., 145
Rayleigh, Lord, 77, 601
Reichardt, H., 142, 373, 375, 536, 541, 619, 620, 628, 691, 695, 696, 718, 730, 733, 747, 749, 752
Reiche, F., 152
Reid, W. H., 214–216, 251, 255, 413
Reiss, L. P., 153
Reynolds, A. R., 577
Reynolds, O., 1, 16, 76, 77
Reynolds, W. C., 604
Rib, L. N., 546
Ribner, H. S., 148, 315
Richardson, E. G., 142
Richardson, L. F., 406, 410
Ricou, F. P., 539, 546
Roache, P. J., 608
Robertson, H. P., 178
Robinson, M. S., 135
Rodriguez, J. M., 427
Roos, F. W., 672
Rose, W. G., 322, 350
Roshko, A., 89
Ross, J. A., 604
Ross, R. C., 384
Rotta, J., 328, 619, 623, 627, 634, 706, 714, 755–757
Rotta, J. C., 754
Rouse, H., 536, 563
Ruden, P., 536, 540

Saffman, P. G., 217, 240, 243, 248, 295, 360, 425–427, 439, 442
Sami, S., 536, 563
Sandborn, V. A., 520, 728
Sato, H., 251, 261, 269
Sauer, F. M., 87
Schapker, R. L., 568
Schlichting, H., 505, 602, 603, 630
Schmitz, G., 150
Schofield, W. H., 637
Schon, J. P., 117
Schraub, F. A., 157
Schubauer, G. B., 599, 605, 609, 626, 639, 643, 653, 731
Schubert, G., 692
Schuh, H., 384
Schultz, B. H., 382
Schultz, D. L., 144
Schultz-Grunow, F., 472, 626, 630, 632
Schuyf, J. P., 149
Schwarz, W. H., 152, 278, 292, 297, 299, 303
Serrin, J., 707, 708
Shapiro, A. H., 534, 536
Sheer, R. E., 655
Sheih, C. M., 243
Shen, S. F., 604
Sherwood, T. K., 448, 747
Shibuya, I., 708
Shlien, D. J., 397, 426
Siddon, T. E., 146–148, 563, 565
Sieder, E. N., 748

Sirkar, K. K., 658, 727
Skramstad, H. K., 120–122, 451, 452, 599, 605
Smith, D. W., 626, 631
Smith, F. B., 412
Smith, J. W., 758
Smith, O. M. O., 155
Snyder, H. A., 153
Snyder, W. H., 426, 470, 471
Spalding, D. B., 384, 539, 546, 622, 704
Squire, H. B., 612
Squyers, A. L., 384
Sternberg, J., 679, 692
Stewart, R. W., 177, 184, 191, 224, 238, 243, 253, 255, 258, 262, 397, 568, 570
Stewart, W. E., 384
Strum, R. C., 145
Stuart, J. T., 604, 614
Stutzman, L. F., 448, 747
Sullivan, P. L., 410
Sunavala, P. D., 545
Sutera, S. P., 762
Szablewski, W., 545

Tan, H. S., 270, 275–277
Tani, I., 613
Tannenbaum, B. S., 258
Tate, C. E., 748
Tatsumi, T., 709
Taylor, G. I., 2, 50–52, 62, 130, 359, 365, 366, 368, 369, 380, 406, 445, 747
Tchen, C. M., 334, 345–347, 350, 353, 461, 463, 466, 470
Tennekes, H., 243, 397, 719
Tetervin, N., 596
Thomas, P., 149
Thring, M. W., 545
Thürlemann, B., 152
Tillmann, W., 144, 626, 638
Toebes, G. H., 89, 144
Tollmien, W., 528, 543, 602
Towle, W. L., 448
Townend, H. C. H., 155, 156, 656
Townsend, A. A., 3, 131–134, 177, 202, 221, 224, 238, 242, 253, 258, 259, 262, 268–270, 273, 275, 293, 370, 371, 376, 414, 415, 438, 443, 445, 446, 455, 456, 505, 506, 508, 510, 513, 515, 517, 566, 568, 572, 574–577, 581, 627, 638, 644, 648, 650, 659, 667, 668, 680, 681, 684, 687, 688, 690, 701, 718, 735, 740
Tritton, D. J., 111
Trottier, G., 142
Trüpel, T., 536
Trutt, R. W., 148
Tsuji, H., 266, 269
Tu, B. J., 669
Tucker, H. J., 577
Tucker, M. J., 129

Uberoi, M. S., 160, 214, 253, 254, 257, 269, 270, 272, 273, 292, 306, 307, 310, 395, 418, 422, 443, 451, 452, 506, 508, 509, 511, 516, 536, 540, 544–546, 548, 550, 673
Ueda, H., 243, 668
Uzkan, T., 604

Van Atta, C. W., 184, 191, 243, 253, 254, 257, 714
Van der Hegge Zijnen, B. G., 90, 105, 142, 158, 166, 167, 269, 451, 452, 536, 539–542
Van Driest, E. R., 622
Van Meel, D. A., 154
Van Muiswinkel, J. C., 609
Vasudeva, B. R., 612
Velte, W., 707
Vermij, H., 154
Vernotte, P., 140
Vogel, W. M., 278, 297
Von Doenhoff, A. E., 596
Von Kármán, T., 2, 3, 58, 178, 185, 196, 221, 227, 233, 244, 248, 252, 264, 368, 369, 614, 618, 619, 748
Von Weizsäcker, G. F., 228, 229
Vrebalovich, T., 254, 257, 269, 270, 272

Walker, J. H., 626, 631
Wallace, J. M., 665
Wallis, S., 253
Walsh, T. J., 423, 427
Webster, C. A. G., 124
Wehrmann, O. H., 141
Weller, A. E., 541, 545
Welling, W. A., 119
Werner, F. D., 150
Wieghardt, K., 597, 630
Williams, G. C., 160, 162
Williams, M. J., 89, 90, 94, 143
Willmarth, W. W., 147, 660, 665, 666, 669, 670, 672, 674, 675, 693, 695
Wills, J. A. B., 142, 143, 673, 675
Wilson, R. E., 654
Winkler, E. M., 654
Winney, H. F., 142
Woertz, B. B., 448, 747
Wood, R. T., 762
Wooldridge, C. E., 669, 672, 675, 693
Wyat, L. A., 142
Wygnanski, I., 536, 538, 541, 543, 546, 552, 555, 556, 560, 565, 570, 577, 581, 686, 687, 709, 713
Wyngaard, J. C., 243

Yaglom, A. M., 242, 243
Yeh, Y., 162
Yen, K. T., 572
Young, C. S., 674, 695

SUBJECT INDEX

Accelerating separation, 411
Acoustic radiation, 311
Active motions, 680, 757
Analogies, 376, 743
Anisotropic turbulence (*see* Nonisotropic turbulence)
Apparent viscosity, 23
Averaging procedure, 4–6
  mean values, 5
Axisymmetric turbulence, 321

Balance equations, 15, 17, 68, 70, 71, 312, 488, 521, 651
Basset term, 463, 468
Boundary layer, 588
  approximation, 486, 489, 589
  buffer layer or region, 588, 619, 629, 659
  laminar, 597, 598
  laminar sublayer (*see* viscous sublayer *below*)
  mean-pressure distribution, 591
  mean-temperature distribution, 743, 754
  mean-velocity distribution, 616, 618, 626, 629, 630
  separation of, 598, 697
Boundary layer:
  shape factor, 596, 599, 624, 625, 633
  shear-stress distribution, 642, 647, 658, 700
  thickness, 594
    displacement, 595, 624, 625, 632
    dissipation, 597
    energy-loss, 596
    momentum-loss, 595, 624, 625, 632
  transition, 597, 599, 601, 602, 654
    effect of free-stream turbulence on, 600
  turbulence distribution, 639, 641
  viscous sublayer, 588, 620
  wall region, 587, 589
Boussinesq's theory, 23, 493
Buffer region, 588, 619, 629, 659, 720, 748, 749
Buoyancy effect, 90
Bursting period, 666, 742

Characteristic length, 487
Characteristic velocity, 487
Closure problem, 197, 704
  direct-interaction approximation, 198
Closure problem:
  physical closure approximation, 198, 250
  quasi-normal approximation, 198
  truncation approximation, 197
Cold length hot-wire anemometer, 99
Compensation, electronic, 114
Compressibility, 91, 159, 310, 471
Conservation of energy, 68
  of mass, 15
  of momentum, 16
  of scalar quantity, 29
  of vorticity, 312
Constant-current method, 94
Constant-stress layer, 617, 626, 643, 681, 718
Constant-temperature method, 94
Constitutive equation of turbulence, 25
Continuum flow, 8
Contraction process, 776
Convection velocity, 556, 565, 577, 673, 675, 688
Convective acceleration, 23
Cooling effect of wire support, 100, 105
Core-region, 588, 717, 730, 747

Corona-discharge anemometer, 150
Correction factor of hot wire, 120
Correction function, 634, 721
Correlation, 30, 33
acceleration, 398
auto-, 184, 191
density, 311
double velocity, 178
Eulerian space, 56, 157
Eulerian space-time, 47, 417, 449
Eulerian time, 45, 57, 184, 556
Lagrangian, 52, 57, 157, 388, 403
measurement, 157
lateral velocity, 34, 39, 56, 122, 653
series expansion of, 411
longitudinal velocity, 34, 39, 122, 157, 184, 653
mixed Eulerian-Lagrangian, 55
one-point double velocity, 56
dynamical behavior, 323
pressure-velocity, 37, 56, 178, 330, 669
pressure-velocity-gradient, 325, 328
scalar, 278
double, 278
series expansion, 279
space-time velocity, 129, 675, 677, 687
time, measurement, 128
triple scalar-velocity, 280
series expansion, 281
triple velocity, 37, 57, 130, 216
measurement, 131
two-point double pressure, 306, 309, 669, 672
two-point double velocity, 36, 178, 555, 669
dynamical behavior, 195, 196, 328
Correlation distance, 43
Correlation ellipse, 129
Correlation product, 33
Correlation tensor, 67, 403
first-order, 38, 178
general expression for isotropic turbulence, 180
second-order, 33, 38, 67, 181, 324
general expression for isotropic turbulence, 182
series expansion, 179, 188, 190
third-order, 38, 189
general expression for isotropic turbulence, 193
Couette flow, 3, 76
Critical point, 602, 603, 606
Crossed-beam method, 161
Cumulant, 198, 706
Cyclic process, 682, 690, 742

Decay of turbulence, 36, 47, 176, 259, 269
final period, 202, 259, 275
initial period, 259, 268, 272
transition period, 259
of turbulent scalar field, 300, 303
Deductive method, 373
Deformation tensor, 18
Degree of turbulence, 4
Density weighted velocity, 21, 70
Detecting element, 84
Developing region, 559
Diffusion, 48
of discrete particles, 460, 462
eddy (*see* Eddy transport or diffusion)
effect of compressibility, 471
gradient type, 361
from line source, 428, 432, 434
from point source, 428, 430, 442
relative, 406, 407, 410, 411
coefficient of, 407, 409
separation velocity, 407
tensor, 411
turbulent or eddy, 3, 29, 48, 50, 227
coefficient of, 227, 467
tensor, 363, 399, 403
Diffusion constant, 49
Diffusion method, 157, 158
Diffusion time, 385, 387
long, 386, 388, 392, 400, 402, 403, 410, 412, 464, 466, 467
short, 386, 392, 393, 398, 399, 402, 403, 407, 439, 466
Dilatation, 15, 573
Direction cosines, 772
Direction sensitivity of hot wire, 123
Discrete particles (*see* Diffusion)
Dissipation, 14, 71, 218, 221, 222, 224, 288
fluctuations, 242
length scale (*see* Scale of turbulence)
isotropic turbulence, expression for, 219
turbulence, 72, 219, 325, 345, 347, 513, 647, 648, 733
Distribution:
joint-probability-density, 456
logarithmic velocity, 616, 627, 630, 719, 720
mean-pressure: boundary layer, 591
jet-flow, 543
tube flow, 727
wake flow, 493
mean-temperature: boundary layer, 743, 754
jet flow, 542
tube flow, 754, 756
wake flow, 503, 508
mean-velocity: boundary layer, 626, 629, 630
jet flow, 521, 537
tube flows, 716, 719, 723, 732
wake flow, 499, 500, 502, 506, 509
power-law velocity, 625, 629, 631, 723
probability density, 401, 440
scalar quantity: boundary layer, 742
jet flow, 521, 541, 542
tube flow, 742
wake flow, 503, 508
shear-stress: boundary layer, 642, 647, 658
Distribution:
sheer stress: jet flow, 540, 548
tube flow, 728
wake flow, 499, 512
of turbulence intensity (*see* Intensity of turbulence)
of turbulence kinetic energy, 510, 520, 642, 647
Double refraction, 159
Drag reduction, 710
Drift velocity, 401
Dynamical viscosity (*see* Viscosity)

Eckert number, 87, 91
Eddy transport or diffusion, 3, 29, 48, 50, 578
coefficient of, 30, 227, 361, 364, 756, 758
tensor (*see* Diffusion)
viscosity, 23–25, 73, 363, 499, 539, 616, 621, 643, 730, 731
second-order tensor, 24
fourth-order tensor, 24
negative value of, 26
Einstein's summation convention, 774
Ejections, 664, 665, 682
Electric conductivity, 152
Electric-discharge anemometer, 149
Electric heating current, 86
Electric resistance, 86
temperature dependence of, 93
Electrochemical probe, 153
Electrokinetic probe, 152
Electromagnetic induction method, 151
Energy of turbulence, 72, 514, 592, 648, 728, 735
dissipation, 72
production, 72
total kinetic, 71, 75
Energy balance, 68, 70, 71, 514, 592, 648, 735
method, 707
Energy containing eddies, 221, 222
Energy (mechanical) equilibrium, 74, 75, 370, 681, 704, 735
Energy relations, 68
Energy spectrum (*see* Spectrum)
Ensemble average, 5
Entrainment, 538, 575, 688
constant, 576, 688
Entrance section, 717
Entropy mode, 112, 314
Entry length, 717
Equilibrium range (*see* Spectrum range)
Equivalent orifice diameter, 545
Ergodic hypothesis, 5
Eulerian correlations (*see* Correlations)
Eulerian description, 360
Eulerian dissipation scales (*see* Scales)
Eulerian integral scales (*see* Scales)
Exchange coefficient, 55, 448

Feedback system, 95, 96
Fickian diffusion equation, 363, 400
Film renewal model, 691
Film temperature, 93, 750
First law of thermodynamics, 70
Flattening (flatness) factor, 146, 242, 243, 513
Flow path, 154
Flow visualization, 154
Fluid jump, 359
Fluid particle, 359
substance (property), 360, 439
volume, 360, 439
Fourier cosine transform, 64
Fourier integral, 62
Free-area ratio of grid, 268, 269
Free boundaries, 13
Free convection effect on hot-wire anemometer, 90
Free jet, 12
Free stream, 587
turbulence, 600, 760
effect on boundary-layer transport, 760
effect on transition, 600
Free turbulence, 2, 483
Friction coefficient (*see* Wall-friction coefficient)
Friction factor, 722, 723, 730
Frozen field, 46

Gaussian distribution, 3
Generalized double-velocity correlation, 55
Generalized velocity, 55
Generation of turbulence (*see* Turbulence, production)
Glow-discharge anemometer (*see* Electric-discharge anemometer)
Grashof number, 87

Half jet, 484
Half-value distance, 485
Harmonic mode, 112, 314
Heat conductivity, 86
Heat-flow resistance, coefficient, 745, 751
Heat-transfer coefficient, 86, 745
Heisenberg's hypothesis, 230, 232, 237, 290, 346
Helmholtz instability, 578, 690
Homogeneous turbulence, 75, 175, 331
Homologous turbulence, 3
Hot-film anemometer, 143
Hot-wire anemometer, 85
cold length, 99
compressibility effect, 91, 111
constant-current method, 94, 113, 123, 126
constant-temperature method, 94, 96, 108, 123, 126
cooling-effect of supports, 100, 105
correction for wire length, 120
deposition on wire, 141
dynamic behavior, 96, 100, 113

Hot-wire anemometer:
electric resistance, 86, 93
temperature dependence, 93, 117
free-convection effect, 90
nonlinearity, 106, 111, 119
nonuniform velocity distribution, effect of, 120
resolution length, 139
resolution power, 139, 140
sensitivity, 103, 112, 113
direction, 123
thermal inertia, 98, 114
compensation for, 114
time constant, 101, 102, 113
V-array, 127
vibrations, 105
X-array, 127
Hydrogen-bubble technique, 155, 156
Hypersonic turbulence, 14

Impermanence, 7
Inactive motions, 573, 680, 682, 757
Induction period, 469
Inductive method, 374, 501
Inner regions (*see* Wall region)
Inrush movement, 664, 665, 683
Instability, 77, 269, 602, 708
energy-balance method, 77
inflexional, 602, 609, 614, 682
secondary, 609, 614, 665, 682
perturbation method, 77, 601
Integral length-scale (*see* Scale)
Integral relations, 588, 593
Integral time-scale (*see* Scale)
Intensity of turbulence, 4
measurement, 123, 158, 510, 547, 639, 641, 657, 724
relative, 4
Interaction:
between main and turbulent motion, 345, 346, 349
between outer and inner region, 682
Interaction function, 233, 258
Interferometer, 159
Intermittency, 12, 242, 373, 380, 517, 553, 685
factor, 242, 381, 518, 552, 565, 572, 644, 646, 667
Irrotational mode, 112, 680
Irrotational zone, 685, 686
Isentropic process, 19
Isotropic turbulence (*see* Turbulence)
Isotropy:
definition of, 177
local, 176, 244

Jet, 483
free, 483
plane, 541
round, 520
mean-pressure distribution, 543
mean-temperature distribution, 542
mean-velocity distribution, 527, 528, 537

Jet:
free: round: shear-stress distribution, 540, 541
spectra, 550
turbulence distribution, 547
half-, 484
wall, 580
Joint-probability distribution (*see* Probability)

Kármán, interpolation formula, 244, 248, 551
for a scalar, 299
Kármán-Howarth equation, 196
(*See also under* Von Kármán)
Kármán similarity hypothesis, 322, 643
Kinetic energy of turbulence, 7
Knudsen number, 86, 87
Kolmogoroff hypothesis, 223, 226
Kolmogoroff length-scale, 121, 223, 391, 399
Kolmogoroff spectrum law, 228, 236
Kolmogoroff time-scale, 399
Kolmogoroff velocity-scale, 223
Kronecker delta, 18, 774

Lagrangian correlation, 388, 403
Lagrangian description, 360
Lagrangian-History Direct-Interaction (L.H.D.I) theory, 224, 249, 256, 409
Lagrangian integral time-scale, 397
Laser-Doppler-shift (L.D.S), 162
Lateral velocity-correlation, 56
Law:
of the wake, 696
of the wall, 615
Light:
absorption, 160
refraction, 159
scattering, 160, 162
Linearizer, 111, 117
Logarithmic velocity-distribution, 616, 618, 718
Loitsianskii's integral, 214, 216, 218, 227, 244, 265, 268
Loitsianskii's invariant, 199, 216, 267, 288
Longitudinal velocity-correlation, 56
Lump of fluid (*see* Fluid lump)

Mach number, 87, 91, 92, 654, 655
of turbulence, 22
Markov process, 50
Material lines, 414
Material surfaces, 414
Mean values:
ensemble, 5
spatial, 5
time, 5
Measurement:
of concentration fluctuations, 139
of double velocity correlation, 127
of eddy-diffusion coefficient, 158
of eddy shear-stress, 126, 158

Measurement:
of Eulerian space correlation, 157, 450
time correlation, 184
of Lagrangian correlation, 157
of mean pressure, 164
of mean velocity, 164, 166, 505
of scale of turbulence, 131
of space-time velocity correlation, 129
of spectrum of turbulence, 131
of temperature fluctuations, 137
of triple velocity correlation, 127, 131
of turbulence in liquids, 143
of turbulence intensity, 123, 126
of turbulent flows, 83
requirements, 84
Mechanical energy, 70, 71, 75
equilibrium (*see* Energy equilibrium)
Memory behavior, 15, 26, 50, 55, 373, 391, 401, 684
Mixed Eulerian-Lagrangian correlation, 55
Mixing length, 359, 385, 507, 555
theory, 361, 362, 370, 380, 416, 500, 504, 527, 532, 619, 622
Molecular effects, 8, 438, 439
Molecular-transport coefficient, 29
Molecular velocity, 8, 49
Momentum-flux balance, 497
Momentum-transfer law, Reichardt's, 375
Momentum-transport theory (*see* Transport)

Navier-Stokes equation, 16, 19
Nonisotropic turbulence, 321
Nonlinearity effect of hot wire, 111
Nusselt number, 86, 745

Objectivity, requirement of, 25
Origin of similarity, 487, 521
Orr-Sommerfeld equation, 601, 693
Outer region, 588, 590, 696, 747
Overhead ratio, 87, 92, 125, 143
Overshoot effect of discrete particle, 470, 471

Partial pressure, 383
Particle of fluid (*see* Fluid particle)
Pathlines, 157
Péclet number, 88
Perturbation method (*see* Instability)
Phase velocity, 614, 675
Phenomenological theories, 484
Piezo-electric crystal, 147
Pitot tube, 85
Potential core, 534, 544
Power-law distribution, 625, 629, 631, 723
Prandtl number, 87, 88, 654, 747, 748, 757
effective, 744
turbulence or eddy, 305, 380, 744, 747, 753
Prandtl's momentum-transport theory (*see* Transport)
Preionization, 150
Pressure, 18
static, 164
total, 166
turbulence, 24
Pressure fluctuation, 305, 315
intensity, 309, 672
Preston tube, 147, 148
Probability, 3, 49, 400
density distribution, 401, 440
joint, 456
Propagation velocity, 614, 711
Propeller anemometer, 148
Property particle (*see* Substance particle)
Pseudo turbulence, 3
Pulse-counting method, 134

Quasi frequency, 7
Quasi periodicity, 7, 14
Quasi permanency, 14

Ranque-Hilsch vortex tube, 473
Rapid-distortion theory, 577
Rayleigh number, 90
Recovery factor, 473
Refraction:
by density variation, 159
double, 159
Reichardt's hypothesis, 375
(*See also* Reichardt's momentum-transfer law)
Reichardt's inductive theory, 374, 501, 504, 529, 533
(*See also* Inductive method)
Reichardt's momentum-transfer law, 375
(*See also* Reichardt's hypothesis)
Relaminarization, 577, 612, 712, 714
Relative intensity (*see* Intensity)
Relaxation time, 392
Resistance thermometer, 137, 138, 303
Resolution power of hot-wire anemometer, 140
Reynolds' analogy, 746, 748, 751
Reynolds number, 87, 599
critical value, 76, 599, 600, 604
similarity, 628, 718
of turbulence, 224, 227, 260
Reynolds stresses, 22
Richardson relative-diffusion law, 410, 412
Rotational, 14, 680
Rough wall, 617, 619, 623, 637
effect on heat transfer, 757
Roughness parameter, 617

Saffman's dynamic invariant, 217, 265, 267, 288
Sand-grain roughness, 634, 722
equivalent, 635, 637, 722
Scale:
dissipation, 279
of eddy diffusion, 54, 58
Scale:
integral, 279, 300
of scalar fluctuations, 279
time, 4
Eulerian dissipation, 45, 47, 57, 61, 65, 121, 133, 210, 224, 418
Eulerian integral, 47, 57, 65, 131
Lagrangian dissipation, 54, 57, 419
Lagrangian integral, 53, 57, 385, 392, 425
relation between Eulerian and Lagrangian, 418, 425
Scale of turbulence, 4
dissipation (micro), 42, 56, 65, 121, 210, 224, 263, 266, 270, 285, 391, 419, 555, 740
integral (macro), 43, 56, 65, 224, 270, 272, 285, 352, 385, 420, 555, 563, 653, 740
measurement, 270, 420
spatial, 4
Schlieren method, 159
Schmidt number, 293, 294, 305, 443, 751
turbulent, 305, 380, 449
Self-noise, 315
Self-preservation, 177, 261, 301, 484, 566
complete, 177, 266, 301, 304
incomplete, 177
partial, 267, 302
Self-preserving (*see* Self-preservation)
Sensitivity of hot-wire anemometer (*see* Hot-wire anemometer)
Separation velocity of two fluid particles (*see* Diffusion)
Shadograph, 159, 160
Shape factor (*see* Boundary layer)
Shear-flow turbulence, 322, 350
Shear layer, 609
concentrated, 609, 682
horizontal, 609, 665, 682
vertical, 613, 665
Shear noise, 315
Shear stress at wall, 144, 147, 672
Shear-stress distribution (*see* Distribution)
Shedding frequency, 761
Shock wave, 315
Similarity, 177, 368, 484, 486, 618, 626
complete, 177, 486, 524, 588
incomplete, 177, 487, 525, 588
Skewness, 451
Skewness factor, 241, 251, 513
Skin-friction coefficient, 90
Smooth wall, 614, 618
Sound mode, 112, 314
Space average, 5
Space scale, 4
(*See also* Scale)
Spectral transfer:
deformation effect, 338, 340
turning effect, 338
among wavenumbers, 336, 338, 339

Spectrum (spectral distribution), 7, 283
energy, 7, 63
higher range, 228
highest range, 228
inertial-convective subrange, 292, 295
inertial-diffusive subrange, 294
inertial subrange, 226, 228, 236
interaction function, 233, 258
Lagrangian, 389
low-range, 228
lower range, 228
lowest range, 228
medium range, 228
one-dimensional, 202, 254, 255, 351, 511, 652, 663, 738
tensor, 204
three-dimensional, 205, 250
dynamic behavior, 211, 334, 336
series expansion, 206, 207
transfer function, 215, 220, 233, 250, 253, 257, 258, 341, 346
integral, 233, 238
series expansion, 215
viscous-convective subrange, 296
viscous range, 238
of scalar fluctuation, 283
one-dimensional, 283, 292, 756
three-dimensional, 284
dynamic behavior, 286
series expansion, 285
transfer function, 286
series expansion, 287
Spectrum range:
of dissipation, 222
of energy-containing eddies, 221, 222, 261, 299
of equilibrium, 223, 261, 341
Spikes, 606, 609
Stagnation temperature, 87, 93, 473
Stanton number, 745
Static-pressure tube, 146
Static temperature, 91
Stationary random function, 4, 62, 176
Streakline, 154, 155, 157
Stress tensor, 18
Strouhall number, 140, 146, 560
Substitution process, 776
Supersonic turbulence, 14
Sweep movements, 664, 683

Taylor's hypothesis, 46, 47, 392, 420, 677
Taylor's vorticity-transport theory (*see* Transport)
Temperature ratio, 89, 118
Tensor, 771
antisymmetric or alternating, 18, 775
correlation (*see* Correlation)
energy-spectrum, 204
general product, 776
relative-diffusion, 411
scalar or dot product, 777
scalar triple product, 778
stress, 24
turbulent or eddy, 363
Tensor:
unit, 18, 775
vector or cross product, 778
Thermal conductivity, 69
Thermal lag, hot-wire anemometer, 103, 114
Thermal sublayer, 748
Thermistor, 145
Thermodynamic pressure, 18
Thomson effect, 86, 125
Time average, 5
Time constant, hot-wire anemometer, 101, 102
constant current, 113
constant temperature, 101
Time lines, 157
Time scale, 4
Tollmien-Schlichting waves, 605, 614
Total-head tube, 85, 145, 166
direction sensitivity, 167
Tracer, 84
Transconductance, 96, 102, 108
Transfer function, 215, 220, 233, 250, 253, 257, 258
Transferable property, 14, 358
Transition, 76, 77, 597, 599, 601, 654, 707, 710
breakdown, 606, 682
natural, 604
reverse, 612
Transition region (*see* Buffer region)
Transport:
in broad sense, 377, 379, 381, 743
by large-scale turbulence, 372, 381, 579
in narrow sense, 377, 379, 743
by small-scale turbulence, 372
by turbulence (*see* Eddy transport or diffusion)
in turbulent flow, 358, 378, 578
analogies, 376, 743
gradient type, 361, 579, 580
(*See also* Diffusion)
momentum, theory, 363, 616
vorticity theory, 365
generalized, 366
modified, 366, 380
Triple velocity correlation (*see* Correlation)
Tube flow, 707
mean-pressure distribution, 727
mean-temperature distribution, 754, 756
mean-velocity distribution, 716, 719, 723, 732
shear-stress distribution, 728
Turbulence:
axisymmetric, 321
burst, 606, 609, 664, 665, 682, 688, 694, 742
characteristic length, 221, 231
characteristic time, 231
characteristic velocity, 221, 231
decay (*see* Decay)
definition, 1
energy (*see* Energy of turbulence)
Turbulence:
fine structure, 176, 241, 242, 269, 742
free, 2
homogeneous, 75, 175, 331
homologous, 3
intensity (*see* Intensity of turbulence)
isotropic, 3, 32, 38, 175, 257
dynamic behavior, 196
(*See also* Isotropy)
large-scale structure, 176, 571, 574, 659
Mach number of (*see* Mach number)
measurement of, in liquids, 143
nonisotropic, 3, 328
production, 2, 72
pseudo, 3
Reynolds number (*see* Reynolds number)
scale of (*see* Scale)
shear-flow, 3, 322
wall, 2, 586
Turbulence level, 4
Turbulence pressure, 24
Turbulence production, 78
Turbulence rate of strain, 230
Turbulence scalar field, 278
dissipation, 288
dynamic behavior, 282
Turbulence slug, 711, 712
Turbulence spot, 605, 609, 712
Turbulence stresses, 24
Turbulence-transport coefficient (*see* Diffusion)
Turbulence viscosity (*see* Eddy transport or diffusion, viscosity)
Turbulence vorticity field, 217
Turbulence zone, 685, 686
Turning effect of spectral transfer (*see* Spectrum)

Ultra microscope, 155
Universal, 223, 228
Universal velocity distribution, 618, 626, 718

V-array of hot-wire anemometer, 127
Vane anemometer, 148
Variational principle, 707
Velocity of sound, 19, 87
Velocity-defect law, 618, 630, 634, 720
Vibrating ribbon technique, 605
Vibrations, hot-wire anemometer, 105
Virtual added mass, 463, 469
Visco-elastic fluid, 26
Viscosity:
decoupling effect, 715
(*See also* Relaminarization)
dual role, 603
dynamic, 19
eddy or turbulence, 23–25, 73
volume, 19
Viscous sublayer, 588, 620, 691, 718, 720
Viscous superlayer, 567
Visualization of flow, 154
emulsions, 155

Visualization of flow:
  hydrogen bubbles, 155
  smoke or dye, 157
  soap bubbles, 156
  suspended particles, 154, 156
Volume particle (*see* Fluid particle)
Volume viscosity (*see* Viscosity)
Von Kármán integral momentum
    balance, 594
Von Kármán universal constant, 619,
    622, 627
  (*See also under* Kármán)
Vorticity, 18
  of mean motion, 346
  root-mean-square, 347
  of turbulence, 346, 347
Vorticity balance equation, 312
Vorticity mode, 112, 314, 573
Vorticity transport theory (*see*
    Transport)

Wake flow, 9, 483, 496, 502
  mean-pressure distribution, 493
  mean-temperature distribution, 503,
    508
  shear-stress distribution, 499, 512
Wall-friction coefficient, 623, 625, 629,
    638, 654
Wall-friction velocity, 615, 624, 681,
    722
Wall-heat transfer temperature, 745
Wall-pressure fluctuations, 668, 674
  intensity, 672
Wall region, 587–589, 615, 659, 717,
    747
Wall roughness (*see* Rough wall)
Wall shear-stress, 144, 147, 672
Wall turbulence, 586, 742
Wave-guide model, 692
Wave packet, 693
Wavenumber, 203
  cutoff value, 235
  of dominant viscous effect, 223
  of energy-containing eddies, 221

X-array of hot-wire anemometer, 127